ile-, *ileum:* ileocecal valve
ili-, ilio-, *ilium,* flank, groin: iliac
in-, *in,* within, or denoting negative effect: inactivate
infra-, *infra,* beneath: infraorbital
inter-, *inter,* between: interventricular
intra-, *intra,* within: intracellular
ipsi-, *ipse,* itself: ipsilateral
iso-, *isos,* equal: isotonic
-itis, *-itis,* inflammation: dermatitis
karyo-, *karyon,* body: megakaryocyte
kerato-, *keros,* horn: keratin
kino-, -kinin, *kinein,* to move: bradykinin
lact-, lacto-, -lactin, lac, milk: prolactin
lapar-, *lapara,* flank or loins: laparoscopy
-lemma, *lemma,* husk: sarcolemma
leuko-, *leukos,* white: leukocyte
liga-, *ligare,* to bind together: ligase
lip-, lipo-, *lipos,* fat: lipase
lith-, *lithos,* stone: otolith
lyso-, -lysis, -lyze, *lysis,* dissolution: hydrolysis
macr-, *makros,* large: macrophage
mal-, *mal,* abnormal: malabsorption
mamilla-, *mamilla,* little breast: mamillary
mast-, masto-, *mustos,* breast: mastoid
mega-, *megas,* big: megakaryocyte
melan-, *melas,* black: melanocyte
men-, *men,* month: menstrual
mero-, *meros,* part: merocrine
meso-, *mesos,* middle: mesoderm
meta-, *meta,* after, beyond: metaphase
micr-, *mikros,* small: microscope
mito-, *mitos,* thread: mitosis
mono-, *monos,* single: monocyte
morpho-, *morphe,* form: morphology
multi-, *multus,* many: multicellular
-mural, *murus,* wall: intramural
myelo-, *myelos,* marrow: myeloblast
myo-, *mys,* muscle: myofilament
narc-, *narkoun,* to numb or deaden: narcotics
nas-, *nasus,* nose: nasolacrimal duct
natri-, *natrium,* sodium: natriuretic
necr-, *nekros,* corpse: necrosis
nephr-, *nephros,* kidney: nephron
neur-, neuro-, *neuron,* nerve: neuromuscular
oculo-, *oculus,* eye: oculomotor
odont-, *odontos,* tooth: odontoid process
-oid, *eidos,* form: odontoid process
oligo-, *oligos,* little, few: oligodendrocyte
-ology, *logos,* the study of: physiology
-oma, -oma, swelling: carcinoma
onco-, *onkos,* mass, tumor: oncology
oo-, *oon,* egg: oocyte
ophthalm-, *ophthalmos,* eye: ophthalmic nerve
-opia, *ops,* eye: optic
orb-, *orbita,* a circle: orbicularis oris
orchi-, *orchis,* testis: orchiectomy
orth-, *orthos,* correct, straight: orthopedist
-osis, -osis, state, condition: neurosis
osteon, osteo-, *os,* bone: osteocyte
oto-, *otikos,* ear: otolith
para-, *para,* beyond: paraplegia
patho-, -path, -pathy, *pathos,* disease: pathology
pedia-, *paidos,* child: pediatrician
per-, *per,* through, throughout: percutaneous
peri-, *peri,* around: perineum
phag-, *phagein,* to eat: phagocyte

-phasia, *phasis,* speech: aphasia
-phil, -philia, *philus,* love: hydrophilic
phleb-, *phleps,* a vein: phlebitis
-phobe, -phobia, *phobos,* fear: hydrophobic
phot-, *phos,* light: photoreceptor
-phylaxis, *phylax,* a guard: prophylaxis
physio-, *physis,* nature: physiology
-plasia, *plasis,* formation: dysplasia
platy-, *platys,* flat: platysma
-plegia, *plege,* a blow, paralysis: paraplegia
-plexy, *plessein,* to strike: apoplexy
pneum-, *pneuma,* air: pneumonia
podo, *podon,* foot: podocyte
-poiesis, *poiesis,* making: hemopoiesis
poly-, *polys,* many: polysaccharide
post-, *post,* after: postganglionic
pre-, *prae,* before: precapillary sphincter
presby-, *presbys,* old: presbyopia
pro-, *pro,* before: prophase
proct-, *proktos,* anus: proctology
pterygo-, *pteryx,* wing: pterygoid
pulmo-, *pulmo,* lung: pulmonary
pulp-, *pulpa,* flesh: pulpitis
pyel-, *pyelos,* trough or pelvis: pyelitis
quadr-, *quadrans,* one quarter: quadriplegia
re-, *re-,* back, again: reinfection
retro-, *retro,* backward: retroperitoneal
rhin-, *rhis,* nose: rhinitis
-rrhage, *rhegnymi,* to burst forth: hemorrhage
-rrhea, *rhein,* flow, discharge: amenorrhea
sarco-, *sarkos,* flesh: sarcomere
scler-, sclero-, *skleros,* hard: sclera
-scope, *skopeo,* to view: microscope
-sect, *sectio,* to cut: transect
semi-, *semis,* half: semilunar valve
-septic, *septikos,* putrid: antiseptic
-sis, state or condition: metastasis
som-, -some, *soma,* body: somatic
spino-, *spina,* spine, vertebral column: spinodeltoid
-stalsis, *staltikos,* contractile: peristalsis
sten-, *stenos,* a narrowing: stenosis
-stomy, *stoma,* mouth, opening: colostomy
stylo-, *stylus,* stake, pole: styloid
sub-, *sub,* below; subcutaneous
super-, *super,* above or beyond: superficial
supra-, *supra,* on the upper side: supraspinous fossa
syn-, *syn,* together: synthesis
tachy-, *tachys,* swift: tachycardia
telo-, *telos,* end: telophase
tetra-, *tettares,* four: tetrad
therm-, thermo-, *therme,* heat: thermoregulation
thorac-, *thorax,* chest: thoracentesis
thromb-, *thrombos,* clot: thrombocyte
-tomy, *temnein,* to cut: appendectomy
tox-, *toxikon,* poison: toxin
trans-, *trans,* through: transport
tri-, *tres,* three: trimester
tropho-, *trophe,* nutrition: trophoblast
-trophy, *trophikos,* nourishing: atrophy
-tropic, *trope,* turning: adrenocorticotropic
tropo-, *tropikos,* turning: troponin
uni-, *unus,* one: unicellular
uro-, -uria, *ouron,* urine: glycosuria
vas-, *vas,* vessel: vascular
zyg-, *zygotos,* yoked: zygote

Essentials of
Anatomy & Physiology

Essentials of
Anatomy & Physiology

THIRD EDITION

Frederic H. Martini, Ph.D.
Edwin F. Bartholomew, M.S.

with

William C. Ober, M.D.
Art Coordinator and Illustrator

Claire W. Garrison, R.N.
Illustrator

Kathleen Welch, M.D.
Clinical Consultant

Ralph T. Hutchings
Biomedical Photographer

Pearson Education, Inc.
Upper Saddle River, New Jersey 07458

Library of Congress Cataloging-in-Publication Data

Martini, Frederic.
  Essentials of anatomy & physiology / Frederic H. Martini, Edwin F. Bartholomew;
with William C. Ober, art coordinator and illustrator; Claire W. Garrison, illustrator;
Kathleen Welch, clinical consultant; Ralph T. Hutchings, biomedical photographer.—
3rd ed.
    p. cm.
  Includes bibliographical references and index.
  ISBN 0-13-061567-6
    1. Human physiology. 2. Human anatomy. I. Title: Essentials of anatomy and
physiology. II. Bartholomew, Edwin F. III. Title.

QP36.M42 2003
612—dc21   2002070051

Editor in Chief, Life and Geosciences: Sheri L. Snavely
Senior Acquisitions Editor: Halee Dinsey
Development Editor / Editor in Chief: Ray Mullaney
Assistant Vice President of Production and Manufacturing:
    David W. Riccardi
Executive Managing Editor: Kathleen Schiaparelli
Assistant Managing Editor: Beth Sweeten
Assistant Managing Editor, Media: Nicole Bush
Assistant Managing Editor, Science Supplements: Dinah Thong
Executive Marketing Manager for Biology: Jennifer Welchans
Marketing Manager: Martha McDonald
Cover Photo: John Post/Panoramic Images
Manufacturing Manager: Trudy Pisciotti

Assistant Manufacturing Manager: Michael Bell
Illustrators: William C. Ober, M.D., Claire W. Garrison, R.N.
Director Creative Services: Paul Belfanti
Director of Design: Carole Anson
Art Director: Kenny Beck
Assisstant to Art Director: Geoffrey Cassar
Managing Editor, Audio Visual Assets and Production: Patricia Burns
Art Editor: Adam Velthaus
Editorial Assistant: Susan Zeigler
Project Manager: Crissy Dudonis
Production Services / Composition: Preparé, Inc.
Photo Editor: Carolyn Gauntt
Photo Researcher: Yvonne Gerin

ISBN 0-13-061567-6

Pearson Education LTD., *London*
Pearson Education Australia PTY, Limited, *Sydney*
Pearson Education Singapore, Pte. Ltd.
Pearson Education North Asia Ltd., *Hong Kong*
Pearson Education Canada, Ltd., *Toronto*
Pearson Educación de Mexico, S.A. de C.V.
Pearson Education—Japan, *Tokyo*
Pearson Education Malaysia, Pte. Ltd.

Text and Illustration Team

 **Frederic H. Martini** received his Ph.D. from Cornell University in comparative and functional anatomy. He has broad interests in vertebrate biology, with special expertise in anatomy, physiology, histology, and embryology. Dr. Martini's publications include journal articles, technical reports, magazine articles, and a book for naturalists on the biology and geology of tropical islands. He is the author of *Fundamentals of Anatomy and Physiology* (5e, Prentice Hall, 2001) and the coauthor of three other undergraduate texts on anatomy or anatomy and physiology.

Dr. Martini has been involved in teaching undergraduate courses in anatomy and physiology (comparative and/or human) since 1970. During the 1980s, he spent his winters teaching courses, including human anatomy and physiology, at Maui Community College and his summers teaching an upper-level field course in vertebrate biology and evolution for Cornell University at the Shoals Marine Laboratory (SML). Dr. Martini now devotes most of his attention to developing new approaches to A&P education, especially the use of appropriate technologies. He is a member of the Human Anatomy and Physiology Society, the American Association of Anatomists, the American Physiological Society, the National Association of Biology Teachers, the Society for Integrative and Comparative Biology, the Society for College Science Teachers, the Western Society of Naturalists, and the National Association of Underwater Instructors.

 **Edwin F. Bartholomew** received his undergraduate degree from Bowling Green State University in Ohio and his M.S. from the University of Hawaii. His interests range widely, from human anatomy and physiology to the marine environment and the "backyard" aquaculture of escargots and ornamental fish. During the last three decades, Mr. Bartholomew has taught human anatomy and physiology at both the secondary and undergraduate levels. In addition, he has taught a wide variety of other science courses (from botany to zoology) at Maui Community College. He is presently teaching at historic Lahainaluna High School, the oldest high school west of the Rockies. He has written journal articles, a weekly newspaper column, and many magazine articles. Working with Dr. Martini, he coauthored *Structure and Function of the Human Body* (Prentice Hall, 1999) and *The Human Body in Health and Disease* (Prentice Hall, 2000). Mr. Bartholomew is a member of the Human Anatomy and Physiology Society, the National Association of Biology Teachers, the National Science Teachers Association, and the American Association for the Advancement of Science.

 **Dr. Kathleen Welch** (clinical consultant) received her M.D. from the University of Washington in Seattle and did her residency at the University of North Carolina in Chapel Hill. For two years she served as Director of Maternal and Child Health at the LBJ Tropical Medical Center in American Samoa and subsequently was a member of the Department of Family Practice at the Kaiser Permanente Clinic in Lahaina, Hawaii. She has been in private practice since 1987. Dr. Welch is a Fellow of the American Academy of Family Practice. She is also a member of the Hawaii Medical Association and the Human Anatomy and Physiology Society.

 **Dr. William C. Ober** (art coordinator and illustrator) received his undergraduate degree from Washington and Lee University and his M.D. from the University of Virginia in Charlottesville. While in medical school, he also studied in the Department of Art as Applied to Medicine at Johns Hopkins University. After graduation Dr. Ober completed a residency in family practice and is currently on the faculty of the University of Virginia in the Department of Sports Medicine. He is also part of the Core Faculty at Shoals Marine Laboratory, where he teaches biological illustration in the summer program. Dr. Ober now devotes his full attention to medical and scientific illustration.

Claire W. Garrison, R.N. (illustrator) practiced pediatric and obstetric nursing for nearly 20 years before turning to medical illustration as a full-time career. Following a five-year apprenticeship, she has worked as Dr. Ober's associate since 1986. Ms. Garrison is also a Core Faculty member at Shoals.

Texts illustrated by Dr. Ober and Ms. Garrison have received national recognition and awards from the Association of Medical Illustrators (Award of Excellence), American Institute of Graphics Arts (Certificate of Excellence), Chicago Book Clinic (Award for Art and Design), Printing Industries of America (Award of Excellence), and Bookbuilders West. They are also recipients of the Art Directors Award.

 **Ralph T. Hutchings** is a biomedical photographer who was associated with the Anatomy Department of the Royal College of Surgeons for 20 years. An engineer by training, Mr. Hutchings has focused for years on photographing the structure of the human body. The result has been a series of color atlases, including the *Color Atlas of Human Anatomy*, *The Color Atlas of Surface Anatomy*, and *The Human Skeleton* (all published by Mosby-Yearbook Publishing, St. Louis, Missouri, USA). Mr. Hutchings makes his home in North London, where he tries to balance the demands of his photographic assignments with his hobbies of early motor cars and airplanes.

Dedication

To Kitty, P. K. , Ivy, and Kate:
We couldn't have done this without you.
Thank you for your encouragement, patience,
and understanding.

Brief Contents

Contents

CHAPTER 18

The Urinary System   547

CHAPTER 19

The Reproductive System   581

CHAPTER 20

Development and Inheritance   613

Preface

THIS TEXTBOOK IS DESIGNED as an introduction to the specialized terms, basic concepts, and principles important to an understanding of the human body. It has two primary goals:

1. **Building a foundation of essential knowledge in human anatomy and physiology.** Constructing this knowledge requires answers to questions such as: What structure is that? How does it work? What happens when it does not work?

2. **Providing a framework for interpreting and applying information that can be used in problem-solving.** This framework is based on the fact that certain themes and patterns appear again and again in the study of anatomy and physiology. These themes and patterns provide the hooks on which to organize and hang the information you will find in the text. Problems that require interpreting and applying information include such general questions as: How does a change in one body system affect the others? How does aging affect body systems?

The landscape of the new millennium in which we now live and work is dynamic, not static. There is a continuing explosion in what we know, how we know what we know, and how we share what we know. Our knowledge base in anatomy and physiology has expanded tremendously. For example, we know much more than ever about the specific roles molecules play in many normal physiological processes and in diseases; we can manipulate the DNA within cells and even clone identical animals; and, an initial survey of the human genome has been completed. These gains in knowledge have led to new methods and procedures to combat disease, reduce suffering, and promote good health. New methods of accessing and sharing information have linked health professionals around the world. Technology has become a means for acquiring and distributing information, and medical professionals have learned to use new technologies to increase their effectiveness as well as to improve the quality of medical care.

Modern technology has enhanced our ability to find specific information. As a result, it has affected how each of us deals with information in general. In many cases, knowing where to look for information is preferable to trying to memorize specific facts. That is certainly the trend in the training of medical and allied health professionals today. Students begin by mastering the terminology and memorizing a substantial core of basic concepts. In the process, they are also given (1) a "mental framework" for organizing new information, (2) the ability to access additional information when needed, by referring to relevant print or electronic data sources, and (3) an understanding of how to apply their knowledge to solve particular problems. The same skills are equally important to people in other career paths. To be effective in almost any job today, you must know how to access and absorb new information, to use (or learn to use) available technology, and to solve problems.

Essentials of Anatomy and Physiology third edition, has been carefully designed to place information in a meaningful context and to help students develop their problem-solving skills. Any one deciding to embark on a career in a medical or allied health field must master the same skills that are needed to succeed in this course. Developing a large technical vocabulary and retaining a large volume of detailed information is not enough. A person must learn how to learn—how to organize new information, how to connect it to what is already known, and then how to apply it as needed. The electronic additions to this edition, including the enclosed CD-ROM and the Companion Website, make it easy for students and instructors to use current technology to access and manage information.

The focus of this text and learning system has been to simplify the processes of teaching and learning anatomy and physiology. Much as we would like to, we cannot create materials that will give any of us more time. But a carefully designed text and supplements package can help both instructors and students make better use of the time they do have. The User's Guide that begins on p. xix outlines our specific teaching framework and shows the many learning aids that are built into this text. These features were developed through feedback from students and instructors on campuses across the United States, Canada, Australia, New Zealand, and Europe. Many students, in person or by phone or e-mail, have told us that this system really works for them when other presentation styles have not.

WHAT'S DIFFERENT ABOUT THE THIRD EDITION?

Each new edition requires some revising and updating. Our first step in planning for the third edition involved integrating the feedback from dozens of reviewers with our own experiences in the classroom. We were then able to identify key areas where students were having trouble following the narrative or grasping important concepts. We then worked with the illustration team to revise the art program while we revised the narrative and the teaching and learning system.

A detailed comparison with the second edition will show significant changes:

- New art design and corresponding revisions in tables and figures.

- The total number of chapters has been reduced from 21 to 20. Previously, the nervous system was spilt into two chapters; but now full coverage is provided in a single chapter.

- Many sections have been revised to focus more closely on the needs of students pursuing careers in allied health in one-semester courses.

- In a number of chapters, including those on the chemical level of organization (2), the skeletal system (6), and the urinary system (18), new illustrations have been added to simplify complex topics. The overall illustration count has been increased by about 7 percent.

- Each of the organ system chapters (i.e., Chapters 5–8 and 10–20) contains new Concept Check questions concerning how the system under consideration interacts with other selected body systems. The questions are keyed to the system integrator figures and are intended to reinforce the concept that the body functions as a unit.

- More clinical information has been added in the form of a list of Relevant Clinical Terms.

- New Concept Check questions have been added to reinforce understanding of the Aging and Systems Integration sections in each body system chapter. As in the second edition, the answers to all Concept Check questions are given at the end of each chapter.

- New Web Explorations. The MediaLab at the end of 11 chapters encourages students to investigate a text topic further on the Web.

- Each new copy of the textbook contains an EAP Interactive Student CD-ROM. It contains visual tools to help understand anatomy and physiology; dissectable 3-D animations, tutorials, interactive clinical case studies; and an audio glossary.

- Each new copy of the textbook includes a copy of the Applications Manual (AM).

INTEGRATED SUPPLEMENTS

For the Instructor

- *300 Full Color Transparencies*   Key illustrations, including artwork, scans, and cadaver photographs, make up this outstanding transparency set. Artwork, labels, and leaders are enlarged for easier visibility and a special production technique is used to saturate all colors making them crisp, vivid, and clear.

- *Instructor's Resource Guide*   This guide makes it easy to see content at a glance in detailed lecture outlines for each chapter. It also provides unique analogies and useful teaching techniques that will engage students by promoting lively classroom discussion and by infusing lectures with interesting and appropriate clinical notes. Tips from other instructors across the country are also included.

- *Instructor's Media Portfolio*   All of the artwork, scans, and photographs can be easily accessed from our Instructor's Media Portfolio. In addition, this fantastic resource provides instructors with both labeled and unlabeled art, images with lecture outlines in PowerPoint, and all of the animations from the student CD-ROM. This easy-to-use tool enables you to build multimedia lectures using PowerPoint.

- *Test Item File and Prentice Hall Custom Text Software*   Access a complete testing package with more than 3000 questions. The testing package parallels the three-level learning system used in the textbook and also includes over 200 pieces of unlabeled text art to allow instructors to create labeling exercises for exams.

- *Prentice Hall Laserdisc for Anatomy and Physiology and Bar Code Manual*   This laserdisc uses videodisc technology to feature high-quality animations based on art from the text.

For the Student

- *Student Tutor CD-ROM*   This interactive CD-ROM offers a wealth of supplemental content and learning exercises to complement the material presented in the text and helps reinforce the concepts students learned in class. Interactive tutorials use text graphics, animations, and audio to help students visualize difficult concepts and encourage self-assessment and promote retention with several types of innovative quizzes. Animations provide step-by-step

views of physiological processes, while 3-D visualizations allow you to explore, rotate, and dissect lifelike three-dimensional models of the human body. Case studies expressly written to complement material in this third edition of the text help students integrate knowledge and practice their analytical and diagnostic skills in real-world situations.

- *WebCT* provides you with high-quality, class-tested material preprogrammed and fully functional in the WebCT environment. Whether used as an online supplement to either a campus-based or a distance-learning course, our preassembled course content gives you a tremendous head start in developing your own online courses.

- *Instructors 1st*   For qualified adopters, Prentice Hall is proud to introduce Instructors 1st—the first integrated service committed to meeting your customization, support, and training needs for your course.

- *Companion Website for Essentials of Anatomy and Physiology* An exciting new edition of Prentice Hall's Companion Website has been developed specifically for students using the third edition of Essentials of Anatomy and Physiology. In addition to multiple-choice, essay, and short-answer questions, this site's self-grading quizzes offer exercises in labeling and concept mapping for each chapter. Numerous interesting, related Websites are referenced and annotated in the Destinations sections, and our NetSearch offers a convenient gateway to hundreds of other sites of interest.

- *Study Guide*   Designed to help students master the topics and concepts covered in the textbook, the study guide includes a variety of review questions, including labeling, concept mapping, and crossword puzzles, that promote an understanding of body systems. It is keyed to each chapter's learning objectives and parallels the three-level learning system in the textbook.

- *Applications Manual*   This supplement provides students with access to interesting and relevant clinical and diagnostic information. It includes introductory sections about the scientific method and the applications of chemistry and cell biology to clinical work, sections about each body system that parallel the textbook organization and provide more detailed clinical information, a full-color Surface Anatomy and Cadaver Atlas, and Critical-Thinking Questions for each body system. The Applications Manual is fully cross-referenced to the textbook to promote the integration of this material into the course.

- *Student Notetaking Guide*   This unique supplement features selected key art from the text in full four-color on the left-hand side of the page with the entire right side left blank for notes. Instead of scrambling to quickly sketch the images depicted on the transparencies, students can spend the time actually listening to the lecture and jotting down the critical notes right beside the artwork.

- *Anatomy and Physiology Video Tutor* This highly praised, 75-minute videotape focuses on the concepts that both instructors and students consistently identify as the most challenging. Physiological processes are demonstrated through the use of top-quality three-dimensional animations and video footage. On-camera narration and the accompanying frame-references study booklet allow for repeated concept review.

- *New York Times "Themes of the Times" Program* Prentice Hall's unique alliance with *The New York Times* enhances your access to current, relevant information and applications. Articles are selected by the text authors and are compiled into a free supplement that helps instructors make the connection between the classroom and the outside world.

- *Life on the Internet—A Student's Guide* This hands-on supplement brings students up to speed on what the Internet is and how to navigate it.

Products for the Laboratory

- *Laboratory Manual for Anatomy and Physiology* by Roberta Meehan

- *Instructor's Manual to accompany the Laboratory Manual,* by Roberta Meehan

- *Anatomy and Physiology Laboratory Manual* by Michael Wood

- *Instructor's Manual to Accompany Anatomy and Physiology Laboratory Manual* by Michael Wood

ACKNOWLEDGMENTS

Every textbook represents a group effort. Foremost on the list are the faculty and reviewers whose advice, comments, and collective wisdom helped shape this edition. Their interest in the subject, their concern for the accuracy and method of presentation, and their experience with students of widely varying abilities and backgrounds made the review process an educational experience. To these individuals, who carefully recorded their comments, opinions, and sources, we express our sincere thanks and best wishes, We would also like to acknowledge the many users, survey respondents, and focus group members whose advice, comments, and collective wisdom helped shape this text into its final form. Their passion for the subject, their concern for accuracy and method of presentation, and their experience with students of widely varying abilities and backgrounds have made the review process much more fruitful.

The following individuals devoted large amounts of time reviewing drafts of *Essentials of Anatomy and Physiology*:

Reviewers for the Third Edition

Reviewers

Professor Jeffrey Kiggins, *Blue Ridge Community College*
Dr. Jennifer Lundmark, *California State University at Sacramento*
Professor Bert Atsma, *Union County College*
Dr. Ann Wright, *Canisius College*
Professor Luanne Clark, *Lansing Community College*
Dr. Louis Miller, *University of Wisconsin-Stout*
Dr. George Spiegel, Jr., *Mid Plains Community College*
Dr. Barbara Christie-Pope, *Cornell College*

Technical Reviewers

We would also like to thank,
Cynthia Herbrandson, *Kellogg Community College*

Lauren Gollahon and Nathan Collie, *Texas Technical University*, and Christine Iltis, *Salt Lake Community College* for their outstanding technical reviews. Their keen eyes and attention to detail helped maintain the highest standards of accuracy and currency in this new edition.

Reviewers for the Second Edition

Reviewers

Teresa Brandon, *New Mexico State Universiy, Dona Ana Branch Community College*
Cliff Fontenot, *Southeastern Louisiana University*
Lori Garrett, *Danville Area Community College*
Judi Lindsley, *Lourdes College*
Janice Meeking, *Mount Royal College, Calgary*
Izak Paul, *Mount Royal College, Calgary*
Carolyn Rivard, *Fanshawe College*
David Thomas, *Fanshawe College*
Donna Van Wynsberghe, *University of Wisconsin–Milwaukee*
Ann Wright, *Canisius College*
Janice Yoder Smith, *Tarrant County Junior College, NW*

Reviewers for the First Edition

Reviewers

Bert Atsma, *Union County College*
Patricia K. Blaney, *Brevard Community College*
Douglas Carmichael, *Pittsburg State University*
Charles Daniels, *Kapiolani Community College*
Darrell Davies, *Kalamazoo Valley Community College*
Connie S. Dempsey, *Stark State College of Technology*
Marty Hitchcock, *Gwinnetta Technical Institute*
Paul Holdaway, *Harper College*
Drusilla B. Jolly, *Forsyth Technical Community College*
Anne L. Lilly, *Santa Rosa Junior College*
Daniel Mark, *Penn Valley Community College*
Christine Martin, *Stark State College of Technology*
Roxine McQuitty, *Milwaukee Area Technical College*
Sherry Medlar, *Southwestern College*
Margaret Merkely, *Delaware Technical & Community College*
Lewis M. Milner, *North Central Technical College*
Red Nelson, *Westark Community College*
William F. Nicholson, *University of Arkansas–Monticello*
Michael Postula, *Parkland College*
Dell Redding, *Evergreen Valley College*
Cecil Sahadath, *Centennial College*
Brian Shmaefsky, *Kingwood College*
Barbara M. Stout, *Stark State College of Technology*
Dave Thomas, *Fanshawe College*
Caryl Tickner, *Stark State College of Technology*
Frank V. Veselovsky, *South Puget Sound Community College*
Connie Vinton-Schoepske, *Hawkeye Community College*

Focus Group Participants

Bert Atsma, *Union County College*
Connie S. Dempsey, *Stark State College of Technology*
Anne L. Lilly, *Santa Rosa Junior College*
Roxine McQuitty, *Milwaukee Area Technical College*
Barbara M. Stout, *Stark State College of Technology*
Dave Thomas, *Fanshawe College*

Technical Reviewers

Brent DeMars, *Lakeland Community College*
Kathleen Flickinger, *Iowa State University*

Our gratitude is also extended to the many faculty and students at campuses across the United States (and out of the country) who made suggestions and comments that helped us improve this edition of *Essentials of Anatomy and Physiology*.

A textbook has two components: narrative and visual. In preparing the narrative, we were ably assisted by Ray Mullaney, Vice President/Editor in Chief of Development, who played a vital role in shaping this text by helping us keep the text organization, general tone, and level of presentation consistent throughout. The accuracy, currency, and clarity of the clinical material in the text and in the Application Manual reflect the detailed clinical reviews performed by Kathleen Welch, M.D.

Virtually without exception, reviewers stressed the importance of accurate, integrated, and visually attractive illustrations in helping students understand essential material. The creative talents brought to this project by our artist team, William Ober, M.D., and Claire Garrison, R.N., are inspiring and very much appreciated. Bill and Claire worked intimately and tirelessly with us, imparting a unity of vision to the book as a whole while making it both clear and beautiful. Their superb art program is greatly enhanced by the incomparable bone and cadaver photographs of Ralph T. Hutchings, formerly of The College of Surgeons of England, and co-author of the best-selling *McMinns Color Atlas of Human Anatomy*.

We are deeply indebted to the Prentice Hall production staff and Preparé, Inc., whose efforts were so vital to the creation of this edition. Special thanks are due to Fran Daniele for her skillful management of the project through the entire production process. Art Director Kenny Beck and Carole Anson, Director of Design, oversaw the development of the beautiful internal text design, as well as an outstanding cover design. We must also express our appreciation to Crissy Dudonis, project manager, for her work on the numerous print and media supplements that accompany this title and to Susan Zeigler, editorial assistant, who greatly supported our efforts in countless ways.

We would like to thank Mike Banino, media consultant, and Sharon Simpson who worked closely with the talented team at VPG on the Student Tutor CD-ROM. Crissy Dudonis also dispayed her creative talents to create the outstanding website that accompanies this text. We greatly appreciate their contributions. Kate Flickinger played an instrumental role assisting with media products and writing the MediaLabs that appear in the text. Mike was joined by a committed media production group led by Nicole Bush (Assistant Managing Editor, Media), whose eye for detail and skill with project management are both envied and appreciated.

We are grateful to Paul Corey, President of the Engineering, Science, and Mathematics Division of Prentice Hall, and Sheri Snavely, Vice President/Editor in Chief of Life and Geosciences, for their continued enthusiastic support of this project, as well as for our other books at Prentice Hall. We would also like to acknowledge the contributions of Marty McDonald, Senior Marketing Manager, and Jennifer Welchans, Executive Marketing Manager, who keep their fingers on the pulse of the market and help us meet the needs of our users. Above all, thanks to our editor, Halee Dinsey, for her patience in nurturing this project, her tireless efforts to coordinate the various components of the package, and her work on the media supplements and Interactive CD-ROM.

Finally, we would like to thank our families for their love and support during the revision process.

No two people could expect to produce a flawless textbook of this scope and complexity. Any errors or oversights are strictly our own rather than those of the reviewers, artists, or editors. In an effort to improve future editions, we ask that readers with pertinent information, suggestions, or comments concerning the organization or content of this textbook send their remarks to us directly, by email, or care of Halee Dinsey, Senior Editor, Applied Biology, Prentice Hall, One Lake Street, Upper Saddle River, NJ, 07458. Any and all comments and suggestions will be deeply appreciated and carefully considered in the preparation of the next edition.

Frederic H. Martini
Haiku, Hawaii
martini@maui.net

Edwin F. Bartholomew
Lahaina, Hawaii
edbarth@maui.net

Emphasizing Concepts

Chapter Outline and Objectives Each chapter opens with an outline that gives you an overview of the concepts. Integrated objectives help you structure your reading and keep your attention focused on the key points.

5

The Integumentary System

Vocabulary Development

cornu	horn; stratum corneum
cutis	skin; cutaneous
derma	skin; dermis
epi-	above or over; epidermis
facere	to make; cornified
germinare	to start growing; stratum germinativum
keros	horn; keratin
kyanos	blue; cyanosis
luna	moon; lunula
melas	black; melanin
onyx	nail; eponychium
papilla	a nipple-shaped mound; dermal papillae

THE BOOGIE BOARD SURFER "shredding" this wave is probably too busy to consider what contributions and sacrifices are normally made by her skin. This remarkable structure absorbs ultraviolet radiation, prevents dehydration, preserves normal body temperature, and tolerates chafing and abrasion. Although few people think of it in these terms, the skin is actually an organ—the largest organ of the body. In this chapter, we will examine its varied functions.

109

The Peripheral Nervous System • The Autonomic Nervous System • Aging and the Nervous System • Integration with Other Systems • Chapter Review

nervous system (CNS), consisting of the *brain* and the *spinal cord*, integrates and coordinates sensory data and motor commands. The CNS is also the seat of higher functions, such as intelligence, memory, and emotion. All communication between the CNS and the rest of the body occurs over the **peripheral nervous system (PNS)**. The peripheral nervous system includes all the neural tissue *outside* the CNS. Its **afferent division** (*af-*, to + *ferre*, to carry) brings sensory information to the CNS, and its **efferent division** (*ef-*, from) carries motor commands to muscles and glands. Within the efferent division, the **somatic nervous system (SNS)** provides control over skeletal muscle contractions, and the **autonomic nervous system (ANS)**, or *visceral motor system*, provides automatic involuntary regulation of smooth muscle, cardiac muscle, and glandular secretions. The ANS includes a *sympathetic division* and a *parasympathetic division*, which commonly have opposite effects. For example, activity of the sympathetic division accelerates the heart rate whereas the parasympathetic division slows the heart rate.

Cellular Organization in Neural Tissue

The nervous system includes all the neural tissue in the body. Neural tissue, introduced in Chapter 4, consists of two kinds

of cells, *neurons* and *neuroglia*. ∞ p. 102 **Neurons** (*neuro-*, nerve) are the basic units of the nervous system. All neural functions involve the communication of neurons with one another and with other cells. The **neuroglia** (noo-ROG-lē-uh or noo-rō-GLĒ-uh; *glia*, glue) regulate the environment around the neurons, provide a supporting framework for neural tissue, and act as phagocytes. Although they are much smaller cells, neuroglia, also called *glial cells*, far outnumber neurons. Unlike most neurons, most glial cells retain the ability to divide.

NEURONS

The General Structure of Neurons

A "model" neuron has (1) a cell body; (2) several branching, sensitive **dendrites**, which receive incoming signals; and (3) an elongate **axon**, which carries outgoing signals toward (4) one or more **synaptic terminals** (Figure 8-2●). At each synaptic terminal, the neuron communicates with another cell. (Because the synaptic terminals between neurons are rounded, they are also called **synaptic knobs**.) Neurons can have a variety of shapes; Figure 8-2● shows a *multipolar neuron*, the most common type of neuron in the CNS.

The cell body of a typical neuron contains a large, round nucleus with a prominent nucleolus. There are usually no centrioles, organelles that in other cells form the spindle fibers that move chromosomes during cell division. Most neurons lose their centrioles during differentiation and become incapable of undergoing mitosis. As a result, most neurons lost to injury or disease cannot be replaced.

The cell body also contains the organelles that provide energy and synthesize organic compounds. The numerous mitochondria, free and fixed ribosomes, and membranes of

● Figure 8-2 The Anatomy of a Multipolar Neuron
The anatomy of a multipolar neuron that innervates a skeletal muscle. (LM × 1400)

Labels: Nissl bodies, Nucleolus, Nucleus, Axon hillock, Collateral, Skeletal muscle, Axon, Mitochondrion, Synaptic terminals, Cell body, Dendrite

22

Concept Links The chain-link icon provides a quick visual signal that new material being presented is related to or builds on earlier discussions.

Vocabulary Development This section lists the important word roots that form the basis of the vocabulary in the chapter.

Figure Reference Locators Red dots serve as place markers, making it easy to return to your spot in the narrative after you've studied an illustration.

Key Terms The most important new terms are highlighted in **bold type** and often include the pronunciation. Key terms are also listed at the end of the chapter for easy review.

CHAPTER REVIEW

Key Terms

Visualizing Structure & Function

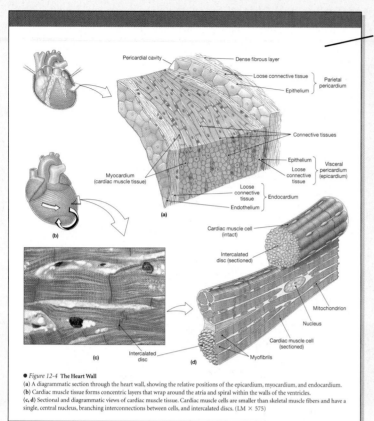

● *Figure 12-4* **The Heart Wall**
(a) A diagrammatic section through the heart wall, showing the relative positions of the epicardium, myocardium, and endocardium.
(b) Cardiac muscle tissue forms concentric layers that wrap around the atria and spiral within the walls of the ventricles.
(c, d) Sectional and diagrammatic views of cardiac muscle tissue. Cardiac muscle cells are smaller than skeletal muscle fibers and have a single, central nucleus, branching interconnections between cells, and intercalated discs. (LM × 575)

Outstanding Anatomy Art and Photos To understand physiology, you must be able to visualize structures in the human body. Macro-to-micro drawings, coupled with histology or electron micrographs, make the details of anatomy easy to understand.

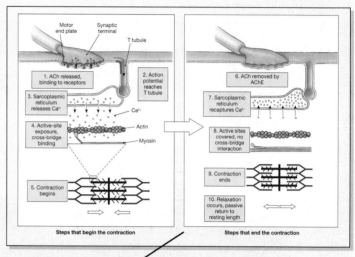

Conceptual Diagrams Show Physiology Physiological processes are easier to understand with flowcharts and diagrams that link structure and function.

Concept Check Questions These questions, located at the ends of major sections in the narrative, will help you assess your understanding of the basic concepts addressed in the previous pages.

CONCEPT CHECK QUESTIONS
Answers on page 222

❶ How would severing the tendon attached to a muscle affect the ability of the muscle to move a body part?

❷ Why does skeletal muscle appear striated when viewed through a microscope?

❸ Where would you expect the greatest concentration of calcium ions in resting skeletal muscle to be?

Answers to Concept Check Questions

Page 182
1. Since tendons attach muscles to bones, severing the tendon would disconnect the muscle from the bone. When the muscle contracted, nothing would happen.
2. Skeletal muscle appears striated when viewed under the microscope because this muscle is composed of the myofilaments actin and myosin, which have an arrangement that produces a banded appearance in the muscle.
3. You would expect to find the greatest concentration of calcium ions in the cisternae of the sarcoplasmic reticulum of the muscle.

Page 184

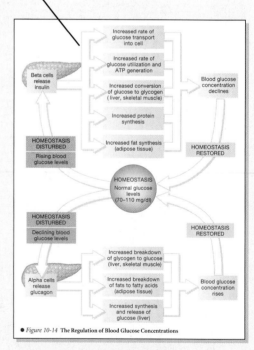

● *Figure 10-14* **The Regulation of Blood Glucose Concentrations**

Answers to Concept Check Questions The answers to concept check questions are provided at the end of each chapter.

Relating Clinical Examples

Several inherited disorders are characterized by the production of abnormal hemoglobin. Two of the best known are *thalassemia* and *sickle-cell anemia (SCA)*.

The various forms of **thalassemia** (thal-ah-SĒ-mē-uh) result from an inability to produce adequate amounts of the globular protein components of hemoglobin. As a result, the rate of RBC production is slowed, and the mature RBCs are fragile and short-lived. The scarcity of healthy RBCs reduces the oxygen-carrying capacity of the blood and leads to problems in growth and development. Individuals with severe thalassemia must undergo frequent *transfusions*—the administration of blood components—to maintain adequate numbers of RBCs in the blood.

Sickle-cell anemia (SCA) results from a mutation affecting the amino acid sequence of one pair of the globular proteins of the hemoglobin molecule. When blood contains an abundance of oxygen, the hemoglobin molecules and the RBCs that carry them appear normal (Figure 11-3a●). But when the defective hemoglobin gives up enough of its stored oxygen, the adjacent hemoglobin molecules interact, and the cells change shape, becoming stiff and markedly curved (Figure 11-3b●). This "sickling" does not affect the oxygen-carrying capabilities of the RBCs, but it makes them more fragile and easily damaged. Moreover, when an RBC that has folded to squeeze into a narrow capillary delivers its oxygen to the surrounding tissue, the cell can become stuck as sickling occurs. This blocks blood flow and nearby tissues become oxygen-starved.

Clinical Note Clinical or health-related topics of particular importance are presented in boxes set off from the main text. These essays cover major diseases, such as lung cancer and AIDS, in addition to other subjects of special interest, such as clinical procedures.

 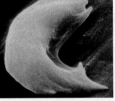

(a) Normal RBC **(b) Sickled RBC**

● *Figure 11-3* **Sickling in Red** ... (a) When fully oxygenated, the ... individual with the sickling tra ... appearance as normal RBCs. (... concentrations the RBCs chan ... becoming more rigid and shar ... (SEM × 5200)

Clinical Discussions Important clinical topics presented in context are set off by an icon and title. These topics have been selected not only for their medical importance, but also to show how an understanding of abnormal conditions can shed light on normal functions, and vice versa.

STRESS AND THE IMMUNE RESPONSE
Interleukin-1 is one of the first cytokines produced as part of the immune response. In addition to promoting inflammation, it also stimulates production of adenocorticotropic hormone (ACTH) by the anterior pituitary gland. This, in turn, leads to the secretion of glucocorticoids by the adrenal cortex. ∞ p. 327 The anti-inflammatory effects of the glucocorticoids may help control the intensity of the immune response. However, the long-term secretion of glucocorticoids, as in chronic stress, can inhibit the immune response and lower resistance to disease. ∞ p. 334 Glucocorticoids depress inflammation, inhibit phagocytes, and reduce interleukin production. The mechanisms involved are still under investigation. It is well known, however, that immune system depression due to chronic stress represents a serious threat to health.

Focus Boxes These essays provide illustrated summaries of important processes or clinical conditions. Topics in the text include visual accommodation problems (p. 293), urine formation (p. 559), and bone fractures (p. 135).

FOCUS

Accommodation Problems

In the normal eye, when the ciliary muscles are relaxed and the lens is flattened, a distant image will be focused on the retinal surface (Figure 9-16a●), a condition called *emmetropia* (*emmetro*-, proper measure). However, irregularities in the shape of the lens or cornea can affect the clarity of the visual image. This condition, called *astigmatism*, can usually be corrected by glasses or special contact lenses.

Figure 9-16● diagrams two other common problems with the accommodation mechanism.

If the eyeball is too deep, the image of a distant object will form in front of the retina, and the retinal picture will be blurry and out of focus (Figure 9-16b●). Vision at close range will be normal, because the lens will be able to round up as needed to focus the image on the retina. As a result, such individuals are said to be "nearsighted." Their condition is more formally termed *myopia* (*myein*, to shut + *ops*, eye). Myopia can be corrected by placing a diverging lens in front of the eye (Figure 9-16c●).

If the eyeball is too shallow, *hyperopia results* (Figure 9-16d●). The ciliary muscles must contract to focus even a distant object on the retina, and at close range the lens cannot provide enough refraction. These individuals are said to be "farsighted" because they can see distant objects most clearly. Older individuals become farsighted as their lenses lose elasticity; this form of hyperopia is called *presbyopia* (*presbys*, old man). Hyperopia can be treated by placing a converging lens in front of the eye (Figure 9-16e●).

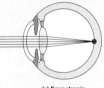 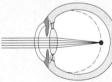 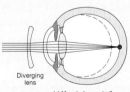

(a) Emmetropia **(b) Myopia** Diverging lens **(c) Myopia (corrected)**

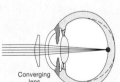

Converging lens **(d) Hyperopia** **(e) Hyperopia (corrected)**

● *Figure 9-16* **Visual Abnormalities** (a) In normal vision, the lens focuses the visual image on the retina. One common problem of accommodation involves (b) an inability to lengthen the focal distance enough to focus the image of a distant object on the retina—myopia. (c) A diverging lens is used to correct myopia. (d) Another accommodation problem is an inability to shorten the focal distance adequately for near objects—hyperopia. (e) A converging lens is used to correct hyperopia.

Related Clinical Terms

ankylosis (ang-ki-LŌ-sis): An abnormal fusion between articulating bones in response to trauma and friction within a joint.
arthritis (ar-THRĪ-tis): Rheumatic diseases that affect synovial joints. Arthritis always involves damage to the articular cartilages, but the specific cause can vary. The diseases of arthritis are usually classified as either *degenerative* or *inflammatory*.
arthroscopic surgery: The surgical modification of a joint by using an *arthroscope* (a fiberoptic instrument used to view the inside of joint cavities).
bursitis: An inflammation of a bursa, causing pain whenever the associated tendon or ligament moves.
carpal tunnel syndrome: An inflammation of the tissues at the anterior ... g compression of ... ndo ...

... tebra ... the ... inn ... bey ... **kyp** ... of t ... bac ... **lor** ... of t ... **lux** ... tio ... ov ... d ... c ... a ... n ...

Related Clinical Terms New to the third edition is a list of relevant clinical terms at the end of each chapter.

Reviewing the Concepts

CHAPTER REVIEW

Key Terms

acidosis 569	filtrate 550	loop of Henle 550	ureters 562
aldosterone 558	glomerular filtration rate 556	micturition 563	urinary bladder.............563
alkalosis 569	glomerulus 550	nephron 550	urine.............555
angiotensin II 560	juxtaglomerular apparatus ...553	peritubular capillaries 554	
buffer system 570	kidney 548	renin.............560	

Key Terms Important terms introduced in the chapter are listed here with a page reference for quick review in context with the relevant material.

Summary Outline

INTRODUCTION **548**

1. The functions of the **urinary system** include (1) eliminating organic waste products, (2) regulating plasma concentrations of ions, (3) regulating blood volume and pressure by adjusting the volume of water lost and releasing hormones, (4) stabilizing blood pH, and (5) conserving nutrients.

THE ORGANIZATION OF THE URINARY SYSTEM **548**

1. The urinary system includes the kidneys, the ureters, the urinary bladder, and the urethra. The kidneys produce urine (a fluid containing water, ions, and soluble compounds); during urination urine is forced out of the body. *(Figure 18-1)*

THE KIDNEYS **548**

1. The left **kidney** extends superiorly slightly more than the right kidney. Both lie in a *retroperitoneal* position. *(Figure 18-2)*

   Superficial and Sectional Anatomy **549**

2. A fibrous renal capsule surrounds each kidney. The **hilus** provides entry for the *renal artery* and exit for the *renal vein* and *ureter.*

3. The ureter is connected to the **renal pelvis**. This chamber branches into two **major calyces**, each connected to four or five **minor calyces**, which enclose the **renal papillae**. Urine production begins in **nephrons**. *(Figure 18-3)*

   The Nephron **550**

4. The nephron (the basic functional unit in the kidney) includes the *renal corpuscle* and a **renal tubule**, which empties into the **collecting system** through a *collecting duct*. From the renal corpuscle, the filtrate travels through the *proximal convoluted tubule*, the *loop of Henle*, and the *distal convoluted tubule*. *(Figures 18-3c,18-4)*

5. Nephrons are responsible for the (1) production of **filtrate**, (2) reabsorption of nutrients, and (3) reabsorption of water and ions.

6. The renal tubule begins at the **renal corpuscle**. It includes a knot of intertwined capillaries called the **glomerulus** surrounded by the **Bowman's capsule**. Blood arrives from the *afferent arteriole* and departs in the *efferent arteriole*. *(Figure 18-5)*

7. At the glomerulus, **podocytes** cover the basement membrane of the capillaries that project into the **capsular space**. The processes of the podocytes are separated by narrow slits. *(Figure 18-5)*

8. The **proximal convoluted tubule (PCT)** actively reabsorbs nutrients, plasma proteins, and electrolytes from the filtrate. They are then released into the surrounding interstitial fluid. *(Figure 18-4)*

9. The **loop of Henle** includes a *descending limb* and an *ascending limb*. The ascending limb is impermeable to water and solutes; the descending limb is permeable to water. *(Figure 18-4)*

574

10. The ascending limb delivers fluid to the **distal convoluted tubule (DCT)**, which actively secretes ions and reabsorbs sodium ions from the urine. The **juxtaglomerular apparatus**, which releases renin and erythropoietin, is located at the start of the DCT. *(Figure 18-5a)*

11. The **collecting ducts** receive urine from nephrons and merge into a **papillary duct** that delivers urine to a minor calyx. The collecting system makes final adjustments to the urine by reabsorbing water or reabsorbing or secreting various ions.

    The Blood Supply to the Kidneys **553**

12. The blood vessels of the kidneys include the **interlobar, arcuate,** and **interlobular arteries**, and the **interlobar, arcuate,** and **interlobular veins**. Blood travels from the a... **peritubular capillaries** and the ... capillaries of the vasa recta and ... tial fluid that surrounds the ne...

BASIC PRINCIPLES OF URINE ...

1. The primary purpose in **urine** ... nation of dissolved solutes, pri... as **urea, creatinine**, and **uric ac**...

2. Urine formation involves **filt**... *(Tables 18–1,18-2)*

   Filtration at the Glomer...

3. Glomerular filtration occurs ... glomerulus into the capsular s... the glomerular capillaries. The ... amount of filtrate produced in t... alters the **filtration** (blood) pr... kidney function.

4. Dropping filtration pressures s... to release **renin**. Renin release ... blood pressure.

   Reabsorption and Secre... **Along the Renal Tubule**...

5. The cells of the PCT norma... of the filtrate produced in the re... ents, sodium and other ions, a... secretes various substances.

6. Water and ions are reclaim... The ascending limb reabsorb... descending limb reabsorbs wate... la encourages the osmotic flow...

Summary Outline This outline provides a detailed summary of all the sections in the chapter—including page references and all corresponding figure and table numbers.

Review Questions

Level 1: Reviewing Facts and Terms

Match each item in column A with the most closely related item in column B. Use letters for answers in the spaces provided.

COLUMN A

____ 1. urination
____ 2. renal capsule
____ 3. hilus
____ 4. medulla
____ 5. nephrons
____ 6. renal corpuscle
____ 7. external urethral sphincter
____ 8. internal urethral sphincter
____ 9. aldosterone
____ 10. podocytes
____ 11. efferent arteriole
____ 12. afferent arteriole
____ 13. vasa recta

COLUMN B

a. site of urine production
b. capillaries around loop of Henle
c. causes sensation of thirst
d. accelerated sodium reabsorption
e. voluntary control
f. glomerular epithelium
g. fibrous covering
h. blood leaves glomerulus
i. dominant cation in ICF
j. renal pyramids
k. blood to glomerulus
l. interstitial fluid
m. contains glomerulus

Level 2: Reviewing Concepts

30. The urinary system regulates blood volume and pressure by:
    (a) adjusting the volume of water lost in the urine
    (b) releasing erythropoietin
    (c) releasing renin
    (d) a, b, and c are correct

33. When pure water is consumed:
    (a) the ECF becomes hypertonic wi...
    (b) the ECF becomes hypotonic wit...
    (c) the ICF becomes hypotonic with...
    (d) water moves from the ICF into t...

34. Increasing or decreasing th...

Level 3: Critical Thinking and Clinical Applications

40. Long-haul trailer truck drivers are on the road for long periods of time between restroom stops. Why might that lead to kidney problems?

41. For the past week, Susan has felt a burning sensation in the urethral area when she urinates. She checks her temperature and finds that

she has a low-grade fever. What unu... present in her urine?

42. *Mannitol* is a sugar that is filtered b... What effect would drinking a soluti... ume of urine produced?

Review Questions Questions are organized in a three-tiered system to help you build your knowledge:

Level 1 questions allow you to test your recall of the chapter's basic information and terminology.

Level 2 questions help you check your grasp of concepts and your ability to integrate ideas presented in different parts of the chapter.

Level 3 questions let you develop your powers of reasoning and analysis by applying chapter material to plausible real-world and clinical situations.

ANSWERS TO CHAPTER REVIEW QUESTIONS

CHAPTER 1

Level 1: Reviewing Facts and Terms

1. g 2. d 3. a 4. j 5. b 6. l 7. n 8. f 9. h 10. e 11. c 12. o 13. k 14. i 15. m 16. c 17. d 18. b 19. c 20. c 21. b

Level 2: Reviewing Concepts

22. responsiveness, adaptability, growth, reproduction, movement, metabolism, absorption, respiration, excretion
23. molecules-cell-tissues-organs-organ systems—organisms
24. Homeostatic regulation refers to adjustments in physiological systems that are responsible for the preservation of homeostasis.
25. In negative feedback, a variation outside normal ranges triggers an au...

follows: 2 in the first level, 8 in the second level, and 6 in the third level. To achieve a full 8 electrons in the third level, the sulfur atom could accept 2 electrons in an ionic bond or share 2 electrons in a covalent bond. Since hydrogen atoms can share 1 electron in a covalent bond, the sulfur atom would form 2 covalent bonds, 1 with each of 2 hydrogen atoms.

27. If a person exhales large amounts of carbon dioxide, the equilibrium will shift to the left and the level of hydrogen ion in the blood will decrease. A decrease in the amount of hydrogen ion will cause the pH to rise.

CHAPTER 3

Level 1: Reviewing Facts and Terms

1. e 2. d 3. h 4. a 5. f 6. b 7. g 8. c 9. m 10. k 11. q

Answers to Chapter Review Questions For easy reference, the chapter review questions are answered in Appendix I, p. A-1.

Integrating Concepts

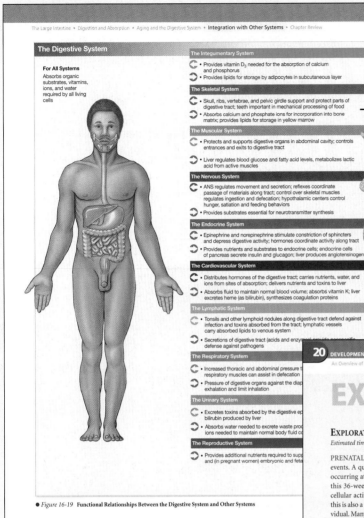

The Digestive System

For All Systems
Absorbs organic substrates, vitamins, ions, and water required by all living cells

The Integumentary System
- Provides vitamin D₃ needed for the absorption of calcium and phosphorus
- Provides lipids for storage by adipocytes in subcutaneous layer

The Skeletal System
- Skull, ribs, vertebrae, and pelvic girdle support and protect parts of digestive tract; teeth important in mechanical processing of food
- Absorbs calcium and phosphate ions for incorporation into bone matrix; provides lipids for storage in yellow marrow

The Muscular System
- Protects and supports digestive organs in abdominal cavity; controls entrances and exits to digestive tract
- Liver regulates blood glucose and fatty acid levels, metabolizes lactic acid from active muscles

The Nervous System
- ANS regulates movement and secretion; reflexes coordinate passage of materials along tract; control over skeletal muscles regulates ingestion and defecation; hypothalamic centers control hunger, satiation and feeding behaviors
- Provides substrates essential for neurotransmitter synthesis

The Endocrine System
- Epinephrine and norepinephrine stimulate constriction of sphincters and depress digestive activity; hormones coordinate activity along tract
- Provides nutrients and substrates to endocrine cells; endocrine cells of pancreas secrete insulin and glucagon; liver produces angiotensinogen

The Cardiovascular System
- Distributes hormones of the digestive tract; carries nutrients, water, and ions from sites of absorption; delivers nutrients and toxins to liver
- Absorbs fluid to maintain normal blood volume; absorbs vitamin K; liver excretes heme (as bilirubin), synthesizes coagulation proteins

The Lymphatic System
- Tonsils and other lymphoid nodules along digestive tract defend against infection and toxins absorbed from the tract; lymphatic vessels carry absorbed lipids to venous system
- Secretions of digestive tract (acids and enzymes) provide nonspecific defense against pathogens

The Respiratory System
- Increased thoracic and abdominal pressure through respiratory muscles can assist in defecation
- Pressure of digestive organs against the diaphragm can affect exhalation and limit inhalation

The Urinary System
- Excretes toxins absorbed by the digestive epithelium; excretes bilirubin produced by liver
- Absorbs water needed to excrete waste products; absorbs ions needed to maintain normal body fluid concentrations

The Reproductive System
- Provides additional nutrients required to support gamete production and (in pregnant women) embryonic and fetal development

● *Figure 16-19* **Functional Relationships Between the Digestive System and Other Systems**

Systems Integrators Found at the end of each body system chapter, systems integrators show how one particular system interacts with each of the body's other systems.

MediaLab New to this edition are the MediaLabs that appear at the end of eleven chapters. These exercises provide an opportunity to explore interesting and thought-provoking situations by supplementing the information presented in the text with additional materials found on the World Wide Web. The theme of the MediaLabs is integration—how the body's systems interact and work together. Many of the exercises deal with the clinical implications of these interactions, focusing on how a problem in one system can affect other systems, and ultimately the whole body.

---

20 DEVELOPMENT AND INHERITANCE
An Overview of Topics in Development • Fertilization • The First Trimester • The Second and Third Trimesters • Labor and Delivery

EXPLORE *MediaLab*

EXPLORATION #1
Estimated time for completion: 10 minutes

PRENATAL DEVELOPMENT is an amazing series of events. A quick look at Table 20-2 reveals how much is occurring at the cellular, tissue, and organ level during this 36-week period. With so much biochemical and cellular activity, you would be correct in thinking that this is also a very sensitive period for the developing individual. Many seemingly innocuous compounds, referred to as teratogens, can have devastating effects on the regulated pattern of development. Alcohol is one of the most common and well-understood teratogens. Alcohol restricts blood flow to the placenta, depriving the developing cells of the fetus of much-needed oxygen. It also crosses the placenta, damaging fetal cells. With chronic exposure, the fetus may develop Fetal Alcohol Syndrome (FAS), Alcohol-Related Neurodevelopmental Disorder (ARND), and Alcohol-Related Birth Defects (ARBD). These three syndromes describe differing degrees of facial abnormalities, growth deficiencies, and central nervous system dysfunction. Prepare a chart of prenatal development similar to Table 20-2. On your chart, indicate when excessive alcohol intake might cause each of these effects. What other detrimental effects would you expect to see? You should be able to list at least one effect for each system. To complete your chart, go to Chapter 20 at the Companion Web site, visit the MediaLab section, and click on the key word "FAS." The information compiled by the ADA, Division of Alcohol and Drug Abuse, will help you add to your chart.

Drinking for two. This beer affects not only the mother's body systems but also the fetus that she is carrying.

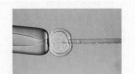

A donor nucleus being injected into a sheep egg for cloning.

EXPLORATION #2
Estimated time for completion: 10 minutes

WITH THE SUCCESSES of the Human Genome Project (see p. 635), the news is full of debates on the issue of human cloning. Two types of cloning are currently being pursued in research laboratories: reproductive cloning and therapeutic cloning. Reproductive cloning, the goal of which is to create a genetically identical copy of an organism, has been successfully performed on a variety of animals, including mice, sheep, and cows. Although reproductive cloning has not yet been attempted on humans, advancements in this technique are occurring rapidly. We may soon have the ability to create genetically identical copies of humans. Therapeutic cloning of human cells has already been successful. The goal of therapeutic cloning is not to create a new individual, but rather to create a viable embryo from which stem cells can be harvested. Figure 20-4● depicts the stage of development at which these cells are removed. Harvesting stem cells destroys the embryo, but provides a population of potentially therapeutically useful tissue cells. A good pictorial explanation of this procedure is provided by *Science* magazine at http://www.sciam.com/explorations/2001/112401ezzell/how.html. As these cloning regimes become mainstream, we are faced with the ethics of using them. What are your thoughts on these issues? Should scientists be able to continue addressing the reproductive cloning of humans? Why or why not? Is therapeutic cloning a medical breakthrough or a moral issue in that viable embryos are destroyed? After preparing a short position paper on this topic, go to Chapter 20 at the Companion Web site, visit the MediaLab section, and click on the key word "cloning." Read through a NOVA staff writer's interview of Dr. Don Wolf, Director of the Andrology/Embryology Laboratory at Oregon Health and Sciences University in Portland. Do you agree with Dr. Wolf's views? Strengthen your position paper with quotes and information from this article.

642

1

An Introduction to Anatomy and Physiology

CHAPTER OUTLINE AND OBJECTIVES

Vocabulary Development

bios ....life; *biology*
cardium ....heart; *pericardium*
dorsum ....back; *dorsal*
homeo- ....unchanging; *homeostasis*
-logy...study of; *biology*
medianus ......situated in the middle; *median*
paries ....wall; *parietal*
pathos ....disease; *pathology*
peri-..around; *perimeter*
pronus ....inclined forward; *prone*
supinus ....lying on the back; *supine*
-stasis ....standing; *homeostasis*
venter ....belly or abdomen; *ventral*

*E*VERY DAY WE PERFORM *a balancing act. An amazingly complex per-formance, it goes on continu-ously with every part of our physical being involved, from individual atoms to the largest organs and organ sys-tems. Injury, disease, and other stress-es, whether they occur inside us or outside, threaten this delicate bal-ance. Adjusting to these changes is crucial, for if the body loses its bal-ance, it may plunge out of control into dysfunction, illness, or even death.*

1 AN INTRODUCTION TO ANATOMY AND PHYSIOLOGY

The Sciences of Anatomy and Physiology • Levels of Organization • An Introduction to Organ Systems • Homeostasis and System Integration

THE WORLD AROUND US contains an enormous diversity of living organisms that vary widely in appearance and lifestyle. Despite their obvious differences, however, all living things perform the same basic functions:

- **Responsiveness**. Organisms respond to changes in their immediate environment; this property is also called *irritability*. You move your hand away from a hot stove, your dog barks at approaching strangers, fish are scared by loud noises, and tiny amoebas glide toward potential prey. Organisms also make longer-term changes as they adjust to their environments. For example, an animal may grow a heavier coat of fur as winter approaches, or it may migrate to a warmer climate. The capacity to make such adjustments is termed *adaptability*.

- **Growth**. Over a lifetime, organisms grow larger, increasing in size through an increase in the size or number of *cells*, the simplest units of life. Familiar organisms, such as dogs, cats, and people, are composed of billions of cells. In such multicellular organisms, the individual cells become specialized to perform particular functions. This specialization is called *differentiation*.

- **Reproduction**. Organisms reproduce, creating subsequent generations of similar organisms.

- **Movement**. Organisms are capable of producing movement, which may be internal (transporting food, blood, or other materials inside the body) or external (moving through the environment).

- **Metabolism**. Organisms rely on complex chemical reactions to provide the energy for responsiveness, growth, reproduction, and movement. They must also synthesize complex chemicals, such as proteins. *Metabolism* refers to all of the chemical operations under way in the body. Normal metabolic operations require the absorption of materials from the environment. To generate energy efficiently, most cells require various nutrients, as well as oxygen, a gas. *Respiration* refers to the absorption, transport, and use of oxygen by cells. Metabolic operations often generate unneeded or potentially harmful waste products that must be eliminated through the process of *excretion*.

For very small organisms, absorption, respiration, and excretion involve the movement of materials across exposed surfaces. But creatures larger than a few millimeters thick seldom absorb nutrients directly from their environment. For example, humans cannot absorb steaks, apples, or ice cream without processing them first. That processing, called *digestion*, occurs in specialized areas where complex foods are broken down into simpler components that can be absorbed easily. Respiration and excretion are also more complicated for large organisms. Humans have specialized structures responsible for gas exchange (lungs) and excretion (kidneys). Although digestion, respiration, and excretion occur in different parts of the body, the cells of the body cannot travel to one place for nutrients, another for oxygen, and a third to get rid of waste products. Instead, individual cells remain where they are but communicate with other areas of the body through an internal transport system, or circulation. For example, your blood absorbs the waste products released by each of your cells and carries those wastes to the kidneys for excretion.

Biology, the study of life, includes many subspecialties. This text considers two biological subjects: **anatomy** (ah-NAT-o-mē) and **physiology** (fiz-ē-OL-o-jē). Over the course of 20 chapters, you will become familiar with the basic anatomy and physiology of the human body.

The Sciences of Anatomy and Physiology

The word *anatomy* has Greek origins, as do many other anatomical terms and phrases. A literal translation would be "a cutting open." **Anatomy** is the study of internal and external structure and the physical relationships between body parts. **Physiology**, another word derived from Greek, is the study of how living organisms perform their vital functions. The two subjects are interrelated. Anatomical information provides clues about probable functions, and physiological mechanisms can be explained only in terms of the underlying anatomy.

The link between structure and function is always present but not always understood. For example, the anatomy of the heart was clearly described in the fifteenth century, but almost 200 years passed before anyone realized that it pumped blood. This text will familiarize you with basic anatomy and give you an appreciation of the physiological processes that make human life possible. The information will enable you to understand many kinds of disease processes and will help you make informed decisions about your personal health.

ANATOMY

Anatomy can be divided into microscopic anatomy or macroscopic (gross) anatomy on the basis of the degree of structural detail under consideration. Other anatomical specialties focus on specific processes or medical applications.

Microscopic Anatomy

Microscopic anatomy deals with structures that cannot be seen without magnification. The boundaries of microscopic anatomy are established by the limits of the equipment used. A light microscope reveals basic details about cell structure, whereas an electron microscope can visualize individual molecules only a few nanometers (nm) across. As we proceed through the text, we will be considering details at all levels, from macroscopic to microscopic. (Readers unfamiliar with the terms used to describe measurements and weights over this size range should consult the tables in Appendix III.)

Microscopic anatomy can be subdivided into specialties that consider features within a characteristic range of sizes. **Cytology** (sī-TOL-o-jē) analyzes the internal structure of individual **cells**. The trillions of living cells in our bodies are composed of chemical substances in various combinations, and our lives depend on the chemical processes occurring in those cells. For this reason we will consider basic chemistry (Chapter 2) before examining cell structure (Chapter 3).

Histology (his-TOL-o-jē) takes a broader perspective and examines **tissues**, groups of specialized cells and cell products that work together to perform specific functions (Chapter 4). Tissues combine to form **organs**, such as the heart, kidney, liver, and brain. Many organs can be examined without a microscope, so at the organ level we cross the boundary into gross anatomy.

Gross Anatomy

Gross anatomy, or *macroscopic anatomy*, considers features visible with the unaided eye. There are many ways to approach gross anatomy. **Surface anatomy** refers to the study of general form and superficial markings. **Regional anatomy** considers all of the superficial and internal features in a specific region of the body, such as the head, neck, or trunk. **Systemic anatomy** considers the structure of major *organ systems*, which are groups of organs that function together in a coordinated manner. For example, the heart, blood, and blood vessels form the *cardiovascular system*, which circulates oxygen and nutrients throughout the body.

PHYSIOLOGY

Physiology examines the function of anatomical structures; it considers the physical and chemical processes responsible for the characteristics of life, or vital functions. Because these vital functions are complex and much more difficult to examine than are most anatomical structures, the science of physiology includes even more specialties than does the science of anatomy.

Human physiology is the study of the functions of the human body. The cornerstone of human physiology is **cell physiology**, the study of the functions of living cells. Cell physiology includes events at the chemical or molecular levels—both chemical processes within cells and between cells. **Special physiology** is the study of the physiology of specific organs. Examples are renal physiology (kidney function) and cardiac physiology (heart function). **System physiology** considers all aspects of the function of specific organ systems. Respiratory physiology and reproductive physiology are examples. **Pathological physiology**, or **pathology** (pah-THOL-o-jē), studies the effects of diseases on organ or system functions. (The Greek word *pathos* means disease.) Modern medicine depends on an understanding of both normal and pathological physiology; not only what is wrong but also how to correct it.

Special topics in physiology address specific functions of the human body as a whole. These specialties focus on physiological interactions between multiple-organ systems. For example, exercise physiology studies the physiological adjustments to exercise. Many other such applied topics are discussed in boxes throughout the book.

CONCEPT CHECK QUESTIONS
Answers on page 25

❶ How are vital functions such as growth, responsiveness, reproduction, and movement dependent on metabolism?

❷ Would a histologist more likely be considered a specialist in microscopic anatomy or in gross anatomy? Why?

Levels of Organization

When considering the human body from the microscopic to macroscopic scales, we are examining several levels of organization. Figure 1-1● presents the relationships among the various levels of organization using the cardiovascular system as an example.

- **Chemical**, or **Molecular**, **Level**. *Atoms*, the smallest stable units of matter, combine to form *molecules* with complex shapes. Even at this simplest level, the specialized shape of a molecule determines its function. This is the chemical, or molecular, level of organization.

- **Cellular Level**. Different molecules interact to form *organelles*, such as the protein filaments found in muscle cells. Each type of organelle has a specific function. For example, interactions among protein filaments produce the contractions of

1 AN INTRODUCTION TO ANATOMY AND PHYSIOLOGY

The Sciences of Anatomy and Physiology • Levels of Organization • An Introduction to Organ Systems • Homeostasis and System Integration

● *Figure 1-1* **Levels of Organization**
Interacting atoms form molecules that combine to form cells, such as heart muscle cells. Groups of similar cells combine to form tissues with specific functions, such as heart muscle tissue. Organs, such as the heart, are composed of different tissues. The heart is one component of the cardiovascular system, which also includes the blood and blood vessels. All of the organ systems combine and interact to create an organism, a living human being.

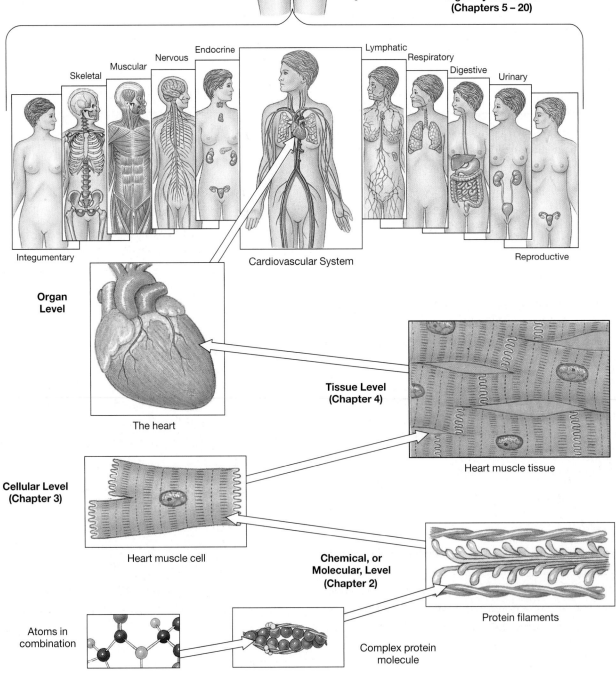

Organism Level

Organ System Level (Chapters 5 – 20)

Skeletal — Muscular — Nervous — Endocrine — Lymphatic — Respiratory — Digestive — Urinary

Integumentary

Cardiovascular System

Reproductive

Organ Level

The heart

Tissue Level (Chapter 4)

Heart muscle tissue

Cellular Level (Chapter 3)

Heart muscle cell

Chemical, or Molecular, Level (Chapter 2)

Protein filaments

Atoms in combination

Complex protein molecule

muscle cells in the heart. Organelles perform the vital functions that keep cells alive. *Cells*, the smallest living units in the body, represent the cellular level of organization.

- **Tissue Level**. A *tissue* is composed of similar cells working together to perform a specific function. Heart muscle cells form *cardiac muscle tissue*, an example of the tissue level of organization.

- **Organ Level**. An *organ* consists of two or more different tissues that work together to perform specific functions. Layers of muscle and other tissues form the wall of the heart, a hollow, three-dimensional organ. This is an example of the organ level of organization.

- **Organ System Level**. Organs interact in *organ systems*. Each time it contracts, the heart pushes blood into a network of blood vessels. Together, the heart, blood, and blood vessels form the *cardiovascular system*, an example of the organ system level of organization.

- **Organism Level**. All of the organ systems of the body work together to maintain life and health. This brings us to the highest level of organization, that of the *organism*—in this case, a human being.

Each level of organization depends on the others, and damage at the cellular, tissue, or organ level can affect the entire system. For example, a chemical change in heart muscle cells can cause abnormal contractions or even stop the heartbeat. Physical damage to the muscle tissue, as in a chest wound, can make the heart ineffective even when most of the heart muscle cells are intact and uninjured. An inherited abnormality in heart structure can make it an ineffective pump, although the muscle cells and muscle tissue are perfectly normal. Because all parts of a system are interdependent, damage to one component will ultimately affect the system as a whole. For example, the heart cannot pump blood effectively after a massive blood loss if there is not enough blood to fill the circulatory system. If the heart cannot pump and blood cannot flow, oxygen and nutrients cannot be distributed. In a very short time, the cardiac muscle tissue begins to break down as individual muscle cells die from oxygen and nutrient starvation. Finally, because all of the body systems are interdependent, disruption of one system will affect all others. If the heart stops pumping blood, the damage will not be restricted to the cardiovascular system; unless something is done quickly, the person will die from the resulting damage to cells, tissues, and organs throughout the body.

An Introduction to Organ Systems

Figure 1-2● introduces the 11 organ systems in the human body and their major functions. These organ systems are (1) the integumentary system, (2) the skeletal system, (3) the muscular system, (4) the nervous system, (5) the endocrine system, (6) the cardiovascular system, (7) the lymphatic system, (8) the respiratory system, (9) the digestive system, (10) the urinary system, and (11) the reproductive system.

Homeostasis and System Integration

Organ systems are interdependent, interconnected, and packaged together in a relatively small space. The cells, tissues, organs, and systems of the body live together in a shared environment, like the inhabitants of a large city. City dwellers breathe the city air and drink the water provided by the local water company; cells in the human body absorb oxygen from the body fluids that surround them. All living cells are in contact with blood or some other body fluid, and any change in the composition of these fluids will affect them in some way. For example, changes in the temperature or salt content of the blood could cause anything from a minor adjustment (heart muscle tissue contracts more often, and the heart rate goes up) to a total disaster (the heart stops beating altogether).

HOMEOSTATIC REGULATION

A variety of physiological mechanisms act to prevent potentially dangerous changes in the environment inside the body. **Homeostasis** (*homeo*, unchanging + *stasis*, standing) refers to the existence of a stable internal environment. To survive, every living organism must maintain homeostasis. The term **homeostatic regulation** refers to the adjustments in physiological systems that preserve homeostasis.

Homeostatic regulation usually involves (1) a **receptor** that is sensitive to a particular environmental change or *stimulus*; (2) a **control center**, or *integration center*, which receives and processes the information from the receptor; and (3) an **effector**, which responds to the commands of the control center and whose activity opposes or enhances the stimulus. You are probably already familiar with several examples of homeostatic regulation, although not in those terms.

1 AN INTRODUCTION TO ANATOMY AND PHYSIOLOGY

The Sciences of Anatomy and Physiology • Levels of Organization • An Introduction to Organ Systems • Homeostasis and System Integration

● *Figure 1-2*
The Organ Systems of the Human Body

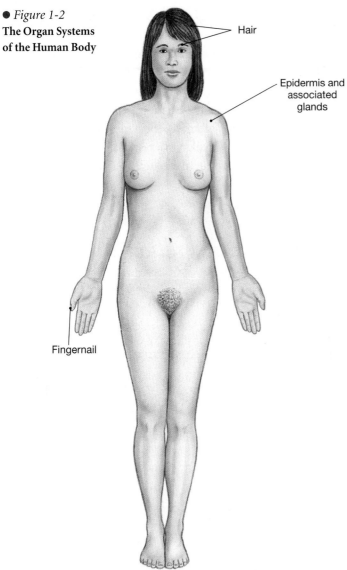

Hair

Epidermis and associated glands

Fingernail

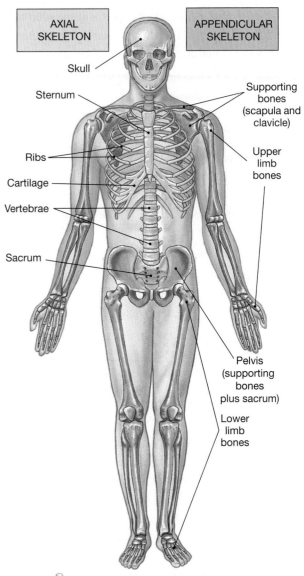

AXIAL SKELETON

APPENDICULAR SKELETON

Skull

Sternum

Ribs

Cartilage

Vertebrae

Sacrum

Supporting bones (scapula and clavicle)

Upper limb bones

Pelvis (supporting bones plus sacrum)

Lower limb bones

(a) THE INTEGUMENTARY SYSTEM

Protects against environmental hazards; helps control body temperature

ORGAN/COMPONENT	PRIMARY FUNCTIONS
CUTANEOUS MEMBRANE	
Epidermis	Covers surface; protects deeper tissues
Dermis	Nourishes epidermis; provides strength; contains glands
HAIR FOLLICLES	Produce hair
Hairs	Provide some protection for head
Sebaceous glands	Secrete oil that lubricates hair shaft and epidermis
SWEAT GLANDS	Produce perspiration for evaporative cooling
NAILS	Protect and stiffen distal tips of digits
SENSORY RECEPTORS	Provide sensations of touch, pressure, temperature, pain
SUBCUTANEOUS LAYER	Stores lipids; attaches skin to deeper structures

(b) THE SKELETAL SYSTEM

Provides support; protects tissues; stores minerals; forms blood

ORGAN/COMPONENT	PRIMARY FUNCTIONS
BONES, CARTILAGES, AND JOINTS	Support, protect soft tissues; bones store minerals
Axial skeleton (skull, vertebrae, ribs, sternum, sacrum, cartilages, and ligaments)	Protects brain, spinal cord, sense organs, and soft tissues of thoracic cavity; supports the body weight over the lower limbs
Appendicular skeleton (limbs and supporting bones and ligaments)	Provides internal support and positioning of the limbs; supports and moves axial skeleton
BONE MARROW	Acts as primary site of blood cell production (red blood cells, white blood cells)

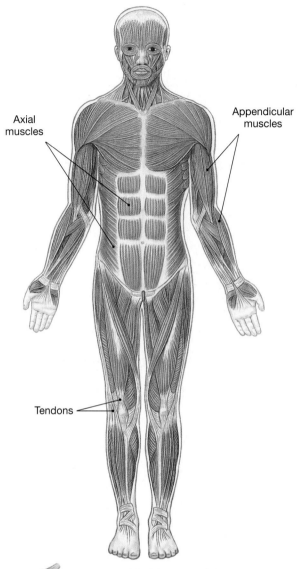

Axial muscles

Appendicular muscles

Tendons

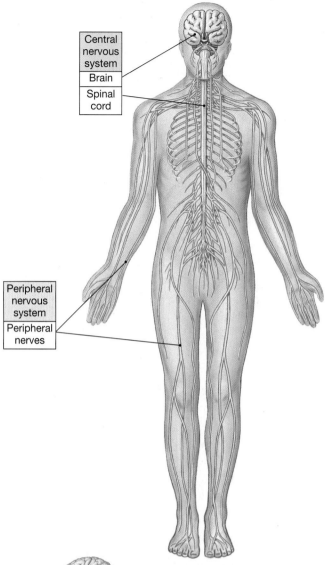

Central nervous system

Brain

Spinal cord

Peripheral nervous system

Peripheral nerves

(c) THE MUSCULAR SYSTEM

Allows for locomotion; provides support; produces heat

(d) THE NERVOUS SYSTEM

Directs immediate responses to stimuli, usually by coordinating the activities of other organ systems

ORGAN/COMPONENT	PRIMARY FUNCTIONS
SKELETAL MUSCLES (700)	Provide skeletal movement; control entrances and exits of digestive tract; produce heat; support skeletal position; protect soft tissues
Axial muscles	Support and position axial skeleton
Appendicular muscles	Support, move, and brace limbs
TENDONS	Harness forces of contraction to perform specific tasks

ORGAN/COMPONENT	PRIMARY FUNCTIONS
CENTRAL NERVOUS SYSTEM (CNS)	Acts as control center for nervous system: processes information; provides short-term control over activities of other systems
Brain	Performs complex integrative functions; controls both voluntary and autonomic activities
Spinal cord	Relays information to and from brain; performs less-complex integrative functions; directs many simple involuntary activities
PERIPHERAL NERVOUS SYSTEM (PNS)	Links CNS with other systems and with sense organs

1 AN INTRODUCTION TO ANATOMY AND PHYSIOLOGY

The Sciences of Anatomy and Physiology • Levels of Organization • An Introduction to Organ Systems • Homeostasis and System Integration

● *Figure 1-2*
Continued

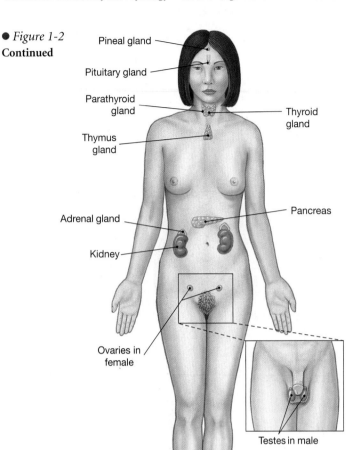

Pineal gland

Pituitary gland

Parathyroid gland

Thymus gland

Thyroid gland

Adrenal gland

Pancreas

Kidney

Ovaries in female

Testes in male

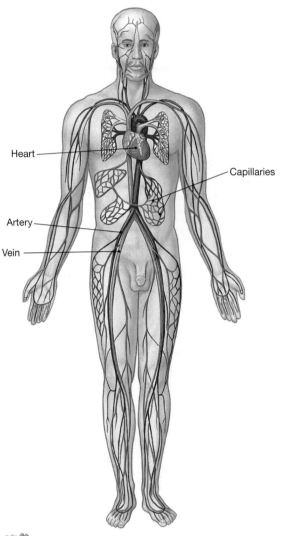

Heart

Capillaries

Artery

Vein

(e) THE ENDOCRINE SYSTEM

Directs long-term changes in activities of other organ systems

ORGAN/COMPONENT	PRIMARY FUNCTIONS
PINEAL GLAND	May control timing of reproduction and set day–night rhythms
PITUITARY GLAND	Controls other endocrine glands; regulates growth and fluid balance
THYROID GLAND	Controls tissue metabolic rate; regulates calcium levels
PARATHYROID GLANDS	Regulate calcium levels (with thyroid)
THYMUS	Controls maturation of lymphocytes
ADRENAL GLANDS	Adjust water balance, tissue metabolism, cardiovascular and respiratory activity
KIDNEYS	Control red blood cell production and elevate blood pressure
PANCREAS	Regulates blood glucose levels
GONADS	
Testes	Support male sexual characteristics and reproductive functions (*see Figure 1-2k*)
Ovaries	Support female sexual characteristics and reproductive functions (*see Figure 1-2l*)

(f) THE CARDIOVASCULAR SYSTEM

Transports cells and dissolved materials, including nutrients, wastes, and gases

ORGAN/COMPONENT	PRIMARY FUNCTIONS
HEART	Propels blood; maintans blood pressure
BLOOD VESSELS	Distribute blood around the body
Arteries	Carry blood from heart to capillaries
Capillaries	Permit diffusion between blood and interstitial fluids
Veins	Return blood from capillaries to the heart
BLOOD	Transports oxygen, carbon dioxide, and blood cells; delivers nutrients and hormones; removes waste products; assists in temperature regulation and defense against disease

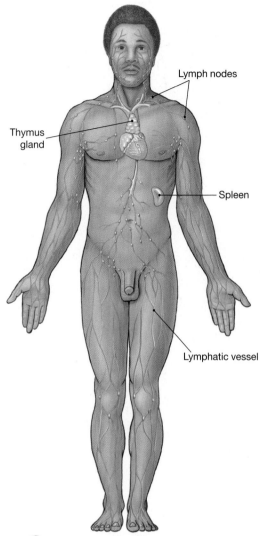

Lymph nodes

Thymus gland

Spleen

Lymphatic vessel

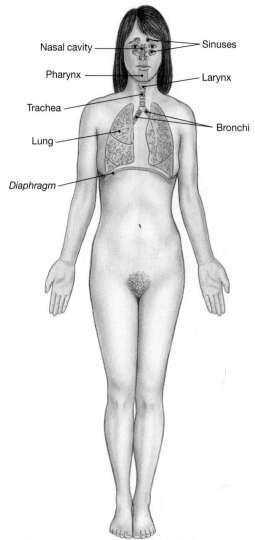

Nasal cavity

Sinuses

Pharynx

Larynx

Trachea

Bronchi

Lung

Diaphragm

(g) THE LYMPHATIC SYSTEM
Defends against infection and disease; returns tissue fluid to the bloodstream

ORGAN/COMPONENT	PRIMARY FUNCTIONS
LYMPHATIC VESSELS	Carry lymph (water and proteins) and lymphocytes from peripheral tissues to veins of the cardiovascular system
LYMPH NODES	Monitor the composition of lymph; engulf pathogens; stimulate immune response
SPLEEN	Monitors circulating blood; engulfs pathogens; stimulates immune response
THYMUS	Controls development and maintenance of one class of lymphocytes (T cells)

(h) THE RESPIRATORY SYSTEM
Delivers air to sites where gas exchange can occur between the air and circulating blood

ORGAN/COMPONENT	PRIMARY FUNCTIONS
NASAL CAVITIES, PARANASAL SINUSES	Filter, warm, humidify air; detect smells
PHARYNX	Conducts air to larynx; a chamber shared with the digestive tract (see Figure 1-2i)
LARYNX	Protects opening to trachea and contains vocal cords
TRACHEA	Filters air, traps particles in mucus; cartilages keep airway open
BRONCHI	(Same functions as trachea)
LUNGS	Responsible for air movement through volume changes during movements of ribs and diaphragm; include airways and alveoli
Alveoli	Act as sites of gas exchange between air and blood

1 AN INTRODUCTION TO ANATOMY AND PHYSIOLOGY

The Sciences of Anatomy and Physiology • Levels of Organization • An Introduction to Organ Systems • **Homeostasis and System Integration**

● *Figure 1-2*
Continued

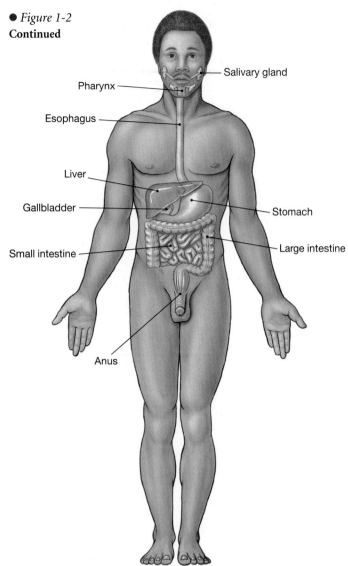

- Salivary gland
- Pharynx
- Esophagus
- Liver
- Gallbladder
- Stomach
- Small intestine
- Large intestine
- Anus

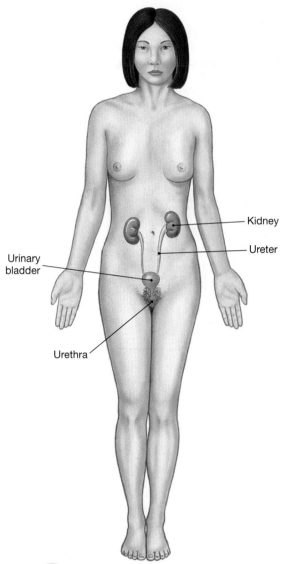

- Kidney
- Ureter
- Urinary bladder
- Urethra

(i) THE DIGESTIVE SYSTEM
Processes food and absorbs nutrients

(j) THE URINARY SYSTEM
Eliminates excess water, salts, and waste products

ORGAN/COMPONENT	PRIMARY FUNCTIONS
SALIVARY GLANDS	Provide buffers and lubrication; produce enzymes that begin digestion
PHARYNX	Conducts solid food and liquids to esophagus; chamber shared with respiratory tract *(see Figure 1-2h)*
ESOPHAGUS	Delivers food to stomach
STOMACH	Secretes acids and enzymes
SMALL INTESTINE	Secretes digestive enzymes, buffers, and hormones; absorbs nutrients
LIVER	Secretes bile; regulates nutrient composition of blood
GALLBLADDER	Stores bile for release into small intestine
PANCREAS	Secretes digestive enzymes and buffers; contains endocrine cells *(see Figure 1-2e)*
LARGE INTESTINE	Removes water from fecal material; stores wastes

ORGAN/COMPONENT	PRIMARY FUNCTIONS
KIDNEYS	Form and concentrate urine; regulate blood pH and ion concentrations; perform endocrine functions *(see Figure 1-2e)*
URETERS	Conduct urine from kidneys to urinary bladder
URINARY BLADDER	Stores urine for eventual elimination
URETHRA	Conducts urine to exterior

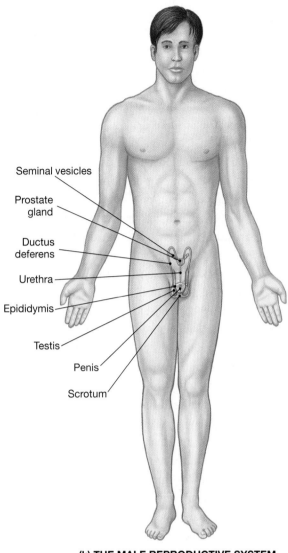

(k) THE MALE REPRODUCTIVE SYSTEM
Produces sex cells and hormones

Seminal vesicles
Prostate gland
Ductus deferens
Urethra
Epididymis
Testis
Penis
Scrotum

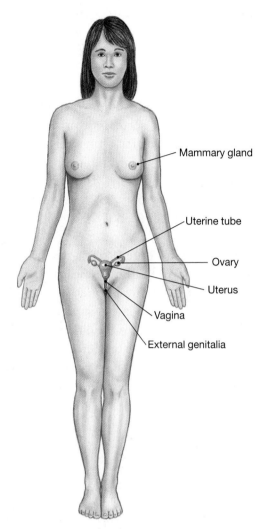

(l) THE FEMALE REPRODUCTIVE SYSTEM
Produces sex cells and hormones

Mammary gland
Uterine tube
Ovary
Uterus
Vagina
External genitalia

ORGAN/COMPONENT	PRIMARY FUNCTIONS
TESTES	Produce sperm and hormones *(see Figure 1-2e)*
ACCESSORY ORGANS	
Epididymis	Acts as site of sperm maturation
Ductus deferens (sperm duct)	Conducts sperm between epididymis and prostate gland
Seminal vesicles	Secrete fluid that makes up much of the volume of semen
Prostate gland	Secretes fluid and enzymes
Urethra	Conducts semen to exterior
EXTERNAL GENITALIA	
Penis	Contains erectile tissue; deposits sperm in vagina of female; produces pleasurable sensations during sexual activities
Scrotum	Surrounds the testes and controls their temperature

ORGAN/COMPONENT	PRIMARY FUNCTIONS
OVARIES	Produce oocytes and hormones *(see Figure 1-2e)*
UTERINE TUBES	Deliver oocyte or embryo to uterus; normal site of fertilization
UTERUS	Site of embryonic development and exchange between maternal and embryonic bloodstreams
VAGINA	Site of sperm deposition; acts as birth canal at delivery; provides passageway for fluids during menstruation
EXTERNAL GENITALIA	
Clitoris	Contains erectile tissue; produces pleasurable sensations during sexual activities
Labia	Contain glands that lubricate entrance to vagina
MAMMARY GLANDS	Produce milk that nourishes newborn infant

1 AN INTRODUCTION TO ANATOMY AND PHYSIOLOGY

The Sciences of Anatomy and Physiology • Levels of Organization • An Introduction to Organ Systems • **Homeostasis and System Integration**

As an example, consider the operation of the thermostat in a house or apartment (Figure 1-3●).

The thermostat is a control center that monitors room temperature. The gauge on the thermostat establishes the set point, the "ideal" room temperature. In our example, the set point is 22°C (about 72°F). The function of the thermostat is to keep room temperature during the year within acceptable limits, usually within a degree or so of the set point. The thermostat receives information from a receptor, a thermometer exposed to air in the room, and it controls two effectors: a heater and an air conditioner. The principle is simple: The heater turns on if the room becomes too cold (winter), and the air conditioner turns on if the room becomes too warm (summer).

When the temperature at the thermometer increases outside of the normal range, the thermostat turns on the air conditioner. The air conditioner then cools the room. When temperature at the thermometer approaches the set point, the thermostat turns off the air conditioner. A comparable pattern of events occurs if the temperature drops below normal levels: Temperature falls, and the thermostat turns on the heater; the temperature then rises, and the thermostat turns off the heater (Figure 1-3a●).

Negative Feedback

Regardless of whether the temperature at the receptor rises or falls, *a variation outside normal limits triggers an automatic response that corrects the situation*. This method of homeostatic regulation is called **negative feedback** because the effector that is activated by the control center opposes or eliminates the stimulus.

Most homeostatic mechanisms in the body involve negative feedback. For example, consider the control of body temperature, a process called *thermoregulation*. Thermoregulation involves altering the relationship between heat loss, which occurs primarily at the body surface, and heat production, which occurs in all active tissues. In the human body, skeletal muscles are the most important generators of body heat.

The cells of the thermoregulatory control center are located in the brain. Temperature receptors are located in the skin, and the cells in the control center are sensitive to local body temperature. The thermoregulatory center has a set point near 37°C (98.6°F) (Figure 1-3b●). If temperature at the thermoregulatory center rises above 37.2°C, activity in the control center targets two different effectors: (1) smooth muscles in the walls of blood vessels supplying the skin and (2) sweat glands. The blood vessels widen, or dilate, increasing blood

flow at the body surface, and the sweat glands accelerate their secretion. The skin then acts like a radiator, losing heat to the environment, and the evaporation of sweat speeds the process. When body temperature returns to normal, the control center becomes inactive, and superficial blood flow and sweat gland activity decrease to normal resting levels.

If temperature at the control center falls below 36.7°C, the control center targets the same two effectors and skeletal muscles. This time blood flow to the skin declines, and sweat gland activity decreases. This combination reduces the rate of heat loss to the environment. Because heat production continues, body temperature gradually rises; once an acceptable temperature has been reached, the thermoregulatory center turns itself "off," and both blood flow and sweat gland activity in the skin increase to normal resting levels. Additional heat may be generated by shivering, which is caused by random contractions of skeletal muscles.

Homeostatic mechanisms using negative feedback usually ignore minor variations, and they maintain a normal range rather than a fixed value. In the example above, body temperature oscillates around the ideal set-point temperature. Thus any measured value, such as body temperature, can vary from moment to moment or day to day for any single individual. The variability between individuals is even greater, for each person has slightly different homeostatic set points. It is therefore impractical to define "normal" homeostatic conditions very precisely. By convention, physiological values are reported either as averages—the average value obtained by sampling a large number of individuals—or as a range that includes 95 percent or more of the sample population. For instance, 5 percent of normal adults have a body temperature outside the "normal" range (below 36.7°C or above 37.2°C). But these temperatures are perfectly normal for them, and the variations have no clinical significance.

Positive Feedback

In **positive feedback**, *the initial stimulus produces a response that reinforces the stimulus*. For example, suppose the thermostat was wired so that when the temperature rose, it would turn on the heater rather than the air conditioner. In that case, the initial stimulus (rising room temperature) would cause a response (heater turns on) that would strengthen the stimulus. The room temperature would continue to rise until some external factor switched off the thermostat, unplugged the heater, or intervened in some other way before the house caught fire and burned down. This kind of escalating process is called a *positive feedback loop*.

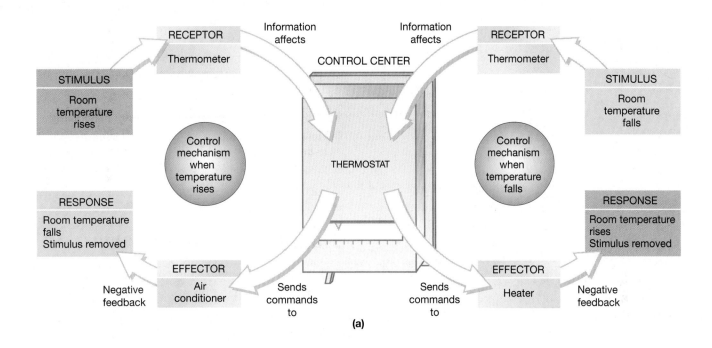

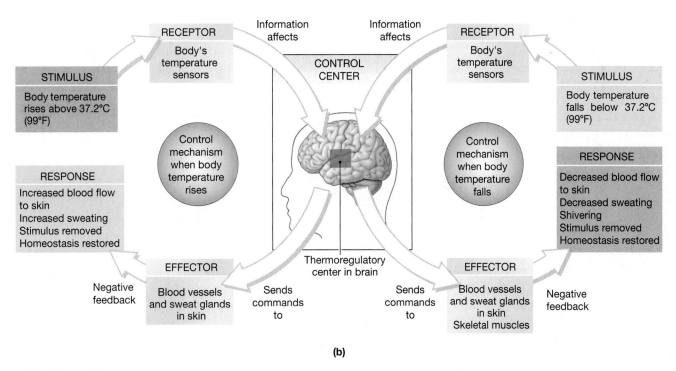

● *Figure 1-3* **Negative Feedback**

In negative feedback, a stimulus produces a response that opposes the original stimulus. (**a**) A thermostat controls heating and cooling systems (a heater and air conditioner) to keep temperatures within acceptable limits. When the room temperature rises or falls, the thermostat (a control center) triggers an effector response that restores normal temperature. (**b**) Body temperature is regulated by a control center in the brain that functions as a thermostat with a set point of 37°C. If body temperature climbs above 37.2°C, heat loss is increased through enhanced blood flow to the skin and increased sweating. If body temperature falls below 36.7°C, heat loss is decreased through decreases in blood flow to the skin and sweating, and heat is produced by shivering.

1 AN INTRODUCTION TO ANATOMY AND PHYSIOLOGY

The Sciences of Anatomy and Physiology • Levels of Organization • An Introduction to Organ Systems • Homeostasis and System Integration

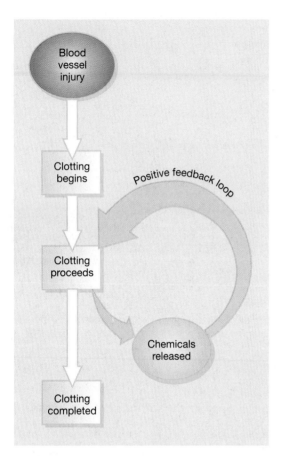

● *Figure 1-4* **Positive Feedback**
In positive feedback, a stimulus produces a response that reinforces the original stimulus. Positive feedback is important in accelerating processes that must proceed to completion rapidly. In this example, positive feedback accelerates blood clotting until bleeding stops.

In the body, positive feedback loops are often involved in the regulation of a dangerous or stressful process that must be completed quickly (Figure 1-4●). For example, the immediate danger from a severe cut is the loss of blood, which can lower blood pressure and reduce the efficiency of the heart. Damage to the blood vessel wall releases chemicals that begin the process of blood clotting. As clotting gets under way, each step releases chemicals that accelerate the process. This escalating process ends with the formation of a blood clot, which patches the vessel wall and stops the bleeding. Blood clotting will be examined more closely in Chapter 11. Labor and delivery, another example of positive feedback in action, will be discussed in Chapter 20.

HOMEOSTASIS AND DISEASE

Physiological mechanisms do a remarkably good job of stabilizing internal conditions, regardless of our ongoing activities. But when homeostatic regulation fails, organ systems begin to malfunction and the individual experiences the symptoms of illness, or **disease**.

CONCEPT CHECK QUESTIONS

Answers on page 25

❶ Why is homeostatic regulation important to humans?

❷ How is positive feedback helpful in producing necessary changes in individuals?

❸ What happens to the body when homeostasis breaks down?

The Language of Anatomy

Early anatomists faced serious communication problems. For example, stating that a bump is "on the back" does not give very precise information about its location. So anatomists created maps of the human body, using prominent anatomical structures as landmarks, specialized directional terms, and reporting distances in centimeters or inches. In effect, anatomy uses a special language that must be learned almost at the start of your study.

A familiarity with Latin and Greek word roots and their combinations makes anatomical terms more understandable. As new terms are introduced in the text, notes on their pronunciation and the relevant word roots will be provided. Additional information on foreign word roots, prefixes, suffixes, and combining forms can be found on the front endpapers of this book.

Latin and Greek terms are not the only foreign words imported into the anatomical vocabulary over the centuries, and the vocabulary continues to expand. Many anatomical structures and clinical conditions were initially named after either the discoverer or, in the case of diseases, the most famous victim. Although most such commemorative names, or *eponyms*, have been replaced by more precise terms, many are still in use.

SURFACE ANATOMY

With the exception of the skin, none of the organ systems can be seen from the body surface. Therefore, you must create your own mental maps and extract information from the terms given in Figures 1-5● and 1-6●. Learning these terms now will make subsequent chapters more understandable.

Anatomical Landmarks

Standard anatomical illustrations show the human form in the **anatomical position**, with the hands at the sides and the palms

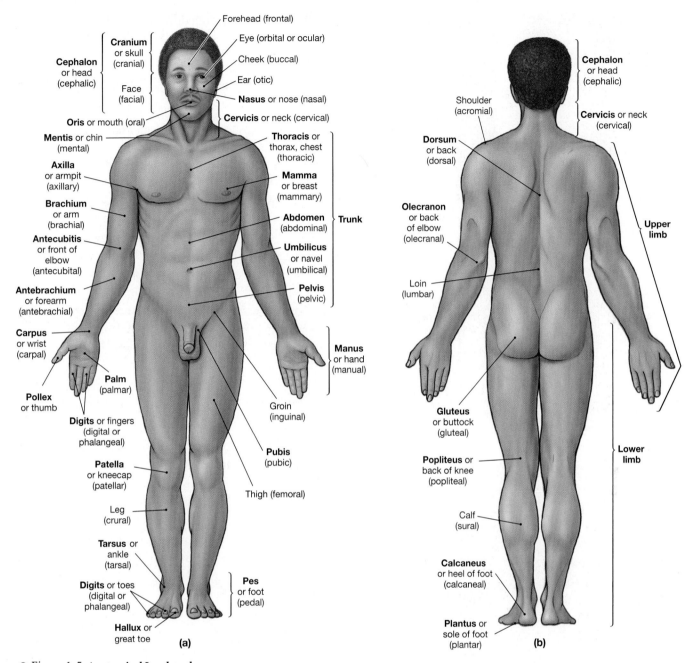

● *Figure 1-5* **Anatomical Landmarks**
Anatomical terms are in boldface type, common names are in plain type, and anatomical adjectives are in parentheses.

facing forward (Figure 1-5●). A person lying down in the anatomical position is said to be **supine** (sū-PĪN) when face up and **prone** when face down.

Important anatomical landmarks are also presented in Figure 1-5●. The anatomical terms are given in boldface, the common names in plain type, and the anatomical adjectives in parentheses. Understanding the terms and their origins

can help you remember the location of a particular structure, as well as its name. For example, the term *brachium* refers to the arm, and later chapters will discuss the brachial artery, brachial nerve, and so forth. You might remember this term more easily if you know that the Latin word *brachium* is also the source of Old English and French words meaning "to embrace."

1 AN INTRODUCTION TO ANATOMY AND PHYSIOLOGY

The Sciences of Anatomy and Physiology • Levels of Organization • An Introduction to Organ Systems • Homeostasis and System Integration

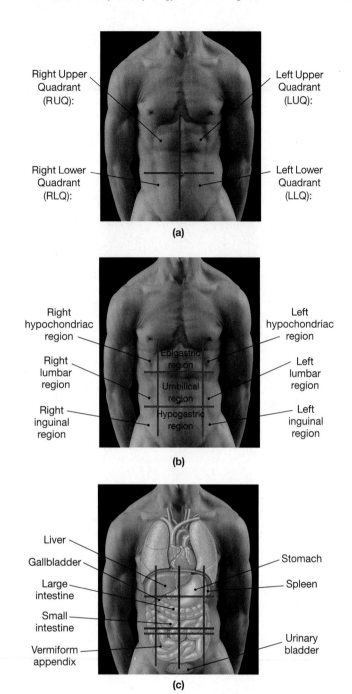

● *Figure 1-6* **Abdominopelvic Quadrants and Regions**
(**a**) Abdominopelvic quadrants divide the area into four sections. These terms, or their abbreviations, are most often used in clinical discussions. (**b**) More precise regional descriptions are provided by reference to the appropriate abdominopelvic region. (**c**) Quadrants or regions are useful because there is a known relationship between superficial anatomical landmarks and underlying organs.

Anatomical Regions

Major regions of the body are listed in Table 1-1 and shown in Figure 1-5●. Anatomists and clinicians often need to use regional terms as well as specific landmarks to describe a general area of interest or injury. Two methods are used to specify

TABLE 1-1 *Regions of the Human Body (see Figure 1-5●)*	
STRUCTURE	**AREA**
Cephalon (head)	Cephalic region
Cervicis (neck)	Cervical region
Thoracis (chest)	Thoracic region
Abdomen	Abdominal region
Pelvis	Pelvic region
Loin (lower back)	Lumbar region
Buttock	Gluteal region
Pubis (anterior pelvis)	Pubic region
Groin	Inguinal region
Axilla (armpit)	Axillary region
Brachium (arm)	Brachial region
Antebrachium (forearm)	Antebrachial region
Manus (hand)	Manual region
Thigh	Femoral region
Leg (anterior)	Crural region
Calf	Sural region
Pes (foot)	Pedal region

locations on the abdominal surface. Clinicians refer to the **abdominopelvic quadrants**: four segments divided by imaginary lines that intersect at the *umbilicus* (navel). This simple method, shown in Figure 1-6a●, is useful for describing aches, pains, and injuries. The location can help a doctor determine the possible cause. For example, tenderness in the right lower quadrant (RLQ) is a symptom of appendicitis, whereas tenderness in the right upper quadrant (RUQ) may indicate gallbladder or liver problems.

Anatomists like to use more precise regional distinctions to describe the location and orientation of internal organs. They recognize nine **abdominopelvic regions** (Figure 1-6b●). Figure 1-6c● shows the relationships between quadrants, regions, and internal organs.

Anatomical Directions

Figure 1-7● and Table 1-2 show the principal directional terms and examples of their use. There are many different terms, and some can be used interchangeably. For example, *anterior* refers to the front of the body, when viewed in the anatomical position; in human beings, this term is equivalent to *ventral*, which refers to the belly. Likewise, *posterior* and *dorsal* are terms that refer to the back of the human body. The terms that appear frequently in later chapters have been emphasized. Remember that *left* and *right* always refer to the left and right sides of the subject, not of the observer.

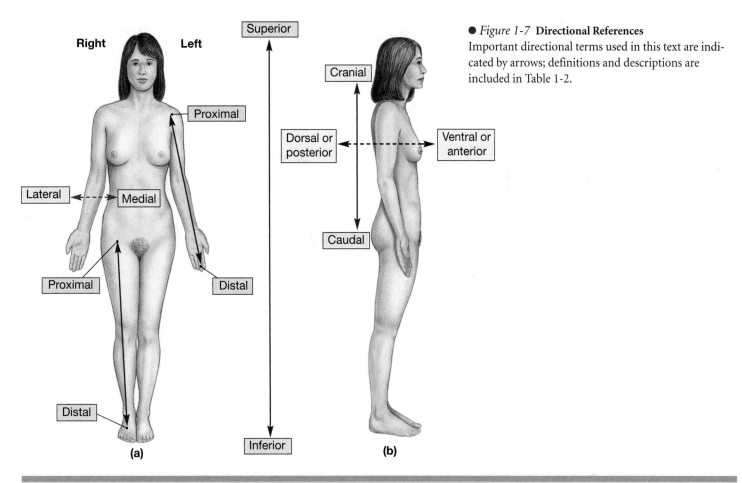

Right　**Left**

Superior

Proximal

Lateral ◄┄┄► Medial

Proximal

Distal

Distal

Inferior

(a)

Cranial

Dorsal or posterior ◄┄┄► Ventral or anterior

Caudal

(b)

● *Figure 1-7* **Directional References**
Important directional terms used in this text are indicated by arrows; definitions and descriptions are included in Table 1-2.

TABLE 1-2 *Directional Terms (see Figure 1-7●)*

TERM	REGION OR REFERENCE	EXAMPLE
ANTERIOR	The front; before	The navel is on the *anterior* (*ventral*) surface of the trunk.
VENTRAL	The belly side (equivalent to anterior when referring to human body)	
POSTERIOR	The back; behind	The shoulder blade is located *posterior* (*dorsal*) to the rib cage.
DORSAL	The back (equivalent to posterior when referring to human body)	The *dorsal* body cavity encloses the brain and spinal cord.
CRANIAL OR CEPHALIC	The head	The *cranial*, or *cephalic*, border of the pelvis is superior to the thigh.
SUPERIOR	Above; at a higher level (in human body, toward the head)	The nose is *superior* to the chin.
CAUDAL	The tail (coccyx in humans)	The hips are *caudal* to the waist.
INFERIOR	Below; at a lower level	The knees are *inferior* to the hips.
MEDIAL	Toward the body's longitudinal axis	The *medial* surfaces of the thighs may be in contact; moving medially from the arm across the chest surface brings you to the sternum.
LATERAL	Away from the body's longitudinal axis	The thigh articulates with the *lateral* surface of the pelvis; moving laterally from the nose brings you to the eyes.
PROXIMAL	Toward an attached base	The thigh is *proximal* to the foot; moving proximally from the wrist brings you to the elbow.
DISTAL	Away from an attached base	The fingers are *distal* to the wrist; moving distally from the elbow brings you to the wrist.
SUPERFICIAL	At, near, or relatively close to the body surface	The skin is *superficial* to underlying structures.
DEEP	Farther from the body surface	The bone of the thigh is *deep* to the surrounding skeletal muscles.

SECTIONAL ANATOMY

A presentation in sectional view is sometimes the only way to illustrate the relationships between the parts of a three-dimensional object. An understanding of sectional views has become increasingly important since the development of procedures that enable us to see inside the living body without resorting to surgery (see Figure 1-11● for representative views).

Planes and Sections

Any slice through a three-dimensional object can be described with reference to three **sectional planes**, indicated in Figure 1-8● and Table 1-3:

1. *Transverse Plane.* The **transverse plane** lies at right angles to the long (head-foot) axis of the body, dividing it into **superior** and **inferior** sections. A cut in this plane is called a **transverse section**, a **horizontal section**, or a *cross section*.
2. *Frontal Plane.* The **frontal plane**, or **coronal plane**, parallels the long axis of the body. The frontal plane extends from side to side, dividing the body into **anterior** and **posterior** sections.

3. *Sagittal Plane.* The **sagittal plane** also parallels the long axis of the body, but it extends from front to back. A sagittal plane divides the body into *left* and *right* sections. A cut that passes along the midline and divides the body into left and right halves is a **midsagittal section**. (Note that a midsagittal section does not cut through the legs.)

Body Cavities

Viewed in sections, the human body is not a solid object, like a rock, in which all of the parts are fused together. Many vital organs are suspended in internal chambers called *body cavities*. These cavities have two essential functions:

1. They protect delicate organs, such as the brain and spinal cord, from accidental shocks and cushion them from the thumps and bumps that occur during walking, jumping, and running.
2. They permit significant changes in the size and shape of visceral organs. For example, because they are situated within body cavities, the lungs, heart, stomach, intestines, urinary bladder, and many other organs can expand and

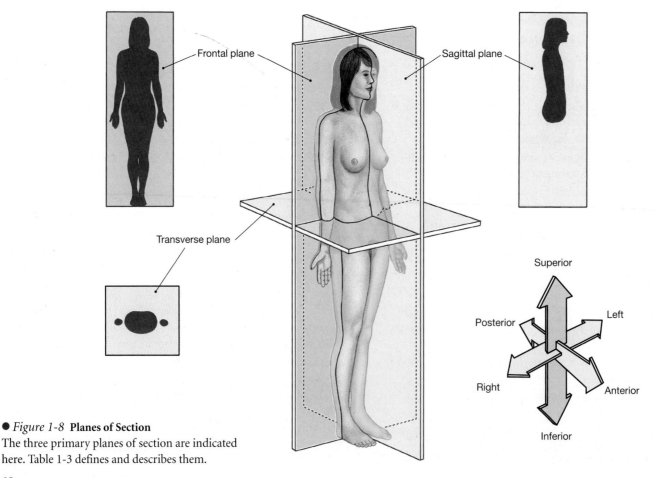

● *Figure 1-8* **Planes of Section**
The three primary planes of section are indicated here. Table 1-3 defines and describes them.

TABLE 1-3 *Terms That Indicate Planes of Section (see Figure 1-8●)*

ORIENTATION OF PLANE	ADJECTIVE	DIRECTIONAL REFERENCE	DESCRIPTION
PARALLEL TO LONG AXIS	Sagittal	Sagittally	A *sagittal section* separates right and left portions. You examine a sagittal section, but you section sagittally.
	Midsagittal		In a *midsagittal section*, the plane passes through the midline, dividing the body in half and separating right and left sides.
	Frontal or coronal	Frontally or coronally	A *frontal*, or *coronal*, *section* separates anterior and posterior portions of the body; *coronal* usually refers to sections passing through the skull.
PERPENDICULAR TO LONG AXIS	Transverse or horizontal	Transversely or horizontally	A *transverse*, or *horizontal*, *section* separates superior and inferior portions of the body.

contract without distorting surrounding tissues and disrupting the activities of nearby organs.

Two body cavities form during embryonic development. A **dorsal body cavity** surrounds the brain and spinal cord, and a much larger **ventral body cavity** surrounds developing organs of the respiratory, cardiovascular, digestive, urinary, and reproductive systems.

Dorsal Body Cavities. The dorsal body cavity is a fluid-filled space whose limits are established by the **cranium**, the bones of the skull that surround the brain, and the spinal vertebrae (Figure 1-9a●). The dorsal body cavity is subdivided into the **cranial cavity**, which encloses the brain, and the **spinal cavity**, which surrounds the spinal cord.

Ventral Body Cavities. As development proceeds, internal organs grow and change their relative positions. These changes lead to the subdivision of the ventral body cavity. The **diaphragm** (DĪ-a-fram), a flat muscular sheet, divides the ventral body cavity into a superior **thoracic cavity**, bounded by the chest wall, and an inferior **abdominopelvic cavity**, enclosed by the abdomen and pelvis. The abdominopelvic cavity has two subdivisions. The **abdominal cavity** extends from the inferior surface of the diaphragm to the level of the superior margins of the pelvis. The **pelvic cavity** is the portion of the ventral body cavity inferior to the abdominal cavity.

Many of the organs in these cavities change size and shape as they perform their functions. For example, the lungs inflate and deflate as you breathe, and your stomach swells at each meal and shrinks between meals. These organs are surrounded by moist internal spaces that permit expansion and limited movement but prevent friction. The internal organs within the thoracic and abdominopelvic cavities are called **viscera**

(VIS-e-ra). A delicate layer called a *serous membrane* lines the walls of these internal cavities and covers the surfaces of the enclosed viscera. Serous membranes secrete a watery fluid that coats the opposing surfaces and reduces friction.

The thoracic cavity contains three internal chambers: a single *pericardial cavity* and a pair of *pleural cavities* (Figure 1-9a,c●). Each of these cavities is lined by shiny, slippery serous membranes. To understand the relationship between a visceral organ and the serous membrane, consider the example in Figure 1-9b●. The heart projects into a space known as the **pericardial cavity**. The relationship resembles that of a fist pushing into a balloon. The wrist corresponds to the base of the heart, and the balloon corresponds to the serous membrane lining the pericardial cavity. The serous membrane is called the **pericardium** (*peri-*, around + *cardium*, heart). The layer covering the heart is the **visceral pericardium**, and the opposing surface is the **parietal pericardium**.

The pericardium lies within the **mediastinum** (mē-dē-as-TĪ-num or mē-dē-AS-ti-num) (Figure 1-9d●). The connective tissue of the mediastinum surrounds the pericardial cavity and heart, the large arteries and veins attached to the heart, and the thymus, trachea, and esophagus.

Each **pleural cavity** surrounds a lung. The spatial relationships between the lungs and the pleural cavities resemble that between the heart and pericardium. The serous membrane lining the pleural cavities is called the **pleura** (PLOO-ra). The *visceral pleura* covers the outer surfaces of a lung, and the *parietal pleura* covers the opposing surface of the mediastinum and the inner body wall.

Most of the visceral organs in the abdominopelvic cavity project into the **peritoneal** (per-i-tō-NĒ-al) **cavity**. The serous membrane lining this cavity is the **peritoneum** (per-i-tō-NĒ-um). Organs such as the stomach, small intestine, and portions of the large intestine are suspended within the peritoneal

1 AN INTRODUCTION TO ANATOMY AND PHYSIOLOGY

The Sciences of Anatomy and Physiology • Levels of Organization • An Introduction to Organ Systems • Homeostasis and System Integration

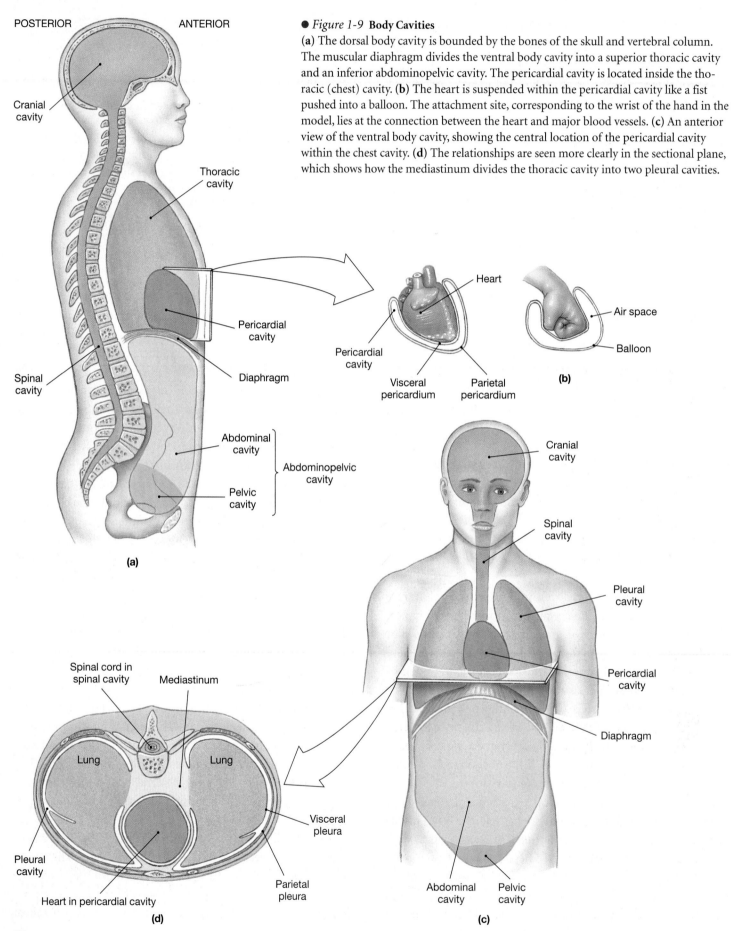

● *Figure 1-9* **Body Cavities**
(**a**) The dorsal body cavity is bounded by the bones of the skull and vertebral column. The muscular diaphragm divides the ventral body cavity into a superior thoracic cavity and an inferior abdominopelvic cavity. The pericardial cavity is located inside the thoracic (chest) cavity. (**b**) The heart is suspended within the pericardial cavity like a fist pushed into a balloon. The attachment site, corresponding to the wrist of the hand in the model, lies at the connection between the heart and major blood vessels. (**c**) An anterior view of the ventral body cavity, showing the central location of the pericardial cavity within the chest cavity. (**d**) The relationships are seen more clearly in the sectional plane, which shows how the mediastinum divides the thoracic cavity into two pleural cavities.

cavity by double sheets of peritoneum, called **mesenteries** (MES-en-ter-ēz). Mesenteries provide support and stability while permitting limited movement.

This chapter provided an overview of the locations and functions of the major components of each organ system. It also introduced the anatomical vocabulary needed to follow more detailed anatomical descriptions in later chapters. Modern methods of visualizing anatomical structures in living individuals are summarized in Figures 1-10● and 1-11● (see "FOCUS: Sectional Anatomy and Clinical Technology," which follows. Many of the figures in later chapters contain images produced by the procedures outlined in this feature).

CONCEPT CHECK QUESTIONS
Answers on page 25

❶ What type of section would separate the two eyes?

❷ If a surgeon makes an incision just inferior to the diaphragm, what body cavity will be opened?

FOCUS

Sectional Anatomy and Clinical Technology

The term **radiological procedures** includes not only those scanning techniques that involve radioisotopes but also methods that employ radiation sources outside the body. Physicians who specialize in the performance and analysis of these procedures are called **radiologists**. Radiological procedures can provide detailed information about internal systems. Figures 1-10● and 1-11● compare the views provided by several different techniques. These figures include examples of X-rays, CT scans, MRI scans, and ultrasound images. Other examples of clinical technology will be found in later chapters.

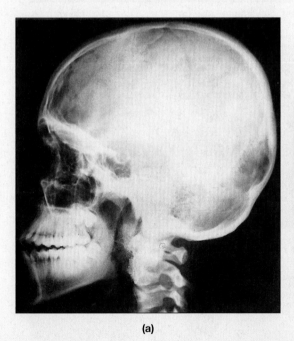

(a)

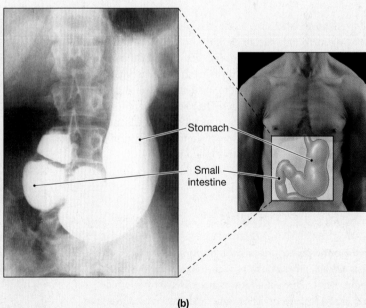

Stomach

Small intestine

(b)

● *Figure 1-10* **X-rays**

(**a**) An X-ray of the skull, taken from the left side. **X-rays** are a form of high-energy radiation that can penetrate living tissues. In the most familiar procedure, a beam of X-rays travels through the body and strikes a photographic plate. All of the projected X-rays do not arrive at the film; some are absorbed or deflected as they pass through the body. The resistance to X-ray penetration is called **radiodensity**. In the human body, the order of increasing radiodensity is as follows: air, fat, liver, blood, muscle, bone. The result is an image with radiodense tissues, such as bone, appearing in white, and less dense tissues in shades of gray to black. The picture is a two-dimensional image of a three-dimensional object; in this image it is difficult to decide whether a particular feature is on the left side (toward the viewer) or on the right side (away from the viewer). (**b**) A **barium-contrast X-ray** of the upper digestive tract. Barium is very dense, and the contours of the gastric and intestinal lining can be seen outlined against the white of the barium solution.

1 AN INTRODUCTION TO ANATOMY AND PHYSIOLOGY

The Sciences of Anatomy and Physiology • Levels of Organization • An Introduction to Organ Systems • Homeostasis and System Integration

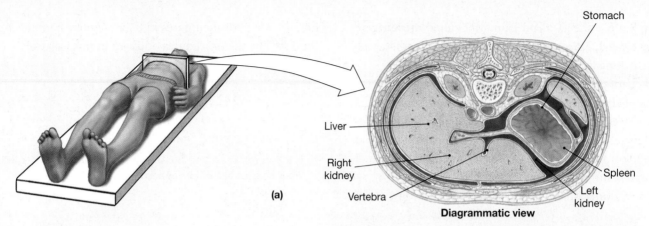

(a)

Diagrammatic view

● *Figure 1-11* **Scanning Techniques**
(a) Diagrammatic views showing the relative position and orientation of the scans shown in parts (b) and (c).

(b) A color-enhanced CT scan of the abdomen. **CT** (computed tomography), formerly called **CAT** (computed axial tomography), uses computers to reconstruct sectional views. A single X-ray source rotates around the body and the X-ray beam strikes a sensor monitored by the computer. The source completes one revolution around the body every few seconds; it then moves a short distance and repeats the process. The result is usually displayed as a sectional view in black and white, but it can be colorized for visual effect. CT scans show three-dimensional relationships and soft tissue structure more clearly than do standard X-rays.

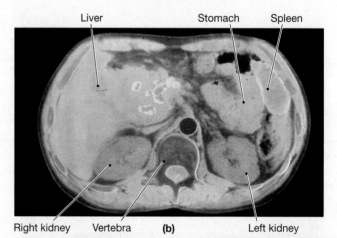

(b)

(c) A color-enhanced **MRI** scan of the same region. Magnetic resonance imaging surrounds part or all of the body with a magnetic field about 3000 times as strong as that of the earth. This field causes particles within the atoms throughout the body to all line up in a uniform direction. Energy from pulses of radio waves are absorbed and released by the different atoms in body tissues. The released energy is used to create an image. Details of soft tissue structure are usually much more clearly detailed than in CT scans. Note the differences in detail between this image, the CT scan, and the ultrasound image.

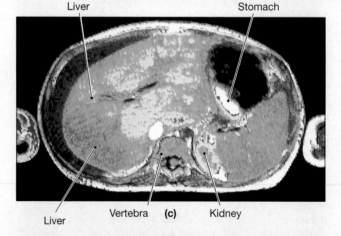

(c)

(d) In **ultrasound** procedures, a small transmitter contacting the skin broadcasts a brief, narrow burst of high-frequency sound and then picks up the echoes. The sound waves are reflected by internal structures, and a picture, or **echogram**, can be assembled from the pattern of echoes. These images lack the clarity of other procedures, but no adverse effects have been reported, and fetal development can be monitored without a significant risk of birth defects. Special methods of transmission and processing permit analysis of the beating heart, without the complications that can accompany dye injections.

(d)

Related Clinical Terms

abdominopelvic quadrant: One of four divisions of the anterior abdominal surface.

abdominopelvic region: One of nine divisions of the anterior abdominal surface.

auscultation (aws-kul-TĀ-shun): Listening to a patient's body sounds by means of a stethoscope.

CT, CAT (computerized [axial] tomography): An imaging technique that uses X-rays to reconstruct the body's three-dimensional structure.

disease: A malfunction of organs or organ systems resulting from a failure in homeostatic regulation.

echogram: An image created by ultrasound.

histology (his-TOL-o-jē): The study of tissues.

MRI (magnetic resonance imaging): An imaging technique that employs a magnetic field and radio waves to portray subtle structural differences.

PET (positron emission tomography) scan: A scan taken after radioisotopes are injected or inhaled; the released radiation is monitored, and computers use the data to visualize physiological activity.

radiology (rā-dē-OL-o-jē)**:** The study of radioactive energy and radioactive substances, and their use in the diagnosis and treatment of disease.

ultrasound: An imaging technique that uses brief bursts of high-frequency sound reflected by internal structures.

X-rays: High-energy radiation that can penetrate living tissues.

CHAPTER REVIEW

Key Terms

Summary Outline

INTRODUCTION 2

1. **Biology** is the study of life; one of its goals is to discover the unity and patterns that underlie the diversity of living organisms.

2. All living things, from single *cells* to large multicellular organisms, perform the same basic functions: they respond to changes in their environment; they grow and reproduce to create future generations; they are capable of producing movement; and they absorb materials from the environment. Organisms absorb and consume oxygen during respiration, and they discharge waste products during excretion. Digestion occurs in specialized areas of the body to break down complex foods. The circulation forms an internal transportation system between areas of the body.

THE SCIENCES OF ANATOMY AND PHYSIOLOGY 2
Anatomy...2

1. **Anatomy** is the study of internal and external structure and the physical relationships between body parts. **Physiology** is the study of how living organisms perform vital functions. All specific functions are performed by specific structures.

2. The boundaries of **microscopic anatomy** are established by the equipment used. **Cytology** analyzes the internal structure of individual **cells**. **Histology** examines **tissues** (groups of cells that have specific functional roles). Tissues combine to form organs, anatomical units with specific functions.

3. **Gross (macroscopic) anatomy** considers features visible without a microscope. It includes **surface anatomy** (general form and superficial markings); **regional anatomy** (superficial and internal features in a specific area of the body); and **systemic anatomy** (structure of major organ systems).

Physiology 3

4. **Human physiology** is the study of the functions of the human body. It is based on **cell physiology**, the study of the functions of living cells. **Special physiology** studies the physiology of specific organs. **System physiology** considers all aspects of the function of specific organ systems. **Pathological physiology (pathology)** studies the effects of diseases on organ or system functions.

LEVELS OF ORGANIZATION 3

1. Anatomical structures and physiological mechanisms are arranged in a series of interacting levels of organization. *(Figure 1-1)*

AN INTRODUCTION TO ORGAN SYSTEMS 5

1. The major organs of the human body are arranged into 11 organ systems. The organ systems of the human body are the *integumentary, skeletal, muscular, nervous, endocrine, cardiovascular, lymphatic, respiratory, digestive, urinary,* and *reproductive systems. (Figure 1-2)*

HOMEOSTASIS AND SYSTEM INTEGRATION 5

1. **Homeostasis** is the tendency for physiological systems to stabilize internal conditions; through **homeostatic regulation** these systems adjust to preserve homeostasis.

Homeostatic Regulation 5

2. Homeostatic regulation usually involves a **receptor** sensitive to a particular stimulus and an **effector** whose activity affects the same stimulus.

3. **Negative feedback** is a corrective mechanism involving an action that directly opposes a variation from normal limits. *(Figure 1-3)*

4. In **positive feedback** the initial stimulus produces a response that reinforces the stimulus. *(Figure 1-4)*

Homeostasis and Disease 14

5. Symptoms of **disease** appear when failure of homeostatic regulation causes organ systems to malfunction.

THE LANGUAGE OF ANATOMY 14
Surface Anatomy..14

1. Standard anatomical illustrations show the body in the **anatomical position**. If the figure is shown lying down, it can be either **supine** (face up) or **prone** (face down). *(Figure 1-5; Table 1-1)*

1
AN INTRODUCTION TO ANATOMY AND PHYSIOLOGY

The Sciences of Anatomy and Physiology • Levels of Organization • An Introduction to Organ Systems • Homeostasis and System Integration

2. **Abdominopelvic quadrants** and **abdominopelvic regions** represent two different approaches to describing anatomical regions of the body. *(Figure 1-6)*

3. The use of special directional terms provides clarity when describing anatomical structures. *(Figure 1-7; Table 1-2)*

Sectional Anatomy 18

4. The three **sectional planes** (**frontal** or **coronal plane**, **sagittal plane**, and **transverse plane**) describe relationships between the parts of the three-dimensional human body. *(Figure 1-8; Table 1-3)*

5. **Body cavities** protect delicate organs and permit changes in the size and shape of visceral organs. The **dorsal body cavity** contains the **cranial cavity** (enclosing the brain) and **spinal cavity** (surrounding the spinal cord). The **ventral body cavity** surrounds developing respiratory, cardiovascular, digestive, urinary, and reproductive organs. *(Figure 1-9a)*

6. The **diaphragm** divides the ventral body cavity into the superior **thoracic** and inferior **abdominopelvic cavities**. The thoracic cavity contains two **pleural cavities** (each containing a lung) and a **pericardial cavity** (which surrounds the heart). The abdominopelvic cavity consists of the **abdominal cavity** and the **pelvic cavity**. *(Figure 1-9b, c, d)* It contains the *peritoneal cavity*, an internal chamber lined by *peritoneum*, a *serous membrane*.

7. Important **radiological procedures** (which can provide detailed information about internal systems) include **X-rays**, **CT scans**, **MRI**, and **ultrasound**. Each technique has its advantages and disadvantages. *(Figures 1-10, 1-11)*

Review Questions

Level 1: Reviewing Facts and Terms

Match each item in column A with the most closely related item in column B. Use letters for answers in the spaces provided.

COLUMN A

_A_ 1. cytology
_D_ 2. physiology
_G_ 3. histology
___ 4. metabolism
_B_ 5. homeostasis
_L_ 6. muscle
_N_ 7. heart
_F_ 8. endocrine
___ 9. temperature regulation
___ 10. blood clot formation
_C_ 11. supine
_O_ 12. prone
_K_ 13. ventral body cavity
_I_ 14. dorsal body cavity
_M_ 15. pericardium

COLUMN B

a. study of tissues
b. constant internal environment
c. face up
d. study of functions
e. positive feedback
f. system
g. study of cells
h. negative feedback
i. brain and spinal cord
j. all chemical activity in body
k. thoracic and abdominopelvic
l. tissue
m. serous membrane
n. organ
o. face down

16. The process by which an organism increases the size and/or number of cells is called:
   (a) reproduction
   (b) adaptation
   (c) growth
   (d) metabolism

17. The terms that apply to the front of the body when in anatomical position are
   (a) posterior, dorsal
   (b) back, front
   (c) medial, lateral
   (d) anterior, ventral

18. A cut through the body that passes perpendicular to the long axis of the body and divides the body into a superior and inferior section is known as a:
   (a) sagittal section
   (b) transverse section
   (c) coronal section
   (d) frontal section

19. The cranial and spinal cavities are found in the:
   (a) ventral body cavity
   (b) thoracic cavity
   (c) dorsal body cavity
   (d) abdominopelvic cavity

20. The diaphragm, a flat muscular sheet, divides the ventral body cavity into a superior _____ cavity and an inferior _____ cavity.

    (a) pleural, pericardial

    (b) abdominal, pelvic

    (c) thoracic, abdominopelvic

    (d) cranial, thoracic

21. The mediastinum is the region between the:

    (a) lungs and heart

    (b) two pleural cavities

    (c) thorax and abdomen

    (d) heart and pericardium

Level 2: Reviewing Concepts

22. What basic functions are performed by all living things?

23. Beginning with the molecular level, list in correct sequence the levels of organization from the simplest level to the most complex level.

24. What is homeostatic regulation, and what is its physiological importance?

25. How does negative feedback differ from positive feedback?

26. Describe the position of the body when it is in the anatomical position.

27. As a surgeon, you perform an invasive procedure that necessitates cutting through the peritoneum. Are you more likely to be operating on the heart or on the stomach?

28. In which body cavity would each of the following organs or systems be found?

    (a) brain and spinal cord

    (b) cardiovascular, digestive, and urinary systems

    (c) heart, lungs

    (d) stomach, intestines

Level 3: Critical Thinking and Clinical Applications

29. A hormone called *calcitonin* from the thyroid gland is released in response to increased levels of calcium ions in the blood. If this hormone exerts negative feedback, what effect will its release have on blood calcium levels?

30. An anatomist wishes to make detailed comparisons of medial surfaces of the left and right sides of the brain. This work requires sections that will show the entire medial surface. Which kind of sections should be ordered from the lab for this investigation?

Answers to Concept Check Questions

Page 3

1. *Metabolism* refers to all of the chemical operations under way in the body. Organisms rely on complex chemical reactions to provide the energy for responsiveness, growth, reproduction, and movement.

2. A *histologist* investigates the structure and properties of tissues. *Histology* is considered a form of microscopic anatomy because it requires the use of a microscope to reveal the cells that make up tissues.

Page 14

1. Physiological systems can function normally only under carefully controlled conditions. Homeostatic regulation prevents potentially disruptive changes in the body's internal environment.

2. Positive feedback is useful in processes that must move quickly to completion once they have begun, such as blood clotting. It is harmful in situations in which a stable condition must be maintained, because it increases any departure from the desired condition. For example, positive feedback in the regulation of body temperature would cause a slight fever to spiral out of control, with fatal results. For this reason, most physiological systems exhibit negative feedback, which tends to oppose any departure from the norm.

3. When homeostasis fails, organ systems function less efficiently or begin to malfunction. The result is the state that we call *disease*. If the situation is not corrected, death can result.

Page 21

1. The two eyes would be separated by a *midsagittal section*.

2. The body cavity inferior to the diaphragm is the *abdominopelvic* (or *peritoneal*) cavity.

The Chemical Level of Organization

CHAPTER OUTLINE AND OBJECTIVES

Vocabulary Development

anabole....................a building up; *anabolism*
endo-.......................................inside; *endergonic*
exo- ....outside; *exergonic*
glyco- ....sugar; *glycogen*
hydro-....water + lysis, breakdown; *hydrolysis*
katabole ....a throwing down; *catabolism*
katalysis ....dissolution; *catalysis*
lipos ....fat; *lipids*
metabole ....change; *metabolism*
sakcharon..sugar
+ mono- ....single; *monosaccharide*
+ di- ....two; *disaccharide*
+ poly- ....many; *polysaccharide*

*C*HEMICALLY, *water is a simple substance: two atoms of hydrogen joined to one of oxygen. Yet when these three atoms are linked by chemical bonds, they produce a substance with many unusual properties. In this chapter you will learn how atoms are bound together to form molecules, the building blocks of living cells. You will also meet the larger molecules that form the structural framework of the body's cells and tissues and that enable our cells to grow, divide, and perform the many other functions that make life possible.*

2 THE CHEMICAL LEVEL OF ORGANIZATION

Atoms and Molecules • Chemical Notation • Chemical Reactions • Inorganic Compounds • Organic Compounds • Chemicals and Living Cells • Chapter Review

OUR STUDY OF THE HUMAN BODY begins at the most basic level of organization, that of individual atoms and molecules. The characteristics of all living and nonliving things—people, elephants, oranges, oceans, rocks, and air—result from the types of atoms involved and the ways those atoms combine and interact. The branch of science that deals with such interactions is **chemistry**. A familiarity with basic chemistry will help us understand how the properties of atoms can affect the anatomy and physiology of the cells, tissues, organs, and organ systems that make up the human body.

Atoms and Molecules

Matter is anything that occupies space and has mass, a property that, on earth, determines its weight. Matter occurs in one of three familiar states: solid (such as a rock), liquid (such as water), or gas (such as the atmosphere). All matter is composed of substances called **elements**, which cannot be broken down by heating or other ordinary physical means. The smallest piece of an element, and the simplest unit of matter, is an **atom**. Although physicists can split atoms apart, chemical reactions cannot change the basic identity of an atom. Atoms are so small that atomic measurements are most conveniently reported in billionths of a meter or *nanometers* (NA-nō-mē-terz) (nm). The very largest atoms approach half of 1 billionth of a meter (0.5 nm) in diameter. One million atoms placed end to end would be no longer than a period on this page.

Atoms contain three major types of subatomic particles: protons, neutrons, and electrons. Protons and neutrons are similar in size and mass, but **protons** (p^+) have a positive electrical charge, and **neutrons** (n^0) are neutral—that is, uncharged. **Electrons** (e^-) are much lighter, only 1/1836th as massive as protons, and have a negative electrical charge. Figure 2-1● is a diagrammatic view of a simple atom of the element *helium*. This atom contains two protons, two neutrons, and two electrons.

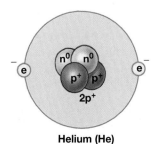

Helium (He)

● *Figure 2-1* **Atomic Structure**
An atom of helium contains two of each subatomic particles: two protons, two neutrons, and two electrons.

THE STRUCTURE OF AN ATOM

All atoms contain protons and electrons, normally in equal numbers. The number of protons in an atom is known as its **atomic number**. A chemical element is a substance that consists entirely of atoms with the same atomic number.

Table 2-1 lists the 13 most abundant elements in the human body. Each element is universally known by its own abbreviation, or chemical symbol. Most of the symbols are easily connected with the names of the elements, but a few, such as Na for sodium, are abbreviations of their original Latin names, in this case *natrium*. (Appendix II, the *periodic table*, gives the chemical symbols and atomic numbers of each element.)

Hydrogen, the simplest element, has an atomic number of 1 because its atom contains one proton. The proton is located in the center of the atom and forms the **nucleus**. Hydrogen

TABLE 2-1 *The Principal Elements in the Human Body*	
ELEMENT (% OF BODY WEIGHT)	**SIGNIFICANCE**
OXYGEN (O) (65)	A component of water and other compounds; oxygen gas is essential for respiration
CARBON (C) (18.6)	Found in all organic molecules
HYDROGEN (H) (9.7)	A component of water and most other compounds in the body
NITROGEN (N) (3.2)	Found in proteins, nucleic acids, and other organic compounds
CALCIUM (Ca) (1.8)	Found in bones and teeth; important for membrane function, nerve impulses, muscle contraction, and blood clotting
PHOSPHORUS (P) (1)	Found in bones and teeth, nucleic acids, and high-energy compounds
POTASSIUM (K) (0.4)	Important for proper membrane function, nerve impulses, and muscle contraction
SODIUM (Na) (0.2)	Important for membrane function, nerve impulses, and muscle contraction
CHLORINE (Cl) (0.2)	Important for membrane function and water absorption
MAGNESIUM (Mg) (0.06)	Required for activation of several enzymes
SULFUR (S) (0.04)	Found in many proteins
IRON (Fe) (0.007)	Essential for oxygen transport and energy capture
IODINE (I) (0.0002)	A component of hormones of the thyroid gland

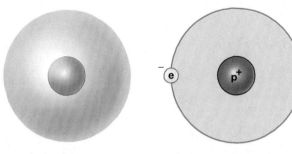

(a) Electron cloud or space-filling model　　**(b) Electron-shell model**

● *Figure 2-2* **Hydrogen Atoms**
(**a**) The electron cloud of a hydrogen atom is formed by the orbiting of an electron around the nucleus. (**b**) A two-dimensional model depicting the electron in an electron shell makes it easier to visualize the atom's components. A typical hydrogen nucleus contains a single proton and no neutrons.

atoms seldom contain neutrons, but when neutrons are present, they are also located in the nucleus. A single electron orbits the nucleus at high speed, forming an **electron cloud** (Figure 2-2a●). To simplify matters, this cloud is usually represented as a spherical **electron shell** (Figure 2-2b●).

Isotopes

The atoms of a single element can differ in terms of the number of neutrons in the nucleus. Such atoms of an element are called **isotopes**. The presence or absence of neutrons generally has no effect on the chemical properties of an atom of a particular element. As a result, isotopes can be distinguished from one another only by their **mass number**—the total number of protons plus neutrons in the nucleus. The nuclei of some isotopes may be unstable. Unstable isotopes are radioactive; that is, they spontaneously emit subatomic particles or radiation in measurable amounts. These *radioisotopes* are sometimes used in diagnostic procedures.

Atomic Weight

Atomic mass numbers are useful because they tell us the number of protons and neutrons in the nuclei of different atoms. However, they do not tell us the *actual* mass of an atom, because they do not take into account the masses of electrons and the slight difference between the masses of a proton and neutron. They also don't tell us the mass of a "typical" atom, since any element consists of a mixture of isotopes. It is therefore useful to know the *average mass* of an atom, and this value is an element's **atomic weight**. Atomic weight takes into account the mass of the subatomic particles and the relative proportions of any isotopes. For example, even though the atomic number of hydrogen is 1, the atomic weight of hydrogen is 1.0079. In this case, the atomic number and atomic weight differ primarily

because a few hydrogen atoms have a mass number of 2 (one proton plus one neutron) and an even smaller number have a mass number of 3 (one proton plus two neutrons). The atomic weights of the elements are included in Appendix II.

Electrons and Electron Shells

Atoms are electrically neutral; every positively charged proton is balanced by a negatively charged electron. These electrons occupy an orderly series of electron shells around the nucleus, and only the electrons in the outer shell can interact with other atoms. *The number of electrons in an atom's outer electron shell determine the chemical properties of that element.*

Atoms with an unfilled outer electron shell are unstable—that is, they will react with other atoms, usually in ways that give them full outer electron shells. An atom with a filled outer shell will not interact with other atoms. The first electron shell (the one closest to the nucleus) is filled when it contains two electrons. A hydrogen atom has one electron in this electron shell (Figure 2-2b●), and hydrogen atoms can react with many other atoms. A helium atom has two electrons in this electron shell (Figure 2-1●). Helium is called an inert gas because the outer electron shell is full and therefore stable. Helium atoms will neither react with one another nor combine with atoms of other elements.

The second electron shell can contain up to eight electrons. Carbon, with an atomic number of 6, has six electrons. In a carbon atom the first shell is filled (two electrons), and the second shell contains four electrons (Figure 2-3a●). In a neon atom (atomic number 10), the second shell is filled; neon is another inert gas (Figure 2-3b●).

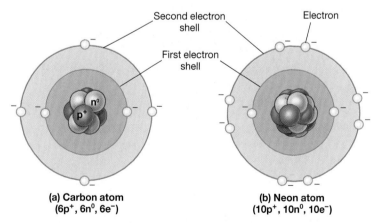

(a) Carbon atom
($6p^+$, $6n^0$, $6e^-$)　　**(b) Neon atom**
($10p^+$, $10n^0$, $10e^-$)

● *Figure 2-3* **Atoms and Electron Shells**
(**a**) The first electron shell can hold only two electrons. In a carbon atom, with six protons and six electrons, the third through sixth electrons occupy the second electron shell. (**b**) The second shell can hold up to eight electrons. A neon atom has 10 protons and 10 electrons; thus both the first and second electron shells are filled. Notice that the nuclei of helium, carbon, and neon contain neutrons as well as protons.

2 THE CHEMICAL LEVEL OF ORGANIZATION

Atoms and Molecules • Chemical Notation • Chemical Reactions • Inorganic Compounds • Organic Compounds • Chemicals and Living Cells • Chapter Review

CHEMICAL BONDS AND CHEMICAL COMPOUNDS

An atom with a full outer electron shell is very stable and not reactive. Atoms with unfilled outer electron shells can achieve stability by sharing, gaining, or losing electrons through chemical reactions. Many of these chemical reactions produce **molecules**, chemical structures each containing more than one atom. Molecules called **compounds** contain atoms of more than one element. A compound is a new chemical substance with properties that can be quite different from those of its component elements. For example, a mixture of hydrogen and oxygen gases is highly flammable, but chemically combining hydrogen and oxygen atoms produces a compound, water, that can put out fires.

Ionic Bonds

Atoms are electrically neutral because the number of protons (each with a +1 charge) is equal to the number of electrons (each with a −1 charge). If an atom loses an electron, it will exhibit a charge of +1 because there will be one proton without a corresponding electron; losing a second electron would

TABLE 2-2 *The Most Common Ions in Body Fluids*

CATIONS	ANIONS
Na^+ (sodium)	Cl^- (chloride)
K^+ (potassium)	HCO_3^- (bicarbonate)
Ca^{2+} (calcium)	HPO_4^{2-} (biphosphate)
Mg^{2+} (magnesium)	SO_4^{2-} (sulfate)

leave the atom with a charge of +2. Similarly, adding one or two extra electrons to the atom will give it a respective charge of −1 or −2. Atoms or molecules that have a (+) or (−) charge are called **ions**. Ions with a positive charge are **cations** (KAT-ī-ons); those with a negative charge are **anions** (AN-ī-ons). Table 2-2 lists several important ions in body fluids.

In an **ionic** (ī-ON-ik) **bond**, anions and cations are held together by the attraction between positive and negative charges. An example of a substance held together by ionic bonds is ordinary table salt (Figure 2-4b●).

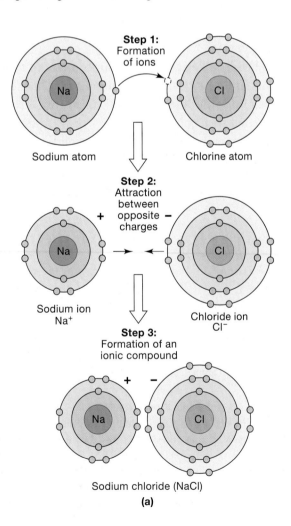

● *Figure 2-4* **Ionic Bonding**
(**a**) **Step 1**: A sodium atom loses an electron, which is accepted by a chlorine atom. **Step 2**: Because the sodium (Na^+) and chloride (Cl^-) ions have opposite charges, they are attracted to one another. **Step 3**: The association of sodium and chloride ions forms the ionic compound sodium chloride. (**b**) Large numbers of sodium and chloride ions form a crystal of sodium chloride.

Step 1: Formation of ions

Sodium atom

Chlorine atom

Step 2: Attraction between opposite charges

Sodium ion
Na^+

Chloride ion
Cl^-

Step 3: Formation of an ionic compound

Sodium chloride (NaCl)
(a)

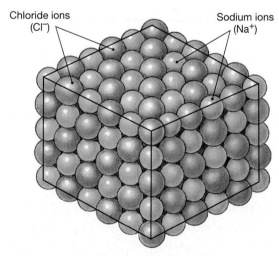

Chloride ions (Cl^-)

Sodium ions (Na^+)

(b) Sodium chloride crystal

The steps in the formation of an ionic bond are illustrated in Figure 2-4a●. In this process, a sodium atom donates an electron to a chlorine atom. This loss of an electron creates a *sodium ion* with a +1 charge and a *chloride ion* with a −1 charge. The two ions do not move apart after the electron transfer, because the positively charged sodium ion is attracted to the negatively charged chloride ion. The combination of oppositely charged ions forms the *ionic compound* **sodium chloride**, the chemical name for table salt.

Covalent Bonds

Another way atoms can complete their outer electron shells is by sharing electrons with other atoms. The result is a molecule held together by **covalent** (kō-VĀ-lent) **bonds**.

For example, individual hydrogen atoms, as diagrammed in Figure 2-2●, are not found in nature. Instead, we find hydrogen molecules (Figure 2-5a●). Molecular hydrogen is a gas present in the atmosphere in very small quantities. The two hydrogen atoms share their electrons, with each electron whirling around both nuclei. The sharing of one pair of electrons creates a **single covalent bond**.

Oxygen, with an atomic number of 8, has two electrons in its first energy level and six in the second. Oxygen atoms (Figure 2-5b●) reach stability by sharing two pairs of electrons, forming a **double covalent bond**. Molecular oxygen is an atmospheric gas that is very important to living organisms; our cells would die without a constant supply of oxygen.

In our bodies, the chemical processes that consume oxygen also produce carbon dioxide (CO_2) as a waste product. The oxygen atoms in a carbon dioxide molecule form double covalent bonds with the carbon atom, as shown in Figure 2-5c●.

Covalent bonds are very strong because the shared electrons tie the atoms together. In most covalent bonds the atoms remain electrically neutral because the electrons are shared equally. Such bonds are called **nonpolar covalent bonds**. Nonpolar covalent bonds between carbon atoms create the stable framework of the large molecules that make up most of the structural components of the human body.

Elements differ in how strongly they hold shared electrons. An unequal sharing creates a **polar covalent bond**. For example, in a molecule of water, an oxygen atom forms covalent bonds with two hydrogen atoms. The oxygen atom has a much stronger attraction for the shared electrons than do the hydrogen atoms, so their electrons spend most of their time with the oxygen atom. Because of the two extra electrons, the oxygen atom develops a slight negative charge. At the same time, the hydrogen atoms develop a slight positive charge, because their electrons are away part of the time.

Hydrogen Bonds

In addition to ionic and covalent bonds, weaker attractive forces act between adjacent molecules and between atoms within a large molecule. Hydrogen bonds are the most important of these attractive forces.

Hydrogen atoms often form polar covalent bonds with the atoms of elements such as oxygen, nitrogen, or both. In the process, the hydrogen atom develops a slight positive charge and its partner develops a weak negative charge. A **hydrogen bond** is the attraction between such a hydrogen atom and a negatively charged atom in another molecule or at another site within the same molecule. Hydrogen bonds do not create molecules, but they can alter molecular shapes or pull molecules

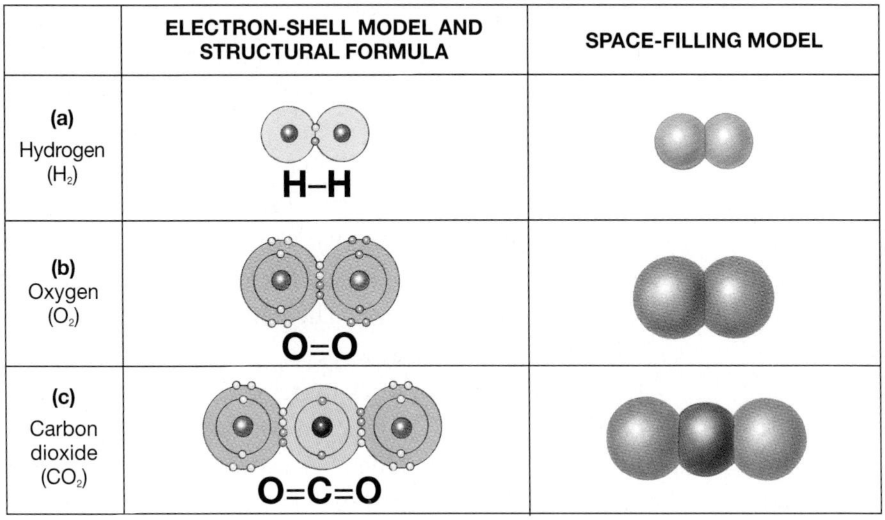

	ELECTRON-SHELL MODEL AND STRUCTURAL FORMULA	SPACE-FILLING MODEL
(a) Hydrogen (H_2)	H–H	
(b) Oxygen (O_2)	O=O	
(c) Carbon dioxide (CO_2)	O=C=O	

● *Figure 2-5* **Covalent Bonds**
(**a**) In a molecule of hydrogen, two hydrogen atoms share their electrons such that each has a filled outer electron shell. This sharing creates a single covalent bond. (**b**) A molecule of oxygen consists of two oxygen atoms that share two pairs of electrons. The result is a double covalent bond. (**c**) In a molecule of carbon dioxide, a central carbon atom forms double covalent bonds with a pair of oxygen atoms.

2 THE CHEMICAL LEVEL OF ORGANIZATION

Atoms and Molecules • **Chemical Notation** • **Chemical Reactions** • Inorganic Compounds • Organic Compounds • Chemicals and Living Cells • Chapter Review

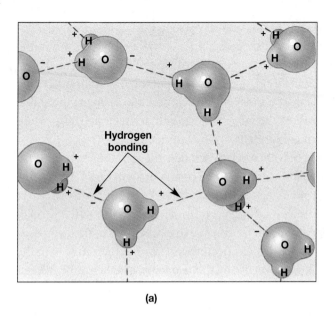

Hydrogen bonding

(a)

(b)

● *Figure 2-6* **Hydrogen Bonds**
(a) The unequal sharing of electrons in a water molecule causes each of its two hydrogen atoms to have a slight positive charge and its oxygen atom to have a slight negative charge. Attraction between a hydrogen atom of one water molecule and the oxygen atom of another is a hydrogen bond (indicated by dashed lines). (b) Hydrogen bonding between water molecules at a free surface restricts evaporation and creates surface tension.

together. For example, hydrogen bonding occurs between water molecules (Figure 2-6●). At the water surface, this attraction slows the rate of evaporation and creates the phenomenon known as surface tension. **Surface tension** acts as a barrier that keeps small objects from entering the water; it is the reason that insects can walk across the surface of a pond or puddle. Surface tension in a layer of tears keeps small dust particles from touching the surface of the eye.

CONCEPT CHECK QUESTIONS

Answers on page 51

① Oxygen and neon are both gases at room temperature. Oxygen combines readily with other elements, but neon does not. Why?

② How is it possible for two samples of hydrogen to contain the same number of atoms but have different weights?

③ Which kind of bond holds atoms in a water molecule together? Which kind of bond attracts water molecules to each other?

Chemical Notation

Complex chemical compounds and reactions are most easily described with a simple form of "chemical shorthand" known as **chemical notation**. The rules of chemical notation are summarized in Table 2-3.

Chemical Reactions

Living cells remain alive by controlling chemical reactions. In every **chemical reaction**, bonds between atoms are broken, and atoms are rearranged into new combinations. In effect, each cell is a chemical factory. For example, growth, mainte-

nance and repair, secretion, and contraction all involve complex chemical reactions. **Metabolism** (meh-TAB-ō-lizm; *metabole*, change) refers to all of the chemical reactions in the body. Cells use chemical reactions to provide the energy to maintain homeostasis and to perform essential functions.

BASIC ENERGY CONCEPTS

Most people are familiar with the concepts of work, energy, and heat. **Work** is movement or a change in the physical structure of matter. **Energy** is the capacity to perform work; movement or physical change cannot occur unless energy is provided. There are two major types of energy: *kinetic energy* and *potential energy*.

Kinetic energy is the energy of motion. When you fall off a ladder, it is kinetic energy that does the damage. **Potential energy** is stored energy. It may result from the position of an object (as when a book sits on a high shelf) or its physical structure (as when a spring is compressed or stretched). Kinetic energy had to be used to lift the book and stretch or compress the spring. The potential energy is converted back into kinetic energy when the book falls or the spring returns to its resting length; the energy released can be used to perform work.

A conversion between potential energy and kinetic energy is not 100 percent efficient. Each time an energy exchange occurs, some of the energy is released as heat. *Heat* is an increase in random molecular motion. The temperature of an object is directly related to its heat content.

Living cells perform work in many forms. For example, when a skeletal muscle contracts, it performs work; potential energy is converted into kinetic energy, and heat is released. The amount of heat is related to the amount of work done. As a result, when you exercise, your body temperature rises.

TABLE 2-3 *Rules of Chemical Notation*

1. The abbreviation of an element indicates one atom of that element:

$$H = \text{an atom of hydrogen}; O = \text{an atom of oxygen}$$

2. A number preceding the abbreviation of an element indicates more than one atom:

$$2\,H = \text{two individual atoms of hydrogen}$$

$$2\,O = \text{two individual atoms of oxygen}$$

3. A subscript following the abbreviation of an element indicates a molecule with that number of atoms:

$$H_2 = \text{a hydrogen molecule composed of two hydrogen atoms}$$

$$O_2 = \text{an oxygen molecule composed of two oxygen atoms}$$

4. In a description of a chemical reaction, the interacting participants are called **reactants**, and the reaction generates one or more **products**. An arrow indicates the direction of the reaction, from reactants (usually on the left) to products (usually on the right). In the reaction below, two atoms of hydrogen combine with one atom of oxygen to produce a single molecule of water.

$$2\,H + O \longrightarrow H_2O$$

5. A superscript plus or minus sign following the abbreviation for an element indicates an ion. A single plus sign indicates an ion with a charge of +1 (loss of one electron). A single minus sign indicates an ion with a charge of −1 (gain of one electron). If more than one electron has been lost or gained, the charge on the ion is indicated by a number preceding the plus or minus.

$$Na^+ = \text{one sodium ion (has lost 1 electron)}$$

$$Cl^- = \text{one chloride ion (has gained 1 electron)}$$

$$Ca^{2+} = \text{one calcium ion (has lost 2 electrons)}$$

6. Chemical reactions neither create nor destroy atoms—they merely rearrange them into new combinations. Therefore, the numbers of atoms of each element must always be the same on both sides of the equation. When this is the case, the equation is balanced.

$$\text{Unbalanced: } H_2 + O_2 \longrightarrow H_2O$$

$$\text{Balanced: } 2\,H_2 + O_2 \longrightarrow 2\,H_2O$$

TYPES OF REACTIONS

Three types of chemical reactions are important to the study of physiology: *decomposition reactions*, *synthesis reactions*, and *exchange reactions*.

Decomposition Reactions

A **decomposition reaction** breaks a molecule into smaller fragments. Such reactions occur during digestion when food molecules are broken into smaller pieces. You could diagram a typical decomposition reaction as:

$$AB \longrightarrow A + B$$

Catabolism (kah-TAB-o-lizm; *katabole*, a throwing down) refers to the breakdown of complex molecules within cells. A chemical bond contains potential energy that is released when that bond is broken. Our cells can capture some of that energy and use it to power essential functions such as growth, movement, and reproduction.

Synthesis Reactions

Synthesis (SIN-the-sis) is the opposite of decomposition. A synthesis reaction assembles larger molecules from smaller components. These relatively simple reactions could be diagrammed as:

$$A + B \longrightarrow AB$$

A and B could be individual atoms that combine to form a molecule, or they could be individual molecules combining to form larger, more complex structures. Synthesis always involves the formation of new chemical bonds, whether the reactants are atoms or molecules.

Anabolism (a-NAB-o-lizm; *anabole*, a building up) is the synthesis of new compounds in the body. Because it takes energy to create a chemical bond, anabolism is usually an "uphill" process. Living cells are constantly balancing their chemical activities, with catabolism providing the energy needed to support anabolism as well as other vital functions.

Exchange Reactions

In an **exchange reaction**, parts of the reacting molecules are shuffled around, as in:

$$AB + CD \longrightarrow AD + CB$$

This equation has two reactants and two products. Although the reactants and products contain the same components

2 THE CHEMICAL LEVEL OF ORGANIZATION

Atoms and Molecules • Chemical Notation • Chemical Reactions • **Inorganic Compounds** • Organic Compounds • Chemicals and Living Cells • Chapter Review

(A, B, C, and D), they are present in different combinations. In an exchange reaction, the reactant molecules AB and CD break apart (a decomposition) before they interact with each other to form AD and CB (a synthesis). An example of such a reaction is the exchange of the components of sodium hydroxide (NaOH) and hydrochloric acid (HCl) to make table salt and water:

$$NaOH + HCl \longrightarrow NaCl + H_2O$$

If breaking the old bonds releases more energy than it takes to create the new ones, the exchange reaction will release energy, usually in the form of heat. Such reactions are said to be **exergonic** (*exo-*, outside). If the energy required for synthesis exceeds the amount released by the associated decomposition reaction, additional energy (usually heat) must be provided. Such reactions are called **endergonic** (*endo-*, inside) because they absorb heat.

REVERSIBLE REACTIONS

Many important biological reactions are freely reversible. Such reactions can be diagrammed as:

$$A + B \longleftrightarrow AB$$

This equation reminds you that there are really two reactions occurring simultaneously, one a synthesis (A + B $\longrightarrow$ AB) and the other a decomposition (AB $\longrightarrow$ A + B). At **equilibrium** (ē-kwi-LIB-rē-um), the two rates are in balance. As fast as a molecule of AB forms, another degrades into A + B. As a result, the number of A, B, and AB molecules present at any given moment does not change. Altering the concentrations of one or more of these molecules will temporarily upset the equilibrium. For example, adding additional molecules of A and B will accelerate the synthesis reaction (A + B $\longrightarrow$ AB). As the concentration of AB rises, however, so does the rate of the decomposition reaction (AB $\longrightarrow$ A + B), until a new equilibrium is established.

CONCEPT CHECK QUESTIONS
Answers on page 51

❶ In living cells, glucose, a six-carbon molecule, is converted into two three-carbon molecules by a reaction that yields energy. How would you classify this reaction?

❷ If the product of a reversible reaction is continuously removed, what do you think the effect will be on the equilibrium?

ACIDS AND BASES

An **acid** is any substance that dissociates to *release* hydrogen ions. (Because a hydrogen ion consists solely of a proton, hydro-

gen ions are often referred to simply as protons, and acids as "proton donors.") A *strong acid* dissociates completely in solution. Hydrochloric acid (HCl) is an excellent example:

$$HCl \longrightarrow H^+ + Cl^-$$

The stomach produces this powerful acid to assist in the breakdown of food.

A **base** is a substance that removes hydrogen ions from a solution. Many common bases are compounds that dissociate in solution to liberate a hydroxide ion (OH^-). Hydroxide ions have a strong affinity for hydrogen ions and quickly react with them to form water molecules. A *strong base* dissociates completely in solution. For example, sodium hydroxide (NaOH) dissociates in solution as:

$$NaOH \longrightarrow Na^+ + OH^-$$

Strong bases have a variety of industrial and household uses; drain openers and lye are two familiar examples.

Weak acids and *weak bases* do not dissociate completely in solution. The human body contains weak bases that are important in counteracting acids produced by cellular metabolism.

pH

The concentration of hydrogen ions in blood or other body fluids is important because hydrogen ions are extremely reactive. In excessive numbers, they will break chemical bonds, change the shapes of complex molecules and disrupt cell and tissue functions. The concentration of hydrogen ions must therefore be precisely regulated.

The concentration of hydrogen ions is usually reported in terms of the **pH** of the solution. The pH value is a number between 0 and 14. Pure water has a pH of 7. A solution with a pH of 7 is called *neutral*, because it contains equal numbers of hydrogen and hydroxide ions. A solution with a pH below 7 is called *acidic* (a-SI-dik), because there are more hydrogen ions than hydroxide ions. A pH above 7 is called *basic*, or *alkaline* (AL-kah-lin), because hydroxide ions outnumber hydrogen ions.

The pH values of some common liquids are indicated in Figure 2-7●. The pH of the blood normally ranges from 7.35 to 7.45. Variations in pH outside this range can damage cells and disrupt normal cellular functions. For example, a blood pH below 7 can produce coma, and a blood pH higher than 7.8 usually causes uncontrollable, sustained muscular contractions.

Buffers and pH

Buffers are compounds that stabilize pH by removing or replacing hydrogen ions. Antacids such as Alka-Seltzer®, Rolaids®,

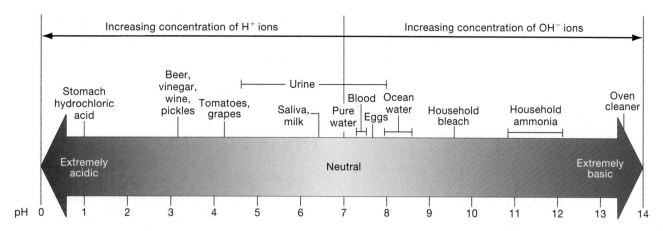

● *Figure 2-7* **pH and Hydrogen Ion Concentration**
An increase or decrease of one unit corresponds to a tenfold change in H⁺ concentration.

and Tums® are buffers that tie up excess hydrogen ions in the stomach. The normal pH of most body fluids ranges from 7.35 to 7.45. A variety of buffers, including sodium bicarbonate, are responsible for regulating pH. We will consider the role of buffers and pH control in Chapter 18.

CONCEPT CHECK QUESTIONS
Answers on page 51

1 What is the difference between an acid and a base?
2 Why would an extreme change in pH of body fluids be undesirable?
3 How does an antacid decrease stomach discomfort?

Inorganic Compounds

The rest of this chapter focuses on nutrients and metabolites. **Nutrients** are the essential elements and molecules obtained from the diet. **Metabolites** (me-TAB-o-līts) include all of the molecules synthesized or broken down by chemical reactions inside our bodies. Like all chemical substances, nutrients and metabolites can be broadly categorized as inorganic or organic. Generally speaking, **inorganic compounds** are small molecules that do not contain carbon and hydrogen atoms. **Organic compounds** are primarily composed of carbon and hydrogen atoms, and they can be much larger and more complex than inorganic compounds.

The most important inorganic substances in the human body are carbon dioxide, oxygen, water, inorganic acids and bases, and salts.

CARBON DIOXIDE AND OXYGEN

Carbon dioxide (CO_2) is produced by cells through normal metabolic activity. It is transported in the blood and released into the air in the lungs. Oxygen (O_2), an atmospheric gas, is absorbed at the lungs, transported in the blood, and consumed by cells throughout the body. The chemical structures of these compounds were introduced earlier in the chapter.

WATER AND ITS PROPERTIES

Water, H_2O, is the single most important constituent of the body, accounting for almost two-thirds of total body weight. A change in body water content can have fatal consequences because virtually all physiological systems will be affected.

Three general properties of water are particularly important to our discussion of the human body:

- *Water is an excellent solvent.* Water dissolves a remarkable variety of inorganic and organic molecules, creating a *solution*. As they dissolve, the molecules break apart, releasing ions or molecules that become uniformly dispersed throughout the solution. The chemical reactions within living cells occur in solution, and the watery component, or plasma, of blood carries dissolved nutrients and waste products throughout the body. Most chemical reactions in the body take place in solution.

- *Water has a very high heat capacity.* It takes a lot of energy to make water boil, and a large amount of energy must be removed before water will freeze. As a result, the water in our cells remains a liquid over a wide range of environmental temperatures. Furthermore, it circulates within the body as the blood transports and redistributes heat. For example, heat absorbed as the blood flows through active muscles will be released when the blood reaches vessels in the relatively cool body surface.

- *Water is an essential reactant in the chemical reactions of living systems.* In such reactions, water is commonly involved in the synthesis and decomposition of various organic

2 THE CHEMICAL LEVEL OF ORGANIZATION

Atoms and Molecules • Chemical Notation • Chemical Reactions • Inorganic Compounds • **Organic Compounds** • Chemicals and Living Cells • Chapter Review

compounds. Water molecules are released during the synthesis of large organic molecules and absorbed during their decomposition.

Solutions

A **solution** consists of a fluid *solvent* and dissolved *solutes*. In biological solutions, the solvent is usually water and the solutes may be inorganic or organic. Inorganic compounds held together by ionic bonds undergo **ionization** (ī-on-i-ZĀ-shun), or *dissociation* (dis-sō-sē-Ā-shun), in solution. As shown in Figure 2-8●, ionic bonds are broken apart as individual ions form hydrogen bonds with water molecules. This process produces a mixture of cations and anions surrounded by so many water molecules that they are unable to re-form their original bonds.

INORGANIC ACIDS AND BASES

The discussion of acids and bases on p. 34 introduced an inorganic acid, hydrochloric acid (HCl), and an inorganic base, sodium hydroxide (NaOH). Several other inorganic acids are found in body fluids, including carbonic acid, sulfuric acid, and phosphoric acid. These acids, produced during normal metabolism, will be considered further in Chapters 15 and 18.

SALTS

A **salt** is an ionic compound consisting of any cation except a hydrogen ion and any anion except a hydroxide ion. Salts are held together by ionic bonds, and in water they dissociate, releasing cations and anions. For example, table salt (NaCl) in solution dissociates into Na^+ and Cl^- ions; these are the most abundant ions in body fluids.

Salts are examples of **electrolytes** (ē-LEK-trō-līts), compounds whose ions will conduct an electrical current in solution. For example, sodium ions (Na^+), potassium ions (K^+), calcium ions (Ca^{2+}), and chloride ions (Cl^-) are released by the dissociation of electrolytes in blood and other body fluids. Alterations in the concentrations of these ions in body fluids will disturb almost every vital function. For example, declining potassium levels will lead to a general muscular paralysis, and rising concentrations will cause weak and irregular heartbeats.

Organic Compounds

Organic compounds always contain the elements carbon and hydrogen and generally oxygen as well. Many organic molecules contain long chains of carbon atoms linked by covalent bonds. These carbon atoms often form additional covalent bonds with hydrogen or oxygen atoms, and less often, with nitrogen, phosphorus, sulfur, iron, or other elements to produce the complex organic molecules characteristic of living organisms.

Although inorganic acids and bases were used as examples earlier in the chapter, there are also important organic acids and bases. For example, *lactic acid* is an organic acid generated by active muscle tissues.

This section focuses on four major classes of large organic molecules: *carbohydrates*, *lipids*, *proteins*, and *nucleic acids*. We will also consider the high-energy compounds that play a crucial role in many of the chemical reactions under way within our cells. In addition, the human body contains small quantities of many other organic compounds whose structures and functions will be considered in later chapters.

CARBOHYDRATES

A **carbohydrate** (kar-bō-HĪ-drāt) is a molecule that contains carbon, hydrogen, and oxygen in a ratio near $1:2:1$. Familiar carbohydrates include the sugars and starches that make up roughly half of the typical U.S. diet. Our tissues can break down most carbohydrates, and although they sometimes have other

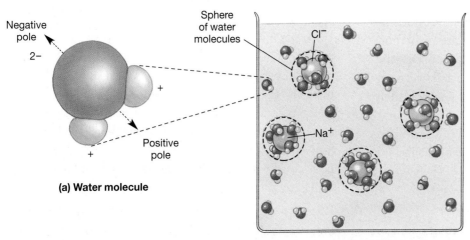

Negative pole
2−
+
Positive pole
+

(a) Water molecule

Sphere of water molecules
Cl^-
Na^+

(b) Sodium chloride in solution

● *Figure 2-8* **Water Molecules and Solutions** (a) In a water molecule, oxygen forms polar covalent bonds with two hydrogen atoms. Because the hydrogen atoms are positioned toward one end of the molecule, the molecule has an uneven distribution of charges. This creates positive and negative poles. (b) Ionic compounds dissociate in water as the polar water molecules disrupt the ionic bonds. The ions remain in solution because the surrounding water molecules prevent the ionic bonds from re-forming.

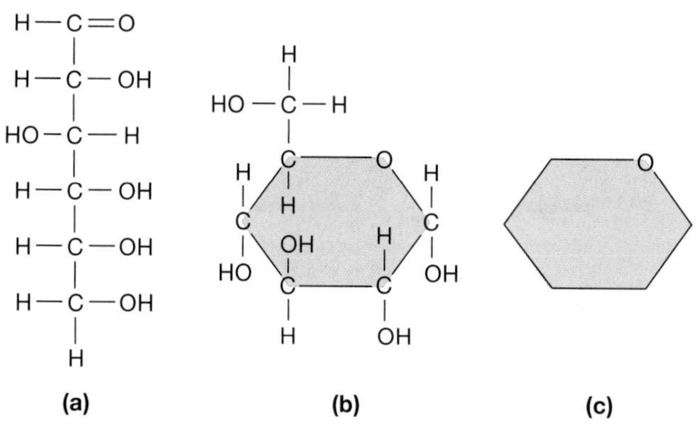

● *Figure 2-9* **Glucose**
(a) The straight-chain structural formula. (b) The ring form that is most common in nature. (c) An abbreviated diagram of the ring form of glucose. In such carbon ring diagrams, symbols of elements other than carbon are shown and the carbon atoms that occupy the points of the ring are not shown.

functions, carbohydrates are most important as sources of energy. Despite their importance as an energy source, however, carbohydrates account for less than 3 percent of our total body weight. The three major types of carbohydrates are *monosaccharides*, *disaccharides*, and *polysaccharides*.

Monosaccharides

A **simple sugar**, or **monosaccharide** (mon-ō-SAK-ah-rīd; *mono-*, single + *sakcharon*, sugar), is a carbohydrate containing from three to seven carbon atoms. Included within this group is **glucose** (GLOO-kōs), $C_6H_{12}O_6$, the most important metabolic "fuel" in the body (Figure 2-9●).

Disaccharides and Polysaccharides

Carbohydrates other than simple sugars are complex molecules composed of monosaccharide building blocks. Through a **dehydration synthesis** reaction, the removal of a water molecule joins two simple sugars to form a **disaccharide** (dī-SAK-ah-rīd; *di-*, two) (Figure 2-10a●). Disaccharides such as **sucrose** (table sugar) have a sugary taste and are quite soluble. Many foods contain disaccharides, but they must be disassembled before they can be broken down to provide useful energy. The breakdown of a disaccharide into its component simple sugars by the addition of a water molecule is an example of **hydrolysis** (hī-DROL-i-sis; *hydro-*, water + *lysis*, breakdown) (Figure 2-10b●). Most junk foods, including candy and soft drinks, abound in simple sugars (often fructose) and disaccharides such as sucrose.

Larger carbohydrate molecules are called **polysaccharides** (pol-ē-SAK-ah-rīdz; *poly-*, many). *Starches* are glucose-based polysaccharides important in our diets. Most of the starches in our diet are synthesized by plants. The human digestive tract can break these molecules into simple sugars, and starches such as those found in potatoes and grains are important energy sources. In contrast, *cellulose*, a component of the cell walls of plants, is a polysaccharide that our bodies cannot digest. The cellulose of foods such as celery contribute to the bulk of digestive wastes but are useless as an energy source.

Glycogen (GLĪ-ko-jen), or *animal starch*, is a branched polysaccharide composed of interconnected glucose molecules (Figure 2-10c●). Like most other large polysaccharides, glycogen will not dissolve in water or other body fluids. Liver and muscle tissues manufacture and store significant amounts of glycogen. When these tissues have a high demand for energy, glycogen molecules are broken down into glucose; when demands are low, the tissues absorb or synthesize glucose and rebuild glycogen reserves.

Table 2-4 summarizes information concerning the carbohydrates.

Some people cannot tolerate sugar for medical reasons; others avoid it because they do not want to gain weight (excess sugars are stored as fat). Many of these people use artificial sweeteners in their foods and beverages. These compounds have a very sweet taste but either cannot be broken down in the body or are used in such small amounts that their breakdown does not contribute to the overall energy balance of the body.

TABLE 2-4	*Carbohydrates in the Body*		
STRUCTURE	**EXAMPLES**	**PRIMARY FUNCTIONS**	**REMARKS**
MONOSACCHARIDES (SIMPLE SUGARS)	Glucose, fructose	Energy source	Manufactured in the body and obtained from food; found in body fluids
DISACCHARIDES	Sucrose, lactose, maltose	Energy source	Sucrose is table sugar, lactose is present in milk; all must be broken down to monosaccharides before absorption
POLYSACCHARIDES	Glycogen	Storage of glucose molecules	Glycogen is in animal cells; other starches and cellulose are in plant cells

2 THE CHEMICAL LEVEL OF ORGANIZATION

Atoms and Molecules • Chemical Notation • Chemical Reactions • Inorganic Compounds • **Organic Compounds** • Chemicals and Living Cells • Chapter Review

(a) During dehydration synthesis two molecules are joined by the removal of a water molecule.

(b) Hydrolysis reverses the steps of dehydration synthesis; a complex molecule is broken down by the addition of a water molecule.

● *Figure 2-10* **The Formation and Breakdown of Complex Sugars and Glycogen**

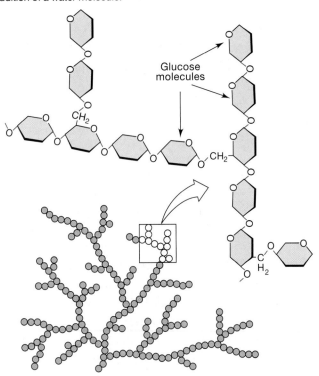

(c) Glycogen, a branching chain of glucose molecules, is stored in muscle cells and liver cells.

LIPIDS

Lipids (*lipos*, fat) also contain carbon, hydrogen, and oxygen, but because they have relatively less oxygen than do carbohydrates, the ratios do not approximate 1 : 2 : 1. In addition, lipids may contain small quantities of other elements, such as phosphorus, nitrogen, or sulfur. Familiar lipids include *fats*, *oils*, and *waxes*. Most lipids are insoluble in water, but special transport mechanisms carry them in the circulating blood.

Lipids form essential structural components of all cells. Lipid deposits also serve as energy sources and reserves. Based on equal weights, lipids provide roughly twice as much energy as carbohydrates when broken down in the body. For this reason there has been great interest in developing *fat substitutes*, such as Olestra®, that provide less energy but have the same taste and texture as lipids.

Lipids normally account for roughly 12 percent of our total body weight. There are many kinds of lipids in the body. The major types are *fatty acids*, *fats*, *steroids*, and *phospholipids* (Table 2-5, p. 39).

Fatty Acids

Fatty acids are long chains of carbon atoms with attached hydrogen atoms that end in a carboxyl group ($-$COOH). Because this group loses a hydrogen ion (H^+) in solution, it is also called a carboxylic acid group. While this group is soluble in water, the rest of the carbon chain is relatively insoluble. Figure 2-11a● shows a representative fatty acid, *lauric acid*.

In a **saturated** fatty acid, such as lauric acid, the four single covalent bonds of each carbon atom permit each neighboring carbon to link to each other and to link to two hydrogen atoms. If any of the carbon-to-carbon bonds are double covalent bonds, then fewer hydrogen atoms are present and the fatty acid is **unsaturated**. The structure of saturated and unsaturated fatty acids is shown in Figure 2-11b●. A **polyunsaturated** fatty acid contains multiple unsaturated bonds.

Both saturated and unsaturated fatty acids can be broken down for energy, but a diet containing large amounts of saturated

TABLE 2-5 *Representative Lipids and Their Functions in the Body*

LIPID TYPE	EXAMPLES	PRIMARY FUNCTIONS	REMARKS
FATTY ACIDS	Lauric acid	Energy sources	Absorbed from food or synthesized in cells; transported in the blood for use in many tissues
FATS	Monoglycerides, diglycerides, triglycerides	Energy source, energy storage, insulation, and physical protection	Stored in fat deposits; must be broken down to fatty acids and glycerol before they can be used as an energy source
STEROIDS	Cholesterol	Structural component of cell membranes, hormones, digestive secretions in bile	All have the same carbon-ring framework
PHOSPHOLIPIDS	Lecithin	Structural components of cell membranes	Composed of fatty acids and nonlipid molecules

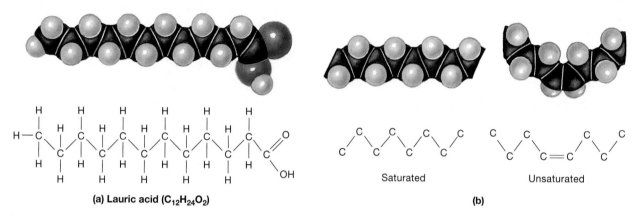

(a) Lauric acid ($C_{12}H_{24}O_2$)

Saturated   Unsaturated

(b)

● *Figure 2-11* **Fatty Acids**
(**a**) Lauric acid shows the basic structure of a fatty acid: a long chain of carbon atoms and a carboxylic acid group (—COOH).
(**b**) A fatty acid is saturated or unsaturated. Unsaturated fatty acids have double covalent bonds; their presence causes a sharp bend in the molecule.

fatty acids increases the risk of heart disease and other circulatory problems. Butter, fatty meat, and ice cream are popular dietary sources of saturated fatty acids. Vegetable oils such as olive oil or corn oil contain a mixture of unsaturated fatty acids.

Fats

Individual fatty acids cannot be strung together in a chain by dehydration synthesis, as simple sugars can. But they can be attached to another compound, **glycerol** (GLI-se-rol), to make a **fat**, through a similar reaction. In a **triglyceride** (trī-GLI-se-rīd), a glycerol molecule is attached to three fatty acids (see Figure 2-12●). Triglycerides are the most common fats in the body. In addition to serving as an energy reserve, fat deposits under the skin serve as insulation, and a mass of fat around a delicate organ, such as a kidney, provides a protective cushion. *Saturated fats,* triglycerides containing saturated fatty acids,

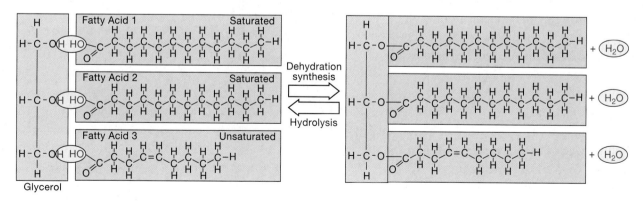

● *Figure 2-12* **Triglyceride Formation**
The formation of a triglyceride involves the attachment of three fatty acids to the carbons of a glycerol molecule. This example shows the attachment of one unsaturated and two saturated fatty acids to a glycerol molecule.

2 THE CHEMICAL LEVEL OF ORGANIZATION

Atoms and Molecules • Chemical Notation • Chemical Reactions • Inorganic Compounds • **Organic Compounds** • Chemicals and Living Cells • Chapter Review

● *Figure 2-13* **A Cholesterol Molecule**
Cholesterol, like all steroids, contains a complex four-ring structure.

are usually solid at room temperature. *Unsaturated fats*, triglycerides containing unsaturated fatty acids, are usually liquid at room temperature.

Steroids

Steroids are large lipid molecules composed of four connected rings of carbon atoms, quite unlike the linear carbon chains of fatty acids. *Cholesterol* is probably the best-known steroid (Figure 2-13●). All of our cells are surrounded by *cell membranes* that contain cholesterol, and some chemical messengers, or *hormones*, are derived from cholesterol. Examples include the sex hormones, such as testosterone and estrogen.

The cholesterol needed to maintain cell membranes and manufacture steroid hormones comes from two sources. One source is the diet; meat, cream, and egg yolks are especially rich in cholesterol. The second is the body itself, for the liver can synthesize large amounts of cholesterol. The ability of the body to synthesize this steroid can make it difficult to control blood cholesterol levels by dietary restriction alone. This difficulty can have serious repercussions because a strong link exists between high blood cholesterol concentrations and heart disease. Current nutritional advice suggests reducing cholesterol intake to under 300 mg per day; this amount represents a 40 percent reduction for the average adult in the United States. The connection between blood cholesterol levels and heart disease will be examined more closely in Chapters 11 and 12.

Phospholipids

Phospholipids (FOS-fō-lip-ids) consist of a glycerol and two fatty acids (a diglyceride) linked to a nonlipid group by a phosphate group (PO_4^{3-}) (Figure 2-14●). The nonlipid portion of a phospholipid is soluble in water, whereas the fatty acid portion is relatively insoluble. Phospholipids are the most abundant lipid components of cell membranes.

PROTEINS

Proteins are the most abundant and diverse organic components of the human body. There are roughly 100,000 different kinds of proteins, and they account for about 20 percent of the total body weight. All proteins contain carbon, hydrogen, oxygen, and nitrogen; smaller quantities of sulfur may also be present.

Protein Function

Proteins perform a variety of functions in seven major categories:

1. *Support.* **Structural proteins** create a three-dimensional supporting framework for the body, providing strength, organization, and support for cells, tissues, and organs.
2. *Movement.* **Contractile proteins** are responsible for muscular contraction; related proteins are responsible for the movement of individual cells.
3. *Transport.* Insoluble lipids, respiratory gases, special minerals, such as iron, and several hormones are carried in the blood attached to special **transport proteins**. Other specialized proteins transport materials from one part of a cell to another.

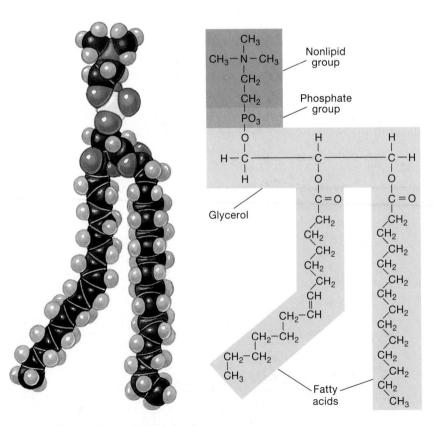

● *Figure 2-14* **A Phospholipid Molecule**
In a phospholipid such as lecithin, a phosphate group links a nonlipid molecule to a glycerol with two fatty acids.

4. *Buffering.* Proteins provide a considerable buffering action, helping to prevent potentially dangerous changes in pH.

5. *Metabolic regulation.* **Enzymes** accelerate chemical reactions in living cells. The sensitivity of enzymes to environmental factors is extremely important in controlling the pace and direction of metabolic operations.

6. *Coordination, communication,* and *control.* Protein **hormones** can influence the metabolic activities of every cell in the body or affect the function of specific organs or organ systems.

7. *Defense.* The tough, waterproof proteins of the skin, hair, and nails protect the body from environmental hazards. In addition, proteins known as **antibodies** protect us from disease, and special clotting proteins restrict bleeding following an injury to the cardiovascular system.

Protein Structure

Proteins are chains of small organic molecules called **amino acids**. The human body contains significant quantities of 20 different amino acids. Each amino acid consists of a central carbon atom bonded to a hydrogen atom, an amino group ($-NH_2$), a carboxylic acid group ($-COOH$), and a variable R group (Figure 2-15a●). The R group may be a straight chain or a ring of atoms. A typical protein contains 1000 amino acids, but the largest protein complexes may have 100,000 or more. The individual amino acids are strung together like beads on a string, with the carboxylic acid group of one amino acid attached to the amino group of another. This connection is called a **peptide bond** (Figure 2-15b●). **Peptides** are molecules made up of amino acids held together by peptide bonds. If a molecule consists of two amino acids, it is called a *dipeptide. Polypeptides* are long chains of amino acids. Proteins are polypeptide chains containing over 100 amino acids.

The shape of a peptide primarily depends on interactions between amino acids at different sites along the peptide chain (Figure 2-16a●). Interactions between the R groups of the amino acids, the formation of hydrogen bonds at different parts of the chain, and interactions between the polypeptide chain and surrounding water molecules contribute to the complex three-dimensional shapes of large proteins. In a *globular protein,* such as *myoglobin,* the peptide chain folds back on itself, creating a rounded mass (Figure 2-16b●). Myoglobin is a protein found in muscle cells. Complex proteins may consist of several protein subunits. Examples are *hemoglobin,* a globular protein found inside red blood cells (Figure 2-16c●), and *keratin* (Figure 2-16d●), the tough, water-resistant protein found in skin, nails, and hair. Keratin is an example of a *fibrous protein.* In fibrous proteins the polypeptide strands are wound together as in a rope. Fibrous proteins are flexible but very strong.

The shape of a protein determines its functional properties, and the primary determinant of shape is the sequence of amino acids. Combining the 20 amino acids in various ways creates an almost limitless variety of proteins. Small differences can have large effects; changing one amino acid in a protein containing 10,000 or more amino acids may make it incapable of performing its normal function.

The shape of a protein—and thus its function—can also be altered by small changes in the ionic composition, temperature, or pH of its surroundings. For example, very high

● *Figure 2-15* **Amino Acids and Peptide Bonds**
(**a**) Each amino acid consists of a central carbon atom to which four different groups are attached: a hydrogen atom, an amino group ($-NH_2$), a carboxylic acid group ($-COOH$), and a variable group generally designated R. (**b**) Peptides form when a dehydration synthesis creates a peptide bond between the carboxyl group of one amino acid and the amino group of another. In this example, glycine and alanine are linked to form a dipeptide.

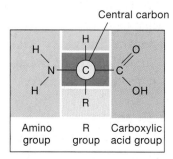

(a) Structure of an amino acid

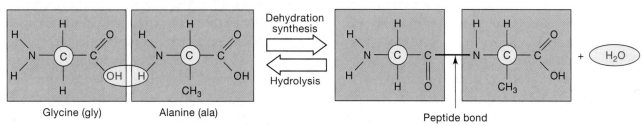

(b) Peptide bond formation

2 THE CHEMICAL LEVEL OF ORGANIZATION

Atoms and Molecules • Chemical Notation • Chemical Reactions • Inorganic Compounds • **Organic Compounds** • Chemicals and Living Cells • Chapter Review

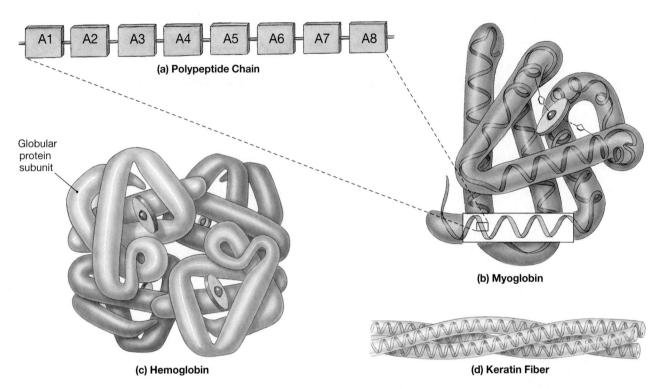

(a) Polypeptide Chain

(b) Myoglobin

(c) Hemoglobin

(d) Keratin Fiber

● *Figure 2-16* **Protein Structure**
(**a**) The shape of a polypeptide is determined by its sequence of amino acids. (**b**) Attraction between R groups plays a large role in forming globular proteins. Myoglobin is a globular protein involved in the storage of oxygen in muscle tissue. (**c**) A single hemoglobin molecule contains four globular subunits, each structurally similar to myoglobin. Hemoglobin transports oxygen in the blood. (**d**) In keratin, three fibrous subunits intertwine like the strands of a rope.

body temperatures (over 43°C, or 110°F) cause death because at these temperatures proteins undergo **denaturation**, a change in their three-dimensional shape. Denatured proteins are nonfunctional, and the loss of structural proteins and enzymes causes irreparable damage to organs and organ systems. You see denaturation in progress each time you fry an egg, because the clear egg white contains abundant dissolved proteins. As the temperature rises, the protein structure changes and eventually the egg proteins form an insoluble white mass.

Enzymes and Chemical Reactions

Most chemical reactions do not occur spontaneously, because they require energy to activate the reactant molecules before a reaction can begin. **Activation energy** is the amount of energy required to start a reaction. Figure 2-17a● diagrams the activation energy needed for a typical reaction. Although many reactions can be activated by changes in temperature or pH, such changes are deadly to cells. For example, to break down a complex sugar in the laboratory, you must boil it in an acid solution. Cells, however, avoid such harsh requirements by using special proteins called *enzymes* to speed up the reactions that support life. Enzymes belong to a class of substances called *catalysts* (KAT-ah-lists; *katalysis*, dissolution), compounds that

accelerate chemical reactions without themselves being permanently changed. A cell makes an enzyme molecule to promote a specific reaction.

Enzymes promote chemical reactions by lowering the activation energy requirements (Figure 2-17b●). Lowering the activation energy affects only the rate of a reaction, not the direction of the reaction or the products that will be formed. An enzyme cannot bring about a reaction that would otherwise be impossible.

Figure 2-18● shows a simple model of enzyme function. The reactants in an enzymatic reaction, called **substrates**, interact to form a specific **product**. Substrate molecules bind to the enzyme at a particular location called the **active site**. This binding depends on the complementary shapes of the two molecules, much as a key fits into a lock. The shape of the active site is determined by the three-dimensional shape of the enzyme molecule. Once the reaction is completed and the products are released, the enzyme is free to catalyze another reaction.

Each enzyme works best at an optimal temperature and pH. As temperatures rise or pH shifts outside normal limits, proteins change shape and enzyme function deteriorates.

The complex reactions that support life proceed in a series of interlocking steps, each step controlled by a different enzyme.

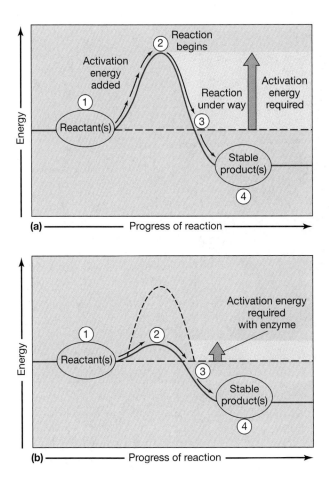

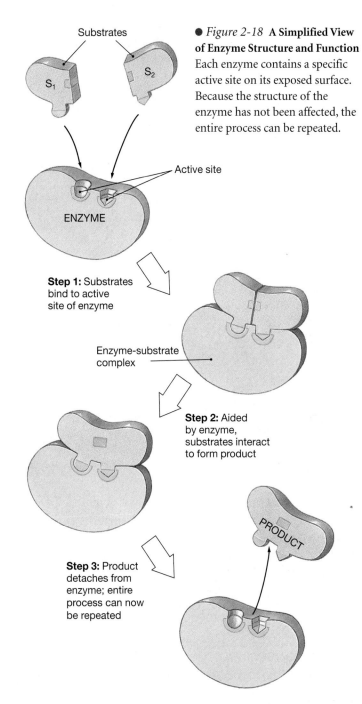

● *Figure 2-18* **A Simplified View of Enzyme Structure and Function** Each enzyme contains a specific active site on its exposed surface. Because the structure of the enzyme has not been affected, the entire process can be repeated.

Step 1: Substrates bind to active site of enzyme

Enzyme-substrate complex

Step 2: Aided by enzyme, substrates interact to form product

Step 3: Product detaches from enzyme; entire process can now be repeated

● *Figure 2-17* **Activation Energy and Enzyme Function** (**a**) Before a reaction can begin, considerable activation energy must be provided. In this diagram, the activation energy represents the energy required to proceed from point 1 to point 2. (**b**) The activation energy requirement of the reaction is much lower in the presence of an appropriate enzyme. This allows the reaction to take place much more rapidly, without the need for extreme conditions that would harm cells.

Such a reaction sequence is called a *pathway*. We will consider important pathways in later chapters.

NUCLEIC ACIDS

Nucleic (noo-KLĀ-ik) **acids** are large organic molecules composed of carbon, hydrogen, oxygen, nitrogen, and phosphorus. Nucleic acids store and process information at the molecular level, inside cells. There are two classes of nucleic acid molecules: **deoxyribonucleic** (dē-ok-sē-rī-bō-noo-KLĀ-ik) **acid**, or **DNA**; and **ribonucleic** (rī-bō-noo-KLĀ-ik) **acid**, or **RNA**.

The DNA in our cells determines our inherited characteristics, such as eye color, hair color, blood type, and so on. It affects all aspects of body structure and function because DNA molecules encode the information needed to build proteins. By directing the synthesis of structural proteins, DNA controls the shape and physical characteristics of our bodies. By con-

trolling the manufacture of enzymes, DNA regulates not only protein synthesis but all aspects of cellular metabolism, including the creation and destruction of lipids, carbohydrates, and other vital molecules.

Several forms of RNA cooperate to manufacture specific proteins using the information provided by DNA. The functional relationships between DNA and RNA will be detailed in Chapter 3.

Structure of Nucleic Acids

A nucleic acid is made up of **nucleotides**. A single nucleotide has three basic components: a *sugar*, a *phosphate group* (PO_4^{3-}),

2 THE CHEMICAL LEVEL OF ORGANIZATION

Atoms and Molecules • Chemical Notation • Chemical Reactions • Inorganic Compounds • **Organic Compounds** • Chemicals and Living Cells • Chapter Review

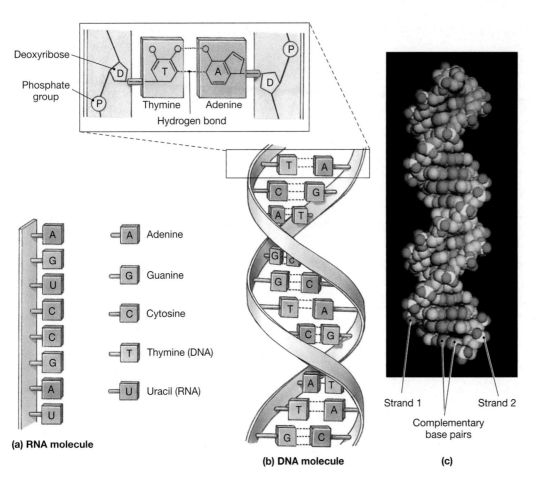

● *Figure 2-19* **Nucleic Acids: RNA and DNA**
Nucleic acids are long chains of nucleotides. Each molecule starts at the phosphate group of the first nucleotide and ends at the sugar-nitrogenous base of the last member of the chain. (**a**) An RNA molecule consists of a single nucleotide chain. Its shape is determined by the sequence of nucleotides and the interactions between them. (**b**) A DNA molecule consists of a pair of nucleotide chains linked by hydrogen bonding between complementary base pairs. (**c**) A three-dimensional model of a DNA molecule shows the double helix formed by the two DNA strands.

and a *nitrogenous (nitrogen-containing) base* (Figure 2-19a●). The sugar is always a five-carbon sugar, either **ribose** (in RNA) or **deoxyribose** (in DNA). There are five nitrogenous bases: **adenine (A)**, **guanine (G)**, **cytosine (C)**, **thymine (T)**, and **uracil (U)**. Both RNA and DNA contain adenine, guanine, and cytosine. Uracil is found only in RNA, and thymine only in DNA.

Important structural differences between RNA and DNA are listed in Table 2-6. A molecule of RNA consists of a single chain of nucleotides (Figure 2-19a●). A DNA molecule consists of two nucleotide chains held together by weak hydrogen bonds between the opposing nitrogenous bases (Figure 2-19b●). Because of their shapes, adenine can bond only with thymine, and cytosine only with guanine. As a

TABLE 2-6 *A Comparison of RNA and DNA*

CHARACTERISTIC	RNA	DNA
SUGAR	Ribose	Deoxyribose
NITROGENOUS BASES	Adenine Guanine Cytosine Uracil	Adenine Guanine Cytosine Thymine
NUMBER OF NUCLEOTIDES IN A TYPICAL MOLECULE	Varies from fewer than 100 nucleotides to about 50,000	Always more than 45 million nucleotides
SHAPE OF MOLECULE	Single strand	Paired strands coiled in a double helix
FUNCTION	Performs protein synthesis as directed by DNA	Stores genetic information that controls protein synthesis

result, adenine-thymine and cytosine-guanine are known as **complementary base pairs**.

The two strands of DNA twist around one another in a **double helix** that resembles a spiral staircase, with the stair steps corresponding to the nitrogenous base pairs. Figure 2-19c● presents this three-dimensional view of a DNA molecule.

HIGH-ENERGY COMPOUNDS

The energy that powers a cell is obtained through the catabolism of organic molecules, such as glucose. Part of the energy released by catabolic reactions is released as heat and part is captured and stored as useful chemical energy in **high-energy bonds**. A high-energy bond is a covalent bond that stores an unusually large amount of energy (as would a tightly wound rubber band attached to the propeller of a model airplane). When that bond is later broken, perhaps in a distant portion of the cell, the energy will be released under controlled conditions (as when the propeller is turned by the unwinding rubber band). In our cells, a high-energy bond usually connects a phosphate group (PO_4^{3-}) to an organic molecule, resulting in a **high-energy compound**. The most important high-energy compound in the body is **adenosine triphosphate**, or **ATP**. Figure 2-20● shows the structure of ATP.

In ATP, a high-energy bond connects a phosphate group to **adenosine diphosphate (ADP)**. Within our cells, the conversion of ADP to ATP represents the primary method of energy storage, and the reverse reaction provides a means for controlled energy release. The arrangement can be summarized as:

$$ADP + phosphate\ group + energy \longleftrightarrow ATP + H_2O$$

Throughout life, our cells continuously generate ATP from ADP and use the energy provided by that ATP to perform vital functions, such as the synthesis of protein molecules or the contraction of muscles. Figure 2-21● shows the cellular relationship between energy flow and the recycling of ADP and ATP.

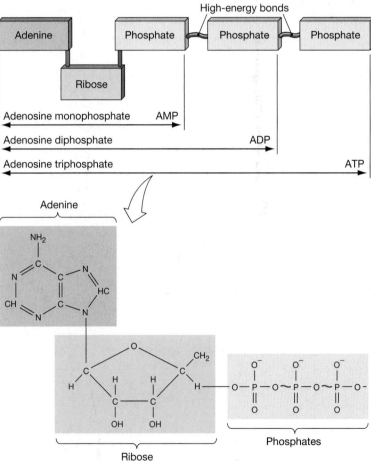

● *Figure 2-20* **The Structure of ATP**
The ATP molecule is made up of an adenosine (adenine and sugar) molecule to which three phosphate groups have been joined. Both the second and third phosphates are bound to the molecule by high-energy bonds. Cells most often store energy by attaching a third phosphate group to ADP. Removing the phosphate group releases the energy for cellular work, including the synthesis of other molecules.

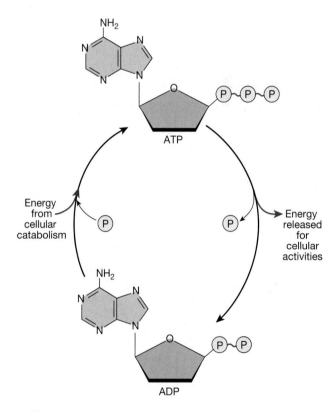

● *Figure 2-21* **Cellular Energy Flow and the Recycling of ATP and ADP**

2 THE CHEMICAL LEVEL OF ORGANIZATION

Atoms and Molecules • Chemical Notation • Chemical Reactions • Inorganic Compounds • Organic Compounds • **Chemicals and Living Cells** • Chapter Review

Chemicals and Living Cells

Figure 2-22● and Table 2-7 review the major chemical components we have discussed in this chapter. But the human body is more than a collection of chemicals. Biochemical building blocks form cells. Each cell behaves like a miniature organism, responding to internal and external stimuli. A lipid membrane separates the cell from its environment, and internal membranes create compartments with specific functions. Proteins form an internal supporting framework and act as enzymes to accelerate and control the chemical reactions that maintain homeostasis. Nucleic acids direct the synthesis of all cellular proteins, including the enzymes that enable the cell to synthesize a wide variety of other substances. Carbohydrates provide energy for vital activities and form part of specialized compounds, in combination with proteins or lipids. The next chapter considers the combination of these compounds within a living, functional cell.

CONCEPT CHECK QUESTIONS
Answers on page 51

❶ A food contains organic molecules with the elements C, H, and O in a ratio of $1:2:1$. What type of compound is this?

❷ Why does boiling a protein affect its structural and functional properties?

❸ How are DNA and RNA similar?

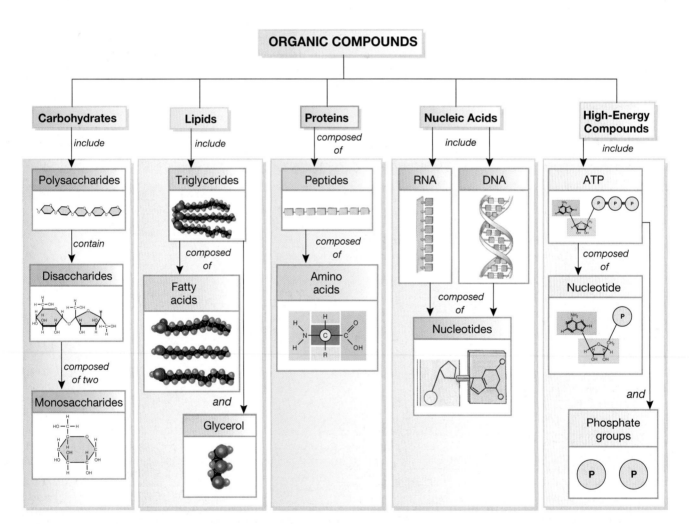

● *Figure 2-22* **A Structural Overview of Organic Compounds in the Body**
Each of the classes of organic compounds is composed of simple structural subunits. Specific compounds within each class are listed above the basic subunits. Fatty acids are the main subunits of all lipids except steroids, such as cholesterol; only one type of lipid, the triglyceride, is represented here.

TABLE 2-7 *The Structure and Function of Biologically Important Compounds*

CLASS	BUILDING BLOCKS	SOURCES	FUNCTIONS
INORGANIC			
Water	Hydrogen and oxygen atoms	Absorbed as liquid water or generated by metabolism	Solvent; transport medium for dissolved materials and heat; cooling through evaporation; medium for chemical reactions; reactant in hydrolysis
Acids, bases, salts	H^+, OH^-, various anions and cations	Obtained from the diet or generated by metabolism	Structural components; buffers; sources of ions
Dissolved gases	Oxygen, carbon, nitrogen, and other atoms	Atmosphere	O_2: required for normal cellular metabolism CO_2: generated by cells as a waste product
ORGANIC			
Carbohydrates	C, H, and O; CHO in a $1:2:1$ ratio	Obtained from the diet or manufactured in the body	Energy source; some structural role when attached to lipids or proteins; energy storage
Lipids	C, H, O, sometimes N or P; CHO not in $1:2:1$ ratio	Obtained from the diet or manufactured in the body	Energy source; energy storage; insulation; structural components; chemical messengers; protection
Proteins	C, H, O, N, often S	20 common amino acids; roughly half can be manufactured in the body, others must be obtained from the diet	Catalysts for metabolic reactions; structural components; movement; transport; buffers; defense; control and coordination of activities
Nucleic acids	C, H, O, N, and P; nucleotides composed of phosphates, sugars, and nitrogenous bases	Obtained from the diet or manufactured in the body	Storage and processing of genetic information
High-energy compounds	Nucleotides joined to phosphates by high-energy bonds	Synthesized by all cells	Storage or transfer of energy

Related Clinical Terms

cholesterol: A steroid, important in the structure of cellular membranes, that in high concentrations increases the risk of heart disease.

familial hypercholesterolemia: A genetic disorder resulting in high cholesterol levels in blood and cholesterol buildup in body tissue, especially the walls of blood vessels.

galactosemia: A metabolic disorder resulting from the lack of an enzyme that converts galactose, a monosaccharide in milk, to glucose within cells. Affected individuals have elevated levels of galactose in the blood and urine. High levels of galactose during childhood can cause abnormalities in nervous system development and liver function, and cataracts.

nuclear imaging: A procedure in which an image is created on a photographic plate or video screen by the radiation emitted by injected radioisotopes.

omega-3 fatty acids: Fatty acids, abundant in fish flesh and fish oils, that have a double bond three carbons away from the end of the hydrocarbon chain. Their presence in the diet has been linked to reduced risks of heart disease and other conditions.

radioisotopes: Isotopes with unstable nuclei, which spontaneously emit subatomic particles or radiation in measurable amounts.

phenylketonuria (PKU): A metabolic disorder resulting from a defect in the enzyme that normally converts the amino acid phenylalanine to tyrosine, another amino acid. If the resulting elevated levels of phenylalanine are not detected in infancy, mental retardation can occur due to damage to the developing nervous system.

radiopharmaceuticals: Drugs that incorporate radioactive atoms; administered to expose specific target tissues to radiation.

tracer: A compound labeled with a radioisotope that can be tracked in the body by the radiation it releases.

2 THE CHEMICAL LEVEL OF ORGANIZATION

Atoms and Molecules • Chemical Notation • Chemical Reactions • Inorganic Compounds • Organic Compounds • Chemicals and Living Cells • **Chapter Review**

CHAPTER REVIEW

Key Terms

Summary Outline

1. **Atoms** are the smallest units of matter; they consist of **protons**, **neutrons**, and **electrons**. *(Figure 2-1)*

2. An **element** consists entirely of atoms with the same number of protons (**atomic number**). Within an atom, an **electron cloud** surrounds the nucleus. *(Figure 2-2; Table 2-1)*

3. The atomic mass of an atom is equal to the total number of protons and neutrons in its nucleus. **Isotopes** are atoms of the same element whose nuclei contain different numbers of neutrons. The **atomic weight** of an element takes into account the abundance of its various isotopes.

4. Electrons occupy a series of **electron shells** around the nucleus. The number of electrons in the outermost electron shell determine an atom's chemical properties. *(Figure 2-3)*

5. An **ionic bond** results from the attraction between **ions**: atoms that have gained or lost electrons. **Cations** are positively charged, and **anions** are negatively charged. *(Figure 2-4; Table 2-2)*

6. Atoms can combine to form a **molecule**; combinations of atoms of different elements form a **compound**. Some atoms share electrons to form a molecule held together by **covalent bonds**.

7. Sharing one pair of electrons creates a single covalent bond; sharing two pairs forms a **double covalent bond**. An unequal sharing of electrons creates a **polar covalent bond**. *(Figure 2-5)*

8. A **hydrogen bond** is the attraction between a hydrogen atom with a slight positive charge and a negatively charged atom in another molecule or within the same molecule. Hydrogen bonds can affect the shapes and properties of molecules. *(Figure 2-6)*

1. Chemical notation allows us to describe reactions between reactants that generate one or more products. *(Table 2-3)*

1. **Metabolism** refers to all the **chemical reactions** in the body. Our cells capture, store, and use energy to maintain homeostasis and support essential functions.

2. **Work** involves movement of an object or a change in its physical structure, and **energy** is the capacity to perform work. There are two major types of energy: kinetic and potential.

3. **Kinetic energy** is the energy of motion. **Potential energy** is stored energy that results from the position or structure of an object. Conversions from potential to kinetic energy are not 100 percent efficient; every energy exchange produces **heat**.

4. A chemical reaction may be classified as a **decomposition**, **synthesis**, or **exchange reaction**. **Exergonic** reactions release heat; **endergonic** reactions absorb heat.

5. Cells gain energy to power their functions by **catabolism**, the breakdown of complex molecules. Much of this energy supports **anabolism**, the synthesis of new organic molecules.

6. Reversible reactions consist of simultaneous synthesis and decomposition reactions. At **equilibrium** the rates of these two opposing reactions are in balance.

7. An **acid** releases hydrogen ions, and a **base** removes hydrogen ions from a solution.

8. The **pH** of a solution indicates the concentration of hydrogen ions it contains. Solutions can be classified as neutral (pH $=$ 7), acidic (pH $<$ 7), or basic (alkaline) (pH $>$ 7) on the basis of pH. *(Figure 2-7)*

9. Buffers maintain pH within normal limits (7.35–7.45 in most body fluids) by releasing or absorbing hydrogen ions.

1. **Nutrients** and **metabolites** can be broadly classified as **organic** or **inorganic compounds**.

2. Living cells in the body generate carbon dioxide and consume oxygen.

3. Water is the most important inorganic component of the body.

4. Water is an excellent solvent, has a high heat capacity, and participates in the metabolic reactions of the body.

5. Many inorganic compounds will undergo **ionization**, or dissociation, in water to form ions. *(Figure 2-8)*

6. Inorganic acids found in the body include hydrochloric acid, carbonic acid, sulfuric acid, and phosphoric acid. Sodium hydroxide is an inorganic base that may form within the body.

7. A **salt** is an ionic compound whose cation is not H^+, and whose anion is not OH^-. Salts are **electrolytes**, compounds that dissociate in water and conduct an electrical current.

ORGANIC COMPOUNDS

1. Organic compounds contain carbon and hydrogen, and usually oxygen as well. Large and complex organic molecules include carbohydrates, lipids, proteins, and nucleic acids.

2. **Carbohydrates** are most important as an energy source for metabolic processes. The three major types are **monosaccharides** (simple sugars), **disaccharides**, and **polysaccharides**. (*Figures 2-9, 2-10; Table 2-4*)

3. **Lipids** are water-insoluble molecules that include fats, oils, and waxes. There are four important classes of lipids: **fatty acids**, **fats**, **steroids**, and **phospholipids**. (*Table 2-5*)

4. **Triglycerides** (**fats**) consist of three fatty acid molecules attached to a molecule of **glycerol**. (*Figure 2-11, 2-12*)

5. Cholesterol is a precursor of steroid hormones and is component of cell membranes. (*Figure 2-13*)

6. Phospholipids are the most abundant components of cell membranes. (*Figure 2-14*)

7. Proteins perform a great variety of functions in the body. Important types of proteins include **structural proteins**, **contractile proteins**, **transport proteins**, **enzymes**, **hormones**, and **antibodies**.

8. Proteins are chains of **amino acids** linked by **peptide bonds**. The sequence of amino acids and the interactions of their R groups influence the final shape of the protein molecule. (*Figures 2-15, 2-16*)

9. The shape of a protein determines its function. Each protein works best at an optimal combination of temperature and pH.

10. **Activation energy** is the amount of energy required to start a reaction. Proteins called **enzymes** control many chemical reactions within our bodies. Enzymes are **catalysts**—substances that accelerate chemical reactions without themselves being permanently changed. (*Figure 2-17*)

11. The reactants in an enzymatic reaction, called **substrates**, interact to form a **product** by binding to the enzyme at the active site. (*Figure 2-18*)

12. **Nucleic acids** store and process information at the molecular level. There are two kinds of nucleic acids: **deoxyribonucleic acid (DNA)** and **ribonucleic acid (RNA)**. (*Figure 2-19; Table 2-6*)

13. Nucleic acids are chains of nucleotides. Each nucleotide contains a sugar, a **phosphate group**, and a **nitrogenous base**. The sugar is always **ribose** or **deoxyribose**. The nitrogenous bases found in DNA are **adenine**, **guanine**, **cytosine**, and **thymine**. In RNA, **uracil** replaces thymine.

14. Cells store energy in **high-energy compounds**. The most important high-energy compound is **ATP (adenosine triphosphate)**. When energy is available, cells make ATP by adding a phosphate group to ADP. When energy is needed, ATP is broken down to ADP and phosphate. (*Figures 2-20, 2-21*)

CHEMICALS AND LIVING CELLS

1. Biochemical building blocks form cells. (*Figure 2-22; Table 2-7*)

Review Questions

Level 1: Reviewing Facts and Terms

Match each item in column A with the most closely related item in column B. Use letters for answers in the spaces provided.

COLUMN A

___ 1. atomic number

___ 2. covalent bond

___ 3. ionic bond

___ 4. catabolism

___ 5. anabolism

___ 6. exchange reaction

___ 7. reversible reaction

___ 8. acid

___ 9. enzyme

___ 10. buffer

___ 11. organic compounds

___ 12. inorganic compounds

COLUMN B

a. synthesis

b. catalyst

c. sharing of electrons

d. $A + B \longleftrightarrow AB$

e. stabilize pH

f. number of protons

g. decomposition

h. carbohydrates, lipids, proteins

i. loss or gain of electrons

j. water, salts

k. H^+ donor

l. $AB + CD \longrightarrow AD + CB$

2 THE CHEMICAL LEVEL OF ORGANIZATION

Atoms and Molecules • Chemical Notation • Chemical Reactions • Inorganic Compounds • Organic Compounds • Chemicals and Living Cells • **Chapter Review**

13. In atoms, protons and neutrons are found:
    (a) only in the nucleus
    (b) outside the nucleus
    (c) inside and outside the nucleus
    (d) in the electron cloud

14. The number and arrangement of electrons in an atom's outer electron shell determines its:
    (a) atomic weight
    (b) atomic number
    (c) electrical properties
    (d) chemical properties

15. The bond between sodium and chlorine in the compound sodium chloride (NaCl) is:
    (a) an ionic bond
    (b) a single covalent bond
    (c) a nonpolar covalent
    (d) a double covalent bond

16. What is the role of enzymes in chemical reactions?

17. List the six most abundant elements in the body.

18. What four major classes of organic compounds are found in the body?

19. List seven major functions performed by proteins.

Level 2: Reviewing Concepts

20. Oxygen has 8 protons, 8 neutrons, and 8 electrons. What is its atomic mass?
    (a) 8
    (b) 16
    (c) 24
    (d) 32

21. Of the following selections, the one that contains only inorganic compounds is
    (a) water, electrolytes, oxygen, carbon dioxide
    (b) oxygen, carbon dioxide, water, sugars
    (c) water, electrolytes, salts, nucleic acids
    (d) carbohydrates, lipids, proteins, vitamins

22. Glucose and fructose are examples of:
    (a) monosaccharides (simple sugars)
    (b) isotopes
    (c) lipids
    (d) a, b, and c are all correct

23. Explain the differences among (1) nonpolar covalent bonds, (2) polar covalent bonds, and (3) ionic bonds.

24. Why does pure water have a neutral pH?

25. A biologist analyzes a sample that contains an organic molecule and finds the following constituents: carbon, hydrogen, oxygen, nitrogen, and phosphorus. On the basis of this information, is the molecule a carbohydrate, a lipid, a protein, or a nucleic acid?

Level 3: Critical Thinking and Clinical Applications

26. The element sulfur has an atomic number of 16 and an atomic mass of 32. How many neutrons are in the nucleus of a sulfur atom? Assuming that sulfur forms covalent bonds with hydrogen, how many hydrogen atoms could bond to one sulfur atom?

27. An important buffer system in the human body involves carbon dioxide (CO_2) and bicarbonate ions (HCO_3^-) as shown:

$$CO_2 + H_2O \longleftrightarrow H_2CO_3 \longleftrightarrow H^+ + HCO_3^-$$

If a person becomes excited and exhales large amounts of CO_2, how will his body's pH be affected?

Answers to Concept Check Questions

Page 32

1. Atoms combine with each other so as to gain a complete set of eight electrons in their outer energy levels. Oxygen atoms do not have a full outer energy level and so will readily react with many other elements to attain this stable arrangement. Neon already has a full outer energy level and thus has little tendency to combine with other elements.

2. Hydrogen can exist as three different isotopes: hydrogen-1, with a mass of 1; deuterium, with a mass of 2; and tritium, with a mass of 3. The heavier sample must contain a higher proportion of one or both of the heavier isotopes.

3. A water molecule is formed by polar covalent bonds. Water molecules are attracted to one another by hydrogen bonds.

Page 34

1. Since this reaction involves a large molecule being broken down into two smaller ones, it is a decomposition reaction. Because energy is released in the process, the reaction can also be classified as exergonic.

2. Removing the product of a reversible reaction would keep its concentration low compared with the concentration of the reactants. Thus the formation of product molecules would continue, but the reverse reaction would slow down, resulting in a shift in the equilibrium toward the product.

Page 35

1. An acid is a solute that releases H^+ ions in a solution; a base is a solute that removes H^+ from a solution.

2. The normal pH range of body fluids is 7.35 to 7.45. Fluctuations in pH outside this range can break chemical bonds, alter the shape of molecules, and affect the functioning of cells, thereby harming cells and tissues.

3. Stomach discomfort is often the result of excess stomach acidity ("acid indigestion"). Antacids contain a weak base that neutralizes the excessive acid.

Page 46

1. A C:H:O ratio of 1:2:1 would indicate that the molecule is a carbohydrate. The body uses carbohydrates chiefly as an energy source.

2. The heat of boiling will break bonds that maintain the protein's three-dimensional shape and/or binding with other proteins or polypeptides. The resulting change in shape will affect the ability of the protein molecule to perform its normal biological functions. These alterations are known as denaturation.

3. DNA and RNA are nucleic acids. Both are composed of sequences of nucleotides. Each nucleotide consists of a five-carbon sugar, a phosphate group (PO_4^{3-}), and a nitrogenous base.

Cell Structure and Function

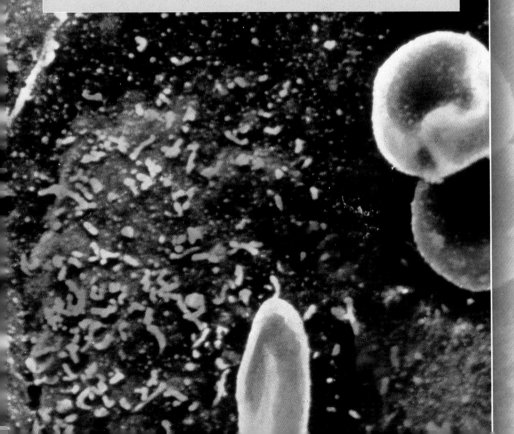

CHAPTER OUTLINE AND OBJECTIVES

Vocabulary Development

aero- ....air; *aerobic*
ana- ....apart; *anaphase*
chondrion ....granule; *mitochondrion*
chroma ....color; *chromosome*
cyto-..cell; *cytoplasm*
endo- ....inside; *endocytosis*
exo- ....outside; *exocytosis*
hemo- ....blood; *hemolysis*
hyper- ....above; *hypertonic*
hypo- ....below; *hypotonic*
inter- ....between; *interphase*
interstitium...................something standing
　................................between; *interstitial fluid*
iso- ....equal; *isotonic*
kinesis............................motion; *cytokinesis*
meta-...................................after; *metaphase*
micro- ....small; *microtubules*
mitos..thread; *mitosis*
osmos ....thrust; *osmosis*
phagein...............................to eat; *phagocyte*
pinein ....to drink; *pinocytosis*
podon ....foot; *pseudopod*
pro- ....before; *prophase*
pseudo- ....false; *pseudopod*
ptosis ....a falling away; *apoptosis*
reticulum ....network; *endoplasmic reticulum*
soma ....body; *lysosome*
telos ....end; *telophase*
tonos ....tension; *isotonic*

MANY FAMILIAR STRUCTURES, *from pyramids to patchwork quilts, are made up of numerous small, similar components. The human body is built on the same principle, for it is made up of several trillion tiny units known as cells. Cells are the smallest entities that can perform all basic life functions. But no single cell performs all of the functions of the human body. Instead, each cell in the body performs its own specialized functions and, in doing so, helps maintain homeostasis. For example, these small red blood cells carry oxygen within the bloodstream, but they cannot divide. In this chapter, we will explore the internal structures of cells and their specific functions.*

3 CELL STRUCTURE AND FUNCTION

Studying Cells • The Cell Membrane • The Cytoplasm • The Nucleus • The Cell Life Cycle • Cell Diversity and Differentiation • Chapter Review

A S ATOMS ARE the building blocks of molecules, **cells** are the building blocks of the human body. Over the years, biologists have developed the **cell theory**, which includes the following four basic concepts:

1. Cells are the building blocks of all plants and animals.
2. Cells are the smallest functioning units of life.
3. Cells are produced by the division of preexisting cells.
4. Each cell maintains homeostasis.

An individual organism maintains homeostasis only through the combined and coordinated actions of many different types of cells. Figure 3-1● gives some examples of the range of cell sizes and shapes found in the human body.

Numbering in the trillions, the cells of the human body form and maintain anatomical structures and perform physiological functions as different as running and thinking. An understanding of how the human body functions thus requires a familiarity with the nature of cells.

Studying Cells

The study of the structure and function of cells is called **cytology** (sī-TOL-ō-jē; *cyto-*, cell + *-logy*, the study of). What we have learned since the 1950s has given us new insights into the phys-

iology of cells and their means of homeostatic control. Acquiring this knowledge depended on developing better ways of viewing cells and applying new experimental techniques not only from biology but also from chemistry and physics.

The two most common methods used to study cell and tissue structure are light microscopy and electron microscopy. Before the 1950s, cells were viewed through light microscopes. Using a series of glass lenses, *light microscopy* can magnify cellular structures about 1000 times. Light microscopy typically involves looking at thin sections sliced from a larger piece of tissue. A photograph taken through a light microscope is called a *light micrograph* (*LM*). Many fine details of intracellular structure are too small to be seen with a light microscope. These details remained a mystery until cell biologists began using *electron microscopy*, a technique that replaced light with a focused beam of electrons. *Transmission electron micrographs* (*TEMs*) are photographs of very thin sections, and they can reveal fine details of cell membranes and intracellular structures. *Scanning electron micrographs* (*SEMs*) provide less magnification but reveal the three-dimensional nature of cell structures. An SEM provides a superficial view of a cell, a portion of a cell, or extracellular structures rather than a detailed sectional view.

You will see examples of light micrographs and both kinds of electron micrographs in figures throughout this text. The

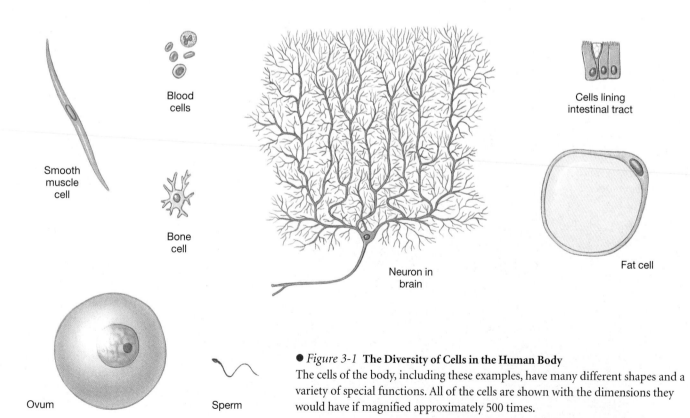

Blood cells

Smooth muscle cell

Bone cell

Neuron in brain

Cells lining intestinal tract

Fat cell

Ovum

Sperm

● *Figure 3-1* **The Diversity of Cells in the Human Body**
The cells of the body, including these examples, have many different shapes and a variety of special functions. All of the cells are shown with the dimensions they would have if magnified approximately 500 times.

abbreviations LM, TEM, and SEM are followed by a number that indicates the total magnification of the image. For example, LM × 160 indicates that the structures in this light micrograph have been magnified 160 times. (See Appendix III for the size ranges and scales included in the study of anatomy and physiology.)

AN OVERVIEW OF CELLULAR ANATOMY

The "typical" cell is like the "average" person, so any description masks enormous individual variations. Our model cell will share features with most cells of the body without being identical to any specific one. Figure 3-2● shows such a composite cell, and Table 3-1 summarizes the structures and functions of its parts.

A body cell is surrounded by a watery medium known as the **extracellular fluid**. A **cell membrane** separates the cell contents, or **cytoplasm**, from the extracellular fluid. The cytoplasm surrounds the **nucleus**—the control center for cellular operations.

The cytoplasm can be subdivided into a fluid, the *cytosol*, and intracellular structures collectively known as *organelles*. Our discussion of the cell begins at its membrane boundary with the external environment, proceeds to the cytosol and the individual organelles, and concludes with the nucleus.

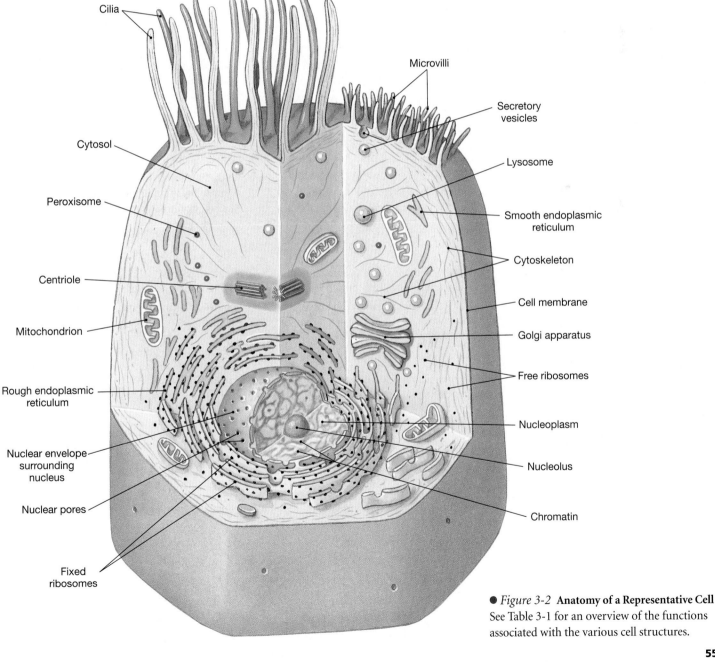

Cilia

Microvilli

Cytosol

Secretory vesicles

Lysosome

Peroxisome

Smooth endoplasmic reticulum

Cytoskeleton

Centriole

Cell membrane

Mitochondrion

Golgi apparatus

Rough endoplasmic reticulum

Free ribosomes

Nuclear envelope surrounding nucleus

Nucleoplasm

Nuclear pores

Nucleolus

Chromatin

Fixed ribosomes

● *Figure 3-2* **Anatomy of a Representative Cell**
See Table 3-1 for an overview of the functions associated with the various cell structures.

3 CELL STRUCTURE AND FUNCTION

Studying Cells • **The Cell Membrane** • The Cytoplasm • The Nucleus • The Cell Life Cycle • Cell Diversity and Differentiation • Chapter Review

TABLE 3-1 *Components of a Representative Cell*

APPEARANCE	STRUCTURE	COMPOSITION	FUNCTION
	Cell membrane	Lipid bilayer, containing phospholipids, steroids, and proteins	Provides isolation, protection, sensitivity, and support; controls entrance/exit of materials
	Cytosol	Fluid component of cytoplasm	Distributes materials by diffusion
	NONMEMBRANOUS ORGANELLES		
	Cytoskeleton: Microtubule Microfilament	Proteins organized in fine filaments or slender tubes	Provides strength and support; enables movement of cellular structures and materials
	Microvilli	Membrane extensions containing microfilaments	Increase surface area to facilitate absorption of extracellular materials
	Cilia	Membrane extensions containing microtubules	Movement of materials over surface
	Centrioles	Two centrioles, at right angles; each composed of microtubules	Essential for movement of chromosomes during cell division
	Ribosomes	RNA + proteins; fixed ribosomes bound to endoplasmic reticulum, free ribosomes scattered in cytoplasm	Protein synthesis
	MEMBRANOUS ORGANELLES		
	Endoplasmic reticulum (ER):	Network of membranous channels extending throughout the cytoplasm	Synthesizes secretory products; provides intracellular storage and transport
	Rough ER	Has ribosomes attached to membranes	Packaging of newly synthesized proteins
	Smooth ER	Lacks attached ribosomes	Synthesizes lipids and carbohydrates
	Golgi apparatus	Stacks of flattened membranes containing chambers	Stores, alters, and packages secretory products; forms lysosomes
	Lysosomes	Vesicles containing powerful digestive enzymes	Remove damaged organelles or pathogens within cells
	Peroxisomes	Vesicles containing degradative enzymes	Catabolism of fats and other organic compounds; neutralizes toxic compounds generated in the process
	Mitochondria	Double membrane, with inner folds (cristae) enclosing important metabolic enzymes	Produce 95% of the ATP required by the cell
	Nucleus	Nucleoplasm containing DNA, nucleotides, enzymes, and proteins; surrounded by double membrane (nuclear envelope)	Controls metabolism; stores and processes genetic information; controls protein synthesis
	Nucleolus	Dense region containing RNA and proteins surrounding segments of DNA	Synthesizes RNA and assembles ribosomal subunits

The Cell Membrane

The outer boundary of the cell is formed by a cell membrane, or *plasma membrane*. Its general functions include:

- *Physical isolation.* The cell membrane is a physical barrier that separates the inside of the cell from the surrounding extracellular fluid. The conditions inside and outside the cell are very different, and those differences must be maintained to preserve homeostasis.

- *Regulation of exchange with the environment.* The cell membrane controls the entry of ions and nutrients, the elimination of wastes, and the release of secretory products.

- *Sensitivity.* The cell membrane is the first part of the cell affected by changes in the extracellular fluid. It also contains a variety of receptors that allow the cell to recognize and respond to specific molecules in its environment.

- *Structural support.* Specialized connections between cell membranes or between membranes and extracellular materials give tissues a stable structure.

MEMBRANE STRUCTURE

The cell membrane is extremely thin and delicate, ranging from 6 to 10 nm in thickness. This membrane contains lipids, proteins, and carbohydrates.

Membrane Lipids

Phospholipids are a major component of cell membranes. p. 40 In a phospholipid, a phosphate group (PO_4^{3-}) serves as a link between a diglyceride (a glycerol molecule bonded to two fatty acid "tails") and a nonlipid "head." The phospholipids in a cell membrane lie in two distinct layers, with the *hydrophilic* (soluble in water) heads on the outside and the *hydrophobic* (insoluble in water) tails on the inside. For this reason, the cell membrane is often called the **phospholipid bilayer** (see Figure 3-3●). Mixed in with the fatty acid tails are cholesterol molecules and small quantities of other lipids.

The hydrophobic lipid tails will not associate with water molecules, and this characteristic allows the cell membrane to act as a selective physical barrier. Lipid-soluble molecules and compounds such as oxygen and carbon dioxide are able to cross the lipid portion of a cell membrane, but ions and water-soluble compounds cannot. Consequently, the cell membrane isolates the cytoplasm from the surrounding extracellular fluid.

Membrane Proteins

Several types of proteins are associated with the cell membrane. The most common of these membrane proteins span the width of the membrane one or more times and are known as *transmembrane proteins*. Other membrane proteins are partially embedded in the membrane's phospholipid bilayer and are loosely bound to its inner or outer surface. Membrane

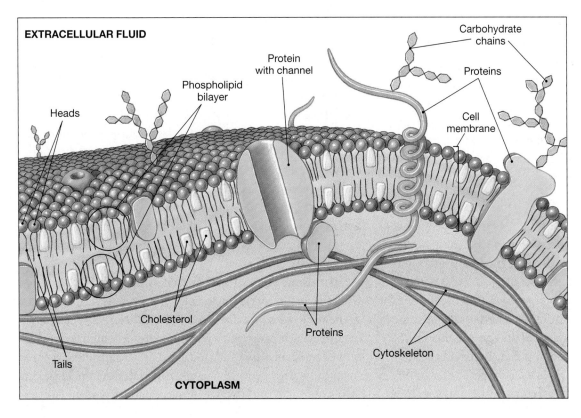

● *Figure 3-3* **The Cell Membrane**

3 CELL STRUCTURE AND FUNCTION

Studying Cells • **The Cell Membrane** • The Cytoplasm • The Nucleus • The Cell Life Cycle • Cell Diversity and Differentiation • Chapter Review

TABLE 3-2 *Types of Membrane Proteins*

CLASS	FUNCTION	EXAMPLE
RECEPTOR PROTEINS	Sensitive to specific extracellular materials that bind to them and trigger a change in a cell's activity.	Binding of the hormone insulin to membrane receptors increases the rate of glucose absorption by the cell.
CHANNEL PROTEINS	Central pore, or channel, permits water and solutes to bypass lipid portion of cell membrane.	Calcium ion movement through channels is involved in muscle contraction and the conduction of nerve impulses.
CARRIER PROTEINS	Bind and transport solutes across the cell membrane. This process may or may not require energy.	Carrier proteins bring glucose into the cytoplasm and also transport sodium, potassium, and calcium ions.
ENZYMES	Catalyze reactions in the extracellular fluid or within the cell.	Dipeptides are broken down into amino acids by enzymes on the membranes of cells lining the intestinal tract.
ANCHORING PROTEINS	Attach the cell membrane to other structures and stabilize its position.	Inside the cell, bound to the network of supporting filaments (the cytoskeleton); outside, attach the cell to extracellular protein fibers or to another cell.
RECOGNITION (IDENTIFIER) PROTEINS	Identify a cell as self or nonself, normal or abnormal, to the immune system.	One group of such recognition proteins is the major histocompatibility complex (MHC) discussed in Chapter 14.

proteins may function as *receptors, channels, carriers, enzymes, anchors,* or *identifiers.* Table 3-2 provides a functional description and example of each class of membrane protein.

Membrane structure is not rigid, and embedded proteins drift from place to place across the surface of the membrane like ice cubes in a punch bowl. In addition, the composition of the cell membrane can change over time, as components of the membrane are added or removed.

Membrane Carbohydrates

Carbohydrates form complex molecules with proteins and lipids on the outer surface of the membrane. The carbohydrate portions of molecules such as *glycoproteins* and *glycolipids* are important as cell lubricants and adhesives, act as receptors for extracellular compounds, and are part of a recognition system that keeps the immune system from attacking its own tissues.

MEMBRANE TRANSPORT

The **permeability** of the cell membrane is the property that determines precisely which substances can enter or leave the cytoplasm. If nothing can cross the cell membrane, it is described as *impermeable.* If any substance can cross without difficulty, the membrane is *freely permeable.* Cell membranes are *selectively permeable,* permitting the free passage of some materials and restricting the passage of others. Whether or not a substance can cross the cell membrane is based on the substance's size, electrical charge, molecular shape, lipid solubility, or some combination of these factors.

Movement across the membrane may be passive or active. **Passive processes** move ions or molecules across the cell membrane without any energy expenditure by the cell. Passive processes include *diffusion, osmosis, filtration,* and *facilitated diffusion.* **Active processes,** discussed later in this chapter, require that the cell expend energy, usually in the form of adenosine triphosphate (ATP).

Diffusion

Ions and molecules are in constant motion, colliding and bouncing off one another and off any obstacles in their paths. **Diffusion** is the net movement of molecules from an area of relatively high concentration (or large number of collisions) to an area of relatively low concentration (or small number of collisions). The difference between the high and low concentrations represents a **concentration gradient**, and diffusion is often described as proceeding "down a concentration gradient" or "downhill." As a result of the process of diffusion, solutes eventually become uniformly distributed, and concentration gradients are eliminated.

Diffusion occurs in air as well as in water. The smell of fresh flowers in a vase can sweeten the air in a large room, just as the molecules of a dissolved sugar cube can sweeten a cup of coffee. Each case begins with an extremely high concentration of molecules in a very localized area. Consider a colored sugar cube dropped into a beaker of water. As the cube dissolves, its sugar and dye molecules establish a steep concen-

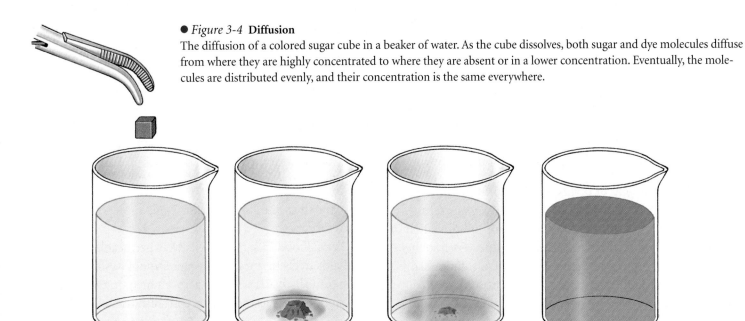

● *Figure 3-4* **Diffusion**
The diffusion of a colored sugar cube in a beaker of water. As the cube dissolves, both sugar and dye molecules diffuse from where they are highly concentrated to where they are absent or in a lower concentration. Eventually, the molecules are distributed evenly, and their concentration is the same everywhere.

tration gradient with the surrounding clear water. Eventually, both of the dissolved molecules spread through the water until they are distributed evenly (Figure 3-4●).

Diffusion is important in body fluids because it tends to eliminate local concentration gradients. For example, every cell in your body generates carbon dioxide, and the intracellular concentration is relatively high. Carbon dioxide concentrations are lower in the surrounding extracellular fluid and lower still in the circulating blood. Because cell membranes are freely permeable to carbon dioxide, it can diffuse down its concentration gradient—traveling from the cell's interior into the extracellular fluid, and from the extracellular fluid into the bloodstream, for eventual delivery to the lungs.

Diffusion Across Cell Membranes. In extracellular fluids of the body, water and dissolved solutes diffuse freely. A cell membrane, however, acts as a barrier that selectively restricts diffusion. Some substances can pass through easily, whereas others cannot penetrate the membrane at all. An ion or molecule can independently diffuse across a cell membrane in one of two ways: (1) by moving across the lipid portion of the membrane or (2) by passing through a membrane channel. Therefore, the primary factors determining whether

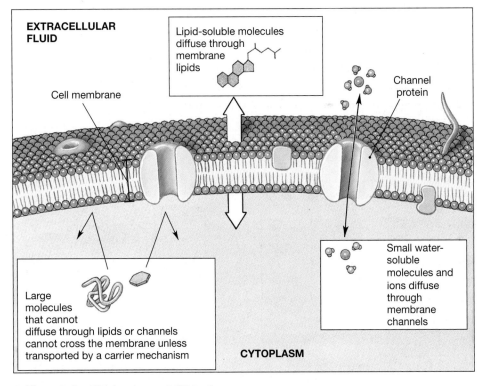

● *Figure 3-5* **Diffusion Across Cell Membranes**
The movement of a substance across the membrane depends on the lipid solubility and size of the substance.

a substance can diffuse across a cell membrane are its lipid solubility and its size relative to the sizes of membrane channels (Figure 3-5●).

Alcohol, fatty acids, and steroids can enter cells easily because they can diffuse through the lipid portions of the membrane. Dissolved gases such as oxygen and carbon dioxide also

3 CELL STRUCTURE AND FUNCTION

Studying Cells • **The Cell Membrane** • The Cytoplasm • The Nucleus • The Cell Life Cycle • Cell Diversity and Differentiation • Chapter Review

enter and leave our cells by diffusion through the phospholipid bilayer.

Ions and most water-soluble compounds are not lipid-soluble, so they must pass through membrane channels to enter the cytoplasm. These channels are very small, about 0.8 nm in diameter. Water molecules can enter or exit freely, as can ions such as sodium and potassium, but even a small organic molecule, such as glucose, is too big to fit through the channels.

Osmosis: A Special Type of Diffusion. The diffusion of water across a membrane is called **osmosis** (oz-MŌ-sis; *osmos*, thrust). Both intracellular and extracellular fluids are solutions that contain a variety of dissolved materials, or **solutes**. Each solute tends to diffuse as if it were the only material in solution. For example, changes in the concentration of potassium ions will have no effect on the rate or direction of sodium ion diffusion. Some ions and molecules diffuse into the cytoplasm, others diffuse out, and a few, such as proteins, are unable to diffuse across a cell membrane. But if we ignore the individual identities and simply count the total number of ions and molecules, we find that the total concentration of ions and molecules on either side of the cell membrane stays the same.

This state of equilibrium persists because *the cell membrane is freely permeable to water*. Whenever a concentration gradient exists, water molecules will diffuse rapidly across the cell membrane until the gradient is eliminated. This movement, which eliminates differences in solute concentrations, occurs in response to a concentration gradient for water molecules.

Dissolved solute molecules occupy space that would otherwise be taken up by water molecules. Thus the higher the solute concentration, the lower the water concentration. As a result, *water molecules will tend to diffuse across a membrane toward the solution containing a higher solute concentration.*

Three characteristics of osmosis are important to remember:

1. Osmosis is the diffusion of water molecules across a membrane.
2. Osmosis occurs across a selectively permeable membrane that is freely permeable to water but not to solutes.
3. In osmosis, water will flow across a membrane toward the solution that has the highest concentration of solutes.

Osmosis and Osmotic Pressure. Figure 3-6● diagrams the process of osmosis. Step 1 shows two solutions (A and B), with different solute concentrations, separated by a selectively permeable membrane. As osmosis occurs, water molecules cross the membrane until the solute concentrations in the two solutions are identical (step 2a). Thus the volume of solution B increases at the expense of solution A. The greater the initial difference in solute concentrations, the stronger the osmotic flow. The **osmotic pressure** of a solution is an indication of the force of water movement *into that solution* as a result of solute con-

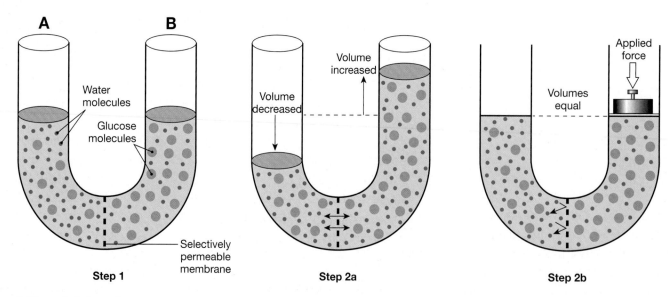

● *Figure 3-6* **Osmosis**
Step 1: Two solutions containing different solute concentrations are separated by a selectively permeable membrane. Water molecules (small blue dots) begin to cross the membrane toward solution B, the solution with the higher concentration of solutes (larger pink circles). **Step 2a:** At equilibrium, the solute concentrations on the two sides of the membrane are equal. The volume of solution B has increased at the expense of that of solution A. **Step 2b:** Osmosis can be prevented by resisting the volume change. The osmotic pressure of solution B is equal to the amount of hydrostatic pressure required to stop the osmotic flow.

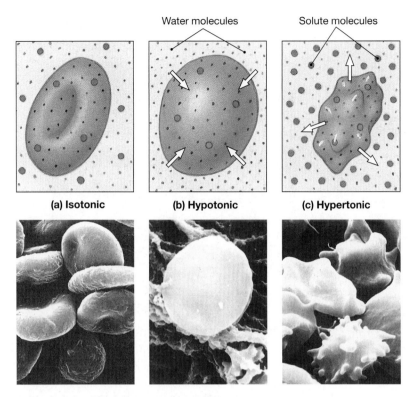

Water molecules   Solute molecules

(a) Isotonic   **(b) Hypotonic**   **(c) Hypertonic**

● *Figure 3-7* **Osmotic Flow Across Cell Membranes**
White arrows indicate the direction of osmotic water movement. (**a**) Because these red blood cells are immersed in an isotonic saline solution, no osmotic flow occurs and the cells have their normal appearance. (**b**) Immersion in a hypotonic saline solution results in the osmotic flow of water into the cells. The swelling may continue until the cell membrane ruptures. (**c**) Exposure to a hypertonic solution results in the movement of water out of the cells. The red blood cells shrivel and become crenated. (SEMs × 833)

centration. As the solute concentration of a solution increases, so does its osmotic pressure. Osmotic pressure can be measured in several ways. For example, a strong enough opposing pressure can prevent the entry of water molecules. Pushing against a fluid generates *hydrostatic pressure*. In step 2b, hydrostatic pressure opposes the osmotic pressure of solution B, so no net osmotic flow occurs.

Solutions of varying solute concentrations are described as *isotonic, hypotonic,* or *hypertonic* with regard to their effects on the shape or tension of the plasma membrane of living cells. Although the effects of various osmotic solutions are difficult to see in most tissues, they are readily observed in red blood cells.

Figure 3-7a● shows the appearance of a red blood cell in an isotonic solution. An **isotonic** (*iso-*, equal + *tonos*, tension) solution is one that will not cause a net movement of water into or out of the cell. In other words, an equilibrium exists, and as one water molecule moves out of the cell another moves in to replace it.

When a cell is placed in a **hypotonic** (*hypo-*, below) solution, water will flow into the cell causing it to swell up like a bal-

loon (Figure 3-7b●). Ultimately the membrane may rupture, or *lyse*. In the case of red blood cells, this event is known as **hemolysis** (*hemo-*, blood + *lysis*, breakdown). A cell in a **hypertonic** (*hyper-*, above) solution will lose water by osmosis. As it does, the cell shrivels and dehydrates. The shrinking of red blood cells is called **crenation** (Figure 3-7c●).

It is often necessary to give large volumes of fluid to patients after severe blood loss or dehydration. One commonly administered fluid is a 0.9 percent (0.9 g/dl) solution of sodium chloride (NaCl). This solution, which approximates the normal *osmotic concentration* (total solute concentration) of the extracellular fluids is called **normal saline**. It is used because sodium and chloride are the most abundant ions in the body's extracellular fluid. Because there is little net movement of either type of ion across cell membranes, normal saline is essentially isotonic with respect to body cells.

Filtration

In **filtration**, hydrostatic pressure forces water across a membrane, and solute molecules may be transported with the water. If they are small enough to fit through the membrane pores, molecules of solute will be carried along with the water. In the body, the heart pushes blood through the circulatory system and generates hydrostatic pressure, or *blood pressure*. Filtration occurs across the walls of small blood vessels, pushing water and dissolved nutrients into the tissues of the body. Filtration across specialized blood vessels in the kidneys is an essential step in the production of urine.

Carrier-Mediated Transport

In **carrier-mediated transport**, membrane proteins bind specific ions or organic substrates and carry them across the cell membrane. These proteins have several characteristics in common with enzymes. For example, they may be used over and over and are very selective about what they will bind and transport; the carrier protein that transports glucose will not carry other simple sugars.

Carrier-mediated transport can be passive (no ATP required) or active (ATP-dependent). In *passive transport*, solutes are typically carried from an area of high concentration to an area of low concentration. *Active transport* mechanisms may follow or oppose an existing concentration gradient.

Many carrier proteins transport one ion or molecule at a time, but some deal with two solutes simultaneously. In *cotransport*, the carrier transports the two substances in the

3
CELL STRUCTURE AND FUNCTION

Studying Cells • **The Cell Membrane** • The Cytoplasm • The Nucleus • The Cell Life Cycle • Cell Diversity and Differentiation • Chapter Review

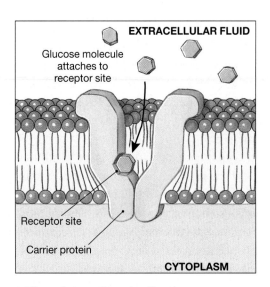
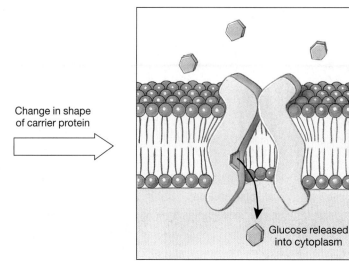

● *Figure 3-8* **Facilitated Diffusion**
In this process, an extracellular molecule, such as glucose, binds to a receptor site on a carrier protein. The binding alters the shape of the protein, which then releases the molecule to diffuse into the cytoplasm.

same direction, either into or out of the cell. In *countertransport*, one substance moves into the cell while the other moves out.

Two major examples of carrier-mediated transport—(1) *facilitated diffusion* and (2) *active transport*—are discussed below.

Facilitated Diffusion. Many essential nutrients, such as glucose or amino acids, are insoluble in lipids and too large to fit through membrane channels. However, these compounds can be passively transported across the membrane by carrier proteins in a process called **facilitated diffusion** (Figure 3-8●). The molecule to be transported first binds to a **receptor site** on the carrier protein. The shape of the protein then changes, moving the molecule to the inside of the cell membrane, where it is released into the cytoplasm.

As in the case of simple diffusion, no ATP is expended in facilitated diffusion, and the molecules move from an area of higher concentration to one of lower concentration. Facilitated diffusion differs from ordinary diffusion, however, because the rate of transport cannot increase indefinitely; only a limited number of carrier proteins are available in the membrane. Once all of them are operating, any further increase in the concentration of the solute in the extracellular fluid will have no effect on the rate of movement into the cell.

Active Transport. In **active transport**, the high-energy bond in ATP provides the energy needed to move ions or molecules across the membrane. The process is complex, and specific enzymes must be present in addition to the carrier molecule. The advantage of active transport is that the cell can import

or export specific materials *regardless of their intracellular or extracellular concentrations.*

All cells contain carrier proteins called **ion pumps** that actively transport the cations sodium (Na^+), potassium (K^+), calcium (Ca^{2+}), and magnesium (Mg^{2+}) across their cell membranes. Specialized cells can transport additional ions such as iodide (I^-), chloride (Cl^-), and iron (Fe^{2+}). Many of these carrier proteins move a specific cation or anion in one direction only, either into or out of the cell. In a few cases, one carrier protein will move more than one ion at a time. If one kind of ion moves in one direction and the other moves in the opposite direction, the carrier molecule is called an **exchange pump**.

A major function of exchange pumps is to maintain cell homeostasis. Sodium and potassium ions are the principal cations in body fluids. Sodium ion concentrations are high in the extracellular fluids, whereas sodium concentrations in the cytoplasm are relatively low. The distribution of potassium in the body is just the opposite—low in the extracellular fluids and high in the cytoplasm. Because of the presence of channel proteins in the membrane that are always open (so-called *leak channels*), sodium ions slowly diffuse into the cell, and potassium ions diffuse out.

Homeostasis within the cell depends on maintaining the sodium and potassium ion concentration gradients with the extracellular fluid. The **sodium-potassium exchange pump** maintains these gradients by ejecting sodium ions and recapturing lost potassium ions. For each ATP molecule consumed, three sodium ions are ejected and two potassium ions are reclaimed by the cell (Figure 3-9●).

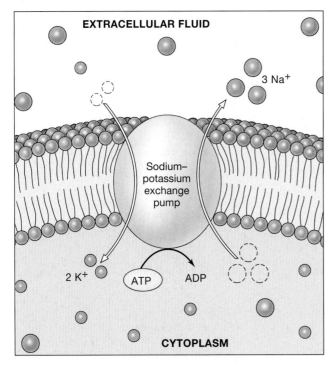

● *Figure 3-9* **The Sodium-Potassium Exchange Pump**
The operation of the sodium-potassium exchange pump is an example of active transport because its operation requires the conversion of ATP to ADP.

Vesicular Transport

In **vesicular transport**, materials move into or out of the cell by means of vesicles, small membranous sacs that form at or fuse with the cell membrane. The two major categories of vesicular transport are *endocytosis* and *exocytosis*.

Endocytosis. Endocytosis (EN-dō-sī-TŌ-sis; *endo-*, inside + *cyte*, cell) is the packaging of extracellular materials in a vesicle at the cell surface for import *into* the cell. This process may involve relatively large volumes of extracellular material. There are three major types of endocytosis: *receptor-mediated endocytosis*, *pinocytosis*, and *phagocytosis*. All three are active processes that require ATP or other sources of energy.

- **Receptor-mediated endocytosis** involves the formation of small vesicles at the membrane surface to import selected substances into the cell. This process produces vesicles that contain a specific target molecule in high concentrations. Receptor-mediated endocytosis begins when molecules in the extracellular fluid bind to receptors on the membrane surface (Figure 3-10●). The receptors bind to specific target molecules, called *ligands* (LĪ-gandz), such as a transport protein or hormone, and then cluster together on the membrane. The area of the membrane with the bound receptors forms a groove or pocket that pinches off to form a vesicle.

Many important substances, such as cholesterol and iron ions (Fe^{2+}), are carried through the body attached to special transport proteins. These proteins are too large to pass through membrane channels, but can enter the cell through this process.

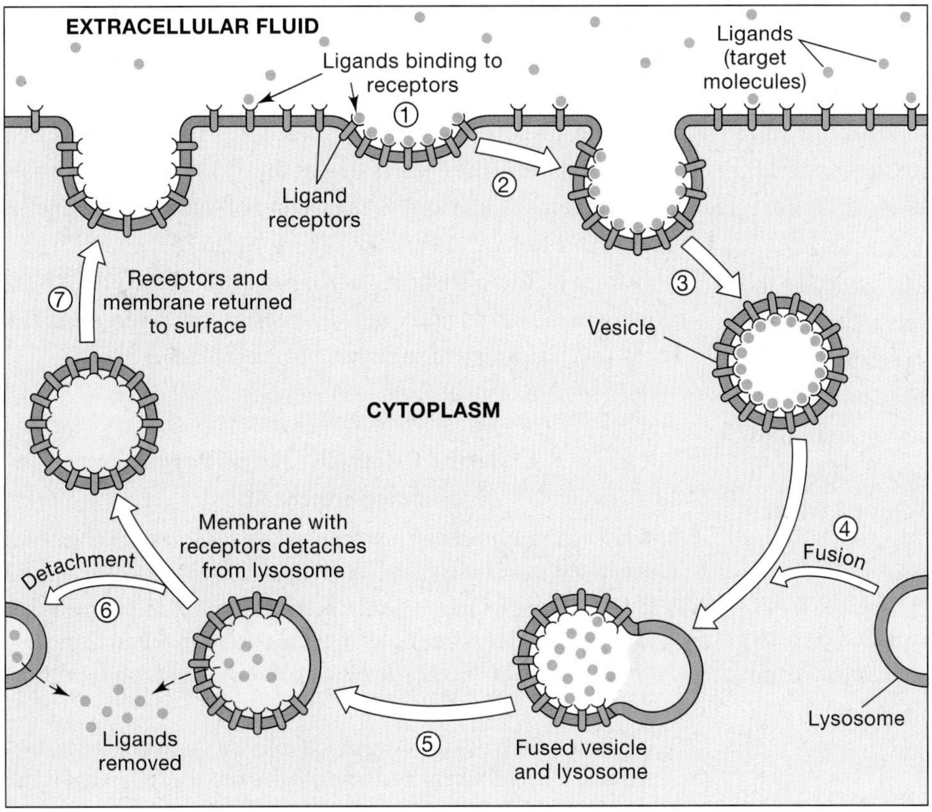

● *Figure 3-10* **Receptor-Mediated Endocytosis**
In this process, ① specific target molecules (ligands) bind to receptor molecules in the cell membrane. ② Membrane areas containing receptors bound with target molecules pinch off to form ③ vesicles that ④ fuse with lysosomes. ⑤ The ligands are freed from the receptors. ⑥ The membrane containing the receptor molecules separates from the lysosome and ⑦ returns to the cell surface.

3   **CELL STRUCTURE AND FUNCTION**

Studying Cells • The Cell Membrane • **The Cytoplasm** • The Nucleus • The Cell Life Cycle • Cell Diversity and Differentiation • Chapter Review

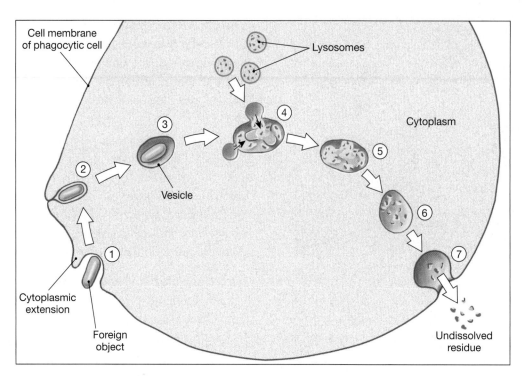

● *Figure 3-11* **Phagocytosis**
① A phagocytic cell first comes in contact with the foreign object and sends cytoplasmic extensions around it. ② The extensions approach one another and then ③ fuse to trap the material within a vesicle. ④ Lysosomes fuse with this vesicle, ⑤-⑥ activating digestive enzymes that gradually break down the structure of the phagocytized material. ⑦ Undissolved residue is then ejected from the cell by exocytosis.

• **Pinocytosis** (pi-nō-si-TŌ-sis; *pinein*, to drink), or "cell drinking," is the formation of small vesicles filled with extracellular fluid. In this process, common to all cells, a deep groove or pocket forms in the cell membrane and then pinches off. Because no receptor proteins are involved, pinocytosis is not as selective a process as receptor-mediated endocytosis.

• **Phagocytosis** (fa-gō-si-TŌ-sis; *phagein*, to eat), or "cell eating," produces vesicles containing solid objects that may be as large as the cell itself (Figure 3-11●). Cytoplasmic extensions called **pseudopodia** (soo-dō-PŌ-dē-a; *pseudo-*, false + *podon*, foot) surround the object, and their membranes fuse to form a vesicle. The vesicle may then fuse with a *lysosome*, a membranous sac of digestive enzymes, whereupon its contents are broken down.

Most cells display pinocytosis, but phagocytosis, especially the entrapment of living or dead cells, is performed only by specialized cells of the immune system. Phagocytic cells will be considered in chapters dealing with blood cells (Chapter 11) and the immune response (Chapter 14).

Exocytosis. Exocytosis (EK-sō-sī-TŌ-sis; *exo-*, outside) is the functional reverse of endocytosis. In this process a vesi-

cle created inside the cell fuses with the cell membrane and discharges its contents into the extracellular environment. The ejected material may be a secretory product, such as a hormone (a compound that circulates in the blood and affects cells in other parts of the body), mucus, or waste products remaining from the recycling of damaged organelles (Figure 3-11●).

Many of the transport mechanisms discussed above are moving materials in and out of the cell at any given moment. These mechanisms are summarized in Table 3-3.

CONCEPT CHECK QUESTIONS

Answers on page 81

❶ What is the difference between active and passive transport processes?

❷ During digestion in the stomach, the concentration of hydrogen (H^+) ions rises to many times the concentration found in the cells of the stomach. What type of transport process could produce this result?

❸ When certain types of white blood cells encounter bacteria, they are able to engulf them and bring them into the cell. What is this process called?

MECHANISM	PROCESS	FACTORS AFFECTING RATE	SUBSTANCES INVOLVED
DIFFUSION	Molecular movement of solutes; direction determined by relative concentrations	Size of gradient, molecular size, charge, lipid solubility, temperature	Small inorganic ions, lipid-soluble materials (all cells)
Osmosis	Movement of water molecules toward solution containing relatively higher solute concentration; requires selectively permeable membrane	Concentration gradient, opposing osmotic or hydrostatic pressure	Water only (all cells)
FILTRATION	Movement of water, usually with solute, by hydrostatic pressure; requires filtration membrane	Amount of pressure, size of pores in filter	Water and small ions (blood vessels)
CARRIER-MEDIATED TRANSPORT			
Facilitated diffusion	Carrier proteins passively transport solutes down a concentration gradient	Size of gradient, temperature and availability of carrier protein	Glucose and amino acids (all cells)
Active transport	Carrier proteins actively transport solutes regardless of any concentration gradients	Availability of carrier proteins, substrate, and ATP	Na^+, K^+, Ca^{2+}, Mg^{2+} (all cells); other solutes by specialized cells
VESICULAR TRANSPORT			
Endocytosis	Creation of vesicles containing fluid or solid material	Stimulus and mechanics incompletely understood; requires ATP	Fluids, nutrients (all cells); debris, pathogens (specialized cells)
Exocytosis	Fusion of vesicles containing fluids and/or solids with the cell membrane	Stimulus and mechanics incompletely understood; requires ATP	Fluids, debris (all cells)

The Cytoplasm

Cytoplasm is a general term for the material inside the cell between the cell membrane and the nucleus. The cytoplasm can be divided into the cytosol and organelles.

THE CYTOSOL

The **cytosol** is the intracellular fluid, which contains dissolved nutrients, ions, soluble and insoluble proteins, and waste products. It differs in composition from the extracellular fluid that surrounds most of the cells in the body:

• The cytosol contains a higher concentration of potassium ions and a lower sodium-ion concentration, whereas extracellular fluid contains a higher concentration of sodium ions and a lower potassium-ion concentration.

• The cytosol contains a high concentration of dissolved proteins, many of them enzymes that regulate metabolic operations. These proteins give the cytosol a consistency that varies between that of thin maple syrup and almost-set gelatin.

• The cytosol contains relatively small quantities of carbohydrates and large reserves of amino acids and lipids. The carbohydrates are broken down to provide energy, and the amino acids are used to manufacture proteins. The lipids are used primarily as an energy source when carbohydrates are unavailable.

The cytosol may also contain insoluble materials known as **inclusions**. Examples include stored nutrients, such as glycogen granules in muscle and liver cells and lipid droplets in fat cells.

ORGANELLES

Organelles (or-gan-ELZ; "little organs") are structures that perform specific functions essential to normal cell structure, maintenance, and metabolism (see Table 3-1). Membrane-enclosed organelles include the *nucleus, mitochondria, endoplasmic reticulum, Golgi apparatus, lysosomes,* and *peroxisomes.* The membrane isolates the organelle from the cytosol so the organelle can manufacture or store secretions, enzymes, or toxins that might otherwise damage the cell. The *cytoskeleton, microvilli,*

3 CELL STRUCTURE AND FUNCTION

Studying Cells • The Cell Membrane • **The Cytoplasm** • The Nucleus • The Cell Life Cycle • Cell Diversity and Differentiation • Chapter Review

centrioles, *cilia*, *flagella*, and *ribosomes* are organelles that are not surrounded by their own individual membranes and their parts are in direct contact with the cytosol.

The Cytoskeleton

The **cytoskeleton** is an internal protein framework of various threadlike filaments and hollow tubules that gives the cytoplasm strength and flexibility (Figure 3-12●). In most cells, the most important cytoskeletal elements are microfilaments and microtubules.

Microfilaments. **Microfilaments** are the thinnest strands, usually composed of the protein **actin**. In most cells, they form a dense layer just inside the cell membrane. Microfilaments attach the cell membrane to the underlying cytoplasm by forming connections with proteins of the cell membrane. In addition, actin microfilaments can interact with thicker filaments made of another protein, **myosin**, to produce active movement of a portion of a cell or to change the shape of the cell.

Microtubules. **Microtubules**, found in all our cells, are hollow tubes built from the globular protein **tubulin**. Microtubules form the primary components of the cytoskeleton, giving the cell strength and rigidity, and anchoring the position of major organelles.

During cell division, microtubules form the *spindle apparatus* that distributes the duplicated chromosomes to opposite ends of the dividing cell. This process will be considered in a later section.

Microvilli

Microvilli are small, finger-shaped projections of the cell membrane supported by microfilaments (see Figures 3-2● and 3-12●). Because they increase the surface area of the membrane, they are common features of cells actively engaged in absorbing materials from the extracellular fluid, such as the cells of the digestive tract and kidneys.

Centrioles, Cilia, and Flagella

In addition to functioning individually in the cytoskeleton, microtubules also interact to form more complex structures known as *centrioles*, *cilia*, and *flagella*.

Centrioles. A **centriole** is a short cylindrical structure composed of microtubules (see Figure 3-2●, p. 55). All animal cells that are capable of dividing contain a pair of centrioles arranged perpendicular to each other. The centrioles create the spindle fibers that move DNA strands during cell division. Mature red blood cells, skeletal muscle cells, cardiac muscle cells, and typical neurons do not have centrioles; as a result, these cells do not divide.

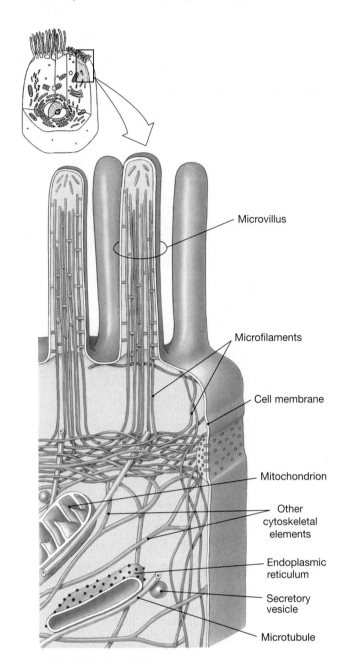

● *Figure 3-12* **The Cytoskeleton**
The cytoskeleton provides strength and structural support for the cell and its organelles. Interactions between cytoskeletal components are also important in moving organelles and changing the shape of the cell.

Cilia. Cilia (SIL-ē-uh; singular *cilium*) are relatively long finger-shaped extensions of the cell membrane (see Figure 3-2●). They are supported internally by a cylindrical array of microtubules. Cilia undergo active movements that require ATP energy. Their movements are coordinated so that their combined efforts move fluids or secretions across the cell surface. For example, cilia lining the respiratory passageways beat in a synchronized manner to move sticky mucus and trapped dust particles toward the throat and away from delicate respiratory surfaces. If the cilia are damaged or immobilized by heavy smoking or a metabolic problem, the cleansing action is lost, and the irritants will no longer be removed. As a result, chronic respiratory infections develop.

Flagella. **Flagella** (fla-JEL-uh; singular *flagellum*, whip) resemble cilia but are much longer. Flagella move a cell through the surrounding fluid, rather than moving the fluid past a stationary cell. The sperm cell is the only human cell that has a flagellum. If the flagella of sperm are paralyzed or otherwise abnormal, the individual will be sterile, because immobile sperm cannot perform fertilization.

Ribosomes

Ribosomes are organelles that manufacture proteins, using information provided by the DNA of the nucleus. Each ribosome consists of ribosomal RNA and protein. Ribosomes are found in all cells, but their number varies depending on the type of cell and its activities. For example, liver cells, which manufacture blood proteins, have much greater numbers of ribosomes than do fat cells, which synthesize triglycerides.

There are two major types of ribosomes: free ribosomes and fixed ribosomes. **Free ribosomes** are scattered throughout the cytoplasm, and the proteins they manufacture enter the cytosol. **Fixed ribosomes** are attached to the *endoplasmic reticulum* (ER), a membranous organelle. Proteins manufactured by fixed ribosomes enter the endoplasmic reticulum, where they are modified and packaged for export.

CONCEPT CHECK QUESTIONS

Answers on page 81

❶ Cells lining the small intestine have numerous fingerlike projections on their free surface. What are these structures, and what is their function?

❷ How would the absence of centrioles affect a cell?

The Endoplasmic Reticulum

The **endoplasmic reticulum** (en-dō-PLAZ-mik re-TIK-ū-lum; *reticulum*, a network), or **ER**, is a network of intracellular membranes that is connected to the membranous *nuclear envelope* surrounding the nucleus (see Figure 3-2●, p. 55). The ER has three major functions:

1. *Synthesis.* Specialized regions of the ER manufacture proteins, carbohydrates, and lipids.
2. *Storage.* The ER can store synthesized molecules or materials absorbed from the cytosol without affecting other cellular operations.

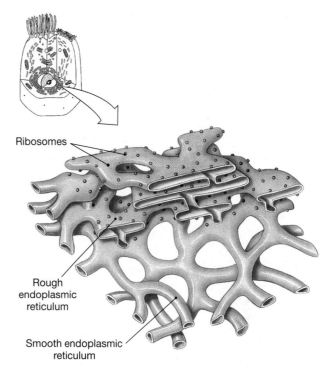

Ribosomes

Rough endoplasmic reticulum

Smooth endoplasmic reticulum

● *Figure 3-13* **The Endoplasmic Reticulum**
This diagrammatic sketch shows the three-dimensional relationships between the rough and smooth endoplasmic reticula.

3. *Transport.* Materials can travel from place to place within the cell without leaving the ER.

There are two types of endoplasmic reticulum, **smooth endoplasmic reticulum (SER)** and **rough endoplasmic reticulum (RER)** (Figure 3-13●). The SER, which lacks ribosomes, is the site where lipids and carbohydrates are produced. The membranes of the RER contain fixed ribosomes, indicating that they participate in protein synthesis.

The SER has a variety of functions, most of which involve the synthesis of lipids and carbohydrates. SER functions include (1) the synthesis of the phospholipids and cholesterol needed for maintenance and growth of the cell membrane, ER, nuclear membrane, and Golgi apparatus in all cells; (2) the synthesis of steroid hormones, such as *testosterone* and *estrogen* (sex hormones) in cells of the reproductive organs; and (3) the synthesis and storage of glycogen in skeletal muscle and liver cells.

The amount of endoplasmic reticulum and the proportion of RER to SER vary depending on the type of cell and its ongoing activities. For example, pancreatic cells that manufacture digestive enzymes contain an extensive RER, and the SER is relatively small. The proportion is just the reverse in the cells that synthesize steroid hormones in the reproductive system.

The lipids and carbohydrates produced by the SER and the proteins produced by the ribosomes of the RER may become incorporated into membranes or enter the inner chambers of

3 CELL STRUCTURE AND FUNCTION

Studying Cells • The Cell Membrane • **The Cytoplasm** • The Nucleus • The Cell Life Cycle • Cell Diversity and Differentiation • Chapter Review

their respective endoplasmic reticulum. These molecules are then packaged into small membrane sacs that pinch off from the tips of the ER. The sacs, called *transport vesicles*, deliver them to the Golgi apparatus, another membranous organelle, where they are processed further.

The Golgi Apparatus

The **Golgi** (GŌL-jē) **apparatus** consists of a set of five or six flattened membrane discs. A single cell may contain several sets, each resembling a stack of dinner plates (see Figure 3-2●, p. 55). The major functions of the Golgi apparatus are (1) the synthesis and packaging of secretions, such as enzymes; (2) the renewal or modification of the cell membrane; and (3) the packaging of special enzymes for use in the cytosol.

The various roles of the Golgi apparatus are diagrammed in Figure 3-14a●. The synthesis of proteins and other substances occurs in the RER, and then transport vesicles move these products to the Golgi apparatus. Enzymes in the Golgi apparatus modify the newly arrived molecules as other vesicles move them closer to the cell surface through succeeding membranes. Ultimately, the modified materials are repackaged in vesicles that leave the Golgi apparatus. The Golgi apparatus creates three classes of vesicles, each with a different fate. One class of vesicles contains secretions that will be discharged from the cell. These are called **secretory vesicles**. The secretion occurs through exocytosis at the cell surface (Figure 3-14b●). A second class of vesicles does not contain secretions, but the vesicle membrane will be incorporated into the cell membrane. Because the Golgi appa-

ratus continuously adds new lipids and proteins to the cell membrane in this way, the properties of the cell membrane can change. For example, receptors can be added or removed, making the cell more or less sensitive to a particular stimulus. A third class of vesicles will remain in the cytoplasm. These vesicles, called *lysosomes*, contain digestive enzymes.

Lysosomes

Lysosomes (LĪ-so-sōmz; *lyso-*, breakdown + *soma*, body) are vesicles filled with digestive enzymes. Lysosomes perform cleanup and recycling functions within the cell. Their enzymes are activated when they fuse with the membranes of damaged organelles, such as mitochondria or fragments of the endoplasmic reticulum. Then the enzymes break down the lysosomal contents. Nutrients reenter the cytosol through passive or active transport processes, and the remaining material is eliminated by exocytosis.

Lysosomes also function in the defense against disease. Cells may engulf bacteria, fluids, and organic debris from their surroundings into vesicles through endocytosis. ∞ p. 63 Lysosomes fuse with vesicles created in this way, and the digestive enzymes then break down the contents and release usable substances such as sugars or amino acids.

Lysosomes perform essential recycling functions inside the cell. For example, when muscle cells are inactive, lysosomes gradually break down their contractile proteins; if the cells become active once again, this destruction ends. However, in damaged or dead cells lysosome membranes disintegrate, releas-

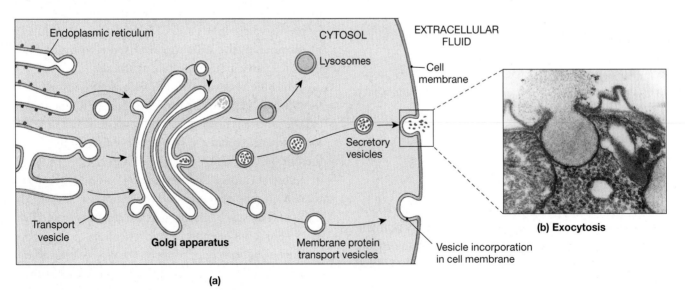

(a)

● *Figure 3-14* **The Golgi Apparatus**
(**a**) Transport vesicles carry molecules manufactured in the endoplasmic reticulum to the Golgi apparatus, where these molecules are modified and transported through succeeding membranes toward the cell surface. At the membrane closest to the cell surface, three types of vesicles develop. Secretory vesicles carry secretions from the Golgi apparatus to the cell surface, another group of vesicles adds additional membrane and protein to the cell membrane, and enzyme-filled lysosomes remain in the cytoplasm. (**b**) Exocytosis of secretions at the cell surface.

ing active enzymes into the cytosol. These enzymes rapidly destroy the proteins and organelles of the cell, a process called **autolysis** (aw-TAH-li-sis; *auto-*, self). Because the breakdown of lysosomal membranes can destroy a cell, lysosomes have been called cellular "suicide packets." We do not know how to control lysosomal activities or why the enclosed enzymes do not digest the lysosomal membranes unless the cell is damaged.

Peroxisomes

Peroxisomes are smaller than lysosomes and carry a different group of enzymes. In contrast to lysosomes, which are produced at the Golgi apparatus, peroxisomes are produced by the growth and subdivision of existing peroxisomes. Peroxisomes break down fatty acids and other organic compounds. During those catabolic reactions, peroxisomes generate hydrogen peroxide (H_2O_2), a potentially dangerous *free radical*. **Free radicals** are ions or molecules that contain unpaired electrons. They are highly reactive and enter additional reactions that can be destructive to vital compounds such as proteins. Other enzymes in peroxisomes break down the hydrogen peroxide into oxygen and water, thus protecting the cell from the damaging effects of free radicals. Found in all cells, peroxisomes are most abundant in metabolically active cells, such as liver cells.

Mitochondria

Mitochondria (mī-tō-KON-drē-uh; singular *mitochondrion*; *mitos*, thread + *chondrion*, granule) are small organelles containing enzymes that regulate the reactions that provide energy for the cell. The number of mitochondria in a particular cell varies with the cell's energy demands. For example, red blood cells lack mitochondria, but these organelles may account for 20 percent of the volume of an active liver cell. Mitochondria have an unusual double membrane: an outer membrane surrounding the entire organelle and an inner membrane containing numerous folds, called *cristae* (Figure 3-15●). Cristae increase the surface area exposed to the fluid contents, or *matrix*, of the mitochondria. Enzymes in the matrix and on the cristae catalyze energy-producing reactions.

Most of the chemical reactions that release energy occur in the mitochondria, but most of the cellular activities that require energy occur in the surrounding cytoplasm. Cells must therefore store energy in a form that can be moved from place to place. Energy is stored and transferred in the high-energy bond of ATP, as discussed in Chapter 2. ∞ p. 45 Living cells break the high-energy phosphate bond under controlled conditions, reconverting ATP to ADP and releasing energy for their use.

Mitochondrial Energy Production. Most cells generate ATP and other high-energy compounds through the breakdown of carbohydrates, especially glucose. Although most of the actual energy production occurs inside mitochondria, the first steps take place in the cytosol. In this reaction sequence, called *glycolysis*, six-carbon glucose molecules are broken down into three-carbon molecules of *pyruvic acid*. These molecules are then absorbed by the mitochondria. If glucose or other carbohydrates are not available, mitochondria can absorb and utilize small carbon chains produced by the breakdown of proteins or lipids.

Because the key reactions involved in mitochondrial activity consume oxygen, the process of mitochondrial energy production is known as **aerobic** (*aero-*, air + *bios*, life) **metabolism**. Aerobic metabolism in mitochondria produces about 95 percent

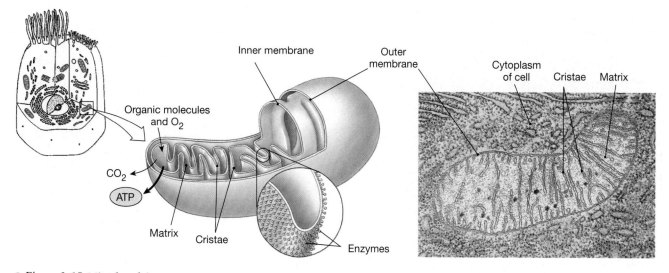

● *Figure 3-15* **Mitochondria**
The three-dimensional organization, and a color enhanced TEM of a typical mitochondrion in section. (TEM × 46,332). Mitochondria absorb short carbon chains and oxygen and generate carbon dioxide and ATP.

3 CELL STRUCTURE AND FUNCTION

Studying Cells • The Cell Membrane • The Cytoplasm • **The Nucleus** • The Cell Life Cycle • Cell Diversity and Differentiation • Chapter Review

of the energy needed to keep a cell alive. Aerobic metabolism is discussed in more detail in Chapters 7 and 17.

Several inheritable disorders result from abnormal mitochondrial activity. The mitochondria involved have defective enzymes that reduce their ability to generate ATP. Cells throughout the body may be affected, but symptoms involving muscle cells, nerve cells, and the light receptor cells in the eye are most common, because these cells have especially high energy demands.

CONCEPT CHECK QUESTIONS
Answers on page 81

❶ Certain cells in the ovaries and testes contain large amounts of smooth endoplasmic reticulum (SER). Why?

❷ Microscopic examination of a cell reveals that it contains many mitochondria. What does this observation imply about the cell's energy requirements?

The Nucleus

The **nucleus** is the control center for cellular operations. A single nucleus stores all the genetic information needed to control the synthesis of the approximately 100,000 different proteins in the human body. The nucleus determines the structural and functional features of the cell by controlling which proteins are synthesized and in what amounts.

Most cells contain a single nucleus, but there are exceptions. For example, skeletal muscle cells have many nuclei, and mature red blood cells have none. Figure 3-16● shows the structure of a typical nucleus. A **nuclear envelope** consisting of a double membrane surrounds the nucleus and its fluid contents, the *nucleoplasm*, from the cytosol. The nucleoplasm contains ions, enzymes, RNA and DNA nucleotides, proteins, small amounts of RNA, and DNA.

Chemical communication between the nucleus and the cytosol occurs through **nuclear pores**. These pores, which cover about 10 percent of the surface of the nucleus, are large enough to permit the movement of ions and small molecules but too small for the passage of proteins or DNA.

Most nuclei contain one to four **nucleoli** (noo-KLĒ-ō-lī; singular *nucleolus*). Nucleoli are organelles that synthesize the components of ribosomes. For this reason, they are most prominent in cells that manufacture large amounts of proteins, such as muscle and liver cells.

CHROMOSOME STRUCTURE

It is the DNA in the nucleus that stores instructions for protein synthesis, and this DNA is contained in **chromosomes** (*chroma*, color). The nuclei of human body cells contain 23 pairs of chromosomes. One member of each pair is derived from the mother and one from the father. The structure of a typical chromosome is shown in Figure 3-17●.

Each chromosome contains DNA strands bound to special proteins called *histones*. At intervals, the DNA strands wind around the histones, coiling up the DNA. The degree of coiling determines whether the chromosome is long and thin or short and fat. Chromosomes in a dividing cell are very tightly coiled, and they can be seen clearly as separate structures in light or electron micrographs. In cells that are not dividing, the DNA is loosely coiled, forming a tangle of fine filaments known as **chromatin**.

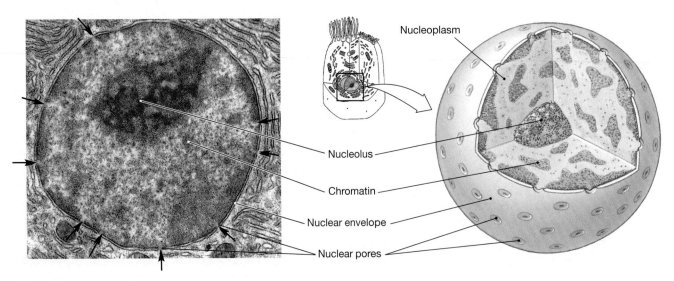

Nucleoplasm

Nucleolus

Chromatin

Nuclear envelope

Nuclear pores

● *Figure 3-16* **The Nucleus**
Electron micrograph and diagrammatic view showing important nuclear structures. The arrows indicate the locations of nuclear pores. (TEM × 4828).

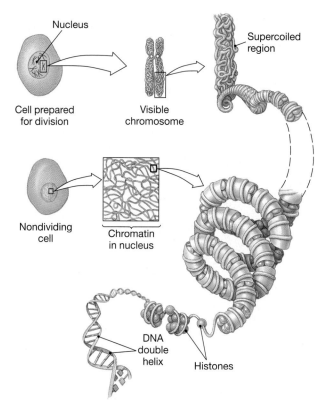

● *Figure 3-17* **Chromosome Structure**
DNA strands wound around histone proteins form coils that may be very tight or rather loose. In cells that are not dividing, the DNA is loosely coiled, forming a tangled network known as chromatin. When the coiling becomes tighter, as it does in preparation for cell division, the DNA becomes visible as distinct structures called chromosomes.

THE GENETIC CODE

The **genetic code** is the method of information storage in the DNA strands of the nucleus. An understanding of the genetic code has enabled us to determine how cells build proteins and how various structural and functional traits, such as hair color or blood type, are inherited from generation to generation.

The basic structure of nucleic acids was described in Chapter 2. ∞ p. 43 A single DNA molecule consists of a pair of strands held together by hydrogen bonding between complementary nitrogenous bases. Information is stored in the sequence of nitrogenous bases (adenine, A; thymine, T; cytosine, C; and guanine, G) along the length of DNA strands.

The genetic code is called a *triplet code* because a sequence of three nitrogenous bases can specify the identity of a single amino acid. A **gene** is the functional unit of heredity, and each one consists of all the triplets needed to produce a specific protein. The number of triplets varies from gene to gene, depending on the size of the protein that will be produced. Each gene also contains special segments responsible for regulating its own activity. In effect these triplets say, "Do (or do not) read this message," "Message starts here," or "Message ends here."

The "read me," "don't read me," and "start" signals form a special region of the DNA called the *promoter*, or *control segment*, at the start of each gene. Each gene ends with a "stop" signal.

⊕ DNA FINGERPRINTING

Every nucleated cell in the body carries a set of 46 chromosomes identical to the set formed at fertilization. Not all the DNA of these chromosomes codes for proteins, however, and long stretches of DNA have no known function. Some segments of this "useless" DNA contain the same nucleotide sequence repeated over and over. The number of segments and the number of repetitions vary from individual to individual. The chance that any two individuals, other than identical twins, have the same pattern is less than one in 9 billion. In other words, it is extremely unlikely that you will ever encounter someone else who has the same pattern of repeating nucleotide sequences present in your DNA.

The identification of individuals can therefore be made on the basis of DNA pattern analysis, just as it can on the basis of a fingerprint. Skin scrapings, blood, semen, hair, or other tissues can be used as a sample source. Information from **DNA fingerprinting** is used to convict (and to acquit) persons accused of committing violent crimes, such as rape or murder.

PROTEIN SYNTHESIS

Each DNA molecule contains thousands of genes and therefore holds the information needed to synthesize thousands of proteins. These genes are normally tightly coiled and bound to histones, which prevent their activation and, in doing so, prevent the synthesis of proteins. Before a specific gene can be activated, enzymes must temporarily break the weak bonds between its nitrogenous bases and remove the histone that guards the promoter. Although the process of gene activation is only partially understood, more is known about protein synthesis. The process of **protein synthesis** is divided into *transcription*, the production of RNA from a single strand of DNA, and *translation*, the assembling of a protein by ribosomes, using the information carried by the RNA molecule. Transcription takes place within the nucleus, and translation occurs in the cytoplasm.

Transcription

Ribosomes, the organelles of protein synthesis, are found in the cytoplasm, whereas the genes remain in the nucleus. This separation between the manufacturing site and the DNA's protein blueprint is overcome by the movement of a molecular messenger, a single strand of RNA known as **messenger RNA (mRNA)**. The process of mRNA formation is called **transcription**. Transcription, which is the process of transcribing or "copying," is an appropriate term because the newly formed mRNA is a copy of the information contained in the gene. Figure 3-18● details this process.

3 **CELL STRUCTURE AND FUNCTION**

Studying Cells • The Cell Membrane • The Cytoplasm • **The Nucleus** • The Cell Life Cycle • Cell Diversity and Differentiation • Chapter Review

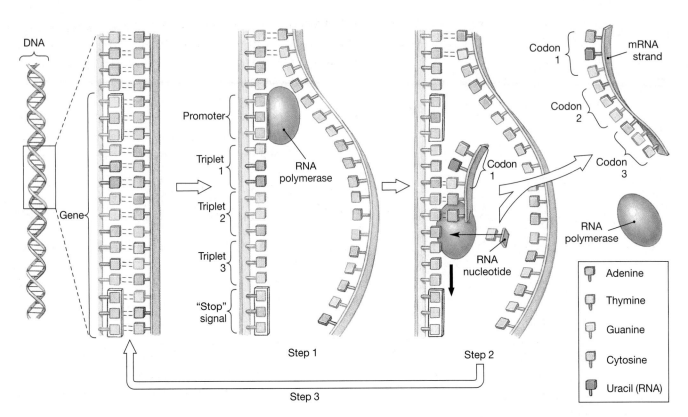

● *Figure 3-18* **Transcription**

A small portion of a single DNA molecule, containing a single gene available for transcription. **Step 1:** The two DNA strands separate, and RNA polymerase binds to the promoter of the gene. **Step 2:** The RNA polymerase moves from one nucleotide to another along the length of the gene. At each site, complementary RNA nucleotides form hydrogen bonds with the DNA nucleotides of the gene. The RNA polymerase then strings the arriving nucleotides together into a strand of mRNA. **Step 3:** On reaching the stop signal at the end of the gene, the RNA polymerase and the mRNA strand detach, and the two DNA strands reattach.

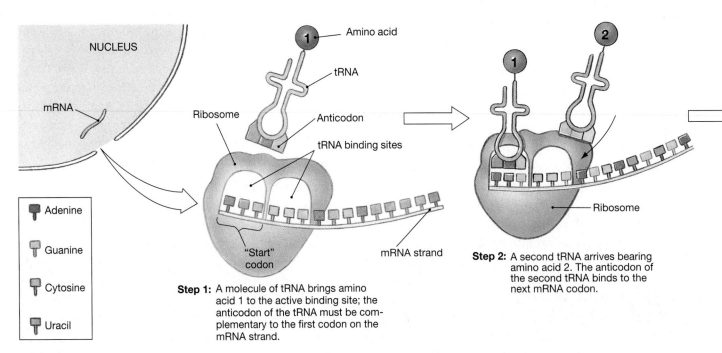

● *Figure 3-19* **Translation**

Once transcription has been completed, the mRNA diffuses into the cytoplasm and interacts with the ribosome.

TABLE 3-4 *Examples of the Triplet Code*

DNA TRIPLET	mRNA CODON	tRNA ANTICODON	AMINO ACID (OR INSTRUCTION)
AAA	UUU	AAA	Phenylalanine
AAT	UUA	AAU	Leucine
ACA	UGU	ACA	Cysteine
CAA	GUU	CAA	Valine
GGG	CCC	GGG	Proline
CGA	GCU	CGA	Alanine
TAC	AUG	UAC	Start codon
ATT	UAA	[none]	Stop codon

Each DNA strand contains thousands of individual genes. Transcription begins when an enzyme, *RNA polymerase*, binds to the promoter of the gene. This enzyme promotes the synthesis of an mRNA strand, using nucleotides complementary to those in the gene. The nucleotides involved are those characteristic of RNA, not DNA; RNA polymerase may attach adenine, guanine, cytosine, or uracil (U), but never thymine. Thus wherever an A occurs in the DNA strand, RNA polymerase will attach a U rather than a T. The mRNA strand thus contains a sequence of nitrogenous bases that are complementary to those of the gene. A sequence of three nitrogenous bases along the new mRNA strand represents a **codon** (KŌ-don) that is complementary to the corresponding triplet along the gene. At the

DNA "stop" signal, the enzyme and the mRNA strand detach, and the complementary DNA strands reassociate.

The mRNA formed in this way may be altered before it leaves the nucleus. For example, some regions (called *introns*) may be removed, and the remaining segments (called *exons*) spliced together. This modification creates a shorter, functional mRNA strand that enters the cytoplasm through one of the nuclear pores.

Translation

Translation is the synthesis of a protein using the information provided by the sequence of codons along the mRNA strand. Every amino acid has at least one unique and specific codon; Table 3-4 includes several examples. During translation, the sequence of codons will determine the sequence of amino acids in the protein.

Translation begins when the newly synthesized mRNA binds with a ribosome. Molecules of **transfer RNA (tRNA)** then deliver amino acids that will be used by the ribosome to assemble a protein. There are more than 20 different types of transfer RNA, at least one for each amino acid used in protein synthesis. Each tRNA molecule contains a complementary triplet of nitrogenous bases, known as an **anticodon**, that will bind to a specific codon on the mRNA.

The translation process is illustrated in Figure 3-19●:

Step 1: Translation begins at the "start" codon of the mRNA strand, with the arrival and binding of the first tRNA. That tRNA carries a specific amino acid.

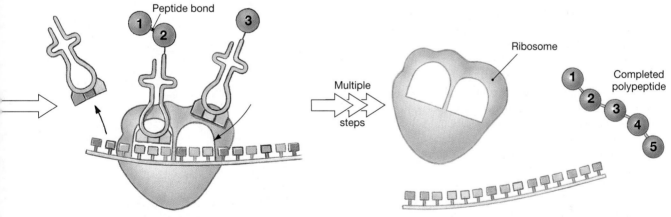

Step 3: Enzymes of the ribosome break the linkage between tRNA-1 and amino acid 1 and join amino acids 1 and 2 with a peptide bond. The ribosome shifts one codon to the right, tRNA-1 departs, and a third tRNA arrives.

Step 4: This process continues until the ribosome reaches the stop codon. The ribosome then breaks the connection between the last tRNA molecule and the polypeptide or protein. The ribosome disengages, leaving the mRNA strand intact.

3 CELL STRUCTURE AND FUNCTION

Studying Cells • The Cell Membrane • The Cytoplasm • The Nucleus • **The Cell Life Cycle** • Cell Diversity and Differentiation • Chapter Review

Step 2: A second tRNA then arrives, carrying a different amino acid, and binds to the second codon. Ribosomal enzymes now remove amino acid 1 from the first tRNA and attach it to amino acid 2 with a peptide bond. p. 41 The first tRNA then detaches from the ribosome and reenters the cytosol, where it can pick up another amino acid molecule and repeat the process. The ribosome now moves one codon farther along the length of the mRNA strand, and a third tRNA arrives, bearing amino acid 3.

Step 3: The ribosomal enzymes remove the dipeptide from the second tRNA and attach it to amino acid 3. The second tRNA is then released, and the ribosome moves one codon farther along the mRNA strand.

Step 4: Amino acids will continue to be added to the growing protein in this way until the ribosome reaches the "stop" codon. The ribosome then detaches, leaving an intact strand of mRNA and a completed polypeptide.

As you may recall from Chapter 2, a protein is a polypeptide containing 100 or more amino acids. p. 41 Translation proceeds swiftly, producing a typical protein (about 1000 amino acids) in around 20 seconds. The protein begins as a simple linear strand, but a more complex structure develops as it grows longer.

✚ MUTATIONS

Mutations are permanent changes in a cell's DNA that affect the nucleotide sequence of one or more genes. The simplest is called a *point mutation*, a change in a single nucleotide that affects one codon. With roughly 3 billion pairs of nucleotides in the DNA of a human cell, a single mistake might seem relatively unimportant. But over 100 inherited disorders have been traced to abnormalities in enzyme or protein structure that reflect single changes in nucleotide sequence. A single change in the amino acid sequence of a structural protein or enzyme can prove fatal. For example, several cancers and two potentially lethal blood disorders, *thalassemia* and *sickle-cell anemia*, result from variations in a single nucleotide. More elaborate mutations such as additions or deletions of nucleotides, can affect multiple codons in one gene, several adjacent genes, or the structure of one or more chromosomes.

Because most mutations occur during DNA replication, they are most likely to involve cells undergoing cell division. A single cell or group of daughter cells may be affected. If the mutations occur early in development, every cell in the body may be affected. For example, a mutation affecting the DNA of an individual's sex cells will be inherited by that individual's children. Our increasing understanding of genetic structure is opening the possibility for diagnosing and correcting some of these problems.

CONCEPT CHECK QUESTIONS
Answers on page 81

① How does the nucleus control the cell's activities?

② What process would be affected by the lack of the enzyme RNA polymerase?

③ During the process of transcription, a nucleotide was deleted from an mRNA sequence that coded for a protein. What effect would this deletion have on the amino acid sequence of the protein?

The Cell Life Cycle

During the time between fertilization of an egg by a sperm and physical maturity, the number of cells making up an individual increases from a single cell to roughly 75 trillion cells. This amazing increase in numbers occurs through a form of cellular reproduction called **cell division**. Even when development has been completed, cell division continues to be essential to survival as it replaces old and damaged cells.

Central to cell division is DNA replication, the accurate duplication of the cell's genetic material, and **mitosis** (mī-TŌ-sis), or nuclear division, which results in the distribution of DNA to the two new daughter cells. Mitosis occurs during the division of the **somatic** (*soma*, body) **cells**, which include the vast majority of the cells in the body. The production of sex cells, sperm and ova (eggs), involves a different form of cell division; this process, called **meiosis** (mī-Ō-sis), is described in Chapter 19.

Figure 3-20● presents the life cycle of a typical cell. Most cells spend only a small part of their life cycle engaged in cell

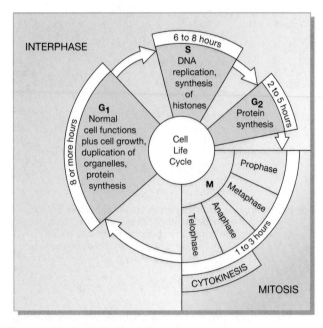

● *Figure 3-20* **The Cell Life Cycle**

division, or mitosis. For most of their lives, cells are in **interphase**, an interval of time between cell divisions when they perform normal functions.

INTERPHASE

A somatic cell in interphase is not necessarily preparing for mitosis. For example, some cells appear to be programmed to self-destruct after a certain length of time as a result of the activation of "suicide genes." The genetically controlled death of cells is called **apoptosis** (ap-op-TŌ-sis or ap-ō-TŌ-sis; *ptosis*, a falling away). Apoptosis is a key process in homeostasis. During fetal development, it causes the loss of tissue, or webbing, between the fingers and toes. Other cells, such as skeletal muscle cells and many nerve cells, never undergo mitosis or cell division.

A cell ready to divide first enters the G_1 phase. In this phase, the cell makes enough organelles and cytosol for two functional cells. These preparations may take hours, days, or weeks to complete, depending on the type of cell and the situation. For example, certain cells in the lining of the digestive tract divide every few days throughout life, whereas specialized cells in other tissues divide only under special circumstances, such as following an injury.

Once these preparations have been completed, the cell enters the S phase and replicates the DNA in its nucleus, a process that takes 6–8 hours. The goal of **DNA replication** is to copy the genetic information in the nucleus so that one set of chromosomes can be given to each of the two cells produced. This process starts when the complementary strands begin to separate and unwind (Figure 3-21●). Molecules of the enzyme *DNA polymerase* then bind to the exposed nitrogenous bases. As a result, complementary nucleotides in the nucleoplasm attach to the exposed nitrogenous bases of the DNA strand and form a pair of identical DNA molecules. Shortly after DNA replication has been completed there is a brief G_2 phase for protein synthesis and the completion of centriole replication. The cell then enters the M phase and mitosis begins.

MITOSIS

Mitosis is a process that separates and encloses the duplicated chromosomes of the original cell into two identical nuclei. Separation of the cytoplasm to form two separate and distinct cells

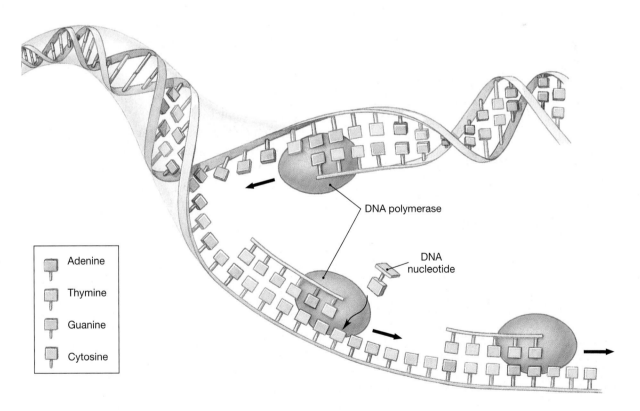

● *Figure 3-21* **DNA Replication**
In replication, the DNA strands unwind and DNA polymerase begins attaching complementary DNA nucleotides along each strand. Two identical copies of the original DNA molecule are produced.

3 CELL STRUCTURE AND FUNCTION

Studying Cells • The Cell Membrane • The Cytoplasm • The Nucleus • The Cell Life Cycle • **Cell Diversity and Differentiation** • Chapter Review

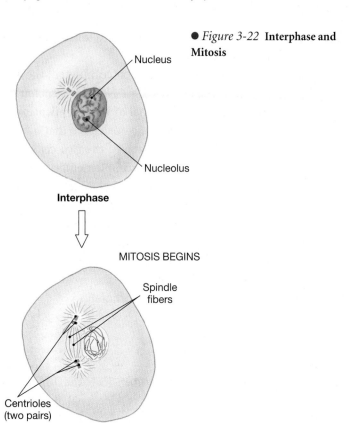

● *Figure 3-22* **Interphase and Mitosis**

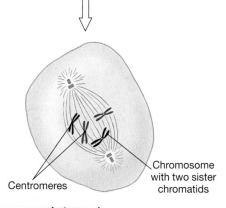

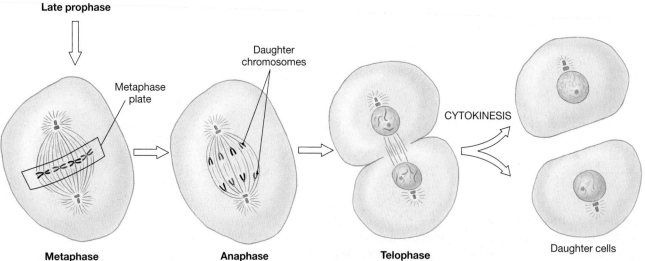

involves a separate but related process known as **cytokinesis** (sī-tō-ki-NĒ-sis; *cyto-*, cell + *kinesis*, motion). Mitosis is divided into four stages: *prophase*, *metaphase*, *anaphase*, and *telophase* (Figure 3-22●).

Stage 1: Prophase

Prophase (PRŌ-fāz; *pro-*, before) begins when the chromosomes coil so tightly that they become visible as individual structures. As a result of DNA replication, there are now two copies of each chromosome. Each copy, called a **chromatid** (KRŌ-ma-tid), is connected to its duplicate at a single point, the **centromere** (SEN-trō-mēr).

As the chromosomes appear, the two pairs of centrioles move toward opposite poles of the nucleus. An array of microtubules, called **spindle fibers**, extend between the centriole pairs. Prophase ends with the disappearance of the nuclear envelope.

Stage 2: Metaphase

Metaphase (MET-a-fāz; *meta-*, after) begins after the disintegration of the nuclear envelope. The spindle fibers now enter the nuclear region and the chromatids become attached to them. Once attachment has been completed, the chromatids move to a narrow central zone called the **metaphase plate**. Metaphase ends when all of the chromatids are aligned in the plane of the metaphase plate.

Stage 3: Anaphase

Anaphase (AN-uh-fāz; *ana-*, apart) begins when the centromere of each chromatid pair splits and the chromatids separate. The two **daughter chromosomes** are now pulled toward opposite ends of the cell. Anaphase ends when the daughter chromosomes arrive near the centrioles at opposite ends of the cell.

Stage 4: Telophase

During **telophase** (TEL-o-fāz; *telos*, end), the cell prepares to return to the interphase state. The nuclear membranes form, the nuclei enlarge, and the chromosomes gradually uncoil. Once the chromosomes have relaxed and the fine filaments of chromatin become visible, nucleoli reappear and the nuclei resemble those of interphase cells.

CYTOKINESIS

Telophase marks the end of mitosis proper, but the daughter cells have yet to complete their physical separation. **Cytokinesis**, the cytoplasmic division of the daughter cells, usually begins in late anaphase. As the daughter chromosomes near the ends of the spindle fibers, the cytoplasm constricts along the plane of the metaphase plate. This process continues throughout telophase and is usually completed sometime after the nuclear membrane has re-formed. The completion of cytokinesis marks the end of cell division.

CELL DIVISION AND CANCER

Within normal tissues, the rate of cell division is balanced with the rate of cell loss. If this balance breaks down, abnormal cell growth and cell division will enlarge the tissue and form a **tumor**, or *neoplasm*. In a **benign tumor**, the abnormal cells remain together in a connective tissue capsule and seldom threaten an individual's life. Surgery can usually remove the tumor if it begins to disturb the functions of the surrounding tissue.

Cells in a **malignant tumor**, however, no longer respond to normal control mechanisms. These cells spread into nearby tissue from the *primary tumor*. Malignant cells may also travel to distant tissues and organs and produce *secondary tumors*. This spreading is called **metastasis** (me-TAS-ta-sis). Metastasis is dangerous and difficult to control.

Cancer is an illness characterized by malignant cells. Such cancer cells lose their resemblance to normal cells and cause organ functions to deteriorate as their numbers increase. Cancer cells use energy less efficiently than normal cells, and they grow and multiply at the expense of normal tissues. The cancer cells steal nutrients from normal cells, and this accounts for the starved appearance of many patients in the late stages of cancer.

CONCEPT CHECK QUESTIONS
Answers on page 81

❶ What major events occur during interphase of cells preparing to undergo mitosis?

❷ List the four stages of mitosis.

❸ What would happen if spindle fibers failed to form in a cell during mitosis?

Cell Diversity and Differentiation

The liver cells, fat cells, and nerve cells of an individual contain the same chromosomes and genes, but in each case a different set of genes has been turned *off*. In other words, these cells differ because different genes are available for transcription. When a gene is deactivated, the cell loses the ability to create a particular protein and thus to perform any functions involving that protein. As more genes are switched off, the cell's functions become more restricted or specialized. This specialization process is called **differentiation**.

Fertilization produces a single cell with an unrestricted genetic potential. A period of repeated cell divisions follows, and differentiation begins as the number of cells increases. Differentiation produces specialized cells with limited capabilities. These cells form organized collections known as *tissues*, each with different functional roles. The next chapter examines the structure and function of tissues, and the role of tissue interactions in the maintenance of homeostasis.

Related Clinical Terms

benign tumor: A mass or swelling in which the cells usually remain within a connective tissue capsule; rarely life-threatening.

cancer: An illness characterized by gene mutations leading to the formation of malignant tumors and metastasis.

carcinogen (kar-SIN-ō-jen): An environmental factor that stimulates the conversion of a normal cell to a cancer cell.

DNA fingerprinting: Identifying an individual on the basis of repeating nucleotide sequences in his or her DNA.

genetic engineering: The research and experiments related to changing the genetic makeup (DNA) of an organism.

malignant tumor: A mass or swelling in which the cells no longer respond to normal control mechanisms but divide rapidly.

metastasis (me-TAS-ta-sis): The spread of malignant cells into distant tissues and organs, where secondary tumors subsequently develop.

normal saline: A solution that approximates the normal osmotic concentration of extracellular fluids.

primary tumor (*primary neoplasm*): The site at which a cancer cell initially develops.

recombinant DNA: DNA created by inserting (splicing) a specific gene from one organism into the DNA strand of another organism.

secondary tumor: A colony of cancerous cells formed by metastasis.

tumor (neoplasm): A mass or swelling produced by abnormal cell growth and division.

3 CELL STRUCTURE AND FUNCTION

Studying Cells • The Cell Membrane • The Cytoplasm • The Nucleus • The Cell Life Cycle • Cell Diversity and Differentiation • **Chapter Review**

CHAPTER REVIEW

Key Terms

Summary Outline

INTRODUCTION 54

1. Modern **cell theory** incorporates several basic concepts: (1) **Cells** are the building blocks of all plants and animals; (2) cells are the smallest functioning units of life; (3) cells are produced by the division of pre-existing cells; and (4) each cell maintains homeostasis. *(Figure 3-1)*

STUDYING CELLS 54

1. Electron microscopes are important tools used in **cytology**, the study of the structure and function of cells.

An Overview of Cellular Anatomy 55

2. A cell is surrounded by **extracellular fluid**. The cell's outer boundary, the **cell membrane**, separates the **cytoplasm**, or cell contents, from the extracellular fluid. *(Figure 3-2; Table 3-1)*

THE CELL MEMBRANE 57

1. The functions of the cell membrane include **(1)** physical isolation; **(2)** control of the exchange of materials with the cell's surroundings; **(3)** sensitivity; and **(4)** structural support.

Membrane Structure 57

2. The cell membrane, or plasma membrane, contains lipids, proteins, and carbohydrates. Its major components, lipid molecules, form a **phospholipid bilayer**. *(Figure 3-3)*

3. Membrane proteins may function as receptors, channels, carriers, enzymes, anchors, or identifiers. *(Table 3-2)*

Membrane Transport...58

4. Cell membranes are **selectively permeable**.

5. **Diffusion** is the net movement of material from an area where its concentration is relatively high to an area where its concentration is lower. Diffusion occurs until the **concentration gradient** is eliminated. *(Figures 3-4, 3-5)*

6. **Osmosis** is the diffusion of water across a membrane in response to differences in concentration. The force of movement is **osmotic pressure**. *(Figures 3-6, 3-7)*

7. In **filtration**, hydrostatic pressure forces water across a membrane. If membrane pores are large enough, molecules of solute will be carried along.

8. **Facilitated diffusion** is a type of **carrier-mediated transport** and requires the presence of carrier proteins in the membrane. *(Figure 3-8)*

9. **Active transport** mechanisms consume ATP but are independent of concentration gradients. Some **ion pumps** are **exchange pumps**. *(Figure 3-9)*

10. In **vesicular transport**, material moves into or out of a cell in membranous sacs. Movement into the cell occurs through **endocytosis**, an active process that includes **receptor-mediated endocytosis**, **pinocytosis** ("cell-drinking"), and **phagocytosis** ("cell-eating"). Movement out of the cell occurs through **exocytosis**. *(Figures 3-10, 3-11; Table 3-3)*

THE CYTOPLASM 65

1. The cytoplasm surrounds the nucleus and contains a fluid **cytosol** and intracellular structures called *organelles*.

The Cytosol 65

2. The **cytosol** differs in composition from the extracellular fluid that surrounds most cells of the body.

Organelles 65

3. Membrane-enclosed **organelles** are surrounded by lipid membranes that isolate them from the cytosol. They include the endoplasmic

reticulum, the nucleus, the Golgi apparatus, lysosomes, and mitochondria. *(Table 3-1)*

4. Organelles that are not membrane-enclosed are always in contact with the cytosol. They include the cytoskeleton, microvilli, centrioles, cilia, flagella, and ribosomes. *(Table 3-1)*

5. The **cytoskeleton** gives the cytoplasm strength and flexibility. Its two main components are **microfilaments** and **microtubules**. *(Figure 3-12)*

6. **Microvilli** are small projections of the cell membrane that increase the surface area exposed to the extracellular environment. *(Figure 3-12)*

7. **Centrioles** direct the movement of chromosomes during cell division.

8. **Cilia** beat rhythmically to move fluids or secretions across the cell surface.

9. **Flagella** move a cell through surrounding fluid rather than moving fluid past a stationary cell.

10. A **ribosome** is an intracellular factory that manufactures proteins. **Free ribosomes** are in the cytoplasm, and **fixed ribosomes** are attached to the endoplasmic reticulum.

11. The **endoplasmic reticulum (ER)** is a network of intracellular membranes. **Rough endoplasmic reticulum (RER)** contains ribosomes and is involved in protein synthesis. **Smooth endoplasmic reticulum (SER)** does not contain ribosomes; it is involved in lipid and carbohydrate synthesis. *(Figure 3-13)*

12. The **Golgi apparatus** forms **secretory vesicles** and new membrane components and *packages lysosomes*. Secretions are discharged from the cell by exocytosis. *(Figure 3-14)*

13. **Lysosomes** are vesicles filled with digestive enzymes. Their functions include ridding the cell of bacteria and debris.

14. **Mitochondria** are responsible for 95 percent of the ATP production within a typical cell. The **matrix**, or fluid contents of a mitochondrion, lie inside the **cristae**, or folds of an inner membrane. *(Figure 3-15)*

THE NUCLEUS 70

1. The **nucleus** is the control center for cellular operations. It is surrounded by a **nuclear envelope**, through which it communicates with the cytosol by way of **nuclear pores**. *(Figure 3-16)*

Chromosome Structure 70

2. The nucleus controls the cell by directing the synthesis of specific proteins using information stored in the DNA of **chromosomes**. *(Figure 3-17)*

The Genetic Code 71

3. The cell's information storage system, the **genetic code**, is called a *triplet code* because a sequence of three nitrogenous bases identifies a single

amino acid. Each **gene** consists of all the triplets needed to produce a specific protein. *(Table 3-4)*

Protein Synthesis 71

4. **Protein synthesis** includes both *transcription*, which occurs in the nucleus, and *translation*, which occurs in the cytoplasm.

5. During **transcription**, a strand of **messenger RNA (mRNA)** is formed and carries protein-making instructions from the nucleus to the cytoplasm. *(Figure 3-18)*

6. During **translation** a functional protein is constructed from the information contained in an mRNA strand. Each triplet of nitrogenous bases along the mRNA strand is a **codon**; the sequence of codons determines the sequence of amino acids in the protein. *(Figure 3-19)*

7. Molecules of **transfer RNA (tRNA)** bring amino acids to the ribosomes involved in translation. *(Figure 3-19)*

THE CELL LIFE CYCLE 74

1. **Cell division** is the reproduction of cells. **Apoptosis** is the genetically controlled death of cells. **Mitosis** is the nuclear division of **somatic cells**. Sex cells (sperm and ova) are produced by **meiosis**.

Interphase 75

2. Most somatic cells are in **interphase** most of the time. Cells preparing for mitosis undergo **DNA replication** in this phase. *(Figures 3-20, 3-21)*

Mitosis 75

3. Mitosis proceeds in four stages: **prophase**, **metaphase**, **anaphase**, and **telophase**. *(Figure 3-22)*

Cytokinesis 77

4. During **cytokinesis**, the cytoplasm divides, producing two identical daughter cells.

Cell Division and Cancer 77

5. Abnormal cell growth and division forms **benign tumors** or **malignant tumors** within a tissue. **Cancer** is a disease characterized by the presence of malignant tumors; over time, cancer cells tend to spread to new areas of the body.

CELL DIVERSITY AND DIFFERENTIATION 77

1. **Differentiation** is the specialization that produces cells with limited capabilities. These specialized cells form organized collections called *tissues*, each of which has specific functional roles.

3 CELL STRUCTURE AND FUNCTION

Studying Cells • The Cell Membrane • The Cytoplasm • The Nucleus • The Cell Life Cycle • Cell Diversity and Differentiation • **Chapter Review**

Review Questions

Level 1: Reviewing Facts and Terms

Match each item in column A with the most closely related item in column B. Use letters for answers in the spaces provided.

COLUMN A

___ 1. filtration

___ 2. osmosis

___ 3. hypotonic solution

___ 4. hypertonic solution

___ 5. isotonic solution

___ 6. facilitated diffusion

___ 7. carrier proteins

___ 8. vesicular transport

___ 9. cytosol

___ 10. cytoskeleton

___ 11. microvilli

___ 12. ribosomes

___ 13. mitochondria

___ 14. lysosomes

___ 15. nucleus

___ 16. chromosomes

___ 17. nucleoli

COLUMN B

a. water out of cell

b. passive carrier-mediated transport

c. endocytosis, exocytosis

d. movement of water

e. hydrostatic pressure

f. normal saline

g. ion pump

h. water into cell

i. manufacture proteins

j. digestive enzymes

k. internal protein framework

l. control center for cellular operations

m. intracellular fluid

n. DNA strands

o. cristae

p. synthesize components of ribosomes

q. increase cell surface area

---

18. The study of the structure and function of cells is called:

    (a) histology

    (b) cytology

    (c) physiology

    (d) biology

19. The proteins in the cell membranes may function as:

    (a) receptors and channels

    (b) carriers and enzymes

    (c) anchors and identifiers

    (d) a, b, and c are correct

20. All of the following membrane transport mechanisms are passive processes except:

    (a) diffusion

    (b) facilitated diffusion

    (c) vesicular transport

    (d) filtration

21. _____ ion concentrations are high in the extracellular fluids, and _____ ion concentrations are high in the cytoplasm.

    (a) calcium, magnesium

    (b) chloride, sodium

    (c) potassium, sodium

    (d) sodium, potassium

22. Structures that perform specific functions within the cell are

    (a) organs

    (b) organisms

    (c) organelles

    (d) chromosomes

23. The construction of a functional protein using the information provided by an mRNA strand is

    (a) translation

    (b) transcription

    (c) replication

    (d) gene activation

24. The term *differentiation* refers to

    (a) the loss of genes from cells

    (b) the acquisition of new functional capabilities by cells

    (c) the production of functionally specialized cells

    (d) the division of genes among different types of cells

25. What are the four general functions of the cell membrane?

26. By what four major transport mechanisms do substances get into and out of cells?

27. What are the three major functions of the endoplasmic reticulum?

28. List the four stages of mitosis in their correct sequence.

Level 2: Reviewing Concepts

29. Diffusion is important in body fluids because this process tends to:

    (a) increase local concentration gradients

    (b) eliminate local concentration gradients

    (c) move substances against their concentration gradients

    (d) create concentration gradients

30. When placed in a _____ solution, a cell will lose water through osmosis. The process results in the _____ of red blood cells.

    (a) hypotonic, crenation

    (b) hypertonic, crenation

    (c) isotonic, hemolysis

    (d) hypotonic, hemolysis

31. Suppose that a DNA segment has the following nucleotide sequence: CTC ATA CGA TTC AAG TTA. Which of the following nucleotide sequences would be found in a complementary mRNA strand?

    (a) GAG UAU GAU AAC UUG AAU

    (b) GAG TAT GCT AAG TTC AAT

    (c) GAG UAU GCU AAG UUC AAU

    (d) GUG UAU GGA UUG AAC GGU

32. How many amino acids are coded in the DNA segment in the previous question?

    (a) 18

    (b) 9

    (c) 6

    (d) 3

33. What are the similarities between facilitated diffusion and active transport? What are the differences?

34. How does the cytosol differ in composition from the extracellular fluid?

35. Differentiate between transcription and translation.

36. List the stages of mitosis, and briefly describe the events that occur in each.

37. What is cytokinesis, and what role does it play in the cell cycle?

Level 3: Critical Thinking and Clinical Applications

38. Experimental evidence shows that the transport of a certain molecule exhibits the following characteristics: (1) The molecule moves along its concentration gradient; (2) at concentrations above a given level, there is no increase in the rate of transport; and (3) cellular energy is not required for transport to occur. Which type of transport process is at work?

39. Two solutions, A and B, are separated by a selectively permeable barrier. Over a period of time, the level of fluid on side A increases. Which solution initially had the higher concentration of solute?

Answers to Concept Check Questions

Page 64

1. Active transport processes require the expenditure of cellular energy in the form of the high-energy bonds of ATP molecules. Passive transport processes (*diffusion, osmosis, filtration,* and *facilitated diffusion*) move ions and molecules across the cell membrane without any energy expenditure by the cell.

2. Energy must be expended to transport H^+ ions against their concentration gradient—that is, from a region where they are less concentrated (the cells lining the stomach) to a region where they are more concentrated (the interior of the stomach). An active transport process must be involved.

3. This is an example of phagocytosis.

Page 67

1. The fingerlike projections on the surface of the intestinal cells are *microvilli*. They increase the cells' surface area so they can absorb nutrients more efficiently.

2. Cells that lack centrioles are unable to divide.

Page 70

1. The SER functions in the synthesis of lipids such as steroids. Ovaries and testes would be expected to have a great deal of SER because these organs produce large amounts of steroid hormones.

2. The function of mitochondria is to produce energy for the cell in the form of ATP molecules. A large number of mitochondria in a cell would indicate a high demand for energy.

Page 74

1. The nucleus of a cell contains DNA that codes for the production of all of the cell's proteins. Some of these proteins are structural proteins that are responsible for the shape and other physical characteristics of the cell. Other proteins are enzymes that govern cellular metabolism, direct the production of cell proteins, and control all of the cell's activities.

2. If a cell lacked the enzyme RNA polymerase it would not be able to transcribe RNA from DNA.

3. The deletion of a base from a coding sequence of DNA during transcription would alter the entire mRNA base sequence after the deletion point. This would result in different codons on the messenger RNA that was transcribed from the affected region, and this, in turn, would result in the incorporation of a different series of amino acids into the protein. Almost certainly the protein product would not be functional.

Page 77

1. Cells that are preparing to undergo mitosis manufacture additional organelles and duplicate sets of their DNA.

2. The four stages of mitosis are prophase, metaphase, anaphase, and telophase.

3. If spindle fibers failed to form during mitosis, the cell would not be able to separate the chromosomes into two sets. If cytokinesis occurred, the result would be one cell with two sets of chromosomes and one cell with none.

The Tissue Level of Organization

Chapter Outline and Objectives

Vocabulary Development

a-...without; *avascular*
apo- ....from; *apocrine*
cardium ....heart; *pericardium*
chondros..................cartilage; *perichondrium*
dendron ....tree; *dendrites*
desmos........................ligament; *desmosome*
glia ....glue; *neuroglia*
histos ....tissue; *histology*
holos ....entire; *holocrine*
hyalos ....glass; *hyaline cartilage*
inter- ....between; *interstitial*
krinein ....to secrete; *exocrine*
lacus ....pool; *lacunae*
meros ....part; *merocrine*
neuro ....nerve; *neuron*
os..bone; *osseous tissue*
peri-................................around; *perichondrium*
phagein ....to eat; *macrophage*
pleura ....rib; *pleural membrane*
pseudes........................false; *pseudostratified*
sistere....................................to set; *interstitial*
soma ....body; *desmosome*
squama ....plate or scale; *squamous*
vas ....vessel; *vascular*

*E*XOTIC CREATURES on the deep-sea floor? Well, no. These "creatures" are much closer to home. These are cells that line the airway leading to your lungs. The "tentacles" are cilia that help remove dirt and harmful microbes from inhaled air. Notice that the surface contains more than one type of cell—some cells have long cilia (green), and others have short microvilli (yellow). Groups of similar cells are specialized to perform a particular set of functions are called tissues. Each of your body's tissues has its own distinctive structure and its own role to play in maintaining homeostasis. We will meet them all in this chapter.

4 THE TISSUE LEVEL OF ORGANIZATION

Epithelial Tissue • Connective Tissues • Membranes • Muscle Tissue • Neural Tissue • Tissue Injuries and Repairs • Tissues and Aging • Chapter Review

N O SINGLE CELL is able to perform the many functions of the human body. Instead, through differentiation, each cell specializes to perform a relatively restricted range of functions. Although there are trillions of individual cells in the human body, there are only about 200 different types of cells. These cell types combine to form **tissues**, collections of specialized cells and cell products that perform a limited number of functions. **Histology** (*histos*, tissue) is the study of tissues. Four basic *tissue types* exist: *epithelial tissue, connective tissue, muscle tissue,* and *neural tissue* (Figure 4-1●).

Epithelial Tissue

Epithelial tissue includes epithelia and *glands*. **Epithelia** (e-pi-THĒ-lē-a; singular, *epithelium*) are layers of cells that cover internal or external surfaces. **Glands** are made up of secreting cells derived from epithelia. Important characteristics of epithelia include the following:

- Cells that are bound closely together. In other tissue types, the cells are often widely separated by extracellular materials.

- A free surface exposed to the environment or to some internal chamber or passageway.

- Attachment to underlying connective tissue by a *basement membrane*.

- The absence of blood vessels. Because of this **avascular** (ā-VAS-kū-lar; *a-*, without + *vas*, vessel) condition, epithelial cells must obtain nutrients from deeper tissues or from their exposed surfaces.

Epithelia cover both external and internal body surfaces. In addition to covering the skin, epithelia line internal passageways that communicate with the outside world, such as the digestive, respiratory, reproductive, and urinary tracts. These epithelia form selective barriers that separate the deep tissues of the body from the external environment.

Epithelia also line internal cavities and passageways, such as the chest cavity, fluid-filled chambers in the brain, eye, and inner ear, and the inner surfaces of blood vessels and the heart. These epithelia prevent friction, regulate the fluid composition of internal cavities, and restrict communication between the blood and tissue fluids.

FUNCTIONS OF EPITHELIA

Epithelia perform four essential functions:

1. *Provide physical protection.* Epithelia protect exposed and internal surfaces from abrasion, dehydration, and destruction by chemical or biological agents. For example, as long as it remains intact, the epithelium of your skin resists impacts and scrapes, restricts water loss, and prevents invasion of underlying structures by bacteria.

2. *Control permeability.* Any substance that enters or leaves the body has to cross an epithelium. Some epithelia are relatively impermeable; others are easily crossed by compounds as large as proteins.

3. *Provide sensation.* Specialized epithelial cells can detect changes in the environment and relay information about such changes to the nervous system. For example, touch receptors in the deepest layers of the epithelium of the skin respond by stimulating neighboring sensory nerves.

4. *Produce specialized secretions.* Epithelial cells that produce secretions are called **gland cells**. In a

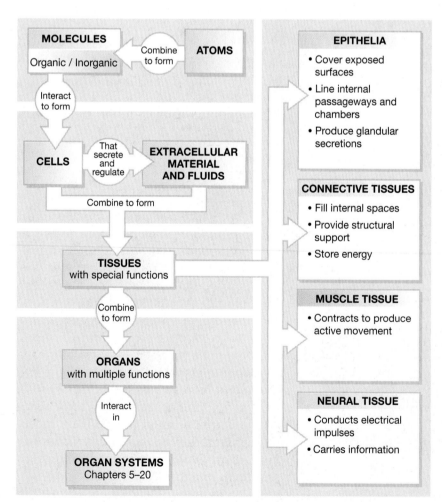

● *Figure 4-1* **An Orientation to the Tissues of the Body**

glandular epithelium, most or all of the cells actively produce secretions. These secretions are classified according to their discharge location:

- **Exocrine** (*exo-*, outside + *krinein*, to secrete) secretions are discharged onto the surface of the skin or other epithelial surface. Enzymes entering the digestive tract, perspiration on the skin, and milk produced by mammary glands are examples.

- **Endocrine** (*endo-*, inside) secretions are released into the surrounding tissues and blood. These secretions, called *hormones*, regulate or coordinate the activities of other tissues, organs, and organ systems. (Hormones are discussed further in Chapter 10.) Endocrine secretions are produced in organs such as the pancreas, thyroid, and pituitary gland.

INTERCELLULAR CONNECTIONS

To be effective in protecting other tissues, epithelial cells must remain firmly attached to one another. If an epithelium is damaged or the connections are broken, it is no longer an effective barrier. For example, when the epithelium of the skin is damaged by a burn or abrasion, disease-causing bacteria can enter underlying tissues and cause an infection. Undamaged epithelia form effective barriers because the epithelial cell membranes are held together by an intercellular cement (composed of a protein-polysaccharide mixture) and by specialized attachment sites, known as *cell junctions*. Three such cell junctions are *tight junctions, gap junctions*, and *desmosomes*.

At a **tight junction**, the outermost lipid layers of adjacent cell membranes are tightly bound together by interlocking membrane proteins (Figure 4-2a●). Tight junctions prevent

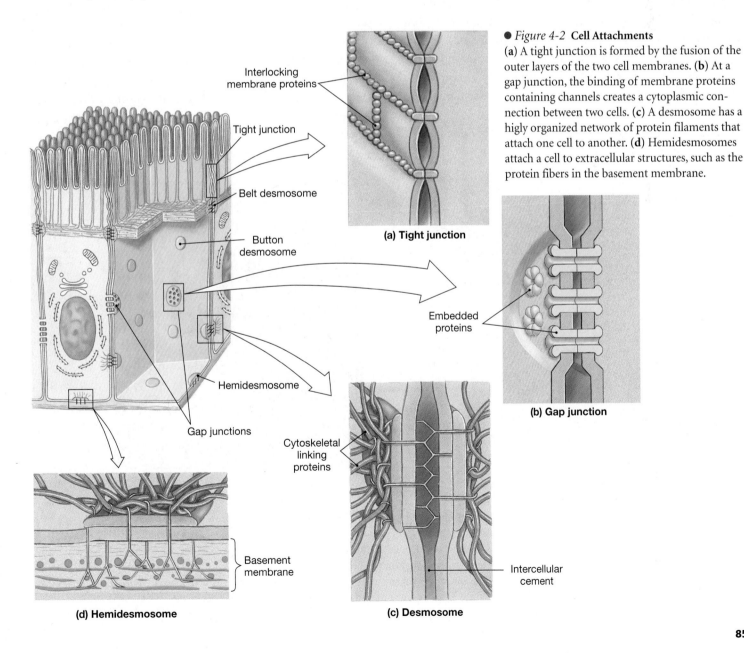

● *Figure 4-2* **Cell Attachments**
(**a**) A tight junction is formed by the fusion of the outer layers of the two cell membranes. (**b**) At a gap junction, the binding of membrane proteins containing channels creates a cytoplasmic connection between two cells. (**c**) A desmosome has a higly organized network of protein filaments that attach one cell to another. (**d**) Hemidesmosomes attach a cell to extracellular structures, such as the protein fibers in the basement membrane.

Interlocking membrane proteins

Tight junction

Belt desmosome

Button desmosome

Hemidesmosome

Gap junctions

(a) Tight junction

Embedded proteins

(b) Gap junction

Cytoskeletal linking proteins

Intercellular cement

Basement membrane

(d) Hemidesmosome

(c) Desmosome

4 THE TISSUE LEVEL OF ORGANIZATION

Epithelial Tissue • Connective Tissues • Membranes • Muscle Tissue • Neural Tissue • Tissue Injuries and Repairs • Tissues and Aging • Chapter Review

the passage of water and solutes between the cells. These junctions are common between epithelial cells exposed to harsh chemicals or powerful enzymes. For example, tight junctions between epithelial cells lining the digestive tract keep digestive enzymes, stomach acids, or waste products from damaging underlying tissues.

At a **gap junction**, two cells are held together by interlocked membrane proteins (Figure 4-2b●). Because these are channel proteins, the result is a narrow passageway that lets small molecules and ions pass from cell to cell. Gap junctions interconnect cells in some epithelia, but they are most abundant in cardiac muscle and smooth muscle tissue, where they are essential to the coordination of muscle contractions.

At **desmosomes** (DEZ-mō-sōmz; *desmos*, ligament + *soma*, body), the cell membranes of two cells are locked together by intercellular cement and a network of fine protein filaments (Figure 4-2c●). A desmosome may form small discs (button desmosomes) or encircle a cell (belt desmosome). *Hemidesmosomes* resemble half of a button desmosome, and attach a cell to the basement membrane (Figure 4-2d●). Desmosomes are very strong, and the connection can resist stretching and twisting. In the skin, these links are so strong that dead cells are usually shed in thick sheets, rather than individually.

THE EPITHELIAL SURFACE

Many epithelia that line internal passageways have microvilli on their exposed surfaces. p. 66 They may vary in number from just a few to so many that they carpet the entire surface (Figure 4-3●). Microvilli are especially abundant on epithelial surfaces where absorption and secretion take place, such as along portions of the digestive and urinary tracts. These epithelial cells specialize in the active and passive transport of materials across their cell membranes. p. 58 A cell with microvilli has at least 20 times the surface area of a cell without them; the greater the surface area of the cell membrane, the more transport proteins will be exposed to the extracellular environment. The diagram in Figure 4-3● also shows elongated microvilli, called *stereocilia*. Stereocilia are found only on specialized cells in the male reproductive tract and on receptor cells in the inner ear.

Some epithelia contain cilia on their exposed surfaces. p. 66 A typical cell within a *ciliated epithelium* has roughly 250 cilia that beat in a coordinated fashion to move materials across the epithelial surface. For example, the ciliated epithelium that lines the respiratory tract moves mucus-trapped irritants away from the lungs and toward the throat.

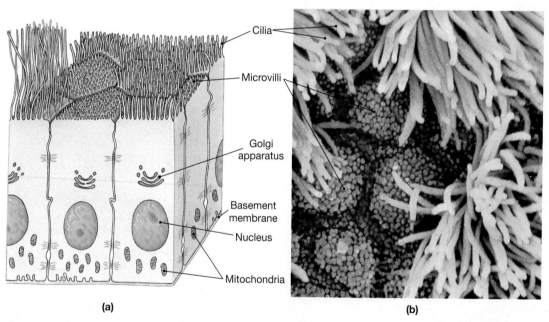

Cilia
Microvilli
Golgi apparatus
Basement membrane
Nucleus
Mitochondria

(a) **(b)**

● *Figure 4-3* **The Surfaces of Epithelial Cells**
(a) The inner and outer surfaces of most epithelia are specialized for specific functions. The free surface often bears microvilli; less commonly, this surface has cilia or (very rarely) stereocilia. (*All three would not normally be on the same group of cells but are depicted here for purposes of illustration.*) Mitochondria are typically concentrated near the base of the cell, probably to provide energy for the cell's transport activities. **(b)** An SEM showing the surface of a ciliated epithelium that lines most of the respiratory tract. The small, bristly areas are microvilli on the exposed surfaces of mucus-producing cells that are scattered among the ciliated epithelial cells. (SEM × 13,469)

The Basement Membrane

Epithelial cells not only must hold onto one another but also must remain firmly connected to the rest of the body. This function is performed by the **basement membrane**, which lies between the epithelium and underlying connective tissues (Figure 4-3a●). There are no cells within the basement membrane, which consists of a network of protein fibers. The epithelial cells adjacent to the basement membrane are firmly attached to these protein fibers by hemidesmosomes. In addition to providing strength and resisting distortion, the basement membrane also provides a barrier that restricts the movement of proteins and other large molecules from the underlying connective tissue into the epithelium.

Epithelial Renewal and Repair

An epithelium must continually repair and renew itself. Epithelial cells may survive for just a day or two, because they are lost or destroyed by exposure to disruptive enzymes, toxic chemicals, pathogenic bacteria, or mechanical abrasion. The only way the epithelium can maintain its structure over time is through the continual division of unspecialized cells known as **stem cells**, or *germinative cells*. These cells are found in the deepest layers of the epithelium, near the basement membrane.

Classifying Epithelia

Epithelia are classified according to the number of cell layers and the shape of the exposed cells. This classification scheme recognizes two types of layering—*simple* and *stratified*—and three cell shapes—*squamous*, *cuboidal*, and *columnar*.

Cell Layers

A **simple epithelium** consists of a single layer of cells covering the basement membrane. Simple epithelia are thin. Since a single layer of cells is fragile and cannot provide much mechanical protection, simple epithelia are found only in protected areas inside the body. They line internal compartments and passageways, including the body cavities and the interior of the heart and blood vessels.

Simple epithelia, shown in Figure 4-4●, are characteristic of regions where secretion or absorption occurs, such as the lining of the digestive and urinary tracts and the gas-exchange surfaces of the lungs. In such places, thinness is an advantage, for it reduces the diffusion time for materials crossing the epithelial barrier.

A **stratified epithelium** provides a greater degree of protection because it has several layers of cells above the basement membrane. Stratified epithelia are usually found in areas subject to mechanical or chemical stresses, such as the surface of the skin and the linings of the mouth and anus.

Cell Shape

In sectional view (perpendicular to the exposed surface and basement membrane), the cells at the surface of the epithelium usually have one of three basic shapes.

1. *Squamous.* In a **squamous epithelium** (SKWĀ-mus; *squama*, a plate or scale), the cells are thin and flat and the nucleus occupies the thickest portion of each cell. Viewed from the surface, the cells look like fried eggs laid side by side.
2. *Cuboidal.* The cells of a **cuboidal epithelium** resemble little hexagonal boxes when seen in three dimensions, but in typical sectional view, they appear square. The nuclei lie near the center of each cell, and they form a neat row.
3. *Columnar.* In a **columnar epithelium**, the cells are also hexagonal, but taller and more slender. The nuclei are crowded into a narrow band close to the basement membrane, and the height of the epithelium is several times the distance between two nuclei.

The two basic epithelial arrangements (simple and stratified) and the three possible cell shapes (squamous, cuboidal, and columnar) enable one to describe almost every epithelium in the body. We will focus here on only a few major types of epithelia; additional examples will be encountered in later chapters.

Simple Squamous Epithelia

A **simple squamous epithelium** is found in protected regions where absorption takes place or where a slick, slippery surface reduces friction (Figure 4-4a●). Examples are portions of the kidney tubules, the exchange surfaces of the lungs, the lining of body cavities, and the lining of blood vessels and the heart.

Simple Cuboidal Epithelia

A **simple cuboidal epithelium** provides limited protection and occurs in regions where secretion or absorption takes place (Figure 4-4b●). These functions are enhanced by larger cells that have more room for the necessary organelles. Simple cuboidal epithelia secrete enzymes and buffers in the pancreas and salivary glands and line the ducts that discharge these secretions. Simple cuboidal epithelia also line portions of the kidney tubules involved in the production of urine.

Simple Columnar Epithelia

A **simple columnar epithelium** provides some protection and may also occur in areas of absorption or secretion (Figure 4-4c●). This type of epithelium lines the stomach, the intestinal tract, and many excretory ducts.

4 THE TISSUE LEVEL OF ORGANIZATION

Epithelial Tissue • Connective Tissues • Membranes • Muscle Tissue • Neural Tissue • Tissue Injuries and Repairs • Tissues and Aging • Chapter Review

Simple Squamous Epithelium

LOCATIONS: Epithelia lining ventral body cavities; lining of heart and blood vessels; portions of kidney tubules (thin sections of loop of Henle); inner lining of cornea; exchange surfaces of lungs

FUNCTIONS: Reduces friction, controls vessel permeability, performs absorption and secretion

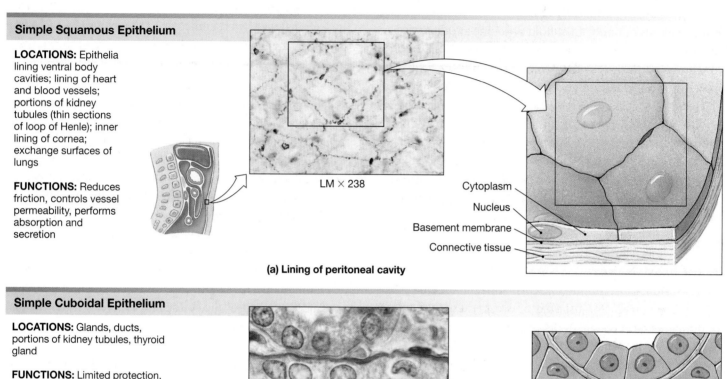

LM × 238

Cytoplasm

Nucleus

Basement membrane

Connective tissue

(a) Lining of peritoneal cavity

Simple Cuboidal Epithelium

LOCATIONS: Glands, ducts, portions of kidney tubules, thyroid gland

FUNCTIONS: Limited protection, secretion and/or absorption

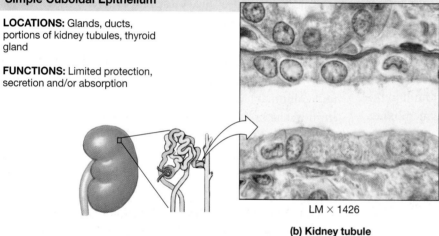

LM × 1426

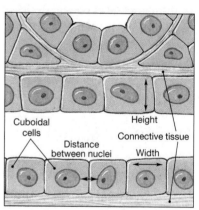

Cuboidal cells

Height

Connective tissue

Distance between nuclei

Width

(b) Kidney tubule

Simple Columnar Epithelium

LOCATIONS: Lining of stomach, intestine, gallbladder, uterine tubes, collecting ducts of kidneys

FUNCTIONS: Protection, secretion, absorption

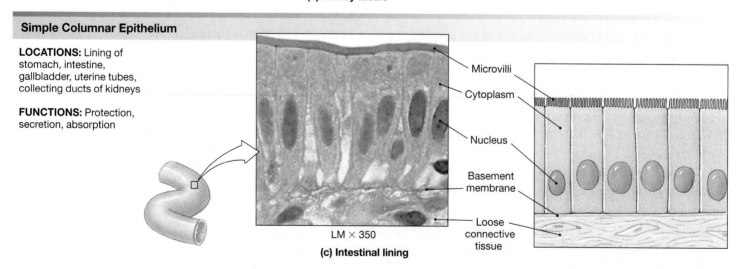

LM × 350

Microvilli

Cytoplasm

Nucleus

Basement membrane

Loose connective tissue

(c) Intestinal lining

● *Figure 4-4* **Simple Epithelia**

(**a**) A superficial view of the simple squamous epithelium that lines the peritoneal cavity. The three-dimensional drawing shows the epithelium in superficial and sectional views. (**b**) A section through the cuboidal epithelial cells of a kidney tubule. The diagrammatic view emphasizes structural details that permit the classification of an epithelium as cuboidal. (**c**) A micrograph showing the characteristics of simple columnar epithelium. In the diagrammatic sketch, note the relationships between the height and width of each cell; the relative size, shape, and location of nuclei; and the distance between adjacent nuclei. Compare with Figure 4-4b.

Pseudostratified Epithelia

Portions of the respiratory tract contain a columnar epithelium that includes a mixture of cell types. Because the nuclei are situated at varying distances from the surface, the epithelium has a layered appearance. But it is not a stratified epithelium, because all of the cells contact the basement membrane. Because it looks stratified but is not, it is known as a **pseudostratified columnar epithelium** (Figure 4-5a●, p. 90). This tissue typically possesses cilia. A ciliated pseudostratified columnar epithelium lines most of the nasal cavity, the trachea (windpipe) and bronchi, and portions of the male reproductive tract.

Transitional Epithelia

A **transitional epithelium** withstands considerable stretching. It lines the ureters and urinary bladder, where large changes in volume occur (Figure 4-5b●). In an empty urinary bladder, the epithelium seems to have many layers, and the outermost cells appear rounded or cuboidal. The layered appearance results from overcrowding; the actual structure of the epithelium can be seen in the full urinary bladder, when the volume of urine has stretched the lining to its natural thickness.

Stratified Squamous Epithelia

A **stratified squamous epithelium** is found where mechanical stresses are severe (Figure 4-5c●). The surface of the skin and the lining of the mouth, tongue, esophagus, and anus are good examples.

EXFOLIATIVE CYTOLOGY

Exfoliative cytology (eks-FŌ-lē-a-tiv; *ex-*, from + *folium*, leaf) is the study of cells shed or collected from epithelial surfaces. The cells are examined for a variety of reasons, including checking for cellular changes that indicate cancer formation and identifying the pathogens involved in an infection. The cells are collected by sampling the fluids that cover the epithelia lining the respiratory, digestive, urinary, or reproductive tract or by removing fluid from one of the ventral body cavities. The sampling procedure is often called a *Pap test*, named after Dr. George Papanicolaou, who pioneered its use. The most familiar Pap test is that for cervical cancer, which involves scraping a small number of cells from the tip of the *cervix*, a portion of the uterus that projects into the vagina.

Amniocentesis is another important test based on exfoliative cytology. In this procedure, exfoliated epithelial cells are collected from a sample of *amniotic fluid*, the fluid that surrounds and protects a developing fetus. Examination of these cells can determine if the fetus has a genetic abnormality, such as *Down syndrome*, that affects the number or structure of chromosomes.

GLANDULAR EPITHELIA

Many epithelia contain gland cells that produce exocrine or endocrine secretions. Exocrine secretions are produced by exocrine glands that discharge their products through a *duct*, or tube, onto some external or internal surface. Endocrine secretions (*hormones*) are produced by ductless glands and released into blood or tissue fluids. Exocrine glands are described in terms of their structure, the *mode of secretion*, and the *type of secretion*. In terms of structure, exocrine glands occur as independent, unicellular glands (called *goblet cells*) or as multicellular glands. Multicellular glands are further classified according to the branching pattern of the duct, and the shape and branching pattern of the secretory portion of the gland. Table 4-1 summarizes the classification of exocrine glands in terms of their mode of secretion or the type of secretion and provides specific examples.

TABLE 4-1 *A Classification of Exocrine Glands*

FEATURE	DESCRIPTION	EXAMPLES
MODE OF SECRETION		
Merocrine	Secretion occurs through exocytosis.	Saliva from salivary glands; mucus in digestive and respiratory tracts; perspiration on the skin; milk in breasts
Apocrine	Secretion occurs through loss of cytoplasm containing secretory product.	Milk in breasts; viscous underarm perspiration
Holocrine	Secretion occurs through loss of entire cell containing secretory product.	Skin oils and waxy coating of hair (produced by sebaceous glands of the skin)
TYPE OF SECRETION		
Serous	Watery solution containing enzymes	Parotid salivary gland
Mucous	Thick, slippery mucus	Sublingual salivary gland
Mixed	Produces more than one type of secretion	Submandibular salivary gland (serous and mucous)

4 THE TISSUE LEVEL OF ORGANIZATION

Epithelial Tissue • Connective Tissues • Membranes • Muscle Tissue • Neural Tissue • Tissue Injuries and Repairs • Tissues and Aging • Chapter Review

Pseudostratified Ciliated Columnar Epithelium

LOCATIONS: Lining of nasal cavity, bronchi, trachea; portions of male reproductive tract

FUNCTIONS: Protection, secretion

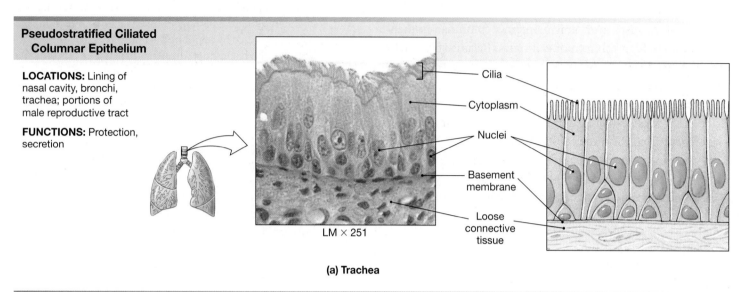

LM × 251

Cilia

Cytoplasm

Nuclei

Basement membrane

Loose connective tissue

(a) Trachea

Transitional Epithelium

LOCATIONS: Urinary bladder, renal pelvis, ureters

FUNCTIONS: Permits expansion and recoil after stretching

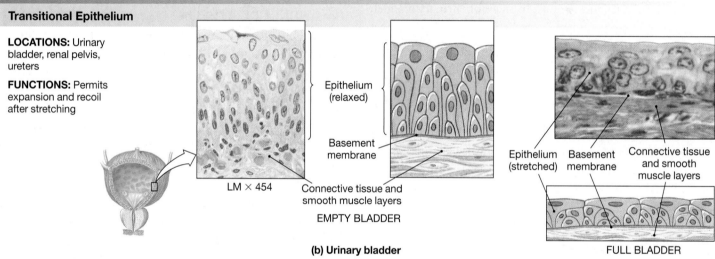

LM × 454

Epithelium (relaxed)

Basement membrane

Connective tissue and smooth muscle layers

EMPTY BLADDER

Epithelium (stretched)

Basement membrane

Connective tissue and smooth muscle layers

FULL BLADDER

(b) Urinary bladder

Stratified Squamous Epithelium

LOCATIONS: Surface of skin; lining of mouth, throat, esophagus, rectum, anus, and vagina

FUNCTIONS: Provides physical protection against abrasion, pathogens, and chemical attack

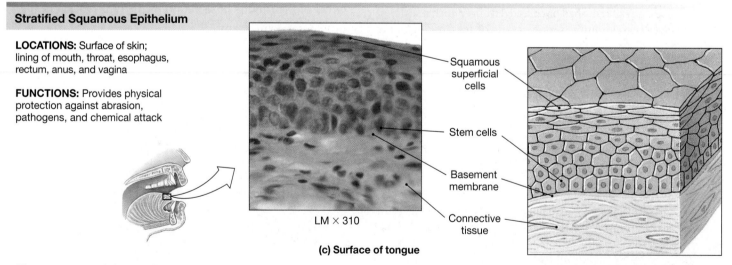

LM × 310

Squamous superficial cells

Stem cells

Basement membrane

Connective tissue

(c) Surface of tongue

● *Figure 4-5* **Stratified Epithelia**
(**a**) The pseudostratified, ciliated, columnar epithelium of the respiratory tract. Note the uneven layering of the nuclei. (**b**) At left, the lining of the empty urinary bladder, showing a transitional epithelium in the relaxed state. At right, the lining of the full bladder, showing the effects of stretching on the arrangement of cells in the epithelium. (**c**) A sectional view of the stratified squamous epithelium that covers the tongue.

Mode of Secretion

A glandular epithelial cell may use one of three methods to release its secretions: (1) *merocrine secretion*, (2) *apocrine secretion*, or (3) *holocrine secretion*.

In **merocrine secretion** (MER-o-krin; *meros*, part + *krinein*, to secrete) the product is released from secretory vesicles by exocytosis. ∞ p. 64 This is the most common mode of secretion (Figure 4-6a●). **Apocrine secretion** (AP-ō-krin; *apo-*, off) involves the loss of both cytoplasm and the secretory product (Figure 4-6b●). The outermost portion of the cytoplasm becomes packed with secretory vesicles before it is shed. (Milk production in the mammary glands involves both merocrine and apocrine secretions). Whereas merocrine and apoc-rine secretions leave the cell intact and able to continue secreting, **holocrine secretion** (HOL-ō-krin; *holos*, entire) does not (Figure 4-6c●). Instead, the entire cell becomes packed with secretions and then bursts apart and dies.

Type of Secretion

There are many kinds of exocrine secretions, all performing a variety of functions. Examples are enzymes entering the digestive tract, perspiration on the skin, and the milk produced by mammary glands.

Exocrine glands may be categorized by the type or types of secretions produced. For example, *serous glands* secrete a watery solution containing enzymes, and *mucous glands* secrete a thick,

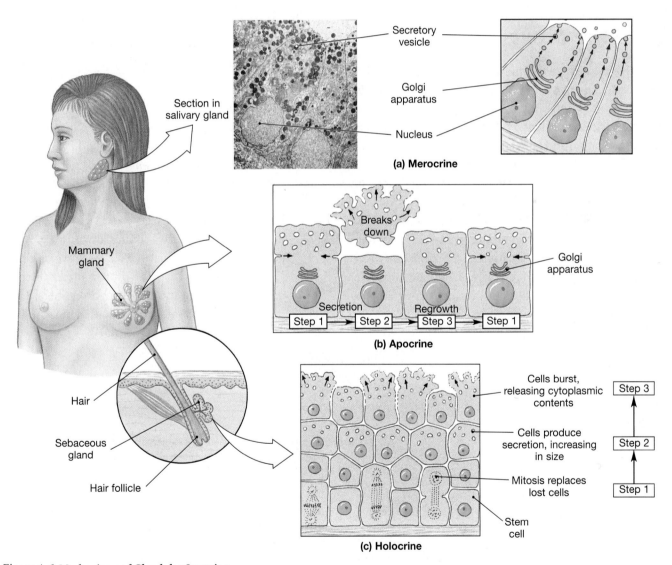

● *Figure 4-6* **Mechanisms of Glandular Secretion**
(**a**) In merocrine secretion, secretory vesicles are discharged at the surface of the gland cell through exocytosis. (**b**) Apocrine secretion involves the loss of cytoplasm. Inclusions, secretory vesicles, and other cytoplasmic components are shed in the process. The gland cell then undergoes a period of growth and repair before releasing additional secretions. (**c**) Holocrine secretion occurs as superficial gland cells break apart. Continued secretion involves the replacement of these cells through the mitotic divisions of underlying stem cells.

4 THE TISSUE LEVEL OF ORGANIZATION

Epithelial Tissue • **Connective Tissues** • Membranes • Muscle Tissue • Neural Tissue • Tissue Injuries and Repairs • Tissues and Aging • Chapter Review

slippery mucus. *Mixed glands* contain more than one type of gland cell and may produce two different exocrine secretions, or both exocrine and endocrine secretions.

CONCEPT CHECK QUESTIONS

Answers on page 107

❶ You look at a tissue under a microscope and see a simple squamous epithelium. Can it be a sample of the skin surface?

❷ Secretory cells associated with hair follicles fill with secretions and then rupture, releasing their contents. What kind of secretion is this?

❸ What physiological functions are enhanced by epithelial cells bearing microvilli and cilia?

Connective Tissues

Connective tissues are the most diverse tissues of the body. Bone, blood, and fat are familiar connective tissues that have very different functions and properties. All connective tissues have three basic components: (1) specialized cells, (2) protein fibers, and (3) a **ground substance**, a fluid that varies in consistency. The extracellular protein fibers and ground substance constitute the **matrix** that surrounds the cells. Whereas epithelial tissue consists almost entirely of cells, the extracellular matrix accounts for most of the volume of connective tissues.

Connective tissues are distributed throughout the body but are never exposed to the outside environment. Many connective tissues are highly vascular (that is, they have many blood vessels) and contain receptors that provide pain, pressure, temperature, and other sensations. Connective tissue functions include:

- *Supporting and protecting.* The minerals and fibers produced by connective tissue cells establish a bony structural framework for the body, protect delicate organs, and surround and interconnect other tissue types.

- *Transporting materials.* Fluid connective tissue provides an efficient means of moving dissolved materials from one region of the body to another.

- *Storing energy reserves.* Fats are stored in connective tissue cells called *adipose cells* until needed.

- *Defending the body.* Specialized connective tissue cells respond to invasions by microorganisms through cell-to-cell interactions and the production of *antibodies*.

CLASSIFYING CONNECTIVE TISSUES

Three classes of connective tissue are recognized on the basis of the physical properties of their matrix (Figure 4-7●).

1. **Connective tissue proper** refers to connective tissues with many types of cells and fibers surrounded by a syrupy ground substance. Examples are the tissue that underlies the skin, fatty tissue, and *tendons* and *ligaments*.

2. **Fluid connective tissues** have a distinctive population of cells suspended in a watery ground substance that contains dissolved proteins. The two fluid connective tissues are *blood* and *lymph*.

3. **Supporting connective tissues** are of two types, *cartilage* and *bone*. These tissues have a less diverse cell population than connective tissue proper and a matrix of dense ground substance and closely packed fibers. The fibrous matrix of bone is said to be calcified because it contains

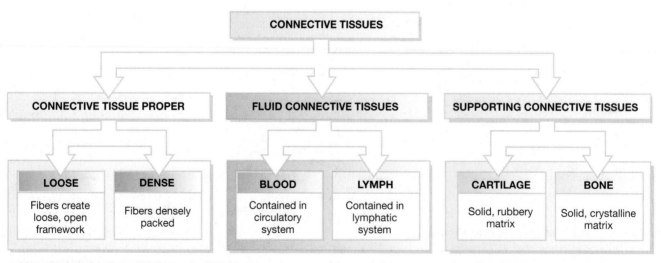

● *Figure 4-7* **Major Types of Connective Tissue**

mineral deposits, primarily calcium salts, which give the bone strength and rigidity.

CONNECTIVE TISSUE PROPER

Connective tissue proper contains a varied cell population, fibers, and a syrupy ground substance (Figure 4-8●). The cells involved with local maintenance, repair, and energy storage (such as fibroblasts and adipocytes) are permanent residents of the connective tissue. Those cells that defend and repair damaged tissues are not always present, and migrate to areas of injured tissue.

The Cell Population

Connective tissue proper includes the following major cell types:

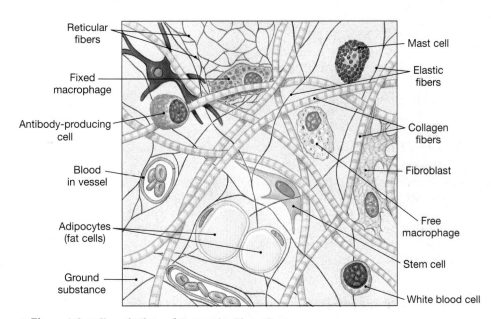

● *Figure 4-8* **Cells and Fibers of Connective Tissue Proper**
A diagrammatic view of common cell types and fibers of connective tissue proper.

- **Fibroblasts** (FĪ-brō-blasts) are the most abundant cells in connective tissue proper. They are responsible for the production and maintenance of the connective tissue fibers and the ground substance.
- **Macrophages** (MAK-rō-fā-jez; *phagein*, to eat) are scattered among the fibers. These cells engulf, or *phagocytize*, damaged cells or pathogens that enter the tissue and release chemicals that mobilize the immune system. ∞ p. 64 Macrophages that spend long periods of time in connective tissue are known as *fixed macrophages*. When an infection occurs, additional macrophages (called *free macrophages*) are drawn to the affected area.
- **Fat cells** are known as adipose cells, or **adipocytes** (AD-i-pō-sīts). A typical adipocyte contains such a large droplet of lipid that the nucleus and other organelles are squeezed to one side of the cell. The number of fat cells varies from one connective tissue to another, from one region of the body to another, and from individual to individual.
- **Mast cells** are small, mobile connective tissue cells often found near blood vessels. The cytoplasm of a mast cell is packed with vesicles filled with chemicals that are released to begin the body's defensive activities after an injury or infection, as discussed later in the chapter.

In addition to mast cells and free macrophages, both phagocytic and antibody-producing white blood cells may move through the connective tissue. Their numbers increase markedly if the tissue is damaged, as does the production of **antibodies**, proteins that destroy invading microorganisms or foreign substances. *Stem cells* also respond to local injury by dividing to produce daughter cells that differentiate into fibroblasts, macrophages, or other connective tissue cells.

Connective Tissue Fibers

The three basic types of fibers—*collagen*, *elastic*, and *reticular*—are formed from protein subunits secreted by fibroblasts:

1. **Collagen fibers** are long, straight, and unbranched. The most common fibers in connective tissue proper, they are strong but flexible.
2. **Elastic fibers** contain the protein *elastin*. They are branched and wavy, and after stretching will return to their original length.
3. **Reticular fibers** (*reticulum*, a network), the least common of the three, are thinner than collagen fibers and commonly form a branching, interwoven framework in various organs.

Ground Substance

Ground substance fills the spaces between cells and surrounds the connective tissue fibers (Figure 4-8●). In normal connective tissue proper, it is clear, colorless, and similar in consistency to maple syrup. This dense consistency slows the movement of bacteria and other pathogens, making them easier for the phagocytes to catch.

4 THE TISSUE LEVEL OF ORGANIZATION

Epithelial Tissue • **Connective Tissues** • Membranes • Muscle Tissue • Neural Tissue • Tissue Injuries and Repairs • Tissues and Aging • Chapter Review

 ## MARFAN'S SYNDROME

Marfan's syndrome is an inherited condition caused by the production of an abnormal form of *fibrillin*, a carbohydrate-protein complex important to normal connective tissue strength and elasticity. Because most organs contain connective tissues, the effects of this defect are widespread. The most visible sign of Marfan's syndrome involves the skeleton; individuals with Marfan's syndrome are tall and have abnormally long arms, legs, and fingers. The most serious consequences involve the cardiovascular system. Roughly 90 percent of the people with Marfan's syndrome have abnormal cardiovascular systems. The most dangerous potential result is that the weakened connective tissues in the walls of major arteries, such as the aorta, may burst, causing a sudden, fatal loss of blood.

Connective tissue proper is categorized as *loose connective tissues* and *dense connective tissues* on the basis of the relative proportions of cells, fibers, and ground substance. Loose connective tissues are the packing material of the body. These tissues fill spaces between organs, provide cushioning, and support epithelia. They also anchor blood vessels and nerves, store lipids, and provide a route for the diffusion of materials. Dense connective tissues are tough, strong, and durable. They resist tension and distortion and interconnect bones and muscles. Dense connective tissue also forms a thick fibrous layer called a *capsule*, that surrounds internal organs, such as the liver, kidneys, and spleen, and that also encloses joint cavities.

Loose Connective Tissue

Loose connective tissue, or *areolar tissue* (*areola*, little space), is the least specialized connective tissue in the adult body (Figure 4-9a●). It contains all of the cells and fibers found in any connective tissue proper, in addition to an extensive circulatory supply.

Loose connective tissue forms a layer that separates the skin from underlying muscles, providing both padding and a considerable amount of independent movement. For example, pinching the skin of the arm does not distort the underlying muscle. The ample blood supply in this tissue carries wandering cells to and from the tissue, and provides for the metabolic needs (oxygen and nutrients) of nearby epithelial tissue.

Adipose Tissue

Adipose tissue, or fat, is a loose connective tissue containing large numbers of fat cells, or adipocytes (Figure 4-9b●). The difference between loose connective tissue and adipose tissue is

one of degree; a loose connective tissue is called adipose tissue when it becomes dominated by fat cells. Adipose tissue provides another source of padding and shock absorption for the body. It also acts as an insulating blanket that slows heat loss through the skin and functions in energy storage.

Adipose tissue is common under the skin of the sides, buttocks, and breasts. It fills the bony sockets behind the eyes, surrounds the kidneys, and dominates extensive areas of loose connective tissue in the pericardial and peritoneal (abdominal) cavities.

 ## ADIPOSE TISSUE AND WEIGHT CONTROL

Adipocytes are metabolically active cells; their lipids are continually being broken down and replaced. When nutrients are scarce, adipocytes deflate like collapsing balloons. This deflation occurs during a weight-loss program. Because the cells are not killed, merely reduced in size, the lost weight can easily be regained in the same areas of the body.

In adults, adipocytes are incapable of dividing. However, an excess of nutrients can cause the division of connective tissue stem cells, which then differentiate into additional fat cells. As a result, areas of loose connective tissue can become adipose tissue after chronic overeating. In a procedure known as *liposuction*, unwanted adipose tissue is surgically removed. Because adipose tissue can regenerate through differentiation of stem cells, liposuction provides only a temporary solution to the problem of excess weight.

Dense Connective Tissues

Dense connective tissues consist mostly of collagen fibers; they may also be called *fibrous*, or *collagenous* (ko-LA-jin-us) tissues. **Tendons** are cords of dense connective tissue that attach skeletal muscles to bones (Figure 4-9c●). Collagen fibers run along the length of the tendon and transfer the pull of the contracting muscle to the bone. **Ligaments** (LIG-a-ments) are bundles of fibers that connect one bone to another. Ligaments often contain elastic fibers as well as collagen fibers and thus can tolerate a modest amount of stretching.

FLUID CONNECTIVE TISSUES

Blood and **lymph** are connective tissues that contain distinctive collections of cells in a fluid matrix. Under normal conditions, the proteins dissolved in this watery matrix do not form large insoluble fibers.

A single cell type, the **red blood cell**, accounts for almost half the volume of blood. Red blood cells transport oxygen in the blood. The watery matrix of blood, called **plasma**, also contains small numbers of **white blood cells**, which are

Loose Connective Tissue

LOCATIONS: Beneath dermis of skin, digestive tract, respiratory and urinary tracts; between muscles; around blood vessels, nerves, and around joints

FUNCTIONS: Cushions organs; provides support but permits independent movement; phagocytic cells provide defense against pathogens

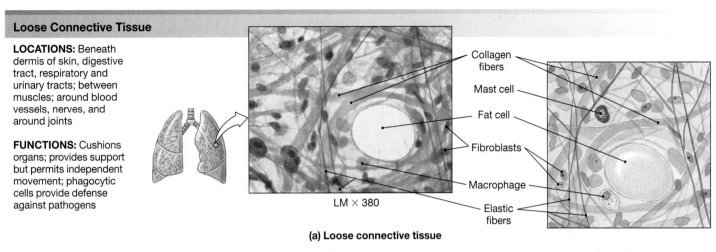

Collagen fibers
Mast cell
Fat cell
Fibroblasts
Macrophage
Elastic fibers

LM × 380

(a) Loose connective tissue

Adipose Tissue

LOCATIONS: Beneath skin, especially at sides, buttocks, breasts; behind eyeballs; around kidneys

FUNCTIONS: Provides padding and cushions shocks; insulates (reduces heat loss); stores energy reserves

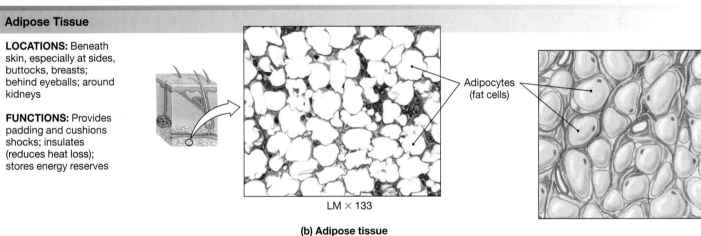

Adipocytes (fat cells)

LM × 133

(b) Adipose tissue

Dense Connective Tissues

LOCATIONS: Between skeletal muscles and skeleton (tendons); between bones (ligaments); covering skeletal muscles; capsules of visceral organs

FUNCTIONS: Provide firm attachment; conduct pull of muscles; reduce friction between muscles; stabilize relative positions of bones; help prevent overexpansion of organs such as the urinary bladder

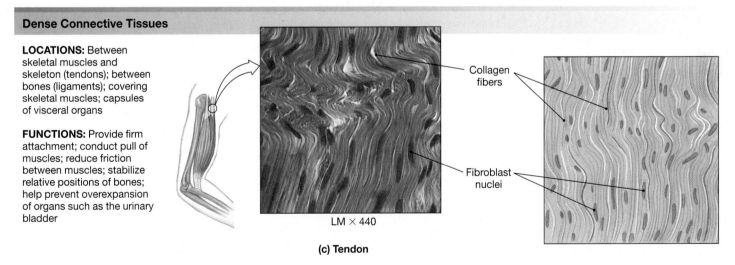

Collagen fibers

Fibroblast nuclei

LM × 440

(c) Tendon

● *Figure 4-9* **Connective Tissue Proper: Loose and Dense Connective Tissues**
(**a**) All of the cells of connective tissue proper are found in loose connective tissue. (**b**) Adipose tissue is loose connective tissue dominated by adipocytes. In standard histological preparations, the tissue looks empty because the lipids in the fat cells dissolve in the alcohol used during tissue processing. (**c**) The dense regular connective tissue in a tendon. Notice the densely packed, parallel bundles of collagen fibers. The fibroblast nuclei are flattened between the bundles.

4 THE TISSUE LEVEL OF ORGANIZATION

Epithelial Tissue • **Connective Tissues** • Membranes • Muscle Tissue • Neural Tissue • Tissue Injuries and Repairs • Tissues and Aging • Chapter Review

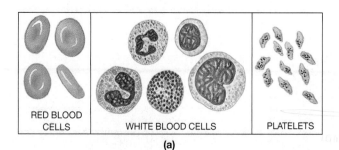

(a)

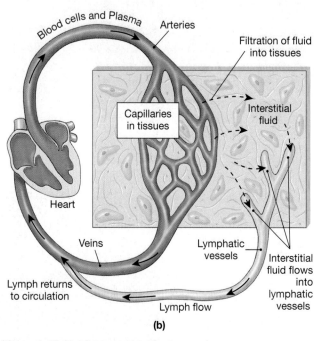

(b)

● *Figure 4-10* **Fluid Connective Tissues**
(a) Specialized cells of the fluid connective tissues. **(b)** Blood travels through the circulatory system, pushed by the contractions of the heart. In capillaries, hydrostatic (blood) pressure forces fluid and dissolved solutes out of the circulatory system. This fluid mixes with the interstitial fluid already in the tissue. Interstitial fluid slowly enters lymphatic vessels; now called lymph, it travels along the lymphatic vessels and reenters the circulatory system at one of the veins that returns blood to the heart.

important components of the immune system, and **platelets**, cell fragments that function in blood clotting (Figure 4-10●).

The extracellular fluid of the body is composed primarily of plasma, **interstitial fluid** (in-ter-STISH-al; *inter*, between + *sistere*, to set), and lymph. Plasma, confined to the vessels of the cardiovascular system, is kept in constant motion by contractions of the heart. A network of **arteries** carries blood away from the heart and toward fine, thin-walled vessels called **capillaries**. **Veins** collect and return blood to the heart, completing the circuit. In the tissues, filtration moves water and small solutes out of the capillaries and into the interstitial fluid, which surrounds the body's cells.

Lymph forms as interstitial fluid enters small passageways, or *lymphatic vessels*, that return it to the cardiovascular system

(Figure 4-10b●). Along the way, cells of the immune system monitor the composition of the lymph and respond to signs of injury and infection. Most of these cells are white blood cells called *lymphocytes*.

SUPPORTING CONNECTIVE TISSUES
Cartilage and bone are called supporting connective tissues because they provide a strong framework that supports the rest of the body. In these connective tissues the matrix contains numerous fibers and, in some cases, deposits of insoluble calcium salts.

Cartilage
The matrix of **cartilage** is a firm gel containing embedded fibers. **Chondrocytes** (KON-drō-sīts), the only cells found within the matrix, live in small pockets known as *lacunae* (la-KOO-nē; *lacus*, pool). Because cartilage is avascular, chondrocytes must obtain nutrients and eliminate waste products by diffusion through the matrix. This lack of blood vessels also limits the repair capabilities of cartilage. Structures of cartilage are covered and set apart from surrounding tissues by a **perichondrium** (per-i-KON-drē-um; *peri-*, around + *chondros*, cartilage), which contains an inner, cellular layer and an outer, fibrous layer.

Types of Cartilage. The three major types of cartilage are *hyaline cartilage, elastic cartilage,* and *fibrocartilage*:

1. **Hyaline cartilage** (HĪ-a-lin; *hyalos*, glass) is the most common type of cartilage (Figure 4-11a●). Tough and somewhat flexible, this type of cartilage connects the ribs to the sternum (breastbone), supports the conducting passageways of the respiratory tract, and covers the surfaces of bones within joints.

2. **Elastic cartilage** contains numerous elastic fibers that make it extremely resilient and flexible (Figure 4-11b●). Elastic cartilage forms the external flap (*pinna*) of the outer ear, the epiglottis, and an airway to the middle ear (the *auditory tube*).

3. **Fibrocartilage** has little ground substance, and its matrix is dominated by collagen fibers (Figure 4-11c●). These fibers are densely interwoven, making this tissue extremely durable and tough. Pads of fibrocartilage lie between the vertebrae of the spinal column, between the pubic bones of the pelvis, and around or within a few joints and tendons. In these positions they resist compression, absorb shocks, and prevent damaging bone-to-bone contact. Cartilages heal poorly, and damaged fibrocartilages in joints such as the knee can interfere with normal movements.

Hyaline Cartilage

LOCATIONS: Between tips of ribs and bones of sternum; covering bone surfaces at synovial joints; supporting larynx (voicebox), trachea, and bronchi; forming part of nasal septum

FUNCTIONS: Provides stiff but somewhat flexible support; reduces friction between bony surfaces

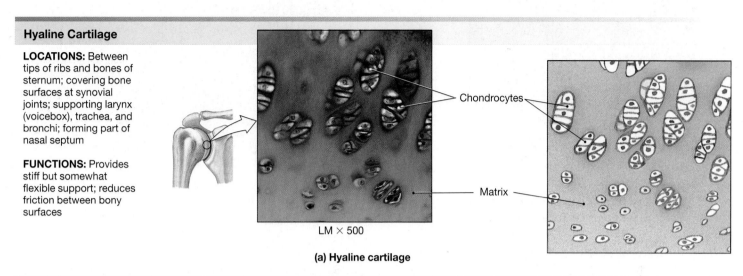

LM × 500

(a) Hyaline cartilage

Chondrocytes

Matrix

Elastic Cartilage

LOCATIONS: Auricle of external ear; auditory canal; epiglottis

FUNCTIONS: Provides support, but tolerates distortion without damage and returns to original shape

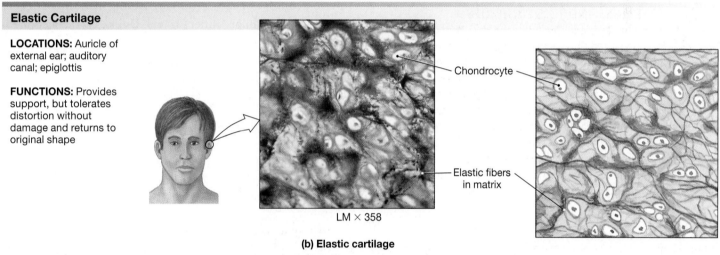

LM × 358

(b) Elastic cartilage

Chondrocyte

Elastic fibers in matrix

Fibrocartilage

LOCATIONS: Intervertebral discs separating vertebrae along spinal column; pads within knee joint; between pubic bones of pelvis

FUNCTIONS: Resists compression; prevents bone-to-bone contact; limits relative movement

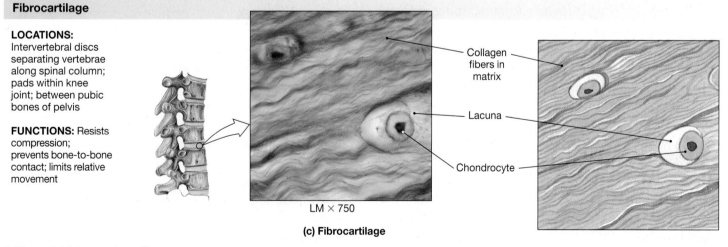

LM × 750

(c) Fibrocartilage

Collagen fibers in matrix

Lacuna

Chondrocyte

● *Figure 4-11* **Types of Cartilage**
(**a**) Hyaline cartilage. Note the translucent matrix and the absence of prominent fibers. (**b**) Elastic cartilage. The closely packed elastic fibers are visible between the chondrocytes. (**c**) Fibrocartilage. The collagen fibers are extremely dense, and the chondrocytes are relatively far apart.

4
THE TISSUE LEVEL OF ORGANIZATION

Epithelial Tissue • Connective Tissues • **Membranes** • Muscle Tissue • Neural Tissue • Tissue Injuries and Repairs • Tissues and Aging • Chapter Review

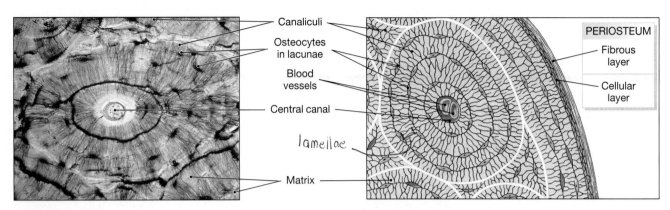

Canaliculi

Osteocytes
in lacunae

Blood
vessels

Central canal

lamellae

Matrix

PERIOSTEUM
Fibrous
layer

Cellular
layer

● *Figure 4-12* **Bone**
The osteocytes in bone are usually organized in groups around a central space that contains blood vessels. For the micrograph, a sample of bone was ground thin enough to become transparent. Bone dust filled the lacunae and the central canal, making them appear dark. (LM × 362).

 ## CARTILAGES AND JOINT INJURIES
Several complex joints, including the knee, contain both hyaline cartilage and fibrocartilage. The hyaline cartilage covers bony surfaces, and fibrocartilage pads in the joint prevent bone contact when movements are underway. Injuries to these joints can produce tears in the fibrocartilage pads, and these tears do not heal. This loss of cushioning places more strain on the cartilages within joints and leads to further joint damage. Eventually, joint mobility is severely reduced. Cartilages heal poorly because they are avascular. Joint cartilages heal even more slowly than other cartilages. Surgery usually results in only a temporary or incomplete repair.

Bone
Because the detailed histology of **bone**, or *osseous tissue* (OS-ē-us; *os*, bone), will be considered in Chapter 6, this discussion will focus on significant differences between cartilage and bone. The volume of ground substance in bone is very small. The matrix of bone consists mainly of hard calcium compounds and flexible collagen fibers. This combination gives bone truly remarkable properties, making it both strong and resistant to shattering. In its overall properties, bone can compete with the best steel-reinforced concrete.

The general organization of bone is shown in Figure 4-12●. Lacunae within the matrix contain bone cells, or **osteocytes** (OS-tē-ō-sīts; *os*, bone + *cyte*, cell). The lacunae surround the blood vessels that branch through the bony matrix. Although diffusion cannot occur through the bony matrix, osteocytes obtain nutrients through cytoplasmic extensions that reach blood vessels and other osteocytes. These extensions run through a branching network within the bony matrix called **canaliculi** (kan-a-LIK-ū-lē; little canals).

Except in joint cavities, where opposing surfaces are covered by hyaline cartilage, each bone is surrounded by a **periosteum** (per-ē-OS-tē-um), a covering made up of fibrous (outer) and cellular (inner) layers. Unlike cartilage, bone is constantly being remodeled throughout life, and complete repairs can be made even after severe damage has occurred. Table 4-2 compares cartilage and bone.

CONCEPT CHECK QUESTIONS
Answers on page 107
❶ Which two types of connective tissue have a fluid matrix?
❷ Chemical analysis of a connective tissue reveals that the tissue contains primarily triglycerides. Which type of connective tissue is this?
❸ Why does cartilage heal so slowly?

Membranes

Some anatomical terms have more than one meaning, depending on the context. One such term is *membrane*. For example, at the cellular level, membranes are lipid bilayers that restrict the passage of ions and other solutes. p. 57 At the tissue level, membranes also form a barrier, such as the basement membranes that separate epithelia from connective tissues. At still another level, epithelia and connective tissues combine to form membranes that cover and protect other structures and tissues. The body has four such membranes: *mucous membranes*, *serous membranes*, the *cutaneous membrane*, and *synovial membranes* (Figure 4-13●).

MUCOUS MEMBRANES
Mucous membranes, or **mucosae** (mū-KŌ-sē), line cavities that communicate with the exterior, including the digestive, respiratory, reproductive, and urinary tracts (Figure 4-13a●).

TABLE 4-2 *A Comparison of Cartilage and Bone*

CHARACTERISTIC	CARTILAGE	BONE
STRUCTURAL FEATURES		
Cells	Chondrocytes in lacunae	Osteocytes in lacunae
Ground substance	Protein-polysaccharide gel	A small volume of liquid surrounding insoluble salts (calcium phosphate and calcium carbonate)
Fibers	Collagen, elastic, reticular fibers (proportions vary)	Collagen fibers predominate
Vascularity	None	Extensive
Covering	Perichondrium	Periosteum
Strength	Limited: bends easily but hard to break	Strong: resists distortion until breaking point is reached
METABOLIC FEATURES		
Oxygen demands	Low	High
Nutrient delivery	By diffusion through matrix	By diffusion through cytoplasm and fluid in canaliculi
Repair capabilities	Limited	Extensive

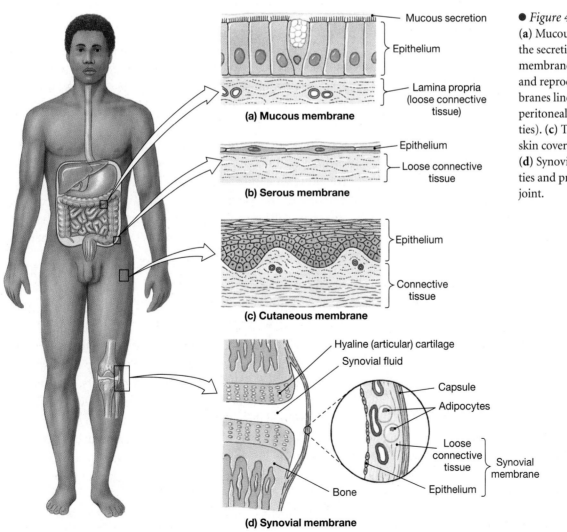

(a) Mucous membrane

- Mucous secretion
- Epithelium
- Lamina propria (loose connective tissue)

(b) Serous membrane

- Epithelium
- Loose connective tissue

(c) Cutaneous membrane

- Epithelium
- Connective tissue

(d) Synovial membrane

- Hyaline (articular) cartilage
- Synovial fluid
- Capsule
- Adipocytes
- Loose connective tissue — Synovial membrane
- Bone
- Epithelium

● *Figure 4-13* **Membranes**
(**a**) Mucous membranes are coated with the secretions of mucous glands. These membranes line the digestive, respiratory, and reproductive tracts. (**b**) Serous membranes line the ventral body cavities (the peritoneal, pleural, and pericardial cavities). (**c**) The cutaneous membrane of the skin covers the outer surface of the body. (**d**) Synovial membranes line joint cavities and produce the fluid within the joint.

4 THE TISSUE LEVEL OF ORGANIZATION

Epithelial Tissue • Connective Tissues • Membranes • **Muscle Tissue** • Neural Tissue • Tissue Injuries and Repairs • Tissues and Aging • Chapter Review

The epithelial surfaces are kept moist at all times, typically by mucous secretions or by exposure to fluids such as urine or semen. The connective tissue portion of a mucous membrane is called the *lamina propria* (PRŌ-prē-uh).

Many mucous membranes are lined by simple epithelia that perform absorptive or secretory functions, such as the simple columnar epithelium of the digestive tract. However, other types of epithelia may be involved. For example, a stratified squamous epithelium covers the mucous membrane of the mouth, and the mucous membrane along most of the urinary tract has a transitional epithelium.

SEROUS MEMBRANES

Serous membranes line the sealed, internal divisions of the ventral body cavity. There are three serous membranes, each consisting of a simple epithelium supported by loose connective tissue (Figure 4-13b●). The **pleura** (PLOO-ra; *pleura*, rib) lines the pleural cavities and covers the lungs. The **peritoneum** (pe-ri-tō-NĒ-um; *peri*, around + *teinein*, to stretch) lines the peritoneal (abdominal) cavity and covers the surfaces of enclosed organs such as the liver and stomach. The **pericardium** (pe-ri-KAR-dē-um) lines the pericardial cavity and covers the heart.

A serous membrane has *parietal* and *visceral* portions that are in close contact at all times. ∞ p. 19 The parietal portion lines the inner surface of the cavity, and the visceral portion covers the outer surface of organs within the body cavity. For example, the visceral pericardium covers the heart, and the parietal pericardium lines the inner surfaces of the pericardial sac that surrounds the pericardial cavity. A watery, *serous fluid* is formed by fluids diffusing from underlying tissues. The serous fluid minimizes the friction between the opposing surfaces of the visceral and parietal membranes.

CUTANEOUS MEMBRANE

The **cutaneous membrane** of the skin covers the surface of the body (Figure 4-13c●). It consists of a stratified squamous epithelium and the underlying connective tissues. In contrast to serous or mucous membranes, the cutaneous membrane is thick, relatively waterproof, and usually dry. The skin is discussed in detail in Chapter 5.

SYNOVIAL MEMBRANES

Bones contact one another at joints, or **articulations** (ar-tik-ū-LĀ-shuns). Joints that allow free movement are surrounded by a fibrous capsule and contain a joint cavity lined by a **synovial** (sin-Ō-vē-al) **membrane** (Figure 4-13d●). Unlike the other three membranes, the synovial membrane consists pri-

marily of loose connective tissue and an incomplete layer of epithelial tissue. In mobile joints, the bony surfaces do not come into direct contact with one another. If they did, abrasion and impacts would damage the opposing surfaces, and smooth movement would be almost impossible. Instead, the ends of the bones are covered with hyaline cartilage and separated by a viscous *synovial fluid* produced by fibroblasts of the **synovial membrane**. The synovial fluid helps lubricate the joint and permits smooth movement.

CONCEPT CHECK QUESTIONS
Answers on page 107

❶ How does a cell membrane differ from a tissue-level membrane?
❷ Serous membranes produce fluids. What is their function?
❸ The lining of the nasal cavity is normally moist, contains numerous goblet cells, and rests on a layer of loose connective tissue. What type of membrane is this?

Muscle Tissue

Muscle tissue is specialized for contraction. Muscle cell contraction involves interaction between filaments of *myosin* and *actin*, proteins found in the cytoskeletons of many cells. ∞ p. 66 In muscle cells, however, the filaments are more numerous and arranged so that their interaction produces a contraction of the entire cell.

There are three types of muscle tissue in the body— *skeletal*, *cardiac*, and *smooth muscle tissues* (Figure 4-14●). The contraction mechanism is the same in all of them, but the organization of their actin and myosin filaments differs. Because each type will be examined in later chapters, notably Chapter 7, this discussion will focus on general characteristics rather than specific details.

SKELETAL MUSCLE TISSUE

Skeletal muscle tissue contains very large, multinucleated cells (Figure 4-14a●). A large skeletal muscle cell may be 100 micrometers (μm; 1 μm = 1/25,000 in.) in diameter and 25 cm (10 in.) long. Because skeletal muscle cells are relatively long and slender, they are usually called *muscle fibers*. Skeletal muscle fibers are incapable of dividing, but new muscle fibers are produced through the divisions of stem cells in adult skeletal muscle tissue. As a result, at least partial repairs can occur after an injury.

Because the actin and myosin filaments are arranged in organized groups, skeletal muscle fibers appear to be marked by a series of bands known as *striations*. Skeletal muscle fibers will not usually contract unless stimulated by nerves. Since the

Skeletal Muscle Tissue

LOCATIONS: Combined with connective tissues and nervous tissue in skeletal muscles

FUNCTIONS: Moves or stabilizes the position of the skeleton; guards entrances and exits to the digestive, respiratory, and urinary tracts; generates heat; protects internal organs

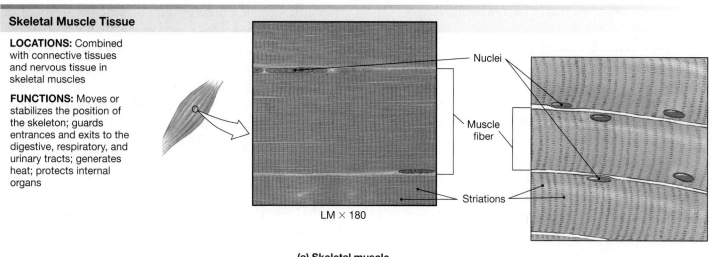

LM × 180

(a) Skeletal muscle

Cardiac Muscle Tissue

LOCATION: Heart

FUNCTIONS: Circulates blood; maintains blood (hydrostatic) pressure

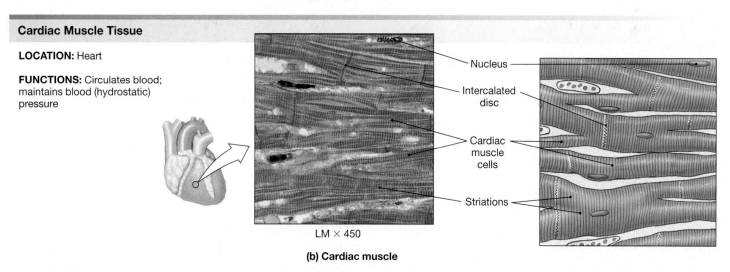

LM × 450

(b) Cardiac muscle

Smooth Muscle Tissue

LOCATIONS: Encircles blood vessels; in the walls of digestive, respiratory, urinary, and reproductive organs

FUNCTIONS: Moves food, urine, and reproductive tract secretions; controls diameter of respiratory passageways; regulates diameter of blood vessels and contributes to regulation of tissue blood flow

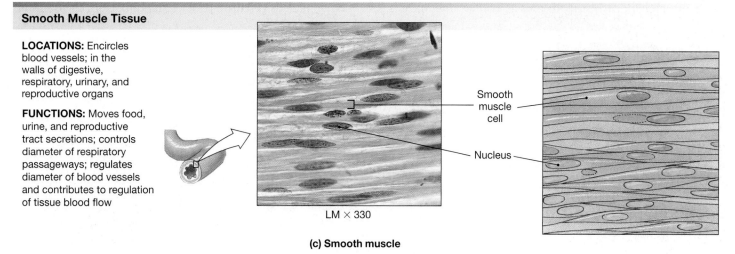

LM × 330

(c) Smooth muscle

● *Figure 4-14* **Muscle Tissue**
(**a**) Skeletal muscle fibers. Note the large fiber size, prominent striations (banding), multiple nuclei, and unbranched arrangement. (**b**) Cardiac muscle cells differ from skeletal muscle fibers in three major ways: size (cardiac muscle cells are smaller), organization (cardiac muscle cells branch), and number of nuclei (a typical cardiac muscle cell has one centrally placed nucleus). Both contain actin and myosin filaments in an organized array that produces striations. (**c**) Smooth muscle cells are small and spindle-shaped, with a central nucleus. They do not branch or have striations.

4 THE TISSUE LEVEL OF ORGANIZATION

Epithelial Tissue • Connective Tissues • Membranes • Muscle Tissue • **Neural Tissue** • Tissue Injuries and Repairs • Tissues and Aging • Chapter Review

nervous system provides voluntary control over its activities, skeletal muscle is described as *striated voluntary muscle.*

CARDIAC MUSCLE TISSUE

Cardiac muscle tissue is found only in the heart (Figure 4-14b●). Cardiac muscle cells are much smaller than skeletal muscle fibers, and each cardiac muscle cell usually has a single nucleus. Striations mark cardiac muscle and, just as in skeletal muscle, are due to the same arrangement of actin and myosin filaments. Cardiac muscle cells are interconnected at **intercalated** (in-TER-ka-lā-ted) **discs**, specialized attachment sites containing gap junctions and desmosomes. The muscle cells branch, forming a network that efficiently conducts the force and stimulus for contraction from one area of the heart to another. The cells of cardiac muscle tissue cannot divide, and because stem cells are also lacking, damaged cardiac muscle tissue cannot regenerate.

Cardiac muscle cells do not rely on nerve activity to start a contraction. Instead, specialized cells, called *pacemaker cells,* establish a regular rate of contraction. Although the nervous system can alter the rate of pacemaker activity, it does not provide voluntary control over individual cardiac muscle cells. Therefore, cardiac muscle is called *striated involuntary muscle.*

SMOOTH MUSCLE TISSUE

Smooth muscle tissue is found in the walls of blood vessels; around hollow organs such as the urinary bladder; and in layers around the respiratory, circulatory, digestive, and reproductive tracts (Figure 4-14c●).

A smooth muscle cell is small and slender, tapering to a point at each end; each smooth muscle cell has one nucleus.

Unlike skeletal and cardiac muscle, the actin and myosin filaments in smooth muscle cells are scattered throughout the cytoplasm, and there are no striations. Smooth muscle cells can divide, so smooth muscle tissue can regenerate after injury.

Smooth muscle cells may contract on their own, or their contractions may be triggered by neural activity. The nervous system usually does not provide voluntary control over smooth muscle contractions, and smooth muscle is known as *nonstriated involuntary muscle.*

Neural Tissue

Neural tissue, which is also known as *nervous tissue* or *nerve tissue*, is specialized for the conduction of electrical impulses from one region of the body to another. Most of the neural tissue (98 percent) is concentrated in the brain and spinal cord, the control centers for the nervous system.

Neural tissue contains two basic types of cells: **neurons** (NOO-ronz; *neuro-*, nerve) and several different kinds of supporting cells, or **neuroglia** (noo-ROG-lē-uh or noo-rō-GLĒ-uh; *glia*, glue). Our conscious and unconscious thought processes reflect the communication between neurons. Neurons communicate through electrical events that affect their cell membranes. The neuroglia provide physical support for neural tissue, maintain the chemical composition of the tissue fluids, and defend the tissue from infection.

The longest cells in your body are neurons, reaching up to a meter (39 in.) long. Most neurons cannot divide under normal circumstances, so they have a very limited ability to repair themselves after injury. A typical neuron has a **cell body**, with a large nucleus (Figure 4-15●). Extending from the cell body

● *Figure 4-15* **Neural Tissue**

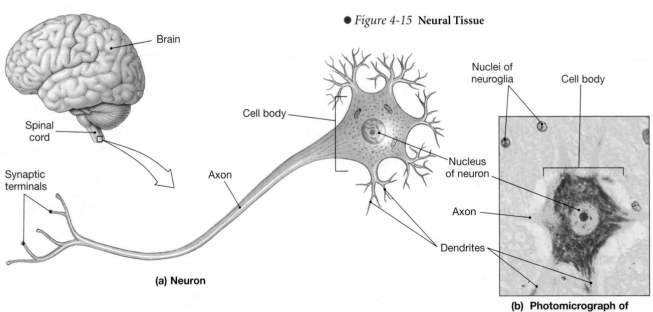

(a) **Neuron**

Brain

Spinal cord

Synaptic terminals

Axon

Cell body

Nuclei of neuroglia

Cell body

Nucleus of neuron

Axon

Dendrites

(b) **Photomicrograph of neuron cell body (LM × 600)**

are various branching projections called **dendrites** (DEN-drīts; *dendron*, tree) and one **axon**. Dendrites receive information, typically from other neurons, and axons carry that information to other cells. Because axons tend to be very long and slender, they are also called *nerve fibers*. Each axon ends at **synaptic terminals** where the neuron communicates with another cell. Chapter 8 considers the properties of neural tissue.

CONCEPT CHECK QUESTIONS
Answers on page 107

1. What type of muscle tissue has small, spindle-shaped cells with single nuclei and no obvious banding pattern?
2. Our voluntary control is restricted to which type of muscle tissue?
3. Why are skeletal muscle cells and axons also called fibers?

Tissue Injuries and Repairs

Tissues in the body are not independent of each other; they combine to form organs with diverse functions. Any injury affects several tissue types simultaneously, and these tissues must respond in a coordinated manner to restore homeostasis.

The restoration of homeostasis following a tissue injury involves two related processes: inflammation and repair. First, the area is isolated from neighboring healthy tissue while damaged cells, tissue components, and any dangerous microorganisms are cleaned up. This phase, which coordinates the activities of several different tissues, is called **inflammation**, or the *inflammatory response*. It produces several familiar sensations, including swelling, warmth, redness, and pain. An **infection** is an inflammation resulting from the presence of pathogens, such as harmful bacteria.

An inflammation can result from many stimuli, including impact, abrasion, chemical irritation, infection by pathogens, such as bacteria or viruses, and extreme temperatures (hot or cold). When these stimuli either kill cells, damage fibers, or injure tissues they trigger the inflammatory response by stimulating connective tissue cells called *mast cells* (see p. 93). The mast cells release chemicals (*histamine* and *heparin*) that cause local blood vessels to *dilate*, or enlarge in diameter, and become more permeable. The increased blood flow to the injured region makes it red and warm to the touch, and the diffusion of blood plasma causes the injured area to swell. The abnormal tissue conditions and the chemicals released by the mast cells also stimulate sensory nerve endings that produce the sensations of pain. These local circulatory changes increase the delivery of nutrients, oxygen, phagocytic white blood cells, blood clotting proteins, and speed up the removal of waste products and tox-

ins. Over a period of hours to days, this coordinated response generally succeeds in eliminating the inflammatory stimulus.

In the second phase, the damaged tissues are replaced or repaired to restore normal function. This repair process is called **regeneration**. During regeneration, fibroblasts produce a dense network of collagen fibers known as *scar tissue* or *fibrous tissue*. Over time, scar tissue is usually remodeled and gradually assumes a more normal appearance. Regeneration is more successful in some tissues than others: epithelia and connective tissue regenerate well; smooth and skeletal muscle tissues do so relatively poorly; and cardiac muscle tissue and neural tissue cannot regenerate at all. Because of different patterns of tissue organization, each organ also has a different ability to regenerate after injury. Your skin, which is made up mostly of epithelia and connective tissues, regenerates rapidly. In contrast, damage to the heart is more serious, because, although its connective tissue can be repaired, lost cardiac muscle cells are replaced by only fibrous connective tissue. Such permanent replacement of normal tissues is called **fibrosis** (fī-BRŌ-sis). Fibrosis may occur in muscle and other tissues in response to injury, disease, or aging.

Inflammation and regeneration are controlled at the tissue level. The two phases overlap; isolation of the area of damaged tissue establishes a framework that guides the cells responsible for reconstruction, and repairs are under way well before cleanup operations have ended. Later chapters will examine inflammation (Chapter 14) and regeneration (Chapter 5) in more detail.

CONCEPT CHECK QUESTIONS
Answers on page 107

1. What sensations are associated with inflammation?
2. What is fibrosis?

Tissues and Aging

Tissues change with age, and there is a decrease in the speed and effectiveness of tissue repairs. Repair and maintenance activities throughout the body slow down, and a combination of hormonal changes and alterations in lifestyle affect the structure and chemical composition of many tissues. Epithelia get thinner and connective tissues more fragile. Individuals bruise more easily and bones become brittle; joint pain and broken bones are common complaints among the elderly. Cardiac muscle fibers and neurons cannot be replaced, and cumulative losses from even relatively minor damage can contribute to major health problems, such as cardiovascular disease or deterioration in mental function.

4

THE TISSUE LEVEL OF ORGANIZATION

Epithelial Tissue • Connective Tissues • Membranes • Muscle Tissue • Neural Tissue • Tissue Injuries and Repairs • Tissues and Aging • **Chapter Review**

In later chapters, we will consider the effects of aging on specific organs and systems. Some of these effects are genetically programmed. For example, as people age, their chondrocytes produce a slightly different form of the gelatinous compound making up the cartilage matrix. This difference in composition probably accounts for the increase in thickness and stiffness of cartilages that we observe in older people.

Other age-related changes in tissue structure have multiple causes. The age-related reduction in bone strength in women, a condition called *osteoporosis*, is often caused by a combination of inactivity, low dietary calcium levels, and a reduction in circulating estrogens (sex hormones). A program of exercise, calcium supplements, and hormonal replacement therapies can generally maintain normal bone structure for many years.

AGING AND CANCER INCIDENCE

Cancer rates increase with age, and roughly 25 percent of all Americans develop cancer at some point in their lives. It has been estimated that 70–80 percent of cancer cases result from chemical exposure, environmental factors, or some combination of the two, and 40 percent of these cancers are caused by cigarette smoke. Each year in the United States, over 500,000 individuals die of cancer, making it second only to heart disease as a cause of death. Cancer development was discussed in Chapter 3. p. 77

Related Clinical Terms

adhesions: Restrictive fibrous connections that can result from surgery, infection, or other injuries to serous membranes.

anaplasia (a-nuh-PLĀ-zē-uh): An irreversible change in the size and shape of tissue cells.

chemotherapy: The administration of drugs that either kill cancerous tissues or prevent mitotic divisions.

dysplasia (dis-PLĀ-zē-uh): A change in the normal shape, size, and organization of tissue cells.

exfoliative cytology: The study of cells shed or collected from epithelial surfaces.

liposuction: A surgical procedure to remove unwanted adipose tissue by sucking it out through a tube.

metaplasia (me-tuh-PLĀ-zē-uh): A structural change that alters the character of a tissue.

necrosis (ne-KRŌ-sis): Tissue degeneration that occurs after cells have been injured or destroyed.

oncologists (on-KOL-o-jists): Physicians who specialize in identifying and treating cancers.

pathologists (pa-THOL-o-jists): Physicians who specialize in the study of disease processes.

pericarditis: An inflammation of the pericardial lining that may lead to the accumulation of pericardial fluid (a *pericardial effusion*).

peritonitis: An inflammation of the peritoneum after infection or injury.

pleural effusion: The accumulation of fluid within the pleural cavities as a result of chronic infection or inflammation of the pleura.

pleuritis (*pleurisy*): An inflammation of the pleural cavities.

regeneration: The repairing of injured tissues that follows inflammation.

remission: A stage in which a tumor stops growing or becomes smaller; the goal of cancer treatment.

CHAPTER REVIEW

Key Terms

Summary Outline

1. **Tissues** are collections of specialized cells and cell products that are organized to perform a relatively limited number of functions. The four **tissue types** are *epithelial tissue, connective tissues, muscle tissue,* and *neural tissue.* **Histology** is the study of tissues. *(Figure 4-1)*

1. An **epithelium** is an **avascular** layer of cells that forms a barrier that covers internal or external surfaces. **Glands** are secretory structures derived from epithelia.

2. Epithelia provide physical protection, control permeability, provide sensations, and produce specialized secretions.

3. Gland cells are epithelial cells that produce secretions. **Exocrine** secretions are released onto body surfaces; **endocrine** secretions, known as *hormones*, are released by gland cells into the surrounding tissues.

4. The individual cells that make up tissues attach to one another or to extracellular protein fibers in three major ways: *tight junctions, gap junctions,* and *desmosomes. (Figure 4-2)*

5. At a **tight junction**, the outer surfaces of the two cell membranes are bound tightly together; these are the strongest intercellular connections.

6. In a **gap junction**, two cells are held together by interlocked membrane proteins, forming a narrow passageway.

7. A **desmosome** has a very thin layer of intercellular cement between the cell membranes, reinforced by a network of protein fibers.

The Epithelial Surface

8. Many epithelial cells have microvilli, and some have stereocilia. The coordinated beating of the cilia on a ciliated epithelium moves materials across the epithelial surface. *(Figure 4-3)*

The Basement Membrane

9. The inner surface of each epithelium is connected to a noncellular **basement membrane**.

Epithelial Renewal and Repair

10. Divisions by **stem cells**, or *germinative cells*, continually replace the short-lived epithelial cells.

Classifying Epithelia

11. Epithelia are classified on the basis of the number of cell layers and the shape of the exposed cells.

12. A **simple epithelium** has a single layer of cells covering the basement membrane; a **stratified epithelium** has several layers. In a **squamous epithelium** the cells are thin and flat. Cells in a **cuboidal epithelium** resemble little hexagonal boxes; those in a **columnar epithelium** are taller and more slender. *(Figures 4-4, 4-5)*

Glandular Epithelia

13. A glandular epithelial cell may release its secretions through *merocrine*, *apocrine*, or *holocrine mechanisms. (Figure 4-6)*

14. In **merocrine secretion**, the most common method of secretion, the product is released through exocytosis. **Apocrine secretion** involves the loss of both secretory product and cytoplasm. Unlike the first two methods, **holocrine secretion** destroys the cell, which becomes packed with secretions and finally bursts.

15. Exocrine secretions may be serous (watery, usually containing enzymes), mucous (thick and slippery), or mixed (containing enzymes and lubricants). *(Table 4-1)*

CONNECTIVE TISSUES

1. All **connective** tissues have specialized cells and a **matrix**, composed of extracellular protein fibers and a **ground substance**.

2. Connective tissues are internal tissues with many important functions: establishing a structural framework; transporting fluids and dissolved materials; protecting delicate organs; supporting, surrounding, and interconnecting tissues; storing energy reserves; and defending the body from microorganisms.

Classifying Connective Tissues

3. **Connective tissue proper** refers to connective tissues that contain varied cell populations and fiber types surrounded by a syrupy ground substance. *(Figure 4-7)*

4. **Fluid connective tissues** have a distinctive population of cells suspended in a watery ground substance containing dissolved proteins. The two types are *blood* and *lymph. (Figure 4-7)*

5. **Supporting connective tissues** have a less diverse cell population than connective tissue proper and a dense matrix that contains closely packed fibers. The two types of supporting connective tissues are cartilage and bone. *(Figure 4-7)*

Connective Tissue Proper

6. Connective tissue proper contains fibers, a viscous ground substance, and a varied cell population.

7. Resident and migrating cells may include **fibroblasts**, **macrophages**, **fat cells**, **mast cells**, and various white blood cells. *(Figure 4-8)*

8. There are three types of fiber in connective tissue: **collagen fibers**, **reticular fibers**, and **elastic fibers**.

9. Connective tissue proper is classified as **loose** or **dense connective tissues**. Loose connective tissues include loose connective tissue, or *areolar tissue*, and **adipose tissue**. *(Figure 4-9a, b)*

10. Most of the volume in dense connective tissue consists of fibers. Dense connective tissues form **tendons** and **ligaments**. *(Figure 4-9c)*

Fluid Connective Tissues

11. **Blood** and **lymph** are connective tissues that contain distinctive collections of cells in a fluid matrix. *(Figure 4-10)*

12. Blood contains **red blood cells**, **white blood cells**, and **platelets**; the watery ground substance is called **plasma**.

13. **Arteries** carry blood from the heart and toward **capillaries**, where water and small solutes move into the **interstitial fluid** of surrounding tissues. **Veins** return blood to the heart.

14. Lymph forms as interstitial fluid enters the **lymphatic vessels**, which return lymph to the cardiovascular system.

Supporting Connective Tissues

15. Cartilage and bone are called supporting connective tissues because they support the rest of the body.

16. The matrix of **cartilage** consists of a firm gel and cells called **chondrocytes**. A fibrous **perichondrium** separates cartilage from surrounding tissues. The three types of cartilage are **hyaline cartilage**, **elastic cartilage**, and **fibrocartilage**. *(Figure 4-11)*

17. Chondrocytes rely on diffusion through the avascular matrix to obtain nutrients.

18. **Bone**, or *osseous tissue*, has a matrix primarily consisting of collagen fibers and calcium salts, which give it unique properties. *(Figure 4-12; Table 4-2)*

19. **Osteocytes** depend on diffusion through **canaliculi** for nutrient intake.

20. Each bone is surrounded by a **periosteum**.

MEMBRANES

1. Membranes form a barrier or an interface. Epithelia and connective tissues combine to form membranes that cover and protect other structures and tissues. There are four types of membranes: *mucous, serous, cutaneous*, and *synovial. (Figure 4-13)*

Mucous Membranes

2. **Mucous membranes** line cavities that communicate with the exterior. Their surfaces are normally moistened by mucous secretions.

Serous Membranes

3. **Serous membranes** line internal cavities and are delicate, moist, and very permeable.

Cutaneous Membrane

4. The **cutaneous membrane** covers the body surface. Unlike serous and mucous membranes, it is relatively thick, waterproof, and usually dry.

Synovial Membranes

5. **Synovial membranes**, located at joints, or articulations, produce *synovial fluid* in joint cavities. Synovial fluid helps lubricate the joint and promotes smooth movement.

MUSCLE TISSUE

1. Muscle tissue is specialized for contraction. The three types of muscle tissue are *skeletal muscle, cardiac muscle*, and *smooth muscle. (Figure 4-14)*

Skeletal Muscle Tissue

2. **Skeletal muscle tissue** contains large cells, or *muscle fibers*, tied together by collagen and elastic fibers. Skeletal muscle fibers are multinucleated and have a striped appearance because of the organization of contractile proteins. Because we can control the contraction of skeletal muscle fibers through the nervous system, skeletal muscle can be considered *striated voluntary muscle*.

4 THE TISSUE LEVEL OF ORGANIZATION

Epithelial Tissue • Connective Tissues • Membranes • Muscle Tissue • Neural Tissue • Tissue Injuries and Repairs • Tissues and Aging • **Chapter Review**

Cardiac Muscle Tissue 102

3. **Cardiac muscle tissue** is found only in the heart. The nervous system does not provide voluntary control over cardiac muscle cells. Thus, cardiac muscle is *striated involuntary muscle*.

Smooth Muscle Tissue..............................102

4. **Smooth muscle tissue** is found in the walls of blood vessels, around hollow organs, and in layers around various tracts. It is classified as *nonstriated involuntary muscle*.

NEURAL TISSUE 102

1. **Neural tissue** is specialized to conduct electrical impulses that convey information from one area of the body to another.

2. Cells in neural tissue are either neurons or neuroglia. **Neurons** transmit information as electrical impulses in their cell membranes. Several kinds of **neuroglia** serve both supporting and defense functions. *(Figure 4-15)*

3. A typical neuron has a **cell body**, **dendrites**, and an **axon**, which ends at **synaptic terminals**.

TISSUE INJURIES AND REPAIRS 103

1. Any injury affects several tissue types simultaneously, and they respond in a coordinated manner. Homeostasis is restored in two processes: *inflammation* and *regeneration*.

2. **Inflammation**, or the *inflammatory response*, isolates the injured area while damaged cells, tissue components, and any dangerous microorganisms are cleaned up.

3. **Regeneration** is the repair process that restores normal function.

TISSUES AND AGING 103

1. Tissues change with age. Repair and maintenance grow less efficient, and the structure and chemical composition of many tissues are altered.

Aging and Cancer Incidence 104

2. Cancer incidence increases with age, with roughly three-quarters of all cases caused by exposure to chemicals or environmental factors.

Review Questions

Level 1: Reviewing Facts and Terms

Match each item in column A with the most closely related item in column B. Use letters for answers in the spaces provided.

COLUMN A

___ 1. histology
___ 2. microvilli
___ 3. gap junction
___ 4. tight junction
___ 5. germinative cells
___ 6. destroys gland cell
___ 7. hormones
___ 8. adipocytes
___ 9. bone-to-bone attachment
___ 10. muscle-to-bone attachment
___ 11. skeletal muscle
___ 12. cardiac muscle

COLUMN B

a. repair and renewal
b. ligament
c. endocrine secretion
d. absorption and secretion
e. fat cells
f. holocrine secretion
g. study of tissues
h. tendon
i. intercellular connection
j. interlocking of membrane proteins
k. intercalated discs
l. striated, voluntary

13. The four basic tissue types found in the body are
   (a) epithelia, connective, muscle, neural
   (b) simple, cuboidal, squamous, stratified
   (c) fibroblasts, adipocytes, melanocytes, mesenchymal
   (d) lymphocytes, macrophages, microphages, adipocytes

14. Long microvilli incapable of movement are called:
   (a) cilia
   (b) flagella
   (c) stereocilia
   (d) a, b, and c are correct

15. The most abundant connections between cells in the superficial layers of the skin are
   (a) intermediate junctions
   (b) gap junctions
   (c) desmosomes
   (d) tight junctions

16. The three cell shapes making up epithelial tissue are
   (a) simple, stratified, transitional
   (b) simple, stratified, pseudostratified
   (c) hexagonal, cuboidal, spherical
   (d) cuboidal, squamous, columnar

17. The tissue that contains the fluid ground substance is
   (a) epithelial
   (b) neural
   (c) muscle
   (d) connective

18. The three major types of cartilage in the body are
   (a) collagen, reticular, elastic
   (b) areolar, adipose, reticular
   (c) hyaline, elastic, fibrocartilage
   (d) keratin, reticular, elastic

19. The primary function of serous membranes in the body is
    (a) to minimize friction between opposing surfaces
    (b) to line cavities that communicate with the exterior
    (c) to perform absorptive and secretory functions
    (d) to cover the surface of the body

20. Large muscle fibers that are multinucleated, striated, and voluntary are found in
    (a) cardiac muscle tissue
    (b) skeletal muscle tissue
    (c) smooth muscle tissue
    (d) a, b, and c are correct

21. Intercalated discs and pacemaker cells are characteristic of
    (a) smooth muscle tissue
    (b) cardiac muscle tissue
    (c) skeletal muscle tissue
    (d) a, b, and c are correct

22. Axons, dendrites, and a cell body are characteristics of cells found in
    (a) neural tissue
    (b) muscle tissue
    (c) connective tissue
    (d) epithelial tissue

23. What are the four essential functions of epithelial tissue?

24. What three types of layering make epithelial tissue recognizable?

25. What three basic components are found in connective tissues?

26. Which fluid connective tissues and supporting connective tissues are found in the human body?

27. Which four kinds of membranes composed of epithelial and connective tissues cover and protect other structures and tissues in the body?

28. What two cell populations make up neural tissue? What is the function of each?

Level 2: Reviewing Concepts

29. In surfaces of the body where mechanical stresses are severe, the dominant epithelium is
    (a) stratified squamous epithelium
    (b) simple cuboidal epithelium
    (c) simple columnar epithelium
    (d) stratified cuboidal epithelium

30. Why does holocrine secretion require continuous cell division?

31. What is the difference between an exocrine and an endocrine secretion?

32. A significant structural feature in the digestive system is the presence of tight junctions located near the exposed surfaces of cells lining the digestive tract. Why are these junctions so important?

33. Why are infections always a serious threat after a severe burn or abrasion?

34. What characteristics make the cutaneous membranes different from the serous and mucous membranes?

Level 3: Critical Thinking and Clinical Applications

35. A biology student accidentally loses the labels of two prepared slides she is studying. One is a slide of animal intestine and the other of animal esophagus. You volunteer to help her sort them out. How would you decide which slide is which?

36. You are asked to develop a scheme that can be used to identify the three different types of muscle tissue in two steps. What would the two steps be?

Answers to Concept Check Questions

Page 92
1. No. A simple squamous epithelium does not provide enough protection against infection, abrasion, and dehydration and is not found in the skin surface.
2. The process described is *holocrine secretion*.
3. The presence of microvilli on the free surface of epithelial cells greatly increases the surface area for absorption. Cilia function to move materials over the surface of epithelial cells.

Page 98
1. The two connective tissues that contain a fluid matrix are blood and lymph.
2. The tissue is adipose (fat) tissue.
3. Cartilage lacks a direct blood supply, which is necessary for rapid healing to occur. Materials that are needed to repair damaged cartilage must diffuse from the blood to the chondroblasts. This diffusion process takes a long time and retards the healing process.

Page 100
1. Cell membranes are composed of lipid bilayers. Tissue membranes consist of a layer of epithelial tissue and a layer of connective tissue.
2. *Serous fluid* minimizes the friction between the serous membranes that cover the surfaces of organs and the surrounding body cavity.
3. This is an example of a mucous membrane.

Page 103
1. Since both cardiac and skeletal muscles are striated (banded), this must be *smooth muscle tissue*.
2. Only skeletal muscle tissue is voluntary.
3. Both skeletal muscle cells and neurons are called fibers because they are relatively long and slender.

Page 103
1. Redness, warmth, swelling, and pain are familiar sensations associated with inflammation.
2. Fibrosis is the permanent replacement of normal tissues by fibrous tissue.

The Integumentary System

CHAPTER OUTLINE AND OBJECTIVES

Vocabulary Development

cornu ....horn; *stratum corneum*
cutis ....skin; *cutaneous*
derma ....skin; *dermis*
epi- ....above or over; *epidermis*
facere ....to make; *cornified*
germinare ....to start growing;
stratum germinativum
keros ....horn; *keratin*
kyanos ....blue; *cyanosis*
luna ....moon; *lunula*
melas ....black; *melanin*
onyx ....nail; *eponychium*
papilla ....a nipple-shaped mound;
dermal papillae

*T*HE BOOGIE BOARD SURFER "shredding" this wave is probably too busy to consider what contributions and sacrifices are normally made by her skin. This remarkable structure absorbs ultraviolet radiation, prevents dehydration, preserves normal body temperature, and tolerates chafing and abrasion. Although few people think of it in these terms, the skin is actually an organ—the largest organ of the body. In this chapter we will examine its varied functions.

5 THE INTEGUMENTARY SYSTEM

Integumentary Structure and Function • Local Control of Homeostasis • Aging and the Integumentary System • Integration with Other Systems

THE INTEGUMENTARY SYSTEM consists of the skin, hair, nails, and various glands. As the most visible organ system of the body, we devote a lot of time to improve its appearance. Washing your face and hands, brushing or trimming your hair, clipping your nails, showering, and applying deodorant are activities that modify the appearance or properties of the skin. And when something goes wrong with your skin, the effects are immediately apparent. You will notice a minor skin condition or blemish at once, whereas more serious problems in other systems are often ignored. Physicians, however, also pay attention to the skin because changes in its color, flexibility, or sensitivity may provide visible signs of a disorder in another body system.

This chapter focuses on the important structural and functional relationships in the skin.

Integumentary Structure and Function

The **integumentary system**, or simply the **integument** (in-TEG-ū-ment), has two major components: the cutaneous membrane and the accessory structures. The **cutaneous membrane**, or *skin*, is an organ composed of the superficial epithelium, or **epidermis** (*epi-*, above), and the underlying connective tissues

of the **dermis**. The **accessory structures** include hair, nails, and a variety of exocrine glands.

The general structure of the integument is shown in Figure 5-1●. Beneath the dermis, the loose connective tissue of the **subcutaneous layer**, or **hypodermis**, attaches the integument to deeper structures, such as muscles or bones. Although often not considered to be part of the integumentary system, this layer will be considered here because its connective tissue fibers are interwoven with those of the dermis.

As we will see, the five major functions of the various parts of the integument are:

1. *Protection.* The skin covers and protects underlying tissues and organs from impacts, chemicals, and infections, and prevents the loss of body fluids.

2. *Temperature maintenance.* The skin maintains normal body temperature by regulating heat gain or loss to the environment.

3. *Synthesis and storage of nutrients.* The epidermis converts a steroid into vitamin D_3, which aids calcium uptake, and the dermis contains large reserves of lipids in adipose tissue.

4. *Sensory reception.* Receptors in the integument detect touch, pressure, pain, and temperature stimuli and relay that information to the nervous system.

5. *Excretion and secretion.* The integument excretes salts, water, and organic wastes, and produces milk, a specialized exocrine secretion.

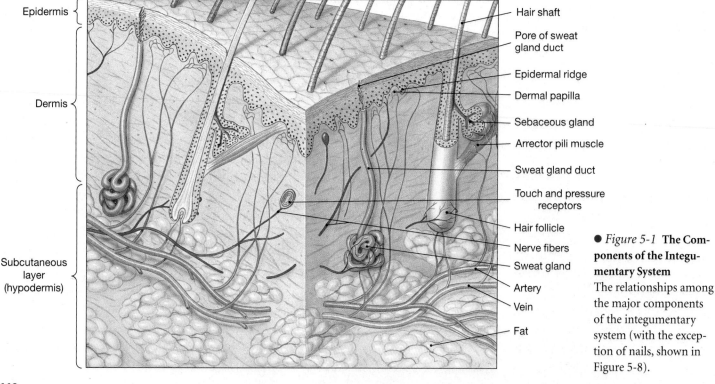

Epidermis {

Dermis {

Subcutaneous layer (hypodermis) {

Hair shaft

Pore of sweat gland duct

Epidermal ridge

Dermal papilla

Sebaceous gland

Arrector pili muscle

Sweat gland duct

Touch and pressure receptors

Hair follicle

Nerve fibers

Sweat gland

Artery

Vein

Fat

● *Figure 5-1* **The Components of the Integumentary System** The relationships among the major components of the integumentary system (with the exception of nails, shown in Figure 5-8).

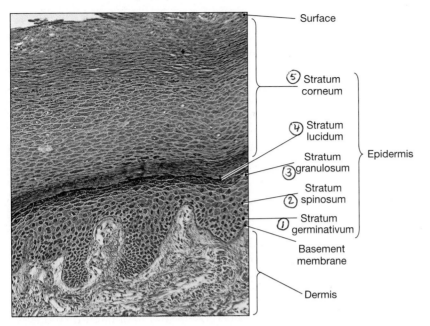

● *Figure 5-2* **The Structure of the Epidermis**
A portion of the epidermis in thick skin, showing all five cell layers. (LM × 150)

Each of these functions will be explored more fully as we discuss the individual components of the integument (Figure 5-2●).

THE EPIDERMIS

The epidermis consists of a stratified squamous epithelium of several different cell layers. ∞ p. 89 In **thick skin**, found on the palms of the hands and soles of the feet, there are five layers. Only four layers make up **thin skin**, which covers the rest of the body. Thin skin may be a mere 0.08 mm thick, whereas thick skin may be as thick as 0.5 mm. The words *thin* and *thick* refer only to the relative thickness of the epidermis, not to the integument as a whole.

Layers of the Epidermis

Figure 5-2● shows the cell layers, or **strata** (singular *stratum*), in a section of thick skin. In order, from the basement membrane toward the free surface, are the *stratum germinativum*, three intermediate layers (the *stratum spinosum*, the *stratum granulosum*, and the *stratum lucidum*), and the *stratum corneum*.

Stratum Germinativum. The deepest epidermal layer is called the **stratum germinativum** (STRA-tum jer-mi-na-TĒ-vum; *stratum*, layer + *germinare*, to start growing), or *stratum basale* (ba-SĀY-lē; *basis*, base). The cells of this layer are firmly attached to the basement membrane by hemidesmosomes. ∞ p. 86 The basement membrane separates the epidermis from the loose connective tissue of the adjacent dermis. The stratum germinativum forms **epidermal ridges**, which extend into the dermis, increasing the area of contact between the two regions. Dermal projections called *dermal papillae* (singular *papilla*; a nipple-shaped mound) extend upward between adjacent ridges (Figure 5-1●). Because there are no blood vessels in the epidermis, epidermal cells must obtain nutrients delivered by dermal blood vessels. The combination of ridges and papillae increases the surface area for diffusion between the dermis and epidermis.

The contours of the skin surface follow the ridge patterns, which vary from small conical pegs (in thin skin) to the complex whorls on the thick skin of the palms and soles. The superficial ridges on the palms and soles (which overlie the dermal papillae) increase the surface area of the skin and increase friction, ensuring a secure grip. Ridge contours are genetically determined; those of each person are unique and do not change over the course of a lifetime. Fingerprints are ridge patterns on the tips of the fingers that can be used to identify individuals; they have been used in criminal investigations for over a century.

Large stem cells, or *germinative cells*, dominate the stratum germinativum, making it the layer where new cells are generated and begin to grow. The divisions of these cells replace cells that are lost or shed at the epithelial surface. The stratum germinativum also contains *melanocytes*, cells whose cytoplasmic processes extend between epithelial cells in this layer, and receptors that provide information about objects touching the skin. Melanocytes synthesize *melanin*, a yellow-brown to black pigment that colors the epidermis.

Intermediate Strata. The cells in these three layers are progressively displaced from the basal layer as they become specialized to form the outer protective barrier of the skin. Each time a stem cell divides, one of the daughter cells enters the next layer, the **stratum spinosum** (spiny layer), where it may continue to divide and add to the thickness of the epithelium. The **stratum granulosum** (grainy layer) consists of cells displaced from the spinosum layer. The cells in this layer have stopped dividing and begin making large amounts of the protein **keratin** (KER-a-tin; *keros*, horn), ∞ p. 41 Keratin is extremely durable and water-resistant. In humans, keratin not only coats the surface of the skin but also forms the basic structure of hair, calluses, and nails. In other animals, it forms structures such as cow horns and hooves, bird feathers, and baleen plates in the mouths of whales. In the thick skin of the palms and soles, a glassy **stratum lucidum** (clear layer) covers the

111

5 THE INTEGUMENTARY SYSTEM

Integumentary Structure and Function • Local Control of Homeostasis • Aging and the Integumentary System • Integration with Other Systems

stratum granulosum. The cells in this layer are flattened, densely packed, and filled with keratin.

Stratum Corneum. The most superficial layer of the epidermis, the **stratum corneum** (KOR-nē-um; *cornu*, horn), consists of 15–30 layers of flattened and dead epithelial cells that have accumulated large amounts of keratin. Such cells are said to be **keratinized** (ker-A-tin-īzed), or **cornified** (KOR-ni-fīd; *cornu*, horn + *facere*, to make). The dead cells in each layer of the stratum corneum remain tightly connected by desmosomes. ∞ p. 86 As a result, the keratinized cells of the stratum corneum are generally shed in large groups or sheets rather than individually.

It takes 2–4 weeks for a cell to move from the stratum germinativum to the stratum corneum. During this time, the cell is displaced from its oxygen and nutrient supply, becomes packed with keratin, and finally dies. The dead cells usually remain in the stratum corneum for an additional 2 weeks before they are shed or washed away. As superficial layers are lost, new layers arrive from the underlying strata. Thus the deeper layers of the epithelium and underlying tissues remain protected by a barrier of dead, durable, and expendable cells. Normally, the surface of the stratum corneum is relatively dry, so it is unsuitable for the growth of many microorganisms.

 DRUG ADMINISTRATION THROUGH THE SKIN

Drugs dissolved in oils or other solvents that are lipid-soluble can be carried across the cell membranes of the epidermis. The movement is slow, particularly through the stratum corneum, but once a drug reaches the underlying tissues, it will be absorbed into the circulation. Drugs can also be administered by packaging them in *liposomes*, artificially produced fat droplets. Liposomes containing DNA fragments have been used experimentally to introduce normal genes into abnormal human cells. For example, genes carried by liposomes have been used to alter the cell membranes of skin cancer cells so that they will be attacked by the immune system.

A useful technique for long-term drug administration involves placing a sticky, drug-containing patch over an area of thin skin. To overcome the slow rate of diffusion, the patch must contain an extremely high concentration of the drug. This procedure is called *transdermal administration*. It has the advantage that a single patch may work for several days, making daily pills unnecessary. Examples of transdermal drugs that are routinely administered include transdermal scopolamine, which affects the nervous system and can control the nausea associated with motion sickness, and transdermal estrogens, which are administered to women to reduce symptoms of menopause.

Skin Color

The color of your skin is caused by the interaction between (1) the epidermal pigmentation and (2) the dermal blood supply.

Pigmentation. The epidermis contains variable amounts of two pigments, carotene and melanin. **Carotene** (KAR-ō-tēn) is an orange-yellow pigment that normally accumulates in epidermal cells. Carotene pigments are found in a variety of orange-colored vegetables, such as carrots and squashes. Eating large amounts of carrots can actually cause the skin of a Caucasian to turn orange. The color change is less striking in the skin of people of other races. Carotene can be converted to vitamin A, which is required for the normal maintenance of epithelial tissues and the synthesis of photoreceptor pigments in the eye. **Melanin** is a brown, yellow-brown, or black pigment produced by melanocytes.

Melanocytes (me-LAN-ō-sīts) manufacture and store melanin and inject that pigment into the epithelial cells of the stratum germinativum and stratum spinosum (Figure 5-3●). This transfer of pigmentation colors the entire epidermis. Melanocyte activity slowly increases in response to sunlight exposure, peaking around 10 days after the initial exposure.

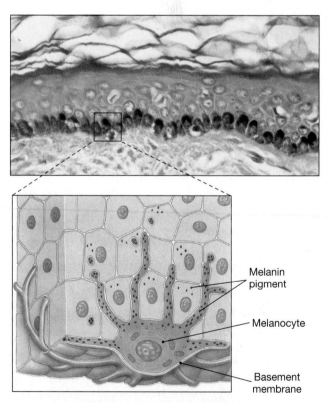

Melanin pigment

Melanocyte

Basement membrane

● *Figure 5-3* **Melanocytes**
The location and orientation of melanocytes in the deepest layer of the epidermis (stratum germinativum) of a dark-skinned person.

Freckles are small pigmented spots that appear on the skin of pale-skinned individuals. Freckles represent areas of larger-than-average melanin production. They tend to be most abundant on surfaces such as the face, that are exposed to the sun.

Sunlight contains significant amounts of **ultraviolet (UV) radiation**. A small amount of UV radiation is beneficial, for it stimulates the synthesis of vitamin D_3 in the epidermis; this process is discussed in a later section. Too much ultraviolet radiation, however, produces immediate effects of mild or even serious burns. Melanin helps prevent skin damage by absorbing ultraviolet radiation before it reaches the deep layers of the epidermis and dermis. Within the epidermal cells, melanin concentrates around the nuclear envelope and absorbs the UV before it can damage the nuclear DNA.

Despite the presence of melanin, long-term damage can result from repeated exposure, even in darkly pigmented individuals. For example, alterations in the underlying connective tissues lead to premature wrinkling, and skin cancers can result from chromosomal damage in stem cells of the stratum germinativum or in melanocytes. One of the major consequences of the global depletion of the ozone layer in the upper atmosphere is likely a sharp increase in the rate of skin cancers, such as *malignant melanoma*. For this reason, limiting UV exposure through a combination of protective clothing and a sunblock with a sun protection factor (SPF) of at least 15 is recommended during outdoor activities. Individuals with fair skin, such as blondes and redheads, are better off with an SPF of 20 to 30.

The ratio between melanocytes and stem cells ranges between 1:4 and 1:20, depending on the region of the body surveyed. The observed differences in skin color between individuals and even races do not reflect different *numbers* of melanocytes, but merely different levels of melanin production. For example, in the inherited condition *albinism*, melanin pigment is not produced by the melanocytes, even though these cells are distributed normally. Individuals with this condition, known as *albinos*, have light-colored skin and hair.

Dermal Circulation. Blood with abundant oxygen is bright red, and blood vessels in the dermis normally give the skin a reddish tint that is most apparent in lightly pigmented individuals. When those vessels are dilated, as during inflammation, the red tones become much more pronounced. When the vessels are temporarily constricted, as when one is frightened, the skin becomes relatively pale. During a sustained reduction in circulatory supply, the blood in the skin loses oxygen and takes on a darker red tone. The skin then takes on a bluish coloration called **cyanosis** (sī-a-NŌ-sis; *kyanos*, blue). In individuals of any skin color, cyanosis is most apparent in areas of thin skin, such as the lips, ears, or beneath the nails. It can be a response to extreme cold or a result of circulatory or respiratory disorders, such as heart failure or severe asthma.

The Epidermis and Vitamin D_3

Although strong sunlight can damage epithelial cells and deeper tissues, limited exposure to sunlight is very beneficial. When exposed to ultraviolet radiation, epidermal cells in the stratum spinosum and stratum germinativum convert a steroid related to cholesterol into **vitamin D_3**. This product is absorbed, modified, and released by the liver and then converted by the kidneys into *calcitriol*, a hormone essential for the absorption of calcium and phosphorus by the small intestine. An inadequate supply of vitamin D_3 leads to abnormal bone growth.

Skin Cancer

Skin cancers are the most common form of cancer. A **basal cell carcinoma** (Figure 5-4a●) is a malignant cancer that originates in the stratum germinativum (basal) layer. This is the most common skin cancer. Less common are **squamous cell carcinomas**. Metastasis seldom occurs in either cancer, and most people survive these cancers. The usual treatment involves surgical removal of the tumor.

Compared with these common and seldom life-threatening cancers, **melanomas** (mel-a-NŌ-mas) (Figure 5-4b●) are extremely dangerous. A melanoma usually begins from a mole but may appear anywhere in the body. In this condition,

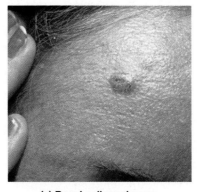
(a) Basal cell carcinoma

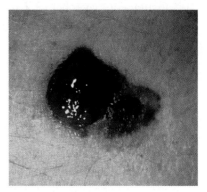
(b) Melanoma

● *Figure 5-4* **Skin Cancers**
(**a**) Basal cell carcinoma. (**b**) Melanoma.

5 THE INTEGUMENTARY SYSTEM

Integumentary Structure and Function • Local Control of Homeostasis • Aging and the Integumentary System • Integration with Other Systems

cancerous melanocytes grow rapidly and metastasize through the lymphatic system. The outlook for long-term survival depends on when the condition is detected and treated. To a large degree, avoiding exposure to UV radiation from the sun, especially during the middle of the day, and the use of a sunblock (not a tanning oil) would prevent all three forms of cancer.

CONCEPT CHECK QUESTIONS
Answers on page 124

❶ Excessive shedding of cells from the outer layer of skin in the scalp causes dandruff. What is the name of this layer of skin?

❷ Some criminals sand the tips of their fingers so as not to leave recognizable fingerprints. Would this practice permanently remove fingerprints? Why or why not?

❸ Why does exposure to sunlight or tanning lamps cause the skin to become darker?

THE DERMIS

The dermis lies beneath the epidermis. It has two major components: a superficial *papillary layer* and a deeper *reticular layer*.

Layers of the Dermis

The **papillary layer**, named after the dermal papillae, consists of loose connective tissue that supports and nourishes the epidermis. This region contains the capillaries and nerves supplying the surface of the skin.

The deeper **reticular layer** consists of an interwoven meshwork of dense, irregular connective tissue. Bundles of collagen fibers leave the reticular layer to blend into those of the papillary layer above, so the boundary line between these layers is indistinct. Collagen fibers of the reticular layer also extend into the subcutaneous layer below. This layer provides support and attachment for the dermis while also allowing flexibility and independent movement.

Other Dermal Components

In addition to protein fibers, the dermis contains a mixed cell population that includes all of the cells of connective tissue proper. ☞ p. 93 Accessory organs of epidermal origin, such as hair follicles and sweat glands, extend into the dermis (see Figure 5-1●, p. 110).

Other systems communicate with the skin through their connections to the dermis. For example, the reticular and papillary layers contain a network of blood vessels (cardiovascular system), lymph vessels (lymphatic system), and nerve fibers (nervous system).

Blood vessels provide nutrients and oxygen and remove carbon dioxide and waste products. Both the blood vessels and the lymph vessels help local tissues defend and repair themselves after an injury or infection. The nerve fibers control blood flow, adjust gland secretion rates, and monitor sensory receptors in the dermis and the deeper layers of the epidermis. These receptors, which provide sensations of touch, pain, pressure, and temperature, will be described in Chapter 9.

THE SUBCUTANEOUS LAYER

An extensive network of connective tissue fibers attaches the dermis to the subcutaneous layer. The boundary between these two layers is indistinct, and although the subcutaneous layer is not actually a part of the integument, it is important in stabilizing the position of the skin in relation to underlying tissues and organs.

The subcutaneous layer (hypodermis) consists of loose connective tissue with many fat cells. These adipose cells provide infants and small children with a layer of "baby fat," which helps them reduce heat loss. Subcutaneous fat also serves as a energy reserve and a shock absorber for the inevitable tumbles during their early years.

As we grow and mature, the distribution of subcutaneous fat changes. Men tend to accumulate subcutaneous fat at the neck, upper arms, along the lower back, and over the buttocks, and women in the breasts, buttocks, hips, and thighs. Both women and men, however, may accumulate distressing amounts of adipose tissue in the abdominal region, producing a prominent "pot belly."

The hypodermis is quite elastic. Below its superficial region with its large blood vessels, the hypodermis contains no vital organs and few capillaries. The lack of vital organs makes *subcutaneous injection* a useful method for administering drugs by means of a *hypodermic needle*.

ACCESSORY STRUCTURES

Accessory structures include hair follicles, sebaceous glands, sweat glands, and nails.

Hair Follicles

Hairs project above the surface of the skin almost everywhere except over the sides and soles of the feet, the palms of the hands, the sides of the fingers and toes, the lips, and portions of the external genital organs. Hairs are nonliving structures produced in organs called **hair follicles**.

The Structure of Hair Follicles. Hair follicles project deep into the dermis and often extend into the underlying subcutaneous layer (Figure 5-5a●). The walls of each follicle contain all the cell layers found in the epidermis. The epithelium at the base

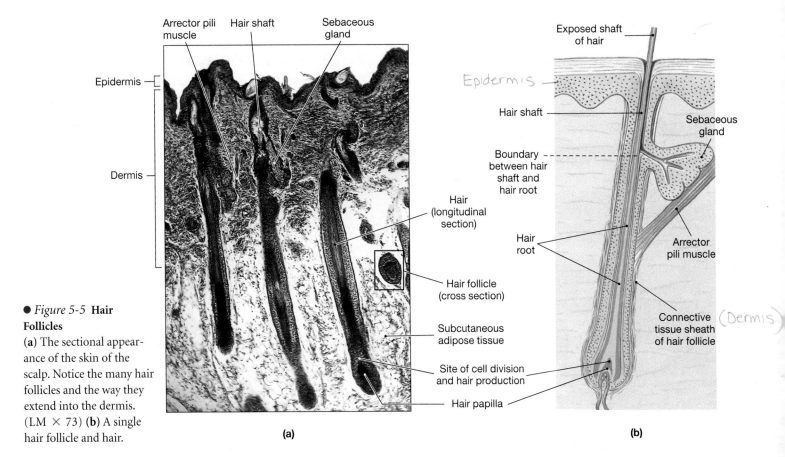

Figure 5-5 Hair Follicles
(a) The sectional appearance of the skin of the scalp. Notice the many hair follicles and the way they extend into the dermis. (LM × 73) (b) A single hair follicle and hair.

Labels in (a): Arrector pili muscle; Hair shaft; Sebaceous gland; Epidermis; Dermis; Hair (longitudinal section); Hair follicle (cross section); Subcutaneous adipose tissue; Site of cell division and hair production; Hair papilla

Labels in (b): Exposed shaft of hair; Epidermis; Hair shaft; Sebaceous gland; Boundary between hair shaft and hair root; Hair root; Arrector pili muscle; Connective tissue sheath of hair follicle (Dermis)

of a follicle surrounds the **hair papilla**, a peg of connective tissue containing capillaries and nerves. Hair is formed by the repeated divisions of epithelial stem cells surrounding the hair papilla. As the daughter cells are pushed toward the surface, the hair lengthens, and the cells undergo keratinization and die. The point at which this occurs marks the boundary of the **hair root** (the portion that anchors the hair into the skin) and the **hair shaft** (the part we see on the surface) (Figure 5-5b●).

Hairs grow and are shed according to a *hair growth cycle* based on the level of activity of hair follicles. In general, a hair in the scalp grows for from 2 to 5 years, at a rate of about 0.3 mm per day, and then its follicle may become inactive for a comparable period of time. When another growth cycle begins, the follicle produces a new hair, and the old hair gets pushed toward the surface to be shed. Variations in growth rate and in the length of the hair growth cycle account for individual differences in the length of uncut hair. Other differences in hair appearance result from the size of the follicles and the shapes of the hairs. For example, straight hairs are round in cross section, whereas curly ones are rather flattened.

Functions of Hair. The 5 million hairs on the human body have important functions. The roughly 100,000 hairs on the head protect the scalp from ultraviolet light, help cushion a light blow to the head, and provide insulating benefits for the skull.

The hairs guarding the entrances to the nostrils and external ear canals help prevent the entry of foreign particles and insects, and eyelashes perform a similar function for the surface of the eye. A sensory nerve fiber is associated with the base of each hair follicle. As a result, you can feel the movement of the shaft of even a single hair. This sensitivity provides an early-warning system that may help prevent injury. For example, you may be able to swat a mosquito before it reaches the skin surface.

Ribbons of smooth muscle, called **arrector pili** (a-REK-tōr PĪ-lē) muscles, extend from the papillary dermis to a connective tissue sheath that surrounds each hair follicle (Figure 5-5b●). When stimulated, the arrector pili pull on the follicles and force the hairs to stand up. Contraction may be caused by emotional states, such as fear or rage, or a response to cold, producing "goose bumps."

Hair Color. Hair color reflects differences in the type and amount of pigment produced by melanocytes at the papilla. Different forms of melanin give dark-brown, yellow-brown, or red coloration to the hair. Although these characteristics are genetically determined, the condition of your hair may also be influenced by hormonal or environmental factors. As pigment production decreases with age, the color of hair lightens. White hair results from the combination of a lack of pigment and the presence of air bubbles within the hair shaft. As the proportion

115

5 THE INTEGUMENTARY SYSTEM

Integumentary Structure and Function • Local Control of Homeostasis • Aging and the Integumentary System • Integration with Other Systems

of white hairs increases, the individual's hair color is described as gray. Because the hair itself is dead and inert, changes in coloration are gradual. Unless bleach is used, it is not possible for hair to "turn white overnight," as some horror stories would have us believe.

 ## HAIR LOSS

Many people are subject to anxiety attacks when they find hairs clinging to their hairbrush instead of to their heads. On the average, about 50 hairs are lost from the head each day, but several factors may affect this rate. Sustained losses of over 100 hairs per day generally indicate that something is wrong. Temporary increases in hair loss can result from drugs, dietary factors, radiation, high fever, stress, or hormonal factors related to pregnancy. In males, changes in the level of circulating sex hormones can affect the scalp, causing a shift in production from normal hair to fine "peach fuzz" hairs. This alteration is called *male pattern baldness*. Some cases of male pattern baldness respond to drug therapies, such as topical application of *minoxidyl (Rogaine™)*.

CONCEPT CHECK QUESTIONS
Answers on page 124

❶ What happens when the arrector pili muscles contract?

❷ A person suffers a burn on the forearm that destroys the epidermis and the deep dermis. When the injury heals, would you expect to find hair growing again in the area of the injury?

❸ Describe the functions of the subcutaneous layer.

Sebaceous Glands

The integument contains two types of exocrine glands: sebaceous glands and *sweat glands*. **Sebaceous (sē-BĀ-shus) glands**, or *oil glands*, are holocrine glands that discharge a waxy, oily secretion into hair follicles or, in some cases, only onto the skin. (Figure 5-6●). The gland cells manufacture large quantities of lipids as they mature, and the lipid is released through holocrine secretion, a process that involves their eventual death. ∞ p. 91 The contraction of the arrector pili muscle that elevates the hair squeezes the sebaceous gland, forcing the oily secretions into the hair follicle and onto the surrounding skin. This secretion, called **sebum** (SĒ-bum), lubricates the hair and skin, and inhibits the growth of bacteria. *Sebaceous follicles* are large sebaceous glands that discharge sebum directly onto the skin. They are located on the face, back, chest, nipples, and male sex organs.

Sebaceous glands are sensitive to changes in the concentrations of sex hormones, and their secretions accelerate at puberty. For this reason, an individual with large sebaceous glands may be especially prone to develop **acne** during adolescence. In acne, sebaceous ducts become blocked and secretions accumulate, causing inflammation and a raised "pimple." The trapped secretions provide a fertile environment for bacterial infection.

Sweat Glands

The skin contains two types of sweat glands, or *sudoriferous glands*: *apocrine sweat glands* and *merocrine sweat glands* (Figure 5-7●). The names refer to their mode of secretion and were discussed in Chapter 4 (see Table 4-1, p. 89).

Apocrine Sweat Glands. **Apocrine sweat glands** secrete their products into hair follicles in the armpits, around the nipples, and in the groin. Although originally thought to use an apoc-

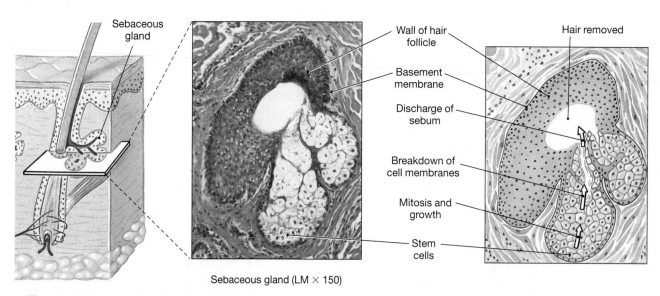

Sebaceous gland (LM × 150)

● *Figure 5-6* Sebaceous Glands and Hair Follicles

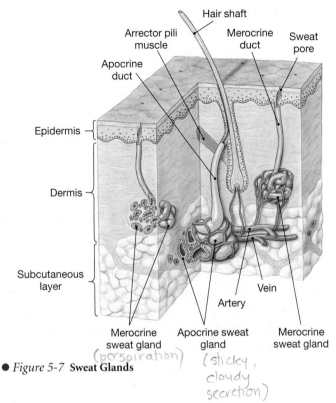

● *Figure 5-7* **Sweat Glands**

[handwritten annotations: (perspiration) (sticky, cloudy secretion)]

into the environment. The role of the skin in thermoregulation (temperature control) was considered in Chapter 1 (see p. 12); Chapter 17 will examine this process in greater detail.

The perspiration, or sweat, produced by merocrine glands is 99 percent water, but it does contain a mixture of electrolytes (chiefly sodium chloride), organic nutrients, and waste products such as urea. The sodium chloride gives sweat its salty taste. When all of the merocrine sweat glands are working at maximum, the rate of perspiration may exceed a gallon (about 4 liters) per hour, and dangerous fluid and electrolyte losses can occur. For this reason, marathon runners and other endurance athletes must drink fluids at regular intervals.

The skin also contains other types of modified sweat glands with specialized secretions. One example is the mammary glands of the breast and the secretion of milk. Another example are the *ceruminous glands* in the passageway of the external ear. Their secretions combine with those of nearby sebaceous glands to form ear wax.

Nails

Nails form on the dorsal surfaces of the fingers and toes, where they protect the exposed tips and help limit their distortion when they are subjected to mechanical stress—for example, when you run or grasp objects. The structure of a nail is shown in Figure 5-8●. The visible **nail body** consists of a dense mass

rine method of secretion, the gland cells are now known to rely on merocrine secretion. However, the name has not changed. At puberty, these glands begin discharging a sticky, cloudy secretion that becomes odorous when broken down by bacteria. In other mammals, this odor is an important form of communication; in our culture, whatever function it has is masked by products such as deodorants. Other products, such as antiperspirants, contain astringent compounds that contract the skin and its sweat gland openings, thus decreasing the quantity of both apocrine and merocrine secretions.

Merocrine Sweat Glands. **Merocrine sweat glands**, or *eccrine* (EK-rin) *sweat glands*, are far more numerous and widely distributed than apocrine glands. The skin of an adult contains 2–5 million eccrine glands. Palms and soles have the highest numbers; it has been estimated that the palm of the hand has about 500 glands per square centimeter (3000 per square inch).

Merocrine sweat glands are coiled tubular glands that discharge their secretions directly onto the surface of the skin. Their secretions, called *perspiration*, cool the surface of the skin and reduce body temperature. When a person sweats in the hot sun, all the merocrine glands are working together. The blood vessels beneath the epidermis are flushed with blood, and the skin assumes a reddish color. The skin surface is warm and wet, and as the moisture evaporates, the skin cools. If body temperature falls below normal, perspiration ceases, blood flow to the skin declines, and the cool, dry surfaces release little heat

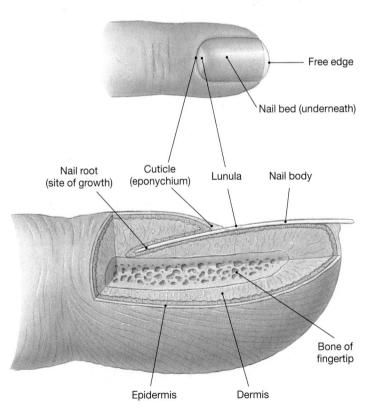

● *Figure 5-8* **The Structure of a Nail**

117

5 THE INTEGUMENTARY SYSTEM

Integumentary Structure and Function • Local Control of Homeostasis • Aging and the Integumentary System • Integration with Other Systems

of dead, keratinized cells. The nail body is recessed beneath the level of the surrounding epithelium. The body of the nail covers an area of epidermis called the **nail bed**. Nail production occurs at the **nail root**, an epithelial fold not visible from the surface. A portion of the stratum corneum of the fold extends over the exposed nail nearest the root, forming the **cuticle**, or *eponychium* (ep-ō-NIK-ē-um; *epi-*, over + *onyx*, nail). Underlying blood vessels give the nail its pink color, but near the root these vessels may be obscured, leaving a pale crescent known as the **lunula** (LOO-nū-la; *luna*, moon).

CONCEPT CHECK QUESTIONS
Answers on page 124

❶ What are the functions of sebaceous secretions?

❷ Deodorants are used to mask the effects of secretions from which type of skin gland?

Local Control of Homeostasis

The integumentary system can respond directly and independently to many local influences or stimuli. For example, when the skin is subjected to mechanical stresses, stem cells in the stratum germinativum divide more rapidly and the depth of the epithelium increases. That is why calluses form on your palms when you perform manual labor. A more dramatic example of local control can be seen after an injury to the skin.

INJURY AND REPAIR

The skin can regenerate effectively even after considerable damage has occurred because stem cells are present in both its epithelial and connective tissue components. Divisions by these stem cells replace lost epidermal and dermal cells, respectively. This process can be slow, and when large surface areas are involved infection and fluid loss complicate the situation. The relative speed and effectiveness of skin repair vary depending on the type of wound. A slender, straight cut, or *incision*, may heal relatively quickly compared with a scrape, or *abrasion*, which involves a much greater area.

Figure 5-9● shows the four stages in the regeneration of the skin after an injury. When damage extends through the epidermis and into the dermis, bleeding generally occurs (Step 1). The blood clot, or **scab**, that forms at the surface tem-

porarily restores the integrity of the epidermis and restricts the entry of additional microorganisms (Step 2). Most of the clot consists of an insoluble network of *fibrin*, a fibrous protein that forms from blood proteins during the clotting response. Cells of the stratum germinativum rapidly divide and begin to migrate along the sides of the wound to replace the missing epidermal cells. Meanwhile, phagocytes patrol the damaged area of the dermis and clear away debris and pathogens.

If the wound covers an extensive area or involves a region covered by thin skin, dermal repairs must be under way before epithelial cells can cover the surface. Fiber-producing cells (*fibroblasts*) and connective tissue stem cells divide to produce mobile cells that invade the deeper areas of injury. Epithelial cells lining damaged blood vessels also begin to divide, and capillaries follow the fibroblasts, providing a circulatory supply. The combination of blood clot, fibroblasts, and an extensive capillary network is called *granulation tissue*. Over time, the clot dissolves and the number of capillaries declines. Fibroblast activity leads to the appearance of collagen fibers and typical ground substance (Step 3).

These repairs do not restore the integument to its original condition, however, because the dermis will contain an abnormally large number of collagen fibers and relatively few blood vessels. Severely damaged hair follicles, sebaceous or sweat glands, muscle cells, and nerves are seldom repaired, and they too are replaced by fibrous tissue. The formation of this rather inflexible, fibrous, noncellular *scar tissue* can be considered a practical limit to the healing process (Step 4).

The process of scar tissue formation is highly variable. For example, surgical procedures performed on a fetus do not leave scars. In some adults, most often those with dark skin, scar tissue formation may continue beyond the requirements of tissue repair. The result is a flattened mass of scar tissue that begins at the injury site and grows into the surrounding dermis. This thickened area of scar tissue, called a **keloid** (KĒ-loyd), is covered by a shiny, smooth epidermal surface. Keloids most commonly develop on the upper back, shoulders, anterior chest, and earlobes. They are harmless, and some cultures intentionally produce keloids as a form of body decoration.

Burns

Burns are relatively common injuries that result from exposure of the skin to heat, radiation, electrical shock, or strong chemical agents. The severity of the burn reflects the depth of

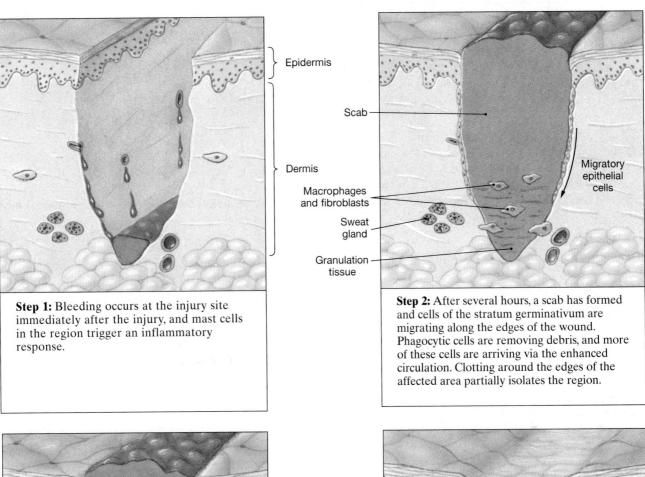

Epidermis

Dermis

Step 1: Bleeding occurs at the injury site immediately after the injury, and mast cells in the region trigger an inflammatory response.

Scab

Migratory epithelial cells

Macrophages and fibroblasts

Sweat gland

Granulation tissue

Step 2: After several hours, a scab has formed and cells of the stratum germinativum are migrating along the edges of the wound. Phagocytic cells are removing debris, and more of these cells are arriving via the enhanced circulation. Clotting around the edges of the affected area partially isolates the region.

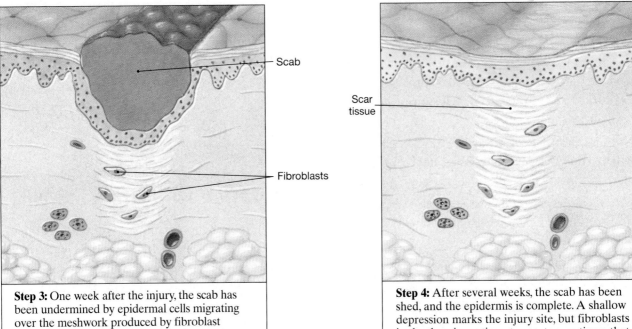

Scab

Fibroblasts

Step 3: One week after the injury, the scab has been undermined by epidermal cells migrating over the meshwork produced by fibroblast activity. Phagocytic activity around the site has almost ended, and the fibrin clot is disintegrating.

Scar tissue

Step 4: After several weeks, the scab has been shed, and the epidermis is complete. A shallow depression marks the injury site, but fibroblasts in the dermis continue to create scar tissue that will gradually elevate the overlying epidermis.

● *Figure 5-9* **Skin Repair**

5 THE INTEGUMENTARY SYSTEM

Integumentary Structure and Function • Local Control of Homeostasis • **Aging and the Integumentary System** • **Integration with Other Systems**

TABLE 5-1	*A Classification of Burns*	
CLASSIFICATION	DAMAGE REPORT	APPEARANCE AND SENSATION
FIRST-DEGREE BURN	*Killed:* superficial cells of epidermis *Injured:* deeper layers of epidermis, papillary dermis	Inflamed; tender
SECOND-DEGREE BURN	*Killed:* superficial and deeper cells of epidermis; dermis may be affected *Injured:* damage may extend into reticular layer of the dermis, but many accessory structures unaffected	Blisters; very painful
THIRD-DEGREE BURN	*Killed:* all epidermal and dermal cells *Injured:* hypodermis and deeper tissues and organs	Charred; no sensation at all

penetration and the total area affected. The most common descriptive terms refer to the depth of penetration, and these are detailed in Table 5-1. The larger the area affected, the greater the impact on integumentary function.

Aging and the Integumentary System

Aging affects all the components of the integumentary system. The major changes including the following:

- *Skin injuries and infections become more common.* Such problems are more likely because the epidermis thins as stem cell activity declines.

- *The sensitivity of the immune system is reduced.* The number of macrophages and other cells of the immune system residing in the skin decreases to around 50 percent of levels seen at maturity (roughly, age 21). This loss further encourages skin damage and infection.

- *Muscles become weaker, and bone strength decreases.* Such changes are related to reduced calcium and phosphate absorption due to a decline in vitamin D_3 production of around 75 percent.

- *Sensitivity to sun exposure increases.* Lesser amounts of melanin are produced because melanocyte activity declines. The skin of Caucasians becomes very pale.

- *The skin becomes dry and often scaly.* Glandular activity declines, reducing sebum production and perspiration (see below).

- *Hair thins and changes color.* Follicles stop functioning or produce thinner, finer hairs. With decreased melanocyte activity, these hairs are gray or white.

- *Sagging and wrinkling of the skin occurs.* The dermis becomes thinner, and the elastic fiber network decreases in size. The integument therefore becomes weaker and less resilient. These effects are most pronounced in areas exposed to the sun.

- *The ability to lose heat decreases.* The blood supply to the dermis is reduced at the same time that sweat glands become less active. This combination makes the elderly less able than younger people to lose body heat. As a result, overexertion or overexposure to high temperatures (such as a sauna or hot tub) can cause dangerously high body temperatures.

- *Skin repairs proceed relatively slowly.* For example, it takes 3–4 weeks to complete the repairs to a blister site in a person age 18–25. The same repairs at age 65–75 take 6–8 weeks. Because repairs are slow, recurrent infections may occur.

CONCEPT CHECK QUESTIONS
Answers on page 124

❶ Why can skin regenerate effectively even after considerable damage has occurred?

❷ Older people do not tolerate the summer heat as well as they did when they were young, and they are more prone to heat-related illness. What accounts for this change?

Integration with Other Systems

Although the integumentary system can function independently, many of its activities are integrated with those of other systems. Figure 5-10● diagrams the major functional relationships. The role of the skin in temperature control, through interactions with the nervous and cardiovascular systems, was detailed in Chapter 1 (see Figure 1-3b●, p. 13).

CONCEPT CHECK QUESTIONS
Answer on page 124

❶ How does vitamin D_3 production in the skin affect the functions of other organ systems in the body?

The Integumentary System

For All Systems
Provides mechanical protection against environmental hazards

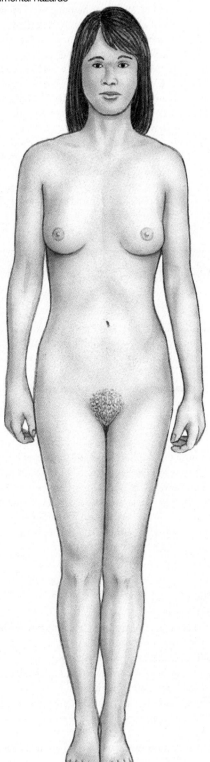

The Skeletal System
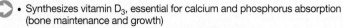
- Provides structural support
- Synthesizes vitamin D_3, essential for calcium and phosphorus absorption (bone maintenance and growth)

The Muscular System
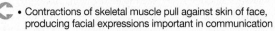
- Contractions of skeletal muscle pull against skin of face, producing facial expressions important in communication
- Synthesizes vitamin D_3, essential for normal calcium absorption (calcium ions play an essential role in muscle contraction)

The Nervous System
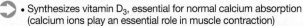
- Controls blood flow and sweat gland activity for thermoregulation; stimulates contraction of arrector pili muscles to elevate hairs
- Receptors in dermis and deep epidermis provide sensations of touch, pressure, vibration, temperature, and pain

The Endocrine System
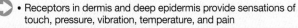
- Sex hormones stimulate sebaceous gland activity; male and female sex hormones influence hair growth, distribution of subcutaneous fat, and apocrine sweat gland activity; adrenal hormones alter dermal blood flow and help mobilize lipids from adipocytes
- Synthesizes vitamin D_3, precursor of calcitriol

The Cardiovascular System

- Provides oxygen and nutrients; delivers hormones and cells of immune system; carries away carbon dioxide, waste products, and toxins; provides heat to maintain normal skin temperature
- Stimulation by mast cells produces localized changes in blood flow and capillary permeability

The Lymphatic System
- Assists in defending the integument by providing additional macrophages and mobilizing lymphocytes
- Provides physical barriers that prevent pathogen entry; macrophages resist infection; mast cells trigger inflammation and initiate the immune reponse

The Respiratory System
- Provides oxygen and eliminates carbon dioxide
- Hairs guard entrance to nasal cavity

The Digestive System
- Provides nutrients for all cells and lipids for storage by adipocytes
- Synthesizes vitamin D_3, needed for absorption of calcium and phosphorus

The Urinary System

- Excretes waste products, maintains normal body fluid pH and ion composition
- Assists in elimination of water and solutes; keratinized epidermis limits fluid loss through skin

The Reproductive System

- Sex hormones affect hair distribution, adipose tissue distribution in subcutaneous layer, and mammary gland development
- Covers external genitalia; provides sensations that stimulate sexual behaviors; mammary gland secretions provide nourishment for newborn infant

● *Figure 5-10* **Functional Relationships Between the Integumentary System and Other Systems**

5 THE INTEGUMENTARY SYSTEM

Integumentary Structure and Function • Local Control of Homeostasis • Aging and the Integumentary System • Integration with Other Systems

Related Clinical Terms

acne: A sebaceous gland inflammation caused by an accumulation of secretions.

biopsy (BĪ-op-sē): The removal and examination of tissue from the body for the diagnosis of disease.

cavernous hemangioma ("port-wine stain"): A birthmark caused by a tumor affecting large vessels in the dermis; generally lasts a lifetime.

contact dermatitis: A dermatitis generally caused by strong chemical irritants. It produces an itchy rash that may spread to other areas as scratching distributes the chemical agent; includes poison ivy.

cyanosis (sī-uh-NŌ-sis): Bluish skin color as a result of reduced oxygenation of the blood in superficial vessels.

dermatitis: An inflammation of the skin that primarily involves the papillary region of the dermis.

dermatology: The branch of medicine concerned with the diagnosis and treatment of diseases of the skin.

granulation tissue: A combination of fibrin, fibroblasts, and capillaries that forms during tissue repair after inflammation.

keloid (KĒ-loyd): A thickened area of scar tissue covered by a shiny, smooth epidermal surface.

lesion (LĒ-zhuns): Changes in skin structure caused by injury or disease.

male pattern baldness: Hair loss in an adult male due to changes in levels of circulating sex hormones.

malignant melanoma (mel-uh-NŌ-muh): A skin cancer originating in malignant melanocytes.

psoriasis (sō-RĪ-uh-sis): A painless condition characterized by rapid stem cell divisions in the stratum germinativum of the scalp, elbows, palms, soles, groin, and nails. Affected areas appear dry and scaly.

sepsis (SEP-sis): A dangerous, widespread bacterial infection; the leading cause of death in burn patients.

ulcer: A localized shedding of an epithelium.

basal cell carcinoma: A malignant cancer that originates in the stratum germinativum; the most common skin cancer. Roughly two-thirds of cases of this cancer appear in areas subjected to chronic UV exposure. Metastasis seldom occurs.

pruritis (proo-RĪ-tis): An irritating itching sensation, common in skin conditions.

squamous cell carcinoma: A form of skin cancer less common than basal cell carcinoma, almost totally restricted to areas of sun-exposed skin. Metastasis seldom occurs.

xerosis (ze-RŌ-sis): "Dry skin," a common complaint of older persons and almost anyone living in an arid climate.

CHAPTER REVIEW

Key Terms

Summary Outline

1. The **integumentary system**, or **integument**, consists of the **cutaneous membrane**, which includes the **epidermis** and **dermis**, and the **accessory structures**. Underneath lies the **subcutaneous layer** (or **hypodermis**). (*Figure 5-1*)

2. **Thin skin** covers most of the body; heavily abraded body surfaces may be covered by **thick skin**.

3. Cell divisions by the stem cells that make up the **stratum germinativum** replace more superficial cells.

4. As epidermal cells age, they pass through the **stratum spinosum**, the **stratum granulosum**, the **stratum lucidum** (in thick skin), and the **stratum corneum**. In the process, they accumulate large amounts of **keratin**. Ultimately, the cells are shed or lost. (*Figure 5-2*)

5. **Epidermal ridges** interlock with the **dermal papillae** of the dermis. Together, they form superficial ridges on the palms and soles that improve our gripping ability.

6. The color of the epidermis depends on two factors: blood supply and pigment composition and concentration. **Melanocytes** protect the stem cells from **ultraviolet (UV) radiation**. (*Figure 5-3*)

7. Epidermal cells synthesize **vitamin D₃** when exposed to sunlight.

8. Skin cancer is the most common form of cancer. **Basal cell carcinoma** and **squamous cell carcinoma** are not as dangerous as a **melanoma**. (*Figure 5-4*)

9. The dermis consists of the **papillary layer** and the deeper **reticular layer**.

10. The papillary layer of the dermis contains blood vessels, lymphatic vessels, and sensory nerves. This layer supports and nourishes the overlying epidermis. The reticular layer consists of a meshwork of collagen and elastic fibers oriented to resist tension in the skin.

11. Components of other systems (cardiovascular, lymphatic, and nervous) that communicate with the skin are in the dermis.

12. The subcutaneous layer stabilizes the skin's position against underlying organs and tissues.

13. **Hairs** originate in complex organs called **hair follicles**. Each hair has a **shaft** composed of dead keratinized cells. (*Figure 5-5*)

14. The **arrector pili** muscles can elevate the hairs.

15. Our hairs grow and are shed according to the *hair growth cycle*. A single hair grows for 2–5 years and is then shed.

16. Typical **sebaceous glands** discharge the waxy **sebum** into hair follicles. *Sebaceous follicles* are sebaceous glands that empty directly onto the skin. (*Figure 5-6*)

17. **Apocrine sweat glands** produce an odorous secretion; the more numerous **merocrine sweat glands** produce perspiration, a watery secretion. (*Figure 5-7*)

18. The **body** of a **nail** covers the **nail bed**. Nail production occurs at the **nail root**. (*Figure 5-8*)

Review Questions

Level 1: Reviewing Facts and Terms

Match each item in column A with the most closely related item in column B. Use letters for answers in the spaces provided.

COLUMN A

___ 1. cutaneous membrane

___ 2. carotene

___ 3. melanocytes

___ 4. epidermal layer containing stem cells

___ 5. smooth muscle

___ 6. epidermal layer of flattened and dead cells

___ 7. bluish skin

___ 8. sebaceous glands

___ 9. merocrine (eccrine) glands

___ 10. vitamin D$_3$

COLUMN B

a. arrector pili

b. cyanosis

c. perspiration

d. sebum

e. stratum corneum

f. skin

g. orange-yellow pigment

h. bone growth

i. pigment cells

j. stratum germinativum

11. The two major components of the integument are
    (a) the cutaneous membrane and the accessory structures
    (b) the epidermis and the hypodermis
    (c) the hair and the nails
    (d) the dermis and the subcutaneous layer

12. The fibrous protein that forms the basic structural component of hair and nails is
    (a) collagen   (b) melanin
    (c) elastin   (d) keratin

13. The two types of exocrine glands in the skin are
    (a) merocrine and sweat glands
    (b) sebaceous and sweat glands
    (c) apocrine and sweat glands
    (d) eccrine and sweat glands

14. The following are all accessory structures of the integumentary system except
    (a) nails   (b) hair
    (c) dermal papillae   (d) sweat glands

15. Sweat glands that communicate with hair follicles in the armpits and produce an odorous secretion are
    (a) apocrine glands
    (b) merocrine glands
    (c) sebaceous glands
    (d) a, b, and c are correct

16. The reason older persons are more sensitive to sun exposure and more likely to get sunburned is that with age
    (a) melanocyte activity declines
    (b) vitamin D$_3$ production declines
    (c) glandular activity declines
    (d) skin thickness decreases

17. Which two skin pigments are found in the epidermis?

18. Which two major layers constitute the dermis, and what components are found in each layer?

19. Which two groups of sweat glands are contained in the integument?

Level 2: Reviewing Concepts

20. During the transdermal administration of drugs, why are fat-soluble drugs more desirable than those that are water-soluble?

21. In our society, a tan body is associated with good health. However, medical research constantly warns about the dangers of excessive exposure to the sun. What are the benefits of a tan?

22. Why is a subcutaneous injection with a hypodermic needle a useful method for administering drugs?

23. Why does skin sag and wrinkle as a person ages?

5 THE INTEGUMENTARY SYSTEM

Integumentary Structure and Function • Local Control of Homeostasis • Aging and the Integumentary System • Integration with Other Systems

Level 3: Critical Thinking and Clinical Applications

24. A new mother notices that her 6-month-old child has a yellow-orange complexion. Fearful that the child may have jaundice (a condition caused by a toxic yellow-orange pigment in the blood), she takes him to her pediatrician. After examining the child, the pediatrician declares him perfectly healthy and advises the mother to watch the child's diet. Why?

25. Vanessa remarks that her 80-year-old grandmother keeps her thermostat set at 80°F and wears a sweater on balmy spring days. When she asks her grandmother why, her grandmother tells her that she is cold. Vanessa can't understand this and asks you for an explanation. What would you tell Vanessa?

Answers to Concept Check Questions

Page 114

1. Cells are constantly shed from the *stratum corneum*.
2. Sanding the tips of one's fingers will not permanently remove fingerprints. Because the ridges of the fingerprints are formed in layers of the skin that are constantly regenerated, the ridges will eventually reappear.
3. When exposed to the ultraviolet radiation in sunlight or tanning lamps, melanocytes in the epidermis (and dermis) synthesize the pigment melanin, darkening the skin.

Page 116

1. Contraction of the arrector pili muscles pulls the hair follicles erect, depressing the area at the base of the hair and making the surrounding skin appear higher. The result is known as "goose bumps" or "goose pimples."
2. Hair is a derivative of the epidermis, but the follicles are in the dermis. Where the epidermis and deep dermis are destroyed, no new hair will grow.
3. The subcutaneous layer stabilizes the position of the skin in relation to underlying tissues and organs; stores fat; and, because its lower region contains few capillaries and no vital organs, provides a useful site for the injection of drugs.

Page 118

1. Sebaceous glands produce a secretion called *sebum*. Sebum lubricates both the hair and skin, and inhibits the growth of bacteria.

2. Apocrine sweat glands produce a secretion containing several kinds of organic compounds. Some of these compounds have an odor, and others produce an odor when metabolized by skin bacteria. Deodorants are used to mask the odor of these secretions.

Page 120

1. Skin can regenerate effectively because stem cells are present in both the epithelial and connective tissues of the skin. When an injury occurs, cells of the stratum germinativum replace epithelial cells and connective tissue stem cells replace cells lost from the dermis.
2. As a person ages, the blood supply to the dermis decreases and merocrine sweat glands become less active. These changes make it more difficult for the elderly to cool themselves in hot weather.

Page 120

1. Vitamin D_3 production in the skin is interrelated with the functions of the endocrine, digestive, skeletal, and muscular systems. Vitamin D_3 production is the starting point in the final synthesis of the hormone, *calcitriol*, by the endocrine system. Calcitriol is essential for the absorption of calcium and phosphorus by the digestive system. The skeletal system is affected because bone maintenance and growth depend on the availability of calcium and phosphorus. The muscular system is affected because calcium is essential for muscle contraction.

EXPLORE *MediaLab*

EXPLORATION #1

Estimated time for completion: 10 minutes

BY NOW YOU ARE FAMILIAR with the chemistry of your body, as well as the organization of cells into tissues and organs. Go to Figure 1-1●, Levels of Organization, and study the relationships depicted. How do you suppose the heart maintains a beat? The answer is in the chemical and cellular levels of organization.

Using your knowledge of cell structure and chemical interactions, can you predict what this relationship might be? Prepare a short description of how the cells of the heart might maintain the heartbeat. Then go to Chapter 5 at the Companion Web site, visit the MediaLab section, and click on the keyword "Heartbeat." Here you will find the *Cells Alive* home page.

Choose cell biology and then "pumping myocytes" to understand how chemistry and cellular biology underlie the beating of your heart. Here you will see a beating cell, read about the chemical interactions that cause that beat, and watch ions travel across the cell membrane. Modify your description above to reflect the information obtained at the website.

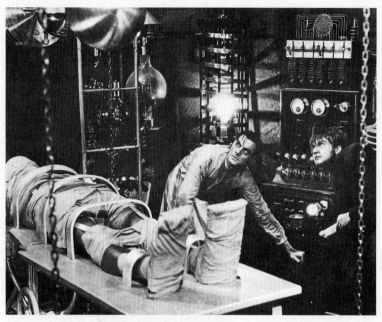

Dr. Frankenstein and his assistant prepare to bring their monster to life in the 1931 movie version of Mary Shelley's *Frankenstein*.

EXPLORATION #2

Estimated time for completion: 10 minutes

DR. FRANKENSTEIN CREATED LIFE from dead tissues in Mary Shelley's famous novel *Frankenstein*. While this practice is *not* being considered in the medical field, there is a serious effort to create new tissues and organs from individual cells and synthetic supporting materials. Imagine you are a histologist working on the regeneration of bone. Your goal is to produce new bone tissue for persons suffering from a severe weakening of bony tissues, a condition called *osteoporosis*. What challenges will you face? How will you locate the needed cells? How will you stimulate and direct their growth? Using your understanding of cellular and tissue biology, compile a list of questions that you must answer before you can complete this project. Once you have finished this list, go to Chapter 5 at the Companion Web site, visit the MediaLab section, and click on the keyword "Neo-organs." Here you will find an April 1999 *Scientific American* article that discusses tissue and organ regeneration in detail. Can you find answers to some of the questions on your list? Include these answers on your list, along with any new information or additional questions that occur to you as you read the article.

The surgeon' hands hold a complete donor heart for a heart transplant.

The Skeletal System

CHAPTER OUTLINE AND OBJECTIVES

Vocabulary Development

ab- ....from; *abduction*
acetabulum ....a vinegar cup; *acetabulum of the hip joint*
ad- ....toward, to; *adduction*
amphi- ....on both sides; *amphiarthrosis*
arthros ....joint; *synarthrosis*
blast ....precursor; *osteoblast*
circum- ....around; *circumduction*
clast ....break; *osteoclast*
clavius ....clavicle; *clavicle*
concha ....shell; *middle concha*
corona ....crown; *coronoid fossa*
cranio- ....skull; *cranium*
cribrum ....sieve; *cribriform plate*
dens ....tooth; *dens*
dia- ....through; *diarthrosis*
duco ....to lead; *adduction*
e- ....out; *eversion*
gennan ....to produce; *osteogenesis*
gomphosis ......a bolting together; *gomphosis*
in- ....into; *inversion*
infra- ....beneath; *infraspinous fossa*
lacrimae ....tears; *lacrimal bones*
lamella ....thin plate; *lamellae of bone*
malleolus .....little hammer; *medial malleolus*
meniscus ....crescent; *menisci*
osteon ....bone; *osteocytes*
penia ....lacking; *osteopenia*
planta ....sole; *plantar*
porosus ....porous; *osteoporosis*
septum ....wall; *nasal septum*
stylos ....pillar; *styloid process*
supra- ....above; *supraspinous fossa*
sutura ....a sewing together; *suture*
teres ....cylindrical; *ligamentum teres*
trabecula ......wall; *trabeculae in spongy bone*
trochlea ....pulley; *trochlea*
vertere ....to turn; *inversion*

*T*HE STEEL GIRDERS the construction workers are standing on will become the internal support system of a new building. Our support system, the skeletal system, is also internal, yet it is made of bone, a remarkable tissue that is strong and (unlike these girders) capable of repairing itself even after serious injury.

6 THE SKELETAL SYSTEM

The Structure of Bone • Bone Development and Growth • Remodeling and Homeostatic Mechanisms • Aging and the Skeletal System

THE SKELETON HAS MANY FUNCTIONS, but the most obvious is supporting the weight of the body. This support is provided by bones, structures as strong as reinforced concrete but considerably lighter. Unlike concrete, bones can be remodeled and reshaped to meet changing metabolic demands and patterns of activity. Bones work together with muscles to maintain body position and to produce controlled, precise movements. With the skeleton to pull against, contracting muscles can make us sit, stand, walk, or run.

The skeletal system includes the bones of the skeleton and the cartilages, ligaments, and other connective tissues that stabilize or connect them. This system has five primary functions:

1. **Support.** The skeletal system provides structural support for the entire body. Individual bones or groups of bones provide a framework for the attachment of soft tissues and organs.

2. **Storage.** The calcium salts of bone represent a valuable mineral reserve that maintains normal concentrations of calcium and phosphate ions in body fluids. In addition, bones store lipids in areas of *yellow marrow* as an energy reserve.

3. **Blood cell production.** Red blood cells, white blood cells, and other blood elements are produced within the *red marrow*, which fills the internal cavities of many bones. The role of the bone marrow in blood cell formation will be discussed in later chapters dealing with the cardiovascular and lymphatic systems (Chapters 11 and 14).

4. **Protection.** Delicate tissues and organs are often surrounded by skeletal elements. The ribs protect the heart and lungs, the skull encloses the brain, the vertebrae shield the spinal cord, and the pelvis cradles delicate digestive and reproductive organs.

5. **Leverage.** Bones of the skeleton function as levers that change the magnitude and direction of the forces generated by skeletal muscles. The resulting movements range from the delicate motion of a fingertip to powerful changes in the position of the entire body.

The Structure of Bone

Bone, or **osseous tissue**, is a supporting connective tissue, and it contains specialized cells and a matrix of extracellular protein fibers and a ground substance. ⟳ p. 96 The distinctive solid, stony character of bone results from the deposition of calcium salts within the matrix. Calcium phosphate, $Ca_3(PO_4)_2$, accounts for almost two-thirds of the weight of bone. The remaining third is dominated by collagen fibers, with osteocytes and other cell types providing only around 2 percent of the mass of a bone.

MACROSCOPIC FEATURES OF BONE

The bones of the human skeleton have four general shapes: long, short, flat, and irregular (Figure 6-1●). **Long bones** are

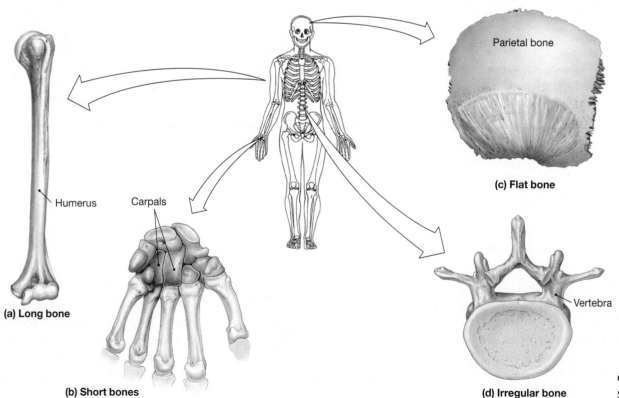

Parietal bone

(c) Flat bone

Humerus

Carpals

(a) Long bone

(b) Short bones

Vertebra

(d) Irregular bone

● *Figure 6-1*
Shapes of Bones

longer than they are wide, whereas **short bones** are of roughly equal dimensions. Examples of long bones are bones of the limbs, such as the bones of the arm (*humerus*) and thigh (*femur*). Short bones include the bones of the wrist (*carpal bones*) and ankles (*tarsal bones*). **Flat bones** are thin and relatively broad, such as the parietal bones of the skull, the ribs, and the shoulder blades (*scapulae*). **Irregular bones** have complex shapes that do not fit easily into any other category. An example is any of the vertebrae of the spinal column.

The typical features of a long bone such as the humerus are shown in Figure 6-2●. A long bone has a central shaft, or **diaphysis** (dī-AF-i-sis), and an expanded portion at each end, or **epiphysis** (ē-PIF-i-sis). The diaphysis surrounds a central *marrow cavity*. The ends, or **epiphyses** (ē-PIF-i-sēz), of adjacent bones articulate with each other and are covered by *articular cartilages*. As will be discussed below, growth in length of an immature long bone occurs at the junctions between the epiphyses and the diaphysis.

The two types of bone tissue are visible in Figure 6-2●. **Compact bone**, or dense bone, is relatively solid, whereas **spongy bone**, or *cancellous* (KAN-sel-us) *bone*, resembles a network of bony rods or struts separated by spaces that are normally filled with **bone marrow**, a loose connective tissue. Both compact and spongy bone are present in the humerus; compact bone forms the diaphysis, and spongy bone fills the epiphyses.

The outer surface of a bone is covered by a **periosteum**, which consists of a fibrous outer layer and a cellular inner layer (Figures 6-2, 6-3●). The periosteum isolates the bone from surrounding tissues, provides a route for circulatory and nervous supplies, and participates in bone growth and repair. The fibers of *tendons* and *ligaments* intermingle with those of the periosteum, attaching skeletal muscles to bones and one bone to another.

Inside the bone, a cellular **endosteum** lines the marrow cavity and other inner surfaces. The endosteum is active during bone growth and whenever repair or remodeling is under way.

MICROSCOPIC FEATURES OF BONE

The general histology of bone was introduced in Chapter 4 (see Figure 4-12●, p. 98). Both compact and spongy bone contain bone cells, or **osteocytes** (OS-tē-ō-sīts; *osteon*, bone), in small pockets called **lacunae** (la-KOO-nē). Lacunae are found between narrow sheets of calcified matrix that are known as **lamellae** (lah-MEL-lē; *lamella*, thin plate). Small channels, called **canaliculi** (ka-na-LIK-ū-lē), radiate through the matrix, interconnecting lacunae and linking them to nearby blood vessels. The canaliculi contain cytoplasmic extensions of the osteocytes. Nutrients from the blood and waste products from osteocytes diffuse through the extracellular fluid that surrounds these cells as well as through their cytoplasmic extensions.

Compact and Spongy Bone

The basic functional unit of compact bone, the **osteon** (OS-tē-on), or *Haversian system*, is shown in Figure 6-3●. Within an osteon, the osteocytes are arranged in concentric layers around a **central canal**, or *Haversian canal*, that contains one or more blood vessels. The lamellae are cylindrical, oriented parallel to the long axis of the central canal. **Perforating canals** provide passageways for linking the blood vessels of the central canals with those of the periosteum or the marrow cavity.

Spongy bone has a different lamellar arrangement and no osteons. Instead, the lamellae form rods or plates called **trabeculae** (tra-BEK-ū-lē; *trabecula*, wall). Frequent branchings of the thin trabeculae create an open network. Canaliculi radiating from the lacunae of spongy bone end at the exposed surfaces of the trabeculae, where nutrients and wastes diffuse between the marrow and osteocytes.

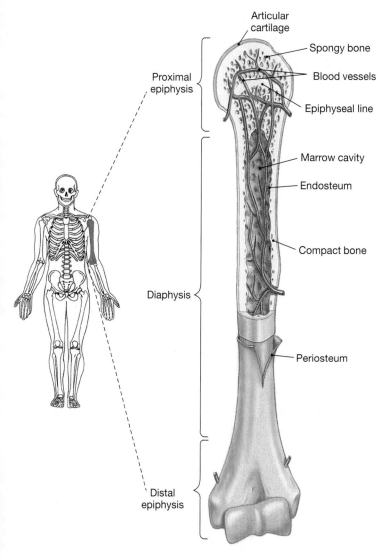

Proximal epiphysis

Diaphysis

Distal epiphysis

Articular cartilage
Spongy bone
Blood vessels
Epiphyseal line
Marrow cavity
Endosteum
Compact bone
Periosteum

● *Figure 6-2* **The Structure of a Long Bone**

129

6

THE SKELETAL SYSTEM

The Structure of Bone • **Bone Development and Growth** • Remodeling and Homeostatic Mechanisms • Aging and the Skeletal System

● *Figure 6-3* **The Structure of a Typical Bone**
(**a**) A thin section through compact bone; in this procedure the intact matrix and central canals appear white, and the lacunae and canaliculi appear black. (LM × 272) (**b**) A diagrammatic view of the structure of a typical bone, the humerus.

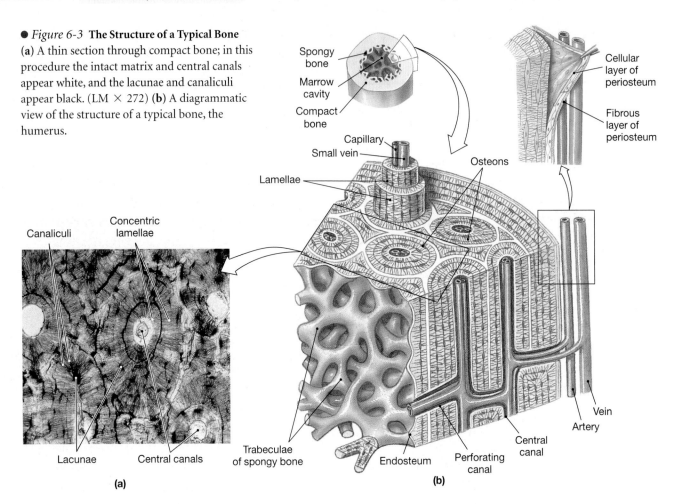

(a)

(b)

A layer of compact bone covers bone surfaces everywhere except inside *joint capsules*, where articular cartilages protect opposing surfaces. Compact bone is usually found where stresses come from a limited range of directions. The limb bones, for example, are built to withstand forces applied at either end. Because the osteons are parallel to the long axis of the shaft, a limb bone does not bend, even when a large force is applied to either end. However, a much smaller force applied to the side of the shaft can break the bone.

Bones do bend in disorders that reduce the amount of calcium salts in the skeleton. As the proportion of calcium to collagen decreases, the bones become very flexible, and they become less able to resist compression and tension. An example of such a condition is *rickets*, a childhood disorder that results from a deficiency of vitamin D_3. Affected individuals develop a bowlegged appearance as the leg bones bend under the weight of the body.

In contrast, spongy bone is found where bones are not heavily stressed or where stresses arrive from many directions. For example, spongy bone is present at the epiphyses of long bones, where stresses are transferred across joints. Spongy bone is also much lighter than compact bone. This reduces the weight of the skeleton and makes it easier for muscles to move the

bones. Finally, the trabecular network of spongy bone supports and protects the cells of red bone marrow, important sites of blood cell formation.

Cells in Bone

Although osteocytes are the most abundant cells in bone, other cell types are also present. These cells, called *osteoclasts* and *osteoblasts*, are associated with the endosteum that lines the inner cavities of both compact and spongy bone and the cellular layer of the periosteum. Three primary cell types occur in bone:

1. **Osteocytes**, mature bone cells. Osteocytes maintain normal bone structure by recycling the calcium salts in the bony matrix around themselves and assisting in repairs.

2. **Osteoclasts** (OS-tē-ō-klasts; *clast*, break), giant cells with 50 or more nuclei. Acids and enzymes secreted by osteoclasts dissolve the bony matrix and release the stored minerals through *osteolysis* (os-tē-OL-i-sis) or *resorption*. This process helps regulate calcium and phosphate concentrations in body fluids.

3. **Osteoblasts** (OS-tē-ō-blasts; *blast*, precursor), the cells responsible for the production of new bone, a process called *osteogenesis* (os-tē-ō-JEN-e-sis; *gennan*, to produce). Osteoblasts produce new bone matrix and promote the deposition of calcium salts in the organic matrix. At any

given moment, osteoclasts are removing matrix and osteoblasts are adding to it. When an osteoblast becomes completely surrounded by calcified matrix, it differentiates into an osteocyte.

CONCEPT CHECK QUESTIONS
Answers on page 173

❶ How would the strength of a bone be affected if the ratio of collagen to calcium increased?

❷ A sample of bone shows concentric layers surrounding a central canal. Is it from the shaft or the end of a long bone?

❸ If the activity of osteoclasts exceeds that of osteoblasts in a bone, how will the mass of the bone be affected?

Bone Development and Growth

The growth of the skeleton determines the size and proportions of your body. Skeletal growth begins about 6 weeks after fertilization, when the embryo is about 12 mm (0.5 in.) long. (At this time, all skeletal elements are made of cartilage.) Bone growth continues through adolescence, and portions of the skeleton usually do not stop growing until roughly age 25. This section considers the process of osteogenesis (bone formation) and growth. The next section examines the maintenance and turnover of mineral reserves in the adult skeleton.

During development, cartilage or connective tissue is replaced by bone. The process of replacing other tissues with bone is called **ossification**. (The process of **calcification**, the deposition of calcium salts, occurs during ossification, but it can also occur in other tissues.) There are two major forms of ossification. In *intramembranous ossification*, bone develops within sheets or membranes of connective tissue. In *endochondral ossification*, bone replaces existing cartilage. Figure 6-4● shows some of the bones formed by these processes in a 16-week-old fetus.

INTRAMEMBRANOUS OSSIFICATION

Intramembranous (in-tra-MEM-bra-nus) **ossification** begins when osteoblasts differentiate within embryonic or fetal connective tissue, or fibrous connective tissue. This type of ossification normally occurs in the deeper layers of the dermis. The osteoblasts differentiate from connective tissue stem cells after the organic components of the matrix secreted by the stem cells becomes calcified. The place where ossification first occurs is called an **ossification center**. As ossification proceeds and new bone branches outward, some osteoblasts become trapped inside bony pockets and change into osteocytes.

Bone growth is an active process, and osteoblasts require oxygen and a reliable supply of nutrients. Blood vessels begin to grow into the area to meet these demands and over time

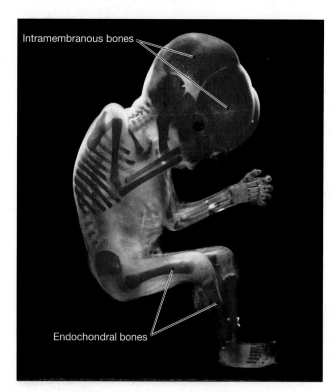

● *Figure 6-4* **Bone Formation in a 16-Week-Old Fetus**

become trapped within the developing bone. At first, the intramembranous bone resembles spongy bone. Further remodeling around the trapped blood vessels can produce osteons typical of compact bone. The flat bones of the skull, the lower jaw, and the collarbones (*clavicles*) form this way.

ENDOCHONDRAL OSSIFICATION

Most of the bones of the skeleton are formed through the **endochondral** (en-dō-KON-dral; *endo*, inside + *chondros*, cartilage) **ossification** of existing hyaline cartilage. The cartilages develop first; they are like miniature cartilage models of the future bone. By the time an embryo is 6 weeks old, the cartilage models of the future limb bones begin to be replaced by true bone. Steps in the growth and ossification of a limb bone are diagrammed in Figure 6-5●.

Step 1: Endochondral ossification starts as chondrocytes within the cartilage model enlarge and the surrounding matrix begins to calcify. The cells die because the diffusion of nutrients slows through the calcified matrix.

Step 2: Bone formation first occurs at the shaft surface. Blood vessels invade the perichondrium and cells of its inner layer differentiate into osteoblasts that begin producing bone matrix. p. 96

Step 3: Blood vessels invade the inner region of the cartilage, and newly differentiated osteoblasts form spongy bone within the center of the shaft at a *primary center of ossification*. Bone

6 THE SKELETAL SYSTEM

The Structure of Bone • Bone Development and Growth • **Remodeling and Homeostatic Mechanisms** • Aging and the Skeletal System

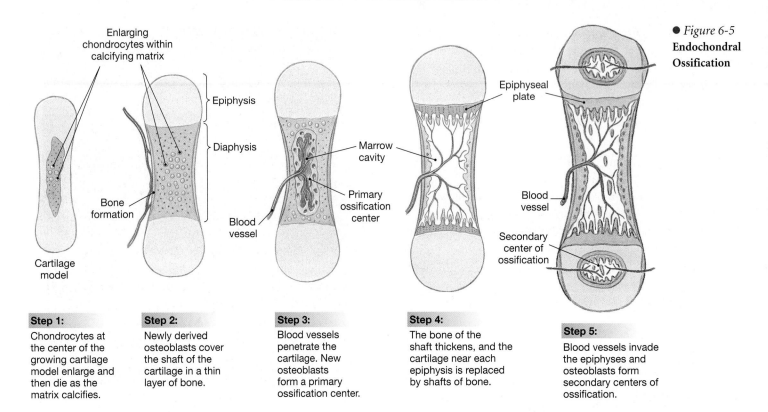

● *Figure 6-5*
Endochondral Ossification

Step 1:
Chondrocytes at the center of the growing cartilage model enlarge and then die as the matrix calcifies.

Step 2:
Newly derived osteoblasts cover the shaft of the cartilage in a thin layer of bone.

Step 3:
Blood vessels penetrate the cartilage. New osteoblasts form a primary ossification center.

Step 4:
The bone of the shaft thickens, and the cartilage near each epiphysis is replaced by shafts of bone.

Step 5:
Blood vessels invade the epiphyses and osteoblasts form secondary centers of ossification.

development proceeds toward either end filling the shaft with spongy bone.

Step 4: As the bone enlarges, osteoclasts break down some of the spongy bone and create a marrow cavity. The cartilage model does not completely fill with bone because the epiphyseal cartilages on the ends continue to enlarge, increasing the length of the developing bone. Although osteoblasts from the shaft continuously invade the epiphyseal cartilages, the bone grows longer because new cartilage is continuously added in front of the advancing osteoblasts. This situation is like a pair of joggers with one in front of the other. As long as they run at the same speed, they can run for miles without colliding.

Step 5: The centers of the epiphyses begin to calcify. As blood vessels and osteoblasts enter these areas, *secondary centers of ossification* form and the epiphyses eventually become filled with spongy bone. A thin cap of the original cartilage model remains exposed to the joint cavity as the **articular cartilage**. At this stage, the bone of the shaft and the bone of each epiphysis are separated by areas of cartilage known as **epiphyseal plates**. As long as the rate of cartilage growth keeps pace with the rate of osteoblast invasion, the epiphyseal plate survives, and the bone continues to grow longer.

When sex hormone production increases at puberty, bone growth accelerates dramatically, and osteoblasts begin to produce bone faster than epiphyseal cartilage expands. As a result,

the epiphyseal plates at each end of the bone get narrower and narrower until they disappear. In adults, the former location of the plate is detected in X-rays as a distinct *epiphyseal line* that remains after epiphyseal growth has ended (see Figure 6-2●, p. 129). The end of epiphyseal growth is called *closure*.

While the bone elongates, it also grows larger in diameter. The diameter enlarges at its outer surface through a process called **appositional growth**. This enlargement occurs as cells of the periosteum develop into osteoblasts and produce additional bony matrix. As new bone is deposited on the outer surface of the shaft, the inner surface is eroded by osteoclasts, and the marrow cavity gradually enlarges (Figure 6-6●).

BONE GROWTH AND BODY PROPORTIONS

The timing of epiphyseal closure varies from bone to bone and individual to individual. The toes may complete their ossification by age 11, whereas portions of the pelvis or the wrist may continue to enlarge until age 25. The epiphyseal plates in the arms and legs usually close by age 18 (women) or 20 (men). Differences in sex hormones account for variations in body size and proportions between men and women.

REQUIREMENTS FOR NORMAL BONE GROWTH

Normal bone growth and maintenance cannot occur without a reliable source of minerals, especially calcium salts. During prenatal development these minerals are absorbed from the mother's bloodstream. The demands are so great that the maternal

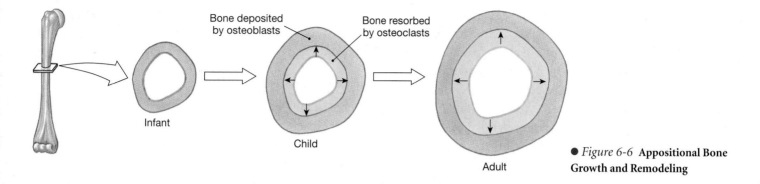

Bone deposited by osteoblasts

Bone resorbed by osteoclasts

Infant

Child

Adult

● *Figure 6-6* **Appositional Bone Growth and Remodeling**

skeleton often loses bone mass during pregnancy. From infancy to adulthood, the diet must provide adequate amounts of calcium and phosphate, and the individual must be able to absorb and transport these minerals to sites of bone formation.

Vitamin D_3 plays an important role in normal calcium metabolism. This vitamin can be obtained from dietary supplements or manufactured by epidermal cells exposed to UV radiation. 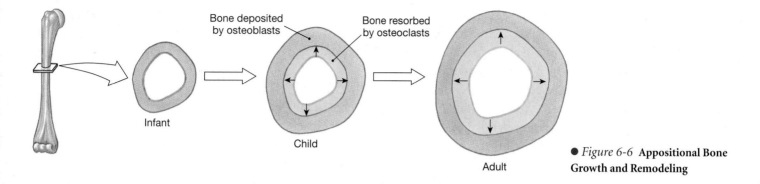 p. 113 After vitamin D_3 has been processed in the liver, the kidneys convert a derivative of this vitamin into *calcitriol*, a hormone that stimulates the absorption of calcium and phosphate ions in the digestive tract. *Rickets* is a condition marked by a softening and bending of bones that occurs in growing children, as a result of vitamin D_3 deficiency.

Vitamin A and *vitamin C* are also essential for normal bone growth and remodeling. For example, a deficiency of vitamin C will cause *scurvy*. One of the primary symptoms of this condition is a reduction in osteoblast activity that leads to weak and brittle bones. In addition to vitamins, hormones such as growth hormone, thyroid hormones, sex hormones, and those involved with calcium metabolism are essential to normal skeletal growth and development.

CONCEPT CHECK QUESTIONS
Answers on page 173

❶ How could X-rays of the femur be used to determine whether a person had reached full height?

❷ In the Middle Ages, choirboys were sometimes castrated (had their testes removed) to prevent their voices from changing. How would castration have affected their height?

❸ Why are pregnant women given calcium supplements and encouraged to drink milk even though their skeletons are fully formed?

Remodeling and Homeostatic Mechanisms

Of the five major functions of the skeleton discussed earlier in this chapter, support and storage depend on the dynamic nature of bone. In adults, osteocytes in lacunae maintain the surrounding matrix, continually removing and replacing the surrounding calcium salts. But osteoclasts and osteoblasts also remain active, even after the epiphyseal plates have closed. Normally their activities are balanced: As one osteon forms through the activity of osteoblasts, another is destroyed by osteoclasts. The turnover rate for bone is quite high, and in adults roughly 18 percent of the protein and mineral components are removed and replaced each year through the process of **remodeling**. Every part of every bone may not be affected, as there are regional and even local differences in the rate of turnover. For example, the spongy bone in the head of the femur may be replaced two or three times each year, whereas the compact bone along the shaft remains largely untouched.

REMODELING AND SUPPORT

Regular mineral turnover gives each bone the ability to adapt to new stresses. Heavily stressed bones become thicker, stronger, and develop more pronounced surface ridges; bones not subjected to ordinary stresses become thin and brittle. Regular exercise is thus an important stimulus that maintains normal bone structure.

Degenerative changes in the skeleton occur after even brief periods of inactivity. For example, using a crutch while wearing a cast takes the weight off the injured leg. After a few weeks, the unstressed leg will lose up to about a third of its bone mass. The bones rebuild just as quickly when they again carry their normal weight.

HOMEOSTASIS AND MINERAL STORAGE

The bones of the skeleton are more than just racks to hang muscles on. They are important mineral reservoirs, and calcium is the most abundant mineral in the human body. A typical human body contains 1–2 kg (2.2–4.4 lb) of calcium, with 99 percent of it deposited in the skeleton.

Calcium ion concentrations must be closely controlled to prevent damage to essential physiological systems. Small variations from normal concentrations will have some effect on

6 THE SKELETAL SYSTEM

The Structure of Bone • Bone Development and Growth • **Remodeling and Homeostatic Mechanisms** • Aging and the Skeletal System

cellular operations, and larger changes can cause a clinical crisis. Neurons and muscle cells are particularly sensitive to changes in the concentration of calcium ions. If the calcium concentration in body fluids increases by 30 percent, neurons and muscle cells become relatively unresponsive. If calcium levels decrease by 35 percent, they become so excitable that convulsions may occur. A 50 percent reduction in calcium concentrations usually causes death. Such gross disturbances in calcium metabolism are relatively rare, because calcium ion concentrations are so closely regulated that daily fluctuations of more than 10 percent are very unusual.

Parathyroid hormone (PTH) and *calcitriol* are hormones that work together to elevate calcium levels in body fluids. Their actions are opposed by *calcitonin*, which depresses calcium levels in body fluids. These hormones and their regulation are discussed further in Chapter 10.

By providing a calcium reserve, the skeleton helps maintain calcium homeostasis in body fluids. This function can directly affect the shape and strength of the bones in the skeleton. When large numbers of calcium ions are mobilized, bones become weaker; when calcium salts are deposited, bones become more massive.

INJURY AND REPAIR

Despite its mineral strength, bone cracks or even breaks if subjected to extreme loads, sudden impacts, or stresses from unusual directions. All such cracks and breaks in bones constitute a **fracture**. Fractures are classified according to their external appearance, the site of the fracture, and the nature of the break. (See "Focus: A Classification of Fractures" on p. 135.)

Bones will usually heal even after they have been severely damaged, as long as the circulatory supply and the cellular components of the endosteum and periosteum survive. Steps in the repair process, which may take from 4 months to well over a year following a fracture, are diagrammed in Figure 6-7●:

Step 1: In even a small fracture, many blood vessels are broken and extensive bleeding occurs. Pooling and clotting of the

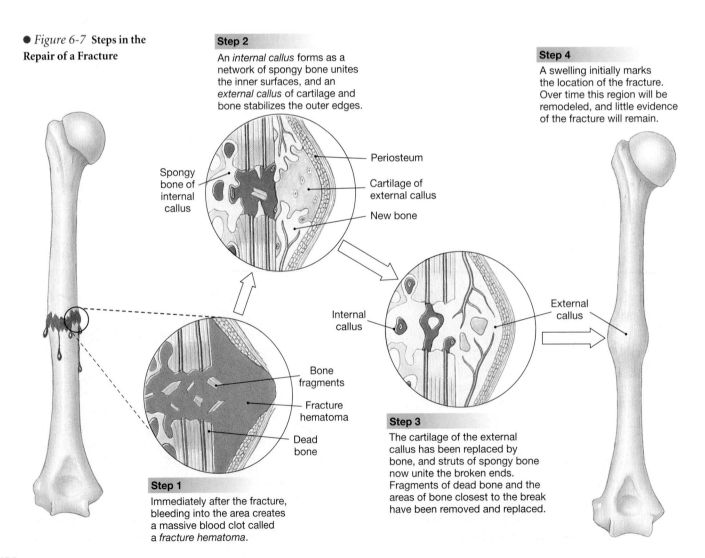

● *Figure 6-7* **Steps in the Repair of a Fracture**

Step 2
An *internal callus* forms as a network of spongy bone unites the inner surfaces, and an *external callus* of cartilage and bone stabilizes the outer edges.

Spongy bone of internal callus

Periosteum

Cartilage of external callus

New bone

Step 4
A swelling initially marks the location of the fracture. Over time this region will be remodeled, and little evidence of the fracture will remain.

Internal callus

External callus

Step 1
Immediately after the fracture, bleeding into the area creates a massive blood clot called a *fracture hematoma*.

Bone fragments

Fracture hematoma

Dead bone

Step 3
The cartilage of the external callus has been replaced by bone, and struts of spongy bone now unite the broken ends. Fragments of dead bone and the areas of bone closest to the break have been removed and replaced.

FOCUS

A Classification of Fractures

Fractures are classified according to their external appearance, the site of the fracture, and the nature of the crack or break in the bone. Important fracture types are indicated below, with representative X-rays. Many fractures fall into more than one category. For example, a Colles' fracture is a transverse fracture, but depending on the injury, it may also be a comminuted fracture that can be either open or closed. Closed, or simple, fractures are completely internal; they do not involve a break in the skin. Open, or compound, fractures project through the skin; they are more dangerous because of the possibility of infection or uncontrolled bleeding.

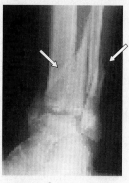

A **Pott's fracture** occurs at the ankle and affects both bones of the leg.

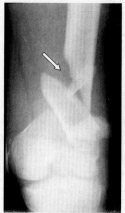

Comminuted fracture of distal femur

Comminuted fractures shatter the affected area into many, small bony fragments.

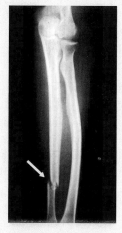

Transverse fractures break a shaft of a bone across its long axis.

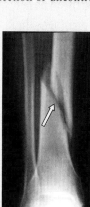

Spiral fracture of tibia

Spiral fractures, produced by twisting stresses, spread along the length of the bone.

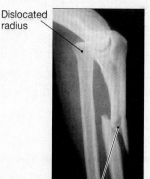

Dislocated radius

Displaced ulnar fracture

Displaced fractures produce new and abnormal arrangements of bony elements.

Nondisplaced fractures retain the normal alignment of the bone elements or fragments.

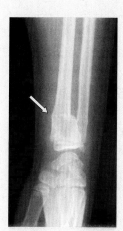

A **Colles' fracture** is a break in the distal portion of the radius, the slender bone of the forearm; it is often the result of reaching out to cushion a fall.

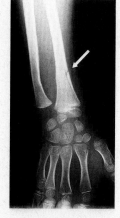

In a **greenstick fracture,** only one side of the shaft is broken, and the other is bent; this type usually occurs in children, whose long bones have yet to ossify fully.

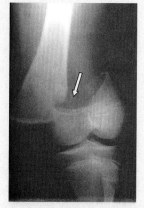

Epiphyseal fracture of femur

Epiphyseal fractures usually occur where the matrix is undergoing calcification and chondrocytes are dying. A clean transverse fracture along this line usually heals well. Fractures between the epiphysis and the epiphyseal plate can permanently halt further longitudinal growth unless carefully treated; often surgery is required.

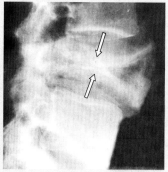

Compression fracture of vertebra

Compression fractures occur in vertebrae subjected to extreme stresses, as when landing on your seat after a fall.

6 THE SKELETAL SYSTEM

The Structure of Bone • Bone Development and Growth • Remodeling and Homeostatic Mechanisms • Aging and the Skeletal System

blood forms a swollen area called a **fracture hematoma** (*hemato-*, blood; + *tumere*, to swell), which closes off the injured blood vessels.

Step 2: Cells of the periosteum and endosteum undergo mitosis and the daughter cells migrate into the fracture zone. There they form localized thickenings—an **external callus** (*callum*, hard skin) and **internal callus**, respectively. At the center of the external callus, cells differentiate into chondrocytes and build blocks of cartilage.

Step 3: Osteoblasts replace the central cartilage of the external callus with spongy bone. When this process is complete, the external and internal calluses form a continuous brace of spongy bone at the fracture site.

Step 4: The remodeling of spongy bone at the fracture site may continue from a period of 4 months to well over a year. When the remodeling is complete, the fragments of dead bone and the spongy bone of the calluses will be gone, and only living compact bone will remain. The repair may be "good as new," with no sign that a fracture occurred, but the bone may be slightly thicker than normal at the fracture site.

Aging and the Skeletal System

The bones of the skeleton become thinner and relatively weaker as a normal part of the aging process. Inadequate ossification is called **osteopenia** (os-tē-ō-PĒ-nē-a; *penia*, lacking), and all people become slightly osteopenic as they age. The reduction in bone mass occurs because between the ages of 30 and 40, osteoblast activity begins to decline while osteoclast activity continues at normal levels. Once the reduction begins, women lose roughly 8 percent of their skeletal mass every decade, whereas men's skeletons deteriorate at about 3 percent per decade. All parts of the skeleton are not equally affected. Epiphyses, vertebrae, and the jaws lose more than their fair share, resulting in fragile limbs, a reduction in height, and the loss of teeth.

 ## OSTEOPOROSIS

Osteoporosis (os-tē-ō-por-Ō-sis; *porosus*, porous) is a condition that produces a reduction in bone mass great enough to compromise normal function. The difference between the "normal" osteopenia of aging and the clinical condition of osteoporosis is a matter of degree. It is estimated that 29 per-

TABLE 6-1 *An Introduction to Skeletal Terms*

GENERAL DESCRIPTION	ANATOMICAL TERM	DEFINITION
Elevations and projections (general)	Process	Any projection or bump
	Ramus	An extension of a bone making an angle to the rest of the structure
Processes formed where tendons or ligaments attach	Trochanter	A large, rough projection
	Tuberosity	A smaller, rough projection
	Tubercle	A small, rounded projection
	Crest	A prominent ridge
	Line	A low ridge
	Spine	A pointed process
Processes formed for articulation with adjacent bones	Head	The expanded articular end of an epiphysis, separated from the shaft by the neck
	Neck	A narrow connection between the epiphysis and diaphysis
	Condyle	A smooth, rounded articular process
	Trochlea	A smooth, grooved articular process shaped like a pulley
	Facet	A small, flat articular surface
Depressions	Fossa	A shallow depression
	Sulcus	A narrow groove
Openings	Foramen	A rounded passageway for blood vessels and/or nerves
	Fissure	An elongate cleft
	Canal	A passageway through the substance of a bone
	Sinus	A chamber within a bone, normally filled with air

cent of women between the ages of 45 and 79 can be considered osteoporotic. The increase in incidence after menopause has been linked to decreases in the production of estrogens (female sex hormones). The incidence of osteoporosis in men of the same age is estimated at 18 percent.

Because bones are more fragile, they break easily and do not repair well. Vertebrae may collapse, distorting the vertebral articulations and putting pressure on spinal nerves. Therapies that boost estrogen levels, dietary changes to elevate calcium levels in the blood, and exercise that stresses bones and stimulates osteoblast activity appear to slow but not completely prevent the development of osteoporosis.

CONCEPT CHECK QUESTIONS

Answers on page 173

❶ Why would you expect the arm bones of a weight lifter to be thicker and heavier than those of a jogger?

❷ Why is osteoporosis more common in older women than older men?

An Overview of the Skeleton

SKELETAL TERMINOLOGY

Each of the bones in the human skeleton not only has a distinctive shape but also has characteristic external features. For example, elevations or projections form where tendons and ligaments attach and where adjacent bones articulate. Depressions and openings indicate sites where blood vessels and nerves lie alongside or penetrate the bone. These external landmarks are called **bone markings**, or *surface features*. The most common terms used to describe bone markings are listed and illustrated in Table 6-1.

SKELETAL DIVISIONS

The skeletal system consists of 206 separate bones and a number of associated cartilages (Figure 6-8●). This system is divided into axial and appendicular divisions (Figure 6-9●). The **axial skeleton** forms the longitudinal axis of the body. This division's 80 bones can be subdivided into (1) the 22 bones of the **skull**, plus associated bones (6 **auditory ossicles** and the

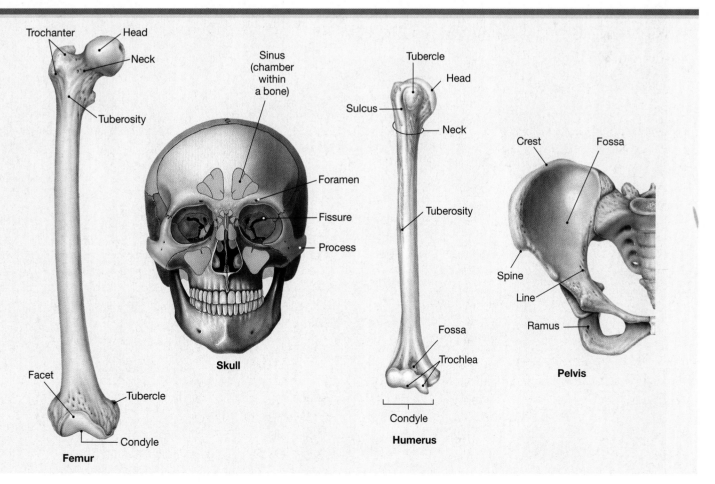

Femur — Trochanter, Head, Neck, Tuberosity, Facet, Tubercle, Condyle

Skull — Sinus (chamber within a bone), Foramen, Fissure, Process

Humerus — Tubercle, Head, Sulcus, Neck, Tuberosity, Fossa, Trochlea, Condyle

Pelvis — Crest, Fossa, Spine, Line, Ramus

6 **THE SKELETAL SYSTEM**

The Structure of Bone • Bone Development and Growth • Remodeling and Homeostatic Mechanisms • Aging and the Skeletal System

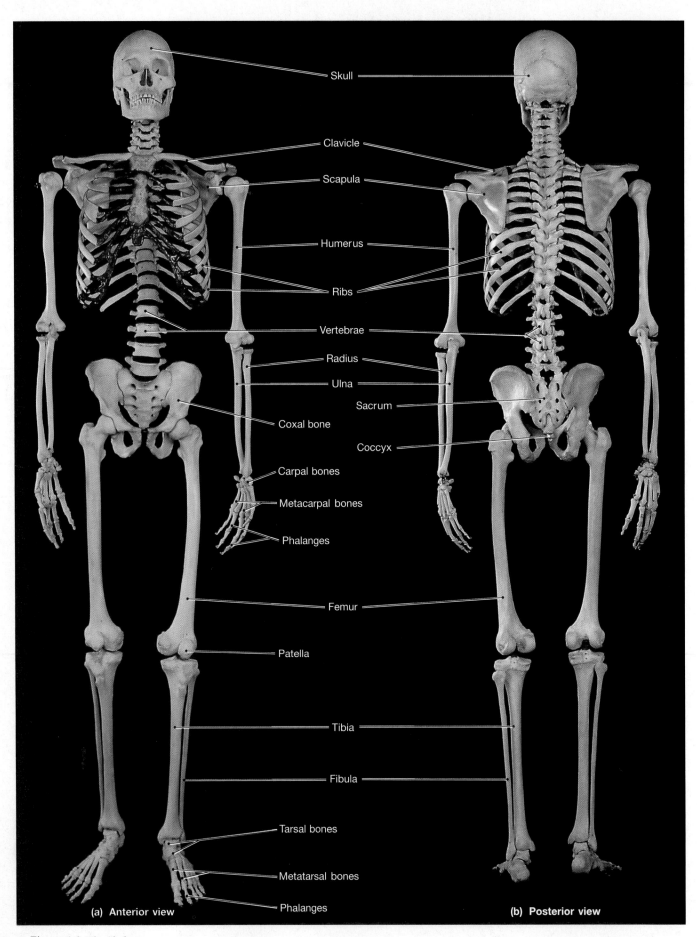

Skull

Clavicle

Scapula

Humerus

Ribs

Vertebrae

Radius

Ulna

Sacrum

Coxal bone

Coccyx

Carpal bones

Metacarpal bones

Phalanges

Femur

Patella

Tibia

Fibula

Tarsal bones

Metatarsal bones

Phalanges

(a) Anterior view

(b) Posterior view

● *Figure 6-8* **The Skeleton**

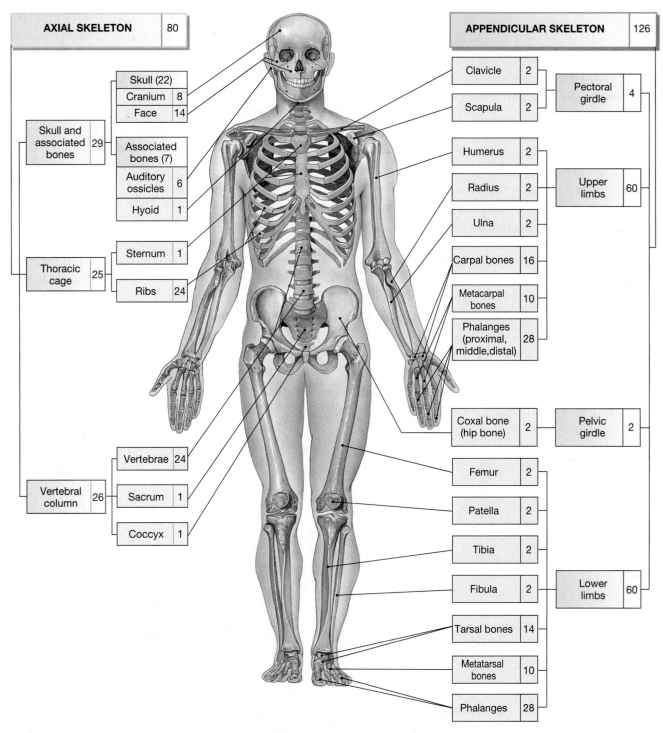

● *Figure 6-9* **The Axial and Appendicular Divisions of the Skeleton**

hyoid bone); (2) the 26 bones of the **vertebral column**; and (3) the **thoracic cage** (*rib cage*), composed of 24 **ribs** and the **sternum**.

The **appendicular skeleton** includes the bones of the limbs and those of the **pectoral** and **pelvic girdles**, which attach the limbs to the trunk. All together there are 126 appendicular bones; 32 are associated with each upper limb, and 31 with each lower limb.

The Axial Division

The **axial skeleton** creates a framework that supports and protects organ systems in the dorsal and ventral body cavities. In addition, it provides an extensive surface area for the attachment of muscles that (1) adjust the positions of the head, neck, and trunk; (2) perform respiratory movements; and (3) stabilize or position elements of the appendicular skeleton.

139

6 **THE SKELETAL SYSTEM**

The Structure of Bone • Bone Development and Growth • Remodeling and Homeostatic Mechanisms • Aging and the Skeletal System

THE SKULL

The bones of the skull protect the brain and support delicate sense organs involved with vision, hearing, balance, olfaction (smell), and gustation (taste). The skull is made up of 22 bones: 8 form the **cranium**, and 14 are associated with the face. Seven additional bones are associated with the skull: 6 *auditory ossicles*, tiny bones involved in sound detection, are encased by the *temporal bones* of the cranium, and the *hyoid bone* has ligamentous connections to the inferior surface of the skull.

The cranium encloses the **cranial cavity**, a fluid-filled chamber that cushions and supports the brain. The outer surface of the cranium provides an extensive area for the attachment of muscles that move the eyes, jaws, and head.

The Bones of the Cranium

The Frontal Bone. The **frontal bone** of the cranium forms the forehead and the superior surface of the **orbits**, the bony recesses that contain the eyes (see Figures 6-10 and 6-11● on pp. 140-141). A **supraorbital foramen** is an opening that pierces the bony ridge above each orbit, forming a passageway for blood vessels and nerves passing to or from the forehead. (Sometimes the ridge has a deep groove, called a *supraorbital notch*, rather than a foramen, but the function is the same.) Above the orbit, the frontal bone contains air-filled internal chambers that com-municate with the nasal cavity. These **frontal sinuses** make the bone lighter and produce mucus that cleans and moistens the nasal cavities (Figure 6-12b●, p. 142). The **infraorbital foramen** is an opening for a major sensory nerve from the face.

The Parietal Bones. On both sides of the skull, a **parietal** (pa-RĪ-e-tal) **bone** is posterior to the frontal bone (Figures 6-11, 6-12●). Together the parietal bones form the roof and the superior walls of the cranium. The parietal bones interlock along the **sagittal suture**, which extends along the midline of the cranium. Anteriorly, the two parietal bones articulate with the frontal bone along the **coronal suture**.

The Occipital Bone. The **occipital bone** forms the posterior and inferior portions of the cranium. Along its superior margin, the occipital bone contacts the two parietal bones at the **lambdoidal** (lam-DOYD-ul) **suture** (Figure 6-10●). Figure 6-11b● presents an inferior view of the skull, showing the orientation of the occipital bone and its relationships with other bones in the floor of the cranium. The occipital bone surrounds the **foramen magnum**, the opening that connects the cranial cavity with the spinal cavity. The spinal cord passes through the foramen magnum to connect with the inferior portion of the brain. On either side of the foramen magnum are the

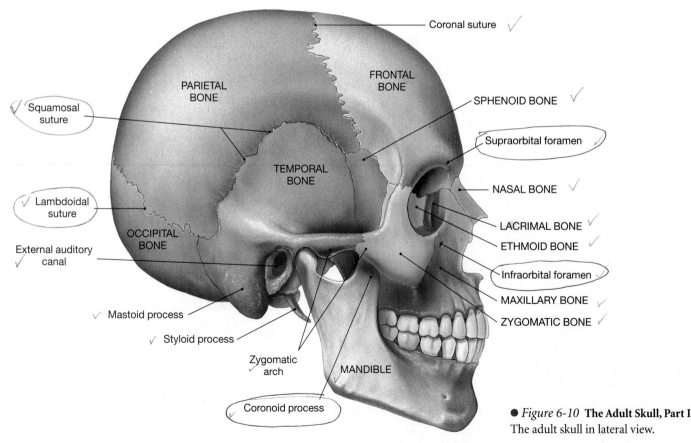

● *Figure 6-10* **The Adult Skull, Part I**
The adult skull in lateral view.

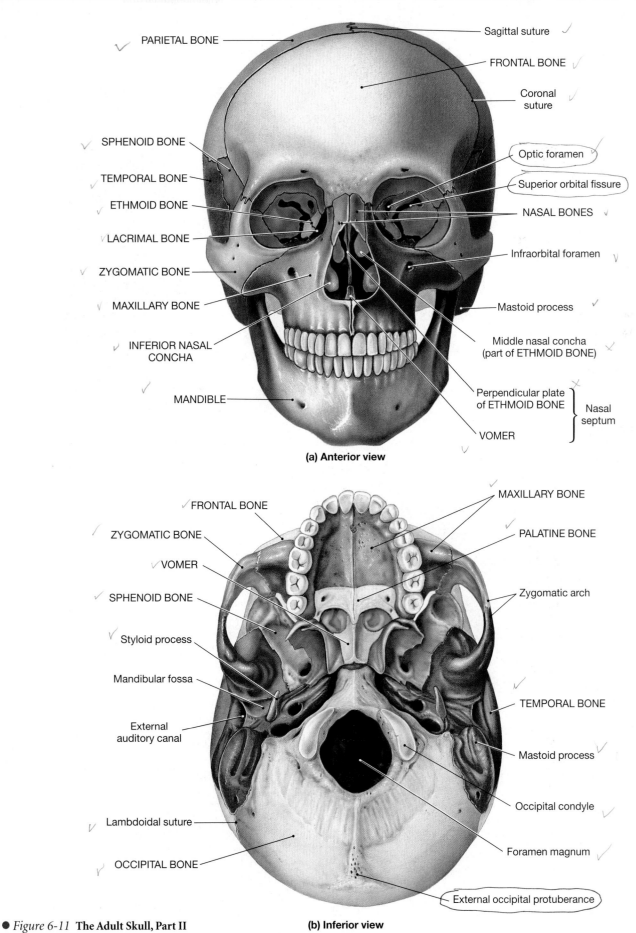

PARIETAL BONE

Sagittal suture

FRONTAL BONE

Coronal suture

SPHENOID BONE

Optic foramen

TEMPORAL BONE

Superior orbital fissure

ETHMOID BONE

NASAL BONES

LACRIMAL BONE

Infraorbital foramen

ZYGOMATIC BONE

MAXILLARY BONE

Mastoid process

INFERIOR NASAL CONCHA

Middle nasal concha (part of ETHMOID BONE)

MANDIBLE

Perpendicular plate of ETHMOID BONE

VOMER

Nasal septum

(a) Anterior view

FRONTAL BONE

MAXILLARY BONE

ZYGOMATIC BONE

PALATINE BONE

VOMER

SPHENOID BONE

Zygomatic arch

Styloid process

Mandibular fossa

External auditory canal

TEMPORAL BONE

Mastoid process

Occipital condyle

Lambdoidal suture

Foramen magnum

OCCIPITAL BONE

External occipital protuberance

(b) Inferior view

● *Figure 6-11* **The Adult Skull, Part II**

6 THE SKELETAL SYSTEM

The Structure of Bone • Bone Development and Growth • Remodeling and Homeostatic Mechanisms • Aging and the Skeletal System

● *Figure 6-12* **Sectional Anatomy of the Skull**
(**a**) A horizontal section through the skull, looking down on the floor of the cranial cavity. (**b**) A sagittal section through the skull. (**c**) A sagittal section through the skull, showing the lateral wall of the right nasal cavity.

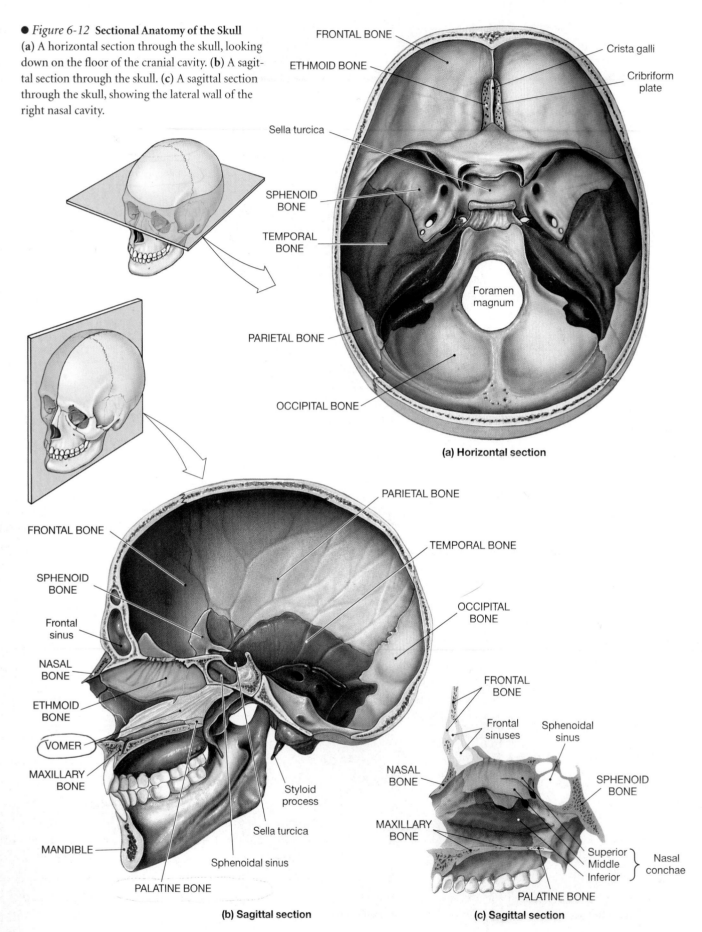

FRONTAL BONE

ETHMOID BONE

Crista galli

Cribriform plate

Sella turcica

SPHENOID BONE

TEMPORAL BONE

Foramen magnum

PARIETAL BONE

OCCIPITAL BONE

(a) Horizontal section

PARIETAL BONE

TEMPORAL BONE

FRONTAL BONE

OCCIPITAL BONE

SPHENOID BONE

Frontal sinus

NASAL BONE

ETHMOID BONE

VOMER

MAXILLARY BONE

Styloid process

Sella turcica

MANDIBLE

Sphenoidal sinus

PALATINE BONE

(b) Sagittal section

FRONTAL BONE

Frontal sinuses

Sphenoidal sinus

NASAL BONE

SPHENOID BONE

MAXILLARY BONE

Superior
Middle
Inferior
} Nasal conchae

PALATINE BONE

(c) Sagittal section

occipital condyles, the sites of articulation between the skull and the vertebral column.

The Temporal Bones. Below the parietal bones and contributing to the sides and base of the cranium are the **temporal bones**. The temporal bones contact the parietal bones along the **squamosal suture** (Figure 6-10●).

The temporal bones display a number of distinctive anatomical landmarks. One of them, the **external auditory canal**, leads to the **tympanum**, or *eardrum*. The eardrum separates the external auditory canal from the *middle ear cavity*, which contains the *auditory ossicles*, or *ear bones*. The structure and function of the tympanum, middle ear cavity, and auditory ossicles will be considered in Chapter 9.

Anterior to the external auditory canal is a transverse depression, the **mandibular fossa**, which marks the point of articulation with the lower jaw (mandible) (Figure 6-11b●). The prominent bulge just posterior and inferior to the entrance to the external auditory canal is the **mastoid process**, which provides a site for the attachment of muscles that rotate or extend the head. Next to the base of the mastoid process is the long, sharp **styloid** (STĪ-loyd; *stylos*, pillar) **process**. The styloid process is attached to ligaments that support the hyoid bone and anchors muscles associated with the tongue and pharynx.

The Sphenoid Bone. The **sphenoid** (SFĒ-noyd) **bone** forms part of the floor of the cranium. It also acts like a bridge, uniting the cranial and facial bones, and it braces the sides of the skull. The general shape of the sphenoid has been compared to that of a giant bat with wings extended. The wings can be seen most clearly on the superior surface (Figure 6-12a●). From the front (Figure 6-11a●) or side (Figure 6-10●), it is covered by other bones. Like the frontal bone, the sphenoid bone also contains a pair of sinuses, called **sphenoidal sinuses** (Figure 6-12b●).

The lateral "wings" of the sphenoid extend to either side from a central depression called the **sella turcica** (TUR-si-kuh) (Turk's saddle) (Figure 6-12a●). It encloses the pituitary gland, an endocrine organ that is connected to the inferior surface of the brain by a narrow stalk of neural tissue.

The Ethmoid Bone. The **ethmoid bone** is anterior to the sphenoid bone. The ethmoid bone consists of two honeycombed masses of bone. It forms part of the cranial floor, contributes to the medial surfaces of the orbit of each eye, and forms the roof and sides of the nasal cavity (Figure 6-11a●). A prominent ridge, the **crista galli**, or "cock's comb," projects above the superior surface of the ethmoid (Figure 6-12a●). Holes in the **cribriform plate** (*cribrum*, sieve) on either side permit passage of sensory nerves traveling from olfactory (smell) receptors in the nasal cavity to the brain.

The lateral portions of the ethmoid bone contain the **ethmoidal sinuses** that drain into the nasal cavity. Projections called the **superior** and **middle nasal conchae** (KONG-kē; *concha*, shell) extend into the nasal cavity toward the *nasal septum* (*septum*, wall) that divides the nasal cavity into left and right portions (Figures 6-11a●, 6-12b,c●). The nasal conchae slow the movement of air through the nasal cavity, allowing time for the air to become warm, moist, and clean before it reaches the delicate portions of the respiratory tract. The **perpendicular plate** of the ethmoid extends inferiorly from the crista galli, passing between the conchae to contribute to the nasal septum (Figure 6-11a●).

CONCEPT CHECK QUESTIONS
Answers on page 173

❶ The mastoid and styloid processes are found on which of the skull bones?

❷ What bone contains the depression called the sella turcica? What is located in the depression?

❸ Which bone of the cranium articulates directly with the vertebral column?

✄ The Bones of the Face

The facial bones protect and support the entrances to the digestive and respiratory tracts. They also provide areas for the attachment of muscles that control our facial expressions and help us manipulate food. Of the 14 facial bones, only the lower jaw, or mandible, is movable.

The Maxillary Bones. The **maxillary** (MAK-si-ler-ē) **bones**, or *maxillae*, articulate with all other facial bones except the mandible. The maxillary bones form (1) the floor and medial portion of the rim of the orbit (Figure 6-11a●); (2) the walls of the nasal cavity; and (3) the anterior roof of the mouth, or *hard palate* (Figure 6-12b●). The maxillary bones contain large **maxillary sinuses**, which lighten the portion of the maxillary bones above the embedded teeth. Infections of the gums or teeth can sometimes spread into the maxillary sinuses, increasing pain and making treatment more complicated.

The Palatine Bones. The paired **palatine bones** form the posterior surface of the *bony palate*, or *hard palate*—the "roof of the mouth" (Figures 6-11b●, 6-12b●). The superior surfaces of the horizontal portion of each palatine bone contributes to the floor of the nasal cavity. The superior tip of the vertical portion of each palatine bone forms a small portion of the inferior wall of the orbit.

6 THE SKELETAL SYSTEM

The Structure of Bone • Bone Development and Growth • Remodeling and Homeostatic Mechanisms • Aging and the Skeletal System

The Vomer. The inferior margin of the **vomer** articulates with the paired palatine bones. The vomer supports a prominent partition that forms part of the *nasal septum*, along with the ethmoid bone (Figures 6-11a●, 6-12b●).

The Zygomatic Bones. On each side of the skull, a **zygomatic** (zī-gō-MA-tik) **bone** articulates with the frontal bone and the maxilla to complete the lateral wall of the orbit (Figures 6-10●, 6-11a●). Along its lateral margin, each zygomatic bone gives rise to a slender bony extension that curves laterally and posteriorly to meet a process from the temporal bone. Together these processes form the **zygomatic arch**, or *cheekbone*.

The Nasal Bones. Forming the bridge of the nose midway between the orbits, the **nasal bones** articulate with the superior frontal bone and the maxillary bones (Figures 6-10●, 6-11a●).

The Lacrimal Bones. The **lacrimal** (*lacrimae*, tears) **bones** are located within the orbit on its medial surface. They articulate with the frontal, ethmoid, and maxillary bones (Figures 6-10●, 6-11a●).

The Inferior Nasal Conchae. The paired **inferior nasal conchae** project from the lateral walls of the nasal cavity (Figure 6-11a●). Their shape helps slow airflow and deflects arriving air toward the olfactory (smell) receptors located near the upper portions of the nasal cavity.

The Nasal Complex. The **nasal complex** includes the bones that form the superior and lateral walls of the nasal cavities and the sinuses that drain into them. The ethmoid bone and vomer form the bony portion of the **nasal septum** that separates the left and right portions of the nasal cavity (Figure 6-11a●). The frontal, sphenoid, ethmoid, and maxillary bones contain air-filled chambers collectively known as the **paranasal sinuses** (Figure 6-13●). In addition to reducing the weight of the skull, the paranasal sinuses help protect the respiratory system. The paranasal sinuses are connected to the nasal cavities and lined by a mucous membrane. The mucous secretions are released into the nasal cavities, and the ciliated epithelium passes the mucus back toward the throat, where it is eventually swallowed. Incoming air is humidified and warmed as it flows across this carpet of mucus. Foreign particles, such as dust and bacteria, become trapped in the sticky mucus and are swallowed. This mechanism helps protect more delicate portions of the respiratory tract.

The Mandible. The broad **mandible** is the bone of the lower jaw. It forms a broad, horizontal curve with vertical processes

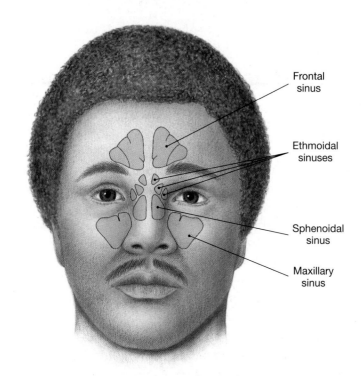

Frontal sinus

Ethmoidal sinuses

Sphenoidal sinus

Maxillary sinus

● *Figure 6-13* **The Paranasal Sinuses**

at either end. The more posterior **condylar process** ends at the *mandibular condyle*, a smoothly curved surface that articulates with the mandibular fossa of the temporal bone on that side. This articulation is quite mobile, and the disadvantage of such mobility is that the jaw can easily be dislocated. The anterior **coronoid** (kō-RŌ-noyd) **process** is the attachment point for the *temporalis muscle*, a powerful muscle that closes the jaws.

The Hyoid Bone

The small, U-shaped **hyoid bone** hangs below the skull, suspended by ligaments from the styloid processes of the temporal bones to the *lesser horns* (Figure 6-14●). The hyoid (1) serves as a base for muscles associated with the *larynx* (voicebox), tongue, and pharynx and (2) supports and stabilizes the position of the larynx.

CONCEPT CHECK QUESTIONS
Answers on page 173

❶ During baseball practice, a ball hits Casey in the eye, fracturing the bones directly above and below the orbit. Which bones were broken?

❷ What are the functions of the paranasal sinuses?

❸ Why would a fracture of the coronoid process of the mandible make it difficult to close the mouth?

❹ What symptoms would you expect to see in a person suffering from a fractured hyoid bone?

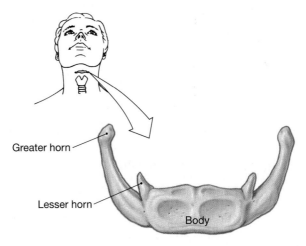

● *Figure 6-14* **The Hyoid Bone**
The stylohyoid ligaments connect the lesser horns to the styloid processes of the temporal bones.

The Skulls of Infants and Children

Many centers of ossification are involved in the formation of the skull. As the fetus develops, the individual centers begin to fuse. This fusion produces a smaller number of composite bones. For example, the sphenoid begins as 14 separate ossification centers but ends as just one. At birth, the fusion has yet to be completed, and there are two frontal bones, four occipital bones, and several sphenoid and temporal elements.

The skull organizes around the developing brain, and as the time of birth approaches, the brain enlarges rapidly. Although the bones of the skull are also growing, they fail to

keep pace, and at birth the cranial bones are connected by areas of fibrous connective tissue known as **fontanels** (fon-tah-NELZ). The fontanels, or "soft spots," are quite flexible and permit distortion of the skull without damage. Such distortion normally occurs during delivery and eases the passage of the infant along the birth canal. Figure 6-15● shows the prominent fontanels and the appearance of the skull at birth.

THE NECK AND TRUNK

The rest of the axial skeleton, the *vertebral column*, is subdivided on the basis of vertebral structure (Figure 6-16●). The **vertebral column**, or **spine**, consists of 26 bones; the **vertebrae** (24), the *sacrum* (SĀ-krum), and the *coccyx* (KOK-siks), or *tailbone*. The **cervical region** of the vertebral column consists of the seven **cervical vertebrae** of the neck (abbreviated as C_1 to C_7). The cervical region begins at the articulation of C_1 with the occipital condyles of the skull and extends inferiorly to the articulation of C_7 with the first thoracic vertebra. The **thoracic region** consists of the 12 **thoracic vertebrae** (T_1 to T_{12}), each of which articulates with one or more pairs of ribs. The **lumbar region** contains the five **lumbar vertebrae** (L_1 to L_5). The first lumbar vertebra articulates with T_{12}, and the fifth lumbar vertebra articulates with the **sacrum**. The sacrum is a single bone formed by the fusion of the five embryonic vertebrae of the **sacral region**. The **coccygeal region** is made up of the small **coccyx**, which also consists of fused vertebrae. The total length of the adult vertebral column averages 71 cm (28 in.).

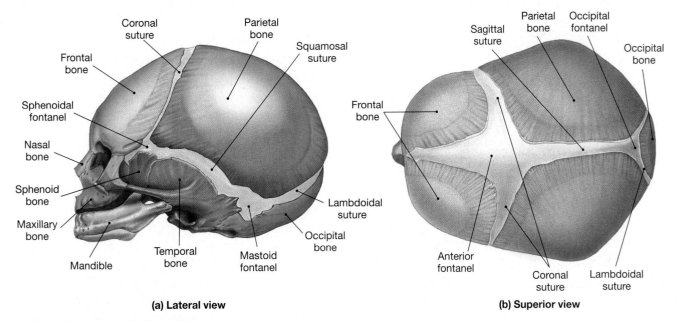

(a) Lateral view

(b) Superior view

● *Figure 6-15* **The Skull of an Infant**
An infant skull contains more individual bones than that of an adult skull. Many of these bones eventually fuse to create the adult skull. The flat bones of the skull are separated by areas of fibrous connective tissue called fontanels, which allow for cranial expansion and distortion during birth. By about age 4, these areas disappear, and skull growth is completed.

6 THE SKELETAL SYSTEM

The Structure of Bone • Bone Development and Growth • Remodeling and Homeostatic Mechanisms • Aging and the Skeletal System

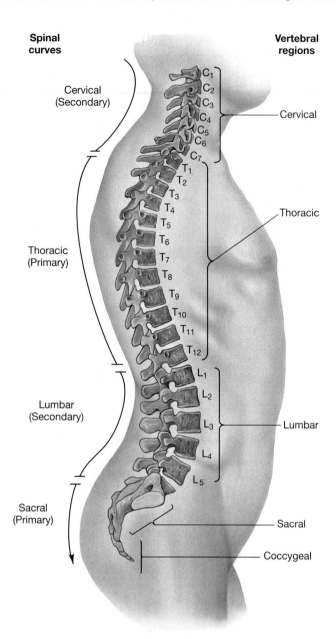

Spinal curves

Cervical (Secondary)

Thoracic (Primary)

Lumbar (Secondary)

Sacral (Primary)

Vertebral regions

Cervical

Thoracic

Lumbar

Sacral

Coccygeal

C₁ C₂ C₃ C₄ C₅ C₆ C₇ T₁ T₂ T₃ T₄ T₅ T₆ T₇ T₈ T₉ T₁₀ T₁₁ T₁₂ L₁ L₂ L₃ L₄ L₅

● *Figure 6-16* **The Vertebral Column**
The major divisions of the vertebral column, showing the four spinal curves.

Spinal Curvature

The vertebral column is not straight and rigid. A side view of the spinal column shows four **spinal curves** (Figure 6-16●). The *thoracic* and *sacral curves* are called **primary curves** because they appear late in fetal development, as the thoracic and abdominal organs enlarge. The *cervical* and *lumbar curves*, known as **secondary curves**, do not appear until months after birth. The cervical curve develops as the infant learns to balance the head upright, and the lumbar curve develops with the ability to stand. When standing, the weight of the body must be transmitted through the spinal column to the pelvic girdle

and ultimately to the legs. Yet most of the body weight lies in front of the spinal column. The secondary curves bring that weight in line with the body axis. All four spinal curves are fully developed by the time a child is 10 years old.

Several abnormal distortions of spinal curvature may appear during childhood and adolescence. Examples are *kyphosis* (kī-FŌ-sis; exaggerated thoracic curvature), *lordosis* (lor-DŌ-sis; exaggerated lumbar curvature), and *scoliosis* (skō-lē-Ō-sis; an abnormal lateral curvature).

Vertebral Anatomy

Figure 6-17● shows representative vertebrae from different regions of the vertebral column. Features shared by all vertebrae include a *vertebral body*, *vertebral arch*, and *articular processes*. The more massive, weight-bearing portion of a vertebra is called the **vertebral body**. The bony faces of the vertebral bodies usually do not contact one another, because an **intervertebral disc** of fibrocartilage lies between them. Intervertebral discs are not found in the sacrum and coccyx, where the vertebrae have fused, or between the first and second cervical vertebrae.

The **vertebral arch** forms the posterior margin of each **vertebral foramen** (plural, *foramina*). Together, the vertebral foramina of successive vertebrae form the **vertebral canal**, which encloses the spinal cord. The vertebral arch has walls, called *pedicles* (PE-di-kulz), and a roof formed by flat *laminae* (LA-mi-nē; singular, *lamina*, a thin plate). *Transverse processes* projecting laterally or dorsolaterally from the pedicles serve as sites for muscle attachment. A *spinous process*, or spinal process, projects posteriorly from where the laminae fuse together. The spinous processes form the bumps that can be felt along the midline of your back.

The articular processes arise at the junction between the pedicles and laminae. Each side of a vertebra has a *superior* and *inferior process*. The articular processes of successive vertebrae contact one another at the **articular facets**. Gaps between the pedicles of successive vertebrae, the *intervertebral foramina*, permit the passage of nerves running to or from the enclosed spinal cord.

Although all vertebrae have many similar characteristics, some regional differences reflect differences in function. The structural differences among the vertebrae are discussed next.

The Cervical Vertebrae

The seven cervical vertebrae extend from the head to the thorax. A typical cervical vertebra is illustrated in Figure 6-17a●. Notice that the body of the vertebra is relatively small compared with the size of the vertebral foramen. At this level the spinal cord still contains most of the axons that connect the brain to the rest of the body. From the first thoracic vertebra

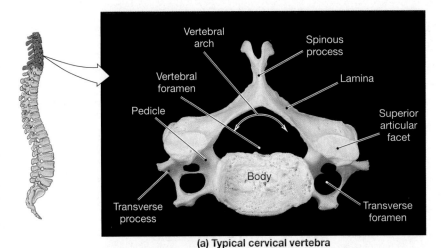

Vertebral arch

Spinous process

Vertebral foramen

Lamina

Pedicle

Superior articular facet

Body

Transverse process

Transverse foramen

(a) Typical cervical vertebra

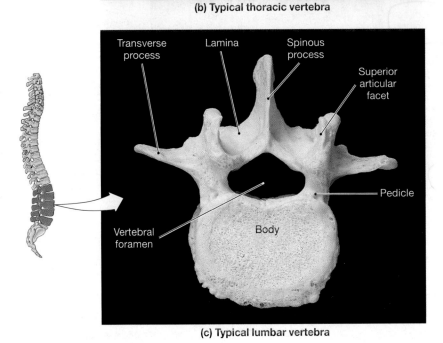

Lamina

Vertebral foramen

Spinous process

Facet for rib articulation

Transverse process

Superior articular facet

Pedicle

Body

Facets for rib articulation

(b) Typical thoracic vertebra

Transverse process

Lamina

Spinous process

Superior articular facet

Pedicle

Vertebral foramen

Body

(c) Typical lumbar vertebra

● *Figure 6-17* **Typical Vertebrae of the Cervical, Thoracic, and Lumbar Regions**
Each vertebra is shown in superior view.

to the sacrum, the diameter of the spinal cord decreases, and so does the size of the vertebral foramen. At the same time, the vertebral bodies gradually enlarge, because they must bear more weight.

Distinctive features of a typical cervical vertebra include: (1) an oval, concave vertebral body; (2) a relatively large vertebral foramen; (3) a stumpy spinous process, usually with a notched tip; and (4) round **transverse foramina** within the transverse processes. These foramina protect important blood vessels supplying the brain.

The first two vertebrae have unique characteristics that allow for specialized movements. The **atlas** (C_1) holds up the head, articulating with the occipital condyles of the skull. It is named after Atlas, who, according to Greek myth, holds the world on his shoulders. The articulation between the occipital condyles and the atlas permits you to nod (as when indicating "yes"). The atlas in turn forms a pivot joint with the **axis** (C_2) through a projection on the axis called the **dens** (*denz*; tooth), or *odontoid process*. This articulation, which permits rotation (as when shaking your head to indicate "no"), is shown in Figure 6-18●.

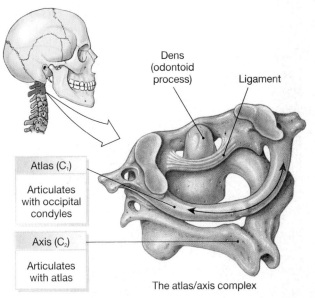

Dens (odontoid process)

Ligament

Atlas (C_1)

Articulates with occipital condyles

Axis (C_2)

Articulates with atlas

The atlas/axis complex

● *Figure 6-18* **The Atlas and Axis**
The articulation between the atlas (C_1) and the axis (C_2).

6

THE SKELETAL SYSTEM

The Structure of Bone • Bone Development and Growth • Remodeling and Homeostatic Mechanisms • Aging and the Skeletal System

The Thoracic Vertebrae

There are 12 thoracic vertebrae (Figure 6-17b●). Distinctive features of a thoracic vertebra include: (1) a characteristic heart-shaped body that is more massive than that of a cervical vertebra; (2) a large, slender spinous process that points inferiorly; and (3) articular surfaces on the body and, in most cases, on the transverse processes for articulation with one or more pairs of ribs.

The Lumbar Vertebrae

The distinctive features of lumbar vertebrae (Figure 6-17c●) include: (1) a vertebral body that is thicker and more oval than that of a thoracic vertebra; (2) a relatively massive, stumpy spinous process that projects posteriorly, providing surface area for the attachment of the lower back muscles; and (3) bladelike transverse processes that lack articulations for ribs.

The lumbar vertebrae are the most massive and least mobile, for they support most of the body weight. As you increase the weight on the vertebrae, the intervertebral discs become increasingly important as shock absorbers. The lumbar discs, which are subjected to the most pressure, are the thickest of all. The lumbar articulations restrict the stresses on the discs by limiting vertebral motion.

The Sacrum and Coccyx

The sacrum consists of the fused elements of five sacral vertebrae. It protects the reproductive, digestive, and excretory organs, and attaches the axial skeleton to the appendicular skeleton by means of paired articulations with the pelvic girdle. The broad surface area of the sacrum provides an extensive area for the attachment of muscles, especially those responsible for leg movement. Figure 6-19● shows the posterior and anterior surfaces of the sacrum.

Because the sacrum resembles a triangle, the narrow, caudal portion is called the **apex**, and the broad superior surface is the **base**. The *articular processes* form articulations with the last lumbar vertebra. The **sacral canal** begins between those processes and extends the length of the sacrum. Nerves and the membranes that line the vertebral canal in the spinal cord continue into the sacral canal. The inferior end of the sacral canal, the *sacral hiatus* (hī-Ā-tus) is covered by connective tissues. A prominent bulge at the anterior tip of the base, the **sacral promontory**, is an important landmark in females during pelvic examinations and during labor and delivery.

Before birth, five vertebrae fuse to form the sacrum, and their spinal processes form a series of elevations along the *median sacral crest*. Four pairs of **sacral foramina** open on either side of the median sacral crest. Along each lateral border of the sacrum, a thickened, flattened area marks the *sacroiliac joint*, the site of articulation with the *coxae* (hip bones).

The coccyx provides an attachment site for a muscle that closes the anal opening. The 3–5 (most often 4) coccygeal vertebrae do not complete their fusion until late in adulthood. In elderly people, the coccyx may also fuse with the sacrum.

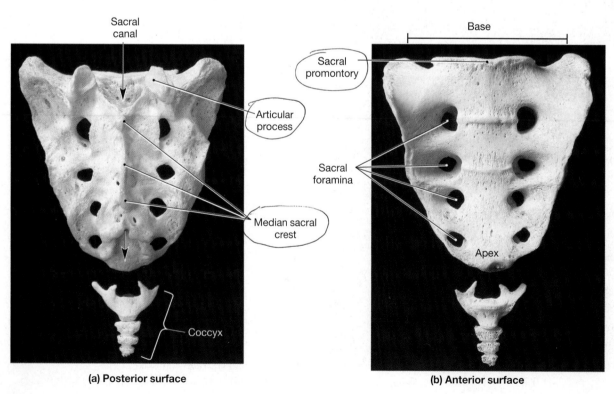

Sacral canal

Articular process

Median sacral crest

Coccyx

(a) Posterior surface

Base

Sacral promontory

Sacral foramina

Apex

(b) Anterior surface

● *Figure 6-19* **The Sacrum and Coccyx**

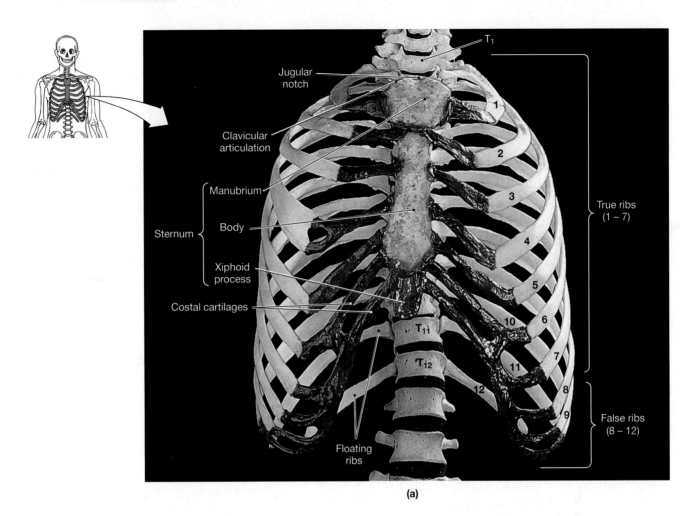

(a)

The Thorax

The skeleton of the chest, or thorax, consists of the thoracic vertebrae, the ribs, and the sternum. The ribs and the sternum form the thoracic cage, or rib cage, and establish the contours of the thoracic cavity (Figure 6-20●). The thoracic cage protects the heart, lungs, and other internal organs and serves as a base for muscles involved with respiration.

The Ribs and Sternum. Ribs, or *costal bones*, are elongate, flattened bones that originate on or between the thoracic vertebrae and end in the wall of the thoracic cavity. There are 12 pairs of ribs. The first seven pairs are called **true ribs**. These ribs reach the anterior body wall and are connected to the sternum by separate cartilaginous extensions, the **costal cartilages**. Ribs 8–12 are called the **false ribs**, because they do not attach directly to the sternum. The costal cartilages of ribs 8–10 fuse together. This fused cartilage merges with the costal cartilage of rib 7 before it reaches the sternum. The last two pairs of ribs are called **floating ribs**, because they have no connection with the sternum.

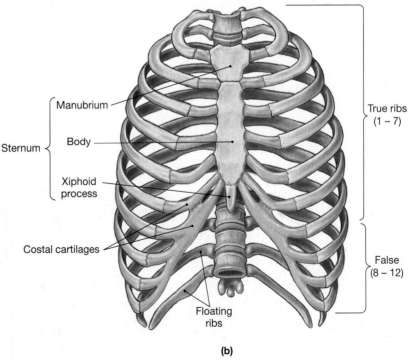

(b)

● *Figure 6-20* **The Thoracic Cage**
(a) An anterior view of the rib cage and sternum from a skeleton.
(b) A diagrammatic view.

6

THE SKELETAL SYSTEM

The Structure of Bone • Bone Development and Growth • Remodeling and Homeostatic Mechanisms • Aging and the Skeletal System

The adult sternum has three parts. The broad, triangular **manubrium** (ma-NOO-brē-um) articulates with the clavicles of the appendicular skeleton and with the cartilages of the first pair of ribs. The *jugular notch* is the shallow indentation on the superior surface of the manubrium. The elongated **body** ends at the slender **xiphoid** (ZĪ-foyd) **process.** Ossification of the sternum begins at six to ten different centers, and fusion is not completed until at least age 25. The xiphoid process is usually the last of the sternal components to ossify and fuse. Impact or strong pressure can drive it into the liver, causing severe damage. Cardiopulmonary resuscitation (CPR) training strongly emphasizes the proper positioning of the hand to reduce the chances of breaking the xiphoid process or ribs.

With their complex musculature, dual articulations at the vertebrae, and flexible connection to the sternum, the ribs are quite mobile. Because they are curved, their movements affect both the width and the depth of the thoracic cage, increasing or decreasing its volume.

CONCEPT CHECK QUESTIONS

Answers on page 173

❶ Joe suffered a hairline fracture at the base of the dens. Which bone is fractured, and where would you find it?

❷ In adults, five large vertebrae fuse to form what single structure?

❸ Why are the bodies of the lumbar vertebrae so large?

❹ How could you distinguish between true ribs and false ribs?

❺ Improper administration of CPR (cardiopulmonary resuscitation) could result in a fracture of which bone?

The Appendicular Division

The appendicular skeleton includes the bones of the upper and lower limbs and the pectoral and pelvic girdles that connect the limbs to the trunk.

THE PECTORAL GIRDLE

Each upper limb articulates with the trunk at the pectoral girdle, or *shoulder girdle.* The pectoral girdle consists of a broad, flat **scapula** (*shoulder blade*) and the slender, curving **clavicle** (*collarbone*). The clavicle articulates with the manubrium of the sternum; this is the only direct connection between the pectoral girdle and the axial skeleton. Skeletal muscles support and position the scapula, which has no bony or ligamentous connections to the thoracic cage.

Movements of the clavicle and scapula position the shoulder joint and provide a base for arm movement. Once the shoulder joint is in position, muscles that originate on the pectoral girdle help to move the arm. The surfaces of the scapula and clavicle are therefore extremely important as sites for muscle attachment.

The Clavicle

The S-shaped clavicle bone, shown in Figure 6-21●, articulates with the manubrium of the sternum at its *sternal end* and with the *acromion* (a-KRŌ-mē-on), a process of the scapula, at its *acromial end.* The smooth superior surface of the clavicle lies just beneath the skin. The rough inferior surface of the acromial end is marked by prominent lines and tubercles, attachment sites for muscles and ligaments.

The clavicles are relatively small and fragile, therefore, fractures are fairly common. For example, you can fracture a clavicle as the result of a simple fall if you land on your hand with your arm outstretched. Fortunately, most fractures of the clavicle heal rapidly without a cast.

The Scapula

The anterior surface of the body of each scapula forms a broad triangle bounded by the **superior, medial,** and **lateral borders** (Figure 6-22●). Muscles that position the scapula attach along these edges. The intersection of the lateral and superior bor-

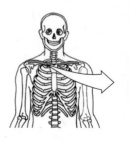

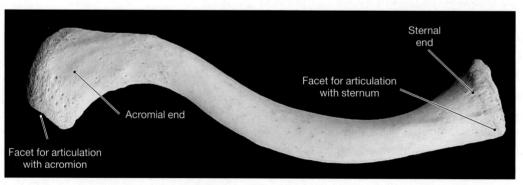

● *Figure 6-21* **The Clavicle**
A superior view of the right clavicle.

● *Figure 6-22* **The Scapula**
Major landmarks on the right scapula.

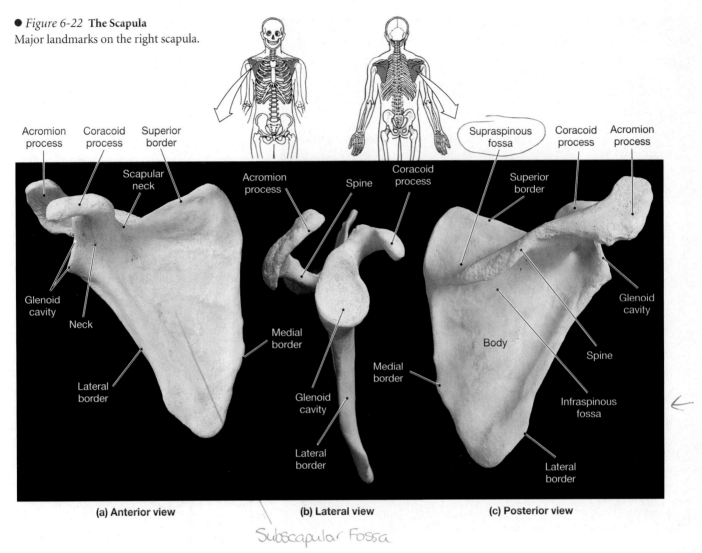

(a) Anterior view

(b) Lateral view

Subscapular Fossa

(c) Posterior view

ders thickens into the shallow, cup-shaped **glenoid cavity**, or *glenoid fossa* (FOS-sah). ⟳ p. 136 At the glenoid cavity, the scapula articulates with the proximal end of the humerus to form the *shoulder joint*. The bone surrounding the glenoid cavity attaches to the body of the scapula at the *scapular neck*.

Figure 6-22b● shows a lateral view of the scapula and the two large processes that extend over the glenoid fossa. The smaller, anterior projection is the **coracoid** (KOR-uh-koyd) **process**. The **acromion** is the larger, posterior projection. If you run your fingers along the superior surface of the shoulder joint, you will feel this process. The acromion articulates with the distal end of the clavicle.

The **scapular spine** divides the posterior surface of the scapula into two regions (Figure 6-22c●). The area superior to the spine is the **supraspinous fossa** (*supra-*, above); the *supraspinatus muscle* attaches here. The region below the spine is the **infraspinous fossa** (*infra-*, beneath), home of the infraspinatus muscle. Both muscles are attached to the humerus, the proximal bone of the upper limb.

THE UPPER LIMB

Each upper limb consists of the arm and forearm. The arm contains a single bone, the **humerus**, which extends from the scapula to the elbow. At its proximal end, the round **head** of the humerus articulates with the scapula. At its distal end, it articulates with the bones of the forearm, the *radius* and *ulna*.

The Humerus

The prominent **greater tubercle** of the humerus is a rounded projection near the lateral surface of the proximal head (Figure 6-23●). It establishes the lateral contour of the shoulder. The **lesser tubercle** lies more anteriorly, separated from the greater tubercle by a deep groove. Muscles are attached to both tubercles, and a large tendon runs along the groove. The *anatomical neck* lies between the tubercles and below the surface of the head. Distal to the tubercles, the narrow *surgical neck* corresponds to the region of growing bone, the *epiphyseal plate*. ⟳ p. 132 It earned its name by being a common fracture site.

151

6 THE SKELETAL SYSTEM

The Structure of Bone • Bone Development and Growth • Remodeling and Homeostatic Mechanisms • Aging and the Skeletal System

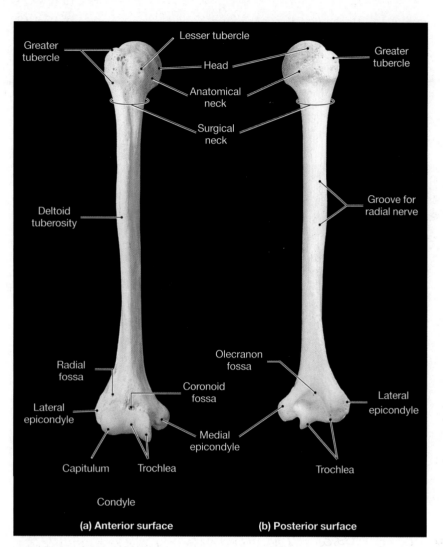

Greater tubercle

Lesser tubercle

Head

Greater tubercle

Anatomical neck

Surgical neck

Deltoid tuberosity

Groove for radial nerve

Radial fossa

Olecranon fossa

Coronoid fossa

Lateral epicondyle

Lateral epicondyle

Capitulum

Medial epicondyle

Trochlea

Trochlea

Condyle

(a) Anterior surface

(b) Posterior surface

● *Figure 6-23* **The Humerus**
Major landmarks on the right humerus.

The proximal shaft of the humerus is round in section. The elevated **deltoid tuberosity** that runs along the lateral border of the shaft is named after the *deltoid muscle* that attaches to it.

Distally, the posterior surface of the shaft flattens and the humerus expands to either side, forming a broad triangle. **Medial** and **lateral epicondyles** project to either side, providing additional surface area for muscle attachment, and the smooth, articular **condyle** dominates the inferior surface of the humerus.

A low ridge crosses the condyle, dividing it into two distinct regions. The **trochlea** is the large medial portion shaped like a spool or pulley (*trochlea*, a pulley). The trochlea extends from the base of the **coronoid** (KOR-ō-noyd; *corona*, crown) **fossa** on the anterior surface to the **olecranon fossa** on the

posterior surface. These depressions accept projections from the surface of the ulna as the elbow reaches its limits of motion. The **capitulum** forms the lateral region of the condyle. A shallow **radial fossa** proximal to the capitulum accommodates a small projection on the radius.

The Radius and Ulna

The **radius** and **ulna** are the bones of the forearm. In the anatomical position, the radius lies along the lateral (thumb) side of the forearm. The ulna forms the medial support of the forearm. Their structure is shown in Figure 6-24a●.

The **olecranon** (ō-LEK-ruh-non) of the ulna is the point of the elbow. On its anterior surface, the **trochlear notch** articulates with the trochlea of the humerus at the elbow joint. The olecranon forms the superior lip of the notch, and the **coronoid process** forms its inferior lip. At the limit of *extension*, with the arm and forearm forming a straight line, the olecranon swings into the olecranon fossa on the posterior surface of the humerus. At the limit of *flexion*, with the arm and forearm forming a V, the coronoid process projects into the coronoid fossa on the anterior surface of the humerus. Lateral to the coronoid process, a smooth **radial notch** accommodates the head of the radius.

A fibrous sheet connects the lateral margin of the ulna to the radius along its length. The ulnar shaft ends at a disc-shaped head whose posterior margin supports a short **styloid process**. The distal end of the ulna is separated from the wrist joint by a pad of cartilage, and only the large distal portion of the radius participates in the wrist joint. The styloid process of the radius assists in the stabilization of the joint by preventing lateral movement of the bones of the wrist (*carpal bones*).

A narrow *neck* extends from the head of the radius to the **radial tuberosity**, which marks the attachment site of the *biceps brachii*, a large muscle on the anterior surface of the arm that flexes the forearm. The disc-shaped head of the radius articulates with the capitulum of the humerus at the elbow joint and with the ulna at the *radial notch*. This proximal articulation with the ulna allows the radius to roll across the ulna, rotating the palm in a movement known as *pronation* (Figure 6-24b●). The reverse movement, which returns the forearm to the anatomical position, is called *supination*.

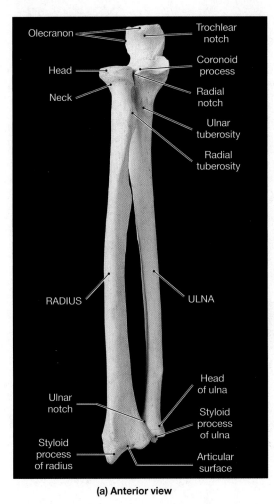

Olecranon
Head
Neck
Trochlear notch
Coronoid process
Radial notch
Ulnar tuberosity
Radial tuberosity
RADIUS
ULNA
Ulnar notch
Head of ulna
Styloid process of ulna
Styloid process of radius
Articular surface

(a) Anterior view

ULNA   RADIUS

(b) Pronation: Anterior view

● *Figure 6-24* **The Radius and Ulna**
(**a**) An anterior view. (**b**) Notice the changes that occur during pronation.

The Wrist and Hand

There are 27 bones in the hand, supporting the wrist, palm, and fingers (Figure 6-25●). The eight bones of the wrist, or *carpus*, form two rows. There are four proximal **carpal bones:** (1) the *scaphoid bone*, (2) the *lunate bone*, (3) the *triquetrum bone*, and (4) the *pisiform* (PIS-i-form) *bone*. There are also four distal carpal bones: (1) the *trapezium*, (2) the *trapezoid bone*, (3) the *capitate bone*, and (4) the *hamate bone*. Joints between the carpal bones permit a limited degree of sliding and twisting.

Five **metacarpal** (met-a-KAR-pal) **bones** articulate with the distal carpal bones and form the palm of the hand. The metacarpal bones in turn articulate with the finger bones, or **phalanges** (fa-LAN-jēz; singular, *phalanx*). Each hand has 14 phalangeal bones. Four of the fingers contain three phalanges

each (proximal, middle, and distal), but the thumb or **pollex** (POL-eks), has only two phalanges (proximal and distal).

THE PELVIC GIRDLE

The pelvic girdle articulates with the thigh bones. Because of the stresses involved in weight bearing and locomotion, the bones of the pelvic girdle and lower limbs are more massive than those of the pectoral girdle and upper limbs. The pelvic girdle is also much more firmly attached to the axial skeleton.

The pelvic girdle consists of two large hip bones, or **coxae**. Each coxa forms by the fusion of three

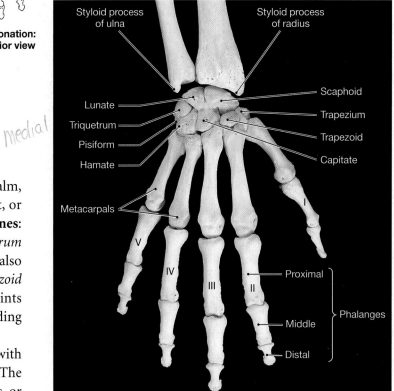

Styloid process of ulna
Styloid process of radius
Lunate
Triquetrum
Pisiform
Hamate
Scaphoid
Trapezium
Trapezoid
Capitate
Metacarpals
medial
lateral
I
V
IV
III
II
Proximal
Middle
Distal
Phalanges

● *Figure 6-25* **Bones of the Wrist and Hand**
A posterior view.

6 THE SKELETAL SYSTEM

The Structure of Bone • Bone Development and Growth • Remodeling and Homeostatic Mechanisms • Aging and the Skeletal System

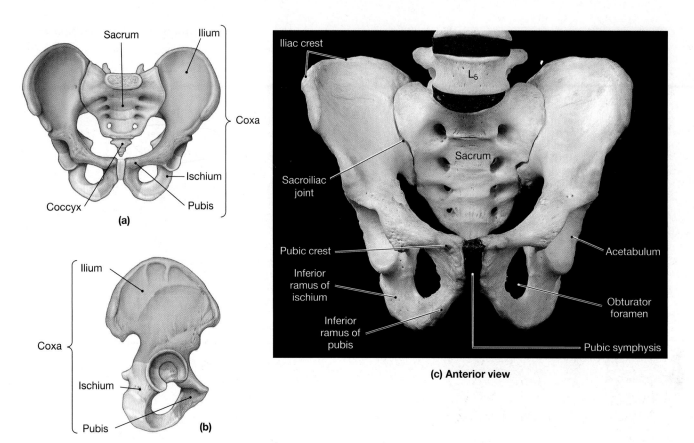

● Figure 6-26 The Pelvis
(a) Anterior view of the components of the pelvis. (b) Lateral view of the components of the pelvis. (c) An anterior view of the pelvis of an adult male.

bones, an **ilium** (IL-ē-um), an **ischium** (IS-kē-um), and a **pubis** (PŪ-bis) (Figure 6-26a, b●). Dorsally, the hipbones articulate with the sacrum at the **sacroiliac joint**. Ventrally the coxae are connected at the *pubic symphysis*, a fibrocartilage pad. At the hip joint on either side, the head of the femur (thighbone) articulates with the curved surface of the **acetabulum** (a-se-TAB-ū-lum; *acetabulum*, a vinegar cup) (Figure 6-26c●).

The Hip

The ilium is the most superior and largest coxal bone. Above the acetabulum, the ilium forms a broad, curved surface that provides an extensive area for the attachment of muscles, tendons, and ligaments. The superior margin of the ilium, the **iliac crest**, marks the sites of attachments of both ligaments and muscles. Near the superior and posterior margin of the acetabulum, the ilium fuses with the ischium. The inferior surface of the ischium, the *ischial tuberosity*, supports the body's weight when sitting.

The fusion of a narrow branch of the ischium with a branch of the pubis completes the encirclement of the **obturator** (OB-tū-rā-tor) **foramen**. This space is closed by a sheet of collagen

fibers whose inner and outer surfaces provide a firm base for the attachment of muscles and visceral structures.

The anterior and medial surface of the pubis contains a roughened area that marks the **pubic symphysis**, an articulation with the pubis of the opposite side. The pubic symphysis limits movement between the two pubic bones.

The Pelvis

The **pelvis** consists of the two hip bones, the sacrum, and the coccyx (see Figure 6-26a●). It is thus a composite structure that includes portions of both the appendicular and axial skeletons. An extensive network of ligaments connects the sacrum with the iliac crests, the ischia, and the pubic bones. Other ligaments tie the ilia to the posterior lumbar vertebrae. These interconnections increase the structural stability of the pelvis.

The shape of the pelvis of a female is somewhat different from that of a male (Figure 6-27●). Some of the differences are the result of variations in body size and muscle mass. For example, in females, the pelvis is generally smoother, lighter in weight, and has less prominent markings. Other differences are adaptations for childbearing and are necessary to support the weight of the developing fetus and to ease passage of the

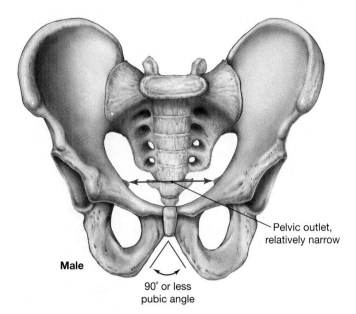

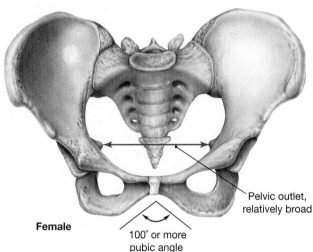

● *Figure 6-27* **Gender Differences in the Anatomy of the Pelvis**

newborn through the pelvis during delivery. Compared to males, females have a relatively broad, low pelvis, enlarged *pelvic outlet* (the opening bounded by the coccyx, the ischial tuberosities, and the pubic symphysis), and a broader *pubic angle* (the inferior angle between the pubic bones).

THE LOWER LIMB

The skeleton of each lower limb consists of a *femur*, the bone of the thigh; a *patella*, the kneecap; a *tibia* and *fibula*, the bones of the leg; and the bones of the ankle and foot.

The Femur

The **femur**, or *thighbone*, is the longest and heaviest bone in the body (Figure 6-28●). The rounded epiphysis, or head, of the femur articulates with the pelvis at the acetabulum. The **greater** and **lesser trochanters** are large, rough projections that extend laterally from the juncture of the neck and the shaft. Both trochanters develop where large tendons attach to the femur. On the posterior surface of the femur, a prominent elevation, the **linea aspera**, marks the attachment of powerful muscles that pull the shaft of the femur toward the midline, a movement called *adduction* (*ad-*, toward + *duco*, to lead).

The proximal shaft of the femur is round in cross section. Moving distally, the shaft becomes more flattened and ends in two large **epicondyles** (**lateral** and **medial**). The inferior surfaces of the epicondyles form the *lateral* and *medial condyles*. The articular condyles merge anteriorly to produce an articular surface with elevated lateral borders.

This is the **patellar surface** over which the **patella** (*kneecap*) glides. ⟶ p. 139 The patella bone forms within the tendon of the *quadriceps femoris*, a group of muscles that straighten the knee.

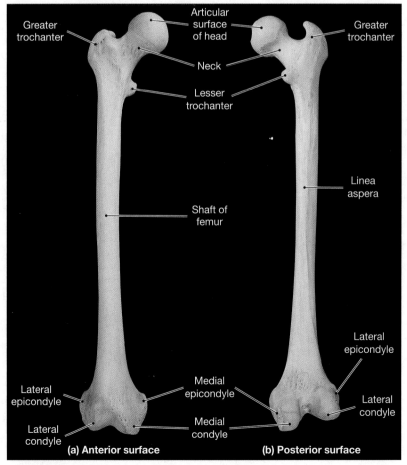

(a) **Anterior surface**   (b) **Posterior surface**

● *Figure 6-28* **The Femur**
Bone markings on the right femur.

6 THE SKELETAL SYSTEM

The Structure of Bone • Bone Development and Growth • Remodeling and Homeostatic Mechanisms • Aging and the Skeletal System

The Tibia and Fibula

The **tibia** (TI-bē-uh), or *shinbone*, is the large medial bone of the leg (Figure 6-29●). The lateral and medial condyles of the femur articulate with the *lateral* and *medial condyles* of the **tibia**. The *patellar ligament* connects the patella to the **tibial tuberosity** just below the knee joint.

A projecting **anterior crest** extends almost the entire length of the anterior tibial surface. The tibia broadens at its distal end into a large process, the **medial malleolus** (ma-LĒ-ō-lus; *malleolus*, hammer). The inferior surface of the tibia forms a joint with the proximal bone of the ankle; the medial malleolus provides medial support for the ankle.

The slender **fibula** (FIB-ū-luh) parallels the lateral border of the tibia. The fibula articulates with the tibia inferior to the lateral tibial condyle. The fibula does not articulate with the femur or help transfer weight to the ankle and foot. However, it is an important surface for muscle attachment, and the distal **lateral malleolus** provides lateral stability to the ankle. A fibrous membrane extending between the two bones helps stabilize their relative positions and provides additional surface area for muscle attachment.

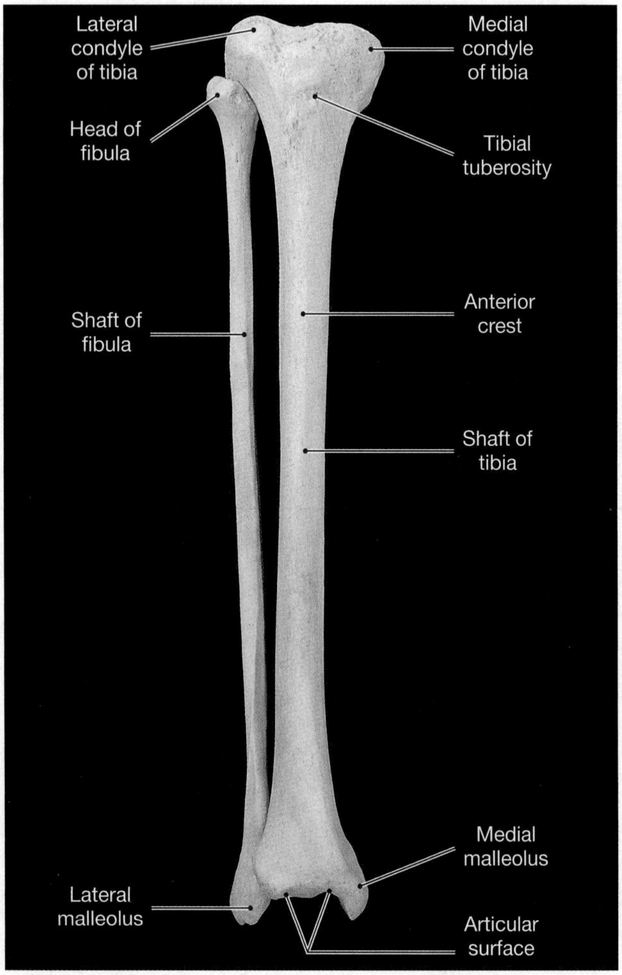

The Ankle and Foot

The ankle, or *tarsus*, includes seven separate **tarsal bones**: (1) the *talus*, (2) the *calcaneus*, (3) the *navicular bone*, (4) the *cuboid bone*, and (5–7) the *first*, *second*, and *third cuneiform bones* (Figure 6-30●). Only the proximal tarsal bone, the **talus**, articulates with the tibia and fibula. The talus then passes the weight to the ground through other bones of the foot.

When you are standing normally, most of your weight is transmitted to the ground through the talus to the large **calcaneus** (kal-KĀ-nē-us), or *heel bone*. The posterior projection of the calcaneus is the attachment site for the *calcaneal tendon*, or *Achilles tendon*, which arises from the calf muscles. These muscles raise the heel and depress the sole, such as when you stand on tiptoes. The rest of the body weight is passed through the cuboid bone and cuneiform bones to the **metatarsal bones**, which support the sole of the foot.

The basic organizational pattern at the metatarsals and phalanges of the foot resembles that of the hand. The metatarsals are numbered by Roman numerals I to V from medial to lateral, and their distal ends form the ball of the foot. The same number of phalanges present in the thumb (2) and fingers (3 each) also make up the great toe, or **hallux**, and other toes.

CONCEPT CHECK QUESTIONS
Answers on page 173

❶ Which three bones make up the hip?

❷ The fibula does not participate in the knee joint or bear weight. When it is fractured, however, walking becomes difficult. Why?

❸ While jumping off the back steps at his house, 10-year-old Joey lands on his right heel and breaks his foot. Which foot bone is most likely broken?

● *Figure 6-29* **The Right Tibia and Fibula**
An anterior view.

Labels: Lateral condyle of tibia; Medial condyle of tibia; Head of fibula; Tibial tuberosity; Shaft of fibula; Anterior crest; Shaft of tibia; Medial malleolus; Lateral malleolus; Articular surface

Articulations

Joints, or **articulations**, exist wherever two bones meet. The characteristic structure of a joint determines the type of movement that may occur. Each joint reflects a workable compromise between the need for strength and the need for movement. When movement is not required, or when relative movement could actually be dangerous, joints can be very strong. For example, joints such as the sutures of the skull are so intricate and extensive that they lock the elements together as if they were a single bone. At other joints, movement is more important than strength. For example, the articulation at the shoul-

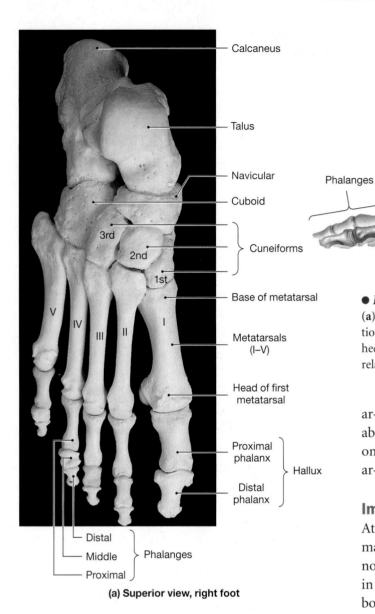

Calcaneus

Talus

Navicular

Cuboid

3rd

2nd

1st

Cuneiforms

Base of metatarsal

Metatarsals (I–V)

Head of first metatarsal

Proximal phalanx

Distal phalanx

Hallux

Distal
Middle
Proximal

Phalanges

(a) Superior view, right foot

Tarsal bones

Tibia

Cuneiform bone

Navicular

Talus

Metatarsal bones

Phalanges

1st

Calcaneus

(b) Medial view, right foot

● *Figure 6-30* **The Bones of the Ankle and Foot**
(**a**) A superior view of the bones of the right foot. Notice the orientation of the tarsal bones, which convey the weight of the body to the heel and the plantar surfaces of the foot. (**b**) Medial view, showing the relative positions and orientation of the tarsal and metatarsal bones.

der permits a range of arm movement that is limited more by the surrounding muscles than by joint structure. The joint itself is relatively weak, and shoulder injuries are rather common.

A CLASSIFICATION OF JOINTS

Joints are classified according to their structure or function. The structural classification is based on the anatomy of the joint. In this framework, joints are classified as **fibrous**, **cartilaginous**, or **synovial** (si-NŌ-vē-al). The first two types reflect the type of connective tissue binding them together. Such joints permit either no or slight movements. Synovial joints are surrounded by fibrous tissue and the ends of bones are covered by cartilage that prevents bone-to-bone contact. Such joints permit free movement. In a functional classification, joints are classified according to the range of motion they permit (Table 6-2). An immovable joint is a **synarthrosis** (sin-

ar-THRŌ-sis; *syn-*, together + *arthros*, joint); a slightly movable joint is an **amphiarthrosis** (am-fē-ar-THRŌ-sis; *amphi-*, on both sides); and a freely movable joint is a **diarthrosis** (dī-ar-THRŌ-sis; *dia-*, through), or *synovial joint*.

Immovable Joints (Synarthroses)

At a synarthrosis, the bony edges are quite close together and may even interlock. A synarthrosis can be fibrous or cartilaginous. Two examples of fibrous immovable joints can be found in the skull. (1) In a **suture** (*sutura*, a sewing together), the bones of the skull are interlocked and bound together by dense connective tissue. (2) In a **gomphosis** (gom-FŌ-sis; *gomphosis*, a bolting together), a ligament binds each tooth in the mouth within a bony socket (*alveolus*).

An epiphyseal plate also represents an articulation between two bones, even though the two are part of the same skeletal element. p. 132 Such a rigid, cartilaginous connection is called a **synchondrosis** (sin-kon-DRŌ-sis; *syn*, together + *chondros*, cartilage).

Slightly Movable Joints (Amphiarthroses)

An amphiarthrosis permits very limited movement, and the bones are usually farther apart than they are at a synarthrosis. Structurally, an amphiarthrosis can be fibrous or cartilaginous.

A **syndesmosis** (sin-dez-MŌ-sis; *desmos*, a band or ligament) is a fibrous joint connected by a ligament. The distal articulation between the two bones of the leg, the tibia and fibula, is an example. A **symphysis** is a cartilaginous joint

157

6 **THE SKELETAL SYSTEM**

The Structure of Bone • Bone Development and Growth • Remodeling and Homeostatic Mechanisms • Aging and the Skeletal System

TABLE 6-2 *A Functional Classification of Articulations*

FUNCTIONAL CATEGORY	STRUCTURAL CATEGORY	DESCRIPTION	EXAMPLE
Synarthrosis (no movement)	**Fibrous** Suture	Fibrous connections plus interlocked surfaces	Between the bones of the skull
	Gomphosis	Fibrous connections plus insertion in bony socket (alveolus)	Between the teeth and jaws
	Cartilaginous Synchondrosis	Interposition of cartilage plate	Epiphyseal plates
Amphiarthrosis (little movement)	**Fibrous** Syndesmosis	Ligamentous connection	Between the tibia and fibula
	Cartilaginous Symphysis	Connection by a fibrocartilage pad	Between right and left halves of pelvis; between adjacent vertebrae of spinal column
Diarthrosis (free movement)	**Synovial**	Complex joint bounded by joint capsule and containing synovial fluid	Numerous; subdivided by range of movement (see Figure 6-35●)

because the bones are separated by a broad disc or pad of fibro-cartilage. The articulations between the spinal vertebrae (at the *intervertebral disc*) and the anterior connection between the two pubic bones are examples of this type of joint.

Freely Movable Joints (Diarthroses)

Diarthroses, or **synovial joints**, permit a wide range of motion. The basic structure of a synovial joint was introduced in Chapter 4 in the discussion of synovial membranes. ⚭ p. 100 Figure 6-31a● shows the structure of a representative synovial joint.

Synovial joints are typically found at the ends of long bones, such as those of the arms and legs. Under normal conditions the bony surfaces do not contact one another, for they are covered with special **articular cartilages**. The joint is surrounded by a fibrous **joint capsule**, or *articular capsule*, and the inner surfaces of the joint cavity are lined with a synovial membrane.

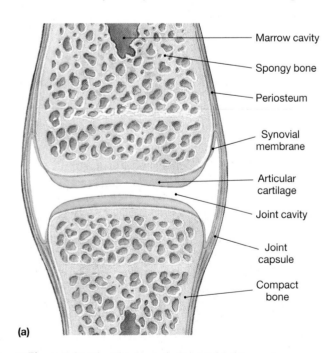

(a)

Marrow cavity
Spongy bone
Periosteum
Synovial membrane
Articular cartilage
Joint cavity
Joint capsule
Compact bone

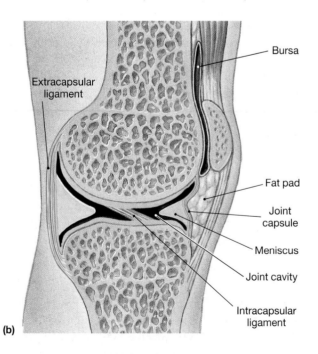

(b)

Bursa
Extracapsular ligament
Fat pad
Joint capsule
Meniscus
Joint cavity
Intracapsular ligament

● *Figure 6-31* **The Structure of a Synovial Joint**
(**a**) A diagrammatic view of a simple articulation. (**b**) A sectional view of the knee joint.

Synovial fluid within the joint cavity provides lubrication that reduces the friction between the moving surfaces in the joint.

In complex joints such as the knee, additional padding lies between the opposing articular surfaces. An example of such shock-absorbing, fibrocartilage pads are the **menisci** (men-IS-kē; *meniscus*, crescent), shown in Figure 6-31b●. Also present in such joints are **fat pads**, which protect the articular cartilages and act as packing material. When the bones move, the fat pads fill in the spaces created as the joint cavity changes shape.

The joint capsule that surrounds the entire joint is continuous with the periostea of the articulating bones. In addition, **ligaments** joining bone to bone may be found outside or inside the joint capsule. Where a tendon or ligament rubs against other tissues, **bursae**, small packets of connective tissue containing synovial fluid, form to reduce friction and act as shock absorbers. Bursae are characteristic of many synovial joints and may also appear around a tendon as a tubular sheath, covering a bone, or within other connective tissues exposed to friction or pressure.

RHEUMATISM AND ARTHRITIS

Rheumatism (ROO-ma-tizm) is a general term describing pain and stiffness that affect the skeletal or muscular systems, or both. There are several major forms of rheumatism. **Arthritis** (ar-THRĪ-tis) includes all of the rheumatic diseases that affect synovial joints. Arthritis always involves damage to the articular cartilages, but the specific cause can vary. For example, arthritis can result from bacterial or viral infection, injury to the joint, metabolic problems, or severe physical stresses.

Osteoarthritis (os-tē-ō-ar-THRĪ-tis), also known as *degenerative arthritis*, or *degenerative joint disease (DJD)*, usually affects individuals age 60 or older. This disease can result from cumulative wear and tear at the joint surfaces or from genetic factors affecting collagen formation. In the U.S. population, 25 percent of women and 15 percent of men over age 60 show signs of this disease. **Rheumatoid arthritis** is an inflammatory condition that affects roughly 2.5 percent of the adult population. At least some cases result when the immune response mistakenly attacks the joint tissues. Allergies, bacteria, viruses, and genetic factors have all been proposed as contributing to or triggering the destructive inflammation.

Regular exercise, physical therapy, and drugs that reduce inflammation, such as aspirin, can slow the progress of the disease. Surgical procedures can realign or redesign the affected joint, and in extreme cases involving the hip, knee, elbow, or shoulder, the defective joint can be replaced by an artificial one.

SYNOVIAL JOINTS: MOVEMENT AND STRUCTURE

Synovial joints are involved in all of our day-to-day movements. In discussions of motion at synovial joints, phrases such as "bend the leg" or "raise the arm" are not sufficiently precise. Anatomists use descriptive terms that have specific meanings.

Types of Movement

Gliding. In **gliding**, two opposing surfaces slide past each other. Gliding occurs between the surfaces of articulating carpal bones, tarsal bones, and between the clavicles and sternum. The movement can occur in almost any direction, but the amount of movement is slight. Rotation is usually prevented by the capsule and associated ligaments.

Angular Motion. Examples of angular motion are *flexion*, *extension*, *adduction*, *abduction*, and *circumduction*. The description of each movement is based on reference to an individual in the anatomical position.

Flexion/Extension. **Flexion** (FLEK-shun) is movement in the anterior-posterior plane that reduces the angle between the articulating elements. **Extension** occurs in the same plane, but it increases the angle between articulating elements (Figure 6-32a●). When you bring your head toward your chest, you flex the intervertebral joints of the neck. When you bend down to touch your toes, you flex the entire vertebral column. Extension reverses these movements.

Flexion at the shoulder joint or hip joint moves the limbs forward (*anterior*), whereas extension moves them back (*posterior*). Flexion of the wrist moves the hand forward, and extension moves it back. In each of these examples, extension can be continued past the anatomical position, in which case **hyperextension** occurs. You can also hyperextend the neck, a movement that enables you to gaze at the ceiling. Hyperextension of other joints is usually prevented by ligaments, bony processes, or soft tissues.

Abduction/Adduction. **Abduction** (*ab-*, from) is movement *away from the midline of the body* in the frontal plane. For example, swinging the upper limb to the side is abduction of the limb (Figure 6-32b●). **Adduction** is movement *toward the midline of the body*. Adduction of the wrist moves the heel of the hand toward the body, whereas abduction moves it farther away. Spreading the fingers or toes apart abducts them, because they move *away* from a central digit (finger or toe), as in Figure 6-32c●. Bringing them together is adduction. Abduction and adduction always refer to movements of the appendicular skeleton, not to those of the axial skeleton.

Circumduction. **Circumduction** (*circum*, around) is another type of angular motion (Figure 6-32d●). An example of circumduction is moving your arm in a loop, as when drawing a large circle on a chalkboard.

Rotation. Rotational movements are also described with reference to a figure in the anatomical position. **Rotation** involves

6 THE SKELETAL SYSTEM

The Structure of Bone • Bone Development and Growth • Remodeling and Homeostatic Mechanisms • Aging and the Skeletal System

● *Figure 6-32*
Angular Movements
The red dots mark the locations of joints involved in the movements.

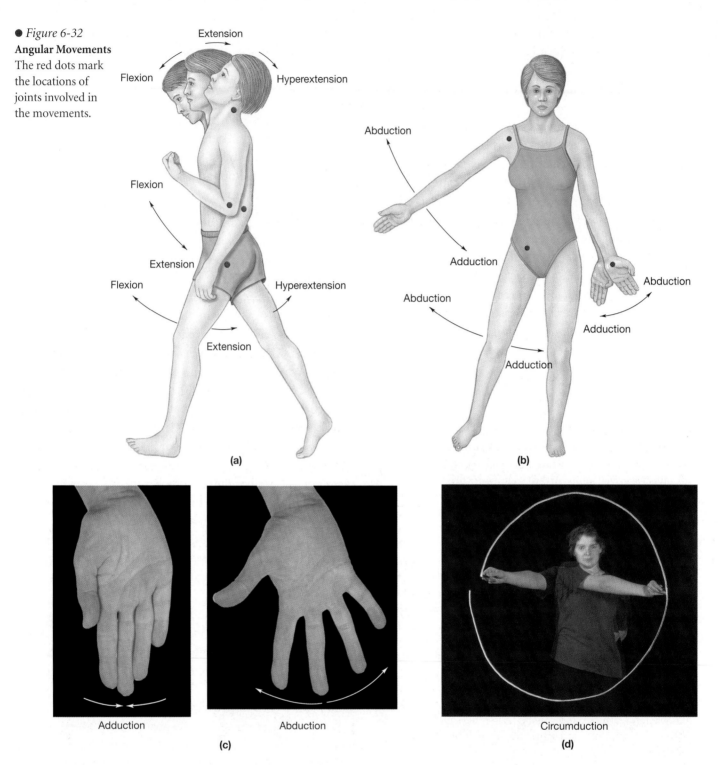

(a)

(b)

Adduction   Abduction

(c)

Circumduction

(d)

turning around the longitudinal axis of the body or limb. For example, you may rotate your head to look to one side, or rotate your arm to screw in a light bulb. Rotational movements are illustrated in Figure 6-33●.

Pronation/Supination. The articulations between the radius and ulna permit the rotation of the distal end of the radius across the anterior surface of the ulna. This rotation moves the wrist and hand from palm-facing-front to palm-facing-back. This motion is called **pronation** (prō-NĀ-shun). The oppos-

ing movement, in which the palm is turned forward, is **supination** (soo-pi-NĀ-shun). ⚬⚬ p. 152

Special Movements. Special terms apply to specific articulations or to unusual types of movement (Figure 6-34●).

• **Inversion** (*in-*, into + *vertere*, to turn) is a twisting motion of the foot that turns the sole inward. The opposite movement is called **eversion** (ē-VER-shun; *e-*, out).

• **Dorsiflexion** is flexion of the ankle and elevation of the sole,

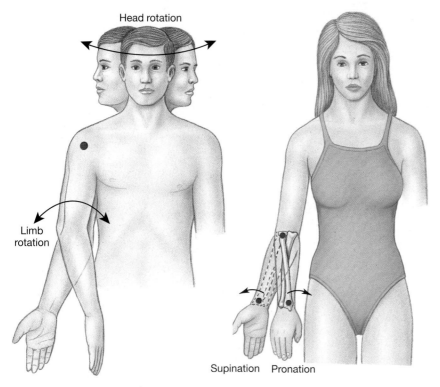

Head rotation

Limb rotation

Supination   Pronation

● *Figure 6-33* **Rotational Movements**

as when you dig in your heel. **Plantar flexion** (*planta*, sole), the opposite movement, extends the ankle and elevates the heel, as when you stand on tiptoe.

· **Opposition** is the special movement of the thumb that enables it to grasp and hold an object.

· **Protraction** occurs when you move a part of the body anteriorly in the horizontal plane. **Retraction** is the reverse move-

ment. You protract your jaw when you grasp your upper lip with your lower teeth, and you protract your clavicles when you cross your arms.

· **Elevation** and **depression** occur when a structure moves in a superior or inferior direction. You depress your mandible when you open your mouth, and elevate it as you close it.

A Structural Classification of Synovial Joints

Synovial joints can be described as *gliding, hinge, pivot, ellipsoidal, saddle,* or *ball-and-socket joints* on the basis of the shapes of the articulating surfaces (Figure 6-35●). Each type of joint permits a different type and range of motion:

· **Gliding joints** have flattened or slightly curved faces (Figure 6-35a●). The relatively flat articular surfaces slide across one another, but the amount of movement is very slight. Although rotation is theoretically possible at such a joint, ligaments usually prevent or restrict such movement. Gliding joints are found at the ends of the clavicles, between the carpal bones, between the tarsal bones, and between the articular facets of adjacent vertebrae.

· **Hinge joints** permit angular movement in a single plane, like the opening and closing of a door (Figure 6-35b●). Examples are the joint between the occipital bone and atlas, in the

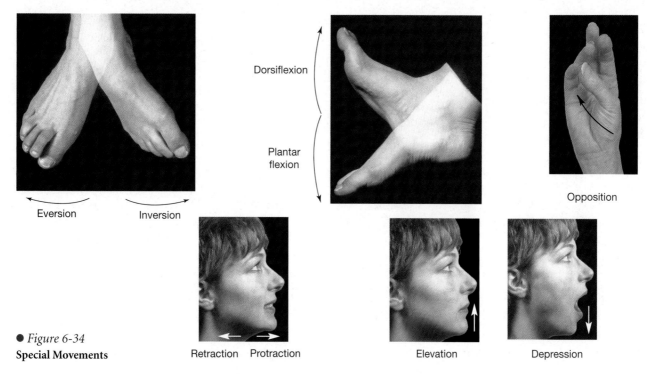

Dorsiflexion

Plantar flexion

Opposition

Eversion   Inversion

● *Figure 6-34*
Special Movements

Retraction   Protraction

Elevation   Depression

161

6

THE SKELETAL SYSTEM

The Structure of Bone • Bone Development and Growth • Remodeling and Homeostatic Mechanisms • Aging and the Skeletal System

● *Figure 6-35* **A Functional Classification of Synovial Joints**

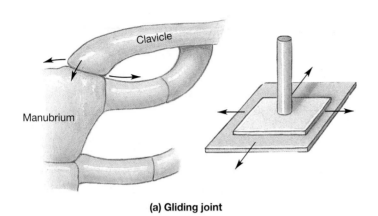

(a) Gliding joint

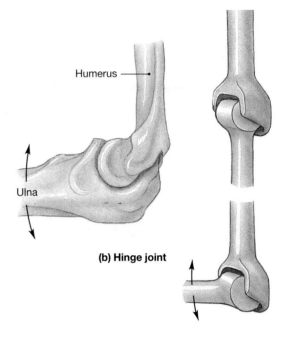

(b) Hinge joint

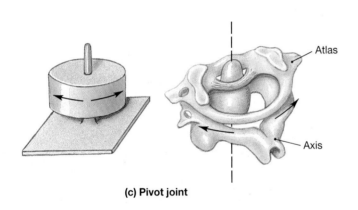

(c) Pivot joint

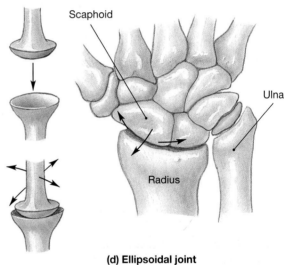

(d) Ellipsoidal joint

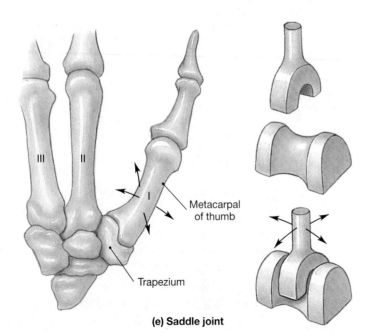

(e) Saddle joint

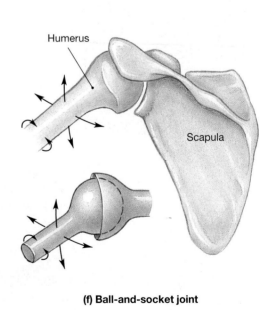

(f) Ball-and-socket joint

axial skeleton, and the elbow, knee, ankle, and interphalangeal joints of the appendicular skeleton.

- **Pivot joints** permit only rotation (Figure 6-35c●). A pivot joint between the atlas and axis enables you to rotate your head to either side, and another between the head of the radius and the proximal shaft of the ulna permits pronation and supination of the palm.

- In an **ellipsoidal joint**, or *condyloid joint*, an oval articular face nestles within a depression on the opposing surface (Figure 6-35d●). Angular motion occurs in two planes, along or across the length of the oval. Ellipsoidal joints connect the radius with the proximal carpal bones, the phalanges of the fingers with the metacarpal bones, and the phalanges of the toes with the metatarsal bones.

- **Saddle joints** have articular faces that resemble saddles (Figure 6-34e●). Each face is concave on one axis and convex on the other, and the opposing faces nest together. This arrangement permits angular motion, including circumduction, but prevents rotation. The carpometacarpal joint at the base of the thumb is the best example of a saddle joint, and twiddling your thumbs will demonstrate the possible movements.

- In a **ball-and-socket joint**, the round head of one bone rests within a cup-shaped depression in another (Figure 6-34f●). All combinations of movements, including circumduction and rotation, can be performed at ball-and-socket joints. Examples are the shoulder and hip joints.

CONCEPT CHECK QUESTIONS
Answers on page 174

❶ In a newborn infant, the large bones of the skull are joined by fibrous connective tissue. Which type of joint is this? These skull bones later grow, interlock, and form immovable joints. Which type of joints are these?

❷ Give the proper term for each of the following types of motion: (a) moving the humerus away from the midline of the body, (b) turning the palms so that they face forward, and (c) bending the elbow.

❸ Which movements are associated with hinge joints?

REPRESENTATIVE ARTICULATIONS

This section considers examples of articulations that demonstrate important functional principles. We will first consider the *intervertebral articulations* of the axial skeleton. We will then proceed to a discussion of the *synovial articulations* of the appendicular skeleton: the shoulder and elbow of the upper limb and the hip and knee of the lower limb.

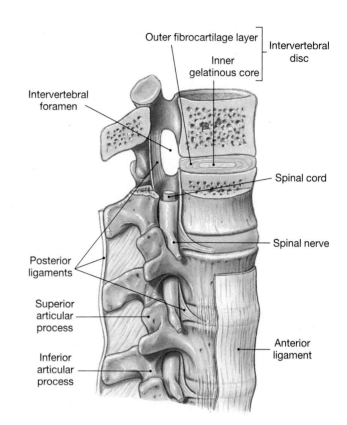

● *Figure 6-36* **Intervertebral Articulations**

Intervertebral Articulations

The vertebrae articulate with one another in two ways: (1) at gliding joints between the superior and inferior *articular processes*, and (2) at *symphyseal joints* between the vertebral bodies (Figure 6-36●). Articulations between the superior and inferior articular processes of adjacent vertebrae permit small movements that are associated with flexion and rotation of the vertebral column. Little gliding occurs between adjacent vertebral bodies.

As noted earlier, the vertebrae are separated and cushioned by pads called *intervertebral discs*. Each intervertebral disc consists of a tough outer layer of fibrocartilage. The collagen fibers of that layer attach the discs to adjacent vertebrae. The fibrocartilage surrounds a soft, elastic and gelatinous core, which gives intervertebral discs resiliency and enables them to act as shock absorbers, compressing and distorting when stressed. This resiliency prevents bone-to-bone contact that might damage the vertebrae or jolt the spinal cord and brain.

Shortly after physical maturity is reached, the gelatinous mass within each disc begins to degenerate, and the "cushion" becomes less effective. Over the same period, the outer fibrocartilage loses its elasticity. If the stresses are sufficient, the inner mass may break through the surrounding fibrocartilage and protrude beyond the intervertebral space. This condition, called a *herniated disc*, further reduces disc function. The term *slipped disc* is often used to describe this problem, although

163

6 THE SKELETAL SYSTEM

The Structure of Bone • Bone Development and Growth • Remodeling and Homeostatic Mechanisms • Aging and the Skeletal System

the disc does not actually slip. The intervertebral discs also make a significant contribution to an individual's height; they account for roughly one-quarter of the length of the spinal column above the sacrum. As we grow older, the water content of each disc decreases; this loss accounts for the characteristic decrease in height with advancing age.

Articulations of the Upper Limb

The shoulder, elbow, and wrist are responsible for positioning the hand, which performs precise and controlled movements. The shoulder has great mobility, the elbow has great strength, and the wrist makes fine adjustments in the orientation of the palm and fingers.

The Shoulder Joint. The shoulder joint permits the greatest range of motion of any joint in the body. Because it is also the most frequently dislocated joint, it provides an excellent demonstration of the principle that strength and stability must be sacrificed to obtain mobility.

Figure 6-37● shows the structure of the shoulder joint. The relatively loose joint capsule extends from the scapular neck to the humerus, and this oversized capsule permits an extensive range of motion. As at other joints, bursae at the shoulder reduce friction where large muscles and tendons pass across the joint capsule. The bursae of the shoulder are especially large and numerous. Several bursae are associated with the capsule, the processes of the scapula, and large shoulder muscles. Inflammation of any of these bursae, a condition called *bursitis*, can restrict motion and produce pain.

Muscles that move the humerus do more to stabilize the shoulder joint than all its ligaments and capsular fibers combined. Powerful muscles originating on the trunk, shoulder girdle, and humerus cover the anterior, superior, and posterior surfaces of the capsule. These muscles form the *rotator cuff*, a group of muscles that swing the arm through an impressive range of motion.

The Elbow Joint. The elbow joint consists of two articulations: the humerus and ulna, and the humerus and radius (Figure 6-38●). The largest and strongest articulation is between the humerus and the ulna. This hinge joint provides stability and limits movement at the elbow joint.

The elbow joint is extremely stable because (1) the bony surfaces of the humerus and ulna interlock; (2) the joint capsule is very thick; and (3) the capsule is reinforced by stout ligaments. Nevertheless, the joint can be damaged by severe impacts or unusual stresses. When you fall on your hand with a partially flexed elbow, powerful contractions of the muscles that extend the elbow can break the ulna at the center of the trochlear notch.

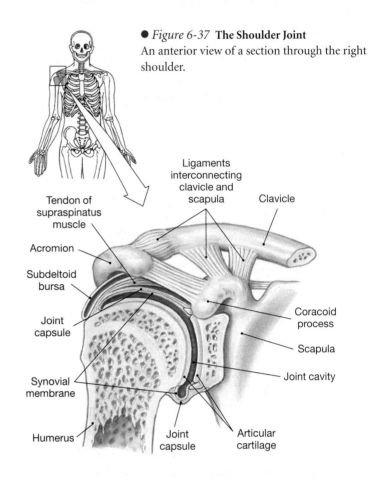

● *Figure 6-37* **The Shoulder Joint**
An anterior view of a section through the right shoulder.

CONCEPT CHECK QUESTIONS
Answers on page 174

❶ Would a tennis player or a jogger be more likely to develop inflammation of the subdeltoid bursa? Why?

❷ Mary falls on her hands with her elbows slightly flexed. After the fall, she can't move her left arm at the elbow. If a fracture exists, which bone is most likely broken?

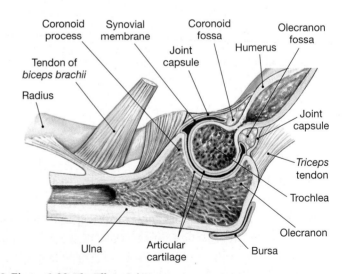

● *Figure 6-38* **The Elbow Joint**
A longitudinal section through the right elbow.

Articulations of the Lower Limb

The joints of the hip, ankle, and foot are sturdier than those at corresponding locations in the upper limb, and they have smaller ranges of motion. The knee has a range of motion comparable to that of the elbow, but it is subjected to much greater forces and therefore is less stable.

The Hip Joint. Figure 6-39● shows the structure of the hip joint. The articulating surface of the acetabulum has a fibrocartilage pad along its edges, a fat pad covered by synovial membrane in its central portion, and a stout central ligament. This combination of coverings and membranes resists compression, absorbs shocks, and stretches and distorts without damage.

Compared with that of the shoulder, the joint capsule of the hip joint, a ball-and-socket diarthrosis, is denser and stronger. It extends from the lateral and inferior surfaces of the pelvic girdle to the femur and encloses both the femoral head and neck. This arrangement helps keep the head from moving away from the acetabulum. Three broad ligaments reinforce the joint capsule, while a fourth, the *ligament of the femoral head* (the *ligamentum teres*), originates inside the acetabulum and attaches to the center of the femoral head. Additional stabilization comes from the bulk of the surrounding muscles.

The combination of an almost complete bony socket, a strong joint capsule, supporting ligaments, and muscular padding makes this an extremely stable joint. Fractures of the femoral neck or between the trochanters are actually more common than hip dislocations. Although flexion, extension, adduction, abduction, and rotation are permitted, the total range of motion is considerably less than that of the shoulder. Hip flexion is the most important normal movement, and the primary limits are imposed by the surrounding muscles. Other directions of movement are restricted by ligaments and the capsule.

 ## HIP FRACTURES

Hip fractures are most often suffered by individuals over the age of 60, when osteoporosis has weakened the thighbones. These injuries may be accompanied by dislocation of the hip or by pelvic fractures. For individuals with osteoporosis, healing of such fractures proceeds very slowly. In addition, the powerful muscles that surround the joint can easily prevent proper alignment of the bone fragments. Trochanteric fractures usually heal well if the joint can be stabilized; steel frames, pins, screws, or some combination of these devices may be needed to preserve alignment and to permit normal healing.

The Knee Joint. The hip joint passes weight to the femur, and at the knee joint, the femur transfers the weight to the tibia. Although the knee functions as a hinge joint, the articulation is far more complex than that of the elbow or even the ankle. The rounded femoral condyles roll across the top of the tibia,

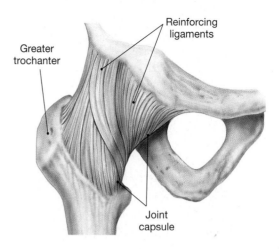

(a) Anterior view

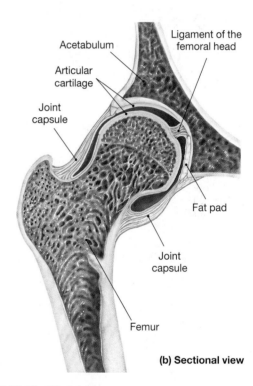

(b) Sectional view

● *Figure 6-39* **The Hip Joint**
(**a**) The right hip joint, which is extremely strong and stable, in part because of the massive capsule and surrounding ligaments. (**b**) A sectional view of the right hip joint.

so the points of contact are constantly changing. Important features of the knee joint are shown in Figure 6-40●.

Structurally, the knee combines three separate joints—two between the femur and tibia (medial condyle to medial condyle and lateral condyle to lateral condyle), and one between the patella and the femur. There is no single unified capsule, nor is there a common synovial cavity. A pair of fibrocartilage pads, the **medial** and **lateral menisci**, lie between the femoral and tibial surfaces. They act as cushions and conform to the shape

6 THE SKELETAL SYSTEM

The Structure of Bone • Bone Development and Growth • Remodeling and Homeostatic Mechanisms • Aging and the Skeletal System

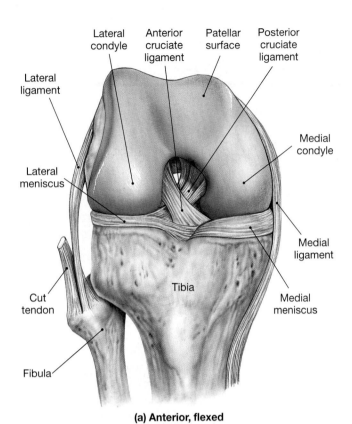

(a) Anterior, flexed

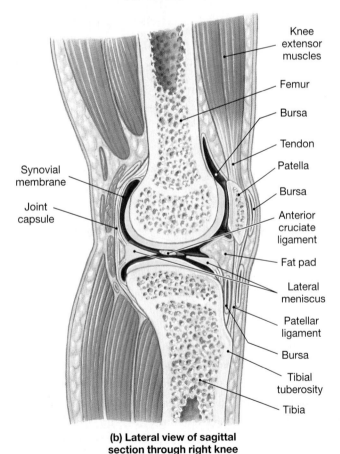

(b) Lateral view of sagittal section through right knee

● *Figure 6-40* **The Knee Joint**
(**a**) The flexed right knee. (**b**) The extended knee.

of the articulating surfaces as the femur changes position. Prominent fat pads provide padding around the margins of the joint and assist the bursae in reducing friction between the patella and other tissues.

Ligaments stabilize the anterior, posterior, medial, and lateral surfaces of this joint, and a complete dislocation of the knee is an extremely rare event. The tendon from the muscles responsible for extending the knee passes over the anterior surface of the joint. The patella lies within this tendon, and the **patellar ligament** continues its attachment on the anterior surface of the tibia. This ligament provides support to the front of the knee joint. Posterior ligaments between the femur and the heads of the tibia and fibula reinforce the back of the knee joint. The lateral and medial surfaces of the knee joint are reinforced by another pair of ligaments. These ligaments stabilize the joint at full extension.

Additional ligaments are found inside the joint capsule (Figure 6-40b●). Inside the joint a pair of ligaments, the *anterior cruciate* and *posterior cruciate*, cross each other as they attach the tibia to the femur. (The term *cruciate* is derived from the Latin word *crucialis*, meaning a cross.) These ligaments limit the anterior and posterior movement of the femur.

CONCEPT CHECK QUESTIONS

Answers on page 174

❶ Why is a complete dislocation of the knee joint an infrequent event?

❷ What symptoms would you expect to see in an individual who has damaged the menisci of the knee joint?

Integration with Other Systems

Although the bones may seem inert, you should now realize that they are quite dynamic structures. The entire skeletal system is intimately associated with other systems. For example, bones are attached to the muscular system, extensively connected to the cardiovascular and lymphatic systems, and largely under the physiological control of the endocrine system. These functional relationships are diagrammed in Figure 6-41●.

CONCEPT CHECK QUESTIONS

Answer on page 174

❶ The bones of the skeletal system contain 99 percent of the calcium in the body. Which other organ systems depend on these reserves of calcium for normal functioning?

The Skeletal System

For All Systems

Provides mechanical support; stores energy reserves; stores calcium and phosphate reserves

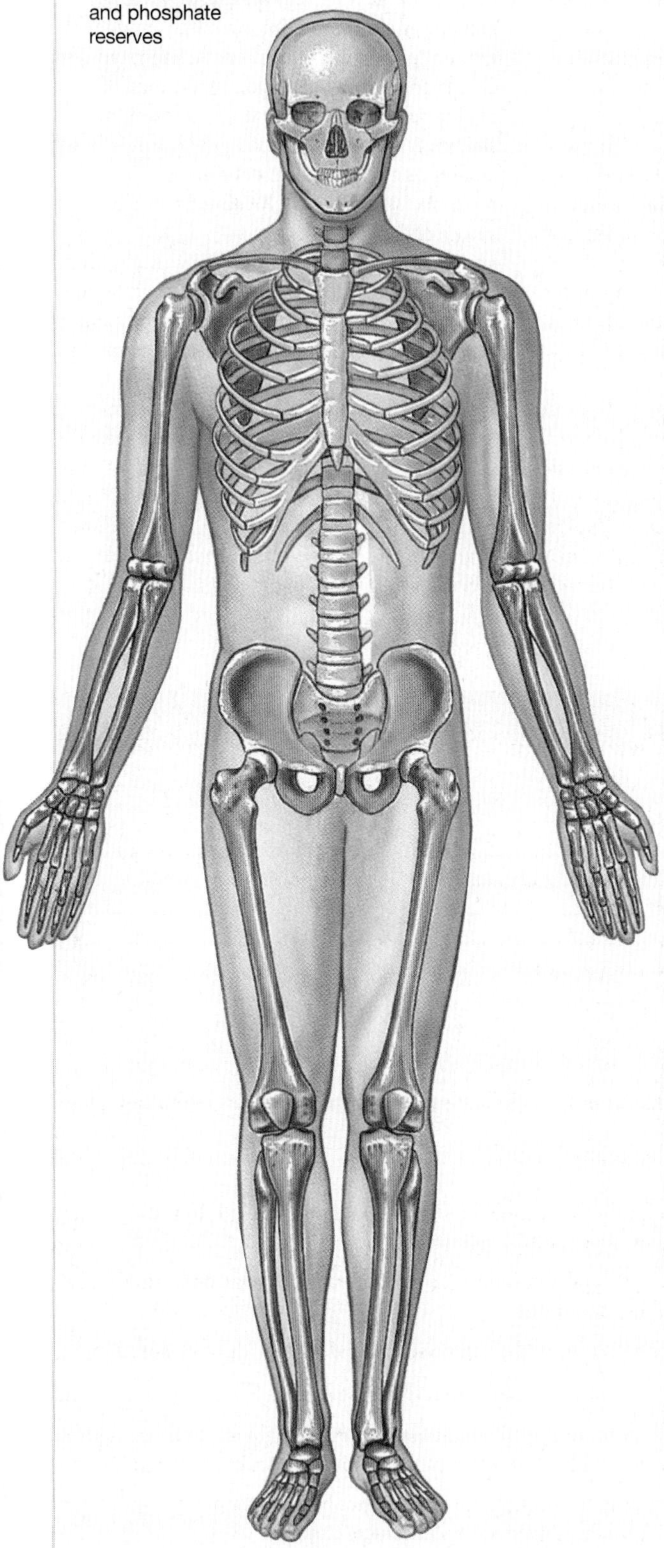

The Integumentary System

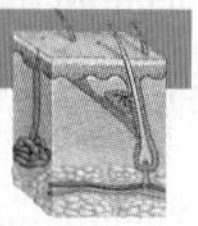

- Synthesizes vitamin D_3, essential for calcium and phosphorus absorption (bone maintenance and growth)
- Provides structural support

The Muscular System

- Stabilizes bone positions; tension in tendons stimulates bone growth and maintenence
- Provides calcium needed for normal muscle contraction; bones act as levers to produce body movements

The Nervous System

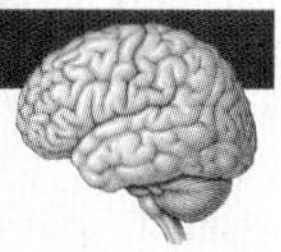

- Regulates bone position by controlling muscle contractions
- Provides calcium for neural function; protects brain, spinal cord; receptors at joints provide information about body position

The Endocrine System

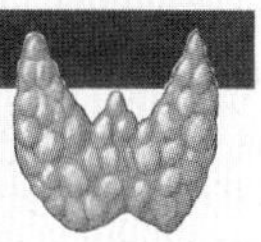

- Skeletal growth regulated by growth hormone, thyroid hormones, and sex hormones; calcium mobilization regulated by parathyroid hormone and calcitonin
- Protects endocrine organs, especially in brain, chest, and pelvic cavity

The Cardiovascular System

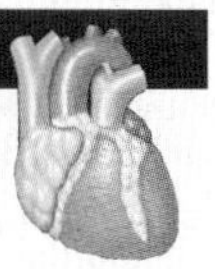

- Provides oxygen, nutrients, hormones, blood cells; removes waste products and carbon dioxide
- Provides calcium needed for cardiac muscle contraction, blood cells produced in bone marrow

The Lymphatic System

- Lymphocytes assist in the defense and repair of bone following injuries
- Lymphocytes and other cells of the immune response are produced and stored in bone marrow

The Respiratory System

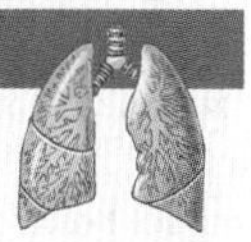

- Provides oxygen and eliminates carbon dioxide
- Movements of ribs important in breathing; axial skeleton surrounds and protects lungs

The Digestive System

- Provides nutrients, calcium, and phosphate
- Ribs protect portions of liver, stomach, and intestines

The Urinary System

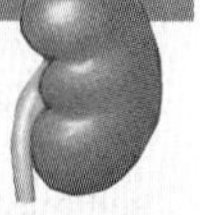

- Conserves calcium and phosphate needed for bone growth; disposes of waste products
- Axial skeleton provides some protection for kidneys and ureters; pelvis protects urinary bladder and proximal urethra

The Reproductive System

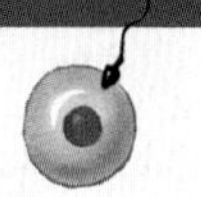

- Sex hormones stimulate growth and maintenance of bones; surge of sex hormones at puberty causes acceleration of growth and closure of epiphyseal cartilages
- Pelvis protects reproductive organs of female, protects portion of ductus deferens and accessory glands in males

● *Figure 6-41* **Functional Relationships Between the Skeletal System and Other Systems**

6 THE SKELETAL SYSTEM

The Structure of Bone • Bone Development and Growth • Remodeling and Homeostatic Mechanisms • Aging and the Skeletal System

Related Clinical Terms

ankylosis (ang-ki-LŌ-sis): An abnormal fusion between articulating bones in response to trauma and friction within a joint.

arthritis (ar-THRĪ-tis): Rheumatic diseases that affect synovial joints. Arthritis always involves damage to the articular cartilages, but the specific cause can vary. The diseases of arthritis are usually classified as either *degenerative* or *inflammatory*.

arthroscopic surgery: The surgical modification of a joint by using an *arthroscope* (a fiberoptic instrument used to view the inside of joint cavities).

bursitis: An inflammation of a bursa, causing pain whenever the associated tendon or ligament moves.

carpal tunnel syndrome: An inflammation of the tissues at the anterior wrist, causing compression of adjacent tendons and nerves. Symptoms are pain and a loss of wrist mobility.

fracture: A crack or break in a bone.

gigantism: A condition of extreme height resulting from an overproduction of growth hormone before puberty.

herniated disc: A condition caused by intervertebral disc compression severe enough to rupture the outer fibrocartilage layer and release of the inner soft, gelatinous core, which may protrude beyond the intervertebral space.

kyphosis (kī-FŌ-sis): An abnormal exaggeration of the thoracic spinal curve that produces a humpback appearance.

lordosis (lor-DŌ-sis): An abnormal lumbar curve of the spine that gives a swayback appearance.

luxation (luks-Ā-shun): A dislocation; a condition in which the articulating surfaces are forced out of position.

orthopedics (or-tho-PĒ-diks): A branch of surgery concerned with disorders of the bones and joints, and their associated muscles, tendons, and ligaments.

osteomyelitis (os-tē-ō-mī-e-LĪ-tis): A painful infection in a bone, generally caused by bacteria.

osteopenia (os-tē-ō-PĒ-nē-uh): Inadequate ossification, leading to thinner, weaker bones.

osteoporosis (os-tē-ō-por-Ō-sis): A reduction in bone mass to a degree that compromises normal function.

rheumatism (ROO-muh-tizm): A general term that indicates pain and stiffness affecting the skeletal system, the muscular system, or both.

rickets: A childhood disorder that reduces the amount of calcium salts in the skeleton; typically characterized by a bowlegged appearance, because the leg bones bend under the body's weight.

scurvy: A condition involving weak, brittle bones as a result of a vitamin C deficiency.

scoliosis (skō-lē-Ō-sis): An abnormal lateral curvature of the spine.

spina bifida (SPĪ-nuh BI-fi-duh): A condition resulting from the failure of the vertebral laminae to unite during development; commonly associated with developmental abnormalities of the brain and spinal cord.

sprain: A condition in which a ligament is stretched to the point at which some of the collagen fibers are torn. The ligament remains functional, and the structure of the joint is not affected.

whiplash: An injury resulting from a sudden change in the body position that can injure the cervical vertebrae.

CHAPTER REVIEW

Key Terms

Summary Outline

INTRODUCTION **128**

1. The skeletal system includes the bones of the skeleton and the cartilages, ligaments, and other connective tissues that stabilize or interconnect bones. Its functions include structural support, storage, blood cell production, protection, and leverage.

THE STRUCTURE OF BONE **128**

1. **Bone**, or **osseous tissue**, is a supporting connective tissue with a solid *matrix*.

Macroscopic Features of Bone **128**

2. General categories of bones are **long bones**, **short bones**, **flat bones**, and **irregular bones**. *(Figure 6-1)*

3. The features of a long bone include a **diaphysis**, **epiphyses**, and a central *marrow cavity*. *(Figure 6-2)*

4. The two types of bone tissue are **compact**, or *dense*, **bone** and **spongy**, or *cancellous*, **bone**.

5. A bone is covered by a **periosteum** and lined with an **endosteum**.

Microscopic Features of Bone **129**

6. Both types of bone contain **osteocytes** in **lacunae**. Layers of calcified matrix are **lamellae**, interconnected by **canaliculi**. *(Figure 6-3)*

7. The basic functional unit of compact bone is the **osteon**, containing osteocytes arranged around a **central canal**.

8. Spongy bone contains **trabeculae**, often in an open network.

9. Compact bone is located where stresses come from a limited range of directions; spongy bone is located where stresses are few or come from many different directions.

10. Cells other than osteocytes are also present in bone. **Osteoclasts** dissolve the bony matrix through the process of *osteolysis.* **Osteoblasts** synthesize the matrix in the process of *osteogenesis.*

BONE DEVELOPMENT AND GROWTH............................131

1. **Ossification** is the process of converting other tissues to bone.

Intramembranous Ossification 131

2. **Intramembranous ossification** begins when stem cells in connective tissue differentiate into osteoblasts and can produce spongy or compact bone. *(Figure 6-4)*

Endochondral Ossification 131

3. **Endochondral ossification** begins by the formation of a cartilage model of a bone that is gradually replaced by bone. *(Figure 6-5)*

4. Bone diameter increases through **appositional growth.** *(Figure 6-6)*

Bone Growth and Body Proportions 132

5. There are differences between bones and between individuals regarding the timing of epiphyseal closure.

Requirements for Normal Bone Growth 132

6. Normal osteogenesis requires a reliable source of minerals, vitamins, and hormones.

REMODELING AND HOMEOSTATIC MECHANISMS 133

1. The organic and mineral components of bone are continuously recycled and renewed through the process of **remodeling.**

Remodeling and Support 133

2. The shapes and thicknesses of bones reflect the stresses applied to them. Mineral turnover allows bone to adapt to new stresses.

Homeostasis and Mineral Storage 133

3. Calcium is the most abundant mineral in the human body, with roughly 99 percent of it located in the skeleton. The skeleton acts as a calcium reserve.

Injury and Repair 134

4. A **fracture** is a crack or break in a bone. Repair of a fracture involves the formation of a **fracture hematoma**, an **external callus**, and an **internal callus.** *(Figure 6-7) (Focus: A Classification of Fractures)*

AGING AND THE SKELETAL SYSTEM.............................136

1. The effects of aging on the skeleton can include **osteopenia** and **osteoporosis.**

AN OVERVIEW OF THE SKELETON 137

Skeletal Terminology..137

1. **Bone markings** can be used to describe and identify specific bones. *(Table 6-1)*

Skeletal Divisions..137

2. The skeletal system consists of the axial skeleton and the appendicular skeleton. The **axial skeleton** can be subdivided into the **skull**, the **auditory ossicles** (ear bones), the **hyoid**, the **thoracic cage** (*rib cage*) composed of the **ribs** and **sternum**, and the **vertebral column.** *(Figures 6-8, 6-9)*

3. The **appendicular skeleton** includes the upper and lower limbs and the **pectoral** and **pelvic girdles.**

THE AXIAL DIVISION 139

The Skull..140

1. The **cranium** encloses the **cranial cavity**, a division of the dorsal body cavity that encloses the brain.

2. The **frontal bone** forms the forehead and superior surface of each **orbit.** *(Figures 6-10, 6-11, 6-12)*

3. The **parietal bones** form the upper sides and roof of the cranium. *(Figures 6-10, 6-12)*

4. The **occipital bone** surrounds the **foramen magnum** and articulates with the sphenoid, temporal, and parietal bones to form the back of the cranium. *(Figures 6-10, 6-11, 6-12)*

5. The **temporal bones** help form the sides and base of the cranium and fuse with the parietal bones along the **squamosal suture.** *(Figures 6-10, 6-11, 6-12)*

6. The **sphenoid bone** acts like a bridge, uniting the cranial and facial bones. *(Figures 6-10, 6-11, 6-12)*

7. The **ethmoid bone** stabilizes the brain and forms the roof and sides of the nasal cavity. Its **cribriform plate** contains perforations for olfactory nerves, and the **perpendicular plate** forms part of the bony *nasal septum.* *(Figures 6-10, 6-11, 6-12)*

8. The left and right **maxillary bones**, or *maxillae*, articulate with all the other facial bones except the *mandible.* *(Figures 6-10, 6-11, 6-12)*

9. The **palatine bones** form the posterior portions of the hard palate and contribute to the walls of the nasal cavity and to the floor of each orbit. *(Figures 6-11, 6-12)*

10. The **vomer** forms the inferior portion of the bony nasal septum. *(Figures 6-11, 6-12)*

11. The **zygomatic bones** help complete the orbit and together with the temporal bones form the **zygomatic arch** (*cheekbone*). *(Figures 6-10, 6-11)*

12. The **nasal bones** articulate with the frontal bone and maxillary bones. *(Figures 6-10, 6-11, 6-12)*

13. The **lacrimal bones** are within the orbit on its medial surface. *(Figures 6-10, 6-11)*

14. The **inferior nasal conchae** inside the nasal cavity aid the **superior** and **middle nasal conchae** of the ethmoid bone to slow incoming air. *(Figures 6-11a, 6-12c)*

15. The **nasal complex** includes the bones that form the superior and lateral walls of the nasal cavity and the sinuses that drain into them. The **nasal septum** divides the nasal cavities. Together the **frontal, sphenoidal, ethmoidal, palatine**, and **maxillary sinuses** make up the **paranasal sinuses.** *(Figures 6-11, 6-12, 6-13)*

16. The **mandible** is the bone of the lower jaw. *(Figures 6-10, 6-11, 6-12)*

17. The **hyoid bone** is suspended below the skull by ligaments from the styloid processes of the temporal bones. *(Figure 6-14)*

18. Fibrous connections of tissue called **fontanels** permit the skulls of infants and children to continue growing. *(Figure 6-15)*

The Neck and Trunk...145

19. There are 7 **cervical vertebrae**, 12 **thoracic vertebrae** (which articulate with ribs), and 5 **lumbar vertebrae** (which articulate with the sacrum). The **sacrum** and **coccyx** consist of fused vertebrae. *(Figure 6-16)*

6 THE SKELETAL SYSTEM

The Structure of Bone • Bone Development and Growth • Remodeling and Homeostatic Mechanisms • Aging and the Skeletal System

20. The spinal column has four **spinal curves**, which accommodate the unequal distribution of body weight and keep it in line with the body axis. *(Figure 6-16)*

21. A typical vertebra has a **body** and a **vertebral arch**; it articulates with other vertebrae at the **articular processes**. Adjacent vertebrae are separated by an **intervertebral disc**. *(Figure 6-17)*

22. Cervical vertebrae are distinguished by the shape of the body and by **transverse foramina** on either side. *(Figures 6-17, 6-18)*

23. Thoracic vertebrae have distinctive heart-shaped bodies. *(Figure 6-17)*

24. The lumbar vertebrae are the most massive and least mobile; they are subjected to the greatest strains, and a *herniated disc* can occur. *(Figure 6-17)*

25. The sacrum protects reproductive, digestive, and excretory organs. At its **apex**, the sacrum articulates with the coccyx. At its **base**, the sacrum articulates with the last lumbar vertebra. *(Figure 6-19)*

26. The skeleton of the thorax consists of the thoracic vertebrae, the ribs, and the sternum. The ribs and sternum form the thoracic cage, or rib cage. *(Figure 6-20)*

27. Ribs 1 to 7 are **true ribs**. Ribs 8 to 12 lack direct connections to the sternum and are called **false ribs**; they include two pairs of **floating ribs**. The medial end of each rib articulates with a thoracic vertebra. *(Figure 6-20)*

28. The sternum consists of a **manubrium**, a **body**, and a **xiphoid process**. *(Figure 6-20)*

THE APPENDICULAR DIVISION 150

1. Each arm articulates with the trunk at the pectoral girdle, or shoulder girdle, which consists of the **scapula** and **clavicle**. *(Figures 6-8, 6-9, 6-21, 6-22)*

2. The clavicle and scapula position the shoulder joint, help move the arm, and provide a base for arm movement and muscle attachment. *(Figures 6-21, 6-22)*

3. Both the **coracoid process** and the **acromion** are attached to ligaments and tendons. The **scapular spine** crosses the posterior surface of the scapular body. *(Figure 6-22)*

4. The **humerus** articulates with the scapula at the shoulder joint. The **greater tubercle** and **lesser tubercle** of the humerus are important sites for muscle attachment. Other prominent landmarks include the **deltoid tuberosity**, the **medial** and **lateral epicondyles**, and the articular **condyle**. *(Figure 6-23)*

5. Distally, the humerus articulates with the radius and ulna. The medial **trochlea** extends from the **coronoid fossa** to the **olecranon fossa**. *(Figure 6-23)*

6. The **radius** and **ulna** are the bones of the forearm. The olecranon fossa accommodates the **olecranon process** during extension of the arm. The coronoid and radial fossae accommodate the **coronoid process** of the ulna. *(Figure 6-24)*

7. The bones of the wrist form two rows of **carpal bones**. The distal carpal bones articulate with the **metacarpal bones** of the palm. The metacarpal bones articulate with the proximal **phalanges**, or finger bones. Four of the fingers contain three phalanges; the **pollex**, or thumb, has only two. *(Figure 6-25)*

8. The pelvic girdle consists of two **coxal bones**. *(Figure 6-26)*

9. The largest coxal bone, the **ilium**, fuses with the **ischium**, which in turn fuses with the **pubis**. The **pubic symphysis** limits movement between the pubic bones. *(Figure 6-26)*

10. The **pelvis** consists of the coxae, the sacrum, and the coccyx. *(Figures 6-26, 6-27)*

11. The **femur**, or *thighbone*, is the longest bone in the body. It articulates with the **tibia** at the knee joint. A ligament from the **patella**, the *kneecap*, attaches at the **tibial tuberosity**. *(Figures 6-28, 6-29)*

12. Other tibial landmarks include the **anterior crest** and the **medial malleolus**. The fibular **head** articulates with the tibia below the knee, and the **lateral malleolus** stabilizes the ankle. *(Figure 6-29)*

13. The ankle includes seven **tarsal bones**; only the **talus** articulates with the tibia and fibula. When we stand normally, most of our weight is transferred to the **calcaneus**, or *heel bone*, and the rest is passed on to the **metatarsal bones**. *(Figure 6-30)*

14. The basic organizational pattern of the metatarsals and phalanges of the foot resembles that of the hand.

ARTICULATIONS 156

1. **Articulations** (joints) exist wherever two bones interact. Immovable joints are **synarthroses**, slightly movable joints are **amphiarthroses**, and those that are freely movable are called **diarthroses**. *(Table 6-2)*

2. Examples of synarthroses are a **suture**, a **gomphosis**, and a **synchondrosis**.

3. Examples of amphiarthroses are a **syndesmosis** and a **symphysis**.

4. The bony surfaces at diarthroses, or **synovial joints**, are covered by **articular cartilages**, lubricated by **synovial fluid**, and enclosed within a **joint capsule**. Other synovial structures include **menisci**, **fat pads**, **bursae**, and various **ligaments**. *(Figure 6-31)*

5. Important terms that describe dynamic motion at synovial joints are **gliding**, **flexion**, **extension**, **hyperextension**, **abduction**, **adduction**, **circumduction**, and **rotation**. *(Figures 6-32, 6-33)*

6. The bones in the forearm permit **pronation** and **supination**. *(Figure 6-33)*

7. Movements of the foot include **inversion** and **eversion**. The ankle undergoes **dorsiflexion** and **plantar flexion**. **Opposition** is the thumb movement that enables us to grasp and hold objects. *(Figure 6-34)*

8. **Protraction** involves moving a part of the body forward; **retraction** involves moving it back. **Depression** and **elevation** occur when we move a structure inferiorly and superiorly, respectively. *(Figure 6-34)*

9. Major types of joints include **gliding joints**, **hinge joints**, **pivot joints**, **ellipsoidal joints**, **saddle joints**, and **ball-and-socket joints**. *(Figure 6-35)*

10. The articular processes of adjacent vertebrae form gliding joints. The vertebral bodies form *symphyseal joints*. They are separated by pads called *intervertebral discs*. *(Figure 6-36)*

11. The shoulder joint is formed by the **glenoid cavity** and the head of the humerus. This joint is extremely mobile and, for that reason, it is also unstable and easily dislocated. *(Figure 6-37)*

12. Bursae at the shoulder joint reduce friction from muscles and tendons during movement. *(Figure 6-37)*

13. The elbow joint permits only flexion and extension. It is extremely stable because of extensive ligaments and the shapes of the articulating elements. *(Figure 6-38)*

14. The hip joint is formed by the union of the **acetabulum** with the head of the femur. This ball-and-socket diarthrosis permits flexion and extension, adduction and abduction, circumduction, and rotation. *(Figure 6-39)*

15. The knee joint is a complicated hinge joint. The patella, or kneecap, is embedded within a tendon that supports the front of the joint. *(Figure 6-40)*

INTEGRATION WITH OTHER SYSTEMS 166

1. The skeletal system is dynamically associated with other systems. *(Figure 6-41)*

Review Questions

Level 1: Reviewing Facts and Terms

Match each item in column A with the most closely related item in column B. Use letters for answers in the spaces provided.

COLUMN A

___ 1. osteocytes
___ 2. diaphysis
___ 3. auditory ossicles
___ 4. cribriform plate
___ 5. osteoblasts
___ 6. C_1
___ 7. C_2
___ 8. hip and shoulder
___ 9. patella
___ 10. calcaneus
___ 11. synarthrosis
___ 12. moving the hand into a palm-front position
___ 13. osteoclasts
___ 14. raising the arm laterally
___ 15. elbow and knee

COLUMN B

a. abduction
b. heelbone
c. ball-and-socket joints
d. bone-dissolving cells
e. hinge joints
f. axis
g. immovable joint
h. bone shaft
i. mature bone cells
j. bone-producing cells
k. atlas
l. olfactory nerves
m. ear bones
n. supination
o. kneecap

16. Skeletal bones store energy reserves as lipids in areas of
    (a) red marrow
    (b) yellow marrow
    (c) the matrix of bone tissue
    (d) the ground substance

17. The two types of osseous tissue are
    (a) compact bone and spongy bone
    (b) dense bone and compact bone
    (c) spongy bone and cancellous bone
    (d) a, b, and c are correct

18. The basic functional units of mature compact bone are
    (a) lacunae
    (b) osteocytes
    (c) osteons
    (d) canaliculi

19. The axial skeleton consists of the bones of the
    (a) pectoral and pelvic girdles
    (b) skull, thorax, and vertebral column
    (c) arm, legs, hand, and feet
    (d) limbs, pectoral girdle, and pelvic girdle

20. The appendicular skeleton consists of the bones of the
    (a) pectoral and pelvic girdles
    (b) skull, thorax, and vertebral column
    (c) arm, legs, hand, and feet
    (d) limbs, pectoral girdle, and pelvic girdle

21. Which of the following contains only bones of the cranium?
    (a) frontal, parietal, occipital, sphenoid
    (b) frontal, occipital, zygomatic, parietal
    (c) occipital, sphenoid, temporal, palatine
    (d) mandible, maxilla, nasal, zygomatic

6

THE SKELETAL SYSTEM

The Structure of Bone • Bone Development and Growth • Remodeling and Homeostatic Mechanisms • Aging and the Skeletal System

22. Of the following bones, which one is unpaired?
    (a) vomer
    (b) maxilla
    (c) palatine
    (d) nasal

23. At the glenoid cavity, the scapula articulates with the proximal end of the
    (a) humerus
    (b) radius
    (c) ulna
    (d) femur

24. While an individual is in the anatomical position, the ulna lies
    (a) medial to the radius
    (b) lateral to the radius
    (c) inferior to the radius
    (d) superior to the radius

25. Each coxal bone of the pelvic girdle consists of three fused bones:
    (a) ulna, radius, humerus
    (b) ilium, ischium, pubis
    (c) femur, tibia, fibula
    (d) hamate, capitate, trapezium

26. Joints that are typically located at the end of long bones are
    (a) synarthroses
    (b) amphiarthroses
    (c) diarthroses
    (d) sutures

27. The function of the synovial fluid is
    (a) to nourish chondrocytes
    (b) to provide lubrication
    (c) to absorb shock
    (d) a, b, and c are correct

28. Abduction and adduction always refer to movements of the
    (a) axial skeleton
    (b) appendicular skeleton
    (c) skull
    (d) vertebral column

29. Standing on tiptoe is an example of a movement called
    (a) elevation
    (b) dorsiflexion
    (c) plantar flexion
    (d) retraction

30. What are the five primary functions of the skeletal system?

31. What is the primary difference between intramembranous ossification and endochondral ossification?

32. What unique characteristic of the hyoid bone makes it different from all the other bones in the body?

33. What two primary functions are performed by the thoracic cage?

34. Which two large scapular processes are associated with the shoulder joint?

Level 2: Reviewing Concepts

35. Why are stresses or impacts to the side of the shaft in a long bone more dangerous than stress applied to the long axis of the shaft?

36. During the growth of a long bone, how is the epiphysis forced farther from the shaft?

37. Why are ruptured intervertebral discs more common in lumbar vertebrae, and dislocations and fractures more common in cervical vertebrae?

38. Why are clavicular injuries common?

39. What is the difference between the pelvic girdle and the pelvis?

40. How do articular cartilages differ from other cartilages in the body?

41. What is the significance of the fact that the pubic symphysis is a slightly movable joint?

Level 3: Critical Thinking and Clinical Applications

42. While playing on her swingset, 10-year-old Sally falls and breaks her right leg. At the emergency room, the physician tells Sally's parents that the proximal end of the tibia, where the epiphysis meets the diaphysis, is fractured. The fracture is properly set and eventually heals. During a routine physical when she is 18, Sally learns that her right leg is 2 cm shorter than her left, probably because of her accident. What might account for this difference?

43. Tess is diagnosed with a disease that affects the membranes surrounding the brain. The physician tells Tess's family that the disease is caused by an airborne virus. Explain how this virus could have entered the cranium.

44. While working at an excavation, an archaeologist finds several small skull bones. She examines the frontal, parietal, and occipital bones and concludes that the skulls are those of children not yet 1 year old. How can she tell their ages from examining the bones?

45. Frank Fireman is fighting a fire in a building when part of the ceiling collapses and a beam strikes him on his left shoulder. He is rescued by his friends, but he has a great deal of pain in his shoulder and cannot move his arm properly, especially in the anterior direction. His clavicle is not broken, and his humerus is intact. What is the probable nature of Frank's injury?

46. Ed "overturns" his ankle while playing tennis. He experiences swelling and pain, but after examination, he is told that there are no torn ligaments and the structure of the ankle is not affected. On the basis of the symptoms and the examination results, what do you think happened to Ed's ankle?

Answers to Concept Check Questions

Page 131

1. If the ratio of collagen to calcium in a bone increased, the bone would be more flexible and less strong.
2. Concentric layers of bone around a central canal are indicative of an osteon. Osteons make up compact bone. Since the ends (epiphyses) of long bones are primarily cancellous (spongy) bone, this sample most likely came from the shaft (diaphysis) of a long bone.
3. Since osteoclasts function in breaking down or demineralizing bone, the bone would have less mineral content and as a result would be weaker.

Page 133

1. Long bones of the body, such as the femur, have a plate of cartilage, called the epiphyseal plate, that separates the epiphysis from the diaphysis as long as the bone is still growing lengthwise. An X-ray would indicate whether the epiphyseal plate is still present. If it is, then growth is still occurring; if not, the bone has reached its adult length.
2. The increase in the male sex hormone testosterone that occurs at puberty contributes to an increased rate of bone growth and the closure of the epiphyseal plates. Since the source of testosterone, the testes, is removed in castration, we would expect these boys to have a longer, though slower, growth period and be taller than they would have been if they had not been castrated.
3. Women who are pregnant need large amounts of calcium to support the needs of the developing fetus for bone growth. If the expectant mother does not include enough calcium in her diet, her body will mobilize the calcium reserves of her own skeleton to provide for the needs of the fetus, resulting in weakened bones and an increased risk of fracture.

Page 137

1. The larger arm muscles of the weight lifter will apply more mechanical stress to the bones of the arms. In response to the stress, the bones will grow thicker. For similar reasons, we would expect the jogger to have heavier thigh bones.
2. The sex hormones known as estrogens play an important role in moving calcium into bones. After menopause, the level of these hormones decreases dramatically. As a result, it is difficult to replace the calcium in bones that is being lost due to normal aging. Males do not show a decrease in sex hormone levels (androgens) until much later in life.

Page 143

1. The mastoid and styloid processes are projections on the temporal bones of the skull.
2. The sella turcica contains the pituitary gland and is located in the sphenoid bone.
3. The occipital condyles of the occipital bone of the cranium articulate with the vertebral column.

Page 144

1. The bone that forms the superior portion of the orbit is the frontal bone, and the maxillary and the zygomatic bones form the inferior portion of the orbit. These would be the three bones fractured by the ball.
2. The paranasal sinuses make some of the heavier skull bones lighter and contain a mucous epithelium that releases mucous secretions into the nasal cavities. The mucus warms, moistens, and filters incoming air.
3. The most powerful muscles that are involved in closing the mouth attach to the mandible at the coronoid process. A fracture of the coronoid process would make it difficult for these muscles to function properly and to close the mouth.
4. Since many muscles that move the tongue and the larynx are attached to the hyoid bone, you would expect a person with a fractured hyoid bone to have difficulty moving the tongue, breathing, and swallowing.

Page 150

1. The dens is located on the second cervical vertebra, or axis, which is in the neck.
2. In adults, the five sacral vertebrae fuse to form a single sacrum.
3. The lumbar vertebrae must support a great deal more weight than do vertebrae that are more superior in the spinal column. The large vertebral bodies allow the weight to be distributed over a larger area.
4. True ribs are each attached directly to the sternum by their own costal cartilage. False ribs either do not attach to the sternum (the floating ribs) or attach by means of a shared, common costal cartilage.
5. Improper compression of the chest during CPR could and frequently does result in a fracture of the sternum or ribs.

Page 153

1. The clavicle attaches the scapula to the sternum and thus restricts the scapula's range of movement. If the clavicle is broken, then the scapula will have a greater range of movement and will be less stable.
2. The two rounded prominences on either side of the elbow are parts of the humerus (the lateral and medial epicondyles).

Page 156

1. The three bones that make up the hip are the ilium, ischium, and the pubis.
2. Although the fibula is not part of the knee joint and does not bear weight, it is an important point of attachment for many leg muscles. When the fibula is fractured, these muscles cannot function properly to move the leg and walking is difficult and painful. The fibula also helps stabilize the ankle joint.
3. Joey has most likely fractured his calcaneus (heel bone).

6 THE SKELETAL SYSTEM

The Structure of Bone • Bone Development and Growth • Remodeling and Homeostatic Mechanisms • Aging and the Skeletal System

Page 163

1. Originally, this is an amphiarthrotic (slightly movable) joint. Structurally, it is a type of syndesmosis. When the skull bones interlock, they form a synarthrotic (immovable) joint. This type of fibrous joint is called a suture.
2. (a) abduction; (b) supination; (c) flexion.
3. Flexion and extension.

Page 164

1. Since the subscapular bursa is located in the shoulder joint, the tennis player would have inflammation of this structure (bursitis). The condition is associated with the intensity of repetitive motion that occurs at the shoulder, such as swinging a tennis racket. The jogger would be more at risk for injuries to the knee joint.
2. Mary has most likely fractured her ulna.

Page 166

1. The knee joint is stabilized by ligaments on its anterior, posterior, medial and lateral surfaces, as well as a pair of ligaments within the joint capsule. As a result, a complete dislocation is rare.
2. Damage to the menisci in the knee joint would result in a decrease in the joint's stability. The individual would have a harder time locking the knee in place while standing and would have to use muscle contractions to stabilize the joint. When the individual stands for long periods, the muscles would fatigue and the knee would "give out." We would also expect the individual to experience pain.

Page 166

1. The skeletal system provides calcium for the muscular system (normal muscle contraction), the nervous system (normal neural function), and the cardiovascular system (normal contraction of cardiac muscle of the heart).

EXPLORE
MediaLab

EXPLORATION #1

Estimated time for completion: 5 minutes

BY NOW YOU ARE AWARE that the systems of the body work in concert to maintain homeostasis. You also know that bone is a dynamic tissue, with a rich blood supply and a continual recycling of the minerals in the matrix. When you review Figure 6-5●, it becomes obvious that bone formation requires a blood supply. Somehow the skeletal system must communicate with the cardiovascular system when creating new bone. Cancerous or malignant tumors also require a blood supply to grow and spread. If we knew precisely how the skeletal and cardiovascular systems communicate, might that information be helpful in controlling growth of cancer cells? Using Figure 6-5● as a reference, predict the nature of the developmental relationship between the skeletal and cardiovascular systems. Once you have your ideas written down, go to the Companion Web site, visit the MediaLab section, and click on the keyword "Bone Formation." Read the article from the Harvard University Archives explaining the link between vascularization and bone growth. Alter your prediction of the interactions between these two systems to include this new information.

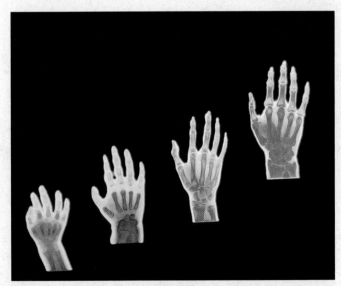

Colored X-ray of bone growth of the hand at different ages (left to right): 1 year, 3 years, 13 years, and 20 years.

Reaching for an item in the cupboard is just one of the common, everyday movements made possible by the shoulder joint.

EXPLORATION #2

Estimated time for completion: 10 minutes

JOINTS ARE AMAZING—supportive, yet flexible, strong, yet easily damaged if abused. The joint with the largest range of motion is the ball and socket joint of the shoulder. It is reinforced by the rotator cuff, four muscle tendons that surround and merge with the joint capsule. Refer to Figures 6-37● and 7-18● for the structure of this joint. Can you predict what might happen to the mobility of this joint if you were to damage it? Using the above two figures as your guides, prepare a list of the structures that might tear were you to fall and land on the outer portion of the arm, near the acromion. What effect would this have on the movements of the arm? If this injury were to occur in early adulthood, what might happen to the repaired joint as the body ages?

Check your predictions by visiting the Companion Web site's MediaLab section and clicking on the keyword "Older Shoulder." Here you will find a case study, "The Older Shoulder," that takes you through an actual rotator cuff injury. Were you able to predict the injury and symptoms experienced by this patient?

The Muscular System

CHAPTER OUTLINE AND OBJECTIVES

Vocabulary Development

aer ....air; *aerobic*
an ....not; *anaerobic*
bi ....two; *biceps*
caput ....head; *biceps*
clavius ....clavicle; *clavicle*
di ....two; *digastricus*
epi- ....on; *epimysium*
ergon ....work; *synergist*
fasciculus ....a bundle; *fascicle*
galea ....a helmet; *galea aponeurotica*
gaster ....stomach; *gastrocnemius*
hyper ....above; *hypertrophy*
iso- ....equal; *isometric*
kneme ....knee; *gastrocnemius*
lemma ....husk; *sarcolemma*
meros ....part; *sarcomere*
metron ....measure; *isometric*
mys ....muscle; *epimysium*
peri- ....around; *perimysium*
platys ....flat; *platysma*
sarkos ....flesh; *sarcolemma*
syn- ....together; *synergist*
tetanos ....convulsive tension; *tetanus*
tonos ....tension; *isotonic*
trope ....a turning; *tropomyosin*
-trophy ....nourishing; *atrophy*

*D*URING A GAME, *basketball players rely on the peak performance of many of the 700 or so skeletal muscles in the body. The contractions of these muscles are responsible for the powerful and delicate movements of the axial and appendicular skeleton.*

7 THE MUSCULAR SYSTEM

Functions of Skeletal Muscle • The Anatomy of Skeletal Muscles • The Control of Muscle Fiber Contraction • Muscle Mechanics • The Energetics of Muscular Activity
Chapter Review • MediaLab

I T IS HARD TO IMAGINE what life would be like without muscle tissue. We would be unable to sit, stand, walk, speak, or grasp objects. Blood would not circulate, because there would be no heartbeat to propel it through the vessels. The lungs could not rhythmically empty and fill, nor could food move through the digestive tract. Muscle tissue, one of the four primary tissue types, consists chiefly of elongated muscle cells that are highly specialized for contraction. The three types of muscle tissue—*skeletal muscle, cardiac muscle,* and *smooth muscle*—were introduced in Chapter 4. ∞ p. 100

This chapter begins with the organization of skeletal muscle tissue, the most abundant muscle tissue in the body. It is followed by an overview of the differences among skeletal, cardiac, and smooth muscle tissue. We will then proceed to the muscular system.

Functions of Skeletal Muscle

Skeletal muscle tissue forms **skeletal muscles**, organs that also contain connective tissues, nerves, and blood vessels. Skeletal muscles are directly or indirectly attached to the bones of the skeleton. The muscular system includes approximately 700 skeletal muscles. These muscles perform the following functions:

1. *Produce movement.* Skeletal muscle contractions pull on tendons and move the bones of the skeleton. These contractions may produce a simple motion, such as extending the arm, or the highly coordinated movements of swimming, skiing, or typing.
2. *Maintain posture and body position.* Continuous muscle contractions maintain body posture. Without this constant action, you could not sit upright without collapsing or stand without toppling over.
3. *Support soft tissues.* The abdominal wall and the floor of the pelvic cavity consist of layers of skeletal muscle. These muscles support the weight of visceral organs and shield internal tissues from injury.
4. *Guard entrances and exits.* Skeletal muscles encircle openings to the digestive and urinary tracts. These muscles provide voluntary control over swallowing, defecation, and urination.
5. *Maintain body temperature.* Muscle contractions require energy, and whenever energy is used in the body, some of it is converted to heat. The heat released by working muscles keeps the body temperature in the range required for normal functioning.

The Anatomy of Skeletal Muscles

The study of the structure of skeletal muscle underlies an understanding of how skeletal muscle contracts. We begin with the organ-level structure of skeletal muscle before describing its cellular-level structure. When naming structural features of muscles and their components, anatomists often used the Greek words *sarkos* (flesh) and *mys* (muscle). These word roots will be encountered in the following discussions.

GROSS ANATOMY

Figure 7-1● illustrates the appearance and organization of a typical skeletal muscle. A skeletal muscle contains connective tissues, blood vessels, nerves, and skeletal muscle tissue. Each cell in skeletal muscle tissue is a single muscle *fiber.*

Connective Tissue Organization

Three layers of connective tissue are part of each muscle: an outer epimysium, a central perimysium, and an inner endomysium (Figure 7-1●). The entire muscle is surrounded by the **epimysium** (ep-i-MIS-ē-um; *epi-,* on + *mys,* muscle), a layer of collagen fibers that separates the muscle from surrounding tissues and organs.

The connective tissue fibers of the **perimysium** (per-i-MIS-ē-um; *peri-,* around) divide the skeletal muscle into a series of compartments, each containing a bundle of muscle fibers called a **fascicle** (FA-sik-ul; *fasciculus,* a bundle). In addition to collagen and elastic fibers, the perimysium contains blood vessels and nerves that supply the fascicles.

Within a fascicle, the **endomysium** (en-dō-MIS-ē-um; *endo-,* inside) surrounds each skeletal muscle fiber and ties adjacent muscle fibers together. Stem cells scattered among the fibers help repair damaged muscle tissue.

At each end of the muscle, the collagen fibers of all three layers come together to form a bundle known as a **tendon**, or a broad sheet called an *aponeurosis.* Tendons are bands of collagen fibers that attach skeletal muscles to bones, and aponeuroses connect different skeletal muscles. ∞ p. 94 The tendon fibers are interwoven into the periosteum of the bone, providing a firm attachment. Any contraction of the muscle will exert a pull on its tendon and in turn on the attached bone.

Nerves and Blood Vessels

The connective tissues of the epimysium and perimysium contain the blood vessels and nerves that supply the muscle fibers. Muscle contraction requires a tremendous amount of energy. An extensive network of blood vessels delivers the oxygen and nutrients needed for the production of ATP in active skeletal muscles.

Skeletal muscles contract only under stimulation from the central nervous system. *Axons,* or nerve fibers, involved with a

● *Figure 7-1* **Gross Anatomy of a Skeletal Muscle**

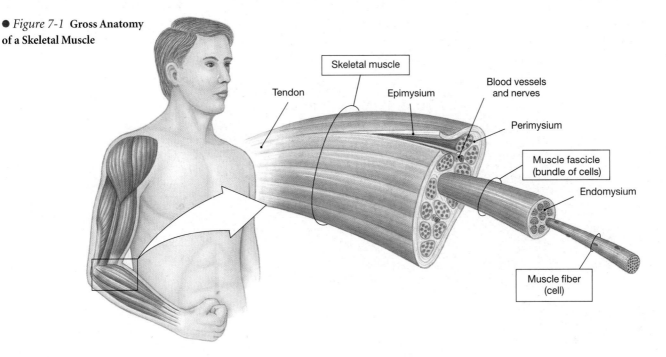

particular skeletal muscle penetrate the epimysium, branch through the perimysium, and enter the endomysium to control individual muscle fibers. Skeletal muscles are often called *voluntary muscles* because we have voluntary control over their contractions. Many of these skeletal muscles may also be controlled at a subconscious level. For example, skeletal muscles involved with breathing, such as the *diaphragm*, usually work outside our conscious awareness.

MICROANATOMY

Skeletal muscle fibers (Figures 7-1 and 7-2●) are quite different from the "typical" cell described in Chapter 3. One obvious difference is their enormous size. For example, a skeletal muscle fiber from a leg muscle could have a diameter of 100 μm and a length equal to that of the entire muscle (up to 30 cm, or 12 in.). In addition, each skeletal muscle fiber is *multinucleate*, containing hundreds of nuclei just beneath the cell membrane.

The Sarcolemma and Transverse Tubules

The cell membrane, or **sarcolemma** (sar-cō-LEM-a; *sarkos*, flesh + *lemma*, husk) of a muscle fiber surrounds the cytoplasm, or **sarcoplasm** (SAR-kō-plazm). Openings scattered across the surface of the sarcolemma lead into a network of narrow tubules called **transverse tubules**, or **T tubules**. Filled with extracellular fluid, the T tubules form passageways through the muscle fiber, like a series of tunnels through a mountain.

The T tubules play a major role in coordinating the contraction of the muscle fiber. A muscle fiber contraction occurs through the orderly interaction of both electrical and chemical events. Electrical impulses conducted by the sarcolemma

trigger a contraction by altering the chemical environment everywhere inside the muscle fiber. The electrical impulses reach the cell interior by traveling along the transverse tubules that extend deep into the sarcoplasm of the muscle fiber.

Myofibrils

Inside the muscle fiber, branches of T tubules encircle cylindrical structures called myofibrils. A myofibril is a cylinder 1–2 μm in diameter and as long as the entire muscle fiber. Each muscle fiber contains hundreds to thousands of myofibrils. Myofibrils are bundles of **myofilaments**, protein filaments consisting primarily of the proteins *actin* and *myosin*. Actin molecules are found in **thin filaments**, and the myosin molecules in thick filaments. Myofibrils are responsible for muscle fiber contraction. Because they are attached to the sarcolemma at each end of the cell, their contraction shortens the entire cell. Scattered among the myofibrils are mitochondria and granules of glycogen, a source of glucose. The breakdown of glucose and the activity of mitochondria provide the ATP needed to power muscular contractions.

The Sarcoplasmic Reticulum

Wherever a T tubule encircles a myofibril, the tubule makes close contact with the membranes of the **sarcoplasmic reticulum (SR)**, a specialized form of smooth endoplasmic reticulum. The sarcoplasmic reticulum forms a tubular network around each myofibril. On either side of a T tubule lie expanded chambers of the SR called *cisternae*. As it encircles a myofibril, a transverse tubule lies sandwiched between a pair of cisternae, forming a *triad* (Figure 7-2a●).

The cisternae contain high concentrations of calcium ions. The calcium ion concentration in the cytoplasm of all cells is

179

7 THE MUSCULAR SYSTEM

Functions of Skeletal Muscle • **The Anatomy of Skeletal Muscles** • The Control of Muscle Fiber Contraction • Muscle Mechanics • The Energetics of Muscular Activity
Chapter Review • MediaLab

● *Figure 7-2* **The Organization of Skeletal Muscles**
(**a**) The structure of a skeletal muscle fiber.
(**b**) The organization of a sarcomere, part of a single myofibril.
(**c**) The sarcomere in part b, stretched such that thick and thin filaments no longer overlap. (This cannot happen in an intact muscle fiber.)
(**d**) The structure of a thin filament.
(**e**) The structure of a thick filament.

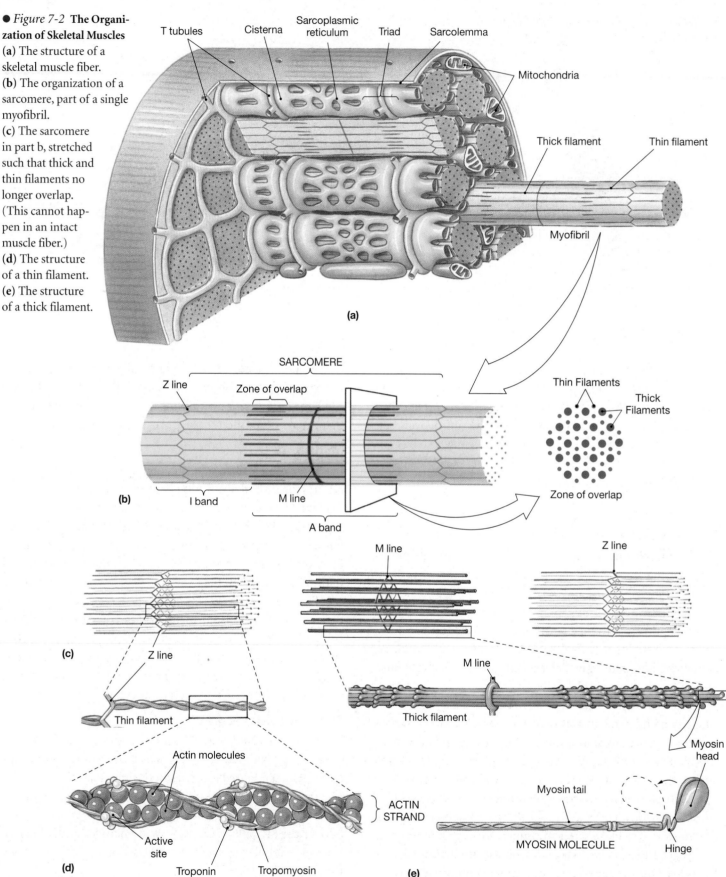

kept very low. Most cells, including skeletal muscle fibers, pump calcium ions across their cell membranes and into the extracellular fluid. Skeletal muscle fibers, however, also actively transport calcium ions into the cisternae of the sarcoplasmic reticulum. A muscle contraction begins when the stored calcium ions are released by the cisternae.

Sarcomeres

Myofilaments are organized into repeating functional units called **sarcomeres** (SAR-kō-mērz; *sarkos*, flesh + *meros*, part) (Figure 7-2b●). Each myofibril consists of approximately 10,000 sarcomeres arranged end to end. *The sarcomere is the smallest functional unit of the muscle fiber. Interactions between the thick and thin filaments of sarcomeres are responsible for muscle contraction.*

The arrangement of thick and thin filaments within a sarcomere produces a banded appearance. All of the myofibrils are arranged parallel to the long axis of the cell, with their sarcomeres lying side by side. As a result, the entire muscle fiber has a banded, or striated, appearance corresponding to the bands of the individual sarcomeres (Figure 7-2a●).

Figure 7-2b● diagrams the external and internal structure of an individual sarcomere. Each sarcomere has a resting length of about 2.6 μm. The thick filaments lie in the center of the sarcomere. Thin filaments at either end of the sarcomere are attached to interconnecting proteins that make up the **Z lines**, the boundaries of each sarcomere. From the Z lines, the thin filaments extend toward the center of the sarcomere, passing among the thick filaments in the zone of overlap. The **M line** is made up of proteins that connect the central portions of each thick filament to its neighbors. The relationships of Z lines and M lines are shown in Figure 7-2c●.

Differences in the size and density of thick and thin filaments account for the banded appearance of the sarcomere. The dark **A band** is the area containing thick filaments. The light region between two successive A bands—including the Z line—is the **I band**. (It may help you to remember that the A band appears d*A*rk and the I band l*I*ght in a light micrograph.)

Thin and Thick Filaments

Each thin filament consists of a twisted strand of actin molecules (Figure 7-2d●). Each actin molecule has an **active site** capable of interacting with myosin. In a resting muscle, the active sites along the thin filaments are covered by strands of the protein **tropomyosin** (trō-pō-MĪ-o-sin; *trope*, turning). The tropomyosin strands are held in position by molecules of **troponin** (TRŌ-pō-nin) that are bound to the actin strand.

Thick filaments are composed of myosin molecules, each with a *tail* and a globular *head*. The myosin molecules are oriented away from the center of the sarcomere, with the heads pro-

jecting outward (Figure 7-2e●). The myosin heads attach to actin molecules during a contraction. This interaction cannot occur unless the troponin changes position, moving the tropomyosin and exposing the active sites.

Calcium is the key that unlocks the active sites and starts a contraction. When calcium ions bind to troponin, the protein changes shape, swinging the tropomyosin away from the active sites. Myosin-actin binding can then occur, and a contraction begins. The cisternae of the sarcoplasmic reticulum are the source of the calcium that triggers muscle contraction.

Sliding Filaments and Cross-Bridges

When a sarcomere contracts, the I bands get smaller, the Z lines move closer together, but the width of the A bands does not change. These observations make sense only if the thin filaments slide toward the center of the sarcomere, alongside the stationary thick filaments (Figure 7-3●). This

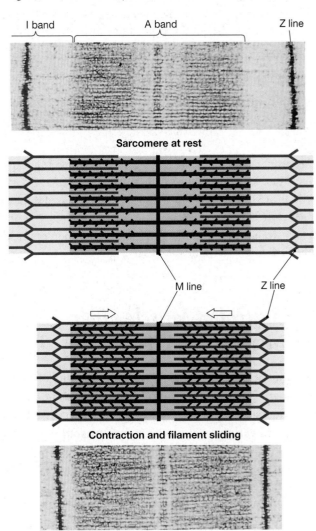

I band   A band   Z line

Sarcomere at rest

M line   Z line

Contraction and filament sliding

● *Figure 7-3* **Changes in the Appearance of a Sarcomere During Contraction of a Skeletal Muscle Fiber**
During a contraction, the A band stays the same width, but the Z lines move closer together and the I band gets smaller.

7 THE MUSCULAR SYSTEM

Functions of Skeletal Muscle • The Anatomy of Skeletal Muscles • **The Control of Muscle Fiber Contraction** • Muscle Mechanics • The Energetics of Muscular Activity
Chapter Review • MediaLab

explanation for sarcomere contraction is called the *sliding filament theory*.

The mechanism responsible for sliding filaments involves the binding of the myosin heads of thick filaments to active sites on the thin filaments. When they connect thick filaments and thin filaments, the myosin heads are called **cross-bridges**. When a cross-bridge binds to an active site, it pivots toward the center of the sarcomere, pulling the thin filament in that direction. The cross-bridge then detaches and returns to its original position, ready to repeat a cycle of "attach, pivot, detach, and return," like a person pulling in a rope one-handed.

 ### THE MUSCULAR DYSTROPHIES

Abnormalities in the genes that code for structural and functional proteins in muscle fibers are responsible for a number of inherited diseases collectively known as the **muscular dystrophies** (DIS-trō-fēz). These conditions, which cause a progressive muscular weakness and deterioration, are the result of abnormalities in the sarcolemma or in the structure of internal proteins. The best-known example is *Duchenne's muscular dystrophy*, which typically develops in males from 3 to 7 years old.

CONCEPT CHECK QUESTIONS

Answers on page 222

❶ How would severing the tendon attached to a muscle affect the ability of the muscle to move a body part?

❷ Why does skeletal muscle appear striated when viewed through a microscope?

❸ Where would you expect the greatest concentration of calcium ions in resting skeletal muscle to be?

The Control of Muscle Fiber Contraction

Skeletal muscle fibers contract only under the control of the nervous system. The communication link between the nervous system and a skeletal muscle fiber occurs at a specialized intercellular connection known as a **neuromuscular junction** (Figure 7-4a●).

THE NEUROMUSCULAR JUNCTION

Each skeletal muscle fiber is controlled by a nerve cell called a *motor neuron*, through a single neuromuscular junction midway along the fiber's length. Figure 7-4b● summarizes key features of this structure. A single axon branches within the perimysium to form a number of fine branches. Each branch ends at an expanded **synaptic terminal**. p. 103

The cytoplasm of the synaptic terminal contains mitochondria and vesicles filled with molecules of **acetylcholine** (as-ē-til-KŌ-lēn), or **ACh**. Acetylcholine is a *neurotransmitter*, a chemical released by a neuron to communicate with other cells. The release of ACh from the synaptic terminal results in changes in the sarcolemma that trigger the contraction of the muscle fiber.

A narrow space, the **synaptic cleft**, separates the synaptic terminal from the sarcolemma. This portion of the membrane, which contains receptors that bind ACh, is known as the **motor end plate**. Both the synaptic cleft and the motor end plate contain the enzyme **acetylcholinesterase** (**AChE**, or *cholinesterase*), which breaks down molecules of ACh.

Neurons control skeletal muscle fibers by stimulating the production of an **action potential**, or electrical impulse, in the sarcolemma. The sequence of events shown in Figure 7-4c● can be summarized as follows:

Step 1: *The release of acetylcholine.* An action potential travels along the axon of a motor neuron. When this impulse reaches the synaptic terminal, vesicles in the synaptic terminal release acetylcholine into the synaptic cleft.

Step 2: *The binding of ACh at the motor end plate.* The ACh molecules diffuse across the synaptic cleft and bind to ACh receptors on the sarcolemma. This event changes the permeability of the membrane to sodium ions. It is this sudden rush of sodium ions into the sarcoplasm that produces an action potential in the sarcolemma. (We will examine the formation and conduction of action potentials more closely in Chapter 8.)

Step 3: *The conduction of action potentials by the sarcolemma.* The action potential spreads over the entire sarcolemma surface. It also travels down all of the transverse tubules toward the cisternae that encircle the sarcomeres of the muscle fiber. The passage of an action potential triggers a sudden, massive release of calcium ions by the cisternae.

As the calcium ion concentration rises, active sites are exposed on the thin filaments, cross-bridge interactions occur, and a contraction begins. Because all of the cisternae in the muscle fiber are affected, this contraction is a combined effort involving every sarcomere on every myofibril. While the contraction process gets under way, the acetylcholine is being broken down by acetylcholinesterase.

THE CONTRACTION CYCLE

In the resting sarcomere, each cross-bridge is bound to a molecule of ADP and phosphate (PO_4^{3-}), the products released by the breakdown of a molecule of ATP. The cross-bridge stores

● *Figure 7-4* **Neural Control of Muscle Contraction**
(**a**) A colorized SEM of a neuromuscular junction.
(**b**) An action potential (in red) arrives at the synaptic
terminal. (**c**) Steps in the chemical communication
between the synaptic terminal and motor end plate.

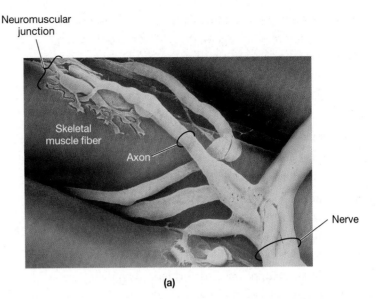

(a)

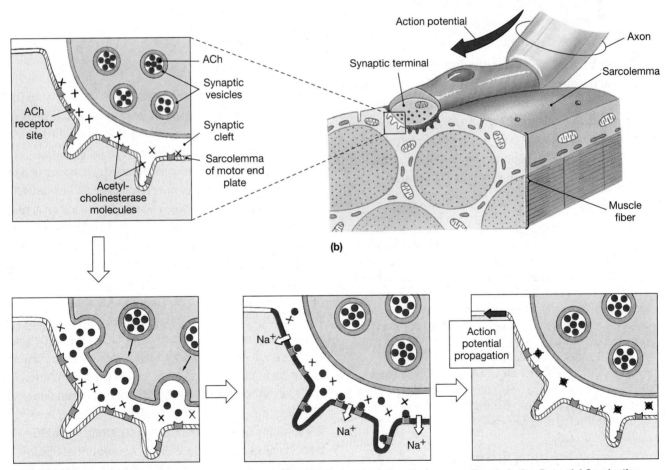

(b)

Step 1: Release of Acetylcholine.
Vesicles in the synaptic terminal release
their contents into the synaptic cleft.

**Step 2: ACh Binding at the Motor End
Plate.**   The binding of ACh to
the receptors changes the membrane
permeability and induces an action
potential in the sarcolemma.

**Step 3: Action Potential Conduction
by the Sarcolemma.**   The action
potential spreads across the membrane
surface and travels down the
transverse tubules, triggering the
release of calcium ions at the cisternae.
While this occurs, AChE removes the
acetylcholine from the synaptic cleft.

(c)

7 THE MUSCULAR SYSTEM

Functions of Skeletal Muscle • The Anatomy of Skeletal Muscles • The Control of Muscle Fiber Contraction • **Muscle Mechanics** • The Energetics of Muscular Activity
Chapter Review • MediaLab

the energy released by the rupture of the high-energy bond. In effect, the resting cross-bridge is "primed" for a contraction, like a cocked pistol or a set mousetrap.

The contraction process involves the following five interlocking steps, which are shown schematically in Figure 7-5● :

Step 1: The exposure of the active site following the binding of calcium ions (Ca^{2+}) to troponin.

Step 2: The attachment of the myosin cross-bridge to the exposed active site on the thin filaments.

Step 3: The pivoting of the attached myosin head toward the center of the sarcomere and the release of ADP and a phosphate group. This step uses the energy that was stored in the myosin molecule at rest.

Step 4: The detachment of the cross-bridges when the myosin head binds another ATP molecule.

Step 5: The reactivation of the detached myosin head as it splits the ATP and captures the released energy. The entire cycle can now be repeated, beginning with step 2.

This cycle is broken when calcium ion concentrations return to normal resting levels, primarily through active transport into the sarcoplasmic reticulum. If a single action potential sweeps across the sarcolemma, calcium ion removal occurs very rapidly and the contraction will be very brief. A sustained contraction will occur only if action potentials occur one after another, and calcium loss from the cisternae continues.

 RIGOR MORTIS

When death occurs, circulation ceases and the skeletal muscles are deprived of nutrients and oxygen. Within a few hours, the skeletal muscle fibers have run out of ATP, and the sarcoplasmic reticulum becomes unable to remove calcium ions from the sarcoplasm. Calcium ions diffusing into the sarcoplasm from the extracellular fluid or leaking out of the sarcoplasmic reticulum then trigger a sustained contraction. Without ATP, the cross-bridges cannot detach from the active sites, and the muscle locks in the contracted position. All of the body's skeletal muscles are involved, and the individual becomes "stiff as a board." This physical state, called **rigor mortis**, lasts until the lysosomal enzymes released by autolysis break down the myofilaments 15–25 hours later.

MUSCLE CONTRACTION: A SUMMARY

Table 7-1 (p. 186) provides a summary of the contraction process, from ACh release to the end of the contraction.

 INTERFERENCE WITH NEURAL CONTROL MECHANISMS

Anything that interferes with the generation of an action potential in the sarcolemma will cause muscular paralysis. Two examples are worth noting.

Cases of **botulism** result from the consumption of contaminated canned or smoked food. The toxin, produced by bacteria, prevents the release of ACh at the synaptic terminals, leading to a potentially fatal muscular paralysis.

The progressive muscular paralysis seen in **myasthenia gravis** results from the loss of ACh receptors at the motor end plate. The primary cause is a misguided attack on the ACh receptors by the immune system. Genetic factors play a role in predisposing individuals to develop this condition.

CONCEPT CHECK QUESTIONS
Answers on page 222

❶ How would a drug that interferes with cross-bridge formation affect muscle contraction?

❷ What would you expect to happen to a resting skeletal muscle if the sarcolemma suddenly became very permeable to calcium ions?

❸ Predict what would happen to a muscle if the motor end plate did not contain acetylcholinesterase.

Muscle Mechanics

Now that we are familiar with muscle contraction of individual muscle fibers, we can examine the performance of skeletal muscles. In this section we will consider the coordinated contractions of an entire population of muscle fibers.

The individual muscle cells in muscle tissue are tied together and surrounded by connective tissue. When muscle cells contract they pull on collagen fibers, producing an active force called **tension**. Tension applied to an object tends to pull the object toward the source of the tension. However, before movement can occur, the applied tension must overcome the object's **resistance**, a passive force that opposes movement. The amount of resistance can depend on an object's weight, shape, friction, and other factors. In contrast, **compression**, or a push applied to an object, tends to force the object away from the source of compression. Muscle cells can use energy to shorten and generate tension but not to lengthen and generate compression.

The amount of tension produced by an individual muscle fiber depends solely on the number of pivoting cross-bridges. If a muscle fiber at a given resting length is stimulated to contract, it will always produce the same amount of tension. There is no mechanism to regulate the amount of tension produced in that contraction by changing the number of contracting sarcomeres. The muscle fiber is either "ON" (producing tension) or "OFF" (relaxed). This is known as the **all-or-none principle**.

An entire skeletal muscle contracts when its component muscle fibers are stimulated. The amount of tension produced in the skeletal muscle *as a whole* is determined by (1) the frequency of stimulation and (2) the number of muscle fibers activated.

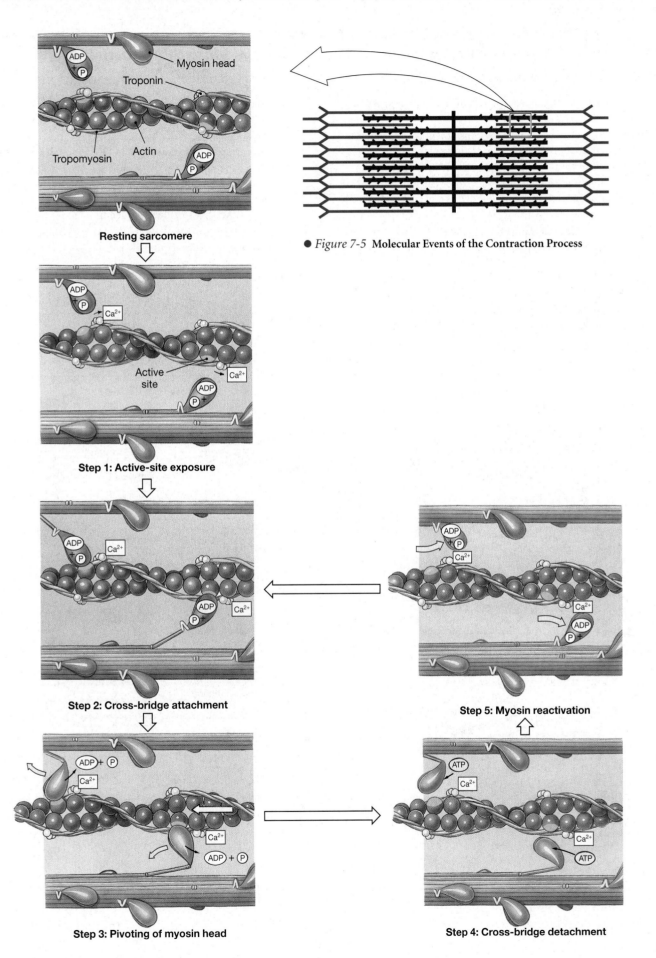

Resting sarcomere

● *Figure 7-5* **Molecular Events of the Contraction Process**

Step 1: Active-site exposure

Step 2: Cross-bridge attachment

Step 3: Pivoting of myosin head

Step 4: Cross-bridge detachment

Step 5: Myosin reactivation

7 THE MUSCULAR SYSTEM

Functions of Skeletal Muscle • The Anatomy of Skeletal Muscles • The Control of Muscle Fiber Contraction • **Muscle Mechanics** • The Energetics of Muscular Activity
Chapter Review • MediaLab

TABLE 7-1 A Summary of the Steps Involved in Skeletal Muscle Contraction

KEY STEPS IN THE INITIATION OF A CONTRACTION INCLUDE:	THIS PROCESS CONTINUES FOR A BRIEF PERIOD, UNTIL:

1. At the neuromuscular junction, ACh released by the synaptic terminal binds to receptors on the sarcolemma.

2. The resulting change in the membrane potential of the muscle fiber leads to the production of an action potential that spreads across its entire surface and along the T tubules.

3. The sarcoplasmic reticulum releases stored calcium ions, increasing the calcium concentration of the sarcoplasm in and around the sarcomeres.

4. Calcium ions bind to troponin, resulting in the movement of tropomyosin and the exposure of active sites on the thin (actin) filaments. Myosin cross bridges form when myosin heads bind to active sites.

5. Repeated cycles of cross-bridge binding, pivoting, and detachment occur, powered by the breakdown of ATP. These events produce filament sliding, and the muscle fiber shortens.

6. Action potential generation ceases as ACh is broken down by acetylcholinesterase.

7. The sarcoplasmic reticulum reabsorbs calcium ions, and the concentration of calcium ions in the sarcoplasm declines.

8. When calcium ion concentrations approach normal resting levels, the troponin and tropomyosin molecules return to their normal positions. These changes cover the active sites and prevent further cross-bridge interaction.

9. Without cross-bridge interactions, further sliding cannot take place and the contraction will end.

10. Muscle relaxation occurs, and the muscle returns passively toward resting length.

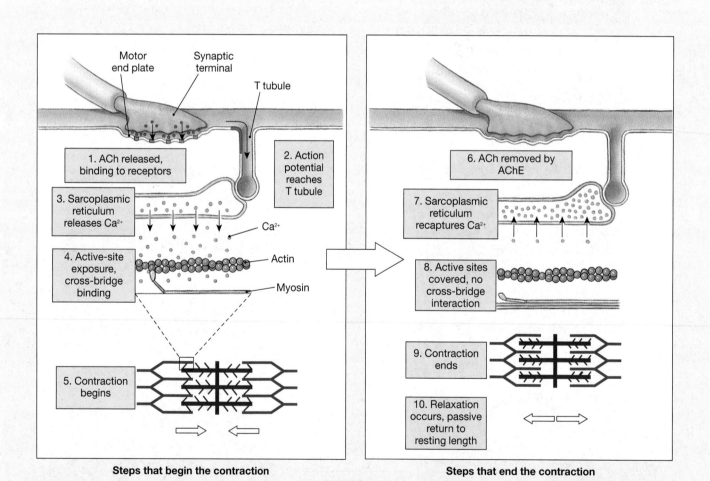

Steps that begin the contraction　　　　**Steps that end the contraction**

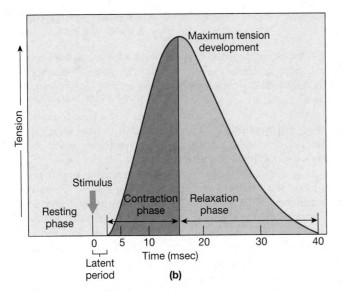

● *Figure 7-6* **The Twitch and Development of Tension**
A myogram showing the time course of a single twitch contraction in the gastrocnemius muscle.

THE FREQUENCY OF MUSCLE FIBER STIMULATION

A **twitch** is a single stimulus-contraction-relaxation sequence in a muscle fiber. Its duration can be as brief as 7.5 msec, as in an eye muscle fiber, or up to 100 msec in fibers of the *soleus*, a small calf muscle. A *myogram* is a graph of tension development in a muscle during a twitch contraction. Figure 7-6● is a myogram of the phases of a 40-msec twitch contraction in the *gastrocnemius muscle*, a prominent calf muscle:

• The **latent period** begins at stimulation and typically lasts about 2 msec. Over this period the action potential sweeps across the sarcolemma, and calcium ions are released by the sarcoplasmic reticulum. No tension is produced by the muscle fiber, because the contractile mechanism is not yet activated.

• In the **contraction phase**, tension rises to a peak. Throughout this period the cross-bridges are interacting with the active sites on the actin filaments. Maximum tension is reached roughly 15 msec after stimulation.

• During the **relaxation phase**, muscle tension falls to resting levels as calcium levels drop and the number of cross-bridges declines. This phase continues for about 25 msec.

A single stimulation produces a single twitch, but twitches in a skeletal muscle do not accomplish anything useful. All normal activities involve sustained muscle contractions. Such contractions result from repeated stimulations.

Summation and Incomplete Tetanus

If a second stimulus arrives before the relaxation phase has ended, a second, more powerful contraction occurs. The addition of one twitch to another in this way is called **summation** (Figure 7-7a●). If you continue to stimulate the muscle, never allowing it to relax completely, tension will peak, as illustrated in Figure 7-7b●. A muscle producing peak tension during rapid cycles of contraction and relaxation is said to be in **incomplete tetanus** (*tetanos*, convulsive tension).

Complete Tetanus

Complete tetanus occurs when the rate of stimulation is increased until the relaxation phase is completely eliminated (Figure 7-7c●). In complete tetanus, the action potentials are arriving so fast that the sarcoplasmic reticulum does not have time to reclaim the calcium ions. The high calcium ion concentration in the cytoplasm prolongs the state of contraction,

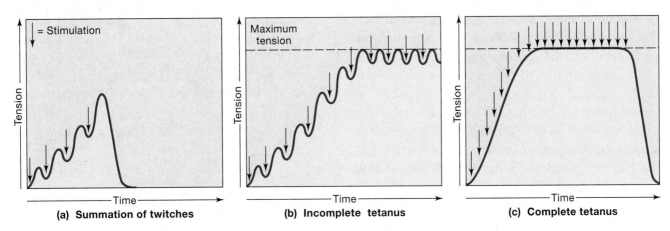

● *Figure 7-7* **The Effects of Repeated Stimulations**
(**a**) During summation, tension rises when successive stimuli arrive before the relaxation phase has been completed. (**b**) Incomplete tetanus occurs if the rate of stimulation increases further. Tension production will rise to a peak, and the periods of relaxation will be very brief. (**c**) In complete tetanus, the frequency of stimulation is so high that the relaxation phase has been completely eliminated and tension plateaus at maximal levels.

7 THE MUSCULAR SYSTEM

Functions of Skeletal Muscle • The Anatomy of Skeletal Muscles • The Control of Muscle Fiber Contraction • Muscle Mechanics • **The Energetics of Muscular Activity**
Chapter Review • MediaLab

✚ **CLINICAL NOTE** *Tetanus*

Children are often told to watch out for rusty nails. Parents should worry most not about the rust or the nail but about infection with a very common bacterium, *Clostridium tetani*. This bacterium can cause **tetanus**. Although they have the same name, the disease tetanus has no relation to the normal muscle response to neural stimulation. The *Clostridium* bacteria, although found virtually everywhere, can thrive only in tissues that contain abnormally low amounts of oxygen. For this reason, a deep puncture wound, such as that from a nail, carries a much greater risk than a shallow, open cut that bleeds freely.

When active in body tissues, these bacteria release a powerful toxin that affects the central nervous system. Motor neurons, which control skeletal muscles throughout the body, are particularly sensitive to it. The toxin suppresses the mechanism that regulates motor neuron activity. The result is a sustained, powerful contraction of skeletal muscles throughout the body. This toxin also has a direct effect on skeletal muscle fibers, producing contraction even in the absence of neural commands.

The incubation period (the time from exposure before symptoms develop) is usually less than 2 weeks. The most common complaints are headache, muscle stiffness, and difficulty in swallowing. Because it soon becomes difficult to open the mouth, this disease is also called *lockjaw*. Widespread muscle spasms usually develop within 2–3 days of the initial symptoms, and they often continue for a week before subsiding. After 2–4 weeks, symptoms in surviving patients disappear with no after-effects.

Although severe tetanus has a 40–60 percent mortality rate, immunization is effective in preventing the disease. There are approximately 500,000 cases of tetanus worldwide each year, but only about 100 of them occur in the United States, thanks to an effective immunization program. ("Tetanus shots" are recommended, with booster shots every 10 years.) Severe symptoms in unimmunized patients can be prevented by early administration of an antitoxin, usually *human tetanus immune globulin*. Such treatment does not reduce symptoms that have already appeared, however.

making it continuous. Virtually all normal muscular contractions involve complete tetanus of the participating muscle fibers.

THE NUMBER OF MUSCLE FIBERS INVOLVED

We have a remarkable ability to control the amount of tension exerted by our skeletal muscles. During a normal movement, our muscles contract smoothly, not jerkily, because activated muscle fibers are responding in complete tetanus. The *total force* exerted by the skeletal muscle depends on how many muscle fibers are activated.

A typical skeletal muscle contains thousands of muscle fibers. Although some motor neurons control a single muscle fiber, most control hundreds or thousands of muscle fibers through multiple synaptic terminals. All of the muscle fibers controlled by a single motor neuron constitute a **motor unit**. The size of a motor unit indicates how fine the control of movement can be. In the muscles of the eye, where precise control is extremely important, a motor neuron may control two or three muscle fibers. We have much less precise control over our leg muscles, where up to 2000 muscle fibers may respond to the call of a single motor neuron. The muscle fibers of each motor unit are intermingled with those of other motor units (Figure 7-8●). This intermingling ensures that the direction of pull on a tendon does not change despite variations in the number of activated motor units. When you decide to perform a specific arm movement, specific groups of motor neurons within the spinal cord are stimulated. The contraction begins with the activation of the smallest motor units in the stimulated muscle. Over time, the number of activated motor units gradually increases. The smooth but steady increase in muscular tension produced by increasing the number of active motor units is called **recruitment**.

Peak tension production occurs when all of the motor units in the muscle are contracting in complete tetanus. Such contractions do not last long, however, because the muscle fibers soon use up their available energy supplies. During a sustained

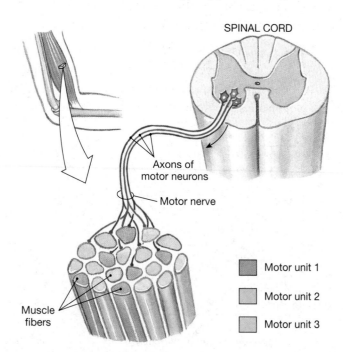

● *Figure 7-8* **Motor Units**
Each motor unit is made up of the muscle fibers controlled by a single motor neuron. Muscle fibers of different motor units are intermingled, so the forces applied to the tendon remain roughly balanced regardless of which muscle groups are stimulated.

tetanic contraction, motor units are activated on a rotating basis, so that some are resting while others are contracting.

Muscle Tone

Some of the motor units within any particular muscle are always active, even when the entire muscle is not contracting. Their contractions do not produce enough tension to cause movement, but they do tense and firm the muscle. This resting tension in a skeletal muscle is called **muscle tone**. A muscle with little muscle tone appears limp and flaccid, whereas one with moderate muscle tone is quite firm and solid.

A skeletal muscle that is not stimulated by a motor neuron on a regular basis will **atrophy** (AT-rō-fē; *a*, without + *-trophy*, nourishing): Its muscle fibers will become smaller and weaker. Individuals paralyzed by spinal injuries or other damage to the nervous system gradually lose muscle tone and volume in the areas affected. Even a temporary reduction in muscle use can lead to muscular atrophy, as is easily seen by comparing limb muscles before and after a cast has been worn. Muscle atrophy is initially reversible, but dying muscle fibers are not replaced, and in extreme atrophy the functional losses are permanent. That is why physical therapy is so important in cases where patients are temporarily unable to move normally.

ISOTONIC AND ISOMETRIC CONTRACTIONS

Muscle contractions may be classified as isotonic or isometric on the basis of the pattern of tension production and overall change in shape. In an **isotonic** (*iso-*, equal + *tonos*, tension) **contraction**, tension rises to a level that is maintained until relaxation occurs. During such contractions, the skeletal muscle's length shortens as the tension in the muscle remains constant. Lifting an object off a desk, walking, and running involve isotonic contractions.

In an **isometric** (*metron*, measure) **contraction**, tension continues to rise but the muscle as a whole does not change in length. Examples of isometric contractions are pushing against a wall and trying to pick up a large car. These are rather unusual movements, but many of the everyday reflexive muscle contractions that keep the body upright when standing or sitting involve isometric contractions of muscles that oppose gravity.

Normal daily activities involve a combination of isotonic and isometric muscular contractions. As you sit reading this text, isometric contractions of postural muscles stabilize your vertebrae and maintain your upright position. When you next turn a page, the movements of your arm, forearm, hand, and fingers are produced by isotonic contractions.

MUSCLE ELONGATION

Recall that there is no active mechanism for muscle fiber elongation; contraction is active, but elongation is passive. After a contraction, a muscle fiber usually returns to its original length through a combination of elastic forces, the movements of opposing muscles, and gravity.

Elastic Forces

Every time a muscle fiber contracts, its organelles are compressed and the extracellular fibers, such as those of the endomysium, are stretched. Their recoil gradually helps return the muscle fiber to its original resting length.

Opposing Muscle Movements

Much more rapid returns to resting length result from the contraction of opposing muscles. For example, contraction of the *biceps brachii* muscle on the anterior part of the arm flexes the elbow; contraction of the *triceps brachii* muscle on the posterior surface of the arm extends the elbow. When the biceps brachii contracts, the triceps brachii is stretched; when the biceps brachii relaxes, contraction of the triceps brachii extends the elbow and stretches the muscle fibers of the biceps brachii to their original length.

Gravity

Gravity may also help lengthen a muscle after a contraction. For example, imagine the biceps brachii muscle fully contracted with the elbow pointing at the ground. When the muscle relaxes, gravity will pull the forearm down and stretch the muscle.

CONCEPT CHECK QUESTIONS

Answers on page 222

❶ What factors are responsible for the amount of tension that a skeletal muscle develops?

❷ A motor unit from a skeletal muscle contains 1500 muscle fibers. Would this muscle be involved in fine, delicate movements or powerful, gross movements? Explain.

❸ Can a skeletal muscle contract without shortening? Explain.

The Energetics of Muscular Activity

Muscle contraction requires large amounts of energy. For example, an active skeletal muscle fiber may require some 600 trillion molecules of ATP each second, not including the energy needed to pump the calcium ions back into the sarcoplasmic reticulum. This enormous amount of energy is not available before a contraction begins in a resting muscle fiber. Instead, a resting muscle fiber contains only enough energy reserves to sustain a contraction until additional ATP can be generated. Throughout the rest of the contraction, the muscle fiber will generate ATP at roughly the same rate as it is used. This section discusses how muscle fibers meet the demand for ATP.

7 THE MUSCULAR SYSTEM

Functions of Skeletal Muscle • The Anatomy of Skeletal Muscles • The Control of Muscle Fiber Contraction • Muscle Mechanics • **The Energetics of Muscular Activity**
Chapter Review • MediaLab

ATP AND CP RESERVES

The primary function of ATP is the transfer of energy from one location to another, not the long-term storage of energy. At rest, a skeletal muscle fiber produces more ATP than it needs. Under these conditions, ATP transfers energy to *creatine*, a small molecule assembled from fragments of amino acids in muscle cells. As Figure 7-9a● shows, the energy transfer creates another high-energy compound, **creatine phosphate (CP):**

$$ATP + creatine \longrightarrow ADP + creatine\ phosphate$$

During a contraction, each cross-bridge breaks down ATP, producing ADP and a phosphate group. The energy stored in creatine phosphate is then used to "recharge" the ADP back to ATP through the reverse reaction:

$$ADP + creatine\ phosphate \longrightarrow ATP + creatine$$

The enzyme that regulates this reaction is **creatine phosphokinase (CPK or CK).** When muscle cells are damaged, CPK leaks into the bloodstream. Thus, a high blood level of CPK usually indicates serious muscle (cardiac or skeletal) damage.

A resting skeletal muscle fiber contains about six times as much creatine phosphate as ATP. But when a muscle fiber is undergoing a sustained contraction, both of these energy reserves will be exhausted in about 15 seconds. The muscle fiber must then rely on other mechanisms to convert ADP to ATP.

ATP GENERATION

As you may recall from Chapter 3, cells in the body generate ATP through *aerobic* (oxygen-requiring) *metabolism* in mitochondria and through *glycolysis* in the cytoplasm. ∞ p. 69 Glycolysis is an **anaerobic** (non-oxygen-requiring) process.

Aerobic Metabolism

Aerobic metabolism normally provides 95 percent of the ATP needed by a resting cell. In this process, mitochondria absorb oxygen, ADP, phosphate ions, and organic substrates from the surrounding cytoplasm. The organic substrates are carbon chains produced by the breakdown of carbohydrates, lipids, or proteins. The substrates are then completely disassembled in the *TCA (tricarboxylic acid) cycle,* a series of chemical reactions described in Chapter 17. In brief, the carbon atoms and oxygen atoms are released as carbon dioxide (CO_2). The hydrogen atoms are shuttled to *respiratory enzymes* in the inner mitochondrial membrane, and they ultimately combine with oxygen to form water (H_2O). Along the way, large amounts of energy are released and used to make ATP. The process is quite efficient; for each organic molecule broken down in the TCA cycle, the cell gains 17 ATP molecules.

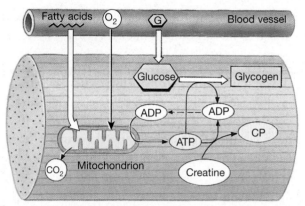

(a) Resting muscle: Fatty acids are catabolized; the ATP produced is used to build energy reserves of ATP, CP, and glycogen.

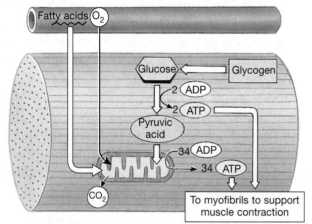

(b) Moderate activity: Glucose and fatty acids are catabolized; the ATP produced is used to power contraction.

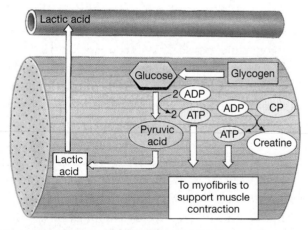

(c) Peak activity: Most ATP is produced through glycolysis, with lactic acid as a by-product. Mitochondrial activity (not shown) now provides only about one-third of the ATP consumed.

● *Figure 7-9* **Muscle Metabolism**
(**a**) A resting muscle generates ATP from the breakdown of fatty acids through aerobic metabolism. The ATP is used to build reserves of ATP, CP, and glycogen. (**b**) A muscle at a modest level of activity meets the ATP demands through the aerobic metabolism of fatty acids and glucose. (**c**) A muscle at peak activity meets the ATP demands primarily through glycolysis, an anaerobic process leading to the production of lactic acid.

Resting skeletal muscle fibers rely almost exclusively on the aerobic metabolism of fatty acids to make ATP (Figure 7-9a●). When the muscle starts contracting, the mitochondria begin breaking down molecules of pyruvic acid instead of fatty acids. The pyruvic acid is provided through glycolysis.

The maximum rate of ATP generation within mitochondria is limited by the availability of oxygen. A sufficient supply of oxygen becomes a problem as the energy demands of the muscle fiber increase. Although oxygen consumption and energy production by mitochondria can increase to 40 times resting levels, the energy demands of the muscle fiber may increase by 120 times. Thus at peak levels of exertion, mitochondrial activity provides only around one-third of the required ATP.

Glycolysis

Glycolysis is the breakdown of glucose to pyruvic acid in the cytoplasm of the cell. This anaerobic reaction provides a net gain of 2 ATP molecules and generates 2 molecules of pyruvic acid from each glucose molecule. The ATP yield of glycolysis is much lower than that of aerobic metabolism; the breakdown of 2 pyruvic acid molecules in mitochondria would generate 34 ATP molecules. However, because it can proceed in the absence of oxygen, *glycolysis can continue to provide ATP when the availability of oxygen limits the rate of mitochondrial ATP production.*

The glucose broken down under these conditions is obtained from glycogen reserves in the sarcoplasm. Glycogen is a polysaccharide chain of glucose molecules. p. 37 Typical skeletal muscle fibers contain large glycogen reserves in the form of insoluble granules. When the muscle fiber begins to run short of ATP and CP, enzymes break the glycogen molecules apart, releasing glucose that can be used to generate more ATP.

Energy Use and the Level of Muscle Activity

In a resting skeletal muscle cell the demand for ATP is low. More than enough oxygen is available for the mitochondria to meet that demand and produce a surplus of ATP. The extra ATP is used to build up reserves of CP and glycogen (Figure 7-9a●). At moderate levels of activity, the demand for ATP increases (Figure 7-9b●). As the rate of mitochondrial ATP production rises, so does the rate of oxygen consumption. As long as sufficient oxygen is available, the demand for ATP can be met by the mitochondria, and the amount of ATP provided by glycolysis remains relatively minor.

At periods of peak activity, oxygen cannot diffuse into the muscle fiber fast enough to enable the mitochondria to produce the required ATP. Mitochondrial activity can provide only about one-third the ATP needed, and glycolysis becomes the primary source of ATP (Figure 7-9c●). The anaerobic process of glycolysis enables the cell to continue generating ATP when mitochondrial activity alone cannot meet the demand. However, this pathway has its drawbacks. For example, when glycolysis produces pyruvic acid faster than it can be used by the mitochondria, pyruvic acid levels rise in the sarcoplasm. Under these conditions, the pyruvic acid is converted to **lactic acid**, a related three-carbon molecule. The conversion of pyruvic acid to lactic acid poses a problem because lactic acid is an organic acid that in body fluids dissociates into a hydrogen ion and *lactate ion*. The accumulation of hydrogen ions can lower the pH and alter the normal functioning of key enzymes. The muscle fiber then cannot continue to contract. Glycolysis is also an inefficient way to generate ATP. Under anaerobic conditions, each glucose generates two pyruvic acids, which are converted to lactic acid. In return, the cell gains 2 ATP molecules. Had those 2 pyruvic acid molecules been catabolized by the mitochondria, the cell would have gained an additional 34 ATP molecules.

MUSCLE FATIGUE

A skeletal muscle fiber is said to be fatigued when it can no longer contract despite continued neural stimulation. **Muscle fatigue** is caused by the exhaustion of energy reserves or the buildup of lactic acid.

If the muscle contractions use ATP at or below the maximum rate of mitochondrial ATP generation, the muscle fiber can function aerobically. Under these conditions, fatigue will not occur until glycogen and other reserves such as lipids and amino acids are depleted. This type of fatigue affects the muscles of long-distance athletes, such as marathon runners, after hours of exertion.

When a muscle produces a sudden, intense burst of activity, the ATP is provided by glycolysis. After a relatively short time (seconds to minutes), the rising lactic acid levels lower the tissue pH, and the muscle can no longer function normally. Athletes running sprints, such as the 100-yard dash, suffer from this type of muscle fatigue.

THE RECOVERY PERIOD

When a muscle fiber contracts, the conditions in the sarcoplasm are changed. For example, energy reserves are consumed, heat is released, and lactic acid may be present. During the **recovery period**, conditions inside the muscle are returned to normal pre-exertion levels. The muscle's metabolic activity focuses on the removal of lactic acid and the replacement of intracellular energy reserves, and the body as a whole loses the heat generated during intense muscular contraction.

7 THE MUSCULAR SYSTEM

Functions of Skeletal Muscle • The Anatomy of Skeletal Muscles • The Control of Muscle Fiber Contraction • Muscle Mechanics • The Energetics of Muscular Activity
Chapter Review • MediaLab

Lactic Acid Recycling

The reaction that converts pyruvic acid to lactic acid is freely reversible. During the recovery period, when lactic acid concentration is high, the lactic acid is converted back to pyruvic acid. This pyruvic acid can then be used (1) to synthesize glucose and (2) to generate ATP through mitochondrial activity. The ATP produced is used to convert creatine to creatine phosphate and to store the newly synthesized glucose as glycogen.

During the recovery period, the body's oxygen demand remains above normal resting levels. The additional oxygen required during the recovery period to restore the normal pre-exertion levels is called an *oxygen debt*. Most of the extra oxygen is consumed by liver cells, as they produce ATP for the conversion of lactic acid back to glucose, and by muscle cells, as they restore their reserves of ATP, creatine phosphate, and glycogen. While the oxygen debt is being repaid, the breathing rate and depth are increased. That is why you continue to breathe heavily for some time even after you stop exercising.

Heat Loss

Muscular activity generates substantial amounts of heat that warms the sarcoplasm, interstitial fluid, and circulating blood. Since muscle makes up a large portion of the total mass of the body, muscle contractions play an important role in the maintenance of normal body temperature. For example, shivering can help keep you warm in a cold environment. But when skeletal muscles are contracting at peak levels, body temperature soon begins to climb. In response, blood flow to the skin increases, promoting heat loss through mechanisms described in Chapters 1 and 5. pp. 12, 117

Muscle Performance

Muscle performance can be considered in terms of **power**, the maximum amount of tension produced by a particular muscle or muscle group, and **endurance**, the amount of time for which the individual can perform a particular activity. Two major factors determine the performance capabilities of a particular skeletal muscle: (1) the types of muscle fibers within the muscle and (2) physical conditioning or training.

TYPES OF SKELETAL MUSCLE FIBERS

There are two contrasting types of skeletal muscle fibers in the human body: fast (or fast-twitch) fibers and slow (or slow-twitch) fibers.

Fast Fibers

Most of the skeletal muscle fibers in the body are called **fast fibers** because they can contract in 0.01 second or less following stimulation. Fast fibers are large in diameter; they contain densely packed myofibrils, large glycogen reserves, and relatively few mitochondria. The tension produced by a muscle fiber is directly proportional to the number of sarcomeres, so fast-fiber muscles produce powerful contractions. However, because these contractions use ATP in massive amounts, their activity is primarily supported by glycolysis, and fast fibers fatigue rapidly.

Slow Fibers

Slow fibers are only about half the diameter of fast fibers, and they take three times as long to contract after stimulation; however, they can continue contracting for extended periods, long after a fast muscle would have become fatigued. Three specializations related to the availability of oxygen and its use make this possible:

1. *Oxygen supply.* Slow muscle tissue contains a more extensive network of capillaries than does typical fast muscle tissue, so oxygen supply is dramatically increased.
2. *Oxygen storage.* Slow muscle fibers contain the red pigment **myoglobin** (MĪ-ō-glō-bin), a globular protein structurally related to hemoglobin, the oxygen-carrying pigment found in blood. p. 41 Because myoglobin also binds oxygen molecules, resting slow muscle fibers contain oxygen reserves that can be mobilized during a contraction.
3. *Oxygen use.* Slow muscle fibers contain a relatively larger number of mitochondria than do fast muscle fibers.

The Distribution of Muscle Fibers and Muscle Performance

The percentage of fast and slow muscle fibers in a particular skeletal muscle can be quite variable. Muscles dominated by fast fibers appear pale, and they are often called **white muscles**. Chicken breasts contain "white meat" because chickens use their wings for only brief intervals, as when fleeing from a predator, and the power for flight comes from fast fibers in their breast muscles. The extensive blood vessels and myoglobin in slow muscle fibers give them a reddish color, and muscles dominated by slow fibers are therefore known as **red muscles**. Chickens walk around all day, and the movements are performed by the slow muscle fibers in the "dark meat" of their legs.

Most human muscles contain a mixture of fiber types, and therefore appear pink. However, there are no slow fibers in muscles of the eye and hand, where swift but brief contractions are required. Many back and calf muscles are dominated by slow fibers; these muscles contract almost continuously to maintain an upright posture. The percentage of fast versus slow fibers in each muscle is genetically determined, but the fatigue resistance of fast muscle fibers can be increased through athletic training.

PHYSICAL CONDITIONING

Physical conditioning and training schedules enable athletes to improve both power and endurance. In practice, the training schedule varies depending on whether the activity is primarily supported by aerobic or anaerobic energy production.

Anaerobic endurance is the length of time muscle contractions can be supported by glycolysis and existing energy reserves of ATP and CP. Examples of activities that require anaerobic endurance are a 50-yard dash or swim, a pole vault, and a weight-lifting competition. Such activities involve contractions of fast muscle fibers. Athletes training to develop anaerobic endurance perform frequent, brief, intensive workouts. The net effect is an enlargement, or **hypertrophy** (hī-PER-trō-fē), of the stimulated muscle as seen in champion weight lifters or bodybuilders. The number of muscle fibers does not change, but the muscle as a whole gets larger because each muscle fiber increases in diameter.

Aerobic endurance is the length of time a muscle can continue to contract while being supported by mitochondrial activities. Aerobic endurance is determined by the availability of substrates for aerobic metabolism from the breakdown of carbohydrates, lipids, or amino acids. Training to improve aerobic endurance usually involves sustained low levels of muscular activity. Examples are jogging, distance swimming, and other exercises that do not require peak tension production. Because glucose is a preferred energy source, aerobic athletes, such as marathon runners, often "load" or "bulk up" on carbohydrates on the day before an event. They may also consume glucose-rich "sports drinks" during a competition.

CONCEPT CHECK QUESTIONS

Answers on page 222

❶ Why would a sprinter experience muscle fatigue before a marathon runner would?

❷ Which activity would be more likely to create an oxygen debt, swimming laps or lifting weights?

❸ Which type of muscle fibers would you expect to predominate in the large leg muscles of someone who excels at endurance activities such as cycling or long-distance running?

Cardiac and Smooth Muscle Tissues

Cardiac muscle tissue and smooth muscle tissue were introduced in Chapter 4. Table 7-2 compares skeletal, cardiac, and smooth muscle tissue in greater detail.

TABLE 7-2   *A Comparison of Skeletal, Cardiac, and Smooth Muscle Tissues*

PROPERTY	SKELETAL MUSCLE FIBER	CARDIAC MUSCLE CELL	SMOOTH MUSCLE CELL
Fiber dimensions (diameter × length)	100 μm × up to 30 cm	10–20 μm × 50–100 μm	5–10 μm × 30–200 μm
Nuclei	Multiple, near sarcolemma	Usually single, centrally located	Single, centrally located
Filament organization	In sarcomeres along myofibrils	In sarcomeres along myofibrils	Scattered throughout sarcoplasm
Control mechanism	Neural, at single neuromuscular junction	Automaticity (pacemaker cells)	Automaticity (pacesetter cells), neural or hormonal control
Ca²⁺ source	Release from sarcoplasmic reticulum	Extracellular fluid and release from sarcoplasmic reticulum	Extracellular fluid and release from sarcoplasmic reticulum
Contraction	Rapid onset; tetanus can occur; rapid fatigue	Slower onset; tetanus cannot occur; resistant to fatigue	Slow onset; tetanus can occur; resistant to fatigue
Energy source	Aerobic metabolism at moderate levels of activity; glycolysis (anaerobic) during peak activity	Aerobic metabolism, usually lipid or carbohydrate substrates	Primarily aerobic metabolism

7 THE MUSCULAR SYSTEM

Functions of Skeletal Muscle • The Anatomy of Skeletal Muscles • The Control of Muscle Fiber Contraction • Muscle Mechanics • The Energetics of Muscular Activity
Chapter Review • MediaLab

CARDIAC MUSCLE TISSUE

Cardiac muscle cells are relatively small and usually have a single, centrally placed nucleus. Cardiac muscle tissue is found only in the heart.

Differences between Cardiac Muscle and Skeletal Muscle

Like skeletal muscle fibers, cardiac muscle cells contain an orderly arrangement of myofibrils and are striated, but significant differences exist in their structure and function. The most obvious structural difference is that cardiac muscle cells are branched, and each cardiac cell contacts several others at specialized sites called **intercalated** (in-TER-ka-lā-ted) **discs** (Figure 7-10a●). These cellular connections contain gap junctions that provide a means for the movement of ions and small molecules and the rapid passage of action potentials from cell to cell, resulting in their simultaneous contraction. Because the myofibrils are also attached to the intercalated discs, the cells "pull together" quite efficiently.

There are also several important functional differences between cardiac muscle and skeletal muscle:

- Cardiac muscle tissue contracts without neural stimulation, a property called *automaticity*. The timing of contractions is normally determined by specialized cardiac muscle cells called **pacemaker cells**.

- Cardiac muscle cell contractions last roughly 10 times as long as those of skeletal muscle fibers.

- The properties of cardiac muscle cell membranes differ from those of skeletal muscle fibers. As a result, cardiac muscle tissue cannot undergo tetanus, or sustained contraction. This property is important because a heart in tetany could not pump blood.

- An action potential not only triggers the release of calcium from the sarcoplasmic reticulum, but it also increases the permeability of the cell membrane to extracellular calcium ions.

- Cardiac muscle cells rely on aerobic metabolism for the energy needed to continue contracting. The sarcoplasm contains large numbers of mitochondria and abundant reserves of myoglobin (to store oxygen).

SMOOTH MUSCLE TISSUE

Smooth muscle cells are similar in size to cardiac muscle cells; they also contain a single, centrally located nucleus within each spindle-shaped cell (Figure 7-10b●). Smooth muscle tissue is found within almost every organ, forming sheets, bundles, or sheaths around other tissues. In the skeletal, muscular, nervous, and endocrine systems, smooth muscles around blood vessels regulate blood flow through vital organs. In the diges-

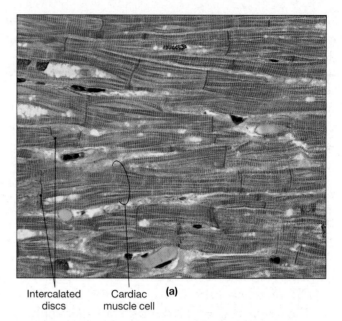

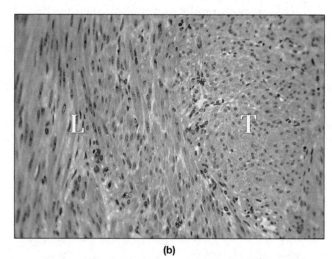

Intercalated discs • Cardiac muscle cell **(a)**

(b)

● *Figure 7-10* **Cardiac and Smooth Muscle Tissues**
(a) An LM of cardiac muscle tissue. Notice the striations and the intercalated discs. **(b)** Many visceral organs contain layers or sheets of smooth muscle fibers. This view shows smooth muscle cells in longitudinal (L) and transverse (T) sections.

tive and urinary systems, rings of smooth muscles, called *sphincters*, regulate movement along internal passageways.

Differences Between Smooth Muscle and Other Muscle Tissues

Structural Differences. Actin and myosin are present in all three muscle types. In skeletal and cardiac muscle cells, these proteins are organized into sarcomeres. The internal organization of a smooth muscle cell is very different from that of skeletal or cardiac muscle cells:

- There are no myofibrils, sarcomeres, or striations in smooth muscle tissue.

- The thin filaments of a smooth muscle cell are anchored within the cytoplasm and to the sarcolemma.
- Adjacent smooth muscle cells are bound together at these anchoring sites, thus transmitting the contractile forces throughout the tissue.

Functional Differences. Smooth muscle tissue differs from other muscle types in several major ways:

- Calcium ions trigger contractions through a different mechanism than that found in other muscle types. Additionally, most of the calcium ions that do trigger contractions enter the cell from the extracellular fluid.
- Smooth muscle cells are able to contract over a greater range of lengths than skeletal or cardiac muscle because the actin and myosin filaments are not rigidly organized. This property is important because layers of smooth muscle are found in the walls of organs such as the urinary bladder and stomach, which undergo large changes in volume.
- Many smooth muscle cells are not innervated by motor neurons, and the muscle cells contract automatically (in response to *pacesetter cells*), or in response to environmental or hormonal stimulation. When smooth muscle fibers are innervated by motor neurons, the neurons involved are not under voluntary control.

CONCEPT CHECK QUESTIONS
Answers on page 222

① How do intercalated discs enhance the functioning of cardiac muscle tissue?

② Extracellular calcium ions are important for the contraction of what type of muscle tissue?

③ Smooth muscle can contract over a wider range of resting lengths than skeletal muscle. Why?

Anatomy of the Muscular System

The **muscular system** includes all of the skeletal muscles that can be controlled voluntarily (Figure 7-11●). The general appearance of each of the nearly 700 skeletal muscles provides clues to its primary function. Muscles involved with locomotion and posture work across joints, producing skeletal movement. Those that support soft tissue form slings or sheets between relatively stable bony elements, whereas those that guard an entrance or exit completely encircle the opening.

ORIGINS, INSERTIONS, AND ACTIONS

Each muscle begins at an **origin**, ends at an **insertion**, and contracts to produce a specific **action**. In general, the origin end remains stationary while the insertion moves. For example, the *gastrocnemius* muscle (in the calf) has its origin on the distal portion of the femur and inserts on the *calcaneus*. Its contraction pulls the insertion closer to the origin, resulting in the action called *plantar flexion*. The determinations of origin and insertion are usually based on movement from the anatomical position.

Almost all skeletal muscles either originate or insert on the skeleton. When they contract, they may produce *flexion, extension, adduction, abduction, protraction, retraction, elevation, depression, rotation, circumduction, pronation, supination, inversion, or eversion.* (You may wish to review Figures 6-32 to 6-34, pp.160–161.)

Actions of muscles may be described in two ways. The first describes muscle actions in terms of the bone affected. Accordingly, the *biceps brachii* muscle is said to perform "flexion of the forearm." The second method, which is increasingly used by specialists of human motion (*kinesiologists*), describes muscle action in terms of the joint involved. Thus, the the biceps brachii muscle is said to perform "flexion at (or of) the elbow." We shall primarily use the second method.

Muscles can be described by their **primary actions**:

- A **prime mover**, or **agonist** (AG-o-nist), is a muscle whose contraction is chiefly responsible for producing a particular movement. The *biceps brachii* muscle is a prime mover that flexes the elbow.

- **Antagonists** (an-TAG-o-nists) are muscles whose actions oppose the movement produced by another muscle. An antagonist may also be a prime mover. For example, the *triceps brachii* muscle is a prime mover that extends the elbow. It is therefore an antagonist of the biceps brachii, and the biceps brachii is an antagonist of the triceps brachii. Agonists and antagonists are functional opposites—if one produces flexion, the other has extension as its primary action.

- A **synergist** (*syn-*, together + *ergon*, work) is a muscle that helps the prime mover to work efficiently. Synergists may provide additional pull near the insertion or stabilize the point of origin. For example, the *deltoid muscle* acts to lift the arm away from the body (abduction). A smaller muscle, the *supraspinatus muscle*, assists the deltoid in starting this movement. **Fixators** are synergists that stabilize the origin of a prime mover by preventing movement at another joint.

● *Figure 7-11* **An Overview of the Major Skeletal Muscles**
(a) Anterior view.

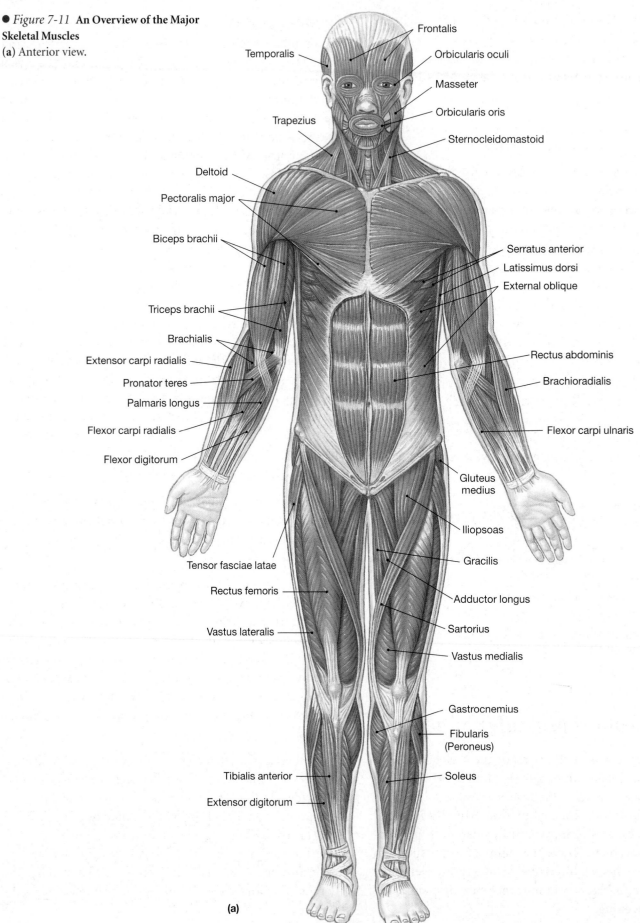

(a)

● *Figure 7-11* *(continued)*
(**b**) Posterior view.

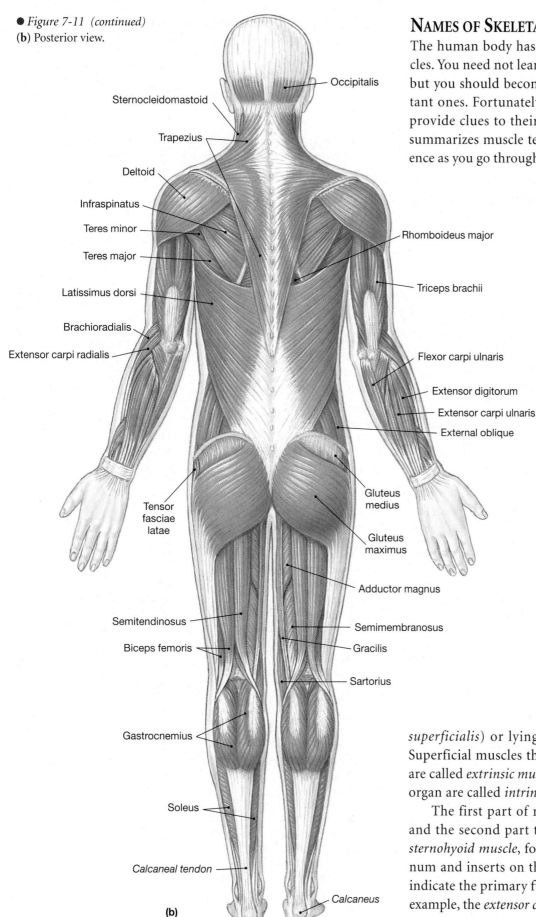

Occipitalis
Sternocleidomastoid
Trapezius
Deltoid
Infraspinatus
Teres minor
Teres major
Latissimus dorsi
Brachioradialis
Extensor carpi radialis
Rhomboideus major
Triceps brachii
Flexor carpi ulnaris
Extensor digitorum
Extensor carpi ulnaris
External oblique
Tensor fasciae latae
Gluteus medius
Gluteus maximus
Adductor magnus
Semitendinosus
Biceps femoris
Semimembranosus
Gracilis
Sartorius
Gastrocnemius
Soleus
Calcaneal tendon
Calcaneus

(**b**)

NAMES OF SKELETAL MUSCLES

The human body has approximately 700 skeletal muscles. You need not learn the name of every one of them, but you should become familiar with the most important ones. Fortunately, the names assigned to muscles provide clues to their identification. Table 7-3, which summarizes muscle terminology, will be a useful reference as you go through the rest of this chapter. (With the exception of the platysma and the diaphragm, the complete name of every muscle includes the word "muscle." For simplicity, we have not included the word "muscle" in figures and tables.)

Some names, often with Greek or Latin roots, refer to the orientation of the muscle fibers. For example, *rectus* means "straight," and *rectus muscles* are parallel muscles whose fibers generally run along the long axis of the body, as in the *rectus abdominis muscle*. In a few cases, a muscle is such a prominent feature that the regional name alone can identify it, such as the *temporalis muscle* of the head. Other muscles are named after structural features. For example, a biceps muscle has two tendons of origin (*bi-*, two + *caput*, head), whereas the *triceps* has three. Table 7-3 also lists names reflecting shape, length, size, and whether a muscle is visible at the body surface (*externus, superficialis*) or lying beneath (*internus, profundus*). Superficial muscles that position or stabilize an organ are called *extrinsic muscles*; those that operate within an organ are called *intrinsic muscles*.

The first part of many names indicates the origin and the second part the insertion of the muscle. The *sternohyoid muscle*, for example, originates at the sternum and inserts on the hyoid bone. Other names may indicate the primary function of the muscle as well. For example, the *extensor carpi radialis muscle* is found along

197

7 **THE MUSCULAR SYSTEM**

Functions of Skeletal Muscle • The Anatomy of Skeletal Muscles • The Control of Muscle Fiber Contraction • Muscle Mechanics • The Energetics of Muscular Activity
Chapter Review • MediaLab

the radial (lateral) border of the forearm, and its contraction produces extension at the wrist joint.

The separation of the skeletal system into axial and appendicular divisions provides a useful guideline for subdividing the muscular system as well:

• The **axial musculature** arises on the axial skeleton. It positions the head and spinal column and also moves the rib cage, assisting in the movements that make breathing possible. It does not play a role in movement or support of the pectoral or pelvic girdles or appendages. This category encompasses roughly 60 percent of the skeletal muscles in the body.

• The **appendicular musculature** stabilizes or moves components of the appendicular skeleton.

CONCEPT CHECK QUESTIONS

Answers on page 222

❶ Which type of muscle would you expect to find guarding the opening between the stomach and the small intestine?

❷ Which muscle would be the antagonist of the biceps brachii?

❸ What does the name *flexor carpi radialis* tell you about this muscle?

TABLE 7-3 *Muscle Terminology*

TERMS INDICATING DIRECTION RELATIVE TO AXES OF THE BODY	TERMS INDICATING SPECIFIC REGIONS OF THE BODY*	TERMS INDICATING STRUCTURAL CHARACTERISTICS OF THE MUSCLE	TERMS INDICATING ACTIONS
Anterior (front)	Abdominis (abdomen)	**Origin**	**General**
Externus (superficial)	Anconeus (elbow)	Biceps (two heads)	Abductor
Extrinsic (outside)	Auricularis (auricle of ear)	Triceps (three heads)	Adductor
Inferioris (inferior)	Brachialis (brachium)	Quadriceps (four heads)	Depressor
Internus (deep, internal)	Capitis (head)		Extensor
Intrinsic (inside)	Carpi (wrist)	**Shape**	Flexor
Lateralis (lateral)	Cervicis (neck)	Deltoid (triangle)	Levator
Medialis/medius	Cleido/clavius (clavicle)	Orbicularis (circle)	Pronator
(medial, middle)	Coccygeus (coccyx)	Pectinate (comblike)	Rotator
Obliquus (oblique)	Costalis (ribs)	Piriformis (pear-shaped)	Supinator
Posterior (back)	Cutaneous (skin)	Platys- (flat)	Tensor
Profundus (deep)	Femoris (femur)	Pyramidal (pyramid)	
Rectus (straight, parallel)	Genio- (chin)	Rhomboideus (rhomboid)	**Specific**
Superficialis (superficial)	Glosso/glossal (tongue)	Serratus (serrated)	Buccinator (trumpeter)
Superioris (superior)	Hallucis (great toe)	Splenius (bandage)	Risorius (laugher)
Transversus (transverse)	Ilio- (ilium)	Teres (long and round)	Sartorius (like a tailor)
	Inguinal (groin)	Trapezius (trapezoid)	
	Lumborum (lumbar region)		
	Nasalis (nose)	**Other Striking Features**	
	Nuchal (back of neck)	Alba (white)	
	Oculo- (eye)	Brevis (short)	
	Oris (mouth)	Gracilis (slender)	
	Palpebrae (eyelid)	Lata (wide)	
	Pollicis (thumb)	Latissimus (widest)	
	Popliteus (behind knee)	Longissimus (longest)	
	Psoas (loin)	Longus (long)	
	Radialis (radius)	Magnus (large)	
	Scapularis (scapula)	Major (larger)	
	Temporalis (temples)	Maximus (largest)	
	Thoracis (thoracic region)	Minimus (smallest)	
	Tibialis (tibia)	Minor (smaller)	
	Ulnaris (ulna)	-tendinosus (tendinous)	
	Uro- (urinary)	Vastus (great)	

*For other regional terms, refer to Figure 1-5, p. 15, which shows anatomical landmarks.

THE AXIAL MUSCLES

The axial muscles fall into four logical groups based on location, function, or both:

1. *The muscles of the head and neck.* These muscles include the muscles responsible for facial expression, chewing, and swallowing.
2. *The muscles of the spine.* This group includes flexors and extensors of the head, neck, and spinal column.

3. *The muscles of the trunk.* The *oblique* and *rectus* muscles form the muscular walls of the thoracic and abdominopelvic cavities.
4. *The muscles of the pelvic floor.* These muscles extend between the sacrum and pelvic girdle and form the muscular *perineum*, which closes the pelvic outlet.

Muscles of the Head and Neck

The muscles of the head and neck are shown in Figures 7-12● and 7-13● and detailed in Table 7-4 (pp. 200–201). The muscles of the face originate on the surface of the skull and insert into the dermis of the skin. When they contract, the skin moves. For example, the **frontalis** muscle of the forehead raises the eyebrows and pulls on the skin of the scalp. The largest group

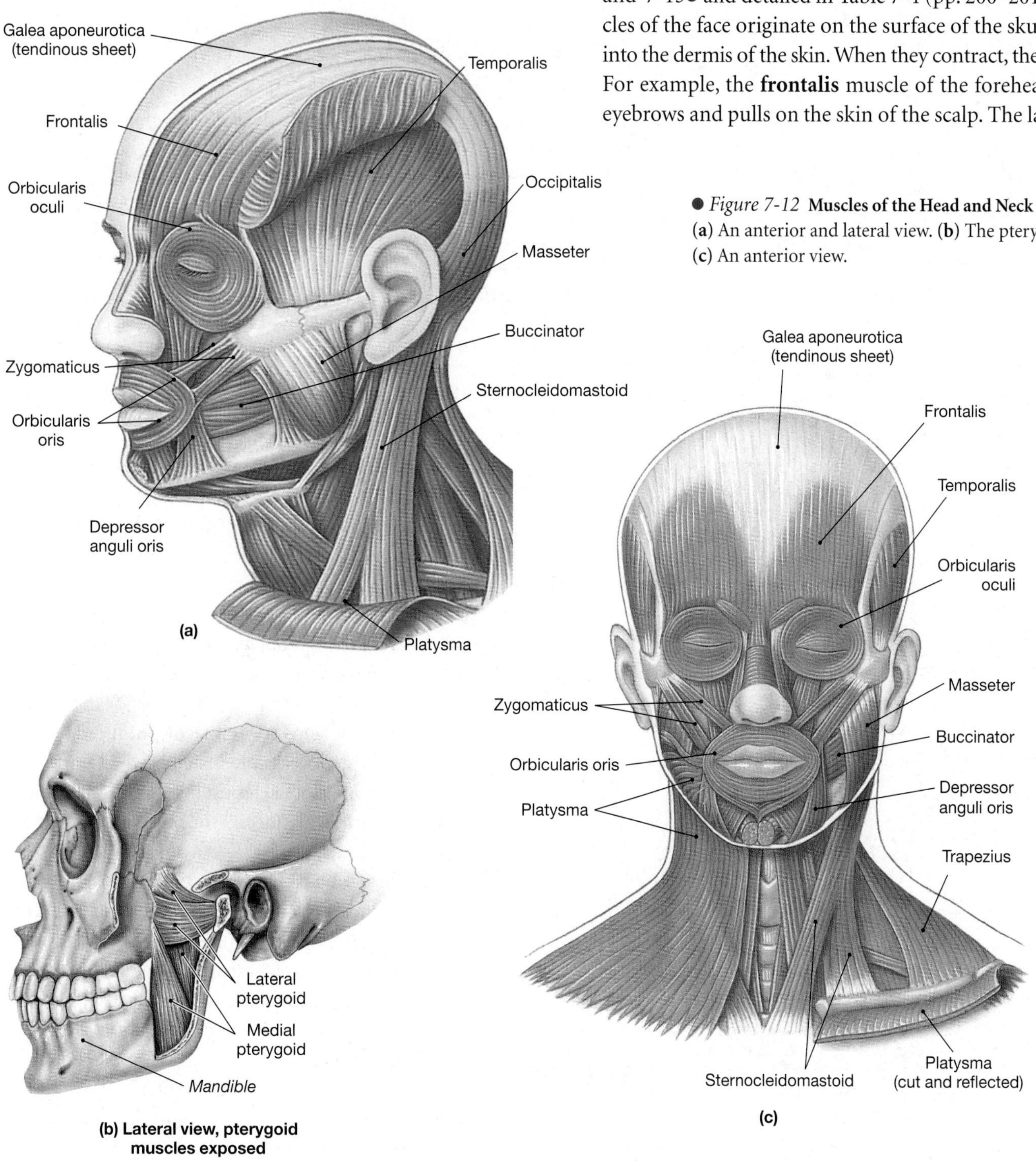

● *Figure 7-12* **Muscles of the Head and Neck**
(**a**) An anterior and lateral view. (**b**) The pterygoid muscles. (**c**) An anterior view.

(a)

Galea aponeurotica (tendinous sheet)
Frontalis
Orbicularis oculi
Zygomaticus
Orbicularis oris
Depressor anguli oris
Temporalis
Occipitalis
Masseter
Buccinator
Sternocleidomastoid
Platysma

(b) Lateral view, pterygoid muscles exposed
Lateral pterygoid
Medial pterygoid
Mandible

(c)
Galea aponeurotica (tendinous sheet)
Frontalis
Temporalis
Orbicularis oculi
Masseter
Buccinator
Depressor anguli oris
Trapezius
Zygomaticus
Orbicularis oris
Platysma
Sternocleidomastoid
Platysma (cut and reflected)

7 **THE MUSCULAR SYSTEM**

Functions of Skeletal Muscle • The Anatomy of Skeletal Muscles • The Control of Muscle Fiber Contraction • Muscle Mechanics • The Energetics of Muscular Activity
Chapter Review • MediaLab

TABLE 7-4 *Muscles of the Head and Neck*

REGION/MUSCLE	ORIGIN	INSERTION	ACTION
MOUTH			
Buccinator	Maxillary bone and mandible	Blends into fibers of orbicularis oris	Compresses cheeks
Orbicularis oris	Maxillary bone and mandible	Lips	Compresses, purses lips
Depressor anguli oris	Anterolateral surface of mandible	Skin at angle of mouth	Depresses corner of mouth
Zygomaticus	Zygomatic bone	Angle of mouth; upper lip	Draws corner of mouth back and up
EYE			
Orbicularis oculi	Medial margin of orbit	Skin around eyelids	Closes eye
SCALP			
Frontalis	Galea aponeurotica	Skin of eyebrow and bridge of nose	Raises eyebrows, wrinkles forehead
Occipitalis	Occipital bone	Galea aponeurotica	Tenses and retracts scalp
LOWER JAW			
Masseter	Zygomatic arch	Lateral surface of mandible	Elevates mandible
Temporalis	Along temporal lines of skull	Coronoid process of mandible	Elevates mandible
Pterygoids	Inferior processes of sphenoid	Medial surface of mandible	Elevate, protract, and/or move mandible to either side

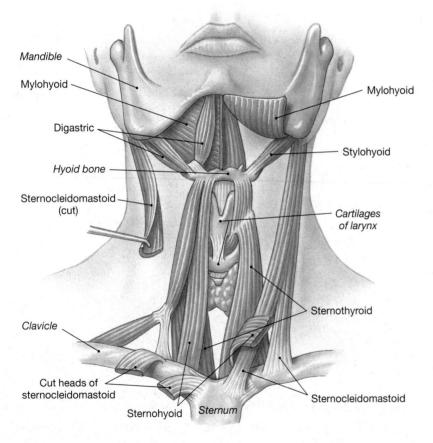

Mandible

Mylohyoid

Digastric

Hyoid bone

Sternocleidomastoid (cut)

Clavicle

Cut heads of sternocleidomastoid

Sternohyoid   *Sternum*

Mylohyoid

Stylohyoid

Cartilages of larynx

Sternothyroid

Sternocleidomastoid

● *Figure 7-13* **Muscles of the Anterior Neck**

of facial muscles is associated with the mouth. The **orbicularis oris** constricts the opening, and other muscles move the lips or the corners of the mouth. The **buccinator** (BUK-si-nā-tor), one of the muscles associated with the mouth, compresses the cheeks, as when pursing the lips and blowing forcefully. (*Buccinator* translates as "trumpet player.") During chewing, contraction and relaxation of the buccinators move food back across the teeth from the space inside the cheeks. The chewing motions are primarily produced by contractions of the **masseter**, assisted by the **temporalis** and the **pterygoid** muscles used in various combinations. In infants, the buccinator provides suction for suckling at the breast.

Smaller groups of muscles control movements of the eyebrows and eyelids, the scalp, the nose, and the external ear. The **epicranius** (ep-i-KRĀ-nē-us), or *scalp*, contains two muscles—the frontalis and the **occipitalis**. These muscles are separated by an *aponeurosis*, or tendinous sheet, called the **galea aponeurotica** (GĀ-lē-uh ap-ō-nū-RO-ti-kuh; *galea*, helmet). The **platysma** (pla-TIZ-ma; *platys*, flat) covers the ventral surface of the neck, extending from the base of the neck to the mandible and the corners of the mouth.

REGION/MUSCLE	ORIGIN	INSERTION	ACTION
NECK			
Platysma	From cartilage of second rib to acromion of scapula	Mandible and skin of cheek	Tenses skin of neck, depresses mandible
Digastric	Mastoid region of temporal bone and inferior surface of mandible	Hyoid bone	Depresses mandible and/or elevates larynx
Mylohyoid	Medial surface of mandible	Median connective tissue band that runs to hyoid bone	Elevates floor of mouth and hyoid, and/or depresses mandible
Sternohyoid	Clavicle and sternum	Hyoid bone	Depresses hyoid bone and larynx
Sternothyroid	Dorsal surface of sternum and 1st rib	Thyroid cartilage of larynx	As above
Stylohyoid	Styloid process of temporal bone	Hyoid bone	Elevates larynx
Sternocleidomastoid	Superior margins of sternum and clavicle	Mastoid region of skull	Together they flex the neck; alone one side bends head toward shoulder and turns face to opposite side

The muscles of the neck control the position of the larynx, depress the mandible, tense the floor of the mouth, and provide a stable foundation for muscles of the tongue and pharynx (Figure 7-13●). These muscles include the following:

- The **digastric**, which has two bellies (*di-*, two + *gaster*, stomach), opens the mouth by depressing the mandible.

- The broad, flat **mylohyoid** provides a muscular floor to the mouth and supports the tongue.

- The **stylohyoid** forms a muscular connection between the hyoid bone and the styloid process of the skull.

- The **sternocleidomastoid** (ster-nō-klī-dō-MAS-toyd) extends from the clavicles and the sternum to the mastoid region of the skull. It can rotate the head or flex the neck.

CONCEPT CHECK QUESTIONS
Answers on page 222

❶ If you were contracting and relaxing your masseter muscle, what would you probably be doing?

❷ Which facial muscle would you expect to be well developed in a trumpet player?

Muscles of the Spine

The muscles of the spine are covered by more superficial back muscles, such as the trapezius and latissimus dorsi (see Figure 7-11b●). The most superior of the spinal muscles are the posterior neck muscles, the superficial **splenius capitis** and the deeper **semispinalis capitis** (Figure 7-14● and Table 7-5). These muscles assist each other in extending the head when their left and right pairs contract together. When they contract on one side, both assist in tilting the head. Because of its more lateral insertion, the contraction of the splenius capitis also acts to rotate the head. The *spinal extensors*, or **erector spinae**, act to maintain an erect spinal column and head. Moving laterally from the spine, these muscles can be subdivided into **spinalis**, **longissimus**, and **iliocostalis** divisions. In the lower lumbar and sacral regions, the border between the longissimus and iliocostalis muscles becomes indistinct, and they are sometimes known as the *sacrospinalis* muscles. When contracting together, these muscles extend the spinal column. When only the muscles on one side contract, the spine is bent laterally (lateral flexion).

The Axial Muscles of the Trunk

The *oblique muscles* and the *rectus muscles* form the muscular walls of the thoracic and abdominopelvic cavities between the

7 **THE MUSCULAR SYSTEM**

Functions of Skeletal Muscle • The Anatomy of Skeletal Muscles • The Control of Muscle Fiber Contraction • Muscle Mechanics • The Energetics of Muscular Activity
Chapter Review • MediaLab

TABLE 7-5 *Muscles of the Spine*

REGION/MUSCLE	ORIGIN	INSERTION	ACTION
SPINAL EXTENSORS			
Splenius capitis	Spinous processes of lower cervical and upper thoracic vertebrae	Mastoid process, base of the skull, and upper cervical vertebrae	The two sides act together to extend the neck; either alone rotates and laterally flexes head to that side
Semispinalis capitis	Spinous processes of lower cervical and upper thoracic vertebrae	Base of skull, upper cervical vertebrae	The two sides act together to extend the neck; either alone laterally flexes head to that side
Spinalis group	Spinous processes and transverse processes of cervical and thoracic vertebrae	Base of skull and spinous processes of cervical and upper thoracic vertebrae	The two sides act together to extend vertebral column; either alone extends neck and laterally flexes head or rotates vertebral column to that side
Longissimus group	Processes of lower cervical, thoracic, and upper lumbar vertebrae	Mastoid processes of temporal bone, transverse processes of cervical vertebrae and inferior surfaces of ribs	The two sides act together to extend vertebral column; either alone rotates and laterally flexes head or vertebral column to that side
Iliocostalis group	Superior borders of ribs and iliac crest	Transverse processes of cervical vertebrae and inferior surfaces of ribs	Extends vertebral column or laterally flexes to that side; moves ribs
SPINAL FLEXOR			
Quadratus lumborum	Iliac crest	Last rib and transverse processes of lumbar vertebrae	Together they depress ribs, flex vertebral column; one side acting alone produces lateral flexion

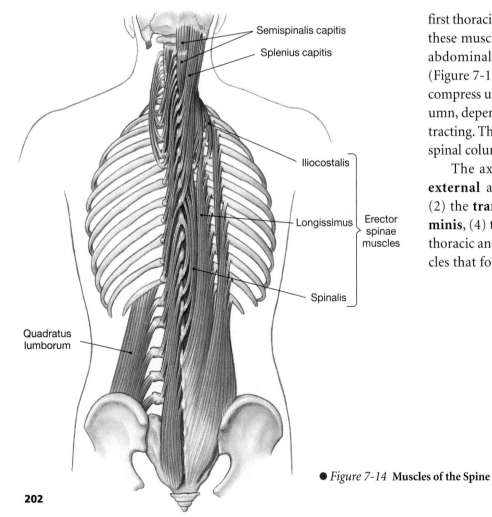

first thoracic vertebra and the pelvis. In the thoracic area, these muscles are partitioned by the ribs, but over the abdominal surface they form broad muscular sheets (Figure 7-15● and Table 7-6). The oblique muscles can compress underlying structures or rotate the spinal column, depending on whether one or both sides are contracting. The rectus muscles are important flexors of the spinal column, opposing the erector spinae.

The axial muscles of the trunk include (1) the **external** and **internal intercostals** and **obliques**, (2) the **transversus abdominis**, (3) the **rectus abdominis**, (4) the muscular **diaphragm** that separates the thoracic and abdominopelvic cavities, and (5) the muscles that form the floor of the pelvic cavity.

● *Figure 7-14* **Muscles of the Spine**

TABLE 7-6 *Axial Muscles of the Trunk*

REGION/MUSCLE	ORIGIN	INSERTION	ACTION
THORACIC REGION			
External intercostals	Inferior border of each rib	Superior border of next rib	Elevate ribs
Internal intercostals	Superior border of each rib	Inferior border of the preceding rib	Depress ribs
Diaphragm	Xiphoid process, cartilages of ribs 4–10, and anterior surfaces of lumbar vertebrae	Central tendinous sheet	Contraction expands thoracic cavity, compresses abdominopelvic cavity
ABDOMINAL REGION			
External oblique	Lower eight ribs	Linea alba and iliac crest	Compresses abdomen, depresses ribs, flexes or laterally flexes vertebral column
Internal oblique	Iliac crest and adjacent connective tissues	Lower ribs, xiphoid of sternum, and linea alba	As above
Transversus abdominis	Cartilages of lower ribs, iliac crest, and adjacent connective tissues	Linea alba and pubis	Compresses abdomen
Rectus abdominis	Superior surface of pubis around symphysis	Inferior surfaces of costal cartilages (ribs 5–7) and xiphoid process	Depresses ribs, flexes vertebral column

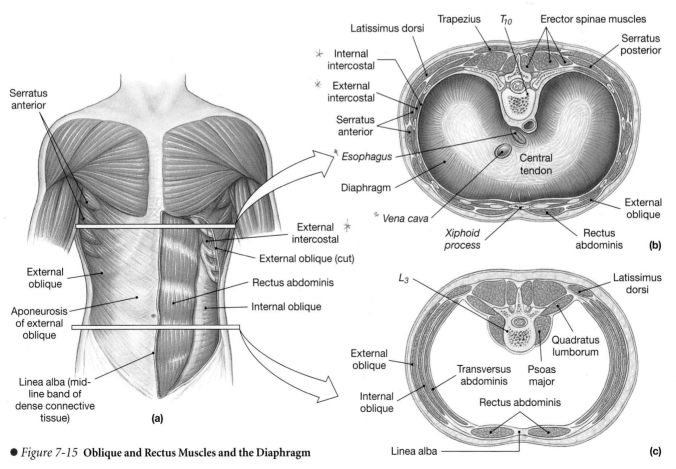

● *Figure 7-15* **Oblique and Rectus Muscles and the Diaphragm**

7 THE MUSCULAR SYSTEM

Functions of Skeletal Muscle • The Anatomy of Skeletal Muscles • The Control of Muscle Fiber Contraction • Muscle Mechanics • The Energetics of Muscular Activity
Chapter Review • MediaLab

Muscles of the Pelvic Floor

The floor of the pelvic cavity is called the **perineum** (Figure 7-16● and Table 7-7). It is formed by a broad sheet of muscles that connects the sacrum and coccyx to the ischium and pubis. These muscles support the organs of the pelvic cavity and control the movement of materials through the urethra and anus.

HERNIAS

When your abdominal muscles contract forcefully, pressure in your abdominopelvic cavity can skyrocket, and those pressures are applied to internal organs. If you exhale at the same time, the pressure is relieved, because your diaphragm can move upward as the lungs collapse. But during vigorous isometric exercises or when lifting a weight while holding your breath, pressure in the abdominopelvic cavity can rise high enough to cause a variety of problems, among them the development of a hernia.

A **hernia** develops when an organ protrudes through an abnormal opening in the surrounding body cavity wall. The most common hernias are inguinal hernias and diaphragmatic hernias. *Inguinal hernias* typically occur in males, at the *inguinal canal*, the site where blood vessels, nerves, and reproductive ducts pass through the abdominal wall to reach the testes. Elevated abdominal pressure can force open the inguinal canal, and push a portion of the intestine into the pocket created. *Diaphragmatic hernias* develop when visceral organs, such as a portion of the stomach, are forced into the left pleural cavity. If herniated structures become trapped or twisted, surgery may be required to prevent serious complications.

CONCEPT CHECK QUESTIONS
Answers on page 222

❶ Damage to the external intercostal muscles would interfere with what important process?

❷ If someone were to hit you in your rectus abdominis muscle, how would your body position then change?

THE APPENDICULAR MUSCLES

The appendicular musculature includes (1) the muscles of the shoulders and upper limbs and (2) the muscles of the pelvic girdle and lower limbs. Because the functions and required ranges of motion are very different, few similarities exist between the two groups. In addition to increasing the mobility of the upper limb, the muscular connections between the pectoral girdle and the axial skeleton must act as shock absorbers. For example, people who are jogging can still perform delicate hand movements because the muscular connections between the axial and appendicular skeleton smooth out the bounces in their stride. In contrast, the pelvic girdle has evolved to transfer weight from the axial to the appendicular skeleton. A muscular connection would reduce the efficiency of the transfer, and the emphasis is on sheer power rather than versatility.

TABLE 7-7 *Muscles of the Perineum*

MUSCLE	ORIGIN	INSERTION	ACTION
Bulbospongiosus:			
Males	Base of penis; fibers cross over urethra	Midline and central tendon of perineum	Compresses base and stiffens penis; ejects urine or semen
Females	Base of clitoris; fibers run on either side of urethral and vaginal openings	Central tendon of perineum	Compresses and stiffens clitoris; narrows vaginal opening
Ischiocavernosus	Inferior medial surface of ischium	Symphysis pubis anterior to base of penis or clitoris	Compresses and stiffens penis or clitoris
Transverse perineus	Inferior medial surface of ischium	Central tendon of perineum	Stabilizes central tendon of perineum
External urethral sphincter:			
Males	Inferior medial surfaces of ischium and pubis	Midline at base of penis; inner fibers encircle urethra	Closes urethra, compresses prostate and bulbourethral glands
Females	As above	Midline; inner fibers encircle urethra	Closes urethra, compresses vagina and greater vestibular glands
External anal sphincter	By tendon from coccyx	Encircles anal opening	Closes anal opening
Levator ani	Ischial spine, pubis	Coccyx	Tenses floor of pelvis, supports pelvic organs, flexes coccyx, elevates and retracts anus

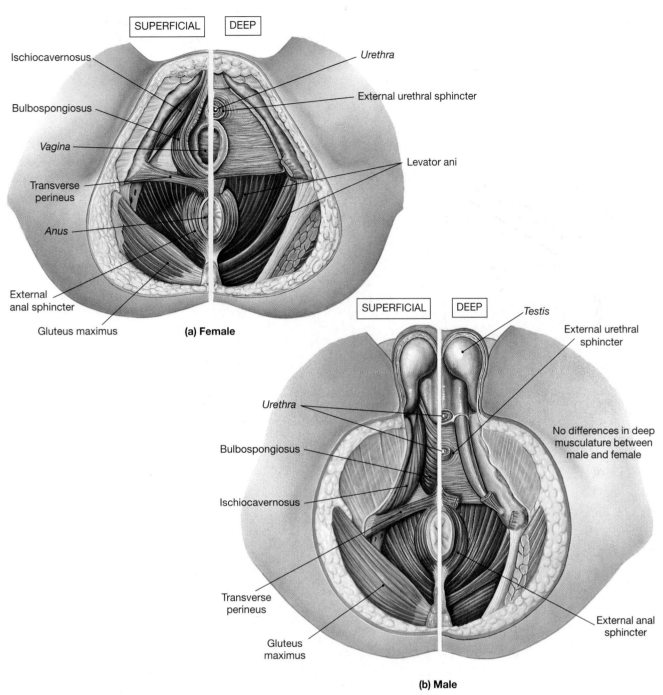

● *Figure 7-16* **Muscles of the Perineum**
(**a**) Female. (**b**) Male.

Muscles of the Shoulders and Upper Limbs

The large, superficial **trapezius** muscles cover the back and portions of the neck, reaching to the base of the skull. These muscles form a broad diamond (Figure 7-17a● and Table 7-8). Its actions are quite varied because specific regions can be made to contract independently. The **rhomboideus** muscles and the **levator scapulae** are covered by the trapezius. Contraction of the rhomboids adducts the scapula, pulling it toward the cen-

ter of the back. The levator scapulae elevates the scapula, as when you shrug your shoulders.

On the chest, the **serratus anterior** originates along the anterior surfaces of several ribs and inserts along the vertebral border of the scapula. When the serratus anterior contracts, it pulls the shoulder anteriorly. The **pectoralis minor** attaches to the coracoid process of the scapula. When it contracts, it depresses and protracts the scapula.

7 THE MUSCULAR SYSTEM

Functions of Skeletal Muscle • The Anatomy of Skeletal Muscles • The Control of Muscle Fiber Contraction • Muscle Mechanics • The Energetics of Muscular Activity
Chapter Review • MediaLab

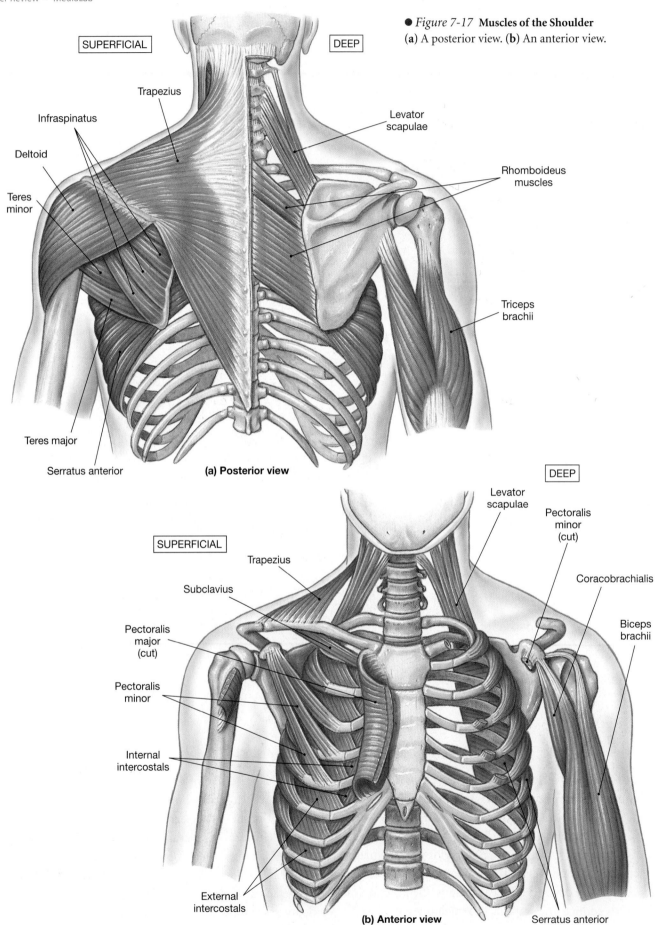

● *Figure 7-17* **Muscles of the Shoulder**
(**a**) A posterior view. (**b**) An anterior view.

SUPERFICIAL

DEEP

Trapezius

Infraspinatus

Deltoid

Teres minor

Levator scapulae

Rhomboideus muscles

Triceps brachii

Teres major

Serratus anterior

(a) Posterior view

DEEP

Levator scapulae

Pectoralis minor (cut)

SUPERFICIAL

Trapezius

Subclavius

Coracobrachialis

Pectoralis major (cut)

Biceps brachii

Pectoralis minor

Internal intercostals

External intercostals

Serratus anterior

(b) Anterior view

TABLE 7-8 *Muscles of the Shoulder*

MUSCLE	ORIGIN	INSERTION	ACTION
Levator scapulae	Transverse processes of first 4 cervical vertebrae	Vertebral border of scapula	Elevates scapula
Pectoralis minor	Anterior surfaces of ribs 3–5	Coracoid process of scapula	Depresses and protracts shoulder; rotates scapula laterally (downward); elevates ribs if scapula is stationary
Rhomboideus muscles	Spinous processes of lower cervical and upper thoracic vertebrae	Vertebral border of scapula	Adducts and rotates scapula laterally (downward)
Serratus anterior	Anterior and superior margins of ribs 1–9	Anterior surface of vertebral border of scapula	Protracts shoulder, abducts and medially rotates scapula (upward)
Subclavius	First rib	Clavicle	Depresses and protracts shoulder
Trapezius	Occipital bone and spinous processes of thoracic vertebrae	Clavicle and scapula (acromion and scapular spine)	Depends on active region and state of other muscles; may elevate, adduct, depress, or rotate scapula and/or elevate clavicle; can also extend or hyperextend neck

Muscles That Move the Arm. The muscles that move the arm (Figure 7-18● and Table 7-9) are easiest to remember when grouped by primary actions:

- The **deltoid** is the major abductor of the arm, and the **supraspinatus** assists at the start of this movement.

- The **subscapularis**, **teres major**, **infraspinatus**, and **teres minor** rotate the arm.

- The **pectoralis major** extends between the chest and the greater tubercle of the humerus; the **latissimus dorsi** extends between the thoracic vertebrae and the lesser tubercle of the humerus. The pectoralis major produces flexion at the shoulder joint, and the latissimus dorsi produces extension. The two muscles also work together to produce adduction and rotation.

These muscles provide substantial support for the shoulder joint. The tendons of the supraspinatus, infraspinatus,

TABLE 7-9 *Muscles That Move the Arm*

MUSCLE	ORIGIN	INSERTION	ACTION
Coracobrachialis	Coracoid process	Medial margin of shaft of humerus	Adduction and flexion at shoulder
Deltoid	Clavicle and scapula (acromion and adjacent scapular spine)	Deltoid tuberosity of humerus	Abduction at shoulder
Latissimus dorsi	Spinous processes of lower thoracic vertebrae, ribs, and lumbar vertebrae	Lesser tubercle, intertubercular groove of humerus	Extension, adduction, and medial rotation at shoulder
Pectoralis major	Cartilages of ribs 2–6, body of sternum, and clavicle	Greater tubercle of humerus	Flexion, adduction, and medial rotation at shoulder
ROTATOR CUFF MUSCLES			
Supraspinatus	Supraspinous fossa of scapula	Greater tubercle of humerus	Abduction at shoulder
Infraspinatus	Infraspinous fossa of scapula	Greater tubercle of humerus	Lateral rotation at shoulder
Subscapularis	Subscapular fossa of scapula	Lesser tubercle of humerus	Medial rotation at shoulder
Teres minor	Lateral border of scapula	Greater tubercle of humerus	Lateral rotation at shoulder
Teres major	Inferior angle of scapula	Intertubercular groove of humerus	Adduction and medial rotation at shoulder

7 THE MUSCULAR SYSTEM

Functions of Skeletal Muscle • The Anatomy of Skeletal Muscles • The Control of Muscle Fiber Contraction • Muscle Mechanics • The Energetics of Muscular Activity
Chapter Review • MediaLab

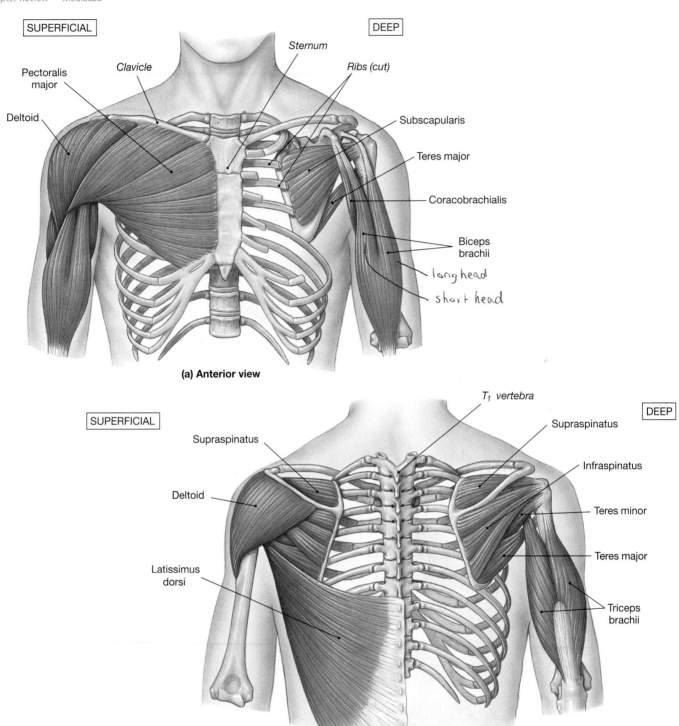

● *Figure 7-18* **Muscles That Move the Arm**
(**a**) An anterior view. (**b**) A posterior view.

(a) Anterior view

(b) Posterior view

subscapularis, and teres minor blend with and support the capsular fibers that enclose the shoulder joint. They are the muscles of the *rotator cuff*, a common site of sports injuries. Powerful, repetitive arm movements, such as pitching a fastball at 96 mph for nine innings, can place intolerable strains on the muscles of the rotator cuff, leading to a *muscle strain* (a tear or break in the muscle), *bursitis*, and other painful injuries.

Muscles That Move the Forearm and Wrist. Although most of the muscles that insert on the forearm and wrist (Figure 7-19●, and Table 7-10, p. 210) originate on the humerus, there are two noteworthy exceptions. The **biceps brachii** and **triceps brachii**, which insert on the bones of the forearm, originate on the scapula. Although their contractions can have a secondary effect on the shoulder, their primary actions are at the elbow. The tri-

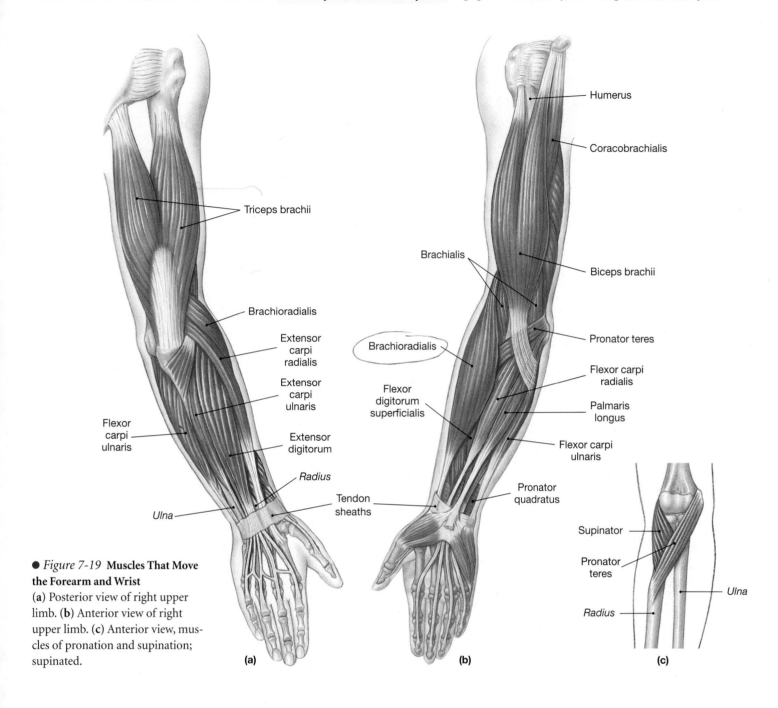

● *Figure 7-19* **Muscles That Move the Forearm and Wrist** (**a**) Posterior view of right upper limb. (**b**) Anterior view of right upper limb. (**c**) Anterior view, muscles of pronation and supination; supinated.

ceps brachii extends the elbow when, for example, you do push-ups. The biceps brachii both flexes the elbow and supinates the forearm. With the forearm pronated (palm facing back), the biceps brachii cannot function effectively. As a result, when picking up a heavy weight in the hand, you always turn the palm forward; the biceps brachii then makes a prominent bulge.

Other important muscles include the following:

- The **brachialis** and **brachioradialis** also flex the elbow, opposed by the triceps brachii.
- The **flexor carpi ulnaris**, the **flexor carpi radialis**, and the **palmaris longus** are superficial muscles that work together

to produce flexion of the wrist. Because they originate on opposite sides of the humerus, the flexor carpi radialis flexes and abducts the wrist while the flexor carpi ulnaris flexes and adducts the wrist.

- The **extensor carpi radialis** muscles and the **extensor carpi ulnaris** have a similar relationship; the former produces extension and abduction at the wrist, the latter extension and adduction.
- The **pronators** and the **supinator** rotate the radius at its proximal and distal articulations with the ulna without either flexing or extending the elbow.

209

7 **THE MUSCULAR SYSTEM**

Functions of Skeletal Muscle • The Anatomy of Skeletal Muscles • The Control of Muscle Fiber Contraction • Muscle Mechanics • The Energetics of Muscular Activity
Chapter Review • MediaLab

TABLE 7-10 *Muscles That Move the Forearm, Wrist, and Hand*

MUSCLE	ORIGIN	INSERTION	ACTION
PRIMARY ACTION AT THE ELBOW			
Flexors			
Biceps brachii	*Short head* from the coracoid process and *long head* from body of scapula	Tuberosity of radius	Flexion at elbow and shoulder; supination
Brachialis	Anterior, distal surface of humerus	Tuberosity of ulna	Flexion at elbow
Brachioradialis	Lateral epicondyle of humerus	Styloid process of radius	As above
Extensor			
Triceps brachii	Superior, posterior, and lateral margins of humerus, and the scapula	Olecranon of ulna	Extension at elbow
Pronators/Supinator			
Pronator quadratus	Medial surface of distal portion of ulna	Anterior and lateral surface of distal portion of radius	Pronation
Pronator teres	Medial epicondyle of humerus and coronoid process of ulna	Distal lateral surface of radius	As above
Supinator	Lateral epicondyle of humerus and ulna	Anterior and lateral surface of radius distal to the radial tuberosity	Supination
PRIMARY ACTION AT THE WRIST			
Flexors			
Flexor carpi radialis	Medial epicondyle of humerus	Bases of 2nd and 3rd metacarpal bones	Flexion and abduction at wrist
Flexor carpi ulnaris	Medial epicondyle of humerus and adjacent surfaces of ulna	Pisiform bone, hamate bone, and base of 5th metacarpal bone	Flexion and adduction at wrist
Palmaris longus	Medial epicondyle of humerus	A tendinous sheet on the palm	Flexion at wrist
Extensors			
Extensor carpi radialis	Distal lateral surface and lateral epicondyle of humerus	Bases of 2nd and 3rd metacarpal bones	Extension and abduction at wrist
Extensor carpi ulnaris	Lateral epicondyle of humerus and adjacent surface of ulna	Base of 5th metacarpal bone	Extension and adduction at wrist
ACTION AT THE HAND			
Extensor digitorum	Lateral epicondyle of humerus	Posterior surfaces of the phalanges	Extension at finger joints and wrist
Flexor digitorum	Medial epicondyle of humerus; anterior surfaces of ulna and radius; medial and posterior surfaces of ulna	Distal phalanges	Flexion at finger joints and wrist

Muscles That Move the Hand and Fingers. The muscles of the forearm perform flexion and extension at the finger joints (Table 7-10). These muscles stop before reaching the hand, and only their tendons cross the wrist. These are relatively large muscles, and keeping them clear of the joints ensures maximum mobility at both the wrist and hand. The tendons that cross the dorsal and ventral surfaces of the wrist are held in place by *tendon sheaths*, wide bands of connective tissue. Inflammation of tendon sheaths can restrict movement and irritate the *median nerve*, a nerve that innervates the palm of the hand. Chronic pain, often associated with weakness in the hand muscles, is the result. This condition is known as *carpal tunnel syndrome*.

Fine control of the hand involves small *intrinsic muscles*, which originate on the carpal and metacarpal bones. No muscles originate on the phalanges, and only tendons extend across the distal joints of the fingers.

CONCEPT CHECK QUESTIONS
Answers on page 222

❶ Which muscle do you use to shrug your shoulders?

❷ Sometimes baseball pitchers will suffer from rotator cuff injuries. Which muscles are involved in this type of injury?

❸ Injury to the flexor carpi ulnaris would impair which two movements?

Muscles of the Pelvis and Lower Limbs
The muscles of the pelvis and the lower limb can be divided into three functional groups: (1) muscles that move the thigh, working across the hip joint; (2) muscles that move the leg, working across the knee joint; and (3) muscles that move the ankles, feet, and toes across the various joints of the foot.

Muscles That Move the Thigh. The muscles that move the thigh are detailed in Figure 7-20● and Table 7-11.

TABLE 7-11 *Muscles That Move the Thigh*

GROUP/MUSCLE	ORIGIN	INSERTION	ACTION
GLUTEAL GROUP			
Gluteus maximus	Iliac crest of ilium, sacrum, and coccyx	Iliotibial tract and gluteal tuberosity of femur	Extension and lateral rotation at hip
Gluteus medius	Anterior iliac crest and lateral surface of ilium	Greater trochanter of femur	Abduction and medial rotation at hip
Gluteus minimus	Lateral surface of ilium	As above	As above
Tensor fasciae latae	Iliac crest and surface of ilium between anterior iliac spines	Iliotibial tract	Flexion, abduction, and medial rotation at hip; tenses fascia lata, which laterally supports the thigh
ADDUCTOR GROUP			
Adductor brevis	Inferior ramus of pubis	Linea aspera of femur	Adduction, flexion, and medial rotation at hip
Adductor longus	Inferior ramus of pubis anterior to adductor brevis	As above	As above
Adductor magnus	Inferior ramus of pubis posterior to adductor brevis	As above	Adduction at hip joint; superior portion produces flexion; inferior portion produces extension
Pectineus	Superior ramus of pubis	Inferior to lesser trochanter of femur	Adduction, flexion, and medial rotation at hip joint
Gracilis	Inferior ramus of pubis	Medial surface of tibia inferior to medial condyle	Flexion at knee; adduction and medial rotation at hip
ILIOPSOAS			
Iliacus	Medial surface of ilium	Femur distal to lesser trochanter; tendon fused with that of psoas major	Flexion at hip and/or lumbar intervertebral joints
Psoas major	Anterior surfaces and transverse processes of T_{12} and lumbar vertebrae	Lesser trochanter in company with iliacus	As above

7 THE MUSCULAR SYSTEM

Functions of Skeletal Muscle • The Anatomy of Skeletal Muscles • The Control of Muscle Fiber Contraction • Muscle Mechanics • The Energetics of Muscular Activity
Chapter Review • MediaLab

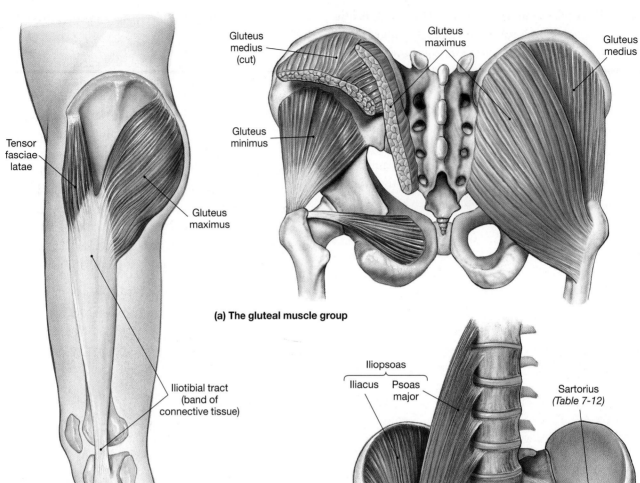

(a) The gluteal muscle group

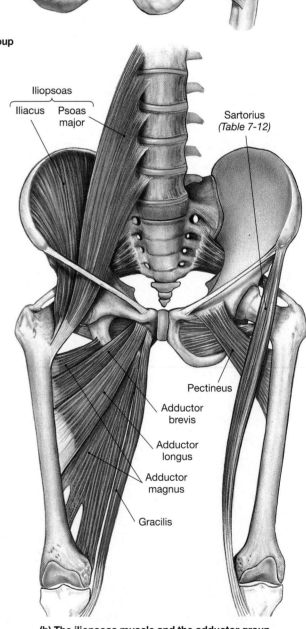

● *Figure 7-20* **Muscles That Move the Thigh**
(**a**) The gluteal muscle group (lateral and posterior views). (**b**) The iliopsoas muscle and the adductor group (anterior view).

• **Gluteal muscles** cover the lateral surfaces of the ilia (Figure 7-20a●). The **gluteus maximus** is the largest and most posterior of the gluteal muscles, which produce extension, rotation, and abduction at the hip.

• The adductors of the thigh include the **adductor magnus**, the **adductor brevis**, the **adductor longus**, the **pectineus** (pek-TIN-ē-us), and the **gracilis** (GRAS-i-lis) (Figure 7-20b●). When an athlete suffers a *pulled groin*, the problem is a strain in one of these adductor muscles.

• The largest hip flexor is the **iliopsoas** (il-ē-ō-SŌ-us) muscle. The iliopsoas is really two muscles, the **psoas major** and the **iliacus** (il-Ē-uh-kus), that share a common insertion at the lesser trochanter of the femur.

Muscles That Move the Leg. The general pattern of muscle distribution in the lower limb is that extensors are found along the anterior and lateral surfaces of the limb, and flexors lie along the posterior and medial surfaces.

(b) The iliopsoas muscle and the adductor group

- The flexors of the knee include three muscles collectively known as the *hamstrings*—the **biceps femoris** (FEM-or-is), the **semimembranosus** (sem-ē-mem-bra-NŌ-sus), and the **semitendinosus** (sem-ē-ten-di-NŌ-sus)—and the **sartorius** (sar-TŌR-ē-us) (Figure 7-21a●). A *pulled hamstring* is a relatively common sports injury caused by a strain affecting one of the hamstring muscles.

- Collectively the *knee extensors* are known as the **quadriceps femoris**. The three **vastus** muscles and the **rectus femoris** insert on the patella, which is attached to the tibial tuberos-

ity by the patellar ligament. (Because the vastus intermedius lies under the other quadriceps femoris muscles, it is not visible in Figure 7-21●.)

- The **popliteus** (pop-LI-tē-us) muscle medially rotates the tibia. When you stand, a slight lateral rotation of the tibia can lock the knee in the extended position. This enables you to stand for long periods with minimal muscular effort, but the locked knee cannot be flexed. The popliteus medially rotates the tibia back into its normal position, unlocking the joint.

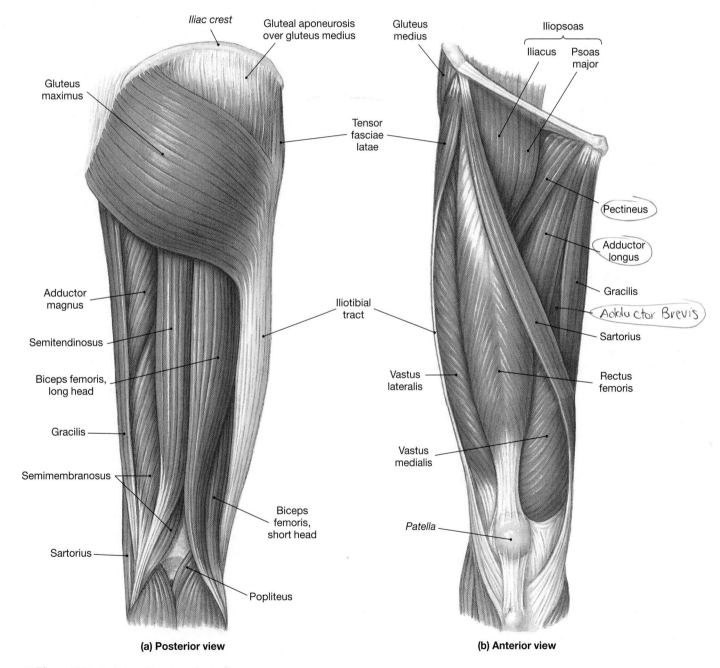

(a) Posterior view

(b) Anterior view

● *Figure 7-21* **Muscles That Move the Leg**
(**a**) A posterior view of the thigh. (**b**) An anterior view of the thigh.

213

7 THE MUSCULAR SYSTEM

Functions of Skeletal Muscle • The Anatomy of Skeletal Muscles • The Control of Muscle Fiber Contraction • Muscle Mechanics • The Energetics of Muscular Activity
Chapter Review • MediaLab

TABLE 7-12 *Muscles That Move the Leg*

MUSCLE	ORIGIN	INSERTION	ACTION
FLEXORS			
Biceps femoris	Ischial tuberosity and linea aspera of femur	Head of fibula, lateral condyle of tibia	Flexion at knee, extension and lateral rotation at hip
Semimembranous	Ischial tuberosity	Posterior surface of medial condyle of tibia	Flexion at knee; flexion and medial rotation at hip
Semitendinosus	As above	Proximal medial surface of tibia	As above
Sartorius	Anterior superior spine of ilium	Medial surface of tibia near tibial tuberosity	Flexion at knee; flexion and lateral rotation at hip
Popliteus	Lateral condyle of femur	Posterior surface of proximal tibial shaft	Rotates tibia medially (or rotates femur laterally)
EXTENSORS			
Rectus femoris	Anterior inferior iliac spine and superior acetabular rim of ilium	Tibial tuberosity by way of patellar ligament	Extension at knee, flexion at hip
Vastus intermedius	Anterior and lateral surface of femur along linea aspera	As above	Extension at knee
Vastus lateralis	Anterior and inferior to greater trochanter of femur and along linea aspera	As above	As above
Vastus medialis	Entire length of linea aspera of femur	As above	As above

The muscles that move the leg are detailed in Table 7-12.

 INTRAMUSCULAR INJECTIONS

Drugs are commonly injected into tissues, rather than directly into the bloodstream. This method enables the physician to introduce a large amount of a drug at one time, yet have it enter the circulation gradually. In an **intramuscular (IM) injection**, the drug is introduced into the mass of a large skeletal muscle. Uptake is usually faster and accompanied by less tissue irritation than when drugs are administered *intradermally* or *subcutaneously* (injected into the dermis or subcutaneous layer, respectively). p. 114 Up to 5 ml of fluid may be injected at one time, and multiple injections are possible.

The most common complications involve accidental injection into a blood vessel or piercing of a nerve. The sudden entry of massive quantities of a drug into the bloodstream can have unpleasant or even fatal consequences, and damage to a nerve can cause motor paralysis or sensory loss. As a result, the site of injection must be selected with care. Bulky muscles that contain few large vessels or nerves make ideal targets, and the gluteus medius or the posterior, lateral, superior portion of the gluteus maximus is often selected. The deltoid muscle of the arm, about 2.5 cm (1 in.) distal to the acromion, is another common site. Probably the most satisfactory from a technical point of view is the vastus lateralis of the thigh, for an injection into this thick muscle will not encounter vessels or nerves. This is the preferred injection site in infants and young children, whose gluteal and deltoid muscles are relatively small. This site is also used in elderly patients or others with atrophied gluteal and deltoid muscles.

Muscles That Move the Foot and Toes. Muscles that move the foot and toes are shown in Figure 7-22● and detailed in Table 7-13, p. 216. Most of the muscles that move the ankle produce the plantar flexion involved with walking and running movements.

- The large **gastrocnemius** (gas-trok-NĒ-mē-us; *gaster*, stomach + *kneme*, knee) of the calf is assisted by the underlying **soleus** (SŌ-lē-us) muscle. These muscles share a common tendon, the **calcaneal tendon**, or *Achilles tendon*.
- A pair of deep **fibularis** muscles (formarly called *peroneus* muscles) produce eversion of the foot as well as plantar flexion at the ankle.
- Inversion of the foot is caused by contraction of the **tibialis** (tib-ē-A-lis) muscles; the large **tibialis anterior** dorsiflexes the ankle, and opposes the gastrocnemius.

Important digital muscles originate on the surface of the tibia, the fibula, or both. Several smaller intrinsic muscles originate on the bones of the tarsus and foot, and their contractions move the toes.

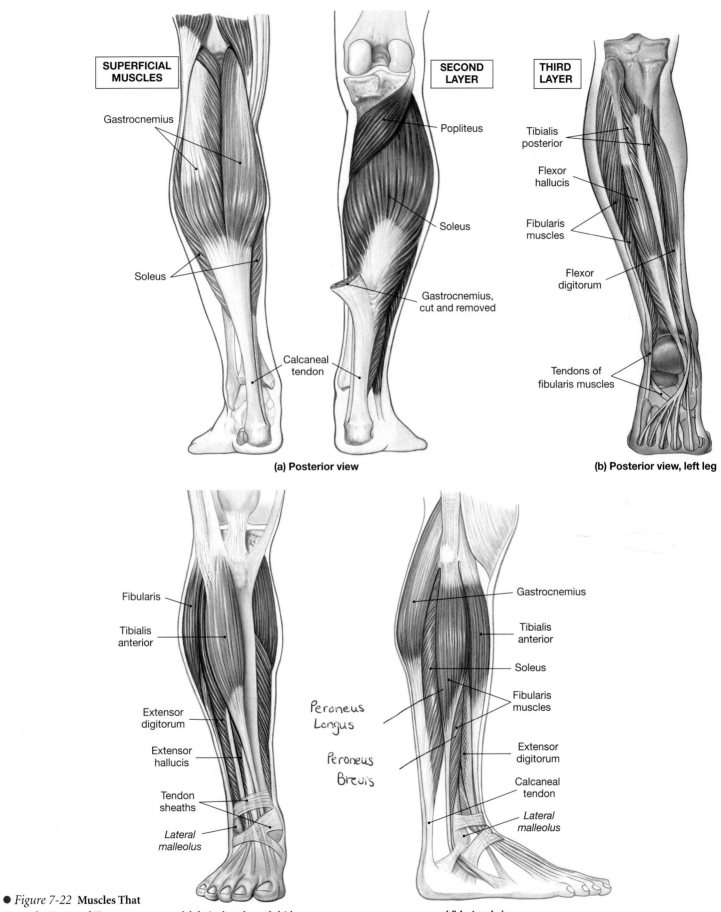

SUPERFICIAL MUSCLES

Gastrocnemius

Soleus

SECOND LAYER

Popliteus

Soleus

Gastrocnemius, cut and removed

Calcaneal tendon

(a) Posterior view

THIRD LAYER

Tibialis posterior

Flexor hallucis

Fibularis muscles

Flexor digitorum

Tendons of fibularis muscles

(b) Posterior view, left leg

Fibularis

Tibialis anterior

Extensor digitorum

Extensor hallucis

Tendon sheaths

Lateral malleolus

(c) Anterior view, right leg

Peroneus Longus

Peroneus Brevis

Gastrocnemius

Tibialis anterior

Soleus

Fibularis muscles

Extensor digitorum

Calcaneal tendon

Lateral malleolus

(d) Lateral view

● *Figure 7-22* **Muscles That Move the Foot and Toes**

215

7

THE MUSCULAR SYSTEM

Functions of Skeletal Muscle • The Anatomy of Skeletal Muscles • The Control of Muscle Fiber Contraction • Muscle Mechanics • The Energetics of Muscular Activity

Chapter Review • MediaLab

TABLE 7-13 *Muscles That Move the Foot and Toes*

MUSCLE	ORIGIN	INSERTION	ACTION
Dorsiflexor			
Tibialis anterior	Lateral condyle and proximal shaft of tibia	Base of 1st metatarsal bone	Dorsiflexion at ankle; inversion of foot
Plantar flexors			
Gastrocnemius	Femoral condyles	Calcaneus by way of calcaneal tendon	Plantar flexion at ankle; inversion and adduction of foot; flexion at knee
Fibularis (formarly **Peroneus**)	Fibula and lateral condyle of tibia	Bases of 1st and 5th metatarsal bones	Eversion of foot and plantar flexion at ankle
Soleus	Head and proximal shaft of fibula, and adjacent shaft of tibia	Calcaneus by way of calcaneal tendon	Plantar flexion at ankle; adduction of foot
Tibialis posterior	Connective tissue membrane and adjacent shafts of tibia and fibula	Tarsal and metatarsal bones	Adduction and inversion of foot; plantar flexion at ankle
ACTION AT THE TOES			
Flexors			
Flexor digitorum	Posterior and medial surface of tibia	Inferior surface of phalanges, toes 2–5	Flexion at joints of toes 2–5
Flexor hallucis	Posterior surface of fibula	Inferior surface, distal phalanx of great toe	Flexion at joints of great toe
Extensors			
Extensor digitorum	Lateral condyle of tibia, anterior surface of fibula	Superior surfaces of phalanges, toes 2–5	Extension at joints of toes 2–5
Extensor hallucis	Anterior surface of fibula	Superior surface, distal phalanx of great toe	Extension at joints of great toe

CONCEPT CHECK QUESTIONS

Answers on page 222

❶ You often hear of athletes suffering a "pulled hamstring." To what does this phrase refer?

❷ How would you expect a torn calcaneal tendon to affect movement of the foot?

Aging and the Muscular System

As the body ages, a general reduction in the size and the power of all muscle tissues occurs. The effects of aging on the muscular system can be summarized as follows:

1. *Skeletal muscle fibers become smaller in diameter.* The reduction in size reflects a decrease in the number of myofibrils. The overall effects are a reduction in muscle strength and endurance and a tendency to fatigue rapidly. Because cardiovascular performance also decreases with age, blood flow to active muscles does not increase with exercise as rapidly as it does in younger people.

2. *Skeletal muscles become less elastic.* Aging skeletal muscles develop increasing amounts of fibrous connective tissue, a process called *fibrosis*. Fibrosis makes the muscle less flexible, and the collagen fibers can restrict movement and circulation.

3. *The tolerance for exercise decreases.* A lower tolerance for exercise as age increases results in part from the tendency for rapid fatigue, and in part from the reduction in thermoregulatory ability described in Chapters 1 and 5. ∞ pp. 12, 117 Individuals over age 65 cannot eliminate heat that their muscles generate as effectively as younger people can, which leads to overheating.

4. *The ability to recover from muscular injuries decreases.* When an injury occurs, repair capabilities are limited, and scar tissue formation is the usual result.

The *rate* of decline in muscular performance is the same in all people, regardless of their exercise patterns or lifestyle. Therefore, to be in good shape late in life, an individual must be in *very* good shape early in life. Regular exercise helps control body weight, strengthens bones, and generally improves the quality of life at all ages. Extremely demanding exercise is not as important as regular exercise. In fact, extreme exercise in the elderly can damage tendons, bones, and joints. Although it has obvious effects on the quality of life, there is no clear evidence that exercise prolongs life expectancy.

CONCEPT CHECK QUESTIONS
Answers on page 222

❶ What major structural change occurs in skeletal muscle fibers as we age and what effect does it have on muscle performance?

Integration with Other Systems

To operate at maximum efficiency, the muscular system must be supported by many other systems. The changes that occur during exercise provide a good example of such interaction. As noted in earlier sections, active muscles consume oxygen and generate carbon dioxide and heat. Responses of other systems include the following:

• *Cardiovascular system.* Blood vessels in the active muscles and the skin dilate, and the heart rate increases. These adjustments speed up the delivery of oxygen and the removal of carbon dioxide at the muscle and bring heat to the skin for radiation into the environment.

• *Respiratory system.* The rate and depth of respiration increase. Air moves into and out of the lungs more quickly, keeping pace with the increased rate of blood flow through the lungs.

• *Integumentary system.* Blood vessels dilate, and sweat gland secretion increases. This combination helps promote evaporation at the skin surface and removes the excess heat generated by muscular activity.

• *Nervous and endocrine systems.* The responses of other systems are directed by controlling the heart rate, the respiratory rate, and sweat gland activity.

Even at rest, the muscular system has extensive interactions with other systems. Figure 7-23● summarizes the range of interactions between the muscular system and other systems of the body.

CONCEPT CHECK QUESTIONS
Answers on page 222

❶ Which organ systems are involved in the recovery period following the contraction of skeletal muscles?

Related Clinical Terms

botulism: A severe, potentially fatal paralysis of skeletal muscles, resulting from the consumption of a bacterial toxin.

carpal tunnel syndrome: An inflammation of the sheath surrounding the flexor tendons of the palm that leads to nerve compression and pain.

compartment syndrome: Ischemia resulting from accumulated blood and fluid trapped within a musculoskeletal compartment.

fibrosis: A process in which muscle tissue is replaced by fibrous connective tissue, making muscles weaker and less flexible.

hernia: A condition involving an organ or a body part that protrudes through an abnormal opening in the wall of a body cavity.

intramuscular (IM) injection: The administration of a drug by injecting it into the mass of a large skeletal muscle.

ischemia (is-KĒ-mē-uh): A deficiency of blood ("blood starvation") in a body part due to compression of regional blood vessels.

muscle cramps: Prolonged, involuntary, painful muscular contractions.

muscular dystrophies (DIS-trō-fēz): A varied collection of congenital diseases that produce progressive muscle weakness and deterioration.

myalgia (mī-AL-jē-uh): Muscular pain; a common symptom of a wide variety of infections.

myasthenia gravis (mī-as-THĒ-nē-uh GRA-vis): A general muscular weakness resulting from a reduction in the number of ACh receptors on the motor end plate.

myoma: A benign tumor of muscle tissue.

myositis (mī-ō-SĪ-tis): An inflammation of muscle tissue.

polio: A viral disease resulting from the destruction of motor neurons and characterized by the paralysis and atrophy of motor units.

rigor mortis: A state following death during which muscles are locked in the contracted position, making the body extremely stiff.

sarcoma: A malignant, cancerous tumor of muscle tissue.

strains: Tears or breaks in muscles.

tendinitis: The inflammation of the connective tissue surrounding a tendon.

tetanus: A disease caused by a bacterial toxin that results in sustained, powerful contractions of skeletal muscles throughout the body.

CHAPTER REVIEW

Key Terms

7 **THE MUSCULAR SYSTEM**

Functions of Skeletal Muscle • The Anatomy of Skeletal Muscles • The Control of Muscle Fiber Contraction • Muscle Mechanics • The Energetics of Muscular Activity

Chapter Review • MediaLab

The Muscular System

For All Systems

Generates heat that maintains normal body temperature

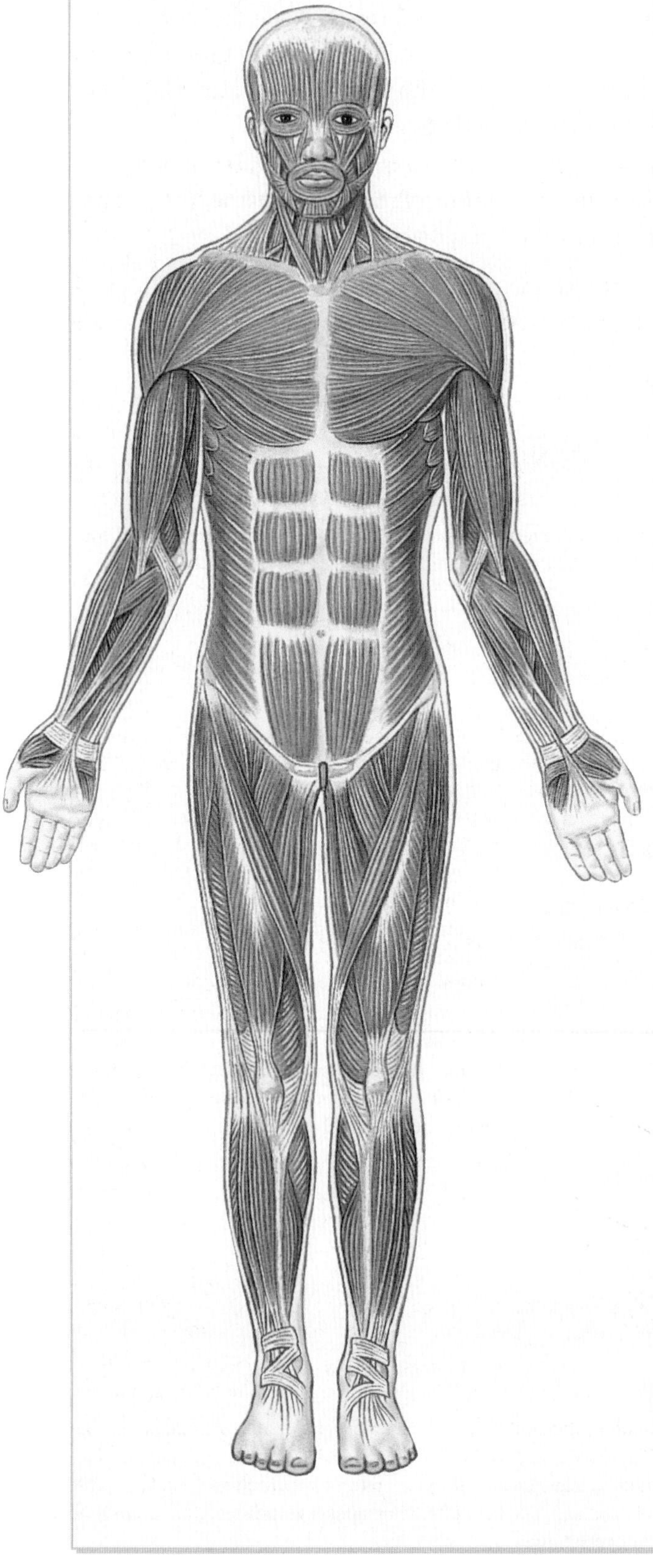

The Integumentary System

- Removes excess body heat; synthesizes vitamin D_3 for calcium and phosphate absorption; protects underlying muscles
- Skeletal muscles pulling on skin of face produce facial expressions

The Skeletal System

- Maintains normal calcium and phosphate levels in body fluids; supports skeletal muscles; provides sites of attachment
- Provides movement and support; stresses exerted by tendons maintain bone mass; stabilizes bones and joints

The Nervous System

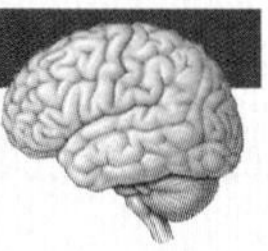

- Controls skeletal muscle contractions; adjusts activities of respiratory and cardiovascular systems during periods of muscular activity
- Muscle spindles monitor body position; facial muscles express emotion; intrinsic laryngeal muscles permit speech

The Endocrine System

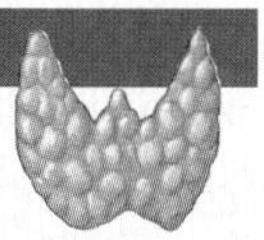

- Hormones adjust muscle metabolism and growth; parathyroid hormone and calcitonin regulate calcium and phosphate ion concentrations
- Skeletal muscles provide protection for some endocrine organs

The Cardiovascular System

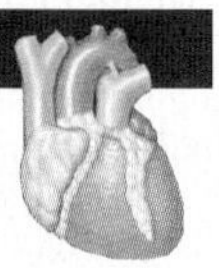

- Delivers oxygen and nutrients; removes carbon dioxide, lactic acid, and heat
- Skeletal muscle contractions assist in moving blood through veins; protects deep blood vessels

The Lymphatic System

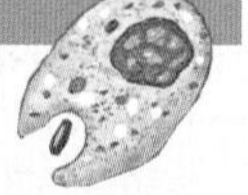

- Defends skeletal muscles against infection and assists in tissue repairs after injury
- Protects superficial lymph nodes and the lymphatic vessels in the abdominopelvic cavity

The Respiratory System

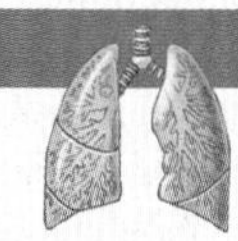

- Provides oxygen and eliminates carbon dioxide
- Muscles generate carbon dioxide; control entrances to respiratory tract, fill and empty lungs, control airflow through larynx, and produce sounds

The Digestive System

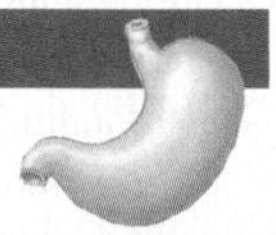

- Provides nutrients; liver regulates blood glucose and fatty acid levels and removes lactic acid from circulation
- Protects and supports soft tissues in abdominal cavity; controls entrances to and exits from digestive tract

The Urinary System

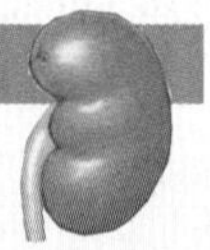

- Removes waste products of protein metabolism; assists in regulation of calcium and phosphate concentrations
- External sphincter controls urination by constricting urethra

The Reproductive System

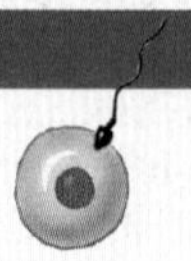

- Reproductive hormones accelerate skeletal muscle growth
- Contractions of skeletal muscles eject semen from male reproductive tract; muscle contractions during sex act produce pleasurable sensations

● *Figure 7-23* **Functional Relationships Between the Muscular System and Other Systems**

Summary Outline

1. The three types of muscle tissue are *skeletal muscle, cardiac muscle*, and *smooth muscle*. The muscular system includes all the skeletal muscle tissue that can be controlled voluntarily.

1. **Skeletal muscles** attach to bones directly or indirectly and perform the following functions: (1) produce skeletal movement, (2) maintain posture and body position, (3) support soft tissues, (4) guard entrances and exits, and (5) maintain body temperature.

1. Each muscle fiber is surrounded by an **endomysium**. Bundles of muscle fibers are sheathed by a **perimysium**, and the entire muscle is covered by an **epimysium**. At the end of the muscle is a **tendon**. *(Figure 7-1)*

2. A muscle cell has a **sarcolemma** (cell membrane), **sarcoplasm** (cytoplasm), and a **sarcoplasmic reticulum**, similar to the smooth endoplasmic reticulum of other cells. **Transverse tubules (T tubules)** and **myofibrils** aid in contraction. Filaments in a myofibril are organized into repeating functional units called **sarcomeres**. *(Figure 7-2a–c)*

3. **Myofilaments** consist of **thin filaments** (*actin*) and **thick filaments** (*myosin*). *(Figure 7-2d,e)*

4. The relationship between the thick and thin filaments changes as the muscle contracts and shortens. The **Z lines** move closer together as the thin filaments slide past the thick filaments. *(Figure 7-3)*

5. The contraction process involves **active sites** on thin filaments and **cross-bridges** of the thick filaments. For each cross-bridge, sliding involves repeated cycles of "attach, pivot, detach, and return." At rest, the necessary interactions are prevented by **tropomyosin** and **troponin** proteins on the thin filaments.

1. Neural control of muscle function involves a link between electrical activity in the sarcolemma and the initiation of a contraction.

2. Each skeletal muscle fiber is controlled by a neuron at a **neuromuscular junction**; the junction includes the **synaptic terminal**, the **synaptic cleft**, and the **motor end plate**. **Acetylcholine (ACh)** and **acetylcholinesterase (AChE)** play a role in the chemical communication between the synaptic terminal and muscle fiber. *(Figure 7-4a,b)*

3. When an **action potential** arrives at the synaptic terminal, acetylcholine is released into the synaptic cleft. The binding of ACh to receptors on the motor end plate leads to the generation of an action potential in the sarcolemma. The passage of an action potential along a transverse tubule triggers the release of calcium ions from the *cisternae* of the sarcoplasmic reticulum. *(Figure 7-4c)*

4. A contraction involves a repeated cycle of "attach, pivot, detach, and return." It begins when calcium ions are released by the sarcoplasmic reticulum. The calcium ions bind to troponin, which changes position and moves tropomyosin away from the active sites of actin. Cross-bridge binding of myosin heads to actin can now occur. After binding, each myosin head pivots at its base, pulling the actin filament toward the center of the sarcomere. *(Figure 7-5)*

5. A summary of the contraction process, from ACh release to the end of the contraction, is shown in *Table 7-1*.

1. There is no mechanism to regulate the amount of tension produced in the contraction of an individual muscle fiber. It is either "ON" (producing tension) or "OFF" (relaxed). This is known as the **all-or-none principle**.

2. Both the number of activated muscle fibers and their rate of stimulation control the tension developed by an entire skeletal muscle.

3. A muscle fiber **twitch** is a cycle of contraction and relaxation produced by a single stimulus. *(Figure 7-6)*

4. Repeated stimulation before the relaxation phase ends can result in the addition of twitches (known as **summation**) and produce **incomplete tetanus** (in which tension will peak because the muscle is never allowed to relax completely), or **complete tetanus** (in which the relaxation phase is completely eliminated). Almost all normal muscular contractions involve the complete tetanus of the participating muscle units. *(Figure 7-7)*

5. The number and size of a muscle's **motor units** indicate how precisely its movements are controlled. *(Figure 7-8)*

6. An increase in muscle tension is produced by increasing the number of motor units through **recruitment**.

7. Resting **muscle tone** stabilizes bones and joints. Inadequate stimulation causes muscles to undergo **atrophy**.

8. Normal activities usually include both **isotonic contractions** (in which the tension in a muscle remains constant as the muscle shortens) and **isometric contractions** (in which the muscle's tension rises but the length of the muscle remains constant).

9. Contraction is an active process, but elongation of a muscle fiber is passive and can occur either through elastic forces or through the contraction of opposing muscles.

1. Muscle contractions require large amounts of ATP energy.

2. ATP is an energy-transfer molecule, not an energy-storage molecule. **Creatine phosphate (CP)** can release stored energy to convert ADP to ATP. A resting muscle cell contains many times more CP than ATP. *(Figure 7-9a)*

3. At rest or moderate levels of activity, aerobic metabolism in mitochondria can provide most of the necessary ATP to support muscle contractions.

4. When a muscle fiber runs short of ATP and CP, enzymes can break down glycogen molecules to release glucose that can be broken down by **glycolysis**. *(Figure 7-9b)*

5. At peak levels of activity the cell relies heavily on the **anaerobic** process of glycolysis to generate ATP, because the mitochondria cannot obtain enough oxygen to meet the existing ATP demands. *(Figure 7-9c)*

6. **Muscle fatigue** occurs when a muscle can no longer contract, because of pH changes due to the buildup of **lactic acid**, a lack of energy, or other problems.

7 THE MUSCULAR SYSTEM

Functions of Skeletal Muscle • The Anatomy of Skeletal Muscles • The Control of Muscle Fiber Contraction • Muscle Mechanics • The Energetics of Muscular Activity

Chapter Review • MediaLab

The Recovery Period 191

7. The **recovery period** begins immediately after a period of muscle activity and continues until conditions inside the muscle have returned to preexertion levels. The *oxygen debt* created during exercise is the amount of oxygen used in the recovery period to restore normal conditions.

MUSCLE PERFORMANCE 192

1. Muscle performance can be considered in terms of **power** (the maximum amount of tension produced by a particular muscle or muscle group) and **endurance** (the duration of muscular activity).

Types of Skeletal Muscle Fibers 192

2. The two types of human skeletal muscle fibers are **fast fibers** and **slow fibers**.

3. Fast fibers are large in diameter, contain densely packed myofibrils, large reserves of glycogen, and few mitochondria. They produce rapid and powerful contractions of relatively short duration.

4. Slow fibers are smaller in diameter and take three times as long to contract after stimulation. Specializations such as an extensive capillary supply, abundant mitochondria, and high concentrations of **myoglobin** enable them to contract for long periods of time.

Physical Conditioning 193

5. **Anaerobic endurance** is the ability to support sustained, powerful muscle contractions through anaerobic mechanisms. Training to develop anaerobic endurance can lead to **hypertrophy** (enlargement) of the stimulated muscles.

6. **Aerobic endurance** is the time over which a muscle can continue to contract while supported by mitochondrial activities.

CARDIAC AND SMOOTH MUSCLE TISSUES 193

Cardiac Muscle Tissue 194

1. Cardiac muscle cells and skeletal muscle fibers differ structurally in terms of (1) size, (2) the number and location of nuclei, (3) their relative dependence on aerobic metabolism when contracting at peak levels, and (4) the presence or absence of **intercalated discs**. *(Figure 7-10a; Table 7-2)*

2. Cardiac muscle cells have *automaticity* and do not require neural stimulation to contract. Their contractions last longer than those of skeletal muscles, and cardiac muscle cannot undergo tetanus.

Smooth Muscle Tissue..194

3. Smooth muscle is nonstriated, involuntary muscle tissue that can contract over a greater range of lengths than skeletal muscle cells. *(Figure 7-10b; Table 7-2)*

4. Many smooth muscle fibers lack direct connections to motor neurons; those that are innervated are not under voluntary control.

ANATOMY OF THE MUSCULAR SYSTEM 195

1. The **muscular system** includes approximately 700 skeletal muscles that can be voluntarily controlled. *(Figure 7-11)*

Origins, Insertions, and Actions 195

2. Each muscle can be identified by its **origin**, **insertion**, and **primary action**. A muscle can be classified by their primary action as a **prime mover**, or **agonist**; a **synergist**; or an **antagonist**.

Names of Skeletal Muscles 197

3. The names of muscles often provide clues to their location, orientation, or function. *(Table 7-3)*

4. The **axial musculature** arises on the axial skeleton; it positions the head and spinal column and moves the rib cage. The **appendicular musculature** stabilizes or moves components of the appendicular skeleton.

The Axial Muscles 199

5. The axial muscles fall into four groups based on location and/or function: muscles of (a) the head and neck, (b) the spine, (c) the trunk, and (d) the pelvic floor.

6. The muscles of the head include the **frontalis, orbicularis oris, buccinator, masseter, temporalis, pterygoids,** and **platysma**. *(Figure 7-12; Table 7-4)*

7. The muscles of the neck include the **digastric, mylohyoid, stylohyoid,** and **sternocleidomastoid**. *(Figures 7-12, 7-13; Table 7-4)*

8. The **splenius capitis** and **semispinalis capitis** are the most superior muscles of the spine. The extensor muscles of the spine, or **erector spinae**, can be classified into the **spinalis, longissimus,** and **iliocostalis** divisions. In the lower lumbar and sacral regions, the longissimus and iliocostalis are sometimes called the *sacrospinalis* muscles. *(Figure 7-14; Table 7-5)*

9. The muscles of the trunk include the **oblique** and **rectus** muscles. The thoracic region muscles include the **intercostal** and **transversus** muscles. Also important to respiration is the **diaphragm**. *(Figure 7-15; Table 7-6)*

10. The muscular floor of the pelvic cavity is called the **perineum**. These muscles support the organs of the pelvic cavity and control the movement of materials through the urethra and anus. *(Figure 7-16; Table 7-7)*

The Appendicular Muscles 204

11. Together, the **trapezius** and the sternocleidomastoid affect the position of the shoulder, head, and neck. Other muscles inserting on the scapula include the **rhomboideus**, the **levator scapulae**, the **serratus anterior**, and the **pectoralis minor**. *(Figure 7-17; Table 7-8)*

12. The **deltoid** and the **supraspinatus** produce abduction of the arm at the shoulder abductors. The **subscapularis, teres major, infraspinatus,** and **teres minor** rotate the arm at the shoulder. *(Figure 7-18; Table 7-9)*

13. The **pectoralis major** flexes the elbow, and the **latissimus dorsi** extends the elbow. Both of these muscles adduct and rotate the arm at the shoulder joint. *(Figure 7-18; Table 7-9)*

14. The primary actions of the **biceps brachii** and the **triceps brachii** (long head) affect the elbow. The **brachialis** and **brachioradialis** flex the elbow. The **flexor carpi ulnaris**, the **flexor carpi radialis**, and the **palmaris longus** cooperate to flex the wrist. They are opposed by the **extensor carpi radialis** and the **extensor carpi ulnaris**. The **pronator** muscles pronate the forearm, opposed by the **supinator** and the biceps brachii. *(Figure 7-19; Table 7-10)*

15. **Gluteal muscles** cover the lateral surfaces of the ilia. They produce extension, abduction, and rotation at the hip. *(Figure 7-20a; Table 7-11)*

16. Adductors of the thigh work across the hip joint; these muscles include the **adductor magnus, adductor brevis, adductor longus, pectineus,** and **gracilis**. *(Figure 7-20b; Table 7-11)*

17. The **psoas major** and the **iliacus** merge to form the **iliopsoas** muscle, a powerful flexor of the hip. *(Figure 7-20b; Table 7-11)*

18. The flexors of the knee, include the hamstrings (**biceps femoris, semimembranosus,** and **semitendinosus**) and **sartorius**. The **popliteus** aids flexion by unlocking the knee. *(Figure 7-21; Table 7-12)*

19. The *knee extensors* are known as the **quadriceps femoris**. This group includes the three **vastus** muscles and the **rectus femoris**. *(Figure 7-21; Table 7-12)*

20. The **gastrocnemius** and **soleus** muscles produce plantar flexion. A pair of **fibularis** muscles produce eversion as well as plantar flexion. The **tibialis anterior** performs dorsiflexion. *(Figure 7-22; Table 7-13)*

21. Control of the phalanges is provided by muscles originating at the tarsal bones and at the metatarsal bones. *(Table 7-13)*

AGING AND THE MUSCULAR SYSTEM 216

1. The aging process reduces the size, elasticity, and power of all muscle tissues. Both exercise tolerance and the ability to recover from muscular injuries decrease.

INTEGRATION WITH OTHER SYSTEMS 217

1. To operate at maximum efficiency, the muscular system must be supported by many other systems. Even at rest, it interacts extensively with other systems. *(Figure 7-23)*

Review Questions

Level 1: Reviewing Facts and Terms

Match each item in column A with the most closely related item in column B. Use letters for answers in the spaces provided.

COLUMN A

____ 1. epimysium
____ 2. fascicle
____ 3. endomysium
____ 4. motor end plate
____ 5. transverse tubule
____ 6. actin
____ 7. myosin
____ 8. extensor of the knee
____ 9. sarcomeres
____ 10. tropomyosin
____ 11. recruitment
____ 12. muscle tone
____ 13. white muscles
____ 14. flexor of the leg
____ 15. red muscles
____ 16. hypertrophy

COLUMN B

a. resting tension
b. contractile units
c. thin filaments
d. surrounds muscle fiber
e. enlargement
f. surrounds muscle
g. slow fibers
h. thick filaments
i. muscle bundle
j. hamstring muscles
k. covers active sites
l. conducts action potentials
m. fast fibers
n. quadriceps muscles
o. multiple motor units
p. binds ACh

17. A skeletal muscle contains:
    (a) connective tissues   (b) blood vessels and nerves
    (c) skeletal muscle tissue (d) a, b, and c are correct

18. The type of contraction in which the tension rises but the resistance does not move is called:
    (a) a wave summation   (b) a twitch
    (c) an isotonic contraction (d) an isometric contraction

19. What are the five functions of skeletal muscle?

20. What five interlocking steps are involved in the contraction process?

21. What forms of energy reserves are found in resting skeletal muscle cells?

22. What two mechanisms are used to generate ATP from glucose in muscle cells?

23. What is the functional difference between the axial musculature and the appendicular musculature?

Level 2: Reviewing Concepts

24. Areas of the body where no slow fibers would be found include the:
    (a) back and calf muscles   (b) eye and hand
    (c) chest and abdomen   (d) a, b, and c are correct

25. Describe the basic sequence of events that occurs at a neuromuscular junction.

26. Why is the multinucleate condition important in skeletal muscle fibers?

27. The muscles of the spine include many dorsal extensors but few ventral flexors. Why?

28. What specific structural characteristic makes voluntary control of urination and defecation possible?

29. What types of movements are affected when the hamstrings are injured?

Level 3: Critical Thinking and Clinical Applications

30. Many potent insecticides contain toxins called *organophosphates* that interfere with the action of the enzyme acetylcholinesterase. Terry is using an insecticide containing organophosphates and is very careless. He does not use gloves or a dust mask and absorbs some of the chemical through his skin. He inhales a large amount as well. What

symptoms would you expect to observe in Terry as a result of the organophosphate poisoning?

31. The time of a murder victim's death is commonly estimated by the flexibility of the body. Explain why this is possible.

32. Jeff is interested in building up his thigh muscles, specifically the quadriceps group. What exercises would you recommend to help Jeff accomplish his goal?

7 THE MUSCULAR SYSTEM

Functions of Skeletal Muscle • The Anatomy of Skeletal Muscles • The Control of Muscle Fiber Contraction • Muscle Mechanics • The Energetics of Muscular Activity
Chapter Review • **MediaLab**

Answers to Concept Check Questions

Page 182

1. Since tendons attach muscles to bones, severing the tendon would disconnect the muscle from the bone. When the muscle contracted, nothing would happen.

2. Skeletal muscle appears striated when viewed under the microscope because this muscle is composed of the myofilaments actin and myosin, which have an arrangement that produces a banded appearance in the muscle.

3. You would expect to find the greatest concentration of calcium ions in the cisternae of the sarcoplasmic reticulum of the muscle.

Page 184

1. Since the ability of a muscle to contract depends on the formation of cross-bridges between the myosin and actin myofilaments, a drug that would interfere with cross-bridge formation would prevent the muscle from contracting.

2. Because the amount of cross-bridge formation is proportional to the amount of available calcium ions, increased permeability of the sarcolemma to calcium ions would lead to an increased intracellular concentration of calcium and a greater degree of contraction. In addition, since relaxation depends on decreasing the amount of calcium in the sarcoplasm, an increase in the permeability of the sarcolemma to calcium could result in a situation in which the muscle would not be able to relax completely.

3. Without acetylcholinesterase, the motor end plate would be continuously stimulated by the acetylcholine, and the muscle would be locked into contraction.

Page 189

1. The amount of tension produced in a skeletal muscle depends on (1) the frequency of stimulation, and (2) the number of activated muscle fibers.

2. A motor unit with 1500 fibers is most likely from a large muscle involved in powerful, gross body movement. Muscles that control fine or precise movements, such as movement of the eye or the fingers, have only a few fibers per motor unit, whereas muscles of the legs, for instance, that are involved in powerful contractions have hundreds of fibers per motor unit.

3. There are two types of muscle contractions, isometric and isotonic. In an isotonic contraction, tension remains constant and the muscle shortens. In isometric contractions, however, the same events of contraction occur, but instead of the muscle's shortening, the tension in the muscle increases.

Page 193

1. The sprinter requires large amounts of energy for a relatively short burst of activity. To supply this demand for energy, the muscles switch to anaerobic metabolism. Anaerobic metabolism is not as efficient in producing energy as is aerobic metabolism, and the process also produces acidic waste products. The lower energy and the waste products contribute to fatigue. Marathon runners, conversely, derive most of their energy from aerobic metabolism, which is more efficient and does not produce the level of waste products that anaerobic respiration does.

2. We would expect activities that require short periods of strenuous activity to produce a greater oxygen debt because this type of activity relies heavily on energy production by anaerobic respiration. Since lifting weights is more strenuous over the short term, we would expect this type of exercise to produce a greater oxygen debt than would swimming laps, which is an aerobic activity.

3. Individuals who are naturally better at endurance activities, such as cycling and marathon running, have a higher percentage of slow muscle fibers, which are physiologically better adapted to this type of activity than are fast fibers, which are less vascular and fatigue faster.

Page 195

1. The cell membranes of cardiac muscle cells are extensively interwoven and are bound tightly to each other at intercalated discs, allowing these muscles to "pull together" efficiently. The intercalated discs also contain gap junctions, which allow ions and small molecules to flow directly from one cell to another. This flow results in the rapid passage of action potentials from cell to cell, so their contraction is simultaneous.

2. Cardiac muscle and smooth muscle require extracellular calcium ions for contraction. In skeletal muscle, the calcium ions come from the sarcoplasmic reticulum.

3. The actin and myosin filaments of smooth muscle are not as rigidly organized as they are in skeletal muscle. This organization allows smooth muscle to contract over a relatively large range of resting lengths.

Page 198

1. The opening between the stomach and the small intestine is guarded by a circular muscle known as a sphincter muscle. The concentric circles of muscle fibers in sphincter muscles are ideally suited for opening and closing holes and for acting as valves in the body.

2. The *triceps brachii* extends the forearm and is an antagonist of the biceps brachii.

3. The name *flexor carpi radialis longus* tells you that this muscle is a long muscle that lies next to the radius and produces flexion at the wrist.

Page 201

1. Contraction of the masseter muscle raises the mandible, whereas relaxation of this muscle depresses the mandible. These movements are important in the process of chewing, or mastication.

2. You would expect the buccinator muscle, which forms the mouth for blowing, to be well-developed in a trumpet player.

Page 204

1. Damage to the external intercostal muscles would interfere with the process of breathing.

2. A blow to the rectus abdominis would cause the muscle to contract forcefully, resulting in flexion of the torso. In other words, you would "double up."

Page 211

1. When you shrug your shoulders, you are contracting your levator scapulae muscles.

2. The rotator cuff muscles include the supraspinatus, infraspinatus, subscapularis, and teres minor. The tendons of these muscles help enclose and stabilize the shoulder joint.

3. Injury to the flexor carpi ulnaris would impair the ability to flex and adduct the wrist.

Page 216

1. The hamstring is a group of three muscles that collectively function in flexing the leg: the biceps femoris, the semimembranous, and the semitendinosus.

2. The Achilles (calcaneal) tendon attaches the soleus and gastrocnemius muscles to the calcaneus (heel bone). When these muscles contract, they extend the foot. A torn Achilles tendon would make extension of the foot difficult and the opposite action, flexion, would be more pronounced as a result of less antagonism from the soleus and gastrocnemius.

Page 217

1. The number of myofibrils in a muscle fiber decreases with age. The overall effects of such a change are a reduction in muscle strength and endurance, and a tendency to fatigue rapidly.

Page 217

1. The recovery of skeletal muscle requires the removal of lactic acid from both muscle tissue and body fluids, and restoration of energy reserves in the muscle tissue. The cardiovascular system absorbs lactic acid from actively contracting muscle and carries it to the liver, an organ of the digestive system. Within the liver, lactic acid forms pyruvic acid. The liver utilizes some of the pyruvic acid to produce ATP and the ATP is then used to combine pyruvic acid molecules together into glucose. The glucose is carried back to the muscles by the cardiovascular system where it may be combined and stored as glycogen.

EXPLORE *MediaLab*

EXPLORATION #1
Estimated time for completion: 10 minutes

WORK THAT BODY! The demand for products that enlarge muscles, develop fit bodies and enhance muscle definition has created a multimillion dollar industry in the United States. Most people associate weight training with muscular development. Reflecting on what you know of the physiology of muscles, what will lifting weights accomplish? Will this training benefit any system other than the muscular system? Prepare a short training program, assuming your goal is to tone your entire body without excessive muscle enlargement. Include a brief description of why each activity is included in your regime, as well as the expected results for a male and female. To get

A pedestrian passes under a bank displaying the temperature on a winter day in Fairbanks, Alaska.

some expert advice on this, visit the Companion Web site's MediaLab section and click on the keyword "weight training." Here you can validate your training program with information from the University of Indiana Health Center.

EXPLORATION #2
Estimated time for completion: 10 minutes

THE LARGEST HEAT producers in the body are the skeletal muscles. The heat that skeletal muscles generate as they break down ATP for both movement and maintenance of posture helps keep our core body temperature at a relatively constant 37°C (98.6°F). The muscular system can also elevate body temperature when we are cold, through the random contraction of muscle cell—shivering uses a lot of ATP and generates large amounts of heat as a result. What happens when the body temperature continues to drop despite the efforts of the muscular system? Review the paragraph on heat loss in this chapter (p. 192), and then look at the discussion on dermal circulation in Chapter 5 (p. 113). Using this information, prepare an outline of how the systems of the body respond to rising and falling temperatures. What symptoms would you expect from someone suffering from hypothermia (below-normal body temperature)? How would you expect those symptoms to be treated? The CW will help you create this outline. For more information on hypothermia, visit the Companion Web site's MediaLab section, and click on the keyword "hypothermia." Use this information from Princeton University to complete your outline, paying close attention to the interplay among the systems of the body.

Working out on a machine that isolates and develops muscles of the arm.

The Nervous System

Vocabulary Development

a- .... without; *aphasia*

af .... to; *afferent*

arachne .... spider; *arachnoid membrane*

astro- .... star; *astrocyte*

ataxia .... a lack of order; *ataxia*

axon .... axis; *axon*

cauda .... tail; *cauda equina*

cephalo- .... head; *diencephalon*

chiasm .... a crossing; *optic chiasm*

choroid .... a vascular coat; *choroid plexus*

colliculus .... a small hill; *superior colliculus*

commissura .. a joining together; *commissure*

cortex .... rind; *neural cortex*

cyte .... cell; *astrocyte*

dia .... through; *diencephalon*

dura .... hard; *dura mater*

equus .... horse; *cauda equina*

ef-, ex- .... from; *efferent*

ferre .... to carry; *afferent*

ganglio .... knot; *ganglion*

glia .... glue; *neuroglia*

hypo- .... below; *hypothalamus*

inter- .... between; *interneurons*

lexis .... diction; *dyslexia*

limbus .... a border; *limbic system*

mamilla .... a little breast; *mamillary bodies*

mater .... mother; *dura mater*

meninx .... membrane; *meninges*

meso- .... middle; *mesencephalon*

neuro- .... nerve; *neuron*

nigra .... black; *substantia nigra*

oligo- .... few; *oligodendrocytes*

phasia .... speech; *aphasia*

pia .... delicate; *pia mater*

plexus .... a network; *choroid plexus*

saltare .... to leap; *saltatory*

syn- .... together; *synapse*

vagus .... wandering; *vagus nerve*

vas .... vessel; *vasomotor*

*I*F YOU THINK OF THE NERVOUS SYSTEM as an organic computer, individual neurons are the "chips" that make it work. This scanning electron micrograph shows a human neuron growing on the surface of a silicon computer chip.

8 THE NERVOUS SYSTEM

The Nervous System • Cellular Organization in Neural Tissue • Neuron Function • Neural Communication • The Central Nervous System

TWO ORGAN SYSTEMS, the *nervous system* and the *endocrine system*, coordinate organ system activities in response to changing environmental conditions. The nervous system responds relatively swiftly but briefly to stimuli, whereas endocrine responses develop more slowly but last much longer. For example, the nervous system adjusts body position and moves your eyes across this page, while the endocrine system adjusts the daily rate of energy use by the entire body and directs long term processes such as growth and maturation. (We will consider the endocrine system in Chapter 10.)

The nervous system is the most complex organ system. As you read these words and think about them, at the involuntary level your nervous system is also monitoring the external environment and your internal systems, and issuing commands as needed to maintain homeostasis. In a few hours, at mealtime or while you are sleeping, the pattern of nervous system activity will be very different. The change from one pattern of activity to another can be almost instantaneous because neural function relies on electrical events that proceed at great speed.

This chapter examines the structure and function of the nervous system, from its cells through their organization into the two major divisions of the nervous system: (1) the *central nervous system* and (2) the *peripheral nervous system*. ∞ p. 7

The Nervous System

The **nervous system** (1) monitors the internal and external environments, (2) integrates sensory information, and (3) coordinates voluntary and involuntary responses of many other organ systems. These functions are performed by neurons, which are supported and protected by surrounding neuroglia.

The functions of the two major anatomical subdivisions of the nervous system are detailed in Figure 8-1●. The **central**

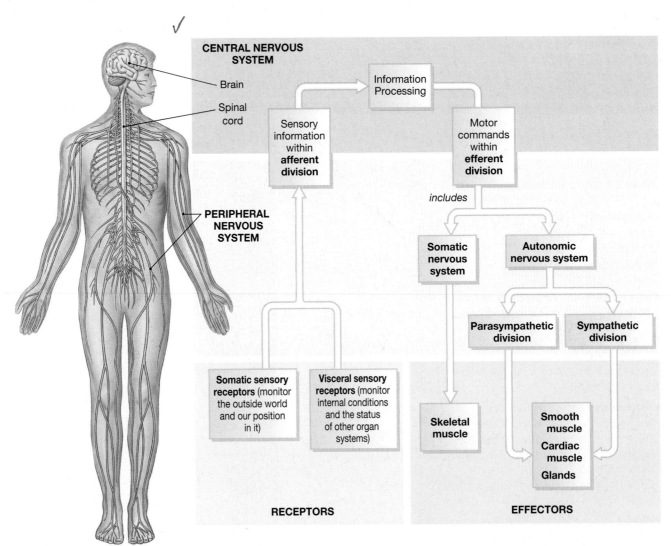

● *Figure 8-1* **A Functional Overview of the Nervous System**

nervous system (CNS), consisting of the *brain* and the *spinal cord*, integrates and coordinates sensory data and motor commands. The CNS is also the seat of higher functions, such as intelligence, memory, and emotion. All communication between the CNS and the rest of the body occurs over the **peripheral nervous system (PNS)**. The peripheral nervous system includes all the neural tissue *outside* the CNS. Its **afferent division** (*af-*, to + *ferre*, to carry) brings sensory information to the CNS, and its **efferent division** (*ef-*, from) carries motor commands to muscles and glands. Within the efferent division, the **somatic nervous system (SNS)** provides control over skeletal muscle contractions, and the **autonomic nervous system (ANS)**, or *visceral motor system*, provides automatic involuntary regulation of smooth muscle, cardiac muscle, and glandular secretions. The ANS includes a *sympathetic division* and a *parasympathetic division*, which commonly have opposite effects. For example, activity of the sympathetic division accelerates the heart rate whereas the parasympathetic division slows the heart rate.

Cellular Organization in Neural Tissue

The nervous system includes all the neural tissue in the body. Neural tissue, introduced in Chapter 4, consists of two kinds of cells, *neurons* and *neuroglia*. ∞ p. 102 **Neurons** (*neuro-*, nerve) are the basic units of the nervous system. All neural functions involve the communication of neurons with one another and with other cells. The **neuroglia** (noo-ROG-lē-uh or noo-rō-GLĒ-uh; *glia*, glue) regulate the environment around the neurons, provide a supporting framework for neural tissue, and act as phagocytes. Although they are much smaller cells, neuroglia, also called *glial cells*, far outnumber neurons. Unlike most neurons, most glial cells retain the ability to divide.

NEURONS
The General Structure of Neurons

A "model" neuron has (1) a cell body; (2) several branching, sensitive **dendrites**, which receive incoming signals; and (3) an elongate **axon**, which carries outgoing signals toward (4) one or more **synaptic terminals** (Figure 8-2●). At each synaptic terminal, the neuron communicates with another cell. (Because the synaptic terminals between neurons are rounded, they are also called **synaptic knobs**.) Neurons can have a variety of shapes; Figure 8-2● shows a *multipolar neuron*, the most common type of neuron in the CNS.

The cell body of a typical neuron contains a large, round nucleus with a prominent nucleolus. There are usually no centrioles, organelles that in other cells form the spindle fibers that move chromosomes during cell division. Most neurons lose their centrioles during differentiation and become incapable of undergoing mitosis. As a result, most neurons lost to injury or disease cannot be replaced.

The cell body also contains the organelles that provide energy and synthesize organic compounds. The numerous mitochondria, free and fixed ribosomes, and membranes of

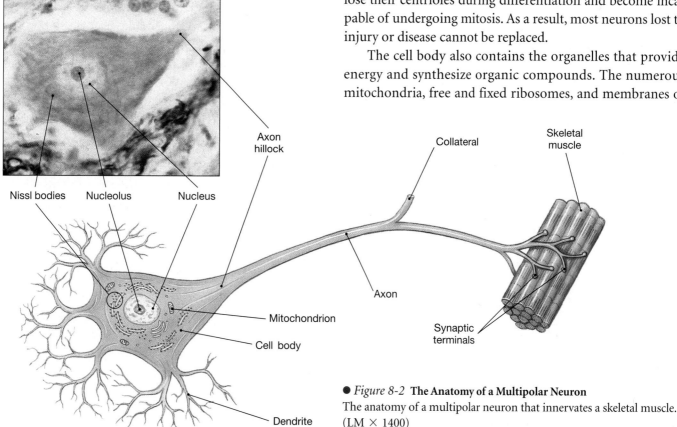

Nissl bodies   Nucleolus   Nucleus

Axon hillock

Mitochondrion

Cell body

Dendrite

Collateral

Skeletal muscle

Axon

Synaptic terminals

● *Figure 8-2* **The Anatomy of a Multipolar Neuron**
The anatomy of a multipolar neuron that innervates a skeletal muscle. (LM × 1400)

227

8 THE NERVOUS SYSTEM

The Nervous System • **Cellular Organization in Neural Tissue** • Neuron Function • Neural Communication • The Central Nervous System

the rough endoplasmic reticulum (RER) give the cytoplasm a coarse, grainy appearance. Clusters of rough ER and free ribosomes, known as **Nissl bodies**, give a gray color to areas containing neuron cell bodies and account for the color of *gray matter* seen in brain and spinal cord dissections.

Projecting from the cell body are a variable number of dendrites and a single large axon. The cell membrane of the dendrites and cell body is sensitive to chemical, mechanical, or electrical stimulation. In a process described later, such stimulation often leads to the generation of an electrical impulse, or *action potential*, that travels along the axon. Action potentials begin at a thickened region of the cell body called the *axon hillock*. The axon may branch along its length, producing branches called *collaterals*. Synaptic terminals are found at the tips of each branch. A synaptic terminal is part of a **synapse**, a site where a neuron communicates with another cell.

Structural Classification of Neurons

The billions of neurons in the nervous system are variable in form. Based on the relationship of the dendrites to the cell body and axon, neurons are classified into three types:

1. A **multipolar neuron** has multiple processes extending away from the cell body (Figure 8-3●). Such neurons are very common within the CNS. (For example, all of the motor neurons that control skeletal muscles are multipolar.)
2. In a **unipolar neuron**, the dendrites and axon are continuous, and the cell body lies off to one side (Figure 8-3b●). In a unipolar neuron, the action potential begins at the base

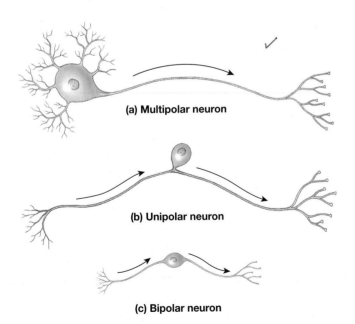

(a) Multipolar neuron

(b) Unipolar neuron

(c) Bipolar neuron

● *Figure 8-3* **A Structural Classification of Neurons**
The neurons are not drawn to scale; bipolar neurons are many times smaller than typical unipolar and multipolar neurons.

of the dendrites and the rest of the process is considered an axon. Most sensory neurons of the peripheral nervous system are unipolar.

3. **Bipolar neurons** have two processes, one dendrite and one axon, with the cell body between them (Figure 8-3c●). Bipolar neurons are rare but occur in special sense organs such as the eye and ear.

Functional Classification of Neurons

Neurons are sorted into three functional groups: (1) *sensory neurons*, (2) *motor neurons*, and (3) *interneurons*.

Sensory Neurons. **Sensory neurons** of the afferent division convey information from both the external and internal environments to other neurons inside the CNS. Sensory neurons form the afferent division of the PNS, and there are approximately 10 million sensory neurons in the human body. These neurons connect a *sensory receptor* in peripheral tissues with the spinal cord or brain. The receptor itself may be simply a dendrite of a sensory neuron or a specialized cell that communicates with the sensory neuron. Receptors may be grouped into three categories based on the information they carry. The **somatic sensory receptors** carry two types of information; one about the outside world and the other about our position within it. The **external receptors** provide information about the external environment in the form of touch, temperature, and pressure sensations and the more complex senses of sight, smell, hearing, and touch, and the **proprioceptors** (prō-prē-ō-SEP-torz; *proprius*, one's own + *capio*, to take) monitor the position and movement of skeletal muscles and joints. The **visceral receptors**, or **internal receptors**, monitor the activities of the digestive, respiratory, cardiovascular, urinary, and reproductive systems and provide sensations of taste, deep pressure, and pain.

Motor Neurons. The half million **motor neurons** of the efferent division carry instructions from the CNS to other tissues, organs, or organ systems. The peripheral targets are called *effectors* because they change their activities in response to the commands issued by the motor neurons. For example, a skeletal muscle is an effector that contracts on neural stimulation. There are two efferent divisions in the PNS, each targeting a separate class of effectors. The **somatic motor neurons** of the somatic nervous system innervate skeletal muscles, and the **visceral motor neurons** of the autonomic nervous system innervate other peripheral effectors, such as cardiac muscle, smooth muscle, and glands.

Interneurons. The 20 billion **interneurons**, or *association neurons*, are located entirely within the brain and the spinal cord.

Interneurons, as the name implies (*inter-*, between), interconnect other neurons. Interneurons are responsible for the distribution of sensory information and the coordination of motor activity. The more complex the response to a given stimulus, the greater the number of interneurons involved.

NEUROGLIA

Neuroglia are found in both the CNS and PNS, but the CNS has the greatest diversity of glial cells. There are four types of glial cells in the central nervous system (Figure 8-4●):

1. **Astrocytes** (AS-trō-sīts; *astro-*, star + *cyte*, cell) are the largest and most numerous neuroglia. Astrocytes secrete chemicals vital to the maintenance of the *blood-brain barrier*,

which isolates the CNS from the general circulation. The secretions cause the capillaries of the CNS to become impermeable to many compounds that could interfere with neuron function. Astrocytes also create a structural framework for the CNS and perform repairs in damaged neural tissues.

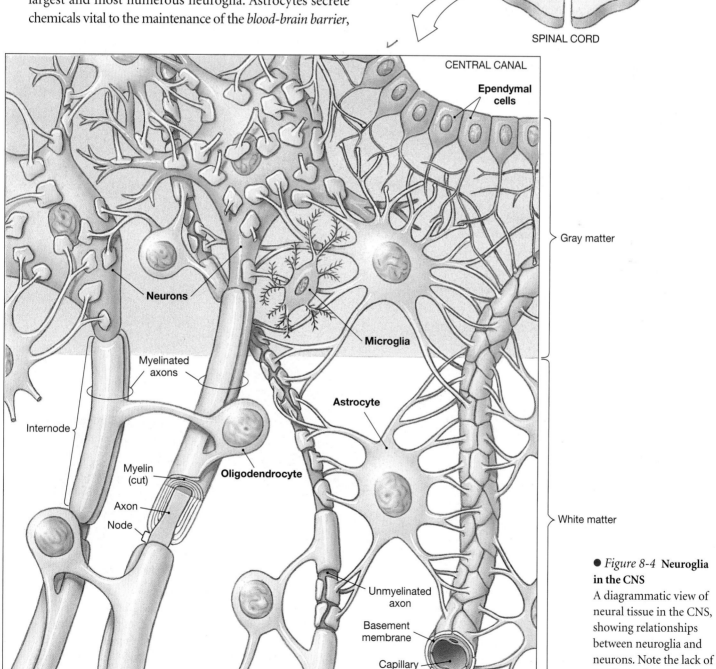

● *Figure 8-4* **Neuroglia in the CNS**
A diagrammatic view of neural tissue in the CNS, showing relationships between neuroglia and neurons. Note the lack of myelin in the gray matter.

229

8 THE NERVOUS SYSTEM

The Nervous System • Cellular Organization in Neural Tissue • **Neuron Function** • Neural Communication • The Central Nervous System

2. **Oligodendrocytes** (o-li-gō-DEN-drō-sīts; *oligo-*, few) have cytoplasmic extensions that can wrap around axons, creating a membranous sheath of insulation called **myelin** (MĪ-e-lin). Many oligodendrocytes are needed to coat an entire axon with myelin. Such an axon is said to be **myelinated**. Myelin increases the speed at which an action potential travels along the axon. Each oligodendrocyte myelinates short segments of several axons. The gaps between adjacent cell processes are called *nodes*, or the *nodes of Ranvier* (RAHN-vē-ā). The areas covered in myelin are called *internodes*. Not every axon in the CNS is myelinated, and those without a myelin coating are said to be **unmyelinated**. Myelin is lipid-rich, and on dissection, areas of the CNS containing myelinated axons appear glossy white. These areas constitute the **white matter** of the CNS, whereas areas of **gray matter** are dominated by neuron cell bodies.

3. **Microglia** (mī-KRŌG-lē-a) are the smallest and rarest of the neuroglia in the CNS. Microglia are phagocytic cells derived from white blood cells that have migrated across capillary walls in the CNS. They perform protective functions such as engulfing cellular waste and pathogens.

4. **Ependymal** (e-PEN-di-mul) **cells** line fluid-filled cavities within the CNS; the *central canal* of the spinal cord and the chambers, or *ventricles*, of the brain. This lining of epithelial cells is called the **ependyma** (e-PEN-di-muh). In some regions of the brain, the ependyma produces cerebrospinal fluid (CSF), and the cilia on ependymal cells in other locations help circulate this fluid within and around the CNS.

The most important glial cells in the peripheral nervous system are **Schwann cells**. Schwann cells cover every axon outside the CNS, whether myelinated or unmyelinated. Whereas an oligodendrocyte in the CNS may myelinate portions of several adjacent axons, a Schwann cell can myelinate only one segment of a single axon (Figure 8-5a●). However, a Schwann cell may surround portions of several different unmyelinated axons (Figure 8-5b●). Whenever a Schwann cell covers an axon, the outer surface of the Schwann cell is called the *neurilemma* (noo-ri-LEM-uh).

Demyelination is the progressive destruction of myelin sheaths, accompanied by inflammation, axon damage, and scarring of neural tissue. The result is a gradual loss of sensation and motor control that leaves affected regions numb and paralyzed. In one demyeli-

nation disorder, **multiple sclerosis** (skler- Ō-sis; *sclerosis*, hardness), or **MS**, axons in the optic nerve, brain, and/or spinal cord are affected. Common symptoms of MS include partial loss of vision and problems with speech, balance, and general motor coordination.

ANATOMICAL ORGANIZATION

Neurons and their axons are not randomly scattered in the CNS and PNS. Instead, they form masses or bundles with distinct anatomical boundaries and are identified by specific terms. We shall use these terms again, so a brief overview here will prove helpful (Figure 8-6●).

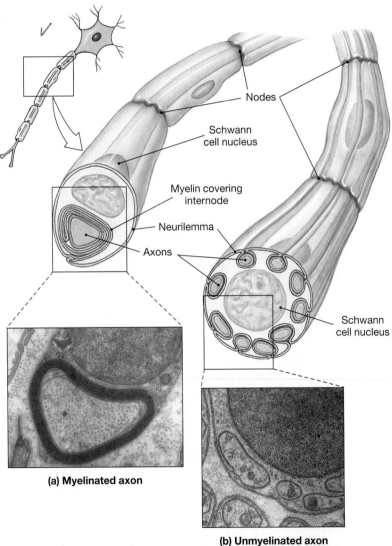

Nodes

Schwann cell nucleus

Myelin covering internode

Neurilemma

Axons

Schwann cell nucleus

(a) Myelinated axon

(b) Unmyelinated axon

● *Figure 8-5* **Schwann Cells and Peripheral Axons**
(**a**) A myelinated axon in the PNS is covered by a series of Schwann cells. Each Schwann cell forms a myelin sheath around a portion of the axon. This arrangement differs from the way myelin forms in the CNS; compare with Figure 8-4. (TEM × 14,048) (**b**) A single Schwann cell can encircle several unmyelinated axons. Every axon in the PNS is completely enclosed by Schwann cells.

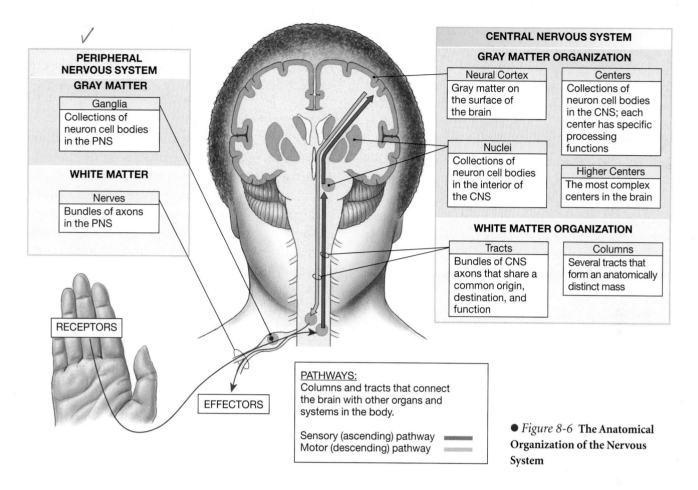

• *Figure 8-6* **The Anatomical Organization of the Nervous System**

In the PNS:

- Neuron cell bodies are located in **ganglia**.
- Axons are bundled together in **nerves**, with *spinal nerves* connected to the spinal cord and *cranial nerves* connected to the brain.

In the CNS:

- A collection of neuron cell bodies with a common function is called a **center**. A center with a discrete boundary is called a **nucleus**. Portions of the brain surface are covered by a thick layer of gray matter called **neural cortex** (*cortex*, rind). The term *higher centers* refers to the most complex integration centers, nuclei, and cortical areas in the brain.
- The white matter of the CNS contains bundles of axons that share common origins, destinations, and functions. These bundles are called **tracts**. Tracts in the spinal cord form larger groups, called **columns**.
- **Pathways** link the centers of the brain with the rest of the body. For example, **sensory** (*ascending*) **pathways** distribute information from sensory receptors to processing centers in the brain, and **motor** (*descending*) **pathways** begin at CNS centers concerned with motor control and end at the skeletal muscles they control.

CONCEPT CHECK QUESTIONS

Answers on page 274

❶ What would damage to the afferent division of the nervous system affect?

❷ Examination of a tissue sample shows unipolar neurons. Are these more likely to be sensory neurons or motor neurons?

❸ Which type of glial cell would you expect to be present in large numbers in brain tissue from a person suffering from an infection of the central nervous system?

Neuron Function

The sensory, integrative, and motor functions of the nervous system are dynamic and ever changing. All of the important communications between neurons and other cells occur through their membrane surfaces. These membrane changes are electrical events that proceed at great speed.

THE MEMBRANE POTENTIAL

The cell membrane of an undisturbed cell has an excess of positive charges on the outside and an excess of negative charges

8 THE NERVOUS SYSTEM

The Nervous System • Cellular Organization in Neural Tissue • **Neuron Function** • Neural Communication • The Central Nervous System

on the inside. Such an uneven distribution of charges is known as a *potential difference*, and the size of the potential difference is measured in units called *volts*. Because the charges are separated by a cell membrane, the potential difference across the cell membrane of a living cell is called a **membrane potential**, or *transmembrane potential*. The membrane potential of an undisturbed cell, also known as its **resting potential**, is very small. For example, the resting potential of a neuron averages about 0.070 volts, versus around 12 volts for a car battery. Because they are so small, membrane potentials are usually reported in *millivolts* (mV, thousandths of volts) rather than volts. The resting potential of a neuron is -70 mV, with the minus sign indicating that the inside of the cell membrane contains an excess of negative charges as compared with the outside.

Factors Responsible for the Membrane Potential

In addition to an imbalance of electrical charges, the intracellular and extracellular fluids differ markedly in ionic composition. For example, the extracellular fluid contains relatively high concentrations of sodium ions (Na^+) and chloride ions (Cl^-), whereas the intracellular fluid contains high concentrations of potassium ions (K^+) and negatively charged proteins (Pr^-).

The cell membrane maintains the differences between the intracellular and extracellular fluids. The proteins within the cytoplasm cannot cross the membrane, and the ions can enter or leave the cell only by passing through membrane channel proteins. ⟳ p. 57 There are many different types of channels in the membrane; some are always open (*leak channels*), and others (*gated channels*) open or close under specific circumstances.

Because the intracellular concentration of potassium ions is relatively high, potassium ions tend to diffuse out of the cell. This movement is driven by the concentration gradient for potassium ions. Similarly, the concentration gradient for sodium ions tends to promote their movement into the cell. However, the cell membrane is much more permeable to potassium ions than to sodium ions. As a result, potassium ions diffuse out of the cell faster than sodium ions enter the cytoplasm. The cell therefore experiences a net loss of positive charges, and as a result, the interior of the cell membrane contains an excess of negative charges, primarily from negatively charged proteins.

The resting potential remains stable over time because the cell membrane contains a carrier protein, the sodium-potassium exchange pump. ⟳ p. 62 At a membrane potential of -70 mV, the rate of sodium entry to potassium loss is precisely balanced by the sodium-potassium exchange pump. Figure 8-7● presents a diagrammatic view of the cell membrane at the resting potential.

Changes in the Membrane Potential

Any stimulus that (1) alters membrane permeability to sodium or potassium or (2) alters the activity of the exchange pump will disturb the resting potential of a cell. Examples of stimuli that can affect the membrane potential include exposure to specific chemicals, mechanical pressure, changes in temperature, or shifts in the extracellular ion concentrations. Any change in the resting potential can have an immediate effect on the cell. For example, in a skeletal muscle fiber, permeability changes in the sarcolemma trigger a contraction. ⟳ p. 182

In most cases, a stimulus opens gated ion channels that are closed when the cell membrane is at the normal resting potential. The opening of these channels accelerates the movement of ions across the cell membrane, and this movement changes the membrane potential. For example, the opening of gated sodium channels will accelerate sodium entry into the cell. As the number of positively charged ions increases on the inner surface of the cell membrane, the membrane potential will shift toward 0 mV. A shift in this direction is called a *depolarization* of the membrane. A stimulus that opens gated potassium ion channels will shift the membrane potential away from 0 mV, because additional potassium ions will leave the cell. Such a change, which may take the membrane potential from -70 mV to -80 mV, is called a *hyperpolarization*.

Information transfer between neurons and other cells involves graded potentials and action potentials. **Graded potentials** affect only a limited portion of the cell membrane. For example, if a chemical stimulus applied to the cell membrane of a neuron opens gated sodium ion channels at a single site, the sodium ions entering the cell will depolarize the membrane at that location. The sodium ions will then diffuse along the inner surface of the membrane in all directions. Because the sodium ions are spreading out, the degree of depolarization decreases with distance away from the point of entry.

Graded potentials occur in the membranes of all cells in response to environmental stimuli. These changes can trigger shifts in cellular function. For example, chemicals released by motor neurons produce graded potentials in the membranes of gland cells, and these potential changes can stimulate or inhibit glandular secretion. However, graded potentials affect

● *Figure 8-7* **The Cell Membrane at the Resting Potential**

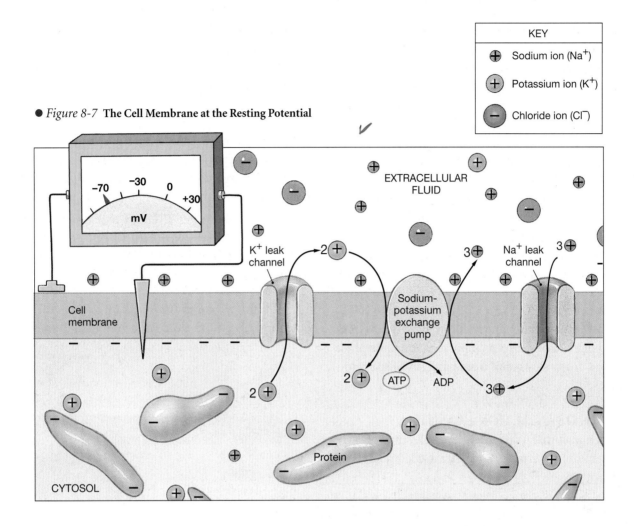

too small an area to have an effect on the activities of relatively enormous cells, such as skeletal muscle fibers or neurons. In these cells, graded potentials can influence operations in distant portions of the cell only if they lead to the production of an **action potential**, an electrical signal that affects the entire membrane surface.

An **action potential** is a propagated change in the permeability of the cell membrane. Skeletal muscle fibers and axons have *excitable membranes* that will conduct action potentials. In a skeletal muscle fiber, the action potential begins at the neuromuscular junction and travels across the entire membrane surface, including the T tubules. ∽ p. 182 The resulting ion movements trigger a contraction. In an axon, an action potential usually begins near the axon hillock and travels along the length of the axon toward the synaptic terminals, where its arrival activates the synapses.

Action potentials are generated by the opening and closing of gated sodium and potassium channels in response to

a local (graded) depolarization. This depolarization acts like pressure on the trigger of a gun. A gun fires only after a certain minimum pressure has been applied to the trigger. It does not matter whether the pressure builds gradually or is exerted suddenly—when the pressure reaches a critical point, the gun will fire. Every time it fires, the bullet that leaves the gun will have the same speed and range, regardless of the forces that were applied to the trigger. In an axon, the graded potential is the pressure on the trigger, and the action potential is the firing of the gun. An action potential will not appear unless the membrane depolarizes sufficiently to a level known as the **threshold**.

Every stimulus, whether minor or extreme, that brings the membrane to threshold will generate an identical action potential. This is called the **all-or-none principle**: A given stimulus either triggers a typical action potential or does not produce one at all. The all-or-none principle applies to all excitable membranes.

8 **THE NERVOUS SYSTEM**

The Nervous System • Cellular Organization in Neural Tissue • Neuron Function • **Neural Communication** • The Central Nervous System

The Generation of an Action Potential

An action potential begins when the cell membrane depolarizes to threshold. Figure 8-8● diagrams the steps involved in the generation of an action potential, beginning with a graded depolarization to threshold (from −70 to −60 mV) and ending with a return to the resting potential (−70 mV).

From the moment that the sodium channels open at threshold until *repolarization* (the return to the resting potential) is completed, the membrane cannot respond normally to further stimulation. This is the **refractory period**. The refractory period limits the number of action potentials that can be generated in an excitable membrane. (The maximum rate of action potential generation is 500–1000 per second.)

Potentially deadly forms of human poisoning result from eating seafood containing *neurotoxins*, poisons that primarily affect neurons. Several neurotoxins, such as *tetrodotoxin* (*TTX*) from puffer fish, either block open sodium channels or prevent their opening. Motor neurons cannot function under these conditions, and death may result from paralysis of the respiratory muscles.

PROPAGATION OF AN ACTION POTENTIAL

An action potential initially involves a relatively small portion of the total membrane surface. But unlike graded potentials, which diminish rapidly with distance, action potentials go on to affect the entire membrane surface. The basic mechanism of action potential propagation along unmyelinated and myelinated axons is shown in Figure 8-9●.

At a given site, for a brief moment at the peak of the action potential, the inside of the cell membrane contains an excess of positive ions. Because opposite charges attract one another, these ions immediately begin spreading along the inner surface of the membrane, drawn to the surrounding negative charges. This *local current* depolarizes adjacent portions of the membrane, and when threshold is reached, action potentials occur at these locations (Figure 8-9a●). Each time a local current develops, the action potential moves forward, not backward, because the previous segment of the axon is still in the refractory period.

The process continues in a chain reaction that soon reaches the most distant portions of the cell membrane. This form of

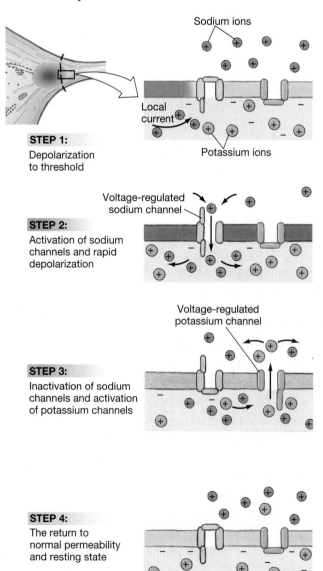

STEP 1:
Depolarization to threshold

Sodium ions

Local current

Potassium ions

STEP 2:
Activation of sodium channels and rapid depolarization

Voltage-regulated sodium channel

STEP 3:
Inactivation of sodium channels and activation of potassium channels

Voltage-regulated potassium channel

STEP 4:
The return to normal permeability and resting state

● *Figure 8-8* **The Generation of an Action Potential**
A sufficiently strong depolarizing stimulus will bring the membrane potential to threshold and trigger an action potential. Each number on the graph indicates a step that is underway at that time.

DEPOLARIZATION   REPOLARIZATION

Threshold

Resting potential

REFRACTORY PERIOD

Transmembrane potential (mV)

Time (msec)

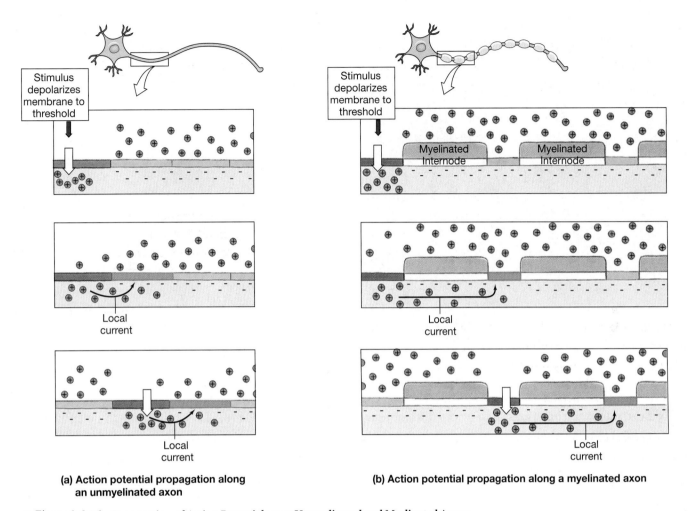

(a) Action potential propagation along an unmyelinated axon

(b) Action potential propagation along a myelinated axon

● *Figure 8-9* **The Propagation of Action Potentials over Unmyelinated and Myelinated Axons**
(**a**) Continuous propagation along an unmyelinated axon. (**b**) Saltatory propagation along a myelinated axon.

action potential transmission is known as **continuous propagation**. You might compare continuous propagation to a person walking heel-to-toe; progress is made in a series of small steps. Continuous propagation occurs along unmyelinated axons at a speed of around 1 meter per second (2 mph).

In a myelinated fiber, the axon is wrapped in layers of myelin. This wrapping is complete except at the nodes, where adjacent glial cells contact one another. Between the nodes, the lipids of the myelin sheath block the flow of ions across the membrane. As a result, continuous propagation cannot occur. Instead, when an action potential occurs at the base of the axon, the local current skips the internode and depolarizes the closest node to threshold. The action potential jumps from node to node, rather than proceeding in a series of small steps (Figure 8-9b●). This process is called **saltatory propagation**, taking its name from *saltare*, the Latin word meaning "to leap." Saltatory propagation carries nerve impulses along an axon at speeds ranging from 18–140 meters per second (40–300 mph).

CONCEPT CHECK QUESTIONS

Answers on page 275

❶ How would a chemical that blocks the sodium channels in a neuron's cell membrane affect the neuron's ability to depolarize?

❷ Two axons are tested for propagation velocities. One carries action potentials at 50 meters per second, the other at 1 meter per second. Which axon is myelinated?

Neural Communication

In the nervous system, information moves from one location to another in the form of action potentials. At the end of an axon, the arrival of an action potential results in the transfer of information to another neuron or effector cell. The information transfer occurs through the release of chemicals called **neurotransmitters** from the synaptic terminal.

When one neuron communicates with another, the synapse may occur on a dendrite, on the cell body, or along the length

235

8 THE NERVOUS SYSTEM

The Nervous System • Cellular Organization in Neural Tissue • Neuron Function • **Neural Communication** • The Central Nervous System

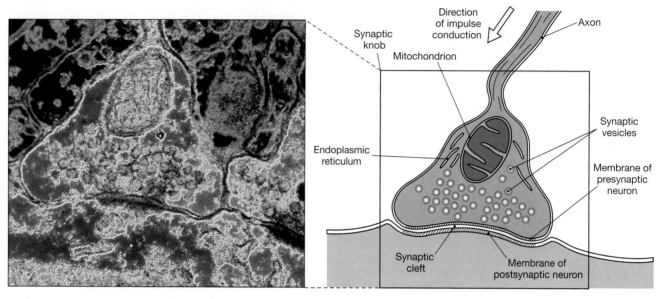

● *Figure 8-10* **The Structure of a Cholinergic Synapse**
The synaptic terminals between neurons are called synaptic knobs because of their rounded shape. (TEM × 222,000)

of the axon. Synapses between a neuron and another cell type are called **neuroeffector junctions**. At a *neuromuscular junction*, the neuron communicates with a muscle cell, as we saw in Chapter 7. ⌒ p. 182 At a *neuroglandular junction*, a neuron controls or regulates the activity of a secretory cell.

STRUCTURE OF A SYNAPSE

Communication between neurons and other cells occurs in only one direction across a synapse—from the synaptic terminal of the **presynaptic neuron** to the **postsynaptic neuron** (Figure 8-10●) or other cell type. The opposing cell membranes are separated by a narrow space called the **synaptic cleft**.

Each synaptic terminal contains mitochondria, synaptic vesicles, and endoplasmic reticulum. Every synaptic vesicle contains several thousand molecules of a specific neurotransmitter, and on stimulation many of these vesicles release their contents into the synaptic cleft. The neurotransmitter then diffuses across the synaptic cleft and binds to receptors on the postsynaptic membrane.

SYNAPTIC FUNCTION AND NEUROTRANSMITTERS

There are many different neurotransmitters. The neurotransmitter **acetylcholine**, or **ACh**, is released at **cholinergic synapses**. Cholinergic synapses are widespread inside and outside of the CNS; the neuromuscular junction described in Chapter 7 is one example. Figure 8-11● shows the major events that occur at a cholinergic synapse after an action potential arrives at the presynaptic neuron:

Step 1: *The arrival of an action potential at the synaptic knob* (Figure 8-11a●). The arriving action potential depolarizes the presynaptic membrane.

Step 2: *The release of neurotransmitter* (Figure 8-11b●). Depolarization of the presynaptic membrane causes the brief opening of calcium channels that allows extracellular calcium ions to enter the synaptic knob. Their arrival triggers the exocytosis of the synaptic vesicles and the release of ACh. The release of ACh stops very soon because the calcium ions are rapidly removed from the cytoplasm by active transport mechanisms.

Step 3: *The binding of ACh and the depolarization of the postsynaptic membrane* (Figure 8-11c●). The binding of ACh to sodium channels causes them to open and allows sodium ions to enter. If the resulting depolarization of the postsynaptic membrane reaches threshold, an action potential is produced.

Step 4: *The removal of ACh by AChE* (Figure 8-11d●). The effects on the postsynaptic membrane are temporary because the synaptic cleft and postsynaptic membrane contain the enzyme acetylcholinesterase (AChE). The AChE removes ACh by breaking it into acetate and choline.

Table 8-1 summarizes the sequence of events that occur at a cholinergic synapse.

Another common neurotransmitter, **norepinephrine** (nōr-ep-i-NEF-rin), or **NE**, is important in the brain and in portions of the autonomic nervous system. It is also called *noradrenaline*, and synapses releasing NE are described as

● *Figure 8-11* **The Function of a Cholinergic Synapse**

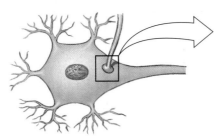

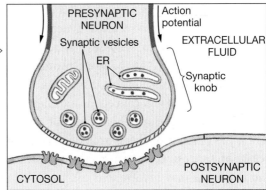

(a) Step 1: Arrival of action potential at synaptic knob

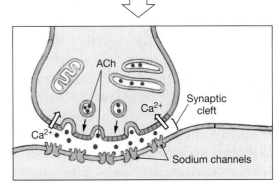

(b) Step 2: Entry of extracellular Ca^{2+} and release of ACh through synaptic vesicle exocytosis

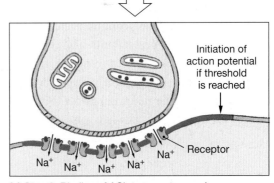

(c) Step 3: Binding of ACh to receptors and depolarization of postsynaptic membrane may bring adjacent segment to threshold

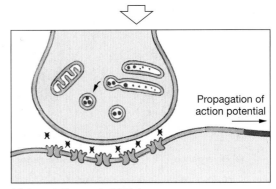

(d) Step 4: Removal of ACh by acetylcholinesterase (AChE)

adrenergic. Both ACh and NE usually have an excitatory effect due to the depolarization of postsynaptic neurons. Like ACh, NE's effect is temporary; it is broken down by an enzyme called *monoamine oxidase.*

Dopamine (DŌ-puh-mēn), **gamma aminobutyric** (GAM-ma a-MĒ-nō-bē-TĒR-ik) **acid**, also known as **GABA**, and **serotonin** (ser-o-TŌ-nin) are CNS neurotransmitters whose effects are usually inhibitory due to the hyperpolarization of postsynaptic neurons. There are at least 50 other neurotransmitters whose functions are not well understood. Included in this group are two gases that have recently been shown to be neurotransmitters: *nitric oxide (NO)* and *carbon monoxide (CO).*

The neurotransmitters released at a synapse may have excitatory or inhibitory effects. Whether an action potential appears in the postsynaptic neuron depends on the balance between the depolarizing and hyperpolarizing stimuli arriving

TABLE 8-1 *The Sequence of Events at a Typical Cholinergic Synapse*

Step 1:
• An arriving action potential depolarizes the synaptic knob and the presynaptic membrane.

Step 2:
• Calcium ions enter the cytoplasm of the synaptic knob.
• ACh release occurs through diffusion and exocytosis of neurotransmitter vesicles.

Step 3:
• ACh diffuses across the synaptic cleft and binds to receptors on the postsynaptic membrane.
• Sodium channels on the postsynaptic surface are activated, producing a graded depolarization.
• ACh release stops because calcium ions are removed from the cytoplasm of the synaptic knob.

Step 4:
• The depolarization ends as ACh is broken down into acetate and choline by AChE.
• The synaptic knob reabsorbs choline from the synaptic cleft and uses it to resynthesize ACh.

8 THE NERVOUS SYSTEM

The Nervous System • Cellular Organization in Neural Tissue • Neuron Function • Neural Communication • **The Central Nervous System**

at any moment. For example, suppose that a neuron will generate an action potential if it receives 10 depolarizing stimuli. That could mean 10 active synapses if all release excitatory neurotransmitters. But if, at the same moment, 10 other synapses release inhibitory neurotransmitters, the excitatory and inhibitory effects will cancel one another out, and no action potential will develop. The activity of a neuron thus depends on the balance between excitation and inhibition. The interactions are extremely complex—synapses at the cell body and dendrites may involve tens of thousands of other neurons, some releasing excitatory neurotransmitters and others releasing inhibitory neurotransmitters.

NEURONAL POOLS

As noted earlier, the human body has about 10 million sensory neurons, 20 billion interneurons, and one-half million motor neurons. These individual neurons represent the simplest level of organization within the central nervous system. The next level is that of the neuronal pool. A **neuronal pool** is a group of interconnected interneurons with specific functions. Each neuronal pool has a limited number of input sources and output destinations, and each may contain excitatory and inhibitory neurons. The output of one pool may stimulate or depress the activity of other pools, or it may exert direct control over motor neurons or peripheral effectors.

Neurons and neuronal pools communicate with one another in several patterns or neural circuits. The two simplest circuit patterns are *divergence* and *convergence*. In **divergence**, information spreads from one neuron to several neurons, or one neuronal pool to multiple neuronal pools (Figure 8-12a●). Considerable divergence occurs when sensory neurons bring

information into the CNS, because the sensory information is distributed to neuronal pools throughout the spinal cord and brain. For example, visual information arriving from the eyes reaches your conscious awareness at the same time it is carried to areas of the brain that control posture and balance at the subconscious level. Divergence is also involved when you step on a sharp object and react with a number of different responses. Stepping on a sharp object stimulates sensory neurons that distribute the information to a number of neuronal pools. As a result, you might withdraw your foot, shift your weight, move your arms, feel the pain, and shout "Ouch!," all at the same time. In **convergence** (Figure 8-12b●), several neurons synapse on the same postsynaptic neuron. Convergence makes possible both voluntary and involuntary control of some body processes. For example, the movements of your diaphragm and ribs are now being involuntarily, or subconsciously, controlled by respiratory centers in the brain. These centers activate or inhibit motor neurons in the spinal cord that control the respiratory muscles. But the same movements can be controlled voluntarily, or consciously, as when you take a deep breath and hold it. Two neuronal pools are involved, both synapsing on the same motor neurons.

CONCEPT CHECK QUESTIONS
Answers on page 275

❶ A neurotransmitter causes potassium channels but not sodium channels to open. What effect would this neurotransmitter produce at the postsynaptic membane?

❷ How would synapse function be affected if the uptake of calcium were blocked at the presynaptic membrane of a cholinergic synapse?

❸ The movements of your diaphragm and ribs may be either controlled consciously, as when you take a deep breath and hold it, or subconsciously, as you sit and read a book. What type of neural circuit or pattern permits both conscious and subconscious control of the same motor neurons?

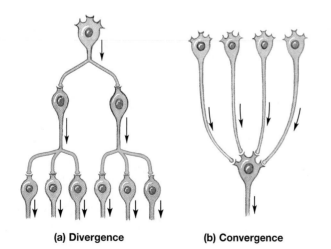

(a) Divergence   **(b) Convergence**

● *Figure 8-12* **The Organization of Neuronal Pools**
(a) Divergence, a mechanism for spreading stimulation to multiple neurons or neuronal pools in the CNS. **(b)** Convergence, a mechanism providing input to a single neuron from multiple sources.

The Central Nervous System

The central nervous system consists of the spinal cord and brain. These masses of neural tissue are extremely delicate and must be protected against shocks, infection, and other dangers. In addition to glial cells within the neural tissue, the CNS is protected by a series of covering layers, the *meninges*.

THE MENINGES

Neural tissue has a very high metabolic rate and requires abundant nutrients and a constant supply of oxygen. At the same time, the CNS must be isolated from a variety of compounds in the blood that could interfere with its complex operations. Delicate neural tissues must also be defended against damaging contact with the surrounding bones. The **meninges** (men-IN-jēz), a series of specialized membranes surrounding the brain and spinal cord, provide physical stability and shock absorption (Figure 8-13●). Blood vessels branching within these layers also deliver oxygen and nutrients to the CNS. At the foramen magnum of the skull, the *cranial meninges* covering the brain are continuous with the *spinal meninges* that surround the spinal cord. There are three meningeal layers: the *dura mater*, the *arachnoid*, and the *pia mater*. The meninges cover the cranial and spinal nerves as they penetrate the skull or pass through the intervertebral foramina, becoming continuous with the connective tissues surrounding the peripheral nerves.

The Dura Mater

The tough, fibrous **dura mater** (DOO-ra MĀ-ter; *dura*, hard + *mater*, mother) forms the outermost covering of the central nervous system. The dura mater surrounding the brain consists of two fibrous layers, with the outermost fused to the periosteum of the skull. The inner and outer layers are separated by a slender gap that contains tissue fluids and blood vessels (Figure 8-13a●). At several locations, the innermost layer of the dura mater extends deep into the cranial cavity, forming folded membranous sheets called *dural folds*. The dural folds act like seat belts in a car and hold the brain in position. Large collecting veins known as *dural sinuses* lie between the two layers of a dural fold.

The outer layer of the dura mater in the spinal cord is not fused to bone. Between the dura mater of the spinal cord and the walls of the vertebral canal lies the **epidural space**, which contains loose connective tissue, blood vessels, and adipose tissue (Figure 8-13b●). Injecting an anesthetic into the epidural space produces a temporary sensory and motor paralysis known as an *epidural block*. This technique has the advantage of affecting only the spinal nerves in the immediate area of the injection. Epidural blocks in the lower lumbar or sacral regions may be used to control pain during childbirth.

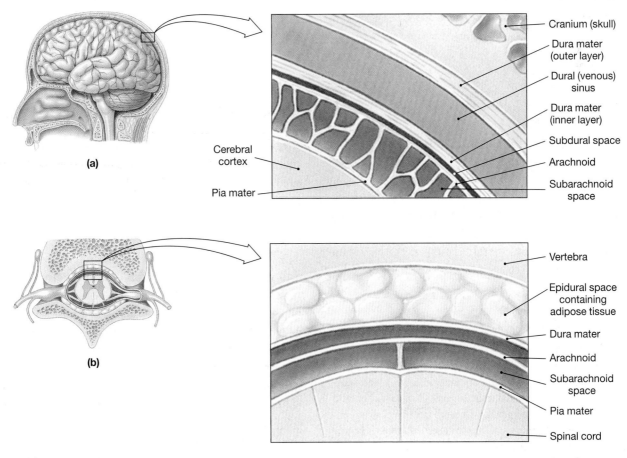

(a)

Cerebral cortex

Pia mater

Cranium (skull)

Dura mater (outer layer)

Dural (venous) sinus

Dura mater (inner layer)

Subdural space

Arachnoid

Subarachnoid space

(b)

Vertebra

Epidural space containing adipose tissue

Dura mater

Arachnoid

Subarachnoid space

Pia mater

Spinal cord

● *Figure 8-13* **The Meninges**
(a) Meninges of the brain. (b) Meninges of the spinal cord.

8

THE NERVOUS SYSTEM

The Nervous System • Cellular Organization in Neural Tissue • Neuron Function • Neural Communication • **The Central Nervous System**

The Arachnoid

A narrow **subdural space** separates the inner surface of the dura mater from the second meningeal layer, the **arachnoid** (a-RAK-noyd; *arachne*, spider). This intervening space contains a small quantity of lymphatic fluid, which reduces friction between the opposing surfaces. The arachnoid is a layer of squamous cells; deep to this epithelial layer lies the **subarachnoid space**, which contains a delicate web of collagen and elastic fibers. The subarachnoid space is filled with *cerebrospinal fluid*, which acts as a shock absorber and transports dissolved gases, nutrients, chemical messengers, and waste products.

EPIDURAL AND SUBDURAL DAMAGE

A severe head injury may damage cerebral blood vessels and cause bleeding into the cranial cavity. If blood is forced between the dura mater and the cranium, the condition is known as an *epidural hemorrhage*. The flow of blood into the lower layer of the dura mater and subdural space is called a *subdural hemorrhage*. These are serious conditions because the blood entering these spaces compresses and distorts the relatively soft tissues of the brain. The symptoms vary depending on whether the damaged vessel is an artery or a vein because arterial blood pressure is higher and can cause more rapid and severe distortion of neural tissue.

The Pia Mater

The subarachnoid space separates the arachnoid from the innermost meningeal layer, the **pia mater** (*pia*, delicate + *mater*, mother). The pia mater is bound firmly to the underlying neural tissue. The blood vessels servicing the brain and spinal cord run along the surface of this layer, within the subarachnoid space. The pia mater of the brain is highly vascular, and large vessels branch over the surface of the brain, supplying the superficial areas of neural cortex. This extensive circulatory supply is extremely important, for the brain has a very high rate of metabolism; at rest, the 1.4 kg (3.1 lb) brain uses as much oxygen as 28 kg (61.6 lb) of skeletal muscle.

MENINGITIS

Meningitis is the inflammation of the meningeal membranes following bacterial or viral infection. Meningitis is dangerous because it can disrupt the normal circulatory and cerebrospinal fluid supplies, damaging or killing neurons and glial cells in the affected areas. Although the initial diagnosis may specify the meninges of the spinal cord (*spinal meningitis*) or brain (*cerebral meningitis*), in later stages the entire meningeal system is usually affected.

THE SPINAL CORD

The spinal cord serves as the major highway for the passage of sensory impulses to the brain and motor impulses from the brain. In addition, the spinal cord integrates information on its own and controls *spinal reflexes*, automatic motor responses ranging from withdrawal from pain to complex reflex patterns involved with sitting, standing, walking, and running.

Gross Anatomy

The spinal cord, detailed in Figure 8-14●, is approximately 45 cm (18 in.) long. It has a **central canal**, a narrow internal passageway filled with cerebrospinal fluid. The posterior surface of the spinal cord has a shallow groove, the *posterior median sulcus*, and the anterior surface has a deeper crease, the *anterior median fissure*. With two exceptions, the diameter of the cord decreases in size as it extends toward the sacral region.

The two exceptions are regions concerned with the sensory and motor control of the limbs. The *cervical enlargement* supplies nerves to the shoulder girdle and upper limbs, and the *lumbar enlargement* provides innervation to the pelvis and lower limbs. Below the lumbar enlargement, the spinal cord becomes tapered and conical. A slender strand of fibrous tissue extends from the inferior tip of the spinal cord to the coccyx, serving as an anchor that prevents superior movement.

The entire spinal cord consists of 31 segments, each identified by a letter and number. Every spinal segment is associated with a pair of **dorsal root ganglia**, which contain the cell bodies of sensory neurons (Figure 8-14b●). The **dorsal roots**, which contain the axons of these neurons, bring sensory information to the spinal cord. A pair of **ventral roots** contain the axons of CNS motor neurons that control muscles and glands. On either side, the dorsal and ventral roots from each segment leave the vertebral column between adjacent vertebrae at the *intervertebral foramen*.

Distal to each dorsal root ganglion, the sensory (dorsal) and motor (ventral) roots are bound together into a single *spinal nerve*. All spinal nerves are classified as *mixed nerves*, because they contain both sensory and motor fibers. The spinal nerves on either side form outside the vertebral canal, where the ventral and dorsal roots unite. p. 146

The adult spinal cord extends only to the level of the first or second lumbar vertebrae. When seen in gross dissection, the long ventral and dorsal roots inferior to the tip of the spinal cord reminded early anatomists of a horse's tail. With that in mind, they called this complex of nerve roots the *cauda equina* (KAW-da ek-WĪ-na; *cauda*, tail + *equus*, horse).

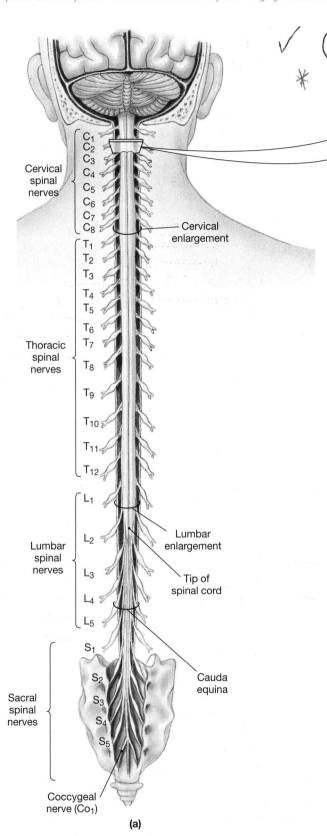

Dorsal root ganglion · Dorsal root · White matter · Posterior median sulcus · Central canal

Spinal nerve · Ventral root · Gray matter · **C₃** · Anterior median fissure

(b)

Sectional Anatomy

The anterior median fissure and the posterior median sulcus mark the division between left and right sides of the spinal cord (Figure 8-14b●). The *gray matter* is dominated by the cell bodies of neurons and glial cells. It forms a rough H, or butterfly shape, around the narrow central canal. Projections of gray matter, called **horns**, extend outward into the *white matter*, which contains large numbers of myelinated and unmyelinated axons.

Figure 8-15a● shows the relationship between the function of a particular nucleus (sensory or motor) and its relative position within the gray matter of the spinal cord. The *posterior gray horns* contain sensory nuclei, whereas the *anterior gray horns* deal with the motor control of skeletal muscles. Nuclei in the *lateral gray horns* contain the visceral motor neurons that control smooth muscle, cardiac muscle, and glands. The *gray commissures* above and below the central canal interconnect the horns on either side of the spinal cord. The gray and white commissures contain axons that cross from one side of the spinal cord to the other.

The white matter on each side can be divided into three regions, or **columns** (Figure 8-15a●). The *posterior white columns* extend between the posterior gray horns and the posterior median sulcus. The *anterior white columns* lie between the anterior gray horns and the anterior median fissure; they are interconnected by the *anterior white commissure*. The white matter between the anterior and posterior columns makes up the *lateral white columns*.

Each column contains tracts whose axons carry either sensory data or motor commands. Small tracts carry sensory or motor signals between segments of the spinal cord, and larger

● *Figure 8-14* **Gross Anatomy of the Spinal Cord**
(a) The superficial anatomy and orientation of the adult spinal cord. The numbers to the left identify the spinal nerves. (b) A cross section through the cervical region of the spinal cord, showing the arrangement of gray matter and white matter.

(Labels on figure (a):) Cervical spinal nerves — C₁ C₂ C₃ C₄ C₅ C₆ C₇ C₈; Cervical enlargement; Thoracic spinal nerves — T₁ T₂ T₃ T₄ T₅ T₆ T₇ T₈ T₉ T₁₀ T₁₁ T₁₂; Lumbar spinal nerves — L₁ L₂ L₃ L₄ L₅; Lumbar enlargement; Tip of spinal cord; Sacral spinal nerves — S₁ S₂ S₃ S₄ S₅; Cauda equina; Coccygeal nerve (Co₁)

(a)

8 THE NERVOUS SYSTEM

The Nervous System • Cellular Organization in Neural Tissue • Neuron Function • Neural Communication • **The Central Nervous System**

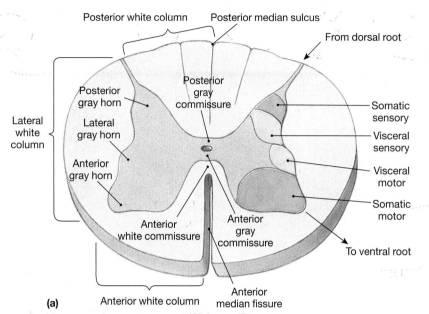

(a)

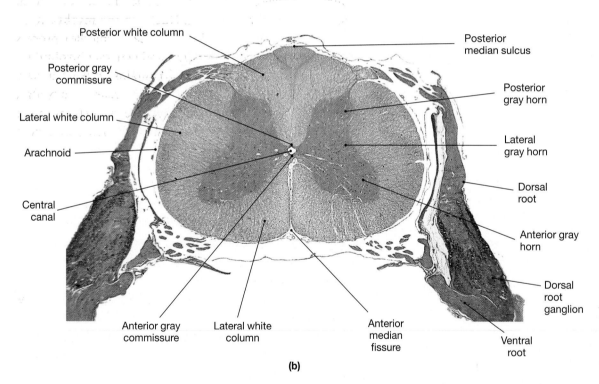

(b)

● *Figure 8-15* **Sectional Anatomy of the Spinal Cord**
(**a**) The left half of this sectional view shows important anatomical landmarks and the major regions of white matter and gray matter. The right half indicates the functional organization of the gray matter in the anterior, lateral, and posterior gray horns. (**b**) A micrograph of a section through the spinal cord, showing major landmarks. Compare with (a).

tracts connect the spinal cord with the brain. **Ascending tracts** carry sensory information toward the brain, and **descending tracts** convey motor commands into the spinal cord.

Injuries affecting the spinal cord or cauda equina produce symptoms of sensory loss or motor paralysis that reflect the specific nuclei, tracts, or spinal nerves involved. A general paralysis can result from severe damage to the spinal cord in an auto crash or other accident, and the damaged tracts

seldom undergo even partial repairs. Extensive damage at the fourth or fifth cervical vertebra will eliminate sensation and motor control of the upper and lower limbs. The extensive paralysis produced is called *quadriplegia. Paraplegia,* the loss of motor control of the lower limbs, may follow damage to the thoracic vertebrae.

The regeneration of spinal cord tissue has long been thought to be an impossibility because mature nervous tissue had not been observed to grow or go through mitosis. Biological cures are being sought by interfering with inhibitory factors in the spinal cord that slow the repair of neurons and by stimulating unspecialized stem cells to grow and divide. Inactive neural stem cells have been found in mature human neural tissues. This is potentially significant because laboratory rats treated with embryonic stem cells at the site of a spinal cord injury have regained limb mobility and strength. Work also continues on electronic methods of restoring some degree of motor control. A technique using computers and wires to stimulate specific muscle groups has let a few paraplegic individuals walk again.

CONCEPT CHECK QUESTIONS
Answers on page 275

❶ Damage to which root of a spinal nerve would interfere with motor function?

❷ A man with polio has lost the use of his leg muscles. In what area of the spinal cord would you expect to locate his virus-infected motor neurons?

❸ Why are spinal nerves also called mixed nerves?

THE BRAIN

The brain is far more complex than the spinal cord, and its responses to stimuli are more versatile. It contains roughly 35 billion neurons organized into hundreds of neuronal pools; it has a complex three-dimensional structure and performs a bewildering array of functions. All of our dreams, passions, plans, and memories are the result of brain activity.

The adult human brain contains almost 98 percent of the neural tissue in the body. A representative adult brain weighs 1.4 kg (3 lb) and has a volume of 1200 cc (71 in.^{3}). There is considerable individual variation, and the brains of males are generally about 10 percent larger than those of females, because of differences in average body size. There is no correlation between brain size and intelligence. Individuals with the smallest brains (750 cc) or largest brains (2100 cc) are functionally normal.

Major Divisions of the Brain

The adult brain has six major regions: (1) the *cerebrum*, (2) the *diencephalon*, (3) the *midbrain*, (4) the *pons*, (5) the *medulla oblongata*, and (6) the *cerebellum*. Major landmarks are indicated in Figure 8-16●.

Viewed from the superior surface, the **cerebrum** (SER-e-brum or se-RĒ-brum) can be divided into large, paired **cerebral hemispheres**. Conscious thoughts, sensations, intellectual functions, memory storage and retrieval, and complex movements originate in the cerebrum. The hollow **diencephalon** (dī-en-SEF-a-lon; *dia-*, through + *cephalo*, head) is connected to the cerebrum. Its sides form the **thalamus** (THAL-a-mus), which contains relay and processing centers for sensory information. A narrow stalk connects the **hypothalamus** (*hypo-*, below), or floor of the diencephalon, to the *pituitary gland*. The hypothalamus contains centers involved with emotions, autonomic function, and hormone production. The pituitary gland, the primary link between the nervous and endocrine systems, is discussed in Chapter 10.

The midbrain, pons, and medulla oblongata form the **brain stem**. The brain stem contains important processing centers and relay stations for information headed to or from the cerebrum or cerebellum. Nuclei in the **midbrain**, or *mesencephalon* (mez-en-SEF-a-lon; *meso-*, middle), process visual and auditory information and generate involuntary motor responses. This region also contains centers that help maintain consciousness. The term **pons** refers to a bridge, and the pons of the brain connects the cerebellum to the brain stem. In addition to tracts and relay centers, this region of the brain also contains nuclei involved with somatic and visceral motor control. The pons is also connected to the **medulla oblongata**, the segment of the brain that is attached to the spinal cord. The medulla oblongata relays sensory information to the thalamus and other brain stem centers; it also contains major centers that regulate autonomic function, such as heart rate, blood pressure, respiration, and digestive activities.

The large cerebral hemispheres and the smaller hemispheres of the **cerebellum** (ser-e-BEL-um) almost completely cover the brain stem. The cerebellum adjusts voluntary and involuntary motor activities on the basis of sensory information and stored memories of previous movements.

CONCEPT CHECK QUESTIONS
Answers on page 275

❶ Describe one major function of each of the six regions of the brain.

❷ The pituitary gland links the nervous and endocrine systems. To which portion of the diencephalon is it attached?

8 THE NERVOUS SYSTEM

The Nervous System • Cellular Organization in Neural Tissue • Neuron Function • Neural Communication • **The Central Nervous System**

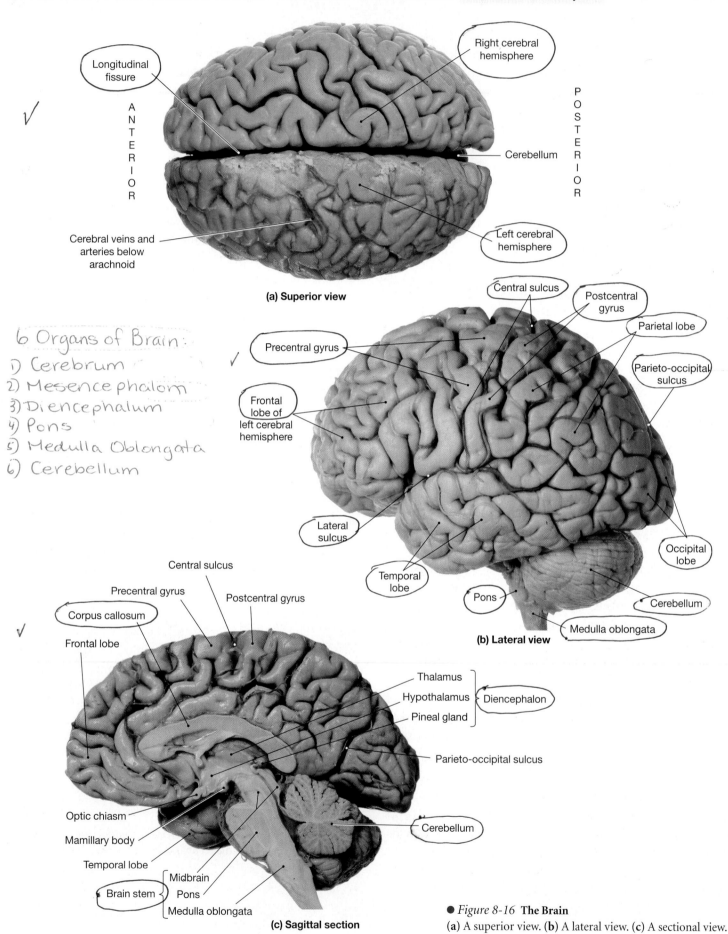

✓

6 Organs of Brain:
1) Cerebrum
2) Mesencephalom
3) Diencephalum
4) Pons
5) Medulla Oblongata
6) Cerebellum

(a) Superior view

Longitudinal fissure

Right cerebral hemisphere

ANTERIOR

POSTERIOR

Cerebellum

Cerebral veins and arteries below arachnoid

Left cerebral hemisphere

Central sulcus

Postcentral gyrus

Parietal lobe

Precentral gyrus

Parieto-occipital sulcus

Frontal lobe of left cerebral hemisphere

Lateral sulcus

Occipital lobe

Temporal lobe

Pons

Cerebellum

Medulla oblongata

(b) Lateral view

Central sulcus

Precentral gyrus

Postcentral gyrus

Corpus callosum

Frontal lobe

Thalamus

Hypothalamus

Diencephalon

Pineal gland

Parieto-occipital sulcus

Optic chiasm

Mamillary body

Temporal lobe

Midbrain

Brain stem

Pons

Medulla oblongata

Cerebellum

(c) Sagittal section

● *Figure 8-16* **The Brain**
(a) A superior view. (b) A lateral view. (c) A sectional view.

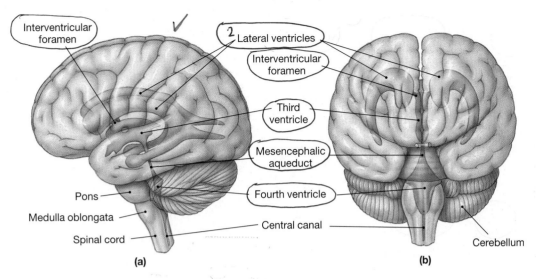

● Figure 8-17 The Ventricles of the Brain (**a**) The orientation and extent of the ventricles as seen through a transparent brain in lateral view. (**b**) An anterior view of the ventricles, showing the relationships between the lateral ventricles and the third ventricle.

The Ventricles of the Brain

The brain and spinal cord contain internal cavities filled with cerebrospinal fluid and lined by cells of the ependyma. ∞ p. 230 The brain has a central passageway that expands to form four chambers, called **ventricles** (VEN-tri-kls) (Figure 8-17●). Each cerebral hemisphere contains a large **lateral ventricle**. There is no direct connection between the lateral ventricles, but an opening, the *interventricular foramen*, allows each of them to communicate with the **third ventricle** of the diencephalon. Instead of a ventricle, the midbrain has a slender canal known as the *mesencephalic aqueduct* (*cerebral aqueduct*), which connects the third ventricle with the **fourth ventricle** of the pons and upper portion of the medulla oblongata. Within the medulla oblongata the fourth ventricle narrows and becomes continuous with the central canal of the spinal cord.

Cerebrospinal Fluid. **Cerebrospinal fluid**, or **CSF**, completely surrounds and bathes the exposed surfaces of the CNS. CSF provides cushioning for delicate neural structures. It also provides support, because the brain essentially floats in the cerebrospinal fluid. A human brain weighs about 1400 g (3.1 lb) in air, but only about 50 g (1.76 oz.) when supported by the cerebrospinal fluid. Finally, the CSF transports nutrients, chemical messengers, and waste products. Except at the *choroid plexus*, where CSF is produced, the ependymal lining is freely permeable and the CSF is in constant chemical communication with the interstitial fluid of the CNS. Because free exchange occurs between the interstitial fluid and CSF, changes in CNS function may produce changes in the composition of the CSF. Samples of the CSF can be obtained through a *lumbar puncture*, or *spinal tap*, providing useful clinical information concerning CNS injury, infection, or disease.

Cerebrospinal fluid is produced at the **choroid plexus** (*choroid*, a vascular coat + *plexus*, a network), a network of permeable capillaries that extends into each of the four ventricles (Figure 8-18a●). The capillaries of the choroid plexus are covered by large ependymal cells that secrete CSF at a rate of about 500 ml/day. The total volume of CSF at any given moment is approximately 150 ml; this means that the entire volume of CSF is replaced roughly every 8 hours. Despite this rapid turnover, the composition of CSF is closely regulated, and the rate of removal normally keeps pace with the rate of production. If it does not, a variety of clinical problems may appear.

Figure 8-18● diagrams the circulation of cerebrospinal fluid. CSF forms at the choroid plexus and circulates between the different ventricles, passes along the central canal, and enters the subarachnoid space. Once inside the subarachnoid space, the CSF circulates around the spinal cord and cauda equina and across the surfaces of the brain. Between the cerebral hemispheres, slender extensions of the arachnoid penetrate the inner layer of the dura mater. These **arachnoid granulations** (Figure 8-18b●) project into the *superior sagittal sinus*, a large cerebral vein. Diffusion across the arachnoid granulations returns excess cerebrospinal fluid to the venous circulation.

The Cerebrum

The cerebrum, the largest region of the brain, is the site where conscious thought and intellectual functions originate. Much of the cerebrum is involved in receiving somatic sensory information and then exerting voluntary or involuntary control over somatic motor neurons. In general, we are aware of these events. However, most sensory processing and all visceral motor (autonomic) control occur elsewhere in the brain, usually outside our conscious awareness.

8 **THE NERVOUS SYSTEM**

The Nervous System • Cellular Organization in Neural Tissue • Neuron Function • Neural Communication • **The Central Nervous System**

FROM OUTSIDE IN

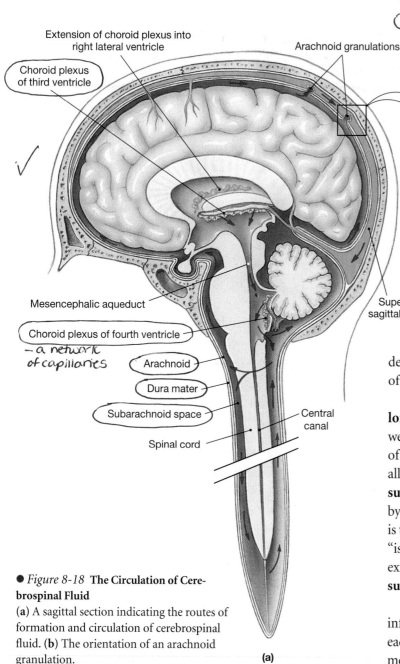

Extension of choroid plexus into right lateral ventricle

Choroid plexus of third ventricle

Arachnoid granulations

(i)

Dura mater (outer layer)   Cranium   Endothelial lining

Fluid movement

Superior sagittal sinus

Arachnoid granulation

Dura mater (inner layer)

2

Cerebral cortex   Subarachnoid space   Subdural space

Pia mater   Arachnoid

4   3

5   **(b)**

Mesencephalic aqueduct

Choroid plexus of fourth ventricle
– a network of capillaries

Arachnoid

Dura mater

Subarachnoid space

Spinal cord

Central canal

Superior sagittal sinus

● *Figure 8-18* **The Circulation of Cerebrospinal Fluid**
(**a**) A sagittal section indicating the routes of formation and circulation of cerebrospinal fluid. (**b**) The orientation of an arachnoid granulation.

(a)

The cerebrum includes gray matter and white matter. Gray matter is found in a superficial layer of neural cortex and in deeper *basal nuclei*. The *central white matter*, composed of myelinated axons, lies beneath the neural cortex and surrounds the basal nuclei.

Structure of the Cerebral Hemispheres. A thick blanket of neural cortex known as the **cerebral cortex** covers the superior and lateral surfaces of the cerebrum (Figure 8-19●). This outer surface forms a series of elevated ridges, or **gyri** (JĪ-rī), separated by shallow depressions, called **sulci** (SUL-sī), or

deeper grooves, called **fissures**. Gyri increase the surface area of the cerebrum and the number of neurons in the cortex.

The two cerebral hemispheres are separated by a deep **longitudinal fissure**. Each hemisphere can be divided into well-defined regions, or **lobes**, named after the overlying bones of the skull (Figures 8-16, p. 244, and 8-19●). Extending laterally from the longitudinal fissure is a deep groove, the **central sulcus**. Anterior to this is the **frontal lobe**, bordered inferiorly by the **lateral sulcus**. The cortex inferior to the lateral sulcus is the **temporal lobe**, which overlaps the **insula** (IN-su-la), an "island" of cortex that is otherwise hidden. The **parietal lobe** extends between the central sulcus and the **parieto-occipital sulcus**. What remains is the **occipital lobe**.

In each lobe, some regions are concerned with sensory information and others with motor commands. Additionally, each hemisphere receives sensory information from, and sends motor commands to, the opposite side of the body. This results in the left cerebral hemisphere controlling the right side of the body, and the right cerebral hemisphere controlling the left side. This crossing over has no known functional significance.

Motor and Sensory Areas of the Cortex. Figure 8-19● shows the major motor and sensory regions of the cerebral cortex. The central sulcus separates the motor and sensory portions of the cortex. The **precentral gyrus** of the frontal lobe forms the anterior margin of the central sulcus, and its surface is the **primary motor cortex**. Neurons of the primary motor cortex direct voluntary movements by controlling somatic motor neurons in the brain stem and spinal cord.

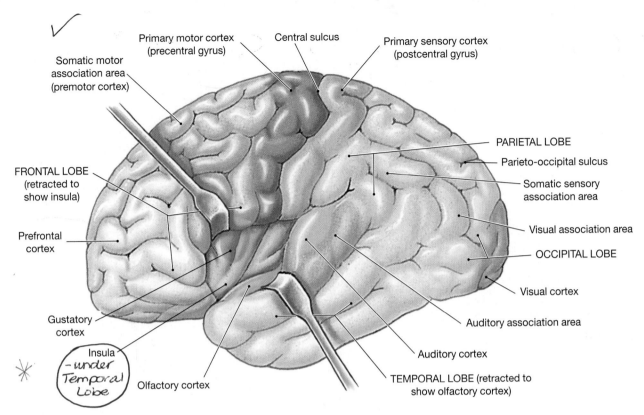

● *Figure 8-19* **The Cerebral Hemispheres**
Major anatomical landmarks on the surface of the left cerebral hemisphere.

The **postcentral gyrus** of the parietal lobe forms the posterior margin of the central sulcus, and its surface contains the **primary sensory cortex.** Neurons in this region receive somatic sensory information from touch, pressure, pain, and temperature receptors. We are consciously aware of these sensations because the brain stem nuclei relay sensory information to the primary sensory cortex.

Sensations of sight, taste, sound, and smell arrive at other portions of the cerebral cortex. The **visual cortex** of the occipital lobe receives visual information, the **gustatory cortex** of the frontal lobe receives taste sensations, and the **auditory cortex** and **olfactory cortex** of the temporal lobe receive information about hearing and smell, respectively.

Association Areas. The sensory and motor regions of the cortex are connected to nearby **association areas**, regions that interpret incoming data or coordinate a motor response. The **somatic sensory association area** monitors activity in the primary sensory cortex. It is this area that allows you to recognize a touch as light as the arrival of a mosquito on your arm. The special senses of smell, sight, and hearing involve separate areas of sensory cortex, and each has its own association area. The **somatic motor association area**, or **premotor cortex**, is responsible for coordinating learned movements. When you perform a voluntary movement, such as picking up a glass or scanning these lines of type, the instructions are relayed to the primary motor cortex by the premotor cortex. The functional distinctions between the motor and sensory association areas are most evident after localized brain damage has occurred. For example, someone with damage to the premotor cortex might understand written letters and words but be unable to read owing to an inability to track along the lines on a printed page. In contrast, someone with a damaged **visual association area** can scan the lines of a printed page but cannot figure out what the letters mean.

Cortical Connections. The various regions of the cerebral cortex are interconnected by the white matter that lies beneath the cerebral cortex. This white matter interconnects areas within a single cerebral hemisphere and links the two hemispheres across the **corpus callosum** (Figure 8-16c●, p. 244). Other bundles of axons link the cerebral cortex with the diencephalon, brain stem, cerebellum, and spinal cord.

Cerebral Processing Centers. There are also "higher-order" integrative centers that receive information through axons from

8 THE NERVOUS SYSTEM

The Nervous System • Cellular Organization in Neural Tissue • Neuron Function • Neural Communication • **The Central Nervous System**

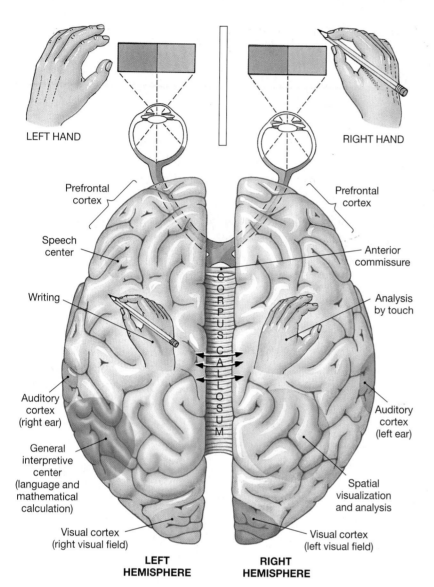

LEFT HAND

RIGHT HAND

Prefrontal cortex

Prefrontal cortex

Speech center

Anterior commissure

Writing

Analysis by touch

CORPUS CALLOSUM

Auditory cortex (right ear)

Auditory cortex (left ear)

General interpretive center (language and mathematical calculation)

Spatial visualization and analysis

Visual cortex (right visual field)

Visual cortex (left visual field)

LEFT HEMISPHERE

RIGHT HEMISPHERE

● *Figure 8-20* **Hemispheric Lateralization**
Functional differences between the left and right cerebral hemispheres.

many different association areas. These regions, shown in Figure 8-20●, control extremely complex motor activities and perform complicated analytical functions. They may be found in both cerebral hemispheres, or they may be restricted to either the left or the right side.

The General Interpretive Area. The **general interpretive area**, or *Wernicke's area*, receives information from all the sensory association areas. This center is present in only one hemisphere, usually the left. Damage to this area affects the ability to interpret what is read or heard, even though the words are understood as individual entities. For example, an individual might understand the meaning of the words "sit" and "here" but be totally bewildered by the instruction "Sit here."

The Speech Center. The general interpretive area is connected to the **speech center** (*Broca's area*), which lies along the edge of the premotor cortex in the same hemisphere as the general interpretive area. The speech center regulates the patterns of breathing and vocalization needed for normal speech. The corresponding regions on the opposite hemisphere are not inactive, but their functions are less well defined. Damage to the speech center can manifest itself in various ways. Some people have difficulty speaking even when they know exactly what words to use; others talk constantly but use all the wrong words.

The Prefrontal Cortex. The **prefrontal cortex** of the frontal lobe coordinates information from the association areas of the entire cortex. In doing so, it performs such abstract intellectual functions as predicting the future consequences of events or actions. Damage to this area leads to problems in estimating time relationships between events. Questions such as "How long ago did this happen?" or "What happened first?" become difficult to answer. The prefrontal cortex also has connections with other cortical areas and with other portions of the brain. Feelings of frustration, tension, and anxiety are generated at the prefrontal cortex as it interprets ongoing events and predicts future situations or consequences. If the connections between the prefrontal cortex and other brain regions are severed, the tensions, frustrations, and anxieties are removed. Early in the 1900s this rather drastic procedure, called a *prefrontal lobotomy*, was used to "cure" a variety of mental illnesses, especially those associated with violent or antisocial behavior.

 APHASIA AND DYSLEXIA

Aphasia (*a-*, without + *phasia*, speech) is a disorder affecting the ability to speak or read. *Global aphasia* results from extensive damage to the general interpretive area or to the associated sensory tracts. Affected individuals cannot speak, read, or understand or interpret the speech of others. Global aphasia often accompanies a severe stroke or tumor that affects a large area of cortex including the speech and language areas. Recovery is possible when the condition results from *edema* (an abnormal accumulation of fluid) or hemorrhage, but the process often takes months. Lesser degrees of aphasia often follow minor strokes, with no initial period of global aphasia. Individuals can understand spoken and written words and may recover completely.

Dyslexia (*lexis*, diction) is a disorder affecting the comprehension and use of words. Developmental dyslexia affects children; estimates indicate that up to 15 percent of children in the United States suffer from some degree of dyslexia. These children have difficulty reading and writing, although their other intellectual functions may be normal or above normal. Their writing looks uneven and unorganized; letters are typically written in the wrong order (*dig* becomes *gid*) or reversed (*E* becomes *Ǝ*). Recent evidence suggests that at least some forms of dyslexia result from problems in processing and sorting visual information.

Hemispheric Lateralization. Each of the two cerebral hemispheres is responsible for specific functions that are not ordinarily performed by the opposite hemisphere. This specialization has been called *hemispheric lateralization*. Figure 8-20● shows the major functional differences between the hemispheres. In most people, the left hemisphere contains the general interpretive and speech centers and is responsible for language-based skills (reading, writing, and speaking). The left hemisphere is also important in performing analytical tasks, such as mathematical calculations and logical decision making. For these reasons, the left hemisphere has been called the dominant hemisphere, or the categorical hemisphere.

The right cerebral hemisphere analyzes sensory information and relates the body to the sensory environment. Interpretive centers in this hemisphere permit you to identify familiar objects by touch, smell, taste, or feel. For example, the right hemisphere plays a dominant role in recognizing faces and in understanding three-dimensional relationships. Interestingly, there may be a link between handedness and sensory/spatial abilities. An unusually high percentage of musicians and artists

are left-handed; the complex motor activities performed by these individuals are directed by the primary motor cortex and association areas on the right hemisphere.

Hemispheric lateralization does not mean that the two hemispheres function independently of each other. As noted above, the white fibers of the corpus callosum link the two hemispheres, including their sensory information and motor commands. The corpus callosum alone contains over 200 million axons, carrying an estimated 4 billion impulses per second!

The Electroencephalogram. The specific areas of the cerebral hemispheres were mapped by direct stimulation in patients undergoing brain surgery. Noninvasive methods are also used to correlate the activity of different cortical regions with function. One of the most common noninvasive methods involves monitoring the electrical activity of the brain.

Neural function depends on electrical events within the cell membrane. The brain contains billions of nerve cells, and their activity generates an electrical field that can be measured by placing electrodes on the brain or the outer surface of the skull. The electrical activity changes constantly as nuclei and cortical areas are stimulated or quiet down. An **electroencephalogram** (**EEG**) is a printed record of this electrical activity over time. The electrical patterns are called **brain waves**, and these can be correlated with the individual's level of consciousness. Electroencephalograms can also provide useful diagnostic information regarding brain disorders. Four types of brain wave patterns are shown in Figure 8-21●.

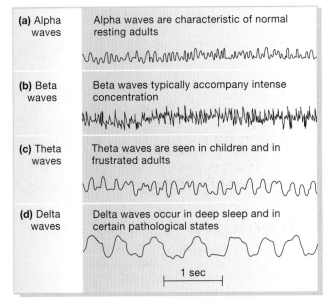

(a) Alpha waves — Alpha waves are characteristic of normal resting adults

(b) Beta waves — Beta waves typically accompany intense concentration

(c) Theta waves — Theta waves are seen in children and in frustrated adults

(d) Delta waves — Delta waves occur in deep sleep and in certain pathological states

1 sec

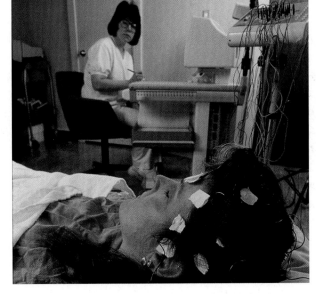

● *Figure 8-21* **Brain Waves**

8

THE NERVOUS SYSTEM

The Nervous System • Cellular Organization in Neural Tissue • Neuron Function • Neural Communication • **The Central Nervous System**

Memory

What was the topic of the last sentence you read? What do your parents look like? What is your Social Security number? What does a red traffic light mean? What does a hot dog taste like? Answering these questions involves accessing memories, stored bits of information gathered through prior experience. **Fact memories** are specific bits of information, such as the color of a stop sign or the smell of a certain perfume. **Skill memories** are learned motor behaviors. For example, you can probably remember how to light a match or open a screw-top jar. With repetition, skill memories become incorporated at the unconscious level. Examples would include the complex motor patterns involved in skiing, playing the violin, and similar activities. Skill memories related to programmed behaviors, such as eating, are stored in appropriate portions of the brain stem. Complex skill memories involve an interplay between the cerebullum and the cerebral cortex.

Two classes of memories exist. **Short-term memories**, or *primary memories*, do not last long, but while they persist, the information can be recalled immediately. Primary memories contain small bits of information, such as a person's name or a telephone number. Repeating a phone number or other bit of information reinforces the original short-term memory and helps ensure its conversion to a long-term memory. **Long-term memories** remain for much longer periods, in some cases for an entire lifetime. The conversion from short-term to long-term memory is called *memory consolidation*. Some long-term memories fade with time and may require considerable effort to recall. Other long-term memories seem to be part of consciousness, such as your name or the contours of your own body. Most long-term memories are stored in the cerebral cortex. Conscious motor and sensory memories are referred to the appropriate association areas. For example, visual memories are stored in the visual association area, and memories of voluntary motor activity are kept in the premotor cortex. Special portions of the occipital and temporal lobes retain the memories of faces, voices, and words. In at least some cases, a specific memory probably reflects the activity of a single neuron. For example, in one portion of the temporal lobe an individual neuron responds to the sound of one word and ignores others.

Amnesia refers to the loss of memory from disease or trauma. The type of memory loss depends on the specific regions of the brain affected. For example, damage to the auditory association areas may make it difficult to remember sounds. Damage to thalamic and limbic structures, especially the *hippocampus*, will affect memory storage and consolidation.

The Basal Nuclei

While your cerebral cortex is consciously active, other centers of your cerebrum, diencephalon, and brain stem are processing sensory information and issuing motor commands at a subconscious level. Many of these activities outside our conscious awareness are directed by the basal nuclei. The **basal nuclei** are masses of gray matter that lie beneath the lateral ventricles and within the central white matter of each cerebral hemisphere (Figure 8-22a●). The **caudate nucleus** has a massive head and slender, curving tail that follows the curve of the lateral ventricle (Figure 8-22b●). Inferior to the head of the caudate nucleus is the **lentiform** (*lens-shaped*) **nucleus**, which consists of a **globus pallidus** (GLŌ-bus PAL-i-dus; pale globe) and **putamen** (pū-TĀ-men). Together, the caudate and

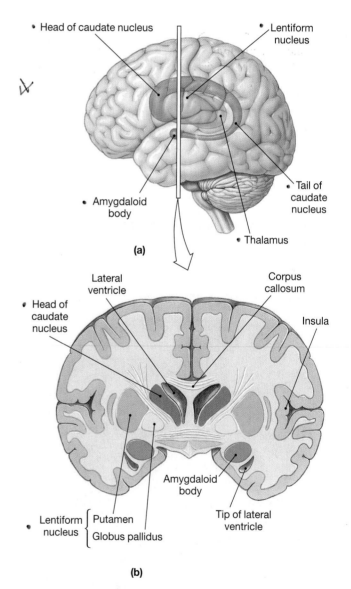

● *Figure 8-22* **The Basal Nuclei**

stimuli, such as a blinding flash of light. The *inferior colliculi* control reflex movements of the head, neck, and trunk to auditory stimuli, such as a loud noise. The midbrain also contains motor nuclei for two of the cranial nerves (N III, IV) involved in the control of eye movements. Descending bundles of nerve fibers on the ventrolateral surface of the midbrain make up the **cerebral peduncles** (*peduncles*, little feet). Some of the descending fibers go to the cerebellum by way of the pons, and others carry voluntary motor commands from the primary motor cortex of each cerebral hemisphere.

The midbrain is also headquarters to one of the most important brain stem components, the **reticular formation**, which regulates many involuntary functions. The reticular formation is a network of interconnected nuclei that extends the length of the brain stem. The reticular formation of the midbrain contains the *reticular activating system (RAS)*. The output of this system directly affects the activity of the cerebral cortex. When the RAS is inactive, so are we; when the RAS is stimulated, so is our state of attention or wakefulness.

The maintenance of muscle tone and posture is due to midbrain nuclei that integrate information from the cerebrum and cerebellum and issue the appropriate involuntary motor commands. Other midbrain nuclei play an important role in regulating the motor output of the basal nuclei. For example, the *substantia nigra* (NĪ-grah; black) inhibit the activity of the basal nuclei by releasing the neurotransmitter dopamine. ∞ p. 237 If the substantia nigra are damaged or the neurons secrete less dopamine, the basal nuclei become more active. The result is a gradual increase in muscle tone and the appearance of symptoms characteristic of *Parkinson's disease*. Persons with Parkinson's disease have difficulty starting voluntary movements, because opposing muscle groups do not relax—they must be overpowered. Once a movement is under way, every aspect must be voluntarily controlled through intense effort and concentration.

The Pons

The pons (Figure 8-24●) links the cerebellum with the midbrain, diencephalon, cerebrum, and spinal cord. One group of nuclei within the pons includes the sensory and motor nuclei for four of the cranial nerves (N V–VIII). Other nuclei are concerned with the involuntary control of the pace and depth of respiration. Tracts passing through the pons link the cerebellum with the brain stem, cerebrum, and spinal cord.

The Cerebellum

The cerebellum (Figure 8-16b,c●, p. 244) is an automatic processing center. Its two important functions are (1) adjusting the postural muscles of the body to maintain balance and (2) programming and fine-tuning movements controlled at the conscious and subconscious levels. These functions are performed indirectly, by regulating activity along motor pathways at the cerebral cortex, basal nuclei, and brain stem. The cerebellum compares the motor commands with proprioceptive information (position sense) and performs adjustments needed to make the movement smooth. The tracts that link the cerebellum with these different regions are the **cerebellar peduncles** (Figure 8-24●). Like the cerebral hemispheres, the cerebellar peduncles are composed of white matter covered by a layer of neural cortex called the *cerebellar cortex*.

The cerebellum can be permanently damaged by trauma or stroke or temporarily affected by drugs such as alcohol. These alterations can produce *ataxia* (a-TAK-sē-a; *ataxia*, a lack of order), a disturbance in balance.

The Medulla Oblongata

The medulla oblongata (Figure 8-24●) connects the brain with the spinal cord. All communication between the brain and spinal cord involves tracts that ascend or descend through the medulla oblongata. These tracts often synapse in the medulla oblongata, in sensory or motor nuclei that act as relay stations and processing centers. In addition to these nuclei, the medulla oblongata contains sensory and motor nuclei associated with five of the cranial nerves (N VIII–XII).

Within the medulla oblongata, a variety of nuclei and centers representing the reticular formation regulate vital autonomic functions. These reflex centers receive inputs from cranial nerves, the cerebral cortex, and the brain stem, and their output controls or adjusts the activities of the cardiovascular and respiratory systems. The **cardiovascular centers** adjust heart rate, the strength of cardiac contractions, and the flow of blood through peripheral tissues. In terms of function, the cardiovascular centers are subdivided into a *cardiac center* regulating the heart rate and a *vasomotor center* controlling peripheral blood flow. The **respiratory rhythmicity centers** set the basic pace for respiratory movements, and their activity is adjusted by the *respiratory centers* of the pons.

CONCEPT CHECK QUESTIONS
Answers on page 275

❶ The thalamus acts as a relay point to all but what type of sensory information?

❷ Which area of the diencephalon would be stimulated by changes in body temperature?

❸ The medulla oblongata is one of the smallest sections of the brain, yet damage there can cause death, whereas similar damage in the cerebrum might go unnoticed. Why?

8 THE NERVOUS SYSTEM

The Nervous System • Cellular Organization in Neural Tissue • Neuron Function • Neural Communication • The Central Nervous System

The Peripheral Nervous System

The peripheral nervous system (PNS) is the link between the neurons of the central nervous system (CNS) and the rest of the body; all sensory information and motor commands are carried by axons of the PNS (see Figure 8-1● p. 226). These axons, bundled together and wrapped in connective tissue, form **peripheral nerves**, or simply nerves. Cranial nerves originate from the brain and spinal nerves connect to the spinal cord. The PNS also contains both the cell bodies and the axons of sensory neurons and motor neurons of the autonomic nervous system. The cell bodies are clustered together in masses called **ganglia** (singular *ganglion*) (see Figure 8-6● p. 231).

THE CRANIAL NERVES

There are 12 pairs of **cranial nerves** that connect to the brain (Figure 8-25●). They are numbered according to their position along the longitudinal axis of the brain. The letter N designates a cranial nerve, and Roman numerals are used to distinguish the individual nerves. For example, N I refers to the first pair of cranial nerves, the olfactory nerves.

Distribution and Function

Functionally, each nerve can be classified as primarily sensory, motor, or mixed (sensory and motor). Many cranial nerves, however, have secondary functions. For example, several cranial nerves (N III, VII, IX, and X) also distribute autonomic fibers to PNS ganglia, just as spinal nerves deliver them to ganglia along the spinal cord. The distribution and functions of the cranial nerves are described below and summarized in Table 8-2 (p. 255).

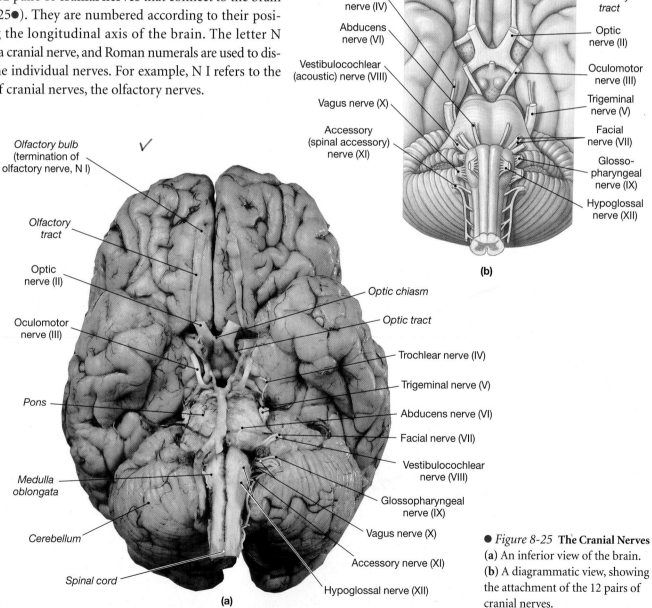

● *Figure 8-25* **The Cranial Nerves**
(**a**) An inferior view of the brain.
(**b**) A diagrammatic view, showing the attachment of the 12 pairs of cranial nerves.

TABLE 8-2 *The Cranial Nerves*

CRANIAL NERVES (NUMBER)	PRIMARY FUNCTION	INNERVATION
Olfactory (I)	Special sensory	Olfactory epithelium
Optic (II)	Special sensory	Retina of eye
Oculomotor (III)	Motor	Inferior, medial, superior rectus, inferior oblique, and intrinsic muscles of eye
Trochlear (IV)	Motor	Superior oblique muscle of eye
Trigeminal (V)	Mixed	*Sensory:* orbital structures, nasal cavity, skin of forehead, eyelids, eyebrows, nose, lips, gums and teeth; cheek, palate, pharynx, and tongue *Motor:* chewing (temporalis, masseter, pterygoids) muscles
Abducens (VI)	Motor	Lateral rectus muscle of eye
Facial (VII)	Mixed	*Sensory:* taste receptors on the anterior $\frac{2}{3}$ of tongue *Motor:* muscles of facial expression, lacrimal (tear) gland, and submandibular and sublingual salivary glands
Vestibulocochlear (Acoustic) (VIII)	Special sensory	Cochlea (receptors for hearing) Vestibule (receptors for motion and balance)
Glossopharyngeal (IX)	Mixed	*Sensory:* posterior $\frac{1}{3}$ of tongue; pharynx and palate (part); receptors for blood pressure, pH, oxygen, and carbon dioxide concentrations *Motor:* pharyngeal muscles, parotid salivary gland
Vagus (X)	Mixed	*Sensory:* pharynx; auricle and external meatus; diaphragm; visceral organs in thoracic and abdominopelvic cavities *Motor:* palatal and pharyngeal muscles and visceral organs in thoracic and abdominopelvic cavities
Accessory (Spinal Accessory) (XI)	Motor	Voluntary muscles of palate, pharynx, and larynx; sternocleidomastoid and trapezius muscles
Hypoglossal (XII)	Motor	Tongue muscles

Few people are able to remember the names, numbers, and functions of the cranial nerves without a struggle. Many people use mnemonic phrases, such as Oh, Once One Takes The Anatomy Final, Very Good Vacations Are Heavenly, in which the first letter of each word represents the cranial nerves.

The Olfactory Nerves (N I). The first pair of cranial nerves, the **olfactory nerves**, are the only cranial nerves attached to the cerebrum. The rest start or end within nuclei of the diencephalon or brain stem. These nerves carry special sensory information responsible for the sense of smell. The olfactory nerves originate in the epithelium of the upper nasal cavity and penetrate the cribriform plate of the ethmoid bone to synapse in the olfactory bulbs of the brain. From the olfactory bulbs, the axons of postsynaptic neurons travel within the olfactory tracts to the olfactory centers of the brain.

The Optic Nerves (N II). The **optic nerves** carry visual information from the eyes. After passing through the **optic foramina** of the orbits, these nerves intersect at the **optic chiasm** ("crossing") (Figure 8-25a●) before they continue as the *optic tracts* to the lateral geniculate nuclei of the thalamus.

The Oculomotor Nerves (N III). The midbrain contains the motor nuclei controlling the third and fourth cranial nerves. Each **oculomotor nerve** innervates four of the six muscles that move an eyeball (the superior, medial, and inferior rectus muscles and the inferior oblique muscle). These nerves also carry autonomic fibers to intrinsic eye muscles that control (1) the amount of light entering the eye and (2) the shape of the lens.

The Trochlear Nerves (N IV). The **trochlear** (TRŌK-lē-ar; *trochlea*, a pulley) **nerves**, the smallest of the cranial nerves, innervate the superior oblique muscles of the eyes. The motor nuclei that control these nerves lie in the midbrain. The name "trochlear" refers to the pulley-shaped, ligamentous sling, through which the tendon of the superior oblique muscle passes to reach its attachment on the eyeball (see Figure 9-8●, p. 286).

The Trigeminal Nerves (N V). The pons contains the nuclei associated with cranial nerve N V. The **trigeminal** (trī-JEM-i-nal) **nerves** are the largest of the cranial nerves. These nerves provide sensory information from the head and face and motor control over the chewing muscles, such as the temporalis and

8 THE NERVOUS SYSTEM

The Nervous System • Cellular Organization in Neural Tissue • Neuron Function • Neural Communication • The Central Nervous System

masseter. The trigeminal has three major branches. The *ophthalmic branch* provides sensory information from the orbit of the eye, the nasal cavity and sinuses, and the skin of the forehead, eyebrows, eyelids, and nose. The *maxillary branch* provides sensory information from the lower eyelid, upper lip, cheek, nose, upper gums and teeth, palate, and portions of the pharynx. The *mandibular branch*, the largest of the three, provides sensory information from the skin of the temples, the lower gums and teeth, the salivary glands, and the anterior portions of the tongue. It also provides motor control over the chewing muscles (the temporalis, masseter, and pterygoid muscles). ∞ p. 200

The Abducens Nerves (N VI).

The **abducens** (ab-DŪ-senz) **nerves** innervate only the lateral rectus, the sixth of the extrinsic eye muscles. Their nuclei are in the pons. The nerves emerge at the border between the pons and the medulla oblongata and reach the orbit of the eye along with the oculomotor and trochlear nerves. The name "abducens" is based on the action of this nerve's innervated muscle, which abducts the eyeball, causing it to rotate laterally, away from the midline of the body.

The Facial Nerves (N VII).

The **facial nerves** are mixed nerves of the face whose sensory and motor roots emerge from the side of the pons. The motor fibers produce facial expressions by controlling the superficial muscles of the scalp and face and muscles near the ear. The sensory fibers monitor proprioceptors in the facial muscles, provide deep pressure sensations over the face, and taste information from receptors along the anterior two-thirds of the tongue. These nerves also carry autonomic fibers that result in control of the tear glands and salivary glands.

The Vestibulocochlear Nerves (N VIII).

The **vestibulocochlear nerves**, also called acoustic nerves, monitor the sensory receptors of the inner ear. The pons and medulla oblongata contain nuclei associated with these nerves. Each vestibulocochlear nerve has two components: (1) a **vestibular nerve** (*vestibulum*, a cavity), which originates at the *vestibule*, the portion of the inner ear concerned with balance sensations, conveys information on position, movement, and balance; and (2) the **cochlear** (KOK-lē-ar; *cochlea*, snail shell) **nerve**, which monitors the receptors of the cochlea, the portion of the inner ear responsible for the sense of hearing.

The Glossopharyngeal Nerves (N IX).

The **glossopharyngeal** (glos-ō-fah-RIN-jē-al; *glossus*, tongue) **nerves** are mixed nerves innervating the tongue and pharynx. The associated sensory and motor nuclei are in the medulla oblongata. The sensory portion of this nerve provides taste sensations from the posterior third of the tongue and monitors blood pressure and dissolved gas concentrations in major blood vessels. The motor portion controls the pharyngeal muscles involved in swallowing. These nerves also carry autonomic fibers that result in control of the parotid salivary glands.

The Vagus Nerves (N X).

The **vagus** (VĀ-gus; *vagus*, wandering) **nerves** provide sensory information from the ear canals, the diaphragm, and taste receptors in the pharynx and from visceral receptors along the esophagus, respiratory tract, and abdominal organs as far away as the last portions of the large intestine. The associated sensory and motor nuclei of the vagus are located in the medulla oblongata. The sensory information provided by N X is vital to the autonomic control of visceral function, but we are not consciously aware of these sensations, because they are seldom relayed to the cerebral cortex. The motor components of the vagus nerves control skeletal muscles of the soft palate, pharynx, and esophagus and affect cardiac muscle, smooth muscle, and glands of the esophagus, stomach, intestines, and gallbladder.

The Accessory Nerves (N XI).

The **accessory nerves**, sometimes called the *spinal accessory nerves*, are mixed nerves that innervate structures in the neck and back. These nerves differ from other cranial nerves in that some of their motor fibers originate in the lateral gray horns of the first five cervical segments of the spinal cord, as well as in the medulla oblongata. The medullary branch innervates the voluntary swallowing muscles of the soft palate and pharynx, and the laryngeal muscles that control the vocal cords and produce speech. The spinal branch controls the sternocleidomastoid and trapezius muscles associated with the pectoral girdle. ∞ pp. 201, 205

The Hypoglossal Nerves (N XII).

The **hypoglossal** (hī-pō-GLOS-al) **nerves** provide voluntary control over the skeletal muscles of the tongue. The nuclei for these motor nerves are located in the medulla oblongata.

CONCEPT CHECK QUESTIONS

Answers on page 275

❶ What symptoms would you associate with damage to the abducens nerve (N VI)?

❷ John is having trouble moving his tongue. His physician tells him it is due to pressure on a cranial nerve. Which cranial nerve is involved?

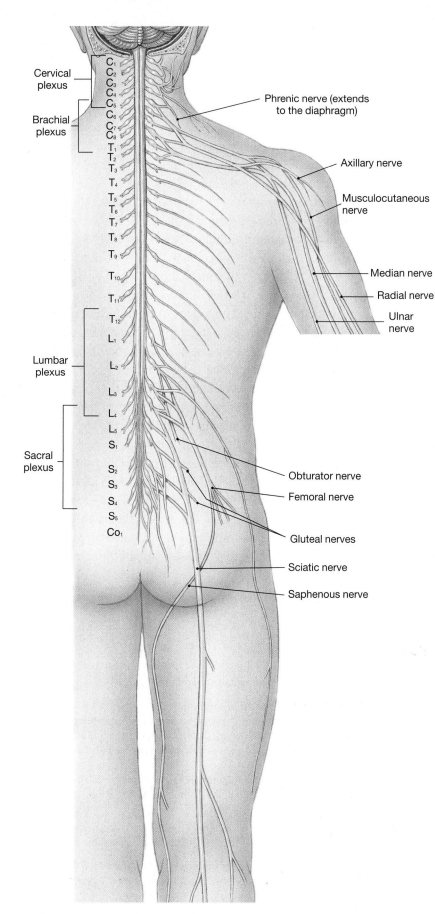

Cervical plexus

Brachial plexus

Phrenic nerve (extends to the diaphragm)

Axillary nerve

Musculocutaneous nerve

Median nerve

Radial nerve

Ulnar nerve

Lumbar plexus

Sacral plexus

Obturator nerve

Femoral nerve

Gluteal nerves

Sciatic nerve

Saphenous nerve

● *Figure 8-26* **Peripheral Nerves and Nerve Plexuses**

THE SPINAL NERVES

The 31 pairs of **spinal nerves** can be grouped according to the region of the vertebral column from which they originate (Figure 8-26●). They include 8 pairs of cervical nerves ($C_1 - C_8$), 12 pairs of thoracic nerves ($T_1 - T_{12}$), 5 pairs of lumbar nerves ($L_1 - L_5$), 5 pairs of sacral nerves ($S_1 - S_5$), and 1 pair of coccygeal nerves (Co_1). Each pair of spinal nerves monitors a specific region of the body surface known as a **dermatome** (Figure 8-27●). Dermatomes are clinically important because damage or infection of a spinal nerve or dorsal root ganglia will produce a characteristic loss of sensation in specific parts of the skin. For example, in *shingles*, a virus that infects dorsal root ganglia causes a painful rash whose distribution corresponds to that of the affected sensory nerves.

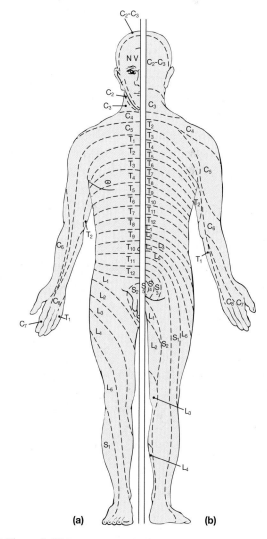

(a)   (b)

● *Figure 8-27* **Dermatomes**
Pattern of dermatomes on the surface of the skin, seen in (a) anterior and (b) posterior views. The face is served by cranial nerves, not spinal nerves.

257

8 THE NERVOUS SYSTEM

The Nervous System • Cellular Organization in Neural Tissue • Neuron Function • Neural Communication • The Central Nervous System

NERVE PLEXUSES

During development, skeletal muscles commonly fuse, forming larger muscles innervated by nerve trunks containing axons derived from several spinal nerves. These compound nerve trunks originate at networks called **nerve plexuses**. The plexuses and major peripheral nerves are shown in Figure 8-26●.

The **cervical plexus** innervates the muscles of the neck and extends into the thoracic cavity to control the diaphragm. The **brachial plexus** innervates the shoulder girdle and upper limb. The **lumbosacral plexus** supplies the pelvic girdle and lower limb. It can be further subdivided into a *lumbar plexus* and a *sacral plexus*. Table 8-3 lists the spinal nerve plexuses and describes some of the major nerves and their distribution.

Peripheral nerve palsies, also known as *peripheral neuropathies*, are characterized by regional losses of sensory and motor function as the result of nerve trauma or compression. You have experienced a mild, temporary palsy if your arm or leg has ever "fallen asleep."

REFLEXES

The central nervous system and peripheral nervous system can be studied separately, but they function together. To consider the ways the CNS and PNS interact, we begin with simple reflex responses to stimulation. A **reflex** is an automatic motor response to a specific stimulus. Reflexes help us preserve homeostasis by making rapid adjustments in the function of our organs or organ systems. The response shows little variability—when a particular reflex is activated it usually produces the same motor response.

Simple Reflexes

A reflex involves sensory fibers delivering information to the CNS and motor fibers carrying motor commands to peripheral effectors. The "wiring" of a single reflex is called a **reflex arc**. Figure 8-28● diagrams the five steps involved in a reflex arc: (1) the arrival of a stimulus and activation of a receptor, (2) the activation of a sensory neuron, (3) information processing, (4) the activation of a motor neuron, and (5) the response by an effector (muscle or gland).

A reflex response usually removes or opposes the original stimulus. In Figure 8-28●, the contracting muscle pulls the hand away from the painful stimulus. This reflex arc is therefore an example of *negative feedback*. p. 12 By opposing potentially harmful changes in the internal or external environment, reflexes play an important role in maintaining homeostasis.

In the simplest reflex arc, a sensory neuron synapses directly on a motor neuron. Such a reflex is called a **monosynaptic reflex**. Because there is only one synapse, monosynaptic reflexes control the most rapid, stereotyped motor responses of the nervous system. The best-known example is the stretch reflex.

The **stretch reflex** provides automatic regulation of skeletal muscle length. The sensory receptors in the stretch reflex are called **muscle spindles**, bundles of small, specialized skeletal muscle fibers scattered throughout the skeletal muscles. The stimulus (increasing muscle length) activates a sensory neuron that triggers an immediate motor response (contraction of the stretched muscle) that counteracts the stimulus.

Stretch reflexes are important in maintaining normal posture and balance and in making automatic adjustments in mus-

TABLE 8-3 *Nerve Plexuses and Major Nerves*

PLEXUS	MAJOR NERVE	DISTRIBUTION
Cervical Plexus (C_1–C_5)	Phrenic nerve	Diaphragm
	Other branches	Muscles of the neck; skin of upper chest, neck, and ears
Brachial Plexus (C_5–T_1)	Axillary nerve	Deltoid and teres minor muscles; skin of shoulder
	Musculocutaneous nerve	Flexor muscles of the arm and forearm; skin on lateral surface of forearm
	Median nerve	Flexor muscles of forearm and hand; skin over lateral surface of hand
	Radial nerve	Extensor muscles of the arm, forearm, and hand; skin over posterolateral surface of the arm
	Ulnar nerve	Flexor muscles of forearm and small digital muscles; skin over medial surface of hand
Lumbosacral Plexus	Obturator nerve	Adductors of hip; skin over medial surface of thigh
Lumbar Plexus (T_{12}–L_4)	Femoral nerve	Adductors of hip, extensors of knee; skin over medial surfaces of thigh and leg
Sacral Plexus (L_4–S_4)	Gluteal nerve	Adductors and extensors of hip; skin over posterior surface of thigh
	Sciatic nerve	Flexors of knee and ankle, flexors and extensors of toes; skin over anterior and posterior surfaces of leg and foot
	Saphenous nerve	Skin over medial surface of leg

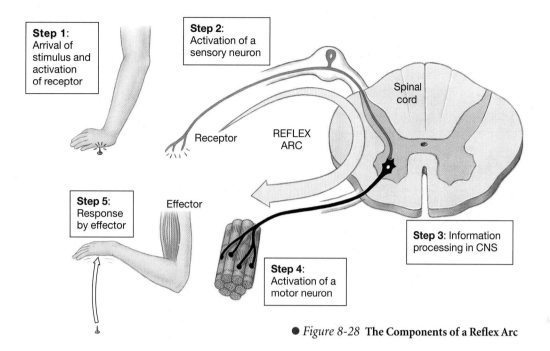

● *Figure 8-28* **The Components of a Reflex Arc**

cle tone. Physicians can use the sensitivity of the stretch reflex to test the general condition of the spinal cord, peripheral nerves, and muscles. For example, in the **knee jerk reflex**, or *patellar reflex*, a sharp rap on the patellar ligament stretches muscle spindles in the quadriceps muscles (Figure 8-29●). With

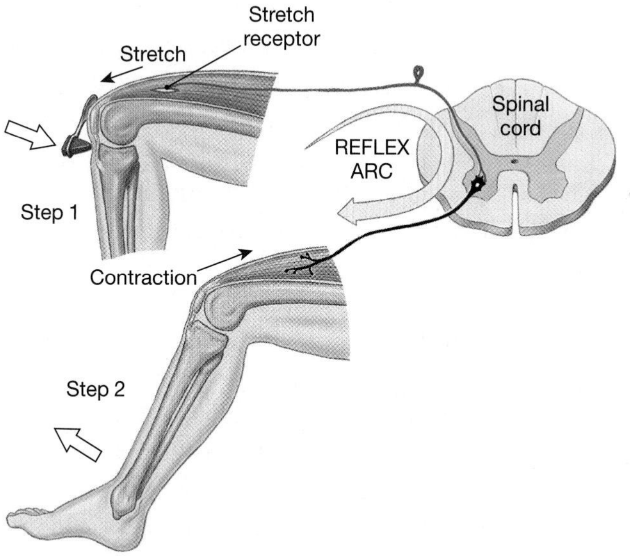

● *Figure 8-29* **The Stretch Reflex**
The patellar reflex is a stretch reflex controlled by stretch receptors (muscle spindles) in the muscles that straighten the knee. **Step 1:** A reflex hammer strikes the muscle tendon, stretching the muscle spindles. This stretching results in a sudden increase in the activity of the sensory neurons. These neurons synapse on motor neurons in the spinal cord. **Step 2:** The activation of spinal motor neurons produces an immediate muscle contraction and a reflexive kick.

so brief a stimulus, the reflexive contraction occurs unopposed and produces a noticeable kick. If this contraction shortens the muscle spindles below their original resting lengths, the sensory nerve endings are compressed, the sensory neuron is inhibited, and the leg drops back.

Complex Reflexes

Many spinal reflexes have at least one interneuron between the sensory (afferent) neuron and the motor (efferent) neuron. Because there are more synapses, these **polysynaptic reflexes** have a longer delay between a stimulus and response. But they can produce far more complicated responses because the interneurons can control several muscle groups simultaneously.

Withdrawal reflexes move stimulated parts of the body away from a source of stimulation. The strongest withdrawal reflexes are triggered by painful stimuli, but these reflexes are also initiated by the stimulation of touch or pressure receptors. The **flexor reflex** is a withdrawal reflex affecting the muscles of a limb. When you step on a tack, a dramatic flexor reflex is produced in the affected limb (Figure 8-30●). When the pain receptors in your foot are stimulated, the sensory neurons activate interneurons in the spinal cord that stimulate motor neurons in the anterior gray horns. The result is a contraction of flexor muscles that yanks your foot off the ground.

When a specific muscle contracts, opposing, or antagonistic, muscles are stretched. For example, the flexor muscles that bend the knee are opposed by extensor muscles, which straighten it out. A potential conflict exists here: contraction of

259

8 THE NERVOUS SYSTEM

The Nervous System • Cellular Organization in Neural Tissue • Neuron Function • Neural Communication • The Central Nervous System

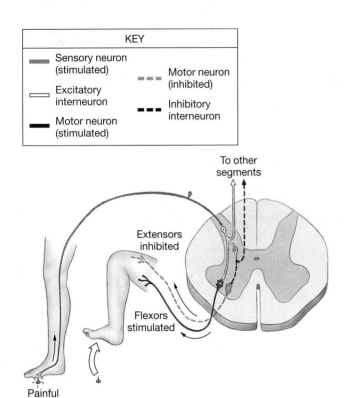

KEY

▬▬ Sensory neuron (stimulated)	▬ ▬ ▬ Motor neuron (inhibited)
☐ Excitatory interneuron	▬ ▬ ▬ Inhibitory interneuron
▬▬ Motor neuron (stimulated)	

To other segments

Extensors inhibited

Flexors stimulated

Painful stimulus

● *Figure 8-30* **The Withdrawal Reflex**

a flexor muscle should trigger a stretch reflex in the extensors that would cause them to contract, opposing the movement that is under way. Interneurons in the spinal cord prevent such competition through **reciprocal inhibition**. When one set of motor neurons is stimulated, those controlling antagonistic muscles are inhibited.

Integration and Control of Spinal Reflexes

Although reflexes are automatic, higher centers in the brain influence these responses by stimulating or inhibiting the interneurons and motor neurons involved. The sensitivity of a reflex can thus be modified. For example, an effort to pull apart clasped hands elevates the general state of stimulation along the spinal cord, leading to an enhancement of spinal reflexes.

Other descending fibers have an inhibitory effect on spinal reflexes. Stroking an infant's foot on the side of the sole produces a fanning of the toes known as the **Babinski sign**, or *positive Babinski reflex*. This response disappears as descending inhibitory synapses develop, so in adults the same stimulus produces a curling of the toes, called a **plantar reflex**, or *negative Babinski reflex*, after about a one-second delay. If either the higher centers or the descending tracts are damaged, the Babinski sign will reappear. As a result, this reflex is often tested if CNS injury is suspected.

When higher centers issue motor commands, they can activate the complex motor patterns already programmed into the spinal cord. The use of preexisting patterns allows a relatively small number of descending fibers to control complex motor functions. For example, the motor patterns for walking, running, and jumping are primarily directed by neuronal pools in the spinal cord. The descending pathways facilitate, inhibit, or fine-tune the established patterns.

CONCEPT CHECK QUESTIONS
Answers on page 275

❶ Which common reflex is used by physicians to test the general condition of the spinal cord, peripheral nerves, and muscles?

❷ Polysynaptic reflexes can produce more complicated responses than can monosynaptic reflexes. Why?

❸ After suffering an injury to his back, Tom exhibits a positive Babinski reflex. What does this reaction imply about Tom's injury?

SENSORY AND MOTOR PATHWAYS

The communication between the central nervous system (CNS), the peripheral nervous system (PNS), and organs and systems occurs over pathways, nerve tracts, and nuclei that relay sensory and motor information. p. 231 The major sensory (ascending) and motor (descending) tracts of the spinal cord are named with regard to the destinations of the axons. If the name of a tract begins with *spino-*, the tract starts in the spinal cord and ends in the brain, and it therefore carries sensory information. If the name of a tract ends in *-spinal*, its axons start in the higher centers and end in the spinal cord, bearing motor commands. The rest of the tract name indicates the associated nucleus or cortical area of the brain.

Table 8-4 lists examples of sensory and motor pathways and their functions.

Sensory Pathways

Sensory receptors monitor conditions in the body or the external environment. A **sensation**, the information gathered by a sensory receptor, arrives in the form of action potentials in an afferent (sensory) fiber. Most of the processing of arriving sensations occurs in centers along the sensory pathways in the spinal cord or brain stem; only about 1 percent of the information provided by afferent fibers reaches the cerebral cortex and our conscious awareness. For example, we usually do not feel the clothes we wear or hear the hum of the engine in our car.

The Posterior Column Pathway. One example of an ascending, sensory pathway is the **posterior column pathway** (Figure 8-31●). It sends highly localized ("fine") touch, pressure, vibra-

TABLE 8-4	*Sensory and Motor Pathways*
PATHWAY	**FUNCTION**
SENSORY	
Posterior column pathway	Delivers highly localized sensations of fine touch, pressure, vibration, and proprioception to the primary sensory cortex
Spinothalamic pathway	Delivers poorly localized sensations of touch, pressure, pain, and temperature to the primary sensory cortex
Spinocerebellar pathway	Delivers proprioceptive information concerning the positions of muscles, bones, and joints to the cerebellar cortex
MOTOR	
Corticospinal pathway	Conscious control of skeletal muscles throughout the body
Medial and lateral pathways	Subconscious regulation of skeletal muscle tone, controls reflexive skeletal muscle responses to equilibrium sensations and to sudden or strong visual and auditory stimuli

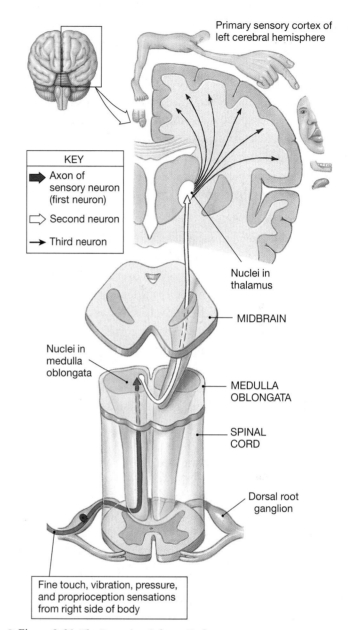

KEY

Axon of sensory neuron (first neuron)

Second neuron

Third neuron

Primary sensory cortex of left cerebral hemisphere

Nuclei in thalamus

MIDBRAIN

Nuclei in medulla oblongata

MEDULLA OBLONGATA

SPINAL CORD

Dorsal root ganglion

Fine touch, vibration, pressure, and proprioception sensations from right side of body

● *Figure 8-31* **The Posterior Column Pathway**
The posterior column pathway delivers fine touch, vibration, and proprioception information to the primary sensory cortex of the cerebral hemisphere on the opposite side of the body. (For clarity, this figure shows only the pathway for sensations originating on the right side of the body.)

tion, and proprioceptive (position) sensations to the cerebral cortex. In the process, the information is relayed from one neuron to another.

Step 1: The sensations reach the CNS through the dorsal roots of spinal nerves. Once inside the spinal cord, the axons ascend within the posterior column pathway to synapse in a sensory nucleus of the medulla oblongata.

Step 2: The axons of the neurons in this nucleus cross over to the opposite side of the brain stem before continuing to the thalamus.

Step 3: The location of the synapse in the thalamus depends on the region of the body involved. The thalamic neuron then relays the information to an appropriate primary sensory cortex.

The sensations arrive organized such that sensory information from the toes reaches one end of the primary sensory cortex and information from the head reaches the other. As a result, the sensory cortex contains a miniature map of the body surface. That map is distorted because the area of sensory cortex devoted to a particular region is proportional not to its size, but to the number of sensory receptors it contains. In other words, it takes many more cortical neurons to process sensory information arriving from the tongue, which has tens of

thousands of taste and touch receptors, than it does to analyze sensations originating on the back, where touch receptors are few and far between.

Motor Pathways

In response to information provided by sensory systems, the CNS issues motor commands that are distributed by the *somatic nervous system (SNS)* and the *autonomic nervous system (ANS)*. The SNS, under voluntary control, issues somatic motor commands that direct the contractions of skeletal muscles. The

261

8 THE NERVOUS SYSTEM

The Nervous System • Cellular Organization in Neural Tissue • Neuron Function • Neural Communication • The Central Nervous System

motor commands of the ANS, which are issued outside our conscious awareness, control the smooth and cardiac muscles, glands, and fat cells.

Three motor pathways provide control over skeletal muscles, the corticospinal pathway, and the medial and lateral pathways. The corticospinal pathway provides conscious, voluntary control over skeletal muscles, whereas the medial and lateral pathways exert more indirect, subconscious control. Refer to Table 8-4 for examples and functions of these motor pathways. We will focus on the corticospinal pathway.

The Corticospinal Pathway. The **corticospinal pathway**, sometimes called the *pyramidal system*, provides conscious, voluntary control of skeletal muscles. Figure 8-32● shows the motor pathway providing voluntary control over the right side of the body. The neurons of the primary motor cortex are organized into a miniature map of the body. It is just as distorted as the sensory map on the primary sensory cortex; the proportions indicate the number of motor units present in that portion of the body. For example, the grossly oversized hands provide an indication of how many different muscles and motor units are involved in writing, grasping, and manipulating objects in our environment.

The corticospinal pathway begins at triangular-shaped *pyramidal cells* of the cerebral cortex. The axons of these cells extend into the brain stem and spinal cord where they synapse on somatic motor neurons (Figure 8-32●). All axons of the corticospinal tracts cross over to reach motor neurons on the opposite side of the body. As a result, the left side of the body is controlled by the right cerebral hemisphere, and the right side is controlled by the left cerebral hemisphere.

The Medial and Lateral Pathways. The **medial and lateral pathways** provide subconscious, involuntary control of muscle tone and movements of the neck, trunk, and limbs. They also coordinate learned movement patterns and other voluntary motor activities (Table 8-4). Together, these pathways were known as the *extrapyramidal system*, because it was thought that they operated independent and parallel to the *pyramidal system* (corticospinal pathway). It is now known that the control of the body's motor functions are integrated between all three motor pathways.

The components of the medial and lateral pathways are spread throughout the brain. They include nuclei in the brain stem (midbrain, pons, and medulla oblongata); relay stations in the thalamus; the basal nuclei of the cerebrum; and the cerebellum. Output from the basal nuclei and cerebellum exerts the highest level of control. For example, they can (1) stimulate or inhibit other nuclei of these pathways or (2) stimulate or inhibit the activities of pyramidal cells

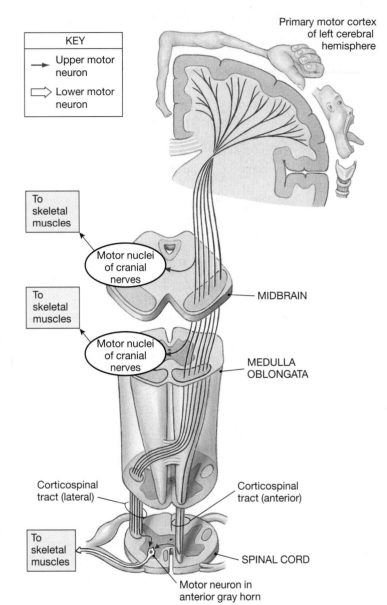

● *Figure 8-32* **The Corticospinal Pathway**
The corticospinal pathway originates at the primary motor cortex. Axons of the pyramidal cells of the primary motor cortex descend to reach motor nuclei in the brain stem and spinal cord. Most of the fibers cross over in the medulla oblongata before descending into the spinal cord as the corticospinal tracts.

in the primary motor cortex. Also, axons from the upper motor neurons in the medial and lateral pathways synapse on the same motor neurons innervated by the corticospinal pathway.

 CEREBRAL PALSY

The term **cerebral palsy** refers to a number of disorders affecting voluntary motor performance, such as movement and posture, that appear during infancy or childhood and persist throughout life. The cause may be trauma associated with premature or unusually stressful birth; maternal exposure to drugs, including alcohol; or a genetic defect that causes improper development of the motor pathways. Problems with labor and delivery result from the compression or interruption of placental circulation or oxygen supplies. If the oxygen concentration in the fetal blood declines significantly for as little as 5–10 minutes, CNS function can be permanently impaired. The cerebral cortex, cerebellum, basal nuclei, hippocampus, and thalamus are likely targets, producing abnormalities in motor skills, posture and balance, memory, speech, and learning abilities.

CONCEPT CHECK QUESTIONS

Answers on page 275

❶ As a result of pressure on her spinal cord, Jill cannot feel touch or pressure on her legs. What sensory pathway is being compressed?

❷ The primary motor cortex of the right cerebral hemisphere controls motor function on which side of the body?

❸ An injury to the superior portion of the motor cortex would affect which part of the body?

The Autonomic Nervous System

Using the pathways already discussed, we can respond to sensory information and exert voluntary control over the activities of our skeletal muscles. Yet our conscious sensations, plans, and responses represent only a tiny fraction of the activities of the nervous system. In practical terms, conscious activities have little to do with our immediate or long-term survival, and the adjustments made by the **autonomic nervous system (ANS)** are much more important. Without the ANS, a simple night's sleep would be a life-threatening event.

Both efferent divisions of the nervous system, the **somatic nervous system (SNS)** and the autonomic nervous system carry motor commands to peripheral effectors. ⟳ p. 227 There are clear anatomical differences between the SNS and ANS (Figure 8-33●). In the SNS, lower motor neurons exert direct control over skeletal muscles (Figure 8-33a●). In the ANS, a second motor neuron always separates the CNS and the peripheral

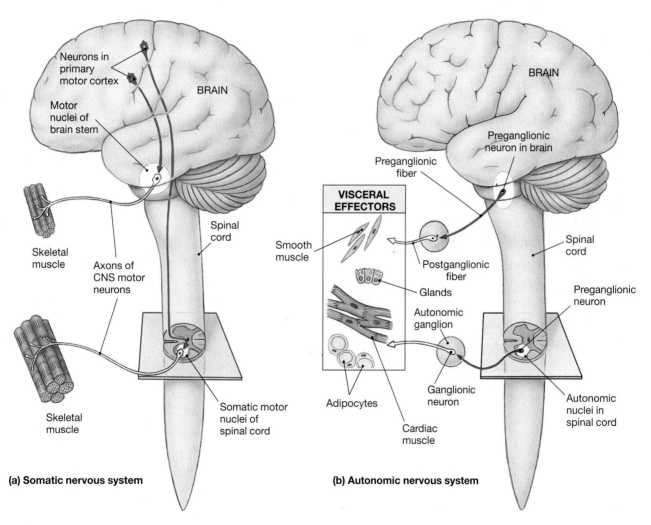

(a) Somatic nervous system

(b) Autonomic nervous system

● *Figure 8-33* **The Somatic and Autonomic Nervous Systems**
(**a**) In the SNS, a motor neuron in the CNS has direct control over skeletal muscle fibers. (**b**) In the ANS, preganglionic neurons synapse on ganglionic neurons that innervate effectors, such as smooth and cardiac muscles, glands, and fat cells.

8

THE NERVOUS SYSTEM

The Nervous System • Cellular Organization in Neural Tissue • Neuron Function • Neural Communication • The Central Nervous System

effector (Figure 8-33b●). The ANS motor neurons in the CNS, known as **preganglionic neurons**, send their axons, called *preganglionic fibers*, to autonomic ganglia outside the CNS. In these ganglia, the axons of the preganglionic neurons synapse on **ganglionic neurons**. The axons, or **postganglionic fibers**, of these neurons leave the ganglia and innervate cardiac muscle, smooth muscles, glands, and fat cells (adipocytes).

The ANS consists of two divisions: (1) the sympathetic division and (2) the parasympathetic division. Preganglionic fibers from the thoracic and lumbar spinal segments synapse in ganglia near the spinal cord; these axons and ganglia are part of the **sympathetic division** of the ANS. The sympathetic division is often called the "fight or flight" system because it usually stimulates tissue metabolism, increases alertness, and prepares the body to deal with emergencies.

Preganglionic fibers originating in the brain and the sacral spinal segments synapse on neurons of **intramural ganglia** (*murus*, wall), located near or within the tissues of visceral organs. These components are part of the **parasympathetic division** of the ANS, often regarded as the "rest and repose" or "rest and digest" system because it conserves energy and promotes sedentary activities, such as digestion.

The sympathetic and parasympathetic divisions affect target organs through the controlled release of specific neurotransmitters by the postganglionic fibers. Whether the result is a stimulation or an inhibition of activity depends on the response of the membrane receptor to the presence of the neurotransmitter. Some general patterns are worth noting:

- All preganglionic autonomic fibers are cholinergic: They release acetylcholine (ACh) at their synaptic terminals. ∽ p. 236 The effects are always excitatory.

- Postganglionic parasympathetic fibers are also cholinergic, but the effects are excitatory or inhibitory, depending on the nature of the target cell receptor.

- Most postganglionic sympathetic terminals release norepinephrine (NE). Neurons that release NE are called *adrenergic*. ∽ p. 237 The effects of NE are usually excitatory.

THE SYMPATHETIC DIVISION

The sympathetic division (Figure 8-34●) consists of:

- *Preganglionic neurons located between segments T_1 and L_2 of the spinal cord.* These neurons are situated in the lateral gray horns, and their short axons enter the ventral roots of these segments.

- *Ganglionic neurons located in ganglia near the vertebral column.* Two types of sympathetic ganglia exist. Paired *sympathetic chain ganglia* on either side of the vertebral column contain neurons that control effectors in the body wall and inside

the thoracic cavity. Unpaired *collateral ganglia*, anterior to the vertebral column, contain ganglionic neurons that innervate tissues and organs in the abdominopelvic cavity.

- *The adrenal medullae.* The center of each adrenal gland, an area known as the adrenal medulla, is a modified ganglion. The ganglionic neurons of the adrenal medullae have very short axons—when stimulated, they release the neurotransmitters norepinephrine (NE) and epinephrine (E) into the general circulation.

Organization

The Sympathetic Chain. From spinal segments T_1 to L_2, sympathetic preganglionic fibers join the ventral root of each spinal nerve. All of these fibers then exit the spinal nerve to enter the sympathetic chain ganglia. For motor commands to the body wall, a synapse occurs at the chain ganglia, and then the postganglionic fibers return to the spinal nerve for distribution. For the thoracic cavity, a synapse also occurs at the chain ganglia, but the postganglionic fibers then form nerves that go directly to their targets (Figure 8-34●).

The Collateral Ganglia. The abdominopelvic tissues and organs receive sympathetic innervation over preganglionic fibers from lower thoracic and upper lumbar segments that pass through the sympathetic chain without synapsing, and instead synapse within separate, unpaired **collateral ganglia**. The nerves traveling to the collateral ganglia are known as *splanchnic nerves*. The fibers leaving the collateral ganglia extend throughout the abdominopelvic cavity. The three collateral ganglia and the organs their postganglionic fibers innervate are diagrammed in Figure 8-34●.

The Adrenal Medullae. Preganglionic fibers entering each adrenal gland proceed to its central region called the **adrenal medulla**. There they synapse on modified neurons that perform an endocrine function. When stimulated, these cells release the neurotransmitters norepinephrine (NE) and epinephrine (E) into surrounding capillaries, which carry them throughout the body. In general, the effects of these neurotransmitters resemble those produced by the stimulation of sympathetic postganglionic fibers.

General Functions

The sympathetic division stimulates tissue metabolism, increases alertness, and prepares the individual for sudden, intense physical activity. Sympathetic innervation distributed by the spinal nerves stimulates sweat gland activity and arrector pili muscles (producing "goose bumps"), reduces circulation to the skin and body wall, accelerates blood flow to skeletal muscles, releases stored lipids from adipose tissue, and dilates the pupils. The activation of the sympathetic nerves to the tho-

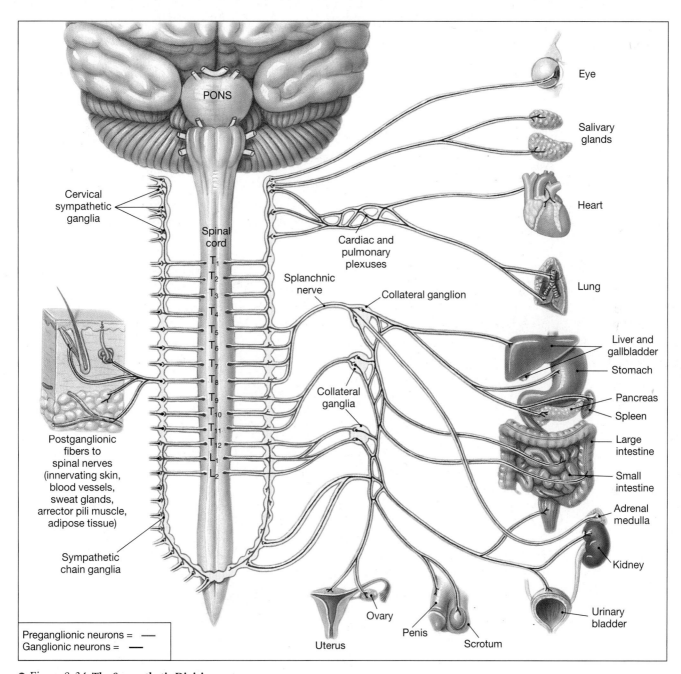

● *Figure 8-34* **The Sympathetic Division**
The distribution of sympathetic fibers is the same on both sides of the body. For clarity, the innervation of somatic structures is shown to the left and the innervation of visceral structures to the right.

racic cavity accelerates the heart rate, increases the force of cardiac contractions, and dilates the respiratory passageways. The postganglionic fibers from the collateral ganglia reduce the blood flow and energy use by visceral organs that are not important to short-term survival, such as the digestive tract, and they stimulate the release of stored energy reserves. The release of NE and E by the adrenal medullae broadens the effects of sympathetic activation to cells not innervated by sympathetic postganglionic fibers and also makes the effects last much longer than those produced by direct sympathetic innervation.

THE PARASYMPATHETIC DIVISION

The parasympathetic division of the ANS includes:

- *Preganglionic neurons in the brain stem and in sacral segments of the spinal cord.* The midbrain, pons, and medulla oblongata contain autonomic nuclei associated with cranial nerves III, VII, IX and X. Other autonomic nuclei lie in the lateral gray horns of spinal cord segments S_2 to S_4.

- *Ganglionic neurons in peripheral ganglia within or adjacent to the target organs.* Preganglionic fibers of the parasympathetic

265

8 **THE NERVOUS SYSTEM**

The Nervous System • Cellular Organization in Neural Tissue • Neuron Function • Neural Communication • The Central Nervous System

division do not diverge as extensively as those of the sympathetic division. Thus, the effects of parasympathetic stimulation are more specific and localized than those of the sympathetic division.

Organization

Figure 8-35● diagrams the pattern of parasympathetic innervation. Preganglionic fibers leaving the brain travel within cranial nerves III (oculomotor), VII (facial), IX (glossopharyngeal), and X (vagus). These fibers synapse in ganglia located in peripheral tissues, and short postganglionic fibers then continue to their targets. The vagus nerves provide preganglionic parasympathetic innervation to ganglia in organs of the thoracic and abdominopelvic cavities as distant as the last segments of the large intestine. The vagus nerves provide roughly 75 percent of all parasympathetic outflow and innervate most of those organs.

Preganglionic fibers in the sacral segments of the spinal cord carry the sacral parasympathetic output. They do not join the spinal nerves and form distinct **pelvic nerves**, which innervate intramural ganglia in the kidney and urinary bladder, the last segments of the large intestine, and the sex organs.

General Functions

Among other things, the parasympathetic division constricts the pupils, increases secretions by the digestive glands, increases

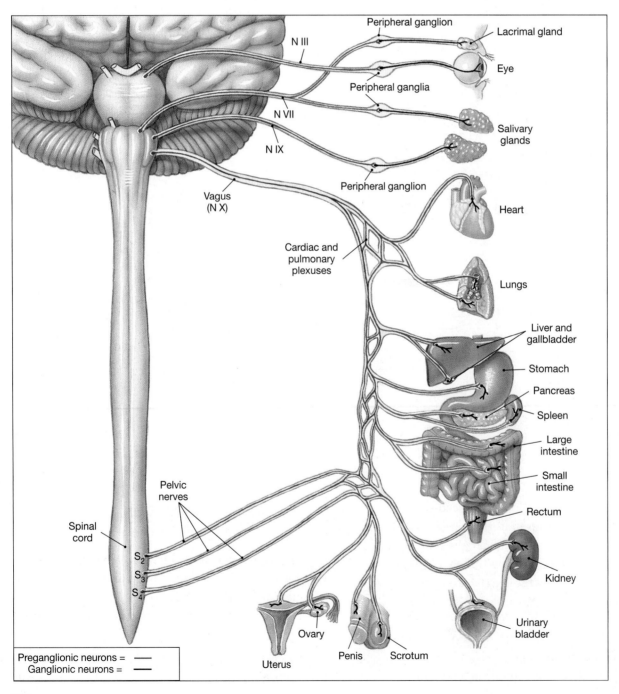

● *Figure 8-35* **The Parasympathetic Division**
The distribution of parasympathetic fibers is the same on both sides of the body.

smooth muscle activity of the digestive tract, stimulates defecation and urination, constricts respiratory passageways, and reduces the heart rate and the force of cardiac contractions. These functions center on relaxation, food processing, and energy absorption. Stimulation of the parasympathetic division leads to a general increase in the nutrient content of the blood. Cells throughout the body respond to this increase by absorbing nutrients and using them to support growth and the storage of energy reserves. The effects of parasympathetic stimulation are usually brief and are restricted to specific organs and sites.

RELATIONSHIPS BETWEEN THE SYMPATHETIC AND PARASYMPATHETIC DIVISIONS

The sympathetic division has widespread impact, reaching visceral and somatic structures throughout the body, whereas the parasympathetic division innervates only visceral structures serviced by the cranial nerves or lying within the abdominopelvic cavity. Although some organs are innervated by one division or the other, most vital organs receive **dual innervation**—that is, instructions from both autonomic divisions. Where dual innervation exists, the two divisions often have opposing effects. Table 8-5 provides examples of different organs and the effects of either single or dual innervation.

CONCEPT CHECK QUESTIONS
Answers on page 275

❶ While out for a brisk walk, Megan is suddenly confronted by an angry dog. Which division of the ANS is responsible for the physiological changes that occur as she turns and runs from the animal?

❷ Why is the parasympathetic division of the ANS sometimes referred to as the anabolic system?

❸ What effect would loss of sympathetic stimulation have on blood flow to a tissue?

❹ What physiological changes would you expect to observe in a patient who is about to undergo a root canal procedure and who is quite anxious about it?

TABLE 8-5 *The Effects of the Sympathetic and Parasympathetic Divisions of the ANS on Various Body Structures*

STRUCTURE	SYMPATHETIC EFFECTS	PARASYMPATHETIC EFFECTS
EYE	Dilation of pupil Focusing for distance vision	Constriction of pupil Focusing for near vision
Tear Glands	None (not innervated)	Secretion
SKIN		
Sweat glands	Increases secretion	None (not innervated)
Arrector pili muscles	Contraction, erection of hairs	None (not innervated)
CARDIOVASCULAR SYSTEM		
Blood vessels	Vasoconstriction and vasodilation	None (not innervated)
Heart	Increases heart rate, force of contraction, and blood pressure	Decreases heart rate, force of contraction, and blood pressure
ADRENAL GLANDS	Secretion of epinephrine and norepinephrine by adrenal medullae	None (not innervated)
RESPIRATORY SYSTEM		
Airways	Increases diameter	Decreases diameter
Respiratory rate	Increases rate	Decreases rate
DIGESTIVE SYSTEM		
General level of activity	Decreases activity	Increases activity
Liver	Glycogen breakdown, glucose synthesis and release	Glycogen synthesis
SKELETAL MUSCLES	Increases force of contraction, glycogen breakdown	None (not innervated)
ADIPOSE TISSUE	Lipid breakdown, fatty acid release	None (not innervated)
URINARY SYSTEM		
Kidneys	Decreases urine production	Increases urine production
Urinary bladder	Constricts sphincter, relaxes urinary bladder	Tenses urinary bladder, relaxes sphincter to eliminate urine
REPRODUCTIVE SYSTEM	Increased glandular secretions; ejaculation in males	Erection of penis (males) or clitoris (females)

8 THE NERVOUS SYSTEM

The Nervous System • Cellular Organization in Neural Tissue • Neuron Function • Neural Communication • The Central Nervous System

Aging and the Nervous System

Age-related anatomical and physiological changes in the nervous system begin shortly after maturity (probably by age 30) and accumulate over time. Although an estimated 85 percent of individuals above age 65 lead relatively normal lives, there are noticeable changes in mental performance and CNS functioning. Common age-related anatomical changes include:

- *A reduction in brain size and weight.* This results primarily from a decrease in the volume of the cerebral cortex. The brains of elderly individuals have narrower gyri and wider sulci than those of young persons.

- *A reduction in the number of neurons.* Brain shrinkage has been linked to a loss of cortical neurons. Evidence exists that the loss of neurons does not occur (at least to the same degree) in brain stem nuclei.

- *A decrease in blood flow to the brain.* With age, fatty deposits gradually accumulate in the walls of blood vessels, and these deposits reduce the rate of blood flow through arteries. (This process, called *arteriosclerosis*, affects arteries throughout the body; it is discussed further in Chapter 13.) The reduction in blood flow may not cause a cerebral crisis, but it does increase the chances that the individual will suffer a stroke.

- *Changes in synaptic organization of the brain.* The number of dendritic branchings and interconnections appears to decrease. Synaptic connections are lost, and the rate of neurotransmitter production declines.

- *Intracellular and extracellular changes in CNS neurons.* Many neurons in the brain accumulate abnormal intracellular deposits of pigments or abnormal proteins that have no known function. The significance of these abnormalities remains to be determined. There is evidence that these changes occur in all aging brains, but when present in excess they seem to be associated with clinical abnormalities.

These anatomical changes are linked to impaired neural function. Memory consolidation, the conversion of short-term memory to long-term memory (such as repeating a telephone number), often becomes more difficult. Other memories, especially those of the recent past, also become harder to access. The sensory systems of the elderly, notably hearing, balance, vision, smell, and taste, become less acute. Light must be brighter, sounds louder, and smells stronger before they are perceived. Reaction times are slowed, and reflexes—even some withdrawal reflexes—weaken or disappear. The precision of motor control decreases, so it takes longer to perform a given motor pattern than it did 20 years earlier.

For roughly 85 percent of the elderly, these changes do not interfere with their abilities to function. But for as yet unknown reasons, many individuals become incapacitated by progressive CNS changes. The most common of such incapacitating conditions is *Alzheimer's disease.*

CONCEPT CHECK QUESTIONS
Answers on page 275

❶ Brain shrinkage is an age-related change of the CNS. What is its major cause?

Integration with Other Systems

The relationships between the nervous system and other organ systems are shown in Figure 8-36●. Many of these interactions will be explored in detail in later chapters.

CONCEPT CHECK QUESTIONS
Answers on page 275

❶ How are the nervous system and the integumentary system functionally related?

✚ CLINICAL NOTE *Alzheimer's Disease*

The most common age-related incapacitating condition of the central nervous system is **Alzheimer's disease**, a progressive disorder characterized by the loss of higher cerebral functions. It is the most common cause of **senile dementia**, or *senility*. The first symptoms usually appear at 50 to 60 years of age, although the disease occasionally affects younger individuals. Alzheimer's disease has widespread impact on the elderly; an estimated 2 million people in the United States, including roughly 15 percent of those over age 65, have some form of the condition, and it causes approximately 100,000 deaths each year.

In its characteristic form, Alzheimer's disease produces a gradual deterioration of mental organization. The afflicted individual loses memories, verbal and reading skills, and emotional control. As memory losses continue to accumulate, problems become more severe. The affected person may forget relatives, a home address, or how to use the telephone. The loss of memory affects both intellectual and motor abilities, and a patient with severe Alzheimer's disease has difficulty performing even the simplest motor tasks. Although by this time individuals with the disease are relatively unconcerned about their mental state or motor abilities, the condition can have devastating emotional effects on the immediate family.

The Nervous System

For All Systems

Monitors pressure, pain, and temperature; adjusts tissue blood flow patterns

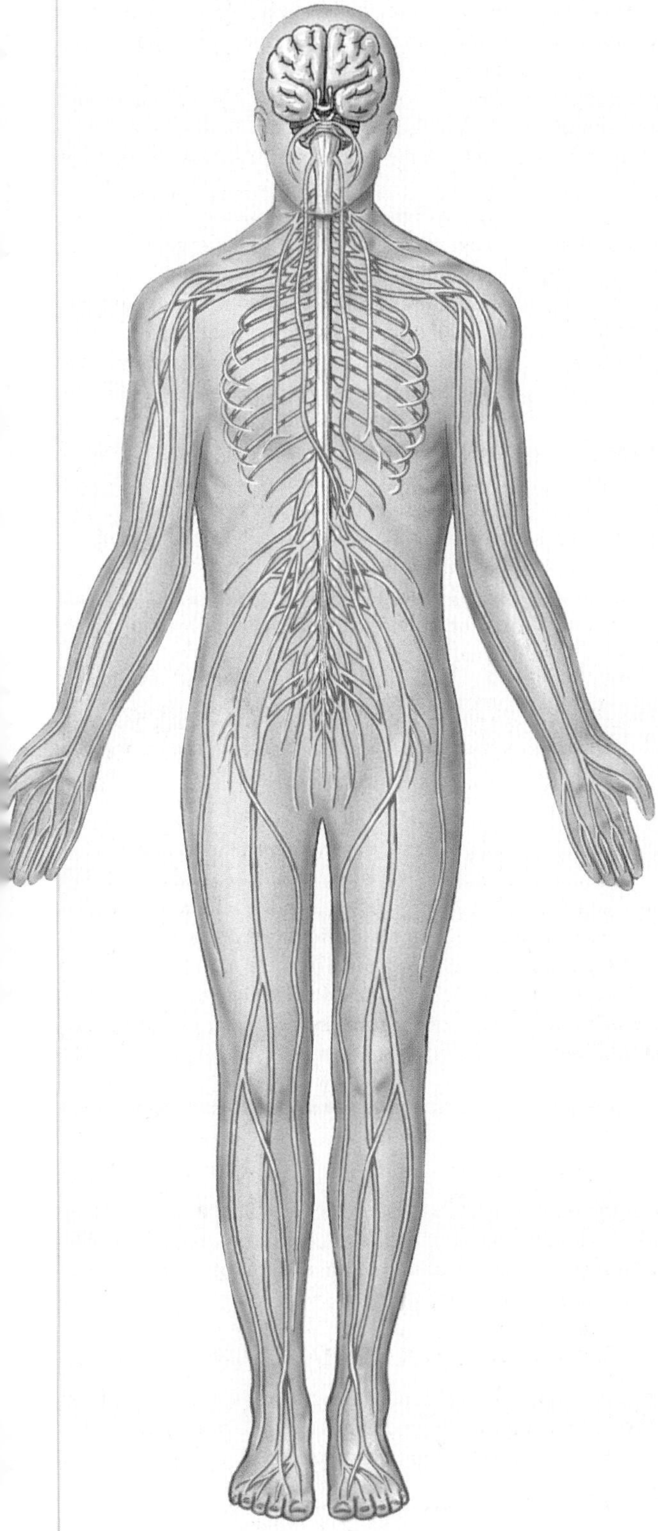

The Integumentary System

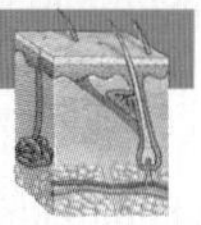

- Provides sensations of touch, pressure, pain, vibration, and temperature; hair provides some protection and insulation for skull and brain; protects peripheral nerves
- Controls contraction of arrector pili muscles and secretion of sweat glands

The Skeletal System

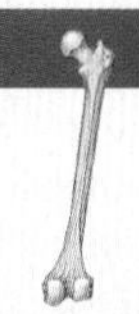

- Provides calcium for neural function; protects brain and spinal cord
- Controls skeletal muscle contractions that produce bone thickening and maintenance and determine bone position

The Muscular System

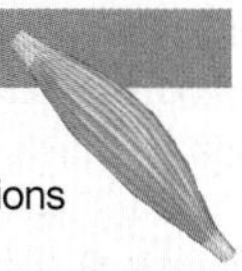

- Facial muscles express emotional state; intrinsic laryngeal muscles permit communication; muscle spindles provide proprioceptive sensations
- Controls skeletal muscle contractions; coordinates respiratory and cardiovascular activities

The Endocrine System

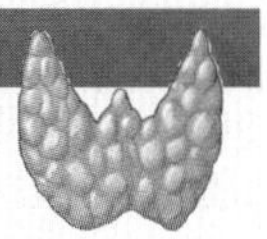

- Many hormones affect CNS neural metabolism; reproductive hormones and thyroid hormone influence CNS development
- Controls pituitary gland and many other endocrine organs; secretes ADH and oxytocin

The Cardiovascular System

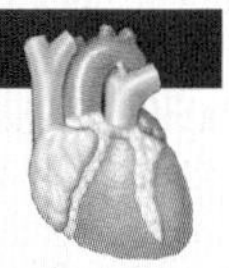

- Endothelial cells of capillaries maintain blood-brain barrier when stimulated by astrocytes; blood vessels (with ependymal cells) produce CSF
- Modifies heart rate and blood pressure; astrocytes stimulate maintenance of blood-brain barrier

The Lymphatic System

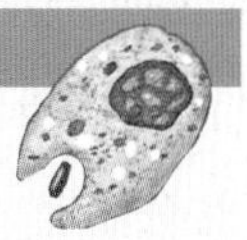

- Defends against infection and assists in tissue repairs
- Release of neurotransmitters and hormones affect sensitivity of immune response

The Respiratory System

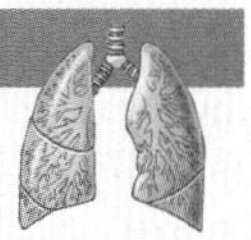

- Provides oxygen and eliminates carbon dioxide
- Controls pace and depth of respiration

The Digestive System

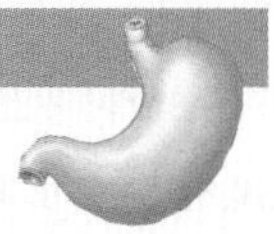

- Provides nutrients for energy production and neurotransmitter synthesis
- Regulates digestive tract movement and secretion

The Urinary System

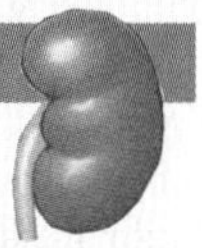

- Eliminates metabolic wastes; regulates body fluid pH and electrolyte concentrations
- Adjusts renal blood pressure; controls urination

The Reproductive System

- Sex hormones affect CNS development and sexual behaviors
- Controls sexual behaviors and sexual function

● *Figure 8-36* **Functional Relationships Between the Nervous System and Other Systems**

8 THE NERVOUS SYSTEM

The Nervous System • Cellular Organization in Neural Tissue • Neuron Function • Neural Communication • The Central Nervous System

Related Clinical Terms

aphasia: A disorder affecting the ability to speak or read.

ataxia: A disturbance of balance that in severe cases leaves the individual unable to stand without assistance. It is caused by problems affecting the cerebellum.

amyotrophic lateral sclerosis (ALS): A progressive, degenerative disorder affecting motor neurons of the spinal cord, brain stem, and cerebral hemispheres.

cerebrovascular accident (CVA), or *stroke:* A condition in which the blood supply to a portion of the brain is blocked off.

dementia: A stable, chronic state of consciousness characterized by deficits in memory, spatial orientation, language, or personality.

demyelination: The destruction of the myelin sheaths around axons in the CNS and PNS.

diphtheria (dif-THĒ-rē-uh): A disease that results from a bacterial infection of the respiratory tract. Among other effects, the bacterial toxins damage Schwann cells and cause PNS demyelination.

epidural block: The injection of anesthetic into the epidural space of the spinal cord to eliminate sensory and motor innervation by spinal nerves in the area of injection.

Hansen's disease *(leprosy):* A bacterial infection that begins in sensory nerves of the skin and gradually progresses to a motor paralysis of the same regions.

Huntington's disease: An inherited disease marked by a progressive deterioration of mental abilities and by motor disturbances.

meningitis: An inflammation of the meninges involving the spinal cord (*spinal meningitis*) and/or brain (*cerebral meningitis*); generally caused by bacterial or viral pathogens.

multiple sclerosis (skler-Ō-sis)(**MS**): A disease marked by recurrent incidents of demyelination affecting axons in the optic nerve, brain, and/or spinal cord.

myelography: A diagnostic procedure in which a radiopaque dye is introduced into the cerebrospinal fluid to obtain an X-ray of the spinal cord and cauda equina.

neurology: The branch of medicine that deals with the study of the nervous system and its disorders.

neurotoxin: A compound that disrupts normal nervous system function by interfering with the generation or propagation of action potentials. Examples include *tetrodotoxin (TTX), saxitoxin (STX), paralytic shellfish poisoning (PSP),* and *ciguatoxin (CTX).*

paraplegia: Paralysis involving a loss of motor control of the lower but not the upper limbs.

paresthesia: An abnormal tingling sensation, usually described as "pins and needles," that accompanies sensory return after a temporary palsy.

Parkinson's disease: A condition characterized by a pronounced increase in muscle tone, resulting from the excitation of neurons in the basal nuclei.

quadriplegia: Paralysis involving the loss of sensation and motor control of the upper and lower limbs.

sciatica (sī-AT-i-kuh): The painful result of compression of the roots of the sciatic nerve.

seizure: A temporary disorder of cerebral function, accompanied by abnormal movements, unusual sensations, and/or inappropriate behavior.

shingles: A condition caused by the infection of neurons in dorsal root ganglia by the virus *Herpes varicella-zoster.* The primary symptom is a painful rash along the sensory distribution of the affected spinal nerves.

spinal shock: A period of depressed sensory and motor function following any severe injury to the spinal cord.

CHAPTER REVIEW

Key Terms

Summary Outline

1. Two organ systems, the nervous and endocrine systems, coordinate organ system activity. The nervous system provides swift but brief responses to stimuli; the endocrine system adjusts metabolic operations and directs long-term changes.

1. The **nervous system** includes all the neural tissue in the body. Its anatomical divisions include the **central nervous system** (**CNS**) (the brain and spinal cord) and the **peripheral nervous system** (**PNS**) (all of the neural tissue outside the CNS).

2. Functionally, it can be divided into an **afferent division**, which brings sensory information to the CNS, and an **efferent division**, which carries motor commands to muscles and glands. The efferent division

includes the **somatic nervous system** (**SNS**) (voluntary control over skeletal muscle contractions) and the **autonomic nervous system** (**ANS**) (automatic, involuntary regulation of smooth muscle, cardiac muscle, and glandular activity). *(Figure 8-1)*

1. There are two types of cells in neural tissue: **neurons**, which are responsible for information transfer and processing, and **neuroglia**, or *glial cells*, which provide a supporting framework and act as phagocytes.

2. **Sensory neurons** form the afferent division of the PNS and deliver information to the CNS. **Motor neurons** stimulate or modify the activity of a peripheral tissue, organ, or organ system. **Interneurons (asso-**

ciation neurons) may be located between sensory and motor neurons; they analyze sensory inputs and coordinate motor outputs.

3. A typical neuron has a cell body, an **axon**, and several branching **dendrites**, and synaptic terminals. *(Figure 8-2)*

4. Neurons may be described as **unipolar, bipolar,** or **multipolar**. *(Figure 8-3)*

Neuroglia 229

5. The four types of neuroglia in the CNS are (1) **astrocytes**, which are the largest and most numerous; (2) **oligodendrocytes**, which are responsible for the **myelination** of CNS axons; (3) **microglia**, phagocytic white blood cells; and (4) **ependymal cells**, with functions related to the *cerebrospinal fluid* (CSF). *(Figure 8-4)*

6. Nerve cell bodies in the PNS are clustered into *ganglia* (singular *ganglion*). Their axons are covered by myelin wrappings of **Schwann cells**. *(Figure 8-5)*

Anatomical Organization 230

7. In the CNS, a collection of neuron cell bodies that share a particular function is called a **center**. A center with a discrete anatomical boundary is called a **nucleus**. Portions of the brain surface are covered by a thick layer of gray matter called the **neural cortex**. The white matter of the CNS contains bundles of axons, or **tracts**, that share common origins, destinations, and functions. Tracts in the spinal cord form larger group, called *columns*. *(Figure 8-6)*

8. **Sensory** *(ascending)* **pathways** carry information from peripheral sensory receptors to processing centers in the brain; **motor** *(descending)* **pathways** extend from CNS centers concerned with motor control to the associated skeletal muscles. *(Figure 8-6)*

NEURON FUNCTION 231

The Membrane Potential 231

1. The **resting potential**, or **membrane potential** of an undisturbed nerve cell, is due to a balance between the rate of sodium ion entry and potassium ion loss and to the sodium-potassium exchange pump. Any stimulus that affects this balance will alter the resting potential of the cell. *(Figure 8-7)*

2. An **action potential** appears when the membrane depolarizes to a level known as the **threshold**. The steps involved include: opening of sodium channels and membrane depolarization; the closing of sodium channels and opening of potassium channels; and the return to normal permeability. *(Figure 8-8)*

Propagation of an Action Potential 234

3. In **continuous propagation**, an action potential spreads across the entire excitable membrane surface in a series of small steps. During **saltatory propagation**, the action potential appears to leap from node to node, skipping the intervening membrane surface. *(Figure 8-9)*

NEURAL COMMUNICATION 235

1. A **synapse** is a site where intercellular communication occurs through the release of chemicals called **neurotransmitters**. A synapse where neurons communicate with other cell types is a **neuroeffector junction**.

Structure of a Synapse 236

2. Neural communication moves from the **presynaptic neuron** to the **postsynaptic neuron** over the **synaptic cleft**. *(Figure 8-10)*

Synaptic Function and Neurotransmitters...............236

3. **Cholinergic synapses** release the neurotransmitter **acetylcholine (ACh)**. ACh is broken down in the synaptic cleft by the enzyme **acetylcholinesterase (AChE)**. *(Figure 8-11; Table 8-1)*

Neuronal Pools...238

4. The roughly 20 billion interneurons are organized into **neuronal pools** (groups of interconnected neurons with specific functions). **Divergence** is the spread of information from one neuron to several neurons or from one neuronal pool to several pools. In **convergence**, several neurons synapse on the same postsynaptic neuron. *(Figure 8-12)*

THE CENTRAL NERVOUS SYSTEM 238

1. The CNS is made up of the *spinal cord* and *brain*.

The Meninges 238

2. Special covering membranes, the **meninges**, protect and support the spinal cord and delicate brain. The *cranial meninges* (the dura mater, arachnoid, and pia mater) are continuous with those of the spinal cord, the *spinal meninges*. *(Figure 8-14)*

3. The **dura mater** covers the brain and spinal cord. The **epidural space** separates the spinal dura mater from the walls of the vertebral canal. The subarachnoid space of the arachnoid layer contains **cerebrospinal fluid** (the CSF), which acts as a shock absorber and a diffusion medium for dissolved gases, nutrients, chemical messengers, and waste products. The **pia mater** is bound to the underlying neural tissue.

The Spinal Cord 240

4. In addition to relaying information to and from the brain, the spinal cord integrates and processes information on its own.

5. The spinal cord has 31 segments, each associated with a pair of **dorsal root ganglia** and their **dorsal roots** and a pair of **ventral roots**. *(Figures 8-14; 8-15b)*

6. The white matter contains myelinated and unmyelinated axons; the gray matter contains cell bodies of neurons and glial cells. The projections of gray matter toward the outer surface of the spinal cord are called **horns**. *(Figure 8-15a)*

The Brain 243

7. There are six regions in the adult brain: cerebrum, diencephalon, midbrain, cerebellum, pons, and medulla oblongata. *(Figure 8-16)*

8. The central passageway of the brain expands to form four chambers called **ventricles**. Cerebrospinal fluid continuously circulates from the ventricles and central canal of the spinal cord into the subarachnoid space of the meninges that surround the CNS. *(Figures 8-17, 8-18)*

9. Conscious thought, intellectual functions, memory, and complex involuntary motor patterns originate in the **cerebrum** *(Figure 8-17)*

10. The cortical surface of the cerebrum contains **gyri** (elevated ridges) separated by **sulci** (shallow depressions) or deeper grooves (**fissures**). The **longitudinal fissure** separates the two **cerebral hemispheres**. The **central sulcus** marks the boundary between the **frontal lobe** and the **parietal lobe**. Other sulci form the boundaries of the **temporal lobe** and the **occipital lobe**. *(Figures 8-16; 8-19)*

11. Each cerebral hemisphere receives sensory information and generates motor commands that concern the opposite side of the body. The **primary motor cortex** of the **precentral gyrus** directs voluntary movements. The **primary sensory cortex** of the **postcentral gyrus** receives somatic sensory information from touch, pressure, pain, and temperature receptors. **Association areas**, such as the **visual association area** and **premotor cortex** (motor association area), control our ability to understand sensory information and coordinate a motor response. *(Figure 8-19)*

12. The left hemisphere is usually the *categorical hemisphere*, which contains the general interpretive and speech centers and is responsible for language-based skills. The right hemisphere or *representational hemisphere*, is concerned with spatial relationships and analyses. *(Figure 8-20)*

13. An **electroencephalogram (EEG)** is a printed record of **brain waves**. *(Figure 8-21)*

14. The **basal nuclei** lie within the central white matter and aid in the coordination of learned movement patterns and other somatic motor activities. *(Figure 8-22)*

15. The **limbic system** includes the *hippocampus*, which is involved in memory and learning, and the *mamillary bodies*, which control reflex

8 THE NERVOUS SYSTEM

The Nervous System • Cellular Organization in Neural Tissue • Neuron Function • Neural Communication • The Central Nervous System

movements associated with eating. The functions of the limbic system involve emotional states and related behavioral drives. *(Figure 8-23)*

16. The **diencephalon** provides the switching and relay centers necessary to integrate the conscious and unconscious sensory and motor pathways. It contains the *pineal gland* and *choroid plexus* (a vascular network that produces cerebrospinal fluid), the **thalamus**, and the **hypothalamus**. *(Figure 8-24)*

17. The thalamus is the final relay point for ascending sensory information. Only a small portion of the arriving sensory information is passed to the cerebral cortex, the rest is relayed to the basal nuclei and centers in the brain stem. *(Figures 8-16c; 8-24)*

18. The hypothalamus contains important control and integrative centers. It can produce emotions and behavioral drives, coordinate activities of the nervous and endocrine systems, secrete hormones, coordinate voluntary and autonomic functions, and regulate body temperature.

19. Three regions make up the **brain stem**. (1) The **midbrain** processes visual and auditory information and generates involuntary somatic motor responses. (2) The **pons** connects the cerebellum to the brain stem and is involved with somatic and visceral motor control. (3) The spinal cord connects to the brain at the **medulla oblongata**, which relays sensory information and regulates autonomic functions. *(Figures 8-16c; 8-24)*

20. The **cerebellum** oversees the body's postural muscles and programs and tunes voluntary and involuntary movements. The **cerebellar peduncles** are tracts that link the cerebellum with the brain stem, cerebrum, and spinal cord. *(Figures 8-16; 8-24)*

21. The medulla oblongata connects the brain to the spinal cord. Its nuclei relay information from the spinal cord and brain stem to the cerebral cortex. Its reflex centers, including the **cardiovascular centers** and the **respiratory rhythmicity centers**, control or adjust the activities of one or more peripheral systems. *(Figure 8-25)*

THE PERIPHERAL NERVOUS SYSTEM 254

1. The **peripheral nervous system (PNS)** links the central nervous system (CNS) with the rest of the body; all sensory information and motor commands are carried by axons of the PNS. The sensory and motor axons are bundled together into **peripheral nerves**, or nerves, and clusters of cell bodies, or **ganglia**.

2. The PNS includes cranial nerves and spinal nerves.

The Cranial Nerves 254

3. There are 12 pairs of **cranial nerves**, which connect to the brain, not to the spinal cord. *(Figure 8-25; Table 8-2)*

4. The **olfactory nerves (N I)** carry sensory information responsible for the sense of smell.

5. The **optic nerves (N II)** carry visual information from special sensory receptors in the eyes.

6. The **oculomotor nerves (N III)** are the primary sources of innervation for four of the six muscles that move the eyeball.

7. The **trochlear nerves (N IV)**, the smallest cranial nerves, innervate the superior oblique muscles of the eyes.

8. The **trigeminal nerves (N V)**, the largest cranial nerves, are mixed nerves with ophthalmic, maxillary, and mandibular branches.

9. The **abducens nerves (N VI)** innervate the sixth extrinsic eye muscle, the lateral rectus.

10. The **facial nerves (N VII)** are mixed nerves that control muscles of the scalp and face. They provide pressure sensations over the face and receive taste information from the tongue.

11. The **vestibulocochlear nerves (N VIII)** contain the vestibular nerves, which monitor sensations of balance, position, and movement, and the cochlear nerves, which monitor hearing receptors.

12. The **glossopharyngeal nerves (N IX)** are mixed nerves that innervate the tongue and pharynx and control swallowing.

13. The **vagus nerves (N X)** are mixed nerves that are vital to the autonomic control of visceral function and have a variety of motor components.

14. The **accessory nerves (N XI)** have a medullary branch, which innervates voluntary swallowing muscles of the soft palate and pharynx, and a spinal branch, which controls muscles associated with the pectoral girdle.

15. The **hypoglossal nerves (N XII)** provide voluntary control over tongue movements.

The Spinal Nerves 257

16. There are 31 pairs of **spinal nerves**: 8 cervical, 12 thoracic, 5 lumbar, 5 sacral, and 1 coccygeal. Each pair monitors a region of the body surface known as a **dermatome** *(Figures 8-26; 8-27)*

Nerve Plexuses 258

17. A complex, interwoven network of nerves is called a **nerve plexus**. The three large plexuses are the **cervical plexus**, the **brachial plexus**, and the **lumbosacral plexus**. The latter can be divided into the lumbar plexus and the sacral plexus. *(Figure 8-26; Table 8-3)*

Reflexes 258

18. A **reflex** is an automatic involuntary motor response to a specific stimulus.

19. A **reflex arc** is the "wiring" of a single reflex. There are five steps involved in a reflex arc: (1) arrival of a stimulus and activation of a receptor, (2) activation of a sensory neuron, (3) information processing, (4) activation of a motor neuron, and (5) response by an effector. *(Figure 8-28)*

20. A **monosynaptic reflex** is the simplest reflex arc, in which a sensory neuron synapses directly on a motor neuron that acts as the processing center. The **stretch reflex** is a monosynaptic reflex that automatically regulates skeletal muscle length and muscle tone. The sensory receptors involved are **muscle spindles**. *(Figure 8-29)*

21. **Polysynaptic reflexes**, which have at least one interneuron between the sensory afferent and the motor efferent, have a longer delay between stimulus and response than does a monosynaptic synapse. Polysynaptic reflexes can also produce more complicated responses. The **flexor reflex** is a withdrawal reflex affecting the muscles of a limb. *(Figure 8-30)*

22. The brain can facilitate or inhibit reflex motor patterns based in the spinal cord.

Sensory and Motor Pathways 260

23. The essential communication between the CNS and PNS occurs over pathways that relay sensory information and motor commands. *(Table 8-4)*

24. A **sensation** arrives in the form of an action potential in an afferent fiber. The **posterior column pathway** carries fine touch, pressure, and proprioceptive sensations. The axons ascend within this pathway and synapse with neurons in the medulla oblongata. These axons then cross over and travel on to the thalamus. The thalamus sorts the sensations according to the region of the body involved and projects them to specific regions of the primary sensory cortex. *(Figure 8-31; Table 8-4)*

25. The **corticospinal pathway** provides conscious skeletal muscle control. The **medial** and **lateral pathways** generally exert subconscious control over skeletal muscles. *(Figure 8-32; Table 8-4)*

THE AUTONOMIC NERVOUS SYSTEM 263

1. The autonomic nervous system (ANS) coordinates cardiovascular, respiratory, digestive, excretory, and reproductive functions.

2. **Preganglionic neurons** in the CNS send axons to synapse on **ganglionic neurons** in **autonomic ganglia** outside the CNS. The axons of the ganglionic neurons (postganglionic fibers) innervate cardiac muscle, smooth muscles, glands, and adipose tissues. *(Figure 8-33)*

3. Preganglionic fibers from the thoracic and lumbar segments form the **sympathetic division** ("fight or flight" system) of the ANS. Preganglionic fibers leaving the brain and sacral segments form the **parasympathetic division** ("rest and repose" or "rest and digest" system).

The Sympathetic Division 264

4. The sympathetic division consists of preganglionic neurons between segments T_1 and L_2, ganglionic neurons in ganglia near the vertebral column, and specialized neurons in the adrenal gland. Sympathetic ganglia are paired **sympathetic chain ganglia** or unpaired **collateral ganglia**. (Figure 8-34)

5. Preganglionic fibers entering the adrenal glands synapse within the **adrenal medullae**. During sympathetic activation these endocrine organs secrete epinephrine and norepinephrine into the bloodstream.

6. In crises, the entire division responds, producing increased alertness, a feeling of energy and euphoria, increased cardiovascular and respiratory activity, and elevation in muscle tone.

The Parasympathetic Division 265

7. The parasympathetic division includes preganglionic neurons in the brain stem and sacral segments of the spinal cord and ganglionic neurons in peripheral ganglia located within or next to target organs. Preganglionic fibers leaving the sacral segments form **pelvic nerves**. (Figure 8-35)

8. The effects produced by the parasympathetic division center on relaxation, food processing, and energy absorption; they are usually brief and restricted to specific sites.

Relationships between the Sympathetic and Parasympathetic Divisions................................267

9. The sympathetic division has widespread impact, reaching visceral and somatic structures throughout the body. The parasympathetic division innervates only visceral structures serviced by cranial nerves or lying within the abdominopelvic cavity. Organs with **dual innervation** receive instructions from both divisions. (Table 8-5)

AGING AND THE NERVOUS SYSTEM 268

1. Age-related changes in the nervous system include (1) a reduction of brain size and weight, (2) a reduction of the number of neurons, (3) decreased blood flow to the brain, (4) changes in synaptic organization of the brain, and (5) intracellular and extracellular changes in CNS neurons.

INTEGRATION WITH OTHER SYSTEMS 268

1. The nervous system monitors pressure, pain, and temperature and adjusts tissue blood flow for all systems. (Figure 8-36)

Review Questions

Level 1: Reviewing Facts and Terms

Match each item in column A with the most closely related item in column B. Use letters for answers in the spaces provided.

COLUMN A

____ 1. neuroglia
____ 2. autonomic nervous system
____ 3. sensory neurons
____ 4. dual innervation
____ 5. ganglia
____ 6. oligodendrocytes
____ 7. ascending tracts
____ 8. descending tracts
____ 9. saltatory propagation
____ 10. continuous propagation
____ 11. dura mater
____ 12. monosynaptic reflex
____ 13. sympathetic division
____ 14. cerebellum
____ 15. somatic nervous system
____ 16. hypothalamus
____ 17. medulla oblongata
____ 18. choroid plexus
____ 19. parasympathetic division
____ 20. motor neurons

COLUMN B

a. coat CNS axons with myelin
b. carry sensory information to the brain
c. occurs along unmyelinated axons
d. outermost covering of brain and spinal cord
e. production of CSF
f. supporting cells
g. controls smooth and cardiac muscle and glands
h. occurs along myelinated axons
i. link between nervous and endocrine systems
j. carry motor commands to spinal cord
k. efferent division of the PNS
l. controls contractions of skeletal muscles
m. masses of neuron cell bodies
n. connects the brain to the spinal cord
o. stretch reflex
p. afferent division of the PNS
q. maintains muscle tone and posture
r. "rest and repose"
s. opposing effects
t. "fight or flight"

21. Regulation by the nervous system provides:

   (a) relatively slow but long-lasting responses to stimuli

   (b) swift, long-lasting responses to stimuli

   (c) swift but brief responses to stimuli

   (d) relatively slow, short-lived responses to stimuli

22. All the motor neurons that control skeletal muscles are

   (a) multipolar neurons

   (b) myelinated bipolar neurons

   (c) unipolar, unmyelinated sensory neurons

   (d) proprioceptors

8 THE NERVOUS SYSTEM

The Nervous System • Cellular Organization in Neural Tissue • Neuron Function • Neural Communication • The Central Nervous System

23. Depolarization of a neuron cell membrane will shift the membrane potential toward:
    (a) 0 mV
    (b) −70 mV
    (c) −90 mV
    (d) a, b, and c are correct

24. The structural and functional link between the cerebral hemispheres and the components of the brain stem is the:
    (a) neural cortex
    (b) medulla oblongata
    (c) midbrain
    (d) diencephalon

25. The ventricles in the brain are filled with:
    (a) blood
    (b) cerebrospinal fluid
    (c) air
    (d) neural tissue

26. Reading, writing, and speaking are dependent on processing in the:
    (a) right cerebral hemisphere
    (b) left cerebral hemisphere
    (c) prefrontal cortex
    (d) postcentral gyrus

27. Establishment of emotional states and related behavioral drives are functions of the:
    (a) limbic system
    (b) pineal gland
    (c) mamillary bodies
    (d) thalamus

28. The final relay point for ascending sensory information that will be projected to the primary sensory cortex is the:
    (a) hypothalamus
    (b) thalamus
    (c) spinal cord
    (d) medulla oblongata

29. Spinal nerves are called mixed nerves because they:
    (a) are associated with a pair of dorsal root ganglia
    (b) exit at intervertebral foramina
    (c) contain sensory and motor fibers
    (d) are associated with a pair of dorsal and ventral roots

30. There is always a synapse between the CNS and the peripheral effector in the:
    (a) ANS
    (b) SNS
    (c) reflex arc
    (d) a, b, and c are correct

31. Approximately 75 percent of parasympathetic outflow is provided by the:
    (a) pelvic nerves
    (b) sciatic nerves
    (c) glossopharyngeal nerves
    (d) vagus nerves

32. State the all-or-none principle of action potentials.

33. Using the mnemonic device "Oh, Once One Takes The Anatomy Final, Very Good Vacations Are Heavenly," list the 12 pairs of cranial nerves and their functions.

34. How does the emergence of sympathetic fibers from the spinal cord differ from the emergence of parasympathetic fibers?

Level 2: Reviewing Concepts

35. A graded potential:
    (a) decreases with distance from the point of stimulation
    (b) spreads passively because of local currents
    (c) may involve either depolarization or hyperpolarization
    (d) a, b, and c are correct

36. The loss of positive ions from the interior of a neuron produces:
    (a) depolarization
    (b) threshold
    (c) hyperpolarization
    (d) an action potential

37. What would happen if the ventral root of a spinal nerve was damaged or transected?

38. Which major part of the brain is associated with respiratory and cardiac activity?

39. Why is response time in a monosynaptic reflex much faster than response time in a polysynaptic reflex?

40. Compare the general effects of the sympathetic and parasympathetic divisions of the ANS.

Level 3: Critical Thinking and Clinical Applications

41. If neurons in the central nervous system lack centrioles and are unable to divide, how can a person develop brain cancer?

42. A police officer has just stopped Bill on suspicion of driving while intoxicated. The officer asks Bill to walk the yellow line on the road and then asks him to place the tip of his index finger on the tip of his nose. How would these activities indicate Bill's level of sobriety? Which part of the brain is being tested by these activities?

43. In some severe cases of stomach ulcers, the branches of the vagus nerve (N X) that lead to the stomach are surgically severed. How might this procedure control the ulcers?

44. Improper use of crutches can produce a condition known as crutch paralysis, which is characterized by a lack of response by the extensor muscles of the arm and a condition known as wrist drop. Which nerve is involved?

45. While playing football, Ramon is tackled hard and suffers an injury to his left leg. As he tries to get up, he finds that he cannot flex his left hip or extend the knee. Which nerve is damaged, and how would this damage affect sensory perception in the left leg?

Answers to Concept Check Questions

Page 231

1. The afferent division of the nervous system is composed of nerves that carry sensory information to the brain and spinal cord. Damage to this division would interfere with a person's ability to experience a variety of sensory stimuli.

2. Sensory neurons of the peripheral nervous system are usually unipolar; thus this tissue is most likely associated with a sensory organ.

3. Microglial cells are small phagocytic cells that are found in increased number in damaged and diseased areas of the CNS.

Page 235

1. Depolarization of the neuron membrane involves the opening of the sodium channels and the rapid influx of sodium ions into the cell. If the sodium channels were blocked, a neuron would not be able to depolarize and propagate an action potential.

2. Action potentials are propagated along myelinated axons by saltatory propagation at speeds much higher than those along unmyelinated axons. An axon with a propagation speed of 10 m/sec must be myelinated.

Page 238

1. A neurotransmitter that opens the potassium channels but not the sodium channels would cause a hyperpolarization at the postsynaptic membrane. The transmembrane potential would be greater and it would be more difficult to bring the membrane to threshold.

2. When an action potential reaches the presynaptic terminal of a cholinergic synapse, calcium channels are opened and the influx of calcium triggers the release of acetylcholine into the synapse to stimulate the next neuron. If the calcium channels were blocked, the acetylcholine would not be released and transmission across the synapse would cease.

3. This is an example of convergence. In this pattern of neural interaction, two neuronal pools synapse on the same motor neurons.

Page 243

1. The ventral root of spinal nerves is composed of visceral and somatic motor fibers. Damage to this root would interfere with motor function.

2. Since the polio virus would be located in the somatic motor neurons, we would find it in the anterior gray horns of the spinal cord, where the cell bodies of these neurons are located.

3. All spinal nerves are classified as mixed nerves because they contain both sensory and motor fibers.

Page 243

1. The six regions in the adult brain and their major functions are (1) the *cerebrum*: conscious thought processes; (2) the *diencephalon*: the thalamic portion contains relay and processing centers for sensory information, and the hypothalamic portion contain centers involved with emotions, autonomic function, and hormone production; (3) the *midbrain*: processes visual and auditory information and generates involuntary motor responses; (4) the *pons*: contains tracts and relay centers that connect the brain stem to the cerebellum; (5) the *medulla oblongata*: contains major centers concerned with the regulation of autonomic function, such as heart rate, blood pressure, respiration, and digestive activities; and (6) the *cerebellum*: adjusts voluntary and involuntary motor activities.

2. The pituitary gland is attached to the hypothalamus, or the floor of the diencephalon.

Page 251

1. Diffusion across the arachnoid granulations is the means by which cerebrospinal fluid reenters the bloodstream. If this process decreased, then excess fluid would start to accumulate in the ventricles and the volume of fluid in the ventricles would increase.

2. The primary motor cortex is located in the precentral gyrus of the frontal lobe of the cerebrum.

3. Damage to the temporal lobes of the cerebrum would interfere with the processing of olfactory (smell) and auditory (sound) sensations.

Page 253

1. All ascending sensory information other than olfactory passes through the thalamus before reaching our conscious awareness.

2. Changes in body temperature would stimulate the hypothalamus, a division of the diencephalon.

3. Even though the medulla oblongata is small, it contains many vital reflex centers, including those that control breathing and regulate the heart and blood pressure. Damage to the medulla oblongata can result in a cessation of breathing or life-threatening changes in heart rate and blood pressure.

Page 256

1. The abducens nerve (N VI) controls lateral movements of the eyes through the lateral rectus muscles. An individual with damage to this nerve would be unable to move his or her eyes laterally, or to the side.

2. The hypoglossal nerve (N XII) controls the voluntary muscles of the tongue.

Page 260

1. Physicians use the sensitivity of stretch reflexes, such as the knee jerk reflex, or patellar reflex, to test the general condition of the spinal cord, peripheral nerves, and muscles.

2. In a monosynaptic reflex, a sensory neuron synapses directly on a motor neuron and produces a rapid, stereotyped movement. More complicated responses occur with polysynaptic reflexes because the interneurons between the sensory and motor neurons may control several muscle groups simultaneously. In addition, some interneurons may stimulate a muscle group or groups, while others may inhibit other muscle groups.

3. A positive Babinski reflex is abnormal for an adult and indicates possible damage of descending tracts in the spinal cord.

Page 263

1. A tract within the posterior column of the spinal cord is responsible for carrying information about touch and pressure from the lower part of the body to the brain.

2. The anatomical basis of opposite-side motor control is that crossing-over occurs, so the pyramidal motor fibers innervate lower motor neurons on the opposite side of the body. Most of the axons of the pyramidal cells cross over to opposite sides in the medulla oblongata.

3. The superior portion of the motor cortex exercises control over the hand, arm, and upper portion of the leg. An injury to this area would affect the ability to control the muscles in those regions of the body.

Page 267

1. The sympathetic division of the autonomic nervous system is responsible for the physiological changes that occur in response to stress and increased activity.

2. The parasympathetic division is sometimes referred to as the anabolic system because parasympathetic stimulation leads to a general increase in the nutrient content of the blood. Cells throughout the body respond to the increase by absorbing the nutrients and using them to support growth and other anabolic activities.

3. Since most blood vessels receive sympathetic stimulation, a decrease in sympathetic stimulation would lead to a relaxation of the muscles in the walls of the vessels and vasodilation (an increase in vessel diameter). These in turn would result in increased blood flow to the tissue.

4. A patient who is anxious about an impending root canal would probably exhibit some or all of the following changes: a dry mouth, increased heart rate, increased blood pressure, increased rate of breathing, cold sweats, an urge to urinate or defecate, change in motility of the digestive tract ("butterflies" in the stomach), and dilated pupils. These changes would be the result of anxiety or stress causing an increase in sympathetic stimulation.

Page 268

1. A reduction in brain size and weight results from a decrease in the volume of the cerebral cortex due to the loss of cortical neurons.

Page 268

1. The nervous system controls contraction of the arrector pili muscles and secretion of the sweat glands within the integumentary system. The integumentary system provides the nervous system with the sensations of touch, pressure, pain, vibration, and temperature through sensory receptors. The integumentary system also protects peripheral nerves and hair provides some protection and insulation for the skull and brain.

The General and Special Senses

Vocabulary Development

akousis...................................hearing; *acoustic*
baro- ....pressure; *baroreceptors*
circa ....about; *circadian*
circum-.............around; *circumvallate papillae*
cochlea ....snail shell; *cochlea*
dies..day; *circadian*
emmetro-..........proper measure; *emmetropia*
incus ....anvil; *incus* (auditory ossicle)
iris..colored circle; *iris*
labyrinthos ....network of canals; *labyrinth*
lacrima...............................tear; *lacrimal gland*
lithos ....a stone; *otolith*
macula...................................spot; *macula lutea*
malleus ..a hammer; *malleus* (auditory ossicle)
myein ....to shut; *myopia*
noceo...............................hurt; *nociceptor*
olfacere ....to smell; *olfaction*
ops..eye; *myopia*
oto- ....ear; *otolith*
presbys ....old man; *presbyopia*
skleros...hard; *sclera*
stapes..............stirrup; *stapes* (auditory ossicle)
tectum ....roof; *tectorial membrane*
tympanon ....drum; *tympanum*
vallum...................wall; *circumvallate papillae*
vitreus ....glassy; *vitreous body*

*O*UR COMPREHENSION OF THE WORLD around us is based on information provided by our senses. For example, our ability to distinguish between sounds as different as a baby's cry or a mother's sigh begins with the movement of these yellow-tinted hairs projecting from sensory cells within the inner ear. This chapter examines the way our receptors provide sensory information and the way the sensory pathways distribute this information to provide us with our senses of smell, taste, vision, equilibrium, and hearing.

9 THE GENERAL AND SPECIAL SENSES

The General Senses • The Special Senses • Smell • Taste • Vision • Equilibrium and Hearing • Aging and the Senses • Chapter Review • MediaLab

OUR KNOWLEDGE OF THE WORLD around us is limited to those characteristics that stimulate our sensory receptors. Although we may not realize it, our picture of the environment is incomplete. Colors invisible to us guide insects to flowers, and sounds and smells we cannot detect are regular information to dolphins, dogs, and cats. Moreover, our senses are sometimes deceptive: In phantom limb pain, a person "feels" pain in a missing limb, and during an epileptic seizure, an individual may experience sights, sounds, or smells that have no physical basis.

All sensory information is picked up by *sensory receptors*, specialized cells or cell processes that monitor internal and external conditions. The simplest receptors are the dendrites of sensory neurons. The branching tips of these dendrites are called **free nerve endings**. Free nerve endings are sensitive to many types of stimuli. For example, the same free nerve endings in the skin may provide the sensation of pain in response to crushing, heat, or a cut. Other receptors are especially sensitive to one kind of stimulus. For example, a touch receptor is very sensitive to pressure but relatively insensitive to chemical stimuli; a taste receptor is sensitive to dissolved chemicals but insensitive to pressure. The most complex receptors, such as the visual receptors of the eye, are protected by accessory cells and layers of connective tissue. Not only are these receptor cells specialized to detect light, they are seldom exposed to any stimulus *except* light.

All sensory information arrives at the CNS in the form of action potentials in a sensory (afferent) fiber. In general, the stronger the stimulus, the higher the frequency of action potentials. The arriving information is called a **sensation**. When sensory information arrives at the CNS, it is routed according to the location and nature of the stimulus. For example, touch, pressure, pain, temperature, and taste sensations arrive at the primary sensory cortex; visual, auditory, and olfactory information reach the visual, auditory, and olfactory regions of the cortex, respectively. The conscious awareness of a sensation is called a *perception*. The CNS interprets the nature of sensory information entirely on the basis of the area of the brain stimulated; it cannot tell the difference between a "true" sensation and a "false" one. For instance, when rubbing your eyes, you may "see" flashes of light. Although the stimulus is mechanical rather than visual, any activity along the optic nerve is projected to the visual cortex and experienced as a visual perception.

Adaptation is a reduction in sensitivity in the presence of a constant stimulus. Familiar examples are stepping into a hot bath or jumping into a cold lake. Shortly afterward, neither temperature seems as extreme as it did initially. Adaptation reduces the amount of information arriving at the cerebral cortex. Most sensory information is routed to centers along the spinal cord or brain stem, potentially triggering involuntary reflexes, such as the withdrawal reflexes. ↪ p. 259 Only about 1 percent of the information provided by afferent fibers reaches the cerebral cortex and our conscious awareness.

Output from higher centers, however, can increase receptor sensitivity or facilitate transmission along a sensory pathway. For example, the *reticular activating system* in the midbrain helps focus attention and thus heightens or reduces awareness of arriving sensations. ↪ p. 253 This adjustment of sensitivity can occur under conscious or unconscious direction. When we "listen carefully," our sensitivity to and awareness of auditory stimuli increase. The reverse occurs when we enter a noisy factory or walk along a crowded city street, as we automatically "tune out" the high level of background noise.

The **general senses** are senses of temperature, pain, touch, pressure, vibration, and **proprioception** (body position). The receptors for the general senses are scattered throughout the body. The **special senses** are smell (**olfaction**), taste (**gustation**), vision, balance (**equilibrium**), and hearing. The receptors for the five special senses are concentrated within specific structures, the sense organs. This chapter explores both the general senses and the special senses.

The General Senses

Receptors for the general senses are scattered throughout the body and are relatively simple in structure. These receptors are classified according to the nature of the stimulus that excites them; important classes include receptors sensitive to pain (*nociceptors*); temperature (*thermoreceptors*); touch, pressure, and position (*mechanoreceptors*); and chemical stimuli (*chemoreceptors*).

PAIN

Pain receptors, or **nociceptors** (nō-sē-SEP-tōrz; *noceo*, hurt) are free nerve endings. They are especially common in the superficial portions of the skin, in joint capsules, within the periostea that cover bones, and around the walls of blood vessels. There are few nociceptors in other deep tissues or in most visceral organs.

Once pain receptors in a region are stimulated, two types of axons carry the painful sensations. Myelinated fibers carry very localized sensations of **fast pain**, or *prickling pain*, such

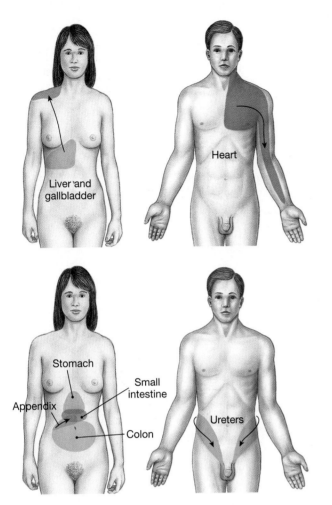

● *Figure 9-1* **Referred Pain**
Pain sensations originating in visceral organs are often perceived as involving specific regions of the body surface innervated by the same spinal nerves.

as that caused by an injection or deep cut. These sensations reach the CNS very quickly, where they often lead to somatic reflexes. They are also relayed to the primary sensory cortex and so receive conscious attention. Slower, unmyelinated fibers carry sensations of **slow pain**, or *burning and aching pain*. Unlike fast pain sensations, slow pain sensations enable you to determine only the general area involved.

Pain sensations from visceral organs are often perceived as originating in more superficial regions, generally regions innervated by the same spinal nerves. The perception of pain coming from parts of the body that are not actually stimulated is called **referred pain**. The precise mechanism responsible for referred pain remains to be determined, but several clinical examples are shown in Figure 9-1●. Cardiac pain, for example, is often perceived as originating in the upper chest and left arm.

Pain receptors continue to respond as long as the painful stimulus remains. However, the perception of the pain can decrease over time because of the inhibition of centers in the thalamus, reticular formation, lower brain stem, and spinal cord.

TEMPERATURE

Temperature receptors, or **thermoreceptors**, are free nerve endings scattered immediately beneath the surface of the skin. They are also located in skeletal muscles, in the liver, and in the hypothalamus. Cold receptors are three or four times as numerous as warm receptors. There are no known structural differences between warm and cold thermoreceptors.

Temperature sensations are relayed along the same pathways that carry pain sensations. They are distributed to the reticular formation, the thalamus, and, to a lesser extent, the primary sensory cortex. Thermoreceptors are very active when the temperature is changing, but they quickly adapt to a stable temperature. When we enter an air-conditioned classroom on a hot summer day or a toasty lecture hall on a brisk fall evening, the temperature seems unpleasant at first, but you quickly become comfortable as adaptation occurs.

TOUCH, PRESSURE, AND POSITION

Mechanoreceptors are receptors sensitive to stimuli that distort their cell membranes. They contain mechanically regulated ion channels, which open or close in response to stretching, compression, twisting, or other distortions of the membrane. There are three classes of mechanoreceptors: (1) *tactile (touch) receptors*, (2) *baroreceptors (pressure)*, and (3) *proprioceptors (position)*.

Tactile Receptors

Tactile receptors provide sensations of touch, pressure, and vibration. The distinctions between these are hazy, for a touch also represents a pressure, and a vibration consists of an oscillating touch/pressure stimulus. **Fine touch and pressure receptors** provide detailed information about a source of stimulation, including its exact location, shape, size, texture, and movement. **Crude touch and pressure receptors** provide poor localization and little additional information about the stimulus.

Tactile receptors range in complexity from free nerve endings to specialized sensory complexes with accessory cells and

9 **THE GENERAL AND SPECIAL SENSES**

The General Senses • The Special Senses • Smell • Taste • Vision • Equilibrium and Hearing • Aging and the Senses • Chapter Review • MediaLab

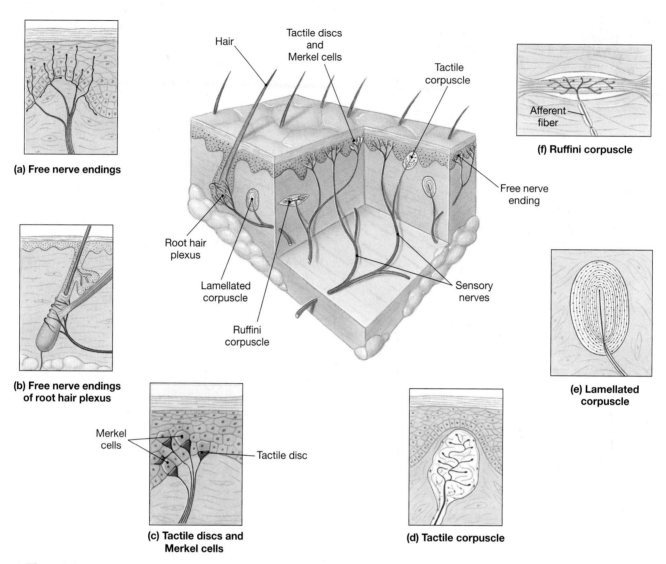

(a) Free nerve endings

(b) Free nerve endings of root hair plexus

(c) Tactile discs and Merkel cells

(d) Tactile corpuscle

(e) Lamellated corpuscle

(f) Ruffini corpuscle

● *Figure 9-2* **Tactile Receptors in the Skin**
(**a**) Free nerve endings. (**b**) A root hair plexus. (**c**) Merkel's discs and Merkel cells. (**d**) A Meissner's corpuscle. (**e**) A Pacinian corpuscle. (**f**) A Ruffini corpuscle.

supporting structures. Figure 9-2● shows six types of tactile receptors in the skin:

1. Free nerve endings are sensitive to touch and pressure and are situated between epidermal cells. No structural differences have been found between these receptors and the free nerve endings that provide temperature or pain sensations.

2. The **root hair plexus** is made up of free nerve endings that monitor the distortion and movement of hairs.

3. **Tactile discs**, or *Merkel's* (MER-kelz) *discs*, are fine touch and pressure receptors. The dendrites of these neurons contact large epithelial cells (*Merkel's cells*) in the lower epidermal layer of the skin.

4. **Tactile corpuscles**, or *Meissner's* (MĪS-nerz) *corpuscles*, are fine touch and pressure receptors. They are abun-

dant in the eyelids, lips, fingertips, nipples, and external genitalia.

5. **Lamellated** (LAM-e-lā-ted) **corpuscles**, or *pacinian* (pa-SIN-ē-an) *corpuscles*, are large receptors sensitive to deep pressure, and to pulsing or high-frequency vibrations. They are common in the skin of the fingers, breasts, and external genitalia. They are also present in joint capsules, mesenteries, the pancreas, and the wall of the urinary bladder.

6. **Ruffini** (roo-FĒ-nē) **corpuscles** are also sensitive to pressure and distortion of the skin, but they are located in its deeper layer, the dermis.

Tactile sensations travel through the posterior column and spinothalamic pathways. ⟳ pp. 260–261 Our sensitivity to tactile sensations can be altered by infection, disease, and damage to sensory neurons or pathways. Mapping tactile responses can sometimes aid clinical assessment. For example, sensory loss along the boundary of a dermatome can help identify the affected spinal nerve or nerves. ∽ p. 257

Baroreceptors

Baroreceptors (bar-ō-rē-SEP-tōrz; *baro-*, pressure) monitor changes in pressure. These receptors consist of free nerve endings that branch within the elastic tissues in the wall of a distensible organ, such as a blood vessel or a portion of the respiratory, digestive, or urinary tract. When the pressure changes, the elastic walls of these vessels or tracts expand or recoil. This movement distorts the dendritic branches and alters the rate of action potential generation. Baroreceptors respond immediately to a change in pressure, but they adapt rapidly, and the output along the afferent fibers gradually returns to "normal."

Figure 9-3● gives examples of baroreceptor distribution and function. Baroreceptors monitor blood pressure in the walls of major blood vessels, including the carotid artery (at the *carotid sinus*) and the aorta (at the *aortic sinus*). The information plays a major role in regulating cardiac function and adjusting blood flow to vital tissues. Baroreceptors in the lungs monitor the degree of lung expansion. This information is relayed to respiratory rhythmicity centers in the brain, which set the pace of respiration. Baroreceptors in the digestive and urinary tracts trigger various visceral reflexes, including those of urination and defecation.

Proprioceptors

Proprioceptors monitor the position of joints, the tension in tendons and ligaments, and the state of muscular contraction. *Free nerve endings* innervate the joint capsules, *Golgi tendon organs* monitor the strain on a tendon, and *muscle spindles* monitor the length of a skeletal muscle. In general, proprioceptors do not adapt to constant stimulation, and each receptor continuously sends information to the CNS. Most of this information is processed subconsciously, and only a small proportion of it reaches your conscience awareness. Your sense of body position results from the integration of information from these three types of proprioceptors, and the receptors of the inner ear.

CHEMICAL DETECTION

In general, **chemoreceptors** respond only to water-soluble and lipid-soluble substances that are dissolved in the surrounding fluid. Adaptation usually occurs over a few seconds following stimulation. Except for the special senses of taste and smell,

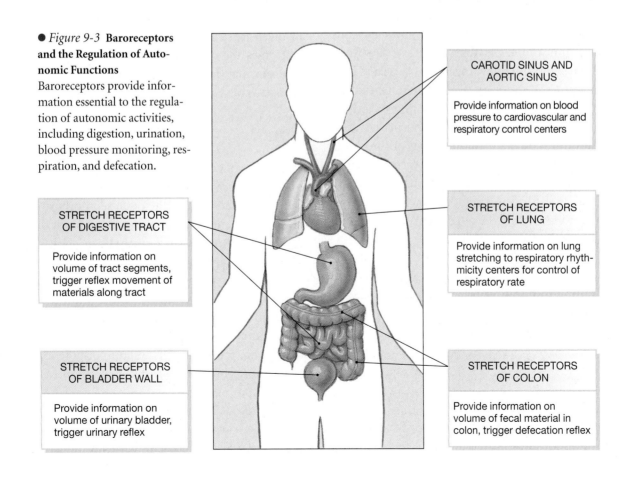

● *Figure 9-3* **Baroreceptors and the Regulation of Autonomic Functions**
Baroreceptors provide information essential to the regulation of autonomic activities, including digestion, urination, blood pressure monitoring, respiration, and defecation.

CAROTID SINUS AND AORTIC SINUS

Provide information on blood pressure to cardiovascular and respiratory control centers

STRETCH RECEPTORS OF LUNG

Provide information on lung stretching to respiratory rhythmicity centers for control of respiratory rate

STRETCH RECEPTORS OF DIGESTIVE TRACT

Provide information on volume of tract segments, trigger reflex movement of materials along tract

STRETCH RECEPTORS OF BLADDER WALL

Provide information on volume of urinary bladder, trigger urinary reflex

STRETCH RECEPTORS OF COLON

Provide information on volume of fecal material in colon, trigger defecation reflex

9 THE GENERAL AND SPECIAL SENSES

The General Senses • **The Special Senses** • **Smell** • **Taste** • Vision • Equilibrium and Hearing • Aging and the Senses • Chapter Review • MediaLab

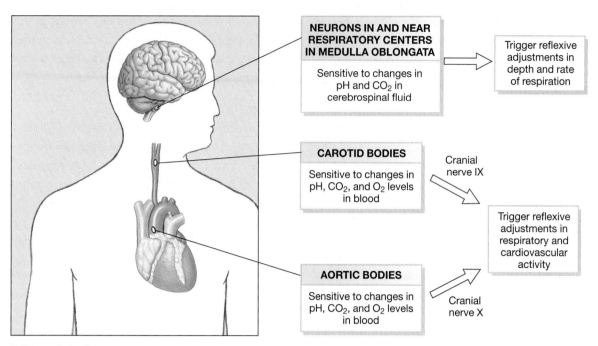

● *Figure 9-4* **Chemoreceptors**
Chemoreceptors are located in the CNS, on the ventrolateral surfaces of the medulla oblongata, and in the aortic and carotid bodies. These receptors are involved in the autonomic regulation of respiratory and cardiovascular function.

there are no well-defined chemosensory pathways in the brain or spinal cord. The chemoreceptors of the general senses send their information to brain stem centers that deal with the autonomic control of respiratory and cardiovascular functions. The locations of important chemoreceptors are shown in Figure 9-4●. Neurons within the respiratory centers of the brain respond to the concentration of hydrogen ions (pH) and carbon dioxide molecules in the cerebrospinal fluid. Chemoreceptors are also located in the **carotid bodies**, near the origin of the internal carotid arteries on each side of the neck, and in the **aortic bodies**, between the major branches of the aortic arch. These receptors monitor the pH and the carbon dioxide and the oxygen concentration of arterial blood. The afferent fibers leaving the carotid and aortic bodies reach the respiratory centers by traveling along the glossopharyngeal (N IX) and vagus (N X) cranial nerves.

CONCEPT CHECK QUESTIONS

Answers on page 310

❶ When you first enter the anatomy and physiology laboratory for dissection, you are very aware of the odor of the preservative, but by the end of the lab period the smell doesn't seem nearly as strong. Why?

❷ When the nociceptors in your hand are stimulated, what sensation do you perceive?

❸ What would happen to an individual if the information from proprioceptors in the legs was blocked from reaching the CNS?

The Special Senses

We next turn our attention to the five *special senses:* smell, taste, vision, equilibrium, and hearing. Although the sense organs involved are structurally more complex than those of the general senses, the same basic principles of receptor function apply. The information these receptors provide is distributed to specific areas of the cerebral cortex and to the brainstem.

Smell

The sense of smell, or *olfaction,* is provided by paired **olfactory organs** (Figure 9-5●). These organs are located in the nasal cavity on either side of the nasal septum just inferior to the cribriform plate of the ethmoid bone. Each olfactory organ consists of an **olfactory epithelium,** which contains the **olfactory receptors,** supporting cells, and *basal cells* (stem cells). The underlying layer of loose connective tissue contains large **olfactory glands** whose secretions produce a pigmented mucus that covers the epithelium. The mucus is produced in a continuous stream that passes across the surface of the olfactory organ, preventing the buildup of potentially dangerous or overpowering stimuli and keeping the area moist and free from dust or other debris.

When you draw air in through your nose, the air swirls within the nasal cavity. A normal, relaxed inhalation carries a small sample (about 2 percent) of the inhaled air to the olfactory organs.

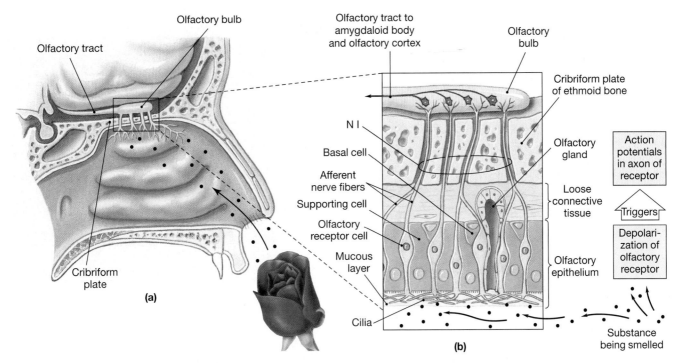

● *Figure 9-5* **The Olfactory Organs**
(a) The structure of the olfactory organ on the left side of the nasal septum. (b) An olfactory receptor is a modified neuron with multiple cilia extending from its free surface.

Sniffing repeatedly increases the flow of air across the olfactory epithelium, intensifying the stimulation of the receptors. Once airborne compounds have reached the olfactory organs, water-soluble and lipid-soluble chemicals must diffuse into the mucus before they can stimulate the olfactory receptors.

The olfactory receptors are highly modified neurons. The exposed tip of each receptor cell provides a base for cilia that extend into the surrounding mucus. Olfactory reception occurs as dissolved chemicals interact with receptors, called *odorant binding proteins*, on the surfaces of the cilia. *Odorants* are chemicals that stimulate olfactory receptors. The binding of an odorant changes the permeability of the receptor membrane, producing action potentials. This information is relayed to the central nervous system, which interprets the smell on the basis of the particular pattern of receptor activity.

Between 10 and 20 million olfactory receptor cells are packed into an area of roughly 5 cm^2. If we take into account the surface area of the exposed cilia, the actual sensory area approaches that of the entire body surface. Nevertheless, our olfactory sensitivities cannot compare with those of other vertebrates such as dogs, cats, or fishes. A German shepherd sniffing for smuggled drugs or explosives has an olfactory receptor surface 72 times greater than that of the nearby customs inspector.

THE OLFACTORY PATHWAYS

The axons leaving the olfactory epithelium collect into 20 or more bundles that penetrate the cribriform plate of the ethmoid bone to reach the **olfactory bulbs**. Axons leaving each olfactory bulb travel along the olfactory tract to reach the olfactory cortex of the cerebrum, the hypothalamus, and portions of the limbic system.

Olfactory stimuli are the only type of sensory information that reaches the cerebral cortex without first synapsing in the thalamus. The extensive limbic and hypothalamic connections help explain the profound emotional and behavioral responses that certain smells can produce. The perfume industry, which understands the practical implications of these connections, spends billions of dollars to develop odors that trigger sexual responses.

Taste

Taste receptors, or *gustatory* (GUS-ta-tōr-ē) *receptors*, are distributed over the surface of the tongue and adjacent portions of the pharynx and larynx. By adulthood, however, the taste receptors of the pharynx and larynx have decreased in importance and abundance. Taste receptors and specialized epithelial cells form sensory structures called **taste buds**. The taste buds are particularly well protected from the excessive mechanical stress due to chewing, for they lie along the sides of epithelial projections called **papillae** (pa-PIL-lē). The greatest number of taste buds are associated with the large *circumvallate papillae*, which form a V that points toward the attached base of the tongue (Figure 9-6●).

Each taste bud contains slender sensory receptors, known as **gustatory cells**, and supporting cells. Each gustatory cell

9

THE GENERAL AND SPECIAL SENSES

The General Senses • The Special Senses • Smell • Taste • **Vision** • Equilibrium and Hearing • Aging and the Senses • Chapter Review • MediaLab

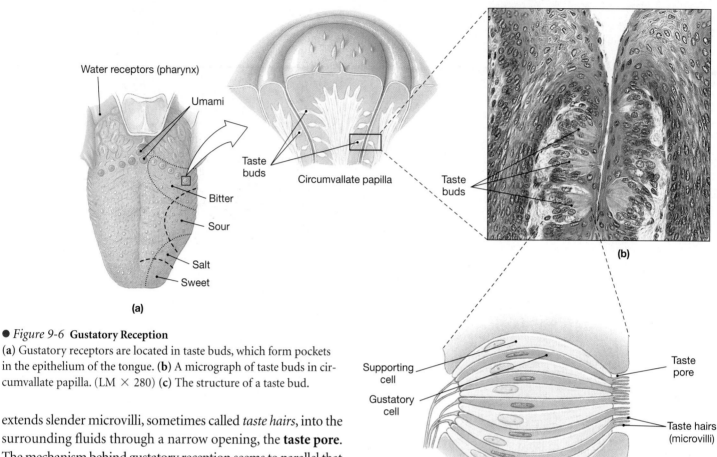

● *Figure 9-6* **Gustatory Reception**
(**a**) Gustatory receptors are located in taste buds, which form pockets in the epithelium of the tongue. (**b**) A micrograph of taste buds in circumvallate papilla. (LM × 280) (**c**) The structure of a taste bud.

extends slender microvilli, sometimes called *taste hairs*, into the surrounding fluids through a narrow opening, the **taste pore**. The mechanism behind gustatory reception seems to parallel that of olfaction. Dissolved chemicals contacting the taste hairs stimulate a change in the membrane potential of the taste cell, which leads to action potentials in the sensory neuron.

You are probably already familiar with the four **primary taste sensations**: sweet, salty, sour, and bitter. Two additional tastes, *umami* and *water*, have been discovered in humans. **Umami** (oo-MAH-mē) is a pleasant taste that is characteristic of beef broth and chicken broth. Most people say water has no flavor, yet **water receptors** are present, especially in the pharynx. Their sensory output is processed in the hypothalamus and affects several systems that deal with water balance and the regulation of blood volume. Each taste bud shows a particular sensitivity to one of these tastes, and a sensory map of the tongue indicates that each tends to be concentrated in a different area (Figure 9-6a●). The threshold for receptor stimulation varies for each of the primary taste sensations, and the taste receptors respond most readily to unpleasant rather than pleasant stimuli.

THE TASTE PATHWAYS

Taste buds are monitored by the facial (N VII), glossopharyngeal (N IX), and vagus (N X) cranial nerves. The sensory fibers of the different nerves synapse within a nucleus in the medulla oblongata, and the axons of the postsynaptic neurons synapse in the thalamus. The information is then projected to the appropriate portions of the primary sensory cortex.

The information received from the taste buds is correlated with other sensory data to assemble a conscious perception of taste. Our perception of the general texture of the food, together with the taste-related sensations of "peppery" or "burning," result from the stimulation of general sensory afferents in the trigeminal nerve (N V). In addition, information from the olfactory receptors plays an overwhelming role in taste perception. You are several thousand times more sensitive to "tastes" when your olfactory organs are fully functional. If you have a cold and airborne molecules cannot reach your olfactory receptors, meals taste dull and unappealing even though your taste buds are responding normally.

CONCEPT CHECK QUESTIONS

Answers on page 310

❶ How does sniffing repeatedly help to identify faint odors?

❷ If you completely dry the surface of the tongue and then place salt or sugar crystals on it, you cannot taste them. Why not?

Vision

We rely more on vision than on any other special sense. Our visual receptors are contained in elaborate structures, the eyes, which enable us not only to detect light but also to create detailed visual images. We will begin our discussion of these complex organs by considering the *accessory structures* of the eye, which provide protection, lubrication, and support.

THE ACCESSORY STRUCTURES OF THE EYE

The **accessory structures** of the eye include the (1) eyelids and associated exocrine glands; (2) the superficial epithelium of the eye; (3) structures associated with the production, secretion, and removal of tears; and (4) the extrinsic eye muscles.

The **eyelids**, or **palpebrae** (pal-PĒ-brē), are a continuation of the skin. They act like windshield wipers: Their blinking movements keep the surface of the eye lubricated and free from dust and debris. They can also close firmly to protect the delicate surface of the eye. The medial and lateral points of attachment of the upper and lower eyelids are called the **medial canthus** (KAN-thus) and the **lateral canthus**, respectively (Figure 9-7a●). The eyelashes are very robust hairs that help prevent foreign particles such as dirt and insects from reaching the surface of the eye.

Several types of exocrine glands protect the eye and its accessory structures. Large sebaceous glands are associated with the eyelashes, as they are with other hairs and hair follicles. ∞ p. 116 In addition, modified sebaceous glands *(tarsal glands)* along the inner margins of the eyelids secrete a lipid-rich substance that keeps them from sticking together. At the medial canthus, the **lacrimal caruncle** (KAR-unk-ul), a soft mass of tissue, contains glands that produce thick secretions that contribute to the gritty deposits occasionally found after a night's sleep. The accessory glands of the eye sometimes become infected by bacteria. An infection in a sebaceous gland of one of the eyelashes, a tarsal gland, or in one of the adjacent sweat glands between the eyelash follicles produces a painful localized swelling known as a **sty**.

The epithelium covering the inner surfaces of the eyelids and the outer surface of the eye is called the **conjunctiva** (kon-junk-TĪ-vuh). The conjunctiva extends to the edges of the transparent **cornea** (KŌR-nē-a), a part of the outer layer of the eye. The cornea is covered by a delicate *corneal epithelium*, that is continuous with the conjunctiva. The conjunctiva contains many free nerve endings and is very sensitive. The painful condition of **conjunctivitis**, or pinkeye, results from damage to and irritation of the conjunctival surface. The most obvious symptom results from dilation of the blood vessels beneath the conjunctival epithelium.

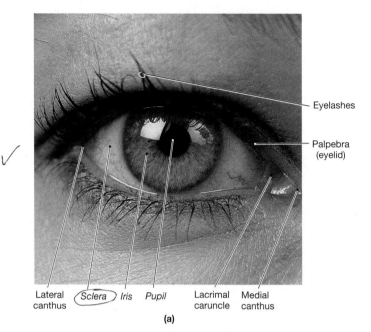

Lateral canthus | Sclera | Iris | Pupil | Lacrimal caruncle | Medial canthus

Eyelashes

Palpebra (eyelid)

(a)

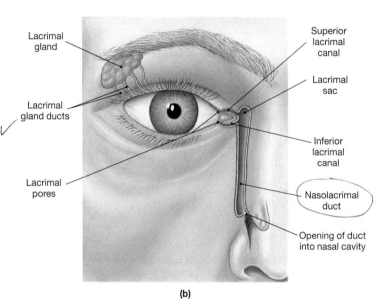

Lacrimal gland

Lacrimal gland ducts

Lacrimal pores

Superior lacrimal canal

Lacrimal sac

Inferior lacrimal canal

Nasolacrimal duct

Opening of duct into nasal cavity

(b)

● *Figure 9-7* **The Accessory Structures of the Eye**
(**a**) The gross and superficial anatomies of the accessory structures.
(**b**) The details of the lacrimal apparatus.

A constant flow of tears keeps the surface of the eyeball moist and clean. Tears reduce friction, remove debris, prevent bacterial infection, and provide nutrients and oxygen to the conjunctival epithelium. The **lacrimal apparatus** produces, distributes, and removes tears (Figure 9-7b●). Above the eyeball is the **lacrimal gland**, or *tear gland*, which has a dozen or more ducts that empty into the pocket between the eyelid and the eye. This gland nestles within a depression in the frontal bone, just inside the orbit and superior and lateral to the eyeball. The lacrimal gland normally provides the key ingredients

9

THE GENERAL AND SPECIAL SENSES

The General Senses • The Special Senses • Smell • Taste • **Vision** • Equilibrium and Hearing • Aging and the Senses • Chapter Review • MediaLab

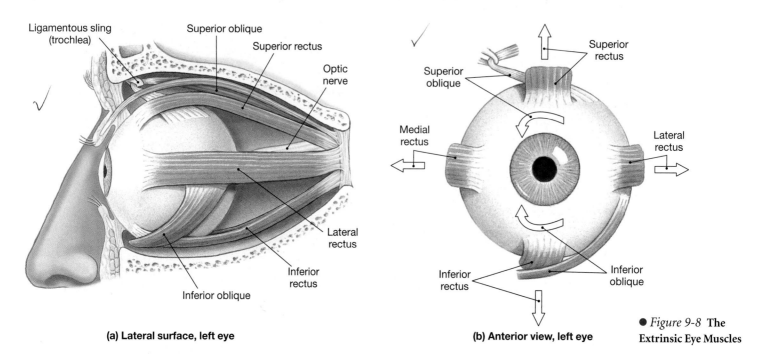

(a) Lateral surface, left eye

(b) Anterior view, left eye

● *Figure 9-8* **The Extrinsic Eye Muscles**

TABLE 9-1 *The Extrinsic Eye Muscles (Figure 9-8)*

MUSCLE	ORIGIN	INSERTION	ACTION	INNERVATION
Inferior rectus	Sphenoid bone around optic canal	Inferior, medial surface of eyeball	Eye looks down	Oculomotor nerve (N III)
Medial rectus	As above	Medial surface of eyeball	Eye rotates medially	As above
Superior rectus	As above	Superior surface of eyeball	Eye looks up	As above
Inferior oblique	Maxillary bone at anterior portion of orbit	Inferior, lateral surface of eyeball	Eye rolls, looks up and to the side	As above
Superior oblique	Sphenoid bone around optic canal	Superior, lateral surface of eyeball	Eye rolls, looks down and to the side	Trochlear nerve (N IV)
Lateral rectus	As above	Lateral surface of eyeball	Eye rotates laterally	Abducens nerve (N VI)

and most of the volume of the tears. Its secretions are watery, slightly alkaline, and contain *lysozyme*, an enzyme that attacks bacteria. The blinking of the eye sweeps the tears across the surface of the eye to the medial canthus. Two small pores direct the tears into the **lacrimal canals**, passageways that end at the lacrimal sac. From this sac, the **nasolacrimal duct** carries the tears to the nasal cavity.

Six **extrinsic eye muscles**, or *oculomotor* (ok-ū-lō-MŌ-ter) *muscles*, originate on the surface of the orbit and control the position of the eye. These muscles are the **inferior rectus**, **lateral rectus**, **medial rectus**, **superior rectus**, **inferior oblique**, and **superior oblique** (Figure 9-8● and Table 9-1).

THE EYE

The eyes are extremely specialized visual organs, more versatile and adaptable than the most expensive cameras, yet light, compact, and durable. Each eye is roughly spherical, with a diameter of nearly 2.5 cm (1 in.), and weighs around 8 g (0.28 oz). The eyeball shares space within the orbit with the extrinsic eye muscles, the lacrimal gland, and the various cranial nerves and blood vessels that service the eye and adjacent areas of the orbit and face. A mass of *orbital fat* provides padding and insulation.

The hollow interior of the eyeball can be divided into two cavities: the posterior cavity and the anterior cavity (Figure 9-9●). The large **posterior cavity** is also called the *vitreous chamber*, because it contains the gelatinous *vitreous body*. The

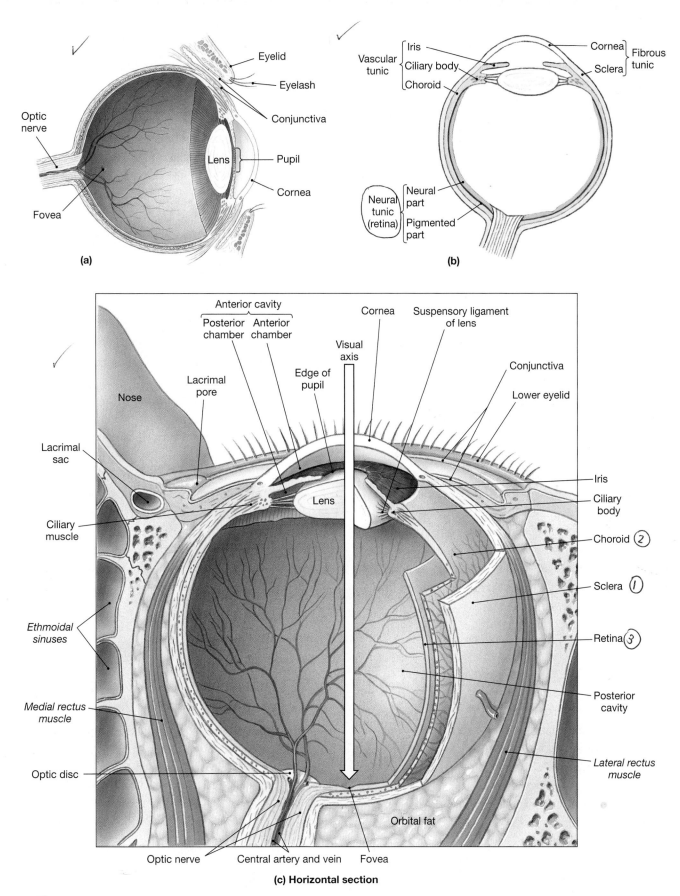

(a)

Eyelid
Eyelash
Conjunctiva
Optic nerve
Lens
Pupil
Cornea
Fovea

(b)

Iris
Cornea
Fibrous tunic
Vascular tunic
Ciliary body
Sclera
Choroid
Neural tunic (retina)
Neural part
Pigmented part

(c) Horizontal section

Anterior cavity
Posterior chamber
Anterior chamber
Cornea
Suspensory ligament of lens
Visual axis
Conjunctiva
Lacrimal pore
Edge of pupil
Lower eyelid
Nose
Lacrimal sac
Lens
Iris
Ciliary muscle
Ciliary body
Choroid ②
Ethmoidal sinuses
Sclera ①
Retina ③
Medial rectus muscle
Posterior cavity
Optic disc
Lateral rectus muscle
Orbital fat
Optic nerve
Central artery and vein
Fovea

● *Figure 9-9* **The Sectional Anatomy of the Eye**
(**a**) A sagittal section through the left eye. (**b**) A horizontal section, showing the three layers, or tunics, of the right eye. (**c**) A horizontal section of the right eye showing various landmarks and features.

9 THE GENERAL AND SPECIAL SENSES

The General Senses • The Special Senses • Smell • Taste • **Vision** • Equilibrium and Hearing • Aging and the Senses • Chapter Review • MediaLab

smaller **anterior cavity** is subdivided into the *anterior chamber* and the *posterior chamber*. The shape of the eye is stabilized in part by the vitreous body and the *aqueous humor*, a fluid which fills the anterior cavity. The wall of the eye contains three distinct layers, or *tunics*: an outer *fibrous tunic*, an intermediate *vascular tunic*, and an inner *neural tunic*.

The Fibrous Tunic

The **fibrous tunic**, the outermost layer of the eye, consists of the *sclera* and *cornea*. The fibrous tunic (1) provides mechanical support and some degree of physical protection, (2) serves as an attachment site for the extrinsic eye muscles, and (3) assists in the focusing process. The **sclera** (SKLER-uh), or "white of the eye," is a layer of dense fibrous connective tissue containing both collagen and elastic fibers (Figure 9-9●). It is thickest over the posterior surface of the eye and thinnest over the anterior surface. The six extrinsic eye muscles insert on the sclera.

The surface of the sclera contains small blood vessels and nerves that penetrate the sclera to reach internal structures. On the anterior surface of the eye, however, these blood vessels lie under the conjunctiva. Because this network of small capillaries does not carry enough blood to lend an obvious color to the sclera, the white color of the collagen fibers is visible.

The transparent **cornea** is continuous with the sclera, but the collagen fibers of the cornea are organized into a series of layers that do not interfere with the passage of light. The cornea has no blood vessels, and its epithelial cells obtain their oxygen and nutrients from the tears that flow across their surfaces. Because the cornea has only a limited ability to repair itself, corneal injuries must be treated immediately to prevent serious vision losses. Restoration of vision after corneal scarring usually requires replacement of the cornea through a corneal transplant.

The Vascular Tunic

The **vascular tunic** contains numerous blood vessels, lymphatic vessels, and the *intrinsic eye muscles*. The functions of this layer include (1) providing a route for blood vessels and lymphatic vessels that supply tissues of the eye, (2) regulating the amount of light entering the eye, (3) secreting and reabsorbing the aqueous humor that circulates within the eye, and (4) controlling the shape of the lens, an essential part of the focusing process.

The vascular tunic includes the *iris*, the *ciliary body*, and the *choroid* (Figure 9-9●). Visible through the

transparent cornea, the **iris** contains blood vessels, melanocytes, and two layers of smooth muscle fibers. When these muscles contract, they change the diameter of the central opening, or **pupil**, of the iris. One group of muscles forms concentric circles around the pupil; their contraction decreases, or constricts, the diameter of the pupil. The second group of muscles extend radially away from the edge of the pupil; their contraction enlarges, or dilates, the pupil (Figure 9-10●). Dilation and constriction are controlled by the autonomic nervous system in response to changes in light intensity. Exposure to bright light produces a reflexive decrease in pupil diameter, under parasympathetic stimulation. A reduction in light levels causes the pupils to dilate, under the control of the sympathetic division.

The color of your eye is determined by the number of melanocytes the iris contains and the pigmented epithelium on its posterior surface. This epithelium contains melanin granules and is part of the neural tunic. When melanocytes are absent, light passes through the iris and bounces off the pigmented epithelium. The eye then appears blue. In order, individuals with gray, brown, or black eyes have increasing numbers of melanocytes in the iris.

Along its outer edge, the iris attaches to the anterior portion of the **ciliary body**, the bulk of which consists of the *ciliary muscle*, a ring of smooth muscle that projects into the interior of the eye. The ciliary body begins at the junction between the cornea and sclera and extends to the scalloped border that also marks the anterior edge of the retina. Posterior to the iris, the surface of the ciliary body is thrown into folds called *ciliary processes*. The **suspensory ligaments** of the lens attach to these processes. These fibers position the lens so that light passing through the pupil passes through the center of the lens.

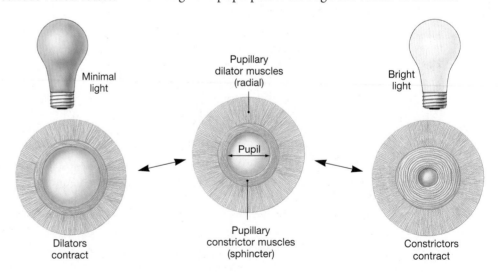

Minimal light

Dilators contract

Pupillary dilator muscles (radial)

Pupil

Pupillary constrictor muscles (sphincter)

Bright light

Constrictors contract

● *Figure 9-10* **The Intrinsic Muscles of the Eye**
Two sets of smooth muscle within the iris control the dilation and constriction of the pupil. In bright light, the concentric muscle fibers constrict the the pupil and reduce the level of incoming light. In dim light, the radial muscle fibers contract and the pupil enlarges, or dilates, to allow the entry of additional light.

The choroid is a layer that separates the fibrous and neural tunics posterior to the ciliary body (Figure 9-9●). The choroid contains a capillary network that delivers oxygen and nutrients to the retina.

The Neural Tunic

The **neural tunic**, or **retina**, is the innermost layer of the eye. It consists of a thin outer pigment layer, called the *pigmented part*, and a thick inner layer, called the *neural part*. The pigmented part absorbs light after it passes through the neural part. The neural part contains (1) the photoreceptors that respond to light, (2) supporting cells and neurons that perform preliminary processing and integration of visual information, and (3) blood vessels supplying tissues that line the posterior cavity. The two layers of the retina are normally very close together but not tightly interconnected. The pigmented part continues over the ciliary body and iris. The neural part forms a cup that establishes the posterior and lateral boundaries of the posterior cavity (Figure 9-9●).

Organization of the Retina. The retina contains several layers of cells (Figure 9-11a●). The outermost layer, closest to the

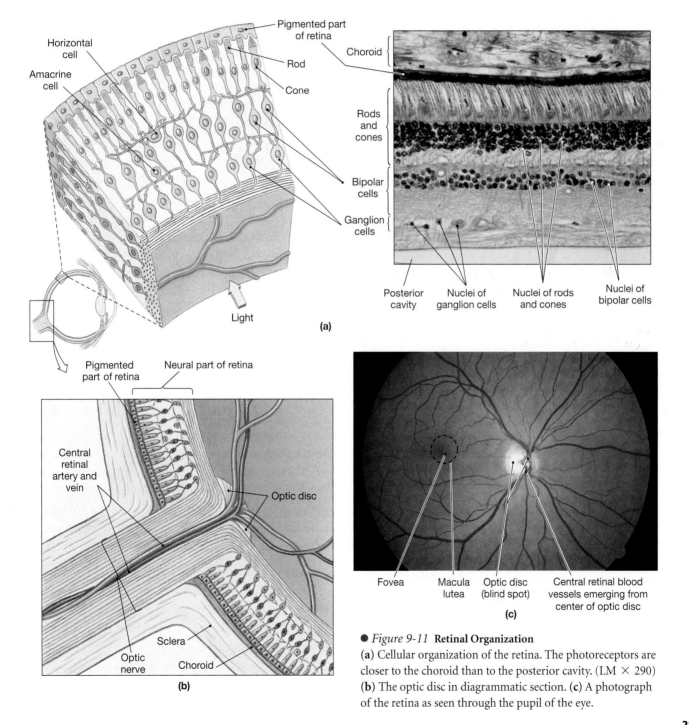

● *Figure 9-11* **Retinal Organization**
(**a**) Cellular organization of the retina. The photoreceptors are closer to the choroid than to the posterior cavity. (LM × 290)
(**b**) The optic disc in diagrammatic section. (**c**) A photograph of the retina as seen through the pupil of the eye.

9 THE GENERAL AND SPECIAL SENSES

The General Senses • The Special Senses • Smell • Taste • **Vision** • Equilibrium and Hearing • Aging and the Senses • Chapter Review • MediaLab

pigment part of the retina, contains the photoreceptors, or cells that detect light. The eye has two types of photoreceptors: **rods** and **cones**. Rods do not discriminate among colors of light. These receptors are very light-sensitive and enable us to see in dimly lit rooms, at twilight, or in pale moonlight. Cones provide us with color vision. Three types of cones are present, and their stimulation in various combinations provides the perception of different colors. Cones give us sharper, clearer images, but they require more intense light than do rods. If you sit outside at sunset (or sunrise), you will probably be able to tell when your visual system shifts from cone-based vision (clear images in full color) to rod-based vision (relatively grainy images in black and white).

Rods and cones are not evenly distributed across the retina. If you think of the retina as a cup, approximately 125 million rods are found on the sides and roughly 6 million cones dominate the bottom. There are no rods in the region where the visual image arrives after passing through the cornea and lens. This area is the **macula lutea** (LOO-tē-uh; yellow spot). The highest concentration of cones is found in the central portion of the macula lutea, an area called the **fovea** (FŌ-vē-uh; shallow depression), or *fovea centralis* (Figure 9-11c●). The fovea is the center of color vision and the site of sharpest vision. When you look directly at an object, its image falls on this portion of the retina.

You are probably already aware of the visual consequences of this distribution. During the day, when there is enough light to stimulate the cones, you see a very good image. In very dim light, cones cannot function. For example, when you try to stare at a dim star, you are unable to see it. But if you look a little to one side rather than directly at the star, you will see it quite clearly. Shifting your gaze moves the image of the star from the fovea, where it does not provide enough light to stimulate the cones, to the sides of the retina, where it stimulates the more sensitive rods.

The rods and cones synapse with roughly 6 million **bipolar cells**. Bipolar cells in turn synapse within the layer of **ganglion cells** that faces the posterior cavity. The axons of the ganglion cells deliver the sensory information to the brain. *Horizontal cells* and *amacrine* (AM-a-krin) cells can regulate communication between photoreceptors and ganglion cells, adjusting the sensitivity of the retina. The effect can be compared to adjusting the "contrast" setting on a television. These cells play an important role in the eye's adjustment to dim or brightly lit environments.

The Optic Disc. Axons from an estimated 1 million ganglion cells converge on the **optic disc**, a circular region just medial to the fovea. The optic disc is the origin of the optic nerve (N II) (Figure 9-11b●). From this point, the axons turn, penetrate the

● *Figure 9-12* **The Optic Disc**
Close your left eye and stare at the cross with your right eye, keeping the cross in the center of your field of vision. Begin with the page a few inches away from your eye and gradually increase the distance. The dot will disappear when its image falls on the blind spot. To check the blind spot in your left eye, close your right eye, stare at the dot, and repeat this sequence.

wall of the eye, and proceed toward the diencephalon. Blood vessels that supply the retina pass through the center of the optic nerve and emerge on the surface of the optic disc (Figure 9-11b, c●). The optic disc has no photoreceptors or other retinal structures. Because light striking this area goes unnoticed, it is commonly called the **blind spot**. You do not notice a blank spot in your visual field, because involuntary eye movements keep the visual image moving and allow your brain to fill in the missing information. A simple experiment, shown in Figure 9-12●, will demonstrate the presence and location of the blind spot.

The Chambers of the Eye

The ciliary body and lens divide the interior of the eye into a small anterior cavity and a larger posterior cavity, or vitreous chamber. The anterior cavity is further subdivided into the **anterior chamber**, which extends from the cornea to the iris, and the **posterior chamber**, between the iris and the ciliary body and lens. The anterior and posterior chambers are filled with **aqueous humor**. This fluid circulates within the anterior cavity, passing from the posterior to the anterior chamber through the pupil (Figure 9-13●). The posterior cavity is filled with a gelatinous substance known as the **vitreous body**, or *vitreous humor*. The vitreous body helps maintain the shape of the eye and also holds the retina against the choroid.

Aqueous Humor. Aqueous humor is secreted by epithelial cells of the ciliary processes into the posterior chamber (Figure 9-13●). Pressure exerted by this fluid helps maintain the shape of the eye, and the circulation of aqueous humor transports nutrients and wastes. In the anterior chamber near the edge of the iris, the aqueous humor enters a passageway, known as the *canal of Schlemm*, that empties into veins in the sclera and returns this fluid to the venous system.

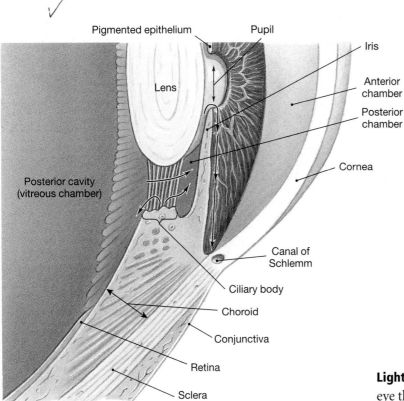

Pigmented epithelium

Pupil

Lens

Posterior cavity (vitreous chamber)

Iris

Anterior chamber

Posterior chamber

Anterior cavity

Cornea

Canal of Schlemm

Ciliary body

Choroid

Conjunctiva

Retina

Sclera

● *Figure 9-13* **Eye Chambers and the Circulation of Aqueous Humor**
The lens is suspended between the vitreous chamber and the posterior chamber. Its position is maintained by suspensory ligaments, which attach the lens to the ciliary body. Aqueous humor secreted at the ciliary body circulates through the posterior and anterior chambers and is reabsorbed after passing along the canal of Schlemm.

Interference with the normal circulation and reabsorption of the aqueous humor leads to an elevation in the pressure inside the eye. If this condition, called **glaucoma**, is left untreated, it can eventually produce blindness by distortion of the retina and the optic disc.

The Lens

The **lens** lies behind the cornea and is held in place by suspensory ligaments that originate on the ciliary body of the choroid. The primary function of the lens is to focus the visual image on the photoreceptors. The lens does so by changing its shape.

The Structure of the Lens. The transparent lens consists of organized concentric layers of cells wrapped in a dense fibrous capsule. The cells making up the lens lack organelles and are filled with transparent proteins. The capsule is elastic; unless an outside force is applied, it will contract and make the lens spherical. However, tension in the suspensory ligaments can overpower the elastic capsule and pull the lens into the shape of a flattened oval.

 ### CATARACTS

The transparency of the lens depends on a precise combination of structural and biochemical characteristics. When that balance is disturbed, the lens loses its transparency, and the abnormal lens is known as a **cataract**. Cataracts can result from drug reactions, injuries, or radiation, but **senile cataracts** are the most common form. As aging proceeds, the lens becomes less elastic, takes on a yellowish hue, and eventual-

ly begins to lose its transparency. As the lens becomes opaque, or "cloudy," the individual needs brighter and brighter reading lights, and visual clarity begins to fade. If the lens becomes completely opaque, the person will be functionally blind, even though the photoreceptors are alive and well. Surgical procedures involve removing the lens, either intact or in pieces, after shattering it with high-frequency sound. The missing lens is replaced by an artificial substitute, and vision is then fine-tuned with glasses or contact lenses.

Light Refraction and Accommodation. As in a camera, in the eye the arriving image must be in focus if it is to provide useful information. "In focus" means that the rays of light arriving from an object strike the sensitive surface of the film (the retina) precisely ordered so as to form a miniature image of the original. If the rays are not perfectly focused, the image will be blurry. Focusing normally occurs in two steps, as light passes through the (1) cornea and (2) lens.

Light is bent, or *refracted*, when it passes from one medium to a medium with a different density. In the human eye, the greatest amount of refraction occurs when light passes from the air into the cornea. When the light enters the relatively dense lens, the lens provides the extra refraction needed to focus the light rays from an object toward a specific **focal point**, a specific point of intersection on the retina. The distance between the center of the lens and the focal point is the *focal distance*. This distance is determined by (1) *the distance of the object from the lens* (the closer the object, the longer the focal distance) and (2) *the shape of the lens* (the rounder the lens, the shorter the focal distance). In the eye, the lens changes shape to keep the focal distance constant, thereby keeping the image focused on the retina. **Accommodation** (Figure 9-14●) is the process of focusing an image on the retina by changing the shape of the lens. During accommodation, the lens either becomes rounder to focus the image of a nearby object on the retina or flattens to focus the image of a distant object.

The lens is held in place by the suspensory ligaments that originate at the ciliary body. Smooth muscle fibers in the ciliary body encircle the lens. As you view a nearby object, your

9 THE GENERAL AND SPECIAL SENSES

The General Senses • The Special Senses • Smell • Taste • **Vision** • Equilibrium and Hearing • Aging and the Senses • Chapter Review • MediaLab

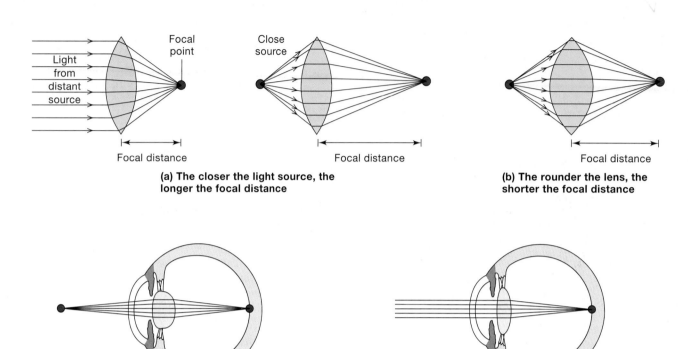

(a) The closer the light source, the longer the focal distance

(b) The rounder the lens, the shorter the focal distance

(c) Ciliary muscle contracted, lens rounded for close vision

(d) Ciliary muscle relaxed, lens flattened for distant vision

● *Figure 9-14* **Focusing and Visual Accommodation**
(**a**) A lens refracts light toward a specific point. The distance from the center of the lens to that point is the focal distance of the lens. Light from a distant source arrives with all of the light waves traveling parallel to one another. Light from a nearby source, however, will still be diverging or spreading out from its source when it strikes the lens. Note the difference in focal distance after refraction. (**b**) For the eye to form a sharp image, the focal distance must equal the distance between the center of the lens and the retina. The lens compensates for variations in the distance between the eye and the object in view by changing its shape. The rounder the lens, the shorter the focal distance. (**c**) When the ciliary muscle contracts, the suspensory ligaments allow the lens to round up. (**d**) When the ciliary muscle relaxes, the ligaments pull against the margins of the lens and flatten it.

ciliary muscles contract and the ciliary body moves toward the lens (Figure 9-14c). This movement reduces the tension in the suspensory ligaments, and the elastic capsule pulls the lens into a more spherical shape. When you view a distant object, your ciliary muscles relax, the suspensory ligaments pull at the circumference of the lens, and the lens becomes relatively flat (Figure 9-14d●).

Image Formation. The image of an object that arrives on the retina is a miniature image of the original, but it is upside down and backward. This makes sense if an object in view can be treated as a large number of individual light sources. Figure 9-15a● shows why an image formed on the retina is upside down. Light from the top of the pole lands at the bottom of the retina and light from the bottom of the pole hits the top of the retina. Figure 9-15b● illustrates why an image formed on the retina is backward. Light from the left side of the fence falls on the right side of the retina and light from the right side of the fence falls on the left side of the retina. The brain com-

pensates for both aspects of image reversal, without our conscious awareness.

✚ VISUAL ACUITY

Clarity of vision, or visual acuity, is rated on the basis of the sight of a "normal" person. A person whose vision is rated 20/20 can see details at a distance of 20 feet as clearly as a "normal" individual would. Vision noted as 20/15 is better than average, for at 20 feet the person is able to see details that would be clear to a normal eye only at a distance of 15 feet. Conversely, a person with 20/30 vision must be 20 feet from an object to discern details that a person with normal vision could make out at a distance of 30 feet.

When visual acuity falls below 20/200, even with the help of glasses or contact lenses, the individual is considered to be legally blind. There are probably fewer than 400,000 legally blind people in the United States; more than half are over 65 years of age. Common causes of blindness include diabetes mellitus, cataracts, glaucoma, corneal scarring, retinal detachment, accidental injuries, and hereditary factors that are as yet poorly understood.

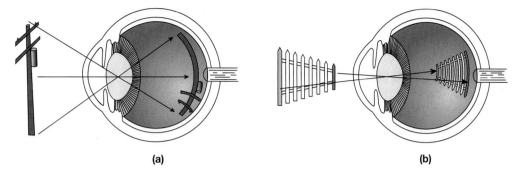

(a) (b)

● *Figure 9-15* **Image Formation**
Light from each portion of an object is focused on a different part of the retina. The resulting image arrives (a) upside down and (b) backward.

 FOCUS

Accommodation Problems

In the normal eye, when the ciliary muscles are relaxed and the lens is flattened, a distant image will be focused on the retinal surface (Figure 9-16a●), a condition called *emmetropia* (*emmetro-*, proper measure). However, irregularities in the shape of the lens or cornea can affect the clarity of the visual image. This condition, called *astigmatism*, can usually be corrected by glasses or special contact lenses.

Figure 9-16● diagrams two other common problems with the accommodation mechanism.

If the eyeball is too deep, the image of a distant object will form in front of the retina, and the retinal picture will be blurry and out of focus (Figure 9-16b●). Vision at close range will be normal, because the lens will be able to round up as needed to focus the image on the retina. As a result, such individuals are said to be "nearsighted." Their condition is more formally termed *myopia* (*myein*, to shut + *ops*, eye). Myopia can be corrected by placing a diverging lens in front of the eye (Figure 9-16c●).

If the eyeball is too shallow, *hyperopia results* (Figure 9-16d●). The ciliary muscles must contract to focus even a distant object on the retina, and at close range the lens cannot provide enough refraction. These individuals are said to be "farsighted" because they can see distant objects most clearly. Older individuals become farsighted as their lenses lose elasticity; this form of hyperopia is called *presbyopia* (*presbys*, old man). Hyperopia can be treated by placing a converging lens in front of the eye (Figure 9-16e●).

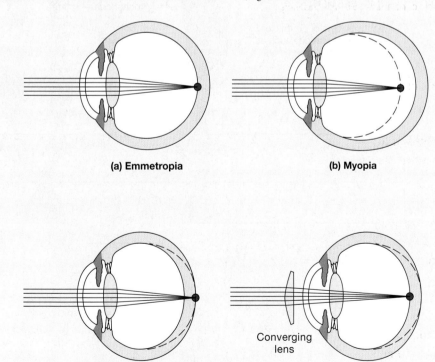

(a) Emmetropia

(b) Myopia

Diverging lens

(c) Myopia (corrected)

(d) Hyperopia

Converging lens

(e) Hyperopia (corrected)

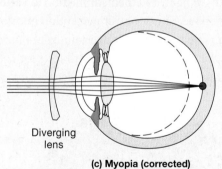

● *Figure 9-16* **Visual Abnormalities**
(a) In normal vision, the lens focuses the visual image on the retina. One common problem of accommodation involves (b) an inability to lengthen the focal distance enough to focus the image of a distant object on the retina—myopia. (c) A diverging lens is used to correct myopia. (d) Another accommodation problem is an inability to shorten the focal distance adequately for near objects—hyperopia. (e) A converging lens is used to correct hyperopia.

9 THE GENERAL AND SPECIAL SENSES

The General Senses • The Special Senses • Smell • Taste • **Vision** • Equilibrium and Hearing • Aging and the Senses • Chapter Review • MediaLab

VISUAL PHYSIOLOGY

The rods and cones of the retina are called **photoreceptors** because they detect *photons*, basic units of visible light. Light is a form of radiant energy. This energy is radiated in waves and described in terms of its wavelength (distance between wave peaks). Our eyes are sensitive to wavelengths that make up the spectrum of **visible light**. This spectrum, seen in a rainbow, can be remembered by the acronym ROY G. BIV (*Red, Orange, Yellow, Green, Blue, Indigo, Violet*). Color depends on the wavelength of the light. The longer the wavelength, the lower the energy content of the light. Photons of red light have the longest wavelength and carry the least energy. Photons from the violet portion of the spectrum have the shortest wavelength and carry the most energy.

Rods and Cones

Rods provide the CNS with information about the presence or absence of photons, without regard to wavelength. As a result, they do not discriminate among colors of light. They are very sensitive, however, and enable us to see in dimly lit rooms, at twilight, and in pale moonlight.

Cones provide information about the wavelength of photons. Because cones are less sensitive than rods, they function only in relatively bright light. We have three types of cones: *blue cones*, *green cones*, and *red cones*. Each type is sensitive to a different range of wavelengths of light, and their stimulation in various combinations accounts for our perception of colors. Persons unable to distinguish certain colors have a form of *color blindness*. The standard tests for color vision involve picking numbers or letters out of a complex image, such as the one in Figure 9-17●. Color blindness occurs because one or more classes of cones are absent or nonfunctional. In the most common condition, the red cones are missing and the individual cannot distinguish red light from green light. Ten percent of men have some color blindness, whereas the incidence among women is only around 0.67 percent. Total color blindness is extremely rare; only 1 person in 300,000 has no cone pigments of any kind.

Photoreceptor Structure

Figure 9-18● compares the structure of rods and cones. The *outer segment* of a photoreceptor contains hundreds to thou-

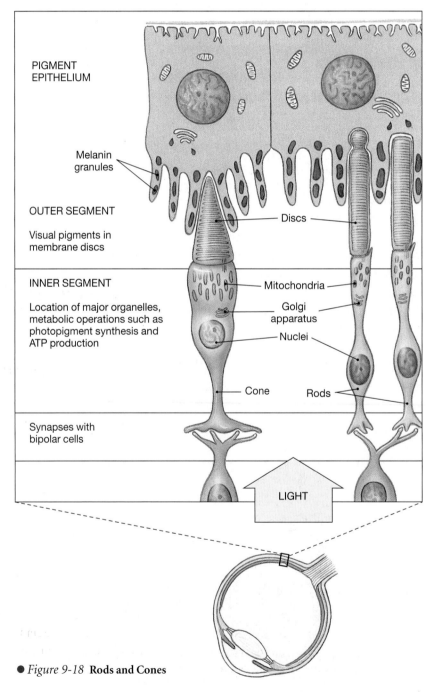

● *Figure 9-17* **A Standard Test for Color Vision**
If you lack one or more populations of cones, you will be unable to distinguish the patterned image (the number 12).

● *Figure 9-18* **Rods and Cones**

sands of flattened membranous discs. The names *rod* and *cone* refer to the shape of the outer segment. The *inner segment* of a photoreceptor synapses with other cells and releases neurotransmitters. In the dark, each photoreceptor continually releases neurotransmitters. The arrival of a photon initiates a chain of events that alters the membrane potential of the photoreceptor and changes the rate of neurotransmitter release.

The discs of the outer segment in both rods and cones contain special organic compounds called **visual pigments**. The absorption of photons by visual pigments is the first key step in the process of photoreception, the detection of light. The visual pigments are derivatives of the compound **rhodopsin** (rō-DOP-sin). Rhodopsin consists of a protein, **opsin**, bound to the pigment **retinal** (RET-i-nal). Retinal is synthesized from **vitamin A**. Retinal is identical in both rods and cones, but a different form of opsin is found in the rods and each of the three types of cones (red, blue, and green).

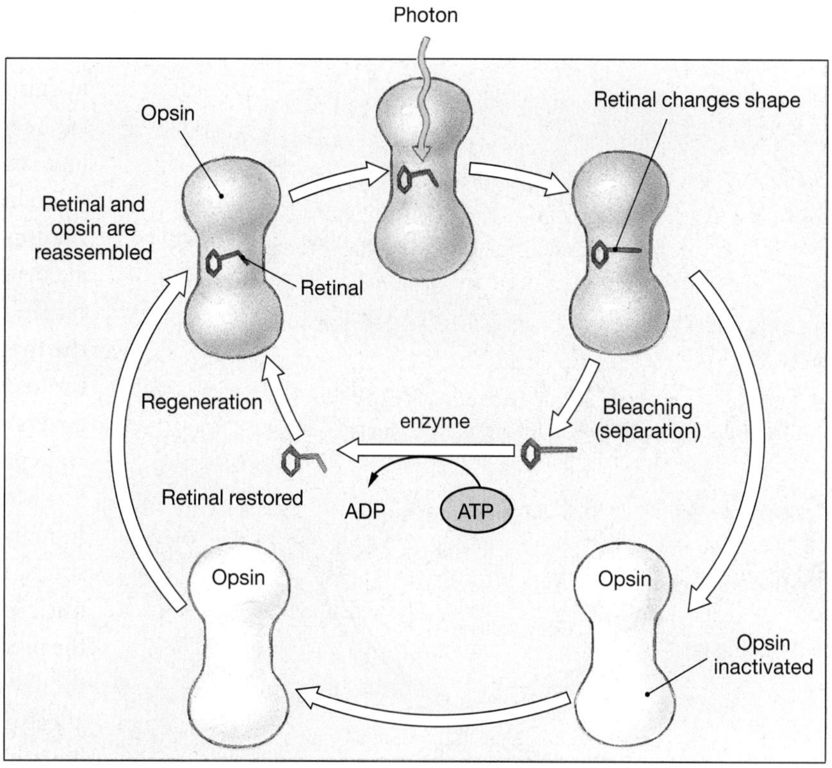

Photoreception

Photoreception begins when a photon strikes a rhodopsin molecule in the outer segment of a photoreceptor. When the photon is absorbed, a change in the shape of the retinal component activates opsin, starting a chain of enzymatic events that alters the rate of neurotransmitter release. This change is the signal that light has struck a photoreceptor at that particular location on the retina.

Shortly after the retinal changes shape, the rhodopsin molecule begins to break down into retinal and opsin, a process known as *bleaching* (Figure 9-19●). The retinal must be converted back to its former shape before it can recombine with opsin. This conversion requires energy in the form of ATP, and it takes time. Bleaching contributes to the lingering visual impression that you have after a camera flash goes off. After an intense exposure to light, a photoreceptor cannot respond to further stimulation until its rhodopsin molecules have been regenerated. As a result, a "ghost" image remains on the retina.

 NIGHT BLINDNESS

The visual pigments of the photoreceptors are synthesized from vitamin A. The body contains vitamin A reserves sufficient for several months, and a significant amount is stored in the cells of the pigmented part of the retina. If dietary sources are inadequate, these reserves are gradually exhausted and the amount of visual pigment in the photoreceptors begins to drop. Daylight vision is affected, but in daytime the

light is usually bright enough to stimulate any remaining visual pigments in the densely packed cone population. As a result, the problem first becomes apparent at night, when the dim light proves insufficient to activate the rods. This condition, known as **night blindness**, can be treated by administration of vitamin A. The body can convert the carotene pigments in many vegetables to vitamin A. Carrots are a particularly good source of carotene, which explains the old adage that carrots are good for your eyes.

THE VISUAL PATHWAY

The visual pathway begins at the photoreceptors and ends at the visual cortex of the cerebral hemispheres. In other sensory pathways we have examined, at most one synapse lies between a receptor and a sensory neuron that delivers information to the CNS. In the visual pathway, the message must cross two synapses (photoreceptor to bipolar cell, and bipolar cell to ganglion cell) before it heads toward the brain. Axons from the entire population of ganglion cells converge on the optic disc, penetrate the wall of the eye, and proceed toward the diencephalon as the optic nerve (N II). The two optic nerves, one from each eye, reach the diencephalon at the optic chiasm (Figure 9-20●). From this point, approximately half of the fibers proceed toward the thalamus on the same side of the brain, while the other half cross over to reach the thalamus on the opposite side. Nuclei in the thalamus act as switching and

● *Figure 9-19* **The Bleaching and Regeneration of Visual Pigments**

9

THE GENERAL AND SPECIAL SENSES

The General Senses • The Special Senses • Smell • Taste • Vision • **Equilibrium and Hearing** • Aging and the Senses • Chapter Review • MediaLab

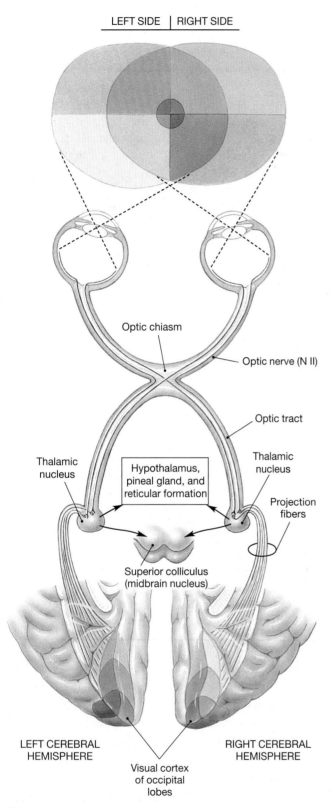

LEFT SIDE | RIGHT SIDE

Optic chiasm

Optic nerve (N II)

Optic tract

Thalamic nucleus

Hypothalamus, pineal gland, and reticular formation

Thalamic nucleus

Projection fibers

Superior colliculus (midbrain nucleus)

LEFT CEREBRAL HEMISPHERE

RIGHT CEREBRAL HEMISPHERE

Visual cortex of occipital lobes

● *Figure 9-20* **The Visual Pathway**
At the optic chiasm, a partial crossover of nerve fibers occurs. As a result, each hemisphere receives visual information from the lateral half of the retina on that side and from the medial half of the retina on the opposite side. Visual association areas integrate this information to develop a composite picture of the entire visual field.

processing centers that relay visual information to reflex centers in the brain stem as well as to the cerebral cortex. The visual information received by the *superior colliculi* (midbrain nuclei in the brain stem) controls constriction or dilation of the pupil and reflexes that control eye movement. ∞ p. 252

The sensation of vision arises from the integration of information arriving at the visual cortex of the cerebrum. The visual cortex of each occipital lobe contains a sensory map of the entire field of vision. As with the primary sensory cortex, the map does not faithfully duplicate the relative areas within the sensory field. For example, the area assigned to the fovea covers about 35 times the surface it would cover if the map were proportionally accurate.

Many centers in the brain stem receive visual information, from the thalamic nuclei or over collateral branches from the optic tracts. For example, some collaterals that bypass the thalamic nuclei synapse in the hypothalamus. Visual inputs there and at the pineal gland establish a daily pattern of activity that is tied to the day-night cycle. This *circadian* (*circa*, about + *dies*, day) *rhythm* affects your metabolic rate, endocrine function, blood pressure, digestive activities, awake-sleep cycle, and other processes.

CONCEPT CHECK QUESTIONS
Answers on page 310

❶ Which layer of the eye would be the first to be affected by inadequate tear production?

❷ When the lens is very round, are you looking at an object that is close to you or far from you?

❸ If a person is born without cones in her eyes, will she be able to see? Explain.

❹ How can a diet deficient in vitamin A affect vision?

Equilibrium and Hearing

The senses of equilibrium and hearing are provided by the *inner ear*, a receptor complex located in the temporal bone of the skull. ∞ p. 143 The basic receptor mechanism for these senses is the same. The receptors, or *hair cells*, are simple mechanoreceptors. The complex structure of the inner ear and the different arrangements of accessory structures account for the abilities of the hair cells to respond to different stimuli and thus to provide the input for two senses:

1. Equilibrium, which informs us of the position of the body in space by monitoring gravity, linear acceleration, and rotation.

2. Hearing, which enables us to detect and interpret sound waves.

THE ANATOMY OF THE EAR

The ear is divided into three anatomical regions: the *external ear*, the *middle ear*, and the *inner ear* (Figure 9-21●). The external ear, the visible portion of the ear, collects and directs sound waves toward the middle ear, a chamber located in a thickened portion of the temporal bone. Structures in the middle ear collect and amplify sound waves and transmit them to appropriate structures of the inner ear. The inner ear contains the sensory organs for hearing and equilibrium.

The External Ear

The **external ear** includes the fleshy **auricle**, or *pinna*, which surrounds the entrance to the **external auditory canal**, or *ear canal*. The auricle, which is supported by elastic cartilage, protects the opening of the canal. It also provides directional sensitivity to the ear: Sounds coming from behind the head are partially blocked by the auricle; sounds coming from the side are collected and channeled into the external auditory canal. (When you "cup your ear" with your hand to hear a faint sound more clearly, you are exaggerating this effect.) **Ceruminous** (se-ROO-mi-nus) **glands** along the external auditory canal secrete a waxy material (*cerumen*) that helps prevent the entry of foreign objects and insects, as do many small, outwardly projecting hairs. The waxy cerumen also slows the growth of microorganisms in the ear canal and reduces the chances of infection. The external auditory canal ends at the **tympanic membrane**, also called the *tympanum* or *eardrum*. The tympanic membrane is a thin sheet that separates the external ear from the middle ear (Figure 9-21●).

The Middle Ear

The **middle ear**, or *tympanic cavity*, is filled with air. It is separated from the external auditory canal by the tympanum, but it communicates with the superior portion of the pharynx, a region known as the *nasopharynx*, and with *air cells* in the mastoid process of the temporal bone. The connection with the nasopharynx is the **auditory tube**, also called the *pharyngotympanic tube* or the *Eustachian tube* (Figure 9-21●). The auditory tube permits the equalization of pressure on either side of the eardrum. Unfortunately, it can also allow microorganisms to travel from the nasopharynx into the tympanic cavity, leading to an unpleasant middle ear infection known as *otitis media*.

The Auditory Ossicles. The middle ear contains three tiny ear bones, collectively called **auditory ossicles**. The ear bones

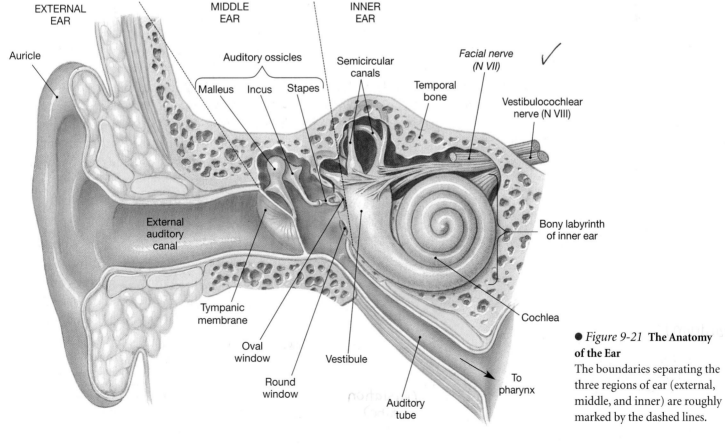

EXTERNAL EAR — MIDDLE EAR — INNER EAR

Auricle
Auditory ossicles
Malleus Incus Stapes
Semicircular canals
Facial nerve (N VII)
Temporal bone
Vestibulocochlear nerve (N VIII)
External auditory canal
Bony labyrinth of inner ear
Tympanic membrane
Oval window
Round window
Vestibule
Cochlea
To pharynx
Auditory tube

● *Figure 9-21* **The Anatomy of the Ear**
The boundaries separating the three regions of ear (external, middle, and inner) are roughly marked by the dashed lines.

9 THE GENERAL AND SPECIAL SENSES

The General Senses • The Special Senses • Smell • Taste • Vision • **Equilibrium and Hearing** • Aging and the Senses • Chapter Review • MediaLab

connect the tympanum with the receptor complex of the inner ear (Figure 9-22●). The three auditory ossicles are the malleus, the incus, and the stapes. The **malleus** (*malleus*, hammer) attaches at three points to the interior surface of the tympanum. The middle bone, the **incus** (*incus*, anvil), attaches the malleus to the inner bone, the **stapes** (*stapes*, stirrup). The base of the stapes almost completely fills the *oval window*, a small opening in the bone that encloses the inner ear.

Vibration of the tympanic membrane converts arriving sound energy into mechanical movements of the auditory ossicles. The ossicles act as levers that conduct the in-out vibrations to the inner ear's fluid-filled inner chamber. The tympanum is larger and heavier than the delicate membrane spanning the oval window, so the amount of movement increases markedly from tympanum to oval window.

This magnification in movement allows us to hear very faint sounds. It can also be a problem, however, when we are exposed to very loud noises. Within the tympanic cavity, two small muscles protect the eardrum and ossicles from violent movements under noisy conditions. The *tensor tympani* (TEN-sor tim-PAN-ē) *muscle* pulls on the malleus, which increases the stiffness of the tympanum and reduces the amount of possible movement. The *stapedius* (sta-PĒ-dē-us) *muscle* pulls on the stapes, thereby reducing its movement at the oval window.

The Inner Ear

The senses of equilibrium and hearing are provided by the receptors of the **inner ear**. The receptors lie within the **membranous labyrinth** (*labyrinthos*, network of canals), a collection of tubes and chambers filled with a fluid called **endolymph** (EN-dō-limf). The **bony labyrinth** is a shell of

dense bone that surrounds and protects the membranous labyrinth. Its inner contours closely follow the contours of the membranous labyrinth (Figure 9-23a●), while its outer walls are fused with the surrounding temporal bone (see Figure 9-21●). Between the bony and membranous labyrinths flows another fluid, the **perilymph** (PER-i-limf).

The bony labyrinth can be subdivided into three parts as seen in Figure 9-23a●.

1. *Vestibule.* The **vestibule** (VES-ti-būl) includes a pair of membranous sacs, the **saccule** (SAK-ūl) and the **utricle** (Ū-tre-kl). Receptors in these sacs provide sensations of gravity and linear acceleration.
2. *Semicircular canals.* The **semicircular canals** enclose slender *semicircular ducts*. Receptors in the semicircular ducts are stimulated by rotation of the head. The combination of vestibule and semicircular canals is called the **vestibular complex**, because the fluid-filled chambers within the vestibule are continuous with those of the semicircular canals.
3. *Cochlea.* The bony **cochlea** (KOK-lē-a; *cochlea*, snail shell) contains the **cochlear duct** of the membranous labyrinth. Receptors in the cochlear duct provide the sense of hearing. The cochlear duct is sandwiched between a pair of perilymph-filled chambers, and the entire complex is coiled around a central bony hub.

The bony labyrinth's walls are dense bone everywhere except at two small areas near the base of the cochlea. The round window is an opening in the bone of the cochlea. A thin membrane spans the opening and separates perilymph in the cochlea from the air in the middle ear. The membrane spanning the oval window is firmly attached to the base of the stapes. When a sound vibrates the tympanic membrane, the movements are conducted over the malleus and incus to the stapes. Movement of the stapes ultimately leads to the stimulation of receptors in the cochlear duct, and we hear the sound.

Receptor Function in the Inner Ear. The receptors of the inner ear are called **hair cells** (Figure 9-23b●). Each hair cell communicates with a sensory neuron by continually releasing small quantities of neurotransmitter. The free surface of this receptor supports 80–100 long microvilli called *stereocilia*. ⌒ p. 86 Hair cells do not actively move their stereocilia. However, when an external force pushes the stereocilia, their movement distorts the cell surface and alters its rate of neurotransmitter release. Displacement of the stereocilia in one direction stimulates the hair cells (and increases neurotransmitter release); displacement in

Temporal bone
Malleus
Incus
Base of stapes in oval window
External auditory canal
Tensor tympani muscle
Stapes
Tympanic membrane (eardrum)
Stapedius muscle
Round window
Auditory tube (eustation tube)

● *Figure 9-22* **The Middle Ear**
The structures of the middle ear.

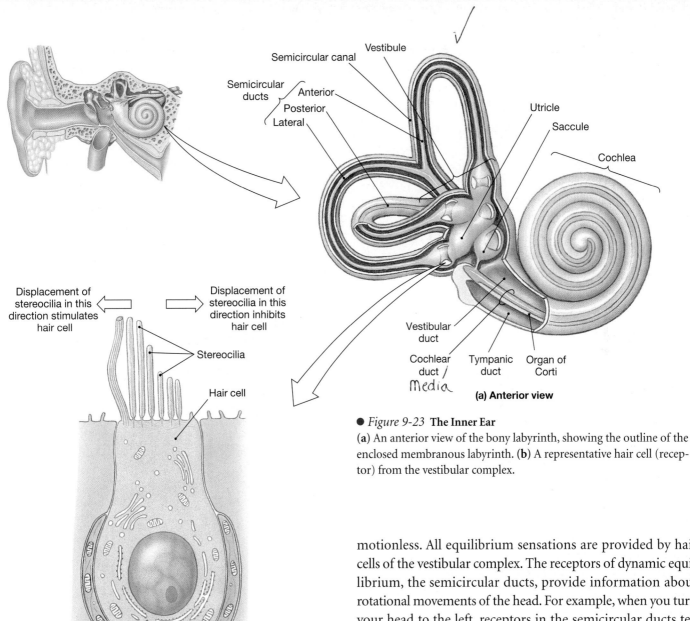

Displacement of stereocilia in this direction stimulates hair cell

Displacement of stereocilia in this direction inhibits hair cell

Stereocilia

Hair cell

Sensory nerve ending

(b) Hair cell

Vestibule

Semicircular canal

Semicircular ducts

Anterior

Posterior

Lateral

Utricle

Saccule

Cochlea

Vestibular duct

Cochlear duct / Media

Tympanic duct

Organ of Corti

(a) Anterior view

● *Figure 9-23* **The Inner Ear**
(**a**) An anterior view of the bony labyrinth, showing the outline of the enclosed membranous labyrinth. (**b**) A representative hair cell (receptor) from the vestibular complex.

motionless. All equilibrium sensations are provided by hair cells of the vestibular complex. The receptors of dynamic equilibrium, the semicircular ducts, provide information about rotational movements of the head. For example, when you turn your head to the left, receptors in the semicircular ducts tell you how rapid the movement is and in which direction. The receptors of static equilibrium, the saccule and the utricle, provide information about your position with respect to gravity. If you stand with your head tilted to one side, hair cells in these receptors will report the angle involved and whether your head tilts forward or backward. These receptors are also stimulated by sudden changes in velocity. For example, when your car accelerates, the saccular and utricular receptors give you the sensation of increasing speed.

The Semicircular Ducts: Rotational Motion
Sensory receptors in the semicircular ducts respond to rotational movements of the head. Figure 9-23a● illustrates the **anterior**, **posterior**, and **lateral semicircular ducts** and their continuity with the utricle. Each semicircular duct contains a swollen region, the *ampulla*, which contains the sensory receptors (Figure 9-24●). Hair cells attached to the wall of the ampulla

the opposite direction inhibits the hair cells (and decreases neurotransmitter release).

EQUILIBRIUM
There are two aspects of equilibrium: (1) **dynamic equilibrium**, which aids us in maintaining our balance when the head and body are moved suddenly, and (2) **static equilibrium**, which maintains our posture and stability when the body is

299

9 THE GENERAL AND SPECIAL SENSES

The General Senses • The Special Senses • Smell • Taste • Vision • **Equilibrium and Hearing** • Aging and the Senses • Chapter Review • MediaLab

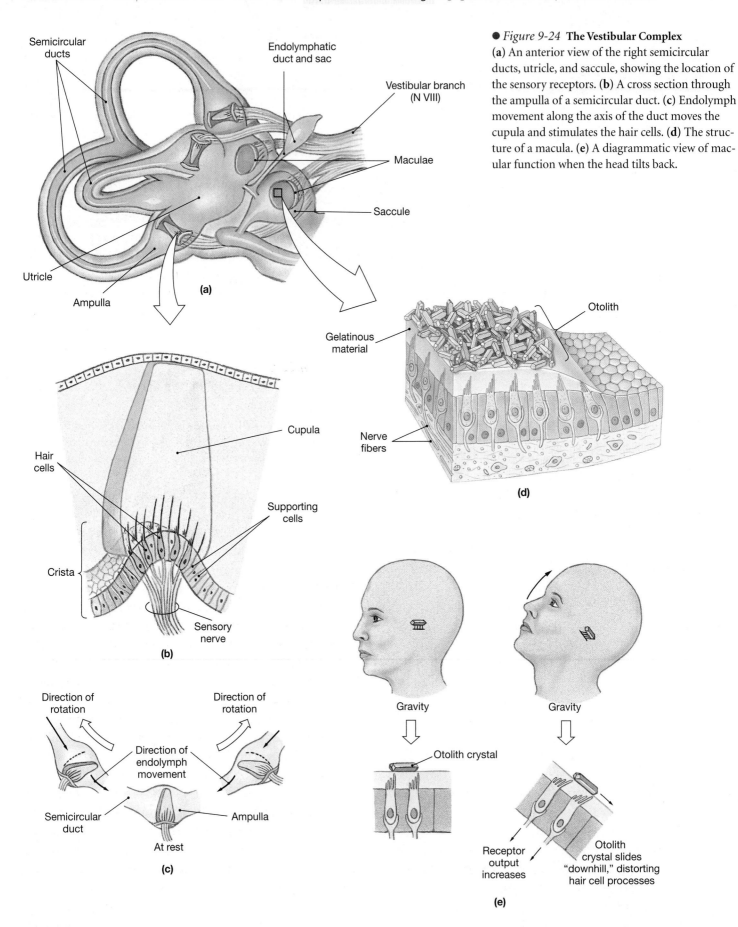

● *Figure 9-24* **The Vestibular Complex** (**a**) An anterior view of the right semicircular ducts, utricle, and saccule, showing the location of the sensory receptors. (**b**) A cross section through the ampulla of a semicircular duct. (**c**) Endolymph movement along the axis of the duct moves the cupula and stimulates the hair cells. (**d**) The structure of a macula. (**e**) A diagrammatic view of macular function when the head tilts back.

Semicircular ducts

Endolymphatic duct and sac

Vestibular branch (N VIII)

Maculae

Saccule

Utricle

Ampulla

(a)

Gelatinous material

Otolith

Nerve fibers

(d)

Hair cells

Cupula

Supporting cells

Crista

Sensory nerve

(b)

Direction of rotation

Direction of rotation

Direction of endolymph movement

Semicircular duct

Ampulla

At rest

(c)

Gravity

Gravity

Otolith crystal

Receptor output increases

Otolith crystal slides "downhill," distorting hair cell processes

(e)

form a raised structure known as a *crista*. The stereocilia of the hair cells are embedded in a gelatinous structure called the *cupula* (KŪ-pū-luh), which nearly fills the ampulla (Figure 9-24b●). When the head rotates in the plane of the canal, movement of the endolymph pushes against this structure and stimulates the hair cells (Figure 9-24c●).

Each semicircular duct responds to one of three possible rotational movements. To distort the cupula and stimulate the receptors, endolymph must flow along the axis of the duct; that flow will occur only when there is rotation in that plane. A horizontal rotation, as in shaking the head "no," stimulates the hair cells of the lateral semicircular duct. Nodding "yes" excites receptors of the anterior duct, and tilting the head from side to side activates receptors in the posterior duct. The three planes monitored by the semicircular ducts correspond to the three dimensions in the world around us, and they provide accurate information about even the most complex movements.

The Vestibule: Gravity and Linear Acceleration

Receptors in the utricle and saccule respond to gravity and linear acceleration. As depicted in Figure 9-24a●, the hair cells of the utricle and saccule are clustered in oval **maculae** (MAK-ū-lē; *macula*, spot). As in the ampullae, the hair cell processes in the maculae are embedded in a gelatinous mass, but the macular receptors lie under a thin layer of densely packed calcium carbonate crystals. The complex as a whole (gelatinous mass and crystals) is called an *otolith* (*oto-*, ear + *lithos*, a stone) (Figure 9-24d●). When the head is in the normal, upright position, the otolith crystals sit atop the macula. Their weight presses down on the macular surface, pushing the sensory hairs downward rather than to one side or another. When the head is tilted, the pull of gravity on the otolith crystals shift their weight to the side, distorting the sensory hairs. The change in receptor activity tells the CNS that the head is no longer level (Figure 9-24e●).

Otolith crytals are relatively dense and heavy, and they are connected to the rest of the body only by the sensory processes of the macular cells. So whenever the rest of the body makes a sudden movement, the otolith crystals lag behind. For example, when an elevator starts downward, we are immediately aware of it because the otolith crystals no longer push so forcefully against the surfaces of the receptor cells. Once they catch up and the elevator has reached a constant speed, we are no longer aware of any movement until the elevator brakes to a halt. As the body slows down, the otolith crystals press harder against the hair cells and we "feel" the force of gravity increase.

A similar mechanism accounts for our perception of linear acceleration in a car that speeds up suddenly. The otoliths lag behind, distorting the sensory hairs and changing the activity in the sensory neurons. A comparable otolith movement occurs when the chin is raised and gravity pulls the otoliths backward. On the basis of visual information, the brain decides whether the arriving sensations indicate acceleration or a change in head position.

Pathways for Equilibrium Sensations

Hair cells of the vestibule and of the semicircular canals are monitored by sensory neurons whose fibers form the **vestibular branch** of the vestibulocochlear nerve, N VIII. These fibers synapse on neurons in the *vestibular nuclei* located at the boundary between the pons and the medulla oblongata. The two vestibular nuclei (1) integrate the sensory information arriving from each side of the head; (2) relay information to the cerebellum; (3) relay information to the cerebral cortex, providing a conscious sense of position and movement; and (4) send commands to motor nuclei in the brain stem and in the spinal cord. These reflexive motor commands are distributed to the motor nuclei for cranial nerves involved with eye, head, and neck movements (N III, IV, VI, and XI). Descending instructions along the *vestibulospinal tracts* of the spinal cord adjust peripheral muscle tone to complement the reflexive movements of the head or neck.

HEARING

The receptors of the cochlear duct provide us with a sense of hearing that enables us to detect the quietest whisper yet remain functional in a crowded, noisy room. The receptors responsible for auditory sensations are hair cells similar to those of the vestibular complex. However, their placement within the cochlear duct and the organization of the surrounding accessory structures shield them from stimuli other than sound. In conveying vibrations from the tympanic membrane to the oval window, the auditory ossicles convert sound energy (pressure waves) in air to pressure pulses in the perilymph of the cochlea. These pressure pulses stimulate hair cells along the cochlear spiral. The *frequency* of the perceived sound is determined by *which part* of the cochlear duct is stimulated. The *intensity* (volume) of the perceived sound is determined by *how many* hair cells are stimulated at that part of the cochlear duct.

9 THE GENERAL AND SPECIAL SENSES

The General Senses • The Special Senses • Smell • Taste • Vision • **Equilibrium and Hearing** • Aging and the Senses • Chapter Review • MediaLab

The Cochlear Duct

In sectional view, the cochlear duct, or *scala media*, lies between a pair of perilymphatic chambers: the **vestibular duct** (*scala vestibuli*) and the **tympanic duct** (*scala tympani*), (Figure 9-25a●). The vestibular and tympanic ducts are interconnected at the tip of the cochlear spiral. The outer surfaces of these ducts are encased by the bony labyrinth everywhere except at the oval window (the base of the vestibular duct) and the round window (the base of the tympanic duct).

The Organ of Corti. The hair cells of the cochlear duct are located in the **organ of Corti** (Figure 9-25b●). This senso-ry structure sits above the **basilar membrane**, which separates the cochlear duct from the tympanic duct. The hair cells are arranged in a series of longitudinal rows, with their stereocilia in contact with the overlying **tectorial membrane** (tek-TŌR-ē-al; *tectum*, roof). This membrane is firmly attached to the inner wall of the cochlear duct. When a portion of the basilar membrane bounces up and down, the stereocilia of the hair cells are distorted as they are pushed up against the tectorial membrane. The basilar membrane moves in response to pressure waves in the perilymph. These waves are produced when sounds arrive at the tympanic membrane.

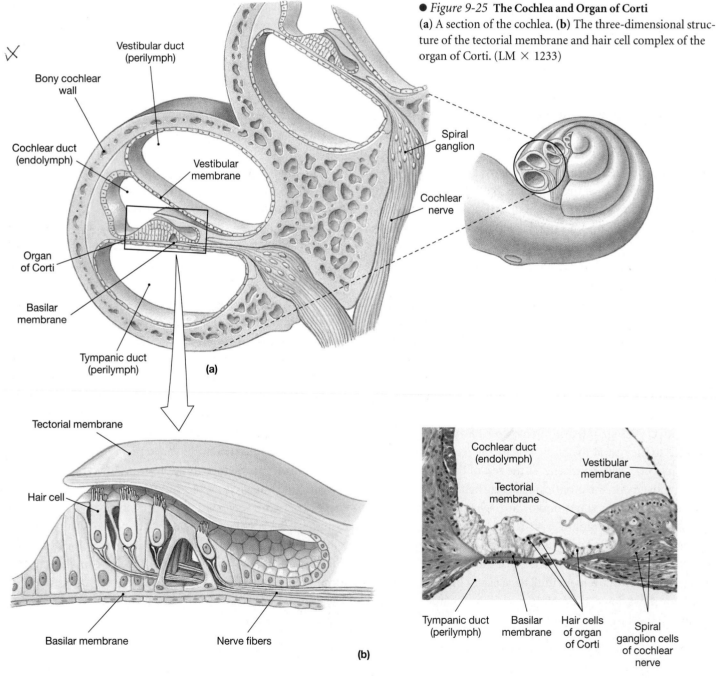

● *Figure 9-25* **The Cochlea and Organ of Corti**
(**a**) A section of the cochlea. (**b**) The three-dimensional structure of the tectorial membrane and hair cell complex of the organ of Corti. (LM × 1233)

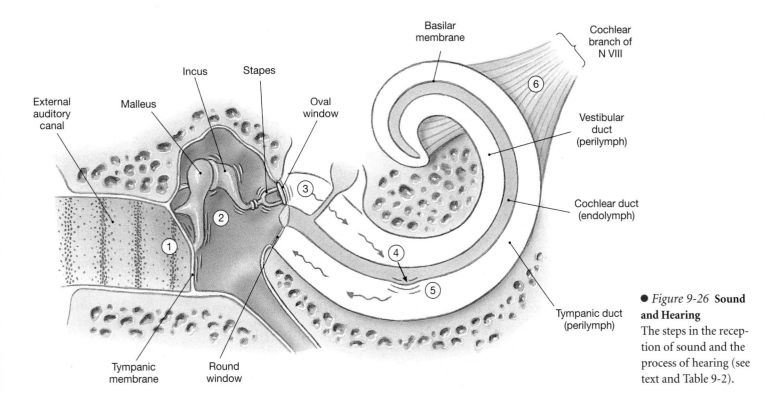

External auditory canal

Malleus

Incus

Stapes

Oval window

Basilar membrane

Cochlear branch of N VIII

Vestibular duct (perilymph)

Cochlear duct (endolymph)

Tympanic duct (perilymph)

Tympanic membrane

Round window

● *Figure 9-26* **Sound and Hearing** The steps in the reception of sound and the process of hearing (see text and Table 9-2).

The Hearing Process

Hearing is the detection of sound, which consists of pressure waves that are conducted through air, water, and solids. Physicists use the term **cycles** rather than waves, and the number of cycles per second (cps), or **hertz (Hz)**, represents the **frequency** of the sound. What we perceive as the **pitch** of a sound (how high or low it is) is our sensory response to its frequency. A sound of high frequency (high pitch) might have a frequency of 15,000 Hz or more; a sound of low frequency (low pitch) could have a frequency of 100 Hz or less. The amount of energy, or power, of a sound determines its *intensity*, or volume. Intensity is reported in **decibels** (DES-i-belz). Some examples of different sounds and their intensities include: a soft whisper (30 decibels); a refrigerator (50 decibels); a gas lawnmower (90 decibels); a chain saw (100 decibels); and a jet plane (140 decibels).

Hearing can be divided into six basic steps, diagrammed in Figure 9-26● and summarized in Table 9-2.

Step 1: *Sound waves arrive at the tympanic membrane.* Sound waves enter the external auditory canal and travel toward the tympanic membrane. Sound waves approaching the side of the head have direct access to the tympanic membrane on that side, whereas sounds arriving from another direction must bend around corners or pass through the auricle or other body tissues.

Step 2: *Movement of the tympanic membrane causes displacement of the auditory ossicles.* The tympanic membrane provides the surface for sound collection. It vibrates to sound waves with frequencies between approximately 20 and 20,000 Hz (in a young child). When the tympanic membrane vibrates, so do the malleus and, through their articulations, the incus and stapes.

Step 3: *The movement of the stapes at the oval window establishes pressure waves in the perilymph of the vestibular duct.* When it moves, the stapes applies pressure to the perilymph of the

---

TABLE 9-2 *Steps in the Production of an Auditory Sensation*

1. Sound waves arrive at the tympanic membrane.
2. The vibration of the tympanic membrane causes movement of the auditory ossicles.
3. The movement of the stapes at the oval window establishes pressure waves in the perilymph of the vestibular duct.
4. The pressure waves distort the basilar membrane on their way to the round window of the tympanic duct.
5. The vibration of the basilar membrane causes the vibration of hair cells against the tectorial membrane.
6. Information about the region and intensity of stimulation is relayed to the CNS over the cochlear branch of N VIII.

9 THE GENERAL AND SPECIAL SENSES

The General Senses • The Special Senses • Smell • Taste • Vision • Equilibrium and Hearing • **Aging and the Senses** • Chapter Review • MediaLab

vestibular duct. Because the rest of the cochlea is sheathed in bone, pressure applied at the oval window can be relieved only at the round window. When the stapes moves inward, the membrane spanning the round window bulges outward. As the stapes moves in and out, vibrating at the frequency of the sound at the tympanic membrane, it creates pressure waves within the perilymph.

Step 4: *The pressure waves distort the basilar membrane on their way to the round window of the tympanic duct.* These pressure waves cause movement in the basilar membrane. The basilar membrane does not have the same structure throughout its length. Near the oval window, it is narrow and stiff, and at its terminal end, it is wider and more flexible. As a result, the location of maximum stimulation varies with the frequency of the sound. High-frequency sounds vibrate the basilar membrane near the oval window. The lower the frequency of the sound, the farther from the oval window the area of maximum distortion will be. The actual *amount* of movement at a given location will depend on the amount of force applied by the stapes. The louder the sound, the greater the movement of the basilar membrane.

Step 5: *Vibration of the basilar membrane causes vibration of hair cells against the tectorial membrane.* The vibration of the affected region of the basilar membrane moves hair cells against the tectorial membrane. The resulting displacement of the hair cells' stereocilia stimulates sensory neurons. The hair cells are arranged in several rows; a very soft sound may stimulate only a few hair cells in a portion of one row. As the volume of a sound increases, not only do these hair cells become more active but additional hair cells—at first in the same row, and then in adjacent rows—are stimulated as well. The number of hair cells responding in a given region of the organ of Corti thus provides information about the volume, or intensity, of the sound.

Step 6: *Information about the region and intensity of stimulation is relayed to the CNS over the cochlear branch of N VIII.* The cell bodies of the sensory neurons that monitor the cochlear hair cells are located at the center of the bony cochlea (Figure 9-25a●) in the *spiral ganglion.* The information is carried to the cochlear nuclei of the medulla oblongata for subsequent distribution to other centers in the brain.

Auditory Pathways

Hair cell stimulation activates sensory neurons whose cell bodies are in the adjacent spiral ganglion. Their afferent fibers form the **cochlear branch** (Figure 9-27●) of the vestibulocochlear nerve (N VIII). These axons enter the medulla oblongata and synapse at the *cochlear nucleus.* From here the information

crosses to the opposite side of the brain and ascends to the *inferior colliculus* of the midbrain. This processing center coordinates a number of responses to acoustic stimuli, including auditory reflexes involving skeletal muscles of the head, face, and trunk. For example, these reflexes automatically change the position of the head in response to a sudden loud noise.

Before reaching the cerebral cortex and our conscious awareness, ascending auditory sensations synapse in the thalamus. Thalamic fibers then deliver the information to the auditory cortex of the temporal lobe. In effect, the auditory cortex contains a map of the organ of Corti. High-frequency sounds activate one portion of the cortex and low-frequency sounds affect another. If the auditory cortex is damaged, the individual will respond to sounds and have normal acoustic reflexes, but sound interpretation and pattern recognition will be difficult or impossible. Damage to the adjacent association area leaves the ability to detect the tones and patterns, but produces an inability to comprehend their meaning.

Auditory Sensitivity

Our hearing abilities are remarkable, though it is difficult to assess the absolute sensitivity of the system. From the softest audible sound to the loudest tolerable blast represents a trillionfold increase in power. Theoretically, if we were to remove the stapes, the receptor mechanism is so sensitive that we could hear the sound of air molecules bouncing off the oval window, responding to displacements as small as one-tenth the diameter of a hydrogen atom. We never utilize the full potential of this system, because body movements and our internal organs produce squeaks, groans, thumps, and other sounds that are tuned out by adaptation. When other environmental noises fade away, the level of adaptation drops and the system becomes increasingly sensitive. If we relax in a quiet room, our heartbeat seems to get louder and louder as the auditory system adjusts to the level of background noise.

 HEARING DEFICITS

Probably more than 6 million people in the United States alone have at least a partial hearing deficit. **Conductive deafness** results from conditions in the outer or middle ear that block the normal transfer of vibration from the tympanic membrane to the oval window. An external auditory canal plugged by accumulated wax or trapped water may cause a temporary hearing loss. Scarring or perforation of the tympanum and immobilization of one or more of the auditory ossicles are more serious causes of conduction deafness.

In **nerve deafness**, the problem lies within the cochlea or somewhere along the auditory pathway. The vibrations are reaching the oval window and entering the perilymph, but the receptors either cannot respond or their response cannot

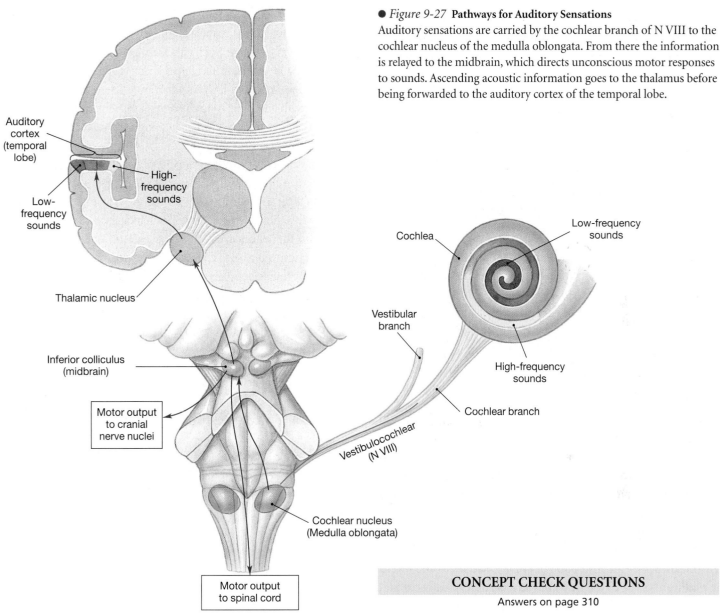

● *Figure 9-27* **Pathways for Auditory Sensations**
Auditory sensations are carried by the cochlear branch of N VIII to the cochlear nucleus of the medulla oblongata. From there the information is relayed to the midbrain, which directs unconscious motor responses to sounds. Ascending acoustic information goes to the thalamus before being forwarded to the auditory cortex of the temporal lobe.

Auditory cortex (temporal lobe)

High-frequency sounds

Low-frequency sounds

Thalamic nucleus

Inferior colliculus (midbrain)

Motor output to cranial nerve nuclei

Motor output to spinal cord

Cochlea

Low-frequency sounds

Vestibular branch

High-frequency sounds

Cochlear branch

Vestibulocochlear (N VIII)

Cochlear nucleus (Medulla oblongata)

---

reach its central destinations. For example, very loud (high-intensity) sounds can produce nerve deafness by breaking stereocilia off the surfaces of the hair cells. (The reflex contraction of the tensor tympani and stapedius muscles in response to a dangerously loud noise occurs in less than 0.1 second, but this may not be fast enough.) Drugs such as the aminoglycoside antibiotics (*neomycin* or *gentamicin*) may diffuse into the endolymph and kill the hair cells. Because hair cells and sensory nerves can also be damaged by bacterial infection, the potential side effects must be balanced against the severity of infection.

Many treatment options are available for conductive deafness; treatment options for nerve deafness are relatively limited. Because many of these problems become progressively worse, early diagnosis improves the chances for successful treatment.

CONCEPT CHECK QUESTIONS

Answers on page 310

❶ If the round window were not able to bulge out with increased pressure in the perilymph, how would sound perception be affected?

❷ How would the loss of stereocilia from the hair cells of the organ of Corti affect hearing?

Aging and the Senses

The general lack of replacement of neurons in the nervous system leads to an inevitable decline in sensory function with age. Although part of this decline can be compensated by an increase in stimuli strength or concentration, the loss of axons to conduct sensory action potentials cannot be increased in a like manner. Other effects of aging also take their toll on the senses:

9

THE GENERAL AND SPECIAL SENSES

The General Senses • The Special Senses • Smell • Taste • Vision • Equilibrium and Hearing • Aging and the Senses • **Chapter Review** • MediaLab

SMELL

Unlike populations of other neurons, the population of olfactory receptor cells is regularly replaced by the division of stem cells in the olfactory epithelium. Despite this process, the total number of receptors declines with age, and the remaining receptors become less sensitive. As a result, elderly individuals have difficulty detecting odors in low concentrations. This drop in the number of receptors accounts for "Grandmother's" tendency to apply perfume in excessive quantities and explains why "Grandfather's" aftershave lotion seems so overdone. They must apply more to be able to smell it.

TASTE

Tasting abilities change with age due to the thinning of mucous membranes and a reduction in the number and sensitivity of taste buds. We begin life with more than 10,000 taste buds, but that number begins declining dramatically by age 50. The sensory loss becomes especially significant because aging individuals also experience a decline in the number of olfactory receptors. As a result, many of the elderly find that their food tastes bland and unappetizing. Children, however, find the same food too spicy.

VISION

Various disorders of vision are associated with normal aging; the most common involve the lens and the neural part of the retina. With age, the lens loses its elasticity and stiffens. As a result, it becomes more difficult to see objects up close, and older individuals become farsighted—a condition called *presbyopia*. ∞ p. 293 For example, the inner limit of clear vision, known as the *near point of vision*, changes from 7–9 cm in children to 15–20 cm in young adults and typically reaches

83 cm by age 60. As noted earlier, the most common cause of the development of a cataract (the loss of transparency in the lens) is advancing age. Such cataracts are called *senile cataracts*. ∞ p. 291 In addition to changes in the near point of vision and some changes in lens transparency, there is a gradual loss of rods with age. This reduction explains why individuals over age 60 need almost twice as much light to read by than those of age 40.

Another contributor to the loss of vision with age is *macular degeneration*, the leading cause of blindness in persons over 50. It is typically associated with the growth and proliferation of blood vessels in the retina. The leakage of blood and plasma from these abnormal vessels causes retinal scarring and a loss of photoreceptors. The vascular growth begins in the macula lutea, the area of the retina correlated with acute vision. Color vision is affected as the cones deteriorate.

HEARING

Hearing is generally affected less by aging than are the other senses. However, because the tympanic membrane loses some of its elasticity, it becomes more difficult to hear high-pitched sounds. The progressive loss of hearing that occurs with aging is called *presbycusis* (prez-bē-KŪ-sis; *presbys*, old man + *akousis*, hearing).

CONCEPT CHECK QUESTIONS

Answers on page 310

❶ How would you explain why the same food may be too spicy for a child, but not an elderly individual?

❷ Explain why we have an increasingly difficult time seeing close objects as we age.

Related Clinical Terms

analgesic: A drug that relieves pain without an individual losing their sensitivity to other stimuli, such as touch or pressure.

anesthesia: A total or partial loss of sensation.

cataract: An abnormal lens that has lost its transparency.

color blindness: A condition in which a person is unable to distinguish certain colors.

conductive deafness: Deafness resulting from conditions in the outer or middle ear that block the transfer of vibrations from the tympanic membrane to the oval window.

glaucoma: A condition characterized by increased fluid pressure within the eye due to the impaired reabsorption of aqueous humor; can result in blindness.

hyperopia, or *farsightedness:* A condition in which nearby objects are blurry but distant objects are clear.

Ménière's disease: A condition in which high fluid pressures rupture the walls of the membranous labyrinth, resulting in acute vertigo (an inappropriate sense of motion) and inappropriate auditory sensations.

myopia, or *nearsightedness:* A condition in which vision at close range is normal but distant objects appear blurry.

nerve deafness: Deafness resulting from problems within the cochlea or along the auditory pathway.

nystagmus: Abnormal eye movements that may appear after the brain stem or inner ear is damaged.

opthalmology (of-thal-MOL-o-jē): The study of the eye and its diseases.

presbyopia: A type of hyperopia that develops with age as lenses become less elastic.

retinitis pigmentosa: A group of inherited retinopathies characterized by the progressive deterioration of photoreceptors, eventually resulting in blindness.

retinopathy (ret-I-NOP-ah-thē): A disease or disorder of the retina.

scotomas (skō-TŌ-muhz): Abnormal blind spots in the field of vision other than the optic disc.

strabismus: Deviations in the alignment of the eyes to each other; one or both eyes turn inward or outward.

Key Terms

Summary Outline

INTRODUCTION 278

1. The **general senses** are temperature, pain, touch, pressure, vibration, and proprioception; receptors for these sensations are distributed throughout the body. Receptors for the **special senses** (smell, taste, vision, balance, and hearing) are located in specialized areas or in sense organs.

2. A *sensory receptor* is a specialized cell that, when stimulated, sends a sensation to the CNS. The simplest receptors are **free nerve endings**; the most complex have specialized accessory structures that isolate them from most all but specific types of stimuli.

3. Sensory information is relayed in the form of action potentials in a sensory (afferent) fiber. In general, the larger the stimulus, the greater the frequency of action potentials. The CNS interprets the nature of the arriving sensory information on the basis of the area of the brain stimulated.

4. **Adaptation** (a reduction in sensitivity in the presence of a constant stimulus) involves changes in receptor sensitivity or inhibition along the sensory pathways.

THE GENERAL SENSES 278

Pain...278

1. **Nociceptors** respond to a variety of stimuli usually associated with tissue damage. The two types of these painful sensations are **fast pain**, or *prickling pain*, and **slow pain**, or *burning and aching pain*.

2. The perception of pain coming from parts of the body that are not actually stimulated is called **referred pain**. *(Figure 9-1)*

Temperature...279

3. **Thermoreceptors** respond to changes in temperature.

Touch, Pressure, and Position 279

4. **Mechanoreceptors** respond to physical distortion, contact, or pressure on their cell membranes; **tactile receptors** to touch, pressure, and vibration; **baroreceptors** to pressure changes in the walls of blood vessels, the digestive and urinary tracts, and the lungs; and **proprioceptors** to positions of joints and muscles.

5. **Fine touch and pressure receptors** provide detailed information about a source of stimulation; **crude touch and pressure receptors** are poorly localized. Important tactile receptors include the *root hair plexus*, *tactile discs*, *tactile corpuscles*, *lamellated corpuscles*, and *Ruffini corpuscles*. *(Figure 9-2)*

6. Baroreceptors in the walls of major arteries and veins respond to changes in blood pressure, and those along the digestive tract help coordinate reflex activities of digestion. *(Figure 9-3)*

7. Proprioceptors monitor the position of joints, tension in tendons and ligaments, and the state of muscular contraction. Proprioceptors include Golgi tendon organs and muscle spindles.

Chemical Detection 281

8. In general, **chemoreceptors** respond to water-soluble and lipid-soluble substances dissolved in the surrounding fluid. They monitor the chemical composition of body fluids. *(Figure 9-4)*

THE SPECIAL SENSES 282

SMELL 282

1. The **olfactory organs** contain the **olfactory epithelium** with **olfactory receptors** (neurons sensitive to chemicals dissolved in the overlying mucus), supporting cells, and *basal* (stem) *cells*. Their surfaces are coated with the secretions of the **olfactory glands**. *(Figure 9-5)*

2. The olfactory receptors are modified neurons.

The Olfactory Pathways 283

3. The olfactory system has extensive limbic and hypothalamic connections.

TASTE 283

1. **Taste (gustatory) receptors** are clustered in **taste buds**, each of which contains **gustatory cells**, which extend *taste hairs* through a narrow **taste pore**. *(Figure 9-6)*

2. Taste buds are associated with **papillae**, epithelial projections on the superior surface of the tongue. *(Figure 9-6)*

3. The **primary taste sensations** are sweet, salt, sour, and bitter; umami and water receptors are also present. *(Figure 9-6)*

The Taste Pathways 284

4. The taste buds are monitored by cranial nerves that synapse within a nucleus of the medulla oblongata.

VISION..285

The Accessory Structures of the Eye 285

1. The **accessory structures** of the eye include the eyelids, the eyelashes, various exocrine glands, and the extrinsic eye muscles.

2. An epithelium called the **conjunctiva** covers most of the the exposed surface of the eye except the transparent **cornea**.

3. The secretions of the **lacrimal gland** bathe the conjunctiva; these secretions contain a *lysozyme* (an enzyme that attacks bacteria). Tears reach the nasal cavity after passing through the **lacrimal canals**, the **lacrimal sac**, and the **nasolacrimal duct**. *(Figure 9-7)*

4. Six **extrinsic eye muscles** control external eye movements: the **inferior** and **superior rectus**, **lateral** and **medial rectus**, and **superior** and **inferior obliques**. *(Figure 9-8; Table 9-1)*

The Eye...286

9 THE GENERAL AND SPECIAL SENSES

The General Senses • The Special Senses • Smell • Taste • Vision • Equilibrium and Hearing • Aging and the Senses • **Chapter Review** • MediaLab

5. The eye has three layers: an outer fibrous tunic, a vascular tunic, and an inner neural tunic. Most of the ocular surface is covered by the **sclera** (a dense fibrous connective tissue), which is continuous with the cornea, both part of the **fibrous tunic.** (*Figure 9-9*)

6. The **vascular tunic** includes the **iris**, the **ciliary body**, and the **choroid.** The iris forms the boundary between the eye's anterior and posterior chambers. The iris regulates the amount of light entering the eye. The ciliary body contains the *ciliary muscle* and the *ciliary processes*, which attach to the **suspensory ligaments** of the **lens.** (*Figures 9-9, 9-10*)

7. The **neural tunic** consists of an outer *pigmented part* and an inner *neural part*; the latter contains visual receptors and associated neurons. (*Figures 9-9, 9-11*)

8. From the photoreceptors, the information is relayed to **bipolar cells**, then to **ganglion cells**, and to the brain by the optic nerve. Horizontal cells and amacrine cells modify the signals passed between other retinal components. (*Figure 9-11*)

9. The ciliary body and lens divide the interior of the eye into a large **posterior cavity** and a smaller **anterior cavity.** The anterior cavity is subdivided into the **anterior chamber**, which extends from the cornea to the iris, and a **posterior chamber** between the iris and the ciliary body and lens. The posterior chamber contains the *vitreous body*, a gelatinous mass that helps stabilize the shape of the eye and supports the retina. (*Figure 9-13*)

10. **Aqueous humor** circulates within the eye and reenters the circulation after diffusing through the walls of the anterior chamber and into veins of the sclera through the *canal of Schlemm.* (*Figure 9-13*)

11. The lens, held in place by the suspensory ligaments, focuses a visual image on the retinal receptors. Light is refracted (bent) when it passes through the cornea and lens. During **accommodation**, the shape of the lens changes to focus an image on the retina. (*Figures 9-14, 9-15, 9-16*)

Visual Physiology

12. Light is radiated in waves with a characteristic wavelength. A *photon* is a single energy packet of visible light. The two types of **photoreceptors** (visual receptors of the retina) are **rods** and **cones.** Rods respond to almost any photon, regardless of its energy content; cones have characteristic ranges of sensitivity. Many cones are densely packed within the **fovea** (the central portion of the **macula lutea**), the site of sharpest vision. (*Figures 9-17, 9-18*)

13. Each photoreceptor contains an outer segment with membranous **discs** containing **visual pigments.** Light absorption occurs in the visual pigments, which are derivatives of **rhodopsin** (**opsin** plus the pigment **retinal**, which is synthesized from **vitamin A**). A photoreceptor responds to light by changing its rate of neurotransmitter release and thereby altering the activity of a bipolar cell. (*Figure 9-19*)

The Visual Pathway

14. The message is relayed from photoreceptors to bipolar cells to ganglion cells within the retina. The axons of ganglion cells converge at the optic disc and leave the eye as the optic nerve. A partial crossover occurs at the optic chiasm before the information reaches a nucleus in the thalamus on each side of the brain. From these nuclei, visual information is relayed to the visual cortex of the occipital lobe, which contains a sensory map of the field of vision. (*Figure 9-20*)

EQUILIBRIUM AND HEARING

1. The senses of equilibrium (**dynamic equilibrium** and **static equilibrium**) and hearing are provided by the receptors of the **inner ear.** Its chambers and canals contain the fluid **endolymph.** The **bony labyrinth** surrounds and protects the **membranous labyrinth**, and the space between them contains the fluid **perilymph.** The bony labyrinth consists of the **vestibule**, the **semicircular canals** (receptors in the vestibule and semicircular canals provide the sense of equilibrium) and the **cochlea** (these receptors provide the sense of hearing). The structures and air spaces of the **external ear** and **middle ear** help capture and transmit sound to the cochlea. (*Figures 9-21, 9-23*)

The Anatomy of the Ear

2. The external ear includes the **auricle** (*pinna*), which surrounds the entrance to the **external auditory canal**, which ends at the **tympanic membrane (eardrum).** (*Figure 9-21*)

3. The middle ear is connected to the nasopharynx by the **auditory tube** (*pharyngotympanic tube* or *Eustachian tube*). The middle ear encloses and protects the **auditory ossicles**, which connect the tympanic membrane with the receptor complex of the inner ear. (*Figures 9-21, 9-22*)

4. The vestibule includes a pair of membranous sacs, the **saccule** and **utricle**, whose receptors provide sensations of gravity and linear acceleration. The semicircular canals contain the **semicircular ducts**, whose receptors provide sensations of rotation. The cochlea contains the **cochlear duct**, an elongated portion of the membranous labyrinth. (*Figure 9-23a*)

5. The basic receptors of the inner ear are **hair cells**, whose surfaces support *stereocilia.* Hair cells provide information about the direction and strength of mechanical stimuli. (*Figure 9-23b*)

Equilibrium

6. The **anterior, posterior,** and **lateral semicircular ducts** are attached to the utricle. Each semicircular duct contains an *ampulla* with sensory receptors. There the stereocilia contact the *cupula*, a gelatinous mass that is distorted when endolymph flows along the axis of the duct. (*Figures 9-23, 9-24a, b, c*)

7. In the saccule and utricle, hair cells cluster within **maculae**, where their cilia contact the *otolith* (densely packed mineral crystals in a gelatinous mass). When the head tilts, the crystals shift, and the resulting distortion in the sensory hairs signals the CNS. (*Figure 9-24d, e*)

8. The vestibular receptors activate sensory neurons whose axons form the **vestibular branch** of the vestibulocochlear nerve (N VIII).

Hearing

9. Sound waves travel toward the tympanic membrane, which vibrates; the auditory ossicles conduct the vibrations to the inner ear. Movement at the oval window applies pressure to the perilymph of the **vestibular duct.** (*Figures 9-25, 9-26; Table 9-2*)

10. Pressure waves distort the **basilar membrane** and push the hair cells of the **organ of Corti** against the **tectorial membrane.** (*Figure 9-26; Table 9-2*)

11. The sensory neurons are located in the **spiral ganglion** of the cochlea. Afferent fibers of sensory neurons form the **cochlear branch** of the vestibulocochlear nerve (N VIII), synapsing at the cochlear nucleus. (*Figure 9-27*)

AGING AND THE SENSES

1. As part of the aging process, there are (1) gradual reductions in smell and taste sensitivity, (2) a tendency toward *presbyopia* and cataract formation in the eyes, and (3) a progressive loss of hearing (*presbycusis*).

Review Questions

Level 1: Reviewing Facts and Terms

Match each item in column A with the most closely related item in column B. Use letters for answers in the spaces provided.

COLUMN A

____ 1. myopia
____ 2. fibrous tunic
____ 3. nociceptors
____ 4. proprioceptors
____ 5. cones
____ 6. accommodation
____ 7. tympanic membrane
____ 8. thermoreceptors
____ 9. rods
____ 10. olfaction
____ 11. fovea
____ 12. hyperopia
____ 13. maculae
____ 14. semicircular ducts

COLUMN B

a. pain receptors
b. free nerve endings
c. sclera and cornea
d. rotational movements
e. provide information on joint position
f. color vision
g. site of sharpest vision
h. active in dim light
i. eardrum
j. change in lens shape to focus retinal image
k. nearsighted
l. farsighted
m. smell
n. gravity and acceleration receptors

---

15. Regardless of the nature of a stimulus, sensory information must be sent to the central nervous system in the form of:
    (a) dendritic processes
    (b) action potentials
    (c) neurotransmitter molecules
    (d) generator potentials

16. A reduction in sensitivity in the presence of constant stimulus is called:
    (a) transduction
    (b) sensory coding
    (c) line labeling
    (d) adaptation

17. Mechanoreceptors that detect pressure changes in the walls of blood vessels and in portions of the digestive, reproductive, and urinary tracts are
    (a) tactile receptors
    (b) baroreceptors
    (c) proprioceptors
    (d) free nerve endings

18. Examples of proprioceptors that monitor the position of joints and the state of muscular contraction are
    (a) Pacinian and Meissner's corpuscles
    (b) carotid and aortic sinuses
    (c) Merkel's discs and Ruffini corpuscles
    (d) Golgi tendon organs and muscle spindles

19. When chemicals dissolve in the nasal cavity, they stimulate:
    (a) gustatory cells
    (b) olfactory hairs
    (c) rod cells
    (d) tactile receptors

20. The taste sensation of sweetness is experienced on the:
    (a) posterior part of the tongue
    (b) anterior part of the tongue
    (c) right and left lateral sides of the tongue
    (d) the middle part of the tongue

21. The purpose of tears produced by the lacrimal apparatus is to:
    (a) keep conjunctival surfaces moist and clean
    (b) reduce friction and remove debris from the eye
    (c) provide nutrients and oxygen to the conjunctival epithelium
    (d) a, b, and c are correct

22. The thickened gel-like substance that helps support the structure of the eyeball is the:
    (a) vitreous body
    (b) aqueous humor
    (c) cupula
    (d) perilymph

23. The retina is considered to be a component of the:
    (a) vascular tunic
    (b) fibrous tunic
    (c) neural tunic
    (d) a, b, and c are correct

24. At sunset your visual system adapts to:
    (a) fovea vision
    (b) rod-based vision
    (c) macular vision
    (d) cone-based vision

25. The malleus, incus, and stapes are the tiny ear bones located in the:
    (a) outer ear
    (b) middle ear
    (c) inner ear
    (d) membranous labyrinth

9

THE GENERAL AND SPECIAL SENSES

The General Senses • The Special Senses • Smell • Taste • Vision • Equilibrium and Hearing • Aging and the Senses • Chapter Review • **MediaLab**

26. Receptors in the saccule and utricle provide sensations of:

    (a) balance and equilibrium

    (b) hearing

    (c) vibration

    (d) gravity and linear acceleration

27. The organ of Corti is located within the _____ of the inner ear.

    (a) utricle   (b) bony labyrinth

    (c) vestibule   (d) cochlea

28. What three types of mechanoreceptors respond to stretching, compression, twisting, or other distortions of the cell membrane?

29. Identify six types of tactile receptors found in the skin and their sensitivities.

30. (a) What structures make up the fibrous tunic of the eye? (b) What are the functions of the fibrous tunic?

31. What structures are part of the vascular tunic of the eye?

32. What six basic steps are involved in the process of hearing?

Level 2: Reviewing Concepts

33. The CNS interprets sensory information entirely on the basis of the:

    (a) strength of the action potential

    (b) number of generator potentials

    (c) area of brain stimulated

    (d) a, b, and c are correct

34. If the auditory cortex is damaged, the individual will respond to sounds and have normal acoustic reflexes, but:

    (a) the sounds may produce nerve deafness

    (b) the auditory ossicle may be immobilized

    (c) sound interpretation and pattern recognition may be impossible

    (d) normal transfer of vibration to the oval window is inhibited

35. Distinguish between the general senses and the special senses in the human body.

36. In what form does the CNS receive a stimulus detected by a sensory receptor?

37. Why are olfactory sensations long-lasting and an important part of our memories and emotions?

38. Jane makes an appointment with the optometrist for a vision test. Her test results are reported as 20/15. What does this test result mean? Is a rating of 20/20 better or worse?

Level 3: Critical Thinking and Clinical Applications

39. You are at a park watching some deer 35 feet away from you. Your friend taps you on the shoulder to ask a question. As you turn to look at your friend, who is standing 2 feet away, what changes will occur regarding your eyes?

40. After attending a Fourth of July fireworks extravaganza, Millie finds it difficult to hear normal conversation, and her ears keep "ringing." What is causing her hearing problems?

41. After riding the express elevator from the twentieth floor to the ground floor, for a few seconds you still feel as if you are descending, even though you have obviously come to a stop. Why?

Answers to Concept Check Questions

Page 282

1. By the end of the lab period, adaptation has occurred. In response to the constant level of stimulation, the receptor neurons have become less active.

2. Since nociceptors are pain receptors, if they are stimulated, you would perceive a painful sensation in your affected hand.

3. Proprioceptors relay information about limb position and movement to the central nervous system, especially the cerebellum. Lack of this information would result in uncoordinated movements, and the individual probably would not be able to walk.

Page 284

1. The perception of faint odors increases because sniffing repeatedly increases the flow of air and the total amount of odorant molecules across the olfactory epithelium.

2. The taste receptors (taste buds) are sensitive only to molecules and ions that are in solution. If you dry the surface of the tongue, there is no moisture for the sugar molecules or salt ions to dissolve in and they will not stimulate the taste receptors.

Page 296

1. The first layer of the eye to be affected by inadequate tear production would be the conjunctiva. Drying of this layer would produce an irritated, scratchy feeling.

2. When the lens is round, you are looking at something close to you.

3. A person with a congenital lack of cone cells in the eye would be able to see as long as he or she had functioning rod cells. Since cone cells function in color vision, such a person would see only black and white.

4. A deficiency or lack of vitamin A in the diet would affect the quantity of retinal the body could produce and thus would interfere with night vision.

Page 305

1. Without the movement of the round window, the perilymph would not be moved by the vibration of the stapes at the oval window, and there would be little or no perception of sound.

2. Loss of stereocilia (as a result of constant exposure to loud noises, for instance) would reduce hearing sensitivity and could eventually result in deafness.

Page 306

1. Tasting abilities change with age as the number and sensitivity of taste buds decrease. The number of taste buds declines dramatically after age 50.

2. With age, the lens loses its elasticity and stiffens. These changes result in a flatter lens that cannot round up to focus the image of a close object on the retina. This condition, called presbyopia, may be corrected with a converging lens.

EXPLORE *MediaLab*

Reading a map in a moving car often produces the symptoms of motion sickness.

EXPLORATION #1

Estimated time for completion: 10 minutes

IMAGINE LIFE WITHOUT the use of your arms and legs. That is a reality for many people who suffer from spinal cord injury. Using Figures 8-31 and 8-32● as guides, diagram a simple sensory neural pathway from the right hand to the brain, and the associated motor pathway back from the brain to the muscles of the same hand. Indicate on your diagram where a spinal cord injury might block transmission. Reflecting on what you know about action potentials and the neuromuscular junction, how might you artificially recreate these pathways? Would an exter-nal electrical stimulus be beneficial? Where would that stimulus need to go? Indicate this on your diagram, then go to the Companios Web site, visit the MediaLab section, and choose the keyword "Freehand" to read what Dr. Peckhman at the Louis Stokes Cleveland VA Medical Center has been working on for 30 years. Are your ideas similar to his? If not, add the "Freehand" ideas to your diagram.

EXPLORATION #2

Estimated time for completion: 10 minutes

"WHILE WE ARE heading down this winding road, could you read that map and tell we where our next turn is?" For a person with severe motion sickness, this simple task might be impossible. Although the exact mechanism remains a mystery, motion sickness is in some way related to conflicting sensory input between the eyes and the inner ear. While this may seem relatively trivial in everyday life, to an Air Force pilot or a sailor, motion sickness can be devastating and extremely dangerous. Prepare a short list of the symptoms of motion sickness that you have experienced yourself or that a friend or relative has described to you. Go to the Companion Web site, visit the MediaLab section, and click on the keywords "Pilot comfort" to read the information about motion sickness given to Air Force pilots. Next to each of your symptoms, indicate the treatment that the Air Force recommended. Why do these treatments differ from those you might use to alleviate motion sickness caused when you try to read a map in a moving vehicle?

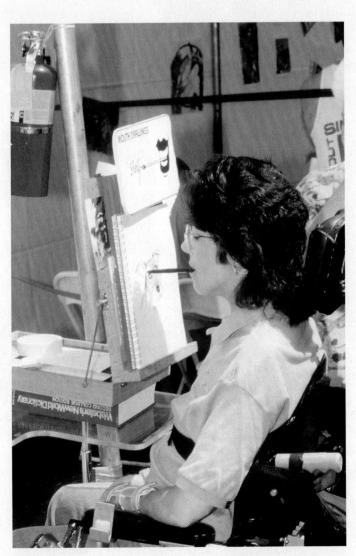

A young woman with quadriplegia making a drawing.

The Endocrine System

CHAPTER OUTLINE AND OBJECTIVES

Vocabulary Development

ad- ....to or toward; *adrenal*
andros..man; *androgen*
angeion ....vessel; *angiotensin*
corpus ....body; *corpus luteum*
diabetes ....to pass through; *diabetes*
diourein ....to urinate; *diuresis*
erythros..............................red; *erythropoietin*
infundibulum ....funnel; *infundibulum*
insipidus ....tasteless; *diabetes insipidus*
krinein..............................to secrete; *endocrine*
lac..milk; *prolactin*
mellitum....................honey; *diabetes mellitus*
natrium ....sodium; *natriuretic*
ouresis ....making water; *polyuria*
oxy- ....quick; *oxytocin*
para....................................beyond; *parathyroid*
poiesis ....making; *erythropoietin*
pro- ....before; *prolactin*
renes ....kidneys; *adrenal*
synairesis ......a drawing together; *synergistic*
teinein ....to stretch; *angiotensin*
thyreos....................an oblong shield; *thyroid*
tokos................................childbirth; *oxytocin*
tropos ....turning; *gonadotropins*

*T*HE DIFFERENCES IN HEIGHT of these teenage girls are clearly evident. Variations in genetically programmed levels of hormones, especially growth hormone, are probably responsible for these differences.

10 THE ENDOCRINE SYSTEM

An Overview of the Endocrine System • The Pituitary Gland • The Thyroid Gland • The Parathyroid Glands • The Adrenal Glands • The Pineal Gland

To FUNCTION EFFECTIVELY, every cell in the body must communicate with its neighbors and with cells and tissues in distant regions of the body. Most of the communication involves the release and receipt of chemical messages. Each living cell is continually "talking" to its neighbors by releasing chemicals into the extracellular fluid. These chemicals let cells know what their neighbors are doing at any given moment, and the result is the coordination of tissue function at the local level.

The nervous system acts like a telephone company, carrying specific "messages" from one location to another inside the body. The source and the destination are quite specific, and the effects are short-lived. This form of communication is ideal for crisis management; if you are in danger of being hit by a speeding bus, the nervous system can coordinate and direct your leap to safety. Once the crisis is over and the neural circuit quiets down, things soon return to normal.

In cellular communication, hormones are like addressed letters, and the circulatory system is the postal service. A hormone released into the circulation will be distributed throughout the body. Each hormone has specific *target cells* that will respond to its presence. These cells possess the receptors needed to bind and "read" the hormonal message. Although a cell may be exposed to the mixture of hormones in circulation at any given moment, it will respond only to those hormones it can bind and read. The other hormones will be treated like junk mail, and ignored.

Because target cells can be anywhere in the body, a single hormone can alter the metabolic activities of multiple tissues and organs simultaneously. These effects may be slow to appear, but they often persist for days. This persistence makes hormones effective in coordinating cell, tissue, and organ activities on a sustained, long-term basis. For example, circulating hormones keep body water content and levels of electrolytes and organic nutrients within normal limits 24 hours a day throughout our entire lives.

While the effects of a single hormone persist, a cell may receive additional instructions from other hormones. The result will be a further modification in cellular operations. Gradual changes in the quantities and identities of circulating hormones can produce complex changes in physical structure and physiological capabilities. Examples are the processes of embryological and fetal development, growth, and puberty.

When viewed from a general perspective, the differences between the nervous and endocrine systems seem relatively clear. In fact, these broad organizational and functional distinctions are the basis for treating them as two separate systems. Yet when considered in detail, the two systems are organized along parallel lines. For example:

- Both systems rely on the release of chemicals that bind to specific receptors on their target cells.
- Both systems use many of the same chemical messengers; for example, norepinephrine and epinephrine are called *hormones* when released into the general circulation, and *neurotransmitters* when released across synapses.
- Both systems are primarily regulated by negative feedback control mechanisms.
- Both systems share a common goal: to coordinate and regulate the activities of other cells, tissues, organs, and systems and to maintain homeostasis.

This chapter introduces the components and functions of the endocrine system and explores the interactions between the nervous and endocrine systems. In later chapters we shall consider specific endocrine organs, hormones, and functions in greater detail.

An Overview of the Endocrine System

The endocrine system includes all of the endocrine cells and tissues of the body. As noted in Chapter 4, **endocrine cells** are glandular secretory cells that release their secretions internally rather than onto an epithelial surface. This feature distinguishes them from *exocrine cells*, which secrete onto epithelial surfaces. ⟶ p. 85 The chemicals released by endocrine cells may affect only adjacent cells, as in the case of *cytokines*, or *local hormones*, such as *prostaglandins*, or they may affect cells throughout the body. **Hormones** are chemical messengers that are released in one tissue and transported by the bloodstream to reach target cells in their tissues.

The components of the endocrine system are introduced in Figure 10-1●. This figure also lists the major hormones produced in each endocrine tissue and organ. Some of these organs, such as the pituitary gland, have endocrine secretion as a primary function; others, such as the pancreas, have many other functions besides endocrine secretion.

THE STRUCTURE OF HORMONES

Hormones can be divided into three groups on the basis of chemical structure: amino acid derivatives, peptide hormones, and lipid derivatives.

1. *Amino acid derivatives.* Some hormones are relatively small molecules that are structurally similar to amino acids.

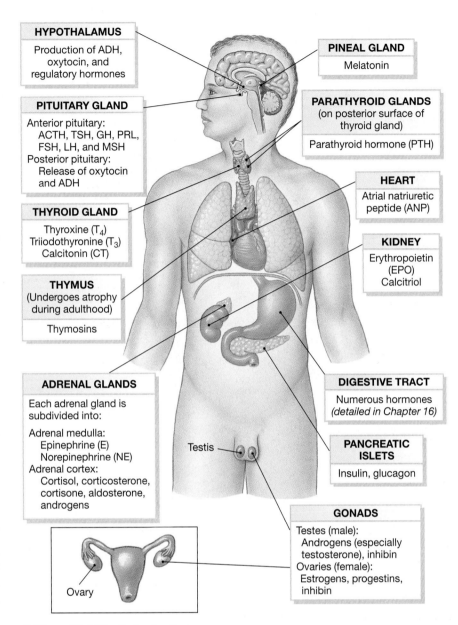

HYPOTHALAMUS

Production of ADH, oxytocin, and regulatory hormones

PITUITARY GLAND

Anterior pituitary:
ACTH, TSH, GH, PRL, FSH, LH, and MSH
Posterior pituitary:
Release of oxytocin and ADH

THYROID GLAND

Thyroxine (T_4)
Triiodothyronine (T_3)
Calcitonin (CT)

THYMUS
(Undergoes atrophy during adulthood)

Thymosins

ADRENAL GLANDS

Each adrenal gland is subdivided into:

Adrenal medulla:
Epinephrine (E)
Norepinephrine (NE)
Adrenal cortex:
Cortisol, corticosterone, cortisone, aldosterone, androgens

PINEAL GLAND

Melatonin

PARATHYROID GLANDS
(on posterior surface of thyroid gland)

Parathyroid hormone (PTH)

HEART

Atrial natriuretic peptide (ANP)

KIDNEY

Erythropoietin (EPO)
Calcitriol

DIGESTIVE TRACT

Numerous hormones
(detailed in Chapter 16)

Testis

PANCREATIC ISLETS

Insulin, glucagon

GONADS

Testes (male):
Androgens (especially testosterone), inhibin
Ovaries (female):
Estrogens, progestins, inhibin

Ovary

● *Figure 10-1* **The Endocrine System**

(Amino acids, the building blocks of proteins, were introduced in Chapter 2.) ⌘ p. 41 This group includes *epinephrine*, *norepinephrine*, the *thyroid hormones*, and *melatonin*.

2. *Peptide hormones.* **Peptide hormones** consist of chains of amino acids. These molecules range from short amino acid chains, such as *ADH* and *oxytocin*, to small proteins such as *growth hormone* and *prolactin*. This is the largest class of hormones and includes all of the hormones secreted by the hypothalamus, pituitary gland, heart, kidneys, thymus, digestive tract, and pancreas.

3. *Lipid derivatives.* There are two classes of lipid-based hormones: *steroid hormones*, derived from cholesterol; and those

derived from *arachidonic acid*, a 20-carbon fatty acid. **Steroid hormones** are lipids structurally similar to cholesterol, a lipid introduced in Chapter 2. ⌘ p. 40 Steroid hormones are released by the reproductive organs and the adrenal glands. The fatty acid-based compounds, which include the **prostaglandins**, coordinate local cellular activities and affect enzymatic processes, such as blood clotting, that occur in extracellular fluids.

THE MECHANISMS OF HORMONAL ACTION

All cellular structures and functions are determined by proteins. Structural proteins determine the general shape and internal structure of a cell, and enzymes direct the cell's metabolic activities. Hormones alter cellular operations by changing the *identities*, *activities*, or *quantities* of important enzymes and structural proteins in various **target cells**. The sensitivity of a target cell is determined by the presence or absence of a specific **receptor** with which the hormone interacts. Hormone receptors are located either on the cell membrane or in the cytoplasm (Figure 10-2●).

Hormones and the Cell Membrane

The receptors for epinephrine, norepinephrine, peptide hormones, and fatty acid-based hormones are in the cell membranes of their respective target cells. Because epinephrine, norepinephrine, and peptide hormones are not lipid soluble, they cannot diffuse through a cell membrane, and they bind to receptor proteins on the outer surface of the cell membrane. Conversely, the fatty acid-based hormones, which are lipid soluble, penetrate the cell membrane and bind to receptor proteins on the inner surface of the cell membrane.

Any hormones that bind to receptors on the outer surface of the cell membrane do not have direct effects on the target cell. Instead, when these hormones, called **first messengers**, bind to an appropriate receptor, they indirectly trigger the appearance of a **second messenger** in the cytoplasm. The link between the first messenger and the second messenger usually involves a *G-protein* enzyme complex. The G-protein is activated when a hormone binds to its receptor on the outer surface of the cell membrane. The second messenger may function as an enzyme activator or inhibitor, but the net result will be a change in the cell's metabolic activities.

10 THE ENDOCRINE SYSTEM

An Overview of the Endocrine System • The Pituitary Gland • The Thyroid Gland • The Parathyroid Glands • The Adrenal Glands • The Pineal Gland

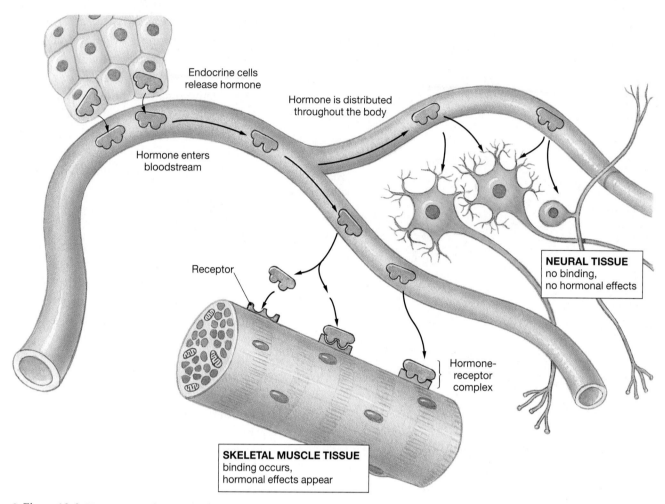

Endocrine cells
release hormone

Hormone is distributed
throughout the body

Hormone enters
bloodstream

Receptor

NEURAL TISSUE
no binding,
no hormonal effects

Hormone-
receptor
complex

SKELETAL MUSCLE TISSUE
binding occurs,
hormonal effects appear

● *Figure 10-2* **Hormones and Target Cells**
For a hormone to affect a target cell, that cell must have receptors that can bind the hormone and initiate a change in cellular activity. This hormone affects skeletal muscle tissue but not neural tissue because only the muscle tissue has the appropriate receptors.

One of the most important second messengers is **cyclic-AMP (cAMP)** (Figure 10-3a●). Its appearance depends on G-protein activation, which activates an enzyme called **adenylate cyclase**. The adenylate cyclase, in turn, converts ATP to a ring-shaped molecule of cAMP. Cyclic-AMP activates *kinase* enzymes, which attach a phosphate group (PO_4^{3-}) to another molecule in a process called *phosphorylation*. The effect on the target cell depends on the nature of the proteins affected. The phosphorylation of membrane proteins can open ion channels, and in the cytoplasm many enzymes can be activated only by phosphorylation. As a result, a single hormone can have one effect in one target tissue and quite different effects in other target tissues. The effects of cAMP are usually very short-lived, because another enzyme present in the cell, *phosphodiesterase (PDE)*, quickly breaks down cAMP.

Cyclic-AMP is one of the most common second messengers, but there are many others. Important examples are *calcium*

ions and *cyclic-GMP*, a derivative of the high-energy compound *guanosine triphosphate (GTP)*.

Hormones and Intracellular Receptors

The steroid hormones and thyroid hormones cross the cell membrane and bind to intracellular receptors (Figure 10-3b●). Steroid hormones diffuse rapidly through the lipid portion of the cell membrane and bind to receptors in the cytoplasm or nucleus. The resulting *hormone-receptor complex* then activates or inactivates specific genes. By this mechanism, steroid hormones can alter the rate of mRNA transcription in the nucleus, and can change the structure or function of the cell. For example, the hormone testosterone stimulates the production of enzymes and proteins in skeletal muscle fibers, increasing muscle size and strength.

Thyroid hormones cross the cell membrane by diffusion or carrier-mediated transport. Once within the cell, these hor-

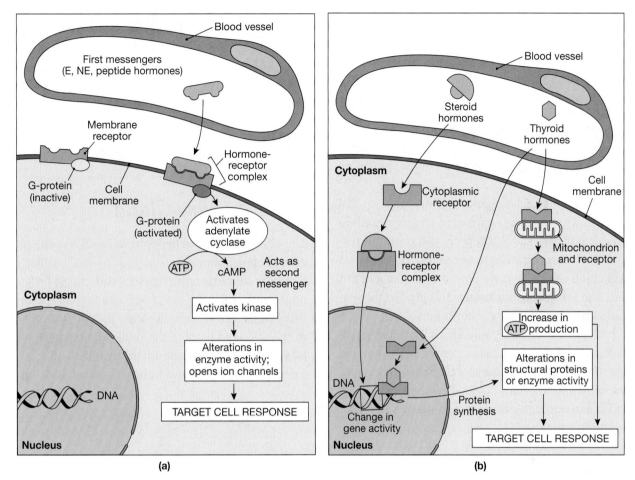

● *Figure 10-3* **Mechanisms of Hormone Action**
(a) Nonsteroidal hormones, such as epinephrine (E), norepinephrine (NE), and the peptide hormones, bind to cell membrane receptors. They exert their effects on target cells through a second messenger, such as cAMP, which alters the activity of enzymes present in the cell. (b) Both steroid hormones and thyroid hormones pass directly through target cell membranes. Steroid hormones bind to receptors in the cytoplasm that then enter the nucleus. Thyroid hormones proceed directly to the nucleus to reach hormonal receptors and mitochondria in the cytoplasm. In the nucleus, both steroid and thyroid hormone-receptor complexes directly affect gene activity and protein synthesis. Thyroid hormones also increase the rate of ATP production in the cell.

mones bind to receptors within the nucleus and on mitochondria. The hormone-receptor complexes in the nucleus activate specific genes or change the rate of transcription. The result is an increase in metabolic activity due to changes in the nature or number of enzymes in the cytoplasm. Thyroid hormones bound to mitochondria increase the mitochondria's rates of ATP production.

THE CONTROL OF ENDOCRINE ACTIVITY

Negative feedback mechanisms provide the basis for the control of endocrine activity. ⚬⚬ p. 12 Endocrine activity may be controlled or regulated by three general mechanisms: (1) directly, by a change in the composition of the extracellular fluid; (2) by changes in the levels of circulating hormones; and (3) by neural stimulation. All three of these regulatory modes are exhibited by the hypothalamus. Its link between the nervous and endocrine systems and its varied roles in the control of endocrine activity are discussed below.

Direct Negative Feedback Control

In direct negative feedback control, endocrine cells respond to a change in the composition of the extracellular fluid by releasing their hormone into the bloodstream. The released hormone stimulates target cells to restore homeostasis. For example, consider the control of calcium levels by two hormones, *parathyroid hormone* and *calcitonin*. When calcium levels in the blood decline, parathyroid hormone is released, and the responses of target cells elevate blood calcium levels. When calcium levels in the blood rise, calcitonin is released, and responses of target cells lower blood calcium levels.

10 THE ENDOCRINE SYSTEM

An Overview of the Endocrine System • **The Pituitary Gland** • The Thyroid Gland • The Parathyroid Glands • The Adrenal Glands • The Pineal Gland

The Hypothalamus and Endocrine Regulation

Coordinating centers in the hypothalamus regulate the activities of the nervous and endocrine systems in three ways (Figure 10-4●):

1. The hypothalamus itself acts as an endocrine organ, releasing the hormones ADH and oxytocin into the circulation at the posterior pituitary. These hormones are released in response to changes in the composition of the circulating blood.

2. The hypothalamus secretes **regulatory hormones**, special hormones that regulate the activities of endocrine cells in the anterior pituitary gland. There are two classes of regulatory hormones: **releasing hormones (RH)** stimulate the production of one or more hormones in the anterior pituitary, and **inhibiting hormones (IH)** prevent the synthesis and secretion of pituitary hormones.

3. The hypothalamus contains autonomic nervous system centers that control the endocrine cells of the adrenal medullae (the interior of the adrenal glands) through sympathetic innervation. ∞ p. 264 When the sympathetic division is activated, the adrenal medullae release hormones into the bloodstream.

CONCEPT CHECK QUESTIONS

Answers on page 340

❶ What is the primary factor that determines each cell's sensitivities to hormones?

❷ How would the presence of a molecule that blocks adenylate cyclase affect the of a hormone that produces its cellular effects by way of cAMP?

❸ Why is cAMP described as a second messenger?

The Pituitary Gland

The **pituitary gland**, or **hypophysis** (hī-POF-i-sis), secretes nine different hormones. All are peptide or small protein hormones that bind to membrane receptors and, except for growth hormone and prolactin, use cAMP as a second messenger. This small, oval gland lies nestled within the *sella turcica*, a depression in the sphenoid bone of the skull (Figure 10-5● p. 319). The pituitary gland hangs beneath the hypothalamus, connected by a slender stalk, the **infundibulum** (in-fun-DIB-ū-lum; funnel).

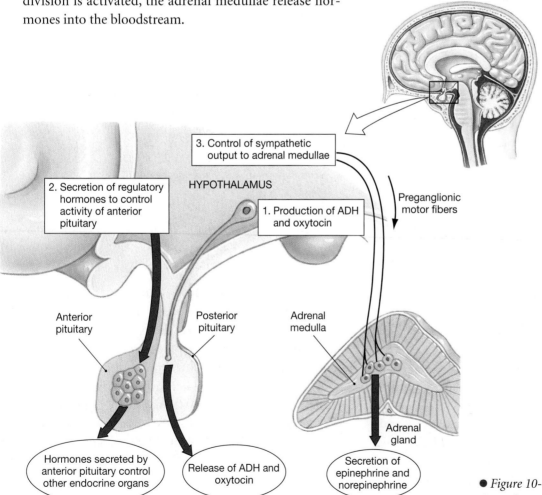

● *Figure 10-4* **Three Methods of Hypothalamic Control over Endocrine Organs**

● *Figure 10-5* **The Anatomy and Orientation of the Pituitary Gland** (LM × 62)

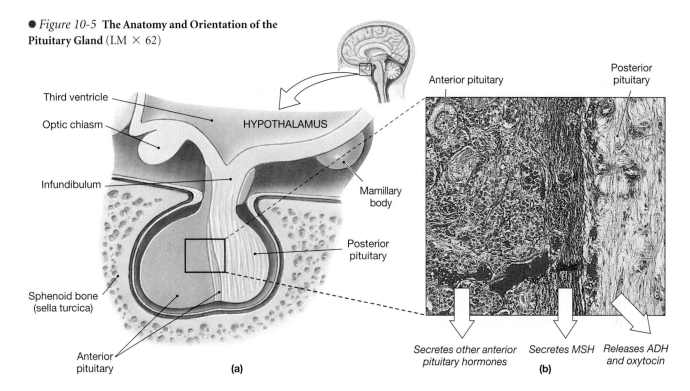

Third ventricle

Optic chiasm

HYPOTHALAMUS

Infundibulum

Mamillary body

Posterior pituitary

Sphenoid bone (sella turcica)

Anterior pituitary

Anterior pituitary

Posterior pituitary

(a)

Secretes other anterior pituitary hormones

Secretes MSH

Releases ADH and oxytocin

(b)

The pituitary gland has a complex structure, with distinct anterior and posterior regions.

THE ANTERIOR PITUITARY GLAND

The **anterior pituitary gland** contains endocrine cells surrounded by an extensive capillary network. The capillary network, part of the *hypophyseal portal system*, provides an entry into the circulatory system for the hormones secreted by endocrine cells of the anterior pituitary.

The Hypophyseal Portal System

The regulatory hormones produced by the hypothalamus regulate the activities of the anterior pituitary. These hormones, released by hypothalamic neurons near the attachment of the infundibulum, enter a network of highly permeable capillaries. Before leaving the hypothalamus, this capillary network unites to form a series of larger vessels that descend to the anterior pituitary and form a second capillary network.

This circulatory arrangement, illustrated in Figure 10-6●, is very unusual. A typical artery usually conducts blood from the heart to a capillary network, and a typical vein carries blood from a capillary network back to the heart. The vessels between the hypothalamus and the anterior pituitary, by contrast, carry blood from one capillary network to another. Blood vessels that link two capillary networks are called *portal vessels*, and the entire complex is termed a **portal system**.

Portal systems ensure that all of the blood entering the portal vessels will reach the intended target cells before return-

ing to the general circulation. Portal vessels are named after their destinations, so this particular network of vessels represents the **hypophyseal** (hī-pō-FI-sē-al) **portal system**.

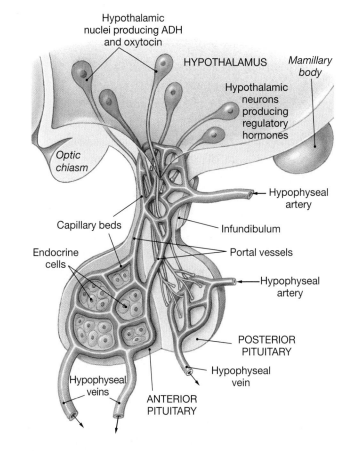

Hypothalamic nuclei producing ADH and oxytocin

HYPOTHALAMUS

Mamillary body

Hypothalamic neurons producing regulatory hormones

Optic chiasm

Hypophyseal artery

Capillary beds

Infundibulum

Endocrine cells

Portal vessels

Hypophyseal artery

POSTERIOR PITUITARY

Hypophyseal vein

Hypophyseal veins

ANTERIOR PITUITARY

● *Figure 10-6* **The Hypophyseal Portal System**

10 THE ENDOCRINE SYSTEM

An Overview of the Endocrine System • **The Pituitary Gland** • The Thyroid Gland • The Parathyroid Glands • The Adrenal Glands • The Pineal Gland

Hypothalamic Control of the Anterior Pituitary

An endocrine cell in the anterior pituitary may be controlled by releasing hormones, inhibiting hormones, or some combination of the two. The regulatory hormones released at the hypothalamus are transported directly to the anterior pituitary by the hypophyseal portal system.

The rate of regulatory hormone secretion by the hypothalamus is regulated through negative feedback mechanisms. The basic regulatory patterns are diagrammed in (Figure 10-7●); these will be referenced in the following description of pituitary hormones. Many of these regulatory hormones are called *tropins* (*tropé*, a turning) because they "turn on" (activate) other endocrine glands or support the functions of other organs.

Hormones of the Anterior Pituitary

Seven hormones are produced by the anterior pituitary gland. Of those, four regulate the production of hormones by other endocrine glands.

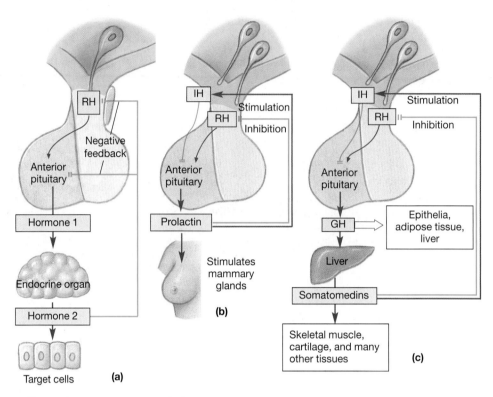

● *Figure 10-7* **Feedback Control of Endocrine Secretion**
(**a**) The typical pattern of regulation when multiple endocrine organs are involved. The hypothalamus produces a releasing hormone (RH) to stimulate hormone production by other glands, and control is through negative feedback. (**b**) In some patterns of feedback control, both a releasing hormone and an inhibiting hormone (IH) are produced by the hypothalamus; when one is stimulated, the other is inhibited. This diagram summarizes the control of prolactin production by the hypothalamus and anterior pituitary. (**c**) The regulation of growth hormone (GH) follows the basic pattern shown in part b, but intermediate steps are involved.

1. **Thyroid-stimulating hormone (TSH)**, or *thyrotropin*, targets the thyroid gland and triggers the release of thyroid hormones. As circulating concentrations of thyroid hormones rise, the production of TSH by the pituitary gland and *thyrotropin releasing hormone (TRH)* by the hypothalamus decline. This pattern of regulatory control is shown in (Figure 10-7a●).

2. **Adrenocorticotropic hormone (ACTH)** stimulates the release of steroid hormones by the adrenal glands. ACTH specifically targets cells producing hormones called *glucocorticoids* (gloo-kō-KŌR-ti-koyds), which affect glucose metabolism. ACTH release occurs under the stimulation of *corticotropin-releasing hormone (CRH)* from the hypothalamus. A rise in glucocorticoid levels causes a decline in the production of ACTH and CRH. This type of negative feedback control is comparable to that for TSH (Figure 10-7a●).

3. The **gonadotropins** (gō-nad-ō-TRŌ-pinz) regulate the activities of the male and female sex organs, or *gonads*. The anterior pituitary produces two gonadotropins, follicle stimulating hormone (FSH) and luteinizing hormone (LH). The production of gonadotropins is stimulated by *gonadotropin releasing hormone (GnRH)* from the hypo-

thalamus. An abnormally low production of gonadotropins produces *hypogonadism*. Children with this condition will not undergo sexual maturation, and adults with hypogonadism cannot produce functional sperm and ova.

4. **Follicle-stimulating hormone (FSH)** promotes follicle (and egg) development in females and stimulates the secretion of estrogens, steroid hormones produced by ovarian cells. In males, FSH production supports sperm production in the testes. A peptide hormone, called *inhibin*, released by the cells of the testes and ovaries is thought to inhibit the release of FSH and GnRH by a feedback control mechanism comparable to that for TSH (Figure 10-7a●).

5. **Luteinizing** (LOO-tē-in-ī-zing) **hormone (LH)** induces *ovulation*, the production of reproductive cells in females. It also promotes the ovarian secretion of estrogens and the **progestins** (such as *progesterone*), which prepare the body for possible pregnancy. In males, LH is sometimes called *interstitial cell-stimulating hormone (ICSH)* because it stimulates the interstitial cells of the testes to produce sex hormones. These sex hormones are called **androgens**

(AN-drō-jenz; *andros*, man); the most important is *testosterone*. GnRH production is inhibited by estrogens, progestins, and androgens, through a feedback control mechanism comparable to that of TSH (Figure 10-7a●).

6. **Prolactin** (prō-LAK-tin; *pro-*, before + *lac*, milk), or **PRL**, works with other hormones to stimulate mammary gland development. In pregnancy and during the period of nursing that follows delivery, PRL also stimulates the production of milk by the mammary glands. Prolactin effects in the human male are poorly understood, but it may help regulate androgen production. The regulation of prolactin release involves interactions between releasing and inhibiting hormones from the hypothalamus. The regulatory pattern is diagrammed in Figure 10-7b●.

7. **Growth hormone (GH)**, also called *human growth hormone (hGH)* or *somatotropin* (*soma*, body), stimulates cell growth and replication by accelerating the rate of protein synthesis. Although virtually every tissue responds to some degree, skeletal muscle cells and chondrocytes (cartilage cells) are particularly sensitive to levels of growth hormone.

   The stimulation of growth by GH involves two mechanisms. The primary mechanism, which is indirect, is best understood. Liver cells respond to the presence of growth hormone by synthesizing and releasing **somatomedins**, or *insulin-like growth factors* (IGF), peptide hormones that bind to receptor sites on a variety of cell membranes. Somatomedins increase the rate of amino acid uptake and their incorporation into new proteins. These effects develop almost immediately after GH release occurs, and they are particularly important after a meal, when the blood contains high concentrations of glucose and amino acids.

   The direct actions of GH usually do not appear until after blood glucose and amino acid concentrations have returned to normal levels. In epithelia and connective tissues, GH stimulates stem cell divisions and the differentiation of daughter cells. GH also has metabolic effects in adipose tissue and in the liver. In adipose tissue, it stimulates the breakdown of stored fats and the release of fatty acids into the blood. In the liver, GH stimulates the breakdown of glycogen reserves and the release of glucose into the circulation. GH thus plays a role in mobilizing energy reserves.

   The production of GH is regulated by *growth hormone-releasing hormone (GH-RH)* and *growth hormone-inhibiting hormone (GH-IH)* from the hypothalamus. Somatomedins stimulate GH-IH and inhibit GH-RH. This regulatory mechanism is summarized in Figure 10-7c●.

8. **Melanocyte-stimulating hormone (MSH)** stimulates the melanocytes of the skin, increasing their production of melanin. MSH is important in the control of skin and hair pigmentation in fishes, amphibians, reptiles, and many mammals other than primates. The MSH-producing cells of the pituitary gland in adult humans are virtually nonfunctional, and the circulating blood usually does not contain MSH. However, MSH is secreted by the human pituitary (1) during fetal development, (2) in very young children, (3) in pregnant women, and (4) in some disease states. The functions of MSH under these circumstances are not known. The administration of a synthetic form of MSH causes darkening of the skin, so MSH has been suggested as a means of obtaining a "sunless tan."

THE POSTERIOR PITUITARY GLAND

The **posterior pituitary gland** contains the axons from two different groups of neurons located within the hypothalamus. One group manufactures antidiuretic hormone (ADH), and the other oxytocin. These products are transported within axons along the infundibulum to the posterior pituitary, as indicated in Figure 10-6●.

Antidiuretic hormone (ADH) is released in response to such stimuli as a rise in the concentration of electrolytes in the blood (increased osmotic pressure) or a fall in blood volume or pressure. The primary function of ADH is to decrease the amount of water lost at the kidneys. With losses minimized, any water absorbed from the digestive tract will be retained, reducing the concentration of electrolytes. ADH also causes the constriction of peripheral blood vessels, which helps to increase blood pressure. The production of ADH is inhibited by alcohol, which explains the increased fluid excretion that follows the consumption of alcoholic beverages.

In females, **oxytocin** (*oxy-*, quick + *tokos*, childbirth) stimulates smooth muscle tissue in the wall of the uterus, promoting labor and delivery, and special contractile cells associated with the mammary glands. Until the last stages of pregnancy, the uterine muscles are insensitive to oxytocin, but they become more sensitive as the time of delivery approaches. The stimulation of uterine muscles by oxytocin helps maintain and complete normal labor and childbirth (discussed in Chapter 20). After delivery, oxytocin also stimulates the contraction of special cells surrounding the secretory cells and ducts of the mammary glands. In the "milk let-down" reflex, oxytocin secreted in response to suckling triggers the release of milk from the breasts.

In the male, oxytocin stimulates the smooth muscle contraction in the walls of the sperm duct and prostate gland. These actions may be important in *emission*, the ejection of prostatic secretions, spermatozoa, and the secretions of other glands into the male reproductive tract before ejaculation occurs.

10 THE ENDOCRINE SYSTEM

An Overview of the Endocrine System • The Pituitary Gland • **The Thyroid Gland** • The Parathyroid Glands • The Adrenal Glands • The Pineal Gland

DIABETES INSIPIDUS

Diabetes (*diabetes*, to pass through) occurs in several forms, all characterized by excessive urine production (*polyuria*). Although diabetes can be caused by physical damage to the kidneys, most forms are the result of endocrine abnormalities. The two most important forms are diabetes mellitus and diabetes insipidus. Diabetes mellitus is described on p. 330.

Diabetes insipidus (*insipidus*, tasteless) develops when the posterior pituitary no longer releases adequate amounts of ADH. Water conservation at the kidneys is impaired, and excessive amounts of water are lost in the urine. As a result, an individual with diabetes insipidus is constantly thirsty, a condition known as *polydipsia* (*dipsa*, thirst), but the fluids consumed are not retained by the body. Mild cases may not require treatment, as long as fluid and electrolyte intake keep pace with urinary losses. In severe diabetes insipidus, the fluid losses can reach 10 liters per day, and a fatal dehydration will occur unless treatment is provided. One innovative treatment method involves administering a synthetic form of ADH, *desmopressin acetate (DDAVP)*, in a nasal spray. The drug enters the bloodstream after diffusing through the nasal epithelium and may alleviate the symptoms of diabetes insipidus.

Figure 10-8● and Table 10-1 (p. 323) summarize important information concerning the hormonal products of the pituitary gland.

CONCEPT CHECK QUESTIONS

Answers on page 340

❶ If a person were dehydrated, how would the level of ADH released by the posterior pituitary change?

❷ A blood sample shows elevated levels of somatomedins. Which pituitary hormone would you expect to be elevated as well?

❸ What effect would elevated levels of cortisol, a hormone from the adrenal gland, have on the pituitary secretion of ACTH?

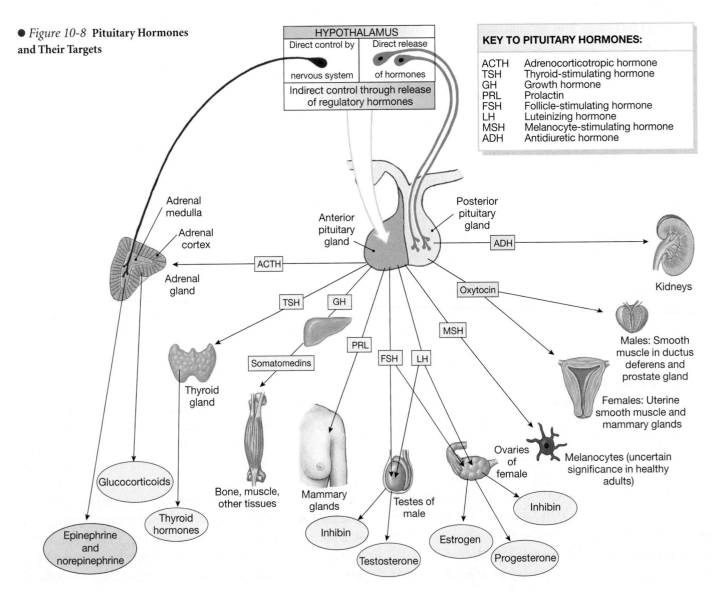

● *Figure 10-8* **Pituitary Hormones and Their Targets**

TABLE 10-1 *The Pituitary Hormones*

REGION	HORMONE	TARGET	HORMONAL EFFECTS
Anterior pituitary	Thyroid-stimulating hormone (TSH)	Thyroid gland	Secretion of thyroid hormones
	Adrenocorticotropic hormone (ACTH)	Adrenal cortex	Glucocorticoid secretion
	Gonadotropins: Follicle-stimulating hormone (FSH)	Follicle cells of ovaries	Estrogen secretion, follicle development
		Sustentacular cells of testes	Sperm maturation
	Luteinizing hormone (LH)	Follicle cells of ovaries	Ovulation, formation of corpus luteum, and progesterone secretion
		Interstitial cells of testes	Testosterone secretion
	Prolactin (PRL)	Mammary glands	Production of milk
	Growth hormone (GH)	All cells	Growth, protein synthesis, lipid mobilization and catabolism
	Melanocyte-stimulating hormone (MSH)	Melanocytes of skin	Increased melanin synthesis in epidermis
Posterior pituitary	Antidiuretic hormone (ADH)	Kidneys	Reabsorption of water, elevation of blood volume and pressure
	Oxytocin	Uterus, mammary glands (females)	Labor contractions, milk ejection
		Sperm duct and prostate gland (males)	Contractions of sperm duct and prostate gland

The Thyroid Gland

The **thyroid gland** lies anterior to the trachea and just below the **thyroid** ("shield-shaped") **cartilage**, which forms most of the anterior surface of the larynx (Figure 10-9a●). The two lobes of the thyroid gland are united by a slender connection, the *isthmus* (IS-mus). An extensive blood supply gives the thyroid gland a deep red color.

THYROID FOLLICLES AND THYROID HORMONES

The thyroid gland contains large numbers of **thyroid follicles**, spheres lined by a simple cuboidal epithelium (Figure 10-9b●). The cavity within each follicle contains a viscous *colloid*, a fluid containing large amounts of suspended proteins and thyroid hormones. A network of capillaries surrounds each follicle, delivering nutrients and regulatory hormones to the glandular cells and picking up their secretory products and metabolic wastes.

Thyroid hormones are manufactured by the follicular epithelial cells and stored within the follicle cavities. Under TSH stimulation from the anterior pituitary, the epithelial cells take hormones from the follicle cavities and release them into the circulation. However, almost all of the released thyroid hormones are unavailable because they become attached to plasma proteins in the bloodstream. Only the remaining unbound thy-

roid hormones, a small percentage of the total relased, are free to diffuse into the target cells of body tissues. As the unbound hormones decrease in concentration, the plasma proteins release additional bound hormone. The bound thyroid hormones are a substantial reserve; in fact, the bloodstream normally contains more than a week's supply of thyroid hormones.

The thyroid hormones are composed of molecules of the amino acid tyrosine, to which three or four iodine atoms have been attached. The hormone **thyroxine** (thī-ROKS-ēn) contains four atoms of iodine; it is also known as *tetraiodothyronine* (tet-ra-ī-ō-dō-THĪ-rō-nēn), or T_4. Thyroxine accounts for roughly 90 percent of all thyroid secretions. **Triiodothyronine**, or T_3, is a related, more potent molecule containing three iodine atoms.

Thyroid hormones readily cross cell membranes, and they affect almost every cell in the body. Inside a cell, they bind to receptor sites on mitochondria and in the nucleus (see Figure 10-3b●). The binding of thyroid hormones to mitochondria increases the rate of ATP production. Thyroid hormone-receptor complexes in the nucleus activate genes coding for the synthesis of enzymes involved in glycolysis and energy production, resulting in an increase in cellular rates of metabolism and oxygen consumption. Because the cell consumes more energy, and energy use is measured in *calories*, the effect is

10 THE ENDOCRINE SYSTEM

An Overview of the Endocrine System • The Pituitary Gland • The Thyroid Gland • **The Parathyroid Glands** • The Adrenal Glands • The Pineal Gland

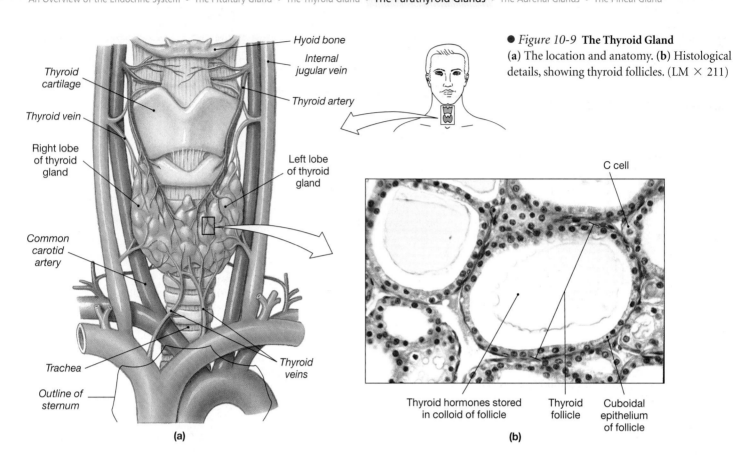

● *Figure 10-9* **The Thyroid Gland**
(a) The location and anatomy. (b) Histological details, showing thyroid follicles. (LM × 211)

called the **calorigenic effect** of thyroid hormones When the metabolic rate increases, more heat is generated and body temperature rises. In growing children, thyroid hormones are also essential to normal development of the skeletal, muscular, and nervous systems.

Normal production of thyroid hormones establishes the background rates of cellular metabolism. These hormones exert their primary effects on active tissues and organs, including skeletal muscles, the liver, the heart, and the kidneys. Overproduction or underproduction of thyroid hormones can therefore cause very serious metabolic problems. In many parts of

the world, inadequate dietary iodine leads to an inability to synthesize thyroid hormones. Under these conditions, TSH stimulation continues, and the thyroid follicles become distended with nonfunctional secretions. The result is an enlarged thyroid gland, or *goiter*. Goiters vary in size, and a large goiter can interfere with breathing and swallowing. This is seldom a problem in the United States because the typical American diet provides roughly three times the minimum daily requirement of iodine, thanks to the addition of iodine to table salt ("iodized salt").

Table 10-2 summarizes the effects of thyroid hormones on major organs and systems.

TABLE 10-2 *Hormones of the Thyroid Gland and Parathyroid Glands*

GLAND/CELLS	HORMONE(S)	TARGETS	HORMONAL EFFECTS
THYROID			
Follicular epithelium	Thyroxine (T_4), triiodothyronine (T_3)	Most cells	Increase energy utilization, oxygen consumption, growth, and development
C cells	Calcitonin (CT)	Bone, kidneys	Decreases calcium concentrations in body fluids (see Figure 10-10)
PARATHYROIDS			
Chief cells	Parathyroid hormone (PTH)	Bone, kidneys	Increases calcium concentrations in body fluids (see Figure 10-10)

THE C CELLS OF THE THYROID GLAND: CALCITONIN

C cells, or *parafollicular cells*, are endocrine cells sandwiched between the follicle cells and their basement membrane. C cells produce the hormone **calcitonin (CT)**. Calcitonin helps regulate calcium ion concentrations in body fluids. The control of calcitonin secretion does not involve the hypothalamus or pituitary gland. As Figure 10-10● illustrates, the C cells release calcitonin when the calcium ion concentration of the blood rises above normal. The target organs are the bones and the kidneys. Calcitonin reduces calcium levels by inhibiting osteoclasts and stimulating calcium excretion at the kidneys. The resulting reduction in the calcium ion concentrations eliminates the stimulus and "turns off" the C cells.

Calcitonin secretion is most important during childhood, when it stimulates active bone growth and calcium deposition in the skeleton. It also acts to reduce the loss of bone mass (1) during prolonged starvation and (2) during late pregnancy when the maternal skeleton competes with the developing fetus for absorbed calcium ions. The role of calcitonin in the healthy non-pregnant adult is uncertain.

Several chapters have dealt with the importance of calcium ions in controlling muscle cell and nerve cell activities. ∞ pp. 182, 236 Calcium ion concentrations also affect the sodium permeabilities of excitable membranes. At high calcium ion concentrations, sodium permeability decreases and membranes become less responsive. Such problems are prevented by the secretion of calcitonin under appropriate conditions. However, under normal conditions, calcium ion levels seldom rise enough to trigger calcitonin secretion. Most homeostatic adjustments are intended to prevent lower than normal calcium ion concentrations. Low calcium concentrations are dangerous because sodium permeabilities then increase, and muscle cells and neurons become extremely excitable. If calcium levels fall too far, convulsions or muscular spasms can result. The parathyroid glands and their hormone secretions respond when calcium levels fall too low and thus prevent such disastrous events.

The Parathyroid Glands

Two tiny pairs of **parathyroid glands** are embedded in the posterior surfaces of the thyroid gland (Figure 10-11●a). The gland

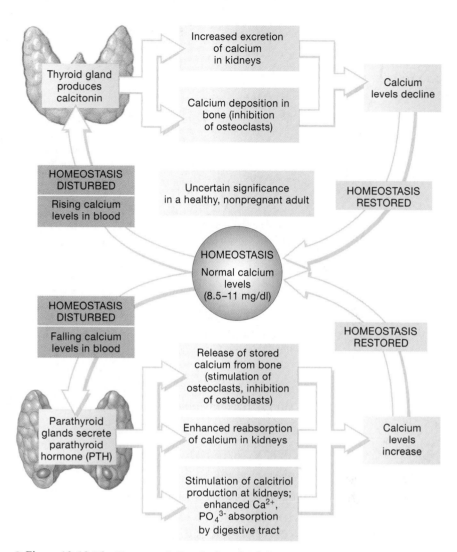

● *Figure 10-10* **The Homeostatic Regulation of Calcium Ion Concentrations**

cells are separated by the connective tissue fibers of the thyroid. The histological appearance of a parathyroid gland is shown in Figure 10-11b●. At least two different cell populations are found in the parathyroid gland. The **chief cells** produce parathyroid hormone; the functions of the other cell type are unknown.

Like the C cells of the thyroid, the chief cells monitor the circulating concentration of calcium ions. When the calcium concentration falls below normal, the chief cells secrete **parathyroid hormone (PTH)**, or *parathormone* (see Figure 10-10●). Although parathyroid hormone acts on the same target organs as calcitonin, it produces the opposite effects. PTH stimulates osteoclasts, inhibits the bone-building functions of osteoblasts, and reduces urinary excretion of calcium ions until blood concentrations return to normal. PTH also stimulates the kidneys to form and secrete *calcitriol*. Calcitriol promotes the absorption of Ca^{2+} and PO_4^{3-} by the digestive tract.

325

10 THE ENDOCRINE SYSTEM

An Overview of the Endocrine System • The Pituitary Gland • The Thyroid Gland • The Parathyroid Glands • **The Adrenal Glands** • The Pineal Gland

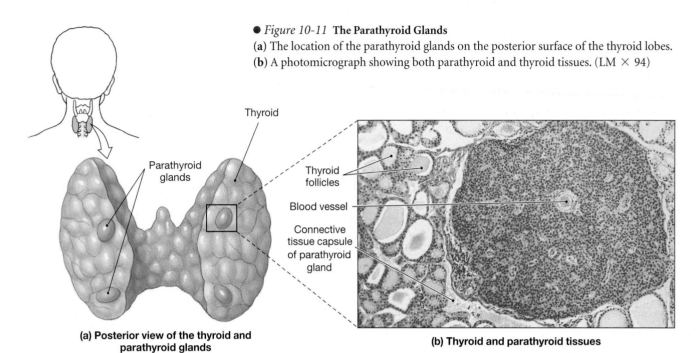

● *Figure 10-11* **The Parathyroid Glands**
(**a**) The location of the parathyroid glands on the posterior surface of the thyroid lobes.
(**b**) A photomicrograph showing both parathyroid and thyroid tissues. (LM × 94)

(a) Posterior view of the thyroid and parathyroid glands

(b) Thyroid and parathyroid tissues

Information concerning the hormones of the thyroid gland and parathyroid glands is summarized in Table 10-2.

CONCEPT CHECK QUESTIONS

Answers on page 340

❶ What symptoms would you expect to see in an individual whose diet lacks iodine?

❷ When a person's thyroid gland is removed, signs of decreased thyroid hormone concentration do not appear until about one week later. Why?

❸ Removal of the parathyroid glands would result in a decrease in the blood of what important mineral?

The Adrenal Glands

A yellow, pyramid-shaped **adrenal gland**, or *suprarenal* (soo-pra-RĒ-nal; *supra-*, above + *renes*, kidneys) *gland*, caps the superior border of each kidney (Figure 10-12a●). Each adrenal gland has two parts: an outer *adrenal cortex* and an inner *adrenal medulla* (Figure 10-12b●).

THE ADRENAL CORTEX

The yellowish color of the **adrenal cortex** is due to the presence of stored lipids, especially cholesterol and various fatty acids. The adrenal cortex produces more than two dozen steroid hormones, collectively called *adrenocortical steroids*, or simply **corticosteroids**. These hormones are vital; if the adrenal glands are destroyed or removed, the individual will die unless corti-

costeroids are administered. Overproduction or underproduction of any of the corticosteroids will have severe consequences because these hormones affect the metabolism of many different tissues.

Corticosteroids

The adrenal cortex contains three distinct zones or regions. Each zone synthesizes specific steroid hormones; the outer zone produces *mineralocorticoids*, the middle zone produces *glucocorticoids*, and the inner zone produces *androgens*. These three classes of hormones are summarized in Table 10-3.

Mineralocorticoids. **Aldosterone** (al-DOS-ter-ōn), the principal **mineralocorticoid**, stimulates the conservation of sodium ions and the elimination of potassium ions. It targets cells that regulate the ionic composition of excreted fluids. It causes the retention of sodium ions at the kidneys, sweat glands, salivary glands, and pancreas, preventing the loss of sodium ions in urine, sweat, saliva, and digestive secretions. The retention of sodium ions is accompanied by a loss of potassium ions. Secondarily, the reabsorption of sodium ions results in the osmotic reabsorption of water at the kidneys, sweat glands, salivary glands, and pancreas. Aldosterone also increases the sensitivity of salt receptors in the tongue resulting in greater interest in consuming salty foods.

Aldosterone secretion occurs in response to a drop in blood sodium content, blood volume, or blood pressure, or to a rise in blood potassium levels. Aldosterone release also occurs in response to the hormone *angiotensin II* (*angeion*, vessel + *teinein*, to stretch). This hormone will be discussed later in the chapter.

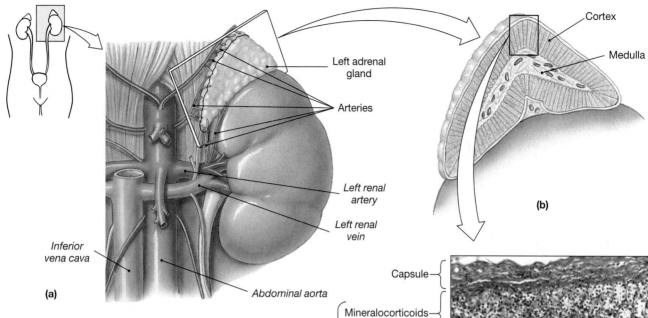

(a)

(b)

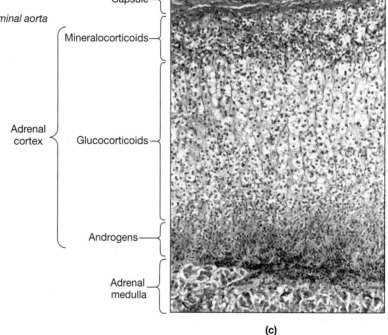

(c)

Glucocorticoids. The steroid hormones collectively known as **glucocorticoids** affect glucose metabolism. **Cortisol** (KOR-ti-sol), also called *hydrocortisone*, **corticosterone** (kor-ti-KOS-te-rōn), and **cortisone** are the three most important glucocorticoids. Glucocorticoid secretion occurs under ACTH stimulation and is regulated by negative feedback (see Figure 10-7a●). These hormones accelerate the rates of glucose synthesis and glycogen formation, especially within the liver. Simultaneously, adipose tissue responds by releasing fatty acids into the blood, and other tissues begin to break down fatty acids instead of glucose. The result is an increase in the level of glucose in the blood. Glucocorticoids also have *anti-inflammatory* effects; they suppress the activities of white blood cells and other components of the immune system. Glucocorticoid creams are often used to control irritating allergic rashes, such as those produced by poison ivy, and injections

● *Figure 10-12* **The Adrenal Gland**
(**a**) A superficial view of the left kidney and adrenal gland. (**b**) An adrenal gland, showing the sectional plane for part (c). (**c**) The major hormones and their relative source regions in the adrenal cortex. (LM × 121)

TABLE 10-3 *The Adrenal Hormones*

REGION	HORMONE	TARGET	EFFECTS
Adrenal cortex	Mineralocorticoids, primarily aldosterone	Kidneys	Increases reabsorption of sodium ions and water from the urine; accelerates urinary loss of potassium ions
	Glucocorticoids: cortisol (hydrocortisone), corticosterone, cortisone	Most cells	Releases of amino acids from skeletal muscles and lipids from adipose tissues; promotes liver formation of glycogen and glucose; promotes peripheral use of lipids; anti-inflammatory effects
	Androgens		Uncertain significance under normal conditions
Adrenal medulla	Epinephrine (E, adrenaline), norepinephrine (NE, noradrenaline)	Most cells	Increased cardiac activity, blood pressure, glycogen breakdown, and blood glucose levels; release of lipids by adipose tissue (see Chapter 8)

10 THE ENDOCRINE SYSTEM

An Overview of the Endocrine System • The Pituitary Gland • The Thyroid Gland • The Parathyroid Glands • The Adrenal Glands • **The Pineal Gland**

of glucocorticoids may be used to control more severe allergic reactions. Because they slow wound healing and suppress immune defenses (such as those against infectious organisms), topical steroids are used to treat superficial rashes but are never applied to open wounds.

Androgens. The adrenal cortex in both sexes produces small quantities of sex hormones called **androgens**. Androgens are produced in large quantities by the testes of males, and the importance of the small adrenal production in both sexes remains uncertain.

THE ADRENAL MEDULLA

The **adrenal medulla** has a reddish brown coloration partly because of the many blood vessels in this area. It contains large, rounded cells similar to those found in other sympathetic ganglia, and these cells are innervated by preganglionic sympathetic fibers. The secretory activities of the adrenal medullae are controlled by the sympathetic division of the ANS. ∞ p. 264

The adrenal medulla contains two populations of secretory cells, one producing **epinephrine** (**E**, or *adrenaline*) and the other **norepinephrine** (**NE**, or *noradrenaline*). These hormones are continuously released at a low rate, but sympathetic stimulation accelerates the rate of discharge dramatically.

Epinephrine makes up 75–80 percent of the secretions from the medulla; the rest is norepinephrine. These hormones accelerate cellular energy utilization and mobilize energy reserves. Receptors for epinephrine and norepinephrine are found on skeletal muscle fibers, in adipose tissues, and in the liver. Secretion by the adrenal medulla triggers a mobilization of glycogen reserves in skeletal muscles and accelerates the breakdown of glucose to provide ATP. This combination increases muscular power and endurance. In adipose tissue, stored fats are broken down to fatty acids, and in the liver, glycogen molecules are converted to glucose. The fatty acids and glucose are then released into the circulation for use by peripheral tissues. The heart also responds to adrenal medulla hormones with an increase in the rate and force of cardiac contractions.

The metabolic changes that follow epinephrine and norepinephrine release peak 30 seconds after adrenal stimulation and linger for several minutes thereafter. As a result, the effects produced by stimulation of the adrenal medullae outlast the other signs of sympathetic activation.

The Pineal Gland

The **pineal gland** lies in the roof of the thalamus. ∞ p. 252 It contains neurons, glial cells, and secretory cells that synthesize the hormone **melatonin** (mel-a-TŌ-nin). Branches from the axons of neurons making up the visual pathways enter the pineal gland and affect the rate of melatonin production. Melatonin production is lowest during daylight hours and highest in the dark of night.

Several functions have been suggested for melatonin in humans:

• *Inhibiting reproductive function.* In a variety of other mammals, melatonin slows the maturation of sperm, eggs, and reproductive organs. The significance of this effect remains uncertain, but there is circumstantial evidence that melatonin may play a role in the timing of human sexual maturation. For example, melatonin levels in the blood decline at puberty, and pineal tumors that eliminate melatonin production will cause premature puberty in young children.

• *Acting as an antioxidant.* Melatonin is a very effective antioxidant that may protect CNS neurons from *free radicals*, such as nitric oxide (NO) or hydrogen peroxide (H_2O_2), that may be generated in active neural tissue. (Free radicals are highly reactive atoms or molecules that contain unpaired electrons in their outer electron shell.) ∞ p. 237

• *Establishing day-night cycles of activity.* Because of the cyclical nature of its rate of secretion, the pineal gland may also be involved with the maintenance of basic *circadian rhythms*, daily changes in physiological processes that follow a regular day-night pattern. Increased melatonin secretion in darkness has been suggested as a primary cause for *seasonal affective disorder (SAD)*. This condition, characterized by changes in mood, eating habits, and sleeping patterns, can develop during the winter in high latitudes, where sunshine is meager or lacking altogether.

CONCEPT CHECK QUESTIONS

Answers on page 340

❶ What effect would elevated cortisol levels have on the level of glucose in the blood?

❷ Increased amounts of light would inhibit the production of which hormone?

The Endocrine Tissues of Other Systems

Many organs that are part of other body systems have secondary endocrine functions. Examples include the pancreas and the intestines (digestive system), the kidneys (urinary system), the heart (cardiovascular system), the thymus (lymphatic system), and the gonads—the testes in the male and ovaries in the female (reproductive system).

THE PANCREAS

The **pancreas** lies in the J-shaped loop between the stomach and small intestine (Figure 10-13●). It is a slender, usually pink organ with a nodular (lumpy) consistency, and it contains both exocrine and endocrine cells. The pancreas is primarily a digestive organ whose exocrine cells make digestive enzyymes. The **exocrine pancreas** is discussed further in Chapter 16. The endocrine cells of the pancreas produce two hormones, *glucagon* and *insulin*.

Cells of the **endocrine pancreas** form clusters known as **pancreatic islets**, or the *islets of Langerhans* (LAN-ger-hanz). The islets are scattered among the exocrine cells and account for only about 1 percent of all pancreatic cells. Each islet contains several cell types. The two most important are **alpha cells**, which produce the hormone **glucagon** (GLOO-ka-gon), and **beta cells**, which secrete **insulin** (IN-su-lin). Glucagon and insulin regulate blood glucose concentrations in the same way parathyroid hormone and calcitonin control blood calcium levels.

Regulation of Blood Glucose Concentrations

Figure 10-14● diagrams the mechanism of hormonal regulation of blood glucose levels. Glucose is the preferred energy source for most cells in the body, and under normal conditions, it is the only energy source for neurons. When blood glucose levels rise above normal, the beta cells of the pancreas release insulin, and this hormone stimulates glucose transport into its target cells. The cell membranes of almost all cells in the body contain insulin receptors; the only exceptions are (1) neurons and red blood cells, which cannot metabolize other nutrients; and (2) epithelial cells of the kidney tubules and intestinal lining, where glucose is reabsorbed (in the kidneys) or obtained from the diet (in the intestines). When glucose is abundant, all cells use it as an energy source and stop breaking down amino acids and lipids.

The ATP generated by the breakdown of glucose molecules is used to build proteins and to increase energy reserves, and most cells increase their rates of protein synthesis in response to insulin. A secondary effect is an increase in the rate of amino acid transport across cell membranes. Insulin also stimulates fat cells to increase their rates of triglyceride (fat) synthesis and storage. In the liver and in skeletal muscles, insulin also accelerates the formation of glycogen. In summary, when glucose is abundant, insulin secretion stimulates glucose utilization to support growth and to establish glycogen and fat reserves.

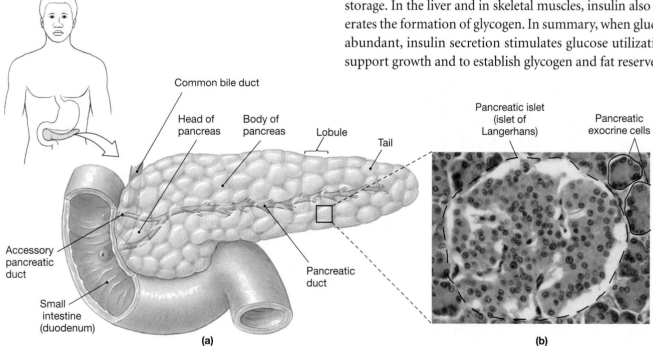

● *Figure 10-13* **The Endocrine Pancreas**
(**a**) The gross anatomy of the pancreas. (**b**) A pancreatic islet surrounded by exocrine-secreting cells. (LM × 276)

10 THE ENDOCRINE SYSTEM

An Overview of the Endocrine System • The Pituitary Gland • The Thyroid Gland • The Parathyroid Glands • The Adrenal Glands • The Pineal Gland

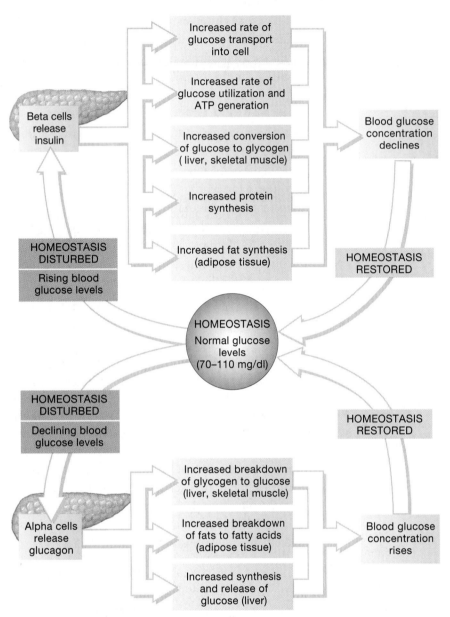

● *Figure 10-14* **The Regulation of Blood Glucose Concentrations**

When glucose levels decline, insulin secretion is suppressed, and so is glucose transport into its target cells. These cells now shift over to other energy sources, such as fatty acids. At the same time, the alpha cells of the pancreas release glucagon, and energy reserves are mobilized. Skeletal muscles and liver cells break down glycogen, adipose tissue releases fatty acids, and proteins are broken down into their component amino acids. The liver takes in the fatty acids and amino acids and converts them to glucose that can be released into the circulation. As a result, blood glucose concentrations rise toward normal levels. The interplay between insulin and glucagon both stabilizes blood glucose levels and prevents competition between neural tissue and other tissues for limited glucose supplies.

Pancreatic alpha and beta cells are sensitive to blood glucose concentrations, and the secretion of glucagon and insulin occur without endocrine or nervous instructions. Yet because the islet cells are very sensitive to variations in blood glucose levels, any hormone that affects blood glucose concentrations will indirectly affect the production of insulin and glucagon. Insulin production is also influenced by autonomic activity. Parasympathetic stimulation enhances insulin release and sympathetic stimulation inhibits it.

➕ DIABETES MELLITUS

Blood glucose levels are usually very closely regulated by insulin and glucagon. Whether glucose is absorbed across the digestive tract or manufactured and released by the liver, very little leaves the body intact once it has entered the circulation. Glucose does get fil-tered out of the blood at the kidneys, but virtually all of it is reabsorbed, so urinary glucose losses are negligible.

Diabetes mellitus (mel-Ī-tus; *mellitum*, honey) is characterized by glucose concentrations that are high enough to overwhelm the reabsorption capabilities of the kidneys. (The term **hyperglycemia** [hī-per-glī-SĒ-mē-ah] refers to the presence of high glucose levels in the blood.) Glucose appears in the urine (*glycosuria*), and urine production becomes excessive (*polyuria*). Other metabolic products, such as fatty acids and other lipids, are also present in abnormal concentrations.

Diabetes mellitus can be caused by genetic abnormalities, and some of the genes responsible have been identified. Mutations that result in inadequate insulin production, the synthesis of abnormal insulin molecules, or the production of defective receptor proteins will produce the same basic symptoms. Diabetes mellitus may also appear as the result of other pathological conditions, injuries, immune disorders, or hormonal imbalances.

There are two major types of diabetes mellitus: *insulin-dependent (type 1) diabetes* and *non-insulin-dependent (type 2) diabetes*. In type 1 diabetes, the primary cause is inadequate insulin production by the beta cells of the pancreatic islets. Type 1 diabetes most often appears in individuals under 40 years of age. Because it frequently appears in childhood, it has been called *juvenile-onset diabetes*. Long-term treatment involves dietary control and the administration of insulin. Type 2 diabetes typically affects obese individuals over 40 years of age, and consequently is often called *maturity-onset diabetes*. In this condition, insulin levels are normal or elevated, but peripheral tissues no longer transport glucose into their cells, often due to a reduction in the number of insulin receptors. Treatment consists of weight loss, dietary restrictions, and drugs (oral hypoglycemic agents) that may elevate insulin production and tissue response.

THE INTESTINES

The intestines, which process and absorb nutrients, release a variety of hormones that coordinate the activities of the digestive system. Although the pace of digestive activities can be affected by the autonomic nervous system, most digestive processes are controlled locally. These hormones will be considered in Chapter 16.

THE KIDNEYS

The kidneys release the steroid hormone *calcitriol*, the peptide hormone *erythropoietin*, and the enzyme *renin*. Calcitriol is important to calcium ion homeostasis. Erythropoietin and renin are involved in the regulation of blood pressure and blood volume.

Calcitriol is secreted by the kidneys in response to the presence of parathyroid hormone (PTH). Its synthesis depends on the availability of vitamin D_3, which may be synthesized in the skin or absorbed from the diet. Vitamin D_3 is absorbed by the liver and converted to an intermediary product that is released into the circulation and absorbed by the kidneys. Calcitriol stimulates the absorption of calcium and phosphate ions across the intestinal lining of the digestive tract.

Erythropoietin (e-rith-rō-poy-Ē-tin; *erythros*, red + *poiesis*, making), or EPO, is released by the kidneys in response to low oxygen levels in kidney tissues. EPO stimulates the production of red blood cells by the bone marrow. The increase in the number of red blood cells elevates blood volume to some degree, and because these cells transport oxygen, the increase in their number improves oxygen delivery to peripheral tissues. EPO will be considered in greater detail when we discuss the formation of blood cells in Chapter 11.

Renin (RĒ-nin) is released by kidney cells in response to a decline in blood volume, blood pressure, or both. Once in the bloodstream, renin starts an enzymatic chain reaction that leads to the formation of **angiotensin II**. Angiotensin II stimulates aldosterone production by the adrenal cortex and ADH at the posterior pituitary gland. This combination restricts salt and water loss at the kidneys. Angiotensin II also stimulates thirst and elevates blood pressure. The *renin-angiotensin system* will be detailed in Chapters 13 and 18.

THE HEART

The endocrine cells in the heart are cardiac muscle cells in the walls of the *right atrium*, the chamber that receives blood from the veins. If the blood volume becomes too great, these cardiac muscle cells are excessively stretched and release the hormone *atrial natriuretic peptide (ANP)* (nā-trē-ū-RET-ik;

natrium, sodium + *ouresis*, making water). In general, the effects of ANP oppose those of angiotensin II. ANP promotes the loss of sodium ions and water at the kidneys, inhibits renin release, and the secretion of ADH and aldosterone. ANP secretion results in a reduction in both blood volume and pressure. We will consider the actions of this hormone further when we discuss the control of blood pressure and volume in Chapter 13.

THE THYMUS

The **thymus** is embedded in the *mediastinum*, usually just posterior to the sternum. In a newborn infant, the thymus is relatively enormous, often extending from the base of the neck to the superior border of the heart. As the child grows, the thymus continues to enlarge slowly, reaching its maximum size just before puberty, at a weight of around 40 g (1.4 oz). After puberty, it gradually diminishes in size; by age 50, the thymus may weigh less than 12 g (0.4 oz).

The thymus produces several hormones, collectively known as the **thymosins** (thī-MŌ-sinz), which play a key role in the development and maintenance of normal immunological defenses. It has been suggested that the gradual decrease in the size and secretory abilities of the thymus may make the elderly more susceptible to disease.

The structure of the thymus and the functions of the thymosins will be further considered in Chapter 14.

THE GONADS
The Testes

In the male, the **interstitial cells** of the paired testes produce the steroid hormones known as androgens. **Testosterone** (tes-TOS-ter-ōn) is the most important androgen. This hormone promotes the production of functional sperm, maintains the secretory glands of the male reproductive tract, and determines secondary sex characteristics such as the distribution of facial hair and body fat. Testosterone also affects metabolic operations throughout the body. It stimulates protein synthesis and muscle growth, and it produces aggressive behavioral responses. During embryonic development, the production of testosterone affects the development of male reproductive ducts, external genitalia, and CNS structures, including hypothalamic nuclei, that will later affect sexual behaviors.

Sustentacular cells in the testes are directly associated with the formation of functional sperm. Under FSH stimulation these cells secrete the hormone **inhibin**, which inhibits the secretion of FSH by the anterior pituitary. Throughout adult life, these two hormones interact to maintain sperm production at normal levels.

331

10 THE ENDOCRINE SYSTEM

An Overview of the Endocrine System • The Pituitary Gland • The Thyroid Gland • The Parathyroid Glands • The Adrenal Glands • The Pineal Gland

The Ovaries

In the ovaries, female sex cells, or *ova*, develop in specialized structures called **follicles**, under stimulation by FSH. Follicle cells surrounding the ova produce **estrogens** (ES-trō-jenz). These steroid hormones support the maturation of the eggs and stimulate the growth of the lining of the uterus. Under FSH stimulation, follicle cells secrete inhibin, which suppresses FSH release through a feedback mechanism comparable to that in males. After ovulation has occurred, the follicular cells reorganize into a **corpus luteum**. The cells of the corpus luteum then begin to release a mixture of estrogens and progestins, especially **progesterone** (prō-JES-ter-ōn). Progesterone accelerates the movement of fertilized eggs along the uterine tubes and prepares the uterus for the arrival of a developing embryo. In combination with other hormones, it also causes an enlargement of the mammary glands.

The production of androgens, estrogens, and progestins is controlled by regulatory hormones released by the anterior pituitary gland. During pregnancy, the placenta itself functions as an endocrine organ, working together with the ovaries and the pituitary gland to promote normal fetal development and delivery.

Table 10-4 summarizes information concerning the reproductive hormones.

ENDOCRINOLOGY AND ATHLETIC PERFORMANCE

The use of hormones to improve athletic performance is banned by the International Olympic Committee, the U.S. Olympic Committee, the National Collegiate Athletic Association, and the National Football League, and condemned by the American Medical Association and the American College of Sports Medicine. A significant number of amateur and professional athletes, however, continue to use hormones. Synthetic forms of testosterone are used most often, but athletes may use any combination of testosterone, growth hormone (GH), erythropoietin (EPO), and complementary drugs such as GHB and clenbuterol. The use of steroids, such as androstenedione, which the body can convert to testosterone, was highlighted during 1998–1999 when baseball slugger Mark McGwire admitted using it to improve his performance. (Performance-enhancing drugs are still legal in professional baseball, although they are banned in some other professional sports.)

LEPTIN: A DIFFERENT HORMONE

Since 1990, several new hormones have been discovered. One of the most interesting is leptin. Unlike typical hormones, which are produced by endocrine cells in one specific location, leptin is secreted by adipose tissues throughout the body. This peptide hormone has several functions, the best known being the feedback control of appetite. When you eat, adipose tissues absorb glucose and lipids and synthesize triglycerides (fats) for storage. At the same time, they release leptin into the bloodstream. Leptin binds to neurons in the hypothalamus that deal with emotion and appetite control. The result is a sense of satiation and the suppression of appetite. Leptin also enhances the synthesis of gonadotropin releasing hormone (GnRH) and gonadotropin hormoness (follicle-stimulating hormone and luteinizing hormone). This effect explains (1) why thin girls commonly enter puberty relatively late, (2) why an increase in body fat content can improve fertility, and (3) why women stop menstruating when their body fat content becomes very low.

TABLE 10-4 *Hormones of the Reproductive System*

STRUCTURE/CELLS	HORMONE	PRIMARY TARGET	EFFECTS
TESTES			
Interstitial cells	Androgens	Most cells	Support functional maturation of sperm, protein synthesis in skeletal muscles, male secondary sex characteristics, and associated behaviors
Sustentacular cells	Inhibin	Anterior pituitary	Inhibits secretion of FSH
OVARIES			
Follicular cells	Estrogens	Most cells	Support follicle maturation, female secondary sex characteristics, and associated behaviors
	Inhibin	Anterior pituitary	Inhibits secretion of FSH
Corpus luteum	Progestins	Uterus, mammary glands	Prepare uterus for implantation; prepare mammary glands for secretory functions

❶ Which pancreatic hormone would cause skeletal muscle and liver cells to convert glucose to glycogen?

❷ What effect would increased levels of glucagon have on the amount of glycogen stored in the liver?

❸ The kidneys secrete the hormones calcitriol and erythropoietin. What are their primary targets and effects?

Patterns of Hormonal Interaction

Although hormones are usually studied individually, the extracellular fluids contain a mixture of hormones whose concentrations change daily and even hourly. When a cell receives instructions from two different hormones at the same time, several results are possible:

- The two hormones may have opposing, or **antagonistic**, effects, as in the case of parathyroid hormone and calcitonin or insulin and glucagon.

- The two hormones may have additive effects. The net result is greater than the effect that each would produce acting alone. In some cases, the net result is greater than the *sum* of their individual effects. An example of such a **synergistic** (sin-er-JIS-tik; *synairesis*, a drawing together) **effect** is the stimulation of mammary gland development by prolactin, estrogens, progestins, and growth hormone.

- One hormone can have a **permissive** effect on another. In such cases, the first hormone is needed for the second to produce its effect. For example, epinephrine by itself has no apparent effect on energy production. It will exert its effect only if thyroid hormones are present in normal concentrations.

- Hormones may also produce different but complementary results in specific tissues and organs. These **integrative** effects are important in coordinating the activities of diverse physiological systems. The differing effects of cacitriol and parathyroid hormone on tissues involved in calcium metabolism are an example.

This next section will discuss how hormones interact to control normal growth, reactions to stress, alterations of behavior, and the effects of aging. More detailed discussions will be found in chapters on cardiovascular function, metabolism, excretion, and reproduction.

HORMONES AND GROWTH

Normal growth requires the cooperation of several endocrine organs. Six hormones—growth hormone, thyroid hormones, insulin, parathyroid hormone, calcitriol, and reproductive hormones—are especially important, although many others have secondary effects on growth rates and patterns:

- *Growth hormone.* Growth hormone helps maintain normal blood glucose concentrations and mobilizes lipid reserves stored in adipose tissues. It is not the primary hormone involved, however, and an adult with a growth hormone deficiency but normal levels of thyroxine, insulin, and glucocorticoids will have no physiological problems. The effects of GH on protein synthesis and cellular growth are most apparent in children, in whom GH supports muscular and skeletal development. Undersecretion or oversecretion of GH can lead to *pituitary dwarfism* or *gigantism*.

- *Thyroid hormones.* Normal growth also requires appropriate levels of thyroid hormones. If these hormones are absent for the first year after birth, the nervous system fails to develop normally, producing mental retardation. If thyroxine concentrations decline later in life but before puberty, normal skeletal development will not continue.

- *Insulin.* Growing cells need adequate supplies of energy and nutrients. Without insulin, produced by the pancreas, the passage of glucose and amino acids across cell membranes will be drastically reduced or eliminated.

- *Parathyroid hormone and calcitriol.* Parathyroid hormone and calcitriol promote the absorption of calcium for building bone. Without adequate levels of both hormones, bones can enlarge but will be poorly mineralized, weak, and flexible. For example, in *rickets*, a condition typically due to inadequate production of calcitriol in growing children, the limb bones are so weak that they bend under the body's weight. ∞ p. 133

- *Reproductive hormones.* The activity of osteoblasts in key locations and the growth of specific cell populations are affected by the presence or absence of sex hormones (androgens in males, estrogens in females). The targets differ for androgens and estrogens, and the differential growth induced by these hormones changes accounts for gender-related differences in skeletal proportions and secondary sex characteristics.

HORMONES AND BEHAVIOR

The hypothalamus regulates many endocrine functions, and its neurons monitor the levels of many circulating hormones.

10 THE ENDOCRINE SYSTEM

An Overview of the Endocrine System • The Pituitary Gland • The Thyroid Gland • The Parathyroid Glands • The Adrenal Glands • The Pineal Gland

✚ CLINICAL NOTE *Hormones and Stress*

Any condition within the body that threatens homeostasis is a form of **stress**. Stresses produced by the action of *stressors* may be (1) physical, such as illness or injury; (2) emotional, such as depression or anxiety; (3) environmental, such as extreme heat or cold; or (4) metabolic, such as acute starvation. Many stresses are opposed by specific homeostatic adjustments. For example, a decline in body temperature will result in responses such as shivering or changes in the pattern of circulation in an attempt to restore normal body temperature.

In addition, the body has a *general* response to stress that can occur while other, more specific, responses are under way. *All stress-causing factors produce the same basic pattern of hormonal and physiological adjustments.* These responses are part of the stress response, also known as the **general adaptation syndrome**, or **GAS**. The GAS has three basic phases: the *alarm phase*, the *resistance phase*, and the *exhaustion phase* (Figure 10-15●).

The **alarm phase** is an immediate response to the stress, under the direction of the sympathetic division of the autonomic nervous system. During this phase, energy reserves are mobilized, mainly in the form of glucose, and the body prepares for any physical activities needed to eliminate the source of the stress or escape from it. Epinephrine is the dominant hormone of the alarm phase, and its secretion accompanies the sympathetic activation that produces the "fight or flight" response discussed in Chapter 8. ∞ p. 264

The temporary adjustments of the alarm phase often remove or overcome the stress. But if a stress lasts longer than a few hours, the individual will enter the **resistance phase** of the GAS. Glucocorticoids (GCs) are the dominant hormones of the resistance phase, but many other hormones are involved. Throughout the resistance phase, energy demands remain higher than normal because of the background production of glucocorticoids, epinephrine, growth hormone, and thyroid hormones.

The endocrine secretions of the resistance phase coordinate three integrated actions to maintain adequate levels of glucose in the blood: (1) the mobilization of lipid and protein reserves, (2) the conservation of glucose for neural tissues, and (3) the synthesis and release of glucose by the liver. When the resistance phase ends, homeostatic regulation breaks down, and the **exhaustion phase** begins. Unless corrective actions are taken almost immediately, the failure of one or more organ systems will prove fatal. Although a single cause, such as heart failure, may be listed as the cause of death, the underlying problem is the inability to support the endocrine and metabolic adjustments of the resistance phase.

● *Figure 10-15* **The General Adaptation Syndrome**

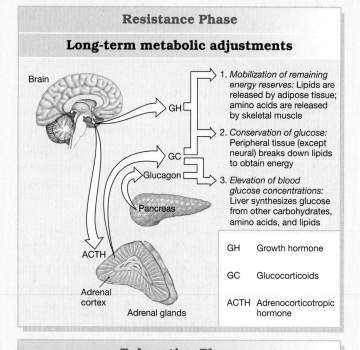

Other portions of the central nervous system are also quite sensitive to hormonal stimulation.

The clearest demonstrations of the effects of specific hormones involve individuals whose endocrine glands are over- secreting or undersecreting. Normal changes in circulating hormone levels can also cause behavioral changes. Chapter 6 noted that one of the triggers for closure of the epiphyses is the increase in sex hormone production at the time of puberty. ∞ p. 132

● *Figure 10-16*
Endocrine Abnormalities

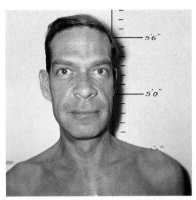

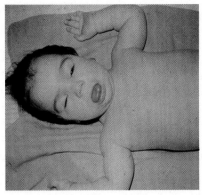

(a) *Acromegaly* results from the overproduction of growth hormone after the epiphyseal cartilages have fused. Bone shapes change, and cartilaginous areas of the skeleton enlarge. Notice the broad facial features and the enlarged lower jaw.

(b) *Cretinism* or *congenital hypothyroidism* results from thyroid hormone insufficiency in infancy.

(c) An enlarged thyroid gland, or *goiter*, can be associated with thyroid hyposecretion due to iodine insufficiency in adults.

✚ ENDOCRINE DISORDERS

Endocrine disorders fall into two basic categories: symptoms of underproduction (inadequate hormonal effects) or symptoms of overproduction (excessive hormonal effects). The observed symptoms may reflect either abnormal hormone production (hyposecretion or hypersecretion) or abnormal cellular sensitivity. The characteristic features of some of these conditions are shown in Figure 10-16●.

In *precocious* (premature) *puberty*, sex hormones are produced at an inappropriate time, perhaps as early as 5 or 6 years of age. The affected children not only begin to develop adult secondary sex characteristics but also undergo significant behavioral changes. The "nice little kid" disappears, and the child becomes aggressive and assertive. These behavioral alterations represent the effects of sex hormones on CNS function. Thus, behaviors that in normal teenagers are usually attributed to external factors, such as peer pressure, actually have some physiological basis as well. In the adult, changes in the mixture of hormones reaching the CNS can have significant effects on intellectual capabilities, memory, learning, and emotional states.

HORMONES AND AGING

The endocrine system shows relatively few functional changes with age. The most dramatic exception is the decline in the concentration of reproductive hormones. Effects of these hormonal changes on the skeletal system were noted in Chapter 6; further discussion will be found in Chapter 20. p. 136

Blood and tissue concentrations of many other hormones, including TSH, thyroid hormones, ADH, PTH,

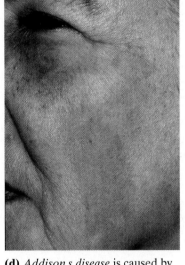

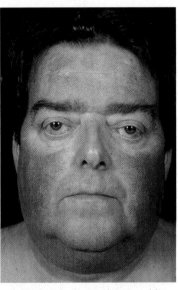

(d) *Addison s disease* is caused by hyposecretion of corticosteroids, especially glucocorticoids. Pigment changes result from stimulation of melanocytes by ACTH, which is structurally similar to MSH.

(e) *Cushing s disease* is caused by hypersecretion of glucocorticoids. Lipid reserves are mobilized, and adipose tissue accumulates in the cheeks and at the base of the neck.

prolactin, and glucocorticoids, remain unchanged with increasing age. Although circulating hormone levels may remain within normal limits, some endocrine tissues become less responsive to stimulation. For example, in elderly people, less GH and insulin are secreted after a carbohydrate-rich meal.

Finally, it should be noted that age-related changes in other tissues affect their abilities to respond to hormonal stimulation. As a result, peripheral tissues may become less responsive to some hormones. This loss of sensitivity has been documented for glucocorticoids and ADH.

10 THE ENDOCRINE SYSTEM

An Overview of the Endocrine System • The Pituitary Gland • The Thyroid Gland • The Parathyroid Glands • The Adrenal Glands • The Pineal Gland

Integration with Other Systems

The relationships between the endocrine system and other systems are summarized in Figure 10-17●. This overview does not consider all of the hormones associated with the digestive system and the control of digestive functions. These hormones will be detailed in Chapter 16.

Related Clinical Terms

Addison's disease: A condition caused by the hyposecretion of glucocorticoids and mineralocorticoids; characterized by an inability to mobilize energy reserves and maintain normal blood glucose levels.

cretinism (KRĒ-tin-ism): A condition caused by hypothyroidism in infancy; marked by inadequate skeletal and nervous development and a metabolic rate as much as 40 percent below normal levels.

Cushing's disease: A condition caused by the hypersecretion of glucocorticoids; characterized by the excessive breakdown and relocation of lipid reserves and proteins.

diabetes insipidus: A disorder that develops when the posterior pituitary no longer releases adequate amounts of ADH or when the kidneys cannot respond to ADH.

diabetes mellitus (mel-Ī-tus): A disorder characterized by glucose concentrations high enough to overwhelm the kidneys' reabsorption capabilities.

diabetic retinopathy, nephropathy, neuropathy: Disorders of the retina, kidneys, and peripheral nerves, respectively, related to diabetes mellitus; the conditions most often afflict middle-aged or older diabetics.

endocrinology (EN-do-kri-NOL-o-jē): The study of hormones, hormone-secreting tissues and glands, and their roles in physiology and disease in the body.

general adaptation syndrome (GAS): The pattern of hormonal and physiological adjustments with which the body responds to all forms of stress.

glycosuria (glī-kō-SOO-rē-a): The presence of glucose in the urine.

goiter: An abnormal enlargement of the thyroid gland.

hyperglycemia: Abnormally high glucose levels in the blood.

hypoglycemia: Abnormally low glucose levels in the blood.

insulin-dependent diabetes mellitus (IDDM), or *type 1 diabetes* or *juvenile-onset diabetes:* A type of diabetes mellitus; the primary cause is inadequate insulin production by the beta cells of the pancreatic islets.

myxedema: In adults, symptoms of hyposecretion of thyroid hormones, including subcutaneous swelling, hair loss, dry skin, low body temperature, muscle weakness, and slowed reflexes.

non-insulin-dependent diabetes mellitus (NIDDM), or *type 2 diabetes,* or *maturity-onset diabetes:* A type of diabetes mellitus in which insulin levels are normal or elevated but peripheral tissues no longer respond normally.

polyuria: The production of excessive amounts of urine; a symptom of diabetes.

thyrotoxicosis: A condition caused by the oversecretion of thyroid hormones (*hyperthyroidism*). Symptoms include increases in metabolic rate, blood pressure, and heart rate; excitability and emotional instability; and lowered energy reserves.

CHAPTER REVIEW

Key Terms

Summary Outline

INTRODUCTION **314**

1. In general, the nervous system performs short-term "crisis management," while the endocrine system regulates longer-term, ongoing metabolic processes. Endocrine cells release **hormones**, chemicals that alter the metabolic activities of many different tissues and organs. (*Figure 10-1*)

AN OVERVIEW OF THE ENDOCRINE SYSTEM **314**

The Structure of Hormones **314**

1. Hormones can be divided into three groups based on chemical structure: amino acid derivatives, peptide hormones, and lipid derivatives.

2. *Amino acid derivatives* are structurally similar to amino acids; they include *epinephrine, norepinephrine, thyroid hormones,* and *melatonin.*

3. **Peptide hormones** are chains of amino acids.

4. There are two classes of *lipid derivatives.* **Steroid hormones**, lipids structurally similar to cholesterol, and fatty acid-based hormones such as **prostaglandins**.

The Mechanisms of Hormonal Action **315**

5. Hormones exert their effects by modifying the activities of **target cells** (peripheral cells that are sensitive to that particular hormone). (*Figure 10-2*)

6. Receptors for amino acid-derived hormones, peptide hormones, and fatty acid-derived hormones are located on the cell membranes of target

The Endocrine System

For All Systems

Adjusts metabolic rates and substrate utilization; regulates growth and development

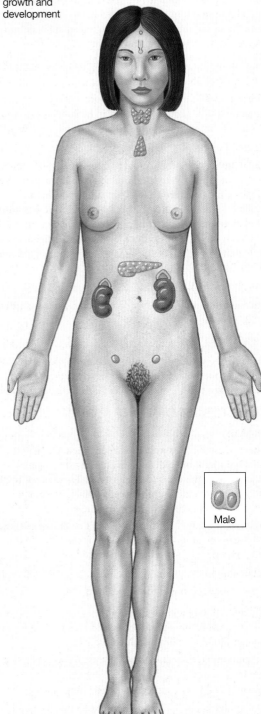

Male

The Integumentary System

- Protects superficial endocrine organs; epidermis synthesizes vitamin D_3

- Sex hormones stimulate sebaceous gland activity, influence hair growth, fat distribution, and apocrine sweat gland activity; PRL stimulates development of mammary glands; adrenal hormones alter dermal blood flow, stimulate release of lipids from adipocytes; MSH stimulates melanocyte activity

The Skeletal System

- Protects endocrine organs, especially in brain, chest, and pelvic cavity

- Skeletal growth regulated by several hormones; calcium mobilization regulated by parathyroid hormone and calcitonin; sex hormones speed growth and closure of epiphyseal plates at puberty and help maintain bone mass in adults

The Muscular System

- Skeletal muscles provide protection for some endocrine organs

- Hormones adjust muscle metabolism, energy production, and growth; regulate calcium and phosphate levels in body fluids; speed skeletal muscle growth

The Nervous System

- Hypothalamic hormones directly control pituitary and indirectly control secretions of other endocrine organs; controls adrenal medullae; secretes ADH and oxytocin

- Several hormones affect neural metabolism; hormones help regulate fluid and electrolyte balance; reproductive hormones influence CNS development and behaviors

The Cardiovascular System

- Circulatory system distributes hormones throughout the body; heart secretes ANP

- Erythropoietin regulates production of RBCs; several hormones elevate blood pressure; epinephrine elevates heart rate and contraction force

The Lymphatic System

- Lymphocytes provide defense against infection and, with other WBCs, assist in repair after injury

- Glucocorticoids have anti-inflammatory effects; thymosins stimulate development of lymphocytes; many hormones affect immune function

The Respiratory System

- Provides oxygen and eliminates carbon dioxide generated by endocrine cells

- Epinephrine and norepinephrine stimulate respiratory activity and dilate respiratory passageways

The Digestive System

- Provides nutrients to endocrine cells; endocrine cells of pancreas secrete insulin and glucagon; liver produces angiotensinogen

- E and NE stimulate constriction of sphincters and depress activity along digestive tract; digestive tract hormones coordinate secretory activities along tract

The Urinary System

- Kidney cells (1) release renin and erythropoietin when local blood pressure declines and (2) produce calcitriol

- Aldosterone, ADH, and ANP adjust rates of fluid and electrolyte reabsorption in kidneys

The Reproductive System

- Steroid sex hormones and inhibin suppress secretory activities in hypothalamus and pituitary

- Hypothalamic regulatory hormones and pituitary hormones regulate sexual development and function; oxytocin stimulates uterine and mammary gland smooth muscle contractions

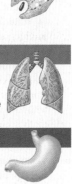

● *Figure 10-17* **Functional Relationships Between the Endocrine System and Other Systems**

10 THE ENDOCRINE SYSTEM

An Overview of the Endocrine System • The Pituitary Gland • The Thyroid Gland • The Parathyroid Glands • The Adrenal Glands • The Pineal Gland

cells; in this case, the hormone acts as a **first messenger** that causes a **second messenger** to appear in the cytoplasm. Thyroid and steroid hormones cross the cell membrane and bind to receptors in the cytoplasm or nucleus. Thyroid hormones also bind to mitochondria, where they increase the rate of ATP production. *(Figure 10-3)*

7. The simplest patterns of endocrine control involve the direct negative feedback of changes in the extracellular fluid on the endocrine cells.

8. The most complex endocrine responses involve the hypothalamus. The hypothalamus regulates the activities of the nervous and endocrine systems by three mechanisms: (1) it acts as an endocrine organ itself by releasing hormones into the circulation; (2) it secretes **regulatory hormones** that control the activities of endocrine cells in the pituitary gland; (3) its autonomic centers exert direct neural control over the endocrine cells of the adrenal medullae. *(Figure 10-4)*

1. The **pituitary gland** (*hypopohysis*) releases nine important peptide hormones; all bind to membrane receptors and most use cyclic-AMP as a second messenger. *(Figure 10-5)*

2. Hypothalamic neurons release regulatory factors into the surrounding interstitial fluids, which then enter highly permeable capillaries.

3. The **hypophyseal portal system** ensures that all of the blood entering the *portal vessels* will reach target cells in the anterior pituitary before returning to the general circulation. *(Figure 10-6)*

4. The rate of regulatory hormone secretion by the hypothalamus is regulated through negative feedback mechanisms. *(Figure 10-7)*

5. The seven hormones of the **anterior pituitary gland** are: (1) **thyroid-stimulating hormone (TSH)**, which triggers the release of thyroid hormones; (2) **adrenocorticotropic hormone (ACTH)**, which stimulates the release of **glucocorticoids** by the adrenal gland; (3) **follicle-stimulating hormone (FSH)**, which stimulates estrogen secretion and egg development in females and sperm production in males; (4) **luteinizing hormone (LH)**, which causes ovulation and **progestin** production in females and **androgen** production in males; (5) **prolactin (PRL)**, which stimulates the development of the mammary glands and the production of milk; (6) **growth hormone (GH)**, which stimulates cell growth and replication by triggering the release of **somatomedins** from liver cells; and (7) **melanocyte-stimulating hormone (MSH)**, which stimulates melanocytes to produce melanin, but MSH is not normally secreted by the nonpregnant human adult.

6. The **posterior pituitary gland** contains the axons of hypothalamic neurons that manufacture **antidiuretic hormone (ADH)** and **oxytocin**. ADH decreases the amount of water lost at the kidneys. In females, oxytocin stimulates smooth muscle cells in the uterus and contractile cells in the mammary glands. In males, it stimulates sperm duct and prostate gland smooth muscle contractions. *(Figure 10-8; Table 10-1)*

1. The **thyroid gland** lies near the **thyroid cartilage** of the larynx and consists of two lobes. *(Figure 10-9)*

2. The thyroid gland contains numerous **thyroid follicles**. Thyroid follicles release several hormones, including **thyroxine (TX or T_4)** and **triiodothyronine (T_3)** *(Table 10-2)*

3. Thyroid hormones exert a **calorigenic effect**, which enables us to adapt to cold temperatures.

4. The **C cells** of the follicles produce **calcitonin (CT)**, which helps regulate calcium ion concentrations in body fluids. *(Table 10-2)*

1. Four **parathyroid glands** are embedded in the posterior surface of the thyroid gland. The **chief cells** of the parathyroid produce **parathyroid hormone (PTH)** in response to lower than normal concentrations of calcium ions. These and the C cells of the thyroid gland maintain calcium ion levels within relatively narrow limits. *(Figures 10-10, 10-11; Table 10-2)*

1. A single **adrenal gland** lies along the superior border of each kidney. Each gland, surrounded by a fibrous capsule, can be subdivided into the superficial adrenal cortex and the inner adrenal medulla. *(Figure 10-12)*

2. The **adrenal cortex** manufactures steroid hormones called *adrenocortical steroids* (**corticosteroids**). The cortex produces: (1) **glucocorticoids**—notably, **cortisol**, **corticosterone**, and **cortisone**, which, in response to ACTH, affect glucose metabolism; (2) **mineralocorticoids**—principally **aldosterone**, which, in response to *angiotensin II*, restricts sodium and water losses at the kidneys, sweat glands, digestive tract, and salivary glands; and (3) androgens of uncertain significance. *(Figure 10-12; Table 10-3)*

3. The **adrenal medulla** produces **epinephrine** and **norepinephrine**. *(Figure 10-12; Table 10-3)*

1. The **pineal gland** synthesizes **melatonin**. Melatonin appears to (1) slow the maturation of sperm, eggs, and reproductive organs; (2) protect neural tissue from free radicals; and (3) establish daily circadian rhythms.

1. The **pancreas** contains both exocrine and endocrine cells. The **exocrine pancreas** secretes an enzyme-rich fluid that travels to the digestive tract. Cells of the **endocrine pancreas** form clusters called **pancreatic islets** (*islets of Langerhans*), containing **alpha cells** (which produce the hormone **glucagon**) and **beta cells** (which secrete **insulin**). *(Figure 10-13)*

2. Insulin lowers blood glucose by increasing the rate of glucose uptake and utilization; glucagon raises blood glucose by increasing the rates of glycogen breakdown and glucose synthesis in the liver. *(Figure 10-14)*

3. The intestines release hormones that coordinate digestive activities.

4. Endocrine cells in the kidneys produce hormones important for calcium metabolism, blood volume, and blood pressure.

5. **Calcitriol** stimulates calcium and phosphate ion absorption along the digestive tract.

6. **Erythropoietin (EPO)** stimulates red blood cell production by the bone marrow.

7. On its release, **renin** functions as an enzyme whose activity leads to the formation of **angiotensin II**, the hormone that stimulates the adrenal production of aldosterone.

8. Specialized muscle cells in the heart produce **atrial natriuretic peptide (ANP)** when blood pressure and/or blood volume becomes excessive.

9. The **thymus** produces several hormones, called **thymosins**, which play a role in developing and maintaining normal immunological defenses.

10. The **interstitial cells** of the paired male testes produce androgens and **inhibin**. The androgen testosterone is the most important sex hormone in the male. *(Table 10-4)*

11. In females, ova (eggs) develop in **follicles**; follicle cells surrounding the eggs produce **estrogens** and inhibin. After ovulation, the cells reorganize into a **corpus luteum** that releases a mixture of estrogens and progestins, especially **progesterone**. If pregnancy occurs, the placenta functions as an endocrine organ. *(Table 10-4)*

1. The endocrine system functions as an integrated unit, and hormones often interact. These interactions may have (1) **antagonistic** (opposing) effects, (2) **synergistic** (additive) effects, (3) **permissive** effects, or (4) **integrative** effects, in which hormones produce different but complementary results.

2. Normal growth requires the cooperation of several endocrine organs. Six hormones are especially important: growth hormone, thyroid hormones, insulin, parathyroid hormone, calcitriol, and reproductive hormones.

3. Many hormones affect the functional state of the nervous system, producing changes in mood, emotional states, and various behaviors.

4. The endocrine system shows relatively few functional changes with advanced age. The most dramatic endocrine change is the decline in the concentration of reproductive hormones.

1. The endocrine system affects all systems by adjusting metabolic rates and regulating growth and development. *(Figure 10-17)*

Review Questions

Level 1: Reviewing Facts and Terms

Match each item in column A with the most closely related item in column B. Use letters for answers in the spaces provided.

COLUMN A

____ 1. thyroid gland
____ 2. pineal gland
____ 3. polyuria
____ 4. parathyroid gland
____ 5. thymus gland
____ 6. adrenal cortex
____ 7. heart
____ 8. endocrine pancreas
____ 9. gonadotropins
____ 10. hypothalamus
____ 11. pituitary gland
____ 12. growth hormone

COLUMN B

a. islets of Langerhans
b. atrophies by adulthood
c. atrial natriuretic peptide
d. cell growth
e. melatonin
f. hypophysis
g. excessive urine production
h. calcitonin
i. secretes regulatory hormones
j. FSH and LH
k. secretes androgens, mineralocorticoids, and glucocorticoids
l. stimulated by low calcium levels

13. Adrenocorticotropic hormone (ACTH) stimulates the release of:
    (a) thyroid hormones by the hypothalamus
    (b) gonadotropins by the adrenal glands
    (c) somatotropins by the hypothalamus
    (d) steroid hormones by the adrenal glands

14. FSH production in males supports:
    (a) maturation of sperm by stimulating sustentacular cells
    (b) development of muscles and strength
    (c) production of male sex hormones
    (d) increased desire for sexual activity

15. The hormone that induces ovulation in women and promotes the ovarian secretion of progesterone is
    (a) interstitial cell-stimulating hormone   (b) estradiol
    (c) luteinizing hormone   (d) prolactin

16. The two hormones released by the posterior pituitary are
    (a) somatotropin and gonadotropin
    (b) estrogen and progesterone
    (c) growth hormone and prolactin
    (d) antidiuretic hormone and oxytocin

17. The primary function of antidiuretic hormone (ADH) is to:
    (a) increase the amount of water lost at the kidneys
    (b) decrease the amount of water lost at the kidneys
    (c) dilate peripheral blood vessels to decrease blood pressure
    (d) increase absorption along the digestive tract

18. The element required for normal thyroid function is
    (a) magnesium   (b) calcium
    (c) potassium   (d) iodine

10 THE ENDOCRINE SYSTEM

An Overview of the Endocrine System • The Pituitary Gland • The Thyroid Gland • The Parathyroid Glands • The Adrenal Glands • The Pineal Gland

19. Reduced fluid losses in the urine due to retention of sodium ions and water is a result of the action of:

    (a) insulin   (b) calcitonin

    (c) aldosterone   (d) cortisone

20. The adrenal medullae produce the hormones:

    (a) cortisol and cortisone

    (b) epinephrine and norepinephrine

    (c) corticosterone and testosterone

    (d) androgens and progesterone

21. What seven hormones are released by the anterior pituitary gland?

22. What effects do calcitonin and parathyroid hormone have on blood calcium levels?

23. (a) What three phases of the general adaptation syndrome (GAS) constitute the body's response to stress?

    (b) What endocrine secretions play dominant roles in the alarm and resistance phases?

Level 2: Reviewing Concepts

24. What is the primary difference in the way the nervous and endocrine systems communicate with their target cells?

25. How can a hormone modify the activities of its target cells?

26. What possible results occur when a cell receives instructions from two different hormones at the same time?

27. How would blocking the activity of phosphodiesterase affect a cell that responds to hormonal stimulation by the cAMP second messenger system?

Level 3: Critical Thinking and Clinical Applications

28. Roger M. has been suffering from extreme thirst; he drinks numerous glasses of water every day and urinates a great deal. Name two disorders that could produce these symptoms. What test could a clinician perform to determine which disorder is present?

29. Julie is pregnant and is not receiving any prenatal care. She has a poor diet consisting mostly of fast food. She drinks no milk, preferring colas instead. How will this situation affect Julie's level of parathyroid hormone?

Answers to Concept Check Questions

Page 318

1. A cell's sensitivity to one or more hormones is determined by the presence or absence of the necessary receptor molecule for a given hormone.

2. Adenylate cyclase is the enzyme that converts ATP to cAMP. A molecule that blocks this enzyme would block the action of any hormone that required cAMP as a second messenger.

3. Epinephrine, norepinephrine, and peptide hormones are first messengers from the endocrine glands. Since these hormones cannot enter their target cells, a second molecule, intracellular cyclic-AMP, acts as the messenger from the endocrine glands.

Page 322

1. Dehydration increases the electrolyte concentration in the blood, which leads to an increase in osmotic pressure. The increase in blood osmotic pressure would stimulate the posterior pituitary gland to release more ADH.

2. Somatomedins are the mediators of growth hormone action. If the level of somatomedins is elevated, we would expect the level of growth hormone to be elevated as well.

3. Increased levels of cortisol would inhibit the cells that control ACTH release from the pituitary gland; therefore the level of ACTH would decrease. This is an example of a negative feedback mechanism.

Page 326

1. An individual who lacked iodine would not be able to form the hormone thyroxine. As a result, we would expect to see the symptoms associated with thyroxine deficiency, such as decreased rate of metabolism, decreased body temperature, poor response to physiological stress, and an increase in the size of the thyroid gland (goiter).

2. Most of the thyroid hormone in the blood is bound to carrier proteins. This represents a large reservoir of thyroxine that guards against rapid fluctuations in the level of this important hormone. Because such a large amount is stored, it takes several days to deplete the supply of hormone, even after the thyroid gland has been removed.

3. The removal of the parathyroid glands would result in a decrease in the blood levels of calcium ion. This decrease could be counteracted by increasing the amount of vitamin D_3 and calcium in the diet.

Page 328

1. One of the functions of cortisol is to decrease the cellular use of glucose while increasing the available glucose by promoting the breakdown of glycogen and the conversion of amino acids to carbohydrates. The net result would be an elevation in the level of glucose in the blood.

2. The pineal gland receives neural input from the optic tracts, and its secretion, melatonin, is influenced by light-dark cycles. Increased amounts of light inhibit the production and release of melatonin from the pineal gland.

Page 333

1. Insulin increases the rate of conversion of glucose to glycogen in skeletal muscle and liver cells.

2. Glucagon stimulates the conversion of glycogen to glucose in the liver. Increased amounts of glucagon would then lead to decreased amounts of liver glycogen.

3. Calcitriol targets cells lining the digestive tract, stimulating their absorption of calcium and phosphate. Erythropoietin (EPO) targets cells in the bone marrow that produce red blood cells. An increase in the number of red blood cells would improve the delivery of oxygen to body tissues.

Page 336

1. The hormonal interaction exemplified by the insulin and glucagon is called antagonistic. In this type of hormonal interaction, two hormones have opposite effects on their target tissues.

2. Growth hormone, thyroid hormone, parathyroid hormone, calcitriol, and the reproductive hormones all play a role in the formation and development of the skeletal system.

EXPLORE *MediaLab*

Female athletes clearing hurdles in an international track event in Moscow.

EXPLORATION #1

Estimated time for completion: 15 minutes

STRESS. IT IS A FACT of life for most of us, causing at the very least some emotional strain and potentially leading to serious concerns. We usually define stress in terms of daily routine aggravations and unforeseen complications. For some individuals, stress is far more pervasive. Imagine how stressful it can be for an infant raised in a sterile, institutional environment for the first 8–12 months of life, who is then transported to an entirely different culture. This is the life history of 65 percent of the internationally adopted infants since 1990. Using the information presented in the Hormones and Stress Clinical Note on page 334 as a guide, prepare a table of the hormones that might be produced in excess in these infants. Indicate what the long-term effects of this early stress might be on the nervous system and emotional state of the individual. Would you expect to see behavioral consequences? After preparing your list, go to Chapter 10 at the Companion Web site, visit the MediaLab section, and click on the key words "Stress Physiology." Here you will find a concise overview of the current research in stress physiology of internationally adopted children. As you read this article, add to your list of stress-related hormones and their detrimental effects.

EXPLORATION #2

Estimated time for completion: 10 minutes

EVERY ATHLETE'S DREAM is to reach peak performance, and perhaps even compete in the Olympic Games. How far will some go in order to attain their goal? Anabolic steroids, the use of which is banned from Olympic competition, are sometimes touted as a quick route to athletic prowess. What are anabolic steroids and how do they alter body functions? Why are they banned from amateur competition? Review the section of Chapter 10 on Hormones and Growth, and take a minute to review Table 10-4, Hormones of the Reproductive System, and Figure 10-7●, Feedback Control of Endocrine Secretion. Prepare a short essay indicating hormones are responsible for the natural gain in strength as we mature. Indicate how these hormones are controlled (What feedback controls exist for these hormones?) and speculate on the effects of an overabundance of these hormones. Then go to Chapter 10 at the Companion Web site, visit the MediaLab section, and click on the key words "Anabolic Steroids." Here you will find a thorough review of the history of anabolic steroids and their properties, side effects, and medicinal uses. After you have read this article, refine your essay using the new information you have gathered.

An American mother holds her adopted Korean infant.

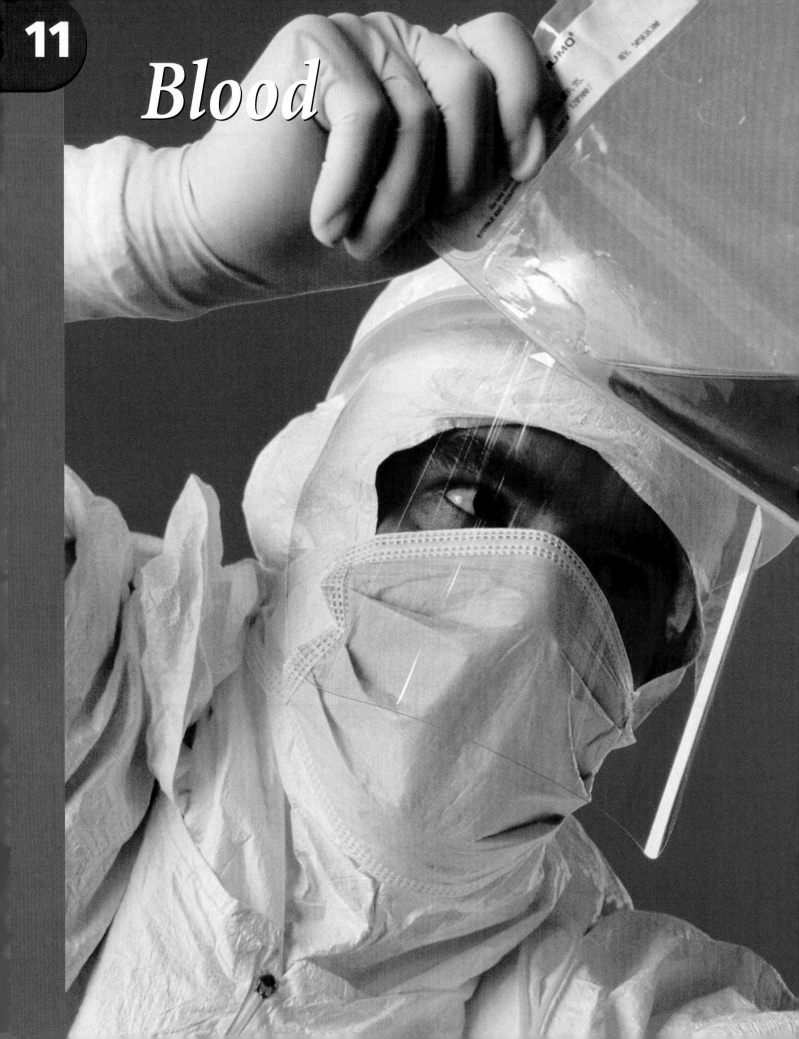

Blood

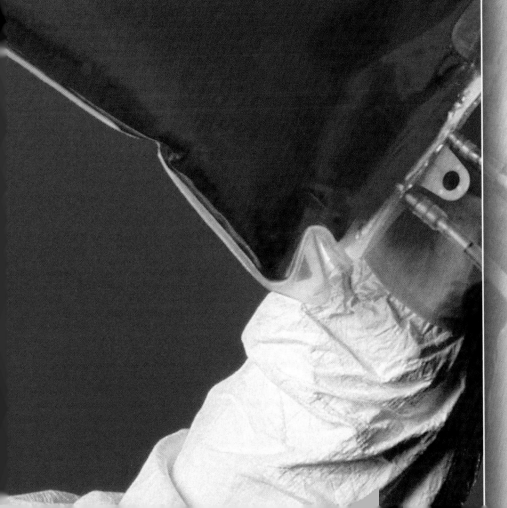

Chapter Outline and Objectives

Vocabulary Development

agglutinins ....gluing; *agglutinization*
embolos ....plug; *embolus*
erythros ....red; *erythrocytes*
haima ....blood; *hemostasis*
hypo- ....below; *hypoxia*
karyon ....nucleus; *megakaryocyte*
leukos ....white; *leukocyte*
megas ....big; *megakaryocyte*
myelos ....marrow; *myeloid*
-osis ....condition; *leukocytosis*
ox- ....presence of oxygen; *hypoxia*
penia ....poverty; *leukopenia*
poiesis ....making; *hemopoiesis*
punctura ....a piercing; *venipuncture*
stasis ....halt; *hemostasis*
thrombos ....clot; *thrombocytes*
vena ....vein; *venipuncture*

"O UT OF SIGHT, *out of mind."* *This old saying applies quite well to blood. As long as we don't see it, we don't think or worry much about it. But looking at vials or bags of it in a hospital—or worse, watching even a drop of our own blood oozing from a cut or scrape—makes most of us feel uneasy; some people even faint at the sight. This anxiety probably comes from being reminded of how dependent we are on our relatively small supply of the precious fluid. A sudden loss of 15 to 20 percent of a person's total blood volume is a clinical emergency—the person will need to get more in a hurry. Fortunately, blood transfusions are now fairly routine matters—but as we will learn in this chapter, not just any blood will do.*

343

THE LIVING BODY is in constant chemical communication with its external environment. Nutrients are absorbed through the lining of the digestive tract, gases move across the delicate epithelium of the lungs, and wastes are excreted in the feces and urine. These chemical exchanges occur at specialized sites or organs, but all parts of the body are linked by the **cardiovascular system**, an internal transport network.

The cardiovascular system can be compared to the cooling system of a car. The basic components are a circulating fluid (blood), a pump (the heart), and conducting pipes (the blood vessels). Although the cardiovascular system is far more complicated and versatile, both mechanical and biological systems can suffer from fluid losses, pump failures, or damaged pipes.

Small embryos don't need cardiovascular systems, because diffusion across their exposed surfaces can exchange materials rapidly enough to meet their demands. By the time the embryo has reached a few millimeters in length, however, developing tissues will consume oxygen and nutrients and generate waste products faster than they can be provided or removed by simple diffusion. At that stage, the cardiovascular system must begin functioning to provide a rapid-transport system for oxygen, nutrients, and waste products. It is the first organ system to become fully operational: The heart begins beating by the end of the third week of embryonic life, when most other systems have barely started their development. When the heart starts beating, the blood begins circulating. The embryo can now make more efficient use of the nutrients obtained from the maternal bloodstream, and its size doubles in the next week.

This chapter considers the nature of the circulating blood. Chapter 12 focuses on the structure and function of the heart, and Chapter 13 examines the organization of blood vessels and the integrated functioning of the cardiovascular system. Chapter 14 considers the lymphatic system, a defense system intimately connected to the cardiovascular system.

The Functions of Blood

The circulating fluid of the body is **blood**, a specialized connective tissue that contains cells suspended in a fluid matrix. ⟳ p. 94 The five major functions of blood are:

1. *The transportation of dissolved gases, nutrients, hormones, and metabolic wastes.* Blood carries oxygen from the lungs to the tissues, and carbon dioxide from the tissues to the lungs. It distributes nutrients that are absorbed at the digestive tract or released from storage in adipose tissue or in the liver. Blood carries hormones from endocrine glands toward their target tissues. It also absorbs the wastes produced by tissue cells and carries these wastes to the kidneys for excretion.

2. *The regulation of the pH and electrolyte composition of interstitial fluids throughout the body.* Blood absorbs and neutralizes the acids generated by active tissues, such as the lactic acid produced by skeletal muscles.

3. *The restriction of fluid losses through damaged vessels or at other injury sites.* Blood contains enzymes and factors that respond to breaks in the vessel walls by initiating the process of *blood clotting.* The blood clot that develops acts as a temporary patch and prevents further changes in blood volume.

4. *Defense against toxins and pathogens.* Blood transports white blood cells, specialized cells that migrate into peripheral tissues to fight infections or remove debris. It also delivers *antibodies,* special proteins that attack invading organisms or foreign compounds.

5. *The stabilization of body temperature.* Blood absorbs the heat generated by active skeletal muscles and redistributes it to other tissues. If the body temperature is already high, that heat will be lost across the surface of the skin. When body temperature is too low, the warm blood is directed to the brain and to other temperature-sensitive organs.

The Composition of Blood

Blood is normally confined to the circulatory system, and it has a characteristic and unique composition (Figure 11-1●). It consists of matrix called *plasma* (PLAZ-muh) and formed elements. **Plasma** contains dissolved proteins rather than the network of insoluble fibers like those in loose connective tissue or cartilage. Because these proteins are in solution, plasma is slightly denser than water. **Formed elements** are blood cells (red or white) and cell fragments (platelets) that are suspended in the plasma. **Red blood cells (RBCs),** or **erythrocytes** (e-RITH-rō-sīts; *erythros,* red), transport oxygen and carbon dioxide. The less numerous **white blood cells (WBCs),** or **leukocytes** (LOO-kō-sīts; *leukos,* white), are cells involved with body's defense mechanisms. **Platelets** are small, membrane-enclosed packets of cytoplasm that contain enzymes and factors important to blood clotting.

Together the plasma and formed elements constitute **whole blood.** The volume of whole blood varies from 5 to 6 liters (5.3–6.4 quarts) in the cardiovascular system of an adult man and from 4 to 5 liters (4.2–5.3 quarts) in that of an adult woman. Whole blood components may be separated, or **fractionated,** for analytical or clinical purposes.

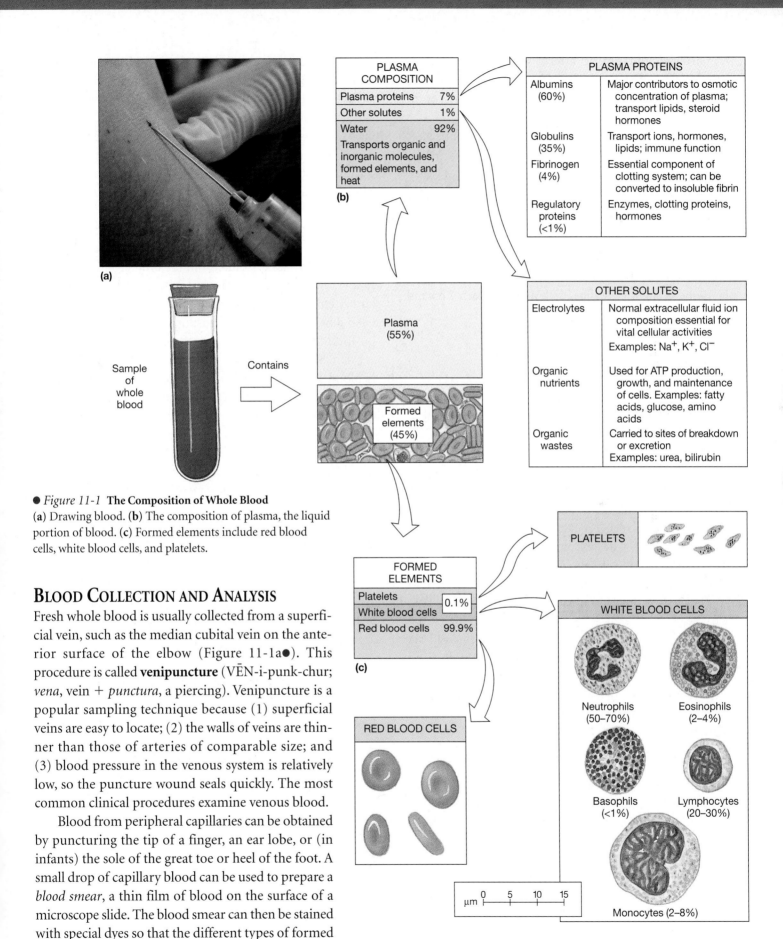

PLASMA COMPOSITION	
Plasma proteins	7%
Other solutes	1%
Water	92%
Transports organic and inorganic molecules, formed elements, and heat	

(b)

PLASMA PROTEINS	
Albumins (60%)	Major contributors to osmotic concentration of plasma; transport lipids, steroid hormones
Globulins (35%)	Transport ions, hormones, lipids; immune function
Fibrinogen (4%)	Essential component of clotting system; can be converted to insoluble fibrin
Regulatory proteins (<1%)	Enzymes, clotting proteins, hormones

OTHER SOLUTES	
Electrolytes	Normal extracellular fluid ion composition essential for vital cellular activities Examples: Na$^+$, K$^+$, Cl$^-$
Organic nutrients	Used for ATP production, growth, and maintenance of cells. Examples: fatty acids, glucose, amino acids
Organic wastes	Carried to sites of breakdown or excretion Examples: urea, bilirubin

Sample of whole blood   Contains

Plasma (55%)

Formed elements (45%)

● *Figure 11-1* **The Composition of Whole Blood**
(**a**) Drawing blood. (**b**) The composition of plasma, the liquid portion of blood. (**c**) Formed elements include red blood cells, white blood cells, and platelets.

PLATELETS

FORMED ELEMENTS	
Platelets	0.1%
White blood cells	
Red blood cells	99.9%

(c)

BLOOD COLLECTION AND ANALYSIS

Fresh whole blood is usually collected from a superficial vein, such as the median cubital vein on the anterior surface of the elbow (Figure 11-1a●). This procedure is called **venipuncture** (VĒN-i-punk-chur; *vena*, vein + *punctura*, a piercing). Venipuncture is a popular sampling technique because (1) superficial veins are easy to locate; (2) the walls of veins are thinner than those of arteries of comparable size; and (3) blood pressure in the venous system is relatively low, so the puncture wound seals quickly. The most common clinical procedures examine venous blood.

Blood from peripheral capillaries can be obtained by puncturing the tip of a finger, an ear lobe, or (in infants) the sole of the great toe or heel of the foot. A small drop of capillary blood can be used to prepare a *blood smear*, a thin film of blood on the surface of a microscope slide. The blood smear can then be stained with special dyes so that the different types of formed elements are easily distinguishable.

WHITE BLOOD CELLS

Neutrophils (50–70%)   Eosinophils (2–4%)

Basophils (<1%)   Lymphocytes (20–30%)

Monocytes (2–8%)

RED BLOOD CELLS

μm   0   5   10   15

An **arterial puncture**, or "arterial stick," may be required for checking the efficiency of gas exchange at the lungs. Samples are usually drawn from the radial artery at the wrist or the brachial artery at the elbow.

Whole blood from any of these sources has the same basic physical characteristics:

- *Temperature.* The temperature of blood is roughly 38°C (100.4°F), slightly higher than normal body temperature.

- *Viscosity.* Blood is five times as viscous as water, because interactions between dissolved proteins, formed elements, and the surrounding water molecules make plasma relatively sticky, cohesive, and resistant to flow.

- *pH.* Blood is slightly alkaline, with a pH between 7.35 and 7.45, averaging 7.4. ⌒⊃ p. 34

Plasma

Plasma makes up approximately 55 percent of the volume of whole blood, with water accounting for 92 percent of the plasma volume (Figure 11-1b●). Together, plasma and interstitial fluid account for most of the volume of extracellular fluid (ECF) in the body.

In many respects, the composition of the plasma resembles that of interstitial fluid (IF). The concentrations of the major plasma ions (electrolytes), for example, are similar to those of the interstitial fluid and differ from those inside cells. The similarity is understandable because there is a constant exchange of water, ions, and small solutes between the interstitial fluid and plasma across the walls of capillaries (see Figure 4-10●, p. 96). The chief differences between plasma and interstitial fluid involve the concentrations of dissolved proteins and the levels of respiratory gases (oxygen and carbon dioxide). The protein concentrations differ because plasma proteins cannot cross the capillary walls of blood vessels. The differences in the levels of respiratory gases are due to the respiratory activities of tissue cells and will be discussed in Chapter 15.

PLASMA PROTEINS

Plasma contains considerable quantities of dissolved proteins. On average, each 100 ml of plasma contains about 7 g of protein, almost five times the concentration in interstitial fluid (Figure 11-1b●). The large size of most blood proteins prevents them from crossing capillary walls, so they remain trapped within the circulatory system. The three primary classes of plasma proteins are *albumins* (al-BŪ-minz), *globulins* (GLOB-ū-linz), and *fibrinogen* (fī-BRIN-ō-jen).

Albumins constitute roughly 60 percent of the plasma proteins. As the most abundant proteins, they are major con-

tributors to the osmotic pressure of the plasma. **Globulins**, accounting for another 35 percent of the plasma proteins, include antibodies and transport proteins. **Antibodies**, also called **immunoglobulins** (i-mū-nō-GLOB-ū-linz), attack foreign proteins and pathogens. **Transport proteins** bind small ions, hormones, or compounds that might otherwise be filtered out of the blood at the kidneys. One example is thyroid-binding globulin, which binds and transports thyroid hormones.

Both albumins and globulins can become attached to lipids, such as triglycerides, fatty acids, or cholesterol. These lipids are not themselves water-soluble, but the protein-lipid combination readily dissolves in plasma. In this way, the cardiovascular system transports insoluble lipids to peripheral tissues. Globulins involved in lipid transport are called *lipoproteins* (lī-pō-PRŌ-tēnz).

The third type of plasma protein, **fibrinogen**, functions in blood clotting. Under certain conditions, fibrinogen molecules interact and combine to form large, insoluble strands of **fibrin** (FĪ-brin). These fibers provide the basic framework for a blood clot. If steps are not taken to prevent clotting in a plasma sample, fibrinogen will convert to fibrin. The fluid left after the clotting proteins are removed is known as **serum**. The liver synthesizes more than 90 percent of the plasma proteins, including all of the albumins, fibrinogen, and most of the globulins. Antibodies (immunoglobulins) are produced by *plasma cells* of the lymphatic system. Because the liver is the primary source of plasma proteins, liver disorders can alter the composition and functional properties of the blood. For example, some forms of liver disease can lead to uncontrolled bleeding, caused by inadequate synthesis of fibrinogen and other plasma proteins involved in the clotting response.

CONCEPT CHECK QUESTIONS

Answers on page 365

❶ Why is venipuncture a common technique for obtaining a blood sample?

❷ What would be the effects of a decrease in the amount of plasma proteins?

Formed Elements

The major types of cells in blood are red blood cells and white blood cells. In addition, blood contains noncellular formed elements—platelets, small packets of cytoplasm that function in the clotting response (Figure 11-1c●).

THE PRODUCTION OF FORMED ELEMENTS

Formed elements are produced through the process of **hemopoiesis** (hēm-ō-poy-Ē-sis), or *hematopoiesis*. Embryonic blood cells appear in the bloodstream during the third week of development. These cells divide repeatedly, increasing in number. The vessels of the yolk sac, an embryonic membrane, are the primary site of blood formation for the first 8 weeks of development. As other organ systems appear, some of the embryonic blood cells move out of the bloodstream and into the liver, spleen, thymus, and bone marrow. These embryonic cells differentiate into **stem cells** that produce blood cells by their divisions. The liver and spleen are the primary sites of hemopoiesis from the second to fifth month of development. As the skeleton enlarges, the bone marrow becomes increasingly important, and it is the primary site after the fifth developmental month. In adults, it is the only site of red blood cell production and the primary site of white blood cell formation.

Hemocytoblasts, or *pleuripotent stem cells*, divide to produce *myeloid stem cells* and *lymphoid stem cells*. Both populations of these stem cells retain the ability to divide, but their daughter cells go on to produce specific types of blood cells. We will consider the fates of the myeloid and lymphoid stem cells as we discuss the formation of each type of formed element.

RED BLOOD CELLS

Red blood cells (RBCs) contain the pigment *hemoglobin*, which binds and transports oxygen and carbon dioxide. Red blood cells are the most abundant blood cells, accounting for 99.9 percent of the formed elements.

The number of RBCs in the blood of a normal individual staggers the imagination. A standard blood test reports the number RBCs per microliter (μl) of whole blood as the *red blood cell count*. One microliter, or one cubic millimeter (mm^3), of whole blood from an adult male contains roughly 5.4 million erythrocytes; a microliter of blood from an adult female contains about 4.8 million. A single drop of whole blood contains some 260 million RBCs. Roughly one-third of the 75 trillion cells in the human body are red blood cells.

The **hematocrit** (he-MA-tō-krit) is the percentage of whole blood occupied by cellular elements. The hematocrit is determined by spinning a blood sample in a centrifuge so that all the formed elements come out of suspension. In adult males, it averages 46% (range: 40–54); in adult females, 42% (range: 37–47). The difference in hematocrit between males and females reflects the fact that androgens (male hormones) stimulate red blood cell production, whereas estrogens (female hormones) have an inhibitory effect. Because whole blood contains roughly 1000 red blood cells for each white blood cell, the hematocrit closely approximates the volume of erythrocytes. For this reason, hematocrit values are often reported as the *volume of packed red cells (VPRC)* or simply the *packed cell volume (PCV)*.

Many factors can alter the hematocrit. For example, dehydration increases the hematocrit by reducing plasma volume, and internal bleeding or problems with RBC formation can decrease the hematocrit through the loss of red blood cells. As a result, the hematocrit alone does not provide specific diagnostic information. However, a change in hematocrit is an indication that other, more specific tests are needed. (Some of those tests will be considered on page 351.)

The Structure of RBCs

Red blood cells are specialized to transport oxygen and carbon dioxide within the bloodstream. As Figure 11-2● shows,

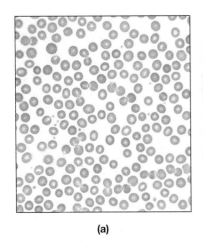
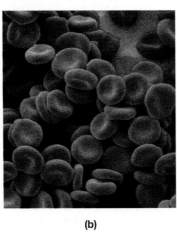
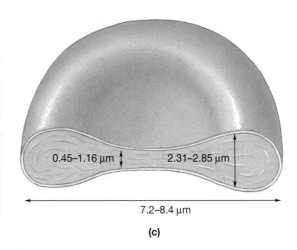

0.45–1.16 μm   2.31–2.85 μm

7.2–8.4 μm

(a)   (b)   (c)

● *Figure 11-2* **The Anatomy of Red Blood Cells**
(a) When viewed in a standard blood smear, red blood cells appear as two-dimensional objects because they are flattened against the surface of the slide. (LM × 320) (b) A scanning electron micrograph of red blood cells reveals their three-dimensional structure. (SEM × 1195) (c) A sectional view of a mature red blood cell, showing average dimensions.

each red blood cell has a thin central region and a thick outer margin. This unusual shape has two important effects on RBC function; (1) it gives each RBC a relatively large surface area to volume ratio that increases the rate of diffusion between its cytoplasm and the surrounding plasma, and (2) it enables RBCs to bend and flex, and squeeze through narrow capillaries.

During their formation, RBCs lose most of their organelles, including mitochondria, ribosomes, and nuclei. Without a nucleus or ribosomes, RBCs can neither undergo cell division nor synthesize structural proteins or enzymes. Without mitochondria, they can obtain energy only through anaerobic metabolism, relying on glucose obtained from the surrounding plasma. This characteristic makes RBCs relatively inefficient in terms of energy, but it ensures that absorbed oxygen will be carried to peripheral tissues, not "stolen" by mitochondria in the cell.

Hemoglobin Structure and Function

A mature red blood cell consists of a cell membrane enclosing a mass of transport proteins. Molecules of **hemoglobin** (HĒ-mō-glō-bin) **(Hb)** account for over 95 percent of the intracellular proteins. Hemoglobin is responsible for the cell's ability to transport oxygen and carbon dioxide.

Two pairs of globular proteins (each pair made up of slightly different chains of polypeptides) combine to form a single molecule of hemoglobin. ∞ p. 41 Each of the four subunits contains a single molecule of an organic pigment called **heme**. Each heme molecule holds an iron ion in such a way that it can interact with an oxygen molecule (O_2). The iron-oxygen interaction is very weak, and the two can easily be separated without damage to either the hemoglobin or the oxygen molecule. RBCs containing hemoglobin with bound oxygen give blood a bright red color. The RBCs give blood a dark red, almost burgundy, color when oxygen is not bound to hemoglobin.

The amount of oxygen bound in each RBC depends on the conditions in the surrounding plasma. When oxygen is abundant in the plasma, the hemoglobin molecules gain oxygen until all of the heme molecules are occupied. As plasma oxygen levels decline, plasma carbon dioxide levels are usually rising. Under these conditions, the hemoglobin molecules release their oxygen reserves and the globin portion of each hemoglobin molecule begins to bind carbon dioxide molecules in a process that is just as reversible as the binding of oxygen to heme.

As red blood cells circulate, they are exposed to varying combinations of oxygen and carbon dioxide concentrations.

At the lungs, where diffusion brings oxygen into the plasma and removes carbon dioxide, the hemoglobin molecules in red blood cells respond by absorbing oxygen and releasing carbon dioxide. In peripheral tissues, the situation is reversed. There active cells consume oxygen and produce carbon dioxide. As blood flows through these areas, oxygen diffuses out of the plasma and carbon dioxide diffuses in. Under these conditions, hemoglobin releases its stored oxygen and binds carbon dioxide.

Normal activity levels can be sustained only when tissue oxygen levels are kept within normal limits. The blood of a person who has a low hematocrit, or whose RBCs have a reduced hemoglobin content, has a reduced oxygen-carrying capacity. This condition is called **anemia**. Anemia causes a variety of symptoms, including premature muscle fatigue, weakness, and a general lack of energy.

RBC Life Span and Circulation

An RBC is exposed to severe physical stresses. A single round-trip of the circulatory system usually takes less than a minute. In that time, a red blood cell is forced along vessels, where it bounces off the walls, collides with other red cells, and is squeezed through tiny capillaries. With all this mechanical battering and no repair mechanisms, a red blood cell has a relatively short life span, only about 120 days. The continuous elimination of RBCs usually goes unnoticed, because new RBCs enter the circulation at a comparable rate. About 1 percent of the circulating RBCs are replaced each day, and approximately 3 million new RBCs enter the circulation each second!

Hemoglobin Recycling. As red blood cells age, they either rupture or are destroyed by phagocytic cells. If a damaged or aged RBC ruptures, its hemoglobin is not recycled. The hemoglobin breaks down and the individual polypeptide chains are filtered by the kidneys and lost in the urine. When large numbers of RBCs break down in the circulation, the urine can turn reddish or brown. This condition, called **hemoglobinuria**, is usually prevented by specialized mechanisms that reclaim and recycle hemoglobin. Only about 10 percent of the RBCs survive long enough to rupture, or **hemolyze** (HĒ-mō-līz), within the bloodstream. Phagocytic cells of the liver, spleen, and bone marrow usually recognize and engulf RBCs before they undergo *hemolysis*. These phagocytes also remove hemoglobin and RBC fragments from the circulation. (Phagocytosis and phagocytic cells were introduced in Chapter 3; additional details are given in Chapter 14.) ∞ p. 64

Abnormal Hemoglobin

Several inherited disorders are characterized by the production of abnormal hemoglobin. Two of the best known are *thalassemia* and *sickle-cell anemia (SCA)*.

The various forms of **thalassemia** (thal-ah-SĒ-mē-uh) result from an inability to produce adequate amounts of the globular protein components of hemoglobin. As a result, the rate of RBC production is slowed, and the mature RBCs are fragile and short-lived. The scarcity of healthy RBCs reduces the oxygen-carrying capacity of the blood and leads to problems in growth and development. Individuals with severe thalassemia must undergo frequent *transfusions*—the administration of blood components—to maintain adequate numbers of RBCs in the blood.

Sickle-cell anemia (SCA) results from a mutation affecting the amino acid sequence of one pair of the globular proteins of the hemoglobin molecule. When blood contains an abundance of oxygen, the hemoglobin molecules and the RBCs that carry them appear normal (Figure 11-3a●). But when the defective hemoglobin gives up enough of its stored oxygen, the adjacent hemoglobin molecules interact, and the cells change shape, becoming stiff and markedly curved (Figure 11-3b●). This "sickling" does not affect the oxygen-carrying capabilities of the RBCs, but it makes them more fragile and easily damaged. Moreover, when an RBC that has folded to squeeze into a narrow capillary delivers its oxygen to the surrounding tissue, the cell can become stuck as sickling occurs. This blocks blood flow and nearby tissues become oxygen-starved.

(a) Normal RBC

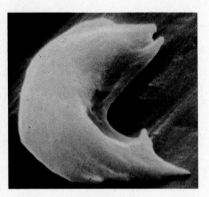

(b) Sickled RBC

● *Figure 11-3* **Sickling in Red Blood Cells**
(a) When fully oxygenated, the cells of an individual with the sickling trait have the same appearance as normal RBCs. **(b)** At lower oxygen concentrations the RBCs change shape, becoming more rigid and sharply curved. (SEM × 5200)

The recycling of hemoglobin and turnover of red blood cells is shown in Figure 11-4●. Once a red blood cell has been engulfed and broken down by a phagocytic cell, each component of a hemoglobin molecule is handled differently:

1. The four globular proteins of each hemoglobin molecule are disassembled into their component amino acids. These amino acids are either metabolized by the cell or released into the circulation for use by other cells.
2. Each heme molecule is stripped of its iron and converted to **biliverdin** (bil-ē-VER-din), a green compound. (Bad bruises often appear greenish because biliverdin forms in the blood-filled tissues.) Biliverdin is then converted to **bilirubin** (bil-ē-ROO-bin), an orange-yellow pigment, and released into circulation. Liver cells absorb the bilirubin and normally release it into the small intestine within the bile. If the bile ducts are blocked (by gallstones, for example), bilirubin then diffuses into peripheral tissues, giving them a yellow color that is most apparent in the skin and eyes. This combination of yellow skin and eyes is called **jaundice** (JAWN-dis). Bilirubin in the large intestine is converted to related pigment molecules. Some of these are absorbed into the bloodstream and excreted into urine. The yellow color of urine and the brown color of feces is due to these bilirubin-derived pigments.
3. Iron extracted from the heme molecules may be stored in the phagocytic cell or released into the bloodstream, where it binds to **transferrin** (tranz-FER-in), a plasma transport protein. Red blood cells developing in the bone marrow absorb the amino acids and transferrins from the circulation and use them to synthesize new hemoglobin molecules. Excess transferrins are removed by the liver and spleen, and the iron is stored in special protein-iron complexes.

In summary, most of the components of an individual red blood cell are recycled following hemolysis or phagocytosis. The entire system is remarkably efficient; although roughly 26 mg of iron is incorporated into hemoglobin molecules each day, a dietary supply of 1–2 mg can keep pace with the incidental losses that occur at the kidney and the digestive tract.

Abnormalities in iron uptake or metabolism can cause serious clinical problems because RBC formation will be

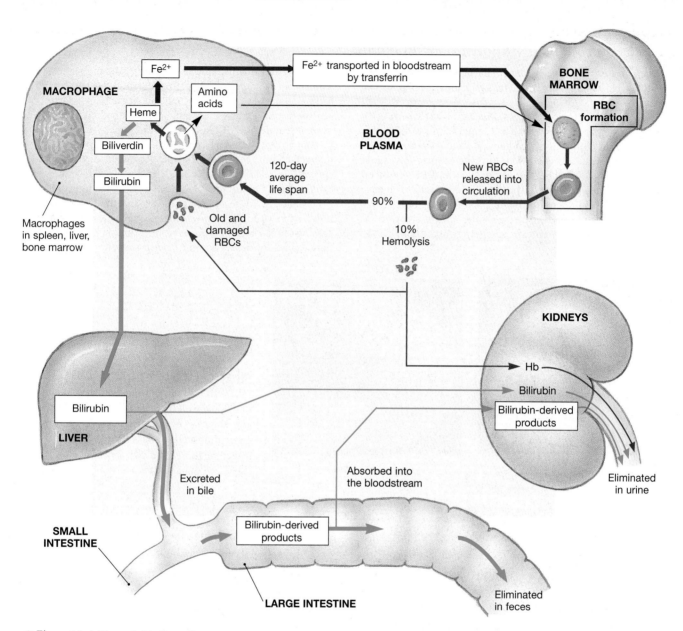

● *Figure 11-4* **Hemoglobin Recycling**
Macrophages in the spleen, liver, and bone marrow break down old and damaged RBCs, recycling the components of the hemoglobin molecules they contain. The breakdown of the protein portion of hemoglobin releases amino acids into the bloodstream for use by other cells, including developing RBCs in the red bone marrow. The iron from the heme units is used in the formation of new hemoglobin or is stored. After removal of the iron, the remaining portions of the heme units are converted to bilirubin. Bilirubin in the circulation is absorbed by the liver and excreted within bile that is released into the intestine. Bilirubin-derived products are excreted in urine or in waste from the digestive tract.

affected. Women are especially dependent on a normal dietary supply of iron, because their iron reserves are smaller than those of men. The body of a normal man contains around 3.5 g of iron in the ionic form Fe^{2+}. Of that amount, 2.5 g is bound to the hemoglobin of circulating red blood cells, and the rest is stored in the liver and bone marrow. In women, the total body iron content averages 2.4 g, with roughly 1.9 g incorporated into red blood cells. Thus a woman's iron reserves consist of only 0.5 g, half that of a typical man. If dietary supplies

of iron are inadequate, hemoglobin production slows down, and symptoms of *iron deficiency anemia* appear. Too much iron can also cause problems because of excessive buildup in the liver and cardiac muscle tissue. Excessive iron deposition in cardiac muscle cells has recently been linked to heart disease.

Red Blood Cell Formation

Red blood cell formation, or **erythropoiesis** (e-rith-rō-poy-Ē-sis), occurs in the red bone marrow. *Red bone marrow*, or **myeloid** tissue (MĪ-e-loyd; *myelos*, marrow), is located in the vertebrae, sternum, ribs, scapulae, pelvis, and proximal limb bones. Other marrow areas contain a fatty tissue known as **yellow bone marrow**. Under extreme stimulation, such as a severe and sustained blood loss, areas of yellow marrow can convert to red marrow, increasing the rate of red blood cell formation.

Stages in RBC Maturation. During its maturation, a red blood cell passes through a series of stages (see Figure 11-8●, p. 358). Hematologists (hē-ma-TOL-o-jists), specialists in blood formation and function, have given specific names to key stages. Divisions of the hemocytoblasts in red bone marrow produce (1) **myeloid stem cells**, which in turn divide to produce red blood cells and several classes of white blood cells, and (2) **lymphoid stem cells**, which divide to produce various classes of lymphocytes. Future RBCs proceed through proerythroblast and erythroblast stages. **Erythroblasts** are very immature red blood cells that are actively synthesizing hemoglobin. After roughly 4 days of differentiation and hemoglobin production, the erythroblast sheds its nucleus and becomes a **reticulocyte** (re-TIK-ū-lō-sīt). After 2 days in the bone marrow, reticulocytes enter the circulation. At this time they can still be detected in a blood smear with stains that specifically combine with RNA. Normally, reticulocytes account for about 0.8 percent of the circulating erythrocytes. After 24 hours in circulation, the reticulocytes complete their maturation and become indistinguishable from other mature RBCs.

The Regulation of Erythropoiesis. For erythropoiesis to proceed normally, the myeloid tissues must receive adequate supplies of amino acids, iron, vitamin B$_{12}$, and other vitamins (B$_6$ and folic acid) required for protein synthesis. Erythropoiesis is stimulated directly by the hormone erythropoietin and indirectly by several hormones, including thyroxine, androgens, and growth hormone. As we noted in the hematocrit discussion, estrogens have an inhibitory effect on erythropoiesis.

Erythropoietin, also called **EPO** or *erythropoiesis-stimulating hormone*, appears in the plasma when peripheral tissues, especially the kidneys, are exposed to low oxygen concentrations. This condition is called **hypoxia** (hī-POKS-ē-a; *hypo-*, below + *ox-*, presence of oxygen). EPO is released during anemia, when blood flow to the kidneys declines, when the oxygen content of air in the lungs declines (owing to disease or high altitudes), and when the respiratory surfaces of the lungs are damaged (Figure 11-5●). Once in the circulation, EPO travels to areas of red bone marrow, where it stimulates stem cells and developing erythrocytes.

Erythropoietin has two major effects: (1) It stimulates increased rates of mitotic divisions in erythroblasts and in the

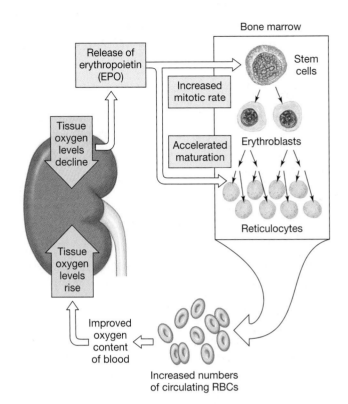

● *Figure 11-5* **The Control of Erythropoiesis**
Tissues deprived of oxygen release EPO, which accelerates division of RBC stem cells and the maturation of erythroblasts. More red blood cells then enter the circulation, improving the delivery of oxygen to peripheral tissues.

stem cells that produce erythroblasts; and (2) it speeds up the maturation of red blood cells, primarily by accelerating the rate of hemoglobin synthesis. Under maximum EPO stimulation, the bone marrow can increase the rate of red blood cell formation tenfold, to around 30 million per second.

This reserve is important to a person recovering from a severe blood loss. If EPO is administered to a healthy person, however, the hematocrit may rise to 65 or more, placing an intolerable strain on the heart. Comparable strains can occur after *blood doping*. In this practice, athletes reinfuse packed RBCs removed at an earlier date. The goal is to increase hematocrit and improve performance, but the result can be sudden death from heart failure.

 BLOOD TESTS AND RBCS

Blood tests provide information about the general health of an individual, usually with a minimum of trouble and expense. Several common blood tests focus on red blood cells, the most abundant formed elements. These tests assess the number, size, shape, and maturity of circulating RBCs. This information provides an indication of the RBC-forming activities and can also be useful in detecting problems, such as internal bleeding, that may not produce obvious symptoms. Table 11-1 lists some important blood tests and related terms.

TABLE 11-1 *Common RBC Tests and Related Terminology*

TEST	DETERMINES	TERMS ASSOCIATED WITH ABNORMAL VALUES	
		ELEVATED	DEPRESSED
HEMATOCRIT (HCT)	Percentage of formed elements in whole blood Normal = 37–54%	Polycythemia (may result from erythrocytosis or leukocytosis)	Anemia
COMPLETE BLOOD COUNT (CBC)			
RBC count	Number of RBCs per μl of whole blood Normal = 4.2–6.3 million/μl	Erythrocytosis	Anemia
Hemoglobin concentration (Hb)	Concentration of hemoglobin in blood Normal = 12–18 g/dl		Anemia
Reticulocyte count (Retic.)	Circulating percentage of reticulocytes Normal = 0.8%	Reticulocytosis	
Mean corpuscular volume (MCV)	Average volume of single RBC Normal = 82–101 μm³ (normocytic)	Macrocytic	Microcytic
Mean corpuscular hemoglobin concentration (MCHC)	Average amount of Hb in one RBC Normal = 27–34 pg/μl (normochromic)	Hyperchromic	Hypochromic

CONCEPT CHECK QUESTIONS

Answers on page 365

❶ How would dehydration affect an individual's hematocrit?

❷ How would the level of bilirubin in the blood be affected by a disease that causes damage to the liver?

❸ How would a decrease in the level of oxygen supplied to the kidneys affect the level of erythropoietin in the blood?

RBCs and Blood Types

Antigens are substances (mostly proteins) that can trigger an *immune response*, a defense mechanism that protects you from infection. Cell membranes contain surface antigens, substances that the body's immune defenses recognize as "normal." In other words, the immune system ignores these substances rather than attacking them as "foreign." A person's blood type is a classification determined by the presence or absence of specific surface antigens in the RBC cell membranes. The surface antigens of RBCs are often called *agglutinogens* (a-gloo-TIN-ō-jenz). The characteristics of RBC surface antigen molecules are genetically determined. Red blood cells have at least 50 kinds of surface antigens. Three of particular importance have been designated antigens **A**, **B**, and **Rh**.

All of the red blood cells of any individual have the same pattern of surface antigens. **Type A** blood has antigen A only, **Type B** has antigen B only, **Type AB** has both A and B, and **Type O** has neither A nor B. The average values for the U.S.

population are Type O, 46 percent; Type A, 40 percent; Type B, 10 percent; and Type AB, 4 percent. Variations in these values result from racial and ethnic genetic differences (Table 11-2).

TABLE 11-2 *Differences in Blood Group Distribution*

POPULATION	PERCENTAGE WITH EACH BLOOD TYPE				
	O	A	B	AB	Rh⁺
U.S. (AVERAGE)	46	40	10	4	85
African American	49	27	20	4	95
Caucasian	45	40	11	4	85
Chinese American	42	27	25	6	100
Filipino American	44	22	29	6	100
Hawaiian	46	46	5	3	100
Japanese American	31	39	21	10	100
Korean American	32	28	30	10	100
NATIVE NORTH AMERICAN	79	16	4	<1	100
NATIVE SOUTH AMERICAN	100	0	0	0	100
AUSTRALIAN ABORIGINE	44	56	0	0	100

The presence or absence of the Rh antigen, sometimes called the Rh factor, is indicated by the terms Rh-positive (present) and Rh-negative (absent). Clinicians usually omit the term Rh and report the complete blood type as O-positive or O-negative, A-positive or A-negative, and so forth.

Antibodies and Cross-Reactions. Blood type is checked before an individual gives or receives blood. The surface antigens on a person's own red blood cells are ignored by that person's immune system. However, plasma contains antibodies, or *agglutinins* (a-GLOO-ti-ninz), that will attack surface antigens on foreign cells (Figure 11-6a●). For example, the plasma of individuals with Type A blood contains circulating anti-B antibodies, which will attack Type B surface antigens. The plasma of Type B individuals contains anti-A antibodies, which will attack Type A surface antigens. The RBCs of an individual with Type O blood lack surface antigens A and B, so the plasma of

such an individual contains both anti-A and anti-B. At the other extreme, Type AB individuals lack antibodies sensitive to either A or B surface antigens.

When an antibody meets its specific antigen, a **cross-reaction** occurs. Initially the RBCs clump together, a process called **agglutination** (a-gloo-ti-NĀ-shun); they may also break up, or *hemolyze* (Figure 11-6b●). Clumps and fragments of RBCs under attack form drifting masses that can plug small vessels in the kidneys, lungs, heart, or brain, damaging or destroying tissues. Such cross-reactions, or *transfusion reactions*, can be avoided by ensuring that the blood type of the donor and that of the recipient are **compatible**. A donor must be chosen whose blood cells will not undergo cross-reaction with the plasma of the recipient. Unless large volumes of whole blood or plasma are transferred, cross-reactions between the donor's plasma and the recipient's blood cells will fail to produce significant agglutination. Packed RBCs, with a minimal

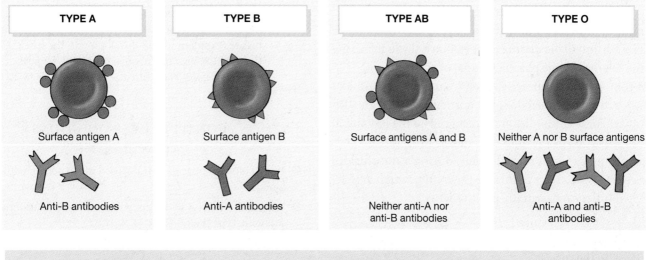

(a)

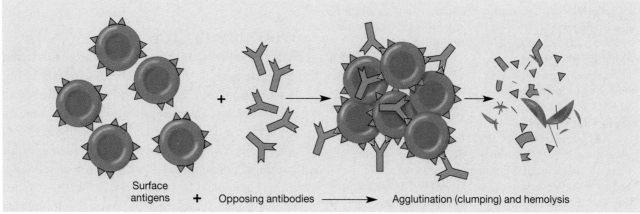

(b)

● *Figure 11-6* **Blood Typing and Cross-Reactions**
(**a**) The blood type depends on the presence of antigens on RBC surfaces. Blood plasma contains antibodies, which will react with foreign antigens. (**b**) A cross-reaction occurs when antibodies encounter their target antigens. The result is extensive clumping (agglutination) and hemolysis of the affected RBCs.

CLINICAL NOTE *Testing for Compatibility*

Testing for compatibility normally involves two steps: a determination of blood type and a *cross-match test*. Although some 50 antigens are known, the standard test for blood type categorizes a blood sample on the basis of the three most likely to produce dangerous cross-reactions. The test involves mixing drops of blood with solutions containing A, B, and Rh antibodies and noting any cross-reactions. For example, if the RBCs clump together when exposed to anti-A and anti-B, the individual has Type AB blood. If no reactions occur, the person must be Type O. The presence or absence of the Rh antigen is also noted, and the individual is classified as Rh-positive or Rh-negative. In the most common type, Type O-positive (O$^+$), the RBCs do not have surface antigens A and B, but they do have the Rh antigen.

Standard blood typing of both donor and recipient can be completed in a matter of minutes. In an emergency, Type O blood can be safely administered. For example, a patient with a severe gunshot wound may require 5 *liters* or more of blood before the damage can be repaired. Under these circumstances, it may not be possible to do more than collect blood, make certain that it is Type O, and administer it to the victim. Because their blood cells are unlikely to produce severe cross-reactions in a recipient, Type O individuals

(especially O-negative) are sometimes called *universal donors*. Type AB individuals are sometimes called *universal recipients*, because they do not have anti-A or anti-B antibodies, which would attack donated RBCs. This term has been dropped from usage—largely because if the recipient's blood type is known to be AB, Type AB blood can be administered.

With at least 48 other possible antigens on the cell surface, however, cross-reactions can occur, even to Type O blood. Whenever time and facilities permit, further testing is performed to ensure complete compatibility. **Cross-match testing** involves exposing the donor's RBCs to a sample of the recipient's plasma under controlled conditions. This procedure reveals the presence of significant cross-reactions involving other antigens and antibodies. It is also possible to replace lost blood with synthetic blood substitutes.

Because blood groups are inherited, blood tests are also used as paternity tests and in crime detection. Results from such tests cannot prove that a particular individual is the criminal or the father involved, but it can prove that he is not involved. For example, a man with Type AB blood cannot be the father of an infant with Type O blood.

---

amount of plasma, are commonly transfused. Even when whole blood is transfused, the plasma will be diluted through mixing with the relatively large plasma volume of the recipient.

In contrast to the surface antigens A and B, the plasma of an Rh-negative individual does not normally contain anti-Rh antibodies. These antibodies are present only if the individual has been sensitized by previous exposure to Rh-positive RBCs. Such exposure can occur accidentally, during a transfusion, but it can also occur in a normal pregnancy when an Rh-negative mother carries an Rh-positive fetus.

HEMOLYTIC DISEASE OF THE NEWBORN

Genes controlling the presence or absence of any surface antigen in the membrane of an RBC are provided by both parents, so a child can have a blood type different from that of either parent. During pregnancy, when fetal and maternal circulatory systems are closely intertwined, the mother's antibodies may cross the placenta, attacking and destroying fetal RBCs. The resulting condition is called **hemolytic disease of the newborn (HDN)**.

The sensitization that causes this condition usually takes place at delivery, when bleeding occurs at the placenta and uterus. This event can trigger the production of anti-Rh antibodies in the mother. Because these antibodies are not produced in significant amounts until after delivery, the first infant is not affected. However, a sensitized mother will respond to a second Rh-positive fetus by producing massive amounts of anti-Rh antibodies. These antibodies attack and hemolyze the fetal RBCs, producing a dangerous anemia, which increases the fetal demand for blood cells. The result-

ing RBCs leave the bone marrow and enter the circulation before completing their development. Because these immature RBCs are erythroblasts, HDN is also known as *erythroblastosis fetalis* (e-rith-rō-blas-TŌ-sis fē-TAL-is).

CONCEPT CHECK QUESTIONS
Answers on page 365

❶ Which blood types can be transfused into a person with Type AB blood?

❷ Why can't a person with Type A blood receive blood from a person with Type B blood?

WHITE BLOOD CELLS

White blood cells, also known as WBCs or **leukocytes**, can easily be distinguished from RBCs because each has a nucleus and lacks hemoglobin. White blood cells help defend the body against invasion by pathogens and remove toxins, wastes, and abnormal or damaged cells. Traditionally, WBCs have been divided into two groups on the basis of their appearance after staining: (1) *granulocytes* (with abundant stained granules) and (2) *agranulocytes* (with few if any stained granules). This categorization is convenient but somewhat misleading because the "granules" in granulocytes are actually secretory vesicles and lysosomes, and the "agranulocytes" contain lysosomes as well—they are just smaller and more difficult to see with the light microscope.

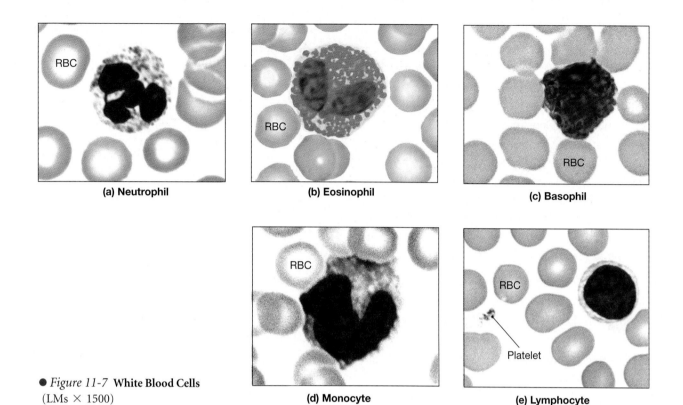

● *Figure 11-7* **White Blood Cells**
(LMs × 1500)

(a) Neutrophil

(b) Eosinophil

(c) Basophil

(d) Monocyte

(e) Lymphocyte

Typical WBCs in the circulating blood are shown in Figure 11-7●. The three types of granulocytes are *neutrophils, eosinophils,* and *basophils;* the two kinds of agranulocytes are *monocytes* and *lymphocytes.* A microliter of blood typically contains 6000–9000 WBCs. Most of the WBCs in the body are located in connective tissues proper or in organs of the lymphatic system. Circulating WBCs represent only a small fraction of the total population.

WBC Circulation and Movement

Unlike RBCs, WBCs do not circulate for extended periods. White blood cells migrate through the loose and dense connective tissues of the body. They use the bloodstream to travel from one organ to another and for rapid transportation to areas of invasion or injury. The WBCs are sensitive to the chemical signs of damage to surrounding tissues. When problems are detected, these cells leave the circulation and enter the damaged area.

Circulating WBCs have four characteristics:

1. They are capable of *amoeboid movement.* Amoeboid movement is a gliding motion accomplished by the flow of cytoplasm into slender cellular processes extended in front of the cell. This mobility allows WBCs to move along the walls of blood vessels and, when outside the bloodstream, through surrounding tissues.

2. They can migrate out of the bloodstream. WBCs can enter surrounding tissue by squeezing between adjacent epithe-

lial cells in the capillary wall. This process is known as *diapedesis* (dī-a-pe-DĒ-sis).

3. They are attracted to specific chemical stimuli. This characteristic, called **positive chemotaxis** (kē-mō-TAK-sis), guides WBCs to invading pathogens, damaged tissues, and other, actively working WBCs.

4. Some of the circulating WBCs, specifically neutrophils, eosinophils, and monocytes, are capable of *phagocytosis.* ∞ p. 64 These cells can engulf pathogens, cell debris, or other materials. Neutrophils and eosinophils are sometimes called *microphages* to distinguish them from the larger macrophages in connective tissue. Macrophages are monocytes that have moved out of the bloodstream and have become actively phagocytic.

General Functions

Neutrophils, eosinophils, basophils, and monocytes contribute to the body's *nonspecific defenses* of the immune system. These defenses respond to a variety of stimuli, but always in the same way—they do not discriminate between one type of threat and another. Lymphocytes, in contrast, are the cells responsible for *specific immunity*: the body's ability to attack invading pathogens or foreign proteins *on a specific, individual basis.* The interactions between WBCs and the relationships between specific and nonspecific defenses will be discussed in Chapter 14.

Neutrophils

Fifty to 70 percent of the circulating white blood cells are **neutrophils** (NOO-trō-filz). This name reflects the fact that their granules are chemically neutral and thus are difficult to stain with either acidic or basic dyes. A mature neutrophil has a very dense, contorted nucleus that may be condensed into a series of lobes resembling beads on a chain (Figure 11-7a●).

Neutrophils are usually the first of the WBCs to arrive at an injury site. They are very active phagocytes, specializing in attacking and digesting bacteria. Most neutrophils have a short life span (about 10 hours). After engulfing one to two dozen bacteria, a neutrophil dies, but its breakdown releases chemicals that attract other neutrophils to the site.

Eosinophils

Eosinophils (ē-ō-SIN-ō-filz) were so named because their granules stain darkly with the red dye eosin (Figure 11-7b●). They usually represent 2–4 percent of circulating WBCs and are similar in size to neutrophils. Their deep red granules and a two-lobed nucleus make them easy to identify. Although they are phagocytes, eosinophils generally ignore bacteria and cellular debris. They are attracted instead to foreign compounds that have reacted with circulating antibodies. Their numbers increase dramatically during an allergic reaction or a parasitic worm infection.

Basophils

Basophils (BĀ-sō-filz) have numerous granules that stain darkly with basic dyes. In a standard blood smear, the granules are a deep purple or blue (Figure 11-7c●). These cells are somewhat smaller than neutrophils or eosinophils and are relatively rare, accounting for less than 1 percent of the WBC population. Basophils migrate to sites of injury and cross the capillary wall to accumulate within the damaged tissues, from where they discharge their granules into the interstitial fluids. The granules contain the chemicals *heparin*, which prevents blood clotting, and *histamine*. The release of histamine by basophils enhances the local inflammation initiated by the mast cells of damaged connective tissues. ∞ p. 103 Other chemicals released by stimulated basophils attract eosinophils and other basophils to the area.

Monocytes

Monocytes (MON-ō-sīts) are nearly twice the size of a typical erythrocyte (Figure 11-7d●). The nucleus is large and commonly oval or shaped like a kidney bean. Monocytes normally account for 2–8 percent of circulating WBCs. They remain in circulation for only about 24 hours before entering peripheral tissues to become a macrophage. In peripheral tissues, they are called *free macrophages*, to distinguish them from the immobile *fixed macrophages* in many connective tissues. ∞ p. 93 They are aggressive phagocytes, often attempting to engulf items as large

as or larger than themselves. For example, osteoclasts are monocytes that have migrated into bone tissue and aid in calcium balance by breaking down bone matrix. ∞ p. 130 Active monocytes release chemicals that attract and stimulate neutrophils, additional monocytes, and other phagocytes and lure fibroblasts to the region. The fibroblasts then begin producing scar tissue, which will wall off the injured area.

Lymphocytes

Typical **lymphocytes** (LIM-fō-sīts) are roughly the same size as red blood cells and contain a relatively large nucleus surrounded by a thin halo of cytoplasm (Figure 11-7e●). Lymphocytes account for 20–30 percent of the WBC population of the blood. Lymphocytes are continuously migrating from the bloodstream, through peripheral tissues, and back to the bloodstream. Circulating lymphocytes represent only a minuscule fraction of the entire WBC population, for at any moment most of the lymphocytes are in other connective tissues and in organs of the lymphatic system. They also act to protect the body and its tissues, but do not rely on phagocytosis. Lymphocytes attack both foreign cells and abnormal cells of the body, and secrete antibodies into the circulation. The antibodies can attack foreign cells or proteins in distant parts of the body.

The Differential Count and Changes in WBC Profiles

A variety of disorders, including pathogenic infection, inflammation, and allergic reactions, cause characteristic changes in the circulating populations of WBCs. By examining a stained blood smear, we can obtain a **differential count** of the WBC population. The values reported indicate the number of each type of cell encountered in a sample of 100 WBCs.

The normal range for each cell type is indicated in Table 11-3. The term **leukopenia** (loo-kō-PĒ-nē-ah; *penia*, poverty) indicates inadequate numbers of WBCs. **Leukocytosis** (loo-kō-sī-TŌ-sis) refers to excessive numbers of WBCs. Extreme leukocytosis (WBC counts of 100,000/μl or more) usually indicates the presence of some form of **leukemia** (loo-KĒ-mē-ah), a cancer of blood-forming tissues. Not all leukemias are characterized by leukocytosis; other indications are the presence of abnormal or immature WBCs. There are many types of leukemia. Treatment helps in some cases; unless treated, all are fatal.

White Blood Cell Formation

Stem cells responsible for the production of white blood cells originate in red bone marrow, with the divisions of the hemocytoblasts. Myeloid stem cell divisions give rise to all the formed elements except lymphocytes. Neutrophils, eosinophils, and basophils complete their development in myeloid (marrow) tissue; monocytes begin their differentiation in the bone mar-

TABLE 11-3 *A Review of the Formed Elements of the Blood*

CELL	ABUNDANCE (AVERAGE PER μl)	FUNCTIONS	REMARKS
RED BLOOD CELLS	5.2 million (range: 4.4–6.0 million)	Transport oxygen from lungs to tissues and carbon dioxide from tissues to lungs	Remain in bloodstream; 120-day life expectancy; amino acids and iron recycled; produced in bone marrow
WHITE BLOOD CELLS	7000 (range: 6000–9000)		
Neutrophils	4150 (range: 1800–7300) Differential count: 50–70%	Phagocytic: Engulf pathogens or debris in tissues, release cytotoxic enzymes and chemicals	Move into tissues after several hours; survive minutes to days, depending on tissue activity; produced in bone marrow
Eosinophils	165 (range: 0–700) Differential count: < 2–4%	Attack antibody-labeled materials through release of cytotoxic enzymes, and/or phagocytosis	Move into tissues after several hours; survive minutes to days, depending on tissue activity; produced in bone marrow
Basophils	44 (range: 0–150) Differential count: < 1%	Enter damaged tissues and release histamine and other chemicals that enhance inflammation	Survival time unknown; assist mast cells of tissues in producing inflammation; produced in bone marrow
Monocytes	456 (range: 200–950) Differential count: 2–8%	Enter tissues to become macrophages; engulf pathogens or debris	Move into tissues after 1–2 days; survive months or longer; primarily produced in bone marrow
Lymphocytes	2185 (range: 1500–4000) Differential count: 20–30%	Cells of lymphatic system, providing defense against specific pathogens or toxins	Survive months to decades; circulate from blood to tissues and back; produced in bone marrow and lymphoid tissues
PLATELETS	350,000 (range: 150,000–500,000)	Hemostasis: Clump together and stick to vessel wall (platelet phase); initiate coagulation cascade	Remain in circulation or in vascular organs; survive 7–12 days; produced by megakaryocytes in bone marrow

row, enter the circulation, and complete their development when they become free macrophages in peripheral tissues. Each of these cell types goes through a characteristic series of developmental stages. Figure 11-8● summarizes the relationships between the various WBC populations and compares the formation of WBCs and RBCs.

Many of the lymphoid stem cells responsible for the production of lymphocytes migrate from the bone marrow to peripheral **lymphoid tissues**, including the thymus, spleen, and lymph nodes. As a result, lymphocytes are produced in these organs as well as in the bone marrow. The process of lymphocyte production is called **lymphopoiesis.**

Various hormones are involved in the regulation of white blood cell populations. The WBCs other than lymphocytes are regulated by hormones called *colony-stimulating factors (CSFs).* Four CSFs have been identified, each targeting single stem cell lines or groups of stem cell lines. Prior to maturity, the thymosins produced by the thymus gland promote the differentiation of one type of lymphocyte, the *T cells*. (In adults, the production of lymphocytes is regulated by exposure to antigens such as foreign proteins, cells, and toxins. We shall describe such control mechanisms in Chapter 14.)

Chemical communication between lymphocytes and other WBCs assists in the coordination of the immune response. For example, active macrophages release chemicals that make lymphocytes more sensitive to antigens and accelerate the development of specific immunity. In turn, active lymphocytes release CSFs, which reinforces nonspecific defenses by stimulating the production of other WBCs.

PLATELETS

Platelets (PLĀT-lets) are another component of the formed elements. Platelets in nonmammalian vertebrates are nucleated cells called **thrombocytes** (THROM-bō-sīts; *thrombos*, clot). Because in humans they are cell fragments rather than individual cells, the term platelet is preferred when referring to our blood.

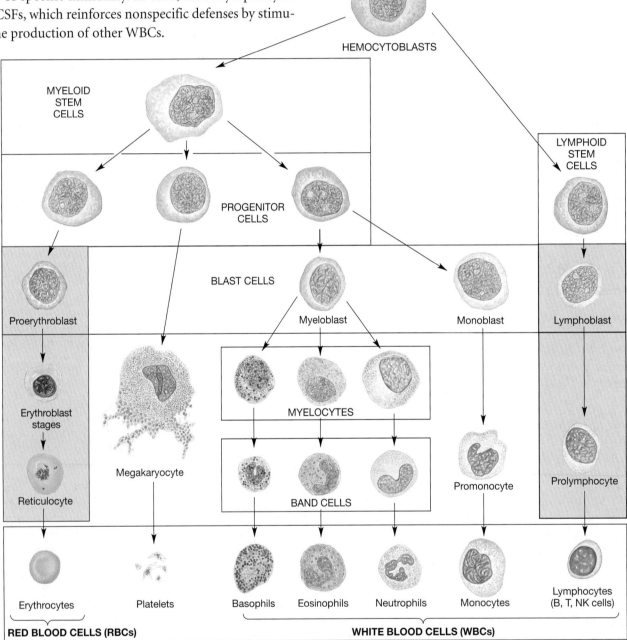

● *Figure 11-8* **The Origins and Differentiation of Formed Elements**
Hemocytoblast divisions give rise to myeloid and lymphoid stem cells. Lymphoid stem cells produce the various lymphocytes. Myeloid stem cells produce progenitor cells that divide to produce the other classes of formed elements. Hematologists use specific terms for the key stages that red and white blood cells pass through to maturity. The stages in RBC development were discussed on p. 351. Basophils, eosinophils and neutrophils proceed from blast cells to myelocytes to band cells in their development. Early developmental stages of monocytes and lymphocytes include blast cells, followed by promonocytes and prolymphocytes, respectively.

Bone marrow contains enormous cells with large nuclei called **megakaryocytes** (meg-a-KĀR-ē-ō-sīts; *megas*, big + *karyon*, nucleus + *-cyte*, cell). Megakaryocytes continuously shed cytoplasm in small membrane-enclosed packets. These packets are the platelets that enter the bloodstream (Figure 11-8●). Platelets initiate the clotting process and help close injured blood vessels. They are one participant in a vascular *clotting system*, detailed in the next section.

Platelets are continuously replaced. An individual platelet circulates for 10–12 days before being removed by phagocytes. Each microliter of circulating blood contains 150,000–500,000 platelets; 350,000/μl is the average concentration. An abnormally low platelet count (80,000/μl or less) is known as *thrombocytopenia* (throm-bō-sī-tō-PĒ-nē-ah), and this condition usually results from excessive platelet destruction or inadequate platelet production. Symptoms include bleeding along the digestive tract, within the skin, and occasionally inside the CNS.

In *thrombocytosis* (throm-bō-sī-TŌ-sis), platelet counts can exceed 1,000,000/μl. Thrombocytosis usually results from accelerated platelet formation in response to infection, inflammation, or cancer.

CONCEPT CHECK QUESTIONS
Answers on page 365

❶ Which type of white blood cell would you expect to find in the greatest numbers in an infected cut?

❷ Which cell type would you expect to find in elevated numbers in a person producing large amounts of circulating antibodies to combat a virus?

❸ How do basophils respond during inflammation?

Hemostasis

Hemostasis (*haima*, blood + *stasis*, halt), the process that stops bleeding, prevents the loss of blood through the walls of damaged vessels. In restricting blood loss, this process also establishes a framework for tissue repairs. Hemostasis is a three step process: (1) the vascular phase, (2) the platelet phase, and (3) the coagulation phase.

Step 1: *The vascular phase.* The walls of blood vessels contain smooth muscle and an inner lining of simple squamous epithelium. The simple squamous epithelium that lines blood vessels (and the inner surfaces of the heart) is known as an **endothelium** (en-dō-THĒ-lē-um). Cutting the wall of a blood vessel triggers a contraction in the smooth muscle fibers in the vessel wall that decreases the vessel's diameter. This local contraction is a *vascular spasm*, which can slow or even stop the loss of blood through the wall of a small vessel. The vascular spasm, which lasts about 30 minutes, is called the **vascular phase** of hemostasis. Changes also occur in the vessel endothelium; the membranes of endothelial cells at the injury site become "sticky," and in small capillaries the cells may stick together and block the opening completely.

Step 2: *The platelet phase.* The platelets begin to attach to the sticky endothelial surfaces and exposed collagen fibers within 15 seconds of the injury. These attachments mark the start of the **platelet phase** of hemostasis. As more platelets arrive, they form a *platelet plug*, a mass that may close the break in the vessel wall, if the damage is not severe or if it is confined to a small blood vessel.

Step 3: *The coagulation phase.* The vascular and platelet phases begin within a few seconds after the injury. The **coagulation** (kō-ag-ū-LĀ-shun) **phase** does not start until 30 seconds or more after the vessel has been damaged. **Coagulation**, or blood clotting, involves a complex sequence of steps leading to the conversion of circulating fibrinogen into the insoluble protein fibrin. As the fibrin network grows, blood cells and additional platelets are trapped within the fibrous tangle, forming a **blood clot** that effectively seals off the damaged portion of the vessel (Figure 11-9●).

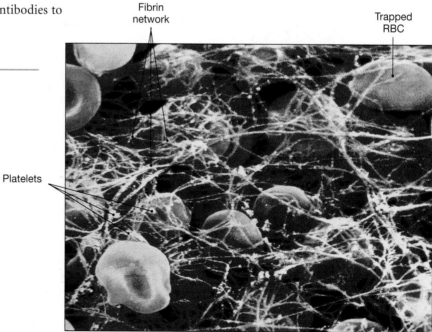

Fibrin network

Trapped RBC

Platelets

● *Figure 11-9* **The Structure of a Blood Clot**
A scanning electron micrograph showing the fibrin network that forms the framework of a clot. (SEM × 4120) Red blood cells trapped in those fibers add to the mass of the blood clot and color it red. Platelets that stick to the fibrin strands gradually contract, shrinking the clot and tightly packing the RBCs.

THE CLOTTING PROCESS

Normal coagulation cannot occur unless the plasma contains the necessary **clotting factors**, which include calcium ions and 11 different plasma proteins. These proteins are converted to active enzymes that direct essential reactions in the clotting response. Most of the circulating clotting proteins are synthesized by the liver.

During the coagulation phase, the clotting proteins interact in sequence such that one protein is converted into an enzyme that activates a second protein and so on, in a chain reaction, or *cascade*. Figure 11-10● provides an overview of the cascades involved in the *extrinsic*, *intrinsic*, and *common pathways* which result in the formation of a blood clot. The extrinsic pathway begins outside the bloodstream; the intrinsic pathway begins inside the bloodstream. These pathways join at the common pathway through the activation of Factor X, a clotting protein produced by the liver.

The Extrinsic, Intrinsic, and Common Pathways

When a blood vessel is damaged, both the extrinsic and intrinsic pathways respond. Clotting usually begins in 15 seconds and is initiated by the shorter and faster extrinsic pathway. The slower, intrinsic pathway reinforces the initial clot, making it larger and more effective.

The **extrinsic pathway** begins with the release of a lipoprotein called **tissue factor** by damaged endothelial cells or peripheral tissues. The greater the damage, the more tissue factor is released and the faster clotting occurs. Tissue factor then combines with calcium ions and one of the clotting proteins (Factor VII) to form an enzyme capable of activating Factor X.

The **intrinsic pathway** begins with the activation of a clotting protein exposed to collagen fibers at the injury site. This pathway proceeds with the assistance of a *platelet factor* released by aggregating platelets. Platelets also release a variety of other factors that speed up the reactions of the intrinsic pathway. After a series of linked reactions, activated clotting proteins also form an enzyme capable of activating Factor X. The **common pathway** begins when enzymes from either the extrinsic or the intrinsic pathway activate Factor X, forming the enzyme prothrombinase. Prothrombinase converts the clotting protein **prothrombin** into the enzyme **thrombin** (THROM-bin). Thrombin then completes the coagulation process by converting fibrinogen to fibrin.

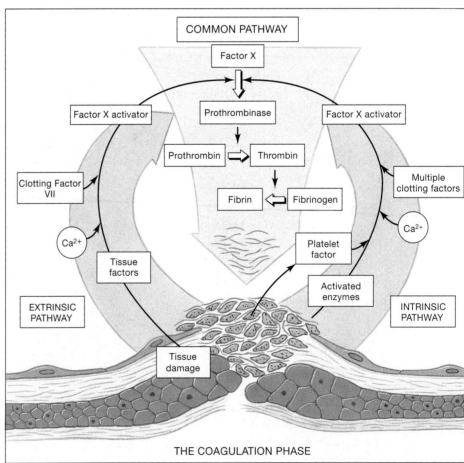

● *Figure 11-10* **The Coagulation Phase of Hemostasis**

The thrombin formed in the common pathway also acts to increase the rate of clot formation. Thrombin stimulates the formation of tissue factor and the release of platelet factor by platelets. The presence of these factors further stimulates both the extrinsic and intrinsic pathways. This positive feedback loop accelerates the clotting process, and speed can be important in reducing blood loss after a severe injury. ∞ p. 12

Calcium ions and a single vitamin, **vitamin K**, affect almost every aspect of the clotting process. For example, all three pathways (intrinsic, extrinsic, and common) require the presence of calcium ions. Adequate amounts of vitamin K must be present for the liver to synthesize four of the clotting factors, including prothrombin. A vitamin K deficiency therefore leads to the breakdown of the common pathway, inactivating the clotting system. Roughly half of the vitamin K is absorbed directly from the diet, and the rest is synthesized by intestinal bacteria.

CLOT RETRACTION AND REMOVAL

Once the fibrin network has appeared, platelets and red blood cells stick to the fibrin strands. The platelets then contract and pull the torn edges of the wound closer together as the entire clot begins to undergo **clot retraction**. This process reduces the size of the damaged area, making it easier for the fibroblasts, smooth muscle cells, and endothelial cells in the area to carry out the necessary repairs.

As the repairs proceed, the clot gradually dissolves. This process, called **fibrinolysis** (fī-bri-NOL-i-sis), begins with the activation of the plasma protein **plasminogen** (plaz-MIN-ō-jen), by thrombin and **tissue plasminogen activator**, or **t-PA**, which is released by damaged tissues. The activation of plasminogen produces the enzyme **plasmin** (PLAZ-min), which begins digesting the fibrin strands and breaking down the clot.

CONCEPT CHECK QUESTIONS
Answers on page 365

❶ A sample of bone marrow has fewer than normal numbers of megakaryocytes. What body process would you expect to be impaired as a result?

❷ Two alternate pathways of interacting clotting proteins lead to coagulation, or blood clotting. How is each pathway initiated?

❸ How would inadequate levels of vitamin K affect blood clotting, or coagulation?

CLINICAL NOTE *Abnormal Hemostasis*

The clotting process involves a complex chain of events, and a disorder that affects any individual clotting factor can disrupt the entire process. As a result, many clinical conditions affect the clotting system.

Excessive Coagulation

If the clotting process is inadequately controlled, clot formation will begin in the circulation, rather than at an injury site. These blood clots do not stick to the wall of the vessel but continue to drift until plasmin digests them or they become stuck in a small blood vessel. A drifting blood clot is a type of **embolus** (EM-bo-lus; *embolos*, plug), an abnormal mass within the bloodstream. (Other common emboli are drifting air bubbles and fat globules.) When an embolus becomes stuck in a blood vessel, it blocks circulation to the area downstream, killing the affected tissues. The blockage is called an **embolism**.

An embolus in the arterial system can get stuck in the capillaries in the brain, causing tissue damage known as a *stroke*. An embolus that forms in the venous system is likely to become lodged in the capillaries in the lung, causing a condition known as a *pulmonary embolism*.

A **thrombus**, or blood clot attached to a vessel wall, begins to form when platelets stick to the wall of an intact blood vessel. Often, the platelets are attracted to areas called *plaques*, where endothelial and smooth muscle cells contain large quantities of lipids. The blood clot gradually enlarges, projecting into the central space, or *lumen*, of the vessel and reducing its diameter. Eventually the vessel can be completely blocked, or a large chunk of the clot can break off, creating an equally dangerous embolus.

Treatment of these conditions must be prompt to prevent irreparable damage to tissues whose vessels have been restricted or blocked by emboli or thrombi. The clots can be surgically removed or attacked by anticoagulants such as the enzymes *streptokinase* (strep-tō-KĪ-nās) or *urokinase* (ū-rō-KĪ-nās) that convert plasminogen to plasmin or by administered *tissue plasminogen activator (t-PA)*, which stimulates plasmin formation.

Inadequate Coagulation

Hemophilia (hēm-ō-FĒL-ē-uh) is one of many inherited disorders characterized by the inadequate production of clotting factors. The incidence of this condition in the general population is about 1 in 10,000, with males accounting for 80–90 percent of those affected. In hemophilia, the production of a single clotting factor (most often Factor VIII, an essential component of the intrinsic clotting pathway) is inadequate; the severity of the condition depends on the degree of underproduction. In severe cases, extensive bleeding accompanies the slightest mechanical stresses, and hemorrhages occur spontaneously at joints and around muscles.

Transfusions of clotting factors can reduce or control the symptoms of hemophilia, but plasma samples from many individuals must be pooled (combined) to obtain adequate amounts of clotting factors. This procedure makes the treatment very expensive and increases the risk of infection with blood-borne infections such as hepatitis or AIDS. Gene-splicing techniques have been used to manufacture Factor VIII. As methods are developed to synthesize other clotting factors, treatment of the various forms of hemophilia will become safer and cheaper.

Related Clinical Terms

anemia (a-NĒ-mē-uh): A condition in which the oxygen-carrying capacity of blood is reduced owing to low hematocrit or low blood hemoglobin concentrations.

embolism: A condition in which a drifting blood clot becomes stuck in a blood vessel, blocking circulation to the area downstream.

erythrocytosis (e-rith-rō-sī-TŌ-sis): A polycythemia affecting only red blood cells.

hematocrit (he-MA-tō-krit): The value that indicates the percentage of whole blood occupied by cellular elements.

hematology (HĒM-ah-tol-o-jē): The study of blood and its disorders.

hematuria (hē-ma-TOO-rē-uh): The presence of red blood cells in urine.

hemolytic disease of the newborn (HDN): A condition in which fetal red blood cells have been destroyed by maternal antibodies.

hemophilia (hē-mō-FĒL-ē-uh): Inherited disorders characterized by the inadequate production of clotting factors.

heterologous marrow transplant: The transplantation of bone marrow from one individual to another to replace bone marrow destroyed during cancer therapy.

hypoxia (hī-POKS-ē-uh): Low tissue oxygen levels.

jaundice (JAWN-dis): A condition characterized by yellow skin and eyes, caused by abnormally high levels of plasma bilirubin.

leukemia (loo-KĒ-mē-uh): A cancer of blood-forming tissues characterized by extremely elevated levels of circulating white blood cells; includes both *myeloid leukemia* (abnormal granulocytes or other cells of the bone marrow) and *lymphoid leukemia* (abnormal lymphocytes).

leukocytosis (loo-kō-sī-TŌ-sis): Excessive numbers of white blood cells in the bloodstream.

leukopenia (loo-kō-PĒ-nē-uh): Inadequate numbers of white blood cells in the bloodstream.

normovolemic (nor-mō-vō-LĒ-mik): Having a normal blood volume.

phlebotomy (fle-BOT-o-mē): The process of withdrawing, or letting, blood from a vein.

polycythemia (po-lē-sī-THĒ-mē-uh): A descriptive term indicating an elevated hematocrit with a normal blood volume.

sickle cell anemia: An anemia resulting from the production of an abnormal form of hemoglobin; causes red blood cells to become sickle-shaped at low oxygen levels.

septicemia: A dangerous infection of the bloodstream that distributes bacteria and bacterial toxins throughout the body.

thalassemia: An inherited blood disorder resulting from the production of an abnormal form of hemoglobin.

thrombus: A blood clot attached to the internal surface of a blood vessel.

transfusion: A procedure in which blood components are given to someone whose blood volume has been reduced or whose blood is defective.

venipuncture (VĒN-i-punk-chur): The puncturing of a vein for any purpose, including the withdrawal of blood or the administration of medication.

CHAPTER REVIEW

Key Terms

Summary Outline

INTRODUCTION **344**

1. The **cardiovascular system** provides a mechanism for the rapid transport of nutrients, waste products, respiratory gases, and cells within the body.

THE FUNCTIONS OF BLOOD..................................**344**

1. **Blood** is a specialized fluid connective tissue. Its functions include (1) transporting dissolved gases, nutrients, hormones, and metabolic wastes; (2) regulating the pH and electrolyte composition of the interstitial fluids; (3) restricting fluid losses through damaged vessels; (4) defending against pathogens and toxins; and (5) regulating body temperature by absorbing and redistributing heat.

THE COMPOSITION OF BLOOD **344**

Blood Collection and Analysis **345**

1. Blood contains **plasma**, **red blood cells (RBCs)**, **white blood cells (WBCs)**, and **platelets**. The plasma and formed elements constitute whole blood, which can be **fractionated** for analytical or clinical purposes. *(Figure 11-1a)*

PLASMA **346**

1. Plasma accounts for about 55 percent of the volume of blood; roughly 92 percent of plasma is water. *(Figure 11-1b)*

2. Compared with interstitial fluid, plasma has a higher dissolved oxygen concentration and more dissolved proteins. The three classes of plasma proteins are *albumins*, *globulins*, and *fibrinogen*.

Plasma Proteins **346**

3. **Albumins** constitute about 60 percent of plasma proteins. **Globulins** constitute roughly 35 percent of plasma proteins; they include **antibodies (immunoglobulins)**, which attack foreign proteins and pathogens, and **transport proteins**, which bind ions, hormones, and other compounds. In the clotting reaction, **fibrinogen** molecules are converted to **fibrin**. The removal of fibrinogen from plasma leaves a fluid called **serum**.

FORMED ELEMENTS **346**

The Production of Formed Elements **347**

1. **Hemopoiesis** is the process of blood cell formation. **Stem cells** called **hemocytoblasts** divide to form all types of blood cells.

Red Blood Cells **347**

2. Red blood cells (**RBCs**), or **erythrocytes**, account for slightly less than half of the blood volume and 99.9 percent of the formed elements. The **hematocrit** value indicates the percentage of whole blood occupied by cellular elements. *(Figure 11-1c; Table 11-1)*

3. RBCs transport oxygen and carbon dioxide within the bloodstream. They are highly specialized cells with a large surface-to-volume ratio. They lack many organelles, and usually degenerate after 120 days in the bloodstream. *(Figure 11-2)*

4. Molecules of **hemoglobin (Hb)** account for over 95 percent of RBC proteins. Hemoglobin is a globular protein formed from four subunits. Each subunit contains a single molecule of **heme** and can reversibly bind an oxygen molecule. Hemoglobin from damaged or dead RBCs is recycled by phagocytes. *(Figure 11-4)*

5. **Erythropoiesis**, the formation of RBCs, occurs mainly in the **red bone marrow (myeloid tissue)** in adults. RBC formation increases under stimulation by **erythropoietin**, or **EPO** (erythropoiesis-stimulating hormone), which occurs when peripheral tissues are exposed to low oxygen concentrations. Stages in RBC development include **erythroblasts** and **reticulocytes**. *(Figures 11-5; 11-8)*

6. **Blood type** is determined by the presence or absence of specific **surface antigens** (*agglutinogens*) in the RBC cell membranes: antigens **A, B,** and **Rh**. Antibodies (agglutinins) in plasma will react with RBCs bearing different surface antigens. Anti-Rh antibodies are synthesized only after an Rh-negative individual becomes sensitized to the Rh surface antigen. *(Figure 11-6; Table 11-2)*

White Blood Cells 354

7. White blood cells (WBCs), or **leukocytes**, defend the body against pathogens and remove toxins, wastes, and abnormal or damaged cells.

8. Leukocytes show *diapedesis* (the ability to move through vessel walls) and **chemotaxis** (an attraction to specific chemicals).

9. *Granulocytes* (granular leukocytes) are subdivided into *neutrophils*, *eosinophils*, and *basophils*. Fifty to 70 percent of circulating WBCs are **neutrophils**, which are highly mobile phagocytes. The much less common **eosinophils** are phagocytes attracted to foreign compounds that have reacted with circulating antibodies. The relatively rare **basophils** migrate to damaged tissues and release histamines, aiding the inflammation response. *(Figure 11-7)*

10. *Agranulocytes* (agranular leukocytes) are subdivided into *monocytes* and *lymphocytes*. **Monocytes** that migrate into peripheral tissues become free macrophages. **Lymphocytes** are the primary cells of the **lymphatic system**. Different classes of lymphocytes attack foreign cells directly, produce antibodies, and destroy abnormal tissue cells. *(Figure 11-7)*

11. Granulocytes and monocytes are produced by **myeloid stem cells** in the bone marrow. **Lymphoid stem cells** responsible for **lymphopoiesis** (production of lymphocytes) also originate in the bone marrow, but many migrate to lymphoid tissues. *(Figure 11-8)*

12. Factors that regulate lymphocyte maturation are not completely understood. *Colony-stimulating factors (CSFs)* are hormones involved in regulating other WBC populations.

Platelets 358

13. **Megakaryocytes** in the bone marrow release packets of cytoplasm (platelets) into the circulating blood. Platelets are essential to the clotting process.

HEMOSTASIS 359

1. **Hemostasis** prevents the loss of blood through the walls of damaged vessels.

2. The initial step of hemostasis, the **vascular phase**, is a period of local contraction of vessel walls resulting from a vascular spasm at the injury site. The **platelet phase** follows as platelets stick to damaged surfaces.

The Clotting Process 360

3. The **coagulation phase** occurs as factors released by endothelial cells or peripheral tissues (**extrinsic pathway**) and platelets (**intrinsic pathway**) interact with **clotting factors** to form a **blood clot**. *(Figures 11-9, 11-10)*

Clot Retraction and Removal 361

4. During **clot retraction**, platelets contract and pull the torn edges closer together. During **fibrinolysis**, the clot gradually dissolves through the action of **plasmin**, the activated form of circulating **plasminogen**.

Review Questions

Level 1: Reviewing Facts and Terms

Match each item in column A with the most closely related item in column B. Use letters for answers in the spaces provided.

COLUMN A

____ 1. interstitial fluid

____ 2. hemopoiesis

____ 3. stem cells

____ 4. hypoxia

____ 5. surface antigens

____ 6. antibodies

____ 7. diapedesis

____ 8. leukopenia

____ 9. agranulocyte

____ 10. leukocytosis

____ 11. granulocyte

____ 12. thrombocytopenia

COLUMN B

a. hemocytoblasts

b. abundant WBCs

c. agglutinogens

d. neutrophil

e. WBC migration

f. extracellular fluid

g. monocyte

h. few WBCs

i. low platelet count

j. agglutinins

k. blood cell formation

l. low oxygen concentration

13. The formed elements of the blood include
    (a) plasma, fibrin, serum
    (b) albumins, globulins, fibrinogen
    (c) WBCs, RBCs, platelets
    (d) a, b, and c are correct

14. Blood temperature is approximately _____, and the blood pH averages _____.
    (a) 98.6°F, 7.0   (b) 104°F, 7.8
    (c) 100.4°F, 7.4   (d) 96.8°F, 7.0

15. Plasma contributes approximately _____ percent of the volume of whole blood, and water accounts for _____ percent of the plasma volume.
    (a) 55, 92   (b) 25, 55
    (c) 92, 55   (d) 35, 72

16. When the clotting proteins are removed from plasma, _____ remains.
    (a) fibrinogen   (b) fibrin
    (c) serum   (d) heme

17. In an adult, the only site of red blood cell production, and the primary site of white blood cell formation, is the:
    (a) liver   (b) spleen
    (c) thymus   (d) red bone marrow

18. The most numerous WBCs found in a differential count of a "normal" individual are
    (a) neutrophils   (b) basophils
    (c) lymphocytes   (d) monocytes

19. Stem cells responsible for the process of lymphopoiesis are located in the:
    (a) thymus and spleen   (b) lymph nodes
    (c) red bone marrow   (d) a, b, and c are correct

20. The first step in the process of hemostasis is
    (a) coagulation   (b) the platelet phase
    (c) fibrinolysis   (d) vascular spasm

21. The complex sequence of steps leading to the conversion of fibrinogen to fibrin is called
    (a) fibrinolysis   (b) clotting
    (c) retraction   (d) the platelet phase

22. What five major functions are performed by the blood?

23. What three primary classes of plasma proteins are found in the blood? What is the major function of each?

24. What type of antibodies does the plasma contain for each of the following blood types?
    (a) Type A   (b) Type B
    (c) Type AB   (d) Type O

25. What three processes facilitate the movement of WBCs to areas of invasion or injury?

26. What contribution from the intrinsic and extrinsic pathways is necessary for the common pathway to begin?

27. Distinguish between an embolus and a thrombus.

Level 2: Reviewing Concepts

28. Dehydration would cause:
    (a) an increase in the hematocrit
    (b) a decrease in the hematocrit
    (c) no effect in the hematocrit
    (d) an increase in plasma volume

29. Erythropoietin directly stimulates RBC formation by:
    (a) increasing rates of mitotic divisions in erythroblasts
    (b) speeding up the maturation of red blood cells
    (c) accelerating the rate of hemoglobin synthesis
    (d) a, b, and c are correct

30. A person with Type A blood has:
    (a) Anti-A in the plasma
    (b) Type B antigens in the plasma
    (c) Type A antigens on the red blood cells
    (d) Anti-B on the red blood cells

31. Hemolytic disease of the newborn can result if:
    (a) the mother is Rh-positive and the father is Rh-negative
    (b) both the father and the mother are Rh-negative
    (c) both the father and the mother are Rh-positive
    (d) an Rh-negative woman carries an Rh-positive fetus

32. How do red blood cells differ from typical cells found in the body?

Level 3: Critical Thinking and Clinical Applications

33. Which of the formed elements would you expect to increase after you have donated a pint of blood?

34. Why do many individuals with advanced kidney disease become anemic?

Answers to Concept Check Questions

Page 346

1. Venipuncture is a common sampling technique because (1) superficial veins are easy to locate; (2) the walls of veins are thinner than those of arteries; and (3) blood pressure in veins is relatively low, so the puncture wound seals quickly.

2. A decrease in the amount of plasma proteins in the blood could cause (1) a decrease in plasma osmotic pressure, (2) a decreased ability to fight infection, and (3) a decrease in the transport and binding of some ions, hormones, and other molecules.

Page 352

1. The hematocrit measures the amount of formed elements (mostly red blood cells) as a percentage of the total blood. The loss of water during dehydration would decrease the plasma volume, resulting in an increased hematocrit.

2. Diseases that damage the liver, such as hepatitis and cirrhosis, would impair the liver's ability to excrete bilirubin. As a result, bilirubin would accumulate in the blood, producing jaundice.

3. A decreased oxygen supply (and therefore a decreased blood flow) to the kidneys would trigger the release of erythropoietin. The elevated erythropoietin would lead to an increase in erythropoiesis (red blood cell formation).

Page 354

1. A person with Type AB blood can accept Type A, Type B, Type AB, or Type O blood.

2. If a person with Type A blood received a transfusion of Type B blood, the transfused RBCs would clump or agglutinate, because the anti-B antibodies in the blood of the Type A person would attack the B antigens on the surfaces of Type B RBCs. The agglutination would potentially block blood flow to various organs and tissues.

Page 359

1. In an infected cut, you would expect to find a large number of neutrophils. Neutrophils are phagocytic white blood cells that are usually the first to arrive at the site of an injury and that are specialized to attack infectious bacteria.

2. The type of white blood cells that produce circulating antibodies are lymphocytes; these would be found in increased numbers.

3. During inflammation, basophils release a variety of chemicals such as histamine and heparin, that exaggerate the inflammation and attract other types of white blood cells.

Page 361

1. Megakaryocytes are the precursors of platelets, which play an important role in hemostasis and the clotting process. A decreased number of megakaryocytes would result in fewer platelets, which in turn would interfere with the ability to clot properly.

2. The faster, extrinsic pathway is initiated by damaged endothelial cells lining the blood vessel and damaged tissues when they release a lipoprotein called tissue factor. The slower, intrinsic pathway is initiated by the activation of a clotting protein exposed to damaged collagen fibers and by the release of platelet factor by aggregating platelets.

3. An inadequate level of vitamin K would result in a decrease in the production of clotting factors necessary for normal blood coagulation.

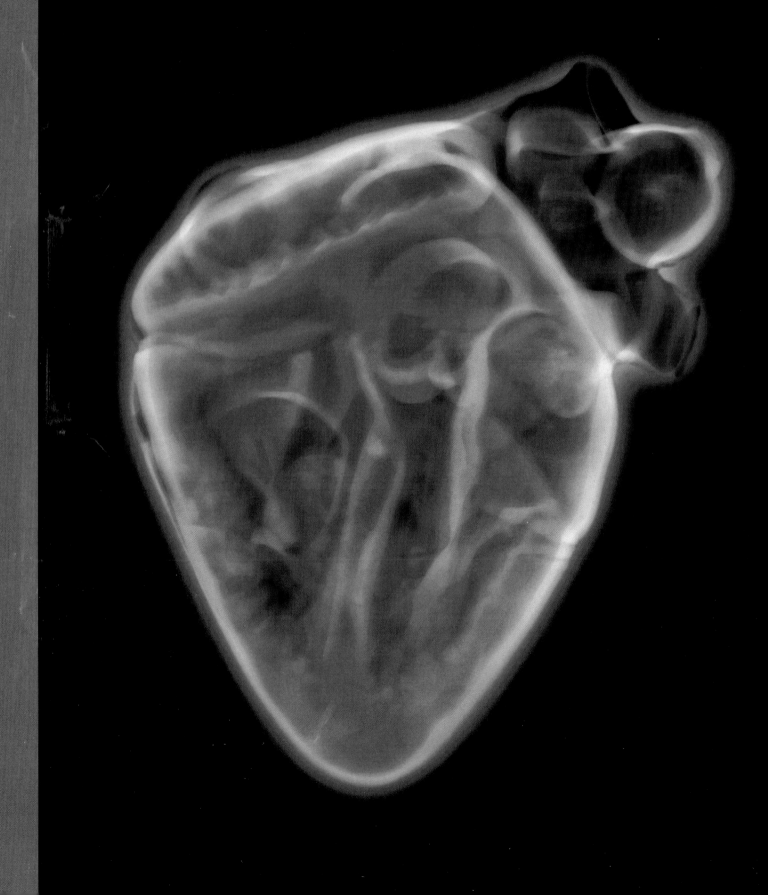

The Heart

Vocabulary Development

anastomosis...................outlet; *anastomoses*

atrion ....hall; *atrium*

auris ....ear; *auricle*

bi-..................................two; *bicuspid*

bradys...................................slow; *bradycardia*

cuspis ....point; *bicuspid valve*

diastole ....expansion; *diastole*

-gram.......................record; *electrocardiogram*

luna ....moon; *semilunar valve*

mitre......................a bishop's hat; *mitral valve*

papilla.......................nipple-shaped elevation; *papillary muscles*

semi-................................half; *semilunar valve*

septum ....wall; *interatrial septum*

systole ....a drawing together; *systole*

tachys ....swift; *tachycardia*

tri- ....three; *tricuspid valve*

ventricle ....little belly; *ventricle*

*H*ARDWORKING MUSCLES *require a steady supply of blood to provide them with nutrients and oxygen. This is especially true of the hardest-working muscle of all: your heart. Any substantial interruption in the flow of blood to this organ has grave consequences—what we commonly call a heart attack. Such attacks typically occur when there is an obstruction in one of the arteries that supply the heart muscle. Physicians can check the status of these vessels by means of scanning techniques that produce computer-enhanced images such as the color-enhanced X-ray shown here.*

12 THE HEART

The Heart and the Circulatory System • The Anatomy and Organization of the Heart • The Heartbeat • Heart Dynamics • Chapter Review

Our BODY CELLS rely on the surrounding interstitial fluid for oxygen, nutrients, and waste disposal. Conditions in the interstitial fluid are kept stable through continuous exchange between the peripheral tissues and circulating blood. If the blood stops moving, its oxygen and nutrient supplies are quickly exhausted, its capacity to absorb wastes is soon saturated, and neither hormones nor white blood cells can reach their intended targets. All cardiovascular functions ultimately depend on the heart. This muscular organ beats approximately 100,000 times each day, pumping roughly 8000 liters of blood—enough to fill forty 55-gallon drums, or nearly 8500 quart-sized milk cartons.

We begin this chapter by examining the structural features that enable the heart to perform so reliably. We will then consider the physiological mechanisms that regulate the activities of the heart to meet the body's ever-changing needs.

the heart beats, the the atria contract first, then the ventricles. The two ventricles contract at the same time and eject equal volumes of blood.

The Anatomy and Organization of the Heart

The heart lies near the anterior chest wall, directly behind the sternum (Figure 12-2a●). It is enclosed by the mediastinum, the connective tissue mass that divides the thoracic cavity into two pleural cavities and also contains the thymus, esophagus, and trachea. ∞ p. 19

The heart is surrounded by the **pericardial** (per-i-KAR-dē-al) **cavity**. The lining of the pericardial cavity is a serous membrane called the **pericardium**. ∞ p. 100 To visualize the relationship between the heart and the pericardial cavity,

The Heart and the Circulatory System

Blood flows through a network of blood vessels that extend between the heart and peripheral tissues. Blood vessels are subdivided into a **pulmonary circuit**, which carries blood to and from exchange surfaces of the lungs, and a **systemic circuit**, which transports blood to and from the rest of the body (Figure 12-1●). Each circuit begins and ends at the heart, and blood travels through these circuits in sequence. Blood returning to the heart from the systemic circuit must complete the pulmonary circuit before reentering the systemic circuit.

Arteries, or *efferent* vessels, carry blood away from the heart; **veins**, or *afferent* vessels, return blood to the heart. **Capillaries** are small, thin-walled vessels between the smallest arteries and veins. The thin walls of the capillaries permit the exchange of nutrients, dissolved gases, and waste products between the blood and surrounding tissues.

Despite its impressive workload, the heart is a small organ, roughly the size of a clenched fist. The heart contains four muscular chambers, two associated with each circuit. The **right atrium** (Ā-trē-um; hall; plural, *atria*) receives blood from the systemic (body) circuit, and the **right ventricle** (VEN-tri-kl; "little belly") discharges blood into the pulmonary (lungs) circuit. The **left atrium** collects blood from the pulmonary circuit, and the **left ventricle** ejects it into the systemic circuit. When

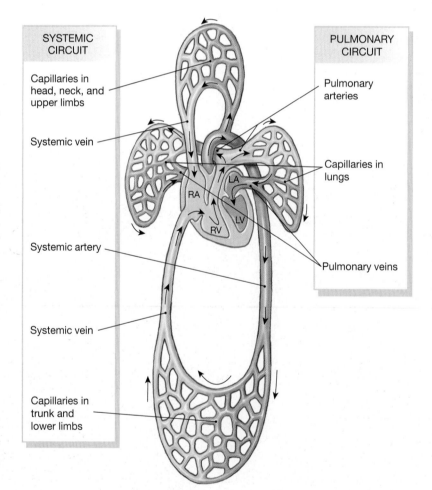

● *Figure 12-1* **An Overview of the Cardiovascular System**
Blood flows through separate pulmonary and systemic circuits, driven by the pumping of the heart. Each circuit begins and ends at the heart and contains arteries, capillaries, and veins.

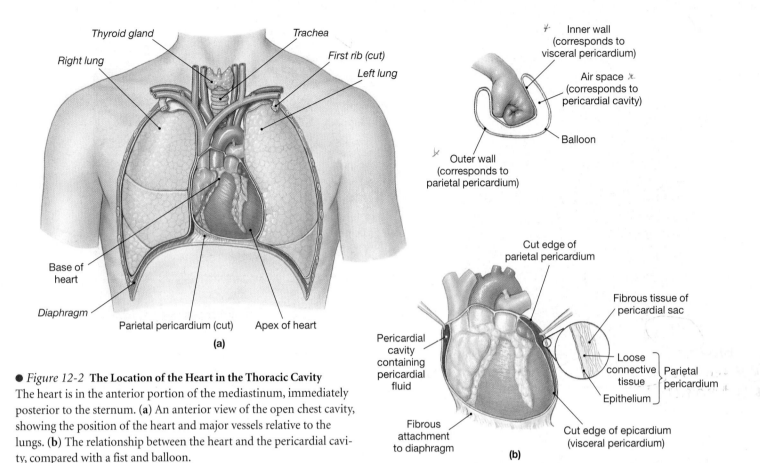

● *Figure 12-2* The Location of the Heart in the Thoracic Cavity
The heart is in the anterior portion of the mediastinum, immediately posterior to the sternum. (**a**) An anterior view of the open chest cavity, showing the position of the heart and major vessels relative to the lungs. (**b**) The relationship between the heart and the pericardial cavity, compared with a fist and balloon.

imagine pushing your fist toward the center of a large balloon (Figure 12-2b●). The balloon represents the pericardium, and your fist represents the heart. Your wrist, where the balloon folds back on itself, corresponds to the **base** of the heart. The air space inside the balloon corresponds to the pericardial cavity.

The pericardium can be subdivided into the visceral pericardium and the parietal pericardium. The **visceral pericardium,** or **epicardium,** covers the outer surface of the heart, and the **parietal pericardium** lines the inner surface of the *pericardial sac,* which surrounds the heart (Figure 12-2b●). The pericardial sac is reinforced by a dense network of collagen fibers that stabilizes the positions of the pericardium, heart, and associated vessels in the mediastinum. The space between the parietal and visceral surfaces is the pericardial cavity. It normally contains a small quantity of *pericardial fluid* secreted by the pericardial membranes. The fluid acts as a lubricant, reducing friction between the opposing surfaces as the heart beats.

THE SURFACE ANATOMY OF THE HEART

Several external features of the heart can be used to identify its four chambers (Figure 12-3a●). The two atria have relatively thin muscular walls and are highly distensible. When an atrium is not filled with blood, its outer portion deflates into a lumpy, wrinkled flap called an **auricle** (AW-ri-kl; *auris,* ear). The **coronary sulcus,** a deep groove, marks the border between the atria and the ventricles. Shallower depressions, the **anterior interventricular sulcus** and **posterior interventricular sulcus,** mark the boundary between the left and right ventricles (Figure 12-3a, b●). The connective tissue of the epicardium at the coronary and interventricular sulci (SUL-sī) usually contains substantial amounts of fat. These sulci, also contain the major arteries and veins that supply blood to the cardiac muscle.

The great veins and arteries of the circulatory system are connected to the superior end of the heart at the base. The inferior, pointed tip of the heart is the **apex** (Ā-peks) (Figure 12-2a●). A typical heart measures approximately 12.5 cm (5 in.) from the attached base to the apex.

The heart sits at an angle to the longitudinal axis of the body. It is also rotated slightly toward the left, so the anterior surface primarily consists of the right atrium and right ventricle (Figure 12-3a●). The wall of the left ventricle forms much of the posterior surface between the base and the apex of the heart (Figure 12-3b●).

12 THE HEART

The Heart and the Circulatory System • **The Anatomy and Organization of the Heart** • The Heartbeat • Heart Dynamics • Chapter Review

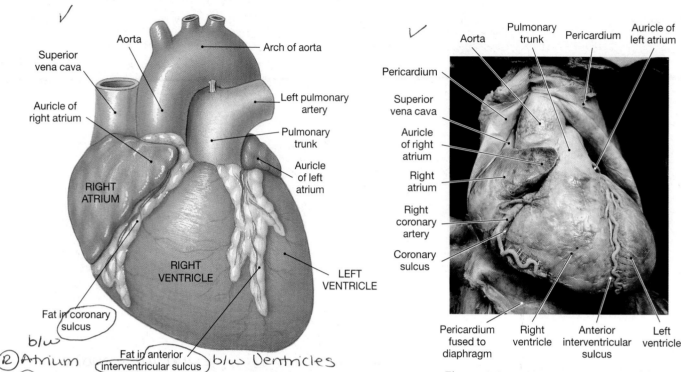

Aorta

Arch of aorta

Superior vena cava

Left pulmonary artery

Auricle of right atrium

Pulmonary trunk

Auricle of left atrium

RIGHT ATRIUM

RIGHT VENTRICLE

LEFT VENTRICLE

Fat in coronary sulcus

blw
(R) Atrium
& (R) Ventricle

Fat in anterior interventricular sulcus

blw Ventricles

(a) Anterior surface

Pulmonary trunk

Aorta

Pericardium

Auricle of left atrium

Pericardium

Superior vena cava

Auricle of right atrium

Right atrium

Right coronary artery

Coronary sulcus

Pericardium fused to diaphragm

Right ventricle

Anterior interventricular sulcus

Left ventricle

● *Figure 12-3* **The Surface Anatomy of the Heart**
(**a**) An anterior view of the heart, showing major anatomical features. (**b**) The posterior surface of the heart.

Left pulmonary artery

Arch of aorta

Right pulmonary artery

Left pulmonary veins

Superior vena cava

Fat in coronary sulcus

LEFT ATRIUM

Right pulmonary veins (superior and inferior)

LEFT VENTRICLE

RIGHT ATRIUM

RIGHT VENTRICLE

Inferior vena cava

Coronary sinus

Fat in posterior interventricular sulcus

(b) Posterior surface

blw ventricles;
on back side.

THE HEART WALL

The wall of the heart contains three distinct layers: the epicardium (visceral pericardium), the myocardium, and the endocardium (Figure 12-4a●). The **epicardium**, which covers the outer surface of the heart, is a serous membrane that consists of an exposed epithelium and an underlying layer of loose connective tissue. The **myocardium**, or muscular wall of the heart, contains cardiac muscle tissue and associated connective tissues, blood vessels, and nerves. The cardiac muscle tissue of the myocardium forms concentric layers that wrap around the atria and spiral into the walls of the ventricles. This arrangement results in squeezing and twisting contractions that increase the pumping efficiency of the heart (Figure 12-4b●). The inner surfaces, including the valves, are covered by the **endocardium** (en-dō-KAR-dē-um), a simple squamous epithelium continuous with the endothelium, or epithelial lining, of the attached blood vessels.

Cardiac Muscle Cells

Typical cardiac muscle cells are shown in Figure 12-4c, d●. These cells are relatively small and contain a single centrally located nucleus. Like skeletal muscle fibers, each cardiac muscle cell contains myofibrils, and contraction involves the shortening of individual sarcomeres. Since cardiac muscle cells are almost totally dependent on aerobic metabolism to obtain the energy needed to continue contracting, they have many mitochondria and abundant reserves of myoglobin (to store oxygen). Energy reserves are maintained in the form of glycogen and lipids.

Each cardiac muscle cell is in contact with several others at specialized sites known as **intercalated** (in-TER-ka-lā-ted)

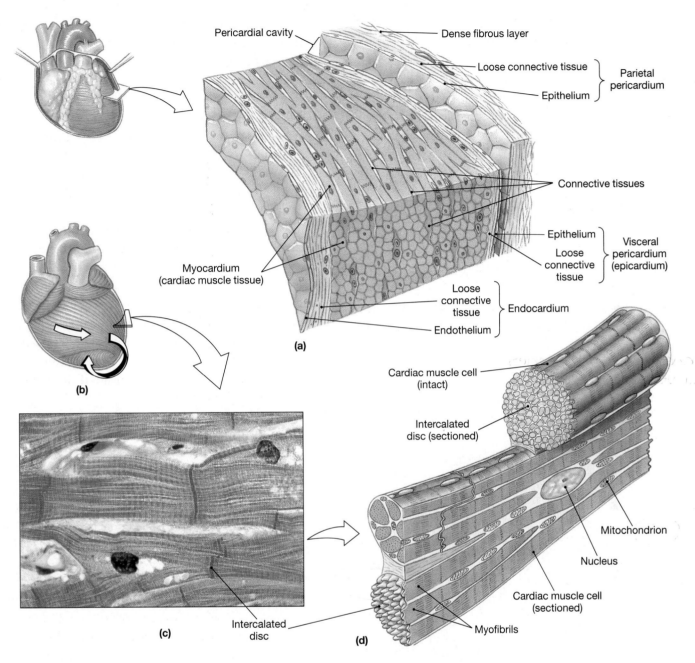

● *Figure 12-4* **The Heart Wall**
(a) A diagrammatic section through the heart wall, showing the relative positions of the epicardium, myocardium, and endocardium.
(b) Cardiac muscle tissue forms concentric layers that wrap around the atria and spiral within the walls of the ventricles.
(c, d) Sectional and diagrammatic views of cardiac muscle tissue. Cardiac muscle cells are smaller than skeletal muscle fibers and have a single, central nucleus, branching interconnections between cells, and intercalated discs. (LM × 575)

discs. ∞ p. 102 At an intercalated disc, the interlocking membranes of adjacent cells are held together by desmosomes and linked by gap junctions. ∞ p. 86 The desmosome connections help convey the force of contraction from cell to cell, increasing their efficiency as they "pull together" during a contraction. The gap junctions at these sites provide for the movement of ions and small molecules, and, as a result, action potentials rapidly travel from cell to cell.

The Fibrous Skeleton

The connective tissues of the heart include large numbers of collagen and elastic fibers that wrap around each cardiac muscle cell and also tie together adjacent cells. Such fibers (1) provide support for cardiac muscle fibers, blood vessels, and nerves of the myocardium, (2) add strength and prevent overexpansion of the heart, and (3) help the heart return to normal shape after contractions.

12 THE HEART

The Heart and the Circulatory System • **The Anatomy and Organization of the Heart** • The Heartbeat • Heart Dynamics • Chapter Review

The **fibrous skeleton** of the heart (see p. 374, Figure 12-6●) consists of dense bands of tough, elastic connective tissue that encircle the bases of the large blood vessels (*pulmonary trunk* and *aorta*) carrying blood away from the heart and each of the heart valves. The fibrous skeleton stabilizes the position of the heart valves and also physically isolates the atrial muscle tissue from the ventricular muscle tissue. This isolation is important to normal heart function because it means that the timing of ventricular contraction relative to atrial contraction can be precisely controlled.

INTERNAL ANATOMY AND ORGANIZATION

The four internal chambers of the heart are shown in Figure 12-5●. The two atria are separated by the **interatrial septum** (*septum*, wall), and the two ventricles are divided by the **interventricular septum**. Each atrium communicates with the ventricle on the same side through an **atrioventicular (AV) valve**, folds of fibrous tissue that ensure a one-way flow of blood from the atria into the ventricles.

The right atrium receives blood from the systemic circuit through two large veins, the superior vena cava (VĒ-na CĀ-va) and the inferior vena cava. The **superior vena cava** delivers blood from the head, neck, upper limbs, and chest. The **inferior vena cava** carries blood from the rest of the trunk, the viscera, and the lower limbs. The *cardiac veins* of the heart return venous blood to the **coronary sinus**, which opens into the right atrium slightly below the connection with the inferior vena cava. From the fifth week of embryonic development until birth, the *foramen ovale*, an oval opening, penetrates the interatrial septum (see Figure 13-24, p. 418). The foramen ovale permits blood flow from the right atrium to the left atrium while the lungs are developing before birth. At birth, the foramen ovale closes, and after 48 hours it is permanently sealed. A small depression, the *fossa ovalis*, persists at this site in the adult heart (Figure 12-5●). Occasionally, the foramen ovale remains open even after birth. Under these conditions, when the left atrium contracts, it pushes blood back into the pulmonary circuit. This leads to heart enlargement and eventual heart failure and death if the condition is not surgically corrected.

Blood travels from the right atrium into the right ventricle through a broad opening bounded by three flaps of fibrous tissue. These flaps, or **cusps**, are part of the **right atrioventricular (AV) valve**, also known as the **tricuspid** (trī-KUS-pid; *tri-*, three + *cuspis*, point) **valve**. Each cusp is braced by connective tissue fibers called **chordae tendineae** (KŌR-dē TEN-di-nē-ē; "tendinous cords"). These fibers are connected to **papillary** (PAP-i-ler-ē) **muscles** on the inner surface of the right ventricle. By tensing the chordae tendineae, these muscles limit the movement of the cusps and ensure proper valve function (Figure 12-5●).

Blood leaving the right ventricle flows into the **pulmonary trunk**, the start of the pulmonary circuit. The **pulmonary semilunar** (*semi-*, half + *luna*, moon; a crescent, or half-moon, shape) **valve** (*pulmonary valve*) guards the entrance to this efferent trunk. Within the pulmonary trunk, blood flows into the **left** and **right pulmonary arteries**. These vessels branch repeatedly in the lungs, before supplying the capillaries where gas exchange occurs. From these respiratory capillaries, oxygenated blood collects into the **left** and **right pulmonary veins**, which deliver it to the left atrium.

Like the right atrium, the left atrium has an external auricle and a valve, the **left atrioventricular (AV) valve**, or **bicuspid** (bī-KUS-pid) **valve**. As the name bicuspid implies, the left AV valve contains a pair, not a trio, of cusps. Clinicians often

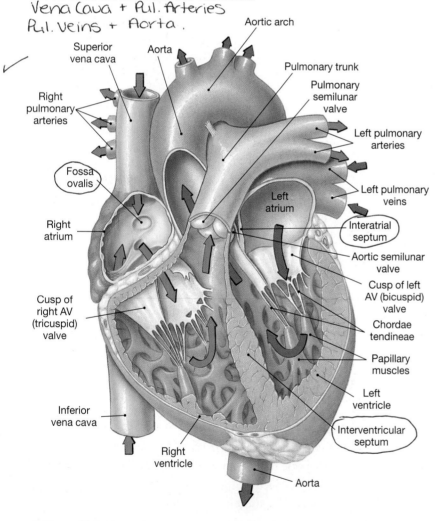

Vena Cava + Pul. Arteries
Pul. Veins + Aorta.

Superior vena cava
Aorta
Aortic arch
Right pulmonary arteries
Pulmonary trunk
Pulmonary semilunar valve
Left pulmonary arteries
Fossa ovalis
Left pulmonary veins
Left atrium
Right atrium
Interatrial septum
Aortic semilunar valve
Cusp of left AV (bicuspid) valve
Cusp of right AV (tricuspid) valve
Chordae tendineae
Papillary muscles
Left ventricle
Inferior vena cava
Interventricular septum
Right ventricle
Aorta

● *Figure 12-5* **The Sectional Anatomy of the Heart**
A diagrammatic frontal section through the heart, showing the major landmarks and the path of blood flow through the atria and ventricles.

use the term **mitral** (MĪ-tral; *mitre*, a bishop's hat) when referring to this valve.

The internal organization of the left ventricle resembles that of the right ventricle. A pair of papillary muscles braces the chordae tendineae that insert on the bicuspid valve. Blood leaving the left ventricle passes through the **aortic semilunar valve** (*aortic valve*) and into the **aorta**, the start of the systemic circuit.

Structural Differences Between the Left and Right Ventricles

The function of an atrium is to collect blood returning to the heart and deliver that blood to the attached ventricle. The demands on the right and left atria are very similar, and the two chambers look almost identical. But the demands on the right and left ventricles are very different, and there are significant structural differences between the two.

The lungs are close to the heart, and the pulmonary arteries and veins are relatively short and wide. Thus the right ventricle normally does not need to push very hard to propel blood through the pulmonary circuit. The wall of the right ventricle is relatively thin, and in sectional view it resembles a pouch attached to the massive wall of the left ventricle. When the right ventricle contracts, it acts like a bellows pump, squeezing the blood against the left ventricle. This mechanism moves blood very efficiently with minimal effort, but it develops relatively low pressures.

A comparable pumping arrangement would not be suitable for the left ventricle, because pushing blood around the systemic circuit takes six to seven times more force than pushing blood around the pulmonary circuit. The left ventricle has an extremely thick muscular wall, and it is round in cross section. When this ventricle contracts, two things happen: (1) The distance between the base and apex decreases, and (2) the diameter of the ventricular chamber decreases. (If you imagine the effects of simultaneously squeezing and rolling up the end of a toothpaste tube, you will get the idea.) As the powerful left ventricle contracts, it also bulges into the right ventricular cavity, an action that helps force blood out of the right ventricle. An individual whose right ventricular musculature has been severely damaged may survive because the contraction of the left ventricle helps push blood through the pulmonary circuit.

The Heart Valves

Details of the structure and function of the heart valves are shown in Figure 12-6●.

The Atrioventricular Valves. The atrioventricular valves prevent the backflow of blood from the ventricles into the atria. The chordae tendineae and papillary muscles play an impor-

tant role in the normal function of the AV valves. When the ventricles are relaxed, the chordae tendineae are loose and the AV valve offers no resistance to the flow of blood from atrium to ventricle (Figure 12-6a●). When the ventricles contract, blood moving back toward the atrium swings the cusps together, closing the valves (Figure 12-6b●). During ventricular contraction, tension in the papillary muscles and chordae tendineae keeps the cusps from swinging into the atrium. This action prevents the backflow, or **regurgitation**, of blood into the atrium each time the ventricle contracts. A small amount of regurgitation often occurs, even in normal individuals. The swirling action creates a soft but distinctive sound called a *heart murmur*.

MITRAL VALVE PROLAPSE

Minor abnormalities in valve shape are relatively common. For example, an estimated 10 percent of normal individuals age 14–30 have some degree of **mitral valve prolapse**. In this condition, the cusps of the left AV valve (the mitral valve) do not close properly. The problem may involve abnormally long (or short) chordae tendineae or malfunctioning papillary muscles. Because the valve does not work perfectly, some regurgitation occurs during the contraction of the left ventricle. Most individuals with mitral valve prolapse are completely asymptomatic; they live normal healthy lives unaware of any circulatory malfunction.

The Semilunar Valves. The pulmonary and aortic semilunar valves prevent the backflow of blood from the pulmonary trunk and aorta into the right and left ventricles, respectively. The semilunar valves do not require muscular braces as the AV valves do, because the arterial walls do not contract and the relative positions of the cusps are stable. When the semilunar valves close, the three symmetrical cusps in each valve support one another like the legs of a tripod, preventing movement back into the ventricles Figure 12-6a●).

VALVULAR HEART DISEASE

Serious valve problems are very dangerous, because they reduce pumping efficiency. If valve function deteriorates such that the heart cannot maintain adequate circulatory flow, symptoms of **valvular heart disease** (**VHD**) appear. Congenital defects may be responsible, but in many cases the condition develops after **carditis**, an inflammation of the heart, occurs.

One relatively common cause of carditis is **rheumatic** (roo-MA-tik) **fever**, an acute childhood reaction to streptococcal bacteria. Valve problems serious enough to reduce cardiac function may not appear until 10–20 years after the initial infection. The resulting disorder is known as *rheumatic heart disease (RHD)*. Rheumatic fever is seldom seen in the United States, due to the widespread availability and use of antibiotics.

12 THE HEART

The Heart and the Circulatory System • The Anatomy and Organization of the Heart • **The Heartbeat** • Heart Dynamics • Chapter Review

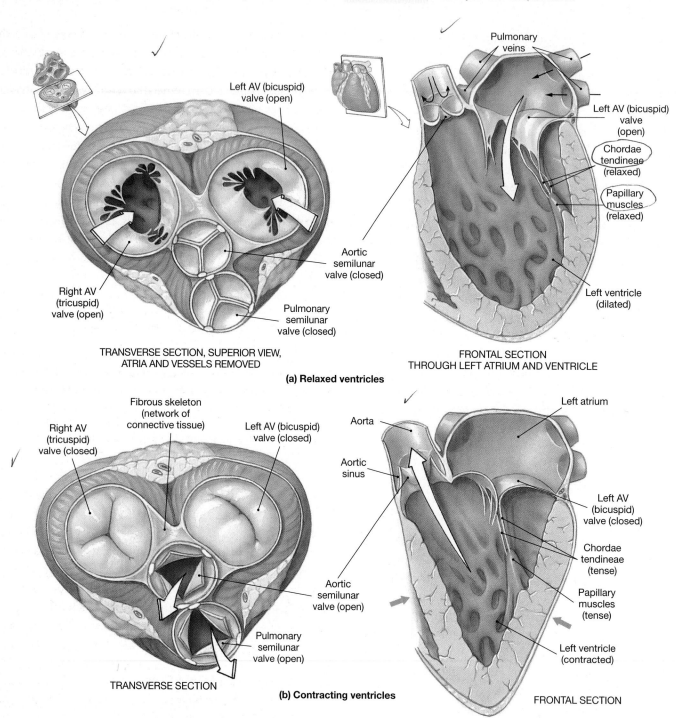

● *Figure 12-6* **The Valves of the Heart**
(**a**) The valve position during ventricular relaxation, when the AV valves are open and the semilunar valves are closed. The chordae tendineae are slack, and the papillary muscles are relaxed. (**b**) The cardiac valves during ventricular contraction, when the AV valves are closed and the semilunar valves are open. In the frontal section, notice that the chordae tendineae and papillary muscles prevent backflow through the left AV valve.

THE BLOOD SUPPLY TO THE HEART

The heart works continuously, and cardiac muscle cells require reliable supplies of oxygen and nutrients. The **coronary circulation** supplies blood to the muscles of the heart. During maximum exertion, the oxygen demand rises considerably, and

the blood flow to the heart may then increase to nine times that of resting levels.

The left and right **coronary arteries** originate at the base of the aorta, at the *aortic sinuses* (Figure 12-7a●). Blood pressure here is the highest anywhere in the systemic circuit. This

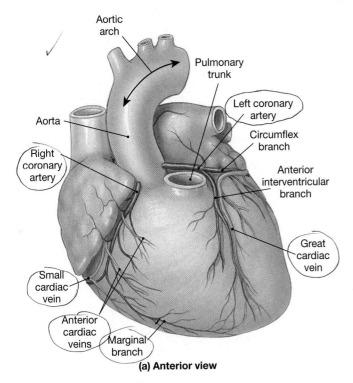

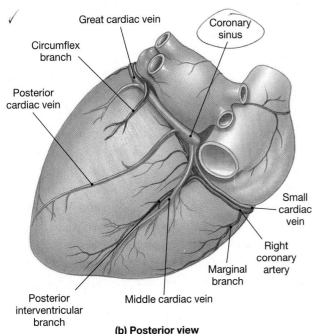

● *Figure 12-7* **The Coronary Circulation**
Coronary vessels supplying the **(a)** anterior and **(b)** posterior surfaces of the heart.

high pressure ensures a continuous flow of blood to meet the demands of active cardiac muscle. The right coronary artery supplies blood to the right atrium and to portions of both ventricles. The left coronary artery supplies blood to the left ventricle, left atrium, and interventricular septum.

Each coronary artery gives rise to two branches. The right coronary artery forms the *marginal* and *posterior interventricular* (*descending*) branches, and the left coronary artery forms the *circumflex* and *anterior interventricular* (*descending*) branches. Small tributaries from these branches of the left and right coronary arteries form interconnections called **anastomoses** (a-nas-to-MŌ-sēz; *anastomosis*, outlet). Because the arteries are interconnected in this way, the blood supply to the cardiac muscle remains relatively constant, regardless of pressure fluctuations in the left and right coronary arteries. The **great** and **middle cardiac veins** carry blood away from the coronary capillaries. They drain into the **coronary sinus**, a large, thin-walled vein in the posterior portion of the coronary sulcus. The coronary sinus opens into the right atrium near the base of the inferior vena cava.

Tissue damage caused by an interruption in blood flow is called an **infarct**. In a **myocardial** (mī-ō-KAR-dē-al) **infarction** (**MI**), or *heart attack*, the coronary circulation becomes blocked and the cardiac muscle cells die from a lack of oxygen. Heart attacks most often result from severe *coronary artery disease*, a condition characterized by the buildup of fatty deposits in the walls of the coronary arteries.

CONCEPT CHECK QUESTIONS

Answers on page 387

❶ Damage to the semilunar valves on the right side of the heart would interfere with blood flow to which vessel?

❷ What prevents the AV valves from opening back into the atria?

❸ Why is the left ventricle more muscular than the right ventricle?

The Heartbeat

In a single **heartbeat**, the entire heart—atria and venticles—contracts in a coordinated manner so that blood flows in the correct direction at the proper time. Each time the heart beats, the contractions of individual cardiac muscle cells in the atria and ventricles must occur in a specific sequence. Two types of cardiac muscle cells are involved in a normal heartbeat: (1) *contractile cells*, which produce the powerful contractions that propel blood; and (2) specialized muscle cells of the *conducting system*, which control and coordinate the activities of the contractile cells.

CONTRACTILE CELLS

Contractile cells form the bulk of the heart's muscle tissue (about 99 percent of all cardiac muscle cells). In both cardiac muscle cells and skeletal muscle fibers an action potential leads to the appearance of Ca²⁺ among the myofibrils and its binding

375

12 THE HEART

The Heart and the Circulatory System • The Anatomy and Organization of the Heart • **The Heartbeat** • Heart Dynamics • Chapter Review

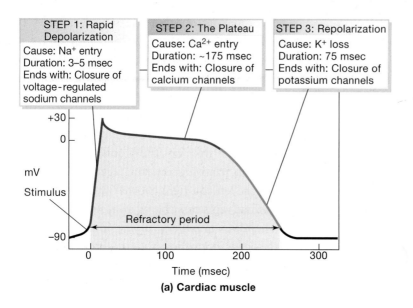

STEP 1: Rapid Depolarization
Cause: Na^+ entry
Duration: 3–5 msec
Ends with: Closure of voltage-regulated sodium channels

STEP 2: The Plateau
Cause: Ca^{2+} entry
Duration: ~175 msec
Ends with: Closure of calcium channels

STEP 3: Repolarization
Cause: K^+ loss
Duration: 75 msec
Ends with: Closure of potassium channels

(a) Cardiac muscle

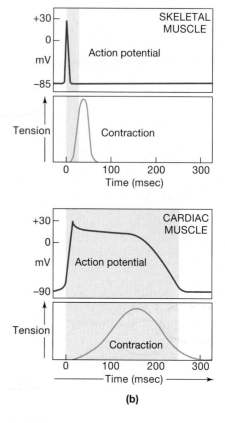

(b)

● *Figure 12-8* **Action Potential and Muscle Cell Contraction in Skeletal and Cardiac Muscle**
(a) An action potential in a ventricular cardiac muscle cell. **(b)** In a skeletal muscle fiber, the action potential is relatively brief and ends as the related twitch contraction begins. In a cardiac muscle cell, both the action potential and twitch contraction are prolonged.

to troponin on the thin filaments to begin a contraction. However, the skeletal and cardiac muscle cells differ in terms of their action potential, the source of Ca^{2+}, and the duration of the resulting contraction. Figure 12-8● shows the 3-step sequence of an action potential in a cardiac muscle cell, and compares the action potential and twitch contractions in skeletal muscle and cardiac muscle.

The first step in triggering the contraction in a contractile cell, as in a skeletal muscle fiber, is the appearance of an action potential in the sarcolemma (cell membrane). 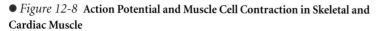 p. 182 In the 10 msec (millisecond) action potential of a skeletal muscle, a rapid depolarization is immediately followed by a rapid repolarization. In the sarcolemma of a cardiac muscle cell, the complete depolarization-repolarization process lasts 250–300 msec, some 25–30 times longer than the duration of an action potential in a skeletal muscle sarcolemma. Until the membrane repolarizes, it cannot respond to further stimulation, and the refractory period of a cardiac muscle cell membrane is relatively long. Thus, a normal cardiac muscle cell is limited to a maximum rate of about 200 contractions per minute.

The action potential is prolonged in a cardiac muscle cell because of the entry of extracellular calcium ions. This portion of the action potential is called the *plateau*. It begins when calcium ions enter the cell through voltage-regulated calcium channels as the sodium channels close and depolarization peaks. The extracellular calcium ions (1) delay repolarization

(their positive charges maintain a transmembrane potential near 0 mV—the plateau) and (2) initiate contraction. Their increased concentration within the cell also triggers the release of Ca^{2+} from reserves in the sarcoplasmic reticulum, which continue the contraction. Extracellular calcium ions thus have both direct and indirect effects on cardiac muscle cell contraction. For this reason, cardiac muscle tissue is sensitive to changes in the Ca^{2+} concentration of the extracellular fluid.

In skeletal muscle fibers, the refractory period ends before the muscle fiber develops peak tension and relaxes. As a result, twitches can build on one another until tension reaches a sustained peak; this state is called *tetanus*. 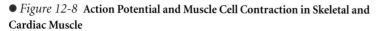 p. 187 In cardiac muscle cells, the refractory period continues until relaxation is under way. A summation of twitches is therefore not possible, and tetanic contractions cannot occur in a normal cardiac muscle cell, regardless of the frequency and intensity of stimulation. This feature is absolutely vital, since a heart in tetany could not pump blood.

THE CONDUCTING SYSTEM

In contrast to skeletal muscle, cardiac muscle tissue contracts on its own in the absence of neural or hormonal stimulation. This property, called *automaticity*, or *autorhythmicity*, also characterizes some types of smooth muscle tissue discussed in Chapter 7. 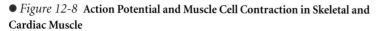 p. 194

In the normal pattern of blood flow, each contraction follows a precise sequence. The atria contract first, followed by the

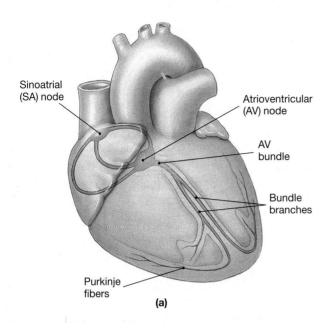

Sinoatrial
(SA) node

Atrioventricular
(AV) node

AV
bundle

Bundle
branches

Purkinje
fibers

(a)

● *Figure 12-9* **The Conducting System of the Heart**
(**a**) The stimulus for contraction is generated by pacemaker cells at the SA node. From there, impulses follow three different paths through the atrial walls to reach the AV node. After a brief delay, the impulses are conducted to the AV bundle (bundle of His) and then on to the left and right bundle branches, the Purkinje fibers, and the ventricular myocardial cells. (**b**) The movement of the contractile stimulus through the heart.

ventricles. Cardiac contractions are coordinated by the **conducting system**, a network of specialized cardiac muscle cells that initiate and distribute electrical impulses. The network is made up of two types of cardiac muscle cells that do not contract: (1) **nodal cells**, which are responsible for establishing the rate of cardiac contraction; and (2) **conducting cells**, which distribute the contractile stimulus to the general myocardium.

Nodal Cells

Nodal cells are unusual because their cell membranes depolarize spontaneously and generate action potentials at regular intervals. Nodal cells are electrically coupled to one another, to conducting cells, and to normal cardiac muscle cells. As a result, when an action potential appears in a nodal cell, it sweeps through the conducting system, reaching all of the cardiac muscle tissue and causing a contraction. In this way, nodal cells determine the heart rate.

Not all nodal cells depolarize at the same rate, and the normal rate of contraction is established by **pacemaker cells**, the nodal cells that reach threshold first. These pacemaker cells are located in the **sinoatrial** (sī-nō-Ā-trē-al) **node (SA node)**, or *cardiac pacemaker*, a tissue mass embedded in the posterior wall of the right atrium near the entrance of the superior vena cava (Figure 12-9a●). Pacemaker cells depolarize rapidly and spontaneously, generating 70–80 action potentials per minute. This results in a heart rate of 70–80 beats per minute (bpm).

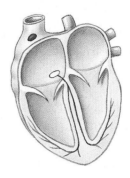

Step 1:
SA node activity and atrial activation begin.

Time = 0

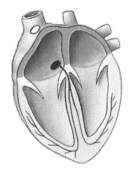

Step 2:
Stimulus spreads across the atrial surfaces, and reaches the AV node.

Elapsed time = 50 msec

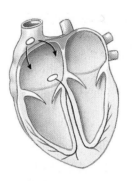

Step 3:
There is a 100 msec delay at the AV node. Atrial contraction begins.

Elapsed time = 150 msec

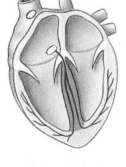

Step 4:
The impulse travels along the interventricular septum, within the AV bundle and the bundle branches, to the Purkinje fibers.

Elapsed time = 175 msec

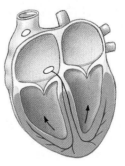

Step 5:
The impulse is distributed by Purkinje fibers and relayed throughout the ventricular myocardium. Atrial contraction is completed, and ventricular contraction begins.

Elapsed time = 225 msec

(b)

12 THE HEART

The Heart and the Circulatory System • The Anatomy and Organization of the Heart • **The Heartbeat** • Heart Dynamics • Chapter Review

Conducting Cells

The stimulus for a contraction is usually generated at the SA node, but it must be distributed so that (1) the atria contract together, before the ventricles; and (2) the ventricles contract together, in a wave that begins at the apex and spreads toward the base. When the ventricles contract in this way, blood is pushed toward the base of the heart, into the aorta and pulmonary trunk.

The conducting system of the heart is illustrated in Figure 12-9a●. The cells of the SA node are electrically connected to those of the larger **atrioventricular** (ā-trē-ō-ven-TRIK-ū-lar) **node (AV node)** by conducting cells in the atrial walls. Although the AV nodal cells also depolarize spontaneously, they generate only 40–60 action potentials per minute. Under normal circumstances, before an AV cell depolarizes to threshold spontaneously, it is stimulated by an action potential generated by the SA node. However, if the AV node does not receive this action potential, it will then become the pacemaker of the heart and establish a heart rate of 40–60 beats per minute.

The AV node is located in the floor of the right atrium near the opening of the coronary sinus. From there the action potentials travel to the **AV bundle**, also known as the *bundle of His* (hiss). This bundle of conducting cells travels along the interventricular septum before dividing into **left** and **right bundle branches**, which radiate across the inner surfaces of the left and right ventricles. At this point, specialized **Purkinje** (pur-KIN-jē) **fibers** (*Purkinje cells*) convey the impulses to the contractile cells of the ventricular myocardium.

Pacemaker cells in the SA node usually generate 60–100 action potentials per minute. It takes an action potential roughly 50 msec to travel from the SA node to the AV node over the conducting pathways (Figure 12-9b●). Along the way, the conducting cells pass the contractile stimulus to cardiac muscle cells of the right and left atria. The action potential then spreads across the atrial surfaces through cell-to-cell contact. The stimulus affects only the atria, because the fibrous skeleton electrically isolates the atria from the ventricles everywhere except at the AV bundle.

At the AV node, the impulse slows down, and another 100 msec passes before it reaches the AV bundle. This delay is important because the atria must be contracting and blood movement must be occurring before the ventricles are stimulated. Once the impulse enters the AV bundle, it flashes down the septum, along the bundle branches, and into the ventricular myocardium along the Purkinje fibers. Within another 75 msec, the stimulus to begin a contraction has reached all of the ventricular muscle cells.

A number of clinical problems result from abnormal pacemaker function. The normal heart rate averages 70–80 bpm. **Bradycardia** (brād-ē-KAR-dē-uh; *bradys*, slow) is a condition in which the heart rate is slower than normal (less than 60 bpm). **Tachycardia** (tak-ē-KAR-dē-uh; *tachys*, swift) indicates a faster than normal heart rate (100 bpm or more). In some cases, an abnormal conducting cell or ventricular muscle cell may begin generating action potentials so rapidly that they override those of the SA or AV node. The origin of such abnormal signals is called an **ectopic** (ek-TOP-ik; out of place) **pacemaker**. The action potentials may completely bypass the conducting system and disrupt the timing of ventricular contractions. Such conditions are commonly diagnosed with the aid of an *electrocardiogram*.

CONCEPT CHECK QUESTIONS

Answers on page 387

❶ Cardiac muscle does not undergo tetanus as skeletal muscle does. How does this characteristic affect the functioning of the heart?

❷ If the cells of the SA node were not functioning, how would the heart rate be affected?

❸ Why is it important for the impulses from the atria to be delayed at the AV node before passing into the ventricles?

THE ELECTROCARDIOGRAM

The electrical events occurring in the heart are powerful enough to be detected by electrodes on the body surface. A recording of these events is an **electrocardiogram** (ē-lek-trō-KAR-dē-ō-gram), also called an **ECG** or **EKG**. Each time the heart beats, a wave of depolarization radiates through the atria, reaches the AV node, travels down the interventricular septum to the apex, turns, and spreads through the ventricular myocardium toward the base.

By comparing the information obtained from electrodes placed at different locations, a clinician can monitor the electrical activity of the heart and check the performance of specific nodal, conducting, and contractile components. For example, when a portion of the heart has been damaged, the affected muscle cells will no longer conduct action potentials, so an ECG will reveal an abnormal pattern of impulse conduction.

The appearance of the ECG tracing varies with the placement of the monitoring electrodes, or *leads*. Figure 12-10● shows the important features of an electrocardiogram as analyzed with the leads in one of the standard configurations:

- The small **P wave** accompanies the depolarization of the atria. The atria begin contracting around 100 msec after the start of the P wave.

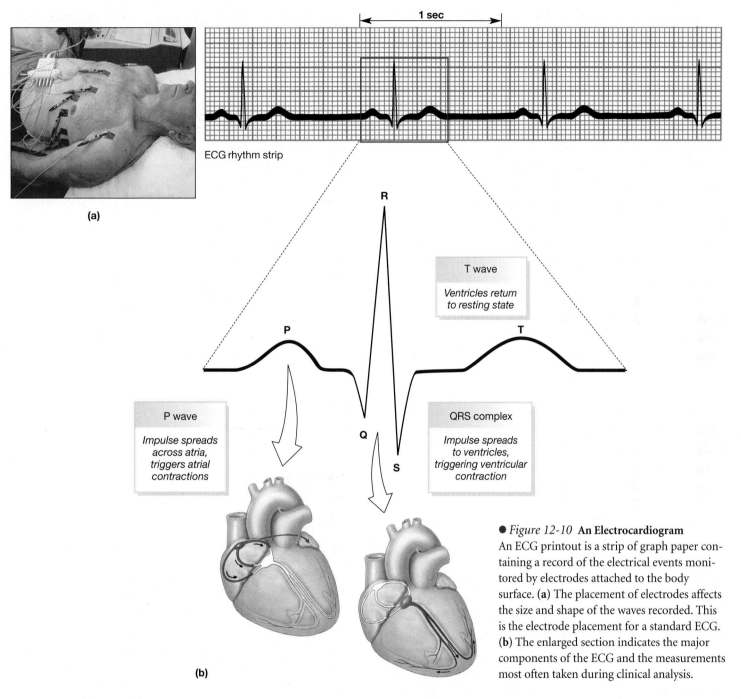

1 sec

ECG rhythm strip

(a)

R

T wave

Ventricles return to resting state

P

T

P wave

Impulse spreads across atria, triggers atrial contractions

Q

S

QRS complex

Impulse spreads to ventricles, triggering ventricular contraction

● *Figure 12-10* **An Electrocardiogram**
An ECG printout is a strip of graph paper containing a record of the electrical events monitored by electrodes attached to the body surface. **(a)** The placement of electrodes affects the size and shape of the waves recorded. This is the electrode placement for a standard ECG. **(b)** The enlarged section indicates the major components of the ECG and the measurements most often taken during clinical analysis.

(b)

- The **QRS complex** appears as the ventricles depolarize. This electrical signal is relatively strong because the mass of the ventricular muscle is much larger than that of the atria. The ventricles begin contracting shortly after the peak of the R wave.

- The smaller **T wave** indicates ventricular repolarization. Atrial repolarization occurs while the ventricles are depolarizing, so the electrical events are masked by the QRS complex.

Analyzing an ECG involves measuring the size of the voltage changes and determining the temporal relationships of the various components. Attention usually focuses on the amount of depolarization occurring during the P wave and the QRS complex. For example, a smaller than normal electrical signal can mean that the mass of the heart muscle has decreased, and excessively strong depolarizations can mean that the heart muscle has become enlarged.

Electrocardiogram analysis is useful in detecting and diagnosing **cardiac arrhythmias** (ā-RITH-mē-az), abnormal patterns of cardiac activity. Momentary arrhythmias are not inherently dangerous, and about 5 percent of the normal population experiences a few abnormal heartbeats each day. Clinical

12 THE HEART

The Heart and the Circulatory System • The Anatomy and Organization of the Heart • The Heartbeat • **Heart Dynamics** • Chapter Review

problems appear when the arrhythmias reduce the heart's pumping efficiency. Serious arrhythmias can indicate damage to the myocardium, injuries to the pacemaker or conduction pathways, exposure to drugs, or variations in the electrolyte composition of the extracellular fluids.

THE CARDIAC CYCLE

The period between the start of one heartbeat and the beginning of the next is a single **cardiac cycle** (Figure 12-11●). The cardiac cycle therefore includes both a period of contraction and one of relaxation. For any one chamber in the heart, the cardiac cycle can be divided into two phases. During contraction, or **systole** (SIS-to-lē), the chamber pushes blood into an adjacent chamber or into an arterial trunk. Systole is followed by the second phase, one of relaxation, or **diastole** (dī-AS-to-lē), when the chamber fills with blood and prepares for the start of the next cardiac cycle.

Fluids tend to move from an area of higher pressure to one of lower pressure. During the cardiac cycle, the pressure within each chamber rises during systole and falls during diastole. An increase in pressure in one chamber will cause the blood to flow to another chamber or vessel where the pressure is lower. The atrioventricular and semilunar valves ensure that blood flows in one direction only during the cardiac cycle.

The correct pressure relationships depend on the careful timing of contractions. The elaborate pacemaking and conduction systems normally provide the required spacing between atrial systole and ventricular systole. If the atria and ventricles were to contract at the same moment, blood could not leave the atria, because the AV valves would be closed. In the normal heart, atrial systole and atrial diastole are slightly out of phase with ventricular systole and diastole. Figure 12-11● shows the duration and timing of systole and diastole for a heart rate of 75 bpm.

The cardiac cycle begins with atrial systole. At the start of a cardiac cycle, the ventricles are partially filled with blood. During atrial systole, the atria contract and the ventricles become completely filled with blood (Figure 12-11a, b●). As

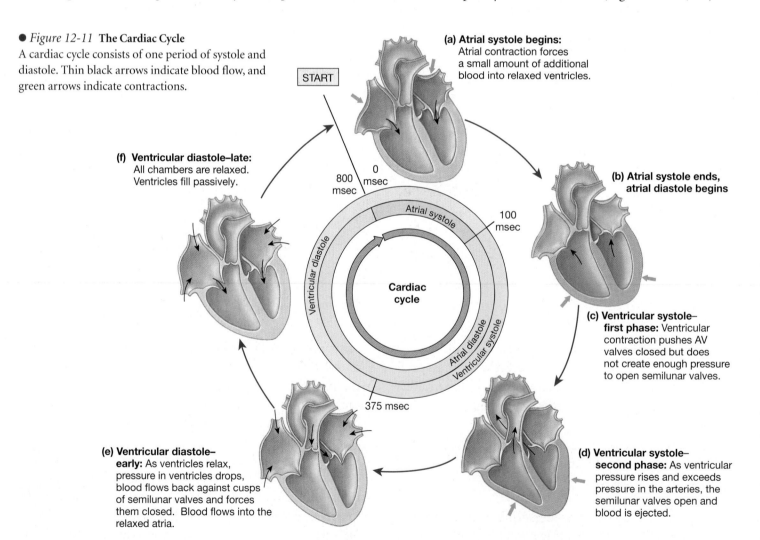

● *Figure 12-11* **The Cardiac Cycle**
A cardiac cycle consists of one period of systole and diastole. Thin black arrows indicate blood flow, and green arrows indicate contractions.

START

(a) Atrial systole begins: Atrial contraction forces a small amount of additional blood into relaxed ventricles.

(f) Ventricular diastole–late: All chambers are relaxed. Ventricles fill passively.

800 msec | 0 msec

Atrial systole | 100 msec

Ventricular diastole

Cardiac cycle

Atrial diastole

Ventricular systole

375 msec

(b) Atrial systole ends, atrial diastole begins

(c) Ventricular systole–first phase: Ventricular contraction pushes AV valves closed but does not create enough pressure to open semilunar valves.

(e) Ventricular diastole–early: As ventricles relax, pressure in ventricles drops, blood flows back against cusps of semilunar valves and forces them closed. Blood flows into the relaxed atria.

(d) Ventricular systole–second phase: As ventricular pressure rises and exceeds pressure in the arteries, the semilunar valves open and blood is ejected.

atrial systole ends, atrial diastole and ventricular systole begin. As pressures in the ventricles rise above those in the atria, the AV valves swing shut. But blood cannot begin moving into the arterial trunks until ventricular pressures exceed the arterial pressures. At this point, the blood pushes open the semilunar valves and flows into the aorta and pulmonary trunk (Figure 12-11c, d●). This blood flow continues for the duration of ventricular systole.

When ventricular diastole begins, ventricular pressures decline rapidly (Figure 12-11e, f●). As they fall below the pressures of the arterial trunks, the semilunar valves close. Ventricular pressures continue to drop; as they fall below atrial pressures, the AV valves open and blood flows from the atria into the ventricles. Both atria and ventricles are now in diastole; blood now flows from the major veins through the relaxed atria and into the ventricles. By the time atrial systole marks the start of another cardiac cycle, the ventricles are roughly 70 percent filled. The relatively minor contribution atrial systole makes to ventricular volume explains why individuals can survive quite normally when their atria have been so severely damaged that they can no longer function. In contrast, damage to one or both ventricles can leave the heart unable to maintain adequate cardiac output. A condition of *heart failure* then exists.

Heart Sounds

Physicians use an instrument called a **stethoscope** to listen to normal and abnormal heart sounds. There are four **heart sounds**. These sounds accompany the action of the heart valves. When you listen to your own heart, you hear the familiar "lubb-dupp" that accompanies each heartbeat. The *first heart sound* ("lubb") is produced as the AV valves close and the semilunar valves open. It marks the start of ventricular systole and lasts a little longer than the second sound. The *second heart sound*, "dupp," occurs at the beginning of ventricular diastole, when the semilunar valves close.

Third and *fourth heart sounds* may be audible as well, but they are usually very faint and are seldom detectable in healthy adults. These sounds are associated with atrial contraction and blood flowing into the ventricles rather than with valve action.

CONCEPT CHECK QUESTIONS

Answers on page 387

❶ Is the heart always pumping blood when pressure in the left ventricle is rising? Explain.

❷ What causes the "lubb-dupp" sounds of the heart that are heard with a stethoscope?

Heart Dynamics

Heart dynamics refers to the movements and forces generated during cardiac contractions. Each time the heart beats, the two ventricles eject equal amounts of blood. The amount ejected by a ventricle during a single beat is the **stroke volume (SV)**. The stroke volume can vary from beat to beat. Hence, physicians are often more interested in the **cardiac output (CO)**, or the amount of blood pumped by each ventricle in 1 minute.

Cardiac output can be calculated by multiplying the average stroke volume by the heart rate (HR):

$$\underset{\substack{\text{cardiac output} \\ \text{(ml/min)}}}{\text{CO}} = \underset{\substack{\text{stroke volume} \\ \text{(ml)}}}{\text{SV}} \times \underset{\substack{\text{heart rate} \\ \text{(bpm)}}}{\text{HR}}$$

For example, if the average stroke volume is 80 ml and the heart rate is 70 beats per minute (bpm), the cardiac output will be:

$$\text{CO} = 80 \text{ ml} \times 70/\text{min}$$
$$= 5600 \text{ ml/min (5.6 liters per minute)}$$

This value represents a cardiac output equivalent to the total volume of blood of an average adult every minute. Cardiac output is highly variable, however, because a normal heart can increase both its rate of contraction and its stroke volume. When necessary, stroke volume in a normal heart can almost double, and the heart rate can increase by 250 percent. When both increase together, the cardiac output can increase by 500 to 700 percent, or up to 30 liters per minute.

FACTORS CONTROLLING CARDIAC OUTPUT

Cardiac output is precisely regulated so that peripheral tissues receive an adequate blood supply under a variety of conditions. The major factors that regulate cardiac output often affect both heart rate and stroke volume simultaneously. These primary factors include *blood volume reflexes*, *autonomic innervation*, and *hormones*. Secondary factors include the concentration of ions in the extracellular fluid and body temperature.

Blood Volume Reflexes

Cardiac muscle contraction is an active process, but relaxation is entirely passive. The force necessary to return cardiac muscle to its precontracted length is provided by the blood pouring into the heart, aided by the elasticity of the fibrous skeleton. As a result, there is a direct relationship between the amount of blood entering the heart and the amount of blood ejected during the next contraction.

12 THE HEART

The Heart and the Circulatory System • The Anatomy and Organization of the Heart • The Heartbeat • **Heart Dynamics** • Chapter Review

Two heart reflexes respond to changes in blood volume. One of these occurs in the right atrium and affects heart rate. The other is a ventricular reflex that affects stroke volume.

The **atrial reflex** (*Bainbridge reflex*) involves adjustments in heart rate that are triggered by an increase in the **venous return**, the flow of venous blood to the heart. When the walls of the right atrium are stretched, the stimulation of stretch receptors in the atrial walls triggers a reflexive increase in heart rate caused by increase in sympathetic activity. As a result of sympathetic stimulation, the cells of the SA node depolarize faster and the heart rate increases.

The amount of blood pumped out of a ventricle each beat depends on the venous return and the filling time. **Filling time** is the duration of ventricular diastole, the period during which blood can flow into the ventricles. Filling time depends primarily on the heart rate: The faster the heart rate, the shorter the available filling time. Venous return changes in response to alterations in cardiac output, peripheral circulation, and other factors that affect the rate of blood flow through the venae cavae.

Over the range of normal activities, the greater the volume of blood entering the ventricles, the more powerful the contraction. The greater the degree of stretching in the walls of the heart, the more force developed when a contraction occurs. In a resting individual, the venous return is relatively low, the walls are not stretched significantly, and the ventricles develop little power. If the venous return suddenly increases, more blood flows into the heart, the myocardium stretches farther, and the ventricles produce greater force on contraction. This general rule of "more in = more out" is often called the **Frank-Starling principle**, in honor of the physiologists who first demonstrated the relationship. The major effect of the Frank-Starling principle is that the output of blood from both ventricles is balanced under a variety of conditions.

Autonomic Innervation

The basic heart rate is established by the pacemaker cells of the SA node, but this rate can be modified by the autonomic nervous system (ANS). As shown in Figure 12-12●, both the sympathetic and parasympathetic divisions of the ANS innervate the heart. Postganglionic sympathetic fibers extend from neurons located in the cervical and upper thoracic ganglia. The vagus nerve (cranial nerve N X) carries parasympathetic preganglionic fibers to small ganglia near the heart. Both ANS divisions innervate the SA and AV nodes as well as the atrial and ventricular cardiac muscle cells.

Autonomic Effects on Heart Rate. Autonomic effects on heart rate primarily reflect the responses of the SA node to

acetylcholine (ACh) and to norepinephrine (NE). Acetylcholine released by parasympathetic motor neurons results in a lowering of the heart rate. Norepinephrine released by sympathetic neurons increases the heart rate. A more sustained rise in heart rate follows the release of epinephrine (E) and norepinephrine by the adrenal medullae during sympathetic activation.

Autonomic Effects on Stroke Volume. Through the release of NE, E, and ACh, the ANS also affects stroke volume by altering the force of myocardial contractions:

- **The Effects of NE and E.** The sympathetic release of NE at synapses in the myocardium and the release of NE and E by the adrenal medullae stimulate cardiac muscle cell metabolism and increase the force and degree of contraction. The result is an increase in stroke volume.

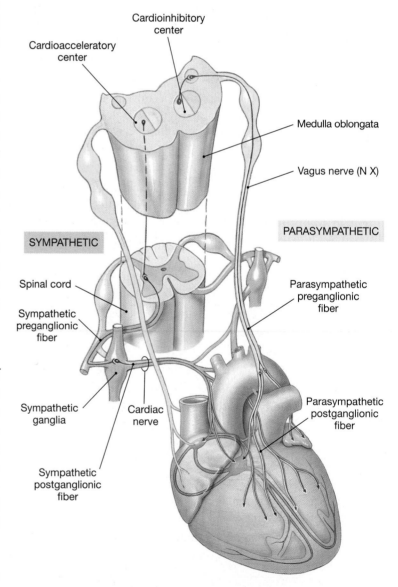

● *Figure 12-12* **Autonomic Innervation of the Heart**

- **The Effects of ACh.** The primary effect of parasympathetic ACh release is inhibition, resulting in a decrease in the force of cardiac contractions. Because parasympathetic innervation of the ventricles is relatively limited, the atria show the greatest reduction in contractile force.

Both autonomic divisions are normally active at a steady background level, releasing ACh and NE both at the nodes and into the myocardium. Thus, cutting the vagus nerves increases the heart rate, and sympathetic blocking agents slow the heart rate. Through dual innervation and adjustments in autonomic tone, the ANS can make very delicate adjustments in cardiovascular function.

Hormones

As noted above, the epinephrine and norepinephrine secreted by the adrenal medullae act to increase both the force of contraction and heart rate. Thyroid hormones and glucagon, which is secreted by the pancreas, produce effects similar to those produced by epinephrine and norepinephrine.

The Coordination of Autonomic Activity. The *cardiac centers* of the medulla oblongata contain the autonomic headquarters for cardiac control. ⟰ p. 253 The stimulation of the **cardioacceleratory center** activates sympathetic motor neurons; the nearby **cardioinhibitory center** controls the parasympathetic motor neurons (Figure 12-12●). Information concerning the status of the cardiovascular system arrives at the cardiac centers over sensory fibers from the vagus nerve and from the sympathetic nerves of the cardiac plexus.

The cardiac centers respond to changes in blood pressure and in the arterial concentrations of dissolved oxygen and carbon dioxide. These properties are monitored by baroreceptors and chemoreceptors innervated by the glossopharyngeal (N IX) and vagus nerves. A decline in blood pressure or oxygen concentrations or an increase in carbon dioxide levels usually indicates that the oxygen demands of peripheral tissues have increased. The cardiac centers then call for an increase in the cardiac output, and the heart works harder.

In addition to making automatic adjustments in response to sensory information, the cardiac centers can be influenced by higher centers, especially centers in the hypothalamus. As a result, changes in emotional state (such as rage, fear, or arousal) have an immediate effect on heart rate.

 ### EXTRACELLULAR IONS, TEMPERATURE, AND CARDIAC OUTPUT

Changes in extracellular calcium ion concentrations primarily affect the strength and duration of cardiac contractions and, thus, stroke volume. If such calcium concentrations are elevated (**hypercalcemia**), cardiac muscle cells become extremely excitable and their contractions become powerful and prolonged. In extreme cases, the heart goes into an extended state of contraction that is usually fatal. When calcium levels are abnormally low (**hypocalcemia**), the contractions become very weak and may cease altogether.

Abnormal extracellular potassium ion concentrations alter the resting potential at the SA node and primarily change the heart rate. When potassium concentrations are high (**hyperkalemia**), cardiac contractions become weak and irregular. When the extracellular concentration of potassium is abnormally low (**hypokalemia**), the heart rate is reduced.

Temperature changes affect metabolic operations throughout the body. For example, a lowered body temperature slows the rate of depolarization at the SA node, lowers the heart rate, and reduces the strength of cardiac contractions. An elevated body temperature accelerates the heart rate and the contractile force—one reason why your heart seems to race and pound when you have a fever.

CONCEPT CHECK QUESTIONS

Answers on page 387

❶ What effect would stimulating the acetylcholine receptors of the heart have on cardiac output?

❷ What effect would an increased venous return have on the stroke volume?

❸ How would increased sympathetic stimulation of the heart affect stroke volume?

Related Clinical Terms

angina pectoris (an-JĪ-nuh PEK-tor-is): A condition in which exertion or stress produces severe chest pain, resulting from temporary ischemia when the heart's workload increases.

balloon angioplasty: A technique for reducing the size of a coronary plaque by compressing it against the arterial walls, using a catheter with an inflatable collar.

bradycardia (brād-ē-KAR-dē-uh): A heart rate that is slower than normal.

cardiac arrhythmias (ā-RITH-mē-az): Abnormal patterns of cardiac electrical activity, indicating abnormal contractions.

cardiac tamponade: A condition, resulting from pericardial irritation and inflammation, in which

12 THE HEART

The Heart and the Circulatory System • The Anatomy and Organization of the Heart • The Heartbeat • Heart Dynamics • **Chapter Review**

fluid collects in the pericardial sac and restricts cardiac output.

cardiology (kar-dē-OL-o-jē): The study of the heart, its functions, and diseases.

carditis (kar-DĪ-tis): A general term indicating inflammation of the heart.

coronary arteriography: The production of an X-ray image of coronary circulation after the introduction of a radiopaque dye into one of the coronary arteries through a catheter; the resulting image is a *coronary angiogram.*

coronary artery bypass graft (CABG): The routing of blood around an obstructed coronary artery (or one of its branches) by a vessel transplanted from another part of the body.

coronary artery disease (CAD): The obstruction of coronary circulation.

coronary ischemia: The restriction of the circulatory supply to the heart, potentially causing tissue damage and a reduction in cardiac efficiency.

coronary thrombosis: A blockage due to the formation of a clot (thrombus) at a plaque in a coronary artery.

defibrillator: A device used to eliminate atrial or ventricular fibrillation and restore normal cardiac rhythm.

echocardiography: Ultrasound analysis of the heart and the blood flow through major vessels.

electrocardiogram (ē-lek-trō-KAR-dē-ō-gram) **(ECG or EKG):** A recording of the electrical activities of the heart over time.

heart block: A condition affecting the normal rhythm of the heart; characterized by impaired communication between the SA node, the AV node, and the ventricular myocardium as a result of damage to conduction pathways due to mechanical distortion, ischemia, infection, or inflammation.

heart failure: A condition in which the heart weakens and peripheral tissues suffer from oxygen and nutrient deprivation.

myocardial (mī-ō-KAR-dē-al) **infarction (MI):** A condition in which the coronary circulation becomes blocked and cardiac muscle cells die from oxygen starvation; also called a *heart attack.*

pericarditis: Inflammation of the pericardium.

rheumatic heart disease (RHD): A disorder in which the heart valves become thickened and stiffen into a partially closed position, affecting the efficiency of the heart.

tachycardia (tak-ē-KAR-dē-uh;): A heart rate that is faster than normal.

valvular heart disease (VHD): A condition caused by abnormal functioning of one of the cardiac valves. The severity of the condition depends on the degree of damage and the valve involved.

CHAPTER REVIEW

Key Terms

Summary Outline

INTRODUCTION **368**

THE HEART AND THE CIRCULATORY SYSTEM **368**

1. The circulatory system can be subdivided into the **pulmonary circuit** (which carries blood to and from the lungs) and the **systemic circuit** (which transports blood to and from the rest of the body). **Arteries** carry blood away from the heart; **veins** return blood to the heart. **Capillaries** are tiny vessels between the smallest arteries and smallest veins. *(Figure 12-1)*

2. The heart has four chambers: the **right atrium**, **right ventricle**, **left atrium**, and **left ventricle**.

THE ANATOMY AND ORGANIZATION OF THE HEART ..**368**

1. The heart is surrounded by the **pericardial cavity** (lined by the **pericardium**). The **visceral pericardium (epicardium)** covers the heart's outer surface, and the **parietal pericardium** lines the inner surface of the *pericardial sac*, which surrounds the heart. *(Figure 12-2)*

The Surface Anatomy of the Heart **369**

2. The **coronary sulcus**, a deep groove, marks the boundary between the atria and ventricles. *(Figure 12-3)*

The Heart Wall..**370**

3. The bulk of the heart consists of the muscular **myocardium**. The **endocardium** lines the inner surfaces of the heart. The **fibrous skeleton** supports the heart's contractile cells and valves. *(Figure 12-4a–b)*

4. **Cardiac muscle cells** are interconnected by **intercalated discs**, which convey the force of contraction from cell to cell and conduct action potentials. *(Figure 12-4c–d)*

Internal Anatomy and Organization **372**

5. The atria are separated by the **interatrial septum**, and the ventricles are divided by the **interventricular septum**. The right atrium receives blood from the systemic circuit via two large veins, the **superior vena cava** and **inferior vena cava**. *(Figure 12-5)*

6. Blood flows from the right atrium into the right ventricle through the **right atrioventricular (AV) valve (tricuspid valve)**. This opening is bounded by three **cusps** of fibrous tissue braced by the tendinous **chordae tendineae**, which are connected to **papillary muscles**.

7. Blood leaving the right ventricle enters the **pulmonary trunk** after passing through the **pulmonary semilunar valve**. The pulmonary trunk divides to form the **left** and **right pulmonary arteries**. The **left** and **right pulmonary veins** return blood to the left atrium. Blood leaving the left atrium flows into the left ventricle through the **left atrioventricular (AV) valve (bicuspid valve or mitral valve)**. Blood leaving the left ventricle passes through the **aortic semilunar valve** and into the systemic circuit via the **aorta**. *(Figure 12-6)*

8. Anatomical differences between the ventricles reflect the functional demands on them. The wall of the right ventricle is relatively thin, while the left ventricle has a massive muscular wall.

9. Valves normally permit blood flow in only one direction, preventing the **regurgitation** (backflow) of blood.

The Blood Supply to the Heart..............................**374**

10. The **coronary circulation** meets the high oxygen and nutrient demands of cardiac muscle cells. The coronary arteries originate at the base of the

aorta. Arterial **anastomoses**, interconnections between arteries, ensure a constant blood supply. The **great** and **middle cardiac veins** carry blood from the coronary capillaries to the **coronary sinus**. *(Figure 12-7)*

THE HEARTBEAT 375

1. Two general classes of cardiac cells are involved in the normal heartbeat: *contractile cells* and cells of the conducting system.

Contractile Cells 375

2. Cardiac muscle cells have a long refractory period, so rapid stimulation produces isolated contractions rather than tetanic contractions. *(Figure 12-8)*

The Conducting System 376

3. The conducting system includes **nodal cells** and **conducting cells**. The conducting system initiates and distributes electrical impulses in the heart. Nodal cells establish the rate of cardiac contraction; **pacemaker cells** are nodal cells that reach threshold first. Conducting cells distribute the contractile stimulus to the general myocardium.

4. Unlike skeletal muscle, cardiac muscle contracts without neural or hormonal stimulation. Pacemaker cells in the **sinoatrial (SA) node** (*cardiac pacemaker*) normally establish the rate of contraction. From the SA node, the stimulus travels to the **atrioventricular (AV) node** and then to the AV bundle, which divides into **bundle branches**. From there **Purkinje fibers** convey the impulses to the ventricular myocardium. *(Figure 12-9)*

The Electrocardiogram 378

5. A recording of electrical activities in the heart is an **electrocardiogram** (**ECG** or **EKG**). Important landmarks of an ECG include the **P wave** (atrial depolarization), **QRS complex** (ventricular depolarization), and **T wave** (ventricular repolarization). *(Figure 12-10)*

The Cardiac Cycle...380

6. The **cardiac cycle** consists of **systole** (contraction) followed by **diastole** (relaxation). Both ventricles contract at the same time, and they eject equal volumes of blood. *(Figure 12-11)*

7. The closing of the heart valves and the rushing of blood through the heart cause characteristic **heart sounds**.

HEART DYNAMICS 381

1. *Heart dynamics* refers to the movements and forces generated during contractions. The amount of blood ejected by a ventricle during a single beat is the **stroke volume (SV)**; the amount of blood pumped each minute is the **cardiac output (CO)**.

Factors Controlling Cardiac Output 381

2. The major factors that affect cardiac output are blood volume reflexes, autonomic innervation, and hormones.

3. Blood volume reflexes are stimulated by changes in **venous return**, the amount of blood entering the heart. The **atrial reflex** accelerates the heart rate when the walls of the right atrium are stretched. Ventricular contractions become more powerful and increase stroke volume when the ventricular walls are stretched (the **Frank-Starling principle**).

4. The basic heart rate is established by the pacemaker cells, but it can be modified by the ANS. *(Figure 12-12)*

5. Acetylcholine (ACh) released by parasympathetic motor neurons lowers the heart rate and stroke volume. Norepinephrine (NE) released by sympathetic neurons increases the heart rate and stroke volume.

6. Epinephrine (E) and norepinephrine, hormones released by the adrenal medullae during sympathetic activation, increase both heart rate and stroke volume. Thyroid hormones and glucagon also act to increase cardiac output.

7. The **cardioacceleratory center** in the medulla oblongata activates sympathetic neurons; the **cardioinhibitory center** governs the activities of the parasympathetic neurons. These cardiac centers receive inputs from higher centers and from receptors monitoring blood pressure and the levels of dissolved gases.

Review Questions

Level 1: Reviewing Facts and Terms

Match each item in column A with the most closely related item in column B. Use letters for answers in the spaces provided.

COLUMN A

___ 1. epicardium

___ 2. right AV valve

___ 3. left AV valve

___ 4. anastomoses

___ 5. myocardial infarction

___ 6. SA node

___ 7. systole

___ 8. diastole

___ 9. cardiac output

___ 10. HR slower than usual

___ 11. HR faster than normal

___ 12. atrial reflex

COLUMN B

a. heart attack

b. cardiac pacemaker

c. tachycardia

d. SV × HR

e. tricuspid valve

f. bradycardia

g. mitral valve

h. interconnections between arteries

i. visceral pericardium

j. increased venous return

k. contractions of heart chambers

l. relaxation of heart chambers

12 THE HEART

The Heart and the Circulatory System • The Anatomy and Organization of the Heart • The Heartbeat • Heart Dynamics • **Chapter Review**

13. The blood supply to the muscles of the heart is provided by the:
    (a) systemic circulation   (b) pulmonary circulation
    (c) coronary circulation   (d) coronary portal system

14. The autonomic centers for cardiac function are located in the:
    (a) myocardial tissue of the heart
    (b) cardiac centers of the medulla oblongata
    (c) cerebral cortex
    (d) a, b, and c are correct

15. The simple squamous epithelium covering the valves of the heart constitutes the:
    (a) epicardium   (b) endocardium
    (c) myocardium   (d) fibrous skeleton

16. The structure that permits blood flow from the right atrium to the left atrium while the lungs are developing is the:
    (a) foramen ovale   (b) interatrial septum
    (c) coronary sinus   (d) fossa ovalis

17. Blood leaves the left ventricle by passing through the:
    (a) aortic semilunar valve   (b) pulmonary semilunar valve
    (c) mitral valve   (d) tricuspid valve

18. The QRS complex of the ECG appears as the:
    (a) atria depolarize   (b) ventricles depolarize
    (c) ventricles repolarize   (d) atria repolarize

19. During diastole in the cardiac cycle, the chambers of the heart:
    (a) relax and fill with blood
    (b) contract and push blood into an adjacent chamber
    (c) experience a sharp increase in pressure
    (d) reach a pressure of approximately 120 mm Hg

20. What role do the chordae tendineae and papillary muscles have in the normal function of the AV valves?

21. What are the principal valves found in the heart, and what is the function of each?

22. Trace the normal pathway of an electrical impulse through the conducting system of the heart.

23. (a) What is the cardiac cycle? (b) What phases and events are necessary to complete the cardiac cycle?

Level 2: Reviewing Concepts

24. Tetanic muscle contractions cannot occur in a normal cardiac muscle cell because:
    (a) cardiac muscle tissue contracts on its own
    (b) there is no neural or hormonal stimulation
    (c) the refractory period lasts until the muscle cell relaxes
    (d) the refractory period ends before the muscle cell reaches peak tension

25. The amount of blood forced out of the heart depends on:
    (a) the degree of stretching at the end of ventricular diastole
    (b) the contractility of the ventricle
    (c) the amount of pressure required to eject blood
    (d) a, b, and c are correct

26. The cardiac output cannot increase indefinitely because:
    (a) available filling time becomes shorter as the heart rate increases
    (b) cardiovascular centers adjust the heart rate
    (c) the rate of spontaneous depolarization decreases
    (d) the ion concentrations of pacemaker cell membranes decrease

27. Describe the association of the four muscular chambers of the heart with the pulmonary and systemic circuits.

28. What are the source and significance of the heart sounds?

29. (a) What effect does sympathetic stimulation have on the heart? (b) What effect does parasympathetic stimulation have on the heart?

Level 3: Critical Thinking and Clinical Applications

30. A patient's ECG tracing shows a consistent pattern of two P waves followed by a normal QRS complex and T wave. What is the cause of this abnormal wave pattern?

31. Karen is taking the medication *verapamil*, a drug that blocks the calcium channels in cardiac muscle cells. What effect would you expect this medication to have on Karen's stroke volume?

Answers to Concept Check Questions

Page 375

1. The semilunar valves on the right side of the heart guard the opening to the pulmonary artery. Damage to these valves would interfere with the blood flow to this vessel.

2. When the ventricles begin to contract, they force the AV valves to close, tensing the chordae tendineae. The chordae tendineae are attached to the papillary muscles. The papillary muscles respond by contracting, counteracting the force that is pushing the valves upward.

3. The wall of the left ventricle is more muscular than that of the right ventricle because the left ventricle must generate enough force to propel blood throughout all of the body's systems except the lungs. The right ventricle must generate only enough force to propel the blood a few centimeters to the lungs.

Page 378

1. Unlike the situation in skeletal muscle fibers, tetanus is not possible. The longer refractory period in cardiac muscle cells results in a relatively long relaxation period, during which the heart's chambers can refill with blood. A heart in tetany could not fill with blood.

2. If the cells of the SA node were not functioning, then the cells of the AV node would become the pacemaker cells. The heart would still continue to beat but at a slower rate.

3. If the impulses from the atria were not delayed at the AV node, they would be conducted through the ventricles so quickly by the bundle branches and Purkinje fibers that the ventricles would begin contracting immediately, before the atria had finished contracting. As a result, the ventricles would not be as full of blood as they could be and the pumping of the heart would not be as efficient, especially during activity.

Page 381

1. When pressure in the left ventricle is rising, the heart is contracting but no blood is leaving the heart. During this initial phase of contraction, both the AV valves and the semilunar valves are closed. The increase in pressure is the result of increased tension as the muscle contracts. When the pressure in the ventricle exceeds the pressure in the aorta, the aortic semilunar valves are forced open and the blood is rapidly ejected from the ventricle.

2. The first sound is produced by the simultaneous closing of the AV valves and the opening of the semilunar valves. The second sound is produced when the semilunar valves close.

Page 383

1. Stimulating the acetylcholine receptors of the heart would cause the heart to slow down. Since the cardiac output is the product of stroke volume times the heart rate, if the heart rate decreases, so will the cardiac output (assuming no change in the stroke volume).

2. The venous return fills the heart with blood, stretching the heart muscle. According to the Frank-Starling principle, the more the heart muscle is stretched, the more forcefully it will contract (to a point). The more forceful the contraction, the more blood the heart will eject with each beat (stroke volume). Therefore, increased venous return will increase the stroke volume if all other factors are constant.

3. Increased sympathetic stimulation of the heart would result in an increased heart rate and an increased force of contraction.

Blood Vessels
and Circulation

Chapter Outline and Objectives

Vocabulary Development

alveolus ....sac; *alveoli*
baro- ....pressure; *baroreceptors*
capillaris ....hairlike; *capillary*
manometer ....device for measuring pressure;
                                  sphygmomanometer
porta ....gate; *portal vein*
pulmo- ....lung; *pulmonary*
pulsus ....stroke; *pulse*
saphenes ....prominent; *saphenous vein*
skleros ....hard; *arteriosclerosis*
sphygmos ....pulse; *sphygmomanometer*
vaso- ....vessel; *vasoconstriction*

*W*HEN WE THINK of the cardiovascular system, we probably think first of the heart
or of the great blood vessels that
leave it and return to it. But the real
work of the cardiovascular system is
done here, in microscopic vessels that
permeate most tissues. This is a network of capillaries, the delicate vessels
that permit diffusion between the
blood and the interstitial fluid.

13 BLOOD VESSELS AND CIRCULATION

The Anatomy of Blood Vessels • Circulatory Physiology • Cardiovascular Regulation • Patterns of Cardiovascular Response • The Blood Vessels

THE LAST TWO CHAPTERS examined the composition of blood and the structure and function of the heart, whose pumping action keeps blood in motion. We will now consider the vessels that carry blood to peripheral tissues and the nature of the exchange that occurs between the blood and interstitial fluids of the body.

Blood leaves the heart in the pulmonary trunk and aorta, each with a diameter of around 2.5 cm (1 in.). These vessels branch repeatedly, forming the major **arteries** that distribute blood to body organs. Within these organs further branching occurs, creating several hundred million tiny arteries, or **arterioles** (ar-TĒ-rē-ōlz). The arterioles provide blood to more than 10 billion **capillaries**. These capillaries, barely the diameter of a single red blood cell, form extensive, branching networks. If all of the capillaries in an average adult's body were placed end to end, they would circle the globe in an unbroken chain 25,000 miles long.

The vital functions of the cardiovascular system occur at the capillary level: *All chemical and gaseous exchange between the blood and interstitial fluid takes place across capillary walls.* Tissue cells rely on capillary diffusion to obtain nutrients and oxygen and to remove metabolic wastes, such as carbon dioxide and urea.

Blood flowing out of the capillary complex first enters the **venules** (VEN-ūlz), the smallest vessels of the venous system. These slender vessels subsequently merge to form small **veins**. Blood then passes through medium-sized and large veins before reaching the venae cavae (in the systemic circuit) or the pul-monary veins (in the pulmonary circuit) (see Figure 12-1●, p. 368).

This chapter examines the structure of the arteries, veins, and capillaries. We then consider their functions and the basic principles of cardiovascular regulation. The final section of the chapter examines the distribution of major blood vessels of the body.

The Anatomy of Blood Vessels

THE STRUCTURE OF VESSEL WALLS

The walls of arteries and veins contain three distinct layers (Figure 13-1●):

1. The **tunica interna** (in-TER-na), or *tunica intima*, is the innermost layer of a blood vessel. It includes the endothelial lining of the vessel and an underlying layer of connective tissue dominated by elastic fibers.

2. The **tunica media**, the middle layer, contains smooth muscle tissue in a framework of collagen and elastic fibers. When these smooth muscles contract, the vessel decreases in diameter, and when they relax, the diameter increases.

3. The outer **tunica externa** (eks-TER-na), or *tunica adventitia* (ad-ven-TISH-ē-uh), forms a sheath of connective tissue around the vessel. Its collagen fibers may intertwine with those of adjacent tissues, stabilizing and anchoring the blood vessel.

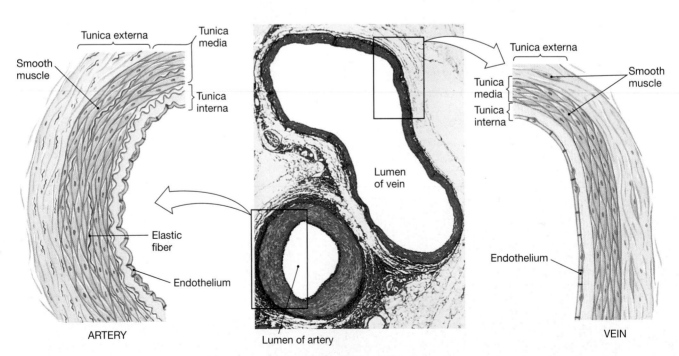

● *Figure 13-1* **A Comparison of a Typical Artery and a Typical Vein** (LM × 74)

Arteries and veins often lie side by side in a narrow band of connective tissue, as in Figure 13-1●. This sectional view clearly shows the greater wall thickness characteristic of arteries. The thicker tunica media of an artery contains more elastic fibers and smooth muscle than does that of a vein. The elastic components permit arterial vessels to resist the pressure generated by the heart as it forces blood into the arterial network. The smooth muscle provides a means of actively controlling changes in vessel diameter. Arterial smooth muscle is under the control of the sympathetic division of the autonomic nervous system. When stimulated, these muscles in the vessel wall contract and the artery constricts, in a process called **vasoconstriction**. Relaxation increases the diameter of the artery and its central opening, or *lumen*, in a process called **vasodilation**.

ARTERIES

In traveling from the heart to the capillaries, blood passes through elastic arteries, muscular arteries, and arterioles. **Elastic arteries** are large, extremely resilient vessels with diameters of up to 2.5 cm (1 in.) (Figure 13-2●). Some examples are the pulmonary trunk and aorta and their major arterial branches. The

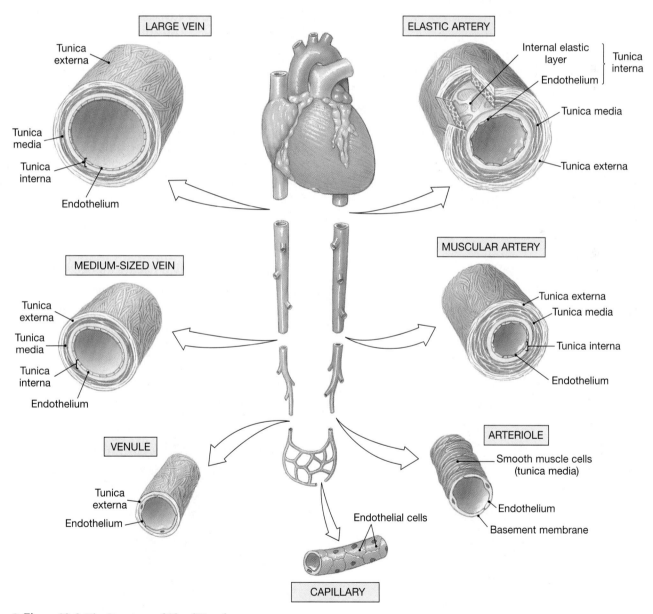

● *Figure 13-2* **The Structure of Blood Vessels**
The walls of arteries and veins are made up of three layers containing different tissues. The inner layer, or tunica interna, includes an endothelium, basement membrane, and underlying connective tissue layer with elastic fibers; the middle layer, or tunica media, contains sheets of smooth muscle fibers within loose connective tissue; the outer layer, or tunica externa, forms a connective tissue covering around the vessel. The wall of a capillary is only one cell thick; it consists of an endothelium and basement membrane.

13 BLOOD VESSELS AND CIRCULATION

The Anatomy of Blood Vessels • Circulatory Physiology • Cardiovascular Regulation • Patterns of Cardiovascular Response • The Blood Vessels

walls of elastic arteries contain a tunica media dominated by elastic fibers rather than smooth muscle cells. As a result, elastic arteries are able to absorb the pressure changes that occur during the cardiac cycle. During ventricular systole, blood pressure rises quickly as additional blood is pushed into the systemic circuit. Over this period, the elastic arteries are stretched, and their diameter increases. During ventricular diastole, arterial blood pressure declines, and the elastic fibers recoil to their original dimensions. The net result is that the expansion of the arteries cushions the rise in pressure during ventricular systole, and the following arterial recoil slows the decline in pressure during ventricular diastole. If the arteries were solid pipes rather than elastic tubes, pressures would rise much higher during systole and would fall much lower during diastole.

Muscular arteries, also known as *medium-sized arteries* or *distribution arteries*, distribute blood to skeletal muscles and internal organs. A typical muscular artery has a diameter of approximately 0.4 cm (0.15 in.). The carotid artery of the neck is one example. The thick tunica media in a muscular artery contains more smooth muscle and fewer elastic fibers than does an elastic artery (Figure 13-2●).

Arterioles, with an internal diameter of about 30 μm, are much smaller than muscular arteries. The tunica media of an arteriole consists of one to two layers of smooth muscle cells. These muscle layers enable muscular arteries and arterioles to change their diameter, thereby altering the blood pressure and the rate of flow through the dependent tissues.

CAPILLARIES

Capillaries are the only blood vessels whose walls permit exchange between the blood and the surrounding interstitial fluid. Because the walls are relatively thin, the diffusion distances are small and exchange can occur quickly. In addition, blood flows through capillaries relatively slowly, allowing sufficient time for the diffusion or active transport of materials across the capillary walls.

A typical capillary consists of a single layer of endothelial cells inside a basement membrane (Figure 13-2●). Neither a tunica externa nor a tunica media is present. The average diameter of a capillary is a 8 μm; very close to that of a red blood cell. In most regions of the body, the endothelium forms a complete lining, and most substances enter or leave the capillary by diffusing through gaps between adjacent endothelial cells. In some areas (notably, the choroid plexus of the brain, the hypothalamus, and filtration sites at the kidneys), small pores in the endothelial cells permit the passage of relatively large molecules, including proteins.

Capillary Beds

Capillaries function as part of an interconnected network called a **capillary bed**, simplified in Figure 13-3a●. On reaching its target area, a single arteriole usually gives rise to dozens of capillaries, which will in turn collect into several *venules*, the smallest vessels of the venous system. The entrance to each capillary is guarded by a

● *Figure 13-3* **The Organization of a Capillary Bed**
(a) Basic features of a typical capillary bed. (b) A micrograph of a capillary network.

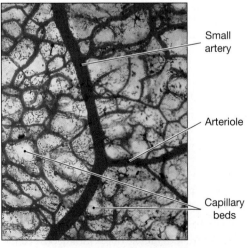

Collateral arteries

Vein

Venule

Arteriole

Capillaries

Small venule

Arteriovenous anastamosis

Precapillary sphincters

Small artery

Arteriole

Capillary beds

(a)

(b)

precapillary sphincter, a band of smooth muscle. Contraction of the smooth muscle fibers narrows the diameter of the capillary entrance and reduces the flow of blood. The relaxation of the sphincter dilates the opening, allowing blood to enter the capillary more rapidly.

Although blood usually flows from the arterioles to the venules at a constant rate, the blood flow within a single capillary can be quite variable. Each precapillary sphincter goes through cycles of activity, alternately contracting and relaxing perhaps a dozen times each minute. As a result of this cyclical change, called **vasomotion** (*vaso-*, vessel), the blood flow within any one capillary occurs in a series of pulses rather than as a steady and constant stream. The net effect is that blood may reach the venules by one route now and by a quite different route later. This process is controlled at the tissue level, as smooth muscle fibers respond to local changes in the concentrations of chemicals and dissolved gases in the interstitial fluid. For example, when dissolved oxygen levels decline within a tissue, the capillary sphincters relax and blood flow to the area increases. Such control at the tissue level is called *autoregulation*.

Under certain conditions, the blood will completely bypass the capillary bed through an **arteriovenous anastomosis** (a-nas-tō-MŌ-sis; "outlet"; plural, *anastomoses*), a vessel that connects an arteriole to a venule (Figure 13-3a●). An anastomosis may also interconnect two arterioles; such a vessel is called an **arterial anastomosis**. An arterial anastomosis provides a second blood supply for capillary beds. This anastomosis thus acts like an insurance policy: If one artery is compressed or blocked, the other can continue to deliver blood to the capillary bed, and the dependent tissues will not be damaged. Arterial anastomoses occur in the brain and in the coronary circulation, as well as many other sites. p. 374

VEINS

Veins collect blood from all tissues and organs and return it to the heart. Veins are classified on the basis of their internal diameters. The smallest, the venules, resemble expanded capillaries, and venules with diameters smaller than 50 µm lack a tunica media altogether. **Medium-sized veins** range from 2 to 9 mm in diameter, comparable in size to muscular arteries (Figure 13-2d●). In these veins, the tunica media contains several smooth muscle layers, and the relatively thick tunica externa has longitudinal bundles of elastic and collagen fibers. **Large veins** include the two venae cavae and their tributaries in the abdominopelvic and thoracic cavities. In these vessels, the thin tunica media is surrounded by a thick tunica externa composed of elastic and collagenous fibers.

Veins have relatively thin walls because they do not have to withstand much pressure. In venules and medium-sized veins, the pressure is so low that it cannot oppose the force of gravity. In the limbs, medium-sized veins contain **valves**, folds of endothelium, that act like the valves in the heart, preventing the backflow

---

✚ CLINICAL NOTE *Arteriosclerosis*

Arteriosclerosis (ar-tē-rē-ō-skle-RŌ-sis; *skleros*, hard) is a thickening and toughening of arterial walls. Complications related to arteriosclerosis account for roughly half of all deaths in the United States. The effects of arteriosclerosis are varied; for example, arteriosclerosis of coronary vessels is responsible for *coronary artery disease (CAD)*, and arteriosclerosis of arteries supplying the brain can lead to strokes. p. 375 There are two major forms of arteriosclerosis, (1) *focal calcification* and (2) *atherosclerosis*. In **focal calcification**, degenerating smooth muscle in the tunica media is replaced with calcium deposits. It occurs as part of the aging process, in association with atherosclerosis, and as a complication of diabetes.

Atherosclerosis (ath-er-ō-skle-RŌ-sis; *athero-*, a pasty deposit) is associated with damage to the endothelial lining and the formation of lipid deposits in the tunica media. Many factors may be involved in the development of atherosclerosis. One major factor is lipid levels in the blood. Atherosclerosis tends to develop in persons whose blood contains elevated levels of plasma lipids, specifically cholesterol. Circulating cholesterol is transported to peripheral tissues in lipoproteins, protein-lipid complexes. (The types of lipoproteins and their interrelationships are discussed in Chapter 17.) When cholesterol-rich lipoproteins remain in circulation for an extended period, circulating monocytes then begin removing them from the bloodstream. Eventually the monocytes become filled with lipid droplets, and they attach themselves to the endothelial walls of blood vessels. These cells then release growth factors that stimulate the divisions of smooth muscle cells near the tunica interna. The vessel wall then thickens and stiffens.

Other monocytes then invade the area, migrating between the endothelial cells. As these changes occur, the monocytes, smooth muscle fibers, and endothelial cells begin phagocytizing lipids as well. The result is a **plaque**, a fatty mass of tissue that projects into the middle of the vessel. At this point, the plaque has a relatively simple structure, and evidence indicates that the process can be reversed if appropriate dietary adjustments are made.

If the conditions persist, the endothelial cells become swollen with lipids and gaps appear in the endothelial lining. Platelets now begin sticking to the exposed collagen fibers, and the combination of platelet adhesion and aggregation leads to the formation of a localized blood clot that will further restrict blood flow through the artery. The structure of the plaque is now relatively complex; plaque growth can be halted, but the structural changes are permanent.

Elderly individuals, especially elderly men, are most likely to develop atherosclerotic plaques. In addition to age and gender, other important risk factors include high blood cholesterol levels, high blood pressure, and cigarette smoking. Other factors that can promote development of atherosclerosis in both men and women include diabetes mellitus, obesity, and stress.

13 **BLOOD VESSELS AND CIRCULATION**

The Anatomy of Blood Vessels • **Circulatory Physiology** • Cardiovascular Regulation • Patterns of Cardiovascular Response • The Blood Vessels

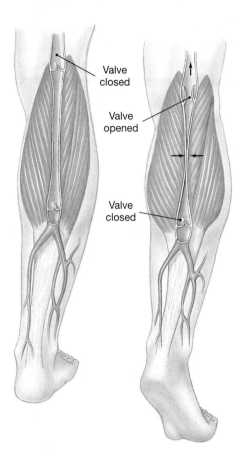

● *Figure 13-4* **The Function of Valves in the Venous System**
Valves in the walls of medium-sized veins prevent the backflow of
blood. Venous compression caused by the contraction of adjacent
skeletal muscles helps maintain venous blood flow.

of blood (Figure 13-4●). As long as the valves function normally,
any movement that compresses a vein will push blood toward the
heart, improving *venous return*. ∞ p. 382 If the walls of the veins
near the valves weaken or become stretched and distorted, the
valves may not work properly. Blood then pools in the veins, and
the vessels become distended. The effects range from mild discomfort and a cosmetic problem, as in superficial *varicose veins* in
the thighs and legs, to painful distortion of adjacent tissues, as in
hemorrhoids, swollen veins in the lining of the anal canal.

CONCEPT CHECK QUESTIONS

Answers on page 425

❶ A cross section of tissue shows several small, thin-walled vessels
with very little smooth muscle tissue in the tunica media. Which
type of vessels are these?

❷ How would the relaxation of precapillary sphincters affect the
blood flow through a tissue?

❸ Why are valves found in veins but not in arteries?

Circulatory Physiology

The components of the cardiovascular system (the blood, heart,
and blood vessels) function together to maintain an adequate
blood flow through all the tissues and organs of the body.
Under normal circumstances, blood flow is equal to cardiac
output. When cardiac output goes up, so does capillary blood
flow; when cardiac output declines, blood flow is reduced. Two
factors—*pressure* and *resistance*—affect the flow rates of blood
through the capillaries.

PRESSURE

Liquids, including blood, cannot be compressed. As a result, a
force exerted against a liquid generates a fluid, or hydrostatic,
pressure that is conducted in all directions. If there is a pressure
difference, a liquid will flow from an area of higher pressure
toward an area of relatively lower pressure. The flow rate is directly proportional to the pressure difference: The greater the difference in pressure, the faster the flow. In the systemic circuit, the
overall pressure difference is measured between the base of the
aorta (as blood leaves the left ventricle) and the entrance to
the right atrium (as it returns). This pressure difference, called the
circulatory pressure, averages around 100 mm Hg (mm Hg = millimeters of mercury, a standardized unit of pressure). This relatively high circulatory pressure is needed primarily to force blood
through the arterioles and into the capillaries.

Circulatory pressure is divided into three components:
(1) *arterial pressure*, (2) *capillary pressure*, and (3) *venous pressure*. Arterial pressure will be referred to as **blood pressure** to
distinguish it from the total circulatory pressure. Capillary
pressure is the pressure within the capillary beds. Venous pressure is the pressure within the venous system.

RESISTANCE

A *resistance* is any force that opposes movement. The resistance of the cardiovascular system opposes the movement of
blood. For blood to flow, the circulatory pressure must be great
enough to overcome the *total peripheral resistance*, the resistance of the entire cardiovascular system. Because the resistance of the venous system is very low, attention focuses on
the **peripheral resistance**, the resistance of the arterial system.

Neural and hormonal control mechanisms regulate blood
pressure, keeping it relatively stable. Adjustments in the peripheral resistance of vessels supplying specific organs allow the
rate of blood flow to be precisely controlled. For example,
Chapter 7 discussed the increase in blood flow to skeletal muscles during exercise. ∞ p. 217 That increase occurs because
there is a drop in the peripheral resistance of the arteries supplying active muscles.

Sources of peripheral resistance include *vascular resistance*, *viscosity*, and *turbulence*. Only vascular resistance can be adjusted by the nervous or endocrine system to regulate blood flow. Viscosity and turbulence, which affect peripheral resistance, are normally constant.

Vascular Resistance

Vascular resistance, the resistance of the blood vessels, is the largest component of peripheral resistance. *The most important factor in vascular resistance is friction between the blood and the vessel walls.* The amount of friction depends on the length of the vessel and its diameter. Vessel length cannot be changed, so vascular resistance is controlled by changing the diameter of blood vessels by contracting or relaxing smooth muscle in the vessel walls.

Most of the vascular resistance occurs in the arterioles, the smallest of the arterial vessels. Arterioles are extremely muscular; for example, the walls of an arteriole with a 30 μm internal diameter can have a 20 μm thick layer of smooth muscle. Local, neural, and hormonal stimuli that stimulate or inhibit the arteriolar smooth muscle tissue can adjust the diameters of these vessels, and a small change in diameter can produce a very large change in resistance.

Viscosity

Viscosity is the resistance to flow caused by interactions among molecules and suspended materials in a liquid. Liquids of low viscosity, such as water, flow even at low pressures, whereas thick, syrupy liquids such as molasses flow only under relatively high pressures. Whole blood has a viscosity about five times that of water, due to the presence of plasma proteins and suspended blood cells. Under normal conditions, the viscosity of the blood remains stable. But disorders that affect the hematocrit or the plasma protein content can change blood viscosity and increase or decrease peripheral resistance. For example, in **anemia**, the hematocrit is reduced due to inadequate production of hemoglobin, RBCs, or both. As a result, both the oxygen-carrying capacity of the blood and blood viscosity are reduced. A reduction in blood viscosity can also result from protein deficiency diseases, in which the liver cannot synthesize normal amounts of plasma proteins.

Turbulence

Blood flow through a vessel is usually smooth, with the slowest flow near the walls and the fastest flow at the center of the vessel. High flow rates, irregular surfaces caused by injury or disease processes, or sudden changes in vessel diameter upset this smooth flow, creating eddies and swirls. This phenomenon, called *turbulence*, slows the flow and increases resistance.

Turbulence normally occurs when blood flows between the heart's chambers and from the heart into the aorta and pulmonary trunk. In addition to increasing resistance, this turbulence generates the *third* and *fourth heart sounds* often heard through a stethoscope. Turbulent blood flow across damaged or misaligned heart valves produces the sound of *heart murmurs*. ∞ p. 373

CIRCULATORY PRESSURE

Blood pressure varies from one vessel to another within the systemic circuit. Systemic pressures are highest in the aorta, peaking at around 120 mm Hg, and lowest at the venae cavae, averaging about 2 mm Hg.

Arterial Blood Pressure

Figure 13-5● graphs the blood pressure throughout the cardiovascular system. It shows that blood pressure in large and small arteries rises and falls, rising during ventricular systole and falling during ventricular diastole. **Systolic pressure** is the peak blood pressure measured during ventricular systole, and **diastolic pressure** is the minimum blood pressure at the end of ventricular diastole. In recording blood pressure, we separate systolic and diastolic pressures by a slash mark, as in "120/80." A *pulse* is a rhythmic pressure oscillation that accompanies each heartbeat. The difference between the systolic and diastolic pressures is the **pulse pressure** (*pulsus*, stroke).

Pulse pressure lessens as the distance from the heart increases. As already discussed, the average pressure declines as a result of friction between the blood and the vessel walls. The pulse pressure fades because arteries are elastic tubes rather than

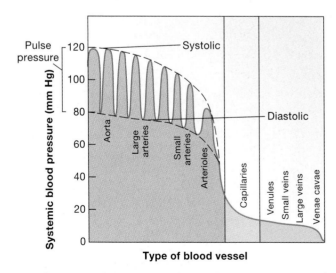

● *Figure 13-5* **Pressures in the Circulatory System**
Notice the general reduction of circulatory pressures within the systemic circuit and the elimination of the pulse pressure in the arterioles, capillaries, and veins.

13 BLOOD VESSELS AND CIRCULATION

The Anatomy of Blood Vessels • Circulatory Physiology • Cardiovascular Regulation • Patterns of Cardiovascular Response • The Blood Vessels

solid pipes. Much like a puff of air expands a partially inflated balloon, the elasticity of the arteries allows them to expand with blood during systole. When diastole begins and blood pressures fall, the arteries recoil to their original dimensions. Because the aortic semilunar valve prevents the return of blood to the heart, the arterial recoil adds an extra push to the flow of blood. The magnitude of this phenomenon, called *elastic rebound*, is greatest near the heart and drops in succeeding arterial sections. By the time blood reaches a precapillary sphincter, there are no pressure oscillations and the blood pressure remains steady at about 35 mm Hg. Along the length of a typical capillary, blood pressure gradually falls from about 35 mm Hg to roughly 18 mm Hg, the pressure at the start of the venous system.

Capillary Pressures and Capillary Exchange

The blood pressure within a capillary bed, or **capillary pressure**, pushes against the capillary walls, just as it does in the arteries. But unlike other blood vessels, capillary walls are quite permeable to small ions, nutrients, organic wastes, dissolved gases, and water. Most of these materials are reabsorbed by the capillaries, but about 3.6 liters (.95 gallons) of water and solutes flow through peripheral tissues each day and enter the **lymphatic vessels** of the lymphatic system, which empty into

the bloodstream (Figure 13-6●). This continuous movement and exchange of water and solutes from the capillaries through the body tissues and back into the bloodstream plays an important role in homeostasis. Capillary exchange has four important functions: (1) maintains constant communication between plasma and interstitial fluid; (2) speeds the distribution of nutrients, hormones, and dissolved gases throughout tissues; (3) assists the movement of insoluble lipids and tissue proteins that cannot cross capillary walls; and (4) flushes bacterial toxins and other chemical stimuli to lymphoid tissues and organs that provide immunity from disease.

Dynamics of Capillary Exchange. The movement of materials across capillary walls occurs by diffusion, filtration, and osmosis. Because Chapter 3 described these processes, only a brief overview is provided here. ⌒ p. 58 Solute molecules tend to diffuse across the capillary lining, driven by their individual concentration gradients. Water-soluble materials, including ions and small organic molecules such as glucose, amino acids, or urea, diffuse through small spaces between adjacent endothelial cells. Larger water-soluble molecules, such as plasma proteins, cannot normally leave the bloodstream. Lipid-soluble materials, including steroids, fatty acids, and dissolved gases, diffuse across the endothelial lining, passing through the membrane lipids.

Water molecules will move when driven by either hydrostatic (fluid) pressure or osmotic pressure. ⌒ pp. 60–61 Hydrostatic pressure is a physical force that pushes water molecules from an area of high pressure to an area of lower pressure. At a capillary, the hydrostatic pressure, or *capillary blood pressure (CBP)*, is greatest at the arteriolar end and least where it empties into a venule. The tendency for water and solutes to move out of the blood is therefore greatest at the start of a capillary, where the capillary pressure is highest, and declines along the length of the capillary as capillary pressure falls.

Osmosis, in contrast, is the movement of water across a selectively permeable membrane separating two solutions of different solute concentrations. Water will move into the solution with the higher solute concentration, and the force of this water movement is called osmotic pressure. Because blood contains more dis-

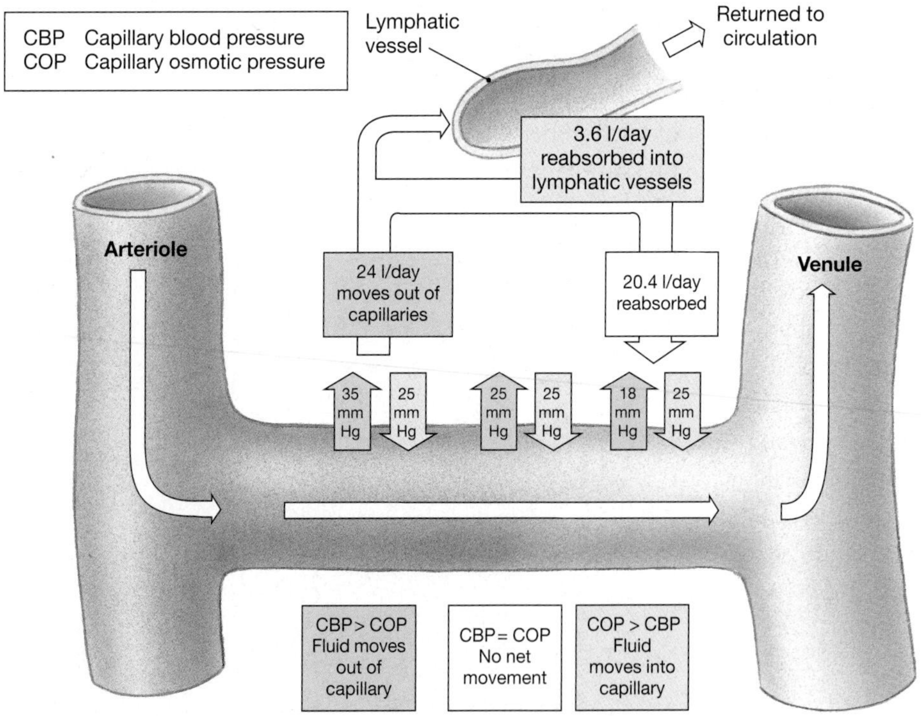

● *Figure 13-6* **Forces Acting Across Capillary Walls**
At the arterial end of the capillary, blood hydrostatic pressure (CBP) is stronger than capillary osmotic pressure (COP), and fluid moves out of the capillary. Near the venule, CBP is lower than COP, and fluid moves into the capillary.

solved proteins than does the interstitial fluid, water tends to move from the interstitial fluid into the blood. ∞ p. 60 Thus, capillary blood pressure (CBP) tends to push water out of the capillary, while capillary osmotic pressure (COP) forces tends to pull it back in (Figure 13-6●).

CAPILLARY DYNAMICS AND BLOOD VOLUME

Any condition that affects hydrostatic or osmotic pressures in the blood or tissues will shift the direction of fluid movement. For example, consider what happens at the capillary level to an accident victim who has lost blood through hemorrhaging or a person lost in the desert, feeling the effects of dehydration. In both cases, the decrease in blood volume causes a drop in blood pressure. In the second case, however, the loss in blood volume is also accompanied by a rise in the blood osmotic pressure, because as water is lost, the blood becomes more concentrated. In either case, there is a net movement of water from the interstitial fluid into the bloodstream, and the blood volume will increase. This process is known as a *recall of fluids*.

Conversely, **edema** (e-DĒ-ma) is an abnormal accumulation of interstitial fluid in the tissues. Although it has many causes, the underlying cause is a disturbance in the normal balance between hydrostatic and osmotic forces at the capillary level. For example, a localized edema often occurs around a bruise. The fluid shift occurs because damaged capillaries at the injury site allow plasma proteins into the interstitial fluid. This leakage decreases the osmotic pressure of the blood and elevates that of the tissues. More water then moves into the tissue, resulting in edema. In the U.S. population, serious cases of edema most often result from an increase in the blood pressure in the arterial system, the venous system, or both. This condition often occurs during **congestive heart failure (CHF)**.

Heart failure occurs when the cardiac output is insufficient to meet the circulatory demand of the body. In congestive heart failure, the left ventricle can no longer keep up with the right ventricle and blood flow becomes congested, or backed up, in the pulmonary circuit. This makes the right ventricle work harder, further elevating pulmonary arterial pressures and forcing blood through the lungs and into the weakened left ventricle. The increased blood pressure in the pulmonary vessels leads to *pulmonary edema*, a buildup of fluids in the lungs.

Venous Pressure

Although blood pressure at the start of the venous system is only about one-tenth that at the start of the arterial system, the blood must still travel through a vascular network as complex as the arterial system before returning to the heart. However, venous pressures are low, and the veins offer little resistance. As a result, once blood enters the venous system, pressure declines very slowly.

As blood travels through the venous system toward the heart, the veins become larger, resistance drops further, and the flow rate increases. Pressures at the entrance to the right atrium fluctuate, but they average around 2 mm Hg. This means that the driving force pushing blood through the venous system is a mere 16 mm Hg (18 mm Hg in the venules −2 mm Hg in the venae cavae = 16 mm Hg) as compared to the 65 mm Hg pressure acting along the arterial system (100 mm Hg at the aorta −35 mm Hg at the capillaries = 65 mm Hg).

When you are lying down, a 16 mm Hg pressure gradient is sufficient to maintain venous flow. But when you are standing, the venous blood from the body below the heart must overcome gravity as it ascends within the inferior vena cava. Two factors help overcome gravity and propel venous blood toward the heart:

1. **Muscular compression.** The contractions of skeletal muscles near a vein compress it, helping push blood toward the heart. The valves in medium-sized veins ensure that blood flow occurs in one direction only (see Figure 13-4●, p. 394).
2. The **respiratory pump.** As you inhale, decreased pressure in the thoracic cavity draws air into the lungs. This drop in pressure also pulls blood into the venae cavae and atria, increasing venous return. On exhalation, the increased pressure that forces air out of the lungs compresses the venae cavae, pushing blood into the right atrium.

During exercise, both factors cooperate to increase venous return and push cardiac output to maximal levels. However, when an individual stands at attention, with knees locked and leg muscles immobile, these mechanisms are impaired and the reduction in venous return leads to a fall in cardiac output. The blood supply to the brain is in turn reduced, sometimes enough to cause *fainting*, a temporary loss of consciousness. The person then collapses; in the horizontal position, both venous return and cardiac output would return to normal.

CONCEPT CHECK QUESTIONS
Answers on page 425

❶ In a normal individual, where would you expect the blood pressure to be greater, in the aorta or in the inferior vena cava? Explain.

❷ While standing in the hot sun, Sally begins to feel light-headed and faints. Explain.

---

13 **BLOOD VESSELS AND CIRCULATION**

The Anatomy of Blood Vessels • Circulatory Physiology • **Cardiovascular Regulation** • Patterns of Cardiovascular Response • The Blood Vessels

✚ **CLINICAL NOTE** *Checking the Pulse and Blood Pressure*

The pulse can be felt within any of the large or medium-sized arteries. The usual procedure involves squeezing an artery with the fingertips against a relatively solid mass, preferably a bone. When the vessel is compressed, the pulse is felt as a pressure against the fingertips.

Figure 13-7a● indicates the locations used to check the pulse. The inside of the wrist is often used because the *radial artery* can easily be pressed against the distal portion of the radius. Other accessible arteries include the *temporal, facial, carotid, brachial, femoral, popliteal, posterior tibial,* and *dorsalis pedis arteries.* Firm pressure exerted at an artery near the base of a limb can reduce or eliminate

arterial bleeding in more distal portions of the limb. These sites are called *pressure points.*

Blood pressure is determined with a *sphygmomanometer* (sfig-mō-ma-NOM-e-ter; *sphygmos,* pulse + *manometer,* device for measuring pressure), as shown in Figure 13-7b●. An inflatable cuff is placed around the arm so the inflation of the cuff squeezes the brachial artery. A stethoscope is placed over the artery distal to the cuff, and the cuff is then inflated. A tube connects the cuff to a pressure gauge that measures the pressure inside the cuff in millimeters of mercury (mm Hg). Inflation continues until cuff pressure is roughly 30 mm Hg above the pressure sufficient to collapse the brachial artery, stop the flow of blood, and eliminate the sound of the pulse.

The investigator then slowly lets the air out of the cuff. When the pressure in the cuff falls below systolic pressure, blood can again enter the artery. At first, blood enters only at peak systolic pressures, and the stethoscope picks up the sound of blood pulsing through the artery. As the pressure falls further, the sound changes because the vessel is remaining open for longer and longer periods. When the cuff pressure falls below diastolic pressure, blood flow becomes continuous and the sound of the pulse becomes muffled or disappears completely. Thus the pressure at which the pulse appears corresponds to the peak systolic pressure; when the pulse fades, the pressure has reached diastolic levels. The distinctive sounds heard during this test are called **sounds of Korotkoff** (sometimes spelled *Korotkov* or *Korotkow*). When the blood pressure is recorded, systolic and diastolic pressures are usually separated by a slashmark, as in "120/80" ("one twenty over eighty") or "110/75." A reading of 120/80 would give a pulse pressure of 40 (mm Hg).

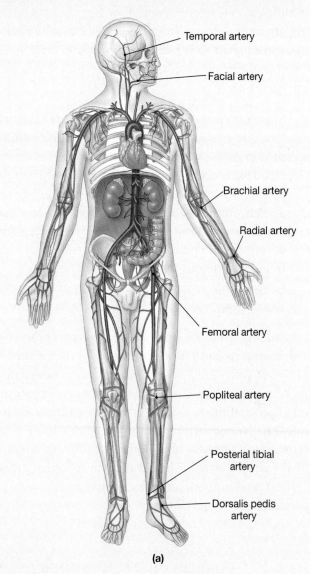

(a)

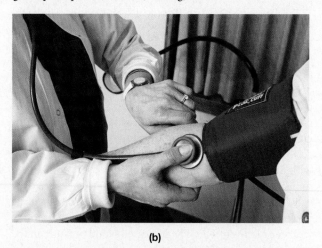

(b)

● *Figure 13-7* **Checking the Pulse and Blood Pressure**
(a) Pressure points used to check the presence and strength of the pulse. (b) The use of a sphygmomanometer to check arterial blood pressure.

Cardiovascular Regulation

Homeostatic mechanisms regulate cardiovascular activity to ensure that tissue blood flow, also called *tissue perfusion,* meets the demand for oxygen and nutrients. The three variable factors that influence tissue blood flow are (1) cardiac output,

(2) peripheral resistance, and (3) blood pressure. We discussed cardiac output in Chapter 12 (∞ p. 381) and considered peripheral resistance and blood pressure earlier in this chapter.

Most cells are relatively close to capillaries. When a group of cells becomes active, the circulation to that region must increase to deliver the necessary oxygen and nutrients and to carry away the waste products and carbon dioxide that they generate. The goal of cardiovascular regulation is to ensure that these blood flow changes occur (1) at an appropriate time, (2) in the right area, and (3) without drastically altering blood pressure and blood flow to vital organs.

Factors involved in the regulation of cardiovascular function include (Figure 13-8●):

- *Local factors.* Local factors change the pattern of blood flow within capillary beds in response to chemical changes in the interstitial fluids. This is an example of *autoregulation* at the tissue level. If autoregulation is unable to normalize tissue conditions, neural mechanisms and endocrine factors are activated.
- *Neural mechanisms.* Neural control mechanisms respond to changes in arterial pressure or blood gas levels at specific sites. When those changes occur, the autonomic nervous system adjusts cardiac output and peripheral resistance to maintain adequate blood flow.
- *Endocrine factors.* The endocrine system releases hormones that enhance short-term adjustments and direct long-term changes in cardiovascular performance.

Short-term responses adjust cardiac output and peripheral resistance to stabilize blood pressure and tissue blood flow. Long-term adjustments involve alterations in blood volume that affect cardiac output and the transport of oxygen and carbon dioxide to and from active tissues.

THE AUTOREGULATION OF BLOOD FLOW

Under normal resting conditions, cardiac output remains stable, and peripheral resistance in individual tissues is adjusted to control local blood flow. When a precapillary sphincter constricts, blood flow decreases; when it relaxes, blood flow increases. Precapillary sphincters respond automatically to alterations in the local environment such as increased or decreased levels of oxygen and carbon dioxide. For example, when oxygen is abundant, the smooth muscles contract and slow down the flow of blood. Oxygen levels decline in active tissues as the cells absorb O_2 for use in aerobic respiration. As tissue oxygen supplies dwindle, carbon dioxide levels rise, and the pH falls. This combination causes the smooth muscle cells in the precapillary sphincter to relax and blood flow to increase. The appearance of specific chemicals in the interstitial fluids can produce the same effect. For example, during inflammation, vasodilation occurs at an injury site because histamine, bacterial toxins, and prostaglandins cause the relaxation of the precapillary sphincters.

Factors that promote the dilation of precapillary sphincters are called *vasodilators*, and those that stimulate the constriction of precapillary sphincters are called *vasoconstrictors*.

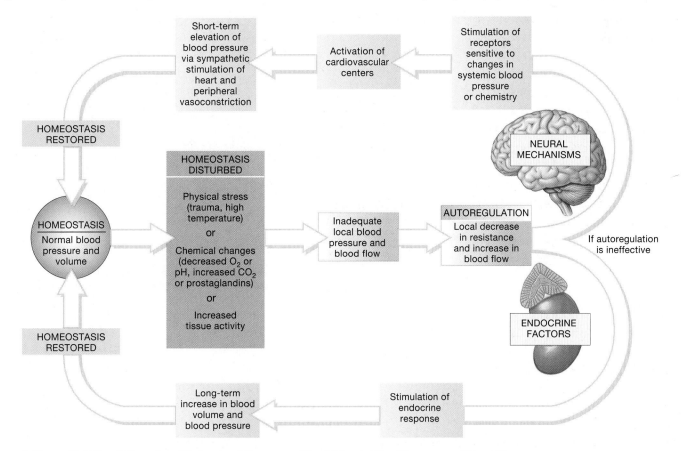

● *Figure 13-8* **Local, Neural, and Endocrine Adjustments That Maintain Blood Pressure and Blood Flow**

13 **BLOOD VESSELS AND CIRCULATION**

The Anatomy of Blood Vessels • Circulatory Physiology • **Cardiovascular Regulation** • Patterns of Cardiovascular Response • The Blood Vessels

Together, such factors control blood flow in a single capillary bed. When present in high concentrations, these factors also affect arterioles, increasing or decreasing blood flow to all of the capillary beds in a given region. Such an event will often trigger a neural response, because significant changes in blood flow to one region of the body will have an immediate effect on circulation to other regions.

The Neural Control of Blood Pressure and Blood Flow

The nervous system adjusts cardiac output and peripheral resistance to maintain adequate blood flow to vital tissues and organs. The *cardiac centers* and *vasomotor centers* of the medulla oblongata are the *cardiovascular (CV) centers* responsible for these regulatory activities. As noted in Chapter 12, each cardiac center includes a *cardioacceleratory center*, which increases cardiac output through sympathetic innervation, and a *cardioinhibitory center*, which reduces cardiac output through parasympathetic innervation. ∞ p. 383

The vasomotor center of the medulla oblongata primarily controls the diameters of the arterioles. Inhibition of the vasomotor center leads to vasodilation, a dilation of arterioles that reduces peripheral resistance. The stimulation of the vasomotor center causes vasoconstriction (the constriction of peripheral arterioles) and, with very strong stimulation, **venoconstriction** (the constriction of peripheral veins), both of which increase peripheral resistance.

The cardiovascular centers detect changes in tissue demand by monitoring arterial blood, especially blood pressure, pH, and dissolved gas concentrations. The *baroreceptor reflexes* respond to changes in blood pressure, and the *chemoreceptor reflexes* respond to changes in chemical composition.

Baroreceptor Reflexes

Baroreceptors monitor the degree of stretch in the walls of expandable organs. ∞ p. 281 The baroreceptors involved with cardiovascular regulation are located in (1) the walls of the **carotid sinuses**, expanded chambers near the bases of the internal carotid arteries of the neck (see Figure 13-17a●, p. 410); (2) the **aortic sinuses**, pockets in the walls of the aorta adjacent to the heart (Figure 12-6b, p. 374); and (3) the wall of the right atrium. These receptors initiate the **baroreceptor reflexes** (*baro-*, pressure), autonomic reflexes that adjust cardiac output and peripheral resistance to maintain normal arterial pressures. Aortic baroreceptors monitor blood pressure within the ascending aorta. There, the aortic reflex adjusts blood pressure in response to changes in pressure. The goal of the aortic reflex is to maintain adequate blood pressure and blood flow through the systemic circuit. Carotid sinus baroreceptors respond to changes in blood pressure at the carotid sinuses. These receptors trigger reflexes that maintain adequate blood flow to the brain. The blood flow to the brain must remain constant, and the carotid sinus reflex is extremely sensitive. The baroreceptor reflexes triggered by changes in blood pressure at these locations are shown in Figure 13-9●.

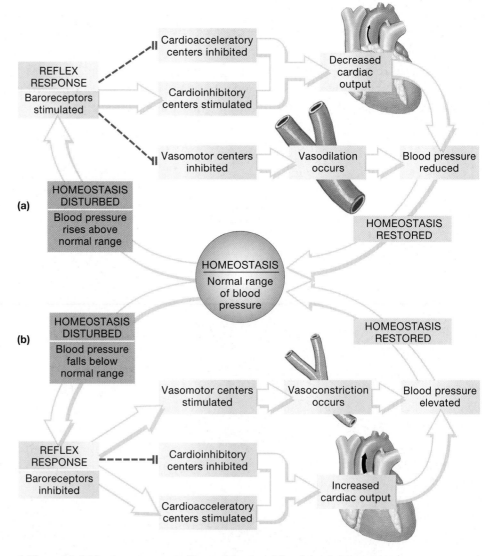

● *Figure 13-9* **The Baroreceptor Reflexes of the Carotid and Aortic Sinuses**

When blood pressure climbs, the increased neural output from the baroreceptors travels to the cardiovascular centers of the medulla oblongata, where it inhibits the cardioacceleratory center, stimulates the cardioinhibitory center, and inhibits the vasomotor center (Figure 13-9a●). The two effects produced are (1) a decrease in cardiac output and (2) widespread peripheral vasodilation. Under the command of the cardioinhibitory center, the vagus nerves release acetylcholine (ACh), which reduces the rate and strength of the cardiac contractions, lowering cardiac output. The inhibition of the vasomotor center leads to the dilation of peripheral arterioles throughout the body. This combination of reduced cardiac output and decreased peripheral resistance then reduces the blood pressure.

This pattern is reversed if blood pressure becomes abnormally low (Figure 13-9b●). In that case, the effects produced by a reduction in baroreceptor output are (1) an increase in cardiac output and (2) widespread peripheral vasoconstriction. Reduced baroreceptor activity stimulates the cardioacceleratory center, inhibits the cardioinhibitory center, and stimulates the vasomotor center. The cardioacceleratory center stimulates sympathetic neurons innervating the sinoatrial (SA) node, atrioventricular (AV) node, and general myocardium. This stimulation increases heart rate and stroke volume, leading to an immediate increase in cardiac output. Vasomotor activity, also carried by sympathetic motor neurons, produces a rapid vasoconstriction, increasing peripheral resistance. These adjustments, increased cardiac output and increased peripheral resistance, work together to elevate blood pressure. Atrial baroreceptors monitor the blood pressure at the end of the systemic circuit—at the venae cavae and the right atrium. The atrial reflex responds to the stretching of the wall of the right atrium. ∞ p. 382 Under normal circumstances, the heart pumps blood into the aorta at the same rate that it is arriving at the right atrium. When blood pressure rises in the atrium, it means that blood is arriving at the heart faster than it is being pumped out. The atrial baroreceptors solve the problem by stimulating the cardioacceleratory center, increasing cardiac output until the

backlog of venous blood is removed. Atrial pressure then returns to normal.

Chemoreceptor Reflexes

The **chemoreceptor reflexes** respond to changes in the carbon dioxide levels, oxygen levels, or pH in the blood and cerebrospinal fluid (Figure 13-10●). The chemoreceptors involved are sensory neurons found in the **carotid bodies**, located in the neck near the carotid sinuses, and the **aortic bodies**, situated near the arch of the aorta. These receptors monitor the composition of the arterial blood. Additional chemoreceptors on the medulla oblongata monitor the composition of the cerebrospinal fluid (CSF).

The activation of chemoreceptors occurs through a drop in pH or in plasma O_2, or a rise in CO_2, levels. Any of these changes leads to a stimulation of the cardioacceleratory and vasomotor centers. This elevates arterial pressure and increases blood flow through peripheral tissues. Chemoreceptor output also affects the respiratory centers in the medulla oblongata. As a result, a rise in blood flow and blood pressure is associated with an elevated respiratory rate. The coordination of cardiovascular and respiratory activity is vital, because accelerating tissue blood flow is useful only if the circulating blood contains adequate oxygen. In addition, a rise in the respiratory

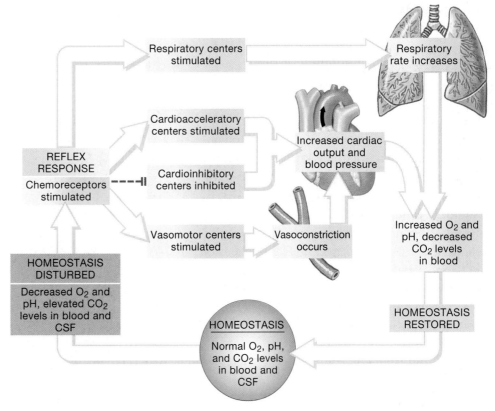

● *Figure 13-10* **The Chemoreceptor Reflexes**

13 BLOOD VESSELS AND CIRCULATION

The Anatomy of Blood Vessels • Circulatory Physiology • Cardiovascular Regulation • **Patterns of Cardiovascular Response** • The Blood Vessels

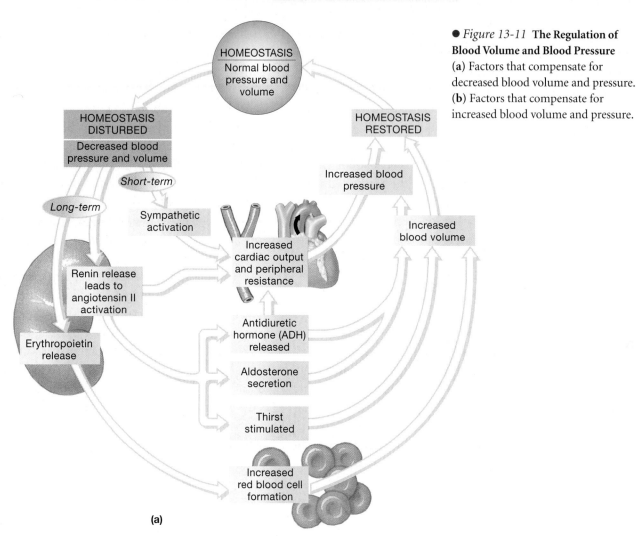

● *Figure 13-11* **The Regulation of Blood Volume and Blood Pressure** (a) Factors that compensate for decreased blood volume and pressure. (b) Factors that compensate for increased blood volume and pressure.

(a)

rate accelerates venous return through the action of the respiratory pump. ∞ p. 397

HORMONES AND CARDIOVASCULAR REGULATION

The endocrine system provides both short-term and long-term regulation of cardiovascular performance. Epinephrine (E) and norepinephrine (NE) from the adrenal medullae act within seconds of their release to stimulate cardiac output and peripheral vasoconstriction. Other regulatory hormones include antidiuretic hormone (ADH), angiotensin II, erythropoietin (EPO), and atrial natriuretic peptide (ANP), introduced in Chapter 10. ∞ pp. 321, 331 Although ADH and angiotensin II affect blood pressure, all four hormones affect the long-term regulation of blood volume, as diagrammed in Figure 13-11●.

Antidiuretic Hormone

Antidiuretic hormone (ADH) is released at the posterior pituitary gland in response to a decrease in blood volume, an increase in the osmotic concentration of the

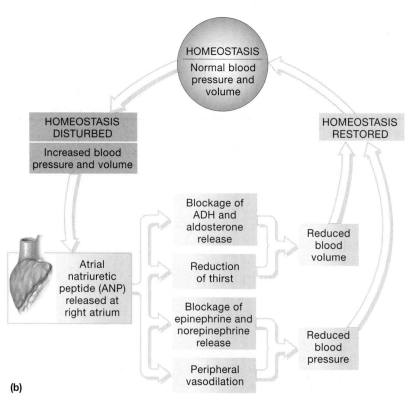

(b)

plasma, or in response to the presence of angiotensin II (Figure 13-11a●). The immediate result is a peripheral vasoconstriction that elevates blood pressure. ADH also has a water-conserving effect on the kidneys.

Angiotensin II

Angiotensin II appears in the blood following the release of the enzyme renin by specialized kidney cells in response to a fall in blood pressure (Figure 13-11a●). Renin starts a chain reaction that ultimately converts an inactive protein, *angiotensinogen*, to the hormone angiotensin II. Angiotensin II stimulates cardiac output and triggers arteriole constriction that, in turn, elevates systemic blood pressure almost at once. It also stimulates the secretion of ADH by the pituitary gland and aldosterone by the adrenal cortex. The effects of these hormones complement one another. The ADH stimulates water conservation at the kidneys, and the aldosterone stimulates the reabsorption of sodium ions and water from the urine. In addition, angiotensin II stimulates thirst, and the presence of ADH and aldosterone ensures that the additional water consumed will be retained, elevating the blood volume.

Erythropoietin

Erythropoietin is released by the kidneys if the blood pressure declines or if the oxygen content of the blood becomes abnormally low (Figure 13-11a●). This hormone stimulates red blood cell production, elevating the blood volume and improving the oxygen-carrying capacity of the blood.

Atrial Natriuretic Peptide

In contrast to the three hormones just described, ANP release is stimulated by *increased* blood pressure (Figure 13-11b●). This hormone is produced by specialized cardiac muscle cells in the atrial walls when they are stretched by excessive venous return. The ANP reduces blood volume and blood pressure by (1) promoting the loss of sodium ions and water at the kidneys, (2) increasing water losses at the kidneys by blocking the release of ADH and aldosterone, (3) reducing thirst, (4) blocking the release of E and NE, and (5) stimulating peripheral vasodilation. As blood volume and blood pressure decline, the stress on the atrial walls is removed and ANP production ceases.

Patterns of Cardiovascular Response

In our day-to-day lives, the individual components of the cardiovascular system—the blood, heart, and blood vessels—operate as an integrated complex. Two common stresses, exercise and blood loss, illustrate the adaptability of this system and its ability to maintain homeostasis. We will also consider the physiological mechanisms involved in shock and heart failure, two important cardiovascular disorders.

EXERCISE AND THE CARDIOVASCULAR SYSTEM

At rest, the cardiac output averages around 5.6 liters per minute. During exercise, both cardiac output and the pattern of blood distribution change markedly. As exercise begins, a number of interrelated changes occur:

- *Extensive vasodilation occurs* as the rate of skeletal muscle oxygen consumption increases. Peripheral resistance drops, blood flow through the capillaries increases, and blood enters the venous system at an accelerated rate.

- *The venous return increases* as skeletal muscle contractions squeeze blood along the peripheral veins and an increased breathing rate pulls blood into the venae cavae (the respiratory pump).

- *Cardiac output rises* as a result of the increased venous return. This increase occurs in direct response to ventricular stretching (the Frank-Starling principle) and in a reflexive response to atrial stretching (the atrial reflex). ∞ p. 382 The increased cardiac output keeps pace with the elevated demand, and arterial pressures are maintained despite the drop in peripheral resistance.

This regulation by venous feedback gradually increases cardiac output to about double resting levels. Over this range, typical of light exercise, the pattern of blood distribution remains relatively unchanged.

At higher levels of exertion, other physiological adjustments occur as the sympathetic nervous system stimulates the cardiac and vasomotor centers. Cardiac output increases, arterial blood pressure increases, and blood flow is severely restricted to "nonessential" organs, such as those of the digestive system, in order to increase the blood flow to active skeletal muscles. When you exercise at maximal levels, your blood essentially races among your skeletal muscles, lungs, and heart. Only the blood supply to the brain is unaffected.

EXERCISE, CARDIOVASCULAR FITNESS, AND HEALTH

Cardiovascular performance improves significantly with training. Trained athletes have larger hearts with greater average stroke volume than do nonathletes. These functional changes are important. Cardiac output is equal to the stroke volume times the heart rate, so for the same cardiac output, an individual with a larger stroke volume has a slower heart rate. A professional athlete at rest can maintain normal blood flow to peripheral tissues at a heart rate as low as 50 bpm (beats per minute), compared with about 80 bpm for a nonathlete.

403

13 BLOOD VESSELS AND CIRCULATION

The Anatomy of Blood Vessels • Circulatory Physiology • Cardiovascular Regulation • Patterns of Cardiovascular Response • **The Blood Vessels**

Regular exercise has several beneficial effects. Even a moderate exercise routine (jogging 5 miles per week, for example) can lower total blood cholesterol levels. High cholesterol is one of the major risk factors for atherosclerosis, which leads to cardiovascular disease and strokes. In addition, a healthy lifestyle with regular exercise, a balanced diet, weight control, and no smoking reduces stress, lowers blood pressure, and slows plaque formation. Large-scale statistical studies indicate that regular moderate exercise can cut the incidence of heart attacks almost in half.

CARDIOVASCULAR RESPONSE TO HEMORRHAGING

In Chapter 11 we considered the local circulatory reaction to a break in the wall of a blood vessel. p. 359 When the clotting response fails to prevent a significant blood loss and arterial blood pressure falls, the entire cardiovascular system begins making adjustments. The immediate goal is to maintain adequate blood pressure and peripheral blood flow. The long-term goal is to restore normal blood volume (Figure 13-11●).

The Short-Term Elevation of Blood Pressure

Short-term responses appear almost as soon as blood pressures start to decline, when the carotid and aortic reflexes increase cardiac output and cause peripheral vasoconstriction. When you donate blood at a blood bank, the amount collected is usually 500 ml, roughly 10 percent of your total blood volume. Such a loss initially causes a drop in cardiac output, but the vasomotor center quickly improves venous return and restores cardiac output to normal levels by a combination of vasoconstriction and mobilization of the **venous reserve** (large reservoirs of slowly moving venous blood in the liver, bone marrow, and skin). This venous compensation can restore normal arterial pressures and peripheral blood flow after losses that amount to 15–20 percent of the total blood volume. With a more substantial blood loss, cardiac output is maintained by increasing the heart rate, often to 180–200 bpm. Sympathetic activation assists by constricting the muscular arteries and arterioles and elevating blood pressure.

Sympathetic activation also causes the secretion of E and NE by the adrenal medullae. At the same time, ADH is released by the posterior pituitary gland, and at the kidneys the fall in blood pressure causes the release of renin, initiating the activation of angiotensin II. The E and NE increase cardiac output, and in combination with ADH and angiotensin II, they cause a powerful vasoconstriction that elevates blood pressure and improves peripheral blood flow.

The Long-Term Restoration of Blood Volume

After a serious hemorrhage, several days may pass before the blood volume returns to normal. When the short-term responses are unable to maintain normal cardiac output and blood pressure, the decline in capillary blood pressure promotes the reabsorption of fluid from the interstitial spaces (recall of fluids). p. 396 Over this period, ADH and aldosterone promote

---

CLINICAL NOTE *Shock*

Shock is an acute circulatory crisis marked by low blood pressure (hypotension) and inadequate peripheral blood flow. Severe and potentially fatal symptoms develop as vital tissues become starved for oxygen and nutrients. Common causes of shock are (1) a fall in cardiac output after hemorrhaging or other fluid losses, (2) damage to the heart, (3) external pressure on the heart, or (4) extensive peripheral vasodilation.

One important form of shock, called **circulatory shock**, is caused by a reduction of about 30 percent of the total blood volume. All such cases share the same six basic symptoms:

1. Hypotension, with systolic pressures below 90 mm Hg.
2. Pale, cool, and moist ("clammy") skin. The skin is pale and cool because of peripheral vasoconstriction; the moisture reflects the sympathetic activation of the sweat glands.
3. Confusion and disorientation, caused by a drop in blood pressure at the brain.
4. A rise in heart rate and a rapid, weak pulse.
5. Cessation of urination, because the reduced blood to the kidneys slows or stops urine production.
6. A drop in blood pH (acidosis), due to lactic acid generation in oxygen-deprived tissues.

When blood volume declines by more than 35 percent, homeostatic mechanisms are unable to cope with the situation, and a vicious cycle begins: Low blood pressure and low venous return lead to decreased cardiac output and myocardial damage, further reducing cardiac output. When the mean arterial blood pressure falls to about 50 mm Hg, carotid sinus baroreceptors trigger a massive activation of sympathetic vasoconstrictors. The resulting vasoconstriction reduces blood flow to peripheral tissues to maintain an adequate blood flow to the brain. The prevention of fatal consequences requires immediate treatment that concentrates on (1) preventing further fluid losses and (2) providing fluids through transfusions.

Several other types of shock also exist. In *cardiogenic shock* and *obstructive shock*, the heart is unable to maintain normal cardiac output. *Septic shock*, *toxic shock syndrome*, and *anaphylactic shock* result from a widespread, uncontrolled vasodilation. All share similar symptoms and consequences with circulatory shock.

fluid retention and reabsorption at the kidneys, preventing further reductions in blood volume. Thirst increases, and additional water is absorbed across the digestive tract. This intake of fluid elevates blood volume and ultimately replaces the interstitial fluids "borrowed" at the capillaries. Erythropoietin targets the bone marrow, stimulating the maturation of red blood cells, which increase the blood volume and improve oxygen delivery to peripheral tissues.

CONCEPT CHECK QUESTIONS

Answers on page 425

❶ How would applying a small pressure to the common carotid artery affect your heart rate? (Each of the two common carotid arteries branches into an internal and external carotid artery.)

❷ Why does blood pressure increase during exercise?

❸ What effect would the vasoconstriction of the renal artery have on blood pressure and blood volume?

The Blood Vessels

You already know that the circulatory system is divided into the **pulmonary circuit** and the **systemic circuit**. ∞ p. 368 The pulmonary circuit is composed of arteries and veins that transport blood between the heart and the lungs. This circuit begins at the right ventricle and ends at the left atrium. From the left ventricle, the arteries of the systemic circuit transport oxygenated blood and nutrients to all organs and tissues, ultimately returning deoxygenated blood to the right atrium. Figure 13-12● summarizes the primary circulatory routes within the pulmonary and systemic circuits.

The distribution of arteries and veins on the left and right sides of the body are usually identical—except near the heart, where large vessels connect to the atria or ventricles. For example, the distribution of the *left* and *right subclavian, axillary, brachial,* and *radial arteries* parallels the *left* and *right subclavian, axillary, brachial,* and *radial veins.* In addition, a single vessel may have several name changes as it crosses specific anatomical boundaries. For example, the *external iliac artery* becomes the *femoral artery* as it leaves the trunk and enters the thigh.

THE PULMONARY CIRCULATION

Blood entering the right atrium has just returned from the peripheral capillary beds, where oxygen was released and carbon dioxide was absorbed. After traveling through the right atrium and ventricle, blood enters the **pulmonary trunk**

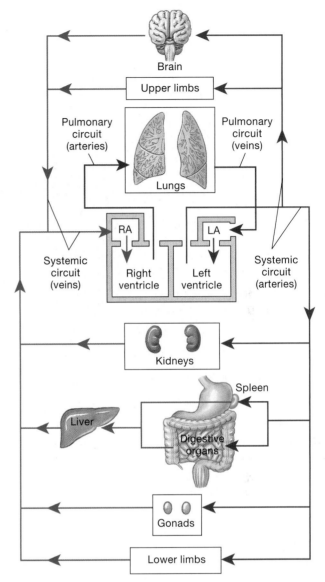

● *Figure 13-12* **An Overview of the Pattern of Circulation** RA stands for right atrium; LA, for left atrium.

(*pulmo-*, lung), the start of the pulmonary circuit (Figure 13-13●). In this circuit, oxygen stores will be replenished, carbon dioxide is released, and the oxygenated blood is returned to the heart for distribution in the systemic circuit.

The arteries of the pulmonary circuit differ from those of the systemic circuit in that they carry deoxygenated blood. (For this reason, color-coded diagrams usually show the pulmonary arteries in blue, the same color as systemic veins.) As it curves over the superior border of the heart, the pulmonary trunk gives rise to the **left** and **right pulmonary arteries**. These large arteries enter the lungs before branching repeatedly, giving rise to smaller and smaller arteries. The smallest branches, the pulmonary arterioles, provide blood to capillary networks that

13 **BLOOD VESSELS AND CIRCULATION**

The Anatomy of Blood Vessels • Circulatory Physiology • Cardiovascular Regulation • Patterns of Cardiovascular Response • **The Blood Vessels**

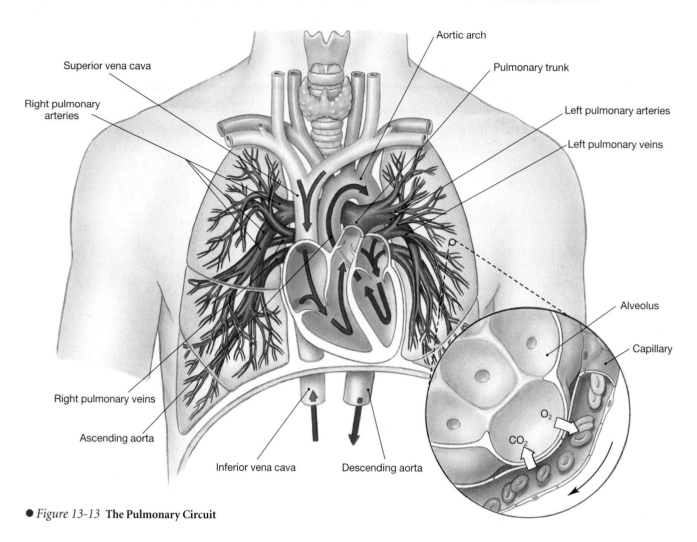

● *Figure 13-13* **The Pulmonary Circuit**

surround small air pockets, or **alveoli** (al-VĒ-ōl-ī; *alveolus*, sac). The walls of alveoli are thin enough for gas exchange to occur between the capillary blood and inspired air, a process detailed in Chapter 15. As it leaves the alveolar capillaries, oxygenated blood enters venules, which in turn unite to form larger vessels carrying blood to the **pulmonary veins**. These four veins, two from each lung, empty into the left atrium, completing the pulmonary circuit.

THE SYSTEMIC CIRCULATION

The systemic circulation supplies the capillary beds in all parts of the body not serviced by the pulmonary circuit.

The systemic circuit, which at any moment contains about 84 percent of the total blood volume, begins at the left ventricle and ends at the right atrium.

Systemic Arteries

Figure 13-14● is an overview of the systemic arteries and their relative locations. The first systemic vessel and largest artery is

the aorta. The **ascending aorta** begins at the aortic semilunar valve of the left ventricle, and the *left* and *right coronary arteries* originate near its base (Figure 12-7●, p. 375). The **aortic arch** curves across the superior surface of the heart, connecting the ascending aorta with the **descending aorta**.

Arteries of the Aortic Arch. Three elastic arteries originate along the aortic arch (Figure 13-15●). These arteries, (1) the **brachiocephalic** (brāk-ē-ō-se-FAL-ik) **trunk**, (2) the **left common carotid**, and the (3) **left subclavia**n (sub-CLĀ-vē-an), deliver blood to the head, neck, shoulders, and upper limbs. The brachiocephalic trunk ascends for a short distance before branching to form the **right common carotid artery** and the **right subclavian artery**. Thus, as the figure makes clear, there is only one brachiocephalic trunk, and the left common carotid and left subclavian arteries arise separately from the aortic arch. In terms of their peripheral distribution, however, the vessels on the left side are mirror images of those on the right side. Because most of the major arteries are paired, with one

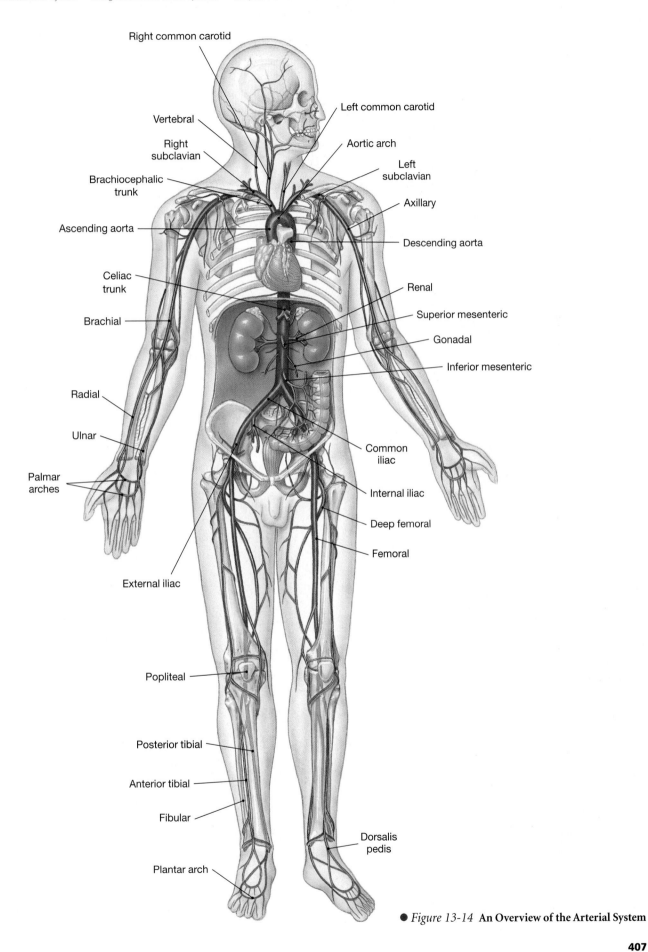

Right common carotid

Left common carotid

Vertebral

Right subclavian

Aortic arch

Left subclavian

Brachiocephalic trunk

Axillary

Ascending aorta

Descending aorta

Celiac trunk

Renal

Superior mesenteric

Brachial

Gonadal

Inferior mesenteric

Radial

Ulnar

Palmar arches

Common iliac

Internal iliac

Deep femoral

Femoral

External iliac

Popliteal

Posterior tibial

Anterior tibial

Fibular

Dorsalis pedis

Plantar arch

● *Figure 13-14* **An Overview of the Arterial System**

13 **BLOOD VESSELS AND CIRCULATION**

The Anatomy of Blood Vessels • Circulatory Physiology • Cardiovascular Regulation • Patterns of Cardiovascular Response • **The Blood Vessels**

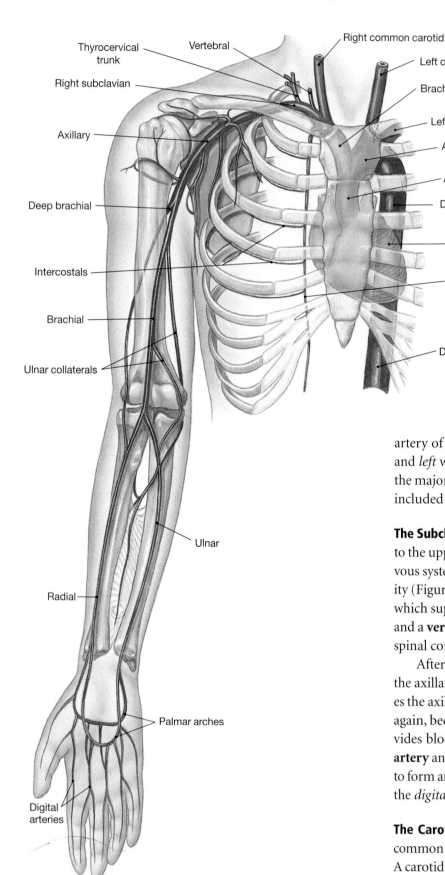

Thyrocervical trunk
Vertebral
Right common carotid
Right subclavian
Left common carotid
Axillary
Brachiocephalic trunk
Left subclavian
Deep brachial
Aortic arch
Ascending aorta
Descending aorta
Intercostals
Heart
Brachial
Internal thoracic
Ulnar collaterals
Descending aorta
Ulnar
Radial
Palmar arches
Digital arteries

● *Figure 13-15* **Arteries of the Chest and Upper Limb**

artery of each pair on either side of the body, the terms *right* and *left* will not be used. Figure 13-14● and 13-15● illustrate the major branches of these arteries, and additional details are included in the flow chart in Figure 13-16●.

The Subclavian Arteries. The subclavian arteries supply blood to the upper limbs, chest wall, shoulders, back, and central nervous system. Before a subclavian artery leaves the thoracic cavity (Figure 13-15●), it gives rise to an **internal thoracic artery**, which supplies the pericardium and anterior wall of the chest, and a **vertebral artery**, which provides blood to the brain and spinal cord.

After passing the first rib, the subclavian gets a new name, the axillary artery (Figure 13-15●). The **axillary artery** crosses the axilla (armpit) to enter the arm, where its name changes again, becoming the **brachial artery**. The brachial artery provides blood to the arm before branching to create the **radial artery** and **ulnar artery** of the forearm. These arteries connect to form anastomoses at the palm, the *palmar arches*, from which the *digital arteries* originate.

The Carotid Artery and the Blood Supply to the Brain. The common carotid arteries ascend deep in the tissues of the neck. A carotid artery can usually be located by pressing gently along either side of the trachea until a strong pulse is felt. Each common carotid artery divides into an **external carotid** and an

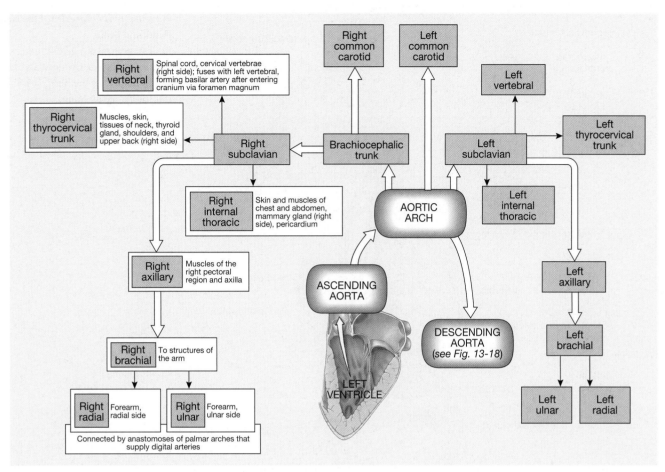

● *Figure 13-16* **A Flow Chart Showing the Arterial Distribution to the Head, Chest, and Upper Limbs**

internal carotid artery (Figure 13-17a●). The external carotid arteries supply blood to the pharynx, esophagus, larynx, and face. The internal carotid arteries enter the skull to deliver blood to the brain and the eyes.

The brain is extremely sensitive to changes in its circulatory supply. An interruption of circulation for several seconds will produce unconsciousness, and after 4 minutes there may be permanent neural damage. Such circulatory crises are rare, because blood reaches the brain through the vertebral arteries and by way of the internal carotid arteries. The vertebral arteries ascend within the transverse foramina of the cervical vertebrae, penetrating the skull at the foramen magnum. Inside the cranium, they fuse to form a large **basilar artery**, which continues along the ventral surface of the brain. This artery gives rise to the vessels in Figure 13-17b●.

Normally, the internal carotid arteries supply the arteries of the anterior half of the cerebrum, and the rest of the brain receives blood from the vertebral arteries. But this circulatory pattern can easily change, because the internal carotids and the basilar artery are interconnected in the **cerebral arterial circle**, or *circle of Willis*, a ring-shaped anastomosis that encircles the infundibulum (stalk) of the pituitary gland. With this arrangement, the brain can receive blood from either the carotids or the vertebral arteries, and the chances for a serious interruption of circulation are reduced.

The Descending Aorta. The descending aorta is continuous with the aortic arch. The diaphragm divides the descending aorta into a superior **thoracic aorta** and an inferior **abdominal aorta** (Figure 13-18●). The thoracic aorta travels within the mediastinum, providing blood to the intercostal arteries, which carry blood to the thoracic spinal cord and the body wall. The thoracic aorta also gives rise to small arteries that end in capillary beds in the esophagus, pericardium, and other mediastinal structures. Near the diaphragm, the **phrenic** (FREN-ik) **arteries** deliver blood to the muscular diaphragm that separates the thoracic and abdominopelvic cavities. The branches of the thoracic aorta are shown in Figure 13-18●.

The abdominal aorta delivers blood to elastic arteries that distribute blood to visceral organs of the digestive system and

13 BLOOD VESSELS AND CIRCULATION

The Anatomy of Blood Vessels • Circulatory Physiology • Cardiovascular Regulation • Patterns of Cardiovascular Response • **The Blood Vessels**

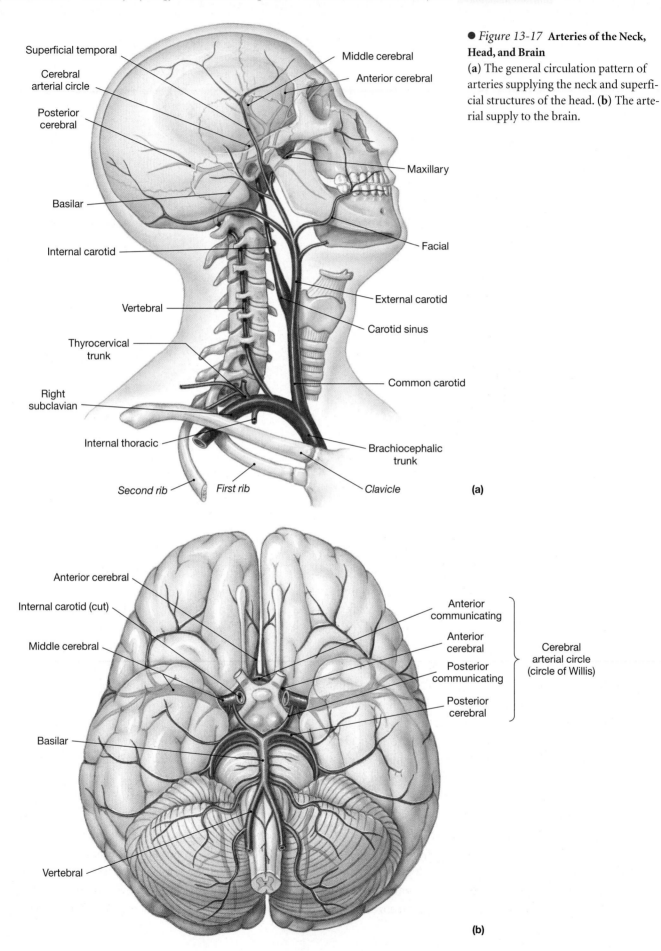

● *Figure 13-17* **Arteries of the Neck, Head, and Brain**
(**a**) The general circulation pattern of arteries supplying the neck and superficial structures of the head. (**b**) The arterial supply to the brain.

(a)

(b)

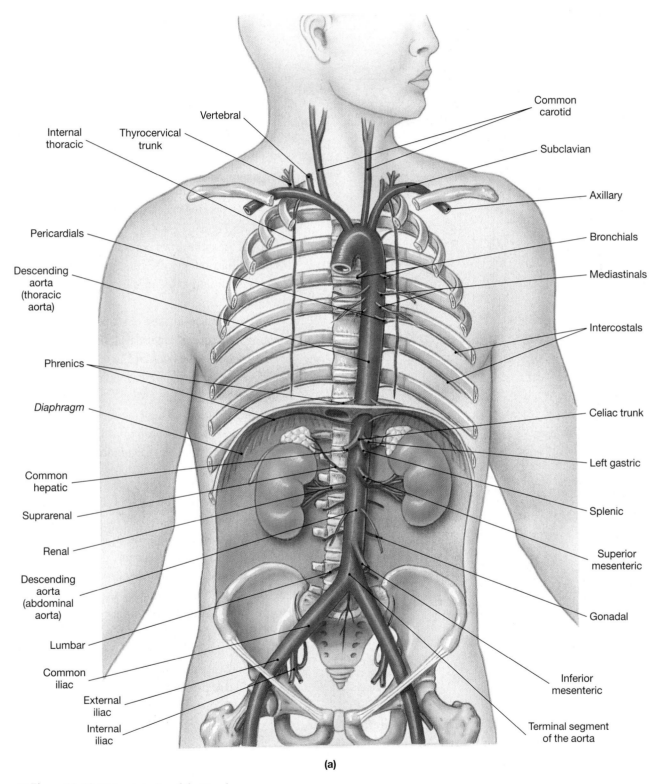

Internal thoracic

Thyrocervical trunk

Vertebral

Common carotid

Subclavian

Axillary

Pericardials

Bronchials

Descending aorta (thoracic aorta)

Mediastinals

Intercostals

Phrenics

Celiac trunk

Diaphragm

Common hepatic

Left gastric

Suprarenal

Splenic

Renal

Superior mesenteric

Descending aorta (abdominal aorta)

Gonadal

Lumbar

Inferior mesenteric

Common iliac

External iliac

Internal iliac

Terminal segment of the aorta

(a)

● *Figure 13-18* **Major Arteries of the Trunk**
(a) A diagrammatic view.

13 BLOOD VESSELS AND CIRCULATION

The Anatomy of Blood Vessels • Circulatory Physiology • Cardiovascular Regulation • Patterns of Cardiovascular Response • **The Blood Vessels**

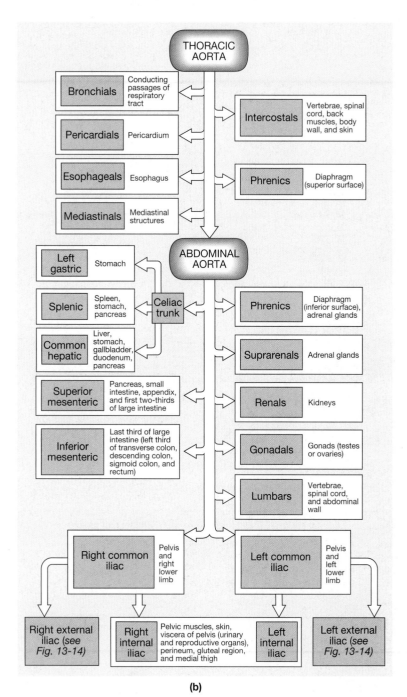

(b)

● *Figure 13-18* (**continued**)
(**b**) A flow chart.

to the kidneys and adrenal glands. The **celiac** (SĒ-lē-ak) **trunk**, **superior mesenteric** (mez-en-TER-ik) **artery**, and **inferior mesenteric artery** arise on the anterior surface of the abdominal aorta and branch in the connective tissues of the mesenteries. These three vessels provide blood to all of the digestive organs in the abdominopelvic cavity. The celiac trunk divides into three branches that deliver blood to the liver, gallbladder, stomach, and spleen. The superior mesenteric artery supplies the pancreas, small intestine, and most of the large intestine.

The inferior mesenteric delivers blood to the last portion of the large intestine and rectum.

Paired **gonadal** (gō-NAD-al) **arteries** originate between the superior and inferior mesenteric arteries. In males they are called *testicular arteries*; in females, *ovarian arteries*.

The **suprarenal arteries** and **renal arteries** arise along the lateral surface of the abdominal aorta and travel behind the peritoneal lining to reach the adrenal glands and kidneys. Small **lumbar arteries** begin on the posterior surface of the aorta and supply the spinal cord and the abdominal wall.

At the level of vertebra L_4, the abdominal aorta divides to form a pair of muscular arteries. These **common iliac** (IL-ē-ak) **arteries** carry blood to the pelvis and lower limbs (see Figure 13-14●). As it travels along the inner surface of the ilium, each common iliac divides to form an **internal iliac artery**, which supplies smaller arteries of the pelvis, and an **external iliac artery**, which enters the lower limb.

Once in the thigh, the external iliac artery branches, forming the **femoral artery** and the **deep femoral artery**. When it reaches the back of the knee, the femoral artery becomes the **popliteal artery**, which almost immediately branches to form the **anterior tibial**, **posterior tibial**, and **fibular arteries**. These arteries are connected by two anastomoses, one on the top of the foot (the *dorsalis pedis*) and one on the bottom (the *plantar arch*).

Systemic Veins

Veins collect blood from each of the tissues and organs of the body by means of a venous network that drains into the right atrium through the superior and inferior vena cavae. Figure 13-19● illustrates the major vessels of the venous system. Complementary arteries and veins often run side by side, and in many cases they have comparable names. For example, the axillary arteries run alongside the axillary veins. In addition, arteries and veins often travel in the company of peripheral nerves that have the same names and innervate the same structures.

One significant difference between the arterial and venous systems concerns the distribution of major veins in the neck and limbs. Arteries in these areas are located deep beneath the skin, protected by bones and surrounding soft tissues. In contrast, the neck and limbs usually have two sets of peripheral veins, one superficial and the other deep. This dual venous drainage helps control body temperature. In hot weather, venous blood flows in superficial veins, where heat loss can occur; in cold weather, blood is routed to the deep veins to minimize heat loss.

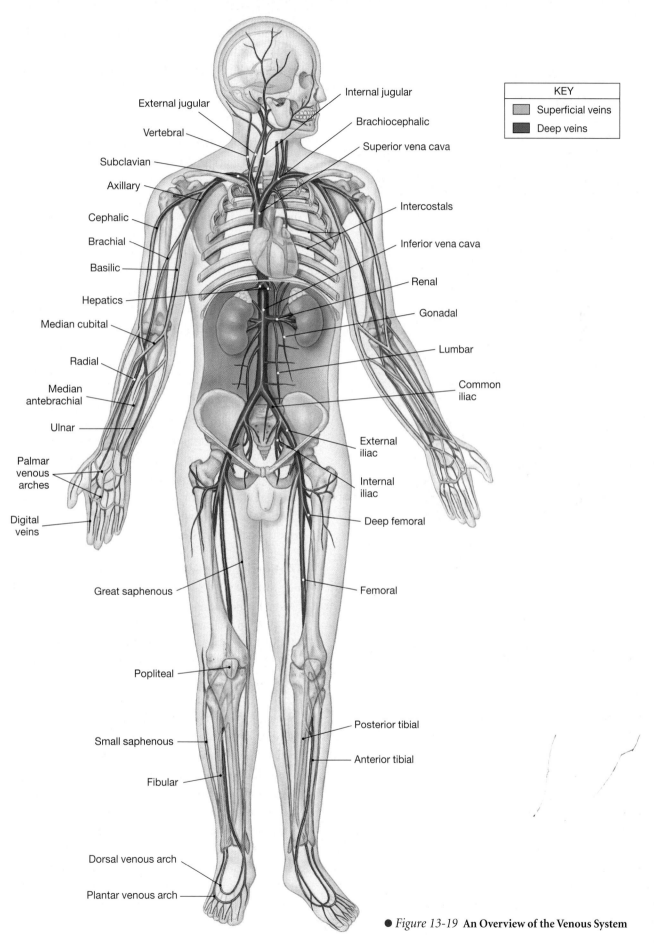

KEY	
	Superficial veins
	Deep veins

External jugular

Vertebral

Subclavian

Axillary

Cephalic

Brachial

Basilic

Hepatics

Median cubital

Radial

Median antebrachial

Ulnar

Palmar venous arches

Digital veins

Internal jugular

Brachiocephalic

Superior vena cava

Intercostals

Inferior vena cava

Renal

Gonadal

Lumbar

Common iliac

External iliac

Internal iliac

Deep femoral

Great saphenous

Femoral

Popliteal

Posterior tibial

Small saphenous

Anterior tibial

Fibular

Dorsal venous arch

Plantar venous arch

● *Figure 13-19* **An Overview of the Venous System**

13 BLOOD VESSELS AND CIRCULATION

The Anatomy of Blood Vessels • Circulatory Physiology • Cardiovascular Regulation • Patterns of Cardiovascular Response • **The Blood Vessels**

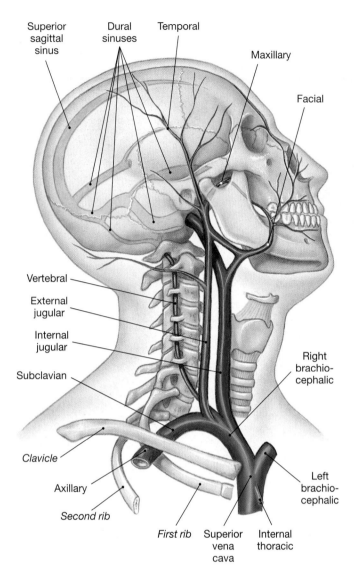

● *Figure 13-20* **Major Veins of the Head and Neck**
Veins draining superficial and deep portions of the head and neck.

The Superior Vena Cava. The **superior vena cava** (**SVC**) receives blood from the head and neck (Figure 13-20●), and the chest, shoulders, and upper limbs (Figure 13-21●).

Venous Return from the Head and Neck. Small veins in the neural tissue of the brain empty into a network of thin-walled channels, the **dural sinuses**. ∞ p. 239 The largest and most important sinus, the **superior sagittal sinus**, is located within the fold of dura mater lying between the cerebral hemispheres. Most of the blood leaving the brain passes through one of the dural sinuses and enters one of the **internal jugular veins**, which penetrate the jugular foramen and descend in the deep tissues of the neck. The more superficial **external jugular veins** collect blood from the superficial structures of the head and neck. These veins travel just beneath the skin,

and a *jugular venous pulse (JVP)* can sometimes be detected at the base of the neck. **Vertebral veins** drain the cervical spinal cord and the posterior surface of the skull, descending within the transverse foramina of the cervical vertebrae along with the vertebral arteries.

Venous Return from the Upper Limbs and Chest. The major veins of the upper body are indicated in Figure 13-21●, and information concerning the venous tributaries of the superior vena cava can be found in (Figure 13-22a● p. 416). A venous network in the palms collects blood from the digital veins. These vessels drain into the **cephalic vein** and the **basilic vein**. The deeper veins of the forearm consist of a **radial vein** and an **ulnar vein**. After crossing the elbow, these veins fuse to form the **brachial vein**. As the brachial vein continues toward the trunk, it joins the basilic vein before entering the axilla as the **axillary vein**. The cephalic vein drains into the axillary vein at the shoulder.

The axillary vein then continues into the trunk; at the level of the first rib it becomes the **subclavian vein**. After traveling a short distance inside the thoracic cavity, the subclavian meets and merges with the external and internal jugular veins of that side. This fusion creates the large **brachiocephalic vein**, also known as the *innominate vein*. Near the heart, the two brachiocephalic veins (one from each side of the body) combine to create the superior vena cava. The SVC receives blood from the thoracic body wall through the **azygos** (AZ-i-gos) **vein** before arriving at the right atrium.

The Inferior Vena Cava. The **inferior vena cava** (**IVC**) collects most of the venous blood from organs below the diaphragm. (A small amount reaches the superior vena cava from the azygos vein.) A flow chart of the tributaries of the IVC is shown in Figure 13-22b●, and the veins of the abdomen are illustrated in Figure 13-21●. Refer to Figure 13-19● for the veins of the lower limbs.

Blood leaving the capillaries in the sole of each foot collects into a network of **plantar veins**. The plantar network provides blood to the **anterior tibial vein**, the **posterior tibial vein**, and the **fibular vein**, the deep veins of the leg. A *dorsal venous arch* drains blood from capillaries on the superior surface of the foot. This arch is drained by two superficial veins, the **great saphenous vein** (sa-FĒ-nus; *saphenes*, prominent) and the **small saphenous vein**. (Surgeons often use segments of the great saphenous vein, the largest superficial vein, as a bypass vessel during *coronary bypass surgery*.) The plantar arch and the dorsal arch interconnect extensively, so the path of blood flow can easily shift from superficial to deep veins.

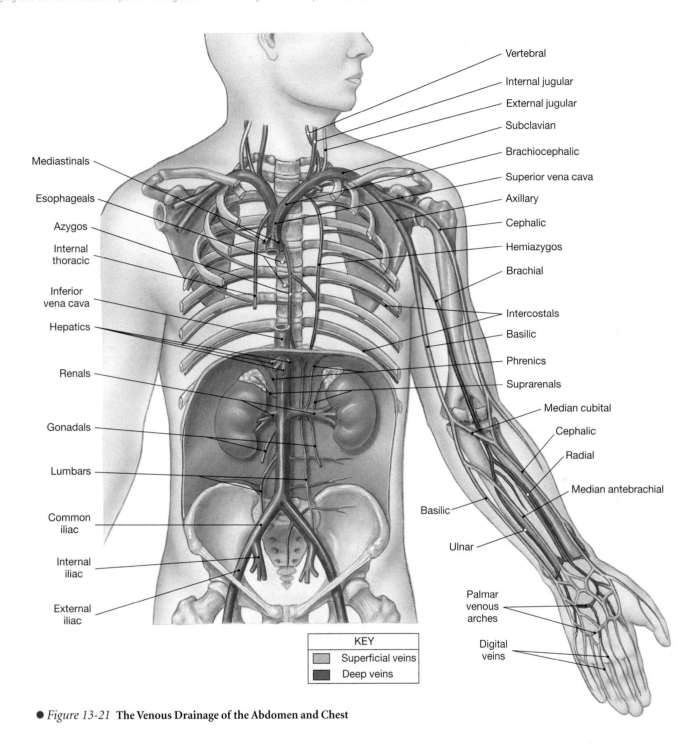

Mediastinals
Esophageals
Azygos
Internal thoracic
Inferior vena cava
Hepatics
Renals
Gonadals
Lumbars
Common iliac
Internal iliac
External iliac

Vertebral
Internal jugular
External jugular
Subclavian
Brachiocephalic
Superior vena cava
Axillary
Cephalic
Hemiazygos
Brachial
Intercostals
Basilic
Phrenics
Suprarenals
Median cubital
Cephalic
Radial
Median antebrachial
Basilic
Ulnar
Palmar venous arches
Digital veins

KEY	
	Superficial veins
	Deep veins

● *Figure 13-21* **The Venous Drainage of the Abdomen and Chest**

Behind the knee, the small saphenous, tibial, and fibular veins unite to form the **popliteal vein**. When the popliteal vein reaches the femur, it becomes the **femoral vein**. Immediately before penetrating the abdominal wall, the femoral, great saphenous, and **deep femoral** veins unite. The large vein that results penetrates the body wall as the **external iliac vein**. As it travels across the inner surface of the ilium, the external iliac fuses with the **internal iliac vein**, which drains the pelvic organs. The resulting **common iliac vein** then meets its counterpart from the opposite side to form the IVC.

Like the aorta, the IVC lies posterior to the abdominopelvic cavity. As it ascends to the heart, it collects blood from several lumbar veins. In addition, the IVC receives blood from the *gonadal, renal, suprarenal, phrenic,* and *hepatic veins* before reaching the right atrium (Figure 13-21●).

13 BLOOD VESSELS AND CIRCULATION

The Anatomy of Blood Vessels • Circulatory Physiology • Cardiovascular Regulation • Patterns of Cardiovascular Response • **The Blood Vessels**

(a)

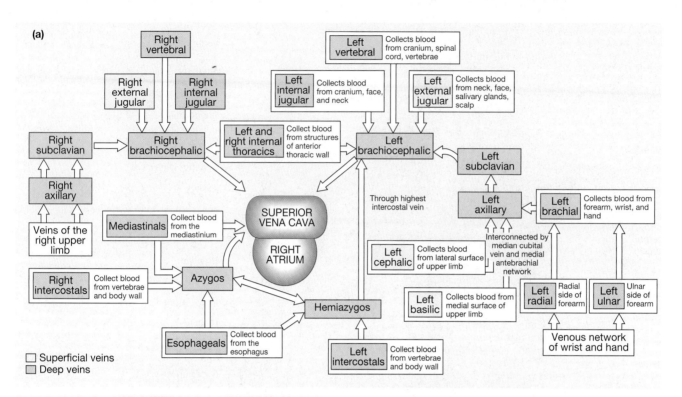

☐ Superficial veins
☐ Deep veins

(b)

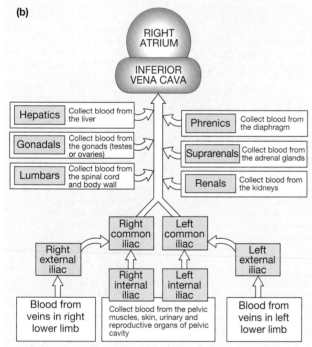

● *Figure 13-22* **A Flow Chart of the Circulation to the Superior and Inferior Venae Cavae**
(a) The tributaries of the SVC. **(b)** The tributaries of the IVC.

The Hepatic Portal System. You may have noticed that the list of veins did not include any names that refer to digestive organs other than the liver. Instead of traveling directly to the inferior vena cava, blood leaving the capillaries supplied by the celi-

ac, superior, and inferior mesenteric arteries flows to the liver through the **hepatic portal system** (*porta*, a gate). A blood vessel connecting two capillary beds is called a *portal vessel*; and the network is a *portal system*. ∞ p. 319 Blood in this system is quite different from that in other systemic veins, because the hepatic portal vessels contain substances absorbed by the digestive tract. For example, levels of blood glucose and amino acids in the hepatic portal vein often exceed those found anywhere else in the circulatory system.

A portal system carries blood from one capillary bed to another and, in the process, prevents its contents from mixing with the entire bloodstream. The liver regulates the concentrations of nutrients, such as glucose or amino acids, in the circulating blood. When digestion is under way, the digestive tract absorbs high concentrations of nutrients, along with various wastes and an occasional toxin. The hepatic portal system delivers these compounds directly to the liver, where liver cells absorb them for storage, metabolic conversion, or excretion. After passing through the liver capillaries, blood collects into the hepatic veins, which empty into the inferior vena cava. Because blood goes to the liver first, the composition of the blood in the general circulation remains relatively stable, regardless of the digestive activities under way.

Figure 13-23● shows the anatomy of the hepatic portal system. The system begins in the capillaries of the digestive organs. Blood from capillaries along the lower portion of the

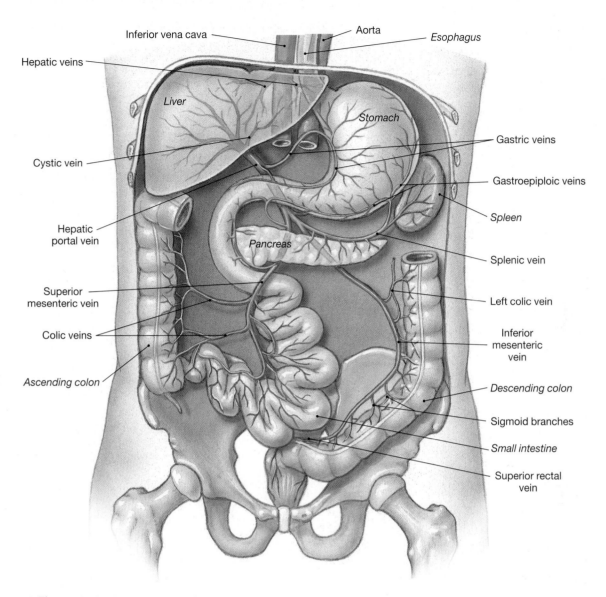

Inferior vena cava

Aorta

Esophagus

Hepatic veins

Liver

Stomach

Gastric veins

Gastroepiploic veins

Cystic vein

Spleen

Hepatic portal vein

Pancreas

Splenic vein

Superior mesenteric vein

Left colic vein

Colic veins

Inferior mesenteric vein

Ascending colon

Descending colon

Sigmoid branches

Small intestine

Superior rectal vein

● *Figure 13-23* **The Hepatic Portal System**

large intestine enters the **inferior mesenteric vein**. As it nears the liver, veins from the spleen, the lateral border of the stomach, and the pancreas fuse with the inferior mesenteric, forming the **splenic vein**. The **superior mesenteric vein** also drains the lateral border of the stomach, through an anastomosis with one of the branches of the splenic vein. In addition, the superior mesenteric collects blood from the entire small intestine and two-thirds of the large intestine. The system ends with the **hepatic portal vein**, which empties into the liver capillaries. The hepatic portal vein forms through the fusion of the superior mesenteric and splenic veins. Of the two, the superior mesenteric normally contributes the greater volume of blood and most of the nutrients. As it proceeds, the hepatic portal

vein receives blood from the **gastric veins**, which drain the medial border of the stomach and the **cystic vein** from the gallbladder.

CONCEPT CHECK QUESTIONS

Answers on page 425

❶ The blockage of which branch of the aortic arch would interfere with the blood flow to the left arm?

❷ Why would the compression of one of the common carotid arteries cause a person to lose consciousness?

❸ Grace is in an automobile accident and ruptures her celiac trunk. Which organs would be affected most directly by this injury?

13 BLOOD VESSELS AND CIRCULATION

The Anatomy of Blood Vessels • Circulatory Physiology • Cardiovascular Regulation • Patterns of Cardiovascular Response • The Blood Vessels

FETAL CIRCULATION

The fetal and adult circulatory systems are significantly different, reflecting different sources of respiratory and nutritional support. The embryonic lungs are collapsed and nonfunctional, and the digestive tract has nothing to digest. All of the embryonic nutritional and respiratory needs are provided by diffusion across the *placenta*, a structure within the uterine wall where the maternal and fetal circulatory systems are in close contact.

Placental Blood Supply

Circulation in a full-term (9-month-old) fetus is diagrammed in Figure 13-24●. Blood flow to the placenta is provided by a pair of **umbilical arteries**, which arise from the internal iliac arteries and enter the umbilical cord. Blood returns from the placenta in a single **umbilical vein**, which brings oxygen and nutrients to the developing fetus. The umbilical vein delivers blood to the developing liver. Some of that blood flows through capillary networks within the liver. The rest bypasses the liver capillaries and reaches the inferior vena cava within the **ductus venosus**. When the placental connection is broken at birth, blood flow ceases along the umbilical vessels, and they soon degenerate.

Circulation in the Heart and Great Vessels

One of the most interesting aspects of circulatory development reflects the differences between the life of an embryo or fetus and that of an infant. Throughout embryonic and fetal life, the lungs are collapsed; yet after delivery, the newborn infant must be able to extract oxygen from inspired air rather than across the placenta.

Although the interatrial and interventricular septa of the heart develop early in fetal life, the interatrial partition remains

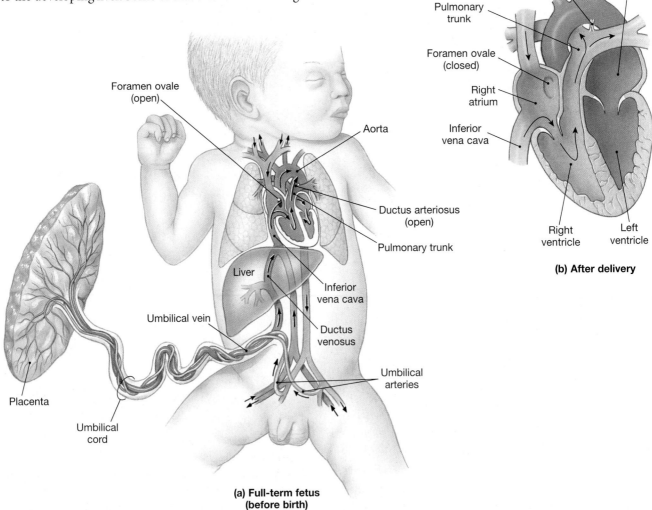

(a) Full-term fetus (before birth)

(b) After delivery

● *Figure 13-24* **Fetal Circulation**
(**a**) Blood flow to and from the placenta. (**b**) Blood flow through the heart of a newborn baby.

functionally incomplete up to the time of birth. The interatrial opening, or **foramen ovale**, is associated with an elongate flap that acts as a valve. Blood can flow freely from the right atrium to the left atrium, but any backflow will close the valve and isolate the two chambers. Thus blood can enter the heart at the right atrium and bypass the pulmonary circuit. A second short-circuit exists between the pulmonary and aortic trunks. This connection, the **ductus arteriosus**, consists of a short, muscular vessel.

With the lungs collapsed, the capillaries are compressed and little blood flows through the lungs. During diastole, blood enters the right atrium and flows into the right ventricle, but it also passes into the left atrium through the foramen ovale. About 25 percent of the blood arriving at the right atrium bypasses the pulmonary circuit in this way. In addition, over 90 percent of the blood leaving the right ventricle passes through the ductus arteriosus and enters the systemic circuit rather than continuing to the lungs.

Circulatory Changes at Birth

At birth, dramatic changes occur in circulatory patterns. When the infant takes its first breath, the lungs expand, and so do the pulmonary vessels. Within a few seconds, the smooth muscles in the ductus arteriosus contract, isolating the pulmonary and aortic trunks, and blood begins flowing through the pulmonary circuit. As pressures rise in the left atrium, the valvular flap closes the foramen ovale and completes the circulatory remodeling (Figure 13-24b●). In adults, the interatrial septum bears a shallow depression, the *fossa ovalis*, that marks the site of the foramen ovale (see Figure 12-5●, p. 372). The remnants of the ductus arteriosus persist as a fibrous cord, the **ligamentum arteriosum**.

If the proper vascular changes do not occur at birth or shortly thereafter, problems will eventually develop because the heart will have to work too hard to provide adequate amounts of oxygen to the systemic circuit. Treatment may involve surgical closure of the foramen ovale, the ductus arteriosus, or both. Other forms of congenital heart defects result from abnormal cardiac development or inappropriate connections between the heart and major arteries and veins.

Aging and the Cardiovascular System

The capabilities of the cardiovascular system gradually decline with age. Major changes affect all parts of the cardiovascular system: blood, heart, and vessels.

In the blood, age-related changes may include (1) decreased hematocrit; (2) the constriction or blockage of peripheral veins by the formation of a *thrombus* (stationary blood clot), which can become detached, pass through the heart, and become wedged in a small artery, most often in the lungs, causing a *pulmonary embolism*; and (3) the pooling of blood in the veins of the legs because valves are not working effectively.

In the heart, age-related changes include (1) a reduction in the maximum cardiac output; (2) changes in the activities of the nodal and conducting cells; (3) a reduction in the elasticity of the fibrous skeleton; (4) a progressive atherosclerosis, which can restrict coronary circulation; and (5) the replacement of damaged cardiac muscle cells by scar tissue.

In blood vessels, age-related changes are often related to arteriosclerosis, a thickening and toughening of the arterial wall. For example, (1) the inelastic walls of arteries become less tolerant of sudden pressure increases, which can lead to a localized dilation, or *aneurysm*, whose subsequent rupture may cause a stroke, infarct, or massive blood loss (depending on the vessel); (2) calcium salts can be deposited on weakened vascular walls, increasing the risk of a stroke or myocardial infarction; and (3) thrombi can form at atherosclerotic plaques.

Integration with Other Systems

The cardiovascular system is both anatomically and functionally linked to all other systems. Figure 13-25● summarizes the physiological relationships between the cardiovascular system and other organ systems. The most extensive communication occurs between the cardiovascular system and the lymphatic system. Not only are the two systems physically interconnected, but cell populations of the lymphatic system use the cardiovascular system as a highway to move from one part of the body to another. Chapter 14 examines the lymphatic system in detail.

CONCEPT CHECK QUESTIONS

Answers on page 425

❶ A blood sample taken from the umbilical cord has a high concentration of oxygen and nutrients and a low concentration of carbon dioxide and waste products. Is this sample from an umbilical artery or from the umbilical vein?

❷ The cardiovascular system is most closely interconnected with what other organ system?

13 BLOOD VESSELS AND CIRCULATION

The Anatomy of Blood Vessels • Circulatory Physiology • Cardiovascular Regulation • Patterns of Cardiovascular Response • The Blood Vessels

The Cardiovascular System

For All Systems

Delivers oxygen, hormones, nutrients, and WBCs; removes carbon dioxide and metabolic wastes; transfers heat

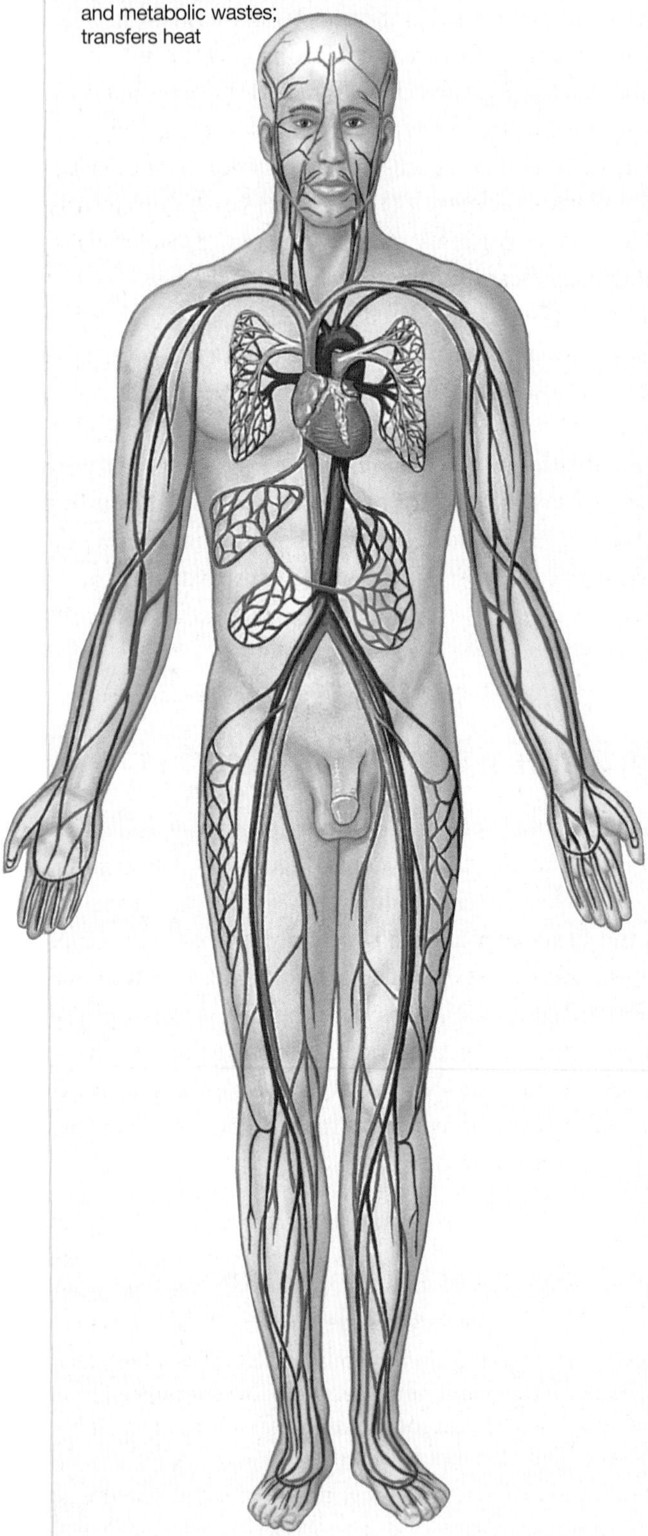

The Integumentary System

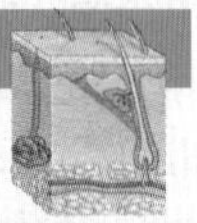

- Stimulation of mast cells produces localized changes in blood flow and capillary permeability
- Delivers immune system cells to injury sites; clotting response seals breaks in skin surface; carries away toxins from sites of infection; provides heat

The Skeletal System

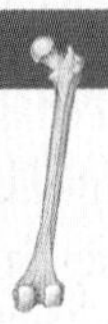

- Provides calcium needed for normal cardiac muscle contraction; protects blood cells developing in bone marrow
- Provides calcium and phosphate for bone deposition; delivers EPO to bone marrow, parathyroid hormone, and calcitonin to osteoblasts and osteoclasts

The Muscular System

- Skeletal muscle contractions assist in moving blood through veins; protects superficial blood vessels, especially in neck and limbs
- Delivers oxygen and nutrients, removes carbon dioxide, lactic acid, and heat during skeletal muscle activity

The Nervous System

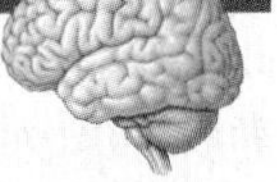

- Controls patterns of circulation in peripheral tissues; modifies heart rate and regulates blood pressure; releases ADH
- Endothelial cells maintain blood-brain barrier; help generate CSF

The Endocrine System

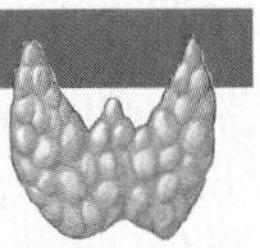

- Erythropoietin regulates production of RBCs; several hormones elevate blood pressure; epinephrine stimulates cardiac muscle, elevating heart rate and contractile force
- Distributes hormones throughout the body; heart secretes ANP

The Lymphatic System

- Defends against pathogens or toxins in blood; fights infections of cardiovascular organs; returns tissue fluid to circulation
- Distributes WBCs; carries antibodies that attack pathogens; clotting response assists in restricting spread of pathogens; granulocytes and lymphocytes produced in bone marrow

The Respiratory System

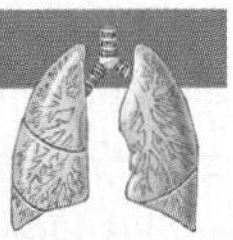

- Provides oxygen to cardiovascular organs and removes carbon dioxide
- RBCs transport oxygen and carbon dioxide between lungs and peripheral tissues

The Digestive System

- Provides nutrients to cardiovascular organs; absorbs water and ions essential to maintenance of normal blood volume
- Distributes digestive tract hormones; carries nutrients, water, and ions away from sites of absorption; delivers nutrients and toxins to liver

The Urinary System

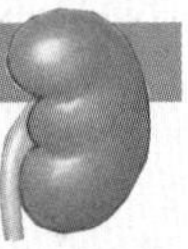

- Releases renin to elevate blood pressure and erythropoietin to accelerate red blood cell production
- Delivers blood to capillaries, where filtration occurs; accepts fluids and solutes reabsorbed during urine production

The Reproductive System

- Maintains healthy vessels and slows development of atherosclerosis with age by estrogens
- Distributes reproductive hormones; provides nutrients, oxygen, and waste removal for developing fetus; local blood pressure changes responsible for physical changes during sexual arousal

● *Figure 13-25* **Functional Relationships Between the Cardiovascular System and Other Systems**

Related Clinical Terms

aneurysm (AN-ū-rizm): A bulge in the weakened wall of a blood vessel, generally an artery.

arteriosclerosis (ar-tē-rē-ō-skle-RŌ-sis): A thickening and toughening of arterial walls.

atherosclerosis (ath-er-ō-skle-RŌ-sis): A type of arteriosclerosis characterized by changes in the endothelial lining and the formation of a plaque.

edema (e-DĒ-muh): An abnormal accumulation of fluid in peripheral tissues.

hypertension: Abnormally high blood pressure; usually defined in adults as blood pressure higher than 150/90.

hypervolemic (hī-per-vō-LĒ-mik): Having an excessive blood volume.

hypotension: Blood pressure so low that circulation to vital organs may be impaired.

hypovolemic (hī-pō-vō-LĒ-mik): Having a low blood volume.

orthostatic hypotension: Low blood pressure on standing, often accompanied by dizziness or fainting; due to a failure of the regulatory mechanisms that increase blood pressure to maintain adequate blood flow to the brain.

phlebitis: The inflammation of a vein.

pulmonary embolism: A circulatory blockage caused by the trapping of a detached thrombus in a pulmonary artery.

shock: An acute circulatory crisis marked by hypotension and inadequate peripheral blood flow.

sphygmomanometer: A device that measures blood pressure by using an inflatable cuff placed around a limb.

thrombus: A stationary blood clot within a blood vessel.

varicose (VAR-i-kōs) **veins:** Sagging, swollen veins distorted by gravity and by the failure of the venous valves.

CHAPTER REVIEW

Key Terms

Summary Outline

INTRODUCTION 390

1. Blood flows through a network of arteries, veins, and capillaries. The vital functions of the cardiovascular system depend on events at the capillary level: All chemical and gaseous exchange between the blood and interstitial fluid takes place across **capillary** walls.

THE ANATOMY OF BLOOD VESSELS 390

1. Arteries and veins form an internal distribution system, propelled by the heart. **Arteries** branch repeatedly, decreasing in size until they become **arterioles**; from the arterioles, blood enters the capillary networks. Blood flowing from the capillaries enters small **venules** before entering larger **veins**.

The Structure of Vessel Walls 390

2. The walls of arteries and veins contain three layers: the **tunica interna**, **tunica media**, and outermost **tunica externa**. *(Figure 13-1)*

Arteries...391

3. The walls of arteries are usually thicker than the walls of veins. The arterial system includes the large **elastic arteries**, medium-sized **muscular arteries**, and smaller arterioles. As we proceed toward the capillaries, the number of vessels increases, but the diameter of the individual vessels decreases and the walls become thinner. *(Figure 13-2)*

Capillaries...392

4. Capillaries are the only blood vessels whose walls permit exchange between blood and interstitial fluid.

5. Capillaries form interconnected networks called **capillary beds**. A **precapillary sphincter** (a band of smooth muscle) adjusts blood flow

into each capillary. Blood flow in a capillary changes as **vasomotion** occurs. *(Figure 13-3)*

Veins 393

6. Venules collect blood from capillaries and merge into **medium-sized veins** and then **large veins**. The arterial system is a high-pressure system; pressure in veins is much lower. **Valves** in these vessels prevent backflow of blood. *(Figure 13-4)*

CIRCULATORY PHYSIOLOGY 394

Pressure 394

1. Flow is proportional to the difference in pressure; blood will flow from an area of higher pressure to one of relatively lower pressure.

Resistance..394

2. For circulation to occur, the *circulatory pressure* must be greater than the *total peripheral resistance* (the resistance of the entire circulatory system). For blood to flow into peripheral capillaries, **blood pressure** (arterial pressure) must be greater than the **peripheral resistance** (the resistance of the arterial system). Neural and hormonal control mechanisms regulate blood pressure.

3. The most important determinant of peripheral resistance is the diameter of arterioles.

Circulatory Pressure................................395

4. The high arterial pressures overcome peripheral resistance and maintain blood flow through peripheral tissues. **Capillary pressures** are normally low; small changes in capillary pressure determine the rate of

13 BLOOD VESSELS AND CIRCULATION

The Anatomy of Blood Vessels • Circulatory Physiology • Cardiovascular Regulation • Patterns of Cardiovascular Response • The Blood Vessels

fluid movement into or out of the bloodstream. Venous pressure, normally low, determines venous return and affects cardiac output and peripheral blood flow.

5. Arterial pressure rises in ventricular systole and falls in ventricular diastole. The difference between the **systolic** and **diastolic pressures** is **pulse pressure**. *(Figure 13-5, 13-7)*

6. At the capillaries, solute molecules diffuse across the capillary lining, and water-soluble materials diffuse through small spaces between endothelial cells. Water will move when driven by either hydrostatic pressure or osmotic pressure. The direction of water movement is determined by the balance between these two opposing pressures. *(Figure 13-6)*

7. Valves, **muscular compression**, and the **respiratory pump** help the relatively low venous pressures propel blood toward the heart. *(Figure 13-4)*

CARDIOVASCULAR REGULATION 398

1. Homeostatic mechanisms ensure that tissue blood flow *(tissue perfusion)* delivers adequate oxygen and nutrients.

2. Blood flow varies directly with cardiac output, peripheral resistance, and blood pressure.

3. Local, autonomic, and endocrine factors influence the coordinated regulation of cardiovascular function. Local factors change the pattern of blood flow within capillary beds in response to chemical changes in interstitial fluids. Central nervous system mechanisms respond to changes in arterial pressure or blood gas levels. Hormones can assist in short-term adjustments (changes in cardiac output and peripheral resistance) and long-term adjustments (changes in blood volume that affect cardiac output and gas transport). *(Figure 13-8)*

The Autoregulation of Blood Flow 399

4. Peripheral resistance is adjusted at the tissues by local factors that result in the dilation or constriction of precapillary sphincters.

The Neural Control of Blood Pressure and Blood Flow 400

5. **Baroreceptor reflexes** are autonomic reflexes that adjust cardiac output and peripheral resistance to maintain normal arterial pressures. Baroreceptor populations include the **aortic** and **carotid sinuses** and *atrial baroreceptors*. *(Figure 13-9)*

6. **Chemoreceptor reflexes** respond to changes in the oxygen or carbon dioxide levels in the blood and cerebrospinal fluid. Sympathetic activation leads to stimulation of the *cardioacceleratory* and *vasomotor centers*; parasympathetic activation stimulates the *cardioinhibitory center*. *(Figure 13-10)*

Hormones and Cardiovascular Regulation 402

7. The endocrine system provides short-term regulation of cardiac output and peripheral resistance with epinephrine and norepinephrine from the adrenal medullae. Hormones involved in long-term regulation of blood pressure and volume are antidiuretic hormone (ADH), angiotensin II, erythropoietin (EPO), and atrial natriuretic peptide (ANP). *(Figure 13-11)*

8. ADH and angiotensin II also promote peripheral vasoconstriction in addition to their other functions. ADH and aldosterone promote water and electrolyte retention and stimulate thirst. EPO stimulates red blood cell production. ANP encourages sodium loss, fluid loss, reduces blood pressure, inhibits thirst, and lowers peripheral resistance.

PATTERNS OF CARDIOVASCULAR RESPONSE 403

Exercise and the Cardiovascular System 403

1. During exercise, blood flow to skeletal muscles increases at the expense of circulation to nonessential organs, and cardiac output rises. Car-

diovascular performance improves with training. Athletes have larger stroke volumes, slower resting heart rates, and greater cardiac reserves than do nonathletes.

Cardiovascular Response to Hemorrhaging 404

2. Blood loss causes an increase in cardiac output, mobilization of venous reserves, peripheral vasoconstriction, and the liberation of hormones that promote fluid retention and the manufacture of red blood cells.

THE BLOOD VESSELS 405

1. The peripheral distributions of arteries and veins are usually identical on both sides of the body except near the heart. *(Figure 13-12)*

The Pulmonary Circulation 405

2. The **pulmonary circuit** includes the **pulmonary trunk**, the **left and right pulmonary arteries**, and the **pulmonary veins**, which empty into the left atrium. *(Figure 13-13)*

The Systemic Circulation 406

3. In the **systemic circuit**, the **ascending aorta** gives rise to the coronary circulation. The **aortic arch** communicates with the **descending aorta**. *(Figures 13-14 to 13-18)*

4. Arteries in the neck and limbs are deep beneath the skin; in contrast, there are usually two sets of peripheral veins, one superficial and one deep. This dual-venous drainage is important for controlling body temperature.

5. The **superior vena cava** (**SVC**) receives blood from the head, neck, chest, shoulders, and arms. *(Figures 13-19 to 13-22)*

6. The **inferior vena cava** (**IVC**) collects most of the venous blood from organs below the diaphragm. *(Figure 13-22)*

7. The **hepatic portal system** directs blood from the other digestive organs to the liver before the blood returns to the heart. *(Figure 13-23)*

Fetal Circulation 418

8. The placenta receives blood from the two **umbilical arteries**. Blood returns to the fetus through the umbilical vein, which delivers blood to the **ductus venosus** in the liver. *(Figure 13-24a)*

9. Prior to delivery, blood bypasses the pulmonary circuit by flowing (1) from the right atrium into the left atrium through the **foramen ovale**, and (2) from the pulmonary trunk into the aortic arch via the **ductus arteriosus**. *(Figure 13-24b)*

AGING AND THE CARDIOVASCULAR SYSTEM 419

1. Age-related changes in the blood can include (1) decreased hematocrit, (2) the constriction or blockage of peripheral veins by a *thrombus* (stationary blood clot), and (3) the pooling of blood in the veins of the legs because the valves are not working effectively.

2. Age-related changes in the heart include (1) a reduction in the maximum cardiac output, (2) changes in the activities of the nodal and conducting cells, (3) a reduction in the elasticity of the fibrous skeleton, (4) a progressive **atherosclerosis** that can restrict coronary circulation, and (5) the replacement of damaged cardiac muscle cells by scar tissue.

3. Age-related changes in blood vessels, often related to **arteriosclerosis**, include (1) a reduced tolerance of inelastic walls of arteries to sudden pressure increases, which can lead to an aneurysm; (2) the deposition of calcium salts on weakened vascular walls, increasing the risk of a stroke or infarct; and (3) the formation of thrombi at atherosclerotic plaques.

INTEGRATION WITH OTHER SYSTEMS 419

1. The cardiovascular system delivers oxygen, nutrients, and hormones to all the body systems. *(Figure 13-25)*

Review Questions

Level 1: Reviewing Facts and Terms

Match each item in column A with the most closely related item in column B. Use letters for answers in the spaces provided.

COLUMN A

____ 1.   diastolic pressure

____ 2.   arterioles

____ 3.   hepatic vein

____ 4.   renal vein

____ 5.   aorta

____ 6.   precapillary sphincter

____ 7.   medulla oblongata

____ 8.   internal iliac artery

____ 9.   external iliac artery

____ 10. baroreceptors

____ 11. systolic pressure

____ 12. saphenous vein

COLUMN B

a.   drains the liver

b.   largest superficial vein in body

c.   carotid sinus

d.   minimum blood pressure

e.   blood supply to leg

f.   blood supply to pelvis

g.   peak blood pressure

h.   vasomotion

i.   largest artery in body

j.   drains the kidney

k.   vasomotor center

l.   smallest arterial vessels

13. Blood vessels that carry blood away from the heart are called:

    (a) veins   (b) arterioles

    (c) venules   (d) arteries

14. The layer of the arteriole wall that provides the properties of contractility and elasticity is the:

    (a) tunica adventitia   (b) tunica media

    (c) tunica interna   (d) tunica externa

15. The two-way exchange of substances between blood and body cells occurs only through:

    (a) arterioles   (b) capillaries

    (c) venules   (d) a, b, and c are correct

16. The blood vessels that collect blood from all tissues and organs and return it to the heart are the:

    (a) veins   (b) arteries

    (c) capillaries   (d) arterioles

17. Blood is compartmentalized within the veins because of the presence of:

    (a) venous reservoirs   (b) muscular walls

    (c) clots   (d) valves

18. The most important factor in vascular resistance is

    (a) the viscosity of the blood

    (b) friction between the blood and the vessel walls

    (c) turbulence due to irregular surfaces of blood vessels

    (d) the length of the blood vessels

19. In a blood pressure reading of 120/80, the 120 represents _____ and the 80 represents _____.

    (a) diastolic pressure; systolic pressure

    (b) pulse pressure; mean arterial pressure

    (c) systolic pressure; diastolic pressure

    (d) mean arterial pressure; pulse pressure

20. Hydrostatic pressure forces water _____ a solution; osmotic pressure forces water _____ a solution.

    (a) into; out of   (b) out of; into

    (c) out of; out of   (d) a, b, and c are incorrect

21. The two factors that assist the relatively low venous pressures in propelling blood toward the heart are

    (a) ventricular systole and valve closure

    (b) gravity and vasomotion

    (c) muscular compression and the respiratory pump

    (d) atrial and ventricular contractions

22. The arteries of the pulmonary circuit differ from those of the systemic circuit in that they carry:

    (a) oxygen and nutrients   (b) deoxygenated blood

    (c) oxygenated blood   (d) oxygen, carbon dioxide, and nutrients

23. The two arteries formed by the division of the brachiocephalic trunk are the:

    (a) aorta and internal carotid   (b) axillary and brachial

    (c) external and internal carotid   (d) common carotid and subclavian

13 **BLOOD VESSELS AND CIRCULATION**

The Anatomy of Blood Vessels • Circulatory Physiology • Cardiovascular Regulation • Patterns of Cardiovascular Response • The Blood Vessels

24. The unpaired arteries supplying blood to the visceral organs include the:
    (a) suprarenal, renal, lumbar
    (b) iliac, gonadal, femoral
    (c) celiac trunk, superior and inferior mesenterics
    (d) a, b, and c are correct

25. The artery generally used to feel the pulse at the wrist is the:
    (a) ulnar   (b) radial
    (c) fibular   (d) dorsalis

26. The vein that drains the dural sinuses of the brain is the:
    (a) cephalic   (b) great saphenous
    (c) internal jugular   (d) superior vena cava

27. The vein that collects most of the venous blood from below the diaphragm is the:
    (a) superior vena cava   (b) great saphenous
    (c) inferior vena cava   (d) azygos

28. (a) What are the primary forces that cause fluid to move out of a capillary and into the interstitial fluid at its arterial end? (b) What are the primary forces that cause fluid to move into a capillary from the interstitial fluid at its venous end?

29. What two effects occur when the baroreceptor response to elevated blood pressure is triggered?

30. What factors affect the activity of chemoreceptors in the carotid and aortic bodies?

31. What circulatory changes occur at birth?

32. What age-related changes take place in the blood, heart, and blood vessels?

Level 2: Reviewing Concepts

33. When dehydration occurs:
    (a) water reabsorption at the kidneys accelerates
    (b) fluids are reabsorbed from the interstitial fluid
    (c) blood osmotic pressure increases
    (d) a, b, and c are correct

34. Increased CO_2 levels in tissues would promote:
    (a) the constriction of precapillary sphincters
    (b) an increase in the pH of the blood
    (c) the dilation of precapillary sphincters
    (d) a decrease of blood flow to tissues

35. Elevated levels of the hormones ADH and angiotensin II will produce:
    (a) increased peripheral vasodilation
    (b) increased peripheral vasoconstriction
    (c) increased peripheral blood flow
    (d) increased venous return

36. Relate the anatomical differences between arteries and veins to their functions.

37. Why do capillaries permit the diffusion of materials whereas arteries and veins do not?

38. Why is blood flow to the brain relatively continuous and constant?

39. An accident victim displays the following symptoms: hypotension; pale, cool, moist skin; confusion and disorientation. Identify her condition and explain why these symptoms occur. If you took her pulse, what would you find?

Level 3: Critical Thinking and Clinical Applications

40. Bob is sitting outside on a warm day and is sweating profusely. His friend Mary wants to practice taking blood pressures, and he agrees to play patient. Mary finds that Bob's blood pressure is elevated, even though he is resting and has lost fluid from sweating. (She reasons that fluid loss should lower blood volume and thus blood pressure.) Mary asks you why Bob's blood pressure is high instead of low. What should you tell her?

41. People with allergies frequently take antihistamines and decongestants to relieve their symptoms. The medication's box warns that the medication should not be taken by individuals being treated for high blood pressure. Why?

42. Gina awakens suddenly to the sound of her alarm clock. Realizing she is late for class, she jumps to her feet, feels light-headed, and falls back on her bed. What probably caused this to happen? Why doesn't this always happen?

Answers to Concept Check Questions

Page 394

1. The blood vessels are veins. Arteries and arterioles have a relatively large amount of smooth muscle tissue in a thick, well-developed tunica media.
2. The relaxation of precapillary sphincters would increase the blood flow to a tissue.
3. Blood pressure in the arterial system pushes blood into the capillaries. Blood pressure on the venous side is very low, and other forces help keep the blood moving. Valves prevent the blood from flowing backward whenever the venous pressure drops.

Page 397

1. In a normal individual, the pressure should be greater in the aorta and least in the venae cavae. Blood, like other fluids, moves along a pressure gradient from high pressure to low pressure. If the pressure were higher in the inferior vena cava, the blood would flow backward.
2. When a person stands for periods of time, blood tends to pool in the lower extremities. The venous return to the heart decreases, and in turn the cardiac output decreases, sending less blood to the brain, causing light-headedness and fainting. A hot day adds to the effect, because body water is lost through sweating.

Page 405

1. Pressure at this site would decrease blood pressure at the carotid sinus, where the carotid baroreceptors are located. This would decrease baroreceptor stimulation and reduce the frequency of action potentials along the glossopharyngeal nerve (IX) to the medulla oblongata. An increase in sympathetic stimulation from the cardioacceleratory center would result in an increase in the heart rate. Stimulation of the vasomotor centers would cause vasoconstriction of arterioles and increase in blood pressure.
2. During exercise, blood pressure increases because (1) blood flow to muscles increases, (2) cardiac output increases, and (3) resistance in visceral tissues increases.
3. The vasoconstriction of the renal artery would decrease both blood flow and blood pressure at the kidney. In response, the kidney would increase the amount of renin that it releases, which in turn would lead to an increase in the level of angiotensin II. The angiotensin II will bring about increased blood pressure and increased blood volume.

Page 417

1. The blockage of the left subclavian artery would interfere with blood flow to the left arm, because that artery is the branch of the aortic arch that sends blood to the left arm.
2. The common carotid arteries carry blood to the head. A compression of one these arteries would decrease blood flow to the brain and lead to the loss of consciousness or even death.
3. Organs served by the celiac trunk include the stomach, spleen, liver, and pancreas; these organs would be affected most directly.

Page 419

1. This blood must have come from the umbilical vein, which carries oxygenated, nutrient-rich blood from the placenta to the fetus.
2. The cardiovascular system and lymphatic system are closely intertwined. The two systems are physically connected (lymphatic vessels drain into veins) and cell populations of the lymphatic system are carried within the cardiovascular system from one part of the body to another.

14

The Lymphatic System and Immunity

Chapter Outline and Objectives

Vocabulary Development

anamnesis....a memory; *anamnestic response*

apo-...away; *apoptosis*

chemo-...........................chemistry; *chemotaxis*

dia- ....through; *diapedesis*

-gen ....to produce; *pyrogen*

humor ....a liquid; *humoral immunity*

immunis ....safe; *immune*

inflammare ....to set on fire; *inflammation*

lympha...water; *lymph*

nodulus ....little knot; *nodule*

pathos ....disease; *pathogen*

pedesis ....a leaping; *diapedesis*

ptosis ....a falling; *apoptosis*

pyr ....fire; *pyrogen*

taxis ....arrangement; *chemotaxis*

*I*N 1876, the German physician Robert Koch proved that bacteria cause disease. He demonstrated that the bacterium shown here, Bacillus anthracis, causes anthrax, a deadly disease of farm animals such as sheep and cattle, and humans. In this chapter, we will examine the mechanisms the body uses to defend itself from general assaults and from specific invaders.

14 THE LYMPHATIC SYSTEM AND IMMUNITY

Organization of the Lymphatic System • Nonspecific Defenses • Specific Defenses: The Immune Response • Patterns of Immune Response

THE WORLD IS NOT ALWAYS KIND to the human body. Accidental bumps, cuts, and scrapes, chemical and thermal burns, extreme cold, and ultraviolet radiation are just a few of the hazards in the physical environment. Making matters worse, an assortment of viruses, bacteria, fungi, and parasites thrive in the environment. Many of these organisms are perfectly capable of not only surviving but also thriving inside our bodies—and potentially causing us great harm. These microorganisms, called **pathogens** (*pathos*, disease + *-gen*, to produce), are responsible for many human diseases. Each has a different mode of life and attacks the body in a characteristic way. For example, viruses, pathogens that lack a cellular structure and consist only of nucleic acid and protein, spend most of their time hiding within cells, the only place where they can reproduce; many bacteria multiply in the interstitial fluids; and the largest parasites, such as roundworms, burrow through internal organs.

Many different organs and systems work together in an effort to keep us alive and healthy. In this ongoing struggle, the **lymphatic system** plays a central role.

This chapter begins by examining the organization of the lymphatic system. We then consider how the lymphatic system interacts with cells and tissues of other systems to defend the body against infection and disease.

Organization of the Lymphatic System

One of the least familiar organ systems, the lymphatic system includes the following three components:

1. *Vessels.* A network of **lymphatic vessels**, often called **lymphatics**, begins in peripheral tissues and ends at connections to the venous system.
2. *Fluid.* A fluid called **lymph** flows through the lymphatic vessels. Lymph resembles plasma but contains a much lower concentration of suspended proteins.
3. *Lymphoid organs.* **Lymphoid organs** are connected to the lymphatic vessels and contain large numbers of lymphocytes. Examples are the lymph nodes, the spleen, and the thymus.

Figure 14-1● provides an overview of the lymphatic system.

FUNCTIONS OF THE LYMPHATIC SYSTEM

The primary functions of the lymphatic system are

• *The production, maintenance, and distribution of lymphocytes.* Lymphocytes are produced and stored within lymphoid organs, such as the spleen, thymus, and bone marrow. These

cells are vital to our ability to resist or overcome infection and disease. They respond to the presence of (1) invading pathogens, such as bacteria or viruses; (2) abnormal body cells, such as virus-infected cells or cancer cells; and (3) foreign proteins, such as the toxins released by some bacteria. Lymphocytes attempt to eliminate these threats or render them harmless by a combination of physical and chemical attack.

Lymphocytes respond to specific threats, such as a bacterial invasion of a tissue, by organizing a defense against

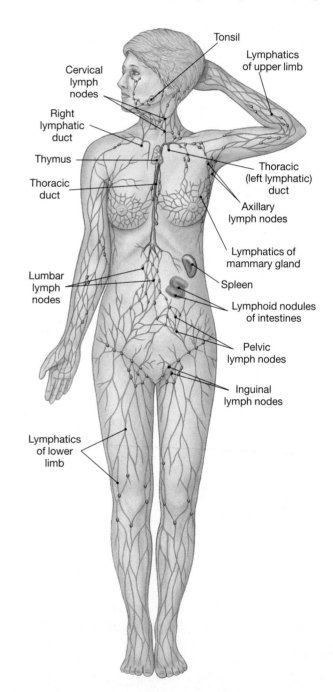

● *Figure 14-1* **The Components of the Lymphatic System**

that specific, or particular, type of bacterium. Such a specific defense of the body is known as an **immune response**. **Immunity** is the ability to resist infection and disease through the activation of specific defenses.

- *The return of fluid and solutes from peripheral tissues to the blood.* The return of tissue fluids through the lymphatic system maintains normal blood volume and eliminates local variations in the composition of the interstitial fluid. The volume of flow is considerable—roughly 3.6 liters per day—and a break in a major lymphatic vessel can cause a rapid and potentially fatal decline in blood volume.

- *The distribution of hormones, nutrients, and waste products from their tissues of origin to the general circulation.* Substances unable to enter the bloodstream directly may do so by way of the lymphatic vessels. For example, lipids absorbed by the digestive tract often fail to enter the bloodstream through capillaries. They reach the bloodstream only after they have traveled along lymphatic vessels (a process described further in Chapter 16).

LYMPHATIC VESSELS

Lymphatic vessels, or lymphatics, carry lymph from peripheral tissues to the venous system in all parts of the body except the central nervous system. The smallest vessels begin as blind pockets in peripheral tissues and are called **lymphatic capillaries** (Figure 14-2a●). The capillaries are lined by an endothelium (simple squamous epithelium). The lack of a basement membrane around the endothelium permits fluid to flow into a lymphatic capillary. The overlapping arrangement of the endothelial cells acts as a one-way valve to prevent backflow into the intercellular spaces.

From the lymphatic capillaries, lymph flows into larger lymphatic vessels that lead toward the trunk of the body. The walls of these lymphatics contain layers comparable to those of veins, and like veins, the larger lymphatic vessels contain valves (Figure 14-2b●). Pressures within the lymphatic system are extremely low, and the valves are essential to maintaining normal lymph flow.

The lymphatic vessels ultimately empty into two large collecting ducts (Figure 14-3●). The **thoracic duct** collects lymph from the lower abdomen, pelvis, and lower limbs and from the left half of the head, neck, and chest. It empties its collected lymph into the venous system near the junction between the left internal jugular vein and the left subclavian vein. The smaller **right lymphatic duct**, which ends at the comparable location on the right side, delivers lymph from the right side of the body above the diaphragm. The blockage of lymphatic drainage from a limb can cause the limb to

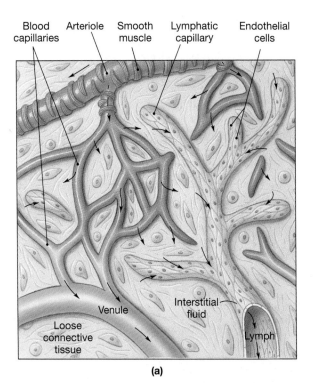

Blood capillaries   Arteriole   Smooth muscle   Lymphatic capillary   Endothelial cells

Venule   Interstitial fluid   Loose connective tissue   Lymph

(a)

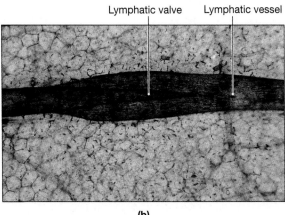

Lymphatic valve   Lymphatic vessel

(b)

● *Figure 14-2* **Lymphatic Capillaries**
(a) A three-dimensional view of the association of blood capillaries, tissue, interstitial fluid, and lymphatic capillaries. Arrows show the directions of interstitial fluid and lymph movement. **(b)** A valve within a small lymphatic vessel. (LM × 43)

swell due to the accumulation of interstitial fluid. This condition is called *lymphedema*.

LYMPHOCYTES

Lymphocytes were introduced in Chapter 11 because they account for roughly 25 percent of the circulating white blood cell population. ⟳ p. 356 But circulating lymphocytes are only a small fraction of the total lymphocyte population. The body contains around 1 trillion (10^{12}) lymphocytes, with a combined weight of over a kilogram (1 kg = 2.2 lb). At any given moment, most of the lymphocytes are found within

429

14 THE LYMPHATIC SYSTEM AND IMMUNITY

Organization of the Lymphatic System • Nonspecific Defenses • Specific Defenses: The Immune Response • Patterns of Immune Response

● *Figure 14-3* **The Lymphatic Ducts and the Venous System**
The thoracic duct carries lymph originating in tissues inferior to the diaphragm and from the left side of the upper body. The right lymphatic duct drains the right half of the body superior to the diaphragm and empties into the right subclavian vein.

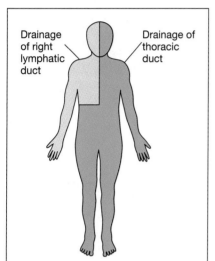

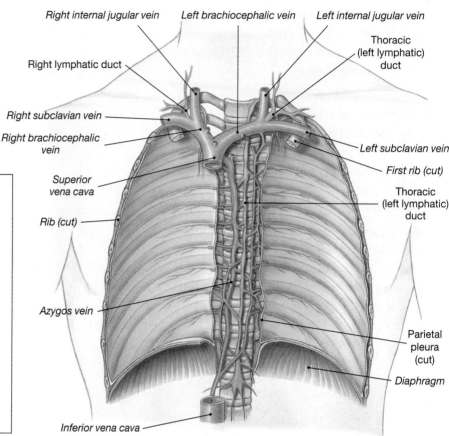

lymphoid organs or other tissues. The bloodstream provides a rapid transport system for lymphocytes moving from one tissue to another.

Types of Lymphocytes

The blood contains three classes of lymphocytes in the blood: *T cells*, *B cells*, and *NK cells*. Each lymphocyte class has distinctive functions.

T Cells. Approximately 80 percent of circulating lymphocytes are **T** (**t**hymus-dependent) **cells**. *Cytotoxic T cells* directly attack foreign cells or body cells infected by viruses. These lymphocytes are the primary cells that provide *cell-mediated immunity*, or *cellular immunity*. *Helper T cells* stimulate the activities of both T cells and B cells, and *suppressor T cells* inhibit both T cells and B cells. Helper and suppressor T cells are also called **regulatory T cells**.

B Cells. **B** (**b**one marrow–derived) **cells** make up 10–15 percent of circulating lymphocytes. B cells can differentiate into **plasma cells**, cells responsible for the production and secretion of **antibodies**. Antibodies are soluble proteins that are often called **immunoglobulins**. ∽ p. 346 Antibodies bind to specific chemical targets called **antigens**. Antigens are usually

pathogens, parts or products of pathogens, or other foreign compounds. When an antigen-antibody complex forms, it starts a chain of events leading to the destruction of the target compound or organism. B cells are said to be responsible for *antibody-mediated immunity*, or *humoral* ("liquid") *immunity*, because antibodies occur in body fluids.

NK Cells. The remaining 5–10 percent of circulating lymphocytes are **NK** (**n**atural **k**iller) **cells**. These lymphocytes will attack foreign cells, normal cells infected with viruses, and cancer cells that appear in normal tissues. Their continual monitoring of peripheral tissues is known as *immunological surveillance*.

The Origin and Circulation of Lymphocytes

Lymphocytes in the blood, bone marrow, spleen, thymus, and peripheral lymphoid tissues are visitors, not residents. Lymphocytes move throughout the body; they wander through a tissue and then enter a blood vessel or lymphatic vessel for transport to another site. In general, lymphocytes have relatively long life spans. Roughly 80 percent survive for four years, and some last 20 years or more. Throughout life, normal lymphocyte populations are maintained through the divisions of stem cells in the bone marrow and lymphoid tissues.

● *Figure 14-4* **The Origin of Lymphocytes**
Hemocytoblast divisions produce lymphoid stem cells with two fates. (**a**) One group remains in the bone marrow, producing daughter cells that mature into B cells and NK cells. (**b**) The second group migrates to the thymus, where subsequent divisions produce daughter cells that mature into T cells. Mature B cells, NK cells, and T cells circulate throughout the body in the bloodstream, and then (**c**) temporarily reside in peripheral tissues.

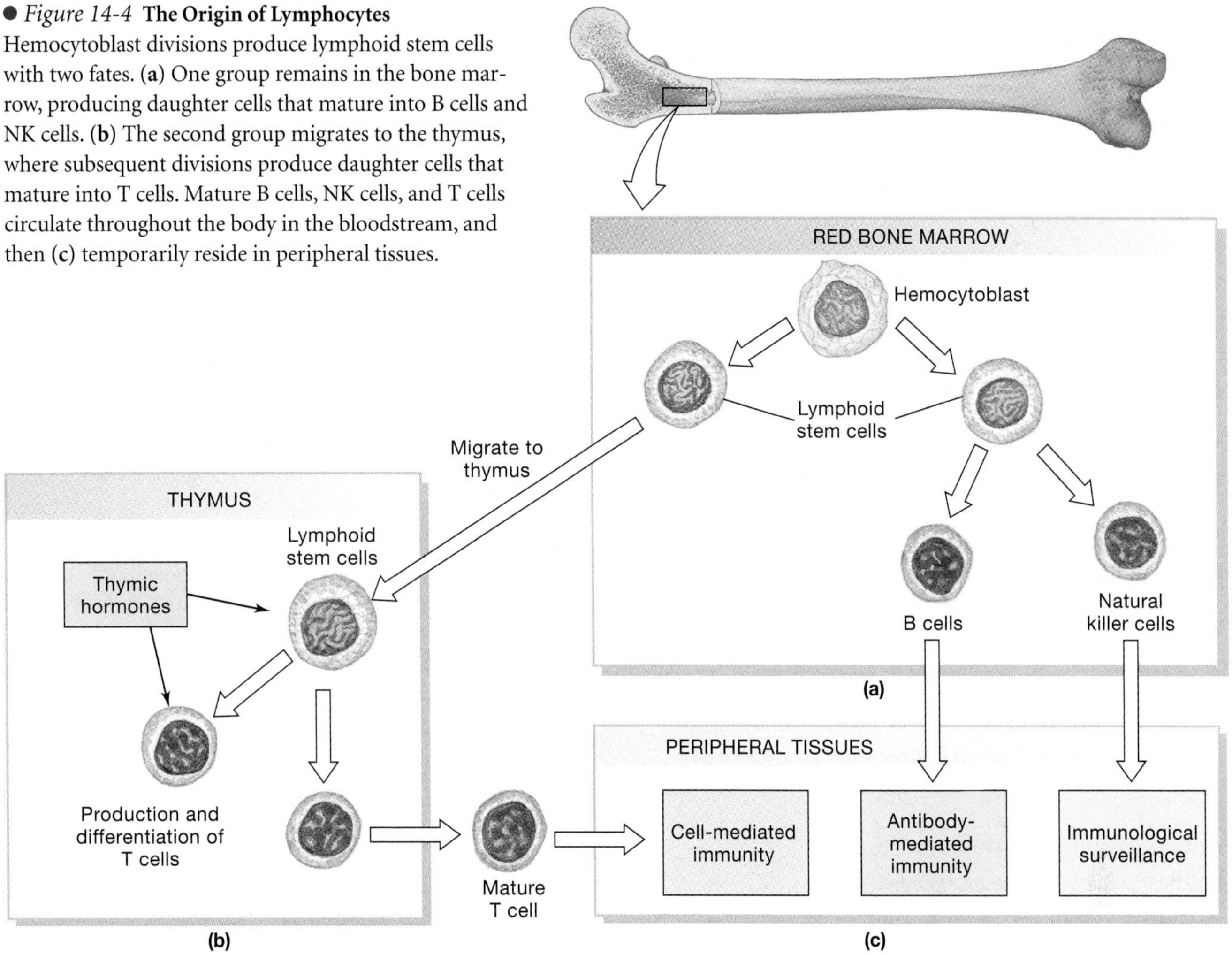

Lymphocyte production, or **lymphopoiesis** (lim-fō-poy-Ē-sis), involves the bone marrow and thymus (Figure 14-4●). As they develop, each B cell and T cell gains the ability to respond to the presence of a specific antigen, and NK cells gain the ability to recognize abnormal cells. *Hemocytoblasts* in the bone marrow produce lymphoid stem cells with two distinct fates. One group remains in the bone marrow and generates functional NK cells and B cells (Figure 14-4a●). The second group of lymphoid stem cells migrates to the thymus. Under the influence of hormones collectively known as the *thymosins*, these cells divide repeatedly, producing large numbers of T cells (Figure 14-4b●). As they mature, all three types of lymphocytes enter the bloodstream and migrate to peripheral tissues (Figure 14-4c●). As these lymphocyte populations migrate through peripheral tissues, they retain the ability to divide and produce daughter cells of the same type. For example, a dividing B cell produces other B cells, not T cells or NK cells. As we shall see, the ability to increase the number of lymphocytes of a specific type is important to the success of the immune response.

LYMPHOID NODULES

Lymphoid tissues are made up of loose connective tissue and lymphocytes. **Lymphoid nodules** are masses of lymphoid tissue that are not surrounded by a fibrous capsule. As a result, their size can increase or decrease, depending on the number of lymphocytes present at any given moment. In large lymphoid nodules, there is often a pale central region, called a *germinal center*, where lymphocytes are actively dividing.

Lymphoid nodules are found beneath the epithelia lining various organs of the respiratory, digestive, and urinary systems. All of these systems are open to the external environment and therefore provide a route of entry into the body for potentially harmful organisms and toxins.

Because our food usually contains foreign proteins and often contains bacteria, lymphoid nodules associated with the digestive tract play a particularly important role in the defense of the body. The **tonsils**, large lymphoid nodules in the walls of the pharynx, guard the entrance to the digestive and

431

14 THE LYMPHATIC SYSTEM AND IMMUNITY

Organization of the Lymphatic System • Nonspecific Defenses • Specific Defenses: The Immune Response • Patterns of Immune Response

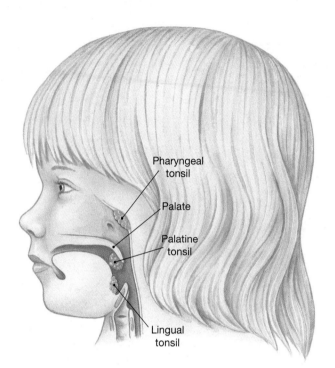

● *Figure 14-5* **The Tonsils**

The tonsils are lymphoid nodules in the wall of the pharynx. The single pharyngeal tonsil (the adenoids) lies above the palatine and lingual tonsils. Each tonsil contains germinal centers where lymphocyte divisions occur.

respiratory tracts (Figure 14-5●). Five tonsils are usually present: a single *pharyngeal tonsil*, or *adenoids*; a pair of *palatine tonsils*; and a pair of *lingual tonsils*. Aggregated lymphoid nodules, or *Peyer's patches*, also lie beneath the epithelial lining of the intestines, and fused lymphoid nodules dominate the walls of the *appendix*, a blind pouch located near the junction of the small and large intestines.

The lymphocytes in a lymphoid nodule are not always able to destroy bacterial or viral invaders, and if pathogens become established in a lymphoid nodule, an infection develops. Two examples are probably familiar to you: *tonsillitis*, an infection of one of the tonsils (usually the pharyngeal tonsil), and *appendicitis*, an infection of the lymphoid nodules in the appendix.

LYMPHOID ORGANS

Lymphoid organs have a stable internal structure and are separated from surrounding tissues by a fibrous capsule. Important lymphoid organs include the *lymph nodes*, the *thymus*, and the *spleen*.

Lymph Nodes

Lymph nodes are small, oval lymphoid organs covered by a fibrous capsule and ranging in diameter from 1–25 mm (up to 1 in.) (Figure 14-6●). Afferent lymphatics deliver lymph to a lymph node, and efferent lymphatics carry the lymph onward, toward the venous system. The lymph node functions like a kitchen water filter: It filters and purifies the lymph before it reaches the venous system. As lymph flows through a lymph node, at least 99 percent of the antigens present in the arriving lymph will be removed. As the antigens are detected and removed, T cells and B cells are stimulated, and an immune response is initiated. Lymph nodes are located in regions where they can detect and eliminate harmful "intruders" before they reach vital organs of the body (see Figure 14-1●, p. 428).

➕ **SWOLLEN GLANDS**

Lymph nodes are often called *lymph glands*, and "swollen glands" usually accompany tissue inflammation or infection. Chronic or excessive enlargement of lymph nodes, a sign called *lymphadenopathy* (lim-fad-e-NOP-a-thē), may occur in response to bacterial or viral infections, endocrine disorders, or cancer.

Since the lymphatic capillaries offer little resistance to the passage of cancer cells, cancer cells often spread along the

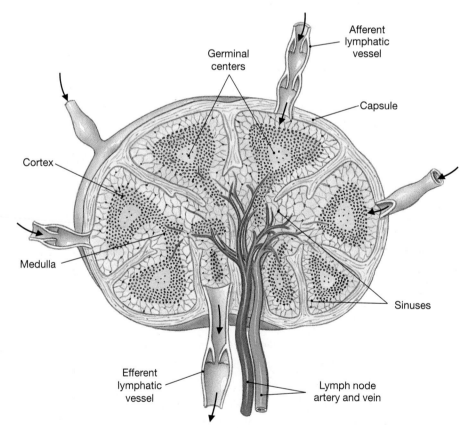

● *Figure 14-6* **The Structure of a Lymph Node**

The arrows indicate the direction of lymph flow. Mature B cells and T cells in the cortex and medulla remove antigens from the lymph and initiate immune responses.

lymphatics and become trapped in the lymph nodes. Thus, an analysis of swollen lymph nodes can provide information on the distribution and nature of the cancer cells, aiding in the selection of appropriate therapies. **Lymphomas** are an important group of lymphatic system cancers.

The Thymus

The **thymus** is a pink gland that lies in the mediastinum behind the sternum (Figure 14-7●). It is the site of T cell production and maturation. The thymus reaches its greatest size (relative to body size) in the first year or two after birth and its maximum absolute size during puberty, when it weighs between 30 and 40 g (1.06 to 1.41 oz). Thereafter, the thymus gradually decreases in size.

The thymus has two lobes, each divided into *lobules* by fibrous partitions, or *septae* (*septum*, a wall). Each lobule consists of a densely packed outer cortex and a paler central *medulla*. Lymphocytes in the cortex are dividing, and as the T cells mature, they migrate into the medulla, eventually entering one of the blood vessels in that region. Other cells within the lobules produce the thymic hormones collectively known as *thymosins*.

The Spleen

The adult **spleen** contains the largest collection of lymphatic tissue in the body. It is around 12 cm (5 in.) long and can weigh about 160 g (5.6 oz). As indicated in Figure 14-8a●, it sits wedged between the stomach, the left kidney, and the muscular diaphragm. The spleen normally has a deep red color because of the blood it contains (14-8b●). The cellular components constitute the **pulp** of the spleen. Areas of *red pulp* contain large quantities of red blood cells, whereas areas of *white pulp* resemble lymphoid nodules (14-8c●). The splenic artery enters the spleen and branches outward toward the capsule into many smaller arteries that are surrounded by white pulp. Capillaries then discharge the blood into a network of reticular fibers that makes up the red pulp. Blood from the red pulp enters *venous sinusoids*, small vessels lined by macrophages, and then flows into small veins and the splenic vein. As blood passes through the spleen, macrophages identify and engulf any damaged or infected cells. The presence of lymphocytes nearby ensures that any microorganisms or other abnormal antigens will stimulate an immune response.

The spleen performs similar functions for the blood that the lymph nodes perform for lymph: (1) removing abnormal

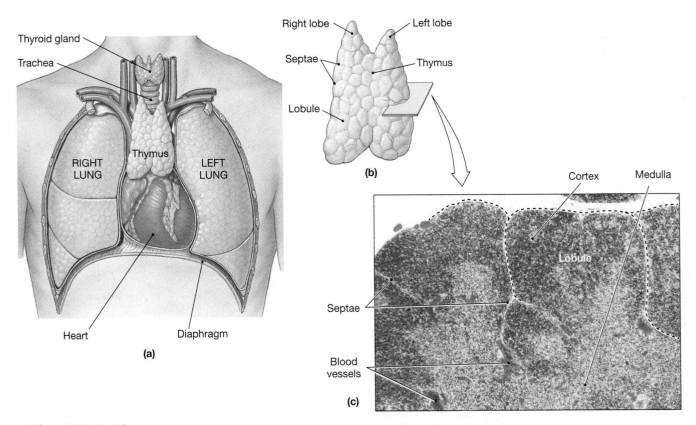

● *Figure 14-7* **The Thymus**
(a) The appearance and position of the thymus on gross dissection; notice its relationship to other organs in the chest. (b) Anatomical landmarks on the thymus. (c) A micrograph of the thymus. (LM × 40)

433

14 THE LYMPHATIC SYSTEM AND IMMUNITY

Organization of the Lymphatic System • **Nonspecific Defenses** • Specific Defenses: The Immune Response • Patterns of Immune Response

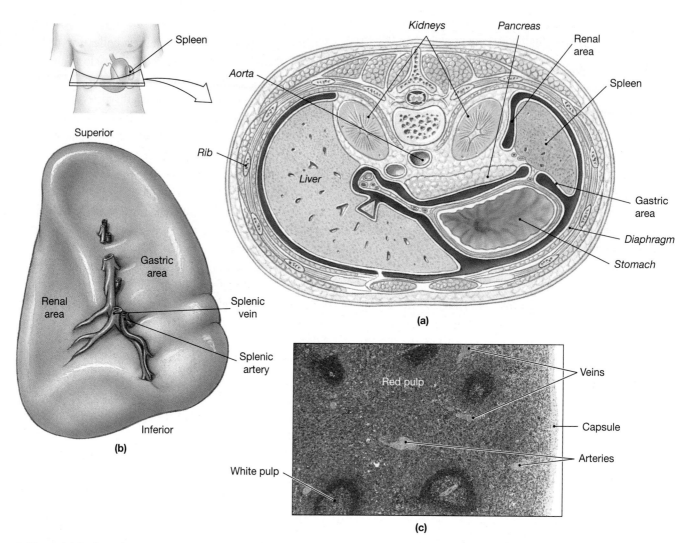

● *Figure 14-8* **The Spleen**
(a) The shape of the spleen roughly conforms to the shapes of adjacent organs. This transverse section through the trunk shows the typical position of the spleen within the abdominopelvic cavity. (b) The external appearance of the intact spleen. (c) The internal appearance of the spleen. The white pulp appears blue because the nuclei of lymphocytes stain very darkly. Red pulp contains an abundance of red blood cells.

blood cells and components, and (2) initiating immune responses by B cells and T cells in response to antigens in the circulating blood. In addition, the spleen stores iron from recycled red blood cells. ◌◌ p. 348

 INJURY TO THE SPLEEN
An impact to the left side of the abdomen can distort or damage the spleen. Such injuries are known risks of contact sports, such as football or hockey, and more-solitary athletic activities, such as skiing or sledding. However, the spleen tears so easily that even a seemingly minor blow to the side may rupture the capsule. The result is serious internal bleeding and eventual circulatory shock. The spleen can also be damaged by infection, inflammation, or invasion by cancer cells.

Because the spleen is relatively fragile, it is very difficult to repair surgically. (Sutures usually tear out before they have

been tensed enough to stop the bleeding.) Treatment for a severely ruptured spleen involves its complete removal, a process called a *splenectomy* (sple-NEK-to-mē). A person without a spleen survives without difficulty, but has a greater risk for bacterial infections than do individuals with a functional spleen.

CONCEPT CHECK QUESTIONS
Answers on page 454

❶ How would blockage of the thoracic duct affect the circulation of lymph?

❷ If the thymus gland failed to produce thymic hormones, which population of lymphocytes would be affected?

❸ Why do the lymph nodes enlarge during some infections?

THE LYMPHATIC SYSTEM AND BODY DEFENSES

The human body has many defense mechanisms, but they can be sorted into two categories:

1. **Nonspecific defenses** do not distinguish between one threat and another. These defenses, which are present at birth, include *physical barriers, phagocytic cells, immunological surveillance, interferons, complement, inflammation,* and *fever.* They provide the body with a defensive capability known as **nonspecific resistance**.

2. **Specific defenses** protect against particular threats. For example, a specific defense may protect against infection by one type of bacterium but ignore other bacteria and viruses. Many specific defenses develop after birth as a result of exposure to environmental hazards. Specific defenses are dependent on the activities of lymphocytes. The body's specific defenses produce a state of protection known as immunity, or **specific resistance**.

Nonspecific and specific resistance do not function in complete isolation from each other; both are necessary to provide adequate resistance to infection and disease.

Nonspecific Defenses

Nonspecific defenses are defenses that deny entrance to, or limit the spread of, microorganisms or other environmental agents to the body (Figure 14-9●). Their response is the same regardless of the type of invading agent.

PHYSICAL BARRIERS

To cause infection or disease, a foreign (antigenic) compound or pathogen must enter the body tissues, which requires crossing an epithelium. The epithelial covering of the skin, described in Chapter 5, has multiple layers, a keratin coating, and a network of desmosomes that lock adja-cent cells together. p. 111 These barriers create very effective protection for underlying tissues.

The exterior surface of the body has several layers of defense. The hairs found in most areas provide some protection against mechanical abrasion (especially on the scalp), and they often prevent hazardous materials or insects from contacting the skin's surface. The epidermal surface also receives

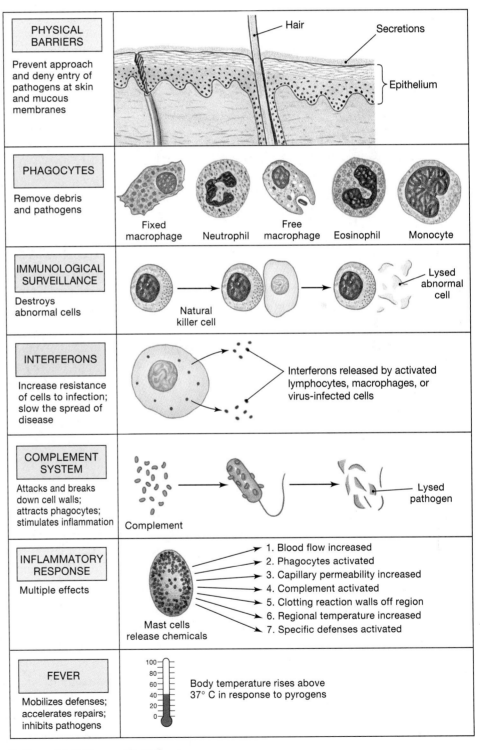

● *Figure 14-9* **Nonspecific Defenses**

14 THE LYMPHATIC SYSTEM AND IMMUNITY

Organization of the Lymphatic System • **Nonspecific Defenses** • Specific Defenses: The Immune Response • Patterns of Immune Response

the secretions of sebaceous glands and sweat glands. These secretions flush the surface, washing away microorganisms and chemical agents. The secretions also contain microbe-killing chemicals, destructive enzymes (*lysozymes*), and antibodies.

The epithelia lining the digestive, respiratory, urinary, and reproductive tracts are more delicate, but they are equally well defended. Mucus bathes most surfaces of the digestive tract, and the stomach contains a powerful acid that can destroy many potential pathogens. Mucus moves across the lining of the respiratory tract, urine flushes the urinary passageways, and glandular secretions do the same for the reproductive tract. Special enzymes, antibodies, and an acidic pH add to the effectiveness of these secretions.

PHAGOCYTES

Phagocytes in peripheral tissues remove cellular debris and respond to invasion by foreign compounds or pathogens. These cells represent the "first line" of cellular defense, often attacking and removing the microorganisms before lymphocytes become aware of the incident. Two general classes of phagocytic cells are found in the human body: *microphages* and *macrophages*.

Microphages are the neutrophils and eosinophils that normally circulate in the blood. These phagocytic cells leave the bloodstream and enter peripheral tissues subjected to injury or infection. As noted in Chapter 11, neutrophils are abundant, mobile, and quick to phagocytize cellular debris or invading bacteria. ⌒ p. 356 The less abundant eosinophils target foreign compounds or pathogens that have been coated with antibodies.

The body also contains several different types of **macrophages**—large, actively phagocytic cells derived from the monocytes of the blood. Almost every tissue in the body shelters resident or visiting macrophages. The relatively diffuse collection of phagocytic cells throughout the body is called the **monocyte-macrophage system**. In some organs, the macrophages have special names. For example, *microglia* are macrophages in the central nervous system and *Kupffer* (KOOP-fer) *cells* are macrophages found in and around blood channels in the liver.

All phagocytic cells function in much the same way, although the items selected for phagocytosis may differ from one cell type to another. Mobile macrophages and microphages also share a number of other functional characteristics in addition to phagocytosis. They can all move through capillary walls by squeezing between adjacent endothelial cells, a process known as *diapedesis* (*dia*, through + *pedesis*, a leaping). They may also be attracted to or repelled by chemicals in the surrounding fluids, a phenomenon called **chemotaxis** (*chemo-*,

chemistry + *taxis*, arrangement). They are particularly sensitive to chemicals released by other body cells or by pathogens.

IMMUNOLOGICAL SURVEILLANCE

Our immune defenses generally ignore normal cells in the body's tissues, but abnormal cells are attacked and destroyed. The constant monitoring of normal tissues is called **immunological surveillance**, and it primarily involves the lymphocytes known as NK (natural killer) cells. NK cells are sensitive to the presence of antigens on cell membranes of abnormal cells. When they encounter such antigens on a bacterium, a cancer cell, or a cell infected with viruses, NK cells secrete proteins that kill the abnormal cell by destroying its cell membrane.

NK cells respond immediately on contact with an abnormal cell. Killing the abnormal cells can slow the spread of a bacterial or viral infection and may eliminate cancer cells before they spread to other tissues. Unfortunately, some cancer cells avoid detection, a process called *immunological escape*. Once immunological escape has occurred, cancer cells can multiply and spread without interference by NK cells.

INTERFERONS

Interferons (in-ter-FĒR-onz) are small proteins released by activated lymphocytes, macrophages, and tissue cells infected with viruses. Normal cells exposed to these molecules respond by producing proteins that interfere with viral replication inside the cell. In addition to slowing the spread of viral infections, interferons stimulate the activities of macrophages and NK cells. Interferons are examples of **cytokines** (SĪ-tō-kīnz), chemical messengers released by tissue cells to coordinate local activities. Most cytokines act within one tissue, but those released by cellular defenders also act as hormones; they affect the activities of cells and tissues throughout the body. Their role in the regulation of specific defenses will be discussed later in the chapter.

COMPLEMENT

Your plasma contains 11 special *complement proteins*, which form the **complement system**. The term complement refers to the fact that this system "complements," or supplements, the action of antibodies. These proteins interact with one another in chain reactions comparable to those of the clotting system. The reaction is begun when a complement binds to an antibody molecule attached to a bacterial cell wall or directly to bacterial cell walls. The bound complement protein then interacts with other complement proteins. Complement activation is known to (1) attract phagocytes, (2) stimulate phagocytosis, (3) destroy cell membranes, and (4) promote inflammation.

INFLAMMATION

Inflammation, or the *inflammatory response*, is a localized tissue response to injury. ∞ p. 103 Inflammation produces local sensations of swelling, redness, heat, and pain. Inflammation can be produced by any stimulus that kills cells or damages loose connective tissue. *Mast cells* within the affected tissue play a pivotal role in this process. ∞ p. 93 When stimulated by mechanical stress or chemical changes in the local environment, mast cells release chemicals, including *histamine* and *heparin*, into the interstitial fluid. These chemicals initiate the process of inflammation. The end result of inflammation is the replacement or repair of damaged tissue.

Figure 14-10● summarizes the events that occur during inflammation of the skin. Comparable events will occur in almost any tissue subjected to physical damage or to infection. The purposes of inflammation are

- To perform a temporary repair at the injury site and prevent the access of additional pathogens.
- To slow the spread of pathogens from the injury site.
- To mobilize a wide range of defenses that can overcome the pathogens and aid permanent tissue repair. The repair process is called *regeneration*.

After an injury, tissue conditions generally become more abnormal before they begin to improve. **Necrosis** (ne-KR-Ō-sis) is the tissue degeneration that occurs after cells have been injured or destroyed. The process begins several hours after the original injury, and is caused by lysosomal enzymes. Lysosomes break down by autolysis, releasing digestive enzymes

that first destroy the injured cells and then attack surrounding tissues. ∞ p. 68 As local inflammation continues, debris and dead and dying cells accumulate at the injury site. This thick fluid mixture is known as **pus**. An accumulation of pus in an enclosed tissue space is called an *abscess*.

FEVER

Fever is the maintenance of a body temperature greater than 37.2°C (99°F). The hypothalamus, which contains nuclei that regulate body temperature, acts as the body's thermostat. ∞ p. 252 Circulating proteins called **pyrogens** (PĪ-rō-jenz; *pyr*, fire + *-gen*, to produce) can reset the thermostat in the hypothalamus and cause a rise in body temperature. Pathogens, bacterial toxins, and antigen-antibody complexes may act as pyrogens or stimulate the release of pyrogens by macrophages.

Within limits, a fever may be beneficial. High body temperatures can accelerate the activities of the immune system, such as phagocytosis. However, high fevers (over 40°C, or 104°F) can damage many physiological systems. For example, a high fever can cause CNS problems, such as nausea, disorientation, hallucinations, or convulsions.

CONCEPT CHECK QUESTIONS

Answers on page 454

❶ What types of cells would be affected by a decrease in the monocyte-forming cells in the bone marrow?

❷ A rise in the level of interferon in the body would suggest what kind of infection?

❸ What effects do pyrogens have in the body?

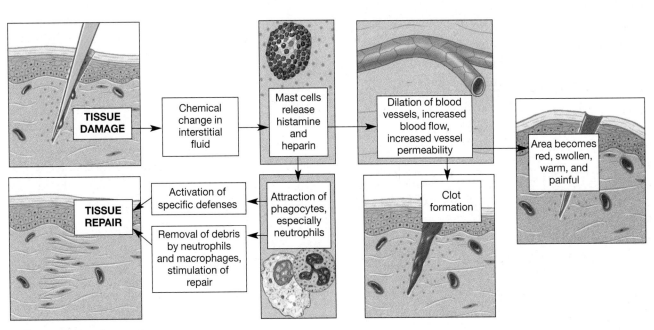

● *Figure 14-10* **Inflammation**

14 THE LYMPHATIC SYSTEM AND IMMUNITY

Organization of the Lymphatic System • Nonspecific Defenses • **Specific Defenses: The Immune Response** • Patterns of Immune Response

Specific Defenses: The Immune Response

The body's specific defenses that produce specific resistance, or immunity, are provided by the coordinated activities of T cells and B cells, which respond to the presence of *specific* antigens. *T cells* provide a defense against abnormal cells and pathogens inside living cells; this process is called **cell-mediated immunity**, or *cellular immunity*. *B cells* provide a defense against antigens and pathogens in body fluids. This process is called **antibody-mediated immunity**, or *humoral immunity*.

FORMS OF IMMUNITY

Immunity is either innate or acquired (Figure 14-11●). **Innate immunity** is genetically determined; it is present at birth and has no relation to previous exposure to the antigen involved. For example, people do not get the same diseases as goldfish.

Acquired immunity is either active or passive. **Active immunity** appears after exposure to an antigen, as a consequence of the immune response. The immune system is capable of defending against an enormous number of antigens. However, the appropriate defenses are mobilized only after an individual's lymphocytes encounter a particular antigen. Active immunity may develop as a result of natural exposure to an antigen in the environment (naturally acquired immunity) or from deliberate exposure to an antigen (induced active immunity).

Naturally acquired immunity normally begins to develop after birth, and it is continually enhanced as the individual encounters "new" pathogens or other antigens. You might compare this process to vocabulary development—a child begins with a few basic common words and learns new ones as needed. The purpose of induced active immunity is to stimulate antibody production under controlled conditions so that the individual will be able to overcome any natural exposure to the same type of pathogen at some time in the future. This is the basic principle behind *immunization*, or *vaccination*, to prevent disease. A **vaccine** is a preparation of antigens derived from a specific pathogen. Vaccines are given orally or by intramuscular or subcutaneous injection.

Passive immunity is produced by the transfer of antibodies from another individual. *Natural passive immunity* results when antibodies produced by the mother protect her baby against infections during development (across the placenta) or in early infancy (through breast milk). In *induced passive immunity*, antibodies are administered to fight infection or prevent disease. For example, antibodies that attack the rabies virus are injected into a person bitten by a rabid animal.

PROPERTIES OF IMMUNITY

There are four general properties of immunity:

1. *Specificity.* A specific defense is activated by a specific antigen, and the immune response targets only that particular antigen. This process is known as *antigen recognition*. **Specificity** occurs because the cell membrane of each T cell and B cell has receptors that will bind only one specific antigen, ignoring all other types of antigens. Either lymphocyte will destroy or inactivate that specific antigen without affecting other antigens or normal tissues.

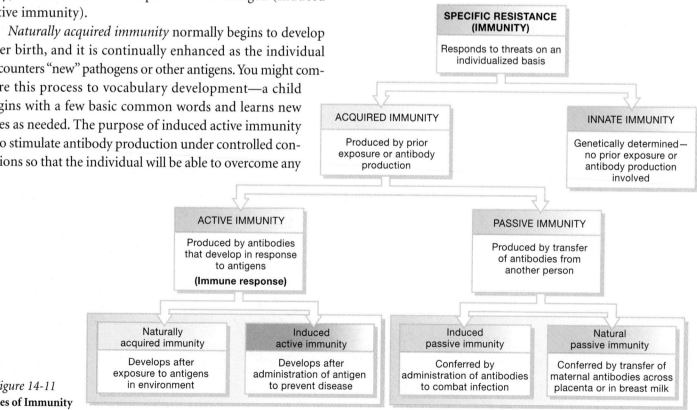

● *Figure 14-11*
Types of Immunity

2. *Versatility.* In the course of a normal lifetime, an individual encounters an enormous number of antigens—perhaps tens of thousands. Your immune system cannot anticipate which antigens it will encounter, so it must be ready to confront *any* antigen at *any* time. **Versatility** results from producing millions of different lymphocyte populations, each with different antigen receptors, and from variability in the structure of synthesized antibodies. In this way, the immune system can produce appropriate and specific responses to each antigen when exposure does occur.

3. *Memory.* The immune system "remembers" antigens that it encounters. As a result of **memory**, the immune response that occurs after a second exposure to an antigen is stronger and lasts longer than the response to the first exposure. During the initial response to an antigen, lymphocytes sensitive to its presence undergo repeated cell divisions. Two kinds of cells are produced: some that attack the invader and others that remain inactive unless they are exposed to the same antigen at a later date. These latter cells are **memory cells**, which enable the immune system to "remember" previously encountered antigens and launch a faster, stronger counterattack if one of them ever appears again.

4. *Tolerance.* **Tolerance** is said to exist when the immune system does not respond to a particular antigen. During their differentiation in the bone marrow (B cells) and thymus (T cells), cells that react to antigens normally present in the body are destroyed. As a result, B and T cells will ignore normal, or **self**, antigens and attack foreign, or **nonself**, antigens.

AN OVERVIEW OF THE IMMUNE RESPONSE

The purpose of the **immune response** is to destroy or inactivate pathogens, abnormal cells, and foreign molecules such as toxins. Figure 14-12● presents an overview of the immune response. When an antigen triggers an immune response, it usually activates T cells first and then B cells. The activation of T cells involves abnormal cells or active phagocytes exposed to the antigen. Once activated, the T cells ultimately attack the antigen and stimulate the activation of B cells. The activated B cells mature into cells that produce antibodies. The antibodies carried in the bloodstream bind to and attack the antigen.

T CELLS AND CELL-MEDIATED IMMUNITY

Before an immune response can begin, T cells must be activated by exposure to an antigen. However, this activation seldom happens by direct lymphocyte-antigen interaction, and foreign compounds or pathogens entering a tissue often fail to stimulate an immediate response.

T cells recognize antigens when they are bound to membrane receptors of other cells. The membrane receptors are called *major histocompatibility complex (MHC) proteins* and are grouped into two classes. Class I MHC proteins are found on the surfaces of all of our cells. These MHC proteins bind small peptide molecules (chains of amino acids) that are normally produced in the cell and carried to the cell membrane. If the cell is healthy and the peptides are normal, T cells ignore them. If the cell contains abnormal (non-self) or viral or bacterial peptides, they will appear in the cell membrane and T cells will be activated. Their activation will lead to destruction of the abnormal cell. The recognition of non-self peptides in transplanted tissue is the primary reason why donated organs are commonly rejected by the recipient. In the case of viruses or bacteria, T cells can be activated by contact with viral or bacterial antigens bound to class I MHC proteins on the surface of infected cells. The activation of T cells by these antigens results in the destruction of the infected cells.

Class II MHC proteins are found only in the membranes of lymphocytes and phagocytic *antigen-presenting cells (APCs)*. Examples of antigen-presenting cells include the monocyte-macrophage group, such as the free and fixed macrophages in connective tissues, and macrophages in the liver (Kupffer cells) and central nervous

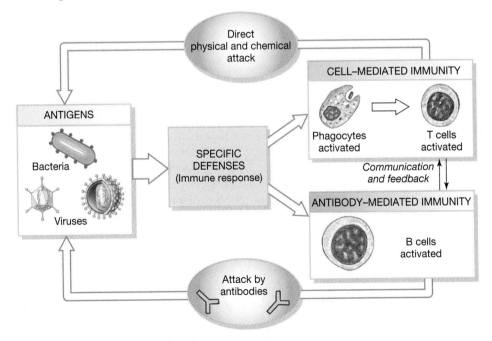

● *Figure 14-12* **An Overview of the Immune Response**

14 THE LYMPHATIC SYSTEM AND IMMUNITY

Organization of the Lymphatic System • Nonspecific Defenses • **Specific Defenses: The Immune Response** • Patterns of Immune Response

system (microglia). The APCs are specialized for activating T cells against foreign cells (including bacteria) and foreign proteins. Phagocytic APCs engulf and break down pathogens or foreign antigens. Fragments of the foreign antigens are displayed on their cell surfaces, bound to class II MHC proteins. T cells that contact this APC macrophage membrane become activated, initiating an immune response.

Inactive T cells have receptors that recognize the Class I or Class II MHC proteins. The receptors also have binding sites that detect specific bound antigens. T cell activation occurs when the MHC protein contains the specific antigen the T cell is programmed to detect. On activation, T cells divide and differentiate into cells with specific functions in the immune response. The major cell types are *cytotoxic T cells, helper T cells, memory T cells,* and *suppressor T cells.*

Cytotoxic T Cells

Cytotoxic T cells are responsible for cell-mediated immunity. They are activated by exposure to antigens bound to Class I MHC proteins. The activated cells undergo cell divisions that produce active cytotoxic cells and memory cells. The cytotoxic cells, also called *killer T cells,* track down and attack the bacteria, fungi, protozoa, or foreign transplanted tissues that contain the target antigen.

A cytotoxic T cell may accomplish its destruction in several ways (Figure 14-13●):

- By rupturing the antigenic cell membrane through the release of a destructive protein called *perforin.*
- By secreting a poisonous *lymphotoxin* (lim-fō-TOK-sin) that kills the target cell.
- By activating genes within the target cell that tell it to die. The process of genetically programmed cell death is called *apoptosis* (ap-op-TŌ-sis; *apo-,* away + *ptosis,* a falling). ∞ p. 75

Helper T Cells

Helper T cells are activated by exposure to antigens bound to Class II MHC proteins on antigen-presenting cells (Figure 14-14●). On activation, they divide to produce active helper T cells and memory cells. Helper T cells release a variety of cytokines that (1) coordinate specific and nonspecific defenses and (2) stimulate cell-mediated immunity and antibody-mediated immunity.

Memory T Cells

During the cell divisions that follow cytotoxic T cell and helper T cell activation, some of the cells develop into **memory T**

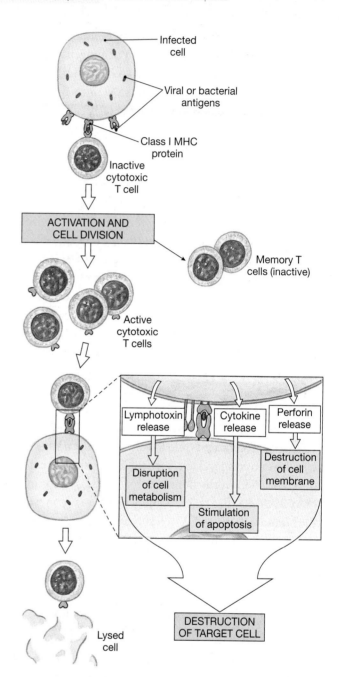

● *Figure 14-13* **The Activation of Cytotoxic T Cells**
For activation, an inactive cytotoxic T cell must encounter an appropriate antigen bound to MHC proteins. Once activated, the T cell undergoes divisions that produce memory T cells and active cytotoxic T cells. When one of these active cells encounters a membrane displaying the target antigen, the cytotoxic T cell will use one of several methods to destroy the cell.

cells. Memory T cells remain "in reserve." If the same antigen appears a second time, these cells will immediately differentiate into cytotoxic T cells and helper T cells, producing a more rapid and effective immune response.

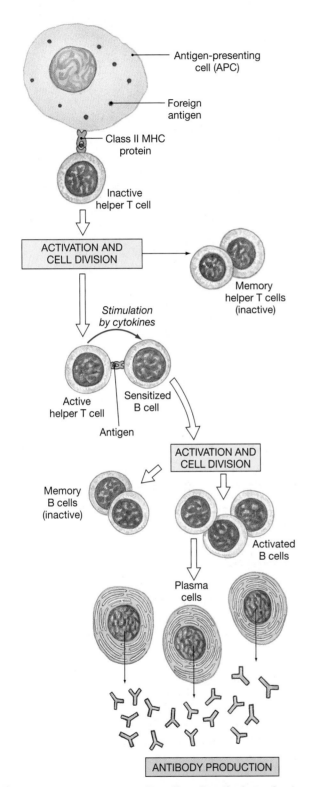

Suppressor T Cells

Activated **suppressor T cells** depress the responses of other T cells and B cells by secreting *suppression factors*. This suppression does not occur immediately, because suppressor T cells take much longer to become activated than other types of T cells. As a result, suppressor T cells act *after* the initial immune response. In effect, these cells "put on the brakes" and limit the degree of immune system activation from a single stimulus.

B CELLS AND ANTIBODY-MEDIATED IMMUNITY

B cells are responsible for launching a chemical attack on antigens by producing appropriate antibodies. B cell activation proceeds in a series of steps diagrammed in Figure 14-14●.

B Cell Activation

The body has millions of populations of B cells. Each B cell carries its antibody molecules in its cell membrane. If the interstitial fluid contains antigens that will bind to those antibodies, then the B cells become *sensitized*. Those antigens then enter the B cell and become displayed on Class II MHC proteins on the surface of the B cell. A helper T cell activated by the same antigen attaches to the MHC protein-antigen complex of the sensitized B cell and secretes cytokines that (1) promote B cell activation, (2) stimulate B cell division, (3) accelerate plasma cell production, and (4) enhance antibody production.

As Figure 14-14● shows, the activated B cell divides several times, producing daughter cells that differentiate into plasma cells and **memory B cells**. Plasma cells begin synthesizing and secreting large numbers of antibodies that have the same target as the antibodies on the surface of the sensitized B cell. Memory B cells perform the same role for antibody-mediated immunity that memory T cells perform for cellular-mediated immunity. They will remain in reserve to deal with subsequent exposure to the same antigens. At that time, they will respond and differentiate into antibody-secreting plasma cells.

Antibody Structure

An antibody molecule consists of two parallel pairs of polypeptide chains: one pair of long *heavy chains* and one pair of shorter *light chains* (Figure 14-15●). Each chain contains *constant* and *variable segments*. The constant segments of the heavy chains form the base of the antibody molecule. B cells produce only five types of constant segments. (These form the basis of the antibody classification scheme described in the following section.) The specificity of the antibody molecule depends on the

● *Figure 14-14* **The Activation of B Cells and Antibody Production**
Helper T cells must be exposed to appropriate antigens bound to Class II MHC proteins for activation. Activated helper T cells encountering a B cell sensitized to the same specific antigen release cytokines, which trigger the activation of the B cell. The activated B cell then goes through cycles of divisions, producing memory B cells and B cells that differentiate into plasma cells. The plasma cells secrete antibodies.

14 THE LYMPHATIC SYSTEM AND IMMUNITY

Organization of the Lymphatic System • Nonspecific Defenses • **Specific Defenses: The Immune Response** • Patterns of Immune Response

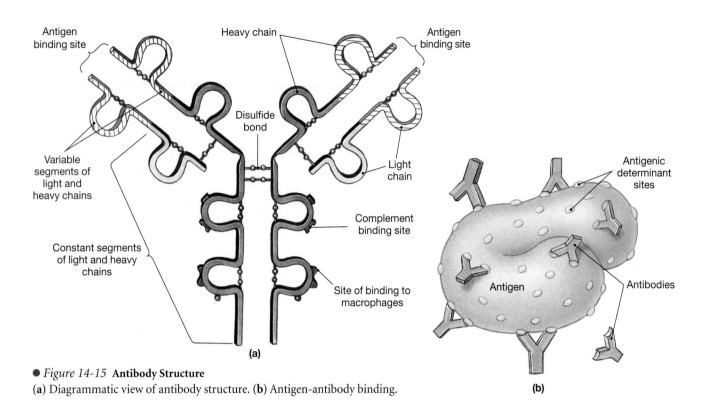

● *Figure 14-15* **Antibody Structure**
(**a**) Diagrammatic view of antibody structure. (**b**) Antigen-antibody binding.

structure of the variable segments of the light and heavy chains. The free tips of the two variable segments contain the **antigen binding sites** of the antibody molecule. Small differences in the amino acid sequence of the variable segments affect the precise shape of the antigen binding sites. The different shapes of these sites account for the differences in specificity between the antibodies produced by different B cells. It has been estimated that the 10 trillion or so B cells of a normal adult can produce an estimated 100 million different antibodies.

When an antibody molecule binds to its proper antigen, an **antigen-antibody complex** is formed. Antibodies do not bind to the entire antigen as a whole. They bind to certain portions of its exposed surface, regions called *antigenic determinant sites*. The specificity of that binding depends on the three-dimensional "fit" between the variable segments of the antibody molecule and the corresponding sites of the antigen. A *complete antigen* has at least two antigenic determinant sites, one for each arm of the antibody molecule. Exposure to a complete antigen can lead to B cell sensitization and an immune response. Most environmental antigens have multiple antigenic determinant sites; entire microorganisms may have thousands.

Classes of Antibodies

Body fluids contain five classes of antibodies, or **immunoglobulins** (**Igs**): *IgG, IgM, IgA, IgE,* and *IgD* (Table 14-1).

Immunoglobulin G, or IgG, is the largest and most diverse class of antibodies. The IgG antibodies are responsible for resistance against many viruses, bacteria, and bacterial toxins. They can also cross the placenta and provide passive immunity to the fetus. Circulating *IgM* antibodies attack bacteria and are responsible for the cross-reactions between incompatible blood types, described in Chapter 11. ∞ p. 353 IgA occurs in exocrine secretions such as mucus, tears, and saliva, and attacks pathogens before they enter the body. IgE with a bound antigen stimulates basophils and mast cells to release chemicals that stimulate inflammation. IgD is attached to B cells and is involved in their activation.

Antibody Function

The function of antibodies is to destroy antigens. The formation of an antigen-antibody complex may cause their elimination in several ways:

1. *Neutralization.* Antibodies can bind to viruses or bacterial toxins, making them incapable of attaching to a cell. This mechanism is called **neutralization**.

2. *Agglutination and precipitation.* When there are a large number of antigens close together, antibodies can bind to two antigenic sites on two different antigens. In this way, antibodies can tie antigens together and create large complexes. When the antigen is a soluble molecule, such as a

TABLE 14-2 *Examples of Cytokines of the Immune Response*

COMPOUND	FUNCTIONS
INTERLEUKINS	
IL-1	Stimulates T cells to produce IL-2, promotes inflammation, causes fever
IL-2, -12	Stimulates T cells and NK cells
IL-3	Stimulates production of blood cells
IL-4, -5, -6, -7, -10, -11	Promote differentiation and growth of B cells, and stimulate plasma cell formation and antibody production
INTERFERONS	Activate other cells to prevent viral entry and replication, stimulate NK cells and macrophages
TUMOR NECROSIS FACTORS (TNFs)	Kill tumor cells, slow tumor growth; stimulate activities of T cells and eosinophils, inhibit parasites and viruses
PHAGOCYTIC REGULATORS	
Monocyte-chemotactic factor (MCF)	Attracts monocytes, activates them to become macrophages
Migration-inhibitory factor (MIF)	Prevents macrophage migration from the area
COLONY-STIMULATING FACTORS (CSFs)	
M-CSF	Stimulates activity in the monocyte-macrophage line
GM-CSF	Stimulates production of both microphages and monocytes

antigen-presenting cells. Those terms are misleading because lymphocytes and macrophages may secrete the same chemical messenger, as well as cells involved with nonspecific defenses and tissue repair. Table 14-2 contains examples of some of the cytokines identified to date.

Interleukins (IL) may be the most diverse and important chemical messengers in the immune system. Interleukins have widespread effects that include increasing T cell sensitivity to antigens, stimulating B cell activity and antibody production, and enhancing nonspecific defenses, such as inflammation or fever. Interferons make the synthesizing cell and its neighbors resistant to viral infection, thereby slowing the spread of the virus. In addition to their antiviral activity, interferons attract and stimulate NK cells and macrophages.

Tumor necrosis factors (TNFs) slow tumor growth and kill sensitive tumor cells. In addition to these effects, tumor necrosis factors stimulate the production of neutrophils, eosinophils, and basophils, promote eosinophil activity, cause fever, and increase T cell sensitivity to interleukins.

Phagocytic regulators include several cytokines that coordinate the specific and nonspecific defenses by adjusting the activities of phagocytic cells. These cytokines include factors that attract free macrophages and microphages to the area and prevent their premature departure.

Colony-stimulating factors (CSFs) are produced by a wide variety of cells including active T cells, cells of the monocyte-macrophage group, endothelial cells, and fibroblasts. CSFs stimulate the production of blood cells in the bone marrow and lymphocytes in lymphoid tissues and organs.

CONCEPT CHECK QUESTIONS
Answers on page 454

1 How can the presence of an abnormal peptide within a cell initiate an immune response?

2 A decrease in the number of cytotoxic T cells would affect what type of immunity?

3 How would a lack of helper T cells affect the antibody-mediated immune response?

4 A sample of lymph contains an elevated number of plasma cells. Would you expect the amount of antibodies in the blood to be increasing or decreasing? Why?

14 **THE LYMPHATIC SYSTEM AND IMMUNITY**

Organization of the Lymphatic System • Nonspecific Defenses • Specific Defenses: The Immune Response • **Patterns of Immune Response**

Patterns of Immune Response

We have discussed the basic chemical and cellular interactions that follow the appearance of a foreign antigen. Figure 14-17● presents a broader, integrated view of the immune response and its relationship to nonspecific defenses.

IMMUNE DISORDERS

Because the immune response is so complex, there are many opportunities for things to go wrong. A variety of clinical conditions result from disorders of immune function. General classes of such disorders include *autoimmune disorders*, *immunodeficiency diseases*, and *allergies*. Immunodeficiency diseases and autoimmune disorders are relatively rare conditions—clear evidence of the effectiveness of the immune system's control mechanisms. Allergies make up a far more common, and usually far less dangerous, class of immune disorders.

Autoimmune Disorders

Autoimmune disorders develop when the immune response mistakenly targets normal body cells and tissues. The immune system usually recognizes and ignores the antigens normally found in the body—self antigens. The recognition system can malfunction, however. When it does, the activated B cells begin to manufacture antibodies against other body cells and tissues. These misguided antibodies are called *autoantibodies*. The resulting symptoms depend on the identity of the antigen attacked. For example, *rheumatoid arthritis* occurs when autoantibodies attack connective tissues around the joints and *insulin-dependent diabetes mellitus (IDDM)* is caused by autoantibodies that attack cells in the pancreatic islets.

Many autoimmune disorders appear to be cases of mistaken identity. For example, proteins associated with the measles, influenza, and other viruses contain amino acid sequences that are similar to those of myelin proteins in nervous tissue. As a result, antibodies that target these viruses may also attack myelin

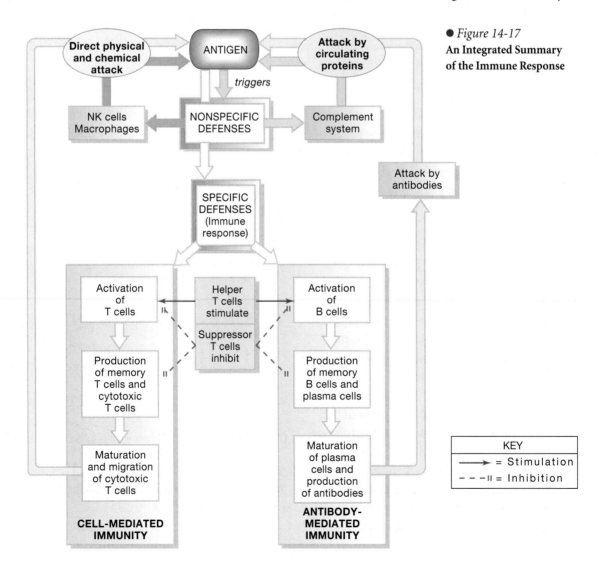

● *Figure 14-17*
An Integrated Summary of the Immune Response

sheaths, producing the neurological complications sometimes associated with a vaccination or viral infection.

Similarly, unusual types of MHC proteins have been linked to at least 50 clinical conditions. These disorders include psoriasis, rheumatoid arthritis, myasthenia gravis, multiple sclerosis, narcolepsy, Graves' disease, Addison's disease, pernicious anemia, systemic lupus erythematosus, and chronic hepatitis.

Immunodefiency Diseases

In an **immunodeficiency disease**, either the immune system fails to develop normally or the immune response is blocked in some way. Individuals with **severe combined immunodefiency disease (SCID)** fail to develop either cellular or antibody-mediated immunity. Such individuals cannot produce an immune response, and even a mild infection can prove fatal. Total isolation offers protection at great cost, with severe restrictions on lifestyle. Bone marrow transplants and gene-splicing techniques have been used to treat some types of SCID.

AIDS, an immunodeficiency disease, was considered on p. 443. AIDS is the result of a viral infection that targets primarily helper T cells. As the number of T cells falls, the normal immune response breaks down.

Allergies

Allergies are inappropriate or excessive immune responses to antigens. The sudden increase in cellular activity or antibody levels can have a number of unpleasant side effects. For example, neutrophils or cytotoxic T cells may destroy normal cells while attacking the antigen, or the antigen-antibody complex may trigger a massive inflammatory response. Antigens that trigger allergic reactions are often called **allergens**.

Four types of allergies are recognized: *immediate hypersensitivity (Type I), cytotoxic reactions (Type II), immune complex disorders (Type III),* and *delayed hypersensitivity (Type IV).* Immediate hypersensitivity is probably the most common type, and it includes "hay fever" and environmental allergies that may affect 15 percent of the U.S. population. The cross-reactions that occur following the transfusion of an incompatible blood type is an example of Type II hypersensitivity. p. 353

Immediate hypersensitivity begins with sensitization. Sensitization leads to the production of large quantities of IgE antibodies. The tendency to produce IgE antibodies in response to an allergen may be genetically determined. Due to the lag time needed to activate B cells, produce plasma cells, and male antibodies, the first exposure to an allergen does not produce allergic symptoms. However, the IgE antibodies that are produced at this time become attached to the cell membranes of basophils and mast cells throughout the body. When exposed to the same allergen later, these cells are stimulated to release histamine, heparin, several cytokines, prostaglandins, and other chemicals into the surrounding tissues. The result is a sudden inflammation of the affected tissues.

The severity of the allergic reaction depends on the person's sensitivity and the location involved. If the allergen exposure occurs at the body surface, the response is usually restricted to that area. If the allergen enters the bloodstream, the response could be lethal.

STRESS AND THE IMMUNE RESPONSE

Interleukin-1 is one of the first cytokines produced as part of the immune response. In addition to promoting inflammation, it also stimulates production of adenocorticotropic hormone (ACTH) by the anterior pituitary gland. This, in turn, leads to the secretion of glucocorticoids by the adrenal cortex. p. 327 The anti-inflammatory effects of the glucocorticoids may help control the intensity of the immune response. However, the long-term secretion of glucocorticoids, as in chronic stress, can inhibit the immune response and lower resistance to disease. p. 334 Glucocorticoids depress inflammation, inhibit phagocytes, and reduce interleukin production. The mechanisms involved are still under investigation. It is well known, however, that immune system depression due to chronic stress represents a serious threat to health.

AGE AND THE IMMUNE RESPONSE

With advancing age, your immune system becomes less effective at combating disease. T cells become less responsive to antigens, so fewer cytotoxic T cells respond to an infection. This effect may, at least in part, be associated with the gradual decrease in size of the thymus and reduced levels of circulating thymic hormones. Because the number of helper T cells is also reduced, B cells are less responsive, and antibody levels do not rise as quickly after antigen exposure. The net result is an increased susceptibility to viral and bacterial infection. For this reason, vaccinations for acute viral diseases, such as the flu (influenza), are strongly recommended for elderly people. The increased incidence of cancer in the elderly reflects the fact that immune surveillance declines, and tumor cells are not eliminated as effectively.

Integration with Other Systems

Figure 14-18● summarizes the interactions between the lymphatic system and other physiological systems. The particularly close relationships among the cells of the immune response and the nervous and endocrine systems are now the focus of intense research. For example, thymic hormones stimulate the

14 THE LYMPHATIC SYSTEM AND IMMUNITY

Organization of the Lymphatic System • Nonspecific Defenses • Specific Defenses: The Immune Response • Patterns of Immune Response

The Lymphatic System

For All Systems

Provides specific defenses against infection; immune surveillance eliminates cancer cells; returns tissue fluid to circulation

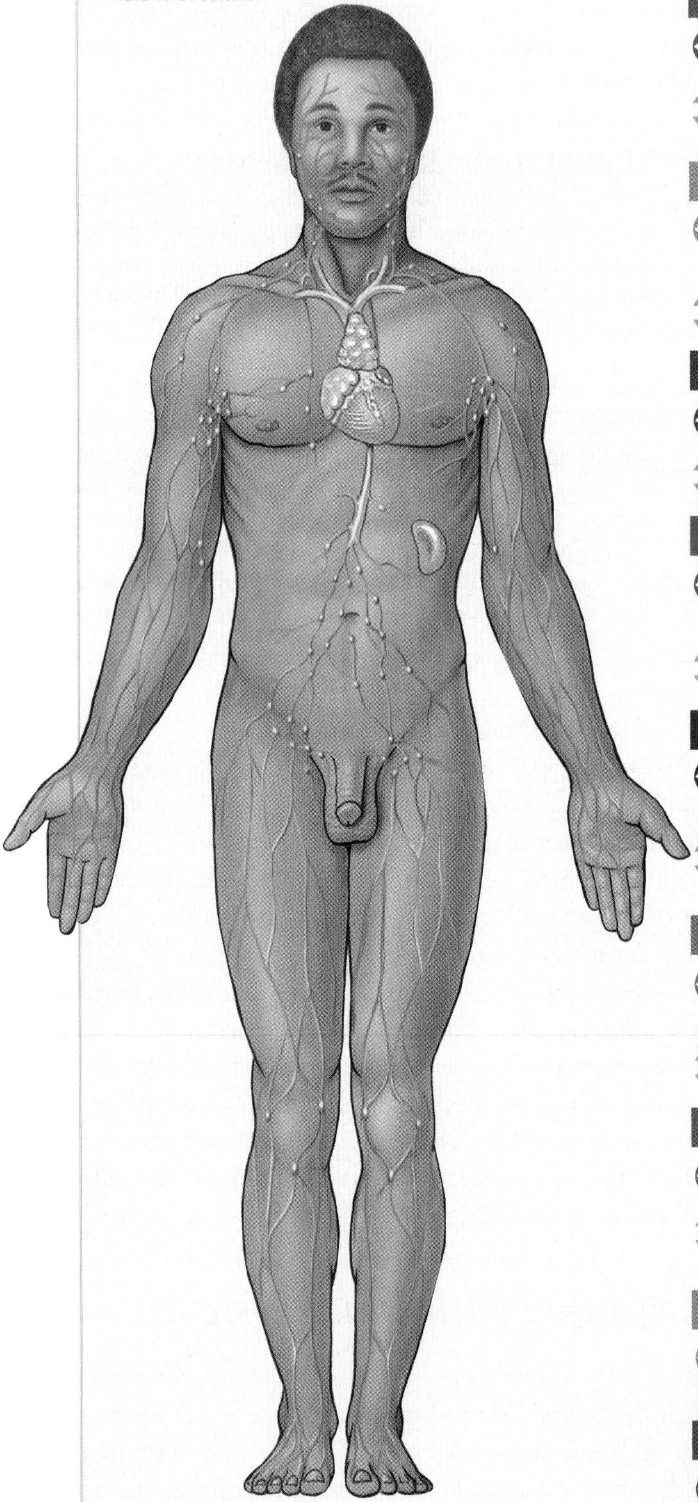

The Integumentary System

- Provides physical barriers to pathogen entry; macrophages in dermis resist infection and present antigens to trigger immune response; mast cells trigger inflammation, mobilize cells of lymphatic system
- Provides IgA for secretion onto integumentary surfaces

The Skeletal System

- Lymphocytes and other cells involved in the immune response are produced and stored in bone marrow
- Assists in repair of bone after injuries; macrophages fuse to become osteoclasts

The Muscular System

- Protects superficial lymph nodes and the lymphatic vessels in the abdominopelvic cavity; muscle contractions help propel lymph along lymphatic vessels
- Assists in repair after injuries

The Nervous System

- Microglia present antigens that stimulate specific defenses; glial cells secrete cytokines; innervation stimulates antigen-presenting cells
- Cytokines affect hypothalamic production of CRH and TRH

The Endocrine System

- Glucocorticoids have anti-inflammatory effects; thymosins stimulate development and maturation of lymphocytes; many hormones affect immune function
- Thymus secretes thymosins; cytokines affect cells throughout the body

The Cardiovascular System

- Distributes WBCs; carries antibodies that attack pathogens; clotting response helps restrict spread of pathogens; granulocytes and lymphocytes produced in bone marrow
- Fights infections of cardiovascular organs; returns tissue fluid to circulation

The Respiratory System

- Alveolar phagocytes present antigens and trigger specific defenses; provides oxygen required by lymphocytes and eliminates carbon dioxide generated during their metabolic activities
- Tonsils protect against infection at entrance to respiratory tract

The Digestive System

- Provides nutrients required by lymphatic tissues; digestive acids and enzymes provide nonspecific defense against pathogens
- Tonsils and lymphoid nodules of the intestine defend against infection and toxins absorbed from the digestive tract; lymphatics carry absorbed lipids to venous system

The Urinary System

- Eliminates metabolic wastes generated by cellular activity; acid pH of urine provides nonspecific defense against urinary tract infection

The Reproductive System

- Lysozymes and bactericidal chemicals in secretions provide nonspecific defense against reproductive tract infections
- Provides IgA for secretion by epithelial glands

● *Figure 14-18* **Functional Relationships Between the Lymphatic System and Other Systems**

production of thyroid hormone-releasing hormone (TRH) by the hypothalamus, leading to the release of thyroid-stimulating hormone (TSH) by the pituitary gland. As a result, circulating thyroid hormone levels increase and stimulate cell and tissue metabolism when an immune response is under way. Conversely, the nervous system can adjust the level of the immune response. For example, some antigen-presenting cells are innervated, and neurotransmitter release exaggerates the local immune response. It is also known that the immune response can decline suddenly after a brief period of emotional distress, such as a heated argument.

 ## MANIPULATING THE IMMUNE RESPONSE

As advances occur in our understanding of the immune system, innovative therapies involving combinations of cytokines, including interleukins, and monoclonal antibodies are appearing. For example, large quantities of identical antibodies, called *monoclonal* (mo-nō-KLŌ-nal) *antibodies*, are now regularly produced from genetically identical cells, under laboratory conditions. Uses of monoclonal antibodies include the clinical analysis of body fluids (as in detecting pregnancy from a urine sample) and providing passive immunity to disease. Because monoclonal antibodies are free of impurities, immunizations with these antibodies do not cause the unpleasant side effects that antibodies from other sources do.

In one treatment procedure, cytotoxic T cells were removed from patients with *malignant melanoma*, a particularly dangerous type of skin cancer. These lymphocytes were able to recognize tumor cells and had migrated to the tumor, but for some reason they appeared to be unable to kill the tumor cells. The extracted lymphocytes were collected and grown in the laboratory, and viruses were used to insert multiple copies of the genes responsible for the production of tumor necrosis factor. The patients were then given periodic infusions of these "supercharged" T cells. To enhance T cell activity further, the researchers also administered doses of interleukin-2. Initial results are promising. It is clear that the ability to manipulate the immune response will revolutionize the treatment of many serious diseases.

CONCEPT CHECK QUESTIONS
Answers on page 454

1 Would the primary response or the secondary response be more affected by a lack of memory B cells for a particular antigen?

2 Which kind of immunity protects a growing fetus, and how does that immunity develop?

3 How does the cardiovascular system support the body's defense mechanisms?

Related Clinical Terms

allergy: An inappropriate or excessive immune response to antigens, triggered by the stimulation of mast cells bound to IgE antibodies.

anaphylactic shock: A drop in blood pressure that may lead to circulatory collapse, resulting from a severe case of anaphylaxis.

anaphylaxis (a-na-fi-LAK-sis): A type of allergy in which a circulating allergen (antigen) affects mast cells throughout the body, producing numerous symptoms very quickly.

appendicitis: An infection and inflammation of the lymphoid nodules in the vermiform appendix.

bone marrow transplantation: The infusion of bone marrow from a compatible donor after the destruction of the host's marrow through radiation or chemotherapy; a treatment option for acute, late-stage lymphoma.

graft-versus-host disease (GVH): A condition that results when T cells in donor tissues, such as

bone marrow, attack the tissues of the recipient.

immunodeficiency disease: A disease in which either the immune system fails to develop normally or the immune response is blocked.

immunology: The study of the functions and disorders of the body's defense mechanisms.

immunosuppression: A reduction in the sensitivity of the immune system.

lymphadenopathy (lim-fad-e-NOP-a-thē): An excessive enlargement of lymph nodes.

lymphedema: A painless accumulation of lymph in a region whose lymphatic drainage has been blocked.

lymphomas: Malignant cancers consisting of abnormal lymphocytes or lymphoid stem cells; examples include *Hodgkin's disease* and *non-Hodgkin's lymphoma*.

mononucleosis: A condition resulting from chronic infection by the *Epstein–Barr virus (EBV)*; symptoms include enlargement of the spleen, fever, sore

throat, widespread swelling of lymph nodes, increased numbers of lymphocytes in the blood, and the presence of circulating antibodies to the virus.

splenomegaly (splen-ō-MEG-a-lē): An enlargement of the spleen.

systemic lupus erythematosus (LOO-pus e-rith-ē-ma-TŌ-sis) **(SLE):** An autoimmune disorder resulting from a breakdown in the antigen recognition mechanism, leading to the production of antibodies that destroy healthy cells and tissues.

tonsillectomy: The removal of an inflamed tonsil.

tonsillitis: An infection of one or more tonsils; symptoms include a sore throat, high fever, and leukocytosis (an abnormally high white blood cell count).

vaccine (vak-SĒN): A preparation of antigens derived from a specific pathogen; administered during *immunization*, or *vaccination*.

14 THE LYMPHATIC SYSTEM AND IMMUNITY

Organization of the Lymphatic System • Nonspecific Defenses • Specific Defenses: The Immune Response • Patterns of Immune Response

CHAPTER REVIEW

Key Terms

Summary Outline

INTRODUCTION 428

1. The cells, tissues, and organs of the **lymphatic system** play a central role in the body's defenses against a variety of **pathogens**, or disease-causing organisms.

2. **Lymphocytes**, the primary cells of the lymphatic system, provide an **immune response** to specific threats to the body. **Immunity** is the ability to resist infection and disease through the activation of specific defenses.

ORGANIZATION OF THE LYMPHATIC SYSTEM 428

1. The lymphatic system includes a network of **lymphatic vessels, lymphatics** that carry **lymph** (a fluid similar to plasma but with a lower concentration of proteins). A series of **lymphoid organs** are connected to the lymphatic vessels. (*Figure 14-1*)

Functions of the Lymphatic System 428

2. The lymphatic system produces, maintains, and distributes lymphocytes (cells that attack invading organisms, abnormal cells, and foreign proteins). The system also helps maintain blood volume and eliminate local variations in the composition of the interstitial fluid.

Lymphatic Vessels 429

3. Lymph flows along a network of lymphatics that originate in the **lymphatic capillaries**. The lymphatic vessels empty into the **thoracic duct** and the **right lymphatic duct**. (*Figures 14-1 to 14-3*)

Lymphocytes 429

4. The three classes of lymphocytes are **T cells** (*t*hymus-dependent), **B cells** (*b*one marrow-derived), and **natural killer (NK) cells**.

5. *Cytotoxic T cells* attack foreign cells or body cells infected by viruses; they provide cellular immunity. **Regulatory T cells** (*helper* and *suppressor T cells*) regulate and coordinate the immune response.

6. B cells can differentiate into **plasma cells**, which produce and secrete antibodies that react with specific chemical targets, or **antigens**. Antibodies in body fluids are also called **immunoglobulins**. B cells are responsible for *antibody-mediated immunity*, or *humoral immunity*.

7. NK cells attack foreign cells, normal cells infected with viruses, and cancer cells. They provide a monitoring service called *immunological surveillance*.

8. Lymphocytes continuously migrate in and out of the blood through the lymphoid tissues and organs. *Lymphopoiesis* (lymphocyte production) involves the bone marrow, thymus, and peripheral lymphoid tissues. (*Figure 14-4*)

Lymphoid Nodules.....................431

9. A **lymphoid nodule** consists of loose connective tissue containing densely packed lymphocytes. **Tonsils** are lymphoid nodules in the pharynx wall. (*Figure 14-5*)

Lymphoid Organs 432

10. Important lymphoid organs include the *lymph nodes*, the *thymus*, and the *spleen*. Lymphoid tissues and organs are distributed in areas especially vulnerable to injury or invasion.

11. **Lymph nodes** are encapsulated masses of lymphoid tissue containing lymphocytes. Lymph nodes monitor the lymph before it drains into the venous system, removing antigens and initiating appropriate immune responses. (*Figure 14-6*)

12. The **thymus** lies behind the sternum. T cells become mature in the thymus. (*Figure 14-7*)

13. The adult **spleen** contains the largest mass of lymphoid tissue in the body. The cellular components form the **pulp** of the spleen. *Red pulp* contains large numbers of red blood cells, and areas of *white pulp* resemble lymphoid nodules. The spleen removes antigens and damaged blood cells from the circulation, initiates appropriate immune responses, and stores iron obtained from recycled red blood cells. (*Figure 14-8*)

The Lymphatic System and Body Defenses.....................435

14. The lymphatic system is a major component of the body's defenses. These fall into two categories: (1) **nonspecific defenses**, which do not discriminate between one threat and another; and (2) **specific defenses**, which protect against threats on an individual basis.

NONSPECIFIC DEFENSES 435

1. Nonspecific defenses prevent the approach, deny the entrance, or limit the spread of living or nonliving hazards. (*Figure 14-9*)

Physical Barriers.....................435

2. Physical barriers include hair, epithelia, and various secretions of the integumentary and digestive systems.

Phagocytes.....................436

3. Two types of cells are **phagocytes: microphages** (neutrophils and eosinophils) and **macrophages** (cells of the *monocyte-macrophage system*).

4. Phagocytes move between cells through *diapedesis*, and they show *chemotaxis* (sensitivity and orientation to chemical stimuli).

5. **Immunological surveillance** involves constant monitoring of normal tissues by NK cells sensitive to abnormal antigens on the surfaces of otherwise normal cells. Cancer cells with tumor-specific antigens on their surfaces are killed.

6. **Interferons**, small proteins released by cells infected with viruses, trigger the production of antiviral proteins that interfere with viral replication inside the cell. Interferons are *cytokines*, chemical messengers released by tissue cells to coordinate local activities.

7. Eleven *complement proteins* make up the **complement system**. They interact with each other in chain reactions to destroy target cell membranes, stimulate inflammation, attract phagocytes, and enhance phagocytosis.

8. **Inflammation** represents a coordinated nonspecific response to tissue injury. *(Figure 14-10)*

9. A **fever** (body temperature greater than 37.2°C or 99°F) can inhibit pathogens and accelerate metabolic processes.

1. Specific defenses are provided by T cells and B cells. T cells provide cell-mediated immunity; B cells provide antibody-mediated immunity.

2. Specific immunity may involve **innate immunity** (genetically determined and present at birth) or **acquired immunity**. The two types of acquired immunity are **active immunity** (which appears following exposure to an antigen) and **passive immunity** (produced by the transfer of antibodies from another person). *(Figure 14-11)*

3. Lymphocytes provide specific immunity, which has four general characteristics: specificity, versatility, memory, and tolerance. **Specificity** occurs because only specific antigens can bind to receptors on T cell and B cell membranes. The immune system is **versatile** in that it can respond to any of the tens of thousands of antigens it encounters. **Memory cells** enable the immune system to "remember" previous target antigens. **Tolerance** refers to the ability of the immune system to ignore some antigens, such as those of body cells.

4. The goal of the **immune response** is to destroy or inactivate pathogens, abnormal cells, and foreign molecules. It is based on the activation of lymphocytes by specific antigens through the process of **antigen recognition**. *(Figure 14-12)*

5. T cells recognize antigens bound to Class I and Class II MHC proteins. Foreign antigens must usually be processed by macrophages and incorporated into their cell membranes bound to *MHC proteins* before they can activate T cells. T cells can also be activated by viral antigens displayed on the surfaces of virus-infected cells.

6. Activated T cells may differentiate into *cytotoxic T cells*, *memory T cells*, *suppressor T cells*, or *helper T cells*.

7. Cell-mediated immunity results from the activation of **cytotoxic**, or *killer*, **T cells**. Activated **memory T cells** remain on reserve to guard against future such attacks. *(Figure 14-13)*

8. **Suppressor T cells** depress the responses of other T and B cells.

9. **Helper T cells** secrete cytokines that help coordinate specific and nonspecific defenses and regulate cellular and humoral immunity. *(Figure 14-14)*

10. B cells, responsible for antibody-mediated immunity, undergo sensitization by a specific antigen before they become activated by helper T cells sensitive to the same antigen.

11. An activated B cell divides and produces plasma cells and **memory B cells**. Antibodies are produced by the plasma cells. *(Figure 14-14)*

12. An antibody molecule consists of two parallel pairs of polypeptide chains containing *fixed segments* and *variable segments*. *(Figure 14-15)*

13. When an antibody molecule binds to an antigen, they form an **antigen-antibody complex**. Antibodies focus on specific *antigenic determinant sites*.

14. Five classes of antibodies exist in body fluids: (1) **immunoglobulin G (IgG)**, responsible for resistance against many viruses, bacteria, and bacterial toxins; (2) **IgM**, the first antibody type secreted after an antigen arrives; (3) **IgA**, found in glandular secretions; (4) **IgE**, which releases chemicals that accelerate local inflammation; and (5) **IgD**, found on the surfaces of B cells. *(Table 14-1)*

15. Antibodies can destroy antigens through **neutralization**, **precipitation**, and **agglutination**, the activation of complement, the attraction of phagocytes, the enhancement of phagocytosis, and the stimulation of inflammation.

16. The antibodies produced by plasma cells on first exposure to an antigen are the agents of the **primary response**. Maximum antibody levels appear during the **secondary response**, which follows subsequent exposure to the same antigen. *(Figure 14-16)*

17. **Interleukins (IL)** increase T cell sensitivity to antigens exposed on macrophage membranes; stimulate B cell activity, plasma cell formation, and antibody production; and enhance nonspecific defenses.

18. Interferons slow the spread of a virus by making the cell that synthesized them, and that cell's neighbors, resistant to viral infections.

19. **Tumor necrosis factors (TNFs)** slow tumor growth and kill tumor cells.

20. Several **phagocytic regulators** adjust the activities of phagocytic cells to coordinate specific and nonspecific defenses. *(Table 14-2)*

1. Foreign antigens may undergo physical or chemical attack by specific and nonspecific defenses. *(Figure 14-17)*

2. **Autoimmune disorders** develop when the immune response mistakenly targets normal body cells and tissues.

14 THE LYMPHATIC SYSTEM AND IMMUNITY

Organization of the Lymphatic System • Nonspecific Defenses • Specific Defenses: The Immune Response • Patterns of Immune Response

3. In an **immunodeficiency disease**, either the immune system does not develop normally or the immune response is somehow blocked.

4. **Allergies** are inappropriate or excessive immune responses to **allergens** (antigens that trigger allergic reactions). The four types of allergies are *immediate hypersensitivity (Type I)*, *cytotoxic reactions (Type II)*, *immune complex disorders (Type III)*, and *delayed hypersensitivity (Type IV)*.

Age and the Immune Response 447

5. With aging, the immune system becomes less effective at combating disease.

INTEGRATION WITH OTHER SYSTEMS 447

1. The lymphatic system has extensive interactions with the nervous and endocrine systems. *(Figure 14-18)*

Review Questions

Level 1: Reviewing Facts and Terms

Match each item in column A with the most closely related item in column B. Use letters for answers in the spaces provided.

COLUMN A

____ 1. humoral immunity

____ 2. lymphoma

____ 3. complement

____ 4. microphages

____ 5. macrophages

____ 6. microglia

____ 7. interferon

____ 8. pyrogens

____ 9. innate immunity

____ 10. active immunity

____ 11. passive immunity

____ 12. apoptosis

COLUMN B

a. induce fever

b. system of circulating proteins

c. CNS macrophages

d. monocytes

e. genetically programmed cell death

f. transfers of antibodies

g. neutrophils, eosinophils

h. secretion of antibodies

i. present at birth

j. cytokine

k. lymphatic system cancer

l. exposure to antigen

---

13. Lymph from the lower abdomen, pelvis, and lower limbs is received by the:
    (a) right lymphatic duct
    (b) inguinal duct
    (c) thoracic duct
    (d) aorta

14. Lymphocytes responsible for providing cell-mediated immunity are called:
    (a) macrophages
    (b) B cells
    (c) plasma cells
    (d) cytotoxic T cells

15. B cells are responsible for:
    (a) cellular immunity
    (b) immunological surveillance
    (c) antibody-mediated immunity
    (d) a, b, and c are correct

16. Lymphoid stem cells that can form all types of lymphocytes occur in the:
    (a) bloodstream
    (b) thymus
    (c) bone marrow
    (d) spleen

17. Lymphatic vessels are found in all portions of the body except the:
    (a) lower limbs
    (b) central nervous system
    (c) head and neck region
    (d) hands and feet

18. The largest collection of lymphoid tissue in the body is contained in the:
    (a) adult spleen
    (b) adult thymus
    (c) bone marrow
    (d) tonsils

19. Red blood cells that are damaged or defective are removed from the circulation by the:
    (a) thymus
    (b) lymph nodes
    (c) spleen
    (d) tonsils

20. Phagocytes move through capillary walls by squeezing between adjacent endothelial cells, a process known as:
    (a) diapedesis
    (b) chemotaxis
    (c) adhesion
    (d) perforation

21. Perforins are destructive proteins associated with the activity of:
    (a) T cells
    (b) B cells
    (c) macrophages
    (d) plasma cells

22. Complement activation:
    (a) stimulates inflammation
    (b) attracts phagocytes
    (c) enhances phagocytosis
    (d) a, b, and c are correct

23. Inflammation:
    (a) aids in temporary repair at an injury site
    (b) slows the spread of pathogens

    (c) facilitates permanent repair
    (d) a, b, and c are correct

24. Memory B cells:
    (a) respond to a threat on first exposure
    (b) secrete large numbers of antibodies into the interstitial fluid
    (c) deal with subsequent injuries or infections that involve the same antigens
    (d) contain binding sites that can activate the complement system

25. Which two large collecting vessels are responsible for returning lymph to the veins of the circulatory system? What areas of the body does each serve?

26. Give a function for each of the following:
    (a) cytotoxic T cells
    (b) helper T cells
    (c) suppressor (regulatory) T cells
    (d) plasma cells
    (e) NK cells
    (f) interferons
    (g) T cells
    (h) B cells
    (i) interleukins

27. What seven defenses, present at birth, provide the body with the defensive capability known as nonspecific resistance?

Level 2: Reviewing Concepts

28. Compared with nonspecific defenses, specific defenses:
    (a) do not discriminate between one threat and another
    (b) are always present at birth
    (c) provide protection against threats on an individual basis
    (d) deny entrance of pathogens to the body

29. T cells and B cells can be activated only by:
    (a) pathogenic microorganisms
    (b) interleukins, interferons, and colony-stimulating factors

    (c) cells infected with viruses, bacterial cells, or cancer cells
    (d) exposure to a specific antigen at a specific site on a cell membrane

30. List and explain the four general properties of immunity.

31. How does the formation of an antibody-antigen complex cause elimination of an antigen?

32. What effects follow activation of the complement system?

Level 3: Critical Thinking and Clinical Applications

33. An investigator at a crime scene discovers some body fluid on the victim's clothing. The investigator carefully takes a sample and sends it to the crime lab for analysis. On the basis of analysis of immunoglobulins, could the crime lab determine whether the sample was blood plasma or semen? Explain.

34. Ted finds out that he has been exposed to the measles. He is concerned that he might have contracted the disease. His physician takes a blood sample and sends it to a lab for antibody titers. The results show an elevated level of IgM antibodies to rubella (measles) virus but very few IgG antibodies to the virus. Did Ted contract the disease?

14 THE LYMPHATIC SYSTEM AND IMMUNITY

Organization of the Lymphatic System • Nonspecific Defenses • Specific Defenses: The Immune Response • Patterns of Immune Response

Answers to Concept Check Questions

Page 434

1. The thoracic duct drains lymph from the area beneath the diaphragm and the left side of the head and thorax. Most of the lymph enters the venous blood by way of this duct. A blockage of this duct would not only impair circulation of lymph through most of the body, but also promote the accumulation of fluid in the extremities (lymphedema).

2. The thymic hormones from the thymus play a role in the differentiation of stem lymphocytes into T lymphocytes. A lack of these hormones would result in an absence of T lymphocytes.

3. During an infection, the lymphocytes and phagocytes in the lymph nodes in the affected region undergo cell division to deal with the infectious agent better. This increase in the number of cells in the nodes causes the nodes to become enlarged or swollen.

Page 437

1. A decrease in the number of monocyte-forming cells in the bone marrow would result in a decreased number of macrophages in the body, since all of the different macrophages are derived from the monocytes. The macrophages would include the microglia of the CNS, the Kuppfer's cells of the liver, and alveolar macrophages in the lungs.

2. A rise in interferon levels would indicate a viral infection. Interferon is released from cells that are infected with viruses. It does not help the infected cell but "interferes" with the virus's ability to infect other cells.

3. Pyrogens stimulate the temperature control area within the hypothalamus. The result is an increase in body temperature, or fever.

Page 445

1. Abnormal peptides within a cell can become attached to MHC proteins and then are displayed on the surface of the cell. Peptides presented in this manner are then recognized by T cells, which can initiate an immune response.

2. Cytotoxic T cells function in cell-mediated immunity. A decrease in the number of cytotoxic T cells would interfere with the ability to kill foreign cells and tissues as well as cells infected by viruses.

3. Helper T cells promote B cell division, the maturation of plasma cells, and the production of antibody by plasma cells. Without helper T cells, the antibody-mediated immune response would be much slower and less efficient.

4. Plasma cells produce and secrete antibodies. We would expect to see increased levels of antibodies in the blood if the number of plasma cells were increased.

Page 449

1. The secondary response would be affected by the lack of memory B cells for a specific antigen. The ability to produce a secondary response depends on the presence of memory B cells and T cells that are formed during the primary response to an antigen. These cells are not involved in the primary response but are held in reserve against future contact with the same antigen.

2. The developing fetus is protected primarily by passive immunity, the product of IgG antibodies that cross the placenta from the mother's circulation. In addition, the fetus may show some degree of active cellular immunity by the third month of development.

3. The cardiovascular system aids the nonspecific and specific body defenses by distributing white blood cells, carrying antibodies, and, through the clotting response, helps restrict the spread of pathogens.

EXPLORE *MediaLab*

EXPLORATION #1

Estimated time for completion: 10 minutes

HYPERTENSION, or high blood pressure, is a common ailment in the United States. Sixty-four percent of American males aged 75 and over suffer from hypertension, as do 75 percent of the females in that age group. There are many risk factors for high blood pressure, including a diet high in sodium, smoking, stressful living, arteriosclerosis, and genetic predisposition. Interestingly, African Americans demonstrate a higher incidence of hypertension than do Caucasians. However, native African populations are not prone to hypertension. Can you explain what might account for this discrepancy? If necessary, review the circulatory physiology discussed in Chapter 13. Combine what you know of blood pressure regulation with what you understand of homeostatic interactions between systems to help formulate an answer. Then go to Chapter 14 at the Companion Web site, visit the MediaLab section, and choose the key word "Hypertension." Here you will find an article published in *Scientific American* in 1999 that will help you understand how genes and the environment interact in this particular instance. Alter your original hypothesis to fit the data presented in the *Scientific American* article.

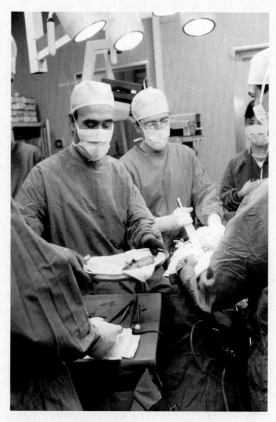

Surgeons performing a kidney transplant.

EXPLORATION #2

Estimated time for completion: 10 minutes

ORGAN TRANSPLANTS are a fact of life in the twenty-first century. In the news, we hear of individuals needing bone marrow transplants and organs such as hearts, livers, kidneys, and lungs. We see documentaries thanking those who donated organs, and we are asked if we would like fill out an organ-donor card when we apply for a driver's license in most states. While this all seems fairly commonplace now, the entire science of organ transplantation would not exist were it not for an understanding of the lymphatic system and the immune response. In order for an organ transplant to be successful, the organ must "match" the recipient. Using Figure 14-11● as a guide, write a short explanation of why tissue matching is such an important part of the entire procedure. How does the lymphatic system relate to transplant surgery? To gain a deeper understanding of tissue typing, go to Chapter 14 at the Companion Web site, visit the MediaLab section, and click on the key words "Tissue Typing." Use the information presented at this site to then expand your explanation of tissue matching.

Relaxing in front of the television surrounded by many of the risk factors that contribute to hypertension.

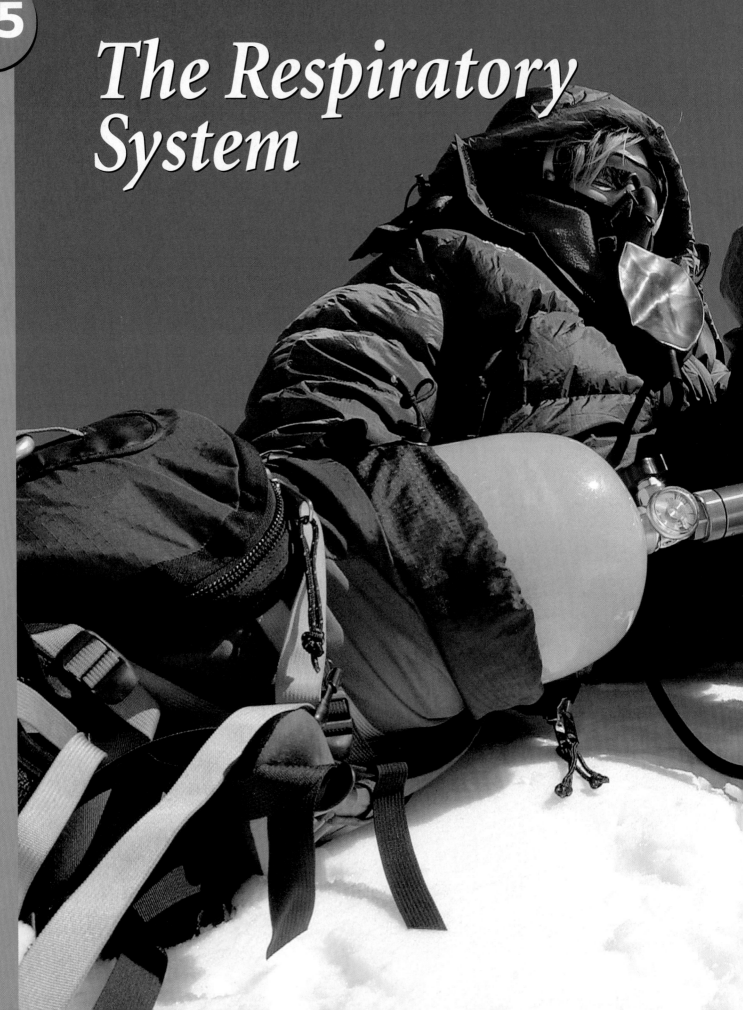

The Respiratory System

CHAPTER OUTLINE AND OBJECTIVES

Vocabulary Development

alveolus ....a hollow cavity; *alveolus, alveolar duct*
ateles ....imperfect; *atelectasis*
bronchus ....windpipe, airway; *bronchus*
cricoid ....ring-shaped; *cricoid cartilage*
ektasis ....expansion; *atelectasis*
-ia ....condition; *pneumonia*
kentesis ....puncture; *thoracentesis*
oris ....mouth; *oropharynx*
pneuma ....air; *pneumothorax*
pneumon ....lung; *pneumonia*
stoma ....mouth; *tracheostomy*
thorac- ....chest; *thoracentesis*
thyroid ....shield-shaped; *thyroid cartilage*

OUR UNDERSTANDING *of the effects of exposure to high altitude have mostly come from studies based on humans. A major problem is the decline in the concentration, or partial pressure, of oxygen at great heights. Here, a climber makes up for the "thin air" by breathing oxygen through a mask.*

15 **THE RESPIRATORY SYSTEM**

The Functions of the Respiratory System • The Organization of the Respiratory System • Respiratory Physiology • The Control of Respiration

L IVING CELLS NEED ENERGY for maintenance, growth, defense, and replication. Our cells obtain that energy through aerobic respiration, a process that requires oxygen and produces carbon dioxide. ∞ p. 69 The respiratory system provides the cells in the body with the means to obtain oxygen and eliminate carbon dioxide. This exchange takes place within our lungs at air-filled pockets called **alveoli** (al-VĒ-o-lī). The gas exchange surfaces of the alveoli are relatively delicate—they must be very thin to encourage rapid diffusion between the air and the blood. The cardiovascular system provides the link between your interstitial fluids and the exchange surfaces of your lungs. The circulating blood carries oxygen from the lungs to peripheral tissues; it also accepts and transports the carbon dioxide generated by those tissues, delivering it to the lungs.

Our discussion of the respiratory system begins by following air as it travels from outside the body to the alveoli of the lungs. We then consider the mechanics of breathing (the physical movement of air into and out of the lungs), and the physiology of respiration, which includes the processes of breathing and gas transport and exchange between the air, blood, and tissues.

The Functions of the Respiratory System

The **respiratory system** has five basic functions: It (1) provides a large area for gas exchange between air and circulating blood; (2) moves air to and from the gas-exchange surfaces of the lungs; (3) protects the respiratory surfaces from dehydration and temperature changes, and provides nonspecific defenses against invading pathogens; (4) produces sounds permitting speech, singing, and nonverbal communication; and (5) provides olfactory sensations to the central nervous system for the sense of smell.

The Organization of the Respiratory System

The respiratory system consists of the nose, nasal cavity, paranasal sinuses, pharynx (throat), larynx (voice box), trachea (windpipe), bronchi and the lungs, including the bronchioles (conducting passageways) and the alveoli (exchange surfaces) (Figure 15-1●).

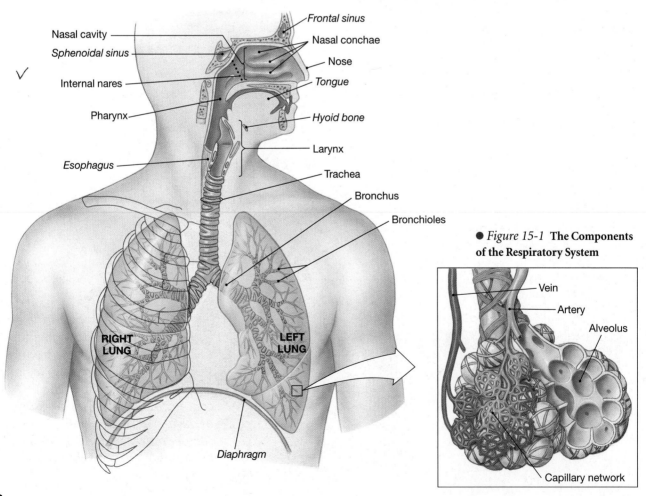

Nasal cavity
Frontal sinus
Sphenoidal sinus
Nasal conchae
Internal nares
Nose
Pharynx
Tongue
Esophagus
Hyoid bone
Larynx
Trachea
Bronchus
Bronchioles

RIGHT LUNG
LEFT LUNG

Diaphragm

● *Figure 15-1* **The Components of the Respiratory System**

Vein
Artery
Alveolus
Capillary network

THE RESPIRATORY TRACT

Your **respiratory tract** consists of the airways that carry air to and from the exchange surfaces of your lungs. The respiratory tract can be divided into a *conducting portion* and a *respiratory portion*. The conducting portion begins at the entrance to the nasal cavity and continues through the pharynx, larynx, trachea, bronchi, and the larger bronchioles. The respiratory portion includes the smallest and most delicate bronchioles and the alveoli, which are the site of gas exchange.

In addition to delivering air to the lungs, the conducting passageways filter, warm, and humidify the air, thereby protecting the alveoli from debris, pathogens, and environmental extremes. By the time the air reaches the alveoli, most foreign particles and pathogens have been removed, and the humidity and temperature are within acceptable limits.

THE NOSE

Air normally enters the respiratory system through the paired **external nares** (NĀ-rēz), or nostrils, which communicate with the **nasal cavity**. The **vestibule** (VES-ti-būl) is the anterior portion of the nasal cavity enclosed by the flexible tissues of the nose. Here coarse hairs extend across the nostrils and guard the nasal cavity from large airborne particles such as sand, dust, and insects.

The maxillary, nasal, frontal, ethmoid, and sphenoid bones form the lateral and superior walls of the nasal cavity. The nasal septum divides the nasal cavity into left and right sides. The anterior portion of the nasal septum is formed of hyaline cartilage. The bony posterior septum includes portions of the vomer and the ethmoid bone. A bony **hard palate**, formed by the palatine and maxillary bones, forms the floor of the nasal cavity and separates the oral and nasal cavities. A fleshy **soft palate** extends behind the hard palate and underlies the **nasopharynx** (nā-zō-FĀR-inks). The nasal cavity opens into the nasopharynx at the **internal nares**.

The *superior, middle,* and *inferior nasal conchae* project toward the nasal septum from the lateral walls of the nasal cavity (Figure 15-2●). To pass from the vestibule to the internal nares, air tends to flow in narrow grooves between adjacent conchae. As the air eddies and swirls, like water flowing over

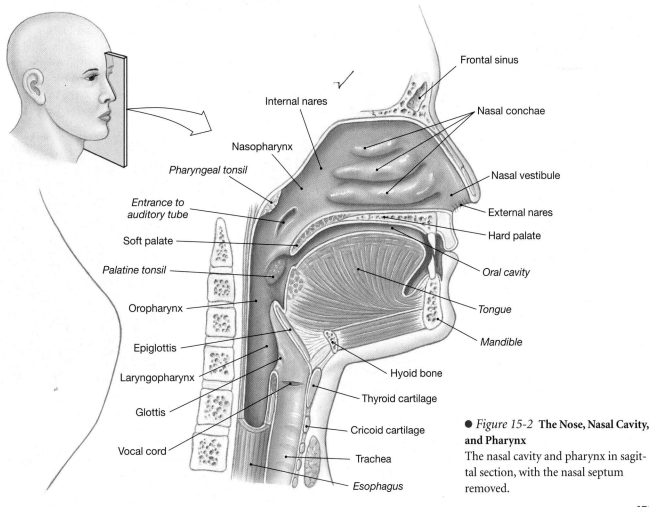

Internal nares
Frontal sinus
Nasal conchae
Nasopharynx
Pharyngeal tonsil
Nasal vestibule
Entrance to auditory tube
External nares
Soft palate
Hard palate
Palatine tonsil
Oral cavity
Oropharynx
Tongue
Epiglottis
Mandible
Laryngopharynx
Hyoid bone
Glottis
Thyroid cartilage
Vocal cord
Cricoid cartilage
Trachea
Esophagus

● *Figure 15-2* **The Nose, Nasal Cavity, and Pharynx**
The nasal cavity and pharynx in sagittal section, with the nasal septum removed.

459

15 THE RESPIRATORY SYSTEM

The Functions of the Respiratory System • **The Organization of the Respiratory System** • Respiratory Physiology • The Control of Respiration

rapids, small airborne particles come in contact with the mucus that coats the lining of the nasal cavity. In addition to promoting filtration, the turbulent flow allows extra time for warming and humidifying the incoming air.

The nasal cavity and much of the rest of the respiratory tract are lined by a protective, mucous membrane. ⟳ p. 98 This membrane is made up of the *respiratory epithelium*, a ciliated epithelium containing many *goblet cells*, and an underlying loose connective tissue layer (the *lamina propria*) containing mucous glands (Figure 15-3●). The goblet cells and mucous glands produce mucus that bathes the exposed surfaces of the nasal cavity and lower respiratory tract. Cilia sweep that mucus and any trapped debris or microorganisms toward the pharynx, where they can be swallowed and exposed to the acids and enzymes of the stomach. The respiratory surfaces of the nasal cavity are also flushed by mucus produced in the *paranasal sinuses* (the *frontal*, *sphenoid*, *ethmoid*, and *maxillary sinuses*), and by tears flowing through the nasolacrimal duct (see Figure 6-13●, p. 144, and Figure 9-7b●, p. 285). Exposure to noxious vapors, large quantities of dust and debris, allergens, or pathogens usually causes a rapid increase in the rate of mucus production, and a "runny nose" develops.

THE PHARYNX

The **pharynx**, or throat, is a chamber shared by the digestive and respiratory systems that extends between the internal nares and the entrances to the larynx and esophagus. Its three subdivisions—the nasopharynx, the oropharynx, and the laryngopharynx—are shown in Figure 15-2●. The nasophar-

ynx is connected to the nasal cavity by the internal nares and extends to the posterior edge of the soft palate. The nasopharynx, lined by a typical respiratory epithelium, contains the entrances to the *auditory tubes* and the *pharyngeal tonsil*. The **oropharynx** extends between the soft palate and the base of the tongue at the level of the hyoid bone. The palatine tonsils lie in the lateral walls of the oropharynx. The narrow **laryngopharynx** (lā-rin-gō-FĀR-inks) includes that portion of the pharynx between the hyoid bone and the entrance to the esophagus. Materials entering the digestive tract pass through both the oropharynx and laryngopharynx. These regions are lined by a stratified squamous epithelium that can resist mechanical abrasion, chemical attack, and pathogenic invasion.

 ## CYSTIC FIBROSIS

Cystic fibrosis (CF) is the most common lethal inherited disease affecting Caucasians of Northern European descent, occurring at a frequency of 1 birth in 2500. It occurs with less frequency in those of southern European ancestry, in the Ashkenazic Jewish population, and in black African Americans. The condition results from a defective gene on chromosome 7. Individuals with CF seldom survive past age 30; death is usually the result of a massive bacterial infection of the lungs and associated heart failure. The most serious symptoms appear because the respiratory mucosa in these individuals produces a dense, viscous mucus that cannot be transported by the cilia of the respiratory tract. Mucus transport stops, and mucus blocks the smaller respiratory passageways. This blockage reduces the diameter of the airways, making breathing difficult, and the inactivation of the normal respiratory defenses leads to frequent bacterial infections.

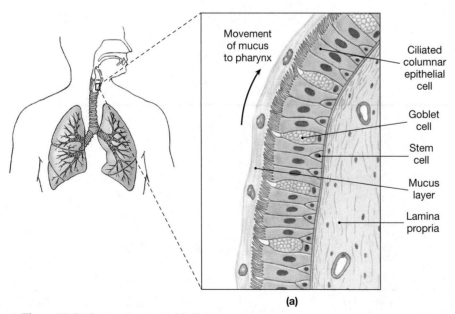

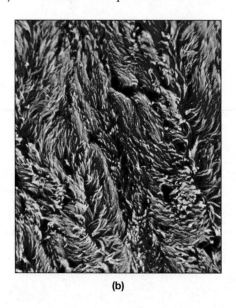

(b)

● *Figure 15-3* **The Respiratory Epithelium**
(a) A sketch showing the sectional appearance of the respiratory epithelium and its role in mucus transport. **(b)** A surface view of the epithelium. The cilia of the epithelial cells form a dense layer that resembles a shag carpet. The movement of these cilia propels mucus across the epithelial surface. (SEM × 1614)

THE LARYNX

Incoming air leaves the laryngopharynx by passing through the **glottis** (GLOT-is), a narrow opening surrounded and protected by the **larynx** (LAR-inks), or *voice box*. The larynx contains nine cartilages that are stabilized by ligaments, skeletal muscles, or both (Figure 15-4●). There are three large cartilages: the *epiglottis*, *thyroid cartilage*, and *cricoid cartilage*.

The elastic **epiglottis** (ep-i-GLOT-is) projects above the glottis. During swallowing, the larynx is elevated and the epiglottis folds back over the glottis, preventing the entry of liquids or solid food into the respiratory tract. The curving **thyroid** (*thyroid*; shield-shaped) **cartilage** forms much of the anterior and lateral surfaces of the larynx. A prominent ridge on the anterior surface of this cartilage forms the "Adam's apple." The thyroid sits atop the **cricoid** (KRĪ-koyd; ring-shaped) **cartilage**, which provides posterior support to the larynx. The thyroid and cricoid cartilages protect the glottis and the entrance to the trachea, and their broad surfaces provide sites for the attachment of important laryngeal muscles and ligaments.

The larynx also contains three pairs of smaller cartilages. The *arytenoid*, *corniculate*, and *cuneiform cartilages* are supported by the cricoid cartilage. Two pairs of ligaments, enclosed by folds of epithelium, extend across the larynx between the thyroid cartilage and these cartilages, considerably reducing the size of the glottis. The upper pair, known as the **false vocal cords**, are relatively inelastic. They help prevent foreign objects

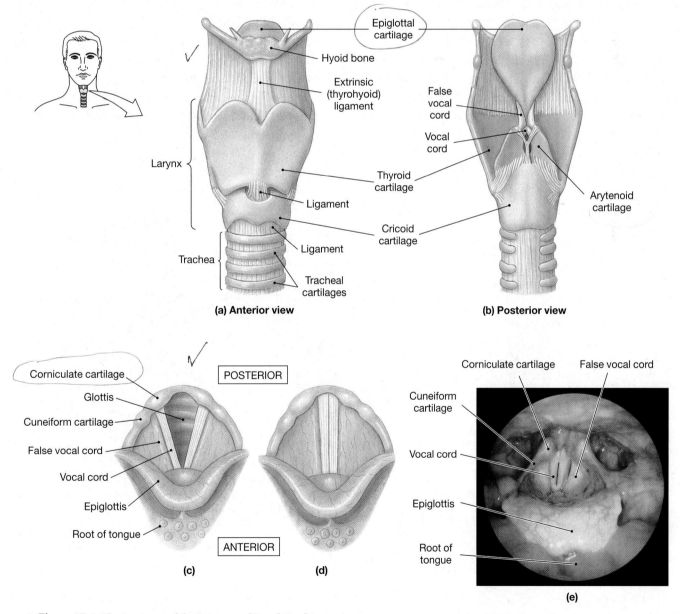

(a) Anterior view

(b) Posterior view

(c)

(d)

(e)

● *Figure 15-4* **The Anatomy of the Larynx and Vocal Cords**
(**a**) An anterior view of the larynx. (**b**) A posterior view of the larynx. (**c**) A superior view of the larynx with the glottis open and (**d**) with the glottis closed. (**e**) A fiber-optic view of the larynx with the glottis almost completely closed.

15 THE RESPIRATORY SYSTEM

The Functions of the Respiratory System • **The Organization of the Respiratory System** • Respiratory Physiology • The Control of Respiration

from entering the glottis, and they protect a more delicate pair of folds. These lower folds, the **true vocal cords**, contain elastic ligaments that extend between the thyroid cartilage and the arytenoid cartilages. Small muscles that insert on these cartilages change their position and alter the tension in these ligaments.

The Vocal Cords and Sound Production

Air passing passing through the glottis vibrates the vocal cords and produces sound waves. The pitch of the sound produced depends on the diameter, length, and tension of the vocal cords. The diameter and length are directly related to larynx size. For example, in a stringed instrument like a harp, short, thin strings vibrate rapidly, producing a high-pitched sound; long, thick strings vibrate more slowly, producing a low-pitched tone. Children have slender, short vocal cords, and their voices tend to be high-pitched. At puberty, the larynx of males enlarges more than that of females. The true vocal cords of an adult male are thicker and longer, and produce lower tones than those of adult females. The amount of tension in the vocal cords is controlled by skeletal muscles that change the position of the arytenoid cartilages. Increased tension in the vocal cords makes the pitch rise; decreased tension lowers the pitch.

The distinctive sound of your voice does not solely depend on the sounds produced by the larynx. Further amplification and resonance occur in the pharynx, the oral cavity, the nasal cavity, and the paranasal sinuses. The final production of distinct words further depends on voluntary movements of the tongue, lips, and cheeks.

THE TRACHEA

The **trachea** (TRĀ-kē-a), or *windpipe*, is a tough, flexible tube with a diameter of about 2.5 cm (1 in.) and a length of approximately 11 cm (4.25 in.) (Figure 15-5●). The trachea begins at the level of the sixth cervical vertebra, where it attaches to the cricoid cartilage of the larynx. It ends in the mediastinum, at the level of the fifth thoracic vertebra, where it branches to form a pair of primary bronchi.

The walls of the trachea are supported by about 20 **tracheal cartilages**. These C-shaped cartilages stiffen the tracheal walls

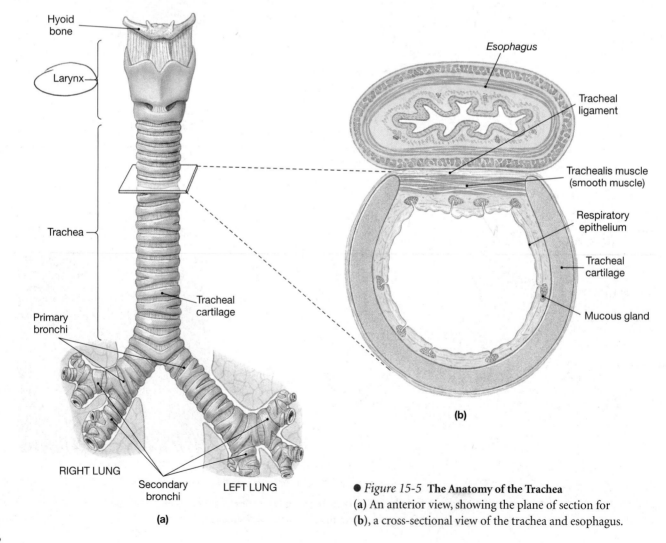

● *Figure 15-5* **The Anatomy of the Trachea**
(a) An anterior view, showing the plane of section for
(b), a cross-sectional view of the trachea and esophagus.

and protect the airway. They also prevent its collapse or over-expansion as pressures change in the respiratory system. The open portions of the C-shaped tracheal cartilages face posteriorly, toward the esophagus. Because the cartilages do not continue around the trachea, the posterior tracheal wall can easily distort, allowing large masses of food to pass along the esophagus. The ends of each tracheal cartilage are connected by an elastic ligament and the *trachealis muscle*, a band of smooth muscle. The diameter of the trachea is adjusted by the contractions of these muscles, which are under autonomic control. Sympathetic stimulation increases the diameter of the trachea and makes it easier to move large volumes of air along the respiratory passageways.

TRACHEAL BLOCKAGE

We sometimes breathe in foreign objects; this process is called *aspiration*. Foreign objects that become lodged in the larynx or trachea are usually expelled by coughing. If the individual can speak or make a sound, the airway is still open and no emergency measures should be taken. If the victim can neither breathe nor speak, an immediate threat to life exists. Unfortunately, many victims become acutely embarrassed by this situation, and rather than seek assistance, they run to the nearest rest room and die there.

In the *Heimlich* (HĪM-lik) *maneuver*, or *abdominal thrust*, a rescuer applies compression to the abdomen just beneath the diaphragm. This action elevates the diaphragm forcefully and may generate enough pressure to remove the blockage. The maneuver must be performed properly to avoid damage to internal organs. Organizations such as the American Red Cross, local fire department, and other charitable groups periodically hold brief training sessions in the proper performance of the Heimlich maneuver.

If blockage results from a swelling of the epiglottis or tissues surrounding the glottis, a professionally qualified rescuer may insert a curved tube through the pharynx and glottis to permit airflow. This procedure is called *intubation*. If a tracheal blockage remains, a *tracheostomy* (trā-kē-OS-to-mē; *stoma*, mouth) may be performed. In this procedure, an incision is made through the anterior tracheal wall and a tube is inserted. The tube bypasses the larynx and permits air to flow directly into the trachea.

THE BRONCHI

The trachea branches within the mediastinum, into the **right** and **left primary bronchi** (BRONG-kī) (Figure 15-5●). The walls of the primary bronchi resemble that of the trachea, including pseudostratified ciliated columnar epithelium and cartilaginous C-shaped rings. The right primary bronchus supplies the right lung, and the left primary bronchus supplies the left lung. The right primary bronchus is larger in diameter and descends toward the lung at a steeper angle than the left primary bronchus. Thus most foreign objects that enter the trachea find their way into the right primary bronchus rather than the left.

The primary bronchi branch into smaller and smaller passageways that ultimately extend to the alveoli in each lung (Figure 15-6a●). These passageways form the **bronchial tree**. As it enters the lung, each primary bronchus gives rise to **secondary bronchi**, which enter the lobes of that lung. The secondary bronchi divide to form 9–10 **tertiary bronchi** in each lung. Each tertiary bronchus branches repeatedly.

The cartilages of the secondary bronchi are quite massive, but farther along the branches of the bronchial tree they become smaller and smaller. When the diameter of the passageway has narrowed to around 1 mm (.04 in.), cartilages disappear completely. This narrow passage is a **bronchiole**.

THE BRONCHIOLES

Bronchioles are to the respiratory system what arterioles are to the circulatory system. Varying the diameter of the bronchioles controls the amount of resistance to airflow and the distribution of air in the lungs. The autonomic nervous system regulates the activity of the smooth muscles in the bronchiole walls, thereby controlling their diameter. Sympathetic activation leads to a relaxation of smooth muscles in the walls of bronchioles, causing a dilation of the respiratory passageways (*bronchodilation*). Parasympathetic stimulation leads to contraction of these smooth muscles and constriction of the respiratory passageways (*bronchoconstriction*). Extreme contraction of the smooth muscles can almost completely block the passageways, making breathing difficult or impossible. Bronchoconstriction also occurs during an *asthma* (AZ-muh) attack or during allergic reactions, in response to inflammation of the bronchioles.

Bronchioles branch further into *terminal bronchioles*, with diameters of 0.3–0.5 mm. Each terminal bronchiole supplies air to a lobule of the lung. A **lobule** (LOB-ūl) is a segment of lung tissue that is bounded by connective tissue partitions and supplied by a single bronchiole, accompanied by branches of the pulmonary arteries and pulmonary veins. Within a lobule, a terminal bronchiole divides to form several *respiratory bronchioles*. These passages, the thinnest branches of the bronchial tree, deliver air to the respiratory surfaces of the lungs. Figure 15-6b● shows the basic structure of a lobule.

THE ALVEOLAR DUCTS AND ALVEOLI

Respiratory bronchioles open into expansive chambers called **alveolar ducts**. These passageways end at **alveolar sacs**, common chambers connected to multiple individual alveoli—the exchange surfaces of the lungs (Figure 15-7a●). Each lung contains

15 THE RESPIRATORY SYSTEM

The Functions of the Respiratory System • **The Organization of the Respiratory System** • Respiratory Physiology • The Control of Respiration

● *Figure 15-6* **The Bronchial Tree and Lobules of the Lung**
(a) The branching pattern of bronchi in the left lung, simplified. (b) The basic structure of a pulmonary lobule. A network of capillaries surrounds each alveolus.

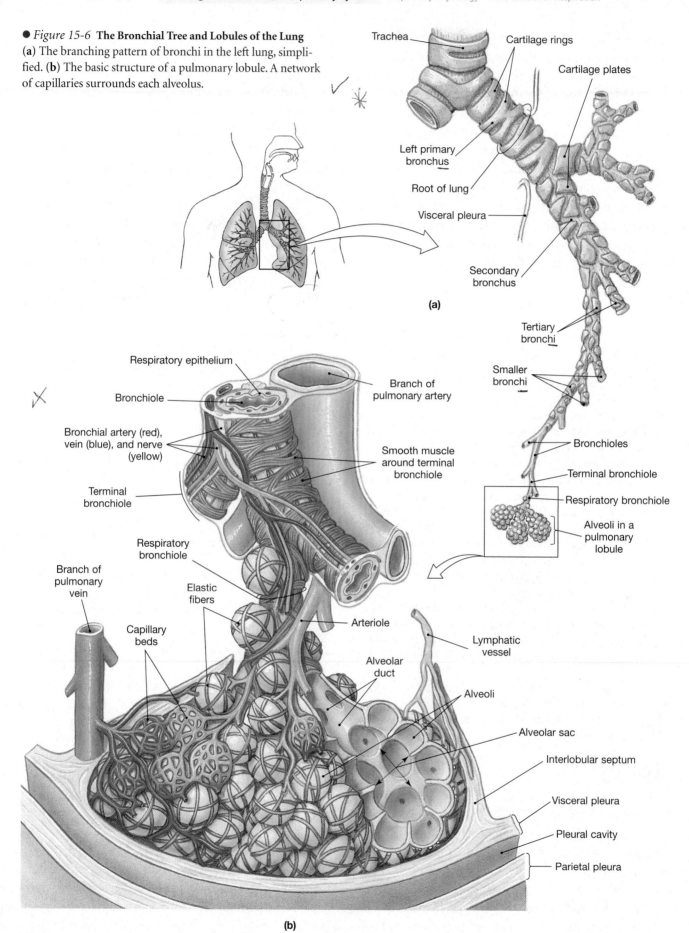

Trachea

Cartilage rings

Cartilage plates

Left primary bronchus

Root of lung

Visceral pleura

Secondary bronchus

Tertiary bronchi

Smaller bronchi

Bronchioles

Terminal bronchiole

Respiratory bronchiole

Alveoli in a pulmonary lobule

(a)

Respiratory epithelium

Bronchiole

Bronchial artery (red), vein (blue), and nerve (yellow)

Terminal bronchiole

Branch of pulmonary artery

Smooth muscle around terminal bronchiole

Respiratory bronchiole

Branch of pulmonary vein

Elastic fibers

Capillary beds

Arteriole

Lymphatic vessel

Alveolar duct

Alveoli

Alveolar sac

Interlobular septum

Visceral pleura

Pleural cavity

Parietal pleura

(b)

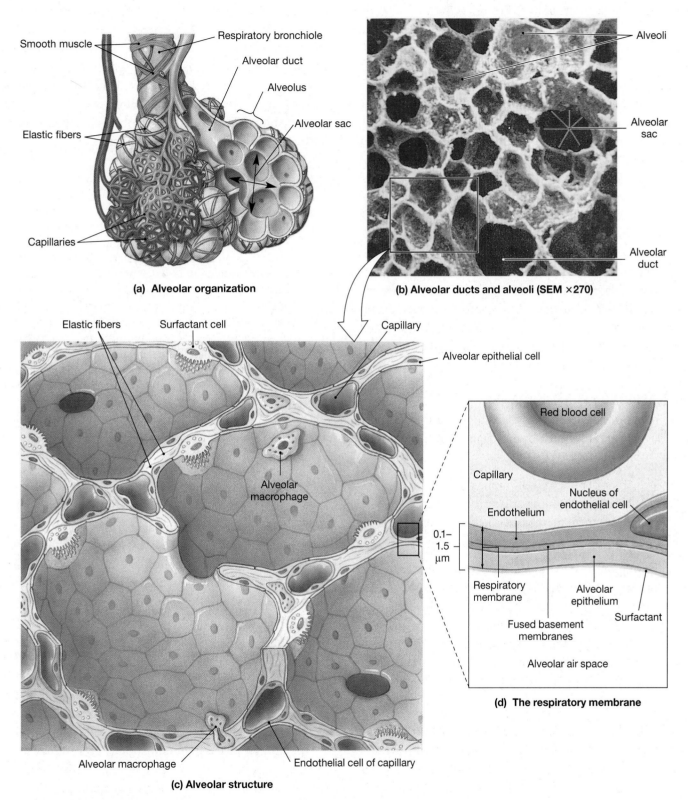

(a) **Alveolar organization**

(b) **Alveolar ducts and alveoli (SEM ×270)**

(c) **Alveolar structure**

(d) **The respiratory membrane**

● *Figure 15-7* **Alveolar Organization**
(**a**) Alveolar organization within a lobule. A network of capillaries surrounds each alveolus. (**b**) An SEM of the lung showing the open, spongy appearance of lung tissue. (**c**) A diagrammatic view of alveolar structure. (**d**) The respiratory membrane is made up of an alveolar epithelial cell, a capillary endothelial cell, and their fused basement membranes.

15 THE RESPIRATORY SYSTEM

The Functions of the Respiratory System • **The Organization of the Respiratory System** • Respiratory Physiology • The Control of Respiration

approximately 150 million alveoli, and their abundance gives the lung an open, spongy appearance (Figure 15-7b●).

To meet our metabolic requirements, the alveolar exchange surfaces of the lungs must be very large, equal to around 140 square meters—roughly the size of a tennis court. The alveolar epithelium primarily consists of an unusually thin and delicate simple squamous epithelium (Figure 15-7c●). Roaming **alveolar macrophages** (*dust cells*) patrol the epithelium, phagocytizing dust or debris that has reached the alveolar surfaces. **Surfactant** (sur-FAK-tant) **cells** produce surfactant, an oily secretion that forms a superficial coating over a thin layer of water on the alveolar epithelium. Surfactant is important because it reduces surface tension within the alveolus. Surface tension results from the attraction between water molecules at an air-water boundary. The alveolar walls are so delicate that without surfactant, the surface tension would be strong enough to collapse them. If surfactant levels are inadequate, as a result of injury or genetic abnormalities, each inhalation must be forceful enough to pop open the alveoli. An individual with this condition, called **respiratory distress syndrome**, is soon exhausted by the effort required to keep inflating and deflating the lungs.

THE RESPIRATORY MEMBRANE

Gas exchange occurs across the **respiratory membrane** of the alveoli. The respiratory membrane (Figure 15-7d●) consists of three components:

1. The squamous epithelial cells lining the alveolus.
2. The endothelial cells lining an adjacent capillary.
3. The fused basement membranes that lie between the alveolar and endothelial cells.

At the respiratory membrane, the total distance separating the alveolar air and the blood can be as little as 0.1 μm. Diffusion across the respiratory membrane proceeds very rapidly, because (1) the distance is small, and (2) both oxygen and carbon dioxide are lipid-soluble. The membranes of the epithelial and endothelial cells thus do not pose a barrier to the movement of oxygen and carbon dioxide between the blood and alveolar air spaces.

Circulation to the Respiratory Membrane

The respiratory exchange surfaces receive blood from arteries of the *pulmonary circuit*. ⟳ p. 405 The pulmonary arteries enter the lungs and branch, following the bronchi and their branches to the lobules. Each lobule receives an arteriole and a venule, and a network of capillaries surrounds each alveolus directly beneath the epithelium. After passing through the pulmonary venules, venous blood enters the pulmonary veins, which deliver it to the left atrium.

Blood pressure in the pulmonary circuit is usually relatively low, with systemic pressures of 30 mm Hg or less. With pressures that low, pulmonary vessels can easily become blocked by small blood clots, fat masses, or air bubbles in the pulmonary arteries. Because the lungs receive the entire cardiac output, any drifting masses in the blood are likely to cause problems almost at once. The blockage of a branch of a pulmonary artery will stop blood flow to a group of lobules or alveoli. This condition is called a **pulmonary embolism**.

 PNEUMONIA

Pneumonia (nū-MŌ-nē-uh; *pneumon*, lung +-*ia*, condition) develops from a pathogenic infection or any other stimulus that causes inflammation of the lobules of the lung. As inflammation occurs, fluids leak into the alveoli and the respiratory bronchioles swell and constrict. Respiratory function deteriorates as a result. When bacteria are involved, they are usually types that are normally found in the mouth and pharynx but have somehow managed to evade the respiratory defenses. Pneumonia becomes more likely when the respiratory defenses have been compromised by other factors, such as epithelial damage from smoking or the breakdown of the immune system in AIDS. The most common pneumonia that develops in AIDS patients results from infection by the fungus *Pneumocystis carinii*. This fungus is normally found in the alveoli, but in healthy individuals the respiratory defenses are able to prevent infection and tissue damage.

THE LUNGS

The left and right **lungs** (Figure 15-8●) occupy the left and right pleural cavities. ⟳ p. 19 Each lung has distinct **lobes** that are separated by deep fissures. The right lung has three lobes (*superior*, *middle*, and *inferior*), and the left lung has two (*superior* and *inferior*). The bluntly rounded *apex* of each lung extends into the base of the neck above the first rib, and the concave base rests on the superior surface of the diaphragm, the muscular sheet that separates the thoracic and abdominopelvic cavities. The curving *costal surface* follows the inner contours of the rib cage. The *mediastinal surface* has a more irregular shape. The mediastinal surface of the left lung bears the *cardiac notch*, an indentation that conforms to the shape of the pericardium, and both lungs bear grooves that mark the passage of vessels traveling to and from the heart.

The lungs have a light and spongy consistency because most of the actual volume of each lung consists of air-filled passageways and alveoli. An abundance of elastic fibers gives the lungs the ability to tolerate large changes in volume.

THE PLEURAL CAVITIES

The thoracic cavity has the shape of a broad cone. Its walls are the rib cage, and its floor is the muscular diaphragm. The two pleural

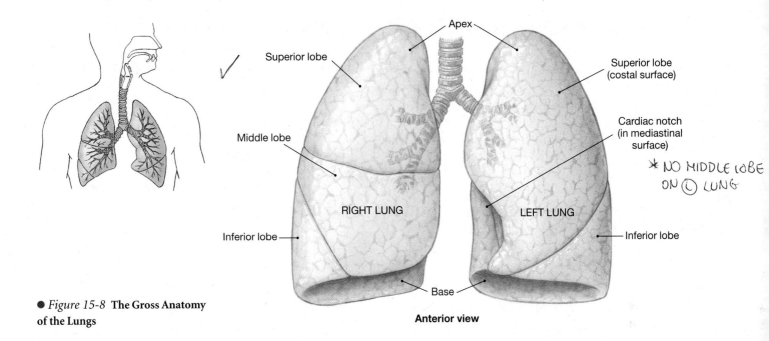

● *Figure 15-8* **The Gross Anatomy of the Lungs**

Anterior view

cavities are separated by the mediastinum (Figure 15-9●). Each lung occupies a single pleural cavity, lined by a serous membrane called the **pleura** (PLOO-ra). The *parietal pleura* covers the inner surface of the body wall and extends over the diaphragm and mediastinum. The *visceral pleura* covers the outer surfaces of the lungs, extending into the fissures between the lobes.

The pleural cavity actually represents a potential space rather than an open chamber, because the parietal and visceral layers are usually in close contact. Both pleural layers secrete *pleural fluid*. Pleural fluid gives a moist, slippery coating that provides lubrication, thereby reducing friction between both pleural surfaces as you breathe. Pleural fluid is sometimes obtained for diagnostic purposes by means of a long needle inserted between the ribs. This procedure is called *thoracentesis*

(thor-a-sen-TĒ-sis; *thorac-*, chest + *kentesis*, puncture). The fluid is examined for the presence of bacteria, blood cells, and other abnormal components.

An injury to the chest wall that penetrates the parietal pleura or damages the alveoli and the visceral pleura can allow air into the pleural cavity. This **pneumothorax** (noo-mō-THŌ-raks; *pneuma*, air) breaks the fluid bond between the pleurae and allows the elastic fibers to contract. The result is a collapsed lung, or *atelectasis* (at-e-LEK-ta-sis; *ateles*, imperfect + *ektasis*, expansion). Treatment involves removing as much of the air as possible before sealing the opening. This procedure restores the fluid bond and reinflates the lung. Lung volume can also be reduced by the accumulation of blood in the pleural cavity. This condition is called a **hemothorax.**

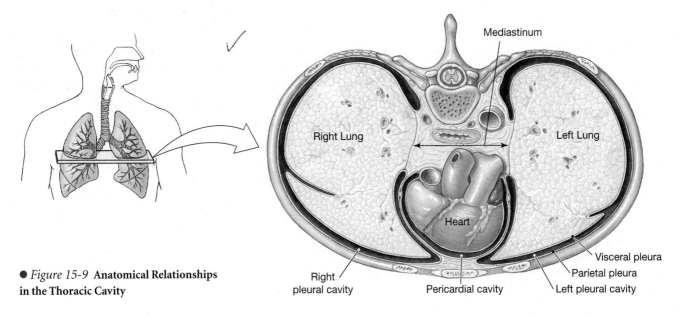

● *Figure 15-9* **Anatomical Relationships in the Thoracic Cavity**

15 THE RESPIRATORY SYSTEM

The Functions of the Respiratory System • The Organization of the Respiratory System • **Respiratory Physiology** • The Control of Respiration

❶ When the tension in the vocal cords increases, what happens to the pitch of the voice?

❷ Why are the cartilages that reinforce the trachea C-shaped instead of complete circles?

❸ What would happen to the alveoli if surfactant were not produced?

Respiratory Physiology

The process of respiration involves three integrated steps:

Step 1: *Pulmonary ventilation*, or breathing, which involves the physical movement of air into and out of the lungs.

Step 2: *Gas exchange*, which involves gas diffusion, across the respiratory membrane that separates the alveolar air from the blood within the alveolar capillaries, and across capillary cell membranes between blood and other tissues.

Step 3: *Gas transport*, which involves the transport of oxygen and carbon dioxide between the alveolar capillaries and the capillary beds in other tissues.

Abnormalities affecting any single step will ultimately affect the gas concentrations of the interstitial fluids. If the oxygen content declines, the affected tissues will become oxygen-starved. **Hypoxia** (hī-POKS-ē-a), or *low tissue oxygen levels*, places severe limits on the metabolic activities of peripheral tissues. If the supply of oxygen gets cut off completely, **anoxia** (a-NOKS-ē-a) results, and cells die very quickly. For example, much of the damage caused by strokes and heart attacks is the result of localized anoxia.

 TUBERCULOSIS

Tuberculosis (too-ber-kū-LŌ-sis), or **TB**, results from a bacterial infection of the lungs, although other organs may be invaded as well. The bacterium, *Mycobacterium tuberculosis*, may colonize the respiratory passageways, the interstitial spaces, the alveoli, or a combination of all three. Symptoms are variable but usually include coughing and chest pain with fever, night sweats, fatigue, and weight loss.

Tuberculosis is a major health problem throughout the world. With roughly 3 million deaths each year from TB, it is the leading cause of death from infectious diseases. An estimated *2 billion* people are infected at this time, and 8 million cases are diagnosed each year. Unlike other deadly diseases, such as AIDS, TB is transmitted through casual contact. Anyone alive and breathing is at risk for this disease; all it takes is exposure to the bacterium.

PULMONARY VENTILATION

Pulmonary ventilation is the physical movement of air into and out of the respiratory tract. A single breath, or *respiratory cycle*, consists of an inhalation, or *inspiration*, and an exhalation, or *expiration*. Breathing functions to maintain adequate **alveolar ventilation**, the movement of air into and out of the alveoli.

 ARTIFICIAL RESPIRATION

Artificial respiration is a technique to provide air to an individual whose respiratory muscles are no longer functioning. In *mouth-to-mouth resuscitation*, a rescuer provides ventilation by exhaling into the mouth or mouth and nose of the victim. After each breath, contact is broken to permit passive exhalation by the victim. Air provided in this way contains adequate oxygen to meet the needs of the victim. Trained rescuers may supply air through an endotracheal tube, which is inserted into the trachea through the glottis. Mechanical ventilators, if available, can be attached to the endotracheal tube. If the victim's cardiovascular system is nonfunctional as well, a technique called *cardiopulmonary resuscitation (CPR)* is required to maintain adequate blood flow and tissue oxygenation.

Pressure and Airflow to the Lungs

As we know from television weather reports, air will flow from an area of higher pressure to an area of lower pressure. The greater the *pressure gradient*, the difference between the high and low pressures, the faster the air will move. This relationship applies both to the movement of atmospheric winds and to the movement of air into and out of the lungs (pulmonary ventilation). Pressure gradients between the atmosphere and the lungs occur when the volume of the lungs changes. Because gases can be compressed, the pressure of a gas in a closed, flexible container (such as a lung) can be altered by increasing or decreasing its volume.

The volume of the lungs depends on the volume of the pleural cavities. The parietal and pleural membranes are separated by only a thin film of pleural fluid, and although the two membranes can slide across each other, they are held together by that fluid film. You encounter the same principle when you set a wet glass on a smooth surface. You can slide the glass quite easily, but when you try to lift it, you encounter considerable resistance from this fluid bond. A comparable fluid bond exists between the parietal pleura and the visceral pleura covering the lungs. As a result, the surface of each lung sticks to the inner wall of the chest and the superior surface of the diaphragm. Movements of the chest wall or the diaphragm thus directly affect the volume of the lungs. The basic principle is shown in Figure 15-10a●. The volume of the thoracic cavity changes when the diaphragm changes position or the rib cage moves:

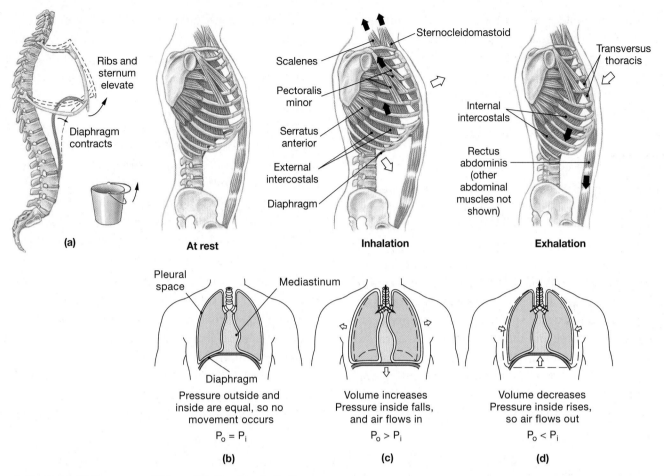

● *Figure 15-10* **Pressure and Volume Relationships in the Lungs**
(**a**) Raising the curved handle of the bucket increases the amount of space between it and the bucket. Similarly, the volume of the thoracic cavity is increased when the ribs are elevated or the diaphragm is depressed when it contracts. (**b**) An anterior view at rest, with no air movement. (**c**) Inhalation: Elevation of the rib cage and depression of the diaphragm increase the size of the thoracic cavity. Pressure decreases, and air flows into the lungs. (**d**) Exhalation: When the rib cage returns to its original position or the diaphragm relaxes, the volume of the thoracic cavity decreases. Pressure rises, and air moves out of the lungs. Accessory muscles may assist such movements of the rib cage to increase the depth and rate of respiration.

• *The diaphragm forms the floor of the thoracic cavity.* The relaxed diaphragm has the shape of a dome and projects upward into the thoracic cavity, compressing the lungs. When the diaphragm contracts, it flattens and increases the volume of the thoracic cavity, expanding the lungs. When the diaphragm relaxes, it returns to its original position and decreases the volume of the thoracic cavity.

• *Owing to the way the ribs and the vertebrae articulate, elevation of the rib cage increases the volume of the thoracic cavity. Lowering the rib cage decreases the volume of the thoracic cavity.* The external intercostal muscles and accessory muscles, such as the sternocleidomastoid, elevate the rib cage. The internal intercostal muscles and accessory muscles, such as the rectus abdominis (and other abdominal muscles) lower the rib cage.

At the start of a breath, pressures inside and outside the lungs are identical and there is no movement of air (Figure 15-10b●). When the thoracic cavity enlarges, the pleural cavities and lungs expand to fill the additional space and the pressure inside the lungs decreases. Air now enters the respiratory passageways because the pressure inside the lungs (P_i) is lower than atmospheric pressure (pressure outside, or P_o) (Figure 15-10c●). Downward movement of the rib cage and upward movement of the diaphragm reverse the process and reduce the volume of the lungs. Pressure inside the lungs now exceeds atmospheric pressure, and air moves out of the lungs (Figure 15-10d●).

The **compliance** of the lungs is an indication of their resilience and ability to expand. The lower the compliance, the greater the force required to fill and empty the lungs. Various disorders affect compliance. For example, the loss of supporting tissues due to alveolar damage, as in *emphysema*, increases compliance (see p. 472). Compliance is reduced if surfactant production is too low to prevent the alveoli from collapsing on exhalation, as in *respiratory distress syndrome* (see p. 466).

15 THE RESPIRATORY SYSTEM

The Functions of the Respiratory System • The Organization of the Respiratory System • **Respiratory Physiology** • The Control of Respiration

Arthritis or other skeletal disorders that affect the joints of the ribs or spinal column will also reduce compliance.

When you are at rest, the muscular activity involved in pulmonary ventilation accounts for 3–5 percent of your resting energy demand. If compliance is reduced, the energy demand increases dramatically and you can become exhausted simply trying to continue breathing.

Modes of Breathing

The respiratory muscles are used in various combinations, depending on the volume of air that must be moved into or out of the system. Respiratory movements are classified as quiet breathing or forced breathing. In *quiet breathing*, inhalation involves muscular contractions, but exhalation is passive. Inhalation results from the contraction of the diaphragm and the external intercostal muscles. Diaphragm contraction normally accounts for around 75 percent of the air movement in normal quiet breathing and the external intercostal muscles account for the remaining 25 percent. These percentages can change, however. For example, pregnant women increasingly rely on movements of the rib cage as expansion of the uterus forces abdominal organs against the diaphragm. In *forced*

breathing*, both inhalation and exhalation are active. Forced breathing involves the accessory muscles during inhalation, and the internal intercostal muscles and abdominal muscles during exhalation.

Lung Volumes and Capacities

As noted earlier, a respiratory cycle is a single cycle of inhalation and exhalation. The **tidal volume** is the amount of air moved into or out of the lungs during a single respiratory cycle. Only a small proportion of the air in the lungs is exchanged during a single quiet respiratory cycle; the tidal volume can be increased by more vigorous inhalation and more complete exhalation. The total volume of the lungs can be divided into *volumes* and *capacities* graphically shown in Figure 15-11●:

• *Expiratory reserve volume.* During a normal quiet respiratory cycle, the tidal volume averages about 500 ml. The amount of air that could be voluntarily expelled at the end of a tidal cycle is about 1000 ml. This volume is the **expiratory reserve volume (ERV)**.

• *Inspiratory reserve volume.* The **inspiratory reserve volume (IRV)** is the amount of air that can be taken in over and above the tidal volume. Because the lungs of males are larger

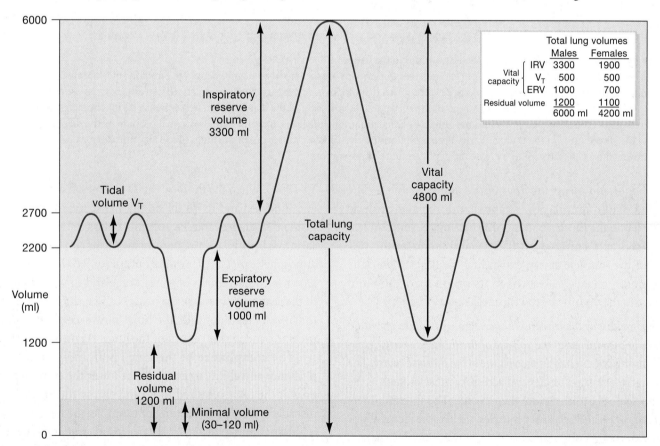

		Total lung volumes	
		Males	Females
Vital capacity	IRV	3300	1900
	V_T	500	500
	ERV	1000	700
Residual volume		1200	1100
		6000 ml	4200 ml

● *Figure 15-11* **Respiratory Volumes and Capacities**
The graph diagrams the relationships between the respiratory volumes and capacities of an average male. The table compares the values for males and females.

than those of females, the IRV of males averages 3300 ml versus 1900 ml in females.

- *Vital capacity.* The sum of the inspiratory reserve volume, the expiratory reserve volume, and the tidal volume is the **vital capacity**—the maximum amount of air that can be moved into and out of the respiratory system in a single respiratory cycle.
- *Residual volume.* Roughly 1200 ml of air remains in the respiratory passageways and alveoli, even after the expiratory reserve volume has been exhausted. Most of this **residual volume** exists because the lungs are held against the thoracic wall, preventing their elastic fibers from contracting further.
- *Minimal volume.* When the chest cavity is opened, as in a pneumothorax, the lungs collapse and the amount of air in the respiratory system is reduced to the **minimal volume**. Some air remains in the lungs, even at minimal volume, because the surfactant coating the alveolar surfaces prevents their collapse.

Not all of the inspired air reaches the alveolar exchange surfaces within the lungs. A typical tidal inspiration pulls around 500 ml of air into the respiratory system. The first 350 ml travels along the conducting passageways and enters the alveolar spaces. The last 150 ml never gets farther than the conducting passageways and does not participate in gas exchange with the blood. The volume of air in the conducting passages is known as the *dead space* of the lungs.

PULMONARY FUNCTION TESTS

Pulmonary function tests monitor several aspects of respiratory function. A *spirometer* (spī-ROM-e-ter) measures parameters such as vital capacity, expiratory reserve, and inspiratory reserve. A *pneumotachometer* determines the rate of air movement. A *peak flow meter* records the maximum rate of forced expiration. These tests are relatively simple to perform, but they have considerable diagnostic significance. For example, in people with asthma the constricted airways tend to close before an exhalation is completed. As a result, pulmonary function tests show a reduction in vital capacity, ERV, and peak flow rate. The functional residual capacity declines, and the narrow respiratory passageways reduce the flow rate as well. Air whistling through the constricted airways produces the characteristic "wheezing" that accompanies an asthmatic attack.

CONCEPT CHECK QUESTIONS
Answers on page 484

❶ Mark breaks a rib, and it punctures the chest wall on his left side. What will happen to his left lung?

❷ In pneumonia, fluid accumulates in the alveoli of the lungs. How would vital capacity be affected?

GAS EXCHANGE

In pulmonary ventilation, the alveoli are supplied with oxygen and carbon dioxide is removed from the bloodstream. The actual process of gas exchange with the external environment occurs between the blood and alveolar air across the respiratory membrane. This process depends on (1) the partial pressures of the gases involved and (2) the diffusion of molecules from a gas into a liquid. p. 58

Mixed Gases and Partial Pressures

The air we breathe is not a single gas but a mixture of gases. Nitrogen molecules (N_2) are the most abundant, accounting for about 78.6 percent of the atmospheric gas molecules. Oxygen molecules (O_2), the second most abundant, constitute roughly 20.8 percent of the atmospheric content. Most of the remaining 0.5 percent consists of water molecules, with carbon dioxide (CO_2) contributing a mere 0.04 percent.

Atmospheric pressure at sea level is approximately 760 mm Hg. Each of the gases contributes to the total pressure in proportion to its relative abundance. The pressure contributed by a single gas is the **partial pressure** of that gas, abbreviated as P. The sum of the partial pressures of all the gases equals the total pressure exerted by the gas mixture. In the case of the atmosphere, this relationship can be summarized as:

$$P_{N_2} + P_{O_2} + P_{H_2O} + P_{CO_2} = 760 \text{ mm Hg}$$

Each gas contributes to the total pressure in proportion to its relative abundance. For example, the partial pressure of oxygen, P_{O_2}, is 20.8 percent of 760 mm Hg, or roughly 159 mm Hg. The partial pressures of other atmospheric gases are given in Table 15-1. These values are important because the partial

TABLE 15-1 *Partial Pressures (mm Hg) and Normal Gas Concentrations in Air*

SOURCE OF SAMPLE	NITROGEN (N₂)	OXYGEN (O₂)	WATER VAPOR (H₂O)	CARBON DIOXIDE (CO₂)
Inspired air (dry)	597 (78.6 percent)	159 (20.8 percent)	3.7 (0.5 percent)	0.3 (0.04 percent)
Alveolar air (saturated)	40 (5.2 percent)	573 (75.4 percent)	100 (13.2 percent)	47 (6.2 percent)
Expired air (saturated)	28 (3.7 percent)	569 (74.8 percent)	116 (15.3 percent)	47 (6.2 percent)

15 THE RESPIRATORY SYSTEM

The Functions of the Respiratory System • The Organization of the Respiratory System • **Respiratory Physiology** • The Control of Respiration

pressure of an individual gas determines its rate of diffusion between the alveolar air and the bloodstream. For instance, the partial pressure of oxygen determines how much oxygen enters solution, but it has no effect on the rate of nitrogen or carbon dioxide diffusion.

Alveolar versus Atmospheric Air

As soon as air enters the respiratory tract, its characteristics begin to change. For example, in passing through the nasal cavity, the air becomes warmer and the amount of water vapor increases. On reaching the alveoli, the incoming air mixes with air that remained in the alveoli after the previous respiratory cycle. The resulting alveolar gas mixture thus contains more carbon dioxide and less oxygen than does atmospheric air. As noted earlier, the last 150 ml of inspired air never gets farther than the conducting passageways and remains in the dead space of the lungs. During expiration, the departing alveolar air mixes with air in the dead space to produce yet another mixture that differs from both atmospheric and alveolar samples. The differences in composition between atmospheric (inspired) and alveolar air can be seen in Table 15-1.

DECOMPRESSION SICKNESS

Decompression sickness is a painful condition that develops when a person is suddenly exposed to a drop in atmospheric pressure. Nitrogen is the gas responsible for the problems experienced, because of its high partial pressure in air. When the pressure drops, nitrogen comes out of solution, forming bubbles like those in a shaken can of soda. The bubbles may form in joint cavities, in the bloodstream, and in the cerebrospinal fluid. Individuals with with decompression sickness typically curl up because of the pain in affected joints; this reaction accounts for the common name for the condition, *the bends*. Decompression sickness most often affects scuba divers, when they return to the surface too quickly after breathing air under greater than normal pressures while submerged. It also can affect people in airplanes subjected to sudden losses of cabin pressure.

Partial Pressures within the Circulatory System

Gas exchange may be divided into external respiration and internal respiration. *External respiration* is the diffusion of gases between the blood and alveolar air across the respiratory membrane. *Internal respiration* is the diffusion of gases between blood and interstitial fluid across the single endothelial cells of capillary walls. Figure 15-12● details the partial pressures of oxygen and carbon dioxide in the alveolar air and capillaries and in the arteries and veins of the pulmonary and systemic circuits. The deoxygenated blood delivered by the pulmonary arteries has a higher P_{CO_2} and a lower P_{O_2} than does alveolar air. Diffusion between the alveolar air and the pulmonary cap-

illaries thus elevates the P_{O_2} of the blood while lowering its P_{CO_2}. By the time it enters the pulmonary venules, the oxygenated blood has reached equilibrium with the alveolar air, so it departs the alveoli with a P_{O_2} of about 100 mm Hg and a P_{CO_2} of roughly 40 mm Hg (Figure 15-12a●).

Normal interstitial fluid has a P_{O_2} of 40 mm Hg and a P_{CO_2} of 45 mm Hg. As a result, oxygen diffuses out of the capillaries and carbon dioxide diffuses in until the capillary partial pressures are the same as those in the adjacent tissues (Figure 15-12b●). At a normal tissue P_{O_2}, the blood entering the venous system still contains around 75 percent of its total oxygen content. The remaining oxygen represents a reserve that can be called on when tissue activity increases or when the oxygen supply is temporarily reduced (such as when you hold your breath). When the blood returns to the alveolar capillaries, it will replace the oxygen released in the tissues at the same time that the excess CO_2 is lost.

EMPHYSEMA

Emphysema (em-fi-SĒ-ma) is a chronic, progressive condition characterized by shortness of breath and an inability to tolerate physical exertion. The underlying problem is the destruction of alveolar surfaces, resulting in an inadequate surface area for oxygen and carbon dioxide exchange. The alveoli gradually expand, their walls become incomplete, and adjacent alveoli merge to form larger, nonfunctional air spaces. Emphysema has been linked to the inhalation of air containing fine particulate matter or toxic vapors, such as those in cigarette smoke. Genetic factors also predispose individuals to this condition. Some degree of emphysema is a normal consequence of aging; an estimated 66 percent of adult males and 25 percent of adult females have detectable areas of emphysema in their lungs.

GAS TRANSPORT

Oxygen and carbon dioxide have limited solubilities in blood plasma. The limited extent to which these gases dissolve are a problem because peripheral tissues need more oxygen and generate more carbon dioxide than the plasma can absorb and transport. The problem is solved by red blood cells (RBCs), which remove dissolved oxygen and carbon dioxide molecules from plasma and bind them (in the case of oxygen) or use them to manufacture soluble compounds (in the case of carbon dioxide). Because these reactions remove dissolved gases from the blood plasma, gases continue to diffuse into the blood and never reach equilibrium. These reactions are also *temporary* and *completely reversible*. When plasma oxygen or carbon dioxide concentrations are high, the excess molecules are removed by the red blood cells; when the plasma concentrations are falling, the red blood cells release their stored reserves.

● *Figure 15-12* **An Overview of Respiration and Respiratory Processes**
(a) External respiration is the diffusion of gases between the blood and alveolar air across the respiratory membrane. Diffusion between the alveolar air and the pulmonary capillaries elevates the P_{O_2} of the blood and lowers its P_{CO_2}. (b) Internal respiration is the diffusion of gases between blood and interstitial fluid across the capillary cell membranes. Diffusion between the systemic capillaries and interstitial fluid lowers the P_{O_2} of the blood and increases its P_{CO_2}.

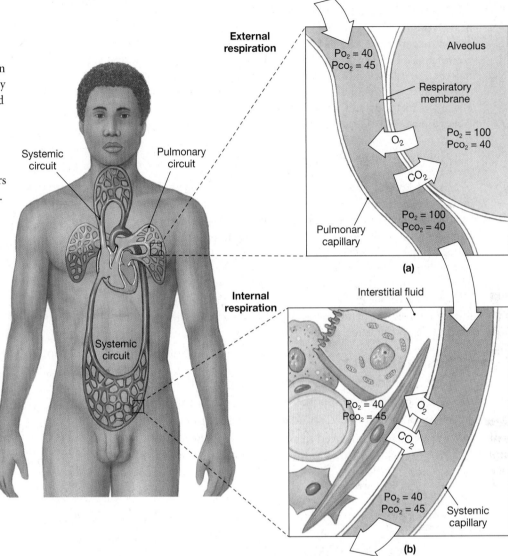

Oxygen Transport

Only around 1.5 percent of the oxygen content of arterial blood consists of oxygen molecules in solution. The rest of the oxygen molecules are bound to hemoglobin (Hb) molecules, specifically to the iron ions in the center of heme units. ∞ p. 348 This is a reversible reaction that can be summarized as follows:

$$Hb + O_2 \longleftrightarrow HbO_2$$

The amount of oxygen bound (or released) by hemoglobin depends primarily on the P_{O_2} in its surroundings. The lower the oxygen content of a tissue, the more oxygen is released by hemoglobin molecules passing through local capillaries. For example, inactive tissues have little demand for oxygen, and the local P_{O_2} is about 40 mm Hg. Under these conditions, hemoglobin releases about 25 percent of its stored oxygen. In contrast, the local P_{O_2} of active tissues may decline to 15–20 mm Hg, about one-half of the P_{O_2} of normal tissue. Hemoglobin

then releases up to 80 percent of its stored oxygen. In practical terms, this means active tissues will receive roughly 3 times as much oxygen as will inactive tissues.

In addition to the effect of P_{O_2}, the amount of oxygen released by hemoglobin is influenced by pH and temperature. Active tissues generate acids that lower the pH of the interstitial fluids. When the pH declines, the hemoglobin molecules release their bound oxygen molecules more readily. Hemoglobin also releases more oxygen when the temperature rises.

All three of these factors (P_{O_2}, pH, and temperature) are important during periods of maximal exertion. When a skeletal muscle works hard, its temperature rises and the local pH and P_{O_2} decline. The combination makes the hemoglobin entering the area release much more oxygen that can be used by active muscle fibers. Without this automatic adjustment, tissue P_{O_2} would fall to very low levels almost immediately and the exertion would come to a premature halt.

15 THE RESPIRATORY SYSTEM

The Functions of the Respiratory System • The Organization of the Respiratory System • Respiratory Physiology • **The Control of Respiration**

✚ CARBON MONOXIDE POISONING

Murder or suicide victims who die in their cars inside a locked garage are popular characters for mystery writers. In real life, entire families each winter are killed by leaky furnaces or space heaters. The cause of death is **carbon monoxide poisoning**. The exhaust of automobiles and other petroleum-burning engines, of oil lamps, and of fuel-fired space heaters contains carbon monoxide (CO) gas. Carbon monoxide competes with oxygen molecules for the binding sites on heme units. Unfortunately, the carbon monoxide usually wins, because at very low partial pressures it has a much stronger affinity for hemoglobin. The bond is extremely strong, and the attachment of a carbon monoxide molecule essentially inactivates that heme unit for respiratory purposes. If carbon monoxide molecules make up just 0.1 percent of the components of inhaled air, enough hemoglobin will be affected that survival is impossible without medical assistance. Treatment may include (1) the administration of pure oxygen, because at sufficiently high partial pressures the oxygen molecules "bump" the carbon monoxide from the hemoglobin; and, if necessary, (2) the transfusion of compatible red blood cells.

Carbon Dioxide Transport

Carbon dioxide is generated by aerobic metabolism in peripheral tissues. After entering the bloodstream, a CO_2 molecule may (1) dissolve in the plasma, (2) bind to the hemoglobin of red blood cells, or (3) be converted to a molecule of carbonic acid (H_2CO_3) (Figure 15-13●). All three processes are completely reversible.

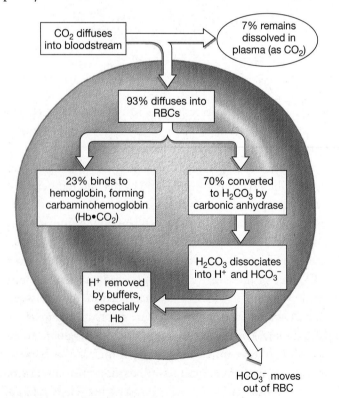

● *Figure 15-13* **Carbon Dioxide Transport in the Blood**

Plasma Transport. Roughly 7 percent of the carbon dioxide absorbed by peripheral capillaries is transported in the form of dissolved gas molecules. The rest diffuses into the red blood cells.

Hemoglobin Binding. Once in the red blood cells, some of the carbon dioxide molecules are bound to the protein "globin" portions of hemoglobin molecules, forming **carbaminohemoglobin** (kar-ba-mē-nō-hē-mō-GLŌ-bin). Such binding does not interfere with the binding of oxygen to iron ions, and, as a result, hemoglobin can transport both oxygen and carbon dioxide simultaneously. Normally, about 23 percent of the carbon dioxide entering the blood in peripheral tissues is transported as carbaminohemoglobin. On arriving at the pulmonary capillaries, the plasma P_{CO_2} declines, and the bound carbon dioxide is released.

Carbonic Acid Formation. The rest of the carbon dioxide molecules, roughly 70 percent of the total, are converted to carbonic acid through the activity of the enzyme carbonic anhydrase. The carbonic acid molecules do not remain intact, however; almost immediately, each of these molecules breaks down into a hydrogen ion and a bicarbonate ion. The entire sequence can be summarized as follows:

$$CO_2 + H_2O \xrightleftharpoons[]{\text{carbonic anhydrase}} H_2CO_3 \rightleftharpoons H^+ + HCO_3^-$$

The reactions occur very rapidly and are completely reversible. Because most of the carbonic acid formed immediately dissociates into bicarbonate and hydrogen ions, we can ignore the intermediary step and summarize the reaction as follows:

$$CO_2 + H_2O \xrightleftharpoons[]{\text{carbonic anhydrase}} H^+ + HCO_3^-$$

In peripheral capillaries, this reaction proceeds vigorously, tying up large numbers of carbon dioxide molecules. The reaction is driven from left to right because carbon dioxide continues to arrive, diffusing out of the interstitial fluids, and the hydrogen ions and bicarbonate ions are continuously being removed. Most of the hydrogen ions get tied up by hemoglobin molecules. Bicarbonate ions diffuse into the surrounding plasma, where they associate with sodium ions to form sodium bicarbonate ($NaHCO_3$).

When venous blood reaches the alveoli, carbon dioxide diffuses out of the plasma and the P_{CO_2} declines. Because all of the carbon dioxide transport mechanisms are reversible, as carbon dioxide diffuses out of the red blood cells, the reactions shown in Figure 15-13● proceed in the opposite direction.

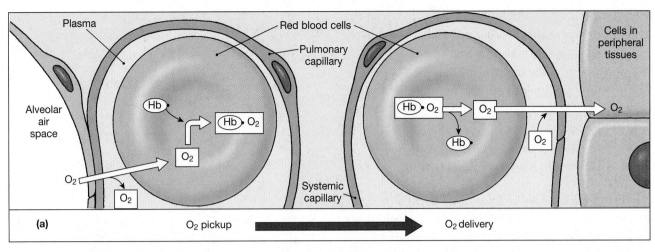

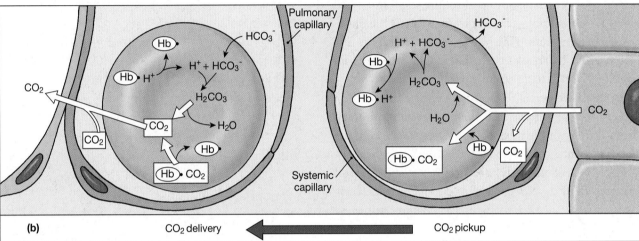

● *Figure 15-14* **A Summary of Gas Transport and Exchange**
(**a**) Oxygen pickup from the alveoli and delivery to peripheral tissues. (**b**) Carbon dioxide pickup from peripheral tissues and delivery to the alveoli.

Hydrogen ions leave the hemoglobin molecules, and bicarbonate ions diffuse into the cytoplasm of the red blood cells, to be converted to water and CO_2.

Figure 15-14● summarizes oxygen and carbon dioxide transport in the respiratory and cardiovascular systems.

CONCEPT CHECK QUESTIONS

Answers on page 484

❶ During exercise, hemoglobin releases more oxygen to the active skeletal muscles than it does when the muscles are at rest. Why?
❷ How would blockage of the trachea affect the blood pH?

The Control of Respiration

Cells continuously absorb oxygen from the interstitial fluids and generate carbon dioxide. Under normal conditions, cellular rates of absorption and generation are matched by capillary rates of delivery and removal, and these rates are identical to those of oxygen absorption and carbon dioxide excretion at the lungs. If these rates become unbalanced, the activities of the cardiovascular and respiratory systems must be adjusted. Equilibrium is restored through homeostatic mechanisms that involve (1) changes in blood flow and oxygen delivery under local control and (2) changes in the depth and rate of respiration under control of the brain's respiratory centers.

THE LOCAL CONTROL OF RESPIRATION

The rate of oxygen delivery in each tissue and the efficiency of oxygen pickup at the lungs are regulated at the local level. If a peripheral tissue becomes more active, the interstitial P_{O_2} falls and the P_{CO_2} rises. This change increases the difference between partial pressures in the tissues and arriving blood, so more oxygen is delivered and more carbon dioxide is carried away. In addition, rising P_{CO_2} levels cause the relaxation of smooth muscles in the walls of arterioles in the area, increasing blood flow.

Local adjustments within the lungs also improve the efficiency of gas transport. For example, as blood flows to alveolar

475

15 THE RESPIRATORY SYSTEM

The Functions of the Respiratory System • The Organization of the Respiratory System • Respiratory Physiology • **The Control of Respiration**

capillaries, it is directed to lobules of the lungs in which P_{O_2} is relatively high. This occurs because precapillary sphincters leading to alveolar capillary beds constrict when the local P_{O_2} is low. (This response is the opposite of that seen in peripheral tissues. ⟳ p. 393) Smooth muscles in the walls of bronchioles are sensitive to the P_{CO_2} of the air they contain. When the P_{CO_2} goes up, the bronchioles increase in diameter. When the P_{CO_2} declines, the bronchioles constrict. Airflow is therefore directed to lobules in which the P_{CO_2} is high.

THE RESPIRATORY CENTERS OF THE BRAIN

Respiratory control involves both involuntary and voluntary components. Your brain's involuntary respiratory centers regulate the respiratory muscles and control the respiratory rate and the depth of breathing. The **respiratory rate** is the number of breaths per minute. This rate in normal adults at rest ranges from 12 to 18 breaths per minute. Children breathe more rapidly, about 18 to 20 breaths per minute.

The **respiratory centers** are three pairs of nuclei in the reticular formation of the pons and medulla oblongata. The **respiratory rhythmicity centers** of the medulla oblongata set the pace for respiration. Each center can be subdivided into a *dorsal respiratory group (DRG)*, which contains an *inspiratory center*, and a *ventral respiratory group (VRG)*, which contains an *expiratory center*. Their output is adjusted by the two pairs of nuclei making up the respiratory centers of the pons. For example, the centers in the pons adjust the respiratory rate and the depth of respiration in response to sensory stimuli, emotional states, or speech patterns.

The Activities of the Respiratory Rhythmicity Centers

The inspiratory center functions in every respiratory cycle. It contains neurons that control the external intercostal muscles and the diaphragm (inspiratory muscles). During quiet breathing, the neurons of the inspiratory center gradually increase stimulation of the inspiratory muscles for two seconds, and then the inspiratory center becomes silent for the next three seconds. During that period of inactivity, the inspiratory muscles relax, and passive exhalation occurs. The inspiratory center will maintain this basic rhythm even in the absence of sensory or regulatory stimuli. The expiratory center functions only during forced breathing, when it activates the accessory muscles involved in inhalation and exhalation. The relationships between the inspiratory and expiratory centers during quiet breathing and forced breathing are diagrammed in Figure 15-15●.

The performance of these respiratory centers can be affected by any factor that alters the metabolic or chemical activities of neural tissues. For example, elevated body temperatures or CNS stimulants, such as amphetamines or caffeine, increase

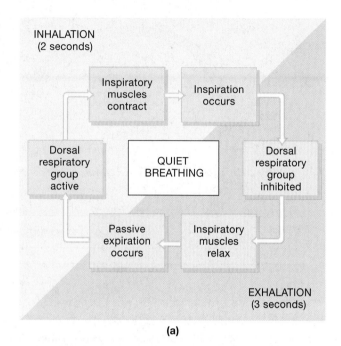

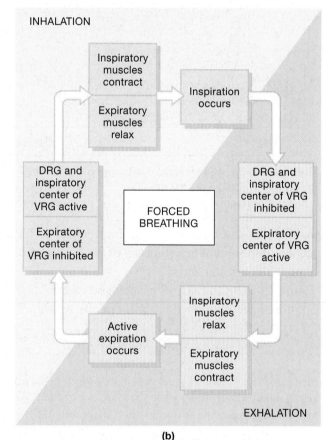

● *Figure 15-15* **Basic Regulatory Patterns**
(**a**) Quiet breathing. (**b**) Forced breathing.

the respiratory rate. Decreased body temperature or CNS depressants, such as barbiturates or opiates, reduce the respiratory rate. Respiratory activities are also strongly influenced by reflexes that are triggered by mechanical or chemical stimuli.

THE REFLEX CONTROL OF RESPIRATION

Normal breathing occurs automatically, without conscious control. The activities of the respiratory centers are modified by sensory information from mechanoreceptors, such as stretch and pressure receptors, and chemoreceptors. Information from these receptors alters the pattern of respiration, producing respiratory reflexes.

Mechanoreceptor Reflexes

Mechanoreceptors respond to changes in the volume of the lungs or to changes in arterial blood pressure. Chapter 9 describes several populations of baroreceptors that are involved in respiratory function. ⌐○○ p. 281

The **inflation reflex** prevents the lungs from overexpanding during forced breathing. The mechanoreceptors involved are stretch receptors that are stimulated when the lungs expand. Sensory fibers leaving these receptors reach the inspiratory and expiratory centers through the vagus nerve. As the volume of the lungs increases, the inspiratory center is gradually inhibited and the expiratory center is stimulated. Thus inspiration stops as the lungs near maximum volume, and active expiration then begins. In contrast, the **deflation reflex** inhibits the expiratory center and stimulates the inspiratory center when the lungs are collapsing. The smaller the volume of the lungs, the greater the inhibition of the expiratory center.

Although neither the inflation nor the deflation reflex is involved in normal quiet breathing, both are important in regulating the forced inhalations and exhalations that accompany strenuous exercise. Together, the inflation and deflation reflexes are known as the *Hering-Breuer reflexes*, after the physiologists who described them in 1865.

The effects of the carotid and aortic baroreceptors on systemic blood pressure are described in Chapter 13. ⌐○○ p. 400 The output from these baroreceptors affects the respiratory centers as well as the cardiac and vasomotor centers. When blood pressure falls, the respiratory rate increases; when blood pressure rises, the respiratory rate declines. This adjustment results from the stimulation or inhibition of the inspiratory and expiratory centers by sensory fibers in the glossopharyngeal (IX) and vagus (X) nerves.

✚ CLINICAL NOTE *Lung Cancer, Smoking, and Diet*

Lung cancer, or *pleuropulmonary neoplasm*, is an aggressive class of malignancies originating in the bronchial passageways or alveoli. These cancers affect the epithelial cells that line conducting passageways, mucous glands, or alveoli. Symptoms usually do not appear until the condition has progressed to the point that the tumor masses are restricting airflow or compressing adjacent structures. Chest pain, shortness of breath, a cough or wheeze, and weight loss are common symptoms. Treatment programs vary with the cellular organization of the tumor and whether metastasis (cancer cell migration) has occurred, but surgery, radiation exposure, or chemotherapy may be involved.

Deaths from lung cancer were rare at the beginning of the twentieth century, but 29,000 such deaths occurred in 1956, 105,000 in 1978, and 154,900 in 2002 in the United States. The rise coincides with the increased rate of smoking in the population. The number of diagnosed cases is also rising, doubling every 15 years. Each year, 22 percent of new cancers detected are lung cancers; in the United States in 2002, an estimated 90,200 men and 79,200 women were diagnosed with this condition. The rate is decreasing for men but increasing markedly among women; in 1989, 101,000 men and 54,000 women were diagnosed with the disease. Lung cancers now account for about 30 percent of all cancer deaths, making this condition the primary cause of cancer death in the U.S. population. Despite advances in the treatment of other forms of cancer, the survival statistics for lung cancer have not changed significantly. Even with early detection, the five-year survival rates are only 30 percent for men and 50 percent for women, and most lung cancer patients die within a year of diagnosis. Detailed statistical and experimental evidence has shown that 85–90 percent of all lung cancers are the direct result of cigarette smoking. Claims to the contrary are simply unjustified and insupportable. The data are far too extensive to detail here, but the incidence of lung cancer for nonsmokers is 3.4 per 100,000 population, while the incidence for smokers ranges from 59.3 per 100,000 for those who smoke between a half-pack and a pack of cigarettes a day to 217.3 per 100,000 for those smoking one to two packs a day.

Smoking changes the quality of the inspired air, making it drier and contaminated with several carcinogenic compounds and particulate matter. The combination overloads the respiratory defenses and damages epithelial cells throughout the respiratory system. Whether lung cancer develops appears to be related to the total cumulative exposure to the carcinogens. The more cigarettes smoked, the greater the risk, whether those cigarettes are smoked over a period of weeks or years. Up to the point at which tumors form, the histological changes induced by smoking are reversible; a normal epithelium will return if the carcinogens are removed. At the same time, the statistical risks decline to significantly lower levels. Ten years after quitting, an ex-smoker stands only a 10 percent greater chance of developing lung cancer than a nonsmoker.

The fact that cigarette smoking often causes cancer is not surprising in view of the toxic chemicals contained in the smoke. What is surprising is that more smokers do not develop lung cancer. Evidence indicates that some smokers have a genetic predisposition for developing one form of lung cancer. Dietary factors may play a role in preventing lung cancer, although the details are controversial. In terms of the influence of such factors on the risk of lung cancer, there is general agreement that (1) vitamin A has no effect; (2) vegetables containing beta-carotene reduce the risk, but pills containing beta-carotene may increase the risk; and (3) a high-cholesterol, high-fat diet increases the risk.

15 THE RESPIRATORY SYSTEM

The Functions of the Respiratory System • The Organization of the Respiratory System • Respiratory Physiology • The Control of Respiration

Chemoreceptor Reflexes

Chemoreceptors respond to chemical changes in the blood and cerebrospinal fluid. Their stimulation leads to an increase in the depth and rate of respiration. Centers in the carotid bodies (adjacent to the carotid sinus) and the aortic bodies (near the aortic arch) are sensitive to the pH, P_{CO_2}, and P_{O_2} in arterial blood. Receptors in the medulla oblongata respond to the pH and P_{CO_2} in the cerebrospinal fluid.

HYPERCAPNIA

The term *hypercapnia* (hī-per-KAP-nē-a) refers to an increase in the P_{CO_2} of arterial blood. Such a change stimulates the chemoreceptors in the carotid and aortic bodies and, since carbon dioxide crosses the blood-brain barrier quite rapidly, also the chemoreceptive neurons of the medulla oblongata. These receptors stimulate the respiratory center to produce **hyperventilation** (hī-per-ven-ti-LĀ-shun), an increase in the rate and depth of respiration. Breathing becomes more rapid, and more air moves into and out of the lungs with each breath. Because more air moves into and out of the alveoli each minute, alveolar concentrations of carbon dioxide decline, accelerating the diffusion of carbon dioxide from the alveolar capillaries. If the arterial P_{CO_2} drops below normal levels, chemoreceptor activity declines and the respiratory rate falls. This **hypoventilation** continues until the carbon dioxide partial pressure returns to normal.

Carbon dioxide levels have a much more powerful effect on respiratory activity than does oxygen. This results because a relatively small increase in arterial P_{CO_2} stimulates CO_2 receptors, but arterial P_{O_2} does not usually decline enough to activate the oxygen receptors. Carbon dioxide levels are therefore responsible for regulating respiratory activity under normal conditions. However, when arterial P_{O_2} does fall, the two types of receptors cooperate. Carbon dioxide is generated during oxygen consumption, so when oxygen concentrations are falling rapidly, carbon dioxide levels are usually increasing. As a result, you cannot hold your breath "until you turn blue." Once the P_{CO_2} rises to critical levels, you will be forced to take a breath. The cooperation between the carbon dioxide and oxygen receptors breaks down only under unusual circumstances. For example, an individual can hold his or her breath longer than normal by taking deep, full breaths, but the practice is very dangerous. The danger lies in the fact that the increased ability is due not to extra oxygen but to the loss of carbon dioxide. If the P_{CO_2} is reduced enough, breath-holding ability may increase to the point that an individual becomes unconscious from oxygen starvation in the brain without ever feeling the urge to breathe.

The chemoreceptors monitoring carbon dioxide levels are also sensitive to pH, and any condition affecting the pH of blood or CSF will affect respiratory performance. For example, the rise in lactic acid levels after exercise causes a drop in pH that helps stimulate respiratory activity.

CONTROL BY HIGHER CENTERS

Higher centers influence respiration through their effects on the respiratory centers of the pons and by the direct control of respiratory muscles. For example, the contractions of respiratory muscles can be voluntarily suppressed or exaggerated; this control is necessary during talking or singing. The depth and pace of respiration also change following the activation of centers involved with rage, eating, or sexual arousal. These changes, directed by the limbic system, occur at an involuntary level. Figure 15-16● summarizes factors involved in the regulation of respiration.

SIDS

Sudden infant death syndrome (SIDS), also known as *crib death*, kills an estimated 10,000 infants each year in the United States alone. Most SIDS deaths occur between midnight and 9:00 A.M., in the late fall or winter, and involve infants 2–4 months old. Eyewitness accounts indicate that the sleeping infant suddenly stops breathing, turns blue, and relaxes. Genetic factors appear to be involved, but controversy remains as to the relative importance of other factors, such as laryngeal spasms, cardiac arrhythmias, upper respiratory tract infections, viral infections, and CNS malfunctions. The age at the time of death corresponds with a period when the respiratory centers are establishing connections with other portions of the brain. It has recently been proposed that SIDS results from a problem in the interconnection process that disrupts the reflexive respiratory pattern.

CONCEPT CHECK QUESTIONS
Answers on page 484

❶ Are peripheral chemoreceptors as sensitive to carbon dioxide levels as to oxygen levels?

❷ Strenuous exercise would stimulate which set of respiratory reflexes?

❸ Johnny tells his mother he will hold his breath until he turns blue and dies. Should she worry?

Respiratory Changes at Birth

Several important differences exist between the respiratory systems of a fetus and a newborn infant. Before delivery, the pulmonary vessels are collapsed, so pulmonary arterial resistance is high. The rib cage is compressed, and the lungs and con-

conducting passageways open, and the surfactant covering the alveolar surfaces prevents their collapse. Subsequent breaths complete the inflation of the alveoli. These physical changes are sometimes used by pathologists to determine whether a newborn infant died before or shortly after delivery. Before the first breath, the lungs are completely filled with fluid, and they will sink if placed in water. After the infant's first breath, even the collapsed lungs contain enough air to keep them afloat.

Aging and the Respiratory System

Many factors interact to reduce the efficiency of the respiratory system in elderly individuals. Three examples are noteworthy:

1. As age increases, elastic tissue deteriorates throughout the body. This breakdown reduces the compliance of the lungs, lowering the vital capacity.
2. Chest movements are restricted by arthritic changes in the rib articulations and by decreased flexibility at the costal cartilages. Along with the changes noted in (1), the stiffening and reduction in chest movement limit pulmonary ventilation. This restriction contributes to the reduction in exercise performance and capabilities with increasing age.
3. Some degree of emphysema is normally found in individuals over age 50. However, the extent varies widely with the lifetime exposure to cigarette smoke and other respiratory irritants. Comparative studies of nonsmokers and those who have smoked for varying lengths of time clearly show the negative effect of smoking on respiratory performance.

Integration with Other Systems

The respiratory system has extensive structural and functional connections to the cardiovascular system. It is functionally linked to all other systems as well (Figure 15-17●).

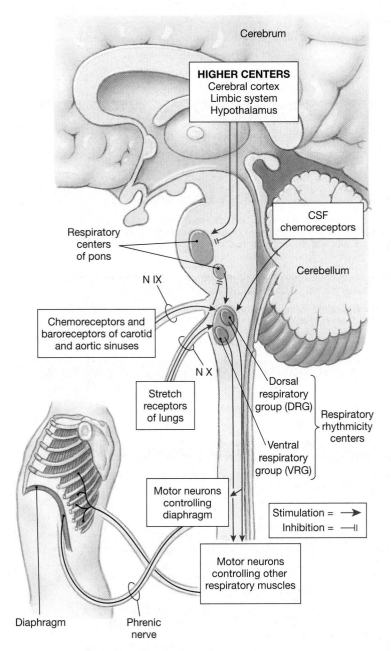

● *Figure 15-16* **The Control of Respiration**

ducting passageways contain only small amounts of fluid and no air. At birth, the newborn infant takes a truly heroic first breath through powerful contractions of the diaphragm and external intercostal muscles. The inspired air enters the passageways with enough force to push the contained fluids out of the way and to inflate the entire bronchial tree and most of the alveolar complexes. The same drop in pressure that pulls air into the lungs pulls blood into the pulmonary circulation.

The exhalation that follows fails to empty the lungs completely, because the rib cage does not return to its former, fully compressed state. Cartilages and connective tissues keep the

---

CONCEPT CHECK QUESTIONS
Answers on page 484

❶ Describe three age-related changes that combine to reduce the efficiency of the respiratory system.
❷ The respiratory system provides all the body systems with oxygen and excretes carbon dioxide. What homeostatic functions does the nervous system perform that support the functional role of the respiratory system?

15 THE RESPIRATORY SYSTEM

The Functions of the Respiratory System • The Organization of the Respiratory System • Respiratory Physiology • The Control of Respiration

The Respiratory System

For All Systems

Provides oxygen and eliminates carbon dioxide

The Integumentary System

• Protects portions of upper respiratory tract; hairs guard entry to external nares

The Skeletal System

• Movements of ribs important in breathing; axial skeleton surrounds and protects lungs

The Muscular System

• Muscular activity generates carbon dioxide; respiratory muscles fill and empty lungs; other muscles control entrances to respiratory tract; intrinsic laryngeal muscles control airflow through larynx and produce sounds

The Nervous System

• Monitors respiratory volume and blood gas levels; controls pace and depth of respiration

The Endocrine System

• Epinephrine and norepinephrine stimulate respiratory activity and dilate respiratory passageways

The Cardiovascular System

• Red blood cells transport oxygen and carbon dioxide between lungs and peripheral tissues
• Bicarbonate ions contribute to buffering capability of blood

The Lymphatic System

• Tonsils protect against infection at entrance to respiratory tract; lymphatic vessels monitor lymph drainage from lungs and mobilize specific defenses when infection occurs
• Alveolar phagocytes present antigens to trigger specific defenses; mucous membrane lining the nasal cavity and upper pharynx traps pathogens, protects deeper tissues

The Digestive System

• Provides substrates, vitamins, water, and ions that are necessary to all cells of the respiratory system
• Increased thoracic and abdominal pressure through contraction of respiratory muscles can assist in defecation

The Urinary System

• Eliminates organic wastes generated by cells of the respiratory system; maintains normal fluid and ion balance in the blood
• Assists in the regulation of pH by eliminating carbon dioxide

The Reproductive System

• Changes in respiratory rate and depth occur during sexual arousal

● *Figure 15-17* **Functional Relationships Between the Respiratory System and Other Systems**

Related Clinical Terms

anoxia (a-NOKS-ē-uh): A lack of oxygen in a tissue.

apnea (AP-nē-uh): The cessation of breathing.

asthma (AZ-ma): An acute respiratory disorder characterized by unusually sensitive, irritated conducting passageways.

atelectasis (at-e-LEK-ta-sis): A collapsed lung.

bronchitis (brong-KĪ-tis): An inflammation of the bronchial lining.

bronchoscope: A fiber-optic bundle small enough to be inserted into the trachea and finer airways; the procedure is called *bronchoscopy.*

chronic obstructive pulmonary disease (COPD): A condition characterized by chronic bronchitis and chronic airways obstruction.

cystic fibrosis (CF): A relatively common lethal inherited disease caused by an abnormal membrane channel protein; a major symptom is that mucous secretions become too thick to be transported easily, leading to respiratory problems.

dyspnea (DISP-nē-uh): Difficulty in breathing.

emphysema (em-fi-SĒ-muh) A chronic, progressive condition characterized by shortness of breath and an inability to tolerate physical exertion.

epistaxis (ep-i-STAK-sis): A nosebleed.

hypercapnia (hī-per-KAP-nē-uh): An abnormally high level of carbon dioxide in the blood.

hypocapnia: An abnormally low level of carbon dioxide in the blood.

hypoxia (hī-POKS-ē-uh): A low oxygen concentration in a tissue.

influenza: A viral infection of the respiratory tract; "the flu."

pleurisy: An inflammation of the pleurae and secretion of excess amounts of pleural fluid.

pneumonia (nū-MŌ-nē-uh): A respiratory disorder characterized by fluid leakage into the alveoli and/or swelling and constriction of the respiratory bronchioles.

pneumothorax (noo-mō-THŌ-raks): The entry of air into the pleural cavity.

pulmonary embolism: A blockage of a branch of a pulmonary artery, with interruption of blood flow to a group of lobules and/or alveoli.

respiratory distress syndrome: A condition resulting from inadequate surfactant production and associated alveolar collapse.

tracheostomy (trā-kē-OS-to-mē): The insertion of a tube directly into the trachea to bypass a blocked or damaged larynx.

CHAPTER REVIEW

Key Terms

Summary Outline

INTRODUCTION **458**

1. To continue functioning, body cells must obtain oxygen and eliminate carbon dioxide.

THE FUNCTIONS OF THE RESPIRATORY SYSTEM **458**

1. The functions of the **respiratory system** include: (1) moving air to and from exchange surfaces where diffusion can occur between air and circulating blood; (2) defending the respiratory system and other tissues from pathogens; (3) permitting vocal communication; and (4) helping control body fluid pH.

THE ORGANIZATION OF THE RESPIRATORY SYSTEM....458

1. The respiratory system includes the nose, nasal cavity, and sinuses as well as the pharynx, larynx, trachea, and conducting passageways leading to the surfaces of the lungs. *(Figure 15-1)*

   The Respiratory Tract 459

2. The **respiratory tract** consists of the conducting passageways that carry air to and from the alveoli.

   The Nose..459

3. Air normally enters the respiratory system via the **external nares**, which open into the **nasal cavity**. The **vestibule** (entrance) is guarded by hairs that screen out large particles. *(Figure 15-2)*

4. The **hard palate** separates the oral and nasal cavities. The **soft palate** separates the superior nasopharynx from the rest of the pharynx. The **internal nares** connects the nasal cavity and nasopharynx. *(Figure 15-2)*

5. Much of the respiratory epithelium is ciliated and produces mucus that traps incoming particles. *(Figure 15-3)*

   The Pharynx 460

6. The **pharynx** (throat) is a chamber shared by the digestive and respiratory systems.

   The Larynx 461

7. Inhaled air passes through the **glottis** en route to the lungs; the **larynx** surrounds and protects the glottis. The **epiglottis** projects into the pharynx. Exhaled air passing through the glottis vibrates the **true vocal cords** and produces sound. *(Figure 15-4)*

   The Trachea 462

8. The wall of the **trachea** ("windpipe") contains C-shaped tracheal cartilages, which protect the airway. The posterior tracheal wall can distort to permit large masses of food to pass. *(Figure 15-5)*

   The Bronchi 463

9. The trachea branches within the mediastinum to form the **right** and **left primary bronchi**. *(Figure 15-5)*

15 THE RESPIRATORY SYSTEM

The Functions of the Respiratory System • The Organization of the Respiratory System • Respiratory Physiology • The Control of Respiration

10. The primary bronchi, **secondary bronchi**, and their branches form the *bronchial tree*. As the **tertiary bronchi** branch within the lung, the amount of cartilage in their walls decreases and the amount of smooth muscle increases.(*Figure 15-5, 15-6a*)

11. Each terminal **bronchiole** delivers air to a single pulmonary **lobule**. Within the lobule, the terminal bronchiole branches into *respiratory bronchioles*. (*Figure 15-6b*)

12. The respiratory bronchioles open into **alveolar ducts**, which end at **alveolar sacs**. Many alveoli are interconnected at each alveolar sac. (*Figure 15-7a, b*)

13. The **respiratory membrane** consists of (1) a simple squamous alveolar epithelium, (2) a capillary endothelium, and (3) their fused basement membranes. **Surfactant cells** produce an oily secretion that keeps the alveoli from collapsing. **Alveolar macrophages** engulf foreign particles. (*Figure 15-7c, d*)

14. The **lungs** are made up of five **lobes**; three in the right lung and two in the left lung. (*Figure 15-8*)

15. Each lung occupies a single pleural cavity lined by a **pleura** (serous membrane). (*Figure 15-9*)

1. Respiratory physiology focuses on a series of integrated processes: *pulmonary ventilation*, or breathing (movement of air into and out of the lungs); *gas exchange*, or diffusion, between the alveoli and circulating blood, and between the blood and interstitial fluids; and *gas transport*, between the blood and interstitial fluids.

2. A single breath, or **respiratory cycle**, consists of an inhalation (*inspiration*) and an exhalation (*expiration*).

3. The relationship between the pressure inside the respiratory tract and atmospheric pressure determines the direction of airflow. (*Figure 15-10*)

4. The diaphragm and the external intercostal muscles are involved in *quiet breathing*, in which exhalation is passive. Accessory muscles become active during the active inspiratory and expiratory movements of *forced breathing*, in which exhalation is active. (*Figure 15-10*)

5. The **vital capacity** includes the **tidal volume** plus the **expiratory reserve volume** and the **inspiratory reserve volume**. The air left in the lungs at the end of maximum expiration is the **residual volume**. (*Figure 15-11*)

6. Gas exchange involves *external respiration*, the diffusion of gases between the blood and alveolar air across the respiratory membrane, and *internal respiration*, the diffusion of gases between blood and interstitial fluid across the single endothelial cells of capillary walls. (*Figure 15-12; Table 15-1*)

7. Blood entering peripheral capillaries delivers oxygen and absorbs carbon dioxide. The transport of oxygen and carbon dioxide in the blood involves reactions that are completely reversible.

8. Over the range of oxygen pressures normally present in the body, a small change in plasma P_{O_2} will mean a large change in the amount of oxygen bound or released by hemoglobin.

9. Aerobic metabolism in peripheral tissues generates carbon dioxide. Roughly 7 percent of the CO_2 transported in the blood is dissolved in the plasma; another 23 percent is bound as **carbaminohemoglobin**; the rest is converted to carbonic acid, which dissociates into a hydrogen ion and a bicarbonate ion. (*Figures 15-13, 15-14*)

1. Large-scale changes in oxygen demand require the integration of cardiovascular and respiratory responses.

2. Arterioles leading to alveolar capillaries constrict when oxygen is low and bronchioles dilate when carbon dioxide is high.

3. The **respiratory centers** include three pairs of nuclei in the reticular formation of the pons and medulla oblongata. These nuclei regulate the respiratory muscles and control the respiratory rate and the depth of breathing. The **respiratory rhythmicity centers** in the medulla oblongata set the basic pace for respiration. (*Figure 15-15*)

4. The **inflation reflex** prevents overexpansion of the lungs during forced breathing; the **deflation reflex** stimulates inspiration when the lungs are collapsing. Chemoreceptor reflexes respond to changes in the pH, P_{O_2}, and P_{CO_2} of the blood and cerebrospinal fluid.

5. Conscious and unconscious thought processes can affect respiration by affecting the respiratory centers or the motor neurons controlling respiratory muscles. (*Figure 15-16*)

1. Before delivery, the fetal lungs are fluid-filled and collapsed. After the first breath, the alveoli normally remain inflated for the life of the individual.

1. The respiratory system is generally less efficient in the elderly because: (1) elastic tissue deteriorates, lowering the vital capacity of the lungs; (2) movements of the chest cage are restricted by arthritic changes and decreased flexibility of costal cartilages; and (3) some degree of emphysema is normal in the elderly.

1. The respiratory system has extensive anatomical connections to the cardiovascular system. (*Figure 15-17*)

Review Questions

Level 1: Reviewing Facts and Terms

Match each item in column A with the most closely related item in column B. Use letters for answers in the spaces provided.

COLUMN A

____ 1.   nasopharynx

____ 2.   laryngopharynx

____ 3.   thyroid cartilage

____ 4.   surfactant cells

____ 5.   dust cells

____ 6.   parietal pleura

____ 7.   visceral pleura

____ 8.   hypoxia

____ 9.   anoxia

____ 10. collapsed lung

____ 11. inhalation

____ 12. exhalation

COLUMN B

a. no O_2 supply to tissues

b. alveolar macrophages

c. produce oily secretion

d. covers inner surface of thoracic wall

e. low O_2 content in tissue fluids

f. inferior portion of pharynx

g. inspiration

h. superior portion of pharynx

i. Adam's apple

j. covers outer surface of lungs

k. expiration

l. atelectasis

13. The structure that prevents the entry of liquids or solid food into the respiratory passageways during swallowing is the:

(a) glottis   (b) arytenoid cartilage

(c) epiglottis   (d) thyroid cartilage

14. The amount of air moved into or out of the lungs during a single respiratory cycle is the:

(a) respiratory minute volume   (b) tidal volume

(c) residual volume   (d) inspiratory capacity

Level 2: Reviewing Concepts

15. When the diaphragm contracts, it tenses and moves inferiorly, causing:

(a) an increase in the volume of the thoracic cavity

(b) a decrease in the volume of the thoracic cavity

(c) decreased pressure on the contents of the abdominopelvic cavity

(d) increased pressure in the thoracic cavity

16. Gas exchange at the respiratory membrane is efficient because:

(a) the differences in partial pressure are substantial

(b) the gases are lipid-soluble

(c) the total surface area is large

(d) a, b, and c are correct

17. What is the functional significance of the decrease in the amount of cartilage and increase in the amount of smooth muscle in the lower respiratory passageways?

18. Why is breathing through the nasal cavity more desirable than breathing through the mouth?

19. How would you justify the statement, "The bronchioles are to the respiratory system what the arterioles are to the cardiovascular system"?

20. How are surfactant cells involved with keeping the alveoli from collapsing?

Level 3: Critical Thinking and Clinical Applications

21. A decrease in blood pressure will trigger a baroreceptor reflex that leads to increased ventilation. What is the possible advantage of this reflex?

22. You spend the night at a friend's house during the winter. Your friend's home is quite old, and the hot-air furnace lacks a humidifier. When you wake up in the morning, you have a fair amount of nasal congestion and decide you might be coming down with a cold. After you take a steamy shower and drink some juice for breakfast, the nasal congestion disappears. Explain.

15 THE RESPIRATORY SYSTEM

The Functions of the Respiratory System • The Organization of the Respiratory System • Respiratory Physiology • The Control of Respiration

Answers to Concept Check Questions

Page 468

1. Increased tension in the vocal cords will cause a higher pitch in the voice.
2. The tracheal cartilages are C-shaped to allow room for esophageal expansion when large portions of food or liquid are swallowed.
3. Without surfactant, surface tension in the thin layer of water that moistens their surfaces would cause the alveoli to collapse.

Page 471

1. Since the rib penetrates the chest wall, atmospheric air will enter the thoracic cavity. This condition is called a pneumothorax. Pressure within the pleural cavity is normally lower than atmospheric pressure. When air enters the pleural cavity, the natural elasticity of the lung may cause it to collapse. The resulting condition is called atelectasis, or a collapsed lung.
2. Since the fluid produced in pneumonia takes up space that would normally be occupied by air, the vital capacity would decrease.

Page 475

1. As skeletal muscles become more active, they generate more heat and more acid waste products and so lower the pH of surrounding fluid. The combination of lower pH and higher temperature causes the hemoglobin to release more oxygen than it would under conditions of lower temperature and higher pH.
2. An obstruction of the trachea would interfere with the body's ability to gain oxygen and eliminate carbon dioxide. Since most carbon dioxide is carried in the blood as bicarbonate ion that is formed from the dissoci-

ation of carbonic acid, an inability to eliminate carbon dioxide would result in an excess of hydrogen ions, thus lowering the body's pH.

Page 478

1. Chemoreceptors are more sensitive to carbon dioxide levels than to oxygen levels. When carbon dioxide dissolves, it produces hydrogen ions, which lower pH and alter cell or tissue activity.
2. Strenuous exercise would stimulate the inflation and deflation reflexes, also known as the Hering-Breuer reflexes. In the inflation reflex, the stimulation of stretch receptors in the lungs results in an inhibition of the inspiratory center and a stimulation of the expiratory center. In contrast, collapse of the lungs initiates the deflation reflex. This reflex results in an inhibition of the expiratory center and a stimulation of the inspiratory center.
3. Johnny's mother shouldn't worry. When Johnny holds his breath, the level of carbon dioxide in his blood will increase. This will lead to increased stimulation of the inspiratory center, forcing Johnny to breathe again.

Page 479

1. Three general age-related changes that reduce the efficiency of the respiratory system are: (1) reduced compliance because of a general deterioration of elastic tissue in the body (and lungs), (2) reduced chest movements due to arthritic changes in rib joints and stiffening of the costal cartilages; and (3) some degree of emphysema because of the gradual destruction of alveolar surfaces.
2. The nervous system controls the pace and depth of respiration, and monitors the respiratory volume of the lungs and levels of blood gases.

The Functions of the Respiratory System • The Organization of the Respiratory System • Respiratory Physiology • The Control of Respiration

484

EXPLORE *MediaLab*

EXPLORATION #1

Estimated time for completion: 10 minutes

IF YOU HAVE EVER EXERCISED you know that there is a limit to what you are comfortably able to accomplish. Less athletic individuals refer to this as being "too tired to continue," and trained athletes speak of "hitting the wall." Is there an actual physiological limit to the amount of aerobic exercise your body can tolerate? In fact there is, and it is referred to as $VO_{2\,max}$. This is a measure of the ability of the body to consume oxygen. Take a minute to reflect on what you know of the interaction between the cardiovascular system and the respiratory system and the possible relationship between these interactions and $VO_{2\,max}$. (Figure 15-12● on p. 473 may help get you started.) Summarize your thoughts on a sheet of paper. Now prepare a list of factors that influence the body's ability to perform work. Predict how you might measure your own $VO_{2\,max}$. What would you need to measure to determine how efficiently you use oxygen? Include this information on your list. To complete this exercise, go to Chapter 15 at the Companion Web site, visit the MediaLab section, and click on the keywords "Oxygen use." You will be taken to a Web site that will help you not only to answer these questions but also to better understand the correlation between the respiratory and cardiovascular systems.

A young asthma sufferer using a home inhaler to dilate his respiratory passageways.

EXPLORATION #2

Estimated time for completion: 10 minutes

BREATHE IN, BREATHE OUT, breathe in, breathe out. It is a cycle we repeat throughout life. The volume of air moved and the timing of air movement into and out of our lungs can help physicians diagnose respiratory ailments and determine appropriate therapeutic treatment. Take a minute to trace Figure 15-11● onto a sheet of paper, including the labels. This will form the basis of a study sheet on pulmonary function. Now read through the Clinical Discussion titled "Pulmonary Function Tests" on page 471. Based on information in this box, predict what the pulmonary function graph might look for an asthmatic and include this on your study sheet. Would you expect the same alterations for someone suffering from an obstructed airway? With proper treatment, how might both of these pulmonary-function graphs appear? The University of Iowa Virtual Hospital provides an excellent simulation of pulmonary function tests that will augment your studies. Go to Chapter 15 at the Companion Web site, visit the MediaLab section, and click the keywords "Pulmonary Function." Read through the first three pages of this site, clicking on the movie links and the descriptions of disorders. Complete your study sheet on pulmonary function by including information on pulmonary volumes and their relation to respiratory health.

A marathon runner catches her breath at the end of a race.

The Digestive System

CHAPTER OUTLINE AND OBJECTIVES

Vocabulary Development

chymos ....juice; *chyme*

deciduus ....falling off; *deciduous*

enteron ....intestine; *myenteric plexus*

frenulum ....small bridle; *lingual frenulum*

gaster ....stomach; *gastric juice*

hepaticus ....liver; *hepatocyte*

hiatus ....gap or opening; *esophageal hiatus*

lacteus ....milky; *lacteal*

nutrients ....nourishing; *nutrient*

odonto-................tooth; *periodontal ligament*

omentum ....fat skin; *greater omentum*

pyle ....gate; *pyloric sphincter*

rugae....................................wrinkles; *rugae*

sigmoides ....Greek letter S; *sigmoid colon*

stalsis ....constriction; *peristalsis*

vermis...................worm; *vermiform appendix*

villus.....................shaggy hair; *intestinal villus*

*T*HEY SAY YOU ARE **what you eat.** *Perhaps that is what Giuseppe Arcimboldo had in mind when he painted this whimsical picture in the sixteenth century. This chapter introduces the organs of the digestive system and how they process foods for use by our cells.*

16 THE DIGESTIVE SYSTEM

An Overview of the Digestive Tract • The Oral Cavity • The Pharynx • The Esophagus • The Stomach • The Small Intestine • The Pancreas • The Liver

FEW PEOPLE GIVE any serious thought to the digestive system unless it malfunctions. Yet we spend hours of conscious effort filling and emptying it. References to this system are part of our everyday language. We "have a gut feeling," "want to chew on" something, or find someone's opinions "hard to swallow." When something does go wrong with the digestive system, even something minor, most people seek relief immediately. For this reason, every hour of television programming contains advertisements that promote toothpaste and mouthwash, diet supplements, antacids, and laxatives.

Chapter 15 discussed the respiratory system, which delivers the oxygen needed to "burn" metabolic fuels. The digestive system provides the fuel and performs many other chemical exchanges. It consists of a muscular tube—the **digestive tract**—and **accessory organs**, including the salivary glands, gallbladder, liver, and pancreas.

Digestive functions involve six related processes:

1. **Ingestion** occurs when foods enter the digestive tract through the mouth.
2. **Mechanical processing** is the physical manipulation of solid foods, first by the tongue and the teeth and then by swirling and mixing motions of the digestive tract.
3. **Digestion** refers to the chemical breakdown of food into small organic fragments that can be absorbed by the digestive epithelium.
4. **Secretion** is the release of water, acids, enzymes, and buffers by the digestive tract and by the accessory organs.
5. **Absorption** is the movement of small organic molecules, electrolytes, vitamins, and water across the digestive epithelium and into the interstitial fluid of the digestive tract.
6. **Excretion** is the removal of waste products from the body. Within the digestive tract, these waste products are compacted and discharged through the process of *defecation* (def-e-KĀ-shun).

The lining of the digestive tract also plays a defensive role by protecting surrounding tissues against the corrosive effects of digestive acids and enzymes and against bacteria that either are swallowed with food or reside in the digestive tract. The digestive epithelium and its secretions provide a nonspecific defense against these bacteria, and bacteria reaching the underlying tissues are attacked by macrophages and other cells of the immune system.

An Overview of the Digestive Tract

The major components of the digestive tract are shown in Figure 16-1●. The digestive tract begins with the oral cavity and continues through the pharynx, esophagus, stomach, small intestine, and large intestine before ending at the rectum and anus. Although these subdivisions of the digestive tract have overlapping functions, each region has certain areas of specialization and shows distinctive histological specializations. After describing the tissues making up the digestive tract wall, we will follow the path of ingested materials from the mouth to the anus.

HISTOLOGICAL ORGANIZATION

The digestive tract has four major layers (Figure 16-2●): the *mucosa*, the *submucosa*, the *muscularis externa*, and the *serosa*.

The Mucosa

The **mucosa**, or inner lining of the digestive tract, is an example of a *mucous membrane*. ∞ p. 98 It consists of an epithelial surface moistened by glandular secretions and an underlying layer of loose connective tissue, the *lamina propria*. Along most of the length of the digestive tract, the mucosa is thrown into folds that increase the surface area available for absorption and permit expansion after a large meal. In the small intestine, the mucosa forms fingerlike projections, called *villi* (*villus*, shaggy hair), that further increase the area for absorption.

The oral cavity, pharynx, esophagus, and anus (where mechanical stresses are most severe) are lined by a stratified squamous epithelium. The remainder of the digestive tract is lined by a simple columnar epithelium, often containing various types of secretory cells. Ducts opening onto the epithelial surfaces carry the secretions of glands located in the lamina propria, in the surrounding submucosa, or within accessory glandular organs. In most regions of the digestive tract, the outer portion of the mucosa contains a narrow band of smooth muscle and elastic fibers. Contractions of this layer, the *muscularis* (mus-kū-LAR-is) *mucosae*, move the mucosal folds and villi.

The Submucosa

The **submucosa** is a second layer of loose connective tissue that surrounds the muscularis mucosae. It contains large blood vessels and lymphatic vessels as well as a network of nerve fibers, sensory neurons, and parasympathetic motor neurons. This neural tissue, the *submucosal plexus*, helps control and coordinate the contractions of smooth muscle layers and also helps regulate the secretion of digestive glands.

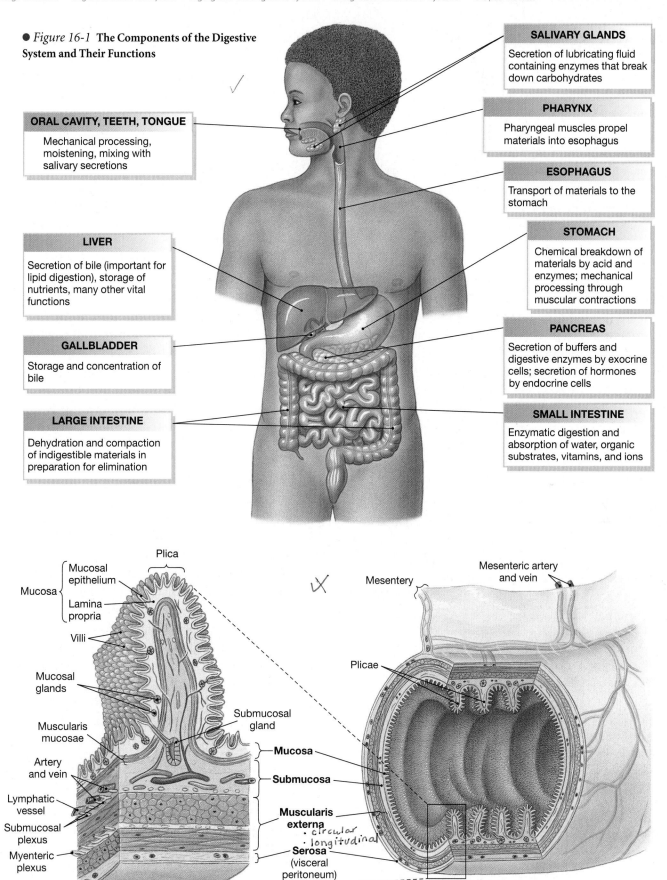

● *Figure 16-1* **The Components of the Digestive System and Their Functions**

SALIVARY GLANDS
Secretion of lubricating fluid containing enzymes that break down carbohydrates

PHARYNX
Pharyngeal muscles propel materials into esophagus

ESOPHAGUS
Transport of materials to the stomach

STOMACH
Chemical breakdown of materials by acid and enzymes; mechanical processing through muscular contractions

PANCREAS
Secretion of buffers and digestive enzymes by exocrine cells; secretion of hormones by endocrine cells

SMALL INTESTINE
Enzymatic digestion and absorption of water, organic substrates, vitamins, and ions

ORAL CAVITY, TEETH, TONGUE
Mechanical processing, moistening, mixing with salivary secretions

LIVER
Secretion of bile (important for lipid digestion), storage of nutrients, many other vital functions

GALLBLADDER
Storage and concentration of bile

LARGE INTESTINE
Dehydration and compaction of indigestible materials in preparation for elimination

Plica
Mucosa
 Mucosal epithelium
 Lamina propria
Villi
Mucosal glands
Muscularis mucosae
Artery and vein
Lymphatic vessel
Submucosal plexus
Myenteric plexus
Submucosal gland

Mucosa
Submucosa
Muscularis externa
 • circular
 • longitudinal
Serosa (visceral peritoneum)

Mesentery
Mesenteric artery and vein
Plicae

● *Figure 16-2* **The Structure of the Digestive Tract**
A representative portion of the digestive tract—the small intestine. The wall of the digestive tract is made up of four layers, the mucosa, submucosa, muscularis externa, and serosa.

16 THE DIGESTIVE SYSTEM

An Overview of the Digestive Tract • **The Oral Cavity** • The Pharynx • The Esophagus • The Stomach • The Small Intestine • The Pancreas • The Liver

The Muscularis Externa

The **muscularis externa** is a collection of smooth muscle cells arranged in an inner circular layer and an outer longitudinal layer. Contractions of these layers in various combinations agitate the contents and propel materials along the digestive tract. These are autonomic reflex movements controlled primarily by another network of nerves, the *myenteric plexus* (*mys*, muscle + *enteron*, intestine), sandwiched between the inner and outer smooth muscle layers. Parasympathetic stimulation increases muscular tone and activity, and sympathetic stimulation promotes muscular inhibition and relaxation.

The Serosa

The **serosa**, a serous membrane, covers the muscularis externa along most portions of the digestive tract inside the peritoneal cavity. This *visceral peritoneum* is continuous with the *parietal peritoneum* that lines the inner surfaces of the body wall. ∽ p. 100 In some areas, the parietal and visceral peritoneum are connected by double sheets of serous membrane called **mesenteries** (MEZ-en-ter-ēz). Portions of the digestive tract are suspended within the peritoneal cavity by mesenteries. The loose connective tissue sandwiched between the epithelial surfaces of the mesenteries provides a passageway for the blood vessels, nerves, and lymphatic vessels servicing the digestive tract. The mesenteries also stabilize the positions of the attached organs and prevent the intestines from becoming entangled during digestive movements or sudden changes in body position.

There is no serosa covering the muscularis externa of the oral cavity, pharynx, esophagus, and rectum. Instead, the muscularis externa is surrounded by a dense network of collagen fibers that firmly attaches these regions of the digestive tract to adjacent structures. This fibrous wrapping is called an *adventitia* (ad-ven-TISH-a).

 ### ASCITES

The peritoneal lining continuously produces peritoneal fluid, which lubricates the opposing parietal and visceral surfaces. About 7 liters of fluid is secreted and reabsorbed each day, although the volume within the cavity at any moment is very small. Liver disease, kidney disease, and heart failure can cause an increase in the rate of fluid movement into the peritoneal cavity. The accumulation of fluid creates a characteristic abdominal swelling called **ascites** (a-SĪ-tēz). The distortion of internal organs by the contained fluid can result in symptoms such as heartburn, indigestion, and low back pain.

THE MOVEMENT OF DIGESTIVE MATERIALS

Smooth muscle tissue is found within almost every organ, forming sheets, bundles, or sheaths around other tissues. *Pacesetter cells* in the smooth muscle of the digestive tract trigger waves of contraction, resulting in rhythmic cycles of activity. The coordinated contractions of the smooth muscle in the walls of the digestive tract play a vital role in moving materials along the tract, through *peristalsis* (*peri-*, around + *stalsis*, constriction), and in mechanical processing, through *segmentation*.

Peristalsis and Segmentation

The muscularis externa propels materials from one part of the digestive tract to another by means of **peristalsis** (per-i-STAL-sis), waves of muscular contractions that move along the length of the digestive tract (Figure 16-3a●). During a peristaltic movement, the circular muscles first contract behind the digestive contents. Then longitudinal muscles contract, shortening adjacent segments. A wave of contraction in the circular muscles then forces the materials in the desired direction.

Regions of the small intestine also undergo **segmentation**, movements that churn and fragment digestive materials (Figure 16-3b●). Over time, this action results in a thorough mixing of the contents with intestinal secretions. Because they do not follow a set pattern, segmentation movements do not propel materials in a particular direction.

The Oral Cavity

The mouth opens into the **oral cavity**, the part of the digestive tract that receives food. The oral cavity (1) analyzes material before swallowing; (2) mechanically processes material through the actions of the teeth, tongue, and surfaces of the palate; (3) lubricates material by mixing it with mucus and salivary secretions; and (4) begins the digestion of carbohydrates and lipids with the help of salivary enzymes.

Figure 16-4● shows the boundaries of the oral cavity, also known as the **buccal** (BUK-al) **cavity**. The **cheeks** form the lateral walls of this chamber; anteriorly they are continuous with the lips, or **labia** (LĀ-bē-a; singular, *labium*). The **vestibule**, a subdivision of the oral cavity, includes the space between the cheeks or lips and the teeth. A pink ridge, the gums, or **gingivae** (JIN-ji-vē; singular, *gingiva*), surrounds the bases of the teeth. The gums cover the tooth-bearing surfaces of the upper and lower jaws.

The **hard palate** and **soft palate** provide a roof for the oral cavity, while the tongue dominates its floor. The free anterior portion of the tongue is connected to the underlying epithelium by a thin fold of mucous membrane, the **lingual frenulum** (FREN-ū-lum; *frenulum*, a small bridle). The dividing line between the oral cavity and pharynx extends between the base of the tongue and the dangling *uvula* (Ū-vū-luh).

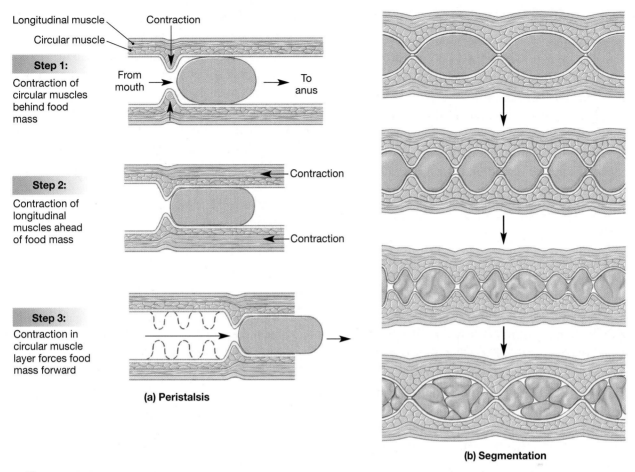

● *Figure 16-3* **Peristalsis and Segmentation**
(**a**) Peristalsis propels materials along the length of the digestive tract. (**b**) Segmentation promotes mixing but does not produce net movement in any direction.

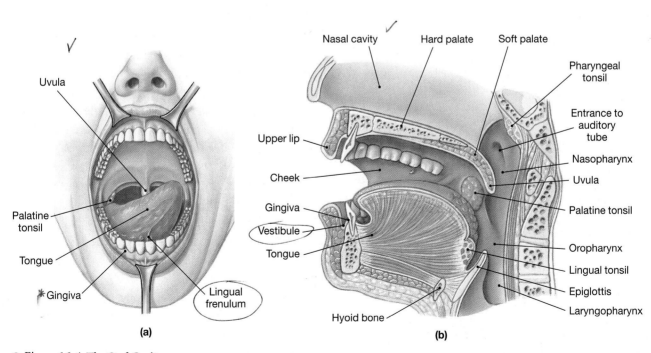

● *Figure 16-4* **The Oral Cavity**
(**a**) An anterior view of the oral cavity, as seen through the open mouth. (**b**) A sagittal section of the oral cavity.

491

16 **THE DIGESTIVE SYSTEM**

An Overview of the Digestive Tract • **The Oral Cavity** • The Pharynx • The Esophagus • The Stomach • The Small Intestine • The Pancreas • The Liver

THE TONGUE

The muscular **tongue** manipulates materials inside the mouth and is occasionally used to bring food into the oral cavity. The primary functions of the tongue are (1) mechanical processing by compression, abrasion, and distortion; (2) manipulation to assist in chewing and to prepare the material for swallowing; and (3) sensory analysis by touch, temperature, and taste receptors. Most of the tongue lies within the oral cavity, but the base of the tongue extends into the pharynx. A pair of prominent lateral swellings at the base of the tongue marks the location of the *lingual tonsils*, lymphoid nodules that help resist infections. p. 431

SALIVARY GLANDS

Figure 16-5● shows the locations and relative sizes of the three pairs of salivary glands (compare to Figure 16-4●). On each side, the large **parotid salivary gland** lies below the zygomatic arch under the skin of the face. p. 144 The **parotid duct** empties into the vestibule at the level of the second upper molar. The **sublingual salivary glands** are located beneath the mucous membrane of the floor of the mouth, and numerous sublingual ducts open along either side of the lingual frenulum. The **submandibular salivary glands** are in the floor of the mouth along the inner surfaces of the mandible; their ducts open into the mouth behind the teeth on either side of the lingual frenulum.

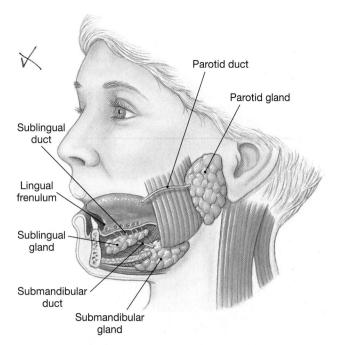

● *Figure 16-5* **The Salivary Glands**
A lateral view, showing the relative positions of the salivary glands and ducts on the left side of the head. The left half of the mandible has been removed to show deeper structures.

These salivary glands produce 1.0–1.5 liters of saliva each day, with a composition of 99.4 percent water, plus an assortment of ions, buffers, waste products, metabolites, and enzymes. At mealtimes, large quantities of saliva lubricate the mouth and dissolve chemicals that stimulate the taste buds. Coating the food with slippery mucus reduces friction and makes swallowing possible. A continuous background level of secretion flushes the oral surfaces, and salivary immunoglobulins (IgA) and lysozymes help control populations of oral bacteria. When salivary secretions are reduced or eliminated, such as by radiation exposure, emotional distress, or other factors, the bacterial population in the oral cavity explodes. This bacterial increase soon leads to recurring infections and the progressive erosion of the teeth and gums.

✚ MUMPS

The *mumps virus* most often targets the salivary glands, especially the parotid salivary glands, although other organs can also become infected. Infection typically occurs at 5–9 years of age. The first exposure stimulates antibody production and usually confers permanent immunity; active immunity can be conferred by immunization. In postadolescent males, the mumps virus may also infect the testes and cause sterility. Infection of the pancreas by the mumps virus may produce temporary or permanent diabetes; other organ systems, including the CNS, may be affected in severe cases. An effective mumps vaccine is available. Widespread distribution of that vaccine has reduced the incidence of the disease in the United States.

Salivary Secretions

Each of the salivary glands produces a slightly different kind of saliva. The parotid glands produce a secretion rich in **salivary amylase**, an enzyme that breaks down starches (complex carbohydrates) into smaller molecules that can be absorbed by the digestive tract. Saliva originating in the submandibular and sublingual salivary glands contains fewer enzymes but more buffers and mucus. During eating, all three salivary glands increase their rates of secretion, and salivary production may reach 7 ml per minute, with about 70 percent of that volume provided by the submandibular glands. The pH also rises, shifting from slightly acidic (pH 6.7) to slightly basic (pH 7.5). Salivary secretions are normally controlled by the autonomic nervous system.

TEETH

Movements of the tongue are important in passing food across the surfaces of the **teeth**. The opposing surfaces of the teeth perform chewing, or **mastication** (mas-ti-KĀ-shun), of food. Mastication breaks down tough connective tissues and plant

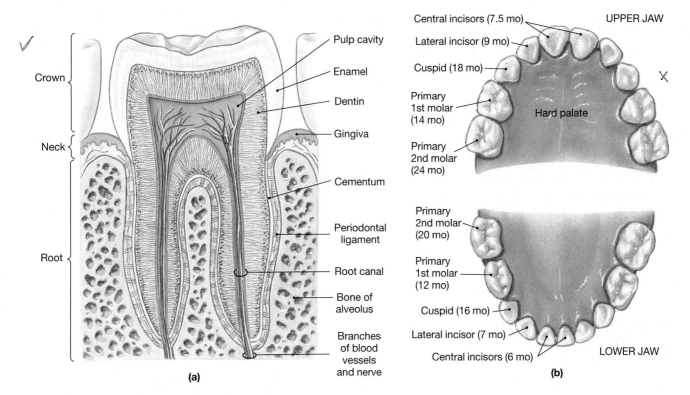

(a)

(b)

● *Figure 16-6* **Teeth: Structural Components and Dental Succession**
(**a**) A diagrammatic section through a typical adult tooth. (**b**) The primary teeth of a child, with the age at eruption given in months. (**c**) The adult teeth, with the age at eruption given in years.

fibers and helps saturate the materials with salivary lubricants and enzymes.

Figure 16-6a● shows the parts of a tooth. The **neck** of the tooth marks the boundary between the **root** and the **crown**. The crown is covered by a layer of **enamel**, which contains a crystalline form of calcium phosphate, the hardest biologically manufactured substance. Adequate amounts of calcium, phosphates, and vitamin D_3 during childhood are essential if the enamel coating is to be complete and resistant to decay.

The bulk of each tooth consists of **dentin** (DEN-tin), a mineralized matrix similar to that of bone. Dentin differs from bone in that it does not contain living cells. Instead, cytoplasmic processes extend into the dentin from cells within the central **pulp cavity**. The pulp cavity receives blood vessels and nerves through a narrow **root canal** at the root of the tooth. The root sits within a bony socket, or *alveolus* (a hollow cavity). Collagen fibers of the **periodontal ligament** (*peri-*, around + *odonto-*, tooth) extend from the dentin of the root to the surrounding bone. A layer of **cementum** (se-MEN-tum) covers the dentin of the root providing protection and firmly anchoring the periodontal ligament. Cementum also resembles bone, but it is softer, and remodeling does not occur following its deposition. Where the tooth penetrates

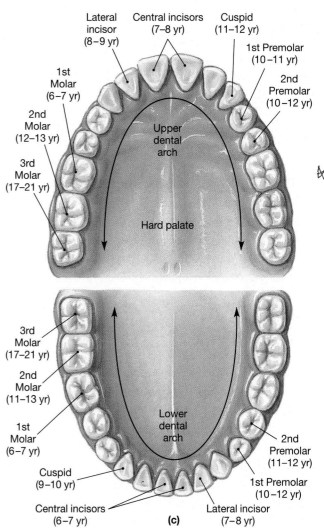

(c)

493

16 THE DIGESTIVE SYSTEM

An Overview of the Digestive Tract • The Oral Cavity • **The Pharynx** • **The Esophagus** • **The Stomach** • The Small Intestine • The Pancreas • The Liver

the gum surface, epithelial cells form tight attachments to the tooth and prevent bacterial access to the easily eroded cementum of the root.

Adult teeth are shown in Figure 16-6c●. Each of the four types of teeth has a specific function: (1) **Incisors** (in-SĪ-zerz), blade-shaped teeth found at the front of the mouth, are useful for clipping or cutting, as when nipping off the tip of a carrot. (2) **Cuspids** (KUS-pidz), or *canines*, are conical with a sharp ridgeline and a pointed tip. They are used for tearing or slashing. You might weaken a tough piece of celery by the clipping action of the incisors, but then take advantage of the shearing action provided by the cuspids. (3) **Bicuspids** (bī-KUS-pidz), or *premolars*, and (4) **molars** have flattened crowns with prominent ridges. They are used for crushing, mashing, and grinding. You can usually shift a tough nut or piece of meat to the premolars and molars for crushing.

Dental Succession

During development, two sets of teeth begin to form. The first to appear are the **deciduous teeth** (de-SID-ū-us; *deciduus*, falling off), also known as *primary teeth*, *milk teeth*, or *baby teeth*. There are usually 20 deciduous teeth (Figure 16-6b●). These teeth will later be replaced by the adult **secondary dentition**, or *permanent dentition*, which permit the processing of a wider variety of foods (Figure 16-6c●). The replacement occurs as the periodontal ligaments and roots of the deciduous teeth erode and either fall out or are pushed aside by the **eruption**, or *emergence*, of the secondary teeth. Adult jaws are larger and can accommodate more than 20 teeth. Three additional teeth appear on each side of the upper and lower jaws as the person ages, bringing the permanent tooth count to 32. The last teeth to appear are the *third molars*, or *wisdom teeth*. Wisdom teeth or any other teeth that develop in locations that do not permit their eruption, are called *impacted teeth*. Impacted teeth can be surgically removed to prevent the formation of abscesses.

The Pharynx

The **pharynx** serves as a common passageway for solid food, liquids, and air. The three major subdivisions of the pharynx were discussed in Chapter 15. p. 460 Food normally passes through the oropharynx and laryngopharynx on its way to the esophagus. The pharyngeal muscles cooperate with muscles of the oral cavity and esophagus to initiate the process of swallowing. The muscular contractions during swallowing force the food mass along the esophagus and into the stomach.

The Esophagus

The **esophagus** is a muscular tube that begins at the pharynx and ends at the stomach (see Figure 16-1●). The upper third of the esophagus contains skeletal muscle, the lower third contains smooth muscle, and a mixture of each makes up the middle third. It is about 25 cm (1 ft) long with a diameter of about 2 cm (0.75 in.). The esophagus lies posterior to the trachea in the neck, passes through the mediastinum in the thoracic cavity, and enters the peritoneal cavity through an opening in the diaphragm, the *esophageal hiatus* (hī-Ā-tus; a gap or opening), before emptying into the stomach.

The esophagus is lined with a stratified squamous epithelium that resists abrasion, hot or cold temperatures, and chemical attack. The secretions of mucous glands lubricate this surface and prevent materials from sticking to the sides of the esophagus during swallowing.

SWALLOWING

Swallowing, or **deglutition**, involves both voluntary actions and involuntary reflexes that transport food from the pharynx to the stomach (Figure 16-7●). Before swallowing can occur, the food must have the proper texture and consistency. Once the material has been shredded or torn by the teeth, moistened with salivary secretions, and approved by the taste receptors, the tongue begins compacting the debris into a small mass, or **bolus**.

Swallowing occurs in three phases. In the **oral phase**, swallowing begins with the compression of the bolus against the hard palate. The tongue then retracts, forcing the bolus into the pharynx and helping to elevate the soft palate, thus preventing the bolus from entering the nasopharynx (Figure 16-7a, b●). The oral phase is the only phase of swallowing that can be consciously controlled.

The **pharyngeal phase** begins when the bolus comes in contact with sensory receptors around the pharynx and the posterior pharyngeal wall and initiates the involuntary *swallowing reflex* (Figure 16-7c, d●) The larynx elevates, and the epiglottis folds to direct the bolus past the closed glottis. In less than a second, the contraction of pharyngeal muscles forces the bolus through the entrance to the esophagus, which is guarded by the *upper esophageal sphincter*.

The **esophageal phase** begins as the bolus enters the esophagus. During this phase, the bolus is pushed toward the stomach by a peristaltic contraction. The approach of the bolus triggers the opening of the *lower esophageal sphincter*, and the bolus enters the stomach (Figure 16-7e–h●).

For a typical bolus, the entire trip from the oral cavity to the esophagus takes about 9 seconds to complete. Fluids may make the journey in a few seconds, arriving ahead of the

ORAL PHASE

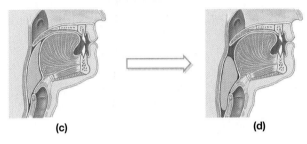

(a) Hard palate, Tongue, Esophagus / Soft palate, Bolus **(b)**

PHARYNGEAL PHASE

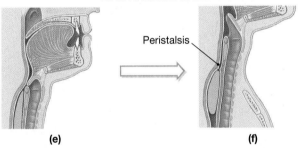

(c) / Epiglottis, Trachea **(d)**

ESOPHAGEAL PHASE

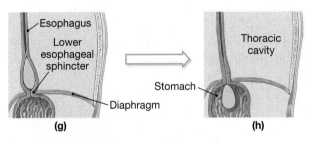

Peristalsis

(e) / **(f)**

Esophagus, Lower esophageal sphincter, Diaphragm / Thoracic cavity, Stomach

(g) / **(h)**

● *Figure 16-7* **The Swallowing Process**
This sequence, based on a series of X-rays, shows the stages of swallowing and the movement of materials from the mouth to the stomach.

peristaltic contractions; a relatively dry or bulky bolus travels much more slowly, and repeated peristaltic waves may be required to drive it into the stomach. A completely dry bolus cannot be swallowed at all, for friction with the walls of the esophagus will make peristalsis ineffective.

✚ ESOPHAGITIS AND DIAPHRAGMATIC (HIATAL) HERNIAS

A weakened or permanently relaxed lower esphageal sphincter can cause inflammation of the esophagus, or **esophagitis** (ē-sof-a-JĪ-tis), as powerful gastric acids enter the lower esoph-

agus. The esophageal epithelium has few defenses from acid and enzyme attack, and inflammation, epithelial erosion, and intense discomfort are the result. Occasional incidents of reflux, or backflow, from the stomach are responsible for the symptoms of "heartburn." This relatively common problem supports a multimillion-dollar industry devoted to producing and promoting antacids.

The esophagus and major blood vessels pass from the thoracic cavity to the abdominopelvic cavity through an opening in the diaphragm called the esophageal hiatus. In a **diaphragmatic hernia**, or **hiatal** (hī-Ā-tal) **hernia**, abdominal organs slide into the thoracic cavity through the *esophageal hiatus*, the opening used by the esophagus. The severity of the condition depends on the location and size of the herniated organ or organs. Hiatal hernias are actually very common, and most go unnoticed. When clinical problems develop, they usually occur because the intruding abdominal organs are exerting pressure on structures or organs in the thoracic cavity.

CONCEPT CHECK QUESTIONS
Answers on page 519

❶ What is the importance of the mesenteries?

❷ Which would be more efficient in propelling intestinal contents from one place to another—peristalsis or segmentation?

❸ What effect would a drug that blocks parasympathetic stimulation of the digestive tract have on peristalsis?

❹ What is occurring when the soft palate and larynx elevate and the glottis closes?

The Stomach

The **stomach**, located within the left upper quadrant of the abdominopelvic cavity, receives the food from the esophagus. The stomach has four primary functions: (1) the temporary storage of ingested food, (2) the mechanical breakdown of resistant materials, (3) breaking chemical bonds in food materials through the action of acids and enzymes, and (4) the production of *intrinsic factor*, a compound necessary for the absorption of vitamin B_{12}. Ingested materials mix with secretions of the glands of the stomach. This mixing produces a viscous, highly acidic, soupy mixture of partially digested food called **chyme** (kīm).

The stomach is a muscular organ with the shape of an expanded J. Figure 16-8a● shows its four regions. The esophagus connects to the smallest part of the stomach, the **cardia** (KAR-dē-a). The bulge of the stomach superior to the cardia is the **fundus** (FUN-dus) of the stomach, and the large area between the fundus and the curve of the J is the **body**. The curve of the J, the **pylorus** (pī-LŌR-us; *pyle*, gate + *ouros*, guard), connects the stomach with the small intestine.

16 THE DIGESTIVE SYSTEM

An Overview of the Digestive Tract • The Oral Cavity • The Pharynx • The Esophagus • **The Stomach** • The Small Intestine • The Pancreas • The Liver

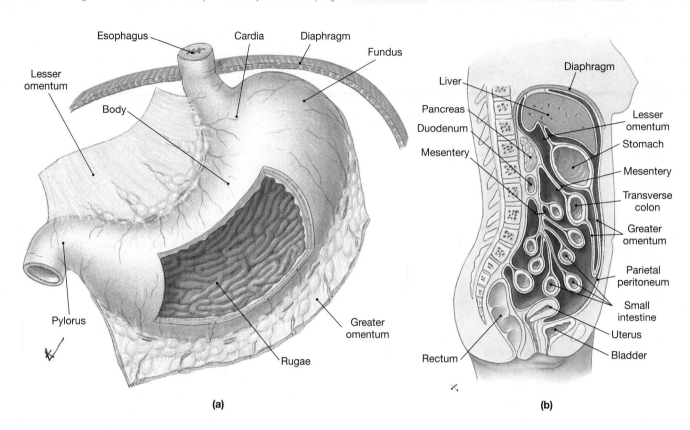

(a)

(b)

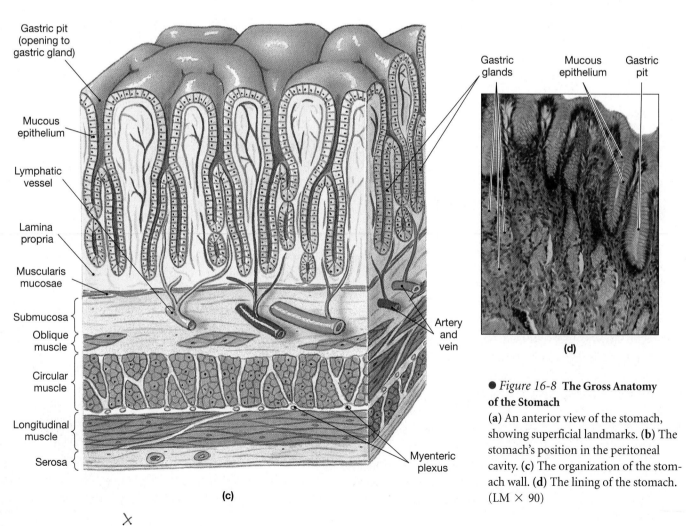

(c)

(d)

● *Figure 16-8* **The Gross Anatomy of the Stomach**
(**a**) An anterior view of the stomach, showing superficial landmarks. (**b**) The stomach's position in the peritoneal cavity. (**c**) The organization of the stomach wall. (**d**) The lining of the stomach. (LM × 90)

A muscular **pyloric sphincter** regulates the flow of chyme between the stomach and small intestine.

The stomach's volume increases when you eat, then decreases as chyme enters the small intestine. When empty, the stomach resembles a muscular tube with a narrow and constricted lumen. When full, it can expand to contain 1–1.5 liters. This degree of expansion is possible because the stomach wall contains thick layers of smooth muscle, and the mucosa of the relaxed stomach contains a number of prominent ridges and folds, called **rugae** (ROO-gē; wrinkles). As the stomach expands, the smooth muscle stretches and the rugae gradually disappear.

Unlike the two-layered muscularis externa of other portions of the digestive tract, that of the stomach contains a longitudinal layer, a circular layer, and an inner oblique layer. This extra layer adds strength and assists in the mixing and churning activities essential to forming chyme.

The visceral peritoneum covering the outer surface of the stomach is continuous with a pair of mesenteries. The **greater omentum** (ō-MEN-tum; *omentum*, a fatty skin) extends below the greater curvature and forms an enormous pouch that hangs over and protects the abdominal viscera (Figure 16-8b●). The much smaller **lesser omentum** extends from the lesser curvature to the liver.

THE GASTRIC WALL

The stomach is lined by a *mucous epithelium* (an epithelium dominated by mucous cells). The alkaline mucus that it secretes covers and protects the stomach lining from acids, enzymes, and abrasive materials. Shallow depressions, called **gastric pits**, open onto the gastric surface (Figure 16-8c●). The mucous cells at the base, or neck, of each gastric pit actively divide and replace superficial cells of the mucous epithelium as well as produce mucus. Each gastric pit communicates with **gastric glands** that extend deep into the underlying lamina propria (Figure 16-8d●). These glands contain two types of secretory cells: *parietal cells* and *chief cells*. Together these cells secrete about 1500 ml of **gastric juice** each day. Gastric glands within the lower stomach, the *pylorus*, also contain endocrine cells that help regulate gastric activity.

Parietal Cells

Parietal cells secrete intrinsic factor and hydrochloric acid (HCl). **Intrinsic factor** facilitates the absorption of vitamin B$_{12}$ across the intestinal lining. Hydrochloric acid lowers the pH of the gastric juice, keeping the stomach contents at a pH of 1.5–2.0. The acidity of the gastric juice kills microorganisms, breaks down cell walls and connective tissues in food, and activates the enzymatic secretions of the chief cells.

Chief Cells

Chief cells secrete **pepsinogen** (pep-SIN-ō-jen), an inactive form of the enzyme **pepsin**. The hydrochloric acid released by the parietal cells converts pepsinogen to pepsin. Pepsin is a proteolytic, or protein-splitting, enzyme. The stomachs of newborn infants (but not of adults) produce *rennin* and *gastric lipase*, enzymes important for the digestion of milk. Rennin coagulates milk proteins, and gastric lipase initiates the digestion of milk fats.

✚ GASTRITIS AND PEPTIC ULCERS

An inflammation of the gastric mucosa is called **gastritis** (gas-TRĪ-tis). This condition can develop after a person has swallowed drugs, including alcohol and aspirin. Gastritis can also appear after severe emotional or physical stress, bacterial infection of the gastric wall, or the ingestion of strongly acid or alkaline chemicals.

A **peptic ulcer** develops when the digestive acids and enzymes manage to erode their way through the defenses of the stomach lining or proximal portions of the small intestine. The specific locations are indicated by the terms **gastric ulcer** (stomach) or **duodenal ulcer** (duodenum of the small intestine). Peptic ulcers result from the excessive production of acid or the inadequate production of the alkaline mucus that defends the epithelium against that acid. Since the late 1970s, drugs such as *cimetidine* (*Tagamet*) have been used to inhibit acid production by parietal cells. Infections involving the bacterium *Helicobacter pylori* are responsible for at least 80 percent of peptic ulcers. Hence treatment for gastric ulcers today commonly involves the administration of antibiotic drugs.

THE REGULATION OF GASTRIC ACTIVITY

The production of acid and enzymes by the stomach mucosa can be controlled by the central nervous system as well as by local hormonal mechanisms. Three stages can be identified, although considerable overlap exists between them (Figure 16-9●):

1. *Cephalic phase.* The sight, smell, taste, or thought of food initiates the **cephalic phase** of gastric secretion. This stage, which is directed by the CNS, prepares your stomach to receive food. Under the control of the vagus nerve, parasympathetic fibers innervate mucous cells, parietal cells, chief cells, and endocrine cells of the stomach. In response to stimulation, the production of gastric juice accelerates, reaching rates of around 500 ml per hour. This phase usually lasts for a relatively brief period before the gastric phase commences.

2. *Gastric phase.* The **gastric phase** begins with the arrival of food in the stomach. The stimulation of stretch receptors in the stomach wall and of chemoreceptors in the

16 THE DIGESTIVE SYSTEM

An Overview of the Digestive Tract • The Oral Cavity • The Pharynx • The Esophagus • The Stomach • **The Small Intestine** • The Pancreas • The Liver

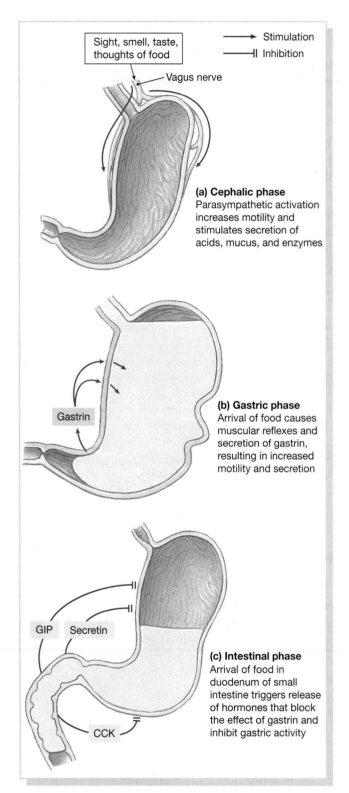

● *Figure 16-9* **The Phases of Gastric Secretion**
(a) The cephalic phase begins when you see, smell, taste, or think of food. (b) The gastric phase begins with the arrival of food in the stomach. (c) The intestinal phase begins when chyme starts to enter the small intestine.

mucosa triggers the release of the hormone **gastrin** into the circulatory system. Proteins, alcohol in small doses, and caffeine are potent stimulators of gastric secretion because they excite the mucosal chemoreceptors. Both parietal and chief cells respond to the presence of gastrin by accelerating their secretory activities. The effect on the parietal cells is the most pronounced, and the pH of the gastric contents drops sharply. This phase may continue for several hours while the ingested materials are processed by the acids and enzymes.

During this period, stomach contractions begin to swirl and churn the gastric contents, mixing the ingested materials with the gastric secretions to form chyme. As digestion proceeds, the contractions begin sweeping down the length of the stomach, and each time the pylorus contracts, a small quantity of chyme squirts through the pyloric sphincter.

3. *Intestinal phase.* The **intestinal phase** begins when chyme starts to enter the small intestine. This phase controls the rate of gastric emptying and ensures that the secretory, digestive, and absorptive functions of the small intestine can proceed efficiently. Most of the regulatory controls are inhibitory, providing a brake for gastric activities. Both endocrine and neural mechanisms are involved. Intestinal hormones, such as *secretin, cholecystokinin (CCK),* and *gastric inhibitory peptide (GIP),* are released when chyme enters the small intestine. These hormones reduce gastric activity and give the small intestine time to deal with the arriving acids.

Inhibitory reflexes that depress gastric activity are stimulated when the proximal portion of the small intestine becomes too full, too acidic, unduly irritated by the chyme, or filled with partially digested proteins, carbohydrates, or fats. For example, as the small intestine distends, inhibitory feedback slows the contractions in the stomach walls, inhibits the parasympathetic nervous system, and stimulates sympathetic innervation. The combination significantly reduces gastric activity.

 STOMACH CANCER
Stomach, or *gastric,* **cancer** is one of the most common lethal cancers, responsible for roughly 14,000 deaths in the United States each year. The incidence is higher in countries such as Japan or Korea, where the typical diet includes large quantities of pickled foods. Because the symptoms can resemble those of gastric ulcers, the condition may not be reported in its early stages. Diagnosis usually involves X-rays of the

stomach at various degrees of distension. The gastric mucosa can also be visually inspected using a flexible instrument called a *gastroscope*. Attachments permit the collection of tissue samples for histological analysis.

The treatment of stomach cancer involves the surgical removal of part or all of the stomach. People can survive even a total *gastrectomy* (gas-TREK-to-mē), because the only absolutely vital function of the stomach is the secretion of intrinsic factor. Protein breakdown can still be performed by the small intestine, although at reduced efficiency, and the loss of such functions as food storage and acid production is not life-threatening.

Digestion in the Stomach

The stomach performs preliminary digestion of proteins by pepsin and, for a variable period, permits the digestion of carbohydrates by salivary amylase. This enzyme remains active until the pH throughout the material in the stomach falls below 4.5, usually within 1–2 hours after a meal.

As the stomach contents become more fluid and the pH approaches 2.0, pepsin activity increases and protein disassembly begins. Protein digestion is not completed in the stomach, but there is usually enough time for pepsin to break down complex proteins into smaller peptide and polypeptide chains before the chyme enters the small intestine.

Although digestion begins in the stomach, little if any nutrient absorption occurs there because (1) the epithelial cells are covered by a blanket of alkaline mucus and are not directly exposed to the chyme, (2) the epithelial cells lack the specialized transport mechanisms found in cells lining the small intestine, (3) the gastric lining is impermeable to water, and (4) digestion has not proceeded to completion by the time chyme leaves the stomach. At this stage, most carbohydrates, lipids, and proteins are only partially broken down.

The Small Intestine

The **small intestine** plays a key role in the digestion and absorption of nutrients. Ninety percent of nutrient absorption occurs in the small intestine, and most of the rest occurs in the large intestine. The small intestine is about 6 meters (20 ft) long and has a diameter ranging from 4 cm at the stomach to about 2.5 cm at the junction with the large intestine. It has three subdivisions: the *duodenum*, the *jejunum*, and the *ileum* (Figure 16-10●):

1. The **duodenum** (doo-AH-de-num or doo-ō-DĒ-num) is the 25 cm (1 ft) closest to the stomach. This portion receives

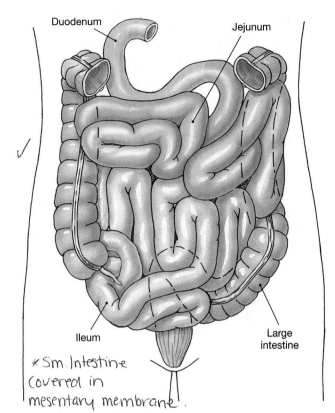

● *Figure 16-10* **The Location and Parts of the Small Intestine** Different colors indicate the three regions of the small intestine: the duodenum, jejunum, and ileum.

chyme from the stomach and exocrine secretions from the pancreas and liver.

2. A rather abrupt bend marks the boundary between the duodenum and the **jejunum** (je-JOO-num). The jejunum, which is supported by a sheet of mesentery, is about 2.5 meters (8 ft) long. The bulk of chemical digestion and nutrient absorption occurs in the jejunum. One rather drastic approach to weight control involves the surgical removal of a significant portion of the jejunum.

3. The jejunum leads to the third segment, the **ileum** (IL-ē-um). It is also the longest, averaging 3.5 m (12 ft) in length. The ileum ends at the *ileocecal valve*, a sphincter that controls the flow of materials from the ileum into the *cecum* of the large intestine.

The small intestine fits in the relatively small peritoneal cavity because it is well packed, and the position of each of the segments is stabilized by mesenteries attached to the dorsal body wall (Figure 16-8b●, p. 496).

16 THE DIGESTIVE SYSTEM

An Overview of the Digestive Tract • The Oral Cavity • The Pharynx • The Esophagus • The Stomach • **The Small Intestine** • The Pancreas • The Liver

THE INTESTINAL WALL

The intestinal lining bears a series of transverse folds called **plicae**, or *plicae circulares* (PLĪ-sē sir-kū-lar-ēs) (Figure 16-11a●). The lining of the intestine is composed of a series of fingerlike projections, the **villi** (Figure 16-11b●). These villi are covered by a simple columnar epithelium carpeted with microvilli. If the small intestine were a simple tube with smooth walls, it would have a total absorptive area of around 3300 cm^2, (3.6 ft^2). Instead, the epithelium contains folds, each fold supports a forest of villi, and each villus is covered by epithelial cells blanketed in microvilli. This arrangement increases the total area for absorption to approximately 2 million cm^2, or more than 2200 ft^2.

Each villus contains a network of capillaries (Figure 16-11c●) that transports respiratory gases and carries absorbed nutrients to the hepatic portal circulation for delivery to the liver.

p. 416 In addition to capillaries and nerve endings, each villus contains a terminal lymphatic called a **lacteal** (LAK-tē-al; *lacteus*, milky). This name refers to the pale, cloudy appearance of lymph that contains large quantities of lipids. Lacteals transport materials that are unable to cross the walls of local capillaries. For example, absorbed fatty acids are assembled into protein-lipid packages that are too large to diffuse into the bloodstream. These packets, called *chylomicrons* (*chylos*, juice), reach the circulation by passage through the lymphatic system.

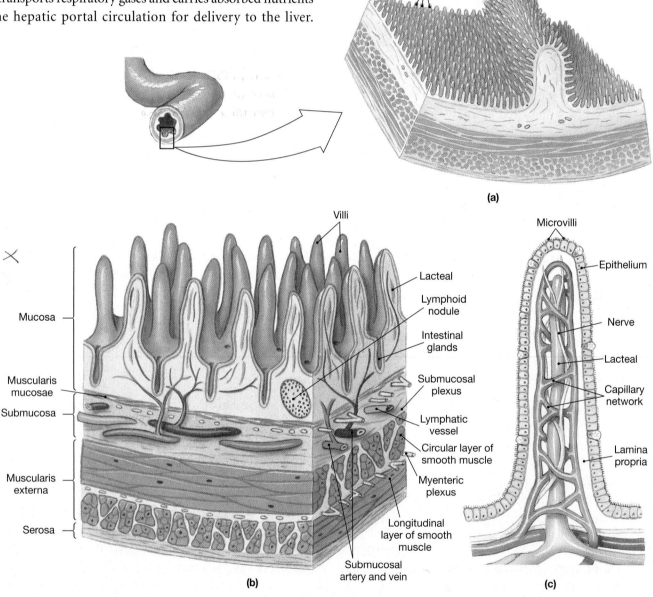

● *Figure 16-11* **The Intestinal Wall**
(a) One plica and multiple villi. (b) The histological structure of the intestinal wall. (c) The internal structure of a single villus.

At the bases of the villi are entrances to intestinal glands that secrete a watery *intestinal juice*. In the duodenum, large intestinal glands secrete an alkaline mucus that helps buffer the acids in chyme. Intestinal glands also contain endocrine cells responsible for the production of intestinal hormones considered in a later section.

Intestinal Movements

Once the chyme is within the small intestine, segmentation contractions mix it with mucous secretions and enzymes before absorption can occur. As absorption occurs, weak peristaltic contractions slowly move the remaining materials along the length of the small intestine. These contractions are local reflexes not under CNS control, and the effects are limited to within a few centimeters of the site of the original stimulus.

More elaborate reflexes, such as the gastroenteric reflex and the gastroileal reflex, coordinate activities along the entire length of the small intestine. Distension of the stomach initiates the *gastroenteric* (gas-trō-en-TER-ik) *reflex*, which immediately accelerates glandular secretion and peristaltic activity in all segments. The increased peristalsis distributes materials along the length of the small intestine and empties the duodenum. The *gastroileal* (gas-trō-IL-ē-al) *reflex* is a response to circulating levels of the hormone gastrin. The entry of food into the stomach triggers the release of gastrin, which relaxes the ileocecal valve at the entrance to the large intestine. Because the valve is relaxed, the increased peristalsis pushes materials from the ileum into the large intestine. On average, it takes about 5 hours for ingested food to pass from the duodenum to the end of the ileum, so the first of the materials to enter the duodenum after breakfast may leave the small intestine at lunch.

INTESTINAL SECRETIONS

Roughly 1.8 liters of watery **intestinal juice** enters the intestinal lumen each day. Intestinal juice moistens the intestinal contents, helps buffer acids, and dissolves both the digestive enzymes provided by the pancreas and the products of digestion. Much of this fluid arrives through osmosis, as water flows out of the mucosa, and the rest is provided by intestinal glands stimulated by the activation of touch and stretch receptors in the intestinal walls.

Hormonal and CNS controls are important in regulating the secretions of the digestive tract and accessory organs. Because the duodenum is the initial region of the intestine to receive chyme, it is the focus of these regulatory mechanisms. Here, the acid content of the chyme must be neutralized and the appropriate enzymes added. The submucosal glands pro-

tect the duodenal epithelium from gastric acids and enzymes. They increase their secretions in response to local reflexes and also to parasympathetic stimulation carried by the vagus nerves. As a result of parasympathetic stimulation, the duodenal glands begin secreting long before chyme reaches the pyloric sphincter. Sympathetic stimulation inhibits their activation, leaving the duodenal lining relatively unprepared for the arrival of the acid chyme. This is probably why duodenal ulcers can be caused by chronic stress or by other factors that promote sympathetic activation.

 VOMITING

The responses of the digestive tract to chemical or mechanical irritation are rather predictable. Gastric secretion accelerates all along the digestive tract, and the intestinal contents are eliminated as quickly as possible. The *vomiting reflex* occurs in response to irritation of the soft palate, pharynx, esophagus, stomach, or proximal portions of the small intestine. These sensations are relayed to the vomiting center of the medulla oblongata, which coordinates the motor responses. In preparation, the pylorus relaxes, and the contents of the duodenum and proximal jejunum are discharged into the stomach by strong peristaltic waves that travel toward the stomach, rather than toward the ileum. Vomiting, or *emesis* (EM-e-sis), then occurs as the stomach regurgitates its contents through the esophagus and pharynx. As regurgitation occurs, the uvula and soft palate block the entrance to the nasopharynx. Increased salivary secretion assists in buffering the stomach acids, thereby preventing erosion of the teeth. In conditions marked by repeated vomiting, severe tooth damage can occur; this is one symptom of the eating disorder *bulimia*. Most of the force of vomiting comes from a powerful expiratory movement that elevates intra-abdominal pressures and presses the stomach against the tensed diaphragm.

Intestinal Hormones

Duodenal endocrine cells produce various peptide hormones that coordinate the secretory activities of the stomach, duodenum, pancreas, and liver. These hormones were introduced in the discussion of gastric activity (see p. 497).

Secretin (sē-KRĒ-tin) is released when the pH falls in the duodenum. This occurs when acid chyme arrives from the stomach. The primary effect of secretin is to increase the secretion of bile and buffers by the liver and pancreas.

Cholecystokinin (kō-lē-sis-tō-KĪ-nin), or **CCK**, is secreted when chyme arrives in the duodenum, especially when it contains lipids and partially digested proteins. This hormone also targets the pancreas and gallbladder. In the pancreas, CCK accelerates the production and secretion of all types of digestive enzymes. At the gallbladder, it causes the ejection of *bile* into the duodenum. The presence of either secretin or

16 THE DIGESTIVE SYSTEM

An Overview of the Digestive Tract • The Oral Cavity • The Pharynx • The Esophagus • The Stomach • The Small Intestine • **The Pancreas** • The Liver

TABLE 16-1 *Important Gastrointestinal Hormones and Their Primary Effects*

HORMONE	STIMULUS	ORIGIN	TARGET	EFFECTS
Gastrin	Vagus nerve stimulation or arrival of food in the stomach	Stomach	Stomach	Stimulates production of acids and enzymes, increases motility
	Arrival of chyme containing large quantities of undigested proteins	Duodenum	Stomach	Stimulates gastric secretion and motion
Secretin	Arrival of chyme in the duodenum	Duodenum	Pancreas	Stimulates production of alkaline buffers
			Stomach	Inhibits gastric secretion and motility
			Liver	Increases rate of bile secretion
Cholecystokinin (CCK)	Arrival of chyme containing lipids and partially digested proteins	Duodenum	Pancreas	Stimulates production of pancreatic enzymes
			Gallbladder	Stimulates contraction of gallbladder
			Duodenum	Causes relaxation of sphincter at base of bile duct
			Stomach	Inhibits gastric secretion and motion
			CNS	Reduces sensation of hunger
Gastric inhibitory peptide (GIP)	Arrival of chyme containing large quantities of fats and glucose	Duodenum	Pancreas	Stimulates release of insulin by pancreatic islets
		Stomach		Inhibits gastric secretion and motility

CCK in high concentrations also reduces gastric motility and secretory rates.

Gastric inhibitory peptide, or **GIP**, is released when fats and carbohydrates, especially glucose, enter the small intestine. GIP inhibits gastric activity and causes the release of insulin from the pancreatic islets.

Functions of the major gastrointestinal hormones are summarized in Table 16-1, and their interactions are diagrammed in Figure 16-12●.

DIGESTION IN THE SMALL INTESTINE

In the stomach, food becomes saturated with gastric juices and exposed to the digestive effects of a strong acid and a proteolytic enzyme, pepsin. Most of the important digestive processes are completed in the small intestine, where the final products of digestion— simple sugars, fatty acids, and amino acids—are absorbed, along with most of the water content. However, the small intestine produces only a few of the enzymes needed to break down the complex materials found in the diet. Most of the enzymes and buffers are contributed by the liver and pancreas, which are discussed in the next section.

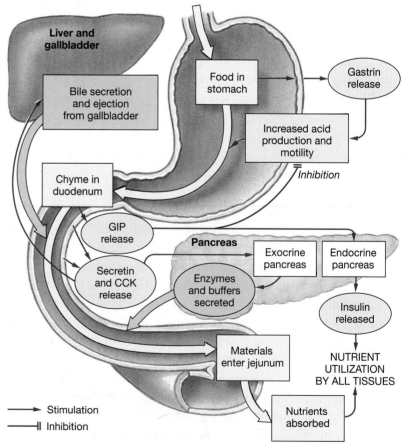

● *Figure 16-12* **The Activities of Major Digestive Tract Hormones**
The primary actions of gastrin, GIP, secretin, and CCK.

CONCEPT CHECK QUESTIONS
Answers on page 519

❶ Which muscle regulates the flow of chyme from the stomach to the small intestine?

❷ When a person suffers from chronic ulcers in the stomach, treatment sometimes involves cutting the branches of the vagus nerve that serve the stomach. Why?

❸ How is the small intestine adapted for the absorption of nutrients?

❹ How would a meal that is high in fat affect the level of cholecystokinin in the blood?

The Pancreas

The **pancreas**, shown in Figure 16-13a●, lies behind your stomach, extending laterally from the duodenum toward the spleen. It is an elongate, pinkish-gray organ with a length of approximately 15 cm (6 in.) and a weight of around 80 g (3 oz). The surface of the pancreas has a lumpy texture, and its tissue is soft and easily torn. Unlike the stomach, the pancreas lies outside the peritoneal cavity with only its anterior surface covered by peritoneum. ∞ p. 19 Organs that lie posterior to, rather than within, the peritoneal cavity are called *retroperitoneal* (*retro*, behind).

HISTOLOGICAL ORGANIZATION

The pancreas is primarily an exocrine organ, producing **pancreatic juice**, a mixture of digestive enzymes and buffers. **Pancreatic islets**, which secrete the hormones insulin and glucagon, account for only around 1 percent of the cellular population of the pancreas. ∞ p. 329 Exocrine cells and their associated ducts account for the rest. The numerous ducts that branch throughout the pancreas begin at saclike pouches called the **pancreatic acini** (AS-i-nī; singular *acinus*, grape) (Figure 16-13b●). Enzymes and buffers are secreted by the *acinar cells* of these pouches and by the cells that line the ducts. The smaller ducts converge to form larger ducts; these ultimately fuse to form the **pancreatic duct**, which carries these secretions to the duodenum. The pancreatic duct penetrates the duodenal wall with the *common bile duct* from the liver and gallbladder.

The pancreatic enzymes do most of the digestive work in the small intestine. Pancreatic enzymes are broadly classified according to their intended targets. **Lipases** (LĪ-pā-sez) attack lipids, **carbohydrases** (kar-bō-HĪ-drā-sez) digest sugars and starches, and **proteases** (prō-tē-ā-sez) (proteolytic enzymes) break proteins apart.

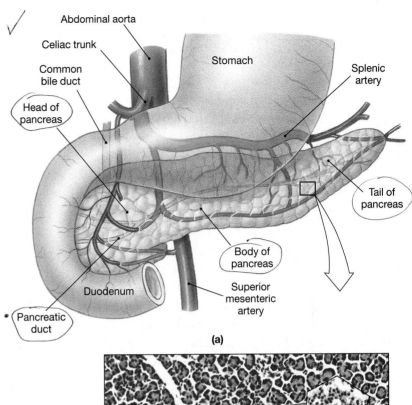

Abdominal aorta
Celiac trunk
Common bile duct
Head of pancreas
Stomach
Splenic artery
Tail of pancreas
Body of pancreas
Superior mesenteric artery
Duodenum
Pancreatic duct

(a)

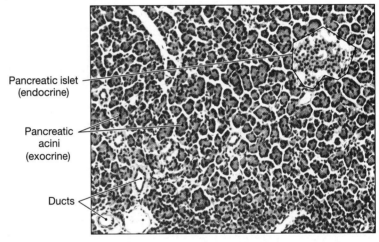

Pancreatic islet (endocrine)

Pancreatic acini (exocrine)

Ducts

(b)

● *Figure 16-13* **The Pancreas**
(a) Gross anatomy. The head of the pancreas is tucked into a curve of the duodenum that begins at the pylorus of the stomach. (b) Cellular organization of the pancreas. (LM × 168)

 PANCREATITIS

Pancreatitis (pan-krē-a-TĪ-tis) is an inflammation of the pancreas. Blockage of the excretory ducts, bacterial or viral infections, circulatory blockage, and drug reactions, especially those involving alcohol, are among the factors that may produce this condition. These stimuli provoke a crisis by injuring exocrine cells in at least a portion of the organ. Lysosomes within the damaged cells then activate the pancreatic enzymes, and autodigestion begins. The proteolytic enzymes digest the surrounding, undamaged cells, activating their enzymes and starting a chain reaction. In most cases, only a portion of the pancreas will be affected, and the condition subsides in a few days. In 10–15 percent of pancreatitis cases,

16 THE DIGESTIVE SYSTEM

An Overview of the Digestive Tract • The Oral Cavity • The Pharynx • The Esophagus • The Stomach • The Small Intestine • The Pancreas • **The Liver**

the process does not subside, and the enzymes may ultimately destroy the pancreas. Loss of the entire pancreas results in *diabetes mellitus* (which requires the administration of insulin) and *nutrient malabsorption* (which requires oral administration of pancreatic enzymes).

THE CONTROL OF PANCREATIC SECRETION

Each day your pancreas secretes about 1000 ml (1 qt) of pancreatic juice. The secretions by the pancreatic exocrine cells are primarily controlled by hormones from the duodenum. When acid chyme arrives in the duodenum, secretin is released, triggering the pancreatic production of an alkaline fluid with a pH of 7.5 to 8.8. Among its other components, this secretion contains buffers, primarily *sodium bicarbonate*, that help increase the pH of the chyme. A different intestinal hormone, CCK, controls the production and secretion of pancreatic enzymes. The specific enzymes involved are **pancreatic amylase**, similar to salivary amylase, **pancreatic lipase**, **nucleases** that break down nucleic acids, and several **proteases** (proteolytic enzymes).

Proteases account for around 70 percent of the total pancreatic enzyme production. The most abundant are **trypsin** (TRIP-sin), **chymotrypsin** (kī-mō-TRIP-sin), and **carboxypeptidase** (kar-bok-sē-PEP-ti-dās). Together they shatter complex proteins into a mixture of short peptide chains and amino acids. The enzymes are quite powerful, and the pancreatic cells protect themselves by secreting them as inactive *proenzymes*, which are activated by other enzymes within the intestinal tract.

The Liver

The **liver** is the largest visceral organ, weighing about 1.5 kg (3.3 lb) and accounting for roughly 2.5 percent of the total body weight. Most of the liver lies in the right hypochondriac and epigastric abdominopelvic regions. ∞ p. 16 This large, firm, reddish-brown organ provides essential metabolic and synthetic functions.

ANATOMY OF THE LIVER

The liver is wrapped in a tough fibrous capsule. Although its overall shape conforms to its surroundings, the liver is divided into four lobes: The large **left** and **right lobes**, and the smaller **caudate** and **quadrate lobes** (Figure 16-14●). A tough connective tissue fold, the *falciform ligament*, marks the division between the left and right lobes. The thickened posterior margin of the falciform ligament is the *round ligament*, a fibrous remnant of the fetal umbilical vein.

Lodged within a recess under the right lobe of the liver is the *gallbladder*, a muscular sac that stores and concentrates bile before it is excreted into the small intestine. The gallbladder and associated structures will be described in a later section.

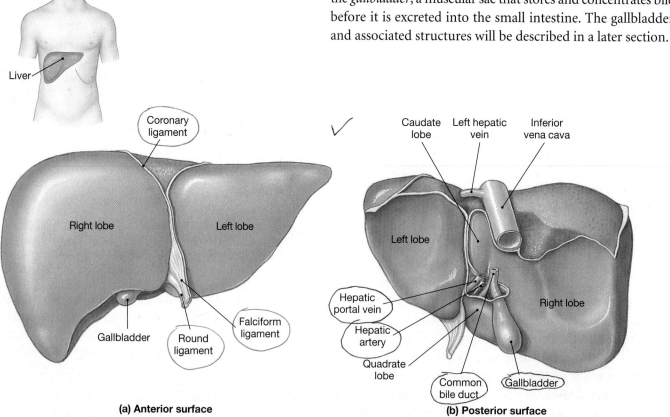

(a) Anterior surface

(b) Posterior surface

● *Figure 16-14* **The Anatomy of the Liver**

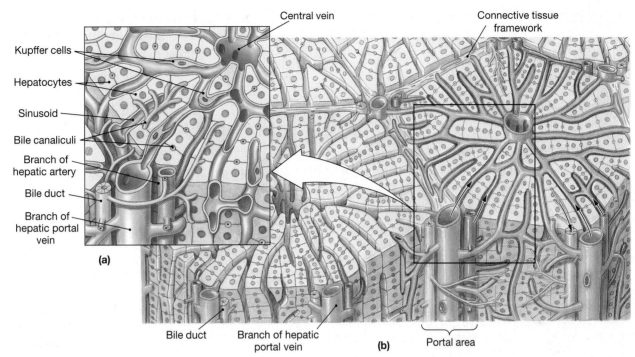

Central vein

Connective tissue framework

Kupffer cells

Hepatocytes

Sinusoid

Bile canaliculi

Branch of hepatic artery

Bile duct

Branch of hepatic portal vein

(a)

Bile duct

Branch of hepatic portal vein

(b)

Portal area

Histological Organization

The lobes of the liver are divided by connective tissue into about 100,000 **liver lobules**, the basic functional unit of the liver. The histological organization and structure of a typical liver lobule is shown in Figure 16-15●.

Liver cells, called *hepatocytes* (he-PAT-ō-sīts), within a lobule are arranged into a series of irregular plates like the spokes of a wheel. The plates are only one cell thick and, where they are exposed, covered with microvilli. *Sinusoids*, specialized and highly permeable capillaries, form passageways between the adjacent plates that empty into the *central vein*. The sinusoidal lining includes a large number of phagocytic *Kupffer* (KOOP-fer) *cells*. These cells, part of the monocyte-macrophage system, engulf pathogens, cell debris, and damaged blood cells.

The circulation of blood to the liver was discussed in Chapter 13 (p. 416) and illustrated in Figure 13-23● (p. 417). Blood enters the sinusoids from branches of the hepatic portal vein and hepatic artery. These two branches and a small branch of the bile duct form a *portal area*, or *hepatic triad*, at each of the six corners of a lobule (Figure 16-15a●). As blood flows through the sinusoids, the liver cells (hepatocytes) adjust circulating levels of nutrients and other solutes by selective absorption (such as glucose) and secretion (such as plasma proteins). Blood then leaves the sinusoids and enters the **central vein** of the lobule. The central veins of all of the lobules ultimately merge to form the *hepatic veins* that empty into the inferior vena cava. Liver diseases, such as the various forms of *hepatitis*, and conditions such as alcoholism can lead to degenerative changes in the liver tissue and constriction of the circulatory supply.

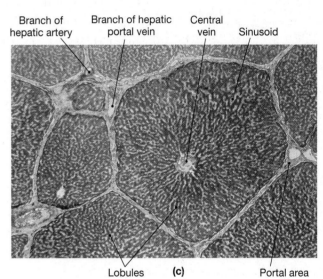

Branch of hepatic artery

Branch of hepatic portal vein

Central vein

Sinusoid

Lobules **(c)** Portal area

● *Figure 16-15* **Liver Histology**
(a) A portion of a single liver lobule. (b) A diagrammatic view of liver structure, showing relationships among lobules. (c) A section through liver lobules from a pig liver. (LM × 38) (The boundaries of human liver lobules are very difficult to see at comparable magnification.)

The hepatocytes also secrete a fluid called **bile**. Bile is released into a network of narrow channels called **bile canaliculi** between adjacent liver cells. These canaliculi extend outward from the central vein, and carry bile toward a network of ever-larger bile ducts within the liver until it eventually leaves the liver through the **common hepatic duct**. The bile in the common hepatic duct may either flow into the **common bile duct**, which empties into the duodenum, or enter the **cystic duct**, which leads to the gallbladder.

16 THE DIGESTIVE SYSTEM

An Overview of the Digestive Tract • The Oral Cavity • The Pharynx • The Esophagus • The Stomach • The Small Intestine • The Pancreas • The Liver

TABLE 16-2 *Major Functions of the Liver*
Digestive and Metabolic Functions
Synthesis and secretion of bile
Storage of glycogen and lipid reserves
Maintenance of normal blood glucose, amino acid, and fatty acid concentrations
Synthesis and interconversion of nutrient types (e.g., transamination of amino acids or conversion of carbohydrates to lipids)
Synthesis and release of cholesterol bound to transport proteins
Inactivation of toxins
Storage of iron reserves
Storage of fat-soluble vitamins
Other Major Functions
Synthesis of plasma proteins
Synthesis of clotting factors
Synthesis of the inactive hormone angiotensinogen
Phagocytosis of damaged red blood cells (by Kupffer cells)
Blood storage (major contributor to venous reserve)
Absorption and breakdown of circulating hormones (insulin, epinephrine) and immunoglobulins
Absorption and inactivation of lipid-soluble drugs

LIVER FUNCTIONS

The liver is responsible for three general functional roles: (1) *metabolic regulation*, (2) *hematological regulation*, and (3) *bile production*. To date, more than 200 liver functions have been identified. (Table 16-2 contains a partial listing.) Only a general overview is provided here.

Metabolic Regulation

The liver is the primary organ involved in regulating the composition of the circulating blood. All blood leaving the absorptive areas of the digestive tract flows through the liver before reaching the general circulation. Thus liver cells can (1) extract absorbed nutrients or toxins from the blood before they reach the rest of the body and (2) monitor and adjust the circulating levels of organic nutrients. Excesses are removed and stored, and deficiencies are corrected by mobilizing stored reserves or synthesizing the necessary compounds. For example, when blood glucose levels rise, the liver removes glucose and synthesizes glycogen. When blood glucose levels fall, the liver breaks down glycogen and releases glucose into the circulation. Circulating toxins and metabolic wastes are also removed for later inactivation or excretion. Additionally, fat-soluble vitamins (A, D, K, and E) are absorbed and stored.

Hematological Regulation

The liver is the largest blood reservoir in the body. In addition to the blood arriving over the hepatic portal vein, the liver receives about 25 percent of the cardiac output. Phagocytic Kupffer cells in the liver remove aged or damaged red blood cells, debris, and pathogens from the circulation. Kupffer cells are antigen-presenting cells that can stimulate an immune response. ∞ p. 439 Equally important, liver cells synthesize the plasma proteins that determine the osmotic concentration of the blood, transport nutrients, and establish the clotting and complement systems. ∞ p. 346

The Production and Role of Bile

Bile is synthesized in the liver and excreted into the lumen of the duodenum. Bile consists mostly of water, ions, *bilirubin* (a pigment derived from hemoglobin), cholesterol, and an assortment of lipids collectively known as **bile salts**. The water and ions in the bile help dilute and buffer acids in chyme as it enters the small intestine. Bile salts are synthesized from cholesterol in the liver and are required for the normal digestion and absorption of fats.

Most dietary lipids are not water-soluble. Mechanical processing along the digestive tract creates large drops containing various lipids that are much too massive to be successfully attacked by digestive enzymes. This problem occurs because pancreatic lipase is not lipid-soluble and can only interact with lipids at the surface of the drop. Bile salts break the drops apart by a process called **emulsification** (ē-mul-si-fi-KĀ-shun), which creates tiny droplets with a superficial coating of bile salts. The formation of tiny droplets increases the surface area available for enzymatic attack. In addition, the layer of bile salts facilitates interaction between the lipids and lipid-digesting enzymes from the pancreas. (We will return to the mechanism of lipid digestion later.)

LIVER DISEASE

Any condition that severely damages the liver represents a serious threat to life. The liver has a limited ability to regenerate itself after injury, but liver function will not fully recover unless a normal circulatory pattern returns. Examples of important types of liver disease include *cirrhosis*, which is characterized by the replacement of lobules by fibrous tissue, and various forms of hepatitis as a result of viral infections. Liver transplants are in some cases used to combat liver disease, but the supply of suitable donor tissue is limited and the success rate is highest in young, otherwise healthy individuals. Clinical trials are now under way to test an artificial liver known as *ELAD* (*extracorporeal liver assist device*), which may prove suitable for the long-term support of persons with chronic liver disease.

THE GALLBLADDER

The **gallbladder** is a muscular organ shaped like a pear (Figure 16-16a●). It has two major functions: bile storage and bile

● *Figure 16-16* **The Gallbladder**
(a) The inferior surface of the liver, showing the position of the gallbladder and ducts that transport bile from the liver to the gallbladder and duodenum. (b) An interior view of the duodenum showing the opening of the duodenal papilla and the location of the hepatopancreatic sphincter.

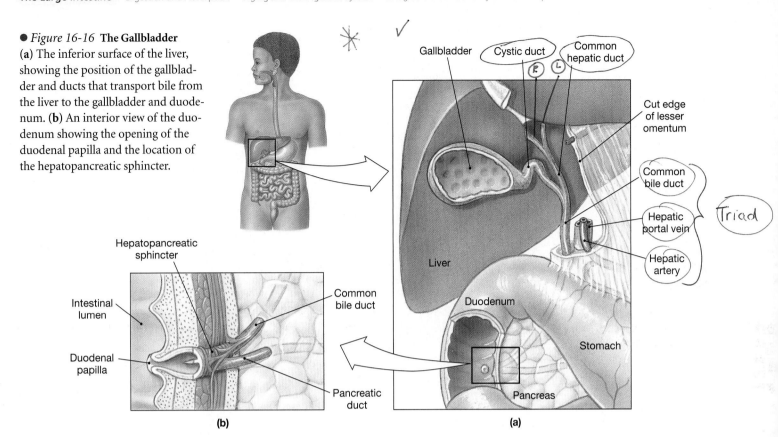

(b)   (a)

modification. Liver cells produce roughly 1 liter of bile each day, but a sphincter at the intestinal end of the common bile duct remains closed until chyme enters the duodenum. In the meantime, while bile cannot flow along the common bile duct, it enters the cystic duct for storage within the expandable gallbladder. When filled to capacity, the gallbladder contains 40–70 ml of bile. The composition of bile gradually changes as it remains in the gallbladder. Water is absorbed, and the bile salts and other components of bile become increasingly concentrated. If they become too concentrated, the bile salts may precipitate, forming *gallstones* that can cause a variety of clinical problems.

Bile secretion occurs continuously, but bile release into the duodenum occurs only under the stimulation of the intestinal hormone cholecystokinin, or CCK. Cholecystokinin relaxes the **hepatopancreatic sphincter** surrounding the shared passageway of the common bile duct and pancreatic duct (Figure 16-16b●). Contractions of the walls of the gallbladder then push bile into the duodenum at the *duodenal papilla*. CCK is released whenever chyme enters the intestine, but the amount secreted increases if the chyme contains large amounts of fat.

CONCEPT CHECK QUESTIONS
Answers on page 519

❶ A narrowing of the ileocecal valve would hamper movement of materials between what two organs?

❷ The digestion of which nutrient would be most impaired by damage to the exocrine pancreas?

❸ How would a decrease in the amount of bile salts in bile affect the digestion and absorption of fat?

The Large Intestine

The horseshoe-shaped **large intestine** begins at the end of the ileum and ends at the anus (Figure 16-17●). The large intestine lies below the stomach and liver and almost completely frames the small intestine. The main functions of the large intestine include (1) the reabsorption of water and compaction of feces, (2) the absorption of important vitamins liberated by bacterial action, and (3) the storing of fecal material prior to defecation.

The large intestine, also called the *large bowel*, has an average length of approximately 1.5 m (5 ft) and a width of 7.5 cm (3 in.). It can be divided into three parts: (1) the pouchlike *cecum*, the first portion; (2) the *colon*, the largest portion; and (3) the *rectum*, the last 15 cm (6 in.) of the large intestine and the end of the digestive tract.

THE CECUM
Material arriving from the ileum first enters an expanded chamber, the **cecum** (SĒ-kum), where compaction begins. A muscular sphincter, the **ileocecal** (il-ē-ō-SĒ-kal) **valve**, guards the

16 THE DIGESTIVE SYSTEM

An Overview of the Digestive Tract • The Oral Cavity • The Pharynx • The Esophagus • The Stomach • The Small Intestine • The Pancreas • The Liver

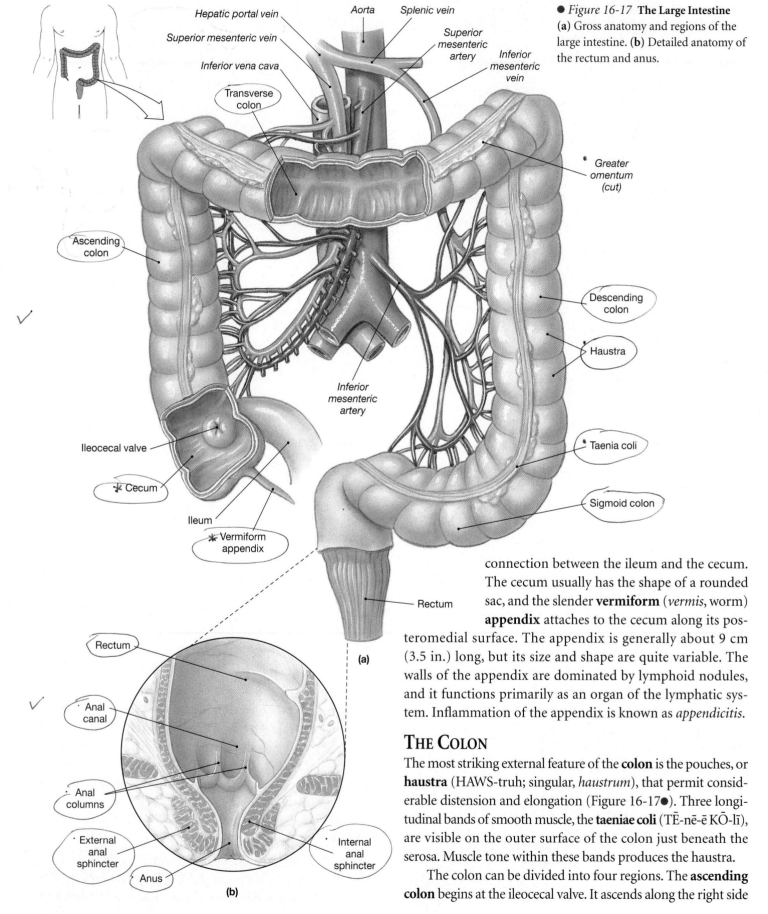

● *Figure 16-17* **The Large Intestine** (a) Gross anatomy and regions of the large intestine. (b) Detailed anatomy of the rectum and anus.

(a)

(b)

connection between the ileum and the cecum. The cecum usually has the shape of a rounded sac, and the slender **vermiform** (*vermis*, worm) **appendix** attaches to the cecum along its posteromedial surface. The appendix is generally about 9 cm (3.5 in.) long, but its size and shape are quite variable. The walls of the appendix are dominated by lymphoid nodules, and it functions primarily as an organ of the lymphatic system. Inflammation of the appendix is known as *appendicitis*.

THE COLON

The most striking external feature of the **colon** is the pouches, or **haustra** (HAWS-truh; singular, *haustrum*), that permit considerable distension and elongation (Figure 16-17●). Three longitudinal bands of smooth muscle, the **taeniae coli** (TĒ-nē-ē KŌ-lī), are visible on the outer surface of the colon just beneath the serosa. Muscle tone within these bands produces the haustra.

The colon can be divided into four regions. The **ascending colon** begins at the ileocecal valve. It ascends along the right side of the peritoneal cavity until it reaches the inferior margin of the liver. It then turns horizontally, becoming the **transverse colon**. The transverse colon continues toward the left side, passing below the stomach and following the curve of the body wall. Near the spleen, it turns inferiorly to form the **descending colon**. The descending colon continues along the left side until it curves and forms the S-shaped **sigmoid** (SIG-moyd; *sigmoides*, the Greek letter S) **colon**. The sigmoid colon empties into the rectum.

 ### COLON CANCER

Colon cancers are relatively common. Approximately 114,300 cases are diagnosed in the United States each year (in addition to 34,700 cases of rectal cancers). In 2002, an estimated 47,900 deaths will occur from colon cancers (and another 8700 deaths from rectal cancers). The mortality rate for these cancers remains high, and the best defense appears to be early detection and prompt treatment. The standard screening test involves checking the feces for blood. This is a simple procedure that can easily be performed on a stool (fecal) sample as part of a routine physical.

THE RECTUM

The **rectum** (REK-tum) forms the end of the digestive tract (see Figure 16-17b●). The last portion of the rectum, the **anal canal**, contains small longitudinal folds called *anal columns*. The distal margins of these columns are joined by transverse folds that mark the boundary between the columnar epithelium of the rectum and a stratified squamous epithelium similar to that found in the oral cavity. Very close to the **anus**, the opening of the anal canal, the epidermis becomes keratinized and identical to that on the skin surface.

The circular muscle layer of the muscularis externa in this region forms the **internal anal sphincter**. The **external anal sphincter** guards the exit of the anorectal canal. This sphincter, which consists of skeletal muscle fibers, is under voluntary control.

THE PHYSIOLOGY OF THE LARGE INTESTINE

The major functions of the large intestine are absorption and preparation of the fecal material for elimination.

Absorption in the Large Intestine

The reabsorption of water is an important function of the large intestine. Although roughly 1500 ml of watery material arrives in the colon each day, some 1300 ml of water is recovered from it and only about 200 ml of feces is ejected. The remarkable efficiency of digestion can best be appreciated by considering the average composition of feces: 75 percent water, 5 percent bacteria, and the rest a mixture of indigestible materials, small quantities of inorganic matter, and the remains of epithelial cells.

In addition to reabsorbing water, the large intestine absorbs a variety of other substances. Examples include useful compounds, such as bile salts and vitamins, organic waste products, such as bilirubin products derived from the breakdown of hemoglobin, and various toxins generated by bacterial action.

Bile Salts. Most of the bile salts remaining in the material reaching the cecum will be reabsorbed and transported to the liver for secretion into bile.

Vitamins. **Vitamins** are organic molecules that are essential to many metabolic reactions. The enzymes controlling many reactions require the binding of an additional ion or molecule, called *cofactors*, before substrates can also bind. *Coenzymes* are nonprotein molecules that function as cofactors, and many vitamins are essential coenzymes.

Bacteria residing within the colon generate three vitamins that supplement our dietary supply:

- Vitamin K, a fat-soluble vitamin needed by the liver to synthesize four clotting factors, including *prothrombin*.
- Biotin, a water-soluble vitamin important in glucose metabolism.
- Vitamin B_5 (*pantothenic acid*), a water-soluble vitamin required in the manufacture of steroid hormones and some neurotransmitters.

Vitamin K deficiencies lead to impaired blood clotting. Intestinal bacteria produce roughly half of our daily vitamin K requirements. Deficiencies of biotin or vitamin B_5 are extremely rare after infancy because the intestinal bacteria produce enough to make up for any shortage in the diet.

Bilirubin Products. Chapter 11 discussed the breakdown of heme and its release as bilirubin in the bile. p. 348 Inside the large intestine, bacteria convert the bilirubin into other products, some of which are absorbed and excreted in the urine producing its yellow color. Others, on exposure to oxygen, are further modified into the pigments that give feces a brown color.

Toxins. Bacterial action breaks down peptides remaining in the feces and generates (1) ammonia, (2) nitrogen-containing compounds that are responsible for the odor of feces, and (3) hydrogen sulfide (H_2S), a gas that produces a "rotten egg" odor. Much of the ammonia and other toxins enter the hepatic portal circulation and are removed by the liver. The liver processes them into relatively nontoxic compounds that will be excreted at the kidneys.

Indigestible carbohydrates are not altered by intestinal enzymes and arrive in the colon intact. These molecules provide a nutrient source for resident bacteria, whose metabolic activities

16 THE DIGESTIVE SYSTEM

An Overview of the Digestive Tract • The Oral Cavity • The Pharynx • The Esophagus • The Stomach • The Small Intestine • The Pancreas • The Liver

are responsible for intestinal gas, or *flatus*, in the large intestine. Meals containing large numbers of indigestible carbohydrates (such as beans) stimulate the bacterial production of gas.

Movements of the Large Intestine

The gastroileal and gastroenteric reflexes move material into the cecum while you eat. Movement from the cecum to the transverse colon is very slow, allowing hours for the reabsorption of water. Movement from the transverse colon through the rest of the large intestine results from powerful peristaltic contractions called *mass movements*, which occur a few times a day. The normal stimulus is distension of the stomach and duodenum. The commands are relayed over the intestinal nerve plexuses. The contractions force fecal materials into the rectum and cause the urge to defecate.

DIVERTICULOSIS

In **diverticulosis** (dī-ver-tik-ū-LŌ-sis), pockets (*diverticula*) form in the mucosa, generally in the sigmoid colon. These pockets get forced outward, probably by the pressures generated during defecation. If the pockets push through weak points in the muscularis externa, they form semi-isolated chambers that are subject to recurrent infection and inflammation. The infections cause pain and occasional bleeding, a condition known as *diverticulitis* (dī-ver-tik-ū-LĪ-tis). Inflammation of other portions of the colon is called *colitis* (ko-LĪ-tis).

Defecation

The rectum is usually empty until a powerful peristaltic contraction forces fecal materials out of the sigmoid colon. Distension of the rectal wall then triggers the **defecation reflex**, with two positive feedback loops:

1. Stretch receptors in the rectal walls stimulate a series of local peristaltic contractions in the colon and rectum, moving feces toward the anus.

2. Parasympathetic motor neurons in the sacral spinal cord, also activated by the stretch receptors, stimulate increased peristalsis throughout the large intestine via motor commands distributed by the pelvic nerves.

The movement of feces through the anal canal requires relaxation of the internal anal sphincter, but when it relaxes, the external sphincter automatically clamps shut. Thus, the actual release of feces requires conscious effort to open the external sphincter voluntarily. If the commands do not arrive, the peristaltic contractions cease until additional rectal expansion triggers the defecation reflex a second time.

In addition to opening the external sphincter, consciously directed activities, such as tensing the abdominal muscles or exhaling while closing the glottis (called the *Valsalva maneuver*),

elevate intra-abdominal pressures and help to force fecal materials out of the rectum. Such pressures also force blood into the network of veins in the lamina propria and submucosa of the anal canal, causing them to stretch. Repeated incidents of straining to force defecation can cause the veins to be permanently distended, producing *hemorrhoids*.

DIARRHEA AND CONSTIPATION

Diarrhea (dī-a-RĒ-a) exists when an individual has frequent, watery bowel movements. Diarrhea results when the mucosa of the colon becomes unable to maintain normal levels of absorption or when the rate of fluid entry into the colon exceeds its maximum reabsorptive capacity. Bacterial, viral, or protozoan infection of the colon or small intestine can cause acute bouts of diarrhea lasting several days. Severe diarrhea is life-threatening due to cumulative fluid and ion losses. In *cholera* (KOL-e-ra), bacteria bound to the intestinal lining release toxins that stimulate a massive fluid secretion across the intestinal epithelium. Without treatment, a person with cholera can die of acute dehydration in a matter of hours.

Constipation is infrequent defecation, generally involving dry and hard feces. Constipation occurs when fecal materials are moving through the colon so slowly that excessive water reabsorption occurs. The feces then become extremely compact, difficult to move, and highly abrasive. Inadequate dietary fiber and fluids, coupled with a lack of exercise, are common causes. Constipation can usually be treated by oral administration of stool softeners, such as Colace™, laxatives, or *cathartics* (ka-THAR-tiks), which promote defecation. These compounds promote water movement into the feces, increase fecal mass, or irritate the lining of the colon to stimulate peristalsis. For example, indigestible fiber adds bulk to the feces, retaining moisture and stimulating stretch receptors that promote peristalsis. The promotion of peristalsis is one benefit of "high-fiber" cereals. Active movement during exercise also assists in the movement of fecal materials through the colon.

Digestion and Absorption

A typical meal contains a mixture of carbohydrates, proteins, lipids, water, electrolytes, and vitamins. Your digestive system handles each of these components differently. Large organic molecules must be broken down through digestion before absorption can occur. Water, electrolytes, and vitamins can be absorbed without preliminary processing, but special transport mechanisms may be involved.

THE PROCESSING AND ABSORPTION OF NUTRIENTS

Food contains large organic molecules, many of them insoluble. The digestive system first breaks down the physical structure of the ingested material and then disassembles the component molecules into smaller fragments. This disassembly produces small organic molecules that can be released into the bloodstream. Once absorbed by cells, they will be used to generate ATP or to synthesize complex carbohydrates, proteins, and lipids. This section will focus on the mechanics of digestion and absorption; the fate of the compounds in the body will be considered in Chapter 17.

Foods are usually complex chains of simpler molecules. In a typical dietary carbohydrate, the basic molecules are simple sugars. In a protein, the building blocks are amino acids, and in lipids they are usually fatty acids. Digestive enzymes break the bonds between the component molecules in a process called

hydrolysis. (The hydrolysis of carbohydrates, lipids, and proteins was detailed in Chapter 2. ∞ pp. 37, 39, 41)

Digestive enzymes differ in their specific targets. Carbohydrases break the bonds between sugars, proteases (proteolytic enzymes) split the linkages between amino acids, and lipases separate the fatty acids from glycerides. Specific enzymes in each class may be even more selective, breaking bonds between specific molecular participants. For example, a carbohydrase might ignore all bonds except those connecting two glucose molecules. Figure 16-18● summarizes the chemical events in the digestion of carbohydrates, lipids,

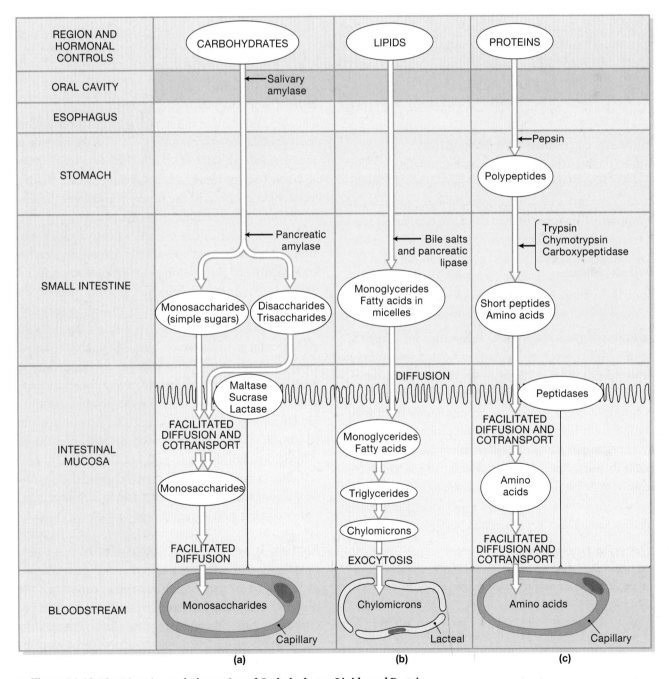

● *Figure 16-18* **The Digestion and Absorption of Carbohydrates, Lipids, and Proteins**

16 THE DIGESTIVE SYSTEM

An Overview of the Digestive Tract • The Oral Cavity • The Pharynx • The Esophagus • The Stomach • The Small Intestine • The Pancreas • The Liver

TABLE 16-3 *Digestive Enzymes and Their Functions*

ENZYME	SOURCE	OPTIMAL PH	TARGET	PRODUCTS
CARBOHYDRASES				
Amylase	Salivary glands, pancreas	6.7–7.5	Complex carbohydrates	Disaccharides and trisaccharides
Maltase, sucrase, lactase	Small intestine	7–8	Maltose, sucrose, lactose	Monosaccharides
PROTEASES				
Pepsin	Chief cells of stomach	1.5–2.0	Proteins, polypeptides	Short polypeptides
Trypsin, chymotrypsin, carboxypeptidase	Pancreas	7–8	Proteins, polypeptides	Short peptide chains
Peptidases	Small intestine	7–8	Dipeptides, tripeptides	Amino acids
LIPASES				
Pancreatic lipase	Pancreas	7–8	Triglycerides	Fatty acids and monoglycerides
Nucleases	Pancreas	7–8	Nucleic acids	Nitrogenous bases and simple sugars

and proteins. Table 16-3 reviews the major digestive enzymes and their functions.

Carbohydrate Digestion and Absorption

Carbohydrate digestion begins in the mouth through the action of salivary amylase (Figure 16-18a●). Amylase breaks down complex carbohydrates into smaller fragments, producing a mixture primarily composed of disaccharides (two simple sugars) and trisaccharides (three simple sugars). Salivary amylase continues to digest the starches and glycogen in the meal for an hour or two before stomach acids render it inactive. In the duodenum, the remaining complex carbohydrates are broken down through the action of pancreatic amylase.

Epithelial Processing and Absorption. Before they are absorbed, disaccharides and trisaccharides are fragmented into simple sugars (monosaccharides) by enzymes found on the surfaces of the intestinal microvilli. The intestinal epithelium then absorbs simple sugars through carrier-mediated transport mechanisms, such as cotransport or facilitated diffusion. ∞ p. 61 For example, glucose uptake occurs with the cotransport of sodium ions. Simple sugars entering an intestinal cell diffuse through the cytoplasm and across the basement membrane to enter the interstitial fluid. They then enter the intestinal capillaries for delivery to the hepatic portal vein and liver.

 ### LACTOSE INTOLERANCE

Lactose is the primary carbohydrate in milk, so by breaking down lactose, the enzyme lactase provides essential services throughout infancy and early childhood. If the intestinal mucosa stops producing lactase by the time of adolescence, the individual becomes lactose-intolerant. Individuals who are lactose intolerant can have a variety of unpleasant digestive problems after eating a meal containing milk and other dairy products.

Lipid Digestion and Absorption

Chapter 2 introduced the structure of fats, or triglycerides, the most abundant dietary lipids. ∞ p. 39 A triglyceride molecule consists of three fatty acids attached to a single molecule of glycerol. Triglycerides and other dietary fats are relatively unaffected by conditions in the stomach and enter the duodenum in the form of large lipid drops. As noted earlier in the chapter, bile salts emulsify these drops into tiny droplets that can be attacked by pancreatic lipase. This enzyme breaks the triglycerides apart, and the resulting mixture of fatty acids and monoglycerides interact with bile salts to form small lipid-bile salt complexes called **micelles** (mī-SELZ) (Figure 16-18b●). When a micelle contacts the intestinal epithelium, the enclosed triglyceride products diffuse across the cell membrane and enter the cytoplasm. The intestinal cells use the arriving fatty acids and monoglycerides to manufacture new triglycerides that are then coated with proteins. This step creates a soluble complex known as a **chylomicron** (kī-lō-MĪ-kron). The chylomicrons are secreted into the interstitial fluids, where they enter the intestinal lacteals through the large gaps between adjacent endothelial cells. From the lacteals they proceed along the lymphatic vessels, through the thoracic duct, and finally enter the circulation at the left subclavian vein.

Protein Digestion and Absorption

Protein digestion first requires that the structure of food be disrupted so that proteolytic enzymes can attack individual protein molecules. This step involves mechanical processing in the oral cavity, through mastication, and chemical processing in the stomach, through the action of hydrochloric acid. Exposure of the ingested food, or bolus, to a strongly acid environment breaks down plant cell walls and the connective tissues in animal products and kills most pathogens. The acidic contents of the stomach also provide the proper environment for the activity of pepsin, the proteolytic enzyme secreted by chief cells of the stomach. Pepsin does not complete protein digestion, but it does reduce the relatively huge proteins of the chyme into smaller polypeptide fragments (Figure 16-18c●).

When chyme enters the duodenum and the pH has risen, pancreatic proteolytic enzymes can now begin working. Trypsin, chymotrypsin, and carboxypeptidase each break peptide bonds between different amino acids and complete the disassembly of the polypeptide fragments into a mixture of short peptide chains and individual amino acids. Other enzymes, on the surfaces of the microvilli, called *peptidases*, complete the process by breaking the peptide chains into their component amino acids, and the amino acids are absorbed into the intestinal epithelial cells through both carrier-mediated transport and cotransport mechanisms. Carrier proteins at the inner surface of the cells release the absorbed amino acids into the interstitial fluid. Once within the interstitial fluids, most of the amino acids diffuse into intestinal capillaries.

WATER AND ELECTROLYTE ABSORPTION

Each day, roughly 2000 ml of water enters the digestive tract in the form of food or drink. The salivary, gastric, intestinal, pancreatic, and bile secretions provide about 7000 ml. Out of that total, only about 150 ml is lost in the fecal wastes. This water conservation occurs passively, following osmotic gradients; water always tends to flow into the solution containing the higher concentration of solutes.

The epithelial cells are continually absorbing dissolved nutrients and ions, and these activities gradually lower the solute concentration of the intestinal contents. As the solute concentration within the intestine decreases, water moves into the surrounding tissues, "following" the solutes and maintaining osmotic equilibrium. The absorption of sodium and chloride ions is the most important factor promoting water movement. Other ions absorbed in smaller quantities are calcium, potassium, magnesium, iodine, bicarbonate, and iron. Calcium absorption occurs under hormonal control, requiring the presence of parathyroid hormone and calcitriol. Regulatory mechanisms governing the absorption or excretion of the other ions are poorly understood.

THE ABSORPTION OF VITAMINS

Vitamins are organic compounds related to lipids and carbohydrates that are required in very small quantities. There are two major groups of vitamins: fat-soluble vitamins and water-soluble vitamins. The four **fat-soluble vitamins**, vitamins A, D, E, and K, enter the duodenum in fat droplets, mixed with dietary lipids. The vitamins remain in association with those lipids when micelles form. The fat-soluble vitamins are then absorbed from the micelles along with the products of lipid digestion. Vitamin K is also produced by the action of resident bacteria in the colon. p. 361

The nine **water-soluble vitamins** function primarily as participants in enzymatic reactions. All but one, vitamin B_{12}, are easily absorbed by the digestive epithelium. Vitamin B_{12} cannot be absorbed by the intestinal mucosa unless it has been bound to intrinsic factor, a protein secreted by the parietal cells of the stomach. p. 497 The bacteria residing in the intestinal tract are an important source for several water-soluble vitamins. In Chapter 17, we consider the functions of vitamins and associated nutritional problems (see Tables 17-2, 17-3; pp. 535–536).

✚ MALABSORPTION SYNDROMES

Malabsorption is a disorder characterized by abnormal nutrient absorption. Difficulties in the absorption of all classes of compounds will result from damage to the accessory glands or the intestinal mucosa. If the accessory organs are functioning normally but their secretions cannot reach the duodenum, the condition is called *biliary obstruction* (bile duct blockage) or *pancreatic obstruction* (pancreatic duct blockage). Alternatively, the ducts may remain open but the glandular cells are damaged and unable to continue normal secretory activities. Two examples, *pancreatitis* and *cirrhosis*, were noted earlier in the chapter.

Even with the normal enzymes in the lumen, absorption will not occur if the mucosa cannot function properly. A genetic inability to manufacture specific enzymes will result in discrete patterns of malabsorption—*lactose intolerance* is a good example. Mucosal damage due to ischemia (an interruption of the blood supply), radiation exposure (such as radiation therapy or contaminated food), toxic compounds, or infection will affect absorption and will deplete nutrient and fluid reserves as a result.

CONCEPT CHECK QUESTIONS
Answers on page 519

❶ What component of a meal would increase the number of chylomicrons in the lacteals?

❷ The absorption of which vitamin would be impaired by the removal of the stomach?

❸ Why is diarrhea potentially life-threatening but constipation is not?

Aging and the Digestive System

Essentially normal digestion and absorption occur in elderly individuals. However, many changes occur in the digestive system that parallel age-related changes already described for other systems:

16 THE DIGESTIVE SYSTEM

An Overview of the Digestive Tract • The Oral Cavity • The Pharynx • The Esophagus • The Stomach • The Small Intestine • The Pancreas • The Liver

- *The rate of epithelial stem cell division declines.* The digestive epithelium becomes more susceptible to damage by abrasion, acids, or enzymes. Peptic ulcers therefore become more likely. In the mouth, esophagus, and anus, the stratified epithelium becomes thinner and more fragile.

- *Smooth muscle tone decreases.* General motility decreases, and peristaltic contractions are weaker. This change slows the rate of intestinal movement and promotes constipation. Sagging pouches (haustra) in the walls of the colon can produce symptoms of diverticulitis. Straining to eliminate compacted fecal materials can stress the less-resilient walls of blood vessels, causing hemorrhoids. Problems are not restricted to the lower digestive tract. For example, weakening of muscular sphincters can lead to esophageal reflux and frequent bouts of "heartburn."

- *The effects of cumulative damage become apparent.* One example is the gradual loss of teeth due to *dental caries* ("cavities") or *gingivitis* (inflammation of the gums). Cumulative damage can involve internal organs as well. Toxins such as alcohol and other injurious chemicals absorbed by the digestive tract are transported to the liver for processing. Liver cells are not immune to these compounds. Chronic exposure can lead to cirrhosis or other types of liver disease.

- *Cancer rates increase.* Cancers are most common in organs where stem cells divide to maintain epithelial cell populations. Rates of colon cancer and stomach cancer rise in the elderly; oral and pharyngeal cancers are particularly common in elderly smokers.

- *Changes in other systems have direct or indirect effects on the digestive system.* For example, the reduction in bone mass and calcium content in the skeleton is associated with erosion of the tooth sockets and eventual tooth loss. The decline in smell and taste sensitivity with age can lead to dietary changes that affect the entire body.

Integration with Other Systems

The digestive system is functionally linked to all other systems, and it has extensive anatomical connections to the nervous, cardiovascular, endocrine, and lymphatic systems. Figure 16-19● summarizes the physiological relationships between the digestive system and other organ systems.

CONCEPT CHECK QUESTIONS

Answers on page 519

❶ The lining of the digestive tract becomes more susceptible to damage by abrasion, acids, or enzymes as an individual ages. What factor is primarily responsible for such changes?

❷ How is the digestive system functionally related to the cardiovascular system?

Related Clinical Terms

ascites (a-SĪ-tēz): Fluid leakage into the peritoneal cavity across the serous membranes of the liver and viscera.

achalasia (ak-a-LĀ-zē-uh): A condition that results when a bolus cannot reach the stomach due to the constriction of the lower esophageal sphincter.

cholecystitis (kŌ-lē-sis-TĪ-tis): An inflammation of the gallbladder due to a blockage of the cystic or common bile duct by gallstones.

cholelithiasis (kō-lē-li-THĪ-a-sis): The presence of gallstones in the gallbladder.

cirrhosis (sir-Ō-sis): A disease characterized by the widespread destruction of hepatocytes by exposure to drugs (especially alcohol), viral infection, ischemia, or blockage of the hepatic ducts.

colectomy (kō-LEK-to-mē): The removal of all or a portion of the colon.

colonoscope (kō-LON-ō-skōp): A fiber-optic instrument inserted into the anus and moved into the colon to look for problems in the rectum, sigmoid colon, and ascending colon.

colostomy (kō-LOS-to-mē): The attachment of the cut end of the colon to an opening in the body wall after a colectomy.

esophagitis (ē-sof-a-JĪ-tis): An inflammation of the esophagus.

gallstones: Deposits of minerals, bile salts, and cholesterol that form if bile becomes too concentrated.

gastroenterology (gas-trŌ-en-ter-OL-o-jē): The study of the digestive system and its diseases and disorders.

gastrectomy (gas-TREK-to-mē): The surgical removal of the stomach, generally to treat advanced stomach cancer.

gastroenteritis (gas-trō-en-ter-Ī-tis): A condition characterized by vomiting and diarrhea and resulting from bacterial toxins, viral infections, or various poisons.

hepatitis (hep-a-TĪ-tis): A virus-induced disease of the liver; the most common forms include *hepatitis A, B,* and *C.*

inflammatory bowel disease (*ulcerative colitis*): A chronic inflammation of the digestive tract, most commonly affecting the colon.

irritable bowel syndrome: A disorder characterized by diarrhea, constipation, or both alternately. When constipation is the primary problem, this condition may be called a *spastic colon* or *spastic colitis.*

laparoscopy (lap-a-ROS-ko-pē): The use of a flexible fiber-optic instrument introduced through the abdominal wall to permit direct visualization of the viscera, tissue sampling, and limited surgical procedures.

liver biopsy: A sample of liver tissue, generally taken through the anterior abdominal wall by means of a long needle.

perforated ulcer: A particularly dangerous ulcer in which the gastric acids erode through the wall of the digestive tract and enter the peritoneal cavity.

periodontal disease: A loosening of the teeth within the bony sockets (alveolar sockets) caused by erosion of the periodontal ligaments by acids produced through bacterial action.

peritonitis (per-i-tō-NĪ-tis): An inflammation of the peritoneal membrane.

polyps (PAH-lips): Small mucosal tumors that grow from the intestinal wall.

The Digestive System

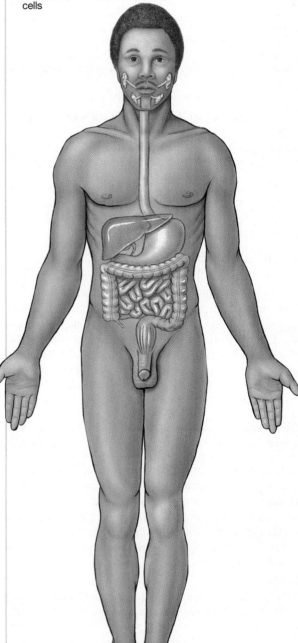

For All Systems

Absorbs organic substrates, vitamins, ions, and water required by all living cells

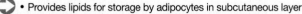

The Integumentary System

- Provides vitamin D_3 needed for the absorption of calcium and phosphorus
- Provides lipids for storage by adipocytes in subcutaneous layer

The Skeletal System

- Skull, ribs, vertebrae, and pelvic girdle support and protect parts of digestive tract; teeth important in mechanical processing of food
- Absorbs calcium and phosphate ions for incorporation into bone matrix; provides lipids for storage in yellow marrow

The Muscular System

- Protects and supports digestive organs in abdominal cavity; controls entrances and exits to digestive tract
- Liver regulates blood glucose and fatty acid levels, metabolizes lactic acid from active muscles

The Nervous System

- ANS regulates movement and secretion; reflexes coordinate passage of materials along tract; control over skeletal muscles regulates ingestion and defecation; hypothalamic centers control hunger, satiation and feeding behaviors
- Provides substrates essential for neurotransmitter synthesis

The Endocrine System

- Epinephrine and norepinephrine stimulate constriction of sphincters and depress digestive activity; hormones coordinate activity along tract
- Provides nutrients and substrates to endocrine cells; endocrine cells of pancreas secrete insulin and glucagon; liver produces angiotensinogen

The Cardiovascular System

- Distributes hormones of the digestive tract; carries nutrients, water, and ions from sites of absorption; delivers nutrients and toxins to liver
- Absorbs fluid to maintain normal blood volume; absorbs vitamin K; liver excretes heme (as bilirubin), synthesizes coagulation proteins

The Lymphatic System

- Tonsils and other lymphoid nodules along digestive tract defend against infection and toxins absorbed from the tract; lymphatic vessels carry absorbed lipids to venous system
- Secretions of digestive tract (acids and enzymes) provide nonspecific defense against pathogens

The Respiratory System

- Increased thoracic and abdominal pressure through contraction of respiratory muscles can assist in defecation
- Pressure of digestive organs against the diaphragm can assist exhalation and limit inhalation

The Urinary System

- Excretes toxins absorbed by the digestive epithelium; excretes some bilirubin produced by liver
- Absorbs water needed to excrete waste products at the kidneys; absorbs ions needed to maintain normal body fluid concentrations

The Reproductive System

- Provides additional nutrients required to support gamete production and (in pregnant women) embryonic and fetal development

● *Figure 16-19* **Functional Relationships Between the Digestive System and Other Systems**

16 THE DIGESTIVE SYSTEM

An Overview of the Digestive Tract • The Oral Cavity • The Pharynx • The Esophagus • The Stomach • The Small Intestine • The Pancreas • The Liver

CHAPTER REVIEW

Key Terms

Summary Outline

1. The digestive system consists of the muscular **digestive tract** and various **accessory organs**.

2. Digestive functions include **ingestion**, **mechanical processing**, **digestion**, **secretion**, **absorption**, **compaction**, and **excretion**.

1. The digestive tract includes the oral cavity, pharynx, esophagus, stomach, small intestine, large intestine, rectum, and anus. *(Figure 16-1)*

2. The epithelium and underlying connective tissue, the *lamina propria*, form the **mucosa** (mucous membrane) of the digestive tract. Next, outward, are the **submucosa**, the **muscularis externa**, and the *adventitia*, a layer of loose connective tissue. In the peritoneal cavity, the muscularis externa is covered by the **serosa**, a serous membrane. *(Figure 16-2)*

3. Double sheets of peritoneal membrane called **mesenteries** suspend the digestive tract.

4. The neurons that innervate the smooth muscle of the muscularis externa are not under voluntary control.

5. The muscularis externa propels materials through the digestive tract by means of the contractions of **peristalsis**. **Segmentation** movements in areas of the small intestine churn digestive materials. *(Figure 16-3)*

1. The functions of the **oral cavity** are (1) analysis of potential foods; (2) mechanical processing using the teeth, tongue, and palatal surfaces; (3) lubrication by mixing with mucus and salivary secretions; and (4) digestion by salivary enzymes.

2. The oral cavity, or **buccal cavity**, is lined by oral mucosa. The **hard palate** and **soft palate** form its roof, and the tongue forms its floor. *(Figure 16-4)*

3. The primary functions of the **tongue** include (1) mechanical processing, (2) manipulation to assist in chewing and swallowing, and (3) sensory analysis.

4. The **parotid, sublingual**, and **submandibular salivary glands** discharge their secretions into the oral cavity. Saliva lubricates the mouth, dissolves chemicals, flushes the oral surfaces, and helps control bacteria. Salivation is usually controlled by the ANS. *(Figure 16-5)*

5. **Mastication** (chewing) occurs through the contact of the opposing surfaces of the **teeth**. The **periodontal ligament** anchors the tooth in a bony socket. **Dentin** forms the basic structure of a tooth. The **crown**

is coated with **enamel**, and the **root** is covered with **cementum**. *(Figure 16-6a)*

6. The 20 primary teeth, or **deciduous teeth**, are replaced by the 32 teeth of the **secondary dentition** during development. *(Figure 16-6b, c)*

1. The **pharynx** serves as a common passageway for solid food, liquids, and air. Pharyngeal muscle contractions during swallowing propel the food mass along the esophagus and into the stomach.

1. The **esophagus** carries solids and liquids from the pharynx to the stomach through an opening in the diaphragm, the *esophageal hiatus*.

2. **Deglutition** (swallowing) can be divided into **oral, pharyngeal**, and **esophageal phases**. Swallowing begins with the compaction of a **bolus** and its movement into the pharynx, followed by the elevation of the larynx, reflection of the epiglottis, and closure of the glottis. After opening of the *upper esophageal sphincter*, peristalsis moves the bolus down the esophagus to the *lower esophageal sphincter*. *(Figure 16-7)*

1. The **stomach** has four major functions: (1) bulk storage of ingested food, (2) the mechanical breakdown of food, (3) the disruption of chemical bonds by acids and enzymes, and (4) the production of intrinsic factor. **Chyme** forms in the stomach as gastric and salivary secretions are mixed with food.

2. The four regions of the stomach are the **cardia, fundus, body**, and **pylorus**. The **pyloric sphincter** guards the exit from the stomach. In a relaxed state the stomach lining contains numerous **rugae** (ridges and folds). *(Figure 16-8)*

3. Within the **gastric glands, parietal cells** secrete **intrinsic factor** and hydrochloric acid. **Chief cells** secrete **pepsinogen**, which acids in the gastric lumen convert to the enzyme **pepsin**. Gastric gland endocrine cells secrete the hormone **gastrin**.

4. Gastric secretion includes: (1) the **cephalic phase**, which prepares the stomach to receive ingested materials; (2) the **gastric phase**, which begins with the arrival of food in the stomach; and (3) the **intestinal phase**, which controls the rate of gastric emptying. *(Figure 16-9)*

1. The **small intestine** includes the **duodenum**, the **jejunum**, and the **ileum**. The *ileocecal valve*, a sphincter, marks the transition between the small and large intestines. *(Figure 16-10)*

The Intestinal Wall..................................500

2. The intestinal mucosa bears transverse folds called **plicae**, or *plicae circulares*, and small projections called intestinal **villi**. They increase the surface area for absorption. Each villus contains a terminal lymphatic called a **lacteal**. *(Figure 16-11)*

3. Some of the smooth muscle cells in the musularis externa of the small intestine contract periodically, without stimulation, to produce brief, localized peristaltic contractions that slowly move materials along the tract. More extensive peristaltic activities are coordinated by the *gastroenteric* and the *gastroileal reflexes*.

Intestinal Secretions 501

4. Intestinal glands secrete **intestinal juice**, mucus, and hormones. Intestinal juice moistens the chyme, helps buffer acids, and dissolves digestive enzymes and the products of digestion.

5. Intestinal hormones include **secretin, cholecystokinin (CCK)**, and **gastric inhibitory peptide (GIP)**. *(Figure 16-12; Table 16-1)*

Digestion in the Small Intestine 502

6. Most of the important digestive and absorptive functions occur in the small intestine. Digestive enzymes and buffers are provided by the pancreas, liver, and gallbladder.

THE PANCREAS 503

1. The **pancreatic duct** penetrates the wall of the duodenum, where it delivers the secretions of the **pancreas**. *(Figure 16-13a)*

Histological Organization 503

2. Exocrine gland ducts branch repeatedly before ending in the **pancreatic acini** (blind pockets). *(Figure 16-13b)*

3. The pancreas has two functions: endocrine (secreting insulin and glucagon into the blood) and exocrine (secreting water, ions, and digestive enzymes into the small intestine). Pancreatic enzymes include **lipases, carbohydrases**, and **proteases**.

The Control of Pancreatic Secretion 504

4. The pancreatic exocrine cells produce a watery **pancreatic juice** in response to hormonal instructions from the duodenum. When chyme arrives in the small intestine, (1) secretin triggers the pancreatic production of a fluid containing buffers, primarily sodium bicarbonate, that help bring the pH of the chyme under control; and (2) CCK is released.

5. CCK stimulates the pancreas to produce and secrete **pancreatic amylase, pancreatic lipase, nucleases**, and several proteolytic protease enzymes—notably **trypsin, chymotrypsin**, and **carboxypeptidase**.

THE LIVER 504

1. The **liver** is the largest visceral organ in the body, with over 200 known functions. *(Figure 16-14)*

Anatomy of the Liver 504

2. The **liver lobule** is the organ's basic functional unit. Blood is supplied to the lobules by the hepatic artery and hepatic portal vein. Within the lobules, blood flows past *hepatocytes* through *sinusoids* to the **central vein**. **Bile canaliculi** carry bile away from the central vein and toward bile ducts. *(Figure 16-15)*

3. The bile ducts from each lobule unite to form the **common hepatic duct**, which meets the **cystic duct** to form the common bile duct, which empties into the duodenum.

Liver Functions.....................................506

4. The liver performs metabolic and hematological regulation and produces **bile**. *(Table 16-2)*

The Gallbladder 506

5. The **gallbladder** stores and concentrates bile for release into the duodenum. Relaxation of the **hepatopancreatic sphincter** by cholecystokinin (CCK) permits bile to enter the duodenum. *(Figure 16-16)*

THE LARGE INTESTINE 507

1. The main functions of the **large intestine** are to (1) reabsorb water and compact the feces, (2) absorb vitamins liberated by bacteria, and (3) store fecal material prior to defecation. *(Figure 16-17a)*

The Cecum 507

2. The **cecum** collects and stores material from the ileum and begins the process of compaction. The **vermiform appendix** is attached to the cecum.

The Colon 508

3. The **colon** has a larger diameter and a thinner wall than the small intestine. It bears **haustra** (pouches) and the **taeniae coli** (longitudinal bands of muscle).

The Rectum 509

4. The **rectum** terminates in the **anal canal**, leading to the **anus**. *(Figure 16-17b)*

The Physiology of the Large Intestine 509

5. The large intestine reabsorbs water and other substances, such as *vitamins, bilirubin products, bile salts*, and *toxins*. Bacteria residing in the large intestine are responsible for intestinal gas, or *flatus*.

6. Distension of the stomach and duodenum stimulates peristalsis, or *mass movements*, of feces from the colon into the rectum. Muscular sphincters control the passage of fecal material to the anus. Distension of the rectal wall triggers the *defecation reflex*. Under normal circumstances, the release of feces cannot occur unless the **external anal sphincter** is voluntarily relaxed.

DIGESTION AND ABSORPTION 510

The Processing and Absorption of Nutrients 510

1. The digestive system breaks down the physical structure of the ingested material and then disassembles the component molecules into smaller fragments through *hydrolysis*. *(Table 16-3)*

2. Amylase breaks down complex carbohydrates into disaccharides and trisaccharides. These are broken down into monosaccharides by enzymes at the epithelial surface and are absorbed by the intestinal epithelium through facilitated diffusion and cotransport. *(Figure 16-18a)*

3. *Triglycerides* are emulsified into large lipid drops. The resulting fatty acids and other lipids interact with bile salts to form **micelles**, from which they diffuse across the intestinal epithelium. The intestinal cells absorb fatty acids and synthesize new triglycerides. These are packaged in **chylomicrons**, which are released into the interstitial fluid and transported to the venous system by lymphatics. *(Figure 16-18b)*

4. Protein digestion involves the gastric enzyme pepsin and the various pancreatic proteases. Peptidases liberate amino acids that are absorbed by the intestinal epithelium and released into the interstitial fluids. *(Figure 16-18c)*

Water and Electrolyte Absorption 513

5. About 2–2.5 liters of water are ingested each day, and digestive secretions provide 6–7 liters. Nearly all is reabsorbed by osmosis.

6. Various processes are responsible for the movement of cations (such as sodium and calcium) and anions (such as chloride and bicarbonate).

The Absorption of Vitamins 513

7. **Fat-soluble vitamins** are enclosed within fat droplets and are absorbed with the products of lipid digestion. The nine **water-soluble vitamins** are important as *cofactors* and *coenzymes* in enzymatic reactions.

16 THE DIGESTIVE SYSTEM

An Overview of the Digestive Tract • The Oral Cavity • The Pharynx • The Esophagus • The Stomach • The Small Intestine • The Pancreas • The Liver

AGING AND THE DIGESTIVE SYSTEM 513

1. Age-related changes include a thinner and more fragile epithelium due to a reduction in epithelial stem cell division and weaker peristaltic contractions as smooth muscle tone decreases.

INTEGRATION WITH OTHER SYSTEMS 514

1. The digestive system has extensive anatomical connections to the nervous, cardiovascular, endocrine, and lymphatic systems. *(Figure 16-19)*

Review Questions

Level 1: Reviewing Facts and Terms

Match each item in column A with the most closely related item in column B. Use letters for answers in the spaces provided.

COLUMN A

___ 1. pyloric sphincter
___ 2. liver cells
___ 3. mucosa
___ 4. mesentery
___ 5. chief cells
___ 6. palate
___ 7. parietal cells
___ 8. parasympathetic stimulation
___ 9. sympathetic stimulation
___ 10. peristalsis
___ 11. bile salts
___ 12. salivary amylase

COLUMN B

a. serous membrane sheet
b. moves materials along digestive tract
c. regulates flow of chyme
d. increases muscular activity of digestive tract
e. starch digestion
f. inhibits muscular activity of digestive tract
g. inner lining of digestive tract
h. roof of oral cavity
i. pepsinogen
j. hydrochloric acid
k. hepatocytes
l. emulsification of fats

13. The enzymatic breakdown of large molecules into their basic building blocks is called:
    (a) absorption   (b) secretion
    (c) mechanical digestion (d) chemical digestion

14. The activities of the digestive system are regulated by:
    (a) hormonal mechanisms (b) local mechanisms
    (c) neural mechanisms   (d) a, b, and c are correct

15. The layer of the peritoneum that lines the inner surfaces of the body wall is the:
    (a) visceral peritoneum (b) parietal peritoneum
    (c) greater omentum   (d) lesser omentum

16. Protein digestion in the stomach results primarily from secretions released by:
    (a) hepatocytes   (b) parietal cells
    (c) chief cells   (d) goblet cells

17. The part of the gastrointestinal tract that plays the primary role in the digestion and absorption of nutrients is the:
    (a) large intestine   (b) small intestine
    (c) stomach   (d) cecum and colon

18. The duodenal hormone that stimulates the production and secretion of pancreatic enzymes is
    (a) pepsinogen   (b) gastrin
    (c) secretin   (d) cholecystokinin

19. The essential physiological service(s) provided by the liver is (are)
    (a) metabolic regulation   (b) hematological regulation
    (c) bile production   (d) a, b, and c are correct

20. Bile release from the gallbladder into the duodenum occurs only under the stimulation of:
    (a) cholecystokinin   (b) secretin
    (c) gastrin   (d) pepsinogen

21. The major function(s) of the large intestine is (are)
    (a) reabsorption of water and compaction of feces
    (b) absorption of vitamins liberated by bacterial action
    (c) storage of fecal material prior to defecation
    (d) a, b, and c are correct

22. The part of the colon that empties into the rectum is the:
    (a) ascending colon   (b) descending colon
    (c) transverse colon   (d) sigmoid colon

23. What are the primary digestive functions?

24. What is the purpose of the transverse or longitudinal folds in the mucosa of the digestive tract?

25. Name and describe the layers of the digestive tract, proceeding from the innermost to the outermost layer.

26. What are the four primary functions of the oral (buccal) cavity?

27. What specific function does each of the four types of teeth perform in the oral cavity?

28. What three subdivisions of the small intestine are involved in the digestion and absorption of food?

29. What are the primary functions of the pancreas, liver, and gallbladder in the digestive process?

30. What are the three major functions of the large intestine?

31. What five age-related changes occur in the digestive system?

Level 2: Reviewing Concepts

32. If the lingual frenulum is too restrictive, an individual:
    (a) has difficulty tasting food
    (b) cannot swallow properly
    (c) cannot control movements of the tongue
    (d) cannot eat or speak normally

33. The gastric phase of secretion is initiated by:
    (a) distension of the stomach
    (b) an increase in the pH of the gastric contents
    (c) the presence of undigested materials in the stomach
    (d) a, b, and c are correct

34. A drop in pH to 4.0 in the duodenum stimulates the secretion of:
    (a) secretin   (b) cholecystokinin
    (c) gastrin   (d) a, b, and c are correct

35. Differentiate between the action and outcome of peristalsis and segmentation.

36. How does the stomach promote and assist in the digestive process?

37. What changes in gastric function occur during the three phases of gastric secretion?

Level 3: Critical Thinking and Clinical Applications

38. Some patients with gallstones develop pancreatitis. How could this occur?

39. Barb suffers from Crohn's disease, a regional inflammation of the intestine that is thought to have some genetic basis, although the actual cause is as yet unknown. When the disease flares up, she experiences abdominal pain, weight loss, and anemia. Which part(s) of the intestine is (are) probably involved, and what is the cause of her symptoms?

Answers to Concept Check Questions

Page 495

1. The mesenteries are double layers of serous membrane that support and stabilize the positions of organs in the abdominopelvic cavity and provide a route for the blood vessels, nerves, and lymphatic vessels associated with the digestive tract.
2. Peristalsis would be more efficient in propelling intestinal contents. Segmentation is essentially a churning action that mixes intestinal contents with digestive fluids.
3. Parasympathetic stimulation increases muscle tone and motility in the digestive tract. A drug that blocks this activity would decrease the rate of peristalsis.
4. The process that is being described is swallowing.

Page 503

1. The pyloric sphincter regulates the flow of chyme into the small intestine.
2. The vagus nerve contains parasympathetic motor fibers that can stimulate gastric secretions even if no food is in the stomach (cephalic phase of gastric digestion). Cutting the branches of the vagus nerve that supply the stomach would prevent this type of secretion and would decrease the chance of ulcer formation.
3. The small intestine has several adaptations that increase surface area to increase its absorptive capacity. First, walls of the small intestine form folds called plicae. The tissue that covers the plicae forms villi, fingerlike projections. The cells that cover the villi have an exposed surface that in turn is covered by small fingerlike projections called microvilli. In addition, the small intestine has a very rich blood supply and lymphatic supply to transport the nutrients that are absorbed.
4. A meal high in fat would increase CCK blood levels.

Page 507

1. A narrowing of the ileocecal valve would interfere with the flow of chyme from the small intestine to the large intestine.
2. Damage to the exocrine pancreas would most affect the digestion of fats (lipids) because that organ is the primary source of lipases.
3. A decrease in the amount of bile salts would decrease the effectiveness of fat digestion and absorption.

Page 513

1. Chylomicrons are formed from the fats that are digested in a meal. A meal that is high in fat would increase the number of chylomicrons in the lacteals.
2. The removal of the upper portion of the stomach would interfere with the absorption of vitamin B_{12}. This vitamin requires intrinsic factor, a molecule produced by the parietal cells in the stomach.
3. Diarrhea is potentially life-threatening because a person could lose fluid and electrolytes faster than these substances can be replaced. This would result in dehydration and possibly death. Although it can be quite uncomfortable, constipation does not interfere with any major body process. The few toxic waste products that are normally eliminated via the digestive system can move into the blood and be eliminated by the kidneys.

Page 514

1. The rate of epithelial stem cell division declines with age. As a result, the digestive epithelium becomes more susceptible to damage by abrasion, acids, or enzymes.
2. Functions of the digestive system that are related to the cardiovascular system include: the absorption of water to maintain blood volume; absorption of vitamin K; host bacteria that produce vitamin K; the liver excretes bilirubin, a product of the breakdown of the heme portion of hemoglobin and, the liver also synthesizes coagulation proteins (blood clotting factors).

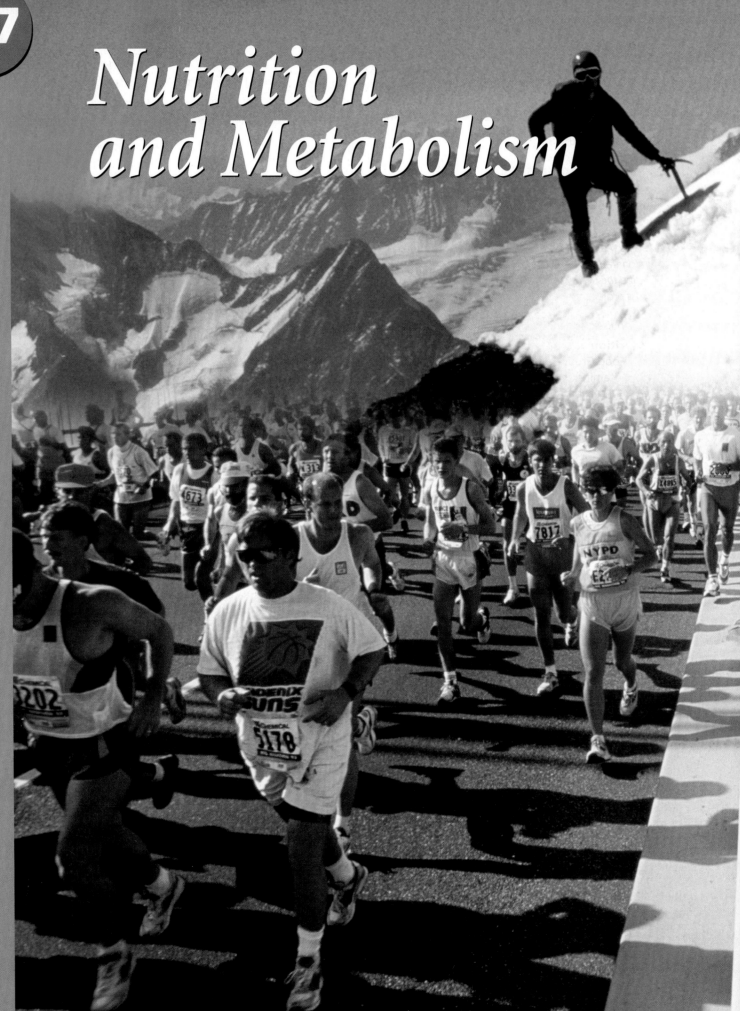

17

Nutrition and Metabolism

CHAPTER OUTLINE AND OBJECTIVES

  ❶ *Define metabolism, and explain why cells need to synthesize new organic structures.*
  ❷ *Describe the basic steps in glycolysis, the TCA cycle, and the electron transport system.*
  ❸ *Describe the pathways involved in lipid metabolism.*
  ❹ *Discuss protein metabolism and the use of proteins as an energy source.*
  ❺ *Discuss nucleic acid metabolism.*

  ❻ *Explain what constitutes a balanced diet and why it is important.*
  ❼ *Discuss the functions of vitamins, minerals, and other important nutrients.*

  ❽ *Describe the significance of the caloric value of foods.*
  ❾ *Define metabolic rate, and discuss the factors involved in determining an individual's metabolic rate.*
  ❿ *Discuss the homeostatic mechanisms that maintain a constant body temperature.*

  ⓫ *Describe the change in nutritional requirements with age.*

Vocabulary Development

anabole ....a building up; *anabolism*
genesis ....an origin; *thermogenesis*
glykus ....sweet; *glycolysis*
katabole ....a throwing down; *catabolism*
lipos ....fat; *lipogenesis*
lysis..................................breakdown; *glycolysis*
neo- ....new; *gluconeogenesis*
therme ....heat; *thermogenesis*
vita ....life; *vitamin*

*H*UMANS ARE REMARKABLY ADAPTABLE ANIMALS; *we can maintain homeostasis from the steamy equator to the frigid North and South Poles, as well as over a wide range of elevations. The key to survival under such varied environmental conditions is the preservation of normal body temperature. In part, this can be accomplished by shedding or adding layers of clothing, but hormonal and metabolic adjustments are also crucial. Thus, the metabolic rate rises in cold climates, generating additional heat that warms the body. This chapter examines several aspects of metabolism, including nutrition, energy use, heat production, and the regulation of body temperature under normal and abnormal conditions.*

521

17 NUTRITION AND METABOLISM

Cellular Metabolism • Diet and Nutrition • Bioenergetics • Aging and Nutritional Requirements • Chapter Review • MediaLab

CELLS ARE CHEMICAL FACTORIES that use chemical reactions to break down organic molecules and to obtain energy, usually in the form of ATP. Reactions within mitochondria provide most of the energy needed by a typical cell. ⚭ p. 69 To carry out these processes, cells in the human body must also obtain oxygen and nutrients. **Nutrients** are essential substances normally obtained from the diet, including water, vitamins, ions, carbohydrates, lipids, and proteins. Oxygen is absorbed at the lungs, but all the other required substances are obtained by absorption at the digestive tract. The cardiovascular system distributes oxygen and nutrients to cells throughout the body.

The energy released in a cell supports growth, cell division, contraction, secretion, and other special functions that vary from cell to cell and tissue to tissue. Because each tissue type contains different populations of cells, the energy and nutrient requirements of any two tissues (such as loose connective tissue and cardiac muscle) are quite different. When cells, tissues, and organs change their patterns or levels of activity, the metabolic needs of the body change. For example, our nutrient requirements vary from moment to moment (resting versus active), hour to hour (asleep versus awake), and year to year (child versus adult).

When organic nutrients, such as carbohydrates or lipids, are abundant, energy reserves are built up. Different tissues and organs are specialized to store excess nutrients; the storage of lipids in adipose tissue is one familiar example. These reserves can then be called on when the diet cannot provide the right quantity or quality of nutrients. The endocrine system, with the assistance of the nervous system, adjusts and coordinates the metabolic activities of the body's tissues and controls the storage and release of nutrient reserves.

The absorption of nutrients from food is called *nutrition*. The mechanisms involved in absorption through the lining of the digestive tract were detailed in Chapter 16. ⚭ p. 510–513 This chapter considers what happens to nutrients after they are inside the body.

Cellular Metabolism

The term **metabolism** refers to all the chemical reactions that occur in the body. ⚭ p. 32 Cellular metabolism, chemical reactions at the cellular level, provide the energy needed to maintain homeostasis and to perform essential functions. Figure 17-1● provides an overview of the processes involved in cellular metabolism. Amino acids, lipids, and simple sugars cross the cell membrane and join other nutrients already in the cytoplasm. All of the cell's metabolic operations rely on the resulting *nutrient pool*.

Catabolism breaks down organic molecules, releasing energy that can be used to synthesize ATP or other high-energy compounds. ⚭ p. 33 Catabolism proceeds in a series of steps. In general, preliminary processing occurs in the cytosol, where enzymes break down large organic molecules into smaller fragments. For example, carbohydrates are broken down into short

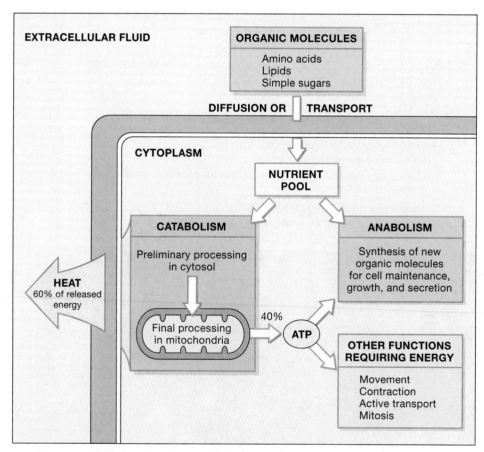

● *Figure 17-1* **Cellular Metabolism**
The cell obtains organic molecules from the extracellular fluid and breaks them down to obtain ATP. Only about 40 percent of the energy released through catabolism is captured in ATP; the rest is radiated as heat. The ATP generated through catabolism provides energy for all vital cellular activities, including anabolism.

carbon chains, triglycerides are split into fatty acids and glycerol, and proteins are broken down to individual amino acids.

Relatively little ATP is formed during these initial steps. However, the simple molecules produced can be absorbed and processed by mitochondria, and the mitochondrial steps release significant amounts of energy. As mitochondrial enzymes break the covalent bonds that hold these molecules together, they capture roughly 40 percent of the energy released. The captured energy is used to convert ADP to ATP, and the rest escapes as heat that warms the interior of the cell and the surrounding tissues.

Anabolism, the synthesis of new organic molecules, involves the formation of new chemical bonds. ∞ p. 33 The ATP produced by mitochondria provides energy to support anabolism as well as other cell functions. Those additional functions, such as contraction, ciliary or cell movement, active transport, and cell division, vary from one cell to another. For example, muscle fibers need ATP to provide energy for contraction, whereas gland cells need ATP to synthesize and transport their secretions.

Cells synthesize new organic components for four basic reasons:

1. *To perform structural maintenance and repairs.* All cells must expend energy to perform ongoing maintenance and repairs, because most structures in the cell, including organelles, are temporary rather than permanent. Their continuous removal and replacement are part of the process of **metabolic turnover**.

2. *To support growth.* Cells preparing for division enlarge and synthesize extra proteins and organelles.

3. *To produce secretions.* Secretory cells must synthesize their products and deliver them to the interstitial fluid.

4. *To build nutrient reserves.* Most cells prepare for a rainy day—a period of emergency, an interval of extreme activity, or a time when the nutrient supply in the bloodstream is inadequate. Cells prepare for such situations by storing nutrients in a form that can be mobilized as needed. For example, muscle cells and liver cells store glucose in the form of glycogen, and adipocytes and liver cells store triglycerides.

The nutrient pool is the source of organic molecules for both catabolism and anabolism. As you might expect, the cell tends to conserve the materials needed to build new compounds and breaks down the rest. The cell continuously replaces membranes, organelles, enzymes, and structural proteins. These anabolic activities require more amino acids than lipids and few carbohydrates. Catabolic activities, however, tend to process these organic molecules in the reverse order. In general, when *a cell with excess carbohydrates, lipids, and amino acids needs energy, it will break down carbohydrates first.* Lipids are broken down when carbohydrates are no longer available, and amino acids are seldom broken down if other energy sources are available.

Mitochondria provide the energy that supports cellular operations. The cell feeds its mitochondria from its nutrient pool, and in return the cell gets the ATP it needs. However, mitochondria will accept only specific organic molecules for processing and energy production. Chemical reactions in the cytoplasm take whichever organic nutrients are available and break them down into smaller fragments acceptable to the mitochondria. The mitochondria then break down the fragments further, generating carbon dioxide, water, and ATP (Figure 17-2●). The mitochondrial activity involves two

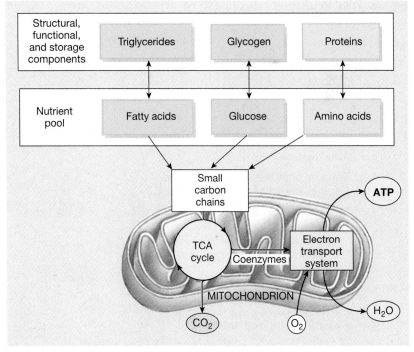

● *Figure 17-2* **Nutrient Use in Cellular Metabolism**
Cells use the nutrient pool to build up reserves and to make cellular structures. Catabolism within mitochondria provide the ATP needed to sustain cell functions. The mitochondria absorb small carbon chains from the breakdown products of carbohydrates (especially glucose), lipids (especially fatty acids from triglycerides), and proteins (amino acids). The small carbon chains are broken down further by means of the tricarboxylic acid (TCA) cycle and the electron transport system (ETS).

17 **NUTRITION AND METABOLISM**

Cellular Metabolism • Diet and Nutrition • Bioenergetics • Aging and Nutritional Requirements • Chapter Review • MediaLab

pathways: the TCA cycle and the electron transfer system. In the next section we shall describe these important catabolic and anabolic reactions that occur in our cells.

CARBOHYDRATE METABOLISM

Carbohydrates, most familiar to us as sugars and starches, are important sources of energy. Most cells generate ATP and other high-energy compounds by breaking down carbohydrates, especially glucose. The complete reaction sequence can be summarized as:

$$C_6H_{12}O_6 \ + \ 6\,O_2 \ \longrightarrow \ 6\,CO_2 \ + \ 6\,H_2O$$

glucose   oxygen   carbon dioxide   water

The breakdown occurs in a series of small steps, and several of the steps release sufficient energy to support the conversion of ADP to ATP. During the complete catabolism of glucose, a typical cell gains 36 ATP molecules.

Although most of the actual energy production occurs inside mitochondria, the preliminary steps take place in the cytosol as a sequence of reactions called *glycolysis*. The steps were outlined in Chapter 7; because those steps do not require

oxygen, they are said to be *anaerobic*. The subsequent reactions, which occur in the mitochondria, consume oxygen and are called *aerobic*. ⟳ p. 190 The mitochondrial activity responsible for ATP production is called **aerobic metabolism**, or **cellular respiration**.

Glycolysis

Glycolysis (glī-KOL-i-sis; *glykus*, sweet + *lysis*, breakdown) is the breakdown of glucose to *pyruvic acid*. In this process, a series of enzymatic steps breaks the six-carbon glucose molecule ($C_6H_{12}O_6$) into two three-carbon molecules of pyruvic acid (CH_3—CO—COOH).

Glycolysis requires (1) glucose molecules, (2) appropriate cytoplasmic enzymes, (3) ATP and ADP, and (4) **NAD** (*n*icotinamide *a*denine *d*inucleotide), a coenzyme that removes hydrogen atoms. *Coenzymes* are organic molecules, usually derived from vitamins, that must be present for an enzymatic reaction to occur. If the cell lacks any of these four participants, glycolysis cannot take place.

The basic steps of glycolysis are summarized in Figure 17-3●. This reaction sequence provides a net gain of two ATP

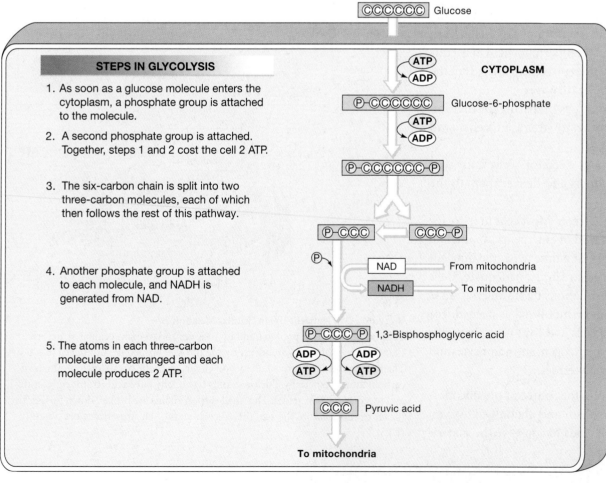

STEPS IN GLYCOLYSIS

1. As soon as a glucose molecule enters the cytoplasm, a phosphate group is attached to the molecule.

2. A second phosphate group is attached. Together, steps 1 and 2 cost the cell 2 ATP.

3. The six-carbon chain is split into two three-carbon molecules, each of which then follows the rest of this pathway.

4. Another phosphate group is attached to each molecule, and NADH is generated from NAD.

5. The atoms in each three-carbon molecule are rearranged and each molecule produces 2 ATP.

● *Figure 17-3*
Glycolysis
Glycolysis breaks down a six-carbon glucose molecule into two three-carbon molecules of pyruvic acid. This process involves a series of enzymatic steps. There is a net gain of two ATPs for each glucose molecule converted to pyruvic acid.

molecules for each glucose molecule converted to two pyruvic acid molecules. A few highly specialized cells, such as red blood cells, lack mitochondria and derive all of their ATP by glycolysis. Skeletal muscle fibers rely on glycolysis for energy production during periods of active contraction, and most cells can survive brief periods of hypoxia (low oxygen levels) by using the ATP glycolysis provides. When oxygen is readily available, however, mitochondrial activity provides most of the ATP required by our cells.

Mitochondrial Energy Production

Glycolysis yields an immediate net gain of two ATP molecules for the cell, but a great deal of additional energy is still locked in the chemical bonds of pyruvic acid. The ability to capture that energy depends on the availability of oxygen. If oxygen supplies are adequate, mitochondria will absorb the pyruvic acid molecules and break them down completely. The hydrogen atoms are removed by coenzymes and will ultimately be the source of most of the energy gain for the cell. The carbon and oxygen atoms are removed and released as carbon dioxide.

Each mitochondrion has two membranes surrounding it. ∞ p. 69 The outer membrane is permeable to pyruvic acid, and a carrier protein in the inner membrane transports the pyruvic acid into the mitochondrial matrix. Once inside the mitochondrion, a pyruvic acid molecule participates in a com-plex reaction involving NAD and another coenzyme, *coenzyme A* (or *CoA*). This reaction yields one molecule of carbon dioxide, one of NADH, and one molecule of **acetyl-CoA** (as-Ē-til-KŌ-ā). Acetyl-CoA consists of a two-carbon *acetyl group* (CH_3CO) bound to coenzyme A. Next, the acetyl group is transferred from CoA to a four-carbon molecule, producing *citric acid*.

The TCA Cycle. The formation of citric acid is the first step in a sequence of enzymatic reactions called the **tricarboxylic** (trī-kar-bok-SIL-ik) **acid (TCA) cycle**, also known as the *citric acid cycle*. This reaction sequence is sometimes known as the *Krebs cycle* in honor of Hans Krebs, the biochemist who described these reactions in 1937. The function of the cycle is to remove hydrogen atoms from organic molecules and transfer them to coenzymes. The electrons in these hydrogen atoms contain energy that can be used by the mitochondria to generate ATP. An overview of the TCA cycle is shown in Figure 17-4●.

At the start of the TCA cycle, the two-carbon acetyl group carried by CoA is attached to a four-carbon molecule to make the six-carbon citric acid molecule. Coenzyme A is then released intact to bind another acetyl group. A complete revolution of the TCA cycle removes the two added carbon atoms, regenerating the four-carbon chain. (This is why the reaction sequence is called a *cycle*.) The two removed carbon atoms generate two

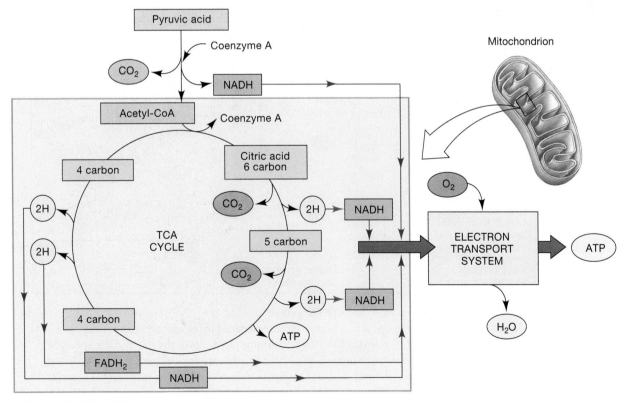

● *Figure 17-4* **The TCA Cycle**
The TCA cycle completes the breakdown of organic molecules begun by glycolysis and other catabolic pathways.

17 NUTRITION AND METABOLISM

Cellular Metabolism • Diet and Nutrition • Bioenergetics • Aging and Nutritional Requirements • Chapter Review • MediaLab

molecules of carbon dioxide (CO_2), a metabolic waste product. The hydrogen atoms of the acetyl group are removed by coenzymes.

The only immediate energy benefit of one revolution of the TCA cycle is the formation of a single molecule of ATP. The real value of the TCA cycle can be seen by following the hydrogen atoms that are removed by coenzymes. The coenzymes (NAD and **FAD,** *flavine adenine dinucleotide*) deliver the hydrogen atoms to the *electron transport system.*

The Electron Transport System. The **electron transport system (ETS)** is embedded in the inner mitochondrial membrane (Figure 17-5●). The ETS consists of coenzymes and an electron transport chain made up of a series of protein-pigment complexes called *cytochromes.* The ETS does not produce ATP directly. Instead, it creates the conditions necessary for ATP production. Coenzymes in the mitochondrial matrix deliver hydrogen atoms to the ETS; the electrons are removed and passed from coenzyme to cytochrome and from cytochrome to cytochrome, losing energy in a series of small steps. At several steps along the way, this energy is used to drive hydrogen ion pumps that move hydrogen ions from the mitochondrial matrix into the space between the outer and inner mitochondrial membranes. This creates a large concentration gradient of hydrogen ions across the inner membrane. The hydrogen ions then diffuse through hydrogen ion channels back into the matrix. The kinetic energy of the passing hydrogen ions is used to attach a phosphate group to ADP, forming ATP. This process is called *chemiosmosis* (kem-ē-oz-MŌ-sis), a term that links the chemical formation of ATP with transport across a membrane. At the end of the electron transport chain, an oxygen atom accepts the electrons and combines with two hydrogen ions (H^+) to form a molecule of water.

The electron transport system is the most important mechanism for the generation of ATP; in fact, it provides roughly 95 percent of the ATP needed to keep our cells alive. Halting or significantly reducing the rate of mitochondrial activity will usually kill a cell. If many cells are affected, the individual may

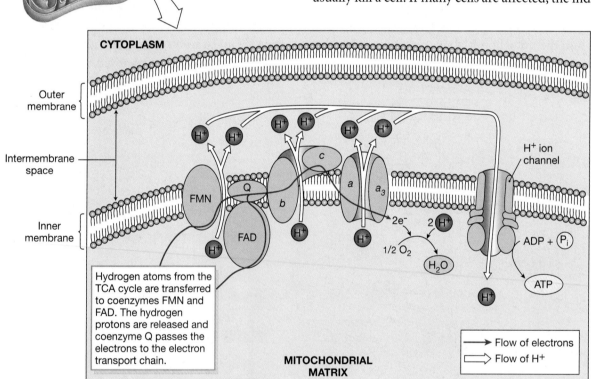

● *Figure 17-5* **The Electron Transport System (ETS) and ATP Formation**
Coenzymes ($NADH$ and $FADH_2$) deliver hydrogen atoms to coenzymes embedded in the inner mitochondrial membrane. The electrons of the hydrogen atoms are transferred to the cytochrome molecules of the ETS and the hydrogen ions remain in the matrix. The cytochrome molecules of the ETS pass the energy-carrying electrons in sequence from one to another. Energy released by the passed electrons is used to pump hydrogen ions from the matrix into the intermembrane space. This creates a difference in the concentrations of hydrogen ions across the inner membrane. The hydrogen ions then diffuse through hydrogen ion channels in the inner membrane and their kinetic energy is used to generate ATP.

die. For example, if the supply of oxygen is cut off, mitochondrial ATP production will cease because the ETS will be unable to get rid of its electrons. With the last reaction in the chain stopped, the entire ETS comes to a halt, like cars at a washed-out bridge. The affected cells quickly die of energy starvation.

Energy Yield of Glycolysis and Cellular Respiration

For most cells, the series of chemical reactions that begin with glucose and end with carbon dioxide and water is the primary method of generating ATP. A cell gains ATP at several steps along the way:

- Through glycolysis in the cytoplasm, the cell gains two molecules of ATP for each glucose molecule broken down to pyruvic acid.

- Inside the mitochondria, the two pyruvic acid molecules derived from each glucose molecule are fully broken down in the TCA cycle. Two revolutions of the TCA cycle, each yielding a molecule of ATP, provide a net gain of two additional molecules of ATP.

- For each molecule of glucose broken down, activity at the electron transport chain in the inner mitochondrial membrane will provide 32 molecules of ATP.

Summing up, for each glucose molecule processed, a typical cell gains 36 molecules of ATP. *All but two of them are produced by the mitochondria.*

 ## CARBOHYDRATE LOADING

Although other nutrients can be broken down to provide substrates, or raw materials, for the TCA cycle, carbohydrates require the least processing and preparation. It is not surprising, therefore, that athletes have tried to devise ways of exploiting these compounds as ready sources of energy.

Eating carbohydrates just before exercising does not improve performance and can actually decrease endurance by slowing the mobilization of existing energy reserves. Runners or swimmers preparing for lengthy endurance contests, such as a marathon or 5 km swim, do not eat immediately before competing, and for 2 hours before the race they limit their intake to water. However, these athletes often eat carbohydrate-rich meals for 3 days before the event. This **carbohydrate loading** increases the carbohydrate reserves of muscle tissue that will be called on during the competition.

You can obtain maximum effects by exercising to exhaustion for 3 days before you start a high-carbohydrate diet; this practice is called *carbohydrate depletion/loading*. Carbohydrate depletion/loading has a number of possible side effects, including muscle and kidney damage. Sports physiologists recommend that athletes use this routine fewer than three times per year.

OTHER CATABOLIC PATHWAYS

Aerobic metabolism is relatively efficient and capable of generating large amounts of ATP. It is the cornerstone of normal cellular metabolism, but it has one obvious limitation—the cell must have adequate supplies of both oxygen and glucose. Cells can survive only for brief periods without oxygen. Low glucose concentrations have a much smaller effect on most cells, because cells can break down other nutrients to provide organic molecules for the TCA cycle, as shown in Figure 17-6●. Many cells can switch from one nutrient source to another as the need arises. For example, many cells can shift from glucose-based to lipid-based ATP production when necessary. When actively contracting, skeletal muscles catabolize glucose, but at rest they rely on fatty acids.

Cells break down proteins for energy only when lipids or carbohydrates are unavailable; this makes sense because the enzymes and organelles that the cell needs to survive are composed of proteins. Nucleic acids are present only in small amounts, and they are seldom catabolized for energy, even when the cell is dying of acute starvation. This restraint makes sense, too, as it is the DNA in the nucleus that determines all of the structural and functional characteristics of the cell. We shall consider the catabolism of other compounds in later sections as we discuss the metabolism of lipids, proteins, and nucleic acids.

Carbohydrate Synthesis

Because some of the steps in glycolysis are not reversible, cells cannot generate glucose by performing glycolysis in reverse, using the same enzymes. Therefore, glycolysis and the

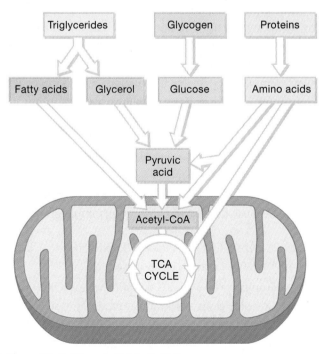

● *Figure 17-6* **Alternate Catabolic Pathways**

17 | NUTRITION AND METABOLISM

Cellular Metabolism • Diet and Nutrition • Bioenergetics • Aging and Nutritional Requirements • Chapter Review • MediaLab

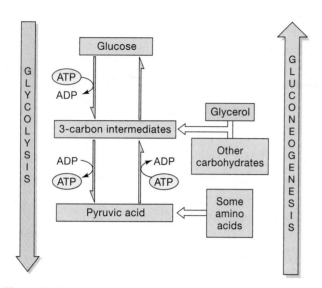

● *Figure 17-7* **Carbohydrate Metabolism**
A flow chart of the major pathways of glycolysis and gluconeogenesis. Some amino acids, other carbohydrates, and glycerol can be converted to glucose. All these reactions occur in the cytosol.

production of glucose involve a different set of regulatory enzymes, and the two processes are independently regulated. Pyruvic acid or other three-carbon molecules can be used to synthesize glucose; as a result, a cell can create glucose from noncarbohydrate precursors, such as lactic acid, glycerol, or amino acids (Figure 17-7●). However, acetyl-CoA cannot be used to make glucose, because the reaction that removes the carbon dioxide molecule (a *decarboxylation*) between pyruvic acid and acetyl-CoA cannot be reversed. The synthesis of glucose from noncarbohydrate (protein or lipid) precursor molecules is called **gluconeogenesis** (gloo-kō-nē-ō-JEN-e-sis; *glykus*, sweet + *neo-*, new + *genesis*, an origin). Fatty acids and many amino acids cannot be used for gluconeogenesis, because their breakdown produces acetyl-CoA.

Glucose molecules created by gluconeogenesis can be used to manufacture other simple sugars, complex carbohydrates, or nucleic acids. In the liver and in skeletal muscle, glucose molecules are stored as **glycogen**. Glycogen is an important energy reserve that can be broken down when the cell cannot obtain enough glucose from the interstitial fluid. Although glycogen molecules are large, glycogen reserves take up very little space because they form compact, insoluble granules.

CONCEPT CHECK QUESTIONS

Answers on page 544

❶ What is the primary role of the TCA cycle in the production of ATP?

❷ Hydrogen cyanide gas is a poison that produces its lethal effect by binding to the last cytochrome molecule in the electron transport system. What effect would this have at the cellular level?

LIPID METABOLISM

Lipid molecules, like carbohydrates, contain carbon, hydrogen, and oxygen, but in different proportions. Because triglycerides are the most abundant lipid in the body, our discussion will focus on pathways for triglyceride breakdown and synthesis. 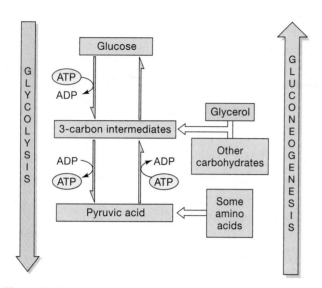 p. 38

Lipid Catabolism

During lipid catabolism, or **lipolysis**, lipids are broken down into pieces that can be converted to pyruvic acid or channeled directly into the TCA cycle (Figure 17-6●). A triglyceride is first split into its component parts through hydrolysis. This step yields one molecule of glycerol and three fatty acid molecules. Glycerol enters the TCA cycle after cytoplasmic enzymes convert it to pyruvic acid. The catabolism of fatty acids, known as *beta-oxidation*, involves a different set of enzymes that break the fatty acids down into two-carbon fragments. The fragments enter the TCA cycle or combine to form *ketone bodies*, short carbon chains discussed in a later section. Beta-oxidation occurs inside mitochondria, so the carbon chains can enter the TCA cycle immediately. The cell gains 144 ATP molecules from the breakdown of one 18-carbon fatty acid molecule—almost 1.5 times the energy obtained from the breakdown of three 6-carbon glucose molecules.

Lipids and Energy Production

Lipids are important as an energy reserve because they can provide large amounts of ATP. They are insoluble, so lipids can be stored in compact droplets in the cytosol. However, if the droplets are large, it is difficult for water-soluble enzymes to get at them. This makes lipid reserves harder to mobilize than carbohydrate reserves. In addition, most lipids are processed inside mitochondria, and mitochondrial activity is limited by the availability of oxygen. The net result is that lipids cannot provide large amounts of ATP in a short amount of time. However, cells with modest energy demands can shift to lipid-based energy production when glucose supplies are limited. Skeletal muscle fibers normally cycle between lipid metabolism and carbohydrate metabolism. When at rest (when energy demands are low), these cells break down fatty acids. When active (when energy demands are large and immediate), skeletal muscle fibers shift to glucose metabolism.

Lipid Synthesis

The synthesis of lipids is known as **lipogenesis** (li-pō-JEN-e-sis; *lipos*, fat). It usually begins with acetyl-CoA, but almost any organic molecule can be used because lipids, amino acids, and carbohydrates can easily be converted to acetyl-CoA (Figure 17-8●). Our cells cannot *build* every fatty acid they can break

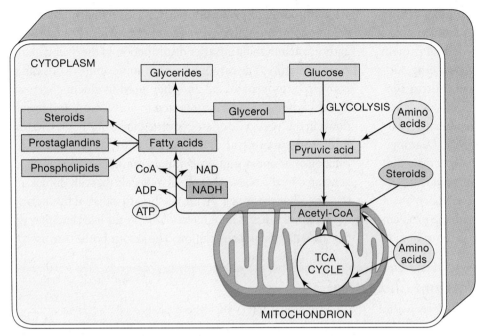

● *Figure 17-8* **Lipid Synthesis**
Pathways of lipid synthesis begin with acetyl-CoA. Molecules of acetyl-CoA can be strung together in the cytosol, yielding fatty acids. Those fatty acids can be used to synthesize glycerides or other lipid molecules. Lipids can also be synthesized from amino acids or carbohydrates.

down. *Linoleic acid* and *linolenic acid*, both 18-carbon unsaturated fatty acids, cannot be synthesized. These are called **essential fatty acids**, because they must be included in your diet. They are synthesized by plants. A diet poor in linoleic acid slows growth and alters the appearance of the skin. These fatty acids are also needed to synthesize prostaglandins and phospholipids for cell membranes.

Lipid Transport and Distribution

All cells need lipids to maintain their cell membranes, and steroid hormones must reach their target cells in many different tissues. Because most lipids are not soluble in water, special transport mechanisms are required to carry them around the body. Most lipids circulate in the bloodstream as lipoproteins. **Lipoproteins** are lipid-protein complexes that contain triglycerides and cholesterol, with an outer coating of phospholipids and proteins. The proteins and phospholipids make the entire complex soluble. Exposed proteins, which bind to specific membrane receptors, determine which cells absorb the associated lipids.

Lipoproteins are classified by size and by their relative proportions of lipid versus protein. One group of lipoproteins, the *chylomicrons*, forms in the intestinal tract. p. 512 Chylomicrons are the largest lipoproteins, and some 95 percent of their weight consists of triglycerides. Chylomicrons carry absorbed lipids from the intestinal tract to the bloodstream.

The chylomicrons transport triglycerides in the bloodstream, from which they are absorbed by skeletal muscle, cardiac muscle, adipose tissue, and the liver.

Two other major groups of lipoproteins are the **low-density lipoproteins (LDLs)** and **high-density lipoproteins (HDLs)**. These lipoproteins are formed in the liver and contain few triglycerides. Their main roles are to shuttle cholesterol between the liver and other tissues. The LDLs deliver cholesterol to peripheral tissues. Because this cholesterol may end up in arterial plaques, LDL cholesterol is often called "bad cholesterol." The HDL cholesterol transports excess cholesterol from peripheral tissues to the liver for storage or excretion in the bile. Because HDL cholesterol is returning cholesterol that will not cause circulatory problems, it is called "good cholesterol."

Free fatty acids (FFA) are lipids that can diffuse easily across cell membranes. A major source of these fatty acids is the breakdown of fat stored in adipose tissue. When released into the blood, the fatty acids bind to albumin, the most abundant plasma protein. Liver cells, cardiac muscle cells, skeletal muscle fibers, and many other body cells can metabolize free fatty acids. They are an important energy source during periods of starvation, when glucose supplies are limited.

PROTEIN METABOLISM

The body can synthesize 100,000 to 140,000 different proteins, each having varied forms, functions, and structures. Yet, each contains some combination of the same 20 amino acids. Under normal conditions, there is a continuous recycling of cellular proteins in the cytosol. Peptide bonds are broken, and the free amino acids are used to manufacture new proteins. If other energy sources are inadequate, mitochondria can break down amino acids in the TCA cycle to generate ATP. Not all amino acids enter the TCA cycle at the same point, so the ATP benefits vary. However, the average ATP yield is comparable to that of carbohydrate catabolism.

Amino Acid Catabolism

The first step in amino acid catabolism is the removal of the amino group, which requires a coenzyme derived from **vitamin B₆** *(pyridoxine)*. The amino group can be removed by transamination or deamination. *Transamination* (trans-am-i-NĀ-shun)

17 NUTRITION AND METABOLISM

Cellular Metabolism • Diet and Nutrition • Bioenergetics • Aging and Nutritional Requirements • Chapter Review • MediaLab

attaches the amino group of an amino acid to another carbon chain, creating a "new" amino acid. Transaminations enable a cell to synthesize many of the amino acids needed for protein synthesis. Cells of the liver, skeletal muscles, heart, lung, kidney, and brain, which are particularly active in protein synthesis, perform many transaminations.

Deamination (dē-am-i-NĀ-shun) is performed in preparing an amino acid for breakdown in the TCA cycle. Deamination is the removal of an amino group in a reaction that generates an ammonia molecule (NH_3). Ammonia is highly toxic, even in low concentrations. The liver, the primary site of deamination, has the enzymes needed to deal with the problem of ammonia generation. Liver cells combine carbon dioxide

with ammonia to produce **urea**, a relatively harmless, water-soluble compound that is excreted in the urine. The fate of the carbon chain remaining after deamination in the liver depends on its structure. The carbon chains of some amino acids can be converted to pyruvic acid and then used in gluconeogenesis. Other carbon chains are converted to acetyl-CoA and broken down in the TCA cycle or are converted to ketone bodies. **Ketone bodies** are organic acids that are also produced when lipid catabolism is under way. *Acetone* is an example of a ketone body generated by the body. It is a small molecule that can diffuse into the alveoli of the lungs, giving the breath a distinctive odor.

Ketone bodies are not metabolized by the liver, and they diffuse into the general circulation. The ketone bodies are used by

---

CLINICAL NOTE *Dietary Fats and Cholesterol*

Elevated cholesterol levels are associated with the development of *atherosclerosis* (Chapter 13) and *coronary artery disease (CAD)* (Chapter 12). pp. 393, 375 Current nutritional advice suggests that you limit cholesterol intake to under 300 mg per day. This amount represents a 40 percent reduction for the average American adult. As a result of rising concerns about cholesterol, such phrases as "low in cholesterol," "contains no cholesterol," and "cholesterol-free" are now widely used in the advertising and packaging of foods. Cholesterol content alone does not tell the entire story; we must consider some basic information about cholesterol and about lipid metabolism in general:

- *Cholesterol has many vital functions in the human body.* It serves as a waterproofing for the epidermis, a lipid component of all cell membranes, a key constituent of bile, and the precursor of several steroid hormones and vitamin D_3. Because cholesterol is so important, dietary restrictions should keep cholesterol levels within acceptable limits; the goal is *not* to eliminate cholesterol from the diet or from the circulating blood.
- *The cholesterol content of the diet is not the only source for circulating cholesterol.* The human body can manufacture cholesterol from the acetyl-CoA obtained through glycolysis or through the breakdown (beta oxidation) of other lipids. If the diet contains an abundance of saturated fats, cholesterol levels in the blood will rise because excess lipids are broken down to acetyl-CoA, which can be used to make cholesterol. This means that a person trying to lower serum cholesterol by dietary control must restrict other lipids—especially saturated fats—as well.
- *Genetic factors affect each individual's cholesterol level.* If you reduce your dietary supply of cholesterol, your body will synthesize more to maintain "acceptable" concentrations in the blood. The acceptable level depends on your genetic programming. Because individuals differ in genetic makeup, their cholesterol levels can vary even on similar diets. In virtually all instances, however, dietary restrictions can lower blood cholesterol significantly.
- *Cholesterol levels vary with age and physical condition.* At age 19, three out of four males have cholesterol levels below 170 mg/dl. Cholesterol levels in females of this age are slightly higher, typically at or below 175 mg/dl. As age increases, the cholesterol values gradually climb; over age 70, typical values are 230 mg/dl (males) and 250 mg/dl (females). Cholesterol levels are considered unhealthy if they are higher than those of 90 percent of the population in that age group. For males, this value ranges from 185 mg/dl at age 19 to 250 mg/dl at age 70. For females, the comparable values are 190 mg/dl and 275 mg/dl.

To determine whether you need to do anything about your cholesterol level, remember three simple rules:

1. Individuals of any age with total cholesterol values below 200 mg/dl probably do not need to change their lifestyle unless they have a family history of coronary artery disease and atherosclerosis.
2. Those with cholesterol levels between 200 and 239 mg/dl should modify their diet, lose weight (if overweight), and have annual checkups.
3. Cholesterol levels over 240 mg/dl warrant drastic changes in dietary lipid consumption, perhaps coupled with drug treatment. Drug therapies are always recommended when the serum cholesterol level exceeds 350 mg/dl. Examples of drugs used to lower cholesterol levels are *cholestyramine, colestipol*, and *lovastatin*.

Most physicians, when ordering a blood test for cholesterol, also request information about circulating triglycerides. When cholesterol levels are high, or when an individual has a family history of atherosclerosis or CAD, further tests and calculations may be performed. The HDL level is measured and the LDL is calculated from an equation relating the levels of cholesterol, HDL, and triglycerides. A high total cholesterol value linked to a high LDL spells trouble. In effect, an unusually large amount of cholesterol is being exported to peripheral tissues. Problems can also exist if the individual has high total cholesterol—or even normal total cholesterol—and low HDL levels (below 35 mg/dl). In this case, excess cholesterol delivered to the tissues cannot easily be returned to the liver for excretion. In either event, the amount of cholesterol in peripheral tissues, and especially in arterial walls, is likely to increase. For years, LDL:HDL ratios were used to predict the risk of developing atherosclerosis. Risk-factor analysis and LDL levels are now thought to be more-accurate indicators. For males with more than one risk factor, many clinicians recommend dietary and drug therapy if LDL levels exceed 130 mg/dl, regardless of the total cholesterol or HDL levels.

other cells, which reconvert them into acetyl-CoA for breakdown in the TCA cycle and the production of ATP. The increased production of ketone bodies that occurs during protein and lipid catabolism by the liver results in high ketone body concentrations in body fluids, a condition called **ketosis** (kē-TŌ-sis).

Several factors make protein catabolism an impractical source of quick energy:

- Proteins are more difficult to break apart than are complex carbohydrates or lipids.

- One of the byproducts, ammonia, is a toxin that can damage cells.

- Proteins form the most important structural and functional components of any cell. Extensive protein catabolism therefore threatens homeostasis at the cellular and systems levels.

Amino Acids and Protein Synthesis

The basic mechanism of protein synthesis was detailed in Chapter 3. ⌾ pp. 71–74 The human body can synthesize roughly half of the different amino acids needed to build proteins. There are ten *essential amino acids*. Eight of them (*isoleucine, leucine, lysine, threonine, tryptophan, phenylalanine, valine,* and *methionine*) cannot be synthesized; the other two (*arginine* and *histidine*) can be synthesized, but in amounts that are insufficient for growing children. The other amino acids, which can be synthesized on demand, are called the *nonessential amino acids.*

Protein deficiency diseases develop when an individual does not consume adequate amounts of all essential amino acids. All amino acids must be available if protein synthesis is to occur. Every transfer RNA molecule must appear at the active ribosome at the proper time bearing its individual amino acid. If that does not happen, the entire process comes to a halt. Regardless of its energy content, if the diet is deficient in essential amino acids, the individual will be malnourished to some degree. Examples of protein deficiency diseases include *marasmus* and *kwashiorkor.* Although over 100 million children worldwide have symptoms of these disorders, neither condition is common in the United States today.

Several inherited metabolic disorders result from an inability to produce specific enzymes involved with amino acid metabolism. Individuals with **phenylketonuria** (fen-il-kē-tō-NOO-rē-a), or **PKU**, cannot convert the amino acid phenylalanine to the amino acid tyrosine, because of a defect in the enzyme phenylalanine hydroxylase. This reaction is an essential step in the synthesis of norepinephrine, epinephrine, and melanin. If PKU is not detected in infancy, central nervous system development is inhibited and severe brain damage results.

 KETOACIDOSIS

A ketone body is also called a *keto acid*, because it dissociates in solution, releasing a hydrogen ion. As a result, the appearance of ketone bodies represents a threat to the pH of blood plasma that must be controlled. During prolonged starvation, ketone levels continue to rise. Eventually, the pH buffering capacities of the blood are exceeded, and a dangerous drop in pH occurs. This acidification of the blood is called **ketoacidosis** (kē-tō-as-i-DŌ-sis). In severe ketoacidosis, the pH may fall below 7.05, low enough to disrupt normal tissue activities and cause coma, cardiac arrhythmias, and death.

In *diabetes mellitus*, most peripheral tissues cannot utilize glucose because of a lack of insulin. ⌾ p. 330 Under these circumstances, the cells survive by catabolizing lipids and proteins. The result is the production of large numbers of ketone bodies. This condition leads to *diabetic ketoacidosis*, the most common and life-threatening form of ketoacidosis.

NUCLEIC ACID METABOLISM

Living cells contain both DNA and RNA. The genetic information contained in the DNA of the nucleus is absolutely essential to the long-term survival of a cell. As a result, the DNA in the nucleus is never catabolized for energy, even if the cell is dying of starvation. But the RNA in the cell is involved in protein synthesis, and RNA molecules are broken down and replaced regularly.

RNA Catabolism

In the breakdown of RNA, the molecule is disassembled into individual nucleotides. Most nucleotides are recycled into new nucleic acids. However, they can be broken down to simple sugars and nitrogen bases. When the nucleotides are broken down, only the sugars, cytosine, and uracil can enter the TCA cycle and be used to generate ATP. Adenine and guanine cannot be catabolized. Instead, they undergo deamination and are excreted as **uric acid**. Like urea, uric acid is a relatively nontoxic waste product, but it is far less soluble than urea. Urea and uric acid are called *nitrogenous wastes*, because they contain nitrogen atoms.

An elevated level of uric acid in the blood is called *hyperuricemia* (hī-per-ū-ri-SĒ-mē-uh). Uric acid saturates the body fluids and, although symptoms may not appear at once, uric acid crystals may begin to form. The condition that then develops is called *gout*. Most cases of hyperuricemia and gout are linked to problems with the excretion of uric acid by the kidneys.

Nucleic Acid Synthesis

Most cells synthesize RNA, but DNA synthesis occurs only in cells that are preparing for mitosis (cell division) or meiosis

17 NUTRITION AND METABOLISM

Cellular Metabolism • **Diet and Nutrition** • Bioenergetics • Aging and Nutritional Requirements • Chapter Review • MediaLab

(gamete production). The process of DNA replication was described in Chapter 3. ∞ p. 75 Messenger RNA (mRNA), transfer RNA (tRNA), and ribosomal RNA (rRNA) are transcribed by different forms of the enzyme RNA polymerase. Messenger RNA is manufactured as needed, when specific genes are activated. A strand of mRNA has a life span measured in minutes or hours. Ribosomal RNA and tRNA are more durable than mRNA strands. For example, the average life of a strand of rRNA is just over 5 days. However, because a typical cell contains roughly 100,000 ribosomes and many times that number of tRNA molecules, their replacement involves a considerable amount of synthetic activity.

CONCEPT CHECK QUESTIONS

Answers on page 544

❶ How would a diet deficient in vitamin B_6 affect protein metabolism?

❷ Elevated levels of uric acid in the blood could indicate that the individual has an increased metabolism of which macromolecule?

❸ Why are high-density lipoproteins (HDLs) considered to be beneficial?

A SUMMARY OF CELLULAR METABOLISM

Figure 17-9● summarizes the major pathways of cellular metabolism. Although this diagram follows the reactions in a "typical" cell, no one cell can perform all of the anabolic and catabolic operations and interconversions required by the body as a whole. As differentiation proceeds, each cell type develops its own complement of enzymes that determines its metabolic capabilities. In the presence of such cellular diversity, homeostasis can be preserved only when the metabolic activities of tissues, organs, and organ systems are coordinated.

Diet and Nutrition

Homeostasis can be maintained indefinitely only if the digestive tract absorbs fluids, organic substrates, minerals, and vitamins at a rate that keeps pace with cellular demands. The absorption of nutrients from food is called **nutrition.**

The individual requirement for each nutrient varies from day to day and from person to person. *Nutritionists* attempt to analyze a diet in terms of its ability to meet the needs of a specific individual. A **balanced diet** contains all of the ingredients

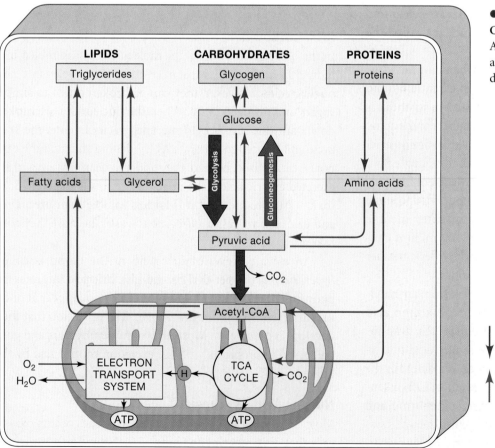

● *Figure 17-9* **A Summary of Pathways of Catabolism and Anabolism**
An overview of major catabolic (red) and anabolic (blue) pathways for lipids, carbohydrates, and proteins.

necessary to maintain homeostasis, including adequate substrates for energy generation, essential amino acids and fatty acids, minerals, and vitamins. In addition, the diet must include enough water to replace losses in urine, feces, and evaporation. A balanced diet prevents **malnutrition**, an unhealthy state resulting from the inadequate or excessive intake of one or more nutrients.

THE BASIC FOOD GROUPS

One method of avoiding malnutrition is to include members of each of the six **basic food groups** in the diet. These are (1) the *milk, yogurt, and cheese group;* (2) the *meat, poultry, fish, dry beans, eggs, and nuts group;* (3) the *vegetable group;* (4) the *fruit group;* (5) the *bread, cereal, rice, and pasta group;* and (6) the *fats, oils, and sweets group.* Each group differs from the others in the typical balance of proteins, carbohydrates, and lipids contained, as well as in the amount and identity of vitamins and minerals. The six groups are arranged in a *food pyramid,* which has the bread, cereal, rice, and pasta group at the bottom (Figure 17-10●). The aim of this arrangement is to emphasize the need to restrict dietary fats, oils, and sugar and to increase the consumption of breads, cereals, rice, and pasta, which are rich in complex carbohydrates (polysaccharides such as starch).

Such groupings can help in planning a balanced diet. However, the wrong selections can lead to malnutrition even if all six groups are represented. What is important is to obtain

● *Figure 17-10* **The Food Pyramid**

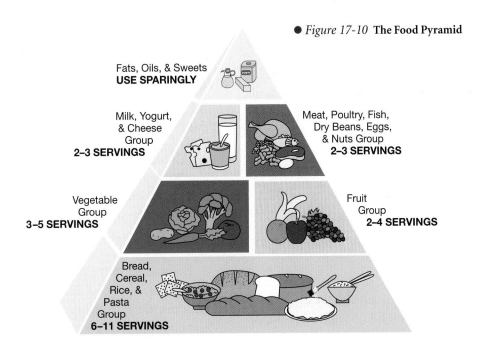

A Guide to Daily Food Choices

NUTRIENT GROUP	PROVIDES	DEFICIENCIES
Fats, oils, sweets	Calories	The majority are deficient in most minerals and vitamins
Milk, yogurt, cheese	Complete proteins; fats; carbohydrates; calcium; potassium; magnesium; sodium; phosphorus; vitamins A, B_{12}, pantothenic acid, thiamine, riboflavin	Dietary fiber, vitamin C
Meat, poultry, fish, dry beans, eggs, nuts	Complete proteins; fats; potassium; phosphorus; iron; zinc; vitamins E, thiamine, B_6	Carbohydrates, dietary fiber, several vitamins
Fruits	Carbohydrates; vitamins A, C, E, folacin; dietary fiber; potassium	Many are low in fats, calories, and protein
Vegetables	Carbohydrates; vitamins A, C, E, folacin; dietary fiber; potassium	Many are low in fats, calories, and protein
Bread, cereal, rice, pasta	Carbohydrates; vitamins E, thiamine, niacin, folacin; calcium; phosphorus; iron; sodium; dietary fiber	Fats

17 NUTRITION AND METABOLISM

Cellular Metabolism • **Diet and Nutrition** • Bioenergetics • Aging and Nutritional Requirements • Chapter Review • MediaLab

nutrients in sufficient *quantity* (adequate to meet energy needs) and *quality* (including essential amino acids, fatty acids, vitamins, and minerals). There is nothing magical about the number six—since 1940, the U.S. government has at various times advocated 11, 7, 4, and 6 food groups. The key is to make intelligent choices about what you eat.

For example, consider the essential amino acids. Your liver cannot synthesize any of these amino acids, and you must obtain them from your diet. Some members of the meat and milk groups, such as beef, fish, poultry, eggs, and milk, contain all of the essential amino acids in sufficient quantities. They are said to contain **complete proteins**. Many plants contain adequate *amounts* of protein, but these are **incomplete proteins**, which are deficient in one or more of the essential amino acids. Vegetarians who restrict themselves to the fruit and vegetable groups (with or without the bread group) must be adept at juggling the constituents of their meals to include a combination of ingredients that will meet all of their amino acid requirements. Even with a proper balance of amino acids, vegetarians face a significant problem, since vitamin B_{12} is obtained only from animal products or from fortified cereals or tofu.

MINERALS, VITAMINS, AND WATER

Minerals, vitamins, and water are essential components of the diet. The body cannot synthesize minerals, and our cells can generate only a small quantity of water and very few vitamins.

Minerals

Minerals are inorganic ions released through the dissociation of electrolytes, such as sodium chloride. Minerals are important for the following reasons:

1. *Ions such as sodium and chloride determine the osmotic concentration of body fluids.* Potassium is important in maintaining the osmotic concentration inside body cells.

2. *Ions in various combinations play major roles in important physiological processes,* as we have discussed. These processes include the maintenance of membrane potentials, the construction and maintenance of the skeleton, muscle contraction, action potential generation, neurotransmitter release, blood clotting, the transport of respiratory gases, buffer systems, fluid absorption, and waste removal, to name but a few examples encountered in earlier chapters.

3. *Ions are essential cofactors in a variety of enzymatic reactions.* For example, the enzyme that breaks down ATP in a contracting skeletal muscle requires the presence of calcium and magnesium ions, and an enzyme required for the conversion of glucose to pyruvic acid needs both potassium and magnesium ions.

The major minerals and a summary of their functions are presented in Table 17-1. Your body contains significant reserves of several important minerals that help reduce the effects of variations in dietary supply. The reserves are often relatively small, however, and chronic dietary reductions can lead to various clinical problems. Alternatively, because storage capabilities are limited, a dietary excess of mineral ions can prove equally dangerous.

Vitamins

Vitamins (*vita*, life) are essential organic nutrients related to lipids and carbohydrates. They can be assigned to either of two groups, depending on their chemical structure and characteristics: fat-soluble vitamins and water-soluble vitamins.

Fat-Soluble Vitamins. Vitamins A, E, and K are **fat-soluble vitamins**. These vitamins are absorbed primarily from the digestive tract along with the lipid contents of micelles. The term *vitamin D* refers to a group of steroids, including vitamin D_3, or *cholecalciferol*. Unlike the other fat-soluble vitamins, which must be obtained by absorption across the digestive tract, vitamin D_3 can usually be synthesized in adequate amounts by the skin when exposed to sunlight. Current information concerning the fat-soluble vitamins is summarized in Table 17-2.

Because they dissolve in lipids, fat-soluble vitamins normally diffuse into cell membranes and other lipids in the body, including the lipid inclusions in the liver and adipose tissue. Your body therefore contains a significant reserve of these vitamins, and normal metabolic operations can continue for several months after dietary sources have been cut off. For this reason, symptoms of **avitaminosis** (ā-vī-ta-min-Ō-sis), or *vitamin deficiency disease*, rarely result from dietary insufficiency of fat-soluble vitamins. However, avitaminosis involving either fat-soluble or water-soluble vitamins can be caused by various factors other than dietary deficiencies. An inability to absorb a vitamin from the digestive tract, inadequate storage, or excessive demand may each produce the same result.

As Table 17-2 points out, *too much* of a vitamin can produce effects just as unpleasant as *too little*. **Hypervitaminosis** (hī-per-vī-ta-min-Ō-sis) occurs when the dietary intake exceeds the ability to store, utilize, or excrete a particular vitamin. This condition most often involves one of the fat-soluble vitamins because the excess is retained and stored in body lipids.

TABLE 17-1 *Minerals and Mineral Reserves*

MINERAL	SIGNIFICANCE	TOTAL BODY CONTENT	PRIMARY ROUTE OF EXCRETION	RECOMMENDED DAILY INTAKE
BULK MINERALS				
Sodium	Major cation in body fluids; essential for normal membrane function	110 g, primarily in body fluids	Urine, sweat, feces	0.5–1.0 g
Potassium	Major cation in cytoplasm; essential for normal membrane function	140 g, primarily in cytoplasm	Urine	1.9–5.6 g
Chloride	Major anion in body fluids	89 g, primarily in body fluids	Urine, sweat	0.7–1.4 g
Calcium	Essential for normal muscle and neuron function, and normal bone structure	1.36 kg, primarily in skeleton	Urine, feces	0.8–1.2 g
Phosphorus	As phosphate in high-energy compounds, nucleic acids, and bone matrix	744 g, primarily in skeleton	Urine, feces	0.8–1.2 g
Magnesium	Cofactor of enzymes, required for normal membrane functions	29 g (skeleton, 17 g; cytoplasm and body fluids, 12 g)	Urine	0.3–0.4 g
TRACE MINERALS				
Iron	Component of hemoglobin, myoglobin, cytochromes	3.9 g, 1.6 g stored (ferritin or hemosiderin)	Urine (traces)	10–18 mg
Zinc	Cofactor of enzyme systems, notably carbonic anhydrase	2 g	Urine, hair (traces)	15 mg
Copper	Required as cofactor for hemoglobin synthesis	127 mg	Urine, feces (traces)	2–3 mg
Manganese	Cofactor for some enzymes	11 mg	Feces, urine (traces)	2.5–5 mg

TABLE 17-2 *The Fat-Soluble Vitamins*

VITAMIN	SIGNIFICANCE	SOURCES	DAILY REQUIREMENT	EFFECTS OF DEFICIENCY	EFFECTS OF EXCESS
A	Maintains epithelia; required for synthesis of visual pigments	Leafy green and yellow vegetables	1 mg	Retarded growth, night blindness, deterioration of epithelial membranes	Liver damage, skin peeling, CNS effects (nausea, anorexia)
D (steroids including cholecalciferol, or D₃)	Required for normal bone growth, calcium and phosphorus absorption at gut and retention at kidneys	Synthesized in skin exposed to sunlight	None*	Rickets, skeletal deterioration	Calcium deposits in many tissues, disrupting functions
E (tocopherols)	Prevents breakdown of vitamin A and fatty acids	Meat, milk, vegetables	12 mg	Anemia, other problems suspected	None reported
K	Essential for liver synthesis of prothrombin and other clotting factors	Vegetables; production by intestinal bacteria	0.07–0.14 mg	Bleeding disorders	Liver dysfunction, jaundice

*Unless sunlight exposure is inadequate for extended periods and alternative sources (fortified milk products) are unavailable.

17 NUTRITION AND METABOLISM

Cellular Metabolism • Diet and Nutrition • **Bioenergetics** • Aging and Nutritional Requirements • Chapter Review • MediaLab

Water-Soluble Vitamins. Most of the water-soluble vitamins are components of coenzymes (Table 17-3). For example, NAD is derived from niacin and coenzyme A from vitamin B$_5$ (pantothenic acid). Water-soluble vitamins are rapidly exchanged between the digestive tract and the circulating blood, and excessive amounts are readily excreted in the urine. For this reason, hypervitaminosis involving water-soluble vitamins is relatively uncommon except among individuals taking large doses of vitamin supplements.

The bacteria that reside in our intestines help prevent deficiency diseases by producing five of the nine water-soluble vitamins, in addition to fat-soluble vitamin K. Your intestinal epithelium can easily absorb all of the water-soluble vitamins except B$_{12}$. The B$_{12}$ molecule is large, and it must be bound to the intrinsic factor secreted by the gastric mucosa before absorption can occur. → p. 497

Water

Daily water requirements average 2500 ml (10 cups), or roughly 40 ml/kg body weight. The specific requirement varies with environmental and metabolic activities. For example, exercise increases metabolic energy requirements and accelerates water losses due to evaporation and perspiration. The temperature rise accompanying a fever has a similar effect; for each degree

TABLE 17-3	*The Water-Soluble Vitamins*				
VITAMIN	SIGNIFICANCE	SOURCES	DAILY REQUIREMENT	EFFECTS OF DEFICIENCY	EFFECTS OF EXCESS
B$_1$ (thiamine)	Coenzyme in decarboxylation reactions (removal of a carbon dioxide molecule)	Milk, meat, bread	1.9 mg	Muscle weakness, CNS and cardiovascular problems including heart disease; called *beriberi*	Hypotension
B$_2$ (riboflavin)	Part of FAD	Milk, meat	1.5 mg	Epithelial and mucosal deterioration	Itching, tingling
Niacin (nicotinic acid)	Part of NAD	Meat, bread, potatoes	14.6 mg	CNS, GI, epithelial, and mucosal deterioration; called *pellagra*	Itching, burning; vasodilation, death after large dose
B$_5$ (pantothenic acid)	Part of acetyl-CoA	Milk, meat	4.7 mg	Retarded growth, CNS disturbances	None reported
B$_6$ (pyridoxine)	Coenzyme in amino acid and lipid metabolism	Meat	1.42 mg	Retarded growth, anemia, convulsions, epithelial changes	CNS alterations, perhaps fatal
Folacin (folic acid)	Coenzyme in amino acid and nucleic acid metabolisms	Vegetables, cereal, bread	0.1 mg	Retarded growth, anemia, gastrointestinal disorders, developmental abnormalities	Few noted except at massive doses
B$_{12}$ (cobalamin)	Coenzyme in nucleic acid metabolism	Milk, meat	4.5 μg	Impaired RBC production causing *pernicious anemia*	Polycythemia (elevated hematocrit)
Biotin	Coenzyme in decarboxylation reactions	Eggs, meat, vegetables	0.1–0.2 mg	Fatigue, muscular pain, nausea, dermatitis	None reported
C (ascorbic acid)	Coenzyme; delivers hydrogen ions, antioxidant	Citrus fruits	60 mg	Epithelial and mucosal deterioration; called *scurvy*	Kidney stones

(°C) the temperature rises above normal, the daily water loss increases by 200 ml. Thus, the advice "drink plenty of fluids" when you are sick has a solid physiological basis.

Most of your daily water ration is obtained by eating or drinking. The food you consume provides roughly 48 percent, and another 40 percent is obtained by drinking fluids. But a small amount of water—called *metabolic water*—is produced in the mitochondria during the operation of the electron transport system. p. 526 The actual amount produced per day varies with the composition of the diet. A typical mixed diet in the United States contains 46 percent carbohydrates, 40 percent lipids, and 14 percent protein. This diet would produce roughly 300 ml of water per day (slightly more than 1 cup), about 12 percent of the average daily water requirement.

DIET AND DISEASE

Diet has a profound influence on general health. We have already considered the effects of too many and too few nutrients, above-normal or below-normal concentrations of minerals, and hypervitaminosis and avitaminosis. More subtle, long-term problems can occur when the diet includes the wrong proportions or combinations of nutrients. The average diet in the United States contains too many calories, and lipids provide too great a proportion of those calories. This diet increases the incidence of obesity, heart disease, atherosclerosis, hypertension, and diabetes in the U.S. population.

CONCEPT CHECK QUESTIONS
Answers on page 544

❶ In terms of servings per day, which of the six food groups is most important?

❷ What is the difference between foods described as complete proteins and those described as incomplete proteins?

❸ How would a decrease in the amount of bile salts in the bile affect the amount of vitamin A in the body?

Bioenergetics

The study of **bioenergetics** examines the acquisition and use of energy by organisms. When chemical bonds are broken, energy is released. Inside cells, some of that energy may be captured as ATP, but much of it is lost to the environment as heat. The unit of energy measurement is the **calorie (c)** (KAL-o-rē), the amount of energy required to raise the temperature of 1 g of water one degree centigrade. One gram of water is not a very practical measure when you are interested in the metabolic operations that keep a 70-kg human alive, however, so the **kilocalorie (kc)** (KIL-o-kal-o-rē), or **Calorie** (with a cap-

ital C), is used instead. One Calorie is the amount of energy needed to raise the temperature of 1 *kilo*gram of water one degree centigrade. Calorie-counting guides that give the caloric value of various foods indicate Calories, not calories.

FOOD AND ENERGY

In cells, organic molecules are broken down to carbon dioxide and water with oxygen. Such oxidation also occurs when something burns, and this process can be experimentally observed and measured. A known amount of food is placed in a chamber, called a **calorimeter** (kal-o-RIM-e-ter), which is filled with oxygen and surrounded by a known volume of water. Once the food is inside, the chamber is sealed and the contents are electrically ignited. When the food is completely burned and only ash remains in the chamber, the number of Calories released can be determined by comparing the water temperatures before and after the test. Such measurements show that the burning, or catabolism, of lipids releases a considerable amount of energy, roughly 9.46 Calories per gram (C/g). In contrast, the catabolism of carbohydrates releases 4.18 C/g, and the catabolism of protein releases 4.32 C/g. Most foods are mixtures of fats, proteins, and carbohydrates, and the values in a "Calorie counter" vary as a result.

METABOLIC RATE

Clinicians can examine your metabolic state and determine how many Calories you are utilizing. The result can be expressed as Calories per hour, Calories per day, or Calories per unit of body weight per day. What is actually measured is the sum of all the varied anabolic and catabolic processes occurring in your body—your **metabolic rate** at that time. The metabolic rate will change according to the activity under way; for instance, sprinting and sleeping measurements are quite different. To reduce such variations, the testing conditions are standardized so as to determine the **basal metabolic rate (BMR)**. Ideally, the BMR is the minimum, resting energy expenditures of an awake, alert person. An average individual has a BMR of 70 C per hour, or about 1680 C per day. Although the test conditions are standardized, other uncontrollable factors influence the BMR, including age, sex, physical condition, body weight, and genetic differences such as ethnic variations.

The actual daily energy expenditure for each individual varies with the activities undertaken. For example, a person leading a sedentary life may have near-basal energy demands, but one hour of swimming can increase the daily caloric requirements by 500 C or more. If the daily energy intake exceeds the total energy demands, the excess will be stored, primarily as triglycerides in adipose tissue. If the daily caloric expenditures exceed the dietary supply, there will be a net reduction in the

17 NUTRITION AND METABOLISM

Cellular Metabolism • Diet and Nutrition • Bioenergetics • **Aging and Nutritional Requirements** • Chapter Review • MediaLab

body's energy reserves and a corresponding loss in weight. This relationship accounts for the significance of Calorie counting and exercise in a weight-control program.

THERMOREGULATION

The BMR (basal metabolic rate) estimates the rate of energy use. Our cells capture only a part of that energy as ATP, and the rest is "lost" as heat. This heat loss serves an important homeostatic purpose. Humans are subject to vast changes in environmental temperatures, but our complex biochemical systems have a major limitation: Enzymes operate over only a relatively narrow range of temperatures. Our bodies have anatomical and physiological mechanisms that keep body temperatures within acceptable limits, regardless of the environmental conditions. This homeostatic process is called **thermoregulation** (*therme*, heat). ↝ p. 12 Failure to control body temperature can result in a series of physiological changes (Figure 17-11●). For example, a body temperature below 36°C (97°F) or above 40°C (104°F) can cause disorientation, and a temperature above 42°C (108°F) can cause convulsions and permanent cell damage.

Mechanisms of Heat Transfer

Heat exchange with the environment involves four basic processes—*radiation, conduction, convection,* and *evaporation*. These mechanisms are illustrated in Figure 17-12●.

1. **Radiation.** Warm objects lose heat energy as infrared radiation. When we feel the sun's heat, we are experiencing radiant heat. Your body loses heat the same way. More than half of the heat you lose occurs by radiation.

2. **Conduction.** Conduction is the direct transfer of energy through physical contact. When you sit on a cold plastic chair in an air-conditioned room, you are immediately aware of this process. Conduction is generally not an effective mechanism of gaining or losing heat.

3. **Convection.** Convection is the result of conductive heat loss to the air that overlies the surface of the body. Warm air rises because it is lighter than cool air. As your body conducts heat to the air next to your skin, that air warms and rises, moving away from your skin surface. Cooler air replaces it, and as it in turn becomes warmed, the pattern repeats.

4. **Evaporation.** When water evaporates, it changes from a liquid to a vapor. This process absorbs roughly 580 calories (0.58 C) per gram of water evaporated. The rate of evaporation occurring at your skin is highly variable. Each hour, 20–25 ml of water crosses epithelia and evaporates from the alveolar surfaces of the lungs and the surface of the skin. This *insensible perspiration* remains relatively constant; at rest it accounts for roughly one-fifth of the average heat loss. The sweat glands responsible for *sensible perspiration* have a tremendous scope of activity, ranging from virtual inactivity to secretory rates of 2–4 liters per

Condition	F°	C°	Thermoregulatory capabilities	Major physiological effects
Heat stroke	114		Severely impaired	Death
		44		Proteins denature, tissue damage accelerates
	110			Convulsions
CNS damage		42	Impaired	Cell damage
	106	40		
Disease-related fevers Severe exercise Active children	102			Disorientation
		38	Effective	Systems normal
Normal range (oral)	98	36		
Early mornings in cold weather	94	34	Impaired	Disorientation
Severe exposure	90	32		Loss of muscle control
	86	30	Severely impaired	Loss of consciousness
Hypothermia for open heart surgery	82	28		Cardiac arrest
	78	26	Lost	Skin turns blue
	74	24		Death

● *Figure 17-11* **Normal and Abnormal Variations in Body Temperature**

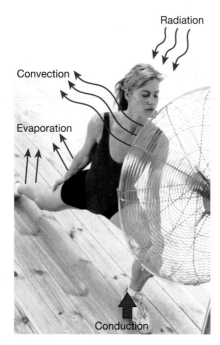

● *Figure 17-12* **Routes of Heat Gain and Loss**

hour. This is equivalent to an entire day's resting water loss in under an hour.

To maintain a constant body temperature, the individual must lose heat as fast as it is generated by metabolic operations. Altering the rates of heat loss and heat gain requires the coordinated activity of many different systems. That activity is coordinated by the **heat-loss center** and **heat-gain center** of the hypothalamus. The heat-loss center adjusts activity in the parasympathetic division of the autonomic nervous system, and the heat-gain center directs its responses through the sympathetic division. The overall effect is to control temperature by influencing two events: the rate of heat production and the rate of heat loss to the environment. These events may be further supported by behavioral changes or modifications, such as the addition or removal of clothing.

Promoting Heat Loss. When the temperature at the heat-loss center exceeds its thermostat setting, three major results occur:

1. Peripheral blood vessels dilate, and warm blood flows to the surface of the body. The skin takes on a reddish color and rises in temperature; heat loss through radiation and convection increases.
2. Sweat glands are stimulated, and as perspiration flows across the skin, heat loss through evaporation accelerates.
3. The respiratory centers are stimulated, and the depth of respiration increases. Often the individual begins respiring through the mouth, increasing heat loss through evaporation from the lungs.

The efficiency of heat loss by evaporation varies with environmental conditions, especially the "relative humidity" of the air. If the air is saturated (100 percent humidity), it holds as much water vapor as it can at that temperature. Under these conditions, evaporation is ineffective as a cooling mechanism. This is why humid, tropical conditions can be so uncomfortable—people perspire continuously but remain warm and wet.

Promoting Heat Gain. The function of the heat-gain center of the brain is to prevent **hypothermia** (hī-pō-THER-mē-uh), or below-normal body temperature. When body temperature falls below acceptable levels, the heat-loss center is inhibited and the heat-gain center is activated. Its activation results in conserving body heat and promoting heat generation. Heat is conserved by decreasing blood flow to the dermis of the skin, thus reducing losses by radiation, convection, and conduction. The skin cools, and with blood flow restricted, it may take on

a bluish or pale coloration. In addition, blood returning from the limbs is shunted into a network of deep veins. In warm weather, blood flows in a superficial venous network. In cold weather, blood is diverted to a network of deep veins that lie beneath an insulating layer of subcutaneous fat.

In addition to conserving heat, the heat-gain center has two mechanisms for increasing the rate of heat production. In *shivering thermogenesis* (ther-mō-JEN-e-sis), muscle tone is gradually increased until stretch receptors stimulate brief, oscillatory contractions of antagonistic skeletal muscles. The shivering stimulates energy consumption by skeletal muscles and the generated heat warms the deep vessels to which the blood has been diverted. Shivering can increase the rate of heat generation by as much as 400 percent.

In *nonshivering thermogenesis*, hormones are released that increase the metabolic activity of all tissues. Epinephrine from the adrenal gland immediately increases the breakdown of glycogen and glycolysis in the liver and in skeletal muscles and the metabolic rate in most tissues. The heat-gain center also stimulates the release of thyroxine by the thyroid gland, accelerating carbohydrate use and the breakdown of all other nutrients. These effects develop gradually over a period of days to weeks.

CONCEPT CHECK QUESTIONS
Answers on page 544

❶ How would the BMR (basal metabolic rate) of a pregnant woman compare with her own BMR while she is in the nonpregnant state?

❷ Under what conditions would evaporative cooling of the body be ineffective?

❸ What effect would the vasoconstriction of peripheral blood vessels have on body temperature on a hot day?

Aging and Nutritional Requirements

Nutritional requirements do not change drastically with age. However, changes in lifestyle, eating habits, and income that often accompany aging can directly affect nutrition and health. For example, current guidelines indicate that dietary calories should come from a mixture of nutrients: Proteins should provide 11–12% of daily calorie intake; carbohydrates, 55–60%; and fats, less than 30%. These percentages do not change with age. Caloric *requirements*, however, do change. For each decade after age 50, caloric requirements decrease by 10 percent. This decrease is associated with changes in metabolic rates, body mass, activity levels, and exercise tolerance.

17 NUTRITION AND METABOLISM

Cellular Metabolism • Diet and Nutrition • Bioenergetics • Aging and Nutritional Requirements • **Chapter Review** • MediaLab

With age, several factors combine to result in an increased need for calcium. Some degree of osteoporosis is a normal consequence of aging. A sedentary lifestyle contributes to the problem. The rate of bone loss decreases if calcium levels are kept elevated. The elderly are also likely to require supplemental vitamin D_3 if they are to absorb the calcium they need. Many elderly people spend most of their time indoors and avoid the sun when outdoors. This behavior slows sun damage to their skin, which is thinner than that of younger people, but it also eliminates vitamin D_3 production by the skin. ∞ p. 113 This vitamin is converted to the hormone calcitriol, which stimulates calcium absorption by the small intestine.

Maintaining a healthy diet becomes more difficult with age as a result of changes in the senses of smell and taste and in the structure of the digestive system. With age, the number and sensitivity of olfactory and gustatory receptors decreases. ∞ p. 306 As a result, food becomes less appetizing, and less food is eaten. Making matters worse, the mucosal lining of the digestive tract becomes thinner as we age, so nutrient absorption becomes less efficient. Thus what food the elderly do eat is not utilized very efficiently. Elderly people on fixed budgets may reduce their consumption of animal protein, which is the primary source of dietary iron. The combination of small quantities plus inefficient absorption makes them prone to iron deficiency, which causes anemia.

Related Clinical Terms

antipyretic drugs: Drugs administered to control or reduce fever.

avitaminosis (ā-vī-ta-min-Ō-sis): A vitamin deficiency disease.

eating disorders: Psychological problems that result in inadequate or excessive food consumption. Examples include anorexia nervosa and bulimia.

gout: A metabolic disorder characterized by the precipitation of uric acid crystals within joint cavities.

heat exhaustion: A malfunction of the thermoregulatory system caused by excessive fluid loss in perspiration.

heat stroke: A condition in which the thermoregulatory center stops functioning and body temperature rises uncontrollably.

hyperuricemia (hī-per-ū-ri-SĒ-mē-uh): Levels of plasma uric acid above 7.4 mg/dl; can result in the condition called *gout*.

hypervitaminosis (hī-per-vī-ta-min-Ō-sis): A disorder caused by the ingestion of excessive quantities of one or more vitamins.

hypothermia (hī-pō-THER-mē-uh): Below-normal body temperature.

ketoacidosis (kē-tō-as-i-DŌ-sis): The acidification of blood due to the presence of ketone bodies.

ketonemia (kē-tō-NĒ-mē-uh): Elevated levels of ketone bodies in blood.

ketonuria (kē-tō-NOO-rē-uh): The presence of ketone bodies in urine.

ketosis (kē-TŌ-sis): Abnormally high concentration of ketone bodies in body fluids.

liposuction: The removal of adipose tissue by suction through an inserted tube.

obesity: A body weight more than 20 percent above the ideal weight for a given individual.

phenylketonuria (fen-il-kē-tō-NOO-rē-uh): An inherited metabolic disorder resulting from an inability to convert phenylalanine to tyrosine.

protein deficiency diseases: Nutritional disorders resulting from a lack of essential amino acids.

pyrexia (pī-REK-sē-uh): An elevated body temperature; a fever is a body temperature maintained at greater than 37.2°C (99°F).

CHAPTER REVIEW

Key Terms

Summary Outline

INTRODUCTION **522**

1. Cells in the body are chemical factories that break down organic molecules and their building blocks to obtain energy.

CELLULAR METABOLISM **522**

1. In general, cells will break down excess carbohydrates first, then lipids, while conserving amino acids. Only about 40 percent of the energy

released through catabolism is captured in ATP; the rest is released as heat. (*Figure 17-1*)

2. Cells synthesize new compounds (1) to perform structural maintenance and repair, (2) to support growth, (3) to produce secretions, and (4) to build nutrient reserves.

3. Cells feed small organic molecules to their mitochondria to obtain ATP to perform cellular functions. (*Figure 17-2*)

4. Most cells generate ATP and other high-energy compounds through the breakdown of carbohydrates.

5. **Glycolysis** and **aerobic metabolism** provide most of the ATP used by typical cells. In glycolysis, each molecule of glucose yields two molecules of pyruvic acid and two molecules of ATP. (*Figure 17-3*)

6. In the presence of oxygen, the pyruvic acid molecules enter the mitochondria, where they are broken down completely in the **tricarboxylic acid (TCA) cycle**. The carbon and oxygen atoms are lost as carbon dioxide, and the hydrogen atoms are passed by *coenzymes* to the *electron transport system*. (*Figure 17-4*)

7. *Cytochromes* pass electrons along the electron transport chain of the **electron transport system (ETS)** to generate ATP and water. (*Figure 17-5*)

8. For each glucose molecule completely broken down by aerobic pathways, a typical cell gains 36 ATP molecules.

9. Cells can break down other nutrients to provide molecules for the TCA cycle if supplies of glucose are limited. (*Figure 17-6*)

10. **Gluconeogenesis**, the synthesis of glucose, enables a cell to create glucose molecules from other carbohydrates, glycerol, or some amino acids. *Glycogen* is an important energy reserve when extracellular glucose is low. (*Figure 17-7*)

11. During **lipolysis** (lipid catabolism), lipids are broken down into pieces that can be converted into pyruvic acid or channeled into the TCA cycle.

12. Triglycerides, the most abundant lipids in the body, are split into glycerol and fatty acids. The glycerol enters the glycolytic pathways, and the fatty acids enter the mitochondria.

13. *Beta-oxidation* is the breakdown of fatty acid molecules into two-carbon fragments. The fragments may be used in the TCA cycle or converted to ketone bodies.

14. Lipids cannot provide large amounts of ATP in a short amount of time. However, cells can shift to lipid-based energy production when glucose reserves are limited.

15. In **lipogenesis**, the synthesis of lipids, almost any organic molecule can be used to form glycerol. **Essential fatty acids** cannot be synthesized and must be included in the diet. (*Figure 17-8*)

16. Lipids circulate as **lipoproteins** (lipid-protein complexes that contain large glycerides and cholesterol) and as **free fatty acids (FFA)** (water-soluble lipids that can diffuse easily across cell membranes).

17. If other energy sources are inadequate, mitochondria can break down amino acids. In the mitochondria, the amino group may be removed by *transamination* or *deamination*. The resulting carbon skeleton may enter the TCA cycle to generate ATP or be converted to ketone bodies.

18. Protein catabolism is impractical as a source of quick energy.

19. Roughly half of the amino acids needed to build proteins can be synthesized. There are 10 *essential amino acids* that must be acquired through the diet.

20. DNA in the nucleus is never catabolized for energy. RNA molecules are broken down and replaced regularly; usually they are recycled as new nucleic acids.

21. No one cell can perform all of the anabolic and catabolic operations necessary to support life. Homeostasis can be preserved only when metabolic activities of different tissues are coordinated. (*Figure 17-9*)

1. **Nutrition** is the absorption of nutrients from food. A *balanced diet* contains all of the ingredients necessary to maintain homeostasis; it prevents **malnutrition**.

2. The six **basic food groups** are milk, yogurt, and cheese; meat, poultry, fish, dry beans, eggs, and nuts; vegetable; fruit; bread, cereal, rice, and pasta; and fats, oils, and sweets. These are arranged in a *food pyramid* with the bread and cereal group forming the base. (*Figure 17-10*)

3. **Minerals** act as cofactors in various enzymatic reactions. They also contribute to the osmotic concentration of body fluids, and they play a role in transmembrane potentials, action potentials, neurotransmitter release, muscle contraction, skeletal construction and maintenance, gas transport, buffer systems, fluid absorption, and waste removal. (*Table 17-1*)

4. **Vitamins** are needed in very small amounts. Vitamins A, D, E, and K are **fat-soluble vitamins**; taken in excess, they can lead to **hypervitaminosis**. **Water-soluble vitamins** are not stored in the body; a lack of adequate dietary supplies can lead to **avitaminosis** (*deficiency disease*). (*Tables 17-2, 17-3*)

5. Daily water requirements average about 40 ml/kg body weight. Water is obtained from food, drink, and metabolic generation.

6. A balanced diet can improve general health. Most Americans consume too many calories, mostly in the form of lipids.

1. The energy content of food is usually expressed as **Calories** per gram (C/g). Less than half of the energy content of glucose or any other organic nutrient can be captured by our cells.

2. The catabolism of lipids releases 9.46 C/g, about twice the amount as equivalent weights of carbohydrates and proteins release.

3. The total of all the anabolic and catabolic processes in the body is the **metabolic rate** of an individual. The **basal metabolic rate (BMR)** is the rate of energy utilization at rest.

4. The homeostatic regulation of body temperature is **thermoregulation**. Heat exchange with the environment involves four processes: **radiation**, **conduction**, **convection**, and **evaporation**. (*Figures 17-11, 17-12*)

5. The hypothalamus acts as the body's thermostat, containing the **heat-loss center** and the **heat-gain center**.

17 NUTRITION AND METABOLISM

Cellular Metabolism • Diet and Nutrition • Bioenergetics • Aging and Nutritional Requirements • **Chapter Review** • MediaLab

6. Mechanisms for increasing heat loss include physiological mechanisms (superficial blood vessel dilation, increased perspiration, and respiration) and behavioral adaptations.

7. Body heat may be conserved by decreased blood flow to the dermis. Heat can be generated by *shivering thermogenesis* and *nonshivering thermogenesis.*

AGING AND NUTRITIONAL REQUIREMENTS 539

1. Caloric requirements drop by 10 percent per decade after age 50. Changes in the senses of smell and taste and in the digestive system dull appetites and decrease the efficiency of nutrient absorption from the digestive tract.

Review Questions

Level 1: Reviewing Facts and Terms

Match each item in column A with the most closely related item in column B. Use letters for answers in the spaces provided.

COLUMN A

____ 1. glucose formation
____ 2. lipid catabolism
____ 3. synthesis of lipids
____ 4. linoleic acid
____ 5. deamination
____ 6. phenylalanine
____ 7. ketoacidosis
____ 8. A, D, E, K
____ 9. B complex and vitamin C
____ 10. calorie
____ 11. uric acid
____ 12. hypothermia

COLUMN B

a. gluconeogenesis
b. essential amino acid
c. below-normal body temperature
d. unit of energy
e. fat-soluble vitamins
f. water-soluble vitamins
g. lipolysis
h. nitrogenous waste
i. essential fatty acid
j. removal of an amino group
k. decrease in pH
l. lipogenesis

13. Cells synthesize new organic components to:
    (a) perform structural maintenance and repairs
    (b) support growth
    (c) produce secretions
    (d) a, b, and c are correct

14. During the complete catabolism of one molecule of glucose, a typical cell gains:
    (a) 4 ATP
    (b) 18 ATP
    (c) 36 ATP
    (d) 144 ATP

15. The breakdown of glucose to pyruvic acid is
    (a) glycolysis
    (b) gluconeogenesis
    (c) cellular respiration
    (d) beta-oxidation

16. Glycolysis yields an immediate net gain of _____ molecules for the cell.
    (a) 1 ATP
    (b) 2 ATP
    (c) 4 ATP
    (d) 36 ATP

17. The electron transport chain yields a total of _____ molecules of ATP in the complete catabolism of one glucose molecule.
    (a) 2
    (b) 4
    (c) 32
    (d) 36

18. The synthesis of glucose from simpler molecules is called:
    (a) glycolysis
    (b) lipolysis
    (c) gluconeogenesis
    (d) beta-oxidation

19. The lipoproteins that transport excess cholesterol from peripheral tissues back to the liver for storage or excretion in the bile are the:
    (a) chylomicrons
    (b) FFA
    (c) LDLs
    (d) HDLs

20. The removal of an amino group in a reaction that generates an ammonia molecule is called:
    (a) ketoacidosis
    (b) transamination
    (c) deamination
    (d) denaturation

21. A complete protein contains:
    (a) the proper balance of amino acids
    (b) all the essential amino acids in sufficient quantities
    (c) a combination of nutrients selected from the food pyramid
    (d) N compounds produced by the body

22. All minerals and most vitamins:
    (a) are fat-soluble
    (b) cannot be stored by the body
    (c) cannot be synthesized by the body
    (d) must be synthesized by the body because they are not present in adequate amounts in the diet

23. The basal metabolic rate represents the:
    (a) maximum energy expenditure when exercising
    (b) minimum, resting energy expenditure of an awake, alert person
    (c) minimum amount of energy expenditure during light exercise
    (d) muscular energy expenditure added to the resting energy expenditure

24. Over half of the heat loss from our bodies is attributable to:
    (a) radiation
    (b) conduction
    (c) convection
    (d) evaporation

25. Define the terms metabolism, anabolism, and catabolism.

26. What is a lipoprotein? What are the major groups of lipoproteins, and how do they differ?

27. Why are vitamins and minerals essential components of the diet?

28. What energy yields in Calories per gram are associated with the catabolism of carbohydrates, lipids, and proteins?

29. What is the basal metabolic rate (BMR)?

30. What four mechanisms are involved in thermoregulation?

Level 2: Reviewing Concepts

31. The function of the TCA cycle is to:
    (a) produce energy during periods of active muscle contraction
    (b) break six-carbon chains into three-carbon fragments
    (c) prepare the glucose molecule for further reactions
    (d) remove hydrogen atoms from organic molecules and transfer them to coenzymes

32. During periods of fasting or starvation, the presence of ketone bodies in the circulation causes:
    (a) an increase in blood pH
    (b) a decrease in blood pH
    (c) a neutral blood pH
    (d) diabetes insipidus

33. What happens during the process of glycolysis? What conditions are necessary for this process to take place?

34. Why is the TCA cycle called a cycle? What substance(s) enter(s) the cycle, and what substance(s) leave(s) it?

35. How are lipids catabolized in the body? How is beta-oxidation involved with lipid catabolism?

36. How can the food pyramid be used as a tool to obtain nutrients in sufficient quantity and quality? Why are the dietary fats, oils, and sugars at the top of the pyramid and the breads, cereals, rices, and pastas at the bottom?

37. How is the brain involved in the regulation of body temperature?

38. Articles in popular magazines sometimes refer to "good cholesterol" and "bad cholesterol." To what types and functions of cholesterol might these terms refer? Explain your answer.

Level 3: Critical Thinking and Clinical Applications

39. Why is an individual who is starving more susceptible to infectious disease than an individual who is well nourished?

40. The drug colestipol™ binds bile salts in the intestine, forming complexes that cannot be absorbed. How would this drug affect cholesterol levels in the blood?

17 NUTRITION AND METABOLISM

Cellular Metabolism • Diet and Nutrition • Bioenergetics • Aging and Nutritional Requirements • Chapter Review • **MediaLab**

Answers to Concept Check Questions

Page 528

1. The primary role of the TCA cycle in ATP production is to transfer electrons from organic substrates to coenzymes. These electrons carry energy that can then be used as an energy source for the production of ATP by the electron transport system.

2. The binding of hydrogen cyanide molecules would prevent the transfer of electrons to oxygen. As a result, cells would be unable to produce ATP in their mitochondria and would die from energy starvation.

Page 532

1. Vitamin B_6 (pyridoxine) is an important coenzyme in the processes of deamination and transamination, the first step in processing amino acids in the cell. A deficiency in this vitamin would interfere with the ability to metabolize proteins.

2. Uric acid is the product of the degradation of the nucleotides adenine and guanine in the body. The macromolecules that contain adenine and guanine are the nucleic acids. An increase in uric acid levels could indicate increased breakdown of nucleic acids.

3. HDLs are considered to be beneficial because they reduce the amount of cholesterol in the bloodstream by transporting it to the liver for storage or for excretion in the bile.

Page 537

1. In terms of servings per day, the bread, cereal, rice, and pasta group is the most important, with between 6 and 11 per day.

2. Foods that contain all of the essential amino acids in nutritionally required amounts are said to contain complete proteins. Foods that are deficient in one or more of the essential amino acids contain incomplete proteins.

3. Bile salts are necessary for the digestion and absorption of fats and fat-soluble vitamins. Vitamin A is a fat-soluble vitamin. A decrease in the amount of bile salts in the bile would result in a decreased ability to absorb the vitamin A from food and would thus lead to a vitamin A deficiency.

Page 539

1. The BMR of a pregnant woman should be higher than the BMR of the woman in a nonpregnant state because of increased metabolism associated with support of the fetus as well as the added effect of fetal metabolism.

2. Evaporation is ineffective as a cooling mechanism under conditions of high relative humidity, when the air is holding large amounts of water vapor.

3. The vasoconstriction of peripheral vessels would decrease blood flow to the skin and decrease the amount of heat that the body can lose. As a result, the body temperature would increase.

EXPLORE *MediaLab*

EXPLORATION #1

Estimated time for completion: 10 minutes

WE ALL KNOW that a healthy diet combined with adequate exercise is the best way to maintain a healthy weight. However, most people like high-Calorie convenience foods with low nutritional value. To counter bad eating habits, many people turn to modern science for help. Look carefully at the weight-control section in your local food store. There you will find everything from diet pills, weight-loss drinks, and fat-burning concoctions to synthetic sweeteners and fats. One of the more publicized synthetic fats is *Olestra™*. Olestra is a compound of sucrose and fatty acids that cannot be broken down by enzymes of the digestive tract. Draw a simple diagram of the digestive tract. Where and how are fats usually digested and absorbed? How might Olestra affect the absorption of lipid-soluble vitamins and drugs? What other effects might you predict? The FDA has a concise article summarizing information on Olestra that will provide relevant background. To get to this report, go to Chapter 17 at the Companion Web site, visit the MediaLab section, and click on the key word "Olestra."

Most of us are concerned about the way we look to others.

EXPLORATION #2

Estimated time for completion: 10 minutes

EATING DISORDERS are surprisingly common, affecting both men and women. Two familiar eating disorders are anorexia nervosa and bulimia. The causes of these disorders are complex and may include emotional and personality disorders, genetic or biologic susceptibility, and a culture in which there is an overabundance of food and an obsession with thinness. Anorexia nervosa is characterized by voluntarily restricted food intake despite progressive starvation and emaciation. Bulimia, which is more common than anorexia, involves cycles of binging and purging. This purging can take the form of self-induced vomiting or abuse of laxatives, diet pills, or diuretics. How serious are these two types of eating disorders? List five potential physiological consequences of eating disorders for each organ system of the body, three that may result from anorexia and two that may result from the binge/purge cycle of bulimia. Pattern your responses after the diagrams of functional relationships between systems that appear at the end of each of the system chapters in the text (see p. 515, for example). For an article that will help you prepare your list, go to Chapter 17 at the Companion Web site, visit the MediaLab section, and click on the key words "Eating Disorders." At this site you can read about the medical and emotional complications associated with eating disorders.

Waiting in line for the world's largest hot dog.

The Urinary System

CHAPTER OUTLINE AND OBJECTIVES

Vocabulary Development

calyx ....a cup of flowers; *minor calyx*
detrudere ......to push down; *detrusor muscle*
fenestra ......a window; *fenestrated capillaries*
glomus ....a ball; *glomerulus*
gonion ....angle; *trigone*
juxta ....near; *juxtaglomerular apparatus*
micturire ....to urinate; *micturition*
nephros ....kidney; *nephron*
papillae ....small, nipple-shaped projections; *renal papillae*
podon ....foot; *podocyte*
rectus ....straight; *vasa recta*
ren ....kidney; *renal artery*
retro- ....behind; *retroperitoneal*
vasa ....vessels; *vasa recta*

U
RINATING IN PUBLIC *is considered bad form. But this whimsical fountain in Brussels, Belgium, depicts a very natural and necessary process. After our digestive system assimilates a meal and our body harvests and uses the nutrients it needs, we must somehow discharge the resulting waste products. The urinary system performs this service while efficiently conserving water and other valuable substances. As we shall see, it also carries out a variety of other vital but less obvious functions (none of them immortalized in sculpture, so far as we know).*

18 THE URINARY SYSTEM

The Organization of the Urinary System • The Kidneys • Basic Principles of Urine Production • Urine Transport, Storage, and Elimination

THE HUMAN BODY CONTAINS trillions of cells bathed in extracellular fluid. In previous chapters we compared these cells to factories that burn nutrients to obtain energy. Imagine what would happen if real factories were built as close together as cells in the body. Each would generate piles of garbage, and the smoke they produced (while depleting the oxygen supply) would drastically reduce air quality. In short, a serious pollution problem would result.

Comparable problems do not develop within the body as long as the activities of the digestive, cardiovascular, respiratory, and urinary systems are coordinated. The digestive tract absorbs nutrients from food and excretes solid wastes, and the liver adjusts the nutrient concentration of the circulating blood. The cardiovascular system delivers these nutrients and oxygen from the respiratory system to peripheral tissues. As blood leaves these tissues, it carries the carbon dioxide and cellular waste products to sites of excretion. The carbon dioxide is eliminated at the lungs, as described in Chapter 15. Most of the organic waste products in the blood are removed by the urinary system.

The **urinary system** performs the vital function of removing the organic waste products generated by cells throughout the body (Figure 18-1●). It also has other essential functions that are often overlooked. A more complete list of its activities includes:

- *Regulating blood volume and blood pressure.* Blood volume and blood pressure are regulated by (1) adjusting the volume of water lost in the urine, (2) releasing erythropoietin, and (3) releasing renin. ⚭ p. 331
- *Regulating plasma concentrations of ions.* The plasma concentrations of sodium, potassium, chloride, and other ions are regulated by controlling the quantities lost in the urine and, for calcium ions, by the synthesis of calcitriol. ⚭ p. 331

- *Helping to stabilize blood pH.* Blood pH is stabilized by controlling the loss of hydrogen ions (H^+) and bicarbonate ions (HCO_3^-) in the urine.
- *Conserving valuable nutrients.* Valuable nutrients (such as glucose and amino acids) are conserved by preventing their excretion in the urine, while organic waste products, especially the nitrogenous wastes *urea* and *uric acid*, are eliminated.

These activities are carefully regulated to keep the composition of the blood within acceptable limits. A disruption of any one of these functions will have immediate and potentially fatal consequences. This chapter examines the organization of the urinary system and describes the major regulatory mechanisms that control urine production and concentration.

The Organization of the Urinary System

The components of the urinary system are indicated in Figure 18-1●. The two *kidneys* produce *urine*, a liquid containing water, ions, and small soluble compounds. Urine leaving the kidneys travels along the paired *ureters* to the *urinary bladder* for temporary storage. When *urination* occurs, contraction of the muscular bladder forces the urine through the *urethra* and out of the body.

The Kidneys

The **kidneys** are located on either side of the vertebral column between the last thoracic and third lumbar vertebrae. The right kidney often sits slightly lower than the left (Figure 18-2a●), and both lie between the muscles of the dorsal body wall and the peritoneal lining (Figure 18-2b●). This position is called

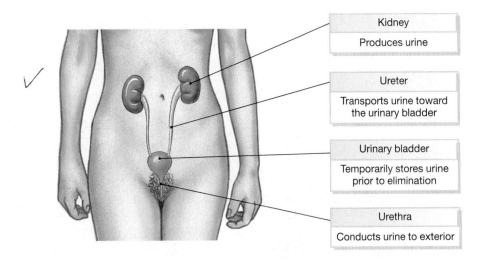

Kidney
Produces urine

Ureter
Transports urine toward the urinary bladder

Urinary bladder
Temporarily stores urine prior to elimination

Urethra
Conducts urine to exterior

● *Figure 18-1* **The Components of the Urinary System**

retroperitoneal (re-trō-per-i-tō-NĒ-al; *retro-*, behind) because the organs are behind the peritoneum.

The position of the kidneys is maintained by (1) the overlying peritoneum, (2) contact with adjacent organs, and (3) supporting connective tissues. Each kidney is covered by a dense, fibrous renal capsule and is packed in a soft cushion of adipose tissue. These connective tissues, along with suspensory collagen fibers, help prevent the jolts and shocks of day-to-day existence from disturbing normal kidney function. Damage to the suspensory fibers may cause the kidney to be displaced. This condition, called a *floating kidney,* is dangerous because the ureters or renal blood vessels may become twisted or kinked during movement.

SUPERFICIAL AND SECTIONAL ANATOMY

Each reddish-brown kidney is shaped like a kidney bean. An indentation called the **hilus** is the point of entry for the renal artery and renal nerve, and exit for the renal vein and ureter. (The adjective "renal" is derived from *ren*, which means kidney in Latin.) The *renal capsule* covers the surface of the kidney and lines the *renal sinus*, an internal cavity.

The kidney is divided into an outer renal **cortex** and an inner renal **medulla** (Figure 18-3a,b●). The medulla contains 6 to 18 conical **renal pyramids**, whose tips, or **papillae**, project into the renal sinus. Renal columns extend from the cortex inward toward the renal sinus between adjacent renal pyramids.

Urine production occurs in the renal pyramids and overlying areas of renal cortex. Ducts within each renal papilla discharge urine into a cup-shaped drain, called a **minor calyx** (KĀ-liks; *calyx*, a cup of flowers; plural *calyces*). Four or five minor calyces (KĀL-i-sēz) merge to form the **major calyces**, both of which combine to form a large, funnel-shaped chamber, the **renal pelvis**. The renal pelvis is connected to the ureter, which drains the kidney.

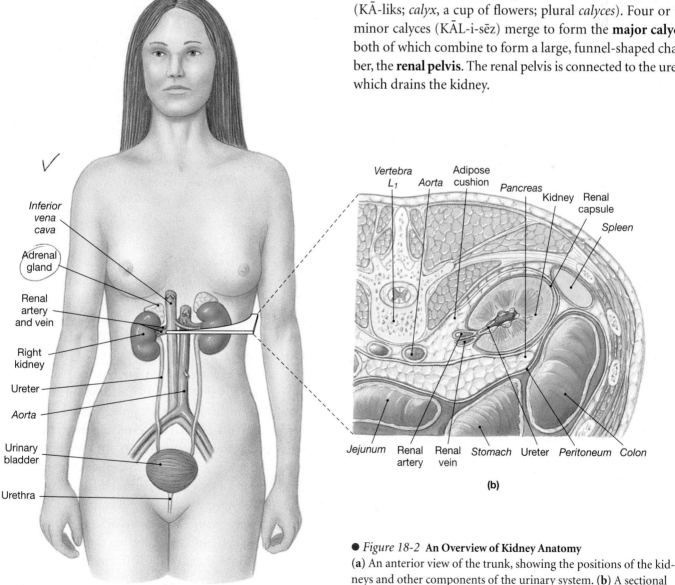

(a)

Inferior vena cava

Adrenal gland

Renal artery and vein

Right kidney

Ureter

Aorta

Urinary bladder

Urethra

Vertebra L₁ *Aorta* Adipose cushion *Pancreas* Kidney Renal capsule *Spleen*

Jejunum Renal artery Renal vein *Stomach* Ureter *Peritoneum* *Colon*

(b)

● *Figure 18-2* **An Overview of Kidney Anatomy**
(**a**) An anterior view of the trunk, showing the positions of the kidneys and other components of the urinary system. (**b**) A sectional view at the level indicated in part (a).

549

18 THE URINARY SYSTEM

The Organization of the Urinary System • **The Kidneys** • Basic Principles of Urine Production • Urine Transport, Storage, and Elimination

● *Figure 18-3* **The Structure of the Kidney**
(**a**) Diagrammatic view of a frontal section through the left kidney. (**b**) Frontal section through the left kidney. (**c**) General appearance and location of a nephron.

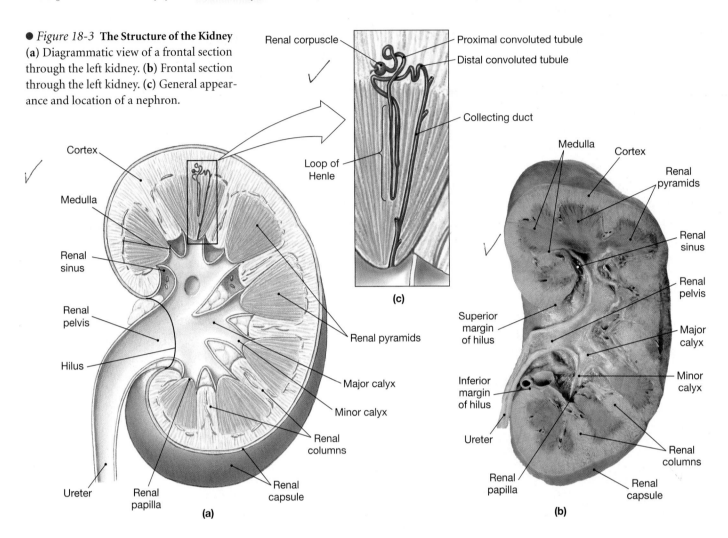

(a)

(b)

(c)

Urine production begins in microscopic, tubular structures called **nephrons** (NEF-ronz) in the renal cortex (Figure 18-3c●). Each kidney has roughly 1.25 million nephrons, with a combined length of about 145 kilometers (85 miles).

THE NEPHRON

A nephron is the basic functional unit in the kidney. Each nephron consists of a *renal tubule* that is roughly 50 mm (2 in.) long. The tubule has two *convoluted* (coiled or twisted) segments separated by a simple U-shaped tube. The convoluted segments are in the cortex, and the U-shaped tube extends partially or completely into the medulla (Figure 18-3c●).

An Overview of the Nephron

A schematic nephron is shown in Figure 18-4●. The nephron begins at a *renal corpuscle* (KOR-pus-l), a cup-shaped chamber (*Bowman's capsule*) that contains a capillary network, or *glomerulus* (glo-MER-ū-lus; *glomus*, a ball). Blood arrives at

the glomerulus by way of an *afferent arteriole* and departs in an *efferent arteriole*. In the renal corpuscle, blood pressure forces fluid and dissolved solutes out of the glomerular capillaries and into the surrounding *capsular space*. This process is called *filtration*. ⬭ p. 61 Filtration produces a protein-free solution known as a **filtrate**.

From the renal corpuscle, the filtrate enters the **renal tubule**, a long passageway that is subdivided into different regions. The major segments are the *proximal convoluted tubule*, the *loop of Henle* (HEN-lē), and the *distal convoluted tubule*. As the filtrate travels along the tubule, its composition gradually changes, and it is then called *tubular fluid*. The changes that occur and the urine that results depend on the specialized activities under way in each segment of the nephron.

Each nephron empties into a *collecting duct*, the start of the **collecting system**. The collecting duct leaves the cortex and descends into the medulla, carrying tubular fluid from many nephrons toward a *papillary duct* that delivers the fluid, now called *urine*, into the renal pelvis.

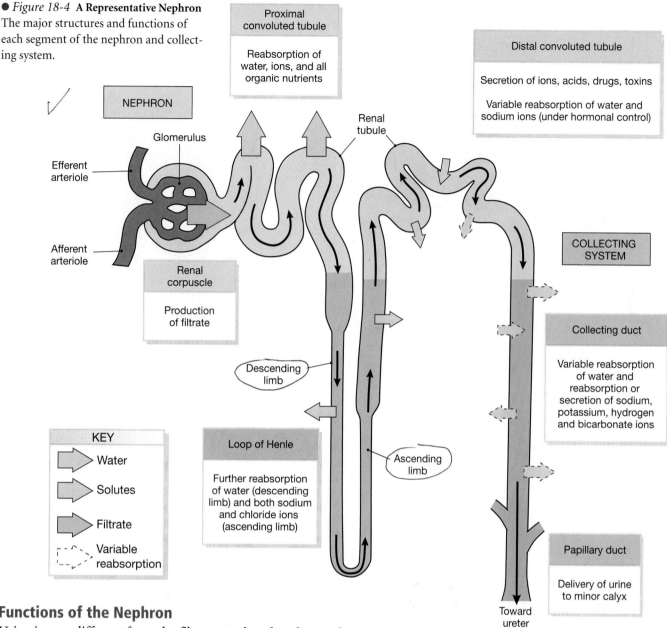

● *Figure 18-4* **A Representative Nephron**
The major structures and functions of each segment of the nephron and collecting system.

Functions of the Nephron

Urine is very different from the filtrate produced at the renal corpuscle. The role of each segment of the nephron in the conversion of filtrate to urine is indicated in Figure 18-4●. The renal corpuscle is the site of filtration. The functional advantage of filtration is that it is passive and does not require an expenditure of energy. The disadvantage of filtration is that any filter with pores large enough to permit the passage of organic waste products is unable to *prevent* the passage of water, ions, and nutrients such as glucose, fatty acids, and amino acids. These substances, along with most of the water, must be reclaimed.

Filtrate enters the renal tubule from the renal corpuscle. The renal tubule is responsible for:

- Reabsorbing all of the useful organic molecules from the filtrate.

- Reabsorbing over 90 percent of the water in the filtrate.
- Secreting into the tubular fluid any waste products that were missed by the filtration process.

Additional water and salts will be removed in the collecting system before the urine is released into the renal sinus. Table 18-1 summarizes the functions of the different regions of the nephron and collecting system.

The Renal Corpuscle

The **renal corpuscle** consists of (1) the capillary knot of the **glomerulus** and (2) the expanded initial segment of the renal tubule, a region known as **Bowman's capsule** (Figure 18-5●).

18 THE URINARY SYSTEM

The Organization of the Urinary System • **The Kidneys** • Basic Principles of Urine Production • Urine Transport, Storage, and Elimination

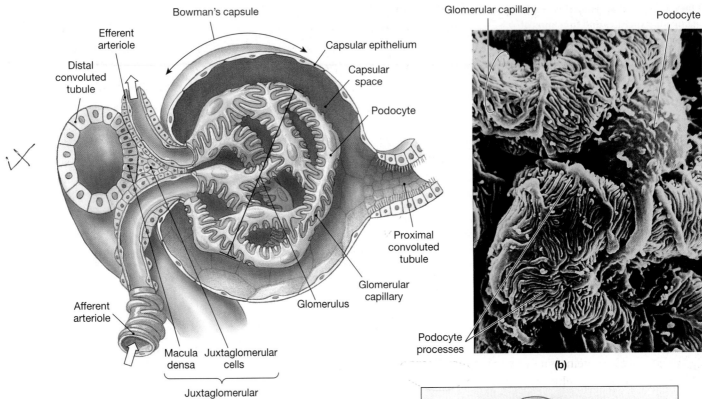

(a)

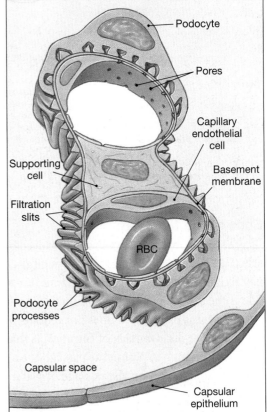

(b)

(c)

● *Figure 18-5* **The Renal Corpuscle**
(a) The renal corpuscle, showing important structural features.
(b) The glomerular surface, showing individual podocytes and their
processes. (SEM × 27,248) (c) A section of a glomerulus, showing the
components of the filtration membrane (capillary endothelium, basement membrane, and filtration slits).

TABLE 18-1	*The Functions of the Nephron and Collecting System in the Kidney*	
REGION	**PRIMARY FUNCTION**
Renal corpuscle	Filtration of plasma to initiate urine formation
Proximal convoluted tubule (PCT)	Reabsorption of ions, organic molecules, vitamins, water
Loop of Henle	Descending limb: reabsorption of water from tubular fluid
	Ascending limb: reabsorption of ions; helps create the medullary concentration gradient
Distal convoluted tubule (DCT)	Reabsorption of sodium ions; secretion of acids, ammonia, drugs
Collecting duct	Reabsorption of water and of sodium and bicarbonate ions
Papillary duct	Conduction of urine to minor calyx

The glomerulus projects into Bowman's capsule much as the heart projects into the pericardial cavity (Figure 18-5a●). A *capsular epithelium* lines the wall of the capsule and a *glomerular epithelium* covers the glomerular capillaries. The two are separated by the **capsular space**, which receives the filtrate and empties into the renal tubule. The glomerular epithelium consists of cells called **podocytes** (PŌ-do-sīts, *podon*, foot). Podocytes have long cellular processes called *pedicels* that wrap around individual capillaries. A thick basement membrane separates the endothelial cells of the capillaries from the podocytes. The glomerular capillaries are said to be *fenestrated* (FEN-e-strā-ted; *fenestra*, a window) because their endothelial cells contain pores (Figure 18-5c●). To enter the capsular space, a solute must be small enough to pass through (1) the pores of the endothelial cells, (2) the fibers of the basement membrane, and (3) the *filtration slits* between the slender processes of the podocytes (Figure 18-5b,c●). The fenestrated capillary, basement membrane, and filtration slits create a **filtration membrane** that prevents the passage of blood cells and most plasma proteins but permits the movement of water, metabolic wastes, ions, glucose, fatty acids, amino acids, vitamins, and other solutes into the capsular space. Most of the valuable solutes will be reabsorbed by the proximal convoluted tubule.

The Proximal Convoluted Tubule

The filtrate next moves into the **proximal convoluted tubule (PCT)** (Figures 18-4●, 18-5a●). The cells lining the PCT actively absorb organic nutrients, plasma proteins, and ions from the tubular fluid. These materials are then released into the interstitial fluid surrounding the renal tubule. As a result of this transport, the solute concentration of the interstitial fluid increases while that of the tubular fluid decreases. Water then moves out of the tubular fluid by osmosis, reducing the volume of tubular fluid.

The Loop of Henle

The last portion of the proximal convoluted tubule bends sharply toward the renal medulla and connects to the **loop of Henle** (Figure 18-4●, p. 551). This loop is composed of a *descending limb* that travels toward the renal pelvis and an *ascending limb* that returns to the cortex. The ascending limb, which is impermeable to water and solutes, actively transports sodium and chloride ions out of the tubular fluid. As a result, the interstitial fluid of the medulla contains an unusually high solute concentration. The descending limb is permeable to water, and as it descends into the medulla, water moves out of the tubular fluid by osmosis.

The Distal Convoluted Tubule

The ascending limb of the loop of Henle ends where it bends and comes in close contact with the glomerulus and its vessels. At this point, the **distal convoluted tubule (DCT)** begins, and it passes between the afferent and efferent arterioles (Figure 18-5a●).

The distal convoluted tubule is an important site for (1) the active secretion of ions, acids, and other materials; and (2) the selective reabsorption of sodium ions from the tubular fluid. In the final portions of the DCT, an osmotic flow of water may assist in concentrating the tubular fluid.

The cells of the DCT closest to the glomerulus are unusually tall, and their nuclei are clustered together. This region, diagrammed in Figure 18-5a●, is called the *macula densa* (MAK-ū-la DEN-sa). The cells of the macula densa are closely associated with unusual smooth muscle fibers, the *juxtaglomerular* (*juxta*, near) *cells* in the wall of the afferent arteriole. Together the macula densa and juxtaglomerular cells form the **juxtaglomerular apparatus,** an endocrine structure that secretes the hormone erythropoietin and the enzyme renin, introduced in Chapter 10. p. 331

The Collecting System

The distal convoluted tubule, the last segment of the nephron, opens into the collecting system. The collecting system consists of collecting ducts and papillary ducts (Figure 18-4●, p. 551). Each **collecting duct** receives tubular fluid from many nephrons, and several collecting ducts merge to form a **papillary duct**, which delivers urine to a minor calyx. In addition to transporting tubular fluid from the nephron to the renal pelvis, the collecting system can make final adjustments to the composition of the urine by reabsorbing water and reabsorbing or secreting sodium, potassium, hydrogen, and bicarbonate ions.

THE BLOOD SUPPLY TO THE KIDNEYS

In healthy individuals, about 1200 ml of blood flows through the kidneys each minute, or some 20–25 percent of the cardiac output. This is a phenomenal amount of blood for organs with a combined weight of less than 300 g (10.5 oz)! Figure 18-6a● diagrams the path of blood flow to each kidney. Each kidney receives blood from a *renal artery* that originates from the abdominal aorta. As the renal artery enters the renal sinus, it divides into branches that supply a series of **interlobar arteries** that radiate outward between the lobes. They then turn, arching along the boundary lines between the cortex and medulla as the **arcuate** (AR-kū-āt) **arteries.** Each of the arcuates gives rise to a number of **interlobular arteries** supplying portions of the adjacent lobe. **Afferent arterioles** branching from each

18 THE URINARY SYSTEM

The Organization of the Urinary System • The Kidneys • **Basic Principles of Urine Production** • Urine Transport, Storage, and Elimination

● *Figure 18-6* **The Blood Supply to the Kidneys**
(**a**) A sectional view, showing the major arteries and veins; compare with Figure 18-3. (**b**) Circulation in the cortex. (**c**) Circulation to an individual nephron.

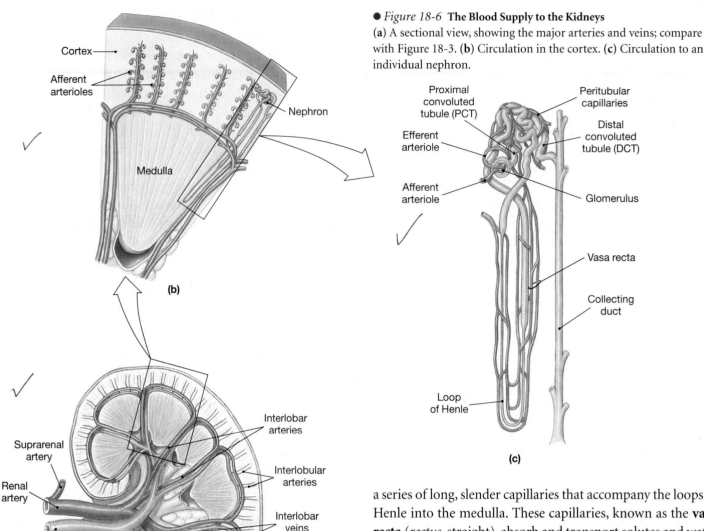

interlobular artery deliver blood to the individual nephrons (Figure 18-6b●).

Blood reaches each glomerulus through an afferent arteriole and leaves in an **efferent arteriole** (Figure 18-6c●). It then travels to the **peritubular capillaries** that surround the proximal and distal convoluted tubules. The peritubular capillaries provide a route for the pickup or delivery of substances that are reabsorbed or secreted by these portions of the nephron.

In nephrons located near the renal medulla, the efferent arterioles and peritubular capillaries are further connected to

a series of long, slender capillaries that accompany the loops of Henle into the medulla. These capillaries, known as the **vasa recta** (*rectus*, straight), absorb and transport solutes and water reabsorbed by the loops of Henle and collecting ducts. Under normal conditions, the removal of solutes and water by the vasa recta balances the rates of solute and water reabsorption in the medulla. These nephrons give the kidneys the ability to produce concentrated urine.

Blood from the peritubular capillaries and vasa recta enters a network of venules and small veins that converge on the **interlobular veins**. In a mirror image of the arterial distribution, blood continues to converge and empty into the **arcuate, interlobar**, and **renal veins** (Figure 18-6a●).

CONCEPT CHECK QUESTIONS

Answer on page 578

❶ How is the position of the kidneys different from most other organs in the abdominal region?

❷ Why don't plasma proteins pass into the capsular space under normal circumstances?

❸ Damage to which part of the nephron would interfere with the control of blood pressure?

Basic Principles of Urine Production

The primary purpose of **urine** production is to maintain homeostasis by regulating the volume and composition of the blood. This process involves the excretion and elimination of dissolved solutes, specifically the following three metabolic waste products:

1. **Urea.** This is the most abundant organic waste, and roughly 21 grams of urea is generated each day. Most of it is produced during the breakdown of amino acids.
2. **Creatinine.** Creatinine is generated in skeletal muscle tissue through the breakdown of creatine phosphate, a high-energy compound that plays an important role in muscle contraction. The body generates roughly 1.8 g of creatinine each day.
3. **Uric acid.** Approximately 480 mg of uric acid is produced each day during the breakdown and recycling of RNA.

Since these waste products must be excreted in solution, their elimination is accompanied by an unavoidable water loss. The kidneys can minimize this water loss by producing a urine that is four to five times more concentrated than normal body fluids. If the kidneys could not concentrate the filtrate produced by glomerular filtration, water losses would lead to fatal dehydration in hours. At the same time, the kidneys ensure that the urine excreted does not contain potentially useful organic substrates, such as sugars or amino acids, that are found in blood plasma.

To accomplish these goals, the kidneys rely on three distinct processes:

1. **Filtration.** In filtration, blood pressure forces water across a filtration membrane. Solute molecules small enough to pass through the membrane are carried by the surrounding water molecules.
2. **Reabsorption.** Reabsorption is the removal of water and solute molecules from the filtrate after it enters the renal tubule. This is a selective process, whereas filtration occurs solely on the basis of size. Solute reabsorption may involve simple diffusion or the activity of carrier proteins in the tubular epithelium. Water reabsorption occurs passively, through osmosis. Reabsorbed water and solutes reenter the circulation at the peritubular capillaries and vasa recta.
3. **Secretion.** Secretion is the transport of solutes across the tubular epithelium and into the filtrate. This process is necessary because filtration does not force all of the dissolved materials out of the plasma, and blood entering the peritubular capillaries may still contain undesirable substances.

TABLE 18-2 *Significant Differences Between Urinary and Plasma Solute Concentrations*

COMPONENT	URINE	PLASMA
IONS (mEq/l)		
Sodium (Na^+)	147.5	138.4
Potassium (K^+)	47.5	4.4
Chloride (Cl^-)	153.3	106
Bicarbonate (HCO_3^-)	1.9	27
METABOLITES AND NUTRIENTS (mg/dl)		
Glucose	0.009	90
Lipids	0.002	600
Amino acids	0.188	4.2
Proteins	0.000	7.5 g/dl
NITROGENOUS WASTES (mg/dl)		
Urea	1800	10–20
Creatinine	150	1–1.5
Ammonia	60	< 0.1
Uric acid	40	3

Together, these processes create a fluid that is very different from other body fluids. Table 18-2 indicates the efficiency of the renal system by comparing the composition of urine and plasma. The kidneys can continue to work efficiently only as long as filtration, reabsorption, and secretion proceed in proper balance. Any disruption in this balance has immediate and potentially disastrous effects on the composition of the circulating blood. If both kidneys fail to perform their assigned roles, death will occur within a few days unless medical assistance is provided.

All segments of the nephron and collecting system participate in the process of urine formation. Most regions perform a combination of reabsorption and secretion, but the balance between the two shifts from one region to another. As indicated in Table 18-1:

- Filtration occurs exclusively in the renal corpuscle, across the capillary walls of the glomerulus.
- Nutrient reabsorption occurs primarily at the proximal convoluted tubule.
- Active secretion occurs primarily at the distal convoluted tubule.
- The loop of Henle and the collecting system interact to regulate the amount of water and the number of sodium and potassium ions lost to the urine.

18 THE URINARY SYSTEM

The Organization of the Urinary System • The Kidneys • **Basic Principles of Urine Production** • Urine Transport, Storage, and Elimination

We will now take a closer look at events under way in each of the segments of the nephron and collecting system.

FILTRATION AT THE GLOMERULUS

Filtration Pressure

Chapter 13 introduced the forces acting across capillary walls. (You may find it helpful to review Figure 13-6● before you proceed.) ∞ p. 396 Blood pressure at the glomerulus tends to force water and solutes out of the bloodstream and into the capsular space. For filtration to occur, this outward force must exceed any opposing pressures, such as the osmotic pressure of the blood. The net force promoting filtration is called the **filtration pressure**. Filtration pressure is higher than capillary blood pressure elsewhere in the body as a result of a difference in the diameters of afferent and efferent arterioles. Because the diameter of the efferent arteriole is slightly smaller, it offers more resistance to blood flow than does the afferent arteriole. As a result, blood "backs up" in the afferent arteriole, increasing the blood pressure in the glomerular capillaries.

The filtration pressure is very low (around 10 mm Hg), and kidney filtration will stop if glomerular blood pressure falls significantly. Reflexive changes in the diameters of the afferent arterioles, the efferent arterioles, and the glomerular capillaries can compensate for minor variations in blood pressure. These changes can occur automatically or in response to sympathetic stimulation. More serious declines in systemic blood pressure can reduce or even stop glomerular filtration. As a result, hemorrhaging, shock, or dehydration can cause a dangerous or even fatal reduction in kidney function. Because the kidneys are more sensitive to blood pressure than are other organs, it is not surprising to find that they control many of the homeostatic mechanisms responsible for regulating blood pressure and blood volume. Examples such as the renin-angiotensin system are considered later in this chapter.

The Glomerular Filtration Rate

Glomerular filtration is the process of filtrate production at the glomerulus. The **glomerular filtration rate (GFR)** is the amount of filtrate produced in the kidneys each minute. Each kidney contains around 6 square meters of filtration surface, and the GFR averages an astounding *125 ml per minute*. This means that almost 20 percent of the fluid delivered to the kidneys by the renal arteries leaves the bloodstream and enters the capsular spaces. In the course of a single day, the glomeruli generate about 180 liters (50 gal) of filtrate, roughly 70 times the total plasma volume. But as the filtrate passes through the renal tubules, over 99 percent of it is reabsorbed. Tubular reabsorption is obviously an extremely important process. An inability to reclaim the water entering the filtrate, as in *diabetes*

insipidus, can quickly cause death by dehydration. (This condition, caused by inadequate ADH secretion, was discussed in Chapter 10.) ∞ p. 322

Glomerular filtration is the vital first step essential to all kidney functions. If filtration does not occur, waste products are not excreted, pH control is jeopardized, and an important mechanism for blood volume regulation is eliminated. Filtration depends on adequate blood flow to the glomerulus and maintaining normal filtration pressures. The regulating factors involved in maintaining a stable GFR are discussed in the section below on the control of kidney function.

REABSORPTION AND SECRETION ALONG THE RENAL TUBULE

Reabsorption and secretion at the kidney involve a combination of diffusion, osmosis, and carrier-mediated transport. In carrier-mediated transport, a specific substrate binds to a carrier protein that facilitates its movement across the cellular membrane. This movement may or may not require energy from ATP molecules. ∞ p. 61

The Proximal Convoluted Tubule

The cells lining the PCT actively reabsorb organic nutrients, plasma proteins, and ions from the filtrate and transport them into the interstitial fluid. As these materials are reabsorbed and transported, osmotic forces pull water across the wall of the PCT and into the surrounding interstitial fluid. The PCT usually reclaims 60–70 percent of the volume of filtrate produced at the glomerulus, along with virtually all of the glucose, amino acids, and other organic nutrients. The PCT also actively reabsorbs ions, including sodium, potassium, calcium, magnesium, bicarbonate, phosphate, and sulfate ions. The ion pumps involved are individually regulated and may be influenced by circulating ion or hormone levels. For example, the presence of parathyroid hormone stimulates calcium ion reabsorption. ∞ p. 325

Although reabsorption represents the primary function of the PCT, a few substances, such as hydrogen ions, can be actively secreted into the tubular fluid. Active secretion can play an important role in the regulation of blood pH, a topic considered in a later section. A few compounds in the tubular fluid, including urea and uric acid, are ignored by the PCT and by other segments of the renal tubule. As water and other nutrients are removed, the concentration of these waste products gradually rises in the tubular fluid.

The Loop of Henle

Roughly 60–70 percent of the volume of the filtrate produced at the glomerulus has been reabsorbed before the tubular fluid reaches the loop of Henle. In the process, the useful organic molecules, along with many mineral ions, have been reclaimed.

The loop of Henle will reabsorb more than half of the remaining water, as well as two-thirds of the sodium and chloride ions remaining in the tubular fluid.

The descending and ascending limbs of the loop of Henle have different permeability characteristics. The descending limb is permeable to water but not to solutes. Thus water can flow in or out by osmosis, but solutes cannot cross the tubular epithelium. The ascending limb is impermeable both to water and to solutes. However, the ascending limb actively pumps sodium and chloride ions out of the tubular fluid and into the interstitial fluid of the renal medulla. Over time, a *concentration gradient* is created in the medulla, with the highest concentration of solutes (roughly four times that of plasma) near the bend in the loop of Henle. Because the descending limb is freely permeable to water, as tubular fluid flows along that limb, water flows out of the tubular fluid and into the interstitial fluid by osmosis.

Roughly half the volume of filtrate that enters the loop of Henle is reabsorbed in the descending limb, and most of the sodium and chloride ions are removed in the ascending limb. With the loss of the sodium and chloride ions, the solute concentration of the filtrate declines to around one-third that of plasma. However, waste products such as urea now make up a significant percentage of the remaining solutes. In essence, most of the water and solutes have been removed, leaving the waste products behind.

The Distal Convoluted Tubule and the Collecting System

By the time the filtrate reaches the distal convoluted tubule, roughly 80 percent of the water and 85 percent of the solutes have already been reabsorbed. The DCT is connected to a collecting duct that drains into the renal pelvis (Figure 18-7●).

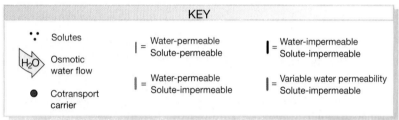

KEY

⠇⠂ Solutes

H_2O Osmotic water flow

● Cotransport carrier

| = Water-permeable Solute-permeable

| = Water-permeable Solute-impermeable

| = Water-impermeable Solute-impermeable

| = Variable water permeability Solute-impermeable

● *Figure 18-7* **The Effects of ADH on the DCT and Collecting Duct** (**a**) Tubule permeabilities and urine production without ADH. (**b**) Tubule permeabilities and urine production with ADH.

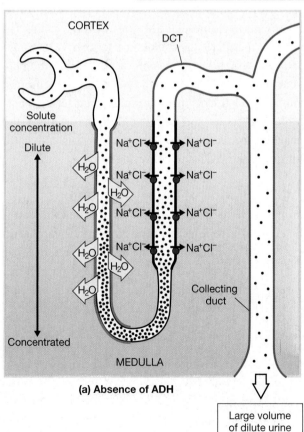

(a) Absence of ADH

Large volume of dilute urine

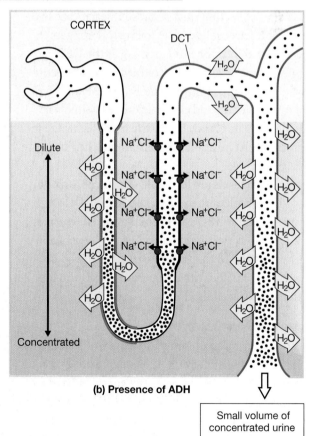

(b) Presence of ADH

Small volume of concentrated urine

18 THE URINARY SYSTEM

The Organization of the Urinary System • The Kidneys • **Basic Principles of Urine Production** • Urine Transport, Storage, and Elimination

As filtrate passes through the DCT and collecting duct, final adjustments are made in its composition and concentration. Composition depends on the type of solutes present; concentration depends on the volume of water in which they are dissolved. Because the DCT and collecting duct are impermeable to solutes, changes in filtrate composition can occur only through active reabsorption or secretion. The DCT is primarily concerned with active secretion.

Throughout most of the DCT, the tubular cells actively transport sodium ions out of the tubular fluid in exchange for potassium ions or hydrogen ions. The DCT and collecting ducts contain ion pumps that regulate the rates of sodium ion reabsorption and potassium ion secretion in response to the hormone **aldosterone**. Aldosterone secretion occurs (1) in response to circulating ACTH from the anterior pituitary and (2) in response to elevated potassium ion concentrations in the extracellular fluid. The higher the aldosterone levels, the more sodium ions are reclaimed, and the more potassium ions are lost.

The amount of water reabsorbed along the DCT and collecting duct is controlled by circulating levels of *antidiuretic hormone (ADH)*. In the absence of ADH, the distal convoluted tubule and collecting duct are impermeable to water. The higher the level of circulating ADH, the greater the water permeability and the more concentrated the urine. Water moves out of the DCT and collecting duct because in each case the tubular fluid contains fewer solutes than the surrounding interstitial fluid. As noted above, the fluid arriving at the DCT has a solute concentration only around one-third that of the cortex, because the ascending limb of the loop of Henle has removed most of the sodium and chloride ions. The fluid passing along the collecting duct travels into the medulla, where it passes through the concentration gradient established by the loop of Henle.

If circulating ADH levels are low, little water reabsorption will occur, and virtually all of the water reaching the DCT will be lost in the urine (Figure 18-7a●). If circulating ADH levels are high, the DCT and collecting duct will be very permeable to water (Figure 18-7b●). In this case, the individual will produce a small quantity of urine with a solute concentration four to five times that of extracellular fluids.

The Properties of Normal Urine

The general characteristics of normal urine are listed in Table 18-3, but the composition of the 1200 ml of urine excreted each day depends on the metabolic and hormonal events under way. Because the composition and concentration of the urine vary independently, an individual can produce a small quantity of concentrated urine or a large quantity of dilute

TABLE 18-3 *General Characteristics of Normal Urine*

pH	6.0 (range: 4.5–8)
Specific gravity (density of urine/density of pure water)	1.003–1.030
Osmotic concentration (Osmolarity) (number of solute particles per liter; for comparison, fresh water ≈ 5 mOsm/l, body fluids ≈ 300 mOsm/l, and sea water ≈ 1000 mOsm/l)	855–1335 mOsm/l
Water content	93–97 percent
Volume	1200 ml/day
Color	Clear yellow
Odor	Varies with composition
Bacterial content	Sterile

urine and still excrete the same amount of dissolved materials. For this reason, physicians often request a 24-hour urine collection rather than a single sample. This enables them to assess both quantity and composition accurately.

THE CONTROL OF KIDNEY FUNCTION

Renal function is regulated in three ways: (1) by local, automatic adjustments in glomerular pressures, through changes in the diameters of the afferent and efferent arterioles; (2) through activities of the sympathetic division of the autonomic nervous system (ANS); and (3) through the effects of hormones. The hormonal mechanisms make complex, long-term adjustments in blood pressure and blood volume that stabilize the GFR, in part by regulating transport mechanisms and water permeabilities of the DCT and collecting duct.

The Local Regulation of Kidney Function

Local, automatic changes in the diameters of the afferent arterioles, the efferent arterioles, and the glomerular capillaries can compensate for minor variations in blood pressure. For example, a reduction in blood flow and a decline in glomerular pressure trigger the dilation of the afferent arteriole and glomerular capillaries and the constriction of the efferent arteriole. This combination keeps glomerular blood pressure and blood flow within normal limits. As a result, glomerular filtration rates remain relatively constant. If blood pressure rises, the afferent arteriole walls become stretched and smooth muscle cells respond by contracting. The reduction in the diameter of the afferent arterioles decreases glomerular blood flow and keeps the GFR within normal limits.

F O C U S

A Summary of Urine Formation

Figure 18-8● summarizes the major steps involved in the reabsorption of water and the production of concentrated urine.

Step 1: Glomerular filtration produces a filtrate resembling blood plasma but containing few plasma proteins. This filtrate has the same osmotic, or solute, concentration as plasma or interstitial fluid.

Step 2: In the proximal convoluted tubule (PCT), 60–70 percent of the water and almost all of the dissolved nutrients are reabsorbed. The osmotic concentration of the tubular fluid remains unchanged.

Step 3: In the PCT and descending limb of the loop of Henle, water moves into the surrounding interstitial fluid, leaving a small fluid volume (roughly 20 percent of the original filtrate) of highly concentrated tubular fluid.

Step 4: The ascending limb is impermeable to water and solutes. The tubular cells actively pump sodium and chloride ions out of the tubular fluid. Because only sodium and chloride ions are removed, urea now accounts for a higher proportion of the solutes in the tubular fluid.

Step 5: The final composition and concentration of the tubular fluid will be determined by the events under way in the DCT and the collecting ducts. These segments are impermeable to solutes, but ions may be actively transported into or out of the filtrate under the control of hormones such as aldosterone.

Step 6: The concentration of urine is controlled by variations in the water permeabilities of the DCT and the collecting ducts. These segments are impermeable to water unless exposed to antidiuretic hormone (ADH). In the absence of ADH, no water reabsorption occurs, and the individual produces a large volume of dilute urine. At high concentrations of ADH, the collecting ducts become freely permeable to water, and the individual produces a small volume of highly concentrated urine.

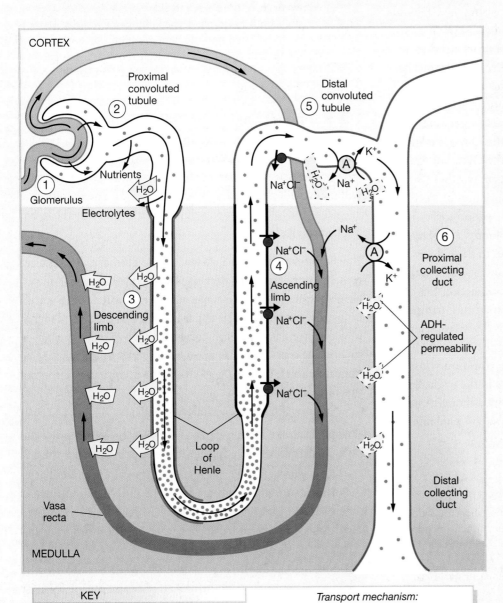

● *Figure 18-8* **The Major Steps in Urine Production**

18 THE URINARY SYSTEM

The Organization of the Urinary System • The Kidneys • **Basic Principles of Urine Production** • Urine Transport, Storage, and Elimination

Sympathetic Activation and Kidney Function

Autonomic regulation of kidney function occurs primarily through the sympathetic division of the ANS. Sympathetic activity primarily serves to shift blood away from the kidneys and lower the glomerular filtration rate (GFR). Sympathetic activation has both direct and indirect effects on kidney function. The direct effect of sympathetic activation is a powerful constriction of the afferent arterioles, decreasing the GFR and slowing production of filtrate. Triggered by a sudden crisis, such as an acute fall in blood pressure or a heart attack, sympathetic activation can override the local regulatory mechanisms that act to stabilize the GFR. As the crisis passes and sympathetic activity decreases, the GFR returns to normal.

When the sympathetic division changes the regional pattern of blood circulation, blood flow to the kidneys is often affected. This change can further reduce the GFR. For example, the dilation of superficial vessels in warm weather shunts blood away from the kidneys, and glomerular filtration declines temporarily. The effect becomes especially pronounced during periods of strenuous exercise. As the blood flow increases to the skin and skeletal muscles, it decreases to the kidneys. At maximal levels of exertion, renal blood flow may be less than one-quarter of normal resting levels. This reduction can create problems for distance swimmers and marathon runners, whose glomerular cells may be damaged by low oxygen levels and the build up of metabolic wastes over the course of a long competition. After such events, protein is commonly lost in the urine, and in some cases, blood appears in the urine. Such problems generally disappear within 48 hours, although a small number of runners experience kidney failure, or renal failure, and permanent impairment of kidney function.

THE TREATMENT OF KIDNEY FAILURE

Kidney failure, or **renal failure**, occurs when the kidneys become unable to perform the excretory functions needed to maintain homeostasis. Many different conditions can result in acute or chronic renal failure. *Acute renal failure* occurs when exposure to toxic drugs, renal ischemia, urinary obstruction, or trauma causes filtration to slow or suddenly stop. In *chronic renal failure*, kidney function deteriorates gradually, and the associated problems accumulate over time. The condition generally cannot be reversed, only prolonged, and symptoms of acute renal failure eventually develop.

Management of chronic kidney failure typically involves restricting water and salt intake and reducing caloric intake to a minimum, with few dietary proteins. This combination reduces strain on the urinary system by (1) minimizing the volume of urine produced and (2) preventing the generation of large quantities of nitrogenous wastes. Acidosis, a blood plasma pH below 7.35, is a common problem in patients with kidney failure; it can be countered with infusions of bicarbonate ions. If drugs, infusions, and dietary controls cannot stabilize the composition of the blood, more drastic measures are taken, such as dialysis or kidney transplantation.

In *dialysis*, the functions of damaged kidneys are partially compensated for by diffusion between the patient's blood and a *dialysis fluid* whose composition is carefully regulated. The process, which takes several hours, must be repeated two or three times each week.

In a *kidney transplant*, the kidney of a healthy donor is surgically inserted into the patient's body and connected to the circulatory system. If the surgical procedure is successful, the transplanted kidney(s) can take over all of the normal kidney functions.

The Hormonal Control of Kidney Function

The major hormones involved in regulating kidney function are angiotensin II, ADH, aldosterone, and ANP. These hormones have been discussed in earlier chapters, so only a brief overview will be provided here. ∞ pp. 321, 326, 331 The secretion of angiotensin II, aldosterone, and ADH is integrated by the *renin-angiotensin system*.

The Renin-Angiotensin System. If the glomerular pressures remain low because of a decrease in blood volume, a fall in systemic pressures, or a blockage in the renal artery or its tributaries, the juxtaglomerular apparatus releases the enzyme renin into the circulation. Renin converts inactive *angiotensinogen* to *angiotensin I,* which a *converting enzyme* activates to **angiotensin II**. Figure 18-9● diagrams the primary effects of this potent hormone.

Angiotensin II has the following effects:

• *In peripheral capillary beds*, it causes a brief but powerful vasoconstriction, elevating blood pressure in the renal arteries.

• *At the nephron*, it triggers the contraction of the efferent arterioles, elevating glomerular pressures and filtration rates.

• *In the CNS*, it triggers the release of ADH, which in turn stimulates the reabsorption of water and sodium ions and causes the sensation of thirst.

• *At the adrenal gland*, it stimulates the secretion of aldosterone by the cortex and epinephrine and norepinephrine (NE) by the adrenal medullae. The result is a sudden, dramatic increase in systemic blood pressure. At the kidneys, aldosterone stimulates sodium reabsorption along the DCT and collecting system.

ADH. Antidiuretic hormone (1) increases the water permeability of the DCT and collecting duct, stimulating the reabsorption of water from the tubular fluid; and (2) causes the sensation of thirst, leading to the consumption of additional water. ADH release occurs under angiotensin II stimulation;

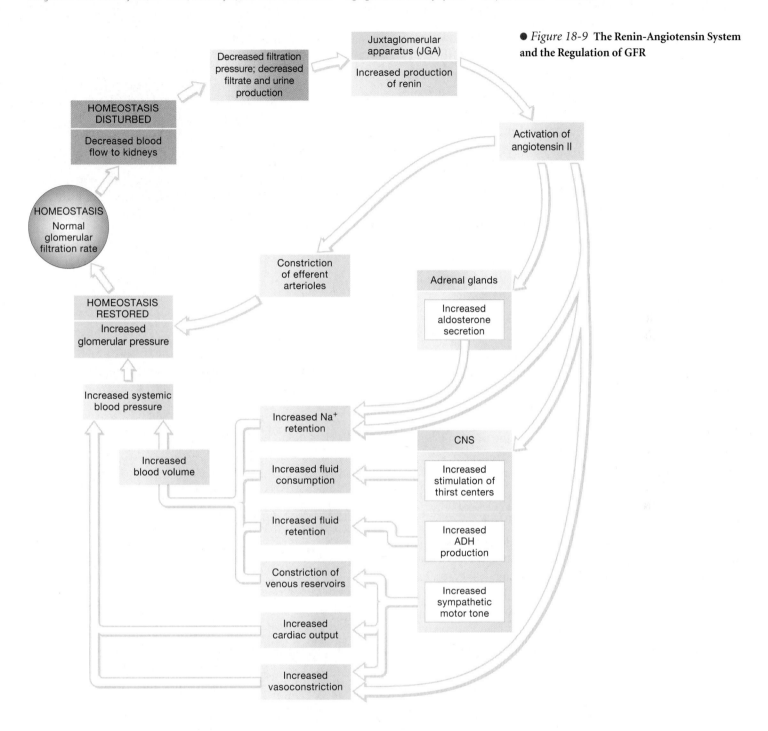

● *Figure 18-9* **The Renin-Angiotensin System and the Regulation of GFR**

it also occurs independently, when hypothalamic neurons are stimulated by a decrease in blood pressure or an increase in the solute concentration of the circulating blood. The nature of the receptors involved has not been determined, but these specialized hypothalamic neurons are called *osmoreceptors*.

Aldosterone. Aldosterone secretion stimulates the reabsorption of sodium ions and the secretion of potassium ions along the DCT and collecting duct. Aldosterone secretion primarily occurs (1) under angiotensin II stimulation and (2) in

response to a rise in the potassium ion concentration of the blood. 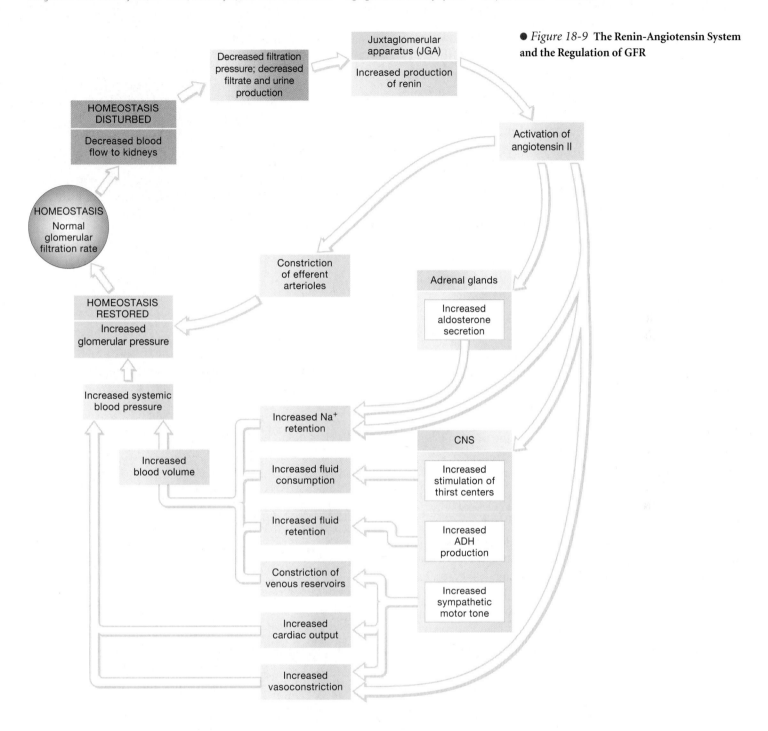 p. 403

Atrial Natriuretic Peptide. The actions of atrial natriuretic peptide (ANP) oppose those of the renin-angiotensin system (Figure 13-11b●, p. 402). This hormone is released by atrial cardiac muscle cells when blood volume and blood pressure are too high. The actions of ANP that affect the kidneys include (1) a decrease in the rate of sodium ion reabsorption in the DCT, leading to increased sodium ion loss in the urine; (2) the

561

18 THE URINARY SYSTEM

The Organization of the Urinary System • The Kidneys • Basic Principles of Urine Production • **Urine Transport, Storage, and Elimination**

dilation of the glomerular capillaries, which results in increased glomerular filtration and urinary water loss; and (3) the inactivation of the renin-angiotensin system through the inhibition of renin, aldosterone, and ADH secretion. The net result is an accelerated loss of sodium ions and an increase in the volume of urine produced. This combination lowers blood volume and blood pressure.

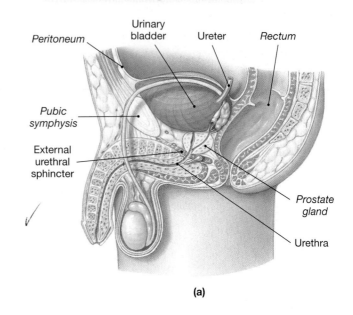

Urinary bladder • Ureter • Rectum

Peritoneum

Pubic symphysis

External urethral sphincter

Prostate gland

Urethra

(a)

CONCEPT CHECK QUESTIONS

Answers on page 578

1. How would a decrease in blood pressure affect the glomerular filtration rate (GFR)?
2. If the nephrons lacked a loop of Henle, what would be the effect on the volume and solute (osmotic) concentration of the urine they produced?

Urine Transport, Storage, and Elimination

Filtrate modification and urine production end when the fluid enters the renal pelvis. The urinary tract (the ureters, urinary bladder, and urethra) is responsible for the transport, storage, and elimination of the urine.

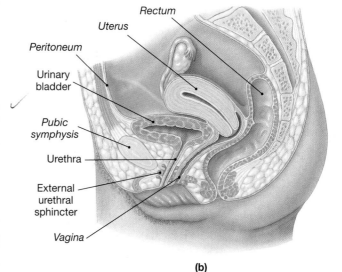

Rectum

Uterus

Peritoneum

Urinary bladder

Pubic symphysis

Urethra

External urethral sphincter

Vagina

(b)

THE URETERS AND URINARY BLADDER

The **ureters** (ū-RĒ-terz) are a pair of muscular tubes that carry urine from the kidneys to the urinary bladder, a distance of about 30 cm (12 in.) (Figures 18-1, p. 548, and 18-10a●). Each ureter begins at the funnel-shaped renal pelvis and ends at the posterior wall of the bladder without entering the peritoneal cavity. Their **ureteral openings** within the urinary bladder are slitlike, a shape that prevents the backflow of urine into the ureters or kidneys when the urinary bladder contracts.

The wall of each ureter contains an inner expandable transitional epithelium, a middle layer of longitudinal and circular bands of smooth muscle, and an outer connective tissue layer continuous with the renal capsule. About every 30 seconds, a peristaltic contraction begins at the renal pelvis and sweeps along the ureter, forcing urine toward the urinary bladder. Occasionally, solids composed of calcium deposits, magnesium salts, or crystals of uric acid form within the kidney,

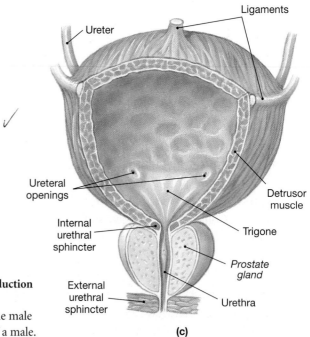

Ligaments

Ureter

Ureteral openings

Detrusor muscle

Internal urethral sphincter

Trigone

Prostate gland

External urethral sphincter

Urethra

(c)

●*Figure 18-10* **Organs Responsible for the Conduction and Storage of Urine**
(**a**) The ureter, urinary bladder, and urethra in the male and (**b**) in the female. (**c**) The urinary bladder of a male.

ureters, or urinary bladder. These solids are called *calculi* (KAL-kū-lī), or *kidney stones*, and their presence results in a painful condition known as *nephrolithiasis* (nef-rō-li-THĪ-a-sis). Kidney stones not only obstruct the flow of urine but may also reduce or prevent filtration in the affected kidney.

The **urinary bladder** is a hollow, muscular organ that stores urine prior to urination. Its dimensions vary depending on the state of distension, but the full urinary bladder can contain up to a liter of urine. The urinary bladder lies in the pelvic cavity, with only its superior surface covered by a layer of peritoneum. It is held in position by peritoneal folds, or *umbilical ligaments*, that extend to the umbilicus (navel), and by bands of connective tissue attached to the pelvic and pubic bones. In males, the base of the urinary bladder lies between the rectum and the pubic symphysis (Figure 18-10a●). In females, the urinary bladder sits inferior to the uterus and anterior to the vagina (Figure 18-10b●).

The triangular area within the urinary bladder that is bounded by the ureteral openings and the entrance to the urethra constitutes the *trigone* (TRĪ-gōn) of the bladder (Figure 18-10c●). The urethral entrance lies at the apex of this triangle, at the lowest point in the bladder. The area surrounding the urethral entrance, called the neck of the urinary bladder, contains a muscular sphincter that also extends along the proximal portions of the urethra. This **internal urethral sphincter** provides involuntary control over the discharge of urine from the bladder.

A *transitional epithelium* continuous with the renal pelvis and the ureters also lines the urinary bladder. This stratified epithelium can tolerate a considerable amount of stretching, as indicated in Figure 4-5b●. ∞ p. 90 The middle layer of the bladder wall consists of inner and outer layers of longitudinal smooth muscle with a circular layer in between. These layers of smooth muscle form the powerful *detrusor* (de-TROO-sor) *muscle* of the bladder. Contraction of this muscle compresses the urinary bladder and expels its contents into the urethra.

 ### URINARY TRACT INFECTIONS

Urinary tract infections (UTIs) result from the colonization of the urinary tract by bacteria or fungi. The intestinal bacterium *Escherichia coli* is most commonly involved. Women are particularly susceptible to UTIs, because the external urethral orifice is so close to the anus. Sexual intercourse can push bacteria into the urethra; because the female urethra is relatively short, the urinary bladder can then become infected.

The condition may have no symptoms, but it can be detected by the presence of bacteria and blood cells in urine. If inflammation of the urethral wall occurs, the condition is termed *urethritis*; inflammation of the lining of the bladder is called *cystitis*. Many infections, including sexually transmitted diseases (STDs) such as *gonorrhea*, cause a combination of urethritis and cystitis. These conditions cause painful urination, a symptom known as *dysuria* (dis-ū-rē-uh), and the urinary

bladder becomes tender and sensitive to pressure. Despite the discomfort that urination produces, the individual feels the urge to urinate frequently. Urinary tract infections generally respond to antibiotic therapies, although reinfections can occur.

In untreated cases, the bacteria may proceed along the ureters to the renal pelvis. The resulting inflammation of the walls of the renal pelvis produces *pyelitis* (pī-e-LĪ-tis). If the bacteria invade the renal cortex and medulla as well, *pyelonephritis* (pī-e-lō-nef-RĪ-tis) results. Signs and symptoms of pyelonephritis include a high fever, intense pain on the affected side, vomiting, diarrhea, and the presence of blood cells and pus in the urine.

THE URETHRA

The urethra extends from the neck of the urinary bladder to the exterior. In females, the **urethra** is very short, extending 2.5–3.0 cm (about 1 in.) from the bladder to its opening in the vestibule anterior to the vagina. In males, the urethra extends from the neck of the urinary bladder to the tip of the penis, about 18–20 cm (7–8 in.) in length. In both genders, as the urethra passes through the muscular floor of the pelvic cavity, a circular band of skeletal muscle forms the **external urethral sphincter**. This sphincter consists of skeletal muscle fibers, and its contractions are under voluntary control.

THE MICTURITION REFLEX AND URINATION

Urine reaches the urinary bladder by the peristaltic contractions of the ureters. The process of **urination**, or **micturition** (mik-tu-RI-shun), is coordinated by the **micturition reflex** (Figure 18-11●). Stretch receptors in the wall of the urinary

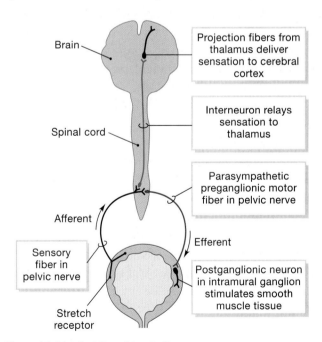

● *Figure 18-11* **The Micturition Reflex**
The basic components of the reflex arc that stimulates smooth muscle contractions in the urinary bladder.

18 THE URINARY SYSTEM

The Organization of the Urinary System • The Kidneys • Basic Principles of Urine Production • Urine Transport, Storage, and Elimination

bladder are stimulated as the bladder fills with urine. Afferent sensory fibers in the pelvic nerves carry the resulting impulses to the sacral spinal cord. Their increased level of activity (1) brings parasympathetic motor neurons in the sacral spinal cord close to threshold and (2) stimulates interneurons that relay sensations to the cerebral cortex. As a result, we become consciously aware of the fluid pressure in the urinary bladder.

The urge to urinate usually occurs when the bladder contains about 200 ml of urine. The micturition reflex begins to function when the stretch receptors have provided adequate stimulation to the parasympathetic motor neurons. At this time, the motor neurons stimulate the smooth muscle in the bladder wall. These commands travel over the pelvic nerves and produce a sustained contraction of the urinary bladder.

This contraction elevates fluid pressures inside the bladder, but urine ejection cannot occur unless both the internal and external sphincters are relaxed. The relaxation of the external sphincter occurs under voluntary control. When the external sphincter relaxes, so does the internal sphincter. If the external sphincter does not relax, the internal sphincter remains closed, and the bladder gradually relaxes. A further increase in bladder volume begins the cycle again, usually within an hour. Each increase in urinary volume leads to an increase in stretch receptor stimulation that makes the sensation more acute. Once the volume of the urinary bladder exceeds 500 ml, the micturition reflex may generate enough pressure to force open the internal sphincter. This opening leads to a reflexive relaxation in the external sphincter, and urination occurs despite voluntary opposition or potential inconvenience. At the end of normal micturition, less than 10 ml of urine remain in the bladder.

INCONTINENCE

Incontinence (in-KON-ti-nens) is the inability to control urination voluntarily. Infants lack voluntary control over urination because the necessary corticospinal connections have yet to be established. Trauma to the internal or external urethral sphincter can contribute to incontinence in otherwise normal adults. For example, childbirth can stretch and damage the sphincter muscles, so some mothers then develop *stress incontinence*. In this condition, elevated intra-abdominal pressures, caused simply by a cough or sneeze, can overwhelm the sphincter muscles, causing urine to leak out. Incontinence may also develop in older individuals, because of a general loss of muscle tone.

Damage to the CNS, the spinal cord, or the nerve supply to the bladder or external sphincter may also produce incontinence. For example, incontinence often accompanies Alzheimer's disease, and it may also result from a stroke or spinal injury. In most cases, the affected individual develops an *automatic bladder*. The micturition reflex remains intact, but voluntary control of the external sphincter is lost, and the person cannot prevent the reflexive emptying of the bladder. Damage to the pelvic nerves can eliminate the micturition reflex entirely, because these nerves carry both afferent and efferent fibers controlling this reflex. The urinary bladder becomes greatly distended with urine. It remains filled to capacity, and the excess trickles into the urethra in an uncontrolled stream. The insertion of a catheter is commonly required to facilitate the discharge of urine.

Integration with Other Systems

The urinary system is not the only organ system concerned with excretion. Along with the urinary system, the integumentary, respiratory, and digestive systems are considered to form an anatomically diverse *excretory system*. Examples of their excretory activities include:

1. *Integumentary system.* Water and electrolyte losses in perspiration affect plasma volume and composition. The effects are most apparent when losses are extreme, as in maximum sweat production. Small amounts of metabolic wastes, including urea, are also excreted in perspiration.
2. *Respiratory system.* The lungs remove the carbon dioxide generated by cells. Small amounts of other compounds, such as acetone and water, evaporate into the alveoli and are eliminated during exhalation.
3. *Digestive system.* Small amounts of metabolic waste products are excreted in liver bile, and a variable amount of water is lost in feces.

These excretory activities affect the composition of body fluids. The respiratory system, for example, is the primary site of carbon dioxide excretion. But the excretory functions of these systems are not regulated as closely as are those of the kidneys, and the effects of integumentary and digestive excretory activities normally are minor compared with those of the urinary system.

Figure 18-12● summarizes the functional relationships between the urinary system and other systems. Many of these relationships will be explored further in the next section, which considers fluid, pH, and electrolyte balance.

The Urinary System

For All Systems

Excretes waste products; maintains normal body fluid pH and ion composition

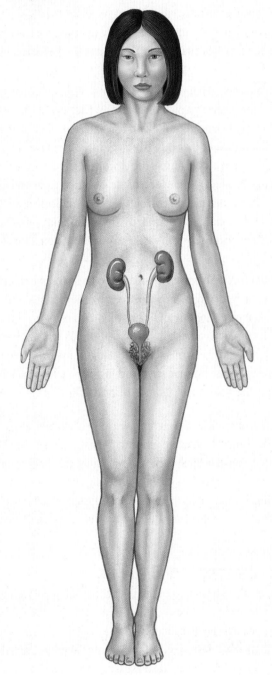

The Integumentary System

- Sweat glands assist in elimination of water and solutes, especially sodium and chloride ions; keratinized epidermis prevents excessive fluid loss through skin surface; epidermis produces vitamin D_3, important for the renal production of calcitriol

The Skeletal System

- Axial skeleton provides some protection for kidneys and ureters; pelvis protects urinary bladder and proximal portion of urethra

- Conserves calcium and phosphate needed for bone growth

The Muscular System

- Sphincter controls urination by closing urethral opening; muscle layers of trunk provide some protection for urinary organs

- Removes waste products of protein metabolism; assists in regulation of calcium and phosphate concentrations

The Nervous System

- Adjusts renal blood pressure; monitors distension of urinary bladder and controls urination

The Endocrine System

- Aldosterone and ADH adjust rates of fluid and electrolyte reabsorption in kidneys

- Kidney cells release renin when local blood pressure declines and erythropoietin (EPO) when renal oxygen levels decline

The Cardiovascular System

- Delivers blood to capillaries, where filtration occurs; accepts fluids and solutes reabsorbed during urine production

- Releases renin to elevate blood pressure and erythropoietin to accelerate red blood cell production

The Lymphatic System

- Provides specific defenses against urinary tract infections

- Eliminates toxins and wastes generated by cellular activities; acid pH of urine provides nonspecific defense against urinary tract infections

The Respiratory System

- Assists in the regulation of pH by eliminating carbon dioxide

- Assists in the elimination of carbon dioxide; provides bicarbonate buffers that assist in pH regulation

The Digestive System

- Absorbs water needed to excrete wastes at kidneys; absorbs ions needed to maintain normal body fluid concentrations; liver removes bilirubin

- Excretes toxins absorbed by the digestive epithelium; excretes bilirubin and nitrogenous wastes produced by the liver; calcitrol production by kidneys aids calcium and phosphate absorption along digestive tract

The Reproductive System

- Accessory organ secretions may have antibacterial action that helps prevent urethral infections in males

- Urethra in males carries semen to the exterior

● *Figure 18-12* **Functional Relationships Between the Urinary System and Other Systems**

18 THE URINARY SYSTEM

The Organization of the Urinary System • The Kidneys • Basic Principles of Urine Production • Urine Transport, Storage, and Elimination

CONCEPT CHECK QUESTIONS

Answers on page 578

❶ What process is responsible for the movement of urine from the kidney to the urinary bladder?

❷ An obstruction of a ureter by a kidney stone would interfere with the flow of urine between what two points?

❸ The ability to control the micturition reflex depends on your ability to control which muscle?

❹ What organ systems contribute to an overall excretory system of the body?

Fluid, Electrolyte, and Acid-Base Balance

The next time you see a small pond, think about the fish it contains. They live out their lives totally dependent on the quality of their isolated environment. If evaporation removes too much of the pond water, oxygen and food supplies run out, and the fish suffocate or starve. Most of the fish in a freshwater pond will die if the water becomes too salty; those in a saltwater pond will be killed if their environment becomes too dilute. The pH of the pond water, too, is a vital factor—water that is too acidic or too alkaline will also kill the fish in the pond.

The cells of our bodies live in a pond whose shores are the exposed surfaces of the skin. Most of the weight of the human body is water. Water accounts for up to 99 percent of the volume outside cells and is an essential ingredient of cytoplasm. All of a cell's operations rely on water as a diffusion medium for the distribution of gases, nutrients, and waste products. If the water content of the body changes, cellular activities are jeopardized. For example, when the water content of the body reaches very low levels, proteins denature, enzymes cease functioning, and cells ultimately die. To survive, we must maintain a normal volume and composition in the **extracellular fluid** (**ECF**; interstitial fluid, plasma, and other body fluids) and **intracellular fluid** (**ICF**, also known as cytosol).

The concentrations of various ions and pH of the body's water are as important as its absolute quantity. If concentrations of calcium or potassium ions in the ECF become too high, cardiac arrhythmias develop, and the individual's life is in jeopardy. A pH outside the normal range can lead to a variety of dangerous problems. Low pH is especially dangerous because hydrogen ions break chemical bonds, change the shapes of complex molecules, disrupt cell membranes, and impair tissue functions.

This section considers the dynamics of exchange between the various body fluids, such as blood plasma and interstitial fluid, and between the body and the external environment. Several different but interrelated types of homeostasis are involved:

- *Fluid balance.* You are in **fluid balance** when the amount of water gained each day is equal to the amount lost to the environment. Maintaining normal fluid balance involves regulating the content and distribution of water in the ECF and ICF. Because your cells and tissues cannot transport water, fluid balance reflects primarily the control of electrolyte balance.

- *Electrolyte balance.* **Electrolytes** are ions released through the dissociation of inorganic compounds; they are so named because they will conduct an electrical current in solution. ∞ p. 36 Each day, your body fluids gain electrolytes from the food and drink you consume and lose electrolytes in urine, sweat, and feces. An **electrolyte balance** exists when there is neither a net gain nor a net loss of any ion in body fluids. Electrolyte balance primarily involves balancing the rates of absorption across the digestive tract with rates of loss at the kidneys.

- *Acid-base balance.* You are in **acid-base balance** when the production of hydrogen ions is equal to their loss. While acid-base balance exists, the pH of body fluids remains within normal limits. Preventing a reduction in pH is a primary problem because your normal metabolic operations generate a variety of acids. The kidneys and lungs play key roles in maintaining the acid-base balance of body fluids.

This section provides an overview that integrates earlier discussions of fluid, electrolyte, and acid-base balance in the body. Few other chapters have such wide-ranging clinical importance: *Treatment of any serious illness affecting the nervous, cardiovascular, respiratory, urinary, or digestive system must always include steps to restore normal fluid, electrolyte, and acid-base balance.*

FLUID AND ELECTROLYTE BALANCE

Water accounts for roughly 60 percent of the total body weight of an adult male and 50 percent of that of an adult female. The gender difference primarily reflects the relatively larger mass of adipose tissue in adult females and the greater average muscle mass in adult males. (Adipose tissue is 10 percent water, whereas skeletal muscle is 75 percent water.) In both genders, intracellular fluid contains more of the total body water than does extracellular fluid.

Figure 18-13● illustrates the distribution of water in the body. Nearly two-thirds of the total body water content is found inside living cells, as the fluid medium of the intracellular fluid

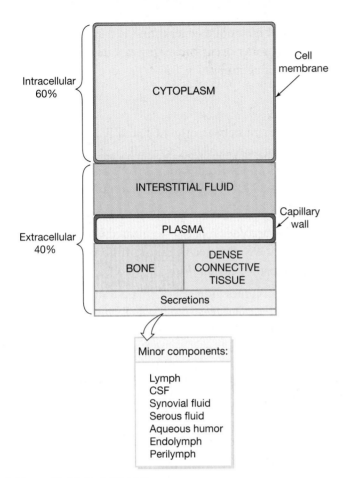

● *Figure 18-13* **Body Fluid Compartments**
The major body fluid compartments in a healthy adult.

(ICF), introduced in Chapter 3. p. 95 The extracellular fluid (ECF) contains the rest of the body water. The largest subdivisions of the ECF are the interstitial fluid in peripheral tissues and the plasma of the circulating blood. Minor components of the ECF include lymph, cerebrospinal fluid (CSF), synovial fluid, serous fluids (pleural, pericardial, and peritoneal fluids), aqueous humor, and the fluids of the inner ear (perilymph and endolymph,). Exchange between the ICF and ECF occurs across cell membranes by osmosis, diffusion, and carrier-mediated processes (see Table 3-3, p. 65).

Exchange among the subdivisions of the ECF occurs primarily across the endothelial lining of capillaries. Fluid may also travel from the interstitial spaces to the plasma by way of the channels of the lymphatic system. The identity and quantity of dissolved electrolytes, proteins, nutrients, and waste products in each ECF subdivision or component are not the same. However, these variations are relatively minor compared with the major differences *between* the ECF and the ICF.

The ICF and ECF are called **fluid compartments** because they commonly behave as distinct entities. As noted in earlier chapters, the principal ions in the ECF are sodium, chloride, and bicarbonate. The ICF contains an abundance of potassium, magnesium, and phosphate ions, plus large numbers of negatively charged proteins. Figure 18-14● compares the ions in the ICF with those in plasma and interstitial fluid, the two main subdivisions of the ECF.

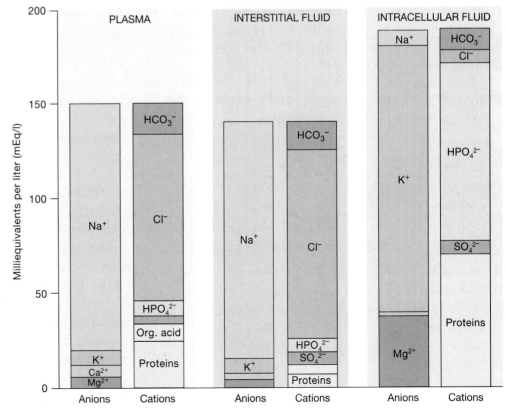

● *Figure 18-14* **Ions in Body Fluids**
Note the differences in the ion concentrations between the the plasma and interstitial fluid of the ECF and the ICF. For information concerning the chemical composition of other body fluids, see Appendix IV.

18 THE URINARY SYSTEM

The Organization of the Urinary System • The Kidneys • Basic Principles of Urine Production • Urine Transport, Storage, and Elimination

Despite these differences in the concentration of specific substances, the intracellular and extracellular osmotic concentrations are identical. Cell membranes are freely permeable to water, and osmosis eliminates any minor concentration differences almost at once. Because changes in solute concentrations lead to immediate changes in water distribution, the regulation of water balance and electrolyte balance are tightly intertwined.

Fluid Balance

Water circulates freely within the extracellular fluid compartment. At capillary beds throughout the body, capillary blood pressure forces water out of the plasma and into the interstitial spaces. Some of that water is reabsorbed along the distal portion of the capillary bed, and the rest circulates into lymphatic vessels for transport to the venous circulation (see Figure 13-6●, p. 396).

Water moves back and forth across the epithelial surfaces lining the peritoneal, pleural, and pericardial cavities and through the synovial membranes lining joint capsules. The flow rate is significant; for example, roughly 7 liters of peritoneal fluid is produced and reabsorbed each day. Water also moves between the blood and the cerebrospinal fluid, the aqueous and vitreous humors of the eye, and the perilymph and endolymph of the inner ear.

Roughly 2500 ml of water is lost each day through urine, feces, and perspiration (Table 18-4). The losses due to perspiration vary, depending on the activities undertaken, but the additional deficits can be considerable, reaching well over 4 liters an hour. ∞ p. 117 Water losses are normally balanced by the gain of fluids through eating (48 percent), drinking (40

TABLE 18-4 *Water Balance*

SOURCE	DAILY INPUT (ml)
Water content of food	1000
Water consumed as liquid	1200
Metabolic water during catabolism	300
Total	2500

METHOD OF ELIMINATION	DAILY OUTPUT (ml)
Urination	1200
Evaporation at skin	750
Evaporation at lungs	400
Loss in feces	150
Total	2500

percent), and metabolic generation (12 percent). *Metabolic generation* of water occurs primarily as a result of mitochondrial ATP production. ∞ p. 537

Fluid Shifts

Water movement between the ECF and ICF is called a *fluid shift*. Fluid shifts occur relatively rapidly, reaching equilibrium within a period of minutes to hours. These shifts occur in response to changes in the osmotic concentration, or *osmolarity*, of the extracellular fluid.

• If the ECF becomes more concentrated (hypertonic) with respect to the ICF, water will move from the cells into the ECF until osmotic equilibrium is restored.

• If the ECF becomes more dilute (hypotonic) with respect to the ICF, water will move from the ECF into the cells, and the volume of the ICF will increase accordingly.

In summary, if the osmolarity of the ECF changes, a fluid shift between the ICF and ECF will tend to oppose the change. Because the volume of the ICF is much greater than that of the ECF, the ICF acts as a "water reserve." In effect, instead of a large change in the composition of the ECF, there are smaller changes in both the ECF and the ICF.

Electrolyte Balance

An individual is in electrolyte balance when the rates of gain and loss are equal for each of the individual electrolytes in the body. Electrolyte balance is important for the following reasons:

• *A gain or loss of electrolytes can cause a gain or loss in water.*

• *The concentrations of individual electrolytes affect a variety of cell functions.* Many examples of ion effects on cell function were described in earlier chapters. For example, the effects of high or low calcium and potassium ion concentrations on cardiac muscle tissue were noted in Chapter 12. ∞ p. 383

Two cations, sodium and potassium, deserve attention because (1) they are major contributors to the osmotic concentrations of the ECF and ICF, and (2) they have direct effects on the normal functioning of living cells. Sodium is the dominant cation within the extracellular fluid. More than 90 percent of the osmotic concentration of the ECF results from the presence of sodium salts, principally sodium chloride (NaCl) and sodium bicarbonate ($NaHCO_3$), so changes in the osmotic concentration of extracellular fluids usually reflect changes in the concentration of sodium ions. Potassium is the dominant cation in the intracellular fluid; extracellular potassium concentrations are normally low. In general:

- *The most common problems with electrolyte balance are caused by an imbalance between sodium gains and losses.*
- *Problems with potassium balance are less common but significantly more dangerous than those related to sodium balance.*

Sodium Balance. The amount of sodium in the ECF represents a balance between sodium ion absorption at the digestive tract and sodium ion excretion at the kidneys and other sites. The rate of uptake varies directly with the amount included in the diet. Sodium losses occur primarily by excretion in urine and through perspiration. The kidneys are the most important site for regulating sodium ion losses. For example, in response to circulating levels of aldosterone the kidneys reabsorb sodium ions (decreases sodium loss) and, in response to atrial natriuretic peptide, the kidneys increase the loss of sodium ions). p. 561

Whenever the rate of sodium intake or output changes, a corresponding gain or loss of water tends to keep the sodium concentration constant. For example, eating a heavily salted meal will not raise the sodium ion concentration of body fluids, because as sodium chloride crosses the digestive epithelium, osmosis brings additional water into the ECF. This is why individuals with high blood pressure are told to restrict their salt intake; dietary salt will be absorbed, and because "water follows salt," the blood volume and blood pressure will increase.

Potassium Balance. Potassium ions are the primary cations of the intracellular fluid. Roughly 98 percent of the potassium content of the body lies within the ICF. The potassium concentration of the ECF, which is relatively low, represents a balance between (1) the rate of potassium ion entry across the digestive epithelium and (2) the rate of loss into urine. The rate of entry is proportional to the amount of potassium in the diet. The rate of loss is strongly affected by aldosterone. Urinary potassium losses are controlled by adjusting the rate of active secretion along the distal convoluted tubules of the kidneys. The ion pumps sensitive to aldosterone reabsorb sodium ions from the filtrate in exchange for potassium ions from the interstitial fluid. High plasma concentrations of potassium ions also stimulate aldosterone secretion directly. When potassium levels rise in the ECF, aldosterone levels climb, and additional potassium ions are lost in the urine. When potassium levels fall in the ECF, aldosterone levels fall and potassium ions are conserved.

1. How would eating a meal high in salt content affect the amount of fluid in the intracellular fluid compartment?
2. What effect would being lost in the desert for a day without water have on your blood osmotic concentration?

ACID-BASE BALANCE

The pH of your body fluids (ECF) represents a balance between the acids, bases, and salts in solution. This pH normally remains within relatively narrow limits, usually from 7.35 to 7.45. Any deviation outside the normal range is extremely dangerous because changes in hydrogen ion concentrations disrupt the stability of cell membranes, alter protein structure, and change the activities of important enzymes. You could not survive for long with a pH below 6.8 or above 7.7.

When the pH falls below 7.35, a state of **acidosis** exists. **Alkalosis** exists if the pH exceeds 7.45. These conditions affect virtually all systems, but the nervous system and cardiovascular system are particularly sensitive to pH fluctuations. For example, severe acidosis (pH below 7.0) can be deadly because (1) CNS function deteriorates, and the individual becomes comatose; (2) cardiac contractions grow weak and irregular, and symptoms of heart failure develop; and (3) peripheral vasodilation produces a dramatic drop in blood pressure, and circulatory collapse can occur.

Although acidosis and alkalosis are both dangerous, in practice, problems with acidosis are much more common than are problems with alkalosis. This is so because several acids, including carbonic acid, are generated by normal cellular activities.

Acids in the Body

Carbonic acid (H_2CO_3) is an important acid in body fluids. At the lungs, carbonic acid breaks down into carbon dioxide and water; the carbon dioxide diffuses into the alveoli. In peripheral tissues, carbon dioxide in solution interacts with water to form molecules of carbonic acid (H_2CO_3). As noted in Chapter 15, the carbonic acid molecules then dissociate to produce hydrogen ions and bicarbonate ions. p. 474 The complete reaction sequence is

$$CO_2 + H_2O \leftrightarrow H_2CO_3 \leftrightarrow H^+ + HCO_3^-$$

| carbon dioxide | water | | carbonic acid | | hydrogen ion | | bicarbonate ion |

569

18 **THE URINARY SYSTEM**

The Organization of the Urinary System • The Kidneys • Basic Principles of Urine Production • Urine Transport, Storage, and Elimination

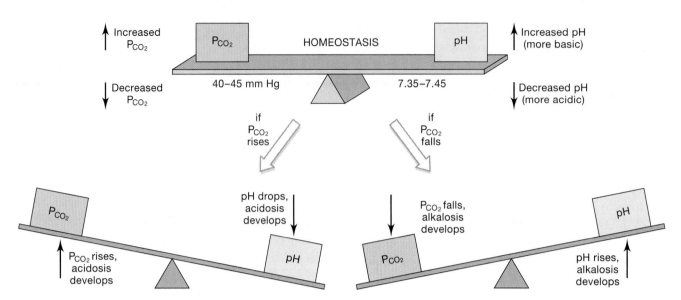

● *Figure 18-15* **Basic Relationships Between Carbon Dioxide and Plasma pH**

This reaction occurs spontaneously in body fluids, but it occurs very rapidly in the presence of *carbonic anhydrase*, an enzyme found in red blood cells, liver and kidney cells, parietal cells of the stomach, and in many other cell types.

Because most of the carbon dioxide in solution is converted to carbonic acid, and most of the carbonic acid dissociates, the partial pressure of carbon dioxide (P_{CO_2}) and the pH are inversely related (Figure 18-15●). When carbon dioxide concentrations rise, additional hydrogen ions and bicarbonate ions are released, and the pH goes down. (Recall that the greater the concentration of hydrogen ions, the lower the pH value.) The P_{CO_2} is the most important factor affecting the pH in body tissues.

At the alveoli, carbon dioxide diffuses into the atmosphere, the number of hydrogen ions and bicarbonate ions drop, and the pH rises. This process, which effectively removes hydrogen ions from solution, will be considered in more detail in the next section.

Organic acids, or metabolic acids, are generated during normal metabolism. Some are generated during the catabolism of amino acids, carbohydrates, or lipids. Examples are lactic acid produced during anaerobic metabolism and ketone bodies produced when breaking down fatty acids. Under normal conditions, metabolic acids are recycled or excreted rapidly, and significant accumulations do not occur.

Buffers and Buffer Systems

The acids discussed above, produced in the course of normal metabolic operations, must be controlled by the buffers and buffer systems in body fluids. *Buffers*, introduced in Chapter 2, are dissolved compounds that can provide or remove hydrogen ions and thereby stabilize the pH of a solution. ∞ p. 34 Buffers include *weak acids* that can donate hydrogen ions and

weak bases that can absorb them. A **buffer system** consists of a combination of a weak acid and its dissociation products: a hydrogen ion (H^+) and an anion. There are three major buffer systems, each with slightly different characteristics and distribution: the *protein buffer system*, the *carbonic acid–bicarbonate buffer system*, and the *phosphate buffer system*.

Protein buffer systems contribute to the regulation of pH in the ECF and ICF. Protein buffer systems depend on the ability of amino acids to respond to alterations in pH by accepting or releasing hydrogen ions. If the pH climbs, the carboxyl group ($-COOH$) of the amino acid can dissociate, releasing a hydrogen ion. If the pH drops, the amino group ($-NH_2$) can accept an additional hydrogen ion, forming an $-NH_3^+$ group.

The plasma proteins and hemoglobin in red blood cells contribute to the buffering capabilities of the blood. Interstitial fluids contain extracellular protein fibers and dissolved amino acids that also help regulate pH. In the ICF of active cells, structural and other proteins provide an extensive buffering capability that prevents destructive pH changes when organic acids, such as lactic acid, are produced by cellular metabolism.

The **carbonic acid–bicarbonate buffer system** is an important buffer system in the ECF. With the exception of red blood cells, your cells generate carbon dioxide 24 hours a day. As detailed earlier, most of the carbon dioxide is converted to carbonic acid, which then dissociates into a hydrogen ion and a bicarbonate ion. The carbonic acid and its dissociation products form the carbonic acid–bicarbonate buffer system; the carbonic acid acts as a weak acid and the bicarbonate ion acts as a weak base. The net effect of this buffer system is that $CO_2 + H_2O \leftrightarrow H^+ + HCO_3^-$. If hydrogen ions are removed, they will be replaced through the combination of water and carbon dioxide; if hydrogen ions are added, most will be removed through the formation of carbon dioxide and water.

The primary role of the carbonic acid–bicarbonate buffer system is to prevent pH changes caused by organic, or metabolic, acids. The hydrogen ions released through the dissociation of these acids combine with bicarbonate ions, producing water and carbon dioxide. The carbon dioxide can then be excreted at the lungs. This buffering system can cope with large amounts of acid because body fluids contain an abundance of bicarbonate ions, known as the *bicarbonate reserve.* When hydrogen ions enter the ECF, the bicarbonate ions that combine with them are replaced by the bicarbonate reserve.

The **phosphate buffer system** consists of an anion, *dihydrogen phosphate* ($H_2PO_4^-$), which is a weak acid. The dihydrogen phosphate ion and its dissociation products form the phosphate buffer system:

$$H_2PO_4^- \leftrightarrow H^+ + HPO_4^{2-}$$

dihydrogen phosphate	hydrogen ion	monohydrogen phosphate

In solution, dihydrogen phosphate ($H_2PO_4^-$) reversibly dissociates into a hydrogen ion and monohydrogen phosphate (HPO_4^{2-}). In the ECF, the phosphate buffer system plays only a supporting role in the regulation of pH, primarily because the concentration of bicarbonate ions far exceeds that of phosphate ions. However, the phosphate buffer system is quite important in buffering the pH of the ICF, where the concentration of phosphate ions is relatively high.

Maintaining Acid-Base Balance

Buffer systems can tie up excess hydrogen ions (H^+), but they provide only a temporary solution. The H^+ have not been eliminated but merely made harmless. For homeostasis to be

---

CLINICAL NOTE *Disturbances of Acid-Base Balance*

The primary source of disturbances of acid-base balance is usually indicated by the name given to the resulting condition. *Respiratory acid-base disorders* result from abnormal carbon dioxide levels in the ECF. These conditions are usually directly related to an imbalance between the rate of carbon dioxide removal at the lungs and its generation in other tissues. *Metabolic acid-base disorders* are caused by the generation of organic acids or by conditions affecting the concentration of bicarbonate ions in the extracellular fluids. The problems of individuals with respiratory disorders are quite different from those individuals with metabolic disorders. Respiratory compensation alone can often restore normal acid-base balance in individuals suffering from respiratory disorders. In contrast, compensation mechanisms for metabolic disorders may be able to stabilize pH, but other aspects of acid-base balance (buffer system function, bicarbonate levels, and P_{CO_2}) remain abnormal until the underlying metabolic problem is corrected.

Respiratory Acidosis
Respiratory acidosis develops when the respiratory system is unable to eliminate all of the CO_2 generated by peripheral tissues. The primary symptom is low plasma pH due to *hypercapnia*, an elevated plasma P_{CO_2}. As carbon dioxide levels climb, hydrogen and bicarbonate ion concentrations rise as well. Other buffer systems can tie up some of the hydrogen ions, but once the combined buffering capacity has been exceeded, the pH begins to fall rapidly.

Respiratory acidosis represents the most frequent challenge to acid-base balance. The usual cause is *hypoventilation*, an abnormally low respiratory rate. Our tissues generate carbon dioxide at a rapid rate, and even a few minutes of hypoventilation can cause acidosis, reducing the pH of the ECF to as low as 7.0. Under normal circumstances, the chemoreceptors monitoring the P_{CO_2} of the plasma and CSF will eliminate the problem by stimulating an increase in the depth and rate of respiration.

Respiratory Alkalosis
Problems with **respiratory alkalosis** are relatively uncommon. This condition develops when respiratory activity lowers plasma

P_{CO_2} to below normal levels, a condition called *hypocapnia*. A temporary hypocapnia can be produced by hyperventilation, when increased respiratory activity leads to a reduction in the arterial P_{CO_2}. Continued hyperventilation can elevate the pH to levels as high as 7.8–8. This condition usually corrects itself, for the reduction in P_{CO_2} removes the stimulation for the chemoreceptors, and the urge to breathe fades until carbon dioxide levels have returned to normal. Respiratory alkalosis caused by hyperventilation seldom persists long enough to cause a clinical emergency.

Metabolic Acidosis
Metabolic acidosis is the second most common type of acid-base imbalance. The most frequent cause is the production of a large number of metabolic acids such as lactic acid or ketone bodies. Metabolic acidosis can also be caused by an impaired ability to excrete hydrogen ions at the kidneys, and any condition accompanied by severe kidney damage can result in metabolic acidosis. Compensation for metabolic acidosis usually involves a combination of respiratory and renal mechanisms. Hydrogen ions interacting with bicarbonate ions form carbon dioxide molecules that are eliminated at the lungs, while the kidneys excrete additional hydrogen ions into the urine and generate bicarbonate ions that are released into the ECF.

Metabolic Alkalosis
Metabolic alkalosis occurs when bicarbonate ion concentrations become elevated. The bicarbonate ions then interact with hydrogen ions in solution, forming carbonic acid, and the reduction in H^+ concentrations then causes symptoms of alkalosis. The secretion of HCl by parietal cells of the stomach commonly results in alkalosis. Cases of severe metabolic alkalosis are relatively rare. Compensation involves a reduction in pulmonary ventilation, coupled with the increased loss of bicarbonates in the urine.

18 THE URINARY SYSTEM

The Organization of the Urinary System • The Kidneys • Basic Principles of Urine Production • Urine Transport, Storage, and Elimination

preserved, the captured H^+ must ultimately be removed from body fluids. The problem is that the supply of buffer molecules is limited; once a buffer binds a H^+, it cannot bind any more H^+. With buffer molecules tied up, the capacity of the ECF to absorb more H^+ is reduced and pH control is impossible.

The maintenance of acid-base balance involves controlling hydrogen ion losses and gains. In this process, the respiratory and renal mechanisms support the buffer systems by (1) secreting or absorbing hydrogen ions; (2) controlling the excretion of acids and bases; and, when necessary; (3) generating additional buffers. It is the *combination* of buffer systems and these respiratory and renal mechanisms that maintain your pH within narrow limits.

Respiratory Contributions to pH Regulation. **Respiratory compensation** is a change in the respiratory rate that helps stabilize pH. Respiratory compensation occurs whenever your pH strays outside normal limits. Respiratory activity has a direct effect on the carbonic acid–bicarbonate buffer system. Increasing or decreasing the rate of respiration alters pH by lowering or raising the partial pressure of CO_2 (P_{CO_2}). Changes in P_{CO_2} have a direct effect on the concentration of hydrogen ions in the plasma. When the P_{CO_2} rises, the pH declines, and when the P_{CO_2} decreases, the pH increases (Figure 18-15●).

Mechanisms responsible for controlling the respiratory rate were discussed in Chapter 15, ∞ p. 476 and only a brief summary will be presented here. A rise in P_{CO_2} stimulates chemoreceptors in the carotid and aortic bodies, and within the CNS; a fall in the P_{CO_2} inhibits them. Stimulation of the chemoreceptors leads to an increase in the respiratory rate. As the rate of respiration increases, more CO_2 is lost at the lungs,

so the P_{CO_2} returns to normal levels. When the P_{CO_2} of the blood or CSF declines, respiratory activity becomes depressed, the breathing rate falls, and the P_{CO_2} in the extracellular fluids rises.

Renal Contributions to pH Regulation. Renal compensation is a change in the rates of hydrogen ion and bicarbonate ion secretion or absorption by the kidneys in response to changes in plasma pH. Under normal conditions, the body generates H^+ through the production of metabolic acids. The H^+ they release must be excreted in the urine to maintain acid-base balance. Glomerular filtration puts hydrogen ions, carbon dioxide, and the other components of the carbonic acid–bicarbonate and phosphate buffer systems into the filtrate. The kidney tubules then modify the pH of the filtrate by secreting hydrogen ions or reabsorbing bicarbonate ions.

Acid-Base Disorders. Together, buffer systems, respiratory compensation, and renal compensation maintain normal acid-base balance. These mechanisms are usually able to control pH very precisely, so that the pH of the extracellular fluids seldom varies more than 0.1 pH units, from 7.35 to 7.45. When buffering mechanisms are severely stressed, the pH wanders outside these limits, producing symptoms of alkalosis or acidosis.

Respiratory acid-base disorders result when abnormal respiratory function causes an extreme rise or fall in CO_2 levels in the ECF. Metabolic acid-base disorders result from the generation of organic acids or by conditions affecting the concentration of bicarbonate ions in the ECF. Table 18-5 summarizes the major classes of acid-base disorders by their characteristic pH values, general causes, and treatments.

TABLE 18-5 *Acid-Base Disorders*

DISORDER	PH (NORMAL = 7.35–7.45)	REMARKS	TREATMENT
Respiratory acidosis	Decreased (below 7.35)	Generally caused by hypoventilation and CO_2 buildup in tissues and blood	Improve ventilation, in some cases, with bronchodilation and mechanical assistance
Metabolic acidosis	Decreased (below 7.35)	Caused by buildup of metabolic acid, impaired H^+ excretion at kidneys, or bicarbonate loss in urine or feces	Administration of bicarbonate (gradual) with other steps as needed to correct primary cause
Respiratory alkalosis	Increased (above 7.45)	Generally caused by hyperventilation and reduction in plasma CO_2 levels	Reduce respiratory rate, allow rise in P_{CO_2}
Metabolic alkalosis	Increased (above 7.45)	Usually caused by prolonged vomiting and associated acid loss	pH below 7.55, no treatment; pH above 7.55 may require administration of ammonium chloride

CONCEPT CHECK QUESTIONS
Answers on page 578

❶ What effect would a decrease in the pH of the body fluids have on the respiratory rate?

❷ How would a prolonged fast affect the body's pH?

❸ Why can prolonged vomiting produce alkalosis?

Aging and the Urinary System

In general, aging is associated with an increased incidence of kidney problems. Age-related changes in the urinary system and aspects of fluid, electrolyte, and acid-base balance include:

1. *A decline in the number of functional nephrons.* The total number of kidney nephrons drops by 30–40 percent between ages 25 and 85.

2. *A reduction in the GFR.* This results from decreased numbers of glomeruli, cumulative damage to the filtration apparatus in the remaining glomeruli, and reductions in renal blood flow. Fewer nephrons and a reduced GFR also reduce the ability to regulate pH through renal compensation.

3. *Reduced sensitivity to ADH and aldosterone.* With age, the distal portions of the nephron and collecting system become less responsive to ADH and aldosterone. Less reabsorption of water and sodium ions occurs, and more potassium ions are lost in the urine.

4. *Problems with the micturition reflex.* Several factors are involved in such problems:

   • The sphincter muscles lose muscle tone and become less effective at voluntarily retaining urine. Loss of muscle tone leads to problems with incontinence, often involving a slow leakage of urine.

   • The ability to control micturition is often lost after a stroke, Alzheimer's disease, or other CNS problems affecting the cerebral cortex or hypothalamus.

   • In males, **urinary retention** may develop secondary to enlargement of the prostate gland. In this condition, swelling and distortion of surrounding prostatic tissues compress the urethra, restricting or preventing the flow of urine.

5. *A gradual decrease of total body water content with age.* Between ages 40 and 60, total body water content averages 60 percent for males and 50 percent for females. After age 60, the values decline to roughly 50 percent for males and 45 percent for females. Among other effects, this decrease results in less dilution of waste products, toxins, and administered drugs.

6. *A net loss in body mineral content in many people over age 60 as muscle mass and skeletal mass decrease.* This loss can, at least in part, be prevented by a combination of exercise and increased dietary mineral supply.

7. *Disorders affecting major systems become more common with increasing age.* Most of these disorders have some impact on either fluid, electrolyte, or acid-base balance.

CONCEPT CHECK QUESTIONS
Answers on page 578

❶ What effect does aging have on the GFR?

Related Clinical Terms

calculi (KAL-kū-lī): Insoluble deposits that form within the urinary tract from calcium salts, magnesium salts, or uric acid.

cystitis: An inflammation of the lining of the urinary bladder.

diuretics (dī-ū-RET-iks): Drugs that promote fluid loss in urine.

dysuria (dis-ū-rē-uh): Painful urination.

glomerulonephritis (glo-mer-ūl-Ō-nef-RĪ-tis): An inflammation of the glomeruli, the filtration units of the nephron.

glycosuria (glī-kŌ-SOO-rē-uh): The presence of glucose in urine.

hematuria: Blood loss in urine.

hypernatremia: A condition of excess sodium in the blood.

hyponatremia: A condition of lower than normal sodium in the blood.

incontinence (in-KON-ti-nens): An inability to control urination voluntarily.

nephritis: An inflammation of the kidneys.

nephrology (ne-FROL-o-jē): The medical specialty concerned with the kidney and its disorders.

proteinuria: Protein loss in urine.

pyelogram (PĪ-el-ō-gram): An image obtained by taking an X-ray of the kidneys after a radiopaque compound has been administered.

pyelonephritis (pī-e-lō-nef-RĪ-tis): An inflammation of the kidney tissues.

renal failure: An inability of the kidneys to excrete wastes in sufficient quantities to maintain homeostasis.

urethritis: An inflammation of the urethra.

urinalysis: A physical and chemical assessment of urine.

urinary tract infection (UTI): An infection that results from the colonization of the urinary tract by bacteria or fungi.

urology (u-ROL-o-jē): Branch of medicine concerned with the urinary system and its disorders, and the male reproductive tract and its disorders.

18 THE URINARY SYSTEM

The Organization of the Urinary System • The Kidneys • Basic Principles of Urine Production • Urine Transport, Storage, and Elimination

---

CHAPTER REVIEW

Key Terms

Summary Outline

1. The functions of the **urinary system** include (1) eliminating organic waste products, (2) regulating plasma concentrations of ions, (3) regulating blood volume and pressure by adjusting the volume of water lost and releasing hormones, (4) stabilizing blood pH, and (5) conserving nutrients.

1. The urinary system includes the kidneys, the ureters, the urinary bladder, and the urethra. The kidneys produce urine (a fluid containing water, ions, and soluble compounds); during urination urine is forced out of the body. *(Figure 18-1)*

1. The left **kidney** extends superiorly slightly more than the right kidney. Both lie in a *retroperitoneal* position. *(Figure 18-2)*

2. A fibrous renal capsule surrounds each kidney. The **hilus** provides entry for the *renal artery* and exit for the *renal vein* and *ureter*.

3. The ureter is connected to the **renal pelvis.** This chamber branches into two **major calyces,** each connected to four or five **minor calyces,** which enclose the **renal papillae.** Urine production begins in **nephrons.** *(Figure 18-3)*

4. The nephron (the basic functional unit in the kidney) includes the *renal corpuscle* and a **renal tubule**, which empties into the **collecting system** through a *collecting duct*. From the renal corpuscle, the filtrate travels through the *proximal convoluted tubule*, the *loop of Henle*, and the *distal convoluted tubule*. *(Figures 18-3c,18-4)*

5. Nephrons are responsible for the (1) production of **filtrate**, (2) reabsorption of nutrients, and (3) reabsorption of water and ions.

6. The renal tubule begins at the **renal corpuscle.** It includes a knot of intertwined capillaries called the **glomerulus** surrounded by the **Bowman's capsule.** Blood arrives from the *afferent arteriole* and departs in the *efferent arteriole.* *(Figure 18-5)*

7. At the glomerulus, **podocytes** cover the basement membrane of the capillaries that project into the **capsular space.** The processes of the podocytes are separated by narrow slits. *(Figure 18-5)*

8. The **proximal convoluted tubule (PCT)** actively reabsorbs nutrients, plasma proteins, and electrolytes from the filtrate. They are then released into the surrounding interstitial fluid. *(Figure 18-4)*

9. The **loop of Henle** includes a *descending limb* and an *ascending limb.* The ascending limb is impermeable to water and solutes; the descending limb is permeable to water. *(Figure 18-4)*

10. The ascending limb delivers fluid to the **distal convoluted tubule (DCT),** which actively secretes ions and reabsorbs sodium ions from the urine. The **juxtaglomerular apparatus,** which releases renin and erythropoietin, is located at the start of the DCT. *(Figure 18-5a)*

11. The **collecting ducts** receive urine from nephrons and merge into a **papillary duct** that delivers urine to a minor calyx. The collecting system makes final adjustments to the urine by reabsorbing water or reabsorbing or secreting various ions.

12. The blood vessels of the kidneys include the **interlobar, arcuate,** and **interlobular arteries**, and the **interlobar, arcuate,** and **interlobular veins.** Blood travels from the **afferent** and **efferent arterioles** to the **peritubular capillaries** and the **vasa recta.** Diffusion occurs between the capillaries of the vasa recta and the tubular cells through the interstitial fluid that surrounds the nephron. *(Figure 18-6)*

1. The primary purpose in **urine** production is the excretion and elimination of dissolved solutes, principally metabolic waste products, such as **urea, creatinine**, and **uric acid**.

2. Urine formation involves **filtration, reabsorption**, and **secretion**. *(Tables 18–1,18-2)*

3. Glomerular filtration occurs as fluids move across the wall of the glomerulus into the capsular space, in response to blood pressure in the glomerular capillaries. The **glomerular filtration rate (GFR)** is the amount of filtrate produced in the kidneys each minute. Any factor that alters the **filtration** (blood) **pressure** will change the GFR and affect kidney function.

4. Dropping filtration pressures stimulate the juxtoglomerular apparatus to release *renin.* Renin release results in increases in blood volume and blood pressure.

5. The cells of the PCT normally reabsorb 60–70 percent of the volume of the filtrate produced in the renal corpuscle. The PCT reabsorbs nutrients, sodium and other ions, and water that enter the filtrate. It also secretes various substances.

6. Water and ions are reclaimed from the filtrate by the loop of Henle. The ascending limb reabsorbs sodium and chloride ions, and the descending limb reabsorbs water. A concentration gradient in the medulla encourages the osmotic flow of water out of the tubular fluid and

into the interstitial fluid. As water is lost by osmosis and the tubular fluid volume decreases, the urea concentration rises.

7. The DCT performs final adjustments by actively secreting or absorbing materials. Sodium ions are actively absorbed in exchange for potassium and hydrogen ions discharged into the tubular fluid. **Aldosterone** secretion increases the rate of sodium reabsorption and potassium loss.

8. The amount of water in the urine of the collecting ducts is regulated by the secretion of *antidiuretic hormone (ADH)*. In the absence of ADH, the DCT, collecting tubule, and collecting duct are impermeable to water. The higher the ADH level in circulation, the more water is absorbed and the more concentrated the urine. *(Figure 18-7)*

9. More than 99 percent of the filtrate produced each day is reabsorbed before reaching the renal pelvis. Yet the water content of normal urine is 93–97 percent. *(Table 18-3)*

10. Each segment of the nephron and collecting system contributes to the production of hypertonic urine. *(Figure 18-8)*

The Control of Kidney Function 558

11. Renal function may be regulated by automatic adjustments in glomerular pressures through changes in the diameters of the afferent and efferent arterioles.

12. Sympathetic activation produces a powerful vasoconstriction of the afferent arterioles, decreasing the GFR and slowing the production of filtrate, and alters the GFR by changing the regional pattern of blood circulation.

13. Hormones that regulate kidney function include angiotensin II, aldosterone, ADH, and atrial natriuretic peptide (ANP). *(Figure 18-9)*

URINE TRANSPORT, STORAGE, AND ELIMINATION 562

1. Filtrate modification and urine production end when the fluid enters the renal pelvis. The rest of the urinary system is responsible for transporting, storing, and eliminating the urine.

The Ureters and Urinary Bladder 562

2. The **ureters** extend from the renal pelvis to the urinary bladder. Peristaltic contractions by smooth muscles in the walls of the ureters move the urine. *(Figures 18-1, 18-10)*

3. Internal features of the **urinary bladder**, a distensible sac for urine storage, include the *trigone*, the neck, and the **internal urethral sphincter**. Contraction of the *detrusor muscle* compresses the bladder and expels the urine into the urethra. *(Figure 18-10)*

The Urethra 563

4. In both genders, as the urethra passes through the muscular pelvic floor, a circular band of skeletal muscles forms the **external urethral sphincter**, which is under voluntary control. *(Figure 18-10)*

The Micturition Reflex and Urination 563

5. The process of **urination** is coordinated by the **micturition reflex**, which is initiated by stretch receptors in the bladder wall. Voluntary urination involves coupling this reflex with the voluntary relaxation of the external urethral sphincter, which allows the opening of the internal urethral sphincter. *(Figure 18-11)*

INTEGRATION WITH OTHER SYSTEMS 564

1. The urinary, integumentary, respiratory, and digestive systems are sometimes considered an anatomically diverse *excretory system*. The systems' components work together to perform all of the excretory functions that affect the composition of body fluids. *(Figure 18-12)*

FLUID, ELECTROLYTE, AND ACID-BASE BALANCE 566

1. The maintenance of normal volume and composition in the extracellular and intracellular fluids is vital to life. Three types of homeostasis are involved: *fluid balance, electrolyte balance*, and *acid-base balance*.

Fluid and Electrolyte Balance 566

2. The **intracellular fluid (ICF)** contains nearly two-thirds of the total body water; the **extracellular fluid (ECF)** contains the rest. Exchange occurs between the ICF and ECF, but the two **fluid compartments** retain their distinctive characteristics. *(Figures 18-13, 18-14)*

3. Water circulates freely within the ECF compartment.

4. Water losses are normally balanced by gains through eating, drinking, and metabolic generation. *(Table 18-4)*

5. Water movement between the ECF and ICF is called a *fluid shift*. If the ECF becomes hypertonic relative to the ICF, water will move from the ICF into the ECF until osmotic equilibrium has been restored. If the ECF becomes hypotonic relative to the ICF, water will move from the ECF into the cells and the volume of the ICF will increase accordingly.

6. Electrolyte balance is important because total electrolyte concentrations affect water balance and because the levels of individual electrolytes can affect a variety of cell functions. Problems with electrolyte balance generally result from an imbalance between sodium gains and losses. Problems with potassium balance are less common but more dangerous.

7. The rate of sodium uptake across the digestive epithelium is directly related to the amount of sodium in the diet. Sodium losses occur mainly in the urine and through perspiration. The rate of sodium reabsorption along the DCT is regulated by aldosterone levels; aldosterone stimulates sodium ion reabsorption.

8. Potassium ion concentrations in the ECF are very low. Potassium excretion increases (1) when sodium ion concentrations decline and (2) as ECF potassium concentrations rise. The rate of potassium excretion is regulated by aldosterone; aldosterone stimulates potassium ion excretion.

Acid-Base Balance 569

9. The pH of normal body fluids ranges from 7.35 to 7.45; variations outside this range produce **acidosis** or **alkalosis**.

10. Carbonic acid is the most important factor affecting the pH of the ECF. In solution, CO_2 reacts with water to form carbonic acid; the dissociation of carbonic acid releases hydrogen ions. An inverse relationship exists between the concentration of CO_2 and pH. *(Figure 18-15)*

11. Organic, or metabolic, acids include products of metabolism such as lactic acid and ketone bodies.

12. A buffer system consists of a weak acid and its anion dissociation product, which acts as a weak base. The three major buffer systems are (1) *protein buffer systems* in the ECF and ICF; (2) the *carbonic acid–bicarbonate buffer system*, most important in the ECF; and (3) the *phosphate buffer system* in the intracellular fluids.

13. In **protein buffer systems**, the component amino acids respond to changes in H^+ concentrations. Blood plasma proteins and hemoglobin in red blood cells help prevent drastic changes in pH.

14. The **carbonic acid–bicarbonate buffer system** prevents pH changes due to organic acids in the ECF.

18 THE URINARY SYSTEM

The Organization of the Urinary System • The Kidneys • Basic Principles of Urine Production • Urine Transport, Storage, and Elimination

15. The **phosphate buffer system** is important in preventing pH changes in the intracellular fluid.

16. In **respiratory compensation**, the lungs help regulate pH by affecting the carbonic acid–bicarbonate buffer system; changing the respiratory rate can raise or lower the P_{CO_2} of body fluids, affecting the buffering capacity.

17. In **renal compensation**, the kidneys vary their rates of hydrogen ion secretion and bicarbonate ion reabsorption depending on the pH of extracellular fluids.

18. Respiratory acid-base disorders result when abnormal respiratory function causes an extreme rise or a fall in CO_2 levels. Metabolic acid-base disorders are caused by the formation of organic acids or conditions affecting the levels of bicarbonate ions. (*Table 18-5*)

AGING AND THE URINARY SYSTEM 573

1. Aging is usually associated with increased kidney problems. Age-related changes in the urinary system include (1) declining number of functional nephrons, (2) reduced GFR, (3) reduced sensitivity to ADH, (4) problems with the micturition reflex (urinary retention may develop in men whose prostate gland is inflamed), (5) declining body water content, (6) a loss of mineral content, and (7) disorders impacting either fluid, electrolyte, or acid-base balance.

Review Questions

Level 1: Reviewing Facts and Terms

Match each item in column A with the most closely related item in column B. Use letters for answers in the spaces provided.

COLUMN A

___ 1. urination
___ 2. renal capsule
___ 3. hilus
___ 4. medulla
___ 5. nephrons
___ 6. renal corpuscle
___ 7. external urethral sphincter
___ 8. internal urethral sphincter
___ 9. aldosterone
___ 10. podocytes
___ 11. efferent arteriole
___ 12. afferent arteriole
___ 13. vasa recta
___ 14. ADH
___ 15. ECF
___ 16. sodium
___ 17. potassium

COLUMN B

a. site of urine production
b. capillaries around loop of Henle
c. causes sensation of thirst
d. accelerated sodium reabsorption
e. voluntary control
f. glomerular epithelium
g. fibrous covering
h. blood leaves glomerulus
i. dominant cation in ICF
j. renal pyramids
k. blood to glomerulus
l. interstitial fluid
m. contains glomerulus
n. exit for ureter
o. dominant cation in ECF
p. involuntary control
q. micturition

18. The filtrate leaving Bowman's capsule empties into the:
    (a) distal convoluted tubule
    (b) loop of Henle
    (c) proximal convoluted tubule
    (d) collecting duct

19. The distal convoluted tubule is an important site for:
    (a) active secretion of ions
    (b) active secretion of acids and other materials
    (c) selective reabsorption of sodium ions from the tubular fluid
    (d) a, b, and c are correct

20. The endocrine structure that secretes renin and erythropoietin is the:
    (a) juxtaglomerular apparatus
    (b) vasa recta
    (c) Bowman's capsule
    (d) adrenal gland

21. The primary purpose of the collecting system is to:
    (a) transport urine from the bladder to the urethra
    (b) selectively reabsorb sodium ions from tubular fluid

(c) transport urine from the renal pelvis to the ureters

(d) make final adjustments to the osmotic concentration and volume of urine

22. A person is in fluid balance when:

(a) the ECF and ICF are isotonic

(b) there is no fluid movement between compartments

(c) the amount of water gained each day is equal to the amount lost to the environment

(d) a, b, and c are correct

23. The primary components of the extracellular fluid are

(a) lymph and cerebrospinal fluid

(b) blood plasma and serous fluids

(c) interstitial fluid and plasma

(d) a, b, and c are correct

24. All the homeostatic mechanisms that monitor and adjust the composition of body fluids respond to changes:

(a) in the ICF

(b) in the ECF

(c) inside the cell

(d) a, b, and c are correct

25. The most common problems with electrolyte balance are caused by an imbalance between gains and losses of:

(a) calcium ions

(b) chloride ions

(c) potassium ions

(d) sodium ions

26. What is the primary function of the urinary system?

27. What are the structural components of the urinary system?

28. What are fluid shifts? What is their function, and what factors can cause them?

29. What three major hormones mediate major physiological adjustments that affect fluid and electrolyte balance? What are the primary effects of each hormone?

Level 2: Reviewing Concepts

30. The urinary system regulates blood volume and pressure by:

(a) adjusting the volume of water lost in the urine

(b) releasing erythropoietin

(c) releasing renin

(d) a, b, and c are correct

31. The balance of solute and water reabsorption in the renal medulla is maintained by the:

(a) segmental arterioles and veins

(b) lobar arteries and veins

(c) vasa recta

(d) arcuate arteries

32. The higher the plasma concentration of aldosterone, the more efficiently the kidney will:

(a) conserve sodium ions

(b) retain potassium ions

(c) stimulate urinary water loss

(d) secrete greater amounts of ADH

33. When pure water is consumed:

(a) the ECF becomes hypertonic with respect to the ICF

(b) the ECF becomes hypotonic with respect to the ICF

(c) the ICF becomes hypotonic with respect to the plasma

(d) water moves from the ICF into the ECF

34. Increasing or decreasing the rate of respiration alters pH by:

(a) lowering or raising the partial pressure of carbon dioxide

(b) lowering or raising the partial pressure of oxygen

(c) lowering or raising the partial pressure of nitrogen

(d) a, b, and c are correct

35. What interacting controls operate to stabilize the glomerular filtration rate (GFR)?

36. Describe the micturition reflex.

37. Differentiate among fluid balance, electrolyte balance, and acid-base balance, and explain why each is important to homeostasis.

38. Why should a person with a fever drink plenty of fluids?

39. Exercise physiologists recommend that adequate amounts of fluid be ingested before, during, and after exercise. Why is adequate fluid replacement during conditions of extensive sweating important?

Level 3: Critical Thinking and Clinical Applications

40. Long-haul trailer truck drivers are on the road for long periods of time between restroom stops. Why might that lead to kidney problems?

41. For the past week, Susan has felt a burning sensation in the urethral area when she urinates. She checks her temperature and finds that she has a low-grade fever. What unusual substances are likely to be present in her urine?

42. *Mannitol* is a sugar that is filtered but not reabsorbed by the kidneys. What effect would drinking a solution of mannitol have on the volume of urine produced?

18 THE URINARY SYSTEM

The Organization of the Urinary System • The Kidneys • Basic Principles of Urine Production • Urine Transport, Storage, and Elimination

Answers to Concept Check Questions

Page 554

1. Unlike most other organs, such as those of the digestive system, the kidneys lie beneath the peritoneal lining in a retroperitoneal position.

2. The filtration slits created by the podocytes are so fine that they will allow only substances smaller than plasma proteins to pass into the capsular space.

3. Damage to the juxtaglomerular apparatus portion of the nephrons would interfere with the normal control of blood pressure.

Page 562

1. Decreases in blood pressure would reduce the blood hydrostatic pressure within the glomerulus and would decrease the GFR.

2. If the nephrons lacked a loop of Henle, the kidneys would not be able to form a concentrated urine.

Page 566

1. Under normal conditions, peristaltic contractions move urine along the minor and major calyces toward the renal pelvis, out of the renal pelvis, and along the ureter to the bladder.

2. An obstruction of the ureters would interfere with the passage of urine from the renal pelvis to the urinary bladder.

3. To control the micturition reflex, you must be able to control the external urinary sphincter, a ring of skeletal muscle that acts as a valve.

4. The body's excretory system is made up of the urinary, the integumentary, the respiratory, and the digestive systems.

Page 569

1. Consuming a meal high in salt would temporarily increase the osmolarity of the ECF. As a result, some of the water in the ICF would shift to the ECF.

2. Fluid loss through perspiration, urine formation, and respiration would increase the osmolarity of body fluids.

Page 573

1. A decrease in the pH of body fluids would have a stimulating effect on the respiratory center in the medulla oblongata. The resulting increase in the rate of breathing would in turn lead to an elimination of more carbon dioxide, which would tend to cause the pH to increase.

2. In a prolonged fast, fatty acids are mobilized and large numbers of ketone bodies are formed. These molecules are acids that lower the body's pH. This situation would eventually lead to ketoacidosis.

3. In vomiting, large amounts of stomach acid are lost from the body. This acid is formed by the parietal cells of the stomach by taking hydrogen ions from the blood. Excessive vomiting would lead to the excessive removal of hydrogen ions from the blood to produce the acid, thus raising the body's pH and causing metabolic alkalosis.

Page 573

1. Aging reduces the GFR because of a decline in the number of nephrons, cumulative damage to the filtration mechanism of the glomeruli, and reduced blood flow to the kidneys.

EXPLORE *MediaLab*

EXPLORATION #1

Estimated time for completion: 10 minutes

URINALYSIS IS QUICKLY becoming a routine procedure. It is performed during routine physical examinations and also sometimes on applicants for certain jobs and on participants in either amateur or professionally sanctioned athletic events. Why is this done? What information can be obtained through analysis of the waste created by the urinary system as it maintains normal fluid balance? Use the information given in Table 18-2 on p. 555 to prepare a short list of the substances that normally appear in urine. Indicate on your list the step during urine formation when the compound is added to the filtrate. Now reflect on the three possible reasons given above for performing a urinalysis. What might physicians, potential employers, or supervisors of athletic events look for in addition to these substances? Add these compounds to your list, indicating which step is most likely responsible for their inclusion in the final urine. To complete your list, go to Chapter 18 at the Companion Web site, visit the MediaLab section, and click on the key word "Urinalysis." Here you will find an excellent tutorial on urinalysis, complete with photographs of actual samples indicating cellular and crystalline inclusions.

A typical meal setting includes table salt (sodium chloride).

EXPLORATION #2

Estimated time for completion: 10 minutes

THE MAIN FUNCTION of the urinary system is to regulate fluid balance and maintain homeostasis of the internal environment. This chapter describes how fluids and electrolytes are managed so as to balance levels between the blood and tissues of the body (pp. 566-572). While the American public may not be familiar with this balancing act, the media often discuss the relationship between sodium and water retention. Reread the section titled "Sodium Balance" in this chapter (p. 569), and then prepare a concept map of how sodium and fluid levels interact. Use either Figure 18-9● (The Renin-Angiotensin System and the Regulation of GFR), or Figure 17-9● (A Summary of Pathways of Catabolism and Anabolism) as your model for the interaction of sodium and extracellular fluids (ECF). Your map should include information on where sodium is absorbed, what hormones are released when sodium levels go up, what hormones are released when sodium levels decline, and how these sodium levels and associated hormones affect the ECF. For help in creating this map, go to Chapter 18 at the Companion Web site, visit the MediaLab section, and click on the key words "Sodium Balance." You will be taken to the Molson Medical home page. To the left, click on the "Ions" pull-down menu, and choose sodium. This will provide you with data that you can use to complete your map.

A personnel director interviewing a job applicant.

The Reproductive System

CHAPTER OUTLINE AND OBJECTIVES

Vocabulary Development

andro-	male; androgen
crypto	hidden; cryptorchidism
diplo	double; diploid
follis	a leather bag; follicle
genesis	generation; oogenesis
gyne	woman; gynecologist
haplo	single; haploid
labium	lip; labium minus
lutea	yellow; corpus luteum
meioun	to make smaller; meiosis
men	month; menopause
metra	uterus; endometrium
myo-	muscle; myometrium
oon	an egg; oocyte
orchis	testis; cryptorchidism
pausis	cessation; menopause
pellucidus	translucent; zona pellucida
rete	a net; rete testis
tetras	four; tetrad

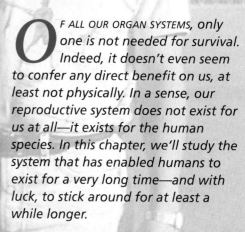

O F ALL OUR ORGAN SYSTEMS, only one is not needed for survival. Indeed, it doesn't even seem to confer any direct benefit on us, at least not physically. In a sense, our reproductive system does not exist for us at all—it exists for the human species. In this chapter, we'll study the system that has enabled humans to exist for a very long time—and with luck, to stick around for at least a while longer.

19 THE REPRODUCTIVE SYSTEM

The Reproductive System of the Male • The Reproductive System of the Female • The Physiology of Sexual Intercourse • Aging and the Reproductive System

A N INDIVIDUAL LIFE SPAN can be measured in decades, but the human species has survived for hundreds of thousands of years through the activities of the reproductive system. The entire process of reproduction seems almost magical; many primitive societies even failed to discover the basic link between sexual activity and childbirth and assumed that cosmic forces were responsible for producing new individuals. Although our society has a much clearer view of the reproductive process, a sense of wonder remains. Sexually mature males and females produce individual reproductive cells that are brought together through sexual intercourse. The fusion of these reproductive cells starts a chain of events leading to the appearance of an infant that will mature as part of the next generation.

This chapter and the next will consider the mechanics of this remarkable process. We will begin by examining the anatomy and physiology of the reproductive system in this chapter. **Gonads** (GŌ-nadz) are reproductive organs that produce hormones and reproductive cells, or **gametes** (GAM-ēts). Gametes are *sperm* in males and *ova* (Ō-va; singular, *ovum*) in females. The other components of the reproductive system store, nourish, and transport the gametes. Chapter 20 begins with **fertilization**, also known as *conception*, in which a male gamete and female gamete unite. All the cells in the body are the mitotic descendants of a **zygote** (ZĪ-gōt), the single cell created at fertilization. The gradual transformation of that single cell into a functional adult occurs through the process of *development*, the topic of Chapter 20. From fertilization to birth, development occurs within specialized organs of the female reproductive system.

The Reproductive System of the Male

The principal structures of the male reproductive system are shown in Figure 19-1●. The male gonads, or *testes* (TES-tēz; singular, *testis*), produce reproductive cells called **sperm**, or **spermatozoa** (sper-ma-tō-ZŌ-a; singular, *spermatozoon*). The spermatozoa leaving each testis travel within the *epididymis* (ep-i-DID-i-mus), the *ductus deferens* (DUK-tus DEF-e-renz), the *ejaculatory* (ē-JAK-ū-la-tō-rē) *duct*, and the *urethra* before leaving the body. Accessory organs—the *seminal* (SEM-i-nal) *vesicles*, the *prostate* (PROS-tāt) *gland*, and the *bulbourethral* (bul-bō-ū-RĒ-thral) *glands*—secrete into the ejaculatory ducts and urethra. The externally visible structures of the reproductive system constitute the *external genitalia* (jen-i-TĀ-lē-a). The external genitalia of the male include the *scrotum* (SKRŌ-tum), which encloses the testes, and the *penis* (PĒ-nis), an erectile organ through which the distal portion of the urethra passes.

THE TESTES

The *primary sex organs* of the male system are the **testes**. The testes hang within the **scrotum**, a fleshy pouch suspended posterior to the base of the penis. The scrotum is subdivided into two chambers, or scrotal cavities, each containing a testis. Each testis has the shape of a flattened egg roughly 5 cm (2 in.) long, 3 cm (1.2 in.) wide, and 2.5 cm (1 in.) thick. An epithelial lining on the inner surface of the scrotum and

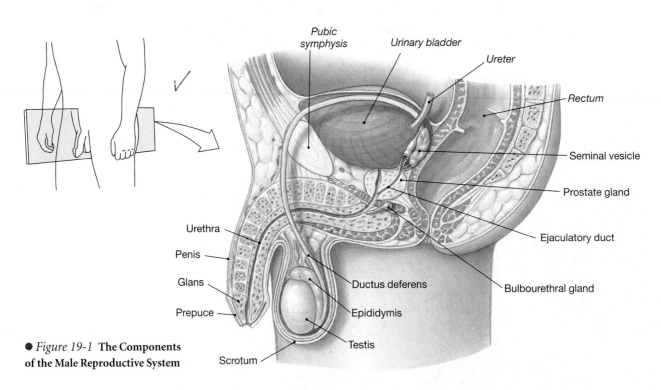

● *Figure 19-1* **The Components of the Male Reproductive System**

Labels: Pubic symphysis · Urinary bladder · Ureter · Rectum · Seminal vesicle · Prostate gland · Ejaculatory duct · Bulbourethral gland · Ductus deferens · Epididymis · Testis · Scrotum · Prepuce · Glans · Penis · Urethra

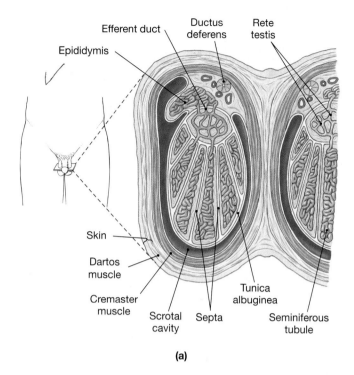

Epididymis

Efferent duct

Ductus deferens

Rete testis

Skin

Dartos muscle

Cremaster muscle

Scrotal cavity

Septa

Tunica albuginea

Seminiferous tubule

(a)

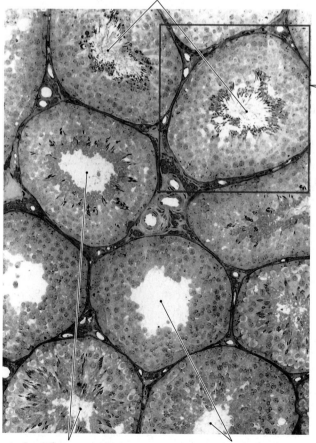

Seminiferous tubules containing nearly mature spermatozoa about to be released into the lumen

Seminiferous tubules containing late spermatids

Seminiferous tubules containing early spermatids

(b)

the outer surface of the testis prevents friction between the opposing surfaces.

The scrotum consists of a thin layer of skin containing smooth muscle (Figure 19-2a●). Sustained contractions of the smooth muscle layer, the *dartos* (DAR-tōs), cause the characteristic wrinkling of the scrotal surface. Beneath the dermis is a layer of skeletal muscle, the **cremaster** (kre-MAS-ter) **muscle**, which can contract to pull the testes closer to the body. Normal sperm development in the testes requires temperatures about 1.1°C (2°F) lower than those elsewhere in the body. When air or body temperatures rise, the cremaster relaxes, and the testes move away from the body. Sudden cooling of the scrotum, as occurs when a man enters a cold swimming pool, results in cremasteric contractions that pull the testes closer to the body and keep testicular temperatures from falling.

Each testis is wrapped in a dense fibrous capsule, the **tunica albuginea** (TŪ-ni-ka al-bū-JIN-ē-a). Collagen fibers from this wrapping extend into the testis, forming partitions, or *septa*, that subdivide the testis into roughly 250 *lobules*. Sperm production occurs in the approximately 800 slender, tightly coiled **seminiferous** (se-mi-NIF-e-rus) **tubules** (Figure 19-2b●) that are distributed among the lobules. Each tubule averages around 80 cm (31 in.) in length, and a typical testis contains nearly half a mile of seminiferous tubules. Sperm production occurs

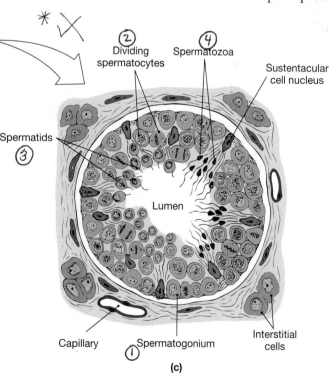

② Dividing spermatocytes

④ Spermatozoa

Sustentacular cell nucleus

Spermatids

③

Lumen

Capillary

① Spermatogonium

Interstitial cells

(c)

● *Figure 19-2* **The Testes and Seminiferous Tubules**
(**a**) Anatomical relationships of the testes. (**b**) A section through a coiled seminiferous tubule. (**c**) Cellular organization of a seminiferous tubule.

583

19 THE REPRODUCTIVE SYSTEM

The Reproductive System of the Male • The Reproductive System of the Female • The Physiology of Sexual Intercourse • Aging and the Reproductive System

within these tubules. Sperm leave the seminiferous tubules through a maze of passageways, known as the **rete** (RĒ-tē; *rete*, a net) **testis**, to enter the epididymis, the beginning of the male reproductive tract.

The spaces between the tubules are filled with loose connective tissue, numerous blood vessels, and large **interstitial cells** that produce male sex hormones, or *androgens* (Figure 19-2c●). The steroid **testosterone** is the most important androgen. (Testosterone and other sex hormones were introduced in Chapter 10.) ∞ p. 331

Sperm cells, or spermatozoa, are produced through **spermatogenesis** (sper-ma-tō-JEN-e-sis). This process begins with stem cells called *spermatogonia* at the outermost layer of cells in the seminiferous tubules and proceeds to the central lumen (Figure 19-2c●). At each step in the process, the daughter cells move closer to the lumen.

Each seminiferous tubule also contains **sustentacular** (sus-ten-TAK-ū-lar) **cells** (or *Sertoli cells*). These large cells extend from the perimeter of the tubule to the lumen. The spermatocytes and spermatogonia lie between and adjacent to the sustentacular cells (Figure 19-2c●).

 ### CRYPTORCHIDISM

In **cryptorchidism** (kript-ŌR-ki-dizm; *crypto*, hidden + *orchis*, testis), one or both of the testes have not descended into the scrotum by the time of birth. Typically, the testes are lodged in the abdominal cavity or within the inguinal canal. This condition occurs in about 3 percent of full-term deliveries and in roughly 30 percent of premature births. In most instances, normal descent occurs a few weeks later, but the condition can be surgically corrected if it persists. Corrective measures should be taken before puberty (sexual maturation) because cryptorchid (abdominal) testes will not produce sperm, and the individual will be *sterile (infertile)* and unable to bear children. If the testes cannot be moved into the scrotum, they will usually be removed, because about 10 percent of those with uncorrected cryptorchid testes eventually develop testicular cancer. This surgical procedure is called a *bilateral orchiectomy* (ōr-kē-EK-to-mē; *ectomy*, excision).

Spermatogenesis

Spermatogenesis involves three processes:

1. *Mitosis.* Stem cells called **spermatogonia** (sper-ma-to-GŌ-nē-a; singular, *spermatogonium*) undergo cell divisions throughout adult life. (See Chapter 3 for a review of mitosis and cell division.) ∞ p. 75 The cell divisions produce daughter cells that are pushed toward the lumen of the seminiferous tubule. These cells differentiate into

spermatocytes (sper-MA-to-sīts) that prepare to begin meiosis.

2. *Meiosis.* Meiosis (mī-Ō-sis; *meioun*, to make smaller) is a special form of cell division that produces gametes. Gametes contain half the number of chromosomes found in other cells. As a result, the fusion of a sperm and an egg yields a single cell with the normal number of chromosomes. In the seminiferous tubules, the meiotic divisions of spermatocytes produce immature gametes called *spermatids*.

3. *Spermiogenesis.* Spermatids are small, relatively unspecialized cells. In *spermiogenesis*, spermatids differentiate into physically mature spermatozoa.

Mitosis and Meiosis. In both males and females, mitosis and meiosis differ significantly in terms of the events that take place in the nucleus. **Mitosis** is part of the process of *somatic* (nonreproductive) cell division, which involves one division that produces two daughter cells, each containing the same number and kind of chromosomes as the original cell. In humans, each somatic cell contains 23 pairs of chromosomes. Each pair consists of one chromosome provided by the father and another by the mother at the time of fertilization. Because each cell contains both members of each chromosome pair, the daughter cells are described as **diploid** (DIP-loyd; *diplo*, double). **Meiosis** involves two cycles of cell division (meiosis I and meiosis II) and produces four cells, or gametes, each containing 23 individual chromosomes. Because these gametes contain only one member of each chromosome pair, gametes are described as **haploid** (HAP-loyd; *haplo*, single).

Figure 19-3● illustrates meiosis and sperm production. (To make the chromosomal events easier to follow, only 3 of the 23 pairs of chromosomes present in spermatogonia are shown.) The mitotic divisions of spermatogonia produce primary spermatocytes. Like spermatogonia, primary spermatocytes are diploid cells, but they divide by meiosis rather than mitosis. As a primary spermatocyte prepares to begin meiosis, all of the chromosomes in the nucleus replicate themselves as if the cell were to undergo mitosis. As prophase of the first meiotic division, meiosis I, occurs, the chromosomes condense and become visible. As in mitosis, each chromosome consists of two duplicate *chromatids* (KRŌ-ma-tidz).

Each primary spermatocyte contains 46 individual chromosomes, the same as any somatic cell in the body. As prophase ends, the corresponding maternal and paternal chromosomes now come together. This event, known as **synapsis**

pair at this stage of meiosis. This exchange increases genetic variation among offspring.

During metaphase of meiosis I, the nuclear envelope disappears and the tetrads line up along the metaphase plate. As anaphase begins, the tetrads break up, and the maternal and paternal chromosomes separate. This is a major difference between mitosis and meiosis: In mitosis, each daughter cell receives one of the two copies of every chromosome, maternal and paternal, whereas in meiosis, each daughter cell receives both copies of *either* the maternal chromosome or the paternal chromosome from each tetrad.

As anaphase proceeds, the maternal and paternal components are randomly distributed. For example, most of the maternal chromosomes may go to one daughter cell, and most of the paternal chromosomes to the other. As a result, telophase I ends with the formation of two daughter cells containing unique combinations of maternal and paternal chromosomes. In the testes, the daughter cells produced by the first meiotic division (meiosis I) are called **secondary spermatocytes**.

Each secondary spermatocyte contains 23 chromosomes. Each of these chromosomes still consists of two duplicate chromatids. The duplicates will separate during **meiosis II**. The interphase separating meiosis I and meiosis II is very brief, and the secondary spermatocyte soon enters prophase of meiosis II. The completion of metaphase II, anaphase II, and telophase II produces four **spermatids** (SPER-ma-tidz), each with 23 chromosomes. In summary, for every primary spermatocyte that enters meiosis, four spermatids are produced (Figure 19-3●).

Spermiogenesis. Each spermatid matures into a single **spermatozoon** (sper-ma-tō-ZŌ-on), or sperm cell, through the process of **spermiogenesis**. The entire process, from spermatogonial division to the release of a physically mature spermatozoon, takes approximately 9 weeks.

Sustentacular cells play a key role in spermatogenesis and spermiogenesis. The developing spermatocytes undergoing meiosis and spermatids are not free in the seminiferous tubules. Instead, they are surrounded by the cytoplasm of sustentacular cells. Because there are no blood vessels inside the seminiferous tubules, all nutrients must enter by diffusion from the surrounding interstitial fluids. The large sustentacular cells control the chemical environment inside the seminiferous tubules and provide nutrients and chemical stimuli that promote the production and differentiation of spermatozoa. They also help regulate spermatogenesis by producing *inhibin*, a hormone introduced in Chapter 10. ⌑ p. 320

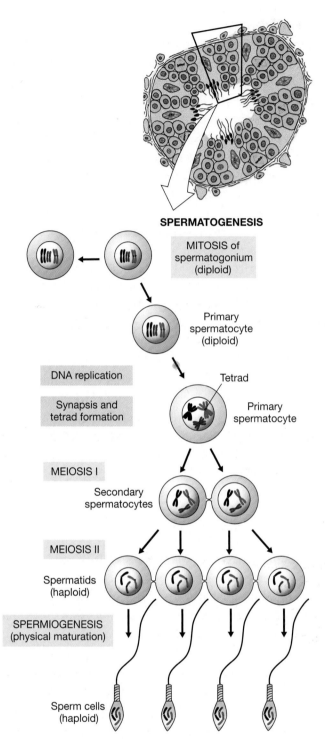

● *Figure 19-3* **Spermatogenesis**
Each diploid primary spermatocyte that undergoes meiosis produces four haploid spermatids. Each spermatid then differentiates into a spermatozoon.

(sin-AP-sis), produces 23 pairs of chromosomes, each member of the pair consisting of two identical chromatids. A matched set of four chromatids is called a **tetrad** (TET-rad; *tetras*, four). An exchange of genetic material, called *crossing-over*, can occur between the chromatids of a chromosome

19 THE REPRODUCTIVE SYSTEM

The Reproductive System of the Male • The Reproductive System of the Female • The Physiology of Sexual Intercourse • Aging and the Reproductive System

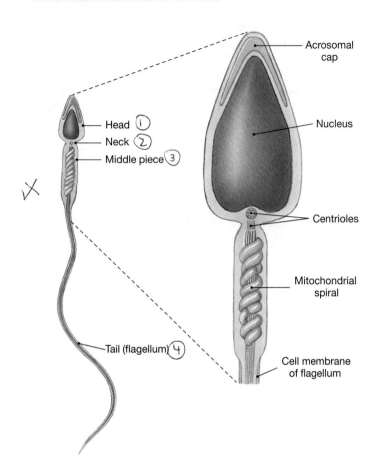

● *Figure 19-4* **Spermatozoon Structure**

Anatomy of a Spermatozoon

A sperm cell has three distinct regions: the head, the middle piece, and the tail (Figure 19-4●). The **head** is a flattened oval filled with densely packed chromosomes. At the tip of the head is the **acrosomal** (ak-rō-SŌ-mal) **cap**, which contains enzymes essential for fertilization. A short **neck** attaches the head to the **middle piece**, which is dominated by a spiral of mitochondria providing the ATP energy for moving the **tail**. The sperm cell's tail, the only example of a *flagellum* in the human body, moves the cell from one place to another. ⟳ p. 67

The entire streamlined structure measures only 60 μm in total length. Unlike most other cells, a mature spermatozoon lacks an endoplasmic reticulum, Golgi apparatus, lysosomes, peroxisomes, and many other intracellular structures. Because the cell does not contain glycogen or other energy reserves, it must absorb nutrients (primarily fructose) from the surrounding fluid.

THE MALE REPRODUCTIVE TRACT

The testes produce physically mature spermatozoa that are, as yet, incapable of fertilizing an ovum. The other portions of the male reproductive system are responsible for the functional maturation, nourishment, storage, and transport of spermatozoa.

The Epididymis

Late in their development, spermatozoa detach from the sustentacular cells and lie within the lumen of the seminiferous tubule. Although they have most of the physical characteristics of mature sperm cells, they are still functionally immature and incapable of coordinated locomotion or fertilization. At this point, cilia-driven fluid currents within the efferent ducts transport them into the **epididymis** (see Figure 19-1●, p. 582). This elongate tubule, almost 7 meters (23 ft) long, is twisted and coiled so as to take up very little space. The epididymis adjusts the composition of the fluid from the seminiferous tubules, acts as a recycling center for damaged spermatozoa, and stores the maturing spermatozoa. Cells lining the epididymis absorb cellular debris from damaged or abnormal spermatozoa and release it for pickup by surrounding blood vessels, as well as absorbing organic nutrients. During the 2 weeks that it takes for a spermatozoon to travel through the epididymis, it completes its physical maturation. Mature spermatozoa then arrive at the ductus deferens.

Although the spermatozoa leaving the epididymus are physically mature, they remain immobile. To become motile (actively swimming) and fully functional, they must undergo **capacitation**. Capacitation occurs when the spermatozoa (1) mix with secretions of the seminal vesicles and (2) are exposed to conditions inside the female reproductive tract. The epididymis secretes a substance that prevents premature capacitation.

The Ductus Deferens

Each **ductus deferens**, or *vas deferens*, is 40–45 cm (16–18 in.) long. It extends toward the abdominal cavity within a connective tissue sheath that also encloses the blood vessels, nerves, and lymphatics serving the testis (Figure 19-5●). The entire complex is called the **spermatic cord**.

After ascending through the *inguinal canal* into the abdominal cavity, the ductus deferens curves downward alongside the urinary bladder toward the prostate gland (Figures 19-1●, p. 582, and 19-5a●). Peristaltic contractions in the muscular walls of the ductus deferens propel spermatozoa and fluid along the length of the duct. The ductus deferens can also store spermatozoa for up to several months. During this period the spermatozoa are in a state of suspended animation, remaining inactive with low metabolic rates.

The junction of the ductus deferens with the duct of the seminal vesicle creates the *ejaculatory duct*, a relatively short (2 cm, or less than 1 in.) passageway. This duct penetrates the muscular wall of the prostate gland and empties into the urethra near the opening of the ejaculatory duct from the other side.

● *Figure 19-5* **The Ductus Deferens** (a) A posterior view of the prostate gland, showing subdivisions of the ductus deferens in relation to surrounding structures. (b) Micrographs showing extensive layering with smooth muscle around the lumen of the ductus deferens. (Reproduced from R. G. Kessel and R. H. Kardon, *Tissues and Organs: A Text-Atlas of Scanning Electron Microscopy*, W. H. Freeman & Co., 1979.) (LM × 34; SEM × 42)

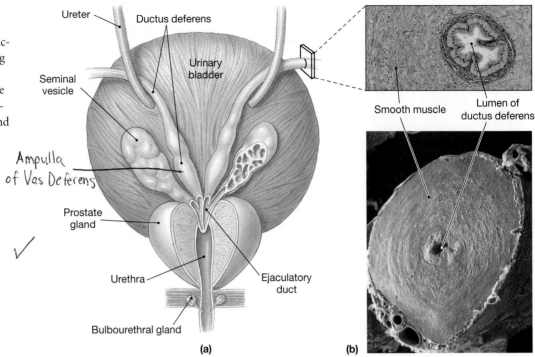

Ureter
Ductus deferens
Urinary bladder
Seminal vesicle
Ampulla of Vas Deferens
Prostate gland
Urethra
Ejaculatory duct
Bulbourethral gland
(a)

Smooth muscle
Lumen of ductus deferens
(b)

The Urethra

The urethra of the male extends 18–20 cm (7–8 in.) from the urinary bladder to the tip of the penis. In males, the urethra is a passageway used by both the urinary and reproductive systems.

THE ACCESSORY GLANDS

The fluids contributed by the seminiferous tubules and the epididymis account for only about 5 percent of the final volume of semen, the fluid that transports and nourishes sperm. The major fraction of seminal fluid is composed of secretions from the *seminal vesicles*, the *prostate gland*, and the *bulbourethral glands*. Primary functions of these glandular organs include: (1) activating the spermatozoa, (2) providing the nutrients spermatozoa need for motility, (3) propelling spermatozoa and fluids along the reproductive tract through peristaltic contractions, and (4) producing buffers that counteract the acidity of the urethral and vaginal contents.

The Seminal Vesicles

Each **seminal vesicle** is a tubular gland with a total length of around 15 cm (6 in.). The body of the gland is coiled and folded into a compact, tapered mass roughly 5 cm by 2.5 cm (2 in. by 1 in.) (Figures 19-1●, p. 582, and 19-5a●).

The seminal vesicles are extremely active secretory glands that contribute about 60 percent of the volume of semen. Their secretions contain (1) fructose, a six-carbon sugar easily metabolized by spermatozoa; (2) prostaglandins, which can stimulate smooth muscle contractions along the male and female reproductive tracts;

and (3) fibrinogen, which after ejaculation forms a temporary clot within the vagina. The secretions are also slightly alkaline, and the alkalinity helps neutralize acids in the secretions of the prostate gland and within the vagina. When mixed with the secretions of the seminal vesicles, previously inactive but mature spermatozoa begin beating their flagella and become highly mobile.

The Prostate Gland

The **prostate gland** is a small, muscular, rounded organ with a diameter of about 4 cm (1.6 in.). The prostate gland surrounds the urethra as it leaves the urinary bladder (Figure 19-5a●). The prostatic fluid produced by the prostate gland is a milky, slightly acidic secretion that contributes about 30 percent of the volume of semen. In addition to several other compounds of uncertain significance, prostatic fluid contains **seminalplasmin** (sem-i-nal-PLAZ-min), an antibiotic that may help prevent urinary tract infections in males. These secretions are ejected into the urethra by peristaltic contractions of the muscular wall.

 PROSTATITIS

Prostatic inflammation, or **prostatitis** (pros-ta-TĪ-tis), can occur in males at any age but most often afflicts older men. Prostatitis can result from bacterial infections but also occurs in the apparent absence of pathogens. Individuals with prostatitis complain of pain in the lower back, perineum, or rectum, sometimes accompanied by painful urination and the discharge of mucous secretions from the urethral opening. Antibiotic therapy is usually effective in treating most cases resulting from bacterial infection.

19 **THE REPRODUCTIVE SYSTEM**

The Reproductive System of the Male • The Reproductive System of the Female • The Physiology of Sexual Intercourse • Aging and the Reproductive System

● *Figure 19-6* **The Penis**
(**a**) The positions of the erectile tissues. (**b**) A frontal section through the penis and associated organs. (**c**) A sectional view through the penis.

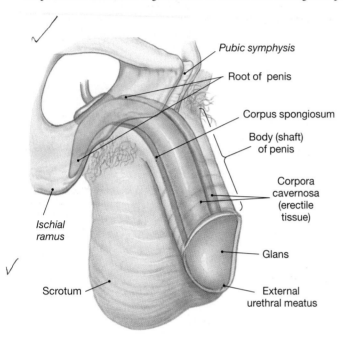

(a) Anterior and lateral view of penis

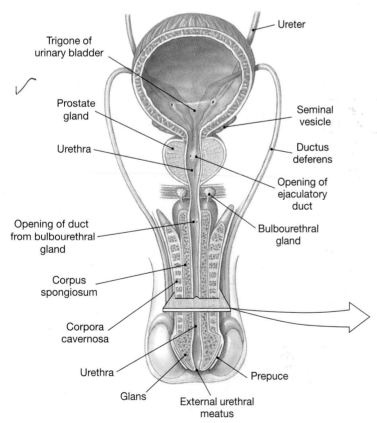

(b) Frontal section

The Bulbourethral Glands

The paired **bulbourethral glands**, or *Cowper's glands*, are round, with diameters approaching 10 mm (less than 0.5 in.) (Figure 19-5a●). These glands secrete a thick, sticky, alkaline mucus that helps neutralize urinary acids in the urethra and has lubricating properties.

SEMEN

Semen (SĒ-men) is the fluid that contains sperm and the secretions of the accessory glands of the male reproductive tract. In a typical **ejaculation** (ē-jak-ū-LĀ-shun), 2–5 ml of semen is expelled from the body. This volume of fluid, called an **ejaculate**, contains:

- *Spermatozoa.* A normal **sperm count** ranges from 20 million to 100 million spermatozoa per milliliter of semen.
- *Seminal fluid.* **Seminal fluid**, the fluid component of semen, is a mixture of glandular secretions with a distinctive ionic and nutrient composition. In terms of total volume, the seminal fluid contains the combined secretions of the seminal vesicles (60 percent), the prostate (30 percent), the sustentacular cells and epididymis (5 percent), and the bulbourethral glands (less than 5 percent).
- *Enzymes.* Several important enzymes are present in the seminal fluid, including: (1) a protease that helps dissolve mucous secretions in the vagina; (2) *seminalplasmin*, a prostatic antibiotic enzyme; (3) a prostatic enzyme that causes the semen to clot in the vagina after ejaculation; and (4) an enzyme that then liquefies the clotted semen.

THE PENIS

The **penis** is a tubular organ through which the urethra passes (see Figure 19-1●). It conducts urine to the exterior and introduces semen into the female vagina during sexual intercourse. As shown in Figure 19-6●, the penis is composed of

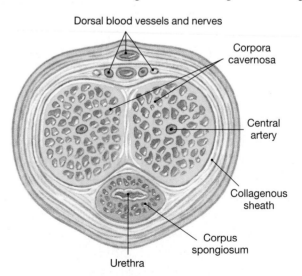

(c) Section through shaft of penis

three regions: (1) the **root**, the fixed portion that attaches the penis to the body wall; (2) the **body (shaft)**, the tubular portion that contains masses of erectile tissue; and (3) the **glans**, the expanded distal end that surrounds the external urethral opening, or *external urethral meatus*.

The skin overlying the penis resembles that of the scrotum. A fold of skin, the **prepuce** (PRĒ-pūs), or *foreskin*, surrounds the tip of the penis. The prepuce attaches to the relatively narrow neck of the penis and continues over the glans. *Preputial glands* in the skin of the neck and inner surface of the prepuce secrete a waxy material called *smegma* (SMEG-ma). Unfortunately, smegma can be an excellent nutrient source for bacteria. Mild inflammation and infections in this region are common, especially if the area is not washed frequently. One way of avoiding trouble is to perform a *circumcision* (ser-kum-SIZH-un), the surgical removal of the prepuce. In Western societies this procedure is generally performed shortly after birth.

Most of the body, or shaft, of the penis consists of three columns of **erectile tissue** (Figure 19-6b●). Erectile tissue consists of a three-dimensional maze of vascular channels incom-

pletely divided by partitions of elastic connective tissue and smooth muscle. On the anterior surface of the penis, two cylindrical **corpora cavernosa** (KŌR-po-ra ka-ver-NŌ-sa) are bound to the pubis and ischium of the pelvis. The corpora cavernosa extend along as far as the glans of the penis. The relatively slender **corpus spongiosum** (spon-jē-Ō-sum) surrounds the urethra and extends all the way to the tip of the penis, where it forms the glans.

In the resting state, there is little blood flow into the erectile tissue because the arterial branches are constricted and the muscular partitions are tense. Parasympathetic stimulation of the penile arteries involves neurons that release nitric oxide (NO). In response to the NO, the smooth muscle in the arterial walls relaxes, resulting in increased blood flow and **erection** of the penis.

HORMONES AND MALE REPRODUCTIVE FUNCTION

Major reproductive hormones were introduced in Chapter 10, and the hormonal interactions in the male are diagrammed in Figure 19-7●. ⟐ pp. 331-332 The anterior pituitary gland

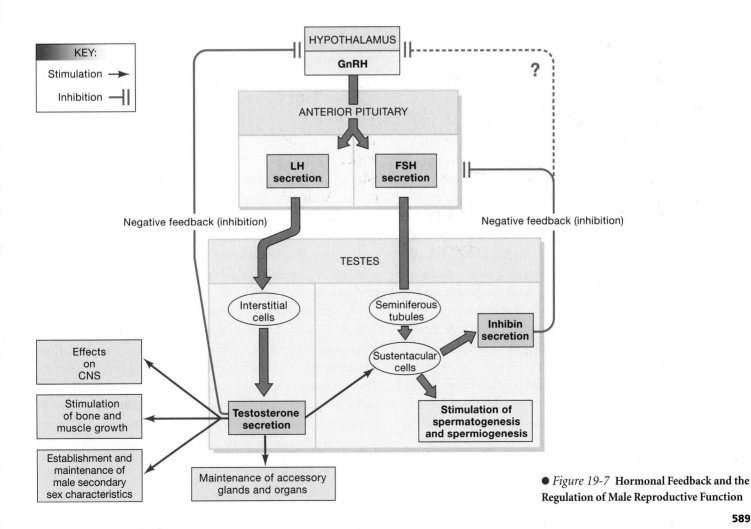

● *Figure 19-7* **Hormonal Feedback and the Regulation of Male Reproductive Function**

19 THE REPRODUCTIVE SYSTEM

The Reproductive System of the Male • **The Reproductive System of the Female** • The Physiology of Sexual Intercourse • Aging and the Reproductive System

releases **follicle-stimulating hormone (FSH)** and **luteinizing hormone (LH)**. (This hormone was called *interstitial cell-stimulating hormone, ICSH*, in males before it was known to be identical to the hormone in females.) The pituitary release of these hormones occurs in the presence of **gonadotropin-releasing hormone (GnRH)**, a peptide synthesized in the hypothalamus and carried to the anterior pituitary by the hypophyseal portal system.

FSH and Spermatogenesis

In the male, FSH targets primarily the sustentacular cells of the seminiferous tubules. Under FSH stimulation, and in the presence of testosterone from the interstitial cells, sustentacular cells promote spermatogenesis and spermiogenesis.

The rate of spermatogenesis is regulated by a negative-feedback mechanism involving GnRH, FSH, and inhibin. Under GnRH stimulation, FSH promotes spermatogenesis along the seminiferous tubules. As spermatogenesis accelerates, however, so does the rate of inhibin secretion by the sustentacular cells of the testes. Inhibin inhibits FSH production in the anterior pituitary and may also suppress secretion of GnRH at the hypothalamus.

The net effect is that when FSH levels become elevated, inhibin production increases until the FSH levels return to normal. If FSH levels decline, inhibin production falls, and the rate of FSH production accelerates.

LH and Androgen Production

In males, LH causes the secretion of testosterone and other androgens by the interstitial cells of the testes. Testosterone, the most important androgen, has numerous functions, such as (1) stimulating spermatogenesis and promoting the functional maturation of spermatozoa; (2) maintaining the accessory glands and organs of the male reproductive tract; (3) determining the secondary sex characteristics, such as facial hair, increased muscle mass and body size, and the quantity and location of adipose tissue; (4) stimulating metabolic operations throughout the body, especially those concerned with protein synthesis and muscle growth; and (5) affecting central nervous system (CNS) function, including the influence of sexual drive (libido) and related behaviors.

Testosterone production begins around the seventh week of development and reaches a peak after roughly 6 months of development. The early surge in testosterone levels stimulates the differentiation of the male duct system and accessory organs. Testosterone secretion accelerates markedly at puberty, initiating sexual maturation and the appearance of secondary sex characteristics. In adult males, negative feedback controls the level of testosterone production. Above-normal testosterone levels inhibit the release of GnRH by the hypothalamus. This inhibition causes a reduction in LH secretion and lowers testosterone levels.

CONCEPT CHECK QUESTIONS
Answers on page 610

❶ On a warm day, would the cremaster muscle be contracted or relaxed? Why?

❷ What will occur if the arteries within the penis dilate?

❸ What effect would low levels of FSH have on sperm production?

The Reproductive System of the Female

A woman's reproductive system produces sex hormones and gametes and also must be able to protect and support a developing embryo and nourish the newborn infant. The principal organs of the female reproductive system are the *ovaries*, the *uterine tubes*, the *uterus* (womb), the *vagina*, and the components of the external genitalia (Figure 19-8●). As in males, a variety of accessory glands secrete into the reproductive tract.

THE OVARIES

The paired ovaries are small, almond-shaped organs near the lateral walls of the pelvic cavity. The ovaries are responsible for (1) the production of female gametes, or **ova** (singular *ovum*); (2) the secretion of female sex hormones, including *estrogens* and *progestins*; and (3) the secretion of inhibin, involved in the feedback control of pituitary FSH production.

A typical **ovary** is a flattened oval that measures approximately 5 cm by 2.5 cm by 8 mm (2 in. by 1 in. by 0.33 in.). It has a pale white or yellowish coloration and a nodular consistency that resembles cottage cheese or lumpy oatmeal. The interior of the ovary is composed of a superficial *cortex* and a deep *medulla*. The production of gametes occurs in the cortex, and the arteries, veins, lymphatics, and nerves within the relatively narrow medulla link the ovary with other body systems.

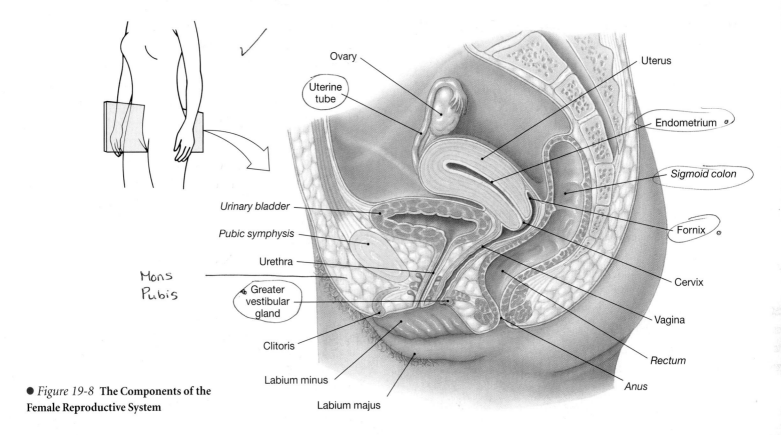

● *Figure 19-8* **The Components of the Female Reproductive System**

Labels: Ovary, Uterine tube, Uterus, Endometrium, Sigmoid colon, Fornix, Urinary bladder, Pubic symphysis, Urethra, Greater vestibular gland, Clitoris, Labium minus, Labium majus, Mons Pubis, Cervix, Vagina, Rectum, Anus

The position of each ovary is stabilized by a mesentery and by a pair of supporting ligaments (see Figure 19-11●, p 594). The mesentery also encloses the uterine tubes and uterus. The ligaments attached to each ovary extend to the uterus and pelvic wall. The latter contain the major blood vessels of each ovary, the *ovarian artery* and *ovarian vein*.

Oogenesis

Ovum production, or **oogenesis** (ō-ō-JEN-e-sis; *oon*, egg), begins before a woman's birth, accelerates at puberty, and ends at *menopause* (*men*, month + *pausis*, cessation). Between puberty and menopause, oogenesis occurs monthly, as part of the *ovarian cycle*. Oogenesis is illustrated in Figure 19-9●; as in Figure 19-3● (p. 585) on spermatogenesis, only 3 of the 23 pairs of chromosomes are illustrated.

In the ovaries, stem cells, or **oogonia** (ō-ō-GŌ-nē-a), complete their mitotic divisions before birth. Between the third and seventh months of fetal development, the daughter cells, or **primary oocytes** (Ō-ō-sīts), prepare to undergo meiosis (Figure 19-9●). They proceed as far as prophase of meiosis I, but at that time the process stops. The primary oocytes then remain in a state of suspended development until the individual reaches puberty, awaiting the hormonal signal to complete meiosis. Not all of the primary oocytes in the ovaries at birth survive until puberty. There are roughly 2 million in the ovaries at birth; by the time of puberty, about 400,000 remain. The rest of the primary oocytes degenerate, a process called *atresia* (a-TRĒ-zē-a).

Although the nuclear events under way during meiosis in the ovary are the same as those in the testis, the process differs in two important details.

1. The cytoplasm of the original oocyte is not evenly distributed during the meiotic divisions. Oogenesis produces one functional ovum, containing most of that cytoplasm, and three nonfunctional **polar bodies** that later disintegrate.
2. The ovary releases a *secondary oocyte*, rather than a mature ovum. The second meiotic division is not completed until *after* fertilization.

The Ovarian Cycle

Oogenesis occurs in the cortex of the ovaries within specialized structures called **ovarian follicles** (ō-VAR-ē-an FOL-i-klz). In the outer cortex, just beneath the capsule, are clusters of primary oocytes, each surrounded by a layer of follicle cells. The combination is known as a **primordial** (prī-MŌR-dē-al) **follicle**. At puberty, an unknown set of signals begins to

591

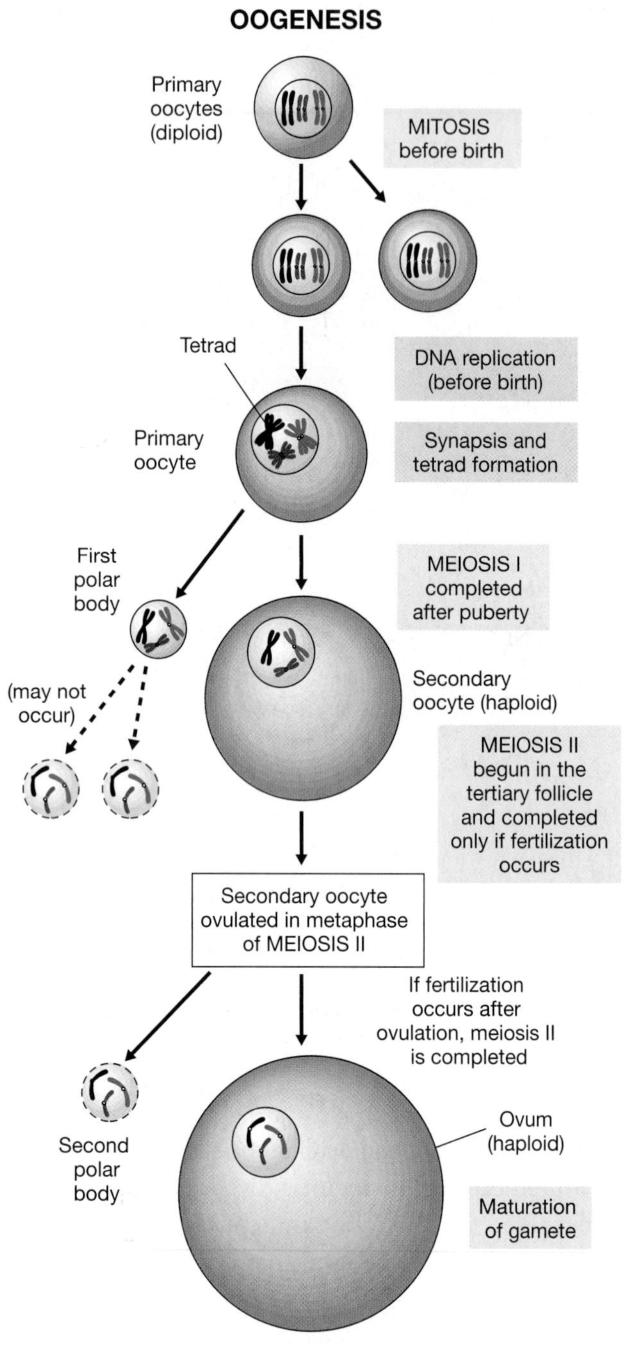

19 **THE REPRODUCTIVE SYSTEM**

The Reproductive System of the Male • **The Reproductive System of the Female** • The Physiology of Sexual Intercourse • Aging and the Reproductive System

OOGENESIS

Primary oocytes (diploid)

MITOSIS before birth

Tetrad

DNA replication (before birth)

Primary oocyte

Synapsis and tetrad formation

First polar body

MEIOSIS I completed after puberty

(may not occur)

Secondary oocyte (haploid)

MEIOSIS II begun in the tertiary follicle and completed only if fertilization occurs

Secondary oocyte ovulated in metaphase of MEIOSIS II

If fertilization occurs after ovulation, meiosis II is completed

Second polar body

Ovum (haploid)

Maturation of gamete

● *Figure 19-9* **Oogenesis**
The production of an ovum.

activate a different group of primordial follicles each month. This monthly process is known as the **ovarian cycle**. Important steps in the ovarian cycle are shown in Figure 19-10●:

Step 1: *Formation of primary follicles.* The cycle begins as the activated primordial follicles develop into **primary follicles**. The follicular cells enlarge, divide, and form several concentric layers of cells around the oocyte.

Microvilli from the surrounding follicular cells intermingle with microvilli originating at the surface of the oocyte. This region is called the **zona pellucida** (ZŌ-na pel-LŪ-si-da; *pellucidus*, translucent). The microvilli increase the surface area available for the transfer of materials from the follicular cells to the developing oocyte.

Step 2: *Formation of secondary follicles.* Although many primordial follicles develop into primary follicles, usually only a few will take the next step under stimulation by rising FSH levels. The transformation begins as the wall of the follicle thickens and the deeper follicular cells begin secreting small amounts of fluid. This *follicular fluid* accumulates in small pockets that gradually expand and separate the inner and outer layers of the follicle. At this stage, the complex is known as a **secondary follicle**. Although the primary oocyte continues to grow slowly, the follicle as a whole now enlarges rapidly because follicular fluid accumulates.

Step 3: *Formation of tertiary follicles.* Eight to 10 days after the start of the ovarian cycle, the ovaries usually contain only a single secondary follicle destined for further development. By days 10 to 14 of the cycle, that follicle has formed a **tertiary follicle**, or mature *Graafian* (GRAF-ē-an) *follicle*, roughly 15 mm in diameter. This complex spans the entire width of the ovarian cortex and stretches the ovarian capsule, creating a prominent bulge in the surface of the ovary. The oocyte, surrounded by a mass of follicular cells, projects into the expanded central chamber of the follicle, the **antrum** (AN-trum).

Until this time, the primary oocyte has been suspended in prophase of meiosis I. As the development of the tertiary follicle ends, rising LH levels prompt the primary oocyte to complete meiosis. The first meiotic division finishes and produces a secondary oocyte and nonfunctional polar body. The secondary oocyte begins meiosis II but stops short of dividing. Meiosis II will not be completed unless fertilization occurs.

Step 4: *Ovulation.* As the time of egg release, or **ovulation** (ōv-ū-LĀ-shun), approaches, the follicular cells surrounding the oocyte lose contact with the follicular wall, and the oocyte floats within the central chamber. This event usually occurs at day 14 of a 28-day cycle. The follicular cells surrounding the oocyte are now known as the *corona radiata* (ko-RŌ-na rā-dē-A-ta). The distended follicular wall then ruptures, releasing the follicular

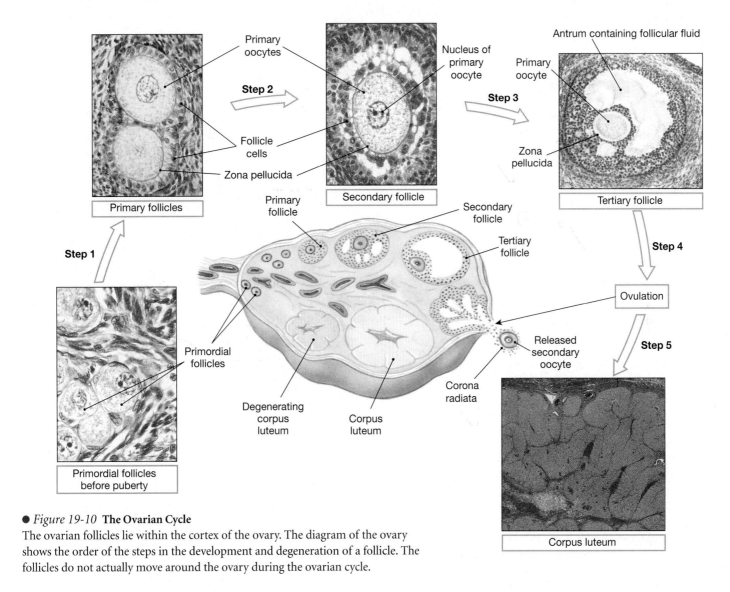

● *Figure 19-10* The Ovarian Cycle
The ovarian follicles lie within the cortex of the ovary. The diagram of the ovary shows the order of the steps in the development and degeneration of a follicle. The follicles do not actually move around the ovary during the ovarian cycle.

contents, including the secondary oocyte, into the pelvic cavity. The sticky follicular fluid keeps the corona radiata attached to the surface of the ovary near the ruptured wall of the follicle. Contact with projections of the uterine tube or with fluid currents established by its ciliated epithelium then sweeps the secondary oocyte into the uterine tube.

Step 5: *Formation and degeneration of the corpus luteum.* The empty follicle collapses, and the remaining follicular cells invade the cavity and multiply to create an endocrine structure known as the **corpus luteum** (LOO-tē-um; *lutea*, yellow). Unless fertilization occurs, the corpus luteum begins to degenerate roughly 12 days after ovulation. The disintegration marks the end of the ovarian cycle, but almost immediately the activation of another set of primordial follicles begins the next ovarian cycle.

THE UTERINE TUBES

Each **uterine tube** (*Fallopian tube*, or *oviduct*) measures roughly 13 cm (5 in.) in length. The end closest to the ovary forms an expanded funnel, or **infundibulum** (in-fun-DIB-ū-lum; *infundibulum*, a funnel), with numerous fingerlike projections that extend into the pelvic cavity (Figure 19-11●). The projections, called **fimbriae** (FIM-brē-ē), and the inner surfaces of the infundibulum are carpeted with cilia that beat toward the broad entrance to the uterine tube. Once inside the uterine tube, the ovum is transported by ciliary movement and peristaltic contractions by smooth muscles in the walls of the uterine tubes. It normally takes 3–4 days for the secondary oocyte to travel from the infundibulum to the uterine cavity. *If fertilization is to occur, the secondary oocyte must encounter spermatozoa during the first 12–24 hours of its passage.* Unfertilized oocytes will degenerate, in the uterine tubes or uterus, without completing meiosis.

19 THE REPRODUCTIVE SYSTEM

The Reproductive System of the Male • **The Reproductive System of the Female** • The Physiology of Sexual Intercourse • Aging and the Reproductive System

● *Figure 19-11*

The Uterus

A posterior view with the left portion of uterus, uterine tube, and ovary shown in section.

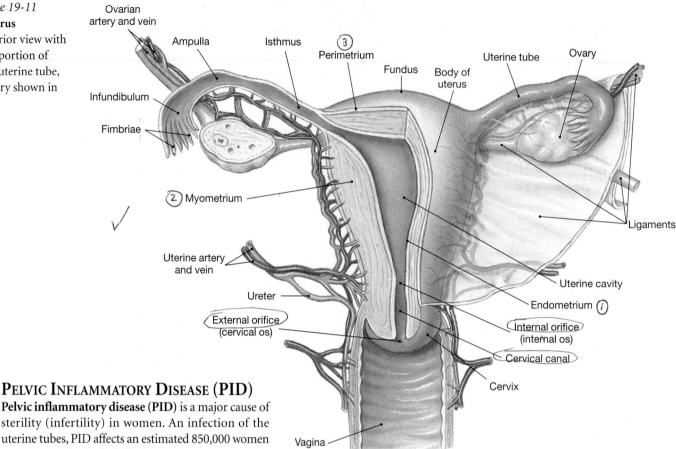

PELVIC INFLAMMATORY DISEASE (PID)

Pelvic inflammatory disease (PID) is a major cause of sterility (infertility) in women. An infection of the uterine tubes, PID affects an estimated 850,000 women each year in the United States. In many cases, sexually transmitted pathogens are involved. As much as 50–80 percent of all first cases may be due to infection by *Neisseria gonorrhoeae*, the organism responsible for symptoms of **gonorrhea** (gon-ō-RĒ-a), a sexually transmitted disease. PID may also result from invasion of the region by bacteria normally found within the vagina. Symptoms of pelvic inflammatory disease include fever, lower abdominal pain, and elevated white blood cell counts. In severe cases, the infection can spread to other visceral organs or produce a generalized peritonitis.

Sexually active women in the 15–24 age group have the highest incidence of PID. Whereas the use of an oral contraceptive decreases the risk of infection, the presence of an intrauterine device (IUD) may increase the risk by 1.4–7.3 times. Treatment with antibiotics may control the condition, but chronic abdominal pain may persist. In addition, damage and scarring of the uterine tubes may cause infertility by preventing the passage of a zygote to the uterus. Recently, another sexually transmitted bacterium, belonging to the genus *Chlamydia*, has been identified as the probable cause of up to 50 percent of all cases of PID. Despite the fact that women with this infection may develop few if any symptoms, scarring of the uterine tubes can still produce infertility.

THE UTERUS

The **uterus** (Ū-ter-us) is a muscular chamber that provides mechanical protection and nutritional support for the devel-

oping embryo (weeks 1–8) and fetus (from week 9 to delivery). In addition, contractions in the muscular wall of the uterus are important in ejecting the fetus at the time of birth.

The typical uterus is a small, pear-shaped organ about 7.5 cm (3 in.) in length with a maximum diameter of 5 cm (2 in.). It weighs 30–40 g (1–1.4 oz.) and is stabilized by various ligaments. In its normal position, the uterus bends anteriorly near its base (see Figure 19-8●, p. 591).

The uterus consists of two regions: the body and the cervix (Figure 19-11●). The **body** is the largest division of the uterus. The *fundus* is the rounded portion of the body superior to the attachment of the uterine tubes. The body ends at a constriction known as the **isthmus**. The **cervix** (SER-viks) is the inferior portion of the uterus. The cervix projects a short distance into the vagina, and the **uterine cavity** opens into the vagina at the **external orifice**, or *cervical os*.

The uterine wall is made up of an inner **endometrium** (en-dō-MĒ-trē-um) and a muscular **myometrium** (mī-ō-MĒ-trē-um; *myo-*, muscle + *metra*, uterus), covered by the **perimetrium**, a layer of visceral peritoneum (Figure 19-11●). The endometrium of the uterus includes the epithelium lining the uterine cavity and the underlying connective tissues. Uterine glands opening onto the endometrial surface extend deep into the connective tissue layer almost all the way to the myometrium. The myometrium consists of a thick mass of interwoven smooth muscle cells. In adult women of reproductive age who have not given birth, the uterine wall is about 1.5 cm (0.5 in.) thick.

The endometrium consists of a superficial *functional zone* and a deeper *basilar zone* that is adjacent to the myometrium. The structure of the basilar layer remains relatively constant over time, but that of the functional zone undergoes cyclical changes in response to sex hormone levels. These alterations produce the characteristic features of the uterine cycle.

The Uterine Cycle

The **uterine cycle**, or *menstrual* (MEN-stroo-al) *cycle*, is a repeating series of changes in the structure of the endometrium. This cycle of events begins with the **menarche** (me-NAR-kē), or first menstrual period at puberty, typically age 11–12. The cycles continue until age 45–55, when **menopause** (MEN-ō-paws), the last menstrual cycle, occurs. In the interim, the regular appearance of menstrual cycles is interrupted only by circumstances, such as illness, stress, starvation, or pregnancy.

The uterine cycle averages 28 days in length, but it can range from 21 to 35 days in normal individuals. It consists of three stages: *menses*, the *proliferative phase*, and the *secretory phase*.

Menses. The menstrual cycle begins with the onset of **menses** (MEN-sēz), a period marked by the degeneration of the superficial layer, or *functional zone*, of the endometrium. The process is triggered by the decline in progesterone and estrogen levels as the corpus luteum disintegrates. The endometrial arteries constrict, reducing blood flow to this region, and the secretory glands, epithelial cells, and other tissues of the functional zone die of oxygen and nutrient deprivation. Eventually the weakened arterial walls rupture, and blood pours into the connective tissues of the functional zone. Blood cells and degenerating tissues break away and enter the uterine cavity, to be lost by passage into the vagina. This sloughing of tissue, which continues until the entire functional zone has been lost, is called **menstruation** (men-stroo-Ā-shun). Menstruation usually lasts 1 to 7 days, and roughly 35 to 50 ml of blood is lost. Painful menstruation, or *dysmenorrhea*, can result from uterine inflammation and contraction or from conditions involving adjacent pelvic structures.

The Proliferative Phase. The **proliferative phase** begins in the days following the completion of menses as the surviving epithelial cells multiply and spread across the surface of the endometrium. This repair process is stimulated by the rising estrogen levels that accompany the growth of another set of ovarian follicles. By the time ovulation occurs, the functional zone is several millimeters thick and its new set of uterine glands are secreting a watery mucus. In addition, the entire functional zone is filled with small arteries that branch from larger trunks in the myometrium.

The Secretory Phase. During the **secretory phase** of the cycle, the uterine glands enlarge, steadily increasing their rates of secretion as the endometrium prepares for the arrival of a developing embryo. This activity is stimulated by the progestins and estrogens from the corpus luteum. This phase begins at the time of ovulation and persists as long as the corpus luteum remains intact. Secretory activities peak about 12 days after ovulation. Over the next day or two the glandular activity declines, and the uterine cycle comes to a close. A new cycle then begins with the onset of menses and the disintegration of the functional zone.

AMENORRHEA

If menarche does not appear by age 16, or if the normal menstrual cycle of an adult becomes interrupted for 6 months or more, the condition of **amenorrhea** (ā-men-ō-RĒ-uh) exists. *Primary amenorrhea* is the failure to initiate menses. This condition may indicate developmental abnormalities, such as nonfunctional ovaries or even the absence of a uterus, or an endocrine or genetic disorder. It can also result from malnutrition; puberty is delayed if leptin levels are too low. p. 332 Transient *secondary amenorrhea* may be caused by severe physical or emotional stresses. In effect, the reproductive system gets "switched off." Factors that cause either type of amenorrhea include drastic weight reduction programs, anorexia nervosa, and severe depression or grief. Amenorrhea has also been observed in marathon runners and other women engaged in training programs that require sustained high levels of exertion and severely reduce body lipid reserves.

THE VAGINA

The **vagina** (va-JĪ-na) is an elastic, muscular tube extending between the uterus and the vestibule, a space bounded by the external genitalia (Figures 19-8●, p. 591, and 19-12●). The vagina has an average length of 7.5–9 cm (3–3.5 in.) but is highly distensible, so its size varies. The cervix of the uterus projects into the vagina. The shallow recess surrounding the cervical protrusion is known as the **fornix** (FOR-niks). The vagina lies parallel to the rectum, and the two are in close

19 THE REPRODUCTIVE SYSTEM

The Reproductive System of the Male • **The Reproductive System of the Female** • The Physiology of Sexual Intercourse • Aging and the Reproductive System

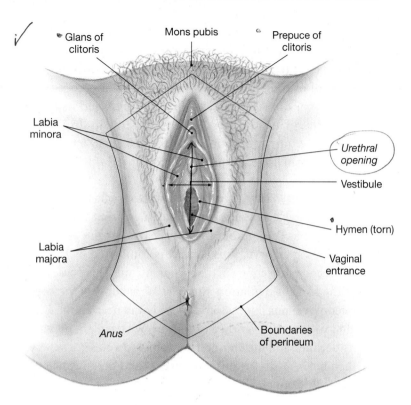

● *Figure 19-12* **The Female External Genitalia**

contact posteriorly. The urethra extends along the superior wall of the vagina from the urinary bladder to the urethral opening, or external urethral meatus.

The vaginal walls contain a network of blood vessels and layers of smooth muscle. The lining is moistened by the mucous secretions of the cervical glands and by the movement of water across the permeable epithelium. The vagina and vestibule are separated by an elastic epithelial fold, the **hymen** (HĪ-men) that partially or completely blocks the entrance to the vagina before the first occasion of sexual intercourse. The two bulbospongiosus muscles pass on either side of the vaginal entrance; their contractions constrict the entrance. p. 205

The vagina (1) serves as a passageway for the elimination of menstrual fluids; (2) receives the penis during sexual intercourse, and holds spermatozoa prior to their passage into the uterus; and (3) during childbirth, forms the lower portion of the birth canal through which the fetus passes during delivery.

The vagina normally contains resident bacteria supported by nutrients in the cervical mucus. As a result of their metabolic activities, the normal pH of the vagina ranges between 3.5 and 4.5, and this acidic environment restricts the growth of many pathogens. An inflammation of the vaginal canal, known as *vaginitis* (va-jin-Ī-tis), is caused by fungal, bacterial, or parasitic organisms. In addition to any discomfort that may result, the condition may affect the survival of sperm, thereby reducing fertility.

THE EXTERNAL GENITALIA

The **perineum**, the muscular floor of the pelvic cavity, includes structures associated with the reproductive system called the **external genitalia**. p. 204 The perineal region enclosing the female external genitalia is the **vulva** (VUL-vuh), or *pudendum* (Figure 19-12●). The vagina opens into the **vestibule**, a central space bounded by the **labia minora** (LĀ-bē-a mi-NOR-a; *labia*, lips; singular *labium minus*). The labia minora are covered with a smooth, hairless skin. The urethra opens into the vestibule just anterior to the vaginal entrance. Anterior to the urethral opening, the **clitoris** (KLI-to-ris) projects into the vestibule. The clitoris is the female equivalent of the penis, derived from the same embryonic structures. Internally it contains erectile tissue comparable to the corpora cavernosa of the penis. A small erectile *glans* sits atop the organ, and extensions of the labia minora encircle the body of the clitoris, forming the *prepuce*.

A variable number of small **lesser vestibular glands** discharge secretions onto the exposed surface of the vestibule, keeping it moist. During **arousal**, a pair of ducts discharges the secretions of the **greater vestibular glands** into the vestibule near the vaginal entrance (Figure 19-8●). These mucous glands resemble the bulbourethral glands of males.

The outer limits of the vulva are formed by the mons pubis and labia majora. The prominent bulge of the **mons pubis** is created by adipose tissue beneath the skin anterior to the pubic symphysis. Adipose tissue also accumulates in the fleshy **labia majora** (singular *labium majus*), which encircle and partially conceal the labia minora and vestibular structures.

THE MAMMARY GLANDS

A newborn infant cannot fend for itself, and several of its key systems have yet to complete their development. While adjusting to an independent existence, the infant gains nourishment from the milk secreted by the maternal **mammary glands**. Milk production, or **lactation** (lak-TĀ-shun), occurs in the mammary glands of the **breasts**, specialized accessory organs of the female reproductive system (Figure 19-13●).

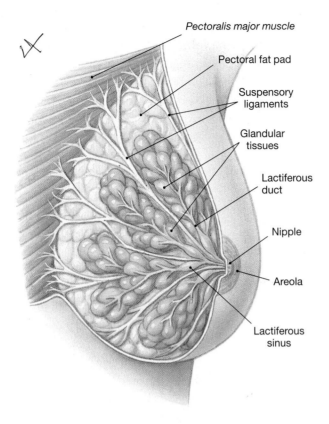

- *Figure 19-13* **The Mammary Glands**
Lobes of glandular tissue within each mammary gland are responsible for the production of breast milk.

On each side, a mammary gland lies in the subcutaneous layer beneath the skin of the chest. Each breast bears a **nipple**, a small conical projection where the ducts of underlying mammary glands open onto the body surface. The reddish-brown skin surrounding each nipple is the **areola** (a-RĒ-ō-la). Large sebaceous glands beneath the areolar surface give it a granular texture.

The glandular tissue of the breast consists of a number of separate lobes, each containing several secretory lobules. Within each lobe, the ducts leaving the lobules converge, giving rise to a single **lactiferous** (lak-TIF-e-rus) **duct**. Near the nipple, that lactiferous duct expands, forming an expanded chamber called a **lactiferous sinus**. Some 15–20 lactiferous sinuses open onto the surface of each nipple. Dense connective tissue surrounds the duct system and forms partitions that extend between the lobes and lobules. These bands of connective tissue, the *suspensory ligaments of the breast*, originate in the dermis of the overlying skin. A layer of loose connective tissue separates the mammary complex from the underlying muscles, and the two can move relatively independently.

BREAST CANCER

Breast cancer is a malignant, metastasizing cancer of the mammary gland. It is the leading cause of death in women between the ages of 35 and 45, but it is most common in women over age 50. For 2002, it is estimated that there will be 39,600 deaths and approximately 203,500 new cases of breast cancer in the United States. An estimated 12 percent of women in the United States will develop breast cancer at some point in their lifetime, and the rate is steadily rising. The incidence is highest among Caucasians, somewhat lower in African Americans, and lowest in Asians and American Indians. Notable risk factors include (1) a family history of breast cancer, (2) a pregnancy after age 30, and (3) early menarche (first menstrual period) or late menopause (last menstrual period). Breast cancers in males are very rare, but about 400 men die of breast cancer each year in the United States.

CONCEPT CHECK QUESTIONS
Answers on page 610

❶ As the result of pelvic inflammatory disease, scar tissue can block the lumen of each uterine tube. How would this blockage affect a woman's ability to conceive?

❷ What is the advantage of the normally acidic pH of the vagina?

❸ Which layer of the uterus sloughs off during menstruation?

❹ Would the blockage of a single lactiferous sinus interfere with delivery of milk to the nipple? Explain.

HORMONES AND THE FEMALE REPRODUCTIVE CYCLE

The activity of the female reproductive tract is under hormonal control by both pituitary and gonadal secretions. But the regulatory pattern in females is much more complicated than in males, for a woman's reproductive system does not just produce functional gametes, it must also coordinate the ovarian and uterine cycles. Circulating hormones, especially estrogen, control the **female reproductive cycle** and coordinate the ovarian and uterine cycles to ensure proper reproductive function. If the two cycles cannot be coordinated normally, infertility results. A woman who fails to ovulate will be unable to conceive, even if her uterus is perfectly normal. A woman who ovulates normally but whose uterus isn't ready to support an embryo will be just as infertile.

Hormones and the Follicular Phase

Follicular development begins under FSH stimulation, and each month some of the primordial follicles begin their development into primary follicles. As the follicular cells enlarge

19 THE REPRODUCTIVE SYSTEM

The Reproductive System of the Male • **The Reproductive System of the Female** • The Physiology of Sexual Intercourse • Aging and the Reproductive System

and multiply, they release steroid hormones collectively known as **estrogens**. The most important estrogen is **estradiol** (es-tra-DĪ-ol). Estrogens have multiple functions including (1) stimulating bone and muscle growth; (2) establishing and maintaining female secondary sex characteristics, such as body hair distribution and the location of adipose tissue deposits; (3) affecting central nervous system (CNS) activity (especially in the hypothalamus, where estrogens increase sexual drive); (4) maintaining functional accessory reproductive glands and organs; and (5) initiating the repair and growth of the endometrium. Figure 19-14● diagrams the hormonal control of ovarian activity, including the effects of estrogens on various aspects of reproductive function.

Figure 19-15a–c● summarizes the hormonal events associated with the ovarian cycle. Its cyclic pattern of hormonal regulation differs between the follicular phase and luteal phase. Early in the follicular phase, estrogen and inhibin levels are low. The estrogens and inhibin have complementary effects on the secretion of FSH and LH. Low levels of estrogen inhibit LH secretion. As follicular development proceeds, the concentration of circulating estrogens and inhibin rises, for the follicular cells are increasing in number and secretory activity. As secondary follicles develop, FSH levels decline due to the negative feedback effects of inhibin, and estrogen levels continue to rise. Despite a slow decline in FSH concentrations, the combination of estrogens, FSH, and LH continues to support follicular development and maturation.

Estrogen concentrations take a sharp upturn in the second week of the ovarian cycle, with the development of a tertiary follicle as it enlarges in preparation for ovulation. The rapid increase in estrogen leads to the secretion of FSH and LH by acting on the hypothalamus and stimulating the production of GnRH. (Figure 19-15a● shows the increases in FSH and LH at this time.) At roughly day 10 of the cycle, the effect of estrogen on LH secretion also changes from inhibition to stimulation. At about day 14, estrogen levels peak, accompanying the maturation of the tertiary follicle. The high estrogen concentration then triggers a massive outpouring, or surge, of LH from the anterior pituitary, which triggers the rupture of the follicular wall and ovulation.

Hormones and the Luteal Phase

The high LH levels that trigger ovulation also stimulate the remaining follicular cells to form a corpus luteum. The yellow color of the corpus luteum results from its lipid reserves. These compounds are used to manufacture steroid hormones known as **progestins** (prō-JES-tinz), predominantly the steroid **progesterone** (prō-JES-ter-ōn). Progesterone is the principal hormone of the luteal phase. It prepares the uterus for pregnancy by stimulating the growth and development of the blood supply and secretory glands of the endometrium. Progesterone also stimulates metabolic activity and elevates basal body temperature.

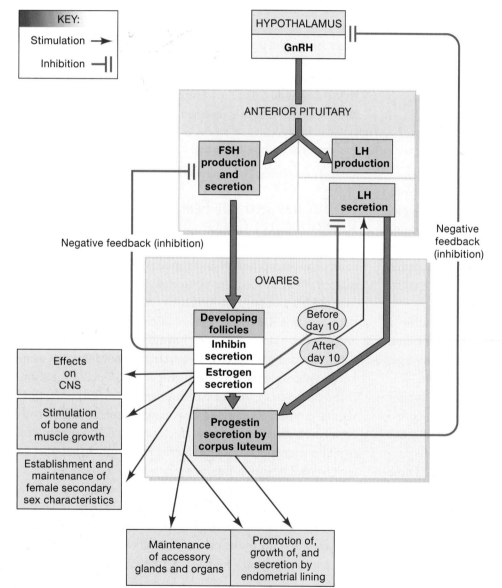

● *Figure 19-14* **Hormonal Regulation of Ovarian Activity**

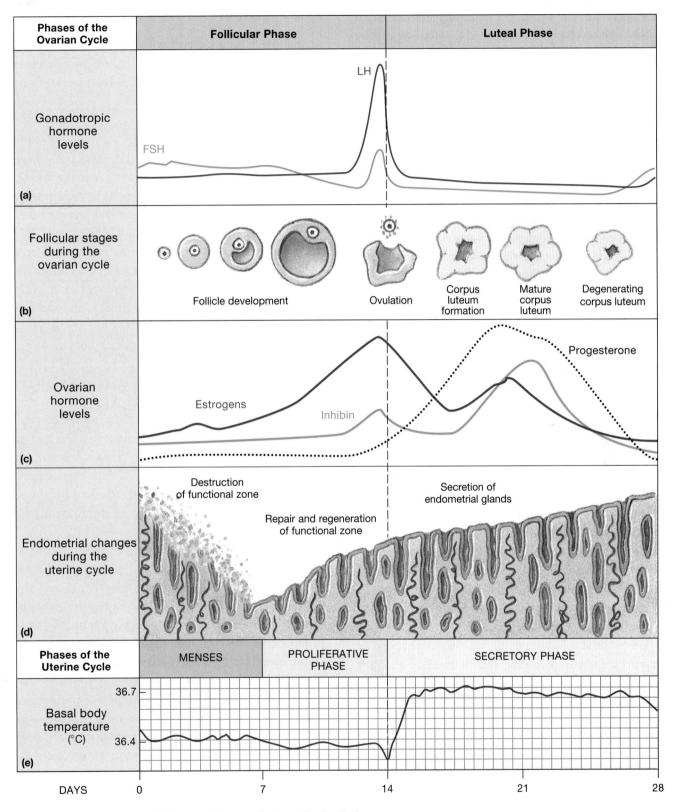

Phases of the Ovarian Cycle	Follicular Phase	Luteal Phase

(a) Gonadotropic hormone levels

LH

FSH

(b) Follicular stages during the ovarian cycle

Follicle development | Ovulation | Corpus luteum formation | Mature corpus luteum | Degenerating corpus luteum

(c) Ovarian hormone levels

Progesterone

Estrogens

Inhibin

(d) Endometrial changes during the uterine cycle

Destruction of functional zone

Repair and regeneration of functional zone

Secretion of endometrial glands

Phases of the Uterine Cycle	MENSES	PROLIFERATIVE PHASE	SECRETORY PHASE

(e) Basal body temperature (°C)

36.7

36.4

DAYS   0   7   14   21   28

● *Figure 19-15* **Hormonal Regulation of the Female Reproductive Cycle**

19 THE REPRODUCTIVE SYSTEM

The Reproductive System of the Male • The Reproductive System of the Female • **The Physiology of Sexual Intercourse** • Aging and the Reproductive System

Luteinizing hormone (LH) levels remain elevated for only 2 days, but that is long enough to stimulate the formation of the functional corpus luteum. Progesterone secretion continues at relatively high levels for the next week. Unless pregnancy occurs, however, the corpus luteum then begins to degenerate. Roughly 12 days after ovulation, the corpus luteum becomes nonfunctional, and progesterone, estrogen, and inhibin levels fall markedly. This decline stimulates the hypothalamus, and GnRH production increases. This increase in turn leads to an increase in the production of both FSH and LH in the anterior pituitary gland, and the entire cycle begins again.

Hormones and the Uterine Cycle

Figure 19-15d● follows changes in the endometrium during a single uterine cycle. The sudden declines in progesterone, estrogen, and inhibin levels that accompany the breakdown of the corpus luteum result in menses. The loss of endometrial tissue continues for several days, until rising estrogen levels stimulate the regeneration of the functional zone of the endometrium.

The follicular phase continues until rising progesterone levels mark the arrival of the luteal phase. The combination of high levels of estrogen and progesterone then causes the enlargement of the endometrial glands as well as an increase in their secretions.

Hormones and Body Temperature

The hormonal fluctuations also cause physiological changes that affect the core body temperature. During the follicular phase, when estrogen is the dominant hormone, the resting, or basal, body temperature measured on awakening in the morning is about 0.3°C (or 0.5°F) lower than it is during the luteal phase, when progesterone dominates. At the time of ovulation, basal temperature declines sharply, making the temperature rise over the following day even more noticeable (Figure 19-15e●). By keeping records of body temperature over a few menstrual cycles, a woman can often determine the precise day of ovulation. This information can be very important for those wishing to avoid or promote a pregnancy, for pregnancy can occur only if an ovum becomes fertilized within a day of its ovulation.

CONCEPT CHECK QUESTIONS

Answers on page 610

❶ What changes would you expect to observe in the ovarian cycle if the LH surge did not occur?

❷ What effect would blockage of progesterone receptors in the uterus have on the endometrium?

❸ What event occurs in the uterine cycle when the levels of estrogen and progesterone decline?

The Physiology of Sexual Intercourse

Sexual intercourse, or **coitus** (KŌ-i-tus), introduces semen into the female reproductive tract. The following sections consider the process as it affects the reproductive systems of males and females.

MALE SEXUAL FUNCTION

Male sexual function is coordinated by reflex pathways involving both divisions of the ANS. During **sexual arousal**, erotic thoughts or the stimulation of sensory nerves in the genital region increase the parasympathetic outflow over the pelvic nerves. This outflow leads to erection of the penis. The skin of the glans of the penis contains numerous sensory receptors, and erection tenses the skin and increases their sensitivity. Subsequent stimulation may initiate the secretion of the bulbourethral glands, lubricating the urethra and the surface of the glans.

During intercourse, the sensory receptors in the penis are rhythmically stimulated, eventually resulting in the coordinated processes of emission and ejaculation. **Emission** occurs under sympathetic stimulation. The process begins with peristaltic contractions of the ductus deferens, which push fluid and spermatozoa through the ejaculatory ducts and into the urethra. The seminal vesicles then contract, followed by waves of contraction in the prostate gland. While these contractions are proceeding, sympathetic commands close the sphincter at the entrance to the urinary bladder, preventing the passage of semen into the bladder.

Ejaculation occurs as powerful, rhythmic contractions appear in the *ischiocavernosus* and *bulbospongiosus* muscles, two superficial skeletal muscles of the pelvic floor. (The positions of these muscles can be seen in Figure 7-16●, p. 205.) Ejaculation is associated with intensely pleasurable sensations, an experience known as male **orgasm** (ŌR-gazm). Several other physiological changes also occur at this time, including temporary increases in heart rate and blood pressure. After ejaculation, blood begins to leave the erectile tissue, and the erection begins to subside. This subsidence, called *detumescence* (de-tū-MES-ens), is mediated by the sympathetic nervous system. An inability to achieve or maintain an erection is called **impotence**.

FEMALE SEXUAL FUNCTION

The phases of female sexual function are comparable to those of male sexual function. During sexual arousal, parasympathetic activation leads to an engorgement of the erectile tissues of the clitoris and increased secretion of cervical mucous glands and the greater vestibular glands. Clitoral erection increases the receptors' sensitivity to stimulation, and the cervical and vestibular glands provide lubrication for the vaginal walls. A network of blood vessels in the vaginal walls becomes filled with blood at this time, and the vaginal surfaces are also moistened by fluid from underlying connective tissues. Parasympathetic stimulation also causes engorgement of blood vessels at the nipples, making them more sensitive to touch and pressure.

During intercourse, rhythmic contact with the clitoris and vaginal walls, reinforced by touch sensations from the breasts and other stimuli (visual, olfactory, and auditory), provides stimulation that can lead to orgasm. Female orgasm is accompanied by peristaltic contractions of the uterine and vaginal walls and, by means of impulses over the pudendal nerves, rhythmic contractions of the bulbospongiosus and ischiocavernosus muscles. The latter contractions give rise to the sensations of orgasm.

 ### SEXUALLY TRANSMITTED DISEASES

Sexually transmitted diseases (STDs) are transferred from individual to individual, primarily or exclusively by sexual intercourse. A variety of bacterial, viral, and fungal infections are included in this category. At least two dozen different STDs are currently recognized. The bacterium *Chlamydia* can cause PID and infertility. Other types of STDs are quite dangerous, and a few, including AIDS, are deadly. The incidence of STDs has been increasing in the United States since 1984; an estimated 15 million cases were diagnosed during 1999. Poverty, coupled with drug use, prostitution, and the appearance of drug-resistant pathogens all contribute to the problem.

Aging and the Reproductive System

Just as the aging process affects our other systems, it also affects the reproductive systems of men and women. These systems become fully functional at puberty. Thereafter, the most striking age-related changes in the female reproductive system occur at menopause. Comparative changes in the male reproductive system occur more gradually and over a longer period of time.

MENOPAUSE

Menopause is usually defined as the time that ovulation and menstruation cease. It typically occurs at age 45–55, but in the years preceding it, the ovarian and menstrual cycles become irregular. A shortage of primordial follicles is the underlying cause of the cycle irregularities. It has been estimated that there are about 2 million potential oocytes at birth and a few hundred thousand at puberty. By age 50, there are often no primordial follicles left to respond to FSH. In *premature menopause*, this depletion occurs before age 40.

Menopause is accompanied by a sharp and sustained rise in the production of GnRH, FSH, and LH, while circulating concentrations of estrogen and progesterone decline. The decline in estrogen levels leads to reductions in the size of the uterus and breasts, accompanied by a thinning of the urethral and vaginal walls. The reduced estrogen concentrations have also been linked to the development of osteoporosis, presumably because bone deposition proceeds at a slower rate. A variety of neural effects are also reported, including "hot flashes," anxiety, and depression, but the hormonal mechanisms involved are not well understood. In addition, the risk of atherosclerosis and other forms of cardiovascular disease increase after menopause.

The majority of women experience only mild symptoms, but some individuals experience unpleasant symptoms during or after menopause. For most of those individuals, hormone replacement therapies involving a combination of estrogens and progestins can prevent osteoporosis and the neural and vascular changes associated with menopause.

THE MALE CLIMACTERIC

Changes in the male reproductive system occur more gradually than do those in the female reproductive system. The period of change is known as the **male climacteric**. Levels of circulating testosterone begin to decline between ages 50 and 60, and levels of FSH and LH increase. Although sperm production continues (men well into their eighties can father children), older men experience a gradual reduction in sexual activity. This decrease may be linked to declining testosterone levels, and some clinicians suggest the use of testosterone replacement therapy to enhance libido (sexual drive) in elderly men.

CONCEPT CHECK QUESTIONS
Answers on page 610

❶ The inability to contract the ischiocavernosus and bulbospongiosus muscles would interfere with which part of the male sexual function?

❷ What changes occur in the female during sexual arousal as the result of increased parasympathetic stimulation?

❸ Why does the level of FSH rise and remain high during menopause?

19 THE REPRODUCTIVE SYSTEM

The Reproductive System of the Male • The Reproductive System of the Female • The Physiology of Sexual Intercourse • Aging and the Reproductive System

Integration with Other Systems

Normal human reproduction is a complex process that requires the participation of multiple body systems. Hormones play a major role in coordinating these events, and Table 19-1 reviews the hormones discussed in this chapter. The reproductive process depends on various physical, physiological, and psychological factors, many of which require cooperation between the reproductive system and other organ systems. Figure 19-16● summarizes these relationships. For example, the male's sperm count must be adequate, the semen must have the correct pH and nutrients, and erection and ejaculation must occur in the proper sequence; the woman's ovarian and uterine cycles must be properly coordinated, ovulation and oocyte transport must occur normally, and her reproductive tract must be suitable for sperm survival, movement, and fertilization. For these steps to occur, the reproductive, digestive, endocrine, nervous, cardiovascular, and urinary systems must all be functioning normally.

Even when all else is normal, and fertilization occurs at the proper time and place, a normal infant will not result unless the zygote, a single cell the size of a pinhead, manages to develop into a full-term fetus that typically weighs about 3 kg. Chapter 20 considers the process of development and the mechanisms that determine both the structure of the body and the distinctive characteristics of each individual.

CONCEPT CHECK QUESTIONS
Answers on page 610

❶ Describe the functional relationships between the integumentary system and the reproductive system.

❷ Describe the functional relationships between the endocrine system and the reproductive system.

TABLE 19-1 *Hormones of the Reproductive System*

HORMONE	SOURCE	REGULATION OF SECRETION	PRIMARY EFFECTS
Gonadotropin-releasing hormone (GnRH)	Hypothalamus	*Males:* inhibited by testosterone *Females:* inhibited by estrogens and/or progestins	Stimulates FSH secretion, LH synthesis
Follicle-stimulating hormone (FSH)	Anterior pituitary gland	*Males:* stimulated by GnRH, inhibited by inhibin *Females:* stimulated by GnRH, inhibited by estrogens and/or progestins	*Males:* stimulates spermatogenesis and spermiogenesis through effects on sustentacular cells *Females:* stimulates follicle development, estrogen production, and egg maturation
Estrogens (primarily estradiol)	Follicular cells of ovaries; corpus luteum	Stimulated by FSH	Stimulates LH secretion, at high levels establishes and maintains secondary sex characteristics and behavior, stimulates repair and growth of endometrium, inhibits secretion of GnRH
Inhibin	Sustentacular cells of testes and follicle cells of ovaries	Stimulated by factors released by developing sperm (*male*) or developing follicles (*female*)	Inhibits secretion of FSH and possibly GnRH
Luteinizing hormone (LH)	Anterior pituitary gland	*Males:* stimulated by GnRH *Females:* production stimulated by GnRH, secretion by estrogens	*Males:* stimulates interstitial cells *Females:* stimulates ovulation, formation of corpus luteum, and progestin secretion
Progestins (primarily progesterone)	Corpus luteum	Stimulated by LH	Stimulates endometrial growth and glandular secretion, inhibits GnRH secretion
Androgens (primarily testosterone)	Interstitial cells of testes	Stimulated by LH	Maintains secondary sex characteristics and sexual behavior, promotes maturation of spermatozoa, inhibits GnRH secretion

The Reproductive System

For All Systems

Secretion of hormones with effects on growth and metabolism

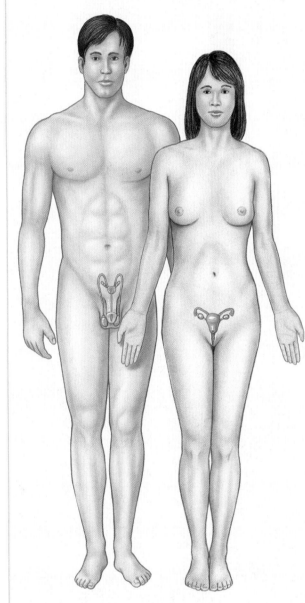

The Integumentary System

- Covers external genitalia; provides sensations that stimulate sexual behaviors; mammary gland secretions provide nourishment for newborn
- Reproductive hormones affect distribution of body hair and subcutaneous fat deposits

The Skeletal System

- Pelvis protects reproductive organs of females, portion of ductus deferens and accessory glands in males
- Sex hormones stimulate growth and maintenance of bones; sex hormones at puberty accelerate growth and closure of epiphyseal plates

The Muscular System

- Contractions of skeletal muscles eject semen from male reproductive tract; muscle contractions during sexual act produce pleasurable sensations in both sexes
- Reproductive hormones, especially testosterone, accelerate skeletal muscle growth

The Nervous System

- Controls sexual behaviors and sexual function
- Sex hormones affect CNS development and sexual behaviors

The Endocrine System

- Hypothalamic regulatory hormones and pituitary hormones regulate sexual development and function; oxytocin stimulates smooth muscle contractions in uterus and mammary glands
- Steroid sex hormones and inhibin inhibit secretory activities of hyopthalamus and pituitary gland

The Cardiovascular System

- Distributes reproductive hormones; provides nutrients, oxygen, and waste removal for fetus; local blood pressure changes responsible for physical changes during sexual arousal
- Estrogens may help maintain healthy vessels and slow development of atherosclerosis

The Lymphatic System

- Provides IgA for secretions by epithelial glands; assists in repairs and defense against infection
- Lysosomes and bactericidal chemicals in secretions provide nonspecific defense against reproductive tract infections

The Respiratory System

- Provides oxygen and removes carbon dioxide generated by tissues of reproductive system
- Changes in respiratory rate and depth occur during sexual arousal, under control of the nervous system

The Digestive System

- Provides additional nutrients required to support gamete production and (in pregnant women) embryonic and fetal development

The Urinary System

- Urethra in males carries semen to exterior; kidneys remove wastes generated by reproductive tissues and (in pregnant women) by a growing embryo and fetus
- Accessory organ secretions may have antibacterial action that helps prevent urethral infections in males

● *Figure 19-16* **Functional Relationships Between the Reproductive System and Other Systems**

19 THE REPRODUCTIVE SYSTEM

The Reproductive System of the Male • The Reproductive System of the Female • The Physiology of Sexual Intercourse • Aging and the Reproductive System

FOCUS

Birth Control Strategies

For physiological, logistical, financial, or emotional reasons, most adults practice some form of conception control during their reproductive years. Well over 50 percent of U.S. women age 15–44 are practicing some method of contraception; in 1996 an estimated 50 million American women were taking oral birth control pills. When the simplest and most obvious method, sexual abstinence, is unsatisfactory for some reason, another method of contraception must be used to avoid unwanted pregnancies. Many methods are available, so the selection process can be quite involved. Because each has specific strengths and weaknesses, the potential risks and benefits must be carefully analyzed on an individual basis.

Sterilization is a surgical procedure that makes one unable to provide functional gametes for fertilization. Either sex partner may be sterilized. In a **vasectomy** (vaz-EK-to-mē), a segment of the ductus deferens is removed, making it impossible for sperm to pass from the epididymis to the distal portions of the reproductive tract. The surgery can be performed in a physician's office in a matter of minutes. The spermatic cords are located as they ascend from the scrotum on either side, and after each cord is opened, the ductus deferens is severed. After a 1 cm section is removed, the cut ends are usually tied shut (Figure 19-17a●). The cut ends cannot reconnect; in time, scar tissue forms a permanent seal. A more recent vasectomy procedure often makes it possible to restore fertility at a later date. In this procedure, the cut ends of the ductus deferens are blocked with silicone plugs that can later be removed. After a vasectomy, the man experiences normal sexual function, for the epididymal and testicular secretions account for only around 5 percent of the volume of the semen. Sperm continue to develop, but they remain within the epididymis until they degenerate. The failure rate for this procedure is 0.08 percent. (*Failure* for a birth control method is defined as a resulting pregnancy.)

The uterine tubes can be blocked through a surgical procedure known as a **tubal ligation** (Figure 19-17b●). The failure rate for this procedure is estimated at 0.45 percent. Because the surgery involves entering the adominopelvic cavity, complications are more likely than with vasectomy. As in a vasectomy, attempts may be made to restore fertility after a tubal ligation.

Oral contraceptives manipulate the female hormonal cycle so that ovulation does not occur.

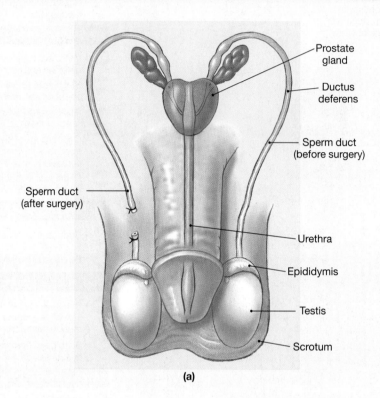

(a)

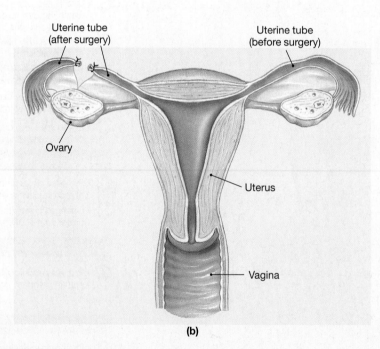

(b)

● *Figure 19-17* **Surgical Sterilization**
(a) In a vasectomy, the removal of a 1 cm section of the male's ductus deferens prevents the passage of sperm cells. (b) In a tubal ligation, the removal of a section of the female's uterine tube prevents the passage of sperm and the movement of an ovum or embryo into the uterus.

The contraceptive pills produced in the 1950s contained relatively large amounts of progestins. The concentrations were adequate to suppress pituitary production of GnRH, so FSH was not released and ovulation did not occur. In most of the oral contraceptive products developed subsequently, small amounts of estrogens have been added. Current *combination pills* differ significantly from the earlier products because the hormonal doses are much lower, with only one-tenth the progestins and less than half the estrogens. The hormones are administered in a cyclic fashion, beginning 5 days after the start of menses and continuing for the next 3 weeks. Then, over the fourth week, the woman takes placebo pills or no pills at all. Low-dosage combination pills are sometimes prescribed for women experiencing irregular uterine cycles, because these pills create a 28-day cycle.

At least 20 brands of combination oral contraceptives are now available, and over 200 million women use them worldwide. In the United States, 25 percent of women under age 45 use the combination pill to prevent conception. The failure rate for the combination oral contraceptives, when used as prescribed, is 0.24 percent over a 2-year period. Birth control pills are not risk-free. Combination pills can worsen problems associated with severe hypertension, diabetes mellitus, epilepsy, gallbladder disease, heart trouble, and acne. Women taking oral contraceptives are also at increased risk for venous thrombosis, strokes, pulmonary embolism, and (for women over 35) heart disease. Pregnancy has similar risks, however.

Two progesterone-only forms of birth control are now available: Depo-provera® and the Norplant® system. *Depo-provera* is injected every 3 months. The Silastic® (silicon rubber) tubes of the Norplant system are saturated with progesterone and inserted under the skin. This method provides birth control for a period of approximately 5 years, but to date the relatively high cost has limited the use of this contraceptive method. Both Depo-provera and the Norplant system can cause irregular menstruation and temporary amenorrhea, but they are easy to use and extremely convenient.

The **condom**, also called a *prophylactic,* or "rubber," covers the body of the penis during intercourse and keeps sperm from reaching the female reproductive tract. Condoms are also used to prevent the transmission of STDs, such as syphilis, gonorrhea, and AIDS. The reported condom failure rate varies from 6 to 17 percent.

Vaginal barriers such as the *diaphragm* and *cervical cap* rely on similar principles. A diaphragm, the most popular form of vaginal barrier in use today, consists of a dome of latex rubber with a small metal hoop supporting the rim. Because vaginas vary in size, women choosing this method must be individually fitted. Before intercourse, the diaphragm is inserted so that it covers the cervical os, and it is usually coated with a small amount of spermicidal jelly or cream, adding to the effectiveness of the barrier. The failure rate of a properly fitted diaphragm is estimated at 5–6 percent. The cervical cap is smaller and lacks the metal rim. It, too, must be fitted carefully, but unlike the diaphragm, it can be left in place for several days. The failure rate (8 percent) is higher than that for diaphragm use.

An **intrauterine device (IUD)** consists of a small plastic loop or a T that can be inserted into the uterine cavity. The mechanism of action remains uncertain, but it is known that IUDs stimulate prostaglandin production in the uterus. The resulting change in the chemical composition of uterine secretions lowers the chances of fertilization and subsequent *implantation* of the zygote in the uterine lining. (Implantation is discussed in Chapter 20.) In the United States, IUDs are in limited use today, but they remain popular in many other countries. The failure rate is estimated at 5–6 percent.

The **rhythm method** involves abstaining from sexual activity on the days ovulation might be occurring. The timing is estimated on the basis of previous patterns of menstruation and in some cases by monitoring changes in basal body temperature. The failure rate for the rhythm method is very high—almost 25 percent.

Sterilization, oral contraceptives, condoms, and vaginal barriers are the primary contraception methods for all age groups. But the proportion of the population using a particular method varies by age group. Sterilization is most popular among older women, who may already have had children. Relative availability also plays a role. For example, a sexually active female under age 18 can buy a condom more easily than she can obtain a prescription for an oral contraceptive.

But many of the differences are attributable to the relationship between risks and benefits for each age group.

When considering the use and selection of contraceptives, many people simply examine the list of potential complications and make the "safest" choice. For example, media coverage of the potential risks associated with oral contraceptives made many women reconsider their use. But complex decisions should not be made on such a simplistic basis, and the risks associated with contraceptive use must be considered in light of their relative efficiencies. Although pregnancy is a natural phenomenon, it has risks, and the mortality rate for pregnant women in the United States averages about 8 deaths per 100,000 pregnancies. That average incorporates a broad range; the rate is 5.4 per 100,000 among women under 20, and 27 per 100,000 for women over 40. Although these risks are small, for pregnant women over age 35, the chances of dying from complications related to pregnancy are almost twice as great as the chances of being killed in an automobile accident and are many times greater than the risks associated with the use of oral contraceptives. For women in developing nations, the comparison is even more striking. The mortality rate for pregnant women in parts of Africa is approximately 1 per 150 pregnancies. In addition to preventing pregnancy, combination birth control pills have also been shown to reduce the risks of ovarian and endometrial cancers and fibrocystic breast disease.

Before age 35, *the risks associated with contraceptive use are lower than the risks associated with pregnancy.* The notable exception involves individuals who take oral contraceptives but also smoke cigarettes. Younger women are more fertile, so despite a lower mortality rate for each pregnancy, they are likely to have more pregnancies. As a result, birth control failures imply a higher risk in the younger age groups.

After age 35, the risks of complications associated with oral contraceptive use increase, but the risks of using other methods remains relatively stable. Women over age 35 (smokers) or 40 (nonsmokers) are therefore typically advised to seek other forms of contraception. Because each contraceptive method has its own advantages and disadvantages, research on contraception control continues.

19 THE REPRODUCTIVE SYSTEM

The Reproductive System of the Male • The Reproductive System of the Female • The Physiology of Sexual Intercourse • Aging and the Reproductive System

Related Clinical Terms

amenorrhea: The failure of menarche to appear before age 16, or a cessation of menstruation for 6 months or more in an adult female of reproductive age.

cervical cancer: A malignant, metastasizing cancer of the cervix that is the most common reproductive cancer in women.

cryptorchidism (kript-ŌR-ki-dizm): The failure of one or both testes to descend into the scrotum by the time of birth.

dysmenorrhea: Painful menstruation.

endometriosis (en-dō-mē-trē-Ō-sis): The growth of endometrial tissue outside the uterus.

gynecology (gī-ne-KOL-o-jē): The study of the female reproductive tract and its disorders.

mammography: The use of X-rays to examine breast tissue.

mastectomy: The surgical removal of part or all of a cancerous mammary gland.

oophoritis (ō-of-ō-RĪ-tis): An inflammation of the ovaries.

orchiectomy (or-kē-EK-to-mē): The surgical removal of a testis.

orchitis (or-KĪ-tis): An inflammation of the testis.

ovarian cancer: A malignant, metastasizing cancer of the ovaries; the most dangerous reproductive cancer in women.

pelvic inflammatory disease (PID): An infection of the uterine tubes.

prostate cancer: A malignant, metastasizing cancer of the prostate gland; the second most common cause of cancer deaths in males.

prostatectomy (pros-ta-TEK-to-mē): The surgical removal of the prostate gland.

testicular torsion: A condition in which the spermatic cord is twisted and blood supply to a testis is obstructed.

vaginitis (va-jin-ī-tis): An inflammation of the vaginal canal.

vasectomy (vaz-EK-to-mē): The surgical removal of a segment of the ductus deferens, making it impossible for spermatozoa to reach the distal portions of the male reproductive tract.

CHAPTER REVIEW

Key Terms

Summary Outline

INTRODUCTION **582**

1. The human reproductive system, including the **gonads** (reproductive organs) produces, stores, nourishes, and transports functional **gametes** (reproductive cells). **Fertilization** is the fusion of male and female gametes to create a **zygote** (fertilized egg).

THE REPRODUCTIVE SYSTEM OF THE MALE **582**

1. In males, the *testes* produce **sperm** cells. The **spermatozoa** travel along the *epididymis*, the *ductus deferens*, the *ejaculatory duct*, and the *urethra* before leaving the body. Accessory organs (notably, the *seminal vesicles*, *prostate gland*, and *bulbourethral glands*) secrete into the ejaculatory ducts and urethra. The scrotum encloses the testes, and the penis is an erectile organ. (*Figure 19-1*)

The Testes..**582**

2. The **testes**, the primary sex organ of males, hang within the **scrotum**. The *dartos* muscle layer gives the scrotum a wrinkled appearance. The **cremaster muscle** pulls the testes closer to the body. The **tunica albuginea** surrounds each testis. Septa subdivide each testis into a series of lobules. **Seminiferous tubules** within each lobule are the sites of sperm production. Between the seminiferous tubules are **interstitial cells**, which secrete sex hormones. (*Figure 19-2*)

3. Seminiferous tubules contain **spermatogonia**, stem cells involved in **spermatogenesis**, and **sustentacular cells**, which sustain and promote the development of spermatozoa. (*Figure 19-3*)

4. Each spermatozoon has a **head**, **middle piece**, and **tail**. (*Figure 19-4*)

The Male Reproductive Tract **586**

5. From the testis, the spermatozoa enter the **epididymis**, an elongate tubule that monitors and adjusts the composition of the tubular fluid and serves as a recycling center for damaged spermatozoa. Spermatozoa leaving the epididymus are functionally mature, yet immobile.

6. The **ductus deferens**, or *vas deferens*, begins at the epididymis and passes through the inguinal canal within the **spermatic cord**. The junction of the base of the seminal vesicle and the ductus deferens creates the **ejaculatory duct**, which penetrates the prostate gland and empties into the urethra. (*Figure 19-5*)

7. The **urethra** extends from the urinary bladder to the tip of the penis and serves as a passageway used by both the urinary and reproductive systems.

8. Each **seminal vesicle** is an active secretory gland that contributes about 60 percent of the volume of semen; its secretions contain fructose, which is easily metabolized by spermatozoa. The **prostate gland** secretes fluid that makes up about 30 percent of seminal fluid. Alkaline mucus secreted by the **bulbourethral glands** has lubricating properties. *(Figure 19-5)*

9. A typical **ejaculation** releases 2–5 ml of semen (an **ejaculate**), which contains 20–100 million sperm per milliliter.

10. The skin overlying the **penis** resembles that of the scrotum. Most of the body of the penis consists of three masses of **erectile tissue**. Beneath the superficial layers are two **corpora cavernosa** and a single **corpus spongiosum**, which surrounds the urethra. Dilation of the erectile tissue with blood produces an **erection**. *(Figure 19-6)*

11. Important regulatory hormones of males include **follicle-stimulating hormone (FSH)**, **luteinizing hormone (LH)**, and **gonadotropin-releasing hormone (GnRH)**. Testosterone is the most important *androgen*. *(Figure 19-7)*

THE REPRODUCTIVE SYSTEM OF THE FEMALE 590

1. Principal organs of the female reproductive system include the *ovaries, uterine tubes, uterus, vagina,* and *external genitalia*. *(Figure 19-8)*

2. The **ovaries** are the primary sex organs of females. Ovaries are the site of **ovum** production, or **oogenesis**, which occurs monthly in **ovarian follicles** as part of the **ovarian cycle**. As development proceeds, **primordial**, **primary**, **secondary**, and **tertiary follicles** form. At **ovulation**, an oocyte and the surrounding follicular walls of the **corona radiata** are released through the ruptured ovarian wall. *(Figures 19-9, 19-10)*

3. Each **uterine tube** has an **infundibulum**, a funnel that opens into the pelvic cavity. For fertilization to occur, the secondary oocyte must encounter spermatozoa during the first 12–24 hours of its passage from the infundibulum to the uterine cavity. *(Figure 19-11)*

4. The **uterus** provides mechanical protection and nutritional support to the developing embryo. It is stabilized by various ligaments. Major anatomical landmarks of the uterus include the **body**, **cervix**, **external orifice**, and **uterine cavity**. The uterine wall consists of an inner **endometrium**, a muscular **myometrium**, and a superficial **perimetrium**. *(Figure 19-11)*

5. A typical 28-day **uterine cycle**, or *menstrual cycle*, begins with the onset of **menses** and the destruction of the *functional zone* of the endometrium. This process of menstruation continues from 1 to 7 days.

6. After menses, the **proliferative phase** begins, and the functional zone undergoes repair and thickens. During the **secretory phase**, the endometrial glands are active and the uterus is prepared for the arrival of an embryo. Menstrual activity begins at **menarche** and continues until **menopause**.

7. The **vagina** is a muscular tube extending between the uterus and the external genitalia. A thin epithelial fold, the **hymen**, partially blocks the entrance to the vagina.

8. The components of the **vulva**, or female external genitalia, include the **vestibule**, the **labia minora**, the **clitoris**, the **labia majora**, and the **lesser** and **greater vestibular glands**. *(Figure 19-12)*

9. A newborn infant gains nourishment from milk secreted by maternal **mammary glands**. **Lactation** is the process of milk production. *(Figure 19-13)*

10. Regulation of the **female reproductive cycle** involves the coordination of the ovarian and uterine cycles by circulating hormones.

11. **Estradiol**, one of the estrogens, is the dominant hormone of the follicular phase of the ovarian cycle. Ovulation occurs in response to a surge in LH secretion. *(Figure 19-14)*

12. The hypothalamic secretion of GnRH triggers the pituitary secretion of FSH and the synthesis of LH. FSH initiates follicular development, and activated follicles and ovarian interstitial cells produce estrogens. **Progesterone**, one of the steroid hormones called **progestins**, is the principal hormone of the *luteal phase*. Hormonal changes regulate the uterine, or menstrual, cycle. *(Figure 19-15)*

THE PHYSIOLOGY OF SEXUAL INTERCOURSE 600

1. During **arousal** in males, erotic thoughts, sensory stimulation, or both lead to parasympathetic activity that produces erection. Stimuli accompanying **coitus** (sexual intercourse) lead to **emission** and ejaculation. Strong muscle contractions are associated with **orgasm**.

2. The phases of female sexual function resemble those of the male, with parasympathetic arousal and muscular contractions associated with orgasm.

AGING AND THE REPRODUCTIVE SYSTEM 601

1. Menopause (the time that ovulation and menstruation cease in women) typically occurs around age 50. Production of GnRH, FSH, and LH rise, while circulating concentrations of estrogen and progesterone decline.

2. During the **male climacteric**, at age 50–60, circulating testosterone levels decline, while levels of FSH and LH rise.

INTEGRATION WITH OTHER SYSTEMS 602

1. Hormones play a major role in coordinating reproduction. *(Table 19-1)*

2. In addition to the endocrine and reproductive systems, reproduction requires the normal functioning of the digestive, nervous, cardiovascular, and urinary systems. *(Figure 19-16)*

19 THE REPRODUCTIVE SYSTEM

The Reproductive System of the Male • The Reproductive System of the Female • The Physiology of Sexual Intercourse • Aging and the Reproductive System

Review Questions

Level 1: Reviewing Facts and Terms

Match each item in column A with the most closely related item in column B. Use letters for answers in the spaces provided.

COLUMN A

____ 1. gametes

____ 2. gonads

____ 3. interstitial cells

____ 4. seminal vesicles

____ 5. prostate gland

____ 6. bulbourethral glands

____ 7. prepuce

____ 8. corpus luteum

____ 9. endometrium

____ 10. myometrium

____ 11. dysmenorrhea

____ 12. menarche

____ 13. clitoris

____ 14. lactation

____ 15. coitus

COLUMN B

a. production of androgens

b. outer muscular uterine wall

c. high concentration of fructose

d. female erectile tissue

e. secretes thick, sticky, alkaline mucus

f. painful menstruation

g. sexual intercourse

h. uterine lining

i. reproductive cells

j. female puberty

k. milk production

l. secretes antibiotic

m. reproductive organs

n. foreskin of penis

o. endocrine structure

16. Perineal structures associated with the reproductive system are collectively known as:
   (a) gonads
   (b) sex gametes
   (c) external genitalia
   (d) accessory glands

17. Meiosis in males produces four spermatids, each of which contains:
   (a) 23 chromosomes
   (b) 23 pairs of chromosomes
   (c) the diploid number of chromosomes
   (d) 46 pairs of chromosomes

18. Erection of the penis occurs when:
   (a) sympathetic activation of penile arteries occurs
   (b) arterial branches are constricted, and muscular partitions are tense
   (c) the vascular channels become engorged with blood
   (d) a, b, and c are correct

19. In males, the primary target of FSH is the:
   (a) sustentacular cells of the seminiferous tubules
   (b) interstitial cells of the seminiferous tubules
   (c) prostate gland
   (d) epididymis

20. The ovaries in females are responsible for:
   (a) the production of female gametes
   (b) the secretion of female sex hormones
   (c) the secretion of inhibin
   (d) a, b, and c are correct

21. Ovum production, or oogenesis, begins:
   (a) before birth
   (b) after birth
   (c) at puberty
   (d) after puberty

22. In females, meiosis is not completed:
    (a) until birth
    (b) until puberty
    (c) unless and until fertilization occurs
    (d) until ovulation occurs

23. If fertilization is to occur, the ovum must encounter spermatozoa during the first _____ of its passage.
    (a) 1 to 5 hours
    (b) 6 to 11 hours
    (c) 12 to 24 hours
    (d) 25 to 36 hours

24. The part of the endometrium that undergoes cyclical changes in response to sexual hormonal levels is the:
    (a) serosa
    (b) basilar zone
    (c) muscular myometrium
    (d) functional zone

25. A sudden surge in LH concentration causes:
    (a) the onset of menses
    (b) the rupture of the follicular wall and ovulation
    (c) the beginning of the proliferative phase
    (d) the end of the uterine cycle

26. At the time of ovulation, the basal body temperature:
    (a) is not affected
    (b) increases noticeably
    (c) declines sharply
    (d) may increase or decrease a few degrees

27. Menopause is accompanied by:
    (a) sustained rises in GnRH, FSH, and LH
    (b) declines in circulating levels of estrogen and progesterone
    (c) thinning of the urethral and vaginal walls
    (d) a, b, and c are correct

28. Which reproductive structures are common to both males and females?

29. Which accessory organs and glands contribute to the composition of semen? What are the functions of each?

30. What are the primary functions of the epididymis in males?

31. What are the primary functions of the ovaries in females?

32. What are the three major functions of the vagina?

Level 2: Reviewing Concepts

33. How does the human reproductive system differ functionally from all other systems in the body?

34. How is meiosis involved in the development of the spermatozoon and the ovum?

35. Using an average uterine cycle of 28 days, describe each of the three phases of the menstrual cycle.

36. Describe the hormonal events associated with the uterine cycle.

37. How does the aging process affect the reproductive systems of men and women?

Level 3: Critical Thinking and Clinical Applications

38. Diane has peritonitis (an inflammation of the peritoneum), which her physician says resulted from a urinary tract infection. Why could this situation occur in females but not in males?

39. Rod injures the sacral region of his spinal cord. Will he still be able to achieve an erection? Explain.

40. Women bodybuilders and women suffering from eating disorders such as anorexia nervosa often cease having menstrual cycles, a condition known as amenorrhea. What does this relationship suggest about the role of body fat and menstruation? What benefit might there be in the discontinuance of menstruation under such circumstances?

19 THE REPRODUCTIVE SYSTEM

The Reproductive System of the Male • The Reproductive System of the Female • The Physiology of Sexual Intercourse • Aging and the Reproductive System

Answers to Concept Check Questions

Page 590

1. The cremaster muscle as well as the dartos muscle would be relaxed on a warm day, so the scrotal sac could descend away from the warmth of the body and could cool the testes.

2. The dilation of the arteries within the penis allows blood flow to increase and the vascular channels become filled with blood, resulting in erection.

3. The hormone FSH is required for maintaining spermatogenesis because FSH stimulates the sustentacular cells to provide nutrients and chemical stimuli to developing sperm. Low levels of FSH would inhibit sustentacular cell function and thus lower sperm production.

Page 597

1. A blockage of the uterine tube would cause sterility.

2. The acidic pH of the vagina helps prevent bacterial, fungal, and parasitic infections in this area.

3. The functional layer of the endometrium sloughs off during menstruation.

4. The blockage of a single lactiferous sinus would not interfere with the movement of milk to the nipple, because each breast usually has 15–20 lactiferous sinuses.

Page 600

1. If the LH surge did not occur during an ovarian cycle, ovulation and the subsequent formation of the corpus luteum would not occur.

2. Progesterone is responsible for the functional maturation and secretion of the endometrium. A blockage of progesterone receptors would inhibit endometrial development (and would make the uterus unprepared for pregnancy).

3. A decline in the levels of estrogen and progesterone signals the end of the uterine cycle and the beginning of menses.

Page 601

1. An inability to contract the ischiocavernosus and bulbospongiosus muscles would interefere with a male's ability to ejaculate and to experience orgasm.

2. As the result of parasympathetic stimulation in females during sexual arousal, the erectile tissues of the clitoris engorge with blood, the secretion of cervical and vaginal glands increases, blood flow to the walls of the vagina increases, and the blood vessels in the nipples engorge.

3. At menopause, circulating estrogen levels begin to drop. Estrogen has an inhibitory effect on FSH (and GnRH). As the level of estrogen declines, the levels of FSH rise and remain elevated.

Page 602

1. The integumentary system covers the external genitalia; sensory receptors in the skin provide sensations that stimulate sexual behaviors; and the mammary glands provide nourishment for the newborn. The reproductive system produces hormones (primarily, testosterone in males and estrogens in females) that affect secondary sex characteristics such as the distribution of hair and subcutaneous fat deposits.

2. The endocrine system regulates sexual development and function by means of hypothalamic regulatory hormones and pituitary hormones; oxytocin in females stimulates smooth muscle contractions in the uterus and mammary glands. The reproductive system produces steroid sex hormones and inhibin that inhibit seretory activities of the hypothalamus and pituitary gland.

EXPLORE *MediaLab*

EXPLORATION #1

Estimated time for completion: 10 minutes

YOU LEARNED in this chapter that reproductive ability depends on hormones, specifically LH and FSH from the pituitary gland and either testosterone or estrogens and progesterone from the reproductive organs. Although this system is quite sophisticated, it is not fool-proof. There are compounds that are able to mimic estrogen in nature as well as under laboratory conditions. A number of these compounds have been identified, and include such commonplace chemicals as CCP (a wood preservative), industrial by-products, and one of the ingredients in insecticides. These compounds, referred to as *environmental estrogens*, are able to react with estrogen receptors in the body, triggering effects within the cell that are similar to those triggered by natural estrogen. In 1995, studies began to indicate that environmental estrogens might cause medical problems for those exposed to them. Referring to Figure 19-14●, predict what effect these compounds might have on a female of reproductive age. What, if any, effect might these environmental estrogens have on a male of reproductive age? In addition to these potential effects, compose a list of questions that you would like answered in regard to these environmental contaminants. For additional information on environmental estrogens, go to Chapter 19 at the Companion Web site, visit the MediaLab section, and click on the key words "Environmental Estrogens."

Pesticides being sprayed onto fruit trees in an orchard.

Pre-adolescent girls sharing a their friendship on a walk. Puberty marks the biological passage from childhood into adulthood.

EXPLORATION #2

Estimated time for completion: 10 minutes

PUBERTY is defined as a period of rapid growth, sexual maturation, and the appearance of secondary sex characteristics. It usually occurs between ages 10 and 15. In most cultures, puberty is associated with many sociological changes and marks the biological passage from childhood into adulthood. In some instances, however, puberty occurs much earlier than age 10. When a child begins to show signs of sexual maturity prior to age 8 in females, or 9 in males, it is referred to as "precocious puberty." In these cases, medical science can help to slow the effects of puberty, allowing the individual more time to mature both physically and socially. Why would this be advisable? What might the consequences of precocious puberty be to the individual, aside from early ability to reproduce? Prepare a list of the events of puberty, and indicate why these might cause problems in a child younger than 9. Using your knowledge of hormones and the endocrine system, what might be the best way to slow this early onset of puberty? Go to Chapter 19 at the Companion Web site, visit the MediaLab section, and click on the key word "Puberty." Modify your list of potential consequences of precocious puberty as well as the treatments available for this condition. If you need clarification of the treatments described at this Web site, review the section on negative feedback control of hormones in Chapter 10 (p. 320).

Development and Inheritance

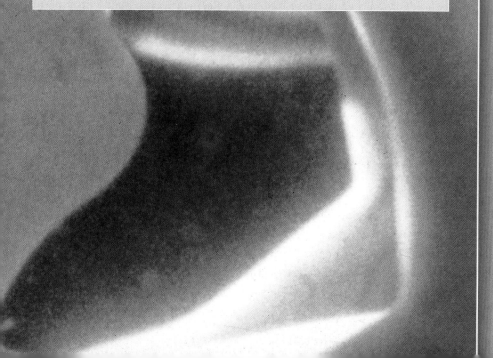

CHAPTER OUTLINE AND OBJECTIVES

Vocabulary Development

allanto- ....sausage; *allantois*
amphi ....two-sided; *amphimixis*
blast.............................precursor; *trophoblast*
gestare ....to bear; *gestation*
heteros...........................other; *heterozygous*
homos ....same; *homozygous*
karyon ....nucleus; *karyotyping*
koiloma............................hollow; *blastocoele*
meros ....part; *blastomere*
meso-..................................middle; *mesoderm*
mixis................................mixing; *amphimixis*
morula ....mulberry; *morula*
phainein ....to display; *phenotype*
praegnans.......................with child; *pregnant*
tropho- ....food; *trophoblast*
typos ....mark; *phenotype*
vitro..glass; *in vitro*

M ANY PHYSIOLOGICAL PROCESSES last only a fraction of a second; others may take hours at most. But some important processes of life are measured in months, years, or decades. A human being develops in the womb for 9 months, grows to maturity in 15 or 20 years, and may live the better part of a century. Birth, growth, maturation, aging, and death are all parts of a single, continuous process. That process does not end with the individual, because human beings can pass at least some of their characteristics on to a new generation that will repeat the same cycle.

20 DEVELOPMENT AND INHERITANCE

An Overview of Topics in Development • Fertilization • The First Trimester • The Second and Third Trimesters • Labor and Delivery

TIME REFUSES TO STAND STILL; today's infant will be tomorrow's adult. The gradual modification of physical and physiological characteristics during the period from conception to physical maturity is called **development**. The changes are truly remarkable—what begins as a single cell slightly larger than the period at the end of this sentence becomes an individual whose body contains trillions of cells organized into tissues, organs, and organ systems. The creation of different cell types in this process is called **differentiation**. Differentiation occurs through selective changes in genetic activity. As development proceeds, some genes are turned off and others are turned on. The types of genes turned off or on vary from one cell type to another.

A basic understanding of human development provides important insights into anatomical structures. In addition, many of the mechanisms of development and growth are similar to those responsible for the repair of injuries. This chapter will focus on major aspects of development and consider highlights of the developmental process. We shall also consider the regulatory mechanisms and how developmental patterns can be modified—for good or ill.

An Overview of Topics in Development

Development involves (1) the division and differentiation of cells and (2) the changes that produce and modify anatomical structures. Development begins at **fertilization**, or **conception**. We can divide development into periods characterized by specific anatomical changes. **Embryological development** considers the events that occur in the first 2 months after fertilization. The study of these events in the developing organism, or **embryo**, is called **embryology** (em-brē-OL-ō-jē). After 2 months, the developing embryo becomes a **fetus**. **Fetal development** begins at the start of the ninth week and continues up to the time of birth. Together, embryological and fetal development are referred to as **prenatal development**, the primary focus of this chapter. **Postnatal development** begins at birth and continues to maturity, when the aging process begins.

Although all human beings go through the same developmental stages, differences in genetic makeup produce distinctive individual characteristics. **Inheritance** refers to the transfer of genetically determined characteristics from generation to generation. **Genetics** is the study of the mechanisms responsible for inheritance. This chapter considers basic genetics as it applies to the appearance of inherited characteristics such as gender, hair color, and various diseases.

Fertilization

Fertilization involves the fusion of two haploid gametes, producing a *zygote* that contains 46 chromosomes, the normal number for a *somatic* (nonreproductive) cell. The roles and contributions of the male and female gametes are very different. The spermatozoon simply delivers the paternal chromosomes to the site of fertilization. In contrast, the female gamete must provide all of the nourishment and genetic programming to support development of the embryo for nearly a week after conception. The volume of the this gamete is therefore much greater than that of the spermatozoon (Figure 20-1a●). Recall from Chapter 19 that ovulation releases a secondary oocyte. p. 591 At fertilization, the diameter of the secondary oocyte is more than twice the length of the spermatozoon. The relationship between their volumes is even more striking—roughly 2000 to 1.

The sperm arriving in the vagina are already motile, as a result of mixing with secretions of the seminal vesicles—the first step of **capacitation**. p. 586 But they cannot fertilize an egg until they have been exposed to conditions in the female reproductive tract. The mechanism responsible for this second step of capacitation is unknown.

Fertilization typically occurs in the upper one-third of the uterine tube within a day of ovulation. Sperm cells can pass from the vagina to the upper portion of a uterine tube within a time range of 2 hours to as little as 30 minutes. In addition to the sperm's own movements, contractions of the uterine musculature and ciliary currents in the uterine tubes aid the passage of sperm to the fertilization site. Of the 200 million spermatozoa introduced into the vagina in a typical ejaculate, only about 10,000 enter the uterine tube, and fewer than 100 actually reach the secondary oocyte. In general, a male with a sperm count below 20 million per milliliter is sterile because too few sperm survive to reach the oocyte. Dozens of sperm cells are required for successful fertilization, because a single sperm cannot penetrate the *corona radiata*, the layer of follicle cells that surrounds the oocyte.

THE OOCYTE AT OVULATION AND ACTIVATION

Ovulation occurs before the oocyte is completely mature. The secondary oocyte leaving the follicle is in metaphase of the second meiotic division (meiosis II). The cell's metabolic operations have been discontinued, and the oocyte drifts in a sort of suspended animation, awaiting the stimulus for further development. If fertilization does not occur, the secondary oocyte disintegrates without completing meiosis.

(a)

Figure 20-1 **Fertilization**
(**a**) An oocyte and sperm cells at the time of fertilization. Notice the difference in size between the gametes. (**b**) Fertilization and the preparations for cleavage.

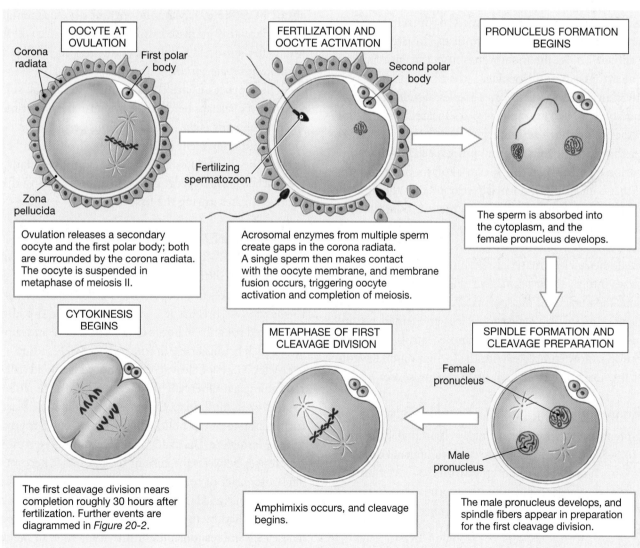

OOCYTE AT OVULATION

Corona radiata

First polar body

Zona pellucida

Ovulation releases a secondary oocyte and the first polar body; both are surrounded by the corona radiata. The oocyte is suspended in metaphase of meiosis II.

FERTILIZATION AND OOCYTE ACTIVATION

Second polar body

Fertilizing spermatozoon

Acrosomal enzymes from multiple sperm create gaps in the corona radiata. A single sperm then makes contact with the oocyte membrane, and membrane fusion occurs, triggering oocyte activation and completion of meiosis.

PRONUCLEUS FORMATION BEGINS

The sperm is absorbed into the cytoplasm, and the female pronucleus develops.

SPINDLE FORMATION AND CLEAVAGE PREPARATION

Female pronucleus

Male pronucleus

The male pronucleus develops, and spindle fibers appear in preparation for the first cleavage division.

METAPHASE OF FIRST CLEAVAGE DIVISION

Amphimixis occurs, and cleavage begins.

CYTOKINESIS BEGINS

The first cleavage division nears completion roughly 30 hours after fertilization. Further events are diagrammed in *Figure 20-2*.

(b)

20 DEVELOPMENT AND INHERITANCE

An Overview of Topics in Development • Fertilization • **The First Trimester** • The Second and Third Trimesters • Labor and Delivery

Fertilization and the events that follow are diagrammed in Figure 20-1b●. The corona radiata protects the oocyte as it passes through the ruptured follicular wall and into the infundibulum of the uterine tube. Although the physical process of fertilization requires only a single sperm in contact with the oocyte membrane, that spermatozoon must first penetrate the corona radiata. The acrosomal cap of the sperm contains several enzymes, including *hyaluronidase* (hī-al-u-RON-a-dās), which breaks down the bonds between adjacent follicle cells. Dozens of sperm cells must release hyaluronidase before an opening forms between the follicular cells.

No matter how many sperm slip through the resulting gap in the corona radiata, only a single spermatozoon will accomplish fertilization and activate the oocyte. That sperm cell first binds to sperm receptors on the zona pellucida. ∞ p. 592 Another enzyme, released by the breakdown of the acrosomal cap, digests a path for the sperm cell through the zona pellucida to the oocyte membrane. On contact, the sperm and oocyte cell membranes begin to fuse. This step triggers **oocyte activation**. As the membranes fuse, the entire sperm enters the cytoplasm of the oocyte.

Activation of the oocyte involves a series of sudden changes in its metabolism. For example, vesicles located just interior to the oocyte membrane undergo exocytosis and release enzymes that prevent an abnormal process called *polyspermy* (fertilization by more than one sperm). Polyspermy produces a zygote that is incapable of normal development. Other important changes include the completion of meiosis II and a rapid increase in the metabolic rate of the oocyte.

After oocyte activation has occurred and meiosis has been completed, the nuclear material remaining within the ovum reorganizes into a *female pronucleus* (Figure 20-1b●). At the same time, the nucleus of the spermatozoon swells, becoming the *male pronucleus*. The male pronucleus and female pronucleus fuse in a process called **amphimixis** (am-fi-MIK-sis). Fertilization is now complete, with the formation of a zygote that has the normal complement of 46 chromosomes.

AN OVERVIEW OF PRENATAL DEVELOPMENT

During prenatal development, a single cell ultimately forms a 3–4 kg (6.6–8.8 lb) infant. The time spent in prenatal development is known as the period of **gestation** (jes-TĀ-shun), or *pregnancy*. Gestation occurs within the uterus over a period of 9 months. For convenience, prenatal development is usually considered to consist of three **trimesters**, each 3 months in duration:

1. The **first trimester** is the period of embryological and early fetal development. During this period, the basic components of each of the major organ systems appear.

2. The **second trimester** is dominated by the development of organs and organ systems and their near completion. The body proportions change, and by the end of the second trimester the fetus looks distinctively human.

3. The **third trimester** is characterized by rapid fetal growth. Early in the third trimester most of the major organ systems become fully functional. An infant born 1 month or even 2 months prematurely has a reasonable chance of survival.

The First Trimester

At the moment of conception, the fertilized ovum is a single cell that has a diameter of about 0.135 mm (0.005 in.) and a weight of approximately 150 mg. By the end of the first trimester, the fetus is almost 75 mm (3 in.) long and weighs about 14 g (0.5 oz). Many important developmental events accompany this increase in size and weight. We will focus on four general processes that occur during this period: *cleavage, implantation, placentation*, and *embryogenesis*.

These processes are complex and vital to the survival of the embryo. Perhaps because the events in the first trimester are so complex, *this is the most dangerous period in prenatal life*. Only about 40 percent of conceptions produce embryos that survive the first trimester. For that reason, pregnant women are warned to take great care to avoid drugs and other disruptive stresses during the first trimester.

CLEAVAGE AND BLASTOCYST FORMATION

Cleavage (KLĒV-ij) is a series of cell divisions that subdivide the cytoplasm of the zygote (Figure 20-2●). The first cleavage division produces two identical cells, called **blastomeres** (BLAS-tō-mērz; *blast*, precursor + *meros*, part). The first division is completed roughly 30 hours after fertilization, and subsequent cleavage divisions occur at intervals of 10–12 hours.

After 3 days of cleavage, the embryo is a solid ball of cells, resembling a mulberry. This stage is called the **morula** (MOR-ū-la; *morula*, mulberry). Over the next 2 days, the blastomeres form a **blastocyst**, a hollow ball with an inner cavity known as the *blastocoele* (BLAS-tō-sēl; *koiloma*, cavity). At this stage, differences between the cells of the blastocyst become visible. The outer layer of cells, separating the outside world from the blastocoele, is called the **trophoblast** (TRŌ-fō-blast). The function is implied by the name: *tropho*, food + *blast*, precursor. These cells are responsible for providing food to the developing embryo. A second group of cells, the **inner cell mass**, lies clustered at one end of the blastocyst. In time, the inner cell mass will form the embryo.

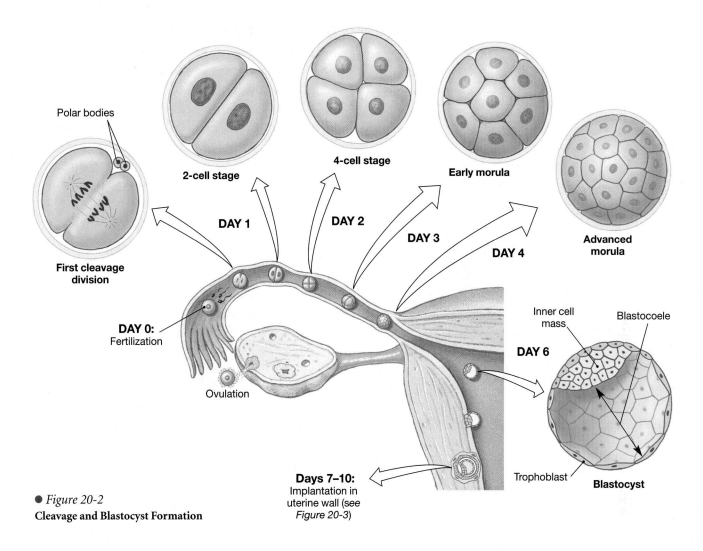

● *Figure 20-2*
Cleavage and Blastocyst Formation

IMPLANTATION

At fertilization, the zygote is still 4 days away from the uterus. It arrives in the uterine cavity as a morula, and over the next 2–3 days, blastocyst formation occurs. During this time, the zona pellucida is shed. The blastocyst is now freely exposed to the fluid contents of the uterine cavity. This glycogen-rich fluid is secreted by the endometrial glands of the uterus and provides nutrients to the blastocyst. When fully formed, the blastocyst contacts the endometrium and implantation occurs. Stages in the implantation process are diagrammed in Figure 20-3●.

Implantation begins as the surface of the blastocyst closest to the inner cell mass touches and adheres to the uterine lining (see day 7, Figure 20-3●). In this area, the superficial cells undergo rapid divisions, making the trophoblast several layers thick. The cells closest to the interior of the blastocyst remain intact and form a layer called the *cellular trophoblast*. Near the endometrial wall, the cell membranes separating the trophoblast cells disappear, creating a layer of cytoplasm containing multiple nuclei (day 8). This outer layer is called the

syncytial (sin-SISH-al) *trophoblast* (day 8). The syncytial trophoblast erodes a path through the uterine epithelium. At first, this erosion creates a gap in the uterine lining, but the division and migration of epithelial cells soon repair the surface. When the repairs are completed, the blastocyst loses contact with the uterine cavity. Further development occurs entirely within the functional zone of the endometrium.

In most cases, implantation occurs in the fundus or elsewhere in the body of the uterus. In an **ectopic pregnancy**, implantation occurs somewhere other than within the uterus, such as in one of the uterine tubes. Approximately 0.6 percent of pregnancies are ectopic pregnancies, which do not produce a viable embryo and can be life-threatening.

As implantation proceeds, the syncytial trophoblast continues to enlarge into the surrounding endometrium (day 9). The erosion of uterine gland cells releases nutrients that are absorbed by the trophoblast and distributed by diffusion to the inner cell mass. These nutrients provide the energy needed to support the early stages of embryo formation. Extensions of the trophoblast grow around endometrial capillaries. As the

617

20 DEVELOPMENT AND INHERITANCE

An Overview of Topics in Development • Fertilization • **The First Trimester** • The Second and Third Trimesters • Labor and Delivery

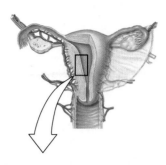

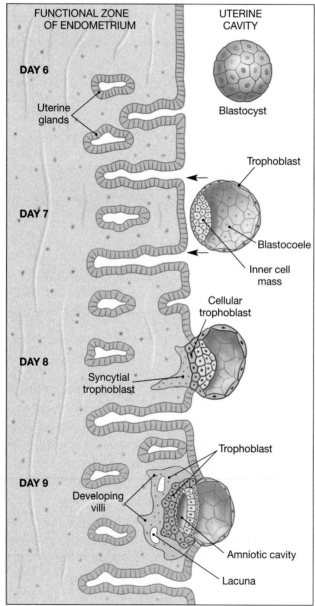

- *Figure 20-3* **Stages in Implantation**

capillary walls are destroyed, maternal blood begins to flow through trophoblastic channels known as *lacunae*. Fingerlike *villi* extend away from the trophoblast into the surrounding endometrium, and these extensions gradually increase in size and complexity as development proceeds.

Formation of the Amniotic Cavity

By the time of implantation, the inner cell mass has separated from the trophoblast. The separation gradually increases, creating a fluid-filled chamber called the **amniotic** (am-nē-OT-ik) **cavity**. The amniotic cavity can be seen in day 9 of Figure 20-3●; additional details from days 10–12 are shown in Figure 20-4●. When the amniotic cavity first appears, the cells of the inner cell mass are organized into an oval sheet that is two cell layers thick; a superficial layer faces the amniotic cavity and a deeper layer that is exposed to the fluid contents of the blastocoele.

 INFERTILITY

Infertility (*sterility*) is usually defined as an inability to achieve pregnancy after 1 year of appropriately timed intercourse. Problems with infertility are relatively common. An estimated 10–15 percent of married couples in the United States are infertile, and another 10 percent are unable to have as many children as they desire. It is thus not surprising that reproductive physiology has become a popular field of medicine and that the treatment of infertility has become a major medical industry. Recent advances in our understanding of reproductive physiology are providing new solutions to fertility problems as varied as low sperm count, abnormal spermatozoa, maternal hormone levels, problems with oocyte production or oocyte transport from the ovary to the uterine tube, blocked uterine tubes, abnormal oocytes, and an abnormal uterine environment. Procedures meant to resolve these reproductive difficulties are known as *assisted reproductive technologies (ART)*.

Gastrulation and Germ Layer Formation

By day 12, a third layer of cells begins forming through the process of **gastrulation** (gas-troo-LĀ-shun) (day 12, Figure 20-4●). During gastrulation, cells on the surface move toward a center line known as the *primitive streak*. These migrating cells leave the surface and move between the two existing layers. This movement creates three distinct embryonic layers of cells with markedly different fates. The superficial layer in contact with the amniotic cavity is called the **ectoderm**, the layer facing the blastocoel is known as the **endoderm**, and the poorly organized layer of migrating cells is the **mesoderm** (*meso-*, middle). Table 20-1 lists the contributions each of these three **germ layers** makes to the body systems described in earlier chapters.

Gastrulation produces an oval, three-layered sheet known as the *embryonic disc*. The embryonic disc will form the body of the embryo, whereas the rest of the blastocyst will be involved in forming the extraembryonic membranes.

The Formation of Extraembryonic Membranes

Germ layers also participate in the formation of four extraembryonic membranes: the *yolk sac*, the *amnion*, the *allantois*, and

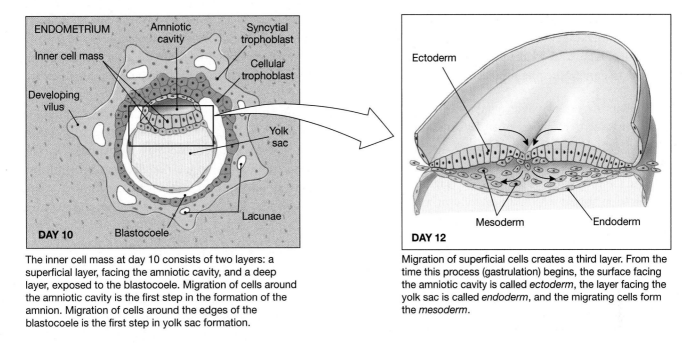

The inner cell mass at day 10 consists of two layers: a superficial layer, facing the amniotic cavity, and a deep layer, exposed to the blastocoele. Migration of cells around the amniotic cavity is the first step in the formation of the amnion. Migration of cells around the edges of the blastocoele is the first step in yolk sac formation.

Migration of superficial cells creates a third layer. From the time this process (gastrulation) begins, the surface facing the amniotic cavity is called *ectoderm*, the layer facing the yolk sac is called *endoderm*, and the migrating cells form the *mesoderm*.

● *Figure 20-4* **The Inner Cell Mass and Gastrulation**

TABLE 20-1 *The Fates of the Primary Germ Layers*

PRIMARY GERM LAYER	DEVELOPMENTAL CONTRIBUTIONS TO THE BODY
Ectoderm	*Integumentary system*: epidermis, hair follicles and hairs, nails, and glands communicating with the skin (apocrine and merocrine sweat glands, mammary glands, and sebaceous glands)
	Skeletal system: pharyngeal cartilages of the embryo develop into portions of sphenoid and hyoid bones, auditory ossicles, and styloid processes of temporal bones
	Nervous system: all neural tissue, including brain and spinal cord
	Endocrine system: pituitary gland and the adrenal medullae
	Respiratory system: mucous epithelium of nasal passageways
	Digestive system: mucous epithelium of mouth and anus, salivary glands
Mesoderm	*Skeletal system*: all components except some pharyngeal cartilage derivatives
	Muscular system: all components
	Endocrine system: adrenal cortex, endocrine tissues of heart, kidneys, and gonads
	Cardiovascular system: all components
	Lymphatic system: all components
	Urinary system: the kidneys, including the nephrons and the initial portions of the collecting system
	Reproductive system: the gonads and the adjacent portions of the duct systems
	Miscellaneous: the lining of the body cavities (pleural, pericardial, and peritoneal) and the connective tissues that support all organ systems
Endoderm	*Endocrine system*: thymus, thyroid gland, and pancreas
	Respiratory system: respiratory epithelium (except nasal passageways) and associated mucous glands
	Digestive system: mucous epithelium (except mouth and anus), exocrine glands (except salivary glands), liver, and pancreas
	Urinary system: urinary bladder and distal portions of the duct system
	Reproductive system: distal portions of the duct system, stem cells that produce gametes

20 DEVELOPMENT AND INHERITANCE

An Overview of Topics in Development • Fertilization • **The First Trimester** • The Second and Third Trimesters • Labor and Delivery

the *chorion*. Although these membranes support embryonic and fetal development, they leave few traces of their existence in adult systems. Figure 20-5● details stages in the development of the extraembryonic membranes.

Yolk Sac. The first extraembryonic membrane to appear is the **yolk sac**. The yolk sac, already present 10 days after fertilization, forms a pouch within the blastocoele (Figure 20-4●). As gastrulation proceeds, mesodermal cells migrate around this pouch and complete the formation of the yolk sac (Figure 20-5a●). Blood vessels soon appear within the mesoderm, and the yolk sac becomes an important site of blood cell formation.

Amnion. The **amnion** (AM-nē-on) is composed of both ectoderm and mesoderm. Ectodermal cells first spread over the inner surface of the amniotic cavity and, soon after, mesodermal cells follow and create a second, outer layer. As the embryo and later the fetus enlarges, the amnion continues to expand, increasing the size of the amniotic cavity. The amnion encloses fluid that surrounds and cushions the developing embryo and fetus (Figure 20-5c–e●).

The Allantois. The **allantois** (a-LAN-tō-is) is a sac of endoderm and mesoderm that extends away from the embryo. The base of the allantois later gives rise to the urinary bladder. The allantois accumulates some of the small amount of urine produced by the kidneys during embryological development.

The Chorion. The **chorion** (KOR-ē-on) is created as migrating mesodermal cells form a layer underneath the trophoblast, separating it from the blastocoele (Figure 20-5a,b●). When implantation first occurs, the nutrients absorbed by the trophoblast can easily reach the inner cell mass by diffusion. But as the embryo and trophoblast enlarge, the distance between them increases, and diffusion alone cannot keep pace with the demands of the embryo. Blood vessels now begin to develop within the mesoderm of the chorion, creating a rapid-transit system that links the embryo with the trophoblast.

PLACENTATION

The **placenta** is a temporary structure in the uterine wall that provides a site for diffusion between the fetal and maternal circulatory systems. **Placentation** (pla-sen-TĀ-shun), or *placenta formation*, takes place when blood vessels form in the chorion around the periphery of the blastocyst (see Figure 20-5a–e●). By the third week of development, the mesoderm extends along each of the trophoblastic villi, forming *chorionic villi* in contact with maternal tissues (Figure 20-5b●). Embryonic blood vessels develop in each villus, and circulation through these

chorionic vessels begins early in the third week, when the heart starts beating. These villi continue to enlarge and branch, forming an intricate network within the endometrium. Blood vessels continue to be eroded, and maternal blood flows slowly through the lacunae. Chorionic blood vessels pass close by, and gases and nutrients diffuse between the embryonic and maternal circulations across the trophoblast layers.

At first, the entire blastocyst is surrounded by chorionic villi. The chorion continues to enlarge, expanding like a balloon within the endometrium, and by the fourth week, the embryo, amnion, and yolk sac are suspended within an expansive, fluid-filled chamber (Figure 20-5c●). The *body stalk*, the connection between the embryo and the chorion, contains the distal portions of the allantois and blood vessels that carry blood to and from the placenta. The narrow connection between the endoderm of the embryo and the yolk sac is called the *yolk stalk*. As the end of the first trimester approaches, the fetus moves farther away from the placenta. The yolk stalk and body stalk begin to fuse, forming an *umbilical stalk*, which contains placental blood vessels (Figure 20-5d●). By week 10, the fetus floats free within the amniotic cavity (Figure 20-5e●). The fetus remains connected to the placenta by the elongate **umbilical cord**, which contains the allantois, umbilical blood vessels, and the yolk sac.

Placental Circulation

Figure 20-6● diagrams circulation at the placenta near the end of the first trimester. Blood flows to the placenta through the paired **umbilical arteries** and returns in a single **umbilical vein**. p. 418 The chorionic villi provide the surface area for the active and passive exchanges of gases, nutrients, and wastes between the fetal and maternal bloodstreams.

Placental Hormones

In addition to its role in the nutrition of the fetus, the placenta acts as an endocrine organ. Hormones are synthesized by the syncytial trophoblast and are released into the maternal circulation. The placental hormones produced include *human chorionic gonadotropin, progesterone, estrogens, human placental lactogen, placental prolactin,* and *relaxin*.

Human chorionic (kō-rē-ON-ik) **gonadotropin (hCG)** appears in the maternal bloodstream soon after implantation has occurred. The presence of hCG in blood or urine samples is a reliable indication of pregnancy. Kits sold for the early detection of a pregnancy are sensitive to the presence of this hormone. In function, hCG resembles luteinizing hormone (LH), because it maintains the corpus luteum and promotes the continued secretion of progesterone. As a result, the endometrial lining remains perfectly functional, and menses does not occur.

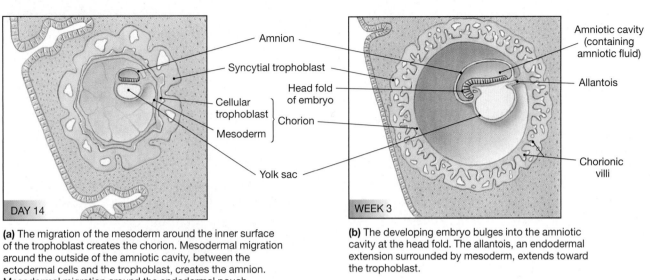

(a) The migration of the mesoderm around the inner surface of the trophoblast creates the chorion. Mesodermal migration around the outside of the amniotic cavity, between the ectodermal cells and the trophoblast, creates the amnion. Mesodermal migration around the endodermal pouch creates the yolk sac.

(b) The developing embryo bulges into the amniotic cavity at the head fold. The allantois, an endodermal extension surrounded by mesoderm, extends toward the trophoblast.

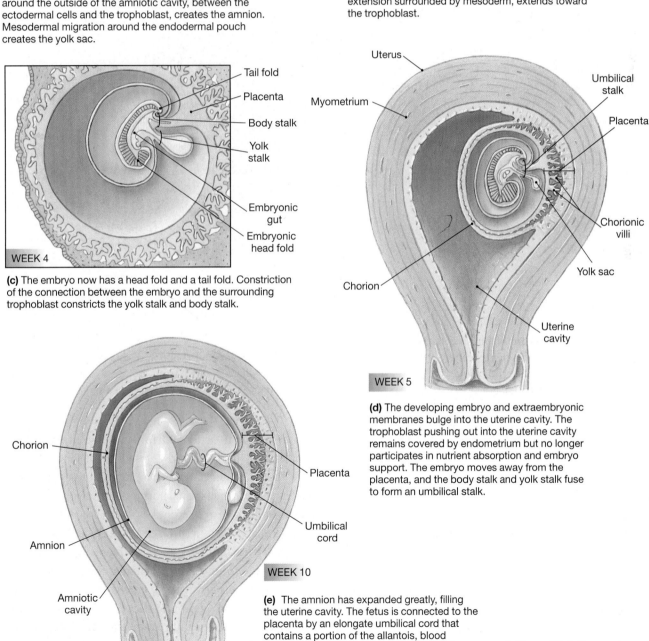

(c) The embryo now has a head fold and a tail fold. Constriction of the connection between the embryo and the surrounding trophoblast constricts the yolk stalk and body stalk.

(d) The developing embryo and extraembryonic membranes bulge into the uterine cavity. The trophoblast pushing out into the uterine cavity remains covered by endometrium but no longer participates in nutrient absorption and embryo support. The embryo moves away from the placenta, and the body stalk and yolk stalk fuse to form an umbilical stalk.

(e) The amnion has expanded greatly, filling the uterine cavity. The fetus is connected to the placenta by an elongate umbilical cord that contains a portion of the allantois, blood vessels, and the remnants of the yolk stalk.

● *Figure 20-5* **Extraembryonic Membranes and Placenta Formation**

621

20 DEVELOPMENT AND INHERITANCE

An Overview of Topics in Development • Fertilization • **The First Trimester** • The Second and Third Trimesters • Labor and Delivery

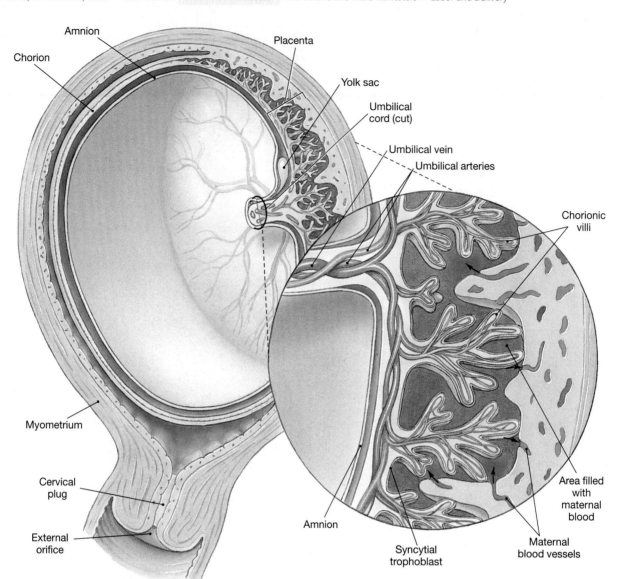

Amnion

Chorion

Placenta

Yolk sac

Umbilical cord (cut)

Umbilical vein

Umbilical arteries

Chorionic villi

Myometrium

Cervical plug

External orifice

Amnion

Area filled with maternal blood

Syncytial trophoblast

Maternal blood vessels

● *Figure 20-6* **The Structure of the Placenta**
For clarity, the uterus is shown after the embryo has been removed and the umbilical cord cut. Blood flows into the placenta through ruptured maternal blood arteries. It then flows around chorionic villi that contain fetal blood vessels. Fetal blood arrives over paired umbilical arteries and leaves over a single umbilical vein. Maternal blood reenters the venous system of the mother through the broken walls of small uterine veins. No mixing of maternal and fetal blood occurs.

In the absence of hCG, the pregnancy would end because another uterine cycle would begin and the endometrial lining would disintegrate.

In the presence of hCG, the corpus luteum persists for 3–4 months before gradually decreasing in size. The decline in the corpus luteum does not trigger the return of menstrual periods, because by the end of the first trimester, the placenta is actively secreting both estrogens and progesterone. After the first trimester, the placenta produces sufficient amounts of progesterone to maintain the endometrial lining and continue the pregnancy. As the end of the third trimester approaches, estrogen production accelerates. The rising estrogen levels play a role in stimulating labor and delivery.

Human placental lactogen (hPL) and **placental prolactin** help prepare the mammary glands for milk production. The conversion of the mammary glands from resting to active status requires the presence of placental hormones (hPL, placental prolactin, estrogen, and progesterone) as well as several maternal hormones (growth hormone, prolactin, and thyroid hormones).

Relaxin is a hormone secreted by the placenta as well as by the corpus luteum. Relaxin (1) increases the flexibility of the pubic symphysis, permitting the pelvis to expand during delivery; (2) causes the dilation of the cervix, making it easier for the fetus to enter the vaginal canal; and (3) suppresses the release of oxytocin by the hypothalamus and delays the onset of labor contractions.

EMBRYOGENESIS

Shortly after gastrulation begins, the body of the embryo begins to separate itself from the rest of the embryonic disc. The body of the embryo and its internal organs now start to form. The process of forming an embryo is called **embryogenesis** (em-brē-ō-JEN-e-sis). It begins as folding and differential growth of the embryonic disc produce a bulge that projects into the amniotic cavity (Figure 20-5b●). This bulge is called the *head fold*. Similar movements lead to the formation of a *tail fold* (Figure 20-5c●). By this time, the orienta-

tion of the embryo can be seen, complete with dorsal and ventral surfaces and left and right sides. The changes in proportions and appearance that occur between the second developmental week and the end of the first trimester are summarized in Figure 20-7●.

The first trimester is a critical period for development, because events during the first 12 weeks establish the basis for **organogenesis**, the process of organ formation. Table 20-2 includes important developmental milestones during the first trimester.

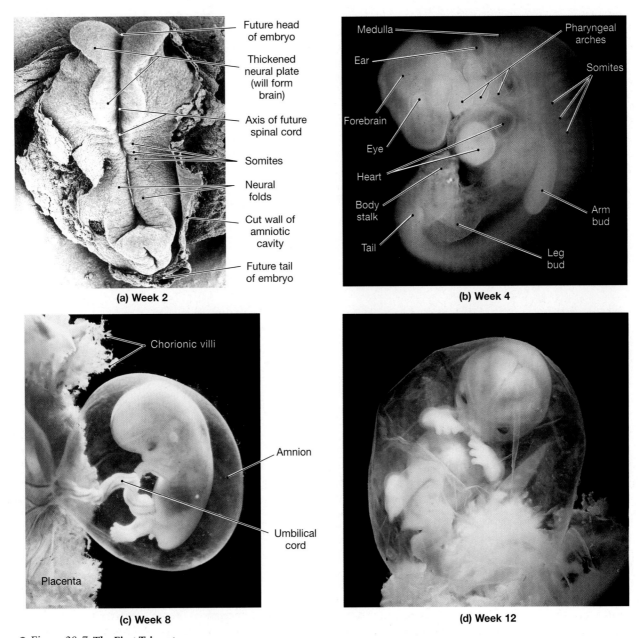

(a) Week 2

(b) Week 4

(c) Week 8

(d) Week 12

● *Figure 20-7* **The First Trimester**
(a) An SEM of the superior surface of a monkey embryo after 2 weeks of development. A human embryo at this stage would look essentially the same. (**b–d**) Fiber-optic views of human embryos at 4, 8, and 12 weeks. For actual sizes, see Figure 20-9●. p. 627.

20 **DEVELOPMENT AND INHERITANCE**

An Overview of Topics in Development • Fertilization • **The First Trimester** • The Second and Third Trimesters • Labor and Delivery

TABLE 20-2 *An Overview of Prenatal and Early Postnatal Development*

GESTATIONAL AGE (MONTHS)	SIZE AND WEIGHT	INTEGUMENTARY SYSTEM	SKELETAL SYSTEM	MUSCULAR SYSTEM	NERVOUS SYSTEM	SPECIAL SENSE ORGANS
1	5 mm 0.02 g		(b) Somite formation	(b) Somite formation	(b) Neural tube	(b) Eye and ear formation
2	28 mm 2.7 g	(b) Nail beds, hair follicles, sweat glands	(b) Axial and appendicular cartilage formation	(c) Rudiments of axial musculature	(b) CNS, PNS organization, growth of cerebrum	(b) Taste buds, olfactory epithelium
3	78 mm 26 g	(b) Epidermal layers appear	(b) Ossification centers spreading	(c) Rudiments of appendicular musculature	(c) Basic spinal cord and brain structure	
4	133 mm 0.15 kg	(b) Hair, sebaceous glands (c) Sweat glands	(b) Articulations (c) Facial and palatal organization	Fetus starts moving	(b) Rapid expansion of cerebrum	(c) Basic eye and ear structure (b) Peripheral receptor formation
5	185 mm 0.46 kg	(b) Keratin production, nail production			(b) Myelination of spinal cord	
6	230 mm 0.64 kg			(c) Perineal muscles	(b) CNS tract formation (c) Layering of cortex	
7	270 mm 1.492 kg	(b) Keratinization, nail formation, hair formation				(c) Eyelids open, retina sensitive to light
8	310 mm 2.274 kg		(b) Epiphyseal plate formation			(c) Taste receptors functional
9	346 mm 3.2 kg					
Postnatal development		Hair changes in consistency and distribution	Formation and growth of epiphyseal plates continue	Muscle mass and control increase	Myelination, layering, CNS tract formation continue	

Note: (b) = beginning to form; (c) = completed

ENDOCRINE SYSTEM	CARDIOVASCULAR AND LYMPHATIC SYSTEMS	RESPIRATORY SYSTEM	DIGESTIVE SYSTEM	URINARY SYSTEM	REPRODUCTIVE SYSTEM
	(b) Heartbeat	(b) Trachea and lung formation	(b) Intestinal tract, liver, pancreas (c) Yolk sac	(c) Allantois	
(b) Thymus, thyroid, pituitary, adrenal glands	(c) Basic heart structure, major blood vessels, lymph nodes and ducts (b) Blood formation in liver	(b) Extensive bronchial branching into mediastinum (c) Diaphragm	(b) Intestinal subdivisions, villi, salivary glands	(b) Kidney formation (adult form)	(b) Mammary glands
(c) Thymus, thyroid gland	(b) Tonsils, blood formation in bone marrow		(c) Gallbladder, pancreas		(b) Gonads, ducts, and genitalia formation
	(b) Migration of lymphocytes to lymphoid organs, blood formation in spleen			(b) Degeneration of embryonic kidneys	
	(c) Tonsils	(c) Nostrils open	(c) Intestinal subdivisions		
(c) Adrenal glands	(c) Spleen, liver, bone marrow	(b) Formation of alveoli	(c) Epithelial organization, glands		
(c) Pituitary gland			(c) Intestinal plicae		(b) Testes descend
		Complete pulmonary branching and alveolar formation		Nephron formation	Descent of testes complete at or near time of delivery
	Cardiovascular changes at birth; immune response gradually becomes operative				

20 DEVELOPMENT AND INHERITANCE

An Overview of Topics in Development • Fertilization • The First Trimester • **The Second and Third Trimesters** • Labor and Delivery

❶ What two important roles do acrosomal enzymes of spermatozoa play in fertilization?

❷ What is the developmental fate of the inner cell mass of the blastocyst?

❸ Sue's pregnancy test indicates elevated levels of the hormone hCG (human chorionic gonadotropin). Is she pregnant?

❹ What are two important functions of the placenta?

The Second and Third Trimesters

By the end of the first trimester (Figure 20-7d●), the rudiments of all the major organ systems have formed. Over the next 3 months, the fetus will grow to a weight of about 0.64 kg (1.4 lb). During this second trimester, the fetus, encircled by the amnion, grows faster than the surrounding placenta. When the outer surface of the amnion contacts the inner surface of the chorion, these layers fuse. Figure 20-8a● shows a 4-month fetus; Figure 20-8b● shows a 6-month fetus. The changes in body form that occur during the first and second trimesters are shown in Figure 20-9●.

During the third trimester, the basic components of all the organ systems appear, and most become ready to fulfill their normal functions. The rate of growth starts to decrease, but in absolute terms, this trimester sees the largest weight gain. In the last 3 months of gestation, the fetus gains about 2.6 kg (5.7 lb), reaching a full-term weight of about 3.2 kg (7 lb). Important events in organ system development during the second and third trimesters are also summarized in Table 20-2.

PREGNANCY AND MATERNAL SYSTEMS

The developing fetus is totally dependent on maternal organ systems for nourishment, respiration, and waste removal. These functions must be performed by maternal systems in addition to their normal operations. For example, the mother must absorb enough oxygen, nutrients, and vitamins for herself and her fetus, and she must eliminate all of the generated wastes. Although this is not a burden over the initial weeks of gestation, the demands become significant as the fetus grows larger. For the mother to survive under these conditions, the maternal systems must make major adjustments. In practical terms, the mother must breathe, eat, and excrete for two.

The major changes that take place in maternal systems include the following:

• *The maternal respiratory rate goes up and the tidal volume increases.* As a result, the mother's lungs obtain the extra

(a)

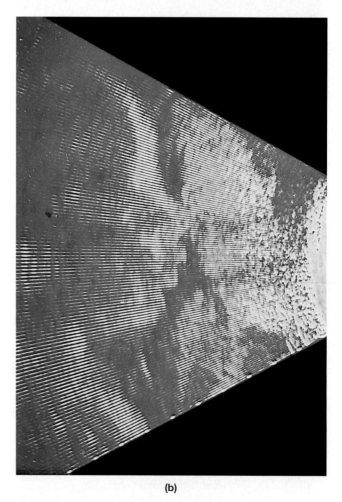
(b)

● *Figure 20-8* **The Second and Third Trimesters**
(a) A 4-month fetus seen through a fiber-optic endoscope.
(b) A 6-month fetus seen through ultrasound.

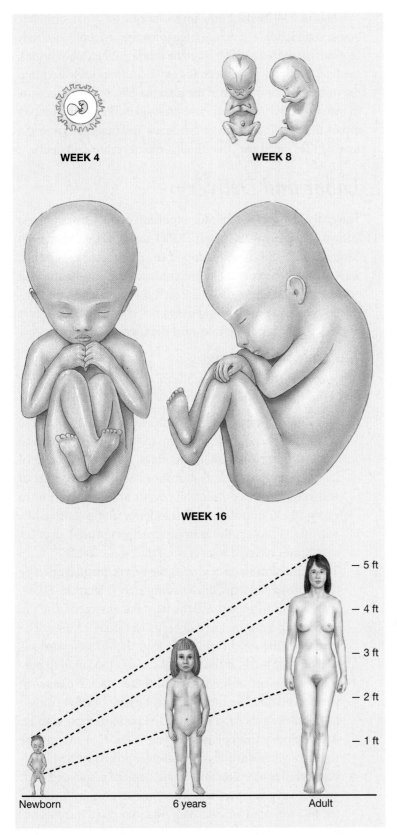

WEEK 4

WEEK 8

WEEK 16

Newborn   6 years   Adult

● *Figure 20-9* **Growth and Changes in Body Form**
The views at fetal ages 4, 8, and 16 weeks are presented at actual size.
Notice the changes in body form and proportions as development
proceeds. These changes do not stop at birth. For example, the head,
which contains the brain and sense organs, is relatively large at birth.

oxygen required and remove the excess carbon dioxide generated by the fetus.

- *The maternal blood volume increases.* This increase occurs
  because (1) blood flowing into the placenta reduces the
  volume in the rest of the systemic circuit, and (2) fetal
  activity lowers the blood P_{O_2} and elevates the P_{CO_2}.
  The combination stimulates the production of renin
  and erythropoietin (EPO) by the kidneys, leading to an
  increase in maternal blood volume (see Figure 13-11a,
  p. 402). By the end of gestation, the maternal blood
  volume has increased by almost 50 percent.

- *The maternal requirements for nutrients and vitamins
  climb 10–30 percent.* Pregnant women must nourish
  both themselves and their fetus and tend to have
  increased sensations of hunger.

- *The maternal glomerular filtration rate increases by
  roughly 50 percent.* This increase, which corresponds to
  the increase in blood volume, accelerates the excretion of
  metabolic wastes generated by the fetus. Because the
  volume of urine produced increases and the weight of
  the uterus presses down on the urinary bladder, pregnant
  women need to urinate frequently.

- *The uterus undergoes a tremendous increase in size.*
  Structural and functional changes in the expanding
  uterus are so important that we will discuss them in a
  separate section.

- *The mammary glands increase in size and secretory
  activity begins.* By the end of the sixth month of
  pregnancy, the mammary glands are fully developed
  and begin producing secretions that are stored in the
  duct system of those glands.

STRUCTURAL AND FUNCTIONAL CHANGES IN THE UTERUS

At the end of gestation, a typical uterus will have grown
from 7.5 cm (3 in.) in length and 60 g (2 oz.) in weight to
30 cm (12 in.) in length and 1100 g (2.4 lb) in weight. It
may then contain almost 5 liters of fluid, so the organ with
contents has a total weight of roughly 10 kg (22 lb). This
remarkable expansion occurs by the enlargement and elongation of existing cells, especially smooth muscle cells, rather
than by an increase in the total number of cells in the uterus.

The tremendous stretching of the myometrium is associated with a gradual increase in the rates of spontaneous
smooth muscle contractions. In the early stages of pregnancy, the contractions are weak, painless, and brief. Evidence indicates that the progesterone released by the placenta

20 **DEVELOPMENT AND INHERITANCE**

An Overview of Topics in Development • Fertilization • The First Trimester • The Second and Third Trimesters • Labor and Delivery

has an inhibitory effect on the uterine smooth muscle, preventing more extensive and powerful contractions.

Three major factors oppose the calming action of progesterone:

1. *Rising estrogen levels.* Estrogens, also produced by the placenta, increase the sensitivity of the uterine smooth muscles and make contractions more likely. Throughout pregnancy, progesterone exerts the dominant effect, but as the time of delivery approaches, estrogen production accelerates and the myometrium becomes more sensitive to stimulation. Estrogens also increase the sensitivity of smooth muscle fibers to oxytocin.

2. *Rising oxytocin levels.* Rising oxytocin levels stimulate an increase in the force and frequency of uterine contractions. Oxytocin release is stimulated by high estrogen levels and by distortion of the cervix.

3. *Prostaglandin production.* Estrogens and oxytocin stimulate the production of prostaglandins in the endometrium. These prostaglandins further stimulate smooth muscle contractions.

After 9 months of gestation, multiple factors interact to produce **labor contractions** in the myometrium of the uterine wall. Once begun, positive feedback ensures that the contractions continue until delivery has been completed.

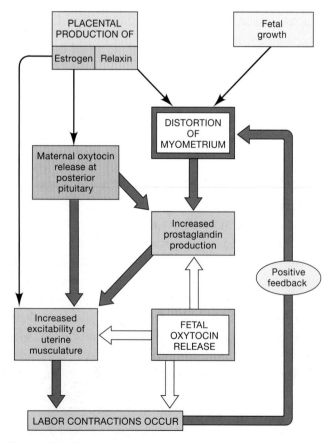

● *Figure 20-10* **Factors Involved in the Initiation of Labor and Delivery**

Figure 20-10● diagrams important factors that stimulate and sustain labor. The actual trigger for the onset of labor may be events in the fetus rather than the mother. When labor begins, the fetal pituitary gland secretes oxytocin that is released into the maternal bloodstream at the placenta. The resulting increase in myometrial contractions and prostaglandin production, on top of the priming effects of estrogens and maternal oxytocin may be the "last straw" that finally initiates labor and delivery.

Labor and Delivery

The goal of labor is the forcible expulsion of the fetus, a process known as **parturition** (par-tū-RISH-un). During labor, each contraction begins near the top of the uterus and sweeps in a wave toward the cervix. These contractions are strong and occur at regular intervals. As parturition approaches, the contractions increase in force and frequency, changing the position of the fetus and moving it toward the cervical canal.

THE STAGES OF LABOR

Labor consists of three stages: the *dilation stage*, the *expulsion stage*, and the *placental stage*.

1. The **dilation stage** begins with the onset of labor, as the cervix dilates and the fetus begins to slide down the cervical canal (Figure 20-11a●). This stage is highly variable in length but typically lasts 8 or more hours. At the start of this stage, labor contractions occur at intervals of once every 10–30 minutes; their frequency increases steadily. Late in the process the amnion usually ruptures, an event sometimes referred to as having the "water break."

2. The **expulsion stage** begins as the cervix, pushed open by the approaching fetus, dilates completely (Figure 20-11b●). Expulsion continues until the fetus has emerged from the vagina, a period that usually lasts less than 2 hours. The arrival of the newborn infant into the outside world is **delivery**, or birth. If the vaginal canal is too small to permit the passage of the fetus and there is acute danger of perineal tearing, the entryway may be temporarily enlarged by making an incision through the perineal musculature. After delivery, this **episiotomy** (e-pēz-ē-OT-o-mē) can be repaired with sutures, a much simpler procedure than dealing with the bleeding and tissue damage associated with an extensive perineal tear. If complications arise during the dilation or expulsion stage, the infant can be removed by **cesarean section**, or "C-section." In such cases, an incision is made through the abdominal wall, and the uterus is opened just enough to allow passage of the infant's head. This procedure is performed during 15–25 percent of the deliveries in the United States—more often than neces-

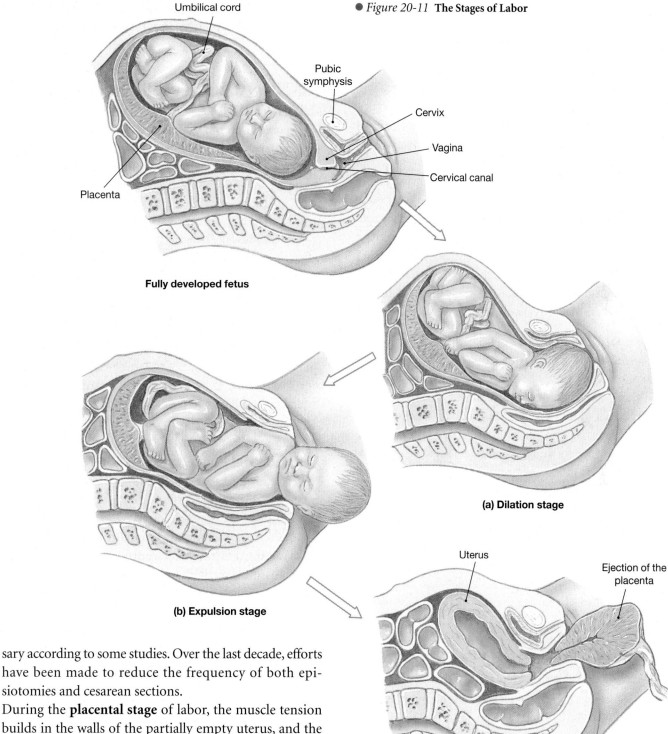

● *Figure 20-11* **The Stages of Labor**

Fully developed fetus

Umbilical cord

Pubic symphysis

Cervix

Vagina

Cervical canal

Placenta

(a) Dilation stage

(b) Expulsion stage

Uterus

Ejection of the placenta

(c) Placental stage

sary according to some studies. Over the last decade, efforts have been made to reduce the frequency of both episiotomies and cesarean sections.

3. During the **placental stage** of labor, the muscle tension builds in the walls of the partially empty uterus, and the organ gradually decreases in size (Figure 20-11c●). This uterine contraction tears the connections between the endometrium and the placenta. Usually within an hour after delivery, the placental stage ends with the ejection of the placenta, or "afterbirth". The disruption of the placenta is accompanied by a loss of blood, perhaps as much as 500–600 ml, but because the maternal blood volume has increased during pregnancy, the loss can be tolerated without difficulty.

20 DEVELOPMENT AND INHERITANCE

An Overview of Topics in Development • Fertilization • The First Trimester • The Second and Third Trimesters • Labor and Delivery

PREMATURE LABOR

Premature labor occurs when labor contractions begin before the fetus has completed normal development. The chances of newborn survival are directly related to the infant's body weight at delivery. Even with massive supportive efforts, newborns weighing less than 400 g (14 oz.) at birth will not survive, primarily because their respiratory, cardiovascular, and urinary systems are unable to support life. As a result, the dividing line between spontaneous abortion and *immature delivery* is usually set at 500 g (17.6 oz.), the normal weight near the end of the second trimester.

Most fetuses born at 25–27 weeks of gestation (birth weight under 600 g) die despite intensive neonatal care; survivors have a high risk of developmental abnormalities. A **premature delivery** usually refers to infants born at 28–36 weeks (birth weight over 1 kg). Given the proper medical care, these infants have a good chance of survival and normal development.

MULTIPLE BIRTHS

Multiple births (twins, triplets, quadruplets, and so forth) can occur for several reasons. The ratio of twin births to single births in the U.S. population is one out of every 89 (1:89). About 70 percent of all twins are *fraternal*, or **dizygotic** (dī-zī-GOT-ik). Fraternal twins develop when two eggs are fertilized at the same time, forming two separate zygotes. Fraternal twins can be of the same or different sexes.

Identical, or **monozygotic**, twins result from the separation of blastomeres early in cleavage or from the splitting of the inner cell mass before gastrulation. In either event, the genetic makeup and sex of the pair are identical because both twins formed from the same pair of gametes. Identical twins occur in about 30 percent of all twin births. Triplets and larger multiples can result from the same processes that produce twins.

CONCEPT CHECK QUESTIONS

Answers on page 641

❶ Why does a mother's blood volume increase during pregnancy?

❷ What effect would a decrease in progesterone have on the uterus during late pregnancy?

❸ During pregnancy, the uterus increases greatly in size and weight. What process do the uterine cells undergo to produce this change?

Postnatal Development

Developmental processes do not cease at delivery. The newborn infant has few of the anatomical, functional, or physiological characteristics of mature adults. In postnatal development, each individual passes through a number of **life stages**—*neonatal, infancy, childhood, adolescence,* and *maturity*—each with a distinctive combination of characteristics and abilities.

THE NEONATAL PERIOD, INFANCY, AND CHILDHOOD

The **neonatal period** extends from the moment of birth to 1 month thereafter. **Infancy** then continues to 2 years of age, and **childhood** lasts until puberty commences. Two major events are under way during these developmental stages:

1. The major organ systems, other than those associated with reproduction, become fully operational and gradually acquire the functional characteristics of adult structures.
2. The individual grows rapidly, and body proportions change significantly.

Pediatrics is a medical specialty focusing on postnatal development from infancy through adolescence. Infants and young children often cannot clearly describe the problems they are experiencing, so pediatricians and parents must be skilled observers. Standardized tests are used to assess developmental progress relative to average values.

The Neonatal Period

Physiological and anatomical changes occur as the fetus completes the transition to the status of a newborn infant, or **neonate**. Before delivery, dissolved gases, nutrients, waste products, hormones, and immunoglobulins were transferred across the placenta. At birth, the newborn infant must become relatively self-sufficient, with respiration, digestion, and excretion performed by its own specialized organs and organ systems. Typical heart rates of 120–140 beats per minute and respiratory rates of 30 breaths per minute in neonates are considerably higher than those of adults. The transition from fetus to neonate can be summarized as follows:

- *The lungs at birth are collapsed and filled with fluid.* Filling them with air involves a powerful inhalation.

- *When the lungs expand, the pattern of cardiovascular circulation changes due to alterations in blood pressure and flow rates.* The circulation changes result in the separation of the pulmonary and systemic circuits.

- *Before birth, the digestive system remains relatively inactive, although it does accumulate a mixture of bile secretions, mucus, and epithelial cells.* This collection of debris is excreted in the first few days of life. Over that period the newborn infant begins to nurse.

- *As waste products build up in the arterial blood, they are excreted at the kidneys.* Glomerular filtration is normal, but the neonate cannot concentrate urine to any significant degree. As a result, urinary water losses are high and neonatal fluid requirements are greater than those of adults.

- *The neonate has little ability to control body temperature, particularly in the first few days after delivery.* As the infant grows larger and its insulating subcutaneous fat "blanket" gets thicker, its metabolic rate also rises. Daily and even hourly alterations in body temperature continue throughout childhood.

Lactation and the Mammary Glands. By the end of the sixth month of pregnancy, the mammary glands are fully developed, and the gland cells begin producing a secretion known as **colostrum** (ko-LOS-trum). Provided to the infant during the first 2 or 3 days of life, colostrum contains more proteins and far less fat than breast milk. Many of the proteins are antibodies that help the infant ward off infections until its own immune system becomes fully functional. As colostrum production declines, the mammary glands convert to milk production. Breast milk consists of a mixture of water, proteins, amino acids, lipids, sugars, and salts. It also contains large quantities of *lysozymes*, enzymes with antibiotic properties.

Mammary gland secretion is triggered when the infant begins to suck on the nipple. The stimulation of tactile receptors there leads to the release of oxytocin at the posterior pituitary. When oxytocin reaches the mammary gland, this hormone causes contraction of contractile cells within the lactiferous ducts and sinuses. The result is the ejection of milk (Figure 20-12●). This *milk let-down reflex* continues to function until weaning, typically 1–2 years after birth. Milk production ceases soon after, and the mammary glands gradually return to a resting state.

Infancy and Childhood

The most rapid growth occurs during prenatal development, and the rate of growth declines after delivery. Postnatal growth during infancy and childhood occurs under the direction of circulating hormones, notably pituitary growth hormone, adrenal steroids, and thyroid hormones. These hormones affect each tissue and organ in specific ways, depending on the sensitivities of the individual cells. As a result, growth does not occur uniformly, and the body proportions gradually change (see Figure 20-9●, p. 627).

ADOLESCENCE AND MATURITY

Adolescence begins at **puberty**, the period of sexual maturation, and ends when growth is completed. Three events interact at the onset of puberty:

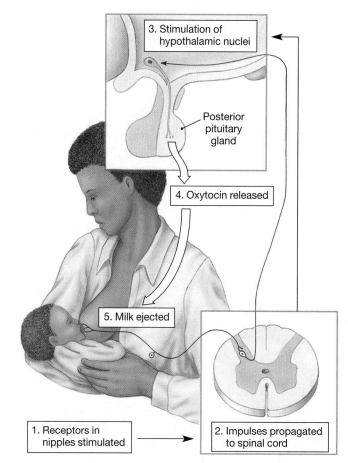

● *Figure 20-12* **The Milk Let-Down Reflex**

1. The hypothalamus increases its production of gonadotropin-releasing hormone (GnRH).
2. The anterior pituitary becomes more sensitive to the presence of GnRH, and the circulating levels of FSH and LH rise rapidly.
3. Ovarian or testicular cells become more sensitive to FSH and LH. These changes initiate (1) gamete formation, (2) the production of sex hormones that stimulate the appearance of secondary sex characteristics and behaviors, and (3) a sudden acceleration in the growth rate, ending in the closure of the epiphyseal cartilages.

The age at which puberty begins varies. In the United States today, puberty generally occurs at about age 14 in boys and 13 in girls, but the normal ranges are broad (9–14 in boys, 8–13 in girls). The combination of sex hormones and growth hormone, adrenal steroids, and thyroxine leads to a sudden acceleration in the growth rate. The timing of the growth spurt varies between the sexes, corresponding to different ages at the onset of puberty. In girls, the growth rate is maximum between ages 10 and 13; boys grow most rapidly between ages 12 and 15. Growth continues at a slower pace until ages 18 to 21. By that time, most of the epiphyseal plates have closed. p. 132

20 DEVELOPMENT AND INHERITANCE

An Overview of Topics in Development • Fertilization • The First Trimester • The Second and Third Trimesters • Labor and Delivery

The boundary between adolescence and maturity is very hazy, for it has physical, emotional, behavioral, and legal implications. Adolescence is often said to be over when growth ends in the late teens or early twenties. The individual is then considered mature. Although growth ends at maturity, physiological changes continue. The gender-specific differences produced at puberty are retained, but further changes occur when sex hormone levels decline at menopause or the male climacteric. ∞ p. 601 All these changes are part of the process of aging, or **senescence**. Aging reduces the efficiency and capabilities of the individual. Even in the absence of other factors, such as disease or injury, aging-related changes at the molecular level ultimately lead to death.

CONCEPT CHECK QUESTIONS

Answers on page 641

❶ An increase in the levels of GnRH, FSH, LH, and sex hormones in the blood mark the onset of which stage of development?

ABORTION

Abortion is the termination of a pregnancy. Most references distinguish among spontaneous, therapeutic, and induced abortions. **Spontaneous abortions**, or *miscarriages*, occur as a result of some developmental or physiological problem. For example, spontaneous abortions can result from chromosomal defects in the embryo or from hormonal problems, such as inadequate LH production by the maternal pituitary gland or placental failure to produce adequate levels of hCG. Spontaneous abortions occur in roughly 15 percent of all pregnancies. **Therapeutic abortions** are performed when continuing the pregnancy represents a threat to the life and health of the mother.

Induced abortions, or *elective abortions*, are performed at the woman's request. Induced abortions remain the focus of considerable controversy. Most induced abortions involve unmarried and/or adolescent women. The ratio between abortions and deliveries for married women averages 1:10, whereas it is nearly 2:1 for unmarried women and adolescents. In most states, induced abortions are legal during the first 3 months after conception; with restrictions, induced abortions may be permitted until the fifth or sixth month of gestation.

Genetics, Development, and Inheritance

Chromosome structure and the functions of genes were introduced in Chapter 3. ∞ pp. 70–71 Chromosomes contain DNA, and genes are segments of DNA. Each gene carries the information needed to direct the synthesis of a specific polypep-

tide. Every somatic cell containing a nucleus in your body carries copies of the original 46 chromosomes present when you were a zygote. Those chromosomes and their component genes represent your **genotype** (JĒN-ō-tīp).

Through development and differentiation, the instructions contained within the genotype are expressed in many ways. No single living cell or tissue makes use of all the information contained within the genotype. For example, in muscle fibers the genes important for excitable membrane formation and contractile proteins are active, while a different set of genes is operating in cells of the pancreatic islets. Collectively, however, the instructions contained within the genotype determine the anatomical and physiological characteristics that make you a unique individual. Those characteristics make up your **phenotype** (FĒN-ō-tīp; *phainein*, to display + *typos*, mark). Specific elements in your phenotype, such as your hair and eye color, skin tone, and foot size are called *phenotypic characters*, or *traits*.

Your genotype is derived from those of your parents, but not in a simple way. You are not an exact copy of either parent, nor are you an easily identifiable mixture of their characteristics. Our discussion of genetics will begin with the basic patterns of inheritance and their implications. Then we will examine the mechanisms responsible for regulating the activities of the genotype during prenatal development.

GENES AND CHROMOSOMES

Every somatic cell contains 46 chromosomes, arranged in 23 pairs. One member of each pair was contributed by the sperm, and the other by the ovum. The members of each pair are known as **homologous** (hō-MOL-o-gus) **chromosomes**. Twenty-two of those pairs are known as **autosomal** (aw-to-SŌ-mal) **chromosomes**. The chromosomes of the twenty-third pair are called the *sex chromosomes* because they differ in the two sexes. Figure 20-13● shows the **karyotype**, or entire set of chromosomes, of a normal male.

Autosomal Chromosomes

The two chromosomes in an autosomal pair have the same structure and carry genes that affect the same traits. Suppose one member of the pair contains three genes in a row, with number 1 determining hair color, number 2 eye color, and number 3 skin pigmentation. The other chromosome will carry genes that affect the same traits, and the genes will be in the same positions and sequence.

The two chromosomes in a pair may not carry the same form of each gene. The various forms of any one gene are called **alleles** (a-LĒLS; *allelon*, of one another). If both chromosomes of a homologous pair carry the same allele of a particular gene,

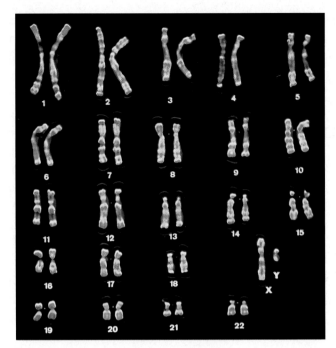

● *Figure 20-13* **Chromosomes of a Normal Male**

you are **homozygous** (hō-mō-ZĪ-gus; *homos*, same) for the trait affected by that gene. That allele will be expressed in your phenotype. For example, if you receive a gene for curly hair from your father and one for curly hair from your mother, you will be homozygous for curly hair. Because the chromosomes of a homologous pair have different origins, one paternal and the other maternal, they need not carry the same alleles. When you have two different alleles of the same gene, you are **heterozygous** (het-er-ō-ZĪ-gus; *heteros*, other) for the trait determined by that gene. In that case, your phenotype will be determined by the interactions between the corresponding alleles:

- An allele that is **dominant** will be expressed in the phenotype *regardless of any conflicting instructions carried by the other allele.*

- An allele that is **recessive** will be expressed in the phenotype only if it is present on both chromosomes of a homologous pair. For example, the albino skin condition is characterized by an inability to synthesize the yellow-brown pigment *melanin.* 🔗 p. 112 A single dominant allele determines normal skin coloration; two recessive alleles must be present to produce albinism.

Predicting Inheritance. Not every allele can be neatly characterized as dominant or recessive. Several that can be are included in Table 20-3. If you consider the traits listed there, you can predict the characteristics of individuals on the basis of the parents' alleles.

TABLE 20-3 *The Inheritance of Selected Phenotypic Characteristics*

DOMINANT TRAITS

One allele determines phenotype; the other is suppressed:
- normal skin pigmentation
- lack of freckles
- brachydactyly (short fingers)
- ability to taste phenylthiocarbamate (PTC)
- free earlobes
- curly hair
- color vision
- presence of Rh factor on red blood cell membranes

Both dominant alleles are expressed (codominance):
- presence of A or B antigens on red blood cell membranes
- structure of serum proteins (albumins, transferrins)
- structure of hemoglobin molecule

RECESSIVE TRAITS
- albinism
- freckles
- normal digits
- attached earlobes
- straight hair
- blond hair
- red hair (expressed only if individual is also homozygous for blond hair)
- lack of A, B surface antigens (Type O blood)
- inability to roll the tongue into a U-shape

SEX-LINKED TRAITS
- color blindness
- hemophilia

POLYGENIC TRAITS
- eye color
- hair colors other than pure blond or red

Dominant traits are traditionally indicated by capitalized abbreviations, and recessives are abbreviated in lowercase letters. For a given trait, the possibilities are indicated by *AA* (homozygous dominant), *Aa* (heterozygous), or *aa* (homozygous recessive). Each gamete involved in fertilization contributes a single allele for a given trait. That allele must be one of the two contained by all cells in the parent's body. Consider, for example, the offspring of an albino mother and a father with normal skin pigmentation. Because albinism is a recessive trait, the maternal alleles are abbreviated *aa*. No matter which of her oocytes gets fertilized, it will carry the recessive *a* allele. The father has normal pigmentation, a dominant trait. He is therefore homozygous or heterozygous for this trait, because both *AA* or *Aa* will produce the same phenotype for normal skin pigmentation.

20 DEVELOPMENT AND INHERITANCE

An Overview of Topics in Development • Fertilization • The First Trimester • The Second and Third Trimesters • Labor and Delivery

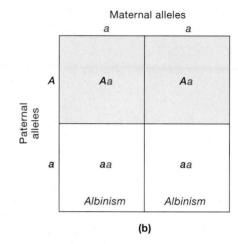

● *Figure 20-14* **Predicting Genotypes and Phenotypes with Punnett Squares**
(a) The offspring of a homozygous dominant father and a homozygous recessive mother will all be heterozygous for that trait. Their phenotype will be the same as that of the father.
(b) The offspring of a heterozygous father and a homozygous recessive mother will either be heterozygous or homozygous for the recessive trait. In this example, half of the offspring will have normal skin coloration and the other half will be albinos. (c) The inheritance of color blindness, a sex-linked trait. A color-blind father and a heterozygous mother will produce daughters with normal vision, but half of the sons will be color-blind.

A simple box diagram known as a *Punnett square* lets us predict the probabilities that children will have particular characteristics by giving the possible combinations of parental alleles they can inherit. In the Punnett squares in Figure 20-14●, the maternal alleles are listed along the horizontal axis and the paternal ones along the vertical axis. The possible combinations are indicated in the small boxes. Figure 20-14a● shows the possible offspring of an *aa* mother and an *AA* father. All of the children must have the genotype *Aa*, and they will all have normal skin pigmentation. Compare these results with those of Figure 20-14b●, for a heterozygous father (*Aa*). The heterozygous male produces two types of gametes, *A* and *a*, and either one may fertilize the oocyte. As a result, the probability is 50 percent that a child of such a father will inherit the genotype *Aa* and so have normal skin pigmentation. The probability of inheriting the genotype *aa*, and thus having the albino phenotype, is also 50 percent.

The Punnett square can also be used to draw conclusions about the identity and genotype of a parent. For example, a man with the genotype *AA* cannot be the father of an albino child (*aa*).

Simple Inheritance. In **simple inheritance**, phenotypes are determined by interactions between a single pair of alleles. The frequency of appearance of an inherited disorder resulting from simple inheritance can be predicted using a Punnett square. Although they are rare disorders in terms of overall numbers, more than 1200 inherited conditions have been identified that reflect the presence of one or two abnormal alleles for a single gene. A partial listing is given in Table 20-4, along with locations with more information.

Polygenic Inheritance. Many phenotypic characters are determined by interactions among several genes. Such interactions are called **polygenic inheritance**. Because multiple alleles are involved, the frequency of occurrence cannot easily be predicted using a simple Punnett square. The risks of developing several important adult disorders, including hypertension and coronary artery disease, fall within this category. Many of the developmental disorders responsible for fetal deaths and congenital malformations result from polygenic inheritance. In these cases, the particular genetic composition of the individual does not by itself determine the onset of the disease. Instead,

TABLE 20-4 *Fairly Common Inherited Disorders*

DISORDER	PAGE IN TEXT
AUTOSOMAL DOMINANTS	
Marfan's syndrome	p. 94
Huntington's disease	p. 270
AUTOSOMAL RECESSIVES	
Deafness	p. 304
Albinism	p. 113
Sickle-cell anemia	p. 349
Cystic fibrosis	p. 460
Phenylketonuria	p. 531
X-LINKED	
Duchenne's muscular dystrophy	p. 182
Hemophilia (one form)	p. 361
Color blindness	p. 294

the conditions regulated by these genes establish a susceptibility to particular environmental influences. This means that not every individual with the genetic tendency for a certain condition will actually develop that condition. It is therefore difficult to track polygenic conditions through successive generations. However, because many inherited polygenic conditions are likely but not guaranteed to occur, steps can be taken to prevent a crisis. For example, you can reduce hypertension by controlling your diet and fluid volume, and you can prevent coronary artery disease by lowering your serum cholesterol levels.

Sex Chromosomes

The **sex chromosomes** determine the biological sex of the individual. Unlike the other 22 chromosomal pairs, the sex chromosomes are not identical in appearance and gene content. There are two different sex chromosomes: an **X chromosome** and a **Y chromosome**. X chromosomes are considerably larger, and have more genes, than do Y chromosomes. The Y chromosome includes dominant alleles that specify that an individual with that chromosome will be male. The normal pair of sex chromosomes in males is *XY*. Females do not have a Y chromosome; their sex chromosome pair is *XX*.

All ova carry an *X* chromosome, because the only sex chromosomes females have are *X* chromosomes. But each sperm carries either an *X* or *Y* sex chromosome. Thus, with a Punnett square you can show that the ratio of males to females in offspring should be 1:1.

The X chromosome also carries genes that affect somatic structures. These characteristics are called **X-linked** because in most cases there are no corresponding alleles on the Y chromosome. They are also known as *sex-linked traits* because the responsible genes are located on the sex chromosomes. The best-known X-linked characteristics are associated with noticeable diseases or defects that are caused by single alleles.

The inheritance of color blindness, a condition discussed in Chapter 9, demonstrates the differences between sex-linked and autosomal inheritance. p. 294 Normal color vision is determined by the presence of a dominant allele, *C*, whereas red-green color blindness results from the absence of *C* and the presence of a recessive allele *c*, on the X chromosome. A woman, with her two X chromosomes, can be either homozygous, *CC*, or heterozygous, *Cc*, and still have normal color vision. She will be unable to distinguish reds from greens only if she carries two recessive alleles, *cc*. But a male has only one X chromosome, so whichever allele that chromosome carries will determine whether he has normal color vision or is red-green color-blind. A Punnett square for an X-linked trait, as in Figure 20-14c●, reveals that the sons produced by a father with normal vision and a heterozygous mother will have a 50 percent chance of being red-green color-blind, whereas the daughters will have normal color vision.

A number of other clinical disorders are X-linked traits, including certain forms of *hemophilia, diabetes insipidus*, and *muscular dystrophy*. In several instances, advances in techniques of molecular genetics have enabled geneticists to locate specific genes on the X chromosome. These techniques provides a relatively direct method of screening for the presence of a particular condition before the symptoms appear, and even before birth.

THE HUMAN GENOME PROJECT

Before the last decade, few of the genes responsible for inherited disorders had been identified or even localized to a specific chromosome. Since then, however, much progress has been made as a result of the **Human Genome Project (HGP)**. Funded by the National Institutes of Health and the Department of Energy, the goal of the HGP is to describe the entire human **genome**—that is, the full set of DNA, or genetic material—chromosome by chromosome, gene by gene, and nucleotide by nucleotide. The project began in October 1990 and was expected to take 15 years. Progress was more rapid than expected. In early 2001, the nucleotide sequence of the human genome was announced to be essentially complete.

The first step in understanding the human genome is to prepare a map of the individual chromosomes. **Karyotyping**

20 DEVELOPMENT AND INHERITANCE

An Overview of Topics in Development • Fertilization • The First Trimester • The Second and Third Trimesters • Labor and Delivery

CLINICAL NOTE *Chromosomal Abnormalities and Genetic Analysis*

Embryos that have abnormal autosomal chromosomes rarely survive. However, *translocation defects* and *trisomy* are two types that do not invariably kill the individual before birth:

In a **translocation defect**, crossing-over occurs between different chromosome pairs. For example, a piece of chromosome 8 may become attached to chromosome 14. The genes moved to their new position may function abnormally, becoming inactive or overactive. A translocation between chromosomes 8 and 14 is responsible for *Burkitt's lymphoma*, a type of lymphatic system cancer.

In **trisomy**, something goes wrong in meiosis. One of the gametes at fertilization carries an extra copy of one chromosome, so the zygote then has three copies of this chromosome rather than two. The nature of the trisomy is indicated by the number of the chromosome involved. Zygotes with extra copies of chromosomes seldom survive. Individuals with trisomy 13 and trisomy 18 may survive until delivery but rarely live longer than a year. The notable exception is trisomy 21.

Trisomy 21, or **Down syndrome**, is the most common viable chromosomal abnormality. Estimates of the frequency of appearance range from 1.5 to 1.9 per 1000 births for the U.S. population. Affected individuals exhibit mental retardation and characteristic physical malformations, including a facial appearance that gave rise to the term *mongolism*, once used to describe this condition. The degree of mental retardation ranges from moderate to severe. Few individuals with this condition lead independent lives. Anatomical problems affecting the cardiovascular system often prove fatal during childhood or early adulthood. Although some individuals survive to moderate old age, many develop Alzheimer's disease while still relatively young (before age 40).

For unknown reasons, there is a direct correlation between maternal age and the risk of having a child with trisomy 21. For a maternal age below 25, the incidence of Down syndrome approaches 1 in 2000 births, or 0.05 percent. For maternal ages 30–34, the odds increase to 1 in 900, and over the next decade, they go from 1 in 290 to 1 in 46, or more than 2 percent. These statistics are becoming increasingly significant, because many women have delayed childbearing until their mid-thirties or later.

Abnormal numbers of sex chromosomes do not produce effects as severe as those induced by extra or missing autosomal chromosomes. In **Klinefelter syndrome**, the individual carries the sex chromosome pattern *XXY*. The phenotype is male, but the extra X chromosome causes reduced androgen production. As a result, the

testes fail to mature, the individuals are sterile, and the breasts are slightly enlarged. The incidence of this condition among newborn males averages 1 in 750 births.

Individuals with **Turner syndrome** have only a single female sex chromosome; their sex chromosome complement is abbreviated *XO*. This kind of chromosomal deletion is known as **monosomy**. The incidence of this condition at delivery has been estimated as 1 in 10,000 live births. At birth, the condition may not be recognized, because the phenotype is normal female. But maturational changes do not appear at puberty. The ovaries are nonfunctional, and estrogen production occurs at negligible levels.

Fragile-X Syndrome causes mental retardation, abnormal facial development, and increased testicular size in affected males. The cause is an abnormal *X* chromosome that contains a *genetic stutter*, an abnormal repetition of a single triplet. The presence of the stutter in some way disrupts the normal functioning of adjacent genes and so produces the symptoms of the disorder.

Many of these conditions can be detected before birth by the analysis of fetal cells. In **amniocentesis**, a sample of amniotic fluid is removed and the fetal cells it contains are analyzed. This procedure permits the identification of more than 20 congenital conditions, including Down syndrome. The needle inserted to obtain a fluid sample is guided into position by using ultrasound. p. 202 Unfortunately, amniocentesis has two major drawbacks:

1. Because the sampling procedure represents a potential threat to the health of the fetus and mother, amniocentesis is performed only when known risk factors are present. Examples of risk factors are a family history of specific conditions, or in the case of Down syndrome, a maternal age over 35.

2. Sampling cannot safely be performed until the volume of amniotic fluid is large enough that the fetus will not be injured during the process. The usual time for amniocentesis is at a gestational age of 14–15 weeks. It may take several weeks to obtain results once samples have been collected, and by the time the results are received, the option of therapeutic abortion may no longer be available.

An alternative procedure known as **chorionic villus sampling** analyzes cells collected from the villi during the first trimester. Although it can be performed at an earlier gestational age, this technique has largely been abandoned because of an associated increased risk of spontaneous abortion (miscarriage).

(KAR-ē-ō-tī-ping; *karyon*, nucleus + *typos*, mark) is the determination of an individual's chromosome complement. Figure 20-13● (p. 633) shows a set of normal human chromosomes. Each chromosome has characteristic banding patterns, and segments can be stained with special dyes. The banding patterns are useful as reference points when more detailed genetic maps are prepared. The banding patterns themselves can be useful, because abnormal banding patterns are characteristic of some genetic disorders and several cancers.

One of the most surprising results of the HGP is that there are apparently far fewer genes than originally expected. Early in the project, estimates of the total number of human genes ranged from 100,000–140,000 genes. The present estimate now lies within a range of 30,000–40,000 genes. A genome is more accurately described by its size in megabases (1 Mb = 1 million base pairs). The human genome consists of some 3,200 Mb. It is the largest genome of any organism determined so far; about 18 times larger than the 180 Mb genome of the common fruit fly. This large difference in genome sizes is not matched by a

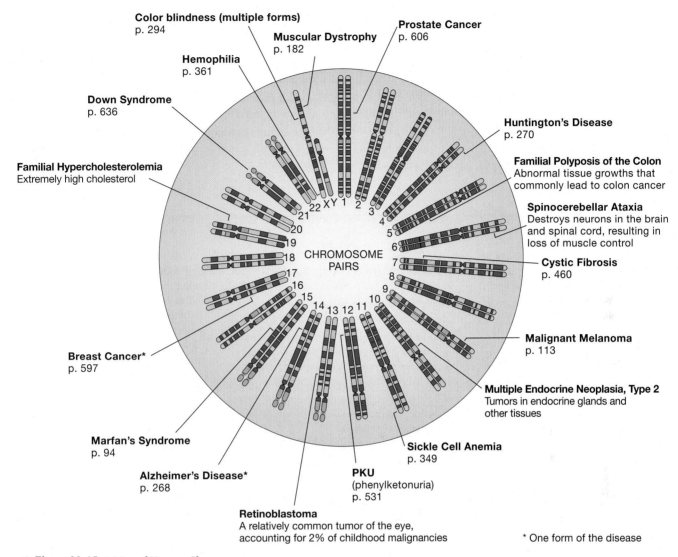

Color blindness (multiple forms)
p. 294

Muscular Dystrophy
p. 182

Prostate Cancer
p. 606

Hemophilia
p. 361

Down Syndrome
p. 636

Huntington's Disease
p. 270

Familial Hypercholesterolemia
Extremely high cholesterol

Familial Polyposis of the Colon
Abnormal tissue growths that
commonly lead to colon cancer

Spinocerebellar Ataxia
Destroys neurons in the brain
and spinal cord, resulting in
loss of muscle control

CHROMOSOME
PAIRS

Cystic Fibrosis
p. 460

Malignant Melanoma
p. 113

Breast Cancer*
p. 597

Multiple Endocrine Neoplasia, Type 2
Tumors in endocrine glands and
other tissues

Marfan's Syndrome
p. 94

Sickle Cell Anemia
p. 349

Alzheimer's Disease*
p. 268

PKU
(phenylketonuria)
p. 531

Retinoblastoma
A relatively common tumor of the eye,
accounting for 2% of childhood malignancies

* One form of the disease

● *Figure 20-15* **A Map of Human Chromosomes**
The banding patterns of typical chromosomes in a male individual and the locations of the genes responsible for specific inherited disorders. The chromosomes are not drawn to scale.

proportionate difference in the number of genes. It is a sobering fact that the common fruit fly has about 13,000 genes, roughly one-third to one-half the number estimated in humans.

At the beginning of 2002, the HGP had sequenced 97.8 percent of the total human genome. It has shown that the DNA nucleotide sequences of any two humans are quite similar, differing by only 0.1 percent. The level of similarity extends far beyond our direct relatives, for differences such as race cannot be deciphered from our genome. In addition, the genes responsible for more than 100 inherited disorders have been identified, including many of the disorders listed in Table 20-4 and shown in Figure 20-15●. Genetic screening can now be done for many of these conditions. Knowing the complete sequence

will help researchers identify the genetic basis of more complex and more common disorders, such as diabetes, asthma, and cancer, in which both genetic and environmental factors interact. Other medical applications of the HGP are based on new techniques that measure the expression, or activity, of thousands of genes at a time. For example, scientists can now compare gene activity in normal and diseased tissue, and also begin to look at why individuals have different responses to the same drugs.

The Human Genome Project has determined the normal genetic composition of a "typical" human being. Yet we are all variations on a basic theme. As we improve our ability to manipulate our own genetic foundations, we will face troubling ethical

20 DEVELOPMENT AND INHERITANCE

An Overview of Topics in Development • Fertilization • The First Trimester • The Second and Third Trimesters • Labor and Delivery

and legal decisions. For example, few people object to the insertion of a "correct" gene into somatic cells to cure a specific disease. But what if we could insert that modified gene into a gamete and change not only that individual but his or her descendants as well? And what if the gene did not correct or prevent a disorder, but "improved" the individual by increasing intelligence, height, or vision or altering some other phenotypic characteristic? These and other difficult questions will not go away, and in the years to come, we will have to find answers acceptable to us all.

CONCEPT CHECK QUESTIONS
Answers on page 641

❶ Curly hair is an autosomal dominant trait. What would be the phenotype of a person who is heterozygous for this trait?

❷ Joe has three daughters and complains that it's his wife's "fault" that he has no sons. What would you tell him?

❸ The human genome consists of approximately 3,200 Mb. What is a genome and how many nucleotide base pairs does 3,200 Mb. represent?

Related Clinical Terms

abruptio placentae (ab-RUP-shē-ō pla-SEN-tē): A tearing away of the placenta from the uterine wall after the fifth gestational month.

amniocentesis: An analysis of fetal cells taken from a sample of amniotic fluid.

Apgar rating: A method of evaluating newborn infants for developmental problems and neurological damage.

breech birth: A delivery during which the legs or buttocks of the fetus enter the vaginal canal first.

chorionic villus sampling: An analysis of cells collected from the chorionic villi during the first trimester.

congenital malformation: A severe structural abnormality, present at birth, that affects major systems.

ectopic pregnancy: A pregnancy in which implantation occurs somewhere other than the uterus.

fetal alcohol syndrome (FAS): A neonatal condition resulting from maternal alcohol consumption; characterized by developmental defects typically involving the skeletal, nervous, and/or cardiovascular systems.

geriatrics: A medical specialty that deals with aging and its medical problems.

infertility: The inability to achieve pregnancy after 1 year of appropriately timed intercourse.

in vitro fertilization: Fertilization outside the body, generally in a petri dish.

pediatrics: A medical specialty focusing on postnatal development from infancy through adolescence.

placenta previa: A condition resulting from implantation in or near the cervix, in which the placenta covers the cervix and prevents normal birth.

teratogens (TER-a-tō-jenz): Stimuli that disrupt normal development by damaging cells, altering chromosome structure, or altering the chemical environment of the embryo.

CHAPTER REVIEW

Key Terms

Summary Outline

1. **Development** is the gradual modification of physical and physiological characteristics from **conception** to maturity. The creation of different cell types is **differentiation**.

1. **Prenatal development** occurs before birth; **postnatal development** begins at birth and continues to maturity, when senescence (aging) begins. **Inheritance** is the transfer of genetically determined characteristics from generation to generation. **Genetics** is the study of the mechanisms of inheritance.

1. **Fertilization** normally occurs in the uterine tube within a day after ovulation. Sperm cannot fertilize an egg until they have undergone **capacitation**.

2. The acrosomal caps of the spermatozoa release *hyaluronidase*, an enzyme that separates cells of the *corona radiata*. Another acrosomal enzyme digests the zona pellucida and exposes the oocyte membrane. When a single spermatozoon contacts that membrane, fertilization occurs and **oocyte activation** follows.

3. During activation, the secondary oocyte completes meiosis, and the penetration of additional sperm is prevented.

4. After activation, the *female pronucleus* and *male pronucleus* fuse in a process called **amphimixis**. *(Figure 20-1)*

5. The 9-month **gestation**, or *pregnancy*, period can be divided into three **trimesters**.

1. The **first trimester** is the most dangerous period of prenatal development. The processes of *cleavage, implantation, placentation*, and *embryogenesis* take place during this critical period.

Cleavage and Blastocyst Formation 616

2. **Cleavage** subdivides the cytoplasm of the zygote through a series of mitotic divisions. The zygote becomes a hollow ball of blastomeres called a **blastocyst**. The blastocyst consists of an outer **trophoblast** and an **inner cell mass**. *(Figure 20-2)*

Implantation 617

3. During **implantation**, the blastocyst burrows into the uterine endometrium. Implantation occurs about 7 days after fertilization. *(Figure 20-3)*

4. As the trophoblast enlarges and spreads, maternal blood flows through open *lacunae*. After **gastrulation**, there is an embryonic disc composed of **endoderm**, **ectoderm**, and **mesoderm**. It is from these three **germ layers** that the body systems differentiate. *(Figure 20-4; Table 20-1)*

5. Germ layers help form four **extraembryonic membranes**: the *yolk sac, amnion, allantois,* and *chorion. (Figure 20-5)*

6. The **yolk sac** is an important site of blood cell formation. The **amnion** encloses fluid that surrounds and cushions the developing embryo. The base of the **allantois** later gives rise to the urinary bladder. Circulation within the vessels of the **chorion** provides a rapid-transport system linking the embryo with the trophoblast.

Placentation 620

7. **Placentation** occurs as blood vessels form around the blastocyst and the **placenta** appears. *Chorionic villi* extend outward into the maternal tissues, forming a branching network through which maternal blood flows. As development proceeds, the **umbilical cord** connects the fetus to the placenta. *(Figure 20-6)*

8. The trophoblast synthesizes **human chorionic gonadotropin (hCG)**, estrogens, progestins, **human placental lactogen (hPL), placental prolactin**, and **relaxin**.

Embryogenesis...623

9. The first trimester is critical because events in the first 12 weeks establish the basis for **organogenesis** (organ formation). *(Figure 20-7; Table 20-2)*

THE SECOND AND THIRD TRIMESTERS 626

1. In the **second trimester**, the organ systems increase in complexity. During the **third trimester**, these organ systems become functional. *(Figures 20-8, 20-9; Table 20-2)*

Pregnancy and Maternal Systems 626

2. The developing fetus is totally dependent on maternal organs for nourishment, respiration, and waste removal. Maternal adaptations include increased blood volume, respiratory rate, tidal volume, nutrient intake, and glomerular filtration.

Structural and Functional Changes in the Uterus ..627

3. Progesterone produced by the placenta has an inhibitory effect on uterine muscles; its calming action is opposed by estrogens, oxytocin, and prostaglandins. At some point, multiple factors interact to produce **labor contractions** in the uterine wall. *(Figure 20-10)*

LABOR AND DELIVERY 628

1. The goal of labor is **parturition**, the forcible expulsion of the fetus.

Stages of Labor 628

2. Labor can be divided into three stages: the **dilation stage, expulsion stage**, and **placental stage**. *(Figure 20-11)*

Premature Labor 630

3. **Premature labor** results in the delivery of a newborn that has not completed normal development.

Multiple Births 630

4. Twin births are either **dizygotic** (fraternal) or **monozygotic** (identical).

POSTNATAL DEVELOPMENT 630

1. Postnatal development involves a series of five **life stages**, including the *neonatal period, infancy, childhood, adolescence*, and *maturity. Senescence* begins at maturity and ends in the death of the individual.

The Neonatal Period, Infancy, and Childhood 630

2. The **neonatal period** extends from birth to 1 month of age. **Infancy** then continues to 2 years of age, and **childhood** lasts until puberty commences. During these stages, major organ systems (other than reproductive) become fully operational and gradually acquire adult characteristics, and the individual grows rapidly.

3. In the transition from fetus to **neonate**, the respiratory, circulatory, digestive, and urinary systems begin functioning independently. The newborn must also begin to thermoregulate.

4. Mammary glands produce protein-rich **colostrum** during the infant's first few days and then convert to milk production. These secretions are released as a result of the *milk let-down reflex. (Figure 20-12)*

Adolescence and Maturity 631

5. **Adolescence** begins at **puberty**: (1) The hypothalamus increases its production of GnRH, (2) circulating levels of FSH and LH rise rapidly, and (3) ovarian or testicular cells become more sensitive to FSH and LH. These changes initiate gametogenesis, production of sex hormones, and a sudden acceleration in growth rate. Adolescence continues until growth is completed.

6. **Maturity**, the end of growth and adolescence, occurs by the early twenties. Postmaturity changes in physiological processes are part of aging, or **senescence**.

GENETICS, DEVELOPMENT, AND INHERITANCE 632

1. Every somatic cell carries copies of the original 46 chromosomes in the zygote; these chromosomes and their genes make up the individual's genotype. The physical expression of the **genotype** is the **phenotype** of the individual.

Genes and Chromosomes 632

2. Every somatic human cell contains 23 pairs of chromosomes; each pair consists of **homologous chromosomes**. Twenty-two pairs are **autosomal chromosomes**. The chromosomes of the twenty-third pair are the sex chromosomes; they differ between the sexes. *(Figure 20-13)*

3. Chromosomes contain DNA, and genes are functional segments of DNA. The various forms of a gene are called **alleles**. If both homologous chromosomes carry the same allele of a particular gene, the individual is **homozygous**; if they carry different alleles, the individual is **heterozygous**.

4. Alleles are either **dominant** or **recessive** depending on how their traits are expressed. *(Table 20-3)*

5. Combining maternal and paternal alleles in a *Punnett square* helps us to predict the characteristics of offspring. *(Figure 20-14)*

6. In **simple inheritance**, phenotypic characters are determined by interactions between a single pair of alleles. **Polygenic inheritance** involves interactions among alleles on several chromosomes. *(Table 20-4)*

7. The two types of **sex chromosomes** are an **X chromosome** and a **Y chromosome**. The normal sex chromosome complement of males is *XY*; that of females is *XX*. The X chromosome carries **X-linked** (sex-linked) **genes**, which affect somatic structures but have no corresponding alleles on the Y chromosome.

The Human Genome Project 635

8. The **Human Genome Project** has estimated that the human genome is made up of 30,000 to 40,000 genes, including some of those responsible for inherited disorders. *(Figure 20-15; Table 20-4)*

20 DEVELOPMENT AND INHERITANCE

An Overview of Topics in Development • Fertilization • The First Trimester • The Second and Third Trimesters • Labor and Delivery

Review Questions

Level 1: Reviewing Facts and Terms

Match each item in column A with the most closely related item in column B. Use letters for answers in the spaces provided.

COLUMN A

___ 1. gestation

___ 2. cleavage

___ 3. gastrulation

___ 4. chorion

___ 5. human chorionic gonadotropin

___ 6. birth

___ 7. episiotomy

___ 8. afterbirth

___ 9. senescence

___ 10. neonate

___ 11. phenotype

___ 12. homozygous recessive

___ 13. heterozygous

___ 14. male genotype

___ 15. female genotype

___ 16. trisomy 21

COLUMN B

a. blastocyst formation

b. ejection of placenta

c. germ-layer formation

d. indication of pregnancy

e. embryo-maternal circulatory exchange

f. visible characteristics

g. time of prenatal development

h. *aa*

i. Down syndrome

j. newborn infant

k. *Aa*

l. *XY*

m. *XX*

n. parturition

o. process of aging

p. perineal musculature incision

---

17. The gradual modification of anatomical structures during the period from conception to maturity is

    (a) development   (b) differentiation

    (c) embryogenesis   (d) capacitation

18. Human fertilization involves the fusion of two haploid gametes, producing a zygote containing:

    (a) 23 chromosomes

    (b) 46 chromosomes

    (c) the normal haploid number of chromosomes

    (d) 46 pairs of chromosomes

19. The secondary oocyte leaving the follicle is in:

    (a) interphase

    (b) metaphase of the first meiotic division

    (c) telophase of the second meiotic division

    (d) metaphase of the second meiotic division

20. The process that establishes the foundation of all major organ systems is

    (a) cleavage   (b) implantation

    (c) placentation   (d) embryogenesis

21. The zygote arrives in the uterine cavity as a:

    (a) morula   (b) trophoblast

    (c) lacuna   (d) blastomere

22. The surface that provides for active and passive exchange between the fetal and maternal bloodstreams is the:

    (a) yolk stalk   (b) chorionic villi

    (c) umbilical veins (d) umbilical arteries

23. Milk let-down is associated with:

    (a) events occurring in the uterus

    (b) placental hormonal influences

    (c) circadian rhythms

    (d) reflex action triggered by suckling

24. If an allele must be present on both the maternal and paternal chromosomes to affect the phenotype, the allele is said to be:

    (a) dominant   (b) recessive

    (c) complementary   (d) heterozygous

25. Summarize the developmental changes that occur during the first, second, and third trimesters.

26. Identify the three stages of labor, and describe the events that characterize each stage.

27. Identify the three life stages that occur between birth and approximately age 10. Describe the characteristics of each stage and when it occurs.

Level 2: Reviewing Concepts

28. Relaxin is a peptide hormone that:
    (a) increases the flexibility of the symphysis pubis
    (b) causes dilation of the cervix
    (c) suppresses the release of oxytocin by the hypothalamus
    (d) a, b, and c are correct

29. During adolescence, the events that interact to promote increased hormone production and sexual maturation result from activity of the:
    (a) hypothalamus   (b) anterior pituitary
    (c) ovaries and testicular cells   (d) a, b, and c are correct

30. In addition to its role in the nutrition of the fetus, what are the primary endocrine functions of the placenta?

31. Discuss the changes that occur in maternal systems during pregnancy. Why are these changes functionally significant?

32. During labor, what physiological mechanisms ensure that uterine contractions continue until delivery has been completed?

33. To what does the phrase "having the water break" refer during the process of labor?

34. What would you conclude about a trait in each of the following situations?
    (a) Children who exhibit this trait have at least one parent who exhibits the same trait.
    (b) Children exhibit this trait even though neither of the parents exhibits it.
    (c) The trait is expressed more frequently in sons than in daughters.
    (d) The trait is expressed equally in both daughters and sons.

35. Explain why more men than women are color-blind. Which type of inheritance is involved?

36. Explain the goals and possible benefits of the Human Genome Project.

Level 3: Critical Thinking and Clinical Applications

37. Hemophilia A, a condition in which the blood does not clot properly, is a recessive trait located on the X chromosome (X^h). A woman heterozygous for the trait marries a normal male. What is the probability that this couple will have hemophiliac daughters? What is the probability that this couple will have hemophiliac sons?

38. Explain why the normal heart and respiratory rates of neonates are so much higher than those of adults, even though adults are so much larger.

39. Sally gives birth to a baby with a congenital deformity of the stomach. She swears that it is the result of a viral infection that she suffered during the third trimester of pregnancy. Do you think this is a possibility? Explain.

Answers to Concept Check Questions

Page 626

1. Fertilization requires a single spermatozoon to contact the cell membrane of the secondary oocyte. The oocyte, however, lies within the corona radiata and zona pellucida. Penetration of the the follicular cells of the corona radiata requires the presence of many spermatozoa. The connections between these cells are broken down by the release of an enzyme (hyaluronidase) from dozens of spermatozoa. The binding of a single spermatozoon to the zona pellucida results in the release of another acrosomal enzyme that digests a path through the zona pellucida to the oocyte membrane.

2. The inner cell mass of the blastocyst eventually develops into the embryo.

3. After fertilization, the developing trophoblasts and, later, the placenta produce and release the hormone hCG. Sue is pregnant.

4. Placental functions include (1) supplying the developing fetus with a route for gas exchange, nutrient transfer, and waste product elimination; and (2) producing hormones that affect maternal systems.

Page 630

1. During pregnancy, blood flow through the placenta reduces the volume of blood in the systemic circuit, and the fetus adds carbon dioxide to the shared circulatory system. The result is the release of renin and EPO, which stimulate an increase in maternal blood volume.

2. Progesterone reduces uterine contractions. A decrease in the progesterone level at any time during the pregnancy can lead to uterine contractions and, in late pregnancy, labor.

3. The expansion of the uterus during gestation is a result of the enlargement of the uterine cells, primarily smooth muscle cells of the myometrium.

Page 632

1. An increase in the levels of GnRH, FSH, LH, and sex hormones in the blood mark the onset of puberty.

Page 638

1. A person who is heterozygous for curly hair would have one dominant gene and one recessive gene. The person's phenotype would be "curly hair."

2. There are two types of sex chromosomes: an X chromosome and a Y chromosome. The normal male chromosome pair is *XY*, and the normal female pair is XX. The ova produced during meiosis by Joe's wife will always carry *X*, and the sperm produced during meiosis by Joe may carry *X* or *Y*. As a result, the sex of Joe's children depends on which type of sperm cell fertilizes the ovum.

3. A genome is the full complement of genetic material (DNA) of an organism. One Mb is equal to one million base pairs, so the 3,200 Mb of the human genome is equal to 3,200,000,000 base pairs.

20 DEVELOPMENT AND INHERITANCE

An Overview of Topics in Development • Fertilization • The First Trimester • The Second and Third Trimesters • Labor and Delivery

EXPLORE *MediaLab*

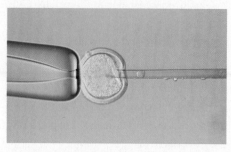

A donor nucleus being injected into a sheep egg for cloning.

EXPLORATION #1

Estimated time for completion: 10 minutes

PRENATAL DEVELOPMENT is an amazing series of events. A quick look at Table 20-2 reveals how much is occurring at the cellular, tissue, and organ level during this 36-week period. With so much biochemical and cellular activity, you would be correct in thinking that this is also a very sensitive period for the developing individual. Many seemingly innocuous compounds, referred to as teratogens, can have devastating effects on the regulated pattern of development. Alcohol is one of the most common and well-understood teratogens. Alcohol restricts blood flow to the placenta, depriving the developing cells of the fetus of much-needed oxygen. It also crosses the placenta, damaging fetal cells. With chronic exposure, the fetus may develop Fetal Alcohol Syndrome (FAS), Alcohol-Related Neurodevelopmental Disorder (ARND), and Alcohol-Related Birth Defects (ARBD). These three syndromes describe differing degrees of facial abnormalities, growth deficiencies, and central nervous system dysfunction. Prepare a chart of prenatal development similar to Table 20-2. On your chart, indicate when excessive alcohol intake might cause each of these effects. What other detrimental effects would you expect to see? You should be able to list at least one effect for each system. To complete your chart, go to Chapter 20 at the Companion Web site, visit the MediaLab section, and click on the key word "FAS." The information compiled by the ADA, Division of Alcohol and Drug Abuse, will help you add to your chart.

Drinking for two. This beer affects not only the mother's body systems but also the fetus that she is carrying.

EXPLORATION #2

Estimated time for completion: 10 minutes

WITH THE SUCCESSES of the Human Genome Project (see p. 635), the news is full of debates on the issue of human cloning. Two types of cloning are currently being pursued in research laboratories: reproductive cloning and therapeutic cloning. Reproductive cloning, the goal of which is to create a genetically identical copy of an organism, has been successfully performed on a variety of animals, including mice, sheep, and cows. Although reproductive cloning has not yet been attempted on humans, advancements in this technique are occurring rapidly. We may soon have the ability to create genetically identical copies of humans. Therapeutic cloning of human cells has already been successful. The goal of therapeutic cloning is not to create a new individual, but rather to create a viable embryo from which stem cells can be harvested. Figure 20-4● depicts the stage of development at which these cells are removed. Harvesting stem cells destroys the embryo, but provides a population of potentially therapeutically useful tissue cells. A good pictorial explanation of this procedure is provided by *Science* magazine at http://www.sciam.com/explorations/2001/112401ezzell/how.html. As these cloning regimes become mainstream, we are faced with the ethics of using them. What are your thoughts on these issues? Should scientists be able to continue addressing the reproductive cloning of humans? Why or why not? Is therapeutic cloning a medical breakthrough or a moral issue in that viable embryos are destroyed? After preparing a short position paper on this topic, go to Chapter 20 at the Companion Web site, visit the MediaLab section, and click on the key word "cloning." Read through a NOVA staff writer's interview of Dr. Don Wolf, Director of the Andrology/Embryology Laboratory at Oregon Health and Sciences University in Portland. Do you agree with Dr. Wolf's views? Strengthen your position paper with quotes and information from this article.

Appendix I

ANSWERS TO CHAPTER REVIEW QUESTIONS

CHAPTER 1

Level 1: Reviewing Facts and Terms

1. g 2. d 3. a 4. j 5. b 6. l 7. n 8. f 9. h 10. e 11. c
12. o 13. k 14. i 15. m 16. c 17. d 18. b 19. c 20. c
21. b

Level 2: Reviewing Concepts

22. responsiveness, adaptability, growth, reproduction, movement, metabolism, absorption, respiration, excretion
23. molecules-cell-tissues-organs-organ systems—organisms
24. Homeostatic regulation refers to adjustments in physiological systems that are responsible for the preservation of homeostasis.
25. In negative feedback, a variation outside normal ranges triggers an automatic response that corrects the situation. In positive feedback, the initial stimulus produces a response that exaggerates the stimulus.
26. The body is erect and the hands are at the sides with the palms facing forward.
27. Stomach. (You would cut the pericardium to access the heart.)
28. (a) dorsal cavity (b) ventral cavity (c) thoracic cavity (d) abdominopelvic cavity

Level 3: Critical Thinking and Clinical Applications

29. Since calcitonin is controlled by negative feedback, it should bring about a decrease in blood calcium level, thus decreasing the stimulus for its release.
30. To see a complete view of the medial surface of each half of the brain, midsagittal sections are needed.

CHAPTER 2

Level 1: Reviewing Facts and Terms

1. f 2. c 3. i 4. g 5. a 6. l 7. d 8. k 9. b 10. e 11. h
12. j 13. a 14. d 15. a
16. Enzymes are specialized protein catalysts that lower the activation energy of chemical reactions. Enzymes speed up chemical reactions but are not used up or changed in the process.
17. carbon, hydrogen, oxygen, nitrogen, calcium, phosphorus
18. carbohydrates, lipids, proteins, nucleic acids
19. support: structural proteins; movement: contractile proteins; transport: transport proteins; buffering; metabolic regulation; coordination and control; defense

Level 2: Reviewing Concepts

20. b 21. a 22. a
23. (1) covalent bond: equal sharing of electrons (2) polar covalent bond: unequal sharing of electrons (3) ionic bond: loss and/or gain of electrons
24. A solution such as pure water with a pH of 7 is neutral because it contains equal numbers of hydrogen and hydroxyl ions.
25. A nucleic acid. Carbohydrates and lipids do not contain the element nitrogen. Although both proteins and nucleic acids contain nitrogen, only nucleic acids contain phosphorus.

Level 3: Critical Thinking and Clinical Applications

26. The number of neutrons in an atom is equal to the atomic mass minus the atomic number. In the case of sulfur, this would be $32 - 16 = 16$ neutrons. Since the atomic number of sulfur is 16, the neutral sulfur atom contains 16 protons and 16 electrons. The electrons would be distributed as follows: 2 in the first level, 8 in the second level, and 6 in the third level. To achieve a full 8 electrons in the third level, the sulfur atom could accept 2 electrons in an ionic bond or share 2 electrons in a covalent bond. Since hydrogen atoms can share 1 electron in a covalent bond, the sulfur atom would form 2 covalent bonds, 1 with each of 2 hydrogen atoms.

27. If a person exhales large amounts of carbon dioxide, the equilibrium will shift to the left and the level of hydrogen ion in the blood will decrease. A decrease in the amount of hydrogen ion will cause the pH to rise.

CHAPTER 3

Level 1: Reviewing Facts and Terms

1. e 2. d 3. h 4. a 5. f 6. b 7. g 8. c 9. m 10. k 11. q
12. i 13. o 14. j 15. l 16. n 17. p 18. b 19. d 20. c 21. d
22. c 23. a 24. c
25. physical isolation; regulation of exchange with the environment; sensitivity; structural support
26. diffusion; filtration; carrier-mediated transport; vesicular transport
27. synthesis of proteins, carbohydrates, and lipids; storage of absorbed molecules; transport of materials
28. prophase; metaphase; anaphase; telophase

Level 2: Reviewing Concepts

29. b 30. b 31. c 32. c
33. similarities: both processes utilize carrier proteins; differences:

Facilitated diffusion	Active transport
passive	active
no ATP expended	ATP expended
concentration gradient	no concentration gradient

34. Cytosol has a high concentration of K^+; interstitial fluid has a high concentration of Na^+. Cytosol also contains a high concentration of suspended proteins, small quantities of carbohydrates, and large reserves of amino acids and lipids. Cytosol may also contain insoluble materials known as inclusions.
35. In transcription, RNA polymerase uses genetic information to assemble a strand of mRNA. In translation, ribosomes use information carried by the mRNA strand to assemble functional proteins.
36. Prophase: Chromatin condenses and chromosomes become visible; centrioles migrate to opposite poles of the cell and spindle fibers develop; nuclear membrane disintegrates. Metaphase: Chromatids attach to spindle fibers and line up along the metaphase plate. Anaphase: Chromatids separate and migrate toward opposite poles of the cell. Telophase: The nuclear membrane re-forms; chromosomes disappear as chromatin relaxes; nucleoli reappear.
37. Cytokinesis is the cytoplasmic movement that separates two daughter cells, thereby completing mitosis.

Level 3: Critical Thinking and Clinical Applications

38. Facilitated transport, which requires a carrier molecule but not cellular energy. The energy for the process is provided by the diffusion gradient for the substance being transported. When all of the carriers are actively involved in transport, the rate of transport plateaus and cannot be increased further.
39. Solution A must have initially had more solutes than solution B. As a result, water moved by osmosis across the semipermeable membrane from side B to side A, increasing the fluid level on side A.

CHAPTER 4

Level 1: Reviewing Facts and Terms

1. g 2. d 3. j 4. i 5. a 6. f 7. c 8. e 9. b 10. h 11. l
12. k 13. a 14. c 15. c 16. d 17. d 18. c 19. a 20. b 21. b
22. a
23. provide physical protection; control permeability; provide sensations; produce specialized secretions
24. simple, stratified, transitional
25. specialized cells, extracellular protein fibers, fluid ground substance
26. fluid connective tissues: blood and lymph; supporting connective tissues: bone and cartilage
27. mucous, serous, cutaneous, synovial
28. Neurons and neuroglia. The neurons transmit electrical impulses. The neuroglia comprise several kinds of supporting cells and play a role in providing nutrients to neurons.

Level 2: Reviewing Concepts

29. a
30. Holocrine secretion destroys the gland cell. During holocrine secretion, the entire cell becomes packed with secretory products and then bursts, releasing the secretion but killing the cell. The gland cells must be replaced by the division of stem cells.
31. Exocrine secretions are secreted onto a surface or outward through a duct. Endocrine secretions are secreted by ductless glands into surrounding tissues. The secretions are called hormones, which usually diffuse into the blood for distribution to other parts of the body.
32. Tight junctions block the passage of water or solutes between cells. In the digestive tract, these junctions keep enzymes, acids and wastes from damaging delicate underlying tissues.
33. The extensive connections between cells formed by tight junctions, intercellular cement, and physical interlocking hold skin cells together and can deny access to chemicals or pathogens that may cover their free surfaces. If the skin is damaged and the connections are broken, infection can easily occur.
34. Cutaneous membranes are thick, relatively waterproof, and usually dry.

Level 3: Critical Thinking and Clinical Applications

35. Since animal intestines are specialized for absorption, you would look for a slide that shows a single layer of epithelium lining the cavity. The cells would be cuboidal or columnar and would probably have microvilli on the surface to increase surface area. With the right type of microscope, you could also see tight junctions between the cells. Since the esophagus receives undigested food, it would have a stratified epithelium consisting of squamous cells to protect it against damage.
36. Step 1: Check for striations. (If striations are present, the choices are skeletal muscle or cardiac muscle. If striations are absent, the tissue is smooth muscle.) Step 2: Check for the presence of intercalated discs. (If the discs are present, the tissue is cardiac muscle. If they are absent, the tissue is skeletal muscle.)

CHAPTER 5

Level 1: Reviewing Facts and Terms

1. f 2. g 3. i 4. j 5. a 6. e 7. b 8. d 9. c 10. h 11. a
12. d 13. b 14. c 15. a 16. a
17. carotene and melanin
18. Papillary layer: consists of loose connective tissue and contains capillaries and sensory neurons; reticular layer: consists of dense irregular connective tissue and bundles of collagen fibers. The reticular and papillary layers of the dermis contain blood vessels, lymphatic vessels, and nerve fibers.
19. apocrine sweat glands and merocrine sweat glands

Level 2: Reviewing Concepts

20. Substances that are lipid soluble pass through the permeability barrier easily, because the barrier is composed primarily of lipids surrounding the epidermal cells. Water-soluble drugs are hydrophobic to the permeability barrier.
21. A tan is a result of the synthesis of melanin in the skin. Melanin helps prevent skin damage by absorbing ultraviolet radiation before it reaches the deep layers of the epidermis and dermis. Within the epidermal cells, melanin concentrates around the outer wall of the nucleus, so it absorbs the UV light before it can damage the nuclear DNA.
22. The subcutaneous layer is not highly vascular and does not contain major organs; thus this method reduces the potential for tissue damage.
23. The dermis becomes thinner and the elastic fiber network decreases in size, weakening the integument and causing loss of resilience.

Level 3: Critical Thinking and Clinical Applications

24. The child probably has a fondness for vegetables high in carotene, such as sweet potatoes, squash, and carrots. It is not uncommon for parents to feed a baby foods that he or she prefers to eat. If the child consumes large amounts of carotene, the yellow-orange pigment will be stored in the skin, producing a yellow-orange skin color.
25. Like most elderly people, Vanessa's grandmother has poor circulation to the skin. As a result, the temperature receptors in the skin do not sense as much warmth as when there is a rich blood supply. The sensory information is relayed to the brain. The brain interprets this as being cool or cold, thus causing Vanessa's grandmother to feel cold.

CHAPTER 6

Level 1: Reviewing Facts and Terms

1. i 2. h 3. m 4. l 5. j 6. k 7. f 8. c 9. o 10. b 11. g
12. n 13. d 14. a 15. e 16. b 17. a 18. c 19. b 20. d
21. a 22. a 23. a 24. a 25. b 26. c 27. d 28. b 29. c
30. support, storage of minerals and lipids, blood cell production, protection, leverage
31. In intramembranous ossification, bone develops from fibrous connective tissue. In endochondral ossification, bone develops from a cartilage model.
32. The hyoid is the only bone in the body that does not articulate with another bone.
33. (1) It protects the heart, lungs, thymus, and other structures in the thoracic cavity. (2) It serves as an attachment point for muscles involved with respiration, the position of the vertebral column, and movements of the pectoral girdle and upper extremities.
34. the acromion and coracoid processes

Level 2: Reviewing Concepts

35. The osteons are parallel to the long axis of the shaft, which does not bend when forces are applied to either end. An impact to the side of the shaft can lead to a fracture.
36. The chondrocytes of the epiphyseal plate enlarge and divide, increasing the thickness of the plate. On the shaft side, the chondrocytes become ossified, "chasing" the expanding epiphyseal plate away from the shaft.
37. The lumbar vertebrae have massive bodies and carry a large amount of weight—both causative factors related to rupturing a disc. The cervical vertebrae are more delicate and have small bodies, increasing the possibility of dislocations and fractures in this region, compared with other regions of the vertebral column.
38. The clavicles are small and fragile, so fractures are quite common. Their position also makes them vulnerable to injury and damage.
39. The pelvic girdle consists of the coxae. The pelvis is a composite structure that includes the coxae of the appendicular skeleton and the sacrum and coccyx of the axial skeleton.
40. Articular cartilages do not have perichondrium, and their matrix contains more water than do other cartilages.
41. The slight movability of the pubic symphysis joint facilitates childbirth, by spreading the pelvis to ease movement of the baby through the birth canal.

Level 3: Critical Thinking and Clinical Applications

42. The fracture might have damaged the epiphyseal plate. Even though the bone healed properly, the damaged plate did not produce as much cartilage as the undamaged plate in the other leg. The result would be a shorter bone on the side of the injury.

43. The virus could have been inhaled through the nose and passed by way of the cribiform plate of the ethmoid bone into the cranium.

44. The large bones of a child's cranium are not yet fused; they are connected by areas of connective tissue called fontanels. By examining the bones, the archaeologist could readily see if sutures had formed yet. By knowing approximately how long it takes for the various fontanels to close and by determining the sizes of the fontanels, she could make a good estimation of the child's age.

45. Frank may have suffered a shoulder dislocation, which is quite a common injury due to the weak nature of the scapulohumeral joint.

46. Ed has a sprained ankle. This condition occurs when ligaments are stretched to the point at which some of the collagen fibers are torn. Stretched ligaments in joints can cause the release of synovial fluid, which results in swelling and pain in the affected area.

CHAPTER 7

Level 1: Reviewing Facts and Terms

1. f 2. i 3. d 4. p 5. l 6. c 7. h 8. n 9. b 10. k 11. o
12. a 13. m 14. j 15. g 16. e 17. d 18. d
19. produce skeletal movement, maintain body posture and body position, support soft tissue, guard entrances and exits, maintain body temperature
20. Step 1: active site exposure. Step 2: cross-bridge attachment. Step 3: pivoting of myosin head (power stroke). Step 4: cross-bridge attachment. Step 5: myosin head activation (cocking)
21. ATP, creatine phosphate, glycogen
22. aerobic metabolism and glycolysis
23. The axial musculature positions the head and spinal column and moves the rib cage, assisting in the movements that make breathing possible. The appendicular musculature stabilizes or moves components of the appendicular skeleton.

Level 2: Reviewing Concepts

24. b
25. Acetylcholine released by the motor neuron at the neuromuscular junction changes the permeability of the cell membrane at the motor end plate. The permeability change allows the influx of positive charge, which in turn triggers an electrical event called an action potential. The action potential spreads across the entire surface of the muscle fiber and into the interior via the T tubules. The cytoplasmic concentration of calcium ions (released from sarcoplasmic reticulum) increases, triggering the start of a contraction. The contraction ends when the ACh has been removed from the synaptic cleft and motor end plate by AChE.
26. Metabolic turnover is rapid in skeletal muscle cells. The genes contained in multiple nuclei direct the production of enzymes and structural proteins required for normal contraction, and the presence of multiple gene copies speeds up the process.
27. The spinal column does not need a massive series of flexors because many of the large trunk muscles flex the spine when they contract. In addition, most of the body weight lies anterior to the spinal column and gravity tends to flex the spine.
28. The urethral and anal sphincter muscles are usually in a constricted state, preventing the passage of urine and feces. The muscles are innervated by nerves that are under conscious control, so the sphincters normally relax to allow the passage of wastes only when the individual decides so.
29. flexion of the leg and extension of the hip

Level 3: Critical Thinking and Clinical Applications

30. Since organophosphates block the action of the enzyme acetylcholinesterase, acetylcholine released into the neuromuscular cleft would not be inactivated. This would allow the acetylcholine to continue to stimulate the muscles, causing a state of persistent contraction (spastic paralysis). If this were to affect the muscles of respiration (which is likely), Terry

would die of suffocation. Prior to death, the most obvious symptoms would be uncontrolled tetanic contractions of the skeletal muscles.

31. In rigor mortis, the muscles lock in the contracted position, making the body extremely stiff. The membranes of the dead muscle cells are no longer selectively permeable and calcium leaks in, triggering contraction. Contraction persists because the dead cells can no longer make ATP, which is necessary for cross-bridge detachment from the active sites. Rigor mortis begins a few hours after death and lasts until 15 to 25 hours later, when the lysosomal enzymes released by autolysis break down the myofilaments.

32. Jeff should do squatting exercises. If he places a weight on his shoulders as he does these, he will notice better results, since the quadriceps muscles would be working against a greater resistance.

CHAPTER 8

Level 1: Reviewing Facts and Terms

1. f 2. g 3. p 4. s 5. m 6. a 7. b 8. j 9. h 10. c 11. d
12. o 13. t 14. q 15. l 16. i 17. n 18. e 19. r 20. k 21. c
22. a 23. a 24. d 25. b 26. b 27. a 28. b 29. c 30. a
31. d
32. The properties of the action potential are independent of the relative strength of the depolarization stimulus.
33. N I: olfactory; N II: optic; N III: oculomotor; N IV: trochlear; N V: trigeminal; N VI: abducens; N VII: facial; N VIII: vestibulocochlear; N IX: glossopharyngeal; N X: vagus; N XI: accessory; N XII: hypoglossal
34. The sympathetic preganglionic fibers emerge from the thoracolumbar area (T_1 through L_2) of the spinal cord. The parasympathetic fibers emerge from the brain stem and the sacral region of the spinal cord (craniosacral).

Level 2: Reviewing Concepts

35. d 36. c
37. Since the ventral roots contain axons of motor neurons, those muscles controlled by the neurons of the damaged root would be paralyzed.
38. Centers in the medulla oblongata are involved in respiratory and cardiac activity.
39. Transmission across a chemical synapse always involves a synaptic delay, but with only one synapse (monosynaptic), the delay between stimulus and response is minimized. In a polysynaptic reflex, the length of delay is proportional to the number of synapses involved.
40.

	Sympathetic	Parasympathetic
mental alertness	increased	decreased
metabolic rate	increased	decreased
digestive/urinary function	inhibited	stimulated
use of energy reserves	stimulated	inhibited
respiratory rate	increased	decreased
heart rate/blood pressure	increased	decreased
sweat glands	stimulated	inhibited

Level 3: Critical Thinking and Clinical Applications

41. Brain tumors result from uncontrolled division of neuroglial cells. Unlike neurons, neuroglial cells are capable of cell division. In addition, cells of the meningeal membranes can give rise to tumors.
42. The officer is testing the function of Bill's cerebellum. Many drugs, including alcohol, have pronounced effects on the function of the cerebellum. A person who is under the influence of alcohol is not able to properly anticipate the range and speed of limb movement because of slow processing and correction by the cerebellum. As a result, Bill would have a difficult time performing simple tasks such as walking a straight line or touching his finger to his nose.
43. Stress-induced stomach ulcers are due to excessive sympathetic stimulation. The sympathetic division causes the vasoconstriction of vessels supplying the digestive organs, leading to an almost total shutdown of blood supply to the stomach. Lack of blood leads to the death of cells and tissues (necrosis), which causes the ulcers.
44. The radial nerve is involved.

45. Ramon damaged the femoral nerve. Since this nerve also supplies the sensory innervation of the skin on the anteromedial surface of the thigh and medial surfaces of the leg and foot, Ramon may also experience numbness in these regions.

CHAPTER 9

Level 1: Reviewing Facts and Terms

1. k 2. c 3. a 4. e 5. f 6. j 7. i 8. b 9. h 10. m 11. g 12. l 13. n 14. d 15. b 16. d 17. b 18. d 19. b 20. b 21. d 22. a 23. c 24. b 25. b 26. a 27. d

28. tactile receptors, baroreceptors, and proprioceptors

29. free nerve endings: sensitive to touch and pressure; root hair plexuses: monitor distortions and movements across the body surface; tactile discs (Merkel's discs): fine touch and pressure receptors; tactile corpuscles (Meissner's corpuscles): fine touch and pressure sensation; lamellated corpuscles (pacinian corpuscles): most sensitive to pulsing or vibrating stimuli (deep pressure); Ruffini corpuscles: -sensitive to pressure and distortion of the skin

30. (a) the sclera and the cornea (b) provides mechanical support and some degree of physical protection; serves as an attachment site for the extrinsic muscles; contains structures that assist in focusing

31. iris, ciliary body, and choroid

32. Step 1: Sound waves arrive at the tympanum. Step 2: Movement of the tympanum causes displacement of the auditory ossicles. Step 3: Movement of the stapes at the oval window establishes pressure waves in the perilymph of the vestibular duct. Step 4: The pressure waves distort the basilar membrane on their way to the round window of the tympanic duct. Step 5: Vibration of the basilar membrane causes hair cells to vibrate against the tectorial membrane. Step 6: Information concerning the region and intensity of stimulation is relayed to the CNS over the cochlear branch of N VIII.

Level 2: Reviewing Concepts

33. c 34. c

35. The general senses include somatic and visceral sensation. The special senses are those whose receptors are confined to the head.

36. Regardless of the type of stimulus, the CNS receives the sensory information in the form of action potentials.

37. The olfactory system has extensive limbic-system connections, accounting for its effect on memories and emotions.

38. A visual acuity of 20/15 means that Jane can discriminate images at a distance of 20 feet, whereas someone with "normal" vision must be 5 feet closer (15 ft.) to see the same details. Jane's visual acuity is better than the acuity of someone with 20/20 vision.

Level 3: Critical Thinking and Clinical Applications

39. As you turn to look at your friend, your medial rectus muscles will contract, directing your gaze more medially. In addition, your pupils will constrict and the lenses will become more spherical.

40. The loud noises from the fireworks have transferred so much energy to the endolymph in the cochlea that the fluid continues to move for a long period of time. As long as the endolymph is moving, it will vibrate the tectorial membrane and stimulate the hair cells. This stimulation produces the "ringing" sensation that Millie perceives. She finds it difficult to hear normal conversation because the vibrations associated with it are not strong enough to overcome the currents already moving through the endolymph, so the pattern of vibrations is difficult to discern against the background "noise."

41. The rapid descent in the elevator causes the maculae in the saccule of the vestibule to slide upward, producing the sensation of downward vertical motion. When the elevator abruptly stops, the maculae do not. Because of their relatively large inertia, it takes a few seconds for them to come to rest in the normal position. As long as the maculae are displaced, the perception of movement will remain.

CHAPTER 10

Level 1: Reviewing Facts and Terms

1. h 2. e 3. g 4. l 5. b 6. k 7. c 8. a 9. j 10. i 11. f 12. d 13. d 14. a 15. c 16. d 17. b 18. d 19. c 20. b

21. thyroid-stimulating hormones (TSH), adrenocorticotropic hormone (ACTH), follicle-stimulating hormone (FSH), luteinizing hormone (LH), prolactin (PRL), growth hormone (GH), melanocyte-stimulating hormone (MSH)

22. The overall effect of calcitonin is to decrease the concentration of calcium ions in body fluids. The overall effect of parathyroid hormone is to cause an increase in the concentration of calcium ions in body fluids.

23. (a) The GAS phase includes alarm, resistance, and exhaustion phases. (b) Epinephrine is the dominant hormone of the alarm phase. Glucocorticoids are the dominant hormones of the resistance phase.

Level 2: Reviewing Concepts

24. In communication by the nervous system, the source and destination are quite specific and the effects are short-lived. In endocrine communication, the effects are slow to appear and often persist for days. A single hormone can alter the metabolic activities of multiple tissues and organs simultaneously.

25. Hormones direct the synthesis of an enzyme (or other protein) that is not already present in the cytoplasm. They also turn an existing enzyme "on" or "off" and increase the rate of synthesis of a particular enzyme or other protein.

26. The two hormones may have opposing, or antagonistic, effects; the hormones may have additive or synergistic effects; one hormone may have a permissive effect on another (the first hormone is needed for the second to produce its effect); the hormones may produce different but complementary effects in specific tissues and organs.

27. Phosphodiesterase is the enzyme that converts cAMP to AMP, thus inactivating it. If this enzyme were blocked, the effect of the hormone would be prolonged.

Level 3: Critical Thinking and Clinical Applications

28. Extreme thirst and frequent urination are characteristics of both diabetes insipidus and diabetes mellitus. To distinguish between the two, glucose levels in the blood and urine could be measured. A high glucose concentration would indicate diabetes mellitus.

29. Julie should exhibit elevated levels of parathyroid hormone in her blood. Her poor diet does not supply enough calcium for her developing fetus. The fetus removes large amounts of calcium from the maternal blood, lowering the mother's blood calcium levels. This would lead to an increase in the blood level of parathyroid hormone and increased mobilization of stored calcium from the maternal skeletal reserves.

CHAPTER 11

Level 1: Reviewing Facts and Terms

1. f 2. k 3. a 4. l 5. c 6. j 7. e 8. h 9. g 10. b 11. d 12. i 13. c 14. c 15. a 16. c 17. d 18. a 19. d 20. d 21. b

22. transportation of dissolved gases, nutrients, hormones, and metabolic wastes; regulation of pH and electrolyte composition of interstitial fluids throughout the body; restriction of fluid losses through damaged vessels or at other injury sites; defense against toxins and pathogens; stabilization of body temperature

23. Albumins maintain osmotic pressure of plasma and important in transport of fatty acids. Transport globulins bind small ions, hormones or compounds that might otherwise be filtered out of the blood at the kidneys or have very low solubility in water, and immunoglobulins attack foreign proteins and pathogens. Fibrinogen functions in blood clotting.

24. (a) anti-B agglutinins (b) anti-A agglutinins (c) neither anti-A nor anti-B agglutinins (d) anti-A and anti-B agglutinins

25. amoeboid movement: extension of a cellular process; diapedesis: squeezing between adjacent endothelial cells in the capillary wall; positive chemotaxis: attraction to specific chemical stimuli

26. The common pathway begins when Factor X activator from either the extrinsic or intrinsic pathway appears in the plasma.

27. An embolus is a drifting blood clot. A thrombus is a blood clot that sticks to the wall of an intact blood vessel.

Level 2: Reviewing Concepts

28. a   29. d   30. c   31. d

32. Red blood cells are biconcave discs that lack mitochondria, ribosomes, and nuclei and contain a large amount of the protein hemoglobin.

Level 3: Critical Thinking and Clinical Applications

33. After donating a pint of blood (or after any other loss of blood), you would expect to see a substantial increase in the number of reticulocytes. Since there is not enough time for large numbers of erythrocytes to mature, the bone marrow releases large numbers of reticulocytes (immature cells) in an effort to maintain a constant number of formed elements.

34. In many cases of kidney disease, the cells responsible for producing erythropoietin are either damaged or destroyed. The reduction in erythropoietin levels leads to reduced erythropoiesis and fewer red blood cells, resulting in anemia.

CHAPTER 12

Level 1: Reviewing Facts and Terms

1. i   2. e   3. g   4. h   5. a   6. b   7. k   8. l   9. d   10. f   11. c
12. j   13. c   14. b   15. b   16. a   17. a   18. b   19. a

20. During ventricular contraction, tension in the papillary muscles and chordae tendineae braces the atrioventricular valve cusps and keeps them from swinging into the atrium. This action prevents the backflow, or regurgitation, of blood into the atrium as the ventricle contracts.

21. The atrioventricular (AV) valves prevent backflow of blood from the ventricles into the atria. The right AV valve is the tricuspid valve, and the left AV valve is the bicuspid (mitral) valve. The pulmonary and aortic semilunar valves prevent the backflow of blood from the pulmonary trunk and aorta into the right and left ventricles.

22. SA node → AV node → AV bundle (bundle of His) → R and L bundle branches → Purkinje fibers (into the mass of ventricular muscle tissue)

23. (a) A single cardiac cycle is the period between the beginning of one heartbeat and the beginning of the next heartbeat. (b) The cycle begins with atrial systole as the atria contract and push blood into the relaxed ventricles. As the atria relax, and atrial diastole begins, the ventricles contract (ventricular systole), forcing blood through the semilunar valves into the pulmonary trunk and aorta. The ventricles then relax (ventricular diastole). For the rest of the cardiac cycle, both the atria and ventricles are in diastole; passive filling occurs.

Level 2: Reviewing Concepts

24. c   25. d   26. a

27. The right atrium receives blood from the systemic circuit and passes it to the right ventricle. The right ventricle discharges blood into the pulmonary circuit. The left atrium collects blood. Contraction of the left ventricle ejects blood into the systemic circuit.

28. Listening to the heart sounds (auscultation) is a simple, effective method of cardiac diagnosis. The first and second heart sounds accompany the action of the heart valves. The first sound ("lubb") marks the start of ventricular contraction and is produced as the AV valves close and the semilunar valves open. The second sound ("dupp") occurs at the beginning of ventricular filling when the semilunar valves close.

29. (a) Sympathetic activation causes the release of norepinephrine by postganglionic fibers and the secretion of norepinephrine and epinephrine by the adrenal medullae. These compounds stimulate cardiac muscle fiber metabolism and increase the force and degree of contraction. (b) Parasympathetic stimulation causes the release of ACh at membrane surfaces, where it produces hyperpolarization and inhibition. The result is a decrease in the heart rate and in the force of cardiac contractions.

Level 3: Critical Thinking and Clinical Applications

30. This patient has second-degree heart block, a condition in which not every signal from the SA node reaches the ventricular muscle. In this case, for every two action potentials generated by the SA node, only one is reaching the ventricles. This accounts for the two P waves but only one QRS complex. This condition frequently results from damage to the internodal pathways of the atria or problems with AV node.

31. Blocking the calcium channels in myocardial cells would lead to a decrease in the force of cardiac contraction. Since the force of cardiac contraction is directly proportional to the stroke volume, you would expect a reduced stroke volume.

CHAPTER 13

Level 1: Reviewing Facts and Terms

1. d   2. l   3. a   4. j   5. i   6. h   7. k   8. f   9. e   10. c   11. g
12. b   13. d   14. b   15. b   16. a   17. d   18. b   19. c   20. b
21. c   22. b   23. d   24. c   25. b   26. c   27. c

28. (a) Fluid leaves the capillary at the arterial end primarily in response to hydrostatic pressure. (b) Fluid returns to the capillary at the venous end primarily in response to osmotic pressure.

29. The cardiac output decreases due to parasympathetic stimulation and inhibition of sympathetic activity, and peripheral vasodilation is widespread due to the inhibition of excitatory neurons in the vasomotor center.

30. changes in the carbon dioxide, oxygen, or pH levels in the blood and cerebrospinal fluid

31. When an infant takes its first breath, the lungs expand, and so do the pulmonary vessels. The smooth muscles in the ductus arteriosus contract, isolating the pulmonary and aortic trunks, and blood begins flowing through the pulmonary circuit. As pressure rises in the left atrium, the valvular flap closes the foramen ovale, completing the circulatory remodeling.

32. blood: decreased hematocrit, formation of thrombin, and valvular malfunction; heart: reduction in maximum cardiac output and changes in the activities of the nodal and conducting fibers, reduction in the elasticity of the fibrous skeleton, progressive atherosclerosis, and replacement of damaged cardiac muscle fibers by scar tissue; blood vessels: progressive inelasticity in arterial walls, deposition of calcium salts on weakened vascular walls, and formation of thrombi at atherosclerotic plaques

Level 2: Reviewing Concepts

33. d   34. c   35. b

36. Artery walls are generally thicker. They contain more smooth muscle and elastic fibers, enabling them to resist and adjust to the pressure generated by the heart. Venous walls are thinner and the pressure in the veins is less than that in the arteries. Arteries constrict more than veins do when not expanded by blood pressure, due to a greater degree of elastic tissue. In addition, the endothelial lining of an artery has a pleated appearance because it forms folds, being unable to contract. The lining of a vein looks like a typical endothelial layer.

37. Capillaries are only one cell thick. Small gaps between adjacent endothelial cells permit the diffusion of water and small solutes into the surrounding interstitial fluid but prevent the loss of blood cells and plasma proteins. Some capillaries also contain pores that permit very rapid exchange of fluids and solutes between the interstitial fluid and the plasma. Conversely, the walls of arteries and veins are several cell layers thick and are not specialized for diffusion.

38. The brain receives arterial blood from four arteries. Because these arteries form anastomoses inside the cranium, interruption of any one vessel will not compromise the circulatory supply to the brain.

39. The accident victim is suffering from shock and acute circulatory crisis characterized by hypotension and inadequate peripheral blood flow. The hypotension results from the loss of blood volume and decreased cardiac output. Her skin is pale and cool due to peripheral vasoconstriction; the moisture results from the sympathetic activation of sweat glands. Falling blood pressure to the brain causes confusion and disorientation. If you took

her pulse, you would find it to be rapid and weak, reflecting the heart's response to reduced blood flow and volume.

Level 3: Critical Thinking and Clinical Applications

40. Three factors contribute to Bob's elevated blood pressure. (1) The loss of water through sweating increases blood viscosity. The number of red blood cells remains about the same, but because there is less plasma volume, the concentration of red cells is increased, thus increasing the blood viscosity. Increased viscosity increases peripheral resistance and contributes to increased blood pressure. (2) To cool Bob's body, blood flow to the skin is increased. This in turn increases venous return, which increases stroke volume and cardiac output (Frank-Starling's law of the heart). The increased cardiac output can also contribute to increased blood pressure. (3) The heat stress that Bob is experiencing leads to increased sympathetic stimulation (the reason for the sweating). Increased sympathetic stimulation of the heart will increased heart rate and stroke volume, thus increasing his cardiac output and blood pressure.

41. Antihistamines and decongestants are drugs that have the same effects on the body as stimulating the sympathetic nervous system. In addition to the desired effects of counteracting the symptoms of the allergy, these medications can produce an increased heart rate, stroke volume, and peripheral resistance, all of which will contribute to elevating blood pressure. If a person has hypertension (high blood pressure), these drugs will aggravate this condition, with potentially hazardous consequences.

42. When Gina rapidly moved from a lying position to a standing position, gravity caused her blood volume to move to the lower parts of her body away from the heart, decreasing venous return. The decreased venous return resulted in a decreased volume of blood at the end of diastole, which lead to a decreased stroke volume and cardiac output. These resulted in decreased blood flow to the brain, where the diminished oxygen supply caused her to be light-headed and feel faint. This usually does not occur, because as soon as the pressure drops due to blood moving inferiorly, the baroreceptor reflex should be triggered. Normally, a rapid change in blood pressure is sensed by baroreceptors in the aortic arch and carotid sinus. Action potentials from these areas are carried to the medulla oblongata, where appropriate responses are integrated. In this case, we would expect an increase in peripheral resistance to compensate for the decreased blood pressure. If this doesn't compensate enough for the drop, then an increase in heart rate and force of contraction would occur. In healthy individuals, these responses occur so quickly that no changes in pressure are noticed after body position is changed.

CHAPTER 14

Level 1: Reviewing Facts and Terms

1. h 2. k 3. b 4. g 5. d 6. c 7. j 8. a 9. i 10. l 11. f 12. e 13. c 14. d 15. c 16. c 17. b 18. a 19. c 20. a 21. a 22. d 23. d 24. c

25. Left thoracic duct: collects lymph from the body below the diaphragm and from the left side of the body above the diaphragm; right thoracic duct: collects lymph from the right side of the body above the diaphragm.

26. (a) responsible for cell-mediated immunity, which defends against abnormal cells and pathogens inside living cells (b) stimulate the activation and function of T cells and B cells (c) inhibit the activation and function of both T cells and B cells (d) produce and secrete antibodies (e) recognize and destroy abnormal cells (f) interfere with viral replication inside the cell and stimulate the activities of macrophages and NK cells (g) provide cell-mediated immunity (h) provide antibody-mediated immunity, which defends against antigens and pathogenic organisms in the body (i) enhance nonspecific defenses and increase T cell sensitivity and stimulate B cell activity

27. physical barriers, phagocytic cells, immunological surveillance, interferon, complement, inflammation, and fever

Level 2: Reviewing Concepts

28. c 29. d

30. Specificity: The immune response is triggered by a specific antigen, and defends against only that antigen. Versatility: The immune system can differentiate from among tens of thousands of antigens it may encounter during a normal lifetime. Memory: The immune response following second exposure to a particular antigen is stronger and lasts longer. Tolerance: Some antigens, such as those on an individual's own cells, do not elicit an immune response.

31. The antigen may be destroyed by neutralization, agglutination and precipitation, activation of complement, attraction of phagocytes, stimulation of inflammation, or prevention of bacterial and viral adhesion.

32. destruction of target cell membranes, stimulation of inflammation, attraction of phagocytes, and enhancement of phagocytosis

Level 3: Critical Thinking and Clinical Applications

33. IgA immunoglobulins are found in body secretions such as tears, saliva, semen, and vaginal secretions but not in blood plasma. Blood plasma contains IgM, IgG, IgD, and IgE of immunoglobulins. By testing for the presence or absence of IgA and IgG antibodies, the lab could determine whether the sample was blood plasma or semen.

34. It appears that Ted has contracted the disease. On initial contact with a virus, the first type of antibody to be produced is IgM. The response is fairly rapid but short-lived. About the time IgM peaks, IgG levels are beginning to rise. IgG plays the more important role in eventually controlling the disease. The fact that Ted's blood sample has an elevated level of IgM antibodies would indicate that he is in the early stages of a primary response to the measles virus.

CHAPTER 15

Level 1: Reviewing Facts and Terms

1. h 2. f 3. i 4. c 5. b 6. d 7. j 8. e 9. a 10. l 11. g 12. k 13. c 14. b

Level 2: Reviewing Concepts

15. a 16. d

17. With less cartilaginous support in the lower respiratory passageways, the amount of tension in the smooth muscles has a greater effect on bronchial diameter and on the resistance to air flow.

18. The nasal cavity is designed to cleanse, moisten, and warm inspired air, whereas the mouth is not. Air entering through the mouth is drier and as a result can irritate the trachea, causing soreness of the throat.

19. The walls of bronchioles, like the walls of arterioles, are dominated by smooth muscle tissue. Varying the diameter (bronchodilation or bronchoconstriction) of the bronchioles provides control over the amount of resistance to air flow and over the distribution of air in the lungs, just as vasodilation and vasoconstriction of the arterioles regulate blood flow/distribution.

20. The surfactant cells produce surfactant, which reduces surface tension in the fluid coating the alveolar surface. The alveolar walls are very delicate; without surfactant, the surface tension would be so high that the alveoli would collapse.

Level 3: Critical Thinking and Clinical Applications

21. An increase in ventilation will increase the movement of venous blood back to the heart. (Recall the respiratory pump.) Increasing venous return would in turn help increase the blood pressure (according to the Frank-Starling law of the heart).

22. While you were sleeping, the air that you were breathing was so dry it absorbed more than the normal amount of moisture as it passed through the nasal cavity. The loss of moisture made the mucous secretions quite viscous and harder for the cilia to move. Your nasal epithelia continued to secrete mucus, but very little of it moved. This ultimately produced the nasal congestion. After the shower and juice, more moisture was transferred to the mucus, loosening it and making it easier to move, thus clearing up the problem.

CHAPTER 16

Level 1: Reviewing Facts and Terms

1. c 2. k 3. g 4. a 5. i 6. h 7. j 8. d 9. f 10. b 11. l 12. e 13. d 14. d 15. b 16. c 17. b 18. d 19. d 20. a 21. d 22. d

23. ingestion, mechanical processing, secretion, digestion, absorption, and excretion

24. The folds increase the surface area available for absorption and may permit expansion of the lumen after a large meal.

25. muscosa: This innermost layer is a mucous membrane consisting of epithelia and loose connective tissue (the lamina propria); submucosa: This layer, which surrounds the mucosa, contains blood vessels, lymphatic vessels, and neural tissue (which helps control and regulate smooth muscle tissue in the lamina propria and glandular secretions into the mucosa); muscularis externa: This layer is made up of two layers of smooth muscle tissue—longitudinal and circular—whose contractions agitate and propel materials along the digestive tract; serosa: This outermost layer is a serous membrane that protects and supports the digestive tract inside the peritoneal cavity.

26. analysis of material before swallowing; mechanical processing through the actions of the teeth, tongue, and palatal surfaces; lubrication by mixing with mucus and salivary secretions; limited digestion of carbohydrates and lipids

27. incisors: clipping or cutting; cuspids: tearing or slashing; bicuspids: crushing, mashing, grinding; molars: crushing and grinding

28. duodenum, jejunum, and ileum

29. The pancreas provides digestive enzymes as well as buffers that assist in the neutralization of acid chyme. The liver and gallbladder provide bile, a solution that contains additional buffers and bile salts that facilitate the digestion and absorption of lipids. The liver is responsible for metabolic regulation, hematological regulation, and bile production. It is the primary organ involved in regulating the composition of the circulating blood.

30. (1) reabsorption of water and compaction of chyme into feces; (2) absorption of important vitamins liberated by bacterial action; (3) storage of fecal material prior to defecation

31. The rate of epithelial stem cell division declines; smooth muscle tone decreases; the effects of cumulative damage become apparent; cancer rates increase; and changes in other systems have direct or indirect effects on the digestive system.

Level 2: Reviewing Concepts

32. d 33. d 34. a

35. Peristalsis consists of waves of muscular contractions that move along the length of the digestive tract. During a peristaltic movement, the circular muscles contract behind the digestive contents. Longitudinal muscles contract next, shortening adjacent segments. A wave of contraction in the circular muscles then forces the materials in the desired direction. Segmentation movements churn and fragment the digestive materials, mixing the contents with intestinal secretions. Because they do not follow a set pattern, segmentation movements do not produce directional movement of materials along the tract.

36. The stomach performs four major functions: the bulk storage of ingested food, the mechanical breakdown of ingested food, the disruption of chemical bonds through the actions of acids and enzymes, and the production of intrinsic factor.

37. The cephalic phase begins with the sight or thought of food. Directed by the CNS, this phase prepares the stomach to receive food. The gastric phase begins with the arrival of food in the stomach. The gastric phase is initiated by distension of the stomach, an increase in the pH of the gastric contents, and the presence of undigested materials in the stomach. The intestinal phase begins when chyme starts to enter the small intestine. This phase controls the rate of gastric emptying and ensures that the secretory, digestive, and absorptive functions of the small intestine can proceed at reasonable efficiency.

Level 3: Critical Thinking and Clinical Applications

38. If the gallstone is small enough, it can pass through the common bile duct and block the pancreatic duct. Enzymes from the pancreas will not be able to reach the small intestine; as they accumulate, they will irritate the duct and ultimately the exocrine pancreas, producing pancreatitis.

39. The small intestine, especially the jejunum and ileum, are probably involved. Regional inflammation is the cause of Barb's pain. The inflamed tissue will not absorb nutrients; this accounts for her weight loss. Among the nutrients that are not absorbed are iron and vitamin B12, which are necessary for formation of hemoglobin and red blood cells; this accounts for her anemia.

CHAPTER 17

Level 1: Reviewing Facts and Terms

1. a 2. g 3. l 4. i 5. j 6. b 7. k 8. e 9. f 10. d 11. h 12. c 13. d 14. c 15. a 16. b 17. c 18. c 19. d 20. c 21. b 22. c 23. b 24. a

25. Metabolism is all of the chemical reactions occurring in the cells of the body. Anabolism is those chemical reactions resulting in the synthesis of complex molecules from simpler reactants. Products of anabolism are used for maintenance/repair, growth, and secretion. Catabolism is the breakdown of complex molecules into their building block molecules, resulting in the release of energy for the synthesis of ATP and related molecules.

26. Lipoproteins are lipid-protein complexes that contain large insoluble glycerides and cholesterol, with a superficial coating of phospholipids and proteins. The major groups are chylomicrons (the largest lipoproteins, which are 95% triglyceride and carry absorbed lipids from the intestinal tract to the circulation), low-density lipoproteins (LDLs, which are mostly cholesterol and function in delivering cholesterol to peripheral tissues; sometimes this cholesterol gets deposited in arteries, hence the designation of LDLs as "bad cholesterol"), and high-density lipoproteins (HDLs, which are known as "good cholesterol," contain equal parts protein and lipid—cholesterol and phospholipids—and function in transporting excess cholesterol back to the liver for storage or excretion in the bile).

27. Most vitamins and all minerals must be provided in the diet because the body cannot synthesize these nutrients.

28. carbohydrates: 4.18 C/g; lipids: 9.46 C/g; proteins: 4.32 C/g

29. The BMR is the minimum, resting energy expenditures of an awake, alert person.

30. radiation (heat loss as infrared waves), conduction (heat loss to surfaces in physical contact, convection (heat loss to the air), and evaporation (heat loss with water becoming gas)

Level 2: Reviewing Concepts

31. d 32. b

33. Glycolysis results in the breakdown of glucose to pyruvic acid through a series of enzymatic steps. 4 ATP and 2 NADH are also produced. Glycolysis requires glucose, specific cytoplasmic enzymes, ATP and ADP, inorganic phosphates, and NAD (nicotinamide adenine dinucleotide), a coenzyme.

34. The TCA reaction sequence is a cycle because the 4-carbon starting compound (oxaloacetic acid) is regenerated at the end. Acetyl-CoA and oxaloacetic acid enter the cycle and CO_2, NADH, ATP, $FADH_2$, and oxaloacetic acid leave the cycle.

35. A triglyceride is hydrolyzed, yielding glycerol and fatty acids. Glycerol is converted to pyruvic acid and enters the TCA cycle. Fatty acids are broken into 2-carbon fragments by beta-oxidation, a process that occurs inside mitochondria. The 2-carbon compounds then enter the TCA cycle.

36. The food pyramid indicates the relative amounts of each food group an individual should consume per day to ensure adequate intake of nutrients and calories. The placement of fats, oils, and sugars at the top of the food pyramid indicates that such foods are to be consumed very sparingly, whereas carbohydrates, represented at the bottom of the pyramid as the bread, cereal, rice, and pasta group, is to be consumed in largest relative quantities.

37. The brain contains the "thermostat" of the body, a region known as the hypothalamus. The hypothalamus regulates the ANS control of such mechanisms as sweating and shivering thermogenesis, through negative feedback homeostatic mechanisms.

38. These terms refer to the HDL and LDL, lipoproteins in the blood that transport cholesterol. HDL ("good cholesterol") transports excess cholesterol to the liver for storage or breakdown, whereas LDL ("bad cholesterol") transports cholesterol to peripheral tissues, which unfortunately may include the arteries. The buildup of cholesterol in the arteries is linked to cardiovascular disease.

Level 3: Critical Thinking and Clinical Applications

39. During starvation, the body must use fat and protein reserves to supply the energy necessary to sustain life. Some of the protein that is metabolized for energy is the gamma globulin fraction of the blood, which is mostly composed of antibodies. This loss of antibodies coupled with a lack of amino acids to synthesize new ones, as well as protective molecules such as interferon and complement proteins, renders an individual more susceptible to contracting a disease and less likely to recover from it.

40. The drug *colestipol* would lead to a decrease in the plasma levels of cholesterol. Bile salts are necessary for the absorption of fats. If the bile salts cannot be absorbed, the amount of fat absorption, namely cholesterol and triglycerides, will decrease. This would in turn lead to a decrease in cholesterol from a dietary source as well as a decrease in fatty acids that could be used to synthesize new cholesterol. In addition, the body will have to replace the bile salts that are being lost with the feces. Since bile salts are formed from cholesterol, this will also contribute to a decline in cholesterol levels.

CHAPTER 18

Level 1: Reviewing Facts and Terms

1. q 2. g 3. n 4. j 5. a 6. m 7. e 8. p 9. d 10. f 11. h 12. k 13. b 14. c 15. l 16. o 17. i 18. c 19. d 20. a 21. d 22. c 23. c 24. b 25. d

26. The urinary system performs vital excretory functions and eliminates the organic waste products generated by cells throughout the body.

27. The urinary system includes the kidneys, ureters, urinary bladder, and urethra.

28. A fluid shift is a water movement between the ECF and ICF; this movement helps prevent drastic variations in the volume of the ECF. Changes in the osmolarity of the ECF can cause fluid shifts. If the ECF becomes more concentrated (hypertonic) with respect to the ICF, water will move from the cells into the ECF until equilibrium is restored. If the ECF becomes more dilute (hypotonic), water will move from the ECF into the cells. Changes in osmolarity can result from water loss (such as excessive perspiration, dehydration, vomiting, or diarrhea), water gain (drinking pure water, administering hypotonic solutions through an IV), or changes in electrolyte concentrations (such as sodium).

29. antidiuretic hormone: stimulates the thirst center and water conservation at the kidneys; aldosterone: determines the rate of sodium absorption along the DCT and collecting system of the kidneys; atrial natriuretic peptide: reduces thirst and blocks the release of ADH and aldosterone

Level 2: Reviewing Concepts

30. d 31. c 32. a 33. b 34. a

35. autoregulation at the local level; hormonal regulation initiated by the kidneys; autonomic regulation, (sympathetic division of the ANS)

36. The urge to urinate usually appears when the bladder contains about 200 ml of urine. The micturition reflex begins to function when the stretch receptors have provided adequate stimulation to the parasympathetic motor neurons. The activity in the motor neurons generates action potentials that reach the smooth muscle in the bladder wall. These efferent impulses travel over the pelvic nerves, producing a sustained contraction of the urinary bladder.

37. Fluid balance is a state in which the amount of water gained each day is equal to the amount lost to the environment. The water content of the body must remain stable, because water is an essential ingredient of cytoplasm and accounts for about 99% of the volume of extracellular fluid. Electrolyte balance exists when there is neither a net gain nor a net loss of any ion in body fluids. The ionic concentrations in body water must remain within normal limits; if levels of calcium or potassium become too high, for instance, cardiac arrhythmias can develop. Acid-base balance exists when the production of hydrogen ions precisely offsets their loss. The pH of body fluids must remain within a relatively narrow range; variations outside this range can be life threatening.

38. "Drink plenty of fluids" is physiologically sound advice because the temperature rise accompanying a fever can also increase water losses. For each degree the temperature rises above normal, the daily water loss increases by 200 ml.

39. Since sweat is usually hypotonic, the loss of a large volume of sweat causes hypertonicity in body fluids. The loss of fluid volume is primarily from the interstitial space, which leads to a reduction in plasma volume and an increase in the hematocrit. Severe dehydration can cause the blood viscosity to increase substantially, resulting in an increased workload on the heart, ultimately increasing the probability of heart failure.

Level 3: Critical Thinking and Clinical Applications

40. Truck drivers may not urinate as frequently as they should. Resisting the urge to urinate can result in urine backing up into the kidneys. This puts pressure on the kidney tissues, which can lead to tissue death and ultimately to kidney failure.

41. Susan may have a urinary tract infection; her urine may contain blood cells and bacteria. She is more likely to have this problem than her husband, because the urethral orifice in females is closer to the anus, and the urethral canal is short and opens near the vagina. Both regions normally harbor bacteria, which can easily reach the urethral entrance (often during sexual intercourse).

42. Because mannitol is filtered but not reabsorbed, drinking a mannitol solution would lead to an increase in the osmolarity of the filtrate. Less water would be absorbed, and an increased volume of urine would be produced.

CHAPTER 19

Level 1: Reviewing Facts and Terms

1. i 2. m 3. a 4. c 5. l 6. e 7. n 8. o 9. h 10. b 11. f 12. j 13. d 14. k 15. g 16. c 17. a 18. c 19. a 20. d 21. a 22. c 23. c 24. d 25. b 26. c 27. d

28. The reproductive system of both males and females includes reproductive organs (gonads) that produce gametes and hormones, ducts that receive and transport gametes, accessory glands and organs that secrete fluids into these or other excretory ducts, and perineal structures collectively known as external genitalia.

29. The accessory organs and glands include the seminal vesicles, prostate gland, and the bulbourethral glands. The major functions of these glands are activating the spermatozoa, providing the nutrients sperm need for motility, propelling sperm and fluids along the reproductive tract, and producing buffers that counteract the acidity of the urethral and vaginal contents.

30. monitors and adjusts the composition of the tubular fluid, acts as a recycling center for damaged spermatozoa, and stores spermatozoa and facilitates their functional maturation

31. produces female gametes (ova); secretes female sex hormones, including estrogens and progestins; and secretes inhibin, involved in the feedback control of pituitary FSH production

32. serves as passageway for the elimination of menstrual fluids; receives the penis during sexual intercourse and holds spermatozoa prior to their passage

into the uterus; and, in childbirth, forms the lower portion of the birth canal through which the fetus passes during delivery

Level 2: Reviewing Concepts

33. The reproductive system is the only physiological system that is not required for the survival of the individual.

34. Meiosis is the two-step nuclear division resulting in the formation of 4 haploid cells from 1 diploid cell. In males, 4 sperm are produced from each diploid cell, whereas in females only 1 ovum (plus 3 polar bodies) is produced from each diploid cell.

35. Menses: This phase is marked by the degeneration and loss of the functional zone of the endometrium; usually lasts 1 to 7 days, and approximately 35–50 ml of blood is lost. Proliferative phase: Growth and vascularization result in the complete restoration of the functional zone; lasts from the end of the menses until the beginning of ovulation around day 14. Secretory phase: The endometrial glands enlarge, accelerating their rates of secretion; the arteries elongate and spiral through the tissues of the functional zone, under the combined stimulatory effects of progestins and estrogens from the corpus luteum; begins at ovulation and persists as long as the corpus luteum remains intact.

36. The corpus luteum degenerates and a decline in progesterone and estrogen levels results in endometrial breakdown of menses. Next, rising levels of FSH, LH, and estrogen stimulate the repair and regeneration of the functional zone of endometrium. During the postovulatory phase, the combination of estrogen and progesterone cause the enlargement of the endometrial glands and an increase in their secretory activity.

37. In women, menopause is defined as the time that ovulation and menstruation cease. Menopause is accompanied by a sharp and sustained rise in the production of GnRH, FSH, and LH, while concentrations of circulating estrogen and progesterone decline. The decline in estrogen levels leads to reductions in the size of the uterus and breasts, accompanied by a thinning of the urethral and vaginal walls. In addition to a variety of neural and cardiovascular effects, reduced estrogen concentrations have been linked to the development of osteoporosis, presumably because bone deposition proceeds at a slower rate. During the male climacteric, circulating testosterone levels begin to decline between ages 50 and 60, coupled with increases in circulating levels of FSH and LH. Although sperm production continues, there is a gradual reduction in sexual activity in older men.

Level 3: Critical Thinking and Clinical Applications

38. There is no direct entry into the abdominopelvic cavity in males as there is in females. In females, the urethral opening is in close proximity to the vaginal orifice; therefore, infectious organisms can exit from the urethral meatus and enter the vagina. They can then proceed through the vagina to the uterus, then into the uterine tubes and finally into the peritoneal cavity.

39. The sacral region of the spinal cord contains the parasympathetic centers that control the genitals. Damage to this area of the spinal cord would interfere with the ability to achieve an erection by way of parasympathetic stimuli. However, erection can also occur by way of sympathetic centers in the lower thoracic region of the spinal cord. Visual, auditory, or cerebral stimuli can result in decreased tone in the arteries serving the penis. This results in increased blood flow and erection. Tactile stimulation of the penis, however, would not generate an erection.

40. It appears that a certain amount of body fat is necessary for menstrual cycles to occur. The ratio of body fat to muscle tissue is somehow monitored by the nervous system; when the ratio falls below a certain set point, menstruation ceases. The actual mechanism appears to be a change in the levels of hypothalamic releasing hormone and pituitary gonadotrophins. Possibly without proper fat reserves, a woman could not have a successful pregnancy. To avoid damage to the female body and death of a fetus, the body prevents pregnancy from occurring by shutting down the ovarian cycle and thus the menstrual cycle. When appropriate energy reserves are available, the cycles begin again.

CHAPTER 20

Level 1: Reviewing Facts and Terms

1. g 2. a 3. c 4. e 5. d 6. n 7. p 8. b 9. o 10. j 11. f 12. h 13. k 14. l 15. m 16. i 17. a 18. b 19. d 20. d 21. a 22. b 23. d 24. b

25. First trimester: The rudiments of all major organ systems appear. Second trimester: The organs and organ systems complete most of their development, and the body proportions change to become more human. Third trimester: The fetus grows rapidly, and most of the major organ systems become fully functional.

26. Dilation stage: This begins with the onset of true labor, as the cervix dilates and the fetus begins to slide down the cervical canal. Late in this stage, the amnion usually ruptures. Expulsion stage: This stage begins as the cervix dilates completely and continues until the fetus has completely emerged from the vagina (delivery). Placental stage: The uterus gradually contracts, tearing the connections between the endometrium and the placenta and ejecting the placenta.

27. Neonatal period (birth to 1 month): The newborn becomes relatively self-sufficient and begins performing respiration, digestion, and excretion for itself. Heart rates and fluid requirements are higher than those of adults. Neonates have little ability to thermoregulate. Infancy (1 month to 2 years): Major organ systems (other than those related to reproduction) become fully operational and start to take on the functional characteristics of adult structures. Daily, even hourly, variations in body temperature continue throughout childhood. Childhood (2 years to puberty): The child continues to grow, and significant changes in body proportions occur.

Level 2: Reviewing Concepts

28. d 29. d

30. The placental hormone human chorionic gonadotropin (HCG) resembles LH, in function, because it maintains the integrity of the corpus luteum and promotes the continued secretion of progesterone, which keeps the endometrial lining perfectly functional. The placental hormone human placental lactogen (HPL) helps prepare the mammary glands for milk production. The conversion from the resting state of the mammary glands to active status requires the presence of HPL and another placental hormone, placental prolactin, as well as several maternal hormones. The placental hormone relaxin is a peptide hormone that increases the flexibility of the symphysis pubis, causes the dilation of the cervix, and suppresses the release of oxytocin by the hypothalamus, delaying the onset of labor contractions.

31. The respiratory rate and tidal volume increase, allowing the lungs to obtain the extra oxygen and to remove the excess carbon dioxide generated by the fetus. Maternal blood volume increases, compensating for blood that will be lost during delivery. Requirements for nutrients and vitamins climb 10–30 percent, reflecting the fact that part of the mother's nutrients must nourish the fetus. The glomerular filtration rate increases by about 50 percent, corresponding to the increased blood volume and accelerates the excretion of metabolic wastes generated by the fetus.

32. Positive feedback mechanisms ensure that labor contractions continue until delivery is complete.

33. The amnion generally ruptures late in the dilation stage of labor.

34. (a) dominant (b) recessive (c) X-linked (d) autosomal

35. The trait of color blindness is carried on the X chromosome. Men are thus more likely to inherit it, because they have only one X chromosome. So whatever that chromosome carries will determine whether he is color blind or has normal vision. Since women have two X chromosomes, they will be color blind only if they are homozygous-recessive. This is an X-linked inheritance.

36. The goal of the Human Genome Project is to determine the normal genetic composition of a "typical" human being. The project will help identify the genes responsible for inherited disorders and their locations on specific chromosomes. Other benefits include gaining an understanding of the genetic basis of diseases such as cancer and why the effectiveness of drugs varies between individuals.

Level 3: Critical Thinking and Clinical Applications

37. None of the couple's daughters will be hemophiliacs, since each will receive a normal allele from her father. There is a 50 percent chance that a son will be hemophiliac since there is a 50 percent chance of receiving either the mother's normal allele or her recessive allele.

38. The fact that the adults are larger is precisely why their rates are lower. Heat is lost across the skin. In infants, the surface-area-to-volume ratio is high, so they lose heat very quickly. To maintain a constant body temperature in the face of the heat loss, cellular metabolism must be high. Cellular metabolism requires oxygen, thus increased metabolism demands an increased respiratory rate. There must also be an increase in cardiac output to move the blood from the lungs to the tissues. Since the range of contraction in the neonate heart is limited, the greatest increase in cardiac output is achieved by increased heart rate.

39. The baby's condition is almost certainly not the result of a virus contracted during the third trimester. The development of organ systems occurs during the first trimester. By the end of the second trimester, almost all of the organ systems are fully formed. During the third trimester, the fetus undergoes tremendous growth but very little new organ formation.

Appendix II

A PERIODIC CHART OF THE ELEMENTS

The periodic table presents the known elements in order of their atomic weights. Each horizontal row represents a single electron shell. The number of elements in that row is determined by the maximum number of electrons that can be stored at that energy level. The element at the left end of each row contains a single electron in its outermost electron shell; the element at the right end of the row has a filled outer electron shell. Organizing the elements in this fashion highlights similarities that reflect the composition of the outer electron shell. These similarities are evident when you examine the vertical columns. All the gases of the right-most column—helium, neon, argon, krypton, xenon, and radon—have full electron shells; each is a gas at normal atmospheric temperature and pressure, and none reacts readily with other elements. These elements, highlighted in blue, are known as the *noble*, or *inert*, *gases*. In contrast, the elements of the left-most column—lithium, sodium, potassium, and so forth—are silvery, soft metals that are so highly reactive that pure forms cannot be found in nature. The fourth and fifth electron levels can hold 18 electrons. Table inserts are used for the *lanthanide* and *actinide series* to save space, as higher levels can store up to 32 electrons. Elements of particular importance to our discussion of human anatomy and physiology are highlighted in pink.

Atomic number — 1
Chemical symbol — H
Element name — Hydrogen
Atomic weight — 1.01

1 H Hydrogen 1.01																	2 He Helium 4.00
3 Li Lithium 6.94	4 Be Beryllium 9.01											5 B Boron 10.81	6 C Carbon 12.01	7 N Nitrogen 14.01	8 O Oxygen 16.00	9 F Fluorine 19.00	10 Ne Neon 20.18
11 Na Sodium 22.99	12 Mg Magnesium 24.31											13 Al Aluminum 26.98	14 Si Silicon 28.09	15 P Phosphorus 30.97	16 S Sulfur 32.07	17 Cl Chlorine 35.45	18 Ar Argon 39.95
19 K Potassium 39.10	20 Ca Calcium 40.08	21 Sc Scandium 44.96	22 Ti Titanium 47.88	23 V Vanadium 50.94	24 Cr Chromium 52.00	25 Mn Manganese 54.94	26 Fe Iron 55.85	27 Co Cobalt 58.93	28 Ni Nickel 58.69	29 Cu Copper 63.55	30 Zn Zinc 65.39	31 Ga Gallium 69.72	32 Ge Germanium 72.61	33 As Arsenic 74.92	34 Se Selenium 78.96	35 Br Bromine 79.90	36 Kr Krypton 83.80
37 Rb Rubidium 85.47	38 Sr Strontium 87.62	39 Y Yttrium 88.91	40 Zr Zirconium 91.22	41 Nb Niobium 92.91	42 Mo Molybdenum 95.94	43 Tc Technetium (98)	44 Ru Ruthenium 101.07	45 Rh Rhodium 102.91	46 Pd Palladium 106.42	47 Ag Silver 107.87	48 Cd Cadmium 112.41	49 In Indium 114.82	50 Sn Tin 118.71	51 Sb Antimony 121.76	52 Te Tellurium 127.60	53 I Iodine 126.90	54 Xe Xenon 131.29
55 Cs Cesium 132.91	56 Ba Barium 137.33	57 La* Lanthanum 138.91	72 Hf Hafnium 178.49	73 Ta Tantalum 180.95	74 W Tungsten 183.85	75 Re Rhenium 186.21	76 Os Osmium 190.2	77 Ir Iridium 192.22	78 Pt Platinum 195.08	79 Au Gold 196.97	80 Hg Mercury 200.59	81 Tl Thallium 204.38	82 Pb Lead 207.2	83 Bi Bismuth 208.98	84 Po Polonium (209)	85 At Astatine (210)	86 Rn Radon (222)
87 Fr Francium (223)	88 Ra Radium 226.03	89 Ac Actinium 227.03	104 Rf Dubnium (261)	105 Db Joliotium (262)	106 Rf Rutherfordium (263)	107 Bh Bohrium (262)	108 Hn Hahnium (265)	109 Mt Meitnerium (266)	110 Unnamed (269)	111 Unnamed (272)	112 Unnamed (272)						

*Lanthanide series

58 Ce Cerium 140.12	59 Pr Praseodymium 140.91	60 Nd Neodymium 144.24	61 Pm Promethium (145)	62 Sm Samarium 150.36	63 Eu Europium 151.96	64 Gd Gadolinium 157.25	65 Tb Terbium 158.93	66 Dy Dysprosium 162.50	67 Ho Holmium 164.93	68 Er Erbium 167.26	69 Tm Thulium 168.93	70 Yb Ytterbium 173.04	71 Lu Lutetium 174.97

Actinide series

90 Th Thorium 232.04	91 Pa Protactinium 231.04	92 U Uranium 238.03	93 Np Neptunium 237.05	94 Pu Plutonium (244)	95 Am Americium (243)	96 Cm Curium (247)	97 Bk Berkelium (247)	98 Cf Californium (251)	99 Es Einsteinium (252)	100 Fm Fermium (257)	101 Md Mendelevium (258)	102 No Nobelium (259)	103 Lr Lawrencium (260)

Appendix III

WEIGHTS AND MEASURES

Accurate descriptions of physical objects would be impossible without a precise method of reporting the pertinent data. Dimensions such as length and width are reported in standardized units of measurement, such as inches or centimeters. These values can be used to calculate the **volume** of an object, a measurement of the amount of space it fills. **Mass** is another important physical property. The mass of an object is determined by the amount of matter it contains. On earth the mass of an object determines its weight.

Most U.S. readers describe length and width in inches, feet, or yards; volumes in pints, quarts, or gallons; and weights in ounces, pounds, or tons. These are units of the **U.S. system** of measurement.

Table 1 summarizes terms used in the U.S. system. For reference, this table also includes a definition of the "household units," popular in recipes. The U.S. system can be very difficult to work with, because there is no logical relationship between the various units. For example, there are 12 inches in a foot, 3 feet in a yard, and 1760 yards in a mile. Without a clear pattern of organization, converting feet to inches or miles to feet can be confusing and time-consuming. The relationships between ounces, pints, quarts, and gallons or ounces, pounds, and tons are no more logical.

In contrast, the **metric system** has a logical organization based on powers of 10, as indicated in Table 2. For example, a **meter** (**m**) is the basic unit for the measurement of size. For measuring larger objects, data can be reported in **dekameters** (*deka*, ten), **hectometers** (*hekaton*, hundred), or **kilometers** (**km**; *chilioi*, thousand); for smaller objects, data can be reported in **decimeters** (0.1 m; *decem*, ten), **centimeters** (**cm** = 0.01 m; *centum*, hundred), **millimeters** (**mm** = 0.001 m; *mille*, thousand), and so forth. Notice that the same prefixes are used to report weights, based on the **gram** (**g**), and volumes, based on the **liter** (**l**).

This text reports data in metric units, usually with U.S. equivalents. Use this opportunity to become familiar with the metric system, because most technical sources report data only in metric units, and most of the rest of the world uses the metric system exclusively. Conversion factors are included in Table 2.

The U.S. system and metric system also differ in their methods of reporting temperature: In the United States, temperature is usually reported in degrees Fahrenheit (°F), whereas scientific literature and individuals in most other countries report temperature in degrees centigrade (or degrees Celsius, °C). The relationship between temperature in degrees Fahrenheit and temperature in degrees centigrade is given in Table 2.

TABLE 1 *The U.S. System of Measurement*

PHYSICAL PROPERTY	UNIT	RELATIONSHIP TO OTHER U.S. UNITS	RELATIONSHIP TO HOUSEHOLD UNITS
Length	inch (in.)	1 in. = 0.083 ft	
	foot (ft)	1 ft = 12 in.	
		= 0.33 yd	
	yard (yd)	1 yd = 36 in.	
		= 3 ft	
	mile (mi)	1 mi = 5280 ft	
		= 1760 yd	
Volume	fluidram (fl dr)	1 fl dr = 0.125 fl oz	
	fluid ounce (fl oz)	1 fl oz = 8 fl dr	= 6 teaspoons (tsp)
		= 0.0625 pt	= 2 tablespoons (tbsp)
	pint (pt)	1 pt = 128 fl dr	= 32 tbsp
		= 16 fl oz	= 2 cups (c)
		= 0.5 qt	
	quart (qt)	1 qt = 256 fl dr	= 4 c
		= 32 fl oz	
		= 2 pt	
		= 0.25 gal	
	gallon (gal)	1 gal = 128 fl oz	
		= 8 pt	
		= 4 qt	
Mass	grain (gr)	1 gr = 0.002 oz	
	dram (dr)	1 dr = 27.3 gr	
		= 0.063 oz	
	ounce (oz)	1 oz = 437.5 gr	
		= 16 dr	
	pound (lb)	1 lb = 7000 gr	
		= 256 dr	
		= 16 oz	
	ton (t)	1 t = 2000 lb	

TABLE 2 *The Metric System of Measurement*

PHYSICAL PROPERTY	UNIT	RELATIONSHIP TO STANDARD METRIC UNITS	CONVERSION TO U.S. UNITS	
Length	nanometer (nm)	1 nm = 0.000000001 m (10^{-9})	= 4×10^{-8} in.	25,000,000 nm = 1 in.
	micrometer (μm)	1 μm = 0.000001 m (10^{-6})	= 4×10^{-5} in.	25,000 mm = 1 in.
	millimeter (mm)	1 mm = 0.001 m (10^{-3})	= 0.0394 in.	25.4 mm = 1 in.
	centimeter (cm)	1 cm = 0.01 m (10^{-2})	= 0.394 in.	2.54 cm = 1 in.
	decimeter (dm)	1 dm = 0.1 m (10^{-1})	= 3.94 in.	0.25 dm = 1 in.
	meter (m)	standard unit of length	= 39.4 in.	0.0254 m = 1 in.
			= 3.28 ft	0.3048 m = 1 ft
			= 1.09 yd	0.914 m = 1 yd
	dekameter (dam)	1 dam = 10 m		
	hectometer (hm)	1 hm = 100 m		
	kilometer (km)	1 km = 1000 m	= 3280 ft	
			= 1093 yd	
			= 0.62 mi	1.609 km = 1 mi
Volume	microliter (μl)	1 μl = 0.000001 l (10^{-6}) = 1 cubic millimeter (mm^3)		
	milliliter (ml)	1 ml = 0.001 l (10^{-3}) = 1 cubic centimeter $(cm^3$ or cc)	= 0.03 fl oz	5 ml = 1 tsp
				15 ml = 1 tbsp
				30 ml = 1 fl oz
	centiliter (cl)	1 cl = 0.01 l (10^{-2})	= 0.34 fl oz	3 cl = 1 fl oz
	deciliter (dl)	1 dl = 0.1 l (10^{-1})	= 3.38 fl oz	0.29 dl = 1 fl oz
	liter (l)	standard unit of volume	= 33.8 fl oz	0.0295 l = 1 fl oz
			= 2.11 pt	0.473 l = 1 pt
			= 1.06 qt	0.946 l = 1 qt
Mass	picogram (pg)	1 pg = 0.000000000001 g (10^{-12})		
	nanogram (ng)	1 ng = 0.000000001 g (10^{-9})		
	microgram (μg)	1 μg = 0.000001 g (10^{-6})	= 0.000015 gr	66,666 mg = 1 gr
	milligram (mg)	1 mg = 0.001 g (10^{-3})	= 0.015 gr	66.7 mg = 1 gr
	centigram (cg)	1 cg = 0.01 g (10^{-2})	= 0.15 gr	6.7 cg = 1 gr
	decigram (dg)	1 dg = 0.1 g (10^{-1})	= 1.5 gr	0.67 dg = 1 gr
	gram (g)	standard unit of mass	= 0.035 oz	28.35 g = 1 oz
			= 0.0022 lb	453.6 g = 1 lb
	dekagram (dag)	1 dag = 10 g		
	hectogram (hg)	1 hg = 100 g		
	kilogram (kg)	1 kg = 1000 g	= 2.2 lb	0.453 kg = 1 lb
	metric ton (mt)	1 mt = 1000 kg	= 1.1 t	
			= 2205 lb	0.907 mt = 1 t

TEMPERATURE	CENTIGRADE	FAHRENHEIT
Freezing point of pure water	0°	32°
Normal body temperature	36.8°	98.6°
Boiling point of pure water	100°	212°
Conversion	°C $\rightarrow$ °F: °F = (1.8 × °C) + 32	°F $\rightarrow$ °C: °C = (°F − 32) × 0.56

The figure below spans the entire range of measurements that we will consider in this book. Gross anatomy traditionally deals with structural organization as seen with the naked eye or with a simple hand lens. A microscope can provide higher levels of magnification and reveal finer details. Before the 1950s, most information was provided by *light microscopy*. A photograph taken through a light microscope is called a **light micrograph (LM)**. Light microscopy can magnify cellular structures about 1000 times and show details as fine as 0.25 μm. The symbol μm stands for *micrometer*; 1 μm = 0.001 mm, or 0.00004 inch. With a light microscope, we can identify cell types, such as muscle cells or neurons, and see large structures within the cell. Because individual cells are relatively transparent, thin sections taken through a cell are treated with dyes that stain specific structures, making them easier to see.

Although special staining techniques can show the general distribution of proteins, lipids, carbohydrates, and nucleic acids in the cell, many fine details of intracellular structure remained a mystery until investigators began using *electron microscopy*. This technique uses a focused beam of electrons, rather than a beam of light, to examine cell structure. In *transmission electron microscopy*, electrons pass through an ultrathin section to strike a photographic plate. The result is a **transmission electron micrograph (TEM)**. Transmission electron microscopy shows the fine structure of cell membranes and intracellular structures. In *scanning electron microscopy*, electrons bouncing off exposed surfaces create a **scanning electron micrograph (SEM)**. Scanning microscopy cannot achieve as much magnification as transmission microscopy, but it provides a three-dimensional perspective of cell structure, whereas transmission microscopy gives a two-dimensional view.

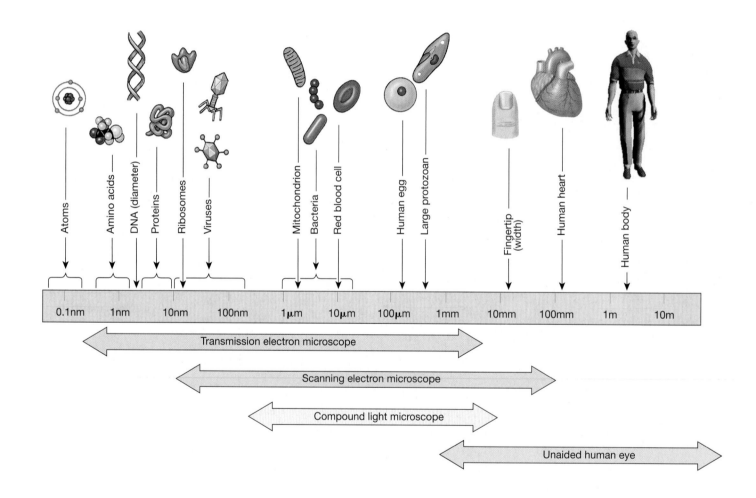

Appendix IV

NORMAL PHYSIOLOGICAL VALUES

Tables 3 and 4 present normal averages or ranges for the chemical composition of body fluids. These values should be considered approximations rather than absolute values, as test results vary from laboratory to laboratory owing to differences in procedures, equipment, normal solutions, and so forth. Sources used in the preparation of these tables are indicated on p. A-16. Blanks indicate tests for which data were not available.

TABLE 3 *The Chemistry of Blood, Cerebrospinal Fluid, and Urine*

TEST	BLOOD[A]	CSF	URINE
		NORMAL RANGES	
pH	S: 7.38–7.44	7.31–7.34	4.6–8.0
OSMOLARITY (mOsm/l)	S: 280–295	292–297	500–800
ELECTROLYTES	(mEq/l unless noted)		(urinary loss per 24-hour period[b])
Bicarbonate	P: 21–28	20–24	
Calcium	S: 4.5–5.3	2.1–3.0	6.5–16.5 mEq
Chloride	S: 100–108	116–122	120–240 mEq
Iron	S: 50–150 µg/l	23–52 µg/l	40–150 µg
Magnesium	S: 1.5–2.5	2–2.5	4.9–16.5 mEq
Phosphorus	S: 1.8–2.6	1.2–2.0	0.8–2 g
Potassium	P: 3.8–5.0	2.7–3.9	35–80 mEq
Sodium	P: 136–142	137–145	120–220 mEq
Sulfate	S: 0.2–1.3		1.07–1.3 g
METABOLITES	(mg/dl unless noted)		(urinary loss per 24-hour period[c])
Amino acids	P/S: 2.3–5.0	10.0–14.7	41–133 mg
Ammonia	P: 20–150 µg/dl	25–80 µg/dl	340–1200 mg
Bilirubin	S: 0.5–1.2	<0.2	0.02–1.9 mg
Creatinine	P/S: 0.6–1.2	0.5–1.9	1.01–2.5
Glucose	P/S: 70–110	40–70	16–132 mg
Ketone bodies	S: 0.3–2.0	1.3–1.6	10–100 mg
Lactic acid	WB: 5–20[d]	10–20	100–600 mg
Lipids (total)	S: 400–1000	0.8–1.7	0–31.8 mg
Cholesterol (total)	S: 150–300	0.2–0.8	1.2–3.8 mg
Triglycerides	S: 40–150	0–0.9	
Urea	P/S: 23–43	13.8–36.4	12.6–28.6
Uric acid	S: 2.0–7.0	0.2–0.3	80–976 mg
PROTEINS	(g/dl)	(mg/dl)	(urinary loss per 24-hour period[c])
Total	S: 6.0–7.8	20–4.5	47–76.2 mg
Albumin	S: 3.2–4.5	10.6–32.4	10–100 mg
Globulins (total)	S: 2.3–3.5	2.8–15.5	7.3 mg (average)
Immunoglobulins	S: 1.0–2.2	1.1–1.7	3.1 mg (average)
Fibrinogen	P: 0.2–0.4	0.65 (average)	

[a] S = serum, P = plasma, WB = whole blood
[b] Because urinary output averages just over 1 liter per day, these electrolyte values are comparable to mEq/l.
[c] Because urinary metabolite and protein data approximate mg/l or g/l, they must be divided by 10 for comparison with CSF or blood concentrations.
[d] Venous blood sample

TABLE 4 *The Composition of Minor Body Fluids*

			NORMAL AVERAGES OR RANGES			
TEST	PERILYMPH	ENDOLYMPH	SYNOVIAL FLUID	SWEAT	SALIVA	SEMEN
pH			7.4	4–6.8	6.4[a]	7.19
SPECIFIC GRAVITY			1.008–1.015	1.001–1.008	1.007	1.028
ELECTROLYTES (mEq/l)						
Potassium	5.5–6.3	140–160	4.0	4.3–14.2	21	31.3
Sodium	143–150	12–16	136.1	0–104	14[a]	117
Calcium	1.3–1.6	0.05	2.3–4.7	0.2–6	3	12.4
Magnesium	1.7	0.02		0.03–4	0.6	11.5
Bicarbonate	17.8–18.6	20.4–21.4	19.3–30.6		6[a]	24
Chloride	121.5	107.1	107.1	34.3	17	42.8
PROTEINS (mg/dl)						
Total	200	150	1.72 g/dl	7.7	386[b]	4.5 g/dl
METABOLITES (mg/dl)						
Amino acids				47.6	40	1.26 g/dl
Glucose	104		70–110	3.0	11	224 (fructose)
Urea				26–122	20	72
Lipids, (total)	12		20.9	[d]	25–500[c]	188

[a] Increases under salivary stimulation
[b] Primarily alpha-amylase, with some lysozymes
[c] Cholesterol
[d] Not present in eccrine secretions

SOURCES

Ballenger, John Jacob. 1977. *Diseases of the Nose, Throat, and Ear.* Philadelphia: Lea and Febiger.

Davidsohn, Israel, and John Bernard Henry, eds. 1969. *Todd-Sanford Clinical Diagnosis by Laboratory Methods*, 14th ed. Philadelphia: W. B. Saunders.

Diem, K., and C. Lenter, eds. 1970. *Scientific Tables*, 7th ed. Basel, Switzerland: Ciba-Geigy.

Halsted, James A. 1976. *The Laboratory in Clinical Medicine: Interpretation and Application.* Philadelphia: W. B. Saunders.

Harper, Harold A. 1987. *Review of Physiological Chemistry.* Los Altos, Calif.: Lange Medical Publications.

Isselbacher, Kurt J., Eugene Braunwauld, Robert G. Petersdorf, Jean D. Wilson, Joseph B. Martin, Anthony S. Fauci, and Dennis L. Kaspar, eds. 1994. *Harrison's Principles of Internal Medicine*, 13th ed. New York: McGraw-Hill.

Lentner, Cornelius, ed. 1981. *Geigy Scientific Tables*, 8th ed. Basel, Switzerland: Ciba-Geigy.

Glossary/Index

Aging, 632
  cancer incidence and, 104
  of cardiovascular system, 419
  of digestive system, 513–514
  hormones and, 335
  of immune system, 447
  of integumentary system, 120
  of muscular system, 216–217
  of nervous system, 268
  nutritional requirements and, 539–540
  of reproductive system, 601
  of respiratory system, 479
  of sensory functions, 305–306
  of skeletal system, 136–137
  of tissues, 103–104
  of urinary system, 573
Agonist: A muscle responsible for a specific movement, 195
Agranulocytes, 354–355
AIDS (acquired immune deficiency syndrome), 443, 447, 466, 601
Air
  alveolar versus atmospheric, 472
  gas concentrations in, 471
Air cells, 297
Alarm phase, of general adaptation syndrome, 334
Albinism: Absence of pigment in hair and skin caused by inability of melanocytes to produce melanin, 113, 633–635
Albumins, 345–346, 529
Alcoholic beverages/alcoholism, 321, 498, 505
Aldosterone: A mineralocorticoid produced by the adrenal cortex; stimulates sodium and water conservation at the kidneys; secreted in response to the presence of angiotensin II, 315, 326–327, 331, 402–403, 558, 560–561, 569, 573
Alkaline solution, 34
Alkalosis (al-kah-LŌ-sis)**:** Condition characterized by a plasma pH of greater than 7.45 and associated with relative deficiency of hydrogen ions or an excess of bicarbonate ions, 569
  metabolic, 571–572
  respiratory, 571–572
Allantois (a-LAN-tō-is)**:** One of the four extraembryonic membranes; it provides blood vessels to the chorion and is therefore essential to placenta formation; the proximal portion becomes the urinary bladder; the distal portion becomes part of the umbilical cord, 618, 620–621
Alleles (a-LĒLZ)**:** Alternate forms of a particular gene, 632–633
Allergen: An antigenic compound that produces a hypersensitivity response, 447
Allergy, 356, 446–447, 449, 463
All-or-none principle, 184, 233
Alpha cells: Cells in the pancreatic islets that secrete glucagon, 329–330
ALS (amyotrophic lateral sclerosis), 270
Alveolar duct, 463–466
Alveolar macrophages, 465–466
Alveolar sac: An air-filled chamber that supplies air to several alveoli, 463–465
Alveolus/alveoli (al-VĒ-o-lī)**:** Blind pockets at the end of the respiratory tree, lined by a simple squamous epithelium and surrounded by a capillary network; site of gas exchange with the blood, 9, 406, 458, 463–466
  alveolar versus atmospheric air, 472
Alveolus (bony socket), 493
Alzheimer's disease, 251, 268, 564, 573, 636–637
Amacrine cells, 289–290
Amenorrhea, 595, 606
Amino acids: Organic compounds whose chemical structure can be summarized as R—CHNH$_2$—COOH, 41, 46, 513

in blood, 555
  catabolism of, 529–531
  essential, 531, 534
  hormones derived from, 314–315
  nonessential, 531
  in protein synthesis, 73–74, 531
  structure of, 41
  in urine, 555
Aminoglycoside, 305
Amino group: NH$_2$.
Ammonia, 530–531, 555
Amnesia: Temporary or permanent memory loss, 250
Amniocentesis, 89, 636, 638
Amnion (AM-nē-on)**:** One of the four extraembryonic membranes; surrounds the developing embryo/fetus, 618, 620–623, 628
Amniotic cavity, 618–619, 621
Amniotic fluid (am-nē-OT-ik)**:** Fluid that fills the amniotic cavity; provides cushioning and support for the embryo/fetus, 89
Amoeboid movement, 355
Amphetamine, 476
Amphiarthrosis (am-fē-ar-THRŌ-sis)**:** An articulation that permits a small degree of independent movement, 157–158
Amphimixis (am-fi-MIK-sis)**:** The fusion of male and female pronuclei following fertilization, 615–616
Ampulla/ampullae (am-PŪL-la)**:** A localized dilation in the lumen of a canal or passageway.
  of semicircular duct, 299–301
Amygdaloid body, 250–251
Amylase: An enzyme that breaks down polysaccharides, produced by the salivary glands and pancreas.
  pancreatic, 504, 511–512
  salivary, 492, 499, 511–512
Amyotrophic lateral sclerosis (ALS), 270
Anabolism (a-NAB-ō-lizm)**:** The synthesis of complex organic compounds from simpler precursors, 33, 522–523, 532
Anaerobic: Without oxygen.
Anaerobic endurance, 193
Anaerobic process, 190–191, 524
Anal canal, 508–510
Anal column, 508–509
Analgesic, 306
Anal sphincter
  external, 204–205, 508–510
  internal, 508–510
Anaphase (AN-a-fāz)**:** Mitotic stage in which the paired chromatids separate and move toward opposite ends of the spindle apparatus.
  meiosis I, 585
  meiosis II, 585
  mitotic, 74, 76
Anaphylactic shock, 404, 449
Anaphylaxis, 449
Anaplasia, 104
Anastomosis (a-nas-to-MŌ-sis)**:** The joining of two tubes, usually referring to a connection between two peripheral blood vessels without an intervening capillary bed, 375, 392–393
Anatomical directions, 16–17
Anatomical landmarks, 14–15
Anatomical neck, of humerus, 151–152
Anatomical position: An anatomical reference position, the body viewed from the anterior surface with the palms facing forward; supine, 14–15
Anatomical regions, 16
Anatomy (a-NAT-ō-mē)**:** The study of the structure of the body, 2

language of, 14–21
  sectional anatomy, 18–21
  surface anatomy, 14–17
  science of, 2–3
Anchoring protein, 58
Androgen (AN-drō-jen)**:** A steroid sex hormone primarily produced by the interstitial cells of the testis and manufactured in small quantities by the adrenal cortex in either sex, 315, 320–321, 326–328, 331–333, 347, 351, 584, 590, 602
Androstenedione, 332
Anemia (a-NĒ-mē-ah)**:** Condition marked by a reduction in the hematocrit and/or hemoglobin content of the blood, 348, 351–352, 362, 395, 535, 540
  iron deficiency, 350
  pernicious, 447, 536
  sickle-cell, 74, 349, 362, 635, 637
Anesthesia, 306
Aneurysm (AN-ū-rizm)**:** A weakening and localized dilation in the wall of a blood vessel, 419, 421
Angina pectoris, 383
Angiogram (AN-jē-ō-gram)**:** An X-ray image of circulatory pathways, 384
Angiotensin I, II: Angiotensin I is a hormone that is converted to angiotensin II; angiotensin II is a hormone that causes an elevation in systemic blood pressure, stimulates secretion of aldosterone, promotes thirst, and causes the release of ADH, 326, 331, 402–404, 560–561
Angiotensinogen: Blood protein produced by the liver that is converted to angiotensin I by the enzyme renin, 403, 560
Angular movement, 159–160
Animal starch. *See* **Glycogen**
Anion (AN-ī-on)**:** An ion bearing a negative charge, 30
Ankle, bones of, 156–157
Ankylosis, 168
Annulus (AN-ū-lus)**:** A cartilage or bone shaped like a ring.
Anoxia (a-NOKS-ē-uh)**:** Tissue oxygen deprivation, 468, 481
ANP. *See* **Atrial natriuretic peptide**
ANS. *See* **Autonomic nervous system**
Antacid, 34–35, 495
Antagonist: A muscle that opposes the movement of an agonist, 195
Antagonistic hormones, 333
Antebrachial region, 15–16
Antebrachial vein, median, 413, 415
Antebrachium, 15–16
Anterior: On or near the front or ventral surface of the body, 16–17
Anterior cavity, of eye, 287–288, 290
Anterior chamber, of eye, 288, 290
Anterior commissure, 248
Anterior crest, of tibia, 156
Anterior cruciate ligament, 166
Anterior fontanel, 145
Anterior interventricular branch, of coronary artery, 375
Anterior median fissure, 240–242
Anterior pituitary, 318–323, 589
  hormones of, 320–321
  hypothalamic control of, 320
Anterior section, 18
Antibiotic: Chemical agent that selectively kills pathogenic microorganisms.
Antibody (AN-ti-bod-ē)**:** A globular protein produced by plasma cells that will bind to specific antigens and promote their destruction or removal from the body, 41, 92–93, 344, 346, 430, 436, 446
  in breast milk, 631

classes of, 442–443
cross-reactions and, 353–354
function of, 442–444
monoclonal, 449
structure of, 441–442
Antibody-mediated immunity. *See* **Humoral immunity**
Anticoagulant: Compound that slows or prevents clot formation by interfering with the clotting system, 361
Anticodon: Triplet of nitrogenous bases on a tRNA molecule that interacts with an appropriate codon on a strand of mRNA, 72–74
Antidiuretic hormone (ADH) (an-tī-dī-ū-RET-ik): Hormone synthesized in the hypothalamus and secreted at the posterior pituitary; causes water retention at the kidneys and an elevation of blood pressure, 252, 315, 318–319, 322–323, 331, 402–404, 557–561, 573
Antigen: A substance capable of inducing the production of antibodies, 352, 430, 446
complete, 442
Antigen-antibody complex: The combination of an antigen and a specific antibody, 442–444
Antigen binding site, 442
Antigenic determinant site: A portion of an antigen that can interact with an antibody molecule, 442
Antigen-presenting cell (APC): A cell that processes antigens and displays them, bound to MHC proteins; essential to the initiation of a normal immune response, 439–440
Antigen recognition, 438
Anti-inflammatory effect, of glucocorticoids, 327
Antioxidant, 328
Antiperspirant, 117
Antipyretic drug, 540
Antrum (AN-trum): A chamber or pocket, 592
Anus: External opening of the anorectal canal, 508–509, 596
Aorta: Large, elastic artery that carries blood away from the left ventricle and into the systemic circuit, 372–373, 390–391, 411, 418, 508, 549
abdominal, 327, 412, 503
ascending, 406–408
descending, 405, 407–412
thoracic, 412
Aortic arch, 370, 375, 406–407
arteries of, 406–409
Aortic body, 282, 401, 478
Aortic reflex: Baroreceptor reflex triggered by increased aortic pressures; leads to a reduction in cardiac output and a fall in systemic pressure.
Aortic semilunar valve, 372–373
Aortic sinus, 281, 374, 400, 479
APC (antigen-presenting cell), 439–440
Apex
of heart, 369
of lung, 466–467
of sacrum, 148
Apgar rating, 638
Aphasia: Inability to speak, 248, 270
Apnea, 481
Apocrine secretion: Mode of secretion in which the glandular cell sheds portions of its cytoplasm, 91
Apocrine sweat gland, 89, 116–117
Aponeurosis/aponeuroses (ap-ō-nū-RŌ-sēz): A broad tendinous sheet that may serve as the origin or insertion of a skeletal muscle, 178
Apoptosis, 75, 440
Appendicitis, 432, 449, 508
Appendicular muscles, 198, 204–216
Appendicular skeleton, 6, 139, 150–166
bones of lower limb, 155–156

bones of upper limb, 151–153
pectoral girdle, 150–151
pelvic girdle, 153–155
Appendix: A blind tube connected to the cecum of the large intestine, 432, 508
Appetite, 332
Appositional growth: Enlargement by the addition of cartilage or bony matrix to the outer surface, 132–133
Aqueous humor: Fluid similar to perilymph or CSF that fills the anterior chamber of the eye, 288, 290–291, 567
Arachidonic acid, 315
Arachnoid (a-RAK-noyd): The middle meninges that encloses CSF and protects the central nervous system, 239–240, 242, 246
Arachnoid granulations, 245–246
Arcuate (AR-kū-āt): Curving.
Arcuate artery, 553–554
Arcuate vein, 554
Areola (a-RĒ-ō-la): Pigmented area that surrounds the nipple of a breast, 597
Areolar: Containing minute spaces, as in areolar connective tissue, 92, 94–95, 99
Arginine, 531
Arm. *See* **Upper limb**
Arousal, 596
Arrector pili (a-REK-tōr PĪ-lī): Smooth muscles whose contractions cause erection of hairs, 110, 115–117, 264–265, 267
Arrhythmias (a-RITH-mē-az): Abnormal patterns of cardiac contractions, 36, 379–380, 383
ART (assisted reproductive technology), 618
Arterial pressure. *See* **Blood pressure**
Arterial puncture, 346
Arteries: Blood vessels that carry blood away from the heart and toward a peripheral capillary, 8, 96, 368, 390. *See also specific arteries*
anatomy of, 391–392
of aortic arch, 406–408
elastic, 391–392
to head, chest, and upper limbs, 409
muscular, 391–392
to neck, head, and brain, 410
structure of vessel walls, 390–391
systemic, 406–412
of trunk, 411–412
Arteriole (ar-TĒ-rē-ōl): A small arterial branch that delivers blood to a capillary network, 390–392, 395, 400
Arteriosclerosis, 268, 393, 421
Arteriovenous anastomosis, 392–393
Arthritis (ar-THRĪ-tis): Inflammation of a joint, 159, 168, 470
Arthroscopic surgery, 168
Articular: Pertaining to a joint.
Articular capsule. *See* **Joint capsule**
Articular cartilage: Cartilage pad that covers the surface of a bone inside a joint cavity, 129, 132, 158–159
Articular facet, 146
Articular process, 146, 148, 163
Articulations. *See* **Joint**
Artificial respiration, 468
Arytenoid cartilage, 461
Ascending tract: A tract carrying information from the spinal cord to the brain, 242
Ascites, 490, 514
Ascorbic acid (vitamin C), 133, 536
Aspiration, 463
Assisted reproductive technology (ART), 618
Association areas: Cortical areas of the cerebrum responsible for integration of sensory inputs and/or motor commands, 247
Association neuron. *See* **Interneuron**

Asthma (AZ-ma): Reversible constriction of smooth muscles around respiratory passageways, frequently caused by an allergic response, 113, 463, 471, 481
Astigmatism: Visual disturbance due to an irregularity in the shape of the cornea, 293
Astrocyte (AS-trō-sīt): One of the glial cells in the CNS; responsible for the blood-brain barrier, 229
Ataxia, 253, 270
Atelectasis, 467, 481
Atherosclerosis (ath-er-ō-skle-RŌ-sis): Formation of fatty plaques in the walls of arteries, leading to circulatory impairment, 393, 404, 419, 421, 530, 537, 601
Athletes
amenorrhea in, 595
blood doping, 351
carbohydrate loading, 527
cardiovascular fitness and, 403–404
endocrinology and athletic performance, 332
Atlas, 147
Atmospheric pressure, 471
Atom: The smallest stable unit of matter, 28–32
Atomic number: The number of protons in the nucleus of an atom, 28
Atomic structure, 28–29
Atomic weight: Roughly, the average total number of protons and neutrons in the atoms of a particular element, 29
ATP (adenosine triphosphate): A high-energy compound consisting of adenosine with three phosphate groups attached; the third is attached by a high-energy bond, 44–46, 69, 522
generation of, 69–70, 523
production in amino acid catabolism, 529–531
production in carbohydrate catabolism, 524–527
production in electron transport system, 526–527
production in lipid catabolism, 528
production in tricarboxylic acid cycle, 525–527
use in active transport, 62
use in muscle contraction, 184, 186, 189–192
use in vision, 295
Atresia, 591
Atria: Thin-walled chambers of the heart that receive venous blood from the pulmonary or systemic circuit, 369
left, 368–370, 372, 405–406, 418–419
right, 331, 368–370, 372, 405, 416, 418–419
Atrial diastole, 380–381
Atrial natriuretic peptide (ANP) (nā-trē-ū-RET-ik): Hormone released by specialized atrial cardiocytes when they are stretched by an abnormally large venous return; promotes fluid loss and reductions in blood pressure and venous return, 315, 331, 402–403, 560–562, 569
Atrial reflex: Reflexive increase in heart rate following an increase in venous return; due to mechanical and neural factors; also called *Bainbridge reflex*, 382, 403
Atrial systole, 380–381
Atrioventricular bundle. *See* **Bundle of His**
Atrioventricular (AV) node (ā-trē-ō-ven-TRIK-ū-lar): Specialized cardiocytes that relay the contractile stimulus to the AV bundle, the bundle branches, the Purkinje fibers, and the ventricular myocardium; located at the boundary between the atria and ventricles, 377–378, 401
Atrioventricular (AV) valve: One of the valves that prevent backflow into the atria during ventricular systole, 372–374, 380
left. *See* **Bicuspid valve**
right. *See* **Tricuspid valve**

Atrophy (AT-rō-fē): Wasting away of tissues from lack of use, ischemia, or nutritional abnormalities.
of skeletal muscle, 189
Auditory: Pertaining to the sense of hearing.
Auditory association area, 247
Auditory cortex, 247–248, 278, 304–305
Auditory ossicles: The bones of the middle ear: malleus, incus, and stapes, 137, 139–140, 143, 297–298, 301, 303
Auditory pathways, 304–305
Auditory sensitivity, 304
Auditory tube, 96, 297–298, 460, 491
Auricle (atrium), 369–370
Auricle (ear): The expanded, projecting portion of the external ear that surrounds the external auditory canal, 96, 297
Auscultation, 23
Autoantibody, 446
Autoimmunity: Immune system sensitivity to normal cells and tissues, resulting in the production of autoantibodies, 446–447
Autolysis: Destruction of a cell due to the rupture of lysosomal membranes in its cytoplasm, 69
Automatic bladder, 564
Automaticity: Spontaneous depolarization to threshold, a characteristic of cardiac pacemaker cells, 194, 376
Autonomic ganglion: A collection of visceral motor neurons outside the CNS.
Autonomic nerve: A peripheral nerve consisting of preganglionic or postganglionic autonomic fibers.
Autonomic nervous system (ANS): Centers, nuclei, tracts, ganglia, and nerves involved in the unconscious regulation of visceral functions; includes components of the CNS and PNS, 226–227, 261–267
innervation of heart, 381–382
parasympathetic division of, 226–227, 264–267
relationship between divisions of, 267
sympathetic division of, 226–227, 264–265, 267, 318
Autoregulation: Alterations in activity that maintain homeostasis in direct response to changes in the local environment; does not require neural or endocrine control.
of blood flow, 393, 399–400
Autorhythmicity. *See* **Automaticity**
Autosomal (aw-to-SŌ-mal): Chromosomes other than the X or Y chromosomes, 632–635
Avascular (ā-VAS-kū-lar): Without blood vessels, 84
Avitaminosis, 534–535, 540
AV node. *See* **Atrioventricular node**
AV valve. *See* **Atrioventricular valve**
Axial muscles, 198–204
Axial skeleton, 6, 137, 139–150
bones of neck and trunk, 145–150
skull, 140–145
Axilla: The armpit, 15–16
Axillary artery, 405, 407–409, 411
Axillary lymph nodes, 428
Axillary nerve, 257–258
Axillary region, 15–16
Axillary vein, 405, 413–416
Axis (vertebrae), 147
Axon: Elongate extension of a neuron that conducts an action potential, 102–103, 178–179, 227–228
action potential, 233
myelinated, 229–230, 234–235
unmyelinated, 230, 234–235
Axon hillock: In a multipolar neuron, the thickened portion of the neural cell body where an action potential begins, 227–228, 233
Axoplasm (AK-sō-plazm): Cytoplasm within an axon.
Azygos vein, 414–416, 430

B

Babinski sign, 260
Baby teeth, 493–494
Bacteria: Single-celled microorganisms, some pathogenic, that are common in the environment.
intestinal, 509, 513, 536
Bainbridge reflex (atrial reflex), 382, 403
Balance, 258
Balanced diet, 532–533
Baldness, male pattern, 116, 122
Ball-and-socket joint, 161–163
Balloon angioplasty, 383
Band cells, 358
Barbiturate, 476
Barium-contrast X-ray, 21
Baroreceptor reflex: A reflexive change in cardiac activity in response to changes in blood pressure, 400–401
Baroreceptors (bar-ō-rē-SEP-torz): Receptors responsible for baroreception, 279, 281, 404, 477, 479
Basal cell carcinoma, 113, 122
Basal cells, 282–283
Basal lamina, 85
Basal metabolic rate (BMR): The resting metabolic rate of a normal fasting subject under homeostatic conditions, 537
Basal nuclei: Nuclei of the cerebrum that are important in the subconscious control of skeletal muscle activity, 246, 250–251
Base: A compound whose dissociation releases a hydroxide ion (OH^-) or removes a hydrogen ion from the solution, 47
functions in body, 47
inorganic, 36
strong, 34
weak, 34, 570
Base (anatomical)
of heart, 369
of sacrum, 148
Basement membrane: A layer of filaments and fibers that attach an epithelium to the underlying connective tissue, 84–85, 87–88, 111
Basic food groups, 533–534
Basic solution, 34
Basilar artery, 409–410
Basilar membrane: Membrane that supports the organ of Corti and separates the cochlear duct from the tympanic duct in the inner ear, 302–304
Basilar zone, of endometrium, 595
Basilic vein, 413–415
Basophils (BĀ-sō-filz): Circulating granulocytes (WBCs) similar in size and function to tissue mast cells, 345, 355–358, 442
B cells: Lymphocytes capable of differentiating into the plasma cells that produce antibodies, 430–431, 438–439, 441–444, 446–447
memory, 441, 444, 446
sensitized, 441
Behavior, hormones and, 333–335
Behavioral drives, 251
Belt desmosome, 85–86
Bends (decompression sickness), 472
Benign tumor, 77
Beriberi, 536
Beta cells: Cells of the pancreatic islets that secrete insulin in response to elevated blood sugar concentrations, 329–330
Beta-oxidation: Fatty acid catabolism that produces molecules of acetyl-CoA, 528
Bicarbonate ions: HCO_3^-; anion components of the carbonic acid-bicarbonate buffer system, 30, 474–475, 569–572
in blood, 555
in body fluids, 567
in urine, 555

Bicarbonate reserve, 571
Biceps brachii muscle, 152, 164, 189, 195–196, 206, 208–210
Biceps femoris muscle, 197, 213–214
Bicuspid (bī-KUS-pid): A premolar tooth, 493–494
Bicuspid valve: The left AV valve; also known as the *mitral valve*, 372–374
Bilateral orchiectomy, 584
Bile: Exocrine secretion of the liver that is stored in the gallbladder and ejected into the duodenum, 349, 505–507
Bile canaliculi, 505
Bile duct, 505
Bile salts: Steroid derivatives in the bile, responsible for the emulsification of ingested lipids, 506–507, 509, 511–512
Biliary obstruction, 513
Bilirubin (bil-ē-ROO-bin): A pigment that is the byproduct of hemoglobin catabolism, 349–350, 506, 509
Biliverdin, 349–350
Bioenergetics, 522, 537–539
Biology, 2
Biopsy: The removal of a small sample of tissue for pathological analysis, 122
liver, 514
Biotin, 509, 536
Biphosphate, 30
Bipolar cells, 289–290, 294–295
Bipolar neuron, 228
Birth control, 604–605
Bitter taste, 284
Bladder: A muscular sac that distends as fluid is stored, and whose contraction ejects the fluid at an appropriate time; used alone, the term usually refers to the urinary bladder.
urinary. *See* **Urinary bladder**
Blast cells, 358
Blastocoele, 616–619
Blastocyst (BLAS-tō-sist): Early stage in the developing embryo, consisting of an outer trophoblast and an inner cell mass, 616–618
Blastomere, 616
Bleaching, of visual pigments, 295
Blindness, 292, 306
Blind spot, 290
Blood, 8, 92, 94–96, 344–365
age-related changes in, 419
carbon dioxide in, 282, 401, 478, 570
collection and analysis of, 345
composition of, 344–346, A-15
formed elements in, 344–359
functions of, 344
oxygen in, 282, 401, 478
pH of, 34–35, 282, 346, 401, 478, 531, 548, 556, 570
plasma. *See* **Plasma**
reservoir in liver, 506
temperature of, 346
viscosity of, 346, 395
Blood-brain barrier: Isolation of the CNS from the general circulation; primarily the result of astrocyte regulation of capillary permeabilities, 229
Blood cells, 54
production of, 128
Blood clot: A network of fibrin fibers and trapped blood cells. *See* **Clot**
Blood clotting. *See* **Clotting system**
Blood doping, 351
Blood flow
autoregulation of, 393, 399–400
to brain, 400, 408–410
to kidneys, 553–554, 560
regulation of, 398–403
to skeletal muscles, 178–179

Blood pressure: A force exerted against the vascular walls by the blood, as the result of the push exerted by cardiac contraction and the elasticity of the vessel walls; usually measured along one of the muscular arteries, with systolic pressure measured during ventricular systole, diastolic pressure during ventricular diastole, 61, 96, 281, 321, 326, 331, 374, 383, 392, 394–396, 556, 558, 560, 569

  baroreceptor reflex and, 400–401
  elevation with hemorrhaging, 404
  low, 401
  measurement of, 398
  neural control of, 400–402
  regulation of, 548

Blood smear, 345
Blood tests, 351–352
Blood type, 352–354
Blood vessels, 8, 390–425. *See also* **Arteries; Capillary; Circulation; Vein**

  aging of, 419
  anatomy of, 390–394
  innervation of, ANS, 267
  structure of vessel walls, 390–391
  in thermoregulation, 12–13

Blood volume, 326, 331, 404, 429, 548, 569
  maternal, 627
  restoration after hemorrhage, 404–405

Blood volume reflexes, 381–382
Blue baby, 372
Blue cones, 294
BMR. *See* **Basal metabolic rate**
Body
  of nail, 117
  of pancreas, 503
  of penis, 588–589
  of sternum, 149–150
  of stomach, 496
  of uterus, 594

Body cavities, 18–21
  dorsal, 19–20
  ventral, 19–21

Body odor, 117
Body position, 178
Body proportions, 132–133
Body stalk, 620–621, 623
Body temperature, 12–13. *See also* **Thermoregulation**
  basal, 599–600
  heart rate and, 383
  hormones and, 599–600
  of neonate, 631
  normal, 12
  respiratory rate and, 476
  sperm development and, 583

Bolus: A compact mass; usually refers to compacted ingested material on its way to the stomach, 494

Bonds, chemical
  covalent, 31
  high-energy, 44–45
  hydrogen, 31–32
  ionic, 30–31, 36

Bone, 6, 92–93, 98. *See also specific bones*
  cells in, 130–131
  compact, 129–130, 136, 158
  compared to cartilage, 99
  development and growth of, 131–133
  of face, 143–144
  flat, 128–129
  fracture of, 134–136, 168
  growth of
    body proportions and, 132–133
    requirements for, 132–133
  irregular, 128–129
  long, 128–129
  macroscopic features of, 128–129

  microscopic features of, 129–131
  mineral storage, 133–134
  organization of, 98
  remodeling of, 133–136
  resorption of, 130
  shapes of, 128
  short, 128–129
  spongy (cancellous), 129–130, 136, 158

Bone markings, 137
Bone marrow, 6, 128–129, 132, 347, 350–351, 357, 359, 431
  transplantation of, 362, 449

Bone mass, 136–137, 325, 514
Bony labyrinth, 298
Bony palate. *See* **Hard palate**
Botulism, 184, 217
Bowel: The intestinal tract.
Bowman's capsule, 551–553
Brachial: Pertaining to the arm, 15–16
Brachial artery, 398, 405, 407–409
Brachialis muscle, 196, 209–210
Brachial plexus: Network formed by branches of spinal nerves C_5 — T_1 en route to innervating the upper limb, 257–258
Brachial region, 15–16
Brachial vein, 405, 413–416
Brachiocephalic artery, 406
Brachiocephalic trunk, 407–410
Brachiocephalic vein, 413–416, 430
Brachioradialis muscle, 196–197, 209–210
Brachium: The arm, 15–16
Brachydactyly, 633
Bradycardia, 378, 383
Brain, 7, 227, 243–253
  basal nuclei, 250–251
  blood flow to, 268, 400, 408–410
  cerebellum, 243–245, 253
  cerebrum, 243–249
  diencephalon, 243–244, 252
  limbic system, 251
  major divisions of, 243–244
  medulla oblongata, 243–245, 253
  memory, 250
  midbrain, 243–244, 252–253
  pons, 243–245, 253
  ventricles of, 230, 245
  weight of, 243, 245, 268

Brain stem: The midbrain, pons, and medulla oblongata (excludes the cerebrum, diencephalon, and cerebellum), 243–244, 253
Brain waves, 249
Breast, 596–597. *See also* **Mammary glands**
  cancer of, 597
Breast milk, 631
Breath-holding, 478
Breathing, modes of, 470, 476
Breech birth, 638
Brevis: Short.
Broca's center: The speech center of the brain, usually found on the neural cortex of the left cerebral hemisphere, 248–249
Bronchial artery, 411–412
Bronchial tree: The trachea, bronchi, and bronchioles, 463–464
Bronchiole, 458, 463–464
  respiratory, 463–465
  terminal, 463–464
Bronchitis (brong-KĪ-tis)**:** Inflammation of the bronchial passageways, 481
Bronchoconstriction, 463
Bronchodilation: Dilation of the bronchial passages; may be caused by sympathetic stimulation, 463
Bronchoscope, 481

Bronchus/bronchi: One of the branches of the bronchial tree between the trachea and bronchioles, 9, 458, 463
  primary, 462–464
  secondary, 462–464
  tertiary, 463–464
Buccal (BUK-al)**:** Pertaining to the cheeks.
Buccal cavity, 490
Buccinator muscle, 199–200
Buffer: A compound that stabilizes the pH of a solution by removing or releasing hydrogen ions, 34–35, 570–571
  in body, 41
Buffer system: Interacting compounds that prevent increases or decreases in the pH of body fluids; includes the carbonic acid–bicarbonate buffer system, the phosphate buffer system, and the protein buffer system, 570–571
Bulbospongiosus muscle, 204–205, 596, 600–601
Bulbourethral glands (bul-bō-ū-RĒ-thral)**:** Mucous glands at the base of the penis that secrete into the penile urethra; also called *Cowper's glands*, 582, 587–588, 600
Bulimia, 501
Bundle branches: Specialized conducting cells in the ventricles that carry the contractile stimulus from the AV bundle to the Purkinje fibers, 377–378
Bundle of His (hiss)**:** Specialized conducting cells in the interventricular septum that carry the contracting stimulus from the AV node to the bundle branches and thence to the Purkinje fibers; also called *AV bundle*, 377–378
Burkitt's lymphoma, 636
Burning pain, 279
Burns, skin, 118, 120
Bursa: A small sac filled with synovial fluid that cushions adjacent structures and reduces friction, 158–159, 166
Bursitis, 164, 168, 208
Buttock, 15–16
Button desmosome, 85–86

C

CABG (coronary artery bypass graft), 384
CAD (coronary artery disease), 383, 393, 530, 634–635
Caffeine, 476, 498
Calcaneal tendon: Large tendon that inserts on the calcaneus; tension on this tendon produces plantar flexion of the foot; also called *Achilles tendon*, 156, 197, 214–215
Calcaneus (kal-KĀ-nē-us)**:** The heelbone, the largest of the tarsal bones, 156–157, 195, 197
Calcification: The deposition of calcium salts within a tissue, 131, 393
Calcitonin (kal-si-TŌ-nin)**:** Hormone secreted by C cells of the thyroid when calcium ion concentrations are abnormally high; restores homeostasis by increasing the rate of bone deposition and the renal rate of calcium loss, 134, 315, 317, 324–325
Calcitriol, 113, 133–134, 315, 325, 331, 333, 513, 540, 548
Calcium, 28, 30, 36, 534–535
  absorption of, 133, 331, 513
  in blood, 317, 548
  in body fluids, 134, 325
  cardiac output and, 383
  in clotting process, 361
  homeostatic regulation of, 325
  mineral storage in bone, 133–134
  in muscle contraction, 181–182, 184–185, 195, 376
  requirement for, 132–133, 540
  as second messenger, 316
  in synaptic function, 236–237

Cation (KAT-ī-on): An ion that bears a positive charge, 30
CAT scan, 21–23
Cauda equina, 240–241
Caudal/caudally: Closest to or toward the tail (coccyx), 17
Caudate nucleus, 250
Cavernous hemangioma, 122
Cavities, body, 18–21
CBC (complete blood count), 352
C cells, 324–325
CCK. See Cholecystokinin
CD4, 443
Cecum (SĒ-kum): An expanded pouch at the start of the large intestine, 499, 507–508
Celiac artery, 407
Celiac trunk, 411–412, 503
Cell: The smallest living unit in the human body, 2–4, 54. See also specific cell types
    chemicals in, 46–47
    components of, 55–56
    sizes and shapes of, 54
    somatic, 74
    study of, 54–55
Cell body, of neuron, 102, 227–228
Cell diversity, 54, 77
Cell division, 74–77
    cancer and, 77
Cell junctions, 85–86
Cell life cycle, 74–77
Cell-mediated immunity: Resistance to disease through the activities of sensitized T cells that destroy antigen-bearing cells by direct contact or through the release of lymphotoxins; also called cellular immunity, 430–431, 438–441, 446
Cell membrane, 40, 55–57, 66, 86, 98
    functions of, 57
    hormones and, 315–317
    movement of substances across, 58–65
    permeability of, 58
    structure of, 57–58
Cell physiology, 3
Cell theory, 54
Cellular immunity. See Cell-mediated immunity
Cellular level of organization, 3–5
Cellular metabolism, 522–532
Cellular respiration, 524
Cellular trophoblast, 617–619, 621
Cellulose, 37
Cementum, 493
Center, neural, 231
Central canal, 98, 129–130, 230, 240, 242, 244, 246
Central nervous system (CNS): The brain and spinal cord, 7, 238–253
    anatomical organization of, 231
    brain, 243–253
    functions of, 226–227
    meninges, 238–240
    spinal cord, 240–243
Central sulcus: Groove in the surface of a cerebral hemisphere, between the primary sensory and primary motor areas of the cortex, 244, 246–247
Central vein, 505
Centriole: A cylindrical intracellular organelle composed of 9 groups of microtubules, 3 in each group; functions in mitosis or meiosis by organizing the microtubules of the spindle apparatus, 55–56, 66, 76, 227
Centromere (SEN-trō-mēr): Localized region where two chromatids remain connected following chromosome replication; site of spindle fiber attachment, 76

Cephalic (anatomical direction), 17
Cephalic phase, of gastric secretion, 497–498
Cephalic region, 15–16
Cephalic vein, 413–416
Cephalon, 15–16
Cerebellar cortex, 253
Cerebellar peduncle, 252–253
Cerebellum (ser-e-BEL-um): Posterior portion of the metencephalon, containing the cerebellar hemispheres; includes the arbor vitae, cerebellar nuclei, and cerebellar cortex, 243–245, 253
Cerebral aqueduct. See Mesencephalic aqueduct
Cerebral arterial circle, 409–410
Cerebral arteries, 410
Cerebral cortex: An extensive area of neural cortex covering the surfaces of the cerebral hemispheres.
    cortical connections, 247
    motor and sensory areas of, 246–247
Cerebral hemispheres: Expanded portions of the cerebrum covered in neural cortex, 243–244
    hemispheric lateralization, 249
    structure of, 246
Cerebral meningitis, 240, 270
Cerebral palsy, 262–263
Cerebral peduncle, 252–253
Cerebrospinal fluid (CSF): Fluid bathing the internal and external surfaces of the CNS; secreted by the choroid plexus, 230, 240, 244–245, 567
    carbon dioxide in, 282, 401
    circulation of, 245–246
    composition of, 245, A-15
    oxygen in, 401
    pH of, 282, 401
    production of, 245
Cerebrovascular accident. See Stroke
Cerebrum (SER-e-brum or ser-Ē-brum): The largest portion of the brain, composed of the cerebral hemispheres; includes the cerebral cortex, the cerebral nuclei, and the white matter, 243–249. See also Cerebral hemispheres
    association areas, 247
    motor and sensory areas of, 246–247
    processing centers, 247–248
Cerumen, 117, 297
Ceruminous gland, 117, 297
Cervical canal, 594, 628
Cervical cap, 605
Cervical curve, 146
Cervical enlargement, 240–241
Cervical gland, 601
Cervical lymph nodes, 428
Cervical os, 594
Cervical plexus, 257–258
Cervical plug, 622
Cervical region, 15–16
Cervical spinal nerves, 241, 257
Cervical vertebrae, 145–147, 242
Cervicis, 15–16
Cervix: The lower part of the uterus, 591, 594
    cancer of, 606
    Pap test, 89
Cesarean section, 628–629
CF (cystic fibrosis), 460, 481, 635, 637
cGMP. See Cyclic-GMP
Channel protein, 57–58, 86
Cheekbone, 140–141, 144
Cheeks, 490–491
Chemical bonds. See Bonds, chemical
Chemical level of organization, 3–4
Chemical notation, 32–33
Chemiosmosis, 526

Chemistry, 28
Chemoreception: Detection of alterations in the concentrations of dissolved compounds or gases.
Chemoreceptor, 278, 281–282, 498
Chemoreceptor reflex, 400–402, 478–479
Chemotaxis (kē-mō-TAK-sis): The attraction of phagocytic cells to the source of abnormal chemicals in tissue fluids, 355, 436
Chemotherapy, 104
Chest
    arteries to, 409
    veins of, 414–415
Chest wall, 468–469
Chewing, 492
CHF (congestive heart failure), 397
Chief cells
    of parathyroid gland, 324–325
    of stomach, 497–498
Childbirth, 321, 596, 628–630
    epidural block, 239
Childhood, 630–631
Chlamydia, 594, 601
Chloride, 30, 36, 534–535
    in blood, 548, 555
    in body fluids, 567
    in urine, 555
Chlorine, 28
Cholecalciferol, 534–535
Cholecystitis, 514
Cholecystokinin (CCK) (kō-lē-sis-tō-KĪ-nin): Duodenal hormone that stimulates the contraction of the gallbladder and the secretion of enzymes by the exocrine pancreas, 498, 501–502, 504, 507
Cholelithiasis, 514
Cholera, 510
Cholesterol: A steroid component of cell membranes and a substrate for the synthesis of steroid hormones and bile salts, 39–40, 47
    in blood, 40, 393, 404, 529–530
    dietary, 40, 530
    functions of, 40, 530
    membrane, 40, 57
    structure of, 40
    synthesis of, 40
Cholestyramine, 530
Cholinergic fiber, 264
Cholinergic synapse (kō-lin-ER-jik): Synapse where the presynaptic membrane releases ACh on stimulation, 236–237
Cholinesterase (kō-li-NES-te-rās): Enzyme that breaks down and inactivates ACh, 182–183, 186, 236–237
Chondrocyte (KON-drō-sīt): Cartilage cell, 96–97, 104, 131–132, 136
Chordae tendineae (KOR-dē TEN-di-nē-ē): Fibrous cords that stabilize the position of the AV valves in the heart, preventing backflow during ventricular systole, 372–374
Chorion/chorionic (KOR-ē-on) (ko-rē-ON-ik): An extraembryonic membrane, consisting of the trophoblast and underlying mesoderm, that forms the placenta, 620–622
Chorionic villi, 620–623
Chorionic villus sampling, 636, 638
Choroid: Middle, vascular layer in the wall of the eye, 287–289
Choroid plexus: The vascular complex in the roof of the third and fourth ventricles of the brain, responsible for CSF production, 245–246, 252
Chromatid (KRŌ-ma-tid): One complete copy of a single chromosome, 76, 584
Chromatin, 55, 70–71
Chromosomal abnormality, 636

Cone: Retinal photoreceptor responsible for color vision, 289–290, 294–295

Congenital (kon-JEN-i-tal): Already present at the birth of an individual.

Congenital heart defect, 419

Congenital malformation, 638

Congestive heart failure (CHF), 397

Conjunctiva (kon-junk-TĪ-va): A layer of stratified squamous epithelium that covers the inner surfaces of the lids and the anterior surface of the eye to the edges of the cornea, 285, 287

Conjunctivitis, 285

Connective tissue: One of the four primary tissue types; provides a structural framework for the body that stabilizes the relative positions of the other tissue types; includes connective tissue proper, cartilage, bone, and blood; always has cell products, cells, and ground substance, 84, 92–99

cells of, 93

classification of, 92

connective tissue fibers, 93

connective tissue proper, 92–94

fluid, 92, 94–96

functions of, 92

ground substance, 93–94

supporting, 92–93, 96–98

Consciousness, 243

Conscious thought, 245

Constant segment, of antibody, 441–442

Constipation, 510, 514

Contact dermatitis, 122

Continuous propagation, action potentials, 235

Contraception, 604–605

Contractile proteins, 40

Contractility: The ability to contract, possessed by skeletal, smooth, and cardiac muscle cells. *See* **Muscle contraction**

Contraction phase, of muscle action, 187

Control center, 5

Control segment, of DNA, 71

Convection, 538–539

Convergence: In the nervous system, the term indicates that the axons from several neurons innervate a single neuron; most common along motor pathways, 238

Convulsion, 325

COP (capillary osmotic pressure), 396–397

COPD (chronic obstructive pulmonary disease), 481

Copper, 535

Coracobrachialis muscle, 206–209

Coracoid process (ko-RA-kōyd): A hook-shaped process of the scapula that projects above the anterior surface of the capsule of the shoulder joint, 151, 164

Cornea (KOR-nē-uh): Transparent portion of the fibrous tunic of the anterior surface of the eye, 285, 287–288

scarring of, 292

transplantation of, 288

Corniculate cartilage, 461

Cornification: The production of keratin by a stratified squamous epithelium; also called *keratinization*, 112

Coronal plane. *See* **Frontal plane**

Coronal suture, 140–141, 145

Corona radiata (ko-RŌ-na rā-dē-A-ta): A layer of follicle cells surrounding a secondary oocyte at ovulation, 592–593, 614–616

Coronary angiogram, 384

Coronary arteriography, 384

Coronary artery, 370, 374–375, 406

Coronary artery bypass graft (CABG), 384

Coronary artery disease (CAD), 383, 393, 530, 634–635

Coronary bypass surgery, 414

Coronary circulation, 374–375

Coronary ischemia, 383

Coronary ligament, 504

Coronary sinus, 372, 375

Coronary sulcus, 369–370

Coronary thrombosis, 383

Coronoid (kō-RŌ-noyd): Hooked or curved.

Coronoid fossa, 152, 164

Coronoid process, 140, 144, 152–153, 164

Corpora cavernosa, 588–589

Corpus/corpora: Body.

Corpus callosum: Bundle of axons linking centers in the left and right cerebral hemispheres, 244, 247, 249–251

Corpus luteum (LOO-tē-um): Progestin-secreting mass of follicle cells that develops in the ovary after ovulation, 332, 593, 595, 598–600, 622

Corpus spongiosum, 588–589

Corpus striatum, 251

Cortex: Outer layer or portion of an organ.

adrenal. *See* **Adrenal cortex**

cerebral. *See* **Cerebral cortex**

of lymph node, 432

of ovary, 590

renal. *See* **Renal cortex**

of thymus, 433

Corti, organ of. *See* **Organ of Corti**

Corticospinal tracts: Descending tracts that carry motor commands from the cerebral cortex to the anterior gray horns of the spinal cord, 261–262

Corticosteroid: A steroid hormone produced by the adrenal cortex, 326–328, 335

Corticosterone, 315, 327

Corticotropin-releasing hormone (CRH), 320

Cortisol (KOR-ti-sol): One of the corticosteroids secreted by the adrenal cortex; a glucocorticoid, 315, 327

Cortisone, 315, 327

Costa/costae: A rib. *See* **Ribs**

Costal cartilage, 149

Costal surface, of lung, 466–467

Cotransport: Membrane transport of a nutrient, such as glucose, in company with the movement of an ion, usually sodium; transport requires a carrier protein but does not involve direct ATP expenditure and can occur regardless of the concentration gradient for the nutrient, 61–62, 511–512

Coughing, 463

Countertransport, 62

Covalent bond (kō-VĀ-lent): A chemical bond between atoms that involves the sharing of electrons, 31

double, 31

nonpolar, 31

polar, 31

single, 31

Cowper's gland, 582, 587–588, 600

Coxa/coxae: A bone of the hip; formed by the fusion of the ilium, ischium, and pubis, 138, 148, 153–154

Coxal bone, 139

CP. *See* **Creatine phosphate**

CPK. *See* **Creatine phosphokinase**

CPR. *See* **Cardiopulmonary resuscitation**

Cranial (anatomical direction), 17

Cranial cavity, 19–20, 140

Cranial meninges, 239

Cranial nerves: The twelve peripheral nerves originating at the brain, 231, 254–256

I. *See* **Olfactory nerve**

II. *See* **Optic nerve**

III. *See* **Oculomotor nerve**

IV. *See* **Trochlear nerve**

V. *See* **Trigeminal nerve**

VI. *See* **Abducens nerve**

VII. *See* **Facial nerve**

VIII. *See* **Vestibulocochlear nerve**

IX. *See* **Glossopharyngeal nerve**

X. *See* **Vagus nerve**

XI. *See* **Accessory nerve**

XII. *See* **Hypoglossal nerve**

Cranium: The brainbox; the skull bones that surround the brain, 19, 139–143

Creatine, 190

Creatine phosphate (CP): A high-energy compound present in muscle cells; during muscular activity, the phosphate group is donated to ADP, regenerating ATP; also called *phosphocreatine* or *phosphorylcreatine*, 190–192, 555

Creatine phosphokinase (CPK, CK), 190

Creatinine: A breakdown product of creatine metabolism, 555

Cremaster muscle, 583

Crenation: Cellular shrinkage due to an osmotic movement of water out of the cytoplasm, 61

Crest (bone), 136–137

Cretinism, 335–336

CRH (corticotropin-releasing hormone), 320

Crib death, 478

Cribriform plate: Portion of the ethmoid bone of the skull that contains the foramina used by the axons of olfactory receptors en route to the olfactory bulbs of the cerebrum, 142–143, 283

Cricoid cartilage (KRĪ-koyd): Ring-shaped cartilage forming the inferior margin of the larynx, 459, 461

Crista/cristae: A ridge-shaped collection of hair cells in the ampulla of a semicircular canal; the crista and cupula form a receptor complex sensitive to movement along the plane of the canal, 69

Crista galli, 142–143

Cross-bridge: Myosin head that projects from the surface of a thick filament and that can bind to an active site of a thin filament in the presence of calcium ions, 181–182, 185–186

Crossing-over, 585, 636

Cross-match test, 354

Cross-reaction, 353–354

Cross section, 18

Crown, of tooth, 493

Crude touch and pressure receptor, 279

Crural region, 15–16

Cryptorchidism, 584, 606

C-section, 628–629

CSF. *See* **Cerebrospinal fluid; Colony-stimulating factor**

CT (computed tomography), 21–23

Cubital vein, 415

Cuboidal epithelium, 87–88

Cuboid bone, 156–157

Cuneiform bones, 156–157

Cuneiform cartilage, 461

Cupula (KŪ-pū-la): A gelatinous mass, in the ampulla of a semicircular canal in the inner ear, whose movement stimulates the hair cells of the crista, 300–301

Curvature, spinal, 146

Cushing's disease, 335–336

Cusp, 372

Cuspid, 493–494

Cutaneous membrane: The epidermis and papillary layer of the dermis, 6, 99–100. *See also* **Skin**

Cuticle: Layer of dead, cornified cells surrounding the shaft of a hair; for nails, 118

Endochondral ossification (en-dō-KON-dral): The conversion of a cartilaginous model to bone; the characteristic mode of formation for skeletal elements other than the bones of the cranium, the clavicles, and sesamoid bones, 131–132

Endocrine cells, 314

Endocrine gland: A gland that secretes hormones into the blood. *See specific glands*

Endocrine pancreas, 329

Endocrine secretions, 85, 89

Endocrine system, 226, 314–341. *See also specific glands and hormones*
  aging of, 335
  athletic performance and, 332
  cardiovascular regulation by, 399, 402–403
  components of, 8
  control of endocrine activity, 317–319
  disorders of, 335
  functions of, 8
  integration with other systems, 336–337
    cardiovascular system, 420
    digestive system, 515
    integumentary system, 121
    lymphatic system, 448
    muscular system, 218
    nervous system, 269
    reproductive system, 603
    respiratory system, 480
    skeletal system, 167
    urinary system, 565
  patterns of hormonal interaction, 333–335
  prenatal and early postnatal development of, 625

Endocrinology, 336

Endocytosis (EN-dō-sī-tō-sis): The movement of relatively large volumes of extracellular material into the cytoplasm via the formation of a membranous vesicle at the cell surface; includes pinocytosis and phagocytosis, 63, 65
  receptor-mediated, 63, 443

Endoderm: One of the three primary germ layers; the layer on the undersurface of the embryonic disc that gives rise to the epithelia and glands of the digestive system, the respiratory system, and portions of the urinary system, 618–619

Endogenous: Produced within the body.

Endolymph (EN-dō-limf): Fluid contents of the membranous labyrinth (the saccule, utricle, semicircular canals, and cochlear duct) of the inner ear, 298, 300–302, 567, A-16

Endometriosis, 606

Endometrium (en-dō-MĒ-trē-um): The mucous membrane lining the uterus, 591, 594–595

Endomysium (en-dō-MĪS-ē-um): A delicate network of connective tissue fibers that surrounds individual muscle cells, 178–179

Endoplasmic reticulum (en-dō-PLAZ-mik re-TIK-ū-lum): A network of membranous channels in the cytoplasm of a cell that function in intracellular transport, synthesis, storage, packaging, and secretion, 65–68
  rough, 55–56, 67
  smooth, 55–56, 67

Endosteum, 129–130

Endothelium (en-dō-THĒ-lē-um): The simple squamous epithelium that lines blood and lymphatic vessels, 359, 390–392, 429, 465–466

Endotracheal tube, 468

Endurance, muscle, 192–193

Energy, 32
  activation, 42–43
  high-energy compounds, 45–47
  types of, 32

Energy storage, in connective tissue, 92

Enzyme: A protein that catalyzes a specific biochemical reaction, 41–43. *See also specific enzymes*
  active site of, 42–43
  membrane, 58
  in seminal fluid, 588

Enzyme-substrate complex, 43

Eosinophil (ē-ō-sin-ō-fil): A granulocyte (WBC) with a lobed nucleus and red-staining granules; participates in the immune response and is especially important during allergic reactions, 345, 355–358, 435–436

Ependyma (ep-EN-di-mah): Layer of cells lining the ventricles and central canal of the CNS, 230

Ependymal cells, 229–230

Epicardium: Serous membrane covering the outer surface of the heart; also called *visceral pericardium*, 19–20, 100, 369–371

Epicranius, 200

Epidermal ridge, 110–111

Epidermis: The epithelium covering the surface of the skin, 6, 110–114
  layers of, 111–112
  skin color, 112–113
  vitamin D_3 and, 113

Epididymis (ep-i-DID-i-mus): Coiled duct that connects the rete testis to the ductus deferens; site of functional maturation of spermatozoa, 11, 582–583, 586

Epidural block, 239, 270

Epidural hemorrhage, 240

Epidural space: Space between the spinal dura mater and the walls of the vertebral foramen; contains blood vessels and adipose tissue; a frequent site of injection for regional anesthesia, 239

Epiglottis (ep-i-GLOT-is): Blade-shaped flap of tissue, reinforced by cartilage, that is attached to the dorsal and superior surface of the thyroid cartilage; it folds over the entrance to the larynx during swallowing, 459, 461, 491, 494

Epilepsy, 251

Epimysium (ep-i-MĪS-ē-um): A dense layer of collagen fibers that surrounds a skeletal muscle and is continuous with the tendons/aponeuroses of the muscle and with the perimysium, 178–179

Epinephrine, 264–265, 314–315, 317–318, 322, 327–328, 333–334, 382, 402–404, 539, 560

Epiphyseal closure, 132

Epiphyseal fracture, 135

Epiphyseal line, 129, 132

Epiphyseal plate (e-pi-FI-sē-al): Cartilaginous region between the epiphysis and diaphysis of a growing bone, 132, 151, 157, 631

Epiphysis (e-PIF-i-sis): The head of a long bone, 129, 132

Episiotomy, 628

Epistaxis, 481

Epithalamus, 252

Epithelium (e-pi-THĒ-lē-um): One of the four primary tissue types; a layer of cells that forms a superficial covering or an internal lining of a body cavity or vessel, 84–92
  basement membrane, 87
  ciliated, 86
  classification of, 87–89
  columnar, 87–90
  cuboidal, 87–88
  exfoliative cytology, 89
  functions of, 84–85
  glandular, 85, 89–92
  intercellular connections, 85–86
  lining mucous membranes, 100
  pseudostratified, 89–90
  renewal and repair of, 87
  simple, 87–88

squamous, 87–90
stratified, 87–90
surface of, 86
transitional, 89–90, 563

EPO. *See* **Erythropoietin**

Eponychium. *See* **Cuticle**

Eponym, 14

Epstein-Barr virus, 449

Equational division: The second meiotic division, 584–585, 591–592

Equilibrium (balance), 278, 296, 299–301
  dynamic, 299
  pathways for equilibrium sensations, 301
  static, 299

Equilibrium (chemical reactions), 34

Erectile tissue, 589

Erection, 589

Erector spinae muscle, 201–203

Eruption, of secondary teeth, 494

ERV. *See* **Expiratory reserve volume**

Erythroblastosis fetalis. *See* **Hemolytic disease of the newborn**

Erythroblasts, 351, 358

Erythrocyte (e-RITH-rō-sīt): A red blood cell; an anucleate blood cell containing large quantities of hemoglobin. *See* **Red blood cells**

Erythrocytosis, 352, 362

Erythropoiesis (e-rith-rō-poy-Ē-sis): The formation of red blood cells, 350–351

Erythropoiesis-stimulating hormone. *See* **Erythropoietin**

Erythropoietin (EPO) (e-rith-rō-poy-Ē-tin): Hormone released by kidney tissues exposed to low oxygen concentrations; stimulates erythropoiesis in bone marrow, 315, 331–332, 351, 402–403, 405, 548

Esophageal artery, 412

Esophageal hiatus, 494–495

Esophageal phase, of swallowing, 494–495

Esophageal sphincter
  lower, 494–495
  upper, 494

Esophageal vein, 415–416

Esophagitis, 495, 514

Esophagus: A muscular tube that connects the pharynx to the stomach, 10, 459, 462, 489, 494–495

Essential amino acids: Amino acids that cannot be synthesized in the body in adequate amounts and must be obtained from the diet, 531, 534

Essential fatty acids: Fatty acids that cannot be synthesized in the body and must be obtained from the diet, 529

Estradiol, 598

Estrogens (ES-trō-jenz): A class of steroid sex hormones that includes estradiol, 40, 67, 137, 315, 322, 332–333, 347, 351, 590, 595, 598–602, 605, 620, 629
  transdermal, 112

Ethmoidal sinus, 143–144, 287, 460

Ethmoid bone, 140–143, 459

Eustachian tube, 96, 297–298, 460, 491

Evaporation: Movement of molecules from the liquid to the gaseous state, 538–539

Eversion (ē-VER-shun): A turning outward, 160, 195

Exchange pump, 62

Exchange reaction, 33–34

Excitable membranes: Membranes that conduct action potentials, a characteristic of muscle and nerve cells, 233

Excretion: Elimination from the body, 2, 110, 488

Exercise
  cardiovascular effects of, 403–404
  effect on bone, 133
  renal blood flow and, 560
  venous return and, 397

Exergonic reaction, 34
Exfoliative cytology, 89, 104
Exhalation, 468–470, 476
Exhaustion phase, of general adaptation syndrome, 334
Exocrine gland: A gland that secretes onto the body surface or into a passageway connected to the exterior, 89
Exocrine pancreas, 329
Exocrine secretions, 85, 89
Exocytosis (EK-sō-sī-tō-sis): The ejection of cytoplasmic materials by fusion of a membranous vesicle with the cell membrane, 63–65, 68, 91, 511
Exon, 73
Expiratory center, 476–477
Expiratory reserve volume (ERV), 470
Expulsion stage, of labor, 628–629
Extension: An increase in the angle between two articulating bones; the opposite of flexion, 152, 159–160, 195
Extensor carpi radialis muscle, 196–197, 209–210
Extensor carpi ulnaris muscle, 197, 209–210
Extensor digitorum muscle, 196–197, 209–210, 215–216
Extensor hallucis muscle, 215–216
External anal sphincter, 204–205, 508–510
External auditory canal: Passageway in the temporal bone that leads to the tympanum, 140–141, 143, 297–298, 303
External callus, 134, 136
External ear: The pinna, external auditory canal, and tympanum, 297
External genitalia
    female, 11, 590, 596
    male, 11, 582
External nares: The nostrils; the external openings into the nasal cavity, 458
External oblique muscle, 196–197, 202–203
External occipital protuberance, 141
External orifice, 594
External receptor, 228
External respiration, 472–473
External urethral sphincter, 204–205, 562–564
Extracellular fluid: All body fluid other than that contained within cells; includes plasma and interstitial fluid, 55, 65, 232, 346, 566–569
Extracorporeal liver assist device (ELAD), 506
Extraembryonic membranes: The yolk sac, amnion, chorion, and allantois, 618–621
Extrapyramidal system, 262
Extrinsic pathway: Clotting pathway that begins with damage to blood vessels or surrounding tissues and ends with the formation of tissue thromboplastin, 360–361
Eye, 285–296
    accessory structures of, 285–286
    anatomy of, 286–293
    color of, 288, 633
    innervation of, ANS, 265–266
    visual physiology, 294–295
Eyeglasses, 292–293
Eyelash, 285
Eyelid, 285, 287
Eye muscles, 253, 286, 288

F

Face, bones of, 143–144
Facet, 136–137
Facial artery, 398, 410
Facial nerve (N VII), 253–256, 266, 284, 297
Facial vein, 414

Facilitated diffusion: Passive movement of a substance across a cell membrane by a protein carrier, 58, 62, 65, 511–512
Fact memory, 250
Factor VII, 360–361
Factor VIII, 361
Factor X, 360
FAD (flavin adenine dinucleotide), 526
Fainting, 397
Falciform ligament, 504
Fallopian tube, 11, 590, 593–594, 614, 616
False ribs, 149
False vocal cords, 461
Familial polyposis of colon, 637
Farsightedness, 293, 306
Fasciae (FASH-ē-ē): Connective tissue fibers, primarily collagenous, that form sheets or bands beneath the skin to attach, stabilize, enclose, and separate muscles and other internal organs.
Fascicle, 178–179
Fasciculus (fa-SIK-ū-lus): A small bundle, usually referring to a collection of nerve axons or muscle fibers.
Fast fiber, 192–193
Fast pain, 278–279
Fat cell. See **Adipocytye**
Fat pad, 158–159
Fats, 38–40. See also **Lipid**
    dietary, 530
    saturated, 39–40, 530
    unsaturated, 40
Fat-soluble vitamins, 506, 513, 534–535
Fat substitute, 38
Fatty acids: Hydrocarbon chains ending in a carboxyl group, 38–40, 46, 512, 529
    breakdown of, 190–191, 528
    essential, 529
    free, 529
    omega-3, 47
    polyunsaturated, 38–39
    saturated, 38–39
    unsaturated, 38–39
Feces: Waste products eliminated by the digestive tract at the anus; contains indigestible residue, bacteria, mucus, and epithelial cells, 509–510, 568
Feedback control, 320
Female reproductive cycle, 597–600
Female reproductive system, 590–600
    aging of, 601
    components of, 11
    functions of, 11
    hormones and, 597–600
Female sexual function, 600–601
Femoral artery, 398, 407, 412
    deep, 407, 412
Femoral nerve, 257–258
Femoral region, 15–16
Femoral vein, 413, 415
    deep, 413, 415
Femur, 129, 138–139, 155, 165
Fenestrated capillary, 553
Fertilization: Fusion of egg and sperm to form a zygote, 582, 593, 614–617
Fetal alcohol syndrome, 638
Fetal circulation, 418–419
Fetus: Developmental stage lasting from the start of the third developmental month to delivery, 75, 614. See also **Development, prenatal**
    bone development in, 131
    oxygen supply to, 263
Fever, 435, 437, 540
Fiber, dietary, 510

Fibrillation (fi-bri-LĀ-shun): Uncoordinated contractions of individual muscle cells that impair or prevent normal function.
Fibrillin, 94
Fibrin (FĪ-brin): Insoluble protein fibers that form the basic framework of a blood clot, 118, 346, 360
Fibrinogen (fī-BRIN-ō-jen): Plasma protein, soluble precursor of the fibrous protein fibrin, 345–346, 360, 587
Fibrinolysis (fī-brin-OL-i-sis): The breakdown of the fibrin strands of a blood clot by a proteolytic enzyme, 361
Fibroblasts (FĪ-brō-blasts): Cells of connective tissue proper that are responsible for the production of extracellular fibers and the secretion of the organic compounds of the extracellular matrix, 93, 95, 103, 118–119, 356
Fibrocartilage: Cartilage containing an abundance of collagen fibers; found around the edges of joints, in the intervertebral discs, the menisci of the knee, etc., 96–98
Fibrosis, 103, 216–217
Fibrous joint, 157
Fibrous protein, 41–42
Fibrous skeleton, of heart, 371–372
Fibrous tissue. See **Dense connective tissue**
Fibrous tunic, 287–288
Fibula: The lateral, relatively small bone of the leg, 138–139, 156
Fibular artery, 407, 412
Fibularis (peroneus) muscles, 196, 214–216
Fibular vein, 413–414
Fight or flight, 264, 334
Filling time, cardiac, 382
Filtrate: Fluid produced by filtration at a glomerulus in the kidney, 550
Filtration: Movement of a fluid across a membrane whose pores restrict the passage of solutes on the basis of size.
    across capillary wall, 396
    by kidney, 61, 550–551, 555–556, 559
    membrane transport, 58, 61, 65
Filtration pressure: Hydrostatic pressure responsible for the filtration process, 556
Filtration slit, 553
Fimbriae (FIM-brē-ē): A fringe; used to describe the fingerlike processes that surround the entrance to the uterine tube, 593–594
Fine touch, 261
Fine touch and pressure receptor, 279
Fingerprints, 111
Fingers, muscles that move, 211
First breath, 419, 479
First-degree burn, 120
First heart sound, 381
First messenger, 315
First trimester, 616–625
Fissure: An elongate groove or opening, 246
Fixator, 195
Fixed macrophages, 93, 356, 435, 439–440
Fixed ribosomes, 55, 67
Flagellum/flagella (fla-JEL-ah): An organelle structurally similar to a cilium but used to propel a cell through a fluid, 66–67, 586
Flat bone, 128–129
Flatus, 510
Flexion (FLEK-shun): A movement that reduces the angle between two articulating bones; the opposite of extension, 152, 159–160, 195
Flexor carpi radialis muscle, 196, 209–210
Flexor carpi ulnaris muscle, 196–197, 209–210
Flexor digitorum muscle, 196, 209–210, 215–216
Flexor hallucis muscle, 215–216

Flexor reflex: A reflex contraction of the flexor muscles of a limb in response to an unpleasant stimulus, 259

Flexure: A bending.

Floating kidney, 549

Floating ribs, 149

Fluid balance, 566–569

Fluid compartments, 567

Fluid connective tissue, 92, 94–96

Fluid shift, 568

FMN, 526

Focal calcification, 393

Focal distance, 291–292

Focal point, 291–292

Folacin, 536

Folic acid, 536

Follicle (FOL-i-kl): A small secretory sac or gland. *See specific types of follicles*

Follicle-stimulating hormone (FSH): A hormone secreted by the anterior pituitary gland; stimulates oogenesis (female) and spermatogenesis (male), 315, 320, 322–323, 331–332, 589–591, 598–599, 602, 631

Follicular cells, ovarian, 598

Follicular fluid, 592–593

Follicular phase, of ovarian cycle, 597–599

Fontanel (fon-tah-NEL): A relatively soft, flexible, fibrous region between two flat bones in the developing skull, 144

Food, energy content of, 537

Food pyramid, 533

Foot

  bones of, 156–157

  muscles that move, 214–217

Foramen: An opening or passage through a bone. *See specific foramina*

Foramen magnum, 140–142, 409

Foramen ovale, 372, 418–419

Forced breathing, 470, 476

Forearm: Distal portion of the arm between the elbow and wrist. *See* **Upper limb**

Forebrain: The cerebrum. *See* **Cerebrum**

Foreskin, 582, 588–589, 596

Formed elements, in blood, 344–359

Fornix, 591, 595

Fossa: A shallow depression or furrow in the surface of a bone, 136–137

Fossa ovalis, 372, 419

Fourth heart sound, 381, 395

Fourth ventricle: An elongate ventricle of the metencephalon (pons and cerebellum) and the myelencephalon (medulla) of the brain; the roof contains a region of choroid plexus, 245–246

Fovea (FŌ-vē-uh): Portion of the retina providing the sharpest vision, with the highest concentration of cones; also called *macula lutea*, 287, 289–290

Fractionated blood, 344

Fracture: A break or crack in a bone, 134–136, 168

  classification of, 135

  of hip, 165

Fracture hematoma, 134, 136

Fragile-X syndrome, 636

Frank-Starling principle, 382, 403

Fraternal twins. *See* **Dizygotic twins**

Freckles, 113, 633

Free earlobes, 633

Free fatty acid, 529

Freely movable joint. *See* **Diarthrosis**

Freely permeable membrane, 58

Free macrophages, 93, 356, 435, 439–440

Free nerve endings, 278–281

Free radicals, 69, 328

Free ribosomes, 55, 67

Frontal bone, 140–142, 145, 459

Frontalis muscle, 196, 199–200

Frontal lobe, 244, 246–247

Frontal plane: A sectional plane that divides the body into anterior and posterior portions; also called *coronal plane*, 18

Frontal section, 18–19

Frontal sinus, 140, 142, 144, 458–460

Fructose: A hexose (simple sugar containing 6 carbons) found in foods and in semen, 37, 587

FSH. *See* **Follicle-stimulating hormone**

Functional zone, of endometrium, 595

Fundus (FUN-dus): The base of an organ.

  of stomach, 495–496

  of uterus, 594

G

GABA. *See* **Gamma aminobutyric acid**

Galactosemia, 47

Galea aponeurotica, 199–200

Gallbladder: Pear-shaped reservoir for the bile secreted by the liver, 10, 265–266, 489, 504–507

Gallstones, 349, 507, 514

Gametes (GAM-ēts): Reproductive cells (sperm or eggs) that contain half the normal chromosome complement, 582

Gametogenesis (ga-mē-tō-JEN-e-sis): The formation of gametes, 584–585, 589, 591

Gamma aminobutyric acid (GABA) (GAM-ma a-MĒ-nō-bū-TIR-ik): A neurotransmitter of the CNS whose effects are usually inhibitory, 237

Ganglion/ganglia: A collection of nerve cell bodies outside the CNS, 231, 254

Ganglion cells, 289–290, 295

Ganglionic neuron: An autonomic neuron whose cell body is in a peripheral ganglion and whose axon is a postganglionic fiber, 264

Gap junctions: Connections between cells that permit the movement of ions and the transfer of graded or conducted changes in the membrane potential from cell to cell, 85–86, 102, 371

GAS (general adaptation syndrome), 334, 336

Gases

  dissolved, 47

  mixed, 471–472

Gas exchange

  alveolar, 406, 458

  pulmonary, 468, 471–472

Gas transport, 468, 472–475

Gastrectomy, 499, 514

Gastric artery, 411–412

Gastric glands: Tubular glands of the stomach whose cells produce acid, enzymes, intrinsic factor, and hormones, 496–497

Gastric inhibitory peptide (GIP), 498, 502

Gastric juice, 497

Gastric phase, of gastric secretion, 497–498

Gastric pit, 496–497

Gastric ulcer, 497

Gastric vein, 417

Gastrin (GAS-trin): Hormone produced by endocrine cells of the stomach, when exposed to mechanical stimuli or vagal stimulation, and of the duodenum, when exposed to chyme containing undigested proteins, 498, 501–502

Gastritis, 497

Gastrocnemius muscle, 187, 195–197, 214–216

Gastroenteric reflex, 501

Gastroenteritis, 514

Gastroenterology, 514

Gastroepiploic vein, 417

Gastroileal reflex, 501

Gastrointestinal tract, bacteria in, 509, 513, 536

Gastroscope, 499

Gastrulation (gas-troo-LĀ-shun): The movement of cells of the inner cell mass that creates the three primary germ layers of the embryo, 618–619

Gated channel, 232

Gene: A portion of a DNA strand that functions as a hereditary unit and is found at a particular location on a specific chromosome, 71–72, 632–638

  suicide, 75

General adaptation syndrome (GAS), 334, 336

General interpretive area, 248–249

General senses, 278–282

Genetic abnormality, 89

Genetic code, 71, 73

Genetic engineering, 77

Genetics: The study of mechanisms of heredity, 614, 632–638

Genetic screening, 637

Genome, 635–638

Genotype (JĒN-ō-tīp): The genetic complement of a particular individual, 632, 634

Gentamicin, 305

Geriatrics, 638

Germinal centers: Pale regions in the interior of lymphoid tissues or nodules, where cell divisions are under way, 431

Germinative cells. *See* **Stem cells**

Germ layer, 618–619

Gestation (jes-TĀ-shun): The period of intrauterine development, 616

GFR. *See* **Glomerular filtration rate**

GHB, 332

Gigantism, 168, 333

Gingivae (JIN-ji-vē): The gums, 490–491, 493

Gingivitis, 514

GIP (gastric inhibitory peptide), 498, 502

Gland: Cells that produce exocrine or endocrine secretions, 84. *See also specific types of glands*

Gland cells, 84, 91

Glandular epithelium, 85, 89–92

  mode of secretion, 91

  type of secretion, 91–92

Glans, of clitoris, 596

Glans penis: Expanded tip of the penis that surrounds the urethral opening; continuous with the corpus spongiosum, 582, 588–589, 600

Glaucoma: Eye disorder characterized by rising intraocular pressures due to inadequate drainage of aqueous humor at the canal of Schlemm, 291–292, 306

Glenoid cavity, 151

Glenoid fossa: A rounded depression that forms the articular surface of the scapula at the shoulder joint.

Glial cells (GLĒ-al): Supporting cells in the neural tissue of the CNS and PNS. *See* **Neuroglia**

Gliding joint, 161–162

Gliding movement, 159

Global aphasia, 248

Globular proteins: Proteins whose tertiary structure makes them rounded and compact, 41–42

Globulins, 345–346

Globus pallidus, 250

Glomerular capillary, 558

Glomerular epithelium, 553

Glomerular filtration rate (GFR): The rate of filtrate formation at the glomerulus, 556, 558, 560, 573, 627

Glomerulonephritis, 573

Glomerulus (glo-MER-ū-lus): A ball or knot; in the kidneys, a knot of capillaries that projects into the enlarged, proximal end of a nephron; the site of filtration, the first step in the production of urine, 550–556, 559

Heat stroke, 538, 540

Heat transfer, mechanisms of, 538–539

Heavy chain, 441–442

Heel bone. *See* **Calcaneus**

Heimlich maneuver, 463

Helicobacter pylori, 497

Helium, 28–29

Helper T cells: Lymphocytes whose secretions and other activities coordinate the cellular and humoral immune responses, 430, 440–441, 443, 446

Hemangioma, cavernous, 122

Hematocrit (hē-MA-tō-krit): Percentage of the volume of whole blood contributed by cells; also called *packed cell volume (PCV)* or *volume of packed red cells (VPRC),* 347, 351–352, 362, 395

Hematological regulation, liver in, 506

Hematologist, 351

Hematology, 362

Hematoma: A tumor or swelling filled with blood, 134, 136

Hematopoiesis. *See* **Hemopoiesis**

Hematuria, 362, 573

Heme (hēm): A special organic compound containing a central iron atom that can reversibly bind oxygen molecules; a component of the hemoglobin molecule, 348–350

Hemiazygos vein, 415–416

Hemidesmosome, 85–87, 111

Hemispheric lateralization, 249

Hemocytoblasts: Stem cells whose divisions produce all of the various populations of blood cells, 347, 351, 358, 431

Hemoglobin (Hb) (HĒ-mō-glō-bin): Protein composed of four globular subunits, each bound to a single molecule of heme; the protein found in red blood cells that gives them the ability to transport oxygen in the blood, 41–42

  abnormal, 349

  buffering capacity of, 570

  carbon dioxide bound to, 348, 474–475

  carbon monoxide bound to, 474

  oxygen bound to, 348, 473–475

  recycling of, 348–350

  structure and function of, 348

Hemoglobin concentration, 352

Hemoglobinuria, 348

Hemolysis: Breakdown (lysis) of red blood cells, 61, 348, 353

Hemolytic disease of the newborn (HDN), 354, 362, 443

Hemophilia, 361–362, 635, 637

Hemopoiesis (hēm-ō-poy-Ē-sis): Blood cell formation and differentiation, 347

Hemorrhage: Blood loss, 556

  epidural, 240

  subdural, 240

Hemorrhaging, 397

  cardiovascular response to, 404–405

Hemorrhoids, 394, 510, 513

Hemostasis, 359–361

  abnormal, 361

  clot retraction and removal, 361

  clotting system, 360–361

Hemothorax, 467

Heparin, 103, 356, 437

Hepatic artery, 504–505

  common, 411–412

Hepatic duct: Duct carrying bile away from the liver lobes and toward the union with the cystic duct.

Hepatic portal system, 416–417, 500

Hepatic portal vein: Vessel that carries blood between the intestinal capillaries and the sinusoids of the liver, 417, 504–505, 508, 512

Hepatic vein, 413, 415–417, 504

Hepatitis, 447, 505–506, 514

Hepatocyte (he-PAT-ō-sit): A liver cell, 505

Hepatopancreatic sphincter, 507

Hering-Breuer reflex, 477

Hernia, 204, 217

  Diaphragmatic, 204, 495

Herniated disc, 163, 168

Hertz, 303

Heterologous marrow transplant, 362

Heterozygous (het-er-ō-ZĪ-gus): Possessing two different alleles at corresponding locations on a chromosome pair; a heterozygous individual's phenotype may be determined by one or both alleles, 633

Hexose: A six-carbon simple sugar.

Hiatal hernia, 204, 495

High-density lipoprotein (HDL): A lipoprotein with a relatively small lipid content, thought to be responsible for the movement of cholesterol from peripheral tissues to the liver, 529–530

High-energy bonds, 45–46

High-energy compound, 45–47

Higher centers, 231

  control of respiration by, 478–479

Hilus (HĪ-lus): A localized region where blood vessels, lymphatics, nerves, and/or other anatomical structures are attached to an organ.

  of kidney, 549–550

Hinge joint, 161–163

Hip joint, 165

  dislocation of, 165

  fracture of, 165

Hippocampus, 250–251

Histamine (HIS-ta-min): Chemical released by stimulated mast cells or basophils to initiate or enhance an inflammatory response, 103, 356, 437

Histidine, 531

Histology (his-TOL-ō-jē): The study of tissues, 3, 23, 84

Histone, 70–71

HIV (human immunodeficiency virus), 443

Hodgkin's disease, 449

Holocrine (HŌ-lō-krin): Form of exocrine secretion where the secretory cell becomes swollen with vesicles and then ruptures, 91

Holocrine gland, 89, 116

Homeostasis (hō-mē-ō-STĀ-sis): The maintenance of a relatively constant internal environment, 5–14, 54

Homeostatic regulation, 5–14

Homologous chromosomes (hō-MOL-o-gus): The members of a chromosome pair, each containing the same gene loci, 632

Homozygous (hō-mō-ZĪ-gus): Having the same allele for a particular character on two homologous chromosomes, 633

Horizontal cells, 289–290

Horizontal section, 18–19

Hormone: A compound secreted by one cell that travels through the circulatory system to affect the activities of cells in another portion of the body, 85, 89, 314–341. *See also specific hormones*

  aging and, 335

  amino acid derivatives, 314–315

  antagonistic, 333

  behavior and, 333–335

  cardiovascular regulation by, 383, 402–403

  cell membrane and, 315–317

  control of endocrine activity, 317–319

  disorders of, 335

  growth and, 333

  of immune system, 444–445

  inhibiting, 318, 320

  integrative effects, 333

  intracellular receptors for, 316–317

  local, 314

  mechanism of action of, 315–317

  patterns of hormonal interaction, 333–335

  peptide, 41, 315, 317

  permissive effects of, 333

  placental, 620–622

  regulatory, 318

  releasing, 318, 320

  of reproductive system, 331–332

  steroid. *See* **Steroid hormones**

  stress and, 334

  structure of, 314–315

  synergistic, 333

Hormone-receptor complex, 316–317

Hormone replacement therapy, 601

Horn (gray matter), 241

Human chorionic gonadotropin (hCG): Placental hormone that maintains the corpus luteum for the first 3 months of pregnancy, 620–622

Human Genome Project, 635–638

Human growth hormone. *See* **Growth hormone**

Human immunodeficiency virus (HIV), 443

Human physiology, 3

Human placental lactogen (hPL): Placental hormone that stimulates the functional development of the mammary glands, 620, 622

Human tetanus immune globulin, 188

Humerus, 129, 138–139, 151–152, 164

Humoral immunity: Immunity resulting from the presence of circulating antibodies produced by plasma cells, 430–431, 438, 441–444, 446

Hunger, 252

Huntington's disease: An inherited disease marked by a progressive deterioration of mental abilities and by motor disturbances, 270, 635, 637

Hyaline cartilage, 96–100, 131–132

Hyaluronidase, 616

Hydrochloric acid, 34, 497, 571

Hydrocortisone, 315, 327

Hydrogen, 28–29, 31

Hydrogen bond: Weak interaction between the hydrogen atom on one molecule and a negatively charged portion of another molecule, 31–32, 41

Hydrogen ion channel, 526

Hydrogen ions, 34–35

Hydrogen peroxide, 69

Hydrolysis (hī-DROL-i-sis): The breakage of a chemical bond through the addition of a water molecule; the reverse of dehydration synthesis, 37–38, 511

Hydrophilic molecule, 57

Hydrophobic molecule, 57

Hydrostatic pressure: Fluid pressure, 60–61, 396

Hydroxide ions, 34–35

Hydroxyl group (hī-DROK-sil): OH⁻.

Hymen, 596

Hyoid bone, 139–140, 144–145, 200, 324, 458–459, 461–462, 491

Hypercalcemia, 383

Hypercapnia, 478, 481, 571

Hypercholesterolemia, familial, 47, 637

Hyperextension, 159

Hyperglycemia, 330, 336

Hyperkalemia, 383

Hypernatremia, 573

Hyperopia, 293, 306

Hyperpolarization of membrane, 232

Hypertension, 393, 421, 537, 634–635

Hyperthyroidism, 336

Hypertonic: A term used when comparing two solutions to refer to the solution with the higher solute or osmotic concentration, 61

Hypertrophy (HĪ-per-trō-fē): Increase in the size of tissue without cell division.
   of skeletal muscle, 193

Hyperuricemia, 531, 540

Hyperventilation (hī-per-ven-ti-LĀ-shun): A rate of respiration sufficient to reduce plasma P_{CO_2} to levels below normal, 478, 571–572

Hypervitaminosis, 534–536, 540

Hypervolemia, 421

Hypocalcemia, 383

Hypocapnia, 571

Hypodermic needle: A needle inserted through the skin to introduce drugs into the subcutaneous layer, 114

Hypodermis. *See* **Subcutaneous layer**

Hypoglossal nerve (N XII), 253–256

Hypoglycemia, 336

Hypogonadism, 320

Hypokalemia, 383

Hyponatremia, 573

Hypophyseal artery, 319

Hypophyseal portal system (hī-pō-FI-sē-al): Network of vessels that carry blood from capillaries in the hypothalamus to capillaries in the anterior pituitary gland (hypophysis), 319

Hypophyseal vein, 319

Hypophysis (hī-POF-i-sis): The anterior pituitary gland. *See* **Pituitary gland**

Hypotension, 404, 421

Hypothalamus: The floor of the diencephalon; region of the brain containing centers involved with the unconscious regulation of visceral functions, emotions, drives, and the coordination of neural and endocrine functions, 243–244, 252, 296, 315, 317–322, 539

Hypothermia, 538–540

Hypothyroidism, congenital, 335

Hypotonic: When comparing two solutions, used to refer to the one with the lower osmotic concentration, 61

Hypoventilation, 478, 571–572

Hypovolemia, 421

Hypoxia (hī-POKS-ē-a): Low tissue oxygen concentrations, 351, 362, 468, 481

I

I band, 180–181

Identical twins. *See* **Monozygotic twins**

Identifier protein, 58

Ig. *See* **Immunoglobulin**

Ileocecal valve (il-ē-ō-SĒ-kal): A fold of mucous membrane that guards the connection between the ileum and the cecum, 499, 507–508

Ileum (IL-ē-um): The last 2.5 m of the small intestine, 499, 508

Iliac artery
   common, 407, 411–412
   external, 407, 411–412
   internal, 411–412

Iliac crest, 154

Iliacus muscle, 211–213

Iliac vein
   common, 413, 415
   external, 413, 415–416
   internal, 413, 415–416

Iliocostalis muscle group, 201–202

Iliopsoas muscle, 196, 211–213

Iliotibial tract, 212–213

Ilium (IL-ē-um): The largest of the three bones whose fusion creates a coxa, 154

Immediate hypersensitivity reaction, 447

Immovable joint, 157–158

Immune complex disorder, 447

Immune disorders, 446–447

Immune response, 352, 358, 429, 438–447
   aging and, 447
   hormones and, 444–445
   immune disorders, 446–447
   manipulation of, 449
   primary, 444
   secondary, 444
   stress and, 447

Immune tolerance, 439

Immunity, 429
   acquired, 438
   active, 438
   cell-mediated, 430–431, 438–441, 446
   forms of, 438
   humoral, 430–431, 438, 441–444, 446
   innate, 438
   passive, 438
   properties of, 438–439

Immunization: Developing immunity by the deliberate exposure to antigens under conditions that prevent the development of illness but stimulate the production of memory B cells, 438, 444, 449

Immunodeficiency: An inability to produce normal numbers and types of antibodies and sensitized lymphocytes.

Immunodeficiency disease, 446–447, 449

Immunoglobulin (Ig) (i-mū-nō-GLOB-ū-lin): A circulating antibody 346, 430

Immunoglobulin A (IgA), 442–443

Immunoglobulin D (IgD), 442–443

Immunoglobulin E (IgE), 442–443

Immunoglobulin G (IgG), 442–444

Immunoglobulin M (IgM), 442–444

Immunological escape, 436

Immunological surveillance, 430–431, 435–436, 447

Immunology, 449

Immunosuppression, 449

Impacted teeth, 494

Impermeable membrane, 58

Implantation (im-plan-TĀ-shun): The erosion of a blastocyst into the uterine wall, 616–620

Impotence, 600

Incision, 118

Incisor, 493–494

Inclusions: Aggregations of insoluble pigments, nutrients, or other materials in the cytoplasm, 65

Incomplete protein, 534

Incomplete tetanus, 187

Incontinence, 564, 573

Incus (IN-kus): The central auditory ossicle, situated between the malleus and the stapes in the middle ear cavity, 297–298, 303

Induced abortion, 632

Induced passive immunity, 438

Infancy, 630–631

Infant skull, 144

Infarct: An area of dead cells resulting from an interruption of circulation, 375

Infection: Invasion and colonization of body tissues by pathogenic organisms, 103

Inferior (anatomical direction), 17

Inferior colliculus, 252–253, 304–305

Inferior oblique muscle, 286

Inferior rectus muscle, 286

Inferior section, 18

Inferior vena cava: The vein that carries blood from the parts of the body below the heart to the right auricle, 327, 370, 372, 397, 406, 414–418, 430, 504, 508, 549

Infertility: Inability to conceive, 601, 638
   female, 594, 618
   male, 584, 618

Inflammation: A nonspecific defense mechanism that operates at the tissue level, characterized by swelling, redness, warmth, pain, and some loss of function, 103, 113, 119, 356, 399, 435, 437, 442, 444

Inflammatory bowel disease, 514

Inflation reflex: A reflex mediated by the vagus nerve that prevents overexpansion of the lungs, 477

Influenza, 481

Infraorbital foramen, 140–141

Infraspinatus fossa, 151

Infraspinatus muscle, 197, 206–208

Infundibulum (in-fun-DIB-ū-lum): A tapering, funnel-shaped structure; in the nervous system, the connection between the pituitary gland and the hypothalamus; in the uterine tube, the entrance bounded by fimbriae that receives the ova at ovulation.
   between pituitary and hypothalamus, 318–319
   of uterine tube, 593–594

Ingestion: The introduction of materials into the digestive tract via the mouth, 488

Inguinal canal: A passage through the abdominal wall that marks the path of testicular descent and that contains the testicular arteries, veins, and ductus deferens, 204, 586

Inguinal hernia, 204

Inguinal lymph nodes, 428

Inguinal region: The area near the junction of the trunk and the thighs that contains the external genitalia, 15–16

Inhalation, 468–470, 476

Inheritance, 614, 632–637
   polygenic, 633–635
   prediction of, 633–634
   simple, 634

Inherited disorder, 70, 74

Inhibin (in-HIB-in): A hormone produced by the sustentacular cells that inhibits the pituitary secretion of FSH, 315, 320, 322, 331–332, 585, 589–590, 598–600, 602

Inhibiting hormones, 318, 320

Injection: Forcing of fluid into a body part or organ, 114, 214, 217

Innate immunity, 438

Inner cell mass: Cells of the blastocyst that will form the body of the embryo, 616–619

Inner ear, 296–299

Inner membrane, of mitochondria, 69, 526

Innervation: The distribution of sensory and motor nerves to a specific region or organ. *See* **Nervous system**

Innominate vein, 414

Inorganic compounds, 35–36
   acids and bases, 36
   in body, 47
   carbon dioxide and oxygen, 35
   salts, 36
   water, 35–36

Insensible perspiration, 538

Insertion: Point of attachment of a muscle that is most movable, 195

Insoluble: Incapable of dissolving in solution.

Inspiratory center, 476–477

Inspiratory reserve volume (IRV): The maximum amount of air that can be drawn into the lungs over and above the normal tidal volume, 470–471

Insula, 246–247, 250

Insulin (IN-su-lin): Hormone secreted by the beta cells of the pancreatic islets; causes a reduction in plasma glucose concentrations, 315, 329–330, 333, 503. *See also* **Diabetes mellitus**

Insulin-like growth factors. *See* **Somatomedins**

Ketone bodies: Organic acids produced during the catabolism of lipids and certain amino acids; acetone is one example, 528, 530–531, 571
Ketonemia, 540
Ketonuria, 540
Ketosis, 531, 540
Kidney: A component of the urinary system; an organ functioning in the regulation of plasma composition, including the excretion of wastes and the maintenance of normal fluid and electrolyte balance, 8, 10, 548–554
blood supply to, 553–554, 560
endocrine functions of, 331
filtration in, 61
floating, 549
hormonal control of function of, 560–562
innervation of, ANS, 265–266, 560
local regulation of function of, 558–560
nephron, 550–553
superficial and sectional anatomy of, 549–550
sympathetic activation and kidney function, 560
transplantation of, 560
urine production, 555–562
Kidney failure, 560, 573
Kidney stones, 536, 563
Killer T cells. *See* **T cells**
Kilocalorie (KIL-o-kal-o-rē): The amount of heat required to raise the temperature of a kilogram of water 1°C, 537
Kinase, 316
Kinesiologist, 195
Kinetic energy, 32
Klinefelter syndrome, 636
Knee extensors, 213
Knee jerk reflex, 259
Knee joint, 158, 165–166
Krebs cycle. *See* **Tricarboxylic acid cycle**
Kupffer cells (KOOP-fer): Stellate reticular cells of the liver; phagocytic cells of the liver sinusoids, 436, 439, 505–506
Kwashiorkor, 531
Kyphosis, 146, 168

L

Labia (LĀ-bē-a): Lips; labia majora and minora are components of the female external genitalia, 11, 490
Labia majora, 591, 596
Labia minora, 591, 596
Labor, 14, 321, 628–630
premature, 630
Labor contractions, 628–629
Labrum: A lip or rim.
Labyrinth: A maze of passageways; usually refers to the structures of the inner ear.
Lacrimal apparatus, 285
Lacrimal bone, 140–141, 144
Lacrimal canal, 285–286
Lacrimal caruncle, 285
Lacrimal gland (LAK-ri-mal): Tear gland on the dorsolateral surface of the eye, 285
Lacrimal pore, 285, 287
Lacrimal sac, 285, 287
Lactase, 511–512
Lactation (lak-TĀ-shun): The production of milk by the mammary glands, 596–597, 631
Lacteal (LAK-tē-al): A terminal lymphatic within an intestinal villus, 500, 511–512
Lactic acid: Compound produced from pyruvic acid under anaerobic conditions, 36, 570–571
in muscle, 191
recycling of, 192
Lactiferous duct (lak-TIF-e-rus): Duct draining one lobe of the mammary gland, 597

Lactiferous sinus: An expanded portion of a lactiferous duct adjacent to the nipple of a breast, 597
Lactose, 37
Lactose intolerance, 512–513
Lacuna (la-KOO-na): A small pit or cavity.
of bone, 98, 129–130
of cartilage, 96–97
of trophoblast, 618
Lambdoidal suture (lam-DOYD-al): Synarthrotic articulation between the parietal and occipital bones of the cranium, 140–141, 144, 145
Lamellae (la-MEL-lē): Concentric layers of bone within an osteon, 129–130
Lamellated corpuscle, 280
Lamina (LA-min-a): A thin sheet or layer.
Lamina propria (LA-mi-na PRO-prē-uh): Loose connective tissue that underlies a mucous epithelium and forms part of a mucous membrane, 99–100, 460, 488–489, 496, 500
Laparoscopy, 514
Large intestine: The terminal portions of the intestinal tract, consisting of the colon, the rectum, and the anorectal canal, 10, 489, 499, 507–510
absorption in, 509–510
cecum, 507–508
colon, 507–509
defecation, 281, 488, 510
functions of, 507
gross anatomy of, 508
innervation of, ANS, 265–266
movements of, 510
rectum, 507, 509
Laryngopharynx (lā-rin-gō-FAR-inks): Division of the pharynx inferior to the epiglottis and superior to the esophagus, 459–460, 491
Larynx (LAR-inks): A complex cartilaginous structure that surrounds and protects the glottis and vocal cords; the superior margin is bound to the hyoid bone, and the inferior margin is bound to the trachea, 9, 144, 458, 461–462
Latent period: The time between the stimulation of a muscle and the start of the contraction phase, 187
Lateral: Pertaining to the side, 17
Lateral canthus, 285
Lateral epicondyle
of femur, 155
of humerus, 152
Lateral malleolus, 156, 215
Lateral meniscus, 165–166
Lateral rectus muscle, 286–287
Lateral sulcus, 244, 246
Lateral ventricle: Fluid-filled chamber within one of the cerebral hemispheres, 245–246, 250
Latissimus dorsi muscle, 196–197, 201, 203, 207
Lauric acid, 38–39
Laxative, 510
LDL (low-density lipoprotein), 529–530
Leads, electrocardiogram, 378
Leak channel, 62, 232
Lecithin, 39–40
Left (anatomical direction), 16–17
Left section, 18
Leg. *See* **Lower limb**
Length, measurement of, A-12, A-13
Lens: The transparent body lying behind the iris and pupil and in front of the vitreous humor, 287–288, 291–292
Lentiform nucleus, 250
Leprosy, 270
Leptin, 332
Lesion, 122
Lesser horn, of hyoid bone, 144–145

Lesser omentum: A small pocket in the mesentery that connects the lesser curvature of the stomach to the liver, 496
Lesser trochanter, 155
Lesser tubercle, of humerus, 151–152
Leucine, 531
Leukemia (loo-KĒ-mē-ah): A malignant disease of the blood-forming tissues, 356, 362
Leukocyte (LOO-kō-sīt): A white blood cell. *See* **White blood cells**
Leukocytosis, 356, 362
Leukopenia, 356, 362
Levator ani muscle, 204
Levator scapulae muscle, 205–207
Leverage, skeletal system and, 128
LH. *See* **Luteinizing hormone**
Libido, 590, 601
Life stages, 630–632
Ligament (LI-ga-ment): Dense band of connective tissue fibers that attach one bone to another, 6, 92, 94, 129
extracapsular, 158–159
intracapsular, 158–159
Ligamentum arteriosum: The fibrous strand in adults that is the remains of the ductus arteriosus of the fetus, 419
Ligamentum teres, 165
Ligand, 63
Light chain, 441–442
Light microscopy, 54–55, A-12
Light refraction, 291–292
Limb bud, 623
Limbic system (LIM-bik): Group of nuclei and centers in the cerebrum and diencephalon that are involved with emotional states, memories, and behavioral drives, 251
Limbus (LIM-bus): The edge of the cornea, marked by the transition from the corneal epithelium to the ocular conjunctiva.
Line (bone), 136–137
Linea alba, 203
Linea aspera, 155
Linear acceleration, perception of, 301
Lingual frenulum: An epithelial fold that attaches the inferior surface of the tongue to the floor of the mouth, 490–492
Lingual tonsils, 432, 491–492
Linoleic acid, 529
Linolenic acid, 529
Lipase (LĪ-pās): A pancreatic enzyme that breaks down triglycerides, 512
gastric, 497
pancreatic, 503–504, 506, 512
Lipid: An organic compound containing carbons, hydrogens, and oxygens in a ratio that does not approximate 1 : 2 : 1; includes fats, oils, and waxes, 36, 38–40, 46
absorption of, 511–512
in blood, 555
caloric content of, 537
catabolism of, 528
digestion of, 506, 511–512
energy production and, 528
functions of, 38–39, 47
membrane, 57, 68
metabolism of, 523, 527–529, 532
structure of, 47
synthesis of, 67, 528–529
transport and distribution of, 529
types of, 38
Lipid-soluble molecules, 59
Lipogenesis, 528–529
Lipolysis: The catabolism of lipids as a source of energy, 528

Lipoprotein (lī-pō-PRŌ-tēn): A compound containing a relatively small lipid bound to a protein, 346, 393, 529

Liposome, 112

Liposuction, 94, 104, 540

Liver: An organ of the digestive system with varied and vital functions, including the production of plasma proteins, the excretion of bile, the storage of energy reserves, the detoxification of poisons, and the interconversion of nutrients, 10, 489, 496, 504–507
  anatomy of, 504–505
  biopsy of, 514
  disease of, 506
  functions of, 506
  histological organization of, 505
  innervation of, ANS, 265–266
  transplantation of, 506

Lobe
  of cerebral hemisphere, 246
  of liver, 504
  of lung, 466–467

Lobotomy, 248

Lobule (LOB-ūl): The basic organizational unit of the liver at the histological level.
  of liver, 505
  of lung, 463–464
  of testis, 583
  of thymus, 433

Local current, 234

Local hormones, 314

Lockjaw, 188, 217

Loin, 15–16

Long bone, 128–129

Longissimus muscle group, 201–202

Longitudinal fissure, 246

Longitudinal muscle, of stomach wall, 496

Long-term memory, 250–251

Loop of Henle, 550–552, 554–557, 559
  ascending limb of, 553, 557
  descending limb of, 553, 557, 559

Loose connective tissue: A loosely organized, easily distorted connective tissue containing several fiber types, a varied population of cells, and a viscous ground substance, 92, 94–95, 99

Lordosis, 146, 168

Lovastatin, 530

Low-density lipoprotein (LDL), 529–530

Lower esophageal sphincter, 494–495

Lower limb, 15–16
  bones of, 139, 155–156
  joints of, 165–166
  lymphatics of, 428
  muscles of, 211–216

Lower motor neuron, 262

Lumbar: Pertaining to the lower back, 15–16

Lumbar artery, 411–412

Lumbar curve, 146

Lumbar enlargement, 240–241

Lumbar lymph nodes, 428

Lumbar plexus, 257–258

Lumbar puncture, 245

Lumbar spinal nerves, 241, 257

Lumbar vein, 413, 415–416

Lumbar vertebrae, 145–148

Lumbosacral plexus, 258

Lumen: The central space within a duct or other internal passageway.
  of blood vessel, 391

Lunate bone, 153

Lungs: Paired organs of respiration, situated in the left and right pleural cavities, 9, 458, 464–467

  airflow to, 468–470
  at birth, 630
  cancer of, 477
  compliance of, 469–470, 479
  innervation of, ANS, 265–266
  pressure and volume relationships in, 469
  volumes and capacities, 470–471

Lunula, 118

Luteal phase, of ovarian cycle, 598–600

Luteinizing hormone (LH) (LOO-tē-in-ī-zing): Anterior pituitary hormone that in females assists FSH in follicle stimulation, triggers ovulation, and promotes the maintenance and secretion of the endometrial glands; in males, stimulates spermatogenesis; formerly known as interstitial cell-stimulating hormone in males, 315, 320–323, 332, 589–590, 592, 598–600, 602, 631

Luxation, 168

Lymph: Fluid contents of lymphatic vessels, similar in composition to interstitial fluid, 92, 94–96, 428–429, 500, 567

Lymphadenopathy, 432, 449

Lymphatic capillary, 429

Lymphatic system, 428–455
  body defenses and, 435
  components of, 9
  functions of, 9, 428–429
  integration with other systems, 447–449
    cardiovascular system, 420
    digestive system, 515
    endocrine system, 337
    integumentary system, 121
    muscular system, 218
    nervous system, 269
    reproductive system, 603
    respiratory system, 480
    skeletal system, 167
    urinary system, 565
  organization of, 428–434
  prenatal and early postnatal development of, 625

Lymphatics: Vessels of the lymphatic system, 9, 96, 396, 428–429, 432, 489, 496

Lymphedema, 429, 449

Lymph nodes: Lymphatic organs that monitor the composition of lymph, 9, 357, 432–433
  swollen glands, 432–433

Lymphoblasts, 358

Lymphocyte (LIM-fō-sīt): A cell of the lymphatic system that participates in the immune response, 93, 96, 345, 355–358, 428–431
  origin and circulation of, 430–431
  types of, 430

Lymphoid leukemia, 362

Lymphoid nodule, 500

Lymphoid nodules, 431–432

Lymphoid organ, 428

Lymphoid organs, 432–434

Lymphoid stem cells, 347, 351, 358

Lymphoid tissue, 357, 431

Lymphokines, 444

Lymphoma, 433, 449

Lymphopoiesis: The production of lymphocytes, 357, 431

Lymphotoxin, 440

Lysine, 531

Lysis (LĪ-sis): The destruction of a cell through the rupture of its cell membrane.

Lysosome (LĪ-so-sōm): Intracellular vesicle containing digestive enzymes, 55–56, 63–65, 68–69, 354, 437

Lysozyme: An enzyme present in some exocrine secretions that has antibiotic properties, 286, 436, 631

M

Macrophage: A phagocytic cell of the monocyte-macrophage system, 93, 95, 119, 355, 436–437, 439–440, 446
  alveolar, 465–466
  fixed, 356, 435, 439–440
  free, 356, 435, 439–440

Macroscopic anatomy. *See* **Gross anatomy**

Macula (MAK-ū-la): A receptor complex in the saccule or utricle that responds to linear acceleration or gravity, 300–301

Macula densa (MAK-ū-la DEN-sa): A group of specialized secretory cells in a portion of the distal convoluted tubule adjacent to the glomerulus and the juxtaglomerular cells; a component of the juxtaglomerular apparatus, 552–553

Macula lutea (LOO-tē-uh): The fovea, 287, 289–290

Macular degeneration, 306

Magnesium, 28, 30, 534–535

Major calyx, 549–550

Major histocompatibility complex (MHC) proteins, 439–441, 447

Malabsorption, 504, 513

Male climacteric, 601

Male pattern baldness, 116, 122

Male reproductive system, 582–590
  accessory glands of, 587
  aging of, 601
  components of, 11
  functions of, 11
  hormones and, 589–590
  male reproductive tract, 586–587

Male sexual function, 600

Malignant tumor, 77
  primary tumor, 77
  secondary tumor, 77

Malleus (MAL-ē-us): The first auditory ossicle, bound to the tympanum and the incus, 297–298, 303

Malnutrition: An unhealthy state produced by inadequate dietary intake of nutrients, calories, and/or vitamins, 533

Maltase, 511–512

Maltose, 37

Mamillary bodies (MAM-i-lar-ē): Nuclei in the hypothalamus concerned with feeding reflexes and behaviors; a component of the limbic system, 244, 251

Mammary glands: Milk-producing glands of the female breast, 11, 91, 117, 428, 596–597, 627, 631

Mammography, 606

Mandible, 140–142, 144–145, 199–200, 459

Mandibular branch, of trigeminal nerve, 256

Mandibular condyle, 144

Mandibular fossa, 141, 143

Manganese, 535

Manual region, 15–16

Manubrium, 149–150

Manus, 15–16

Marasmus, 531

Marfan's syndrome, 94, 635, 637

Marginal branch, of coronary artery, 375

Marrow: A tissue that fills the internal cavities in a bone; may be dominated by hemopoietic cells (red marrow) or adipose tissue (yellow marrow). *See* **Bone marrow**

Marrow cavity, 129, 158

Mass, measurement of, A-12, A-13

Masseter muscle, 196, 199–200

Mass movements, 510
Mass number, 29
Mast cell: A connective tissue cell that when stimulated releases histamine, serotonin, and heparin, initiating the inflammatory response, 93, 95, 103, 119, 435, 437, 442
Mastectomy, 606
Mastication (mas-ti-KĀ-shun): Chewing, 492
Mastoid fontanel, 145
Mastoid process, 140–141, 143, 297
Mastoid sinus: Air-filled space in the mastoid process of the temporal bone.
Maternal age, Down syndrome and, 636
Matrix: The extracellular fibers and ground substance of a connective tissue, 92
Matrix, mitochondrial, 69, 525
Matter, 28
Maturity, 630–631
Maxillary artery, 410
Maxillary bone, 140–143, 145, 459
Maxillary branch, of trigeminal nerve, 256
Maxillary sinus (MAK-si-ler-ē): One of the paranasal sinuses; an air-filled chamber lined by a respiratory epithelium that is located in a maxillary bone and opens into the nasal cavity, 144, 460
Maxillary vein, 414
MCF (monocyte-chemotactic-factor), 445
Mean corpuscular hemoglobin concentration (MCHC), 352
Mean corpuscular volume (MCV), 352
Measurement, A-12–A-14
Mechanical processing, of food, 488
Mechanical ventilator, 468
Mechanoreception: Detection of mechanical stimuli, such as touch, pressure, or vibration, 278–279
Mechanoreceptor reflex, 477
Medial: Toward the midline of the body, 17
Medial and lateral pathways, 261–262
Medial canthus, 285
Medial epicondyle
 of femur, 155
 of humerus, 152
Medial malleolus, 156
Medial meniscus, 165–166
Medial rectus muscle, 286–287
Median nerve, 211, 257–258
Median sacral crest, 148
Mediastinal artery, 411–412
Mediastinal surface, of lung, 466–467
Mediastinal vein, 415–416
Mediastinum (mē-dē-as-TĪ-num): Central tissue mass that divides the thoracic cavity into two pleural cavities; includes the aorta and other great vessels, the esophagus, trachea, thymus, the pericardial cavity and heart, and a host of nerves, small vessels, and lymphatics, 19–20
Medical imaging, 21–22
Medium-sized artery, 391–392
Medium-sized vein, 393–394
Medulla: Inner layer or core of an organ.
 adrenal. *See* **Adrenal medulla**
 of lymph node, 432
 of ovary, 590
 renal, 549–550
 of thymus, 433
Medulla oblongata: The most caudal of the brain regions, 243–245, 252–253
Medullary cavity: The space within a bone that contains the marrow.
Megakaryocytes (meg-a-KĀR-ē-ō-sīts): Bone marrow cells responsible for the formation of platelets, 358–359

Meiosis (mī-Ō-sis): Cell division that produces gametes with half the normal somatic chromosome complement, 74, 584–585, 591–592, 615–616
Meissner's corpuscles, 280
Melanin (ME-la-nin): Yellow-brown pigment produced by the melanocytes of the skin, 111–113, 120, 294, 321, 633
Melanocyte (me-LAN-ō-sīt): Specialized cell in the deeper layers of the stratified squamous epithelium of the skin, responsible for the production of melanin, 111–113, 115, 120, 321, 335
Melanocyte-stimulating hormone (MSH) (me-LAN-ō-sīt): Hormone of the anterior pituitary that stimulates melanin production, 315, 321–323
Melanoma, malignant, 113–114, 122, 449, 637
Melatonin (mel-a-TŌ-nin): Hormone secreted by the pineal gland; inhibits secretion of MSH and GnRH, 252, 315, 328
Membrane potential: The potential difference, in millivolts, measured across the cell membrane; a potential difference that results from the uneven distribution of positive and negative ions across a cell membrane, 231–234
Membranes
 cell. *See* **Cell membrane**
 extraembryonic, 618–621
 tissue, 98–100
  cutaneous membranes, 99–100
  mucous membranes, 98–100
  serous membranes, 99–100
  synovial, 99–100
Membranous labyrinth: Endolymph-filled tubes of the inner ear that enclose the receptors of the inner ear, 298
Memory (brain function), 243, 250
 fact, 250
 long-term, 250–251
 short-term, 250
 skill, 250
Memory, immune, 439–440, 444, 446
Memory B cells, 441, 444, 446
Memory cells, 439–440, 444
Memory consolidation, 250, 268
Memory loss, 268
Memory T cells, 440, 446
Menarche (me-NAR-kē): The first menstrual period, which normally occurs at puberty, 595
Ménière's disease, 306
Meninges (men-IN-jēz): Three membranes that surround the surfaces of the CNS—the dura mater, the pia mater, and the arachnoid, 238–240
Meningitis, 240, 270
Meniscus (men-IS-kus): A fibrocartilage pad between opposing surfaces in a joint, 158–159, 165–166
Menopause, 112, 137, 591, 595, 601
Menses (MEN-sēz): The monthly loss of blood and endometrial tissue from the female reproductive tract; also called *menstruation*, 595
Menstrual cycle, 595, 599–600
Menstruation, 595
Merkel's cells, 280
Merocrine (MER-ō-krin): A method of secretion in which the cell ejects materials through exocytosis, 91, 116
Merocrine sweat glands. *See* **Eccrine glands**
Mesencephalic aqueduct: Passageway that connects the third ventricle (diencephalon) with the fourth ventricle (metencephalon), 245–246
Mesencephalon, 243
Mesenchymal cells, 93
Mesenchyme: Embryonic/fetal connective tissue.

Mesenteric artery, 489
 inferior, 407, 411–412, 508
 superior, 407, 411–412, 503
Mesenteric vein, 489
 inferior, 417, 508
 superior, 417, 508
Mesentery (MEZ-en-ter-ē): A double layer of serous membrane that supports and stabilizes the position of an organ in the abdominopelvic cavity and provides a route for the associated blood vessels, nerves, and lymphatics, 21, 489–490, 496, 591
Mesoderm: The middle germ layer, between the ectoderm and endoderm of the embryo, 618–619
Messenger RNA (mRNA): RNA formed at transcription to direct protein synthesis in the cytoplasm, 71–73, 532
Metabolic acid, 570
Metabolic acidosis, 571–572
Metabolic alkalosis, 571–572
Metabolic rate, 324, 537–538
Metabolic turnover: The continuous breakdown and replacement of organic materials within cells, 523
Metabolic water, 537, 568
Metabolism (meh-TAB-ō-lizm): The sum of all of the biochemical processes underway in the body at a given moment; includes anabolism and catabolism, 32, 522–545
 aerobic, 69–70, 524
 of carbohydrates, 523–528, 532
 cellular, 522–532
 of lipids, 523, 527–529, 532
 liver in metabolic regulation, 506
 of proteins, 523, 527, 529–532
Metabolites (me-TAB-ō-līts): Compounds produced in the body as the result of metabolic reactions, 35
 in body fluids, A-15, A-16
Metacarpals: The five bones of the palm of the hand, 138–139, 153
Metaphase (MET-a-faz): A stage of mitosis wherein the chromosomes line up along the equatorial plane of the cell.
 meiosis I, 585
 meiosis II, 585
 mitotic, 74, 76
Metaphase plate, 76, 585
Metaplasia, 104
Metastasis, 77
Metatarsal: One of the five bones of the foot that articulate with the tarsals (proximally) and the phalanges (distally), 138–139, 156–157
Methionine, 531
Metric system, A-12, A-13
MHC (major histocompatibility complex) proteins: A surface antigen that is important to foreign antigen recognition and that plays a role in the coordination and activation of the immune response, 439–441, 447
Micelle (mī-SEL): A spherical aggregation of bile salts, monoglycerides, and fatty acids in the lumen of the intestinal tract, 512–513
Microfilament, 56, 66
Microglia (mī-KRŌG-lē-uh): Phagocytic glial cells in the CNS, derived from the monocytes of the blood, 229–230, 436, 440
Microphages: Neutrophils and eosinophils, 355, 436
Microscopic anatomy, 3
Microtubules: Microscopic tubules that are part of the cytoskeleton and are found in cilia, flagella, the centrioles, and spindle fibers, 56, 66
Microvilli: Small, fingerlike extensions of the exposed cell membrane of an epithelial cell, 55–56, 65–66, 86, 88, 500

Micturition (mik-tu-RI-shun): Urination. *See* **Urination**

Micturition reflex, 563–564, 573

Midbrain: The mesencephalon, 243–244, 252–253

Middle ear: Space between the external ear and internal ear that contains auditory ossicles, 297–298

Middle ear cavity, 143

Middle piece, of sperm, 586

Midsagittal plane: A plane passing through the midline of the body that divides it into left and right halves.

Midsagittal section, 18–19

Migration-inhibitory factor (MIF), 445

Milk, 89, 91, 117

Milk let-down reflex, 321, 631

Milk teeth, 493–494

Millivolt, 232

Mineralocorticoid: Corticosteroids produced by the adrenal cortex; steroids, such as aldosterone, that affect mineral metabolism, 326–327

Minerals, 534–535, 573

Minimal volume, 470–471

Minor calyx, 549–550

Minoxidyl, 116

Mitochondrion (mī-tō-KON-drē-on): An intracellular organelle responsible for generating most of the ATP required for cellular operations, 55–56, 65–66, 69–70, 190–191, 523

energy production in, 69–70, 525–527

inherited disorders of, 70

Mitosis (mī-TŌ-sis): The division of a single cell that produces two identical daughter cells; the primary mechanism of tissue growth, 74–77, 584–585, 592

Mitral valve (MĪ-tral): The left AV, or bicuspid, valve of the heart, 372–374

Mitral valve prolapse, 373

Mixed gases, 471–472

Mixed gland, 89, 92

Mixed nerve, 240

M line, 180–181

Molar (tooth), 493–494

Molecular weight: The sum of the atomic weights of the atoms in a molecule.

Molecule: A chemical structure containing two or more atoms that are held together by chemical bonds, 3, 30

Mongolism. *See* **Down syndrome**

Monoamine oxidase, 237

Monoblasts, 358

Monoclonal antibody, 449

Monocyte-chemotactic factor (MCF), 445

Monocyte-macrophage system, 436, 439–440

Monocytes (MON-ō-sīts): Phagocytic agranulocytes (white blood cells) in the circulating blood, 345, 355–358, 393, 435

Monoglyceride (mo-nō-GLI-se-rīd): A lipid consisting of a single fatty acid bound to a molecule of glycerol, 39

Monohydrogen phosphate, 571

Monokines, 444

Mononucleosis, 449

Monosaccharide (mon-ō-SAK-ah-rīd): A simple sugar, such as glucose or ribose, 37, 46, 512

Monosomy, 636

Monosynaptic reflex: A reflex in which the sensory afferent neuron synapses directly on the motor efferent neuron, 258

Monozygotic twins: Twins produced by the splitting of a single fertilized egg (zygote); also called *identical twins*, 630

Mons pubis, 596

Morula (MOR-ū-la): A mulberry-shaped collection of cells produced through the mitotic divisions of a zygote, 616–617

Motion sickness, 112

Motor end plate, 182–183

Motor neuron, 182, 188, 228, 259–260

Motor pathways, 231, 261–263

Motor unit: All of the muscle cells controlled by a single motor neuron, 188

Mouth, 459, 488–494, 513

Mouth-to-mouth resuscitation, 468

Movement, 2, 178

M phase, 74

MRI (magnetic resonance imaging), 21–23

mRNA. *See* **Messenger RNA**

MS (multiple sclerosis), 230, 270, 447

MSH. *See* **Melanocyte-stimulating hormone**

Mucosa (mū-KŌ-sa): A mucous membrane; the epithelium plus the lamina propria. *See also* **Mucous membrane**

of digestive tract, 488–490

of small intestine, 500

of stomach, 496–497

Mucous gland, 89, 91–92, 460, 462

Mucous membrane, 98–100, 488

Mucous secretions, 99–100

Mucus: Lubricating secretion produced by unicellular and multicellular glands along the digestive, respiratory, urinary, and reproductive tracts, 436, 460

Multinucleate cell, 179

Multiple birth, 630

Multiple endocrine neoplasia, type 2, 637

Multiple sclerosis (MS), 230, 270, 447

Multipolar neuron: A neuron with many dendrites and a single axon; the typical form of a motor neuron, 227–228

Mumps, 492

Muscle: A contractile organ composed of muscle tissue, blood vessels, nerves, connective tissues, and lymphatics. *See also specific muscles*

antagonist, 195

involuntary, 102

primary actions of, 195

prime mover, 195

red, 192

synergist, 195

voluntary, 102, 179

white, 192

Muscle contraction, 179–181

cardiac muscle, 194, 375–376

contraction cycle, 182–184

energetics of, 189–192

isometric, 189

isotonic, 189

mechanics of, 184–189

frequency of muscle fiber stimulation, 187–188

number of muscle fibers involved, 188–189

neuromuscular junction, 182–183, 186

opposing muscle movements, 189

recovery period, 191

skeletal muscle, 182–184

sliding filament theory of, 181–182, 185–186

smooth muscle, 195

summary of, 186

thermoregulation and, 192

Muscle cramps, 217

Muscle elongation, 189

Muscle fatigue, 191, 216

Muscle fiber, 100–101, 178–180, 183

types of, 192–193

Muscle spindle, 258, 281

Muscle tissue: A tissue characterized by the presence of cells capable of contraction; includes skeletal, cardiac, and smooth muscle tissues, 84, 100–102

cardiac, 100–102

skeletal, 100–102

smooth, 100–102

Muscle tone, 189, 253, 258–259

Muscular artery, 391–392

Muscular dystrophy, 182, 217, 635, 637

Muscularis externa (mus-kū-LAR-is): Concentric layers of smooth muscle responsible for peristalsis.

of digestive tract, 489–490

of small intestine, 500

Muscularis mucosae: Layer of smooth muscle beneath the lamina propria; responsible for moving the mucosal surface.

of digestive tract, 488–489

of small intestine, 500

of stomach, 496

Muscular spasm, 325

Muscular system, 178–223. *See also specific muscles*

aging of, 216–217

anatomy of, 195–217

appendicular muscles, 198, 204–216

of pelvis and lower limbs, 211–216

of shoulders and upper limbs, 205–211

axial muscles, 198–204

of head and neck, 199–201

of pelvic floor, 199, 204

of spine, 199, 201–202

of trunk, 199, 201–203

cardiac muscle, 193–194

components of, 7

functions of, 7

integration with other systems, 217–218

cardiovascular system, 420

digestive system, 515

endocrine system, 337

integumentary system, 121

lymphatic system, 448

nervous system, 269

reproductive system, 603

respiratory system, 480

skeletal system, 167

urinary system, 565

prenatal and early postnatal development of, 624

skeletal muscle. *See* **Skeletal muscle**

smooth muscle, 193–195

Musculocutaneous nerve, 257–258

Mutation, 74

Myalgia, 217

Myasthenia gravis, 184, 217, 447

Mycobacterium tuberculosis, 468

Myelin (MĪ-e-lin): Insulating sheath around an axon consisting of multiple layers of glial cell membrane; significantly increases conduction rate along the axon, 230, 235, 446

Myelinated axon, 229–230, 234–235

Myelination: The formation of myelin.

Myelin sheath, 230

Myeloblasts, 358

Myelocytes, 358

Myelography, 270

Myeloid leukemia, 362

Myeloid stem cells, 347, 351, 356, 358

Myeloid tissue, 350–351

Myenteric plexus (mī-en-TER-ik): Parasympathetic motor neurons and sympathetic postganglionic fibers located between the circular and longitudinal layers of the muscularis externa, 489–490, 496, 500

Mylohyoid muscle, 200–201

Myocardial infarction, 375, 384, 468

Myocardium: The cardiac muscle tissue of the heart, 370–371

Myofibril: Organized collections of myofilaments in skeletal and cardiac muscle cells, 179–181

Myofilament: Protein filament consisting of the proteins actin and myosin; bundles of myofilaments form myofibrils, 179

Myoglobin (MĪ-ō-glō-bin): An oxygen-binding pigment especially common in slow skeletal and cardiac muscle fibers, 41–42, 192

Myogram: A recording of the tension produced by muscle fibers on stimulation, 187

Myoma, 217

Myometrium (mī-ō-MĒ-trē-um): The thick layer of smooth muscle in the wall of the uterus, 594–595, 621–622, 627–628

Myopia, 293, 306

Myosin: Protein component of the thick myofilaments, 66, 100–102, 179–185, 194

Myositis, 217

Myxedema, 336

N

NAD (nicotinamide adenine dinucleotide), 524–526, 536

Nail: Keratinous structure produced by epithelial cells of the nail root, 6, 111, 117–118

Nail bed, 118

Nail body, 117

Nail root, 118

Nanometer, 28

Narcolepsy, 447

Nares, external (NĀ-rēz): The entrance from the exterior to the nasal cavity, 458

Nares, internal: The entrance from the nasal cavity to the nasopharynx, 458–459

Nasal bone, 140–142, 144–145, 459

Nasal cavity: A chamber in the skull bounded by the internal and external nares, 9, 282–283, 458–459, 491

Nasal complex, 144

Nasal conchae, 458–459
    inferior, 141–142, 144, 459–460
    middle, 141–143, 459–460
    superior, 142–143, 459–460

Nasal septum, 141, 144, 459

Nasolacrimal duct: Passageway that transports tears from the nasolacrimal sac to the nasal cavity, 285–286, 460

Nasolacrimal sac: Chamber that receives tears from the lacrimal ducts.

Nasopharynx (nā-zō-FAR-inks): Region posterior to the internal nares, superior to the soft palate, and ending at the oropharynx, 297, 459–460, 491

Natural killer (NK) cells, 430–431, 435–436, 446

Naturally acquired immunity, 438

Natural passive immunity, 438

Navicular bone, 156–157

NE. See **Norepinephrine**

Near point of vision, 306

Nearsightedness, 293, 306

Neck
    arteries of, 410
    bones of, 145–150
    muscles of, 199–201
    veins of, 414

Neck of bone, 136–137
    of femur, 155
    of radius, 152–153
    of scapula, 151

Neck of sperm, 586

Neck of tooth, 493

Necrosis, 104, 437

Negative feedback: Corrective mechanism that opposes or reverses a variation from normal limits and restores homeostasis, 12–13, 258, 314
    direct, 318

Neomycin, 305

Neon, 29

Neonatal period, 630–631

Neonate: A newborn infant, 630–631

circulation in, 419
    respiratory changes at birth, 478–479, 630

Neoplasm: A tumor, or mass of abnormal tissue, 77

Nephritis, 573

Nephrolithiasis, 563

Nephrology, 573

Nephron (NEF-ron): Basic functional unit of the kidney, 550–554, 573

Nephropathy, diabetic, 336

Nerve, 231. See also specific nerves
    of skeletal muscles, 178–179

Nerve deafness, 304–306

Nerve impulse: An action potential in a neuron cell membrane. See **Action potential**

Nerve plexus, 258

Nervous system, 226–275. See also **Central nervous system; Peripheral nervous system**
    aging of, 268
    anatomical organization of, 230–231
    cardiovascular regulation, 399–402
    components of, 7
    functions of, 7, 226–227
    integration with other systems, 268–269
        cardiovascular system, 420
        digestive system, 515
        endocrine system, 337
        integumentary system, 121
        lymphatic system, 448
        muscular system, 218
        reproductive system, 603
        respiratory system, 480
        skeletal system, 167
        urinary system, 565
    prenatal and early postnatal development of, 624

Nervous tissue. See **Neural tissue**

Neural cortex: An area where gray matter is found at the surface of the CNS, 231

Neural fold, 623

Neural plate, 623

Neural retina, 289

Neural tissue, 84, 102–103. See also **Neuroglia; Neuron**

Neural tunic, 287–290

Neurilemma, 230

Neuroeffector junction: A synapse between a motor neuron and a peripheral effector, such as a muscle, gland cell, or fat cell, 236

Neuroglandular junction, 236

Neuroglia (noo-RŌG-lē-ah): Cells of the CNS and PNS that support and protect the neurons, 227, 229–230

Neurology, 270

Neuromuscular junction: A specific type of neuroeffector junction, 182–183, 186, 236

Neuron (NOO-ron): A cell in neural tissue specialized for intercellular communication by (1) changes in membrane potential and (2) synaptic connections, 54, 102, 230
    aging of, 268
    bipolar, 228
    communication, 235–238
    convergence, 238
    divergence, 238
    functional classification of, 228–229
    function of, 231–235
    ganglionic, 264
    interneuron. See **Interneuron**
    motor, 182, 188, 228, 259–260
    multipolar, 227–228
    postganglionic, 264–266
    postsynaptic, 236
    preganglionic, 264–266
    presynaptic, 236
    sensory, 228, 259–261, 278
    structural classification of, 228
    structure of, 227–228

unipolar, 228

Neuronal pool, 238

Neuropathy, diabetic, 336

Neurotoxin, 234, 270

Neurotransmitter: Chemical compound released by one neuron to affect the membrane potential of another, 183, 235–238, 264, 314
    excitatory, 237–238
    inhibitory, 237–238

Neutralization, 442

Neutral solution, 34

Neutron: A fundamental particle that does not carry a positive or negative charge, 28

Neutrophil (NOO-trō-fil): A phagocytic microphage that is very numerous and usually the first of the mobile phagocytic cells to arrive at an area of injury or infection, 345, 355–358, 435–437

Niacin, 536

Night blindness, 295, 535

Nipple: An elevated epithelial projection on the surface of the breast, containing the openings of the lactiferous sinuses, 597

Nissl bodies: The ribosomes, Golgi, rough endoplasmic reticulum, and mitochondria of the cytoplasm of a typical nerve cell, 227–228

Nitric oxide, 237, 589

Nitrogen, 28
    in air, 471
    decompression sickness, 472
    partial pressure of, 472

Nitrogenous base, 44, 71

Nitrogenous wastes: Organic waste products of metabolism that contain nitrogen; include urea, uric acid, and creatinine, 531

NK cells, 430–431, 435–436, 446

Nociception (nō-sē-SEP-shun): Pain perception, 278–279

Node of Ranvier: Area between adjacent glial cells where the myelin covering of an axon is incomplete, 230

Nondisplaced fracture, 135

Nonessential amino acid, 531

Non-Hodgkin's lymphoma, 449

Nonpolar covalent bond, 31

Nonself antigen, 439

Nonshivering thermogenesis, 539

Nonspecific defenses, 355, 435–437, 446

Nonspecific resistance, 435

Noradrenaline. See **Norepinephrine**

Norepinephrine (NE) (nor-ep-i-NEF-rin): A neurotransmitter in the PNS and CNS and a hormone secreted by the adrenal medulla; also called noradrenaline, 236–237, 264–265, 314–318, 322, 327–328, 334, 382, 402–404, 560

Normal saline, 61, 77

Normovolemia, 362

Norplant system, 605

Nose, 458–460

Nuclear envelope, 55, 67, 70

Nuclear imaging, 47

Nuclear membrane, 77

Nuclear pore, 55, 70, 73

Nuclease, 504, 512

Nucleic acid (nū-KLĀ-ik): A chain of nucleotides containing a five-carbon sugar, a phosphate group, and one of four nitrogenous bases that regulate the synthesis of proteins and make up the genetic material in cells, 36, 43–44. See also **DNA; RNA**
    catabolism of, 531
    functions of, 47
    metabolism of, 531–532
    structure of, 47
    synthesis of, 531–532

Nucleolus (noo-KLĒ-ō-lus): Dense region in the nucleus that represents the site of RNA synthesis, 55–56, 70, 76

Nucleoplasm, 55, 70

Nucleotide: Compound consisting of a nitrogenous base, a simple sugar, and a phosphate group, 43–44, 46

Nucleus, atomic, 28–29

Nucleus, cell: Cellular organelle that contains DNA, RNA, and proteins, 55–56, 65, 70–74

Nucleus, cerebral: A mass of gray matter in the CNS, 231

Nucleus pulposus (pul-PŌ-sus): Gelatinous central region of an intervertebral disc.

Nutrient: An organic compound that can be broken down in the body to produce energy, 35, 522
   absorption of, 510–513
   processing of, 510–513
   use in cellular metabolism, 522–532

Nutrient pool, 522–523

Nutrient reserves, 523

Nutrition, 522–545
   aging and nutritional requirements, 539–540
   basic food groups, 533–534
   dietary fats and cholesterol, 530
   minerals, 534–535
   vitamins, 534–536
   water requirements, 536–537

Nutritionist, 532

Nystagmus, 306

O

Obesity: Body weight 10–20 percent above standard values as the result of body fat accumulation, 393, 537, 540

Oblique muscle, of stomach wall, 496

Oblique muscles, 201–202

Obstructive shock, 404

Obturator foramen, 154

Obturator nerve, 257–258

Occipital bone, 140–143, 145

Occipital condyle, 141, 143

Occipitalis muscle, 197, 199–200

Occipital lobe, 244, 246–247

Ocular: Pertaining to the eye.

Oculomotor nerve (ok-ū-lō-MŌ-ter): Cranial nerve III, which controls the extrinsic oculomotor muscles other than the superior oblique and the lateral rectus, 253–255, 266

Odontoid process, 147

Odorant, 283

Odorant binding protein, 283

Oil gland. *See* **Sebaceous glands**

Oils, 38

Olecranon: The proximal end of the ulna that forms the prominent point of the elbow, 152–153, 164

Olecranon fossa, 152, 164

Olestra, 38

Olfaction: The sense of smell, 255, 278, 282–283, 458

Olfactory bulb (ol-FAK-tor-ē): Expanded ends of the olfactory tracts; the sites where the axons of NI synapse on CNS interneurons that lie beneath the frontal lobe of the cerebrum, 254, 283

Olfactory cortex, 247, 251, 278

Olfactory epithelium, 282–283

Olfactory gland, 282

Olfactory nerve (N I), 254–255

Olfactory organ, 282–284

Olfactory receptors, 282–283, 306

Olfactory tract, 251, 254, 283

Oligodendrocytes (o-li-gō-DEN-drō-sīts): CNS glial cells responsible for maintaining cellular organization in the gray matter and providing a myelin sheath in areas of white matter, 229–230

Omega-3 fatty acid, 47

Oncogene (ON-kō-jēn): A gene that can turn a normal cell into a cancer cell.

Oncologists (on-KOL-ō-jists): Physicians specializing in the study and treatment of tumors, 104

Oocyte (Ō-ō-sīt): A cell whose meiotic divisions will produce a single ovum and three polar bodies.
   activation of, 616
   fertilization, 614–616
   at ovulation, 614–616
   primary, 591–593
   secondary, 591–593, 614

Oogenesis (ō-ō-JEN-e-sis): Ovum production, 591

Oogonia (ō-ō-GŌ-nē-uh): Stem cells in the ovaries whose divisions give rise to oocytes, 591

Oophoritis, 606

Ooplasm: The cytoplasm of the ovum.

Open fracture, 135

Ophthalmic branch, of trigeminal nerve, 256

Ophthalmology, 306

Opiate, 476

Opportunistic infection, 443

Opposition (movement), 161

Opsin: A protein that is a structural component of the visual pigment rhodopsin, 295

Opsonin, 444

Opsonization, 444

Optic chiasm (OP-tik KĪ-asm): Crossing point of the optic nerves, 244, 254–255, 295–296, 319

Optic disc, 287, 289–290, 295

Optic foramen, 141, 255

Optic nerve (N II): Nerve that carries signals from the eye to the optic chiasm, 254–255, 287, 289–290, 295–296

Optic tract: Tract over which nerve impulses from the retina are transmitted between the optic chiasm and the thalamus, 252, 254–255, 296

Oral cavity, 459, 488–494
   cancer of, 513
   functions of, 490

Oral contraceptives, 604–605

Oral phase, of swallowing, 494–495

Orbicularis muscle, 196

Orbicularis oculi muscle, 199–200

Orbicularis oris muscle, 196, 199–200

Orbit: Bony cavity of the skull that contains the eyeball, 140

Orbital fat, 286–287

Orchiectomy, 584, 606

Orchitis, 606

Organelle (or-gan-EL): An intracellular structure that performs a specific function or group of functions, 3–5, 55–56, 65–70

Organic compound: A compound containing carbon, hydrogen, and usually oxygen, 35–47
   in body, 46–47

Organism, 5

Organism level of organization, 4–5

Organization, levels of, 3–5

Organ level of organization, 4–5

Organ of Corti: Receptor complex in the cochlear duct that includes the inner and outer hair cells, supporting cells and structures, and the tectorial membrane; provides the sensation of hearing, 299, 302, 304

Organogenesis: The formation of organs during embryological and fetal development, 623–625

Organs: Combinations of tissues that perform complex functions, 3

Organ system, 3–11. *See also specific systems*
   integration of systems, 5–14

Orgasm
   female, 601
   male, 600

Origin: Point of attachment of a muscle that is least movable, 195

Oropharynx: The middle portion of the pharynx, bounded superiorly by the nasopharynx, anteriorly by the oral cavity, and inferiorly by the laryngopharynx, 459–460, 491

Orthopedics, 168

Orthostatic hypotension, 421

Osmolarity (oz-mō-LAR-i-tē): Osmotic concentration; the total concentration of dissolved materials in a solution, regardless of their specific identities, expressed in terms of moles, 568

Osmoreceptor: A receptor sensitive to changes in the osmotic concentration of the plasma, 561

Osmosis (oz-MŌ-sis): The movement of water across a semipermeable membrane toward a solution containing a relatively high solute concentration, 58, 60–61, 65, 396

Osmotic pressure: The force of osmotic water movement; the pressure that must be applied to prevent osmotic movement across a membrane, 60–61, 396

Osseous tissue: Bone tissue, strong connective tissue containing specialized cells and a mineralized matrix of crystalline calcium phosphate and calcium carbonate. *See* **Bone**

Ossicles: Small bones.

Ossification: The formation of bone, 131
   endochondral, 131–132
   intramembranous, 131

Ossification center, 131, 144
   primary, 131–132
   secondary, 132

Osteoarthritis, 159

Osteoblast: A cell that produces the fibers and matrix of bone, 130–133, 136

Osteoclast (OS-tē-ō-klast): A cell that dissolves the fibers and matrix of bone, 130–133, 136

Osteocyte (OS-tē-ō-sīt): A bone cell responsible for the maintenance and turnover of the mineral content of the surrounding bone, 98, 128–131, 133

Osteogenesis, 130

Osteolysis (os-tē-OL-i-sis): The breakdown of the mineral matrix of bone, 130

Osteomyelitis, 168

Osteon (OS-tē-on): The basic histological unit of compact bone, consisting of osteocytes organized around a central canal and separated by concentric lamellae, 129–130

Osteopenia, 136, 168

Osteoporosis, 104, 136–137, 165, 168, 540, 601

Otic: Pertaining to the ear.

Otitis media, 297

Otolith: A complex formed by the combination of a gelatinous matrix and aggregations of calcium carbonate crystals; located above one of the maculae of the vestibule, 300–301

Outer membrane, of mitochondria, 69, 525–526

Oval window: Opening in the bony labyrinth where the stapes attaches to the membranous wall of the vestibular duct, 297–298, 301, 303–304

Ovarian artery, 412, 591, 594

Ovarian cycle (ō-VAR-ē-an): Monthly chain of events that leads to ovulation, 591–593, 597–600

Ovarian follicle, 332, 591–593

Ovarian vein, 591, 594

Ovaries: The female gonads, reproductive glands that produce gametes, 8, 11, 590–594
   cancer of, 606
   as endocrine organ, 315, 332

innervation of, ANS, 265–266
Oviduct, 11, 590, 593–594, 614, 616
Ovulation (ōv-ū-LĀ-shun): The release of a secondary oocyte, surrounded by cells of the corona radiata, following the rupture of the wall of a tertiary follicle, 320, 332, 592–593, 595, 598–600, 614–617
Ovum/ova (Ō-vum): The functional product of meiosis II, produced after fertilization of a secondary oocyte, 54, 582, 592
Oxygen, 28, 31, 35
  in air, 471
  in blood, 282, 383, 401, 478
  bound to hemoglobin, 473–475
  in cerebrospinal fluid, 401
  diffusion out of capillaries, 472
  gas exchange in lungs, 471–472
  partial pressure within circulatory system, 472
  transport in blood, 472–475
  use in electron transport system, 526
Oxygen debt, 192
Oxytocin (oks-i-TŌ-sin): Hormone produced by hypothalamic cells and secreted into capillaries at the posterior pituitary gland; stimulates smooth muscle contractions of the uterus or mammary glands in females, but has no known function in males, 252, 315, 318–319, 322–323, 628

P

Pacemaker cells: Cells of the SA node that set the pace of cardiac contraction, 102, 194, 377
Pacesetter cells, 195, 490
Pacinian corpuscle (pa-SIN-ē-an): Receptor sensitive to vibration, 280
Packed cell volume, 347
Pain, 103, 278–279
  fast, 278–279
  referred, 279
  slow, 279
Palate: Horizontal partition separating the oral cavity from the nasal cavity and nasopharynx; includes an anterior bony (hard) palate and a posterior fleshy (soft) palate. *See* **Hard palate; Soft palate**
Palatine bone, 141–142, 144, 459
Palatine tonsils, 432, 459–460, 491
Palmar arch artery, 407–408
Palmaris longus muscle, 196, 209–210
Palmar venous arch, 413, 415
Palpate: To examine by touch.
Palpebrae (pal-PĒ-brē): Eyelids, 285, 287
Pancreas: Digestive organ containing exocrine and endocrine tissues; exocrine portion secretes pancreatic juice, and endocrine portion secretes hormones, including insulin and glucagon, 8, 10, 489, 496, 503–504
  control of secretions of, 504
  digestive secretions of, 503
  endocrine, 315, 329–331
  exocrine, 329
  histological organization of, 503
  innervation of, ANS, 265–266
Pancreatic acini, 503
Pancreatic duct: A tubular duct that carries pancreatic juice from the pancreas to the duodenum, 329, 503, 507
Pancreatic islets: Aggregations of endocrine cells in the pancreas, 329, 503
Pancreatic juice: A mixture of buffers and digestive enzymes that is discharged into the duodenum under the stimulation of the enzymes secretin and cholecystokinin, 503
Pancreatic obstruction, 513
Pancreatitis, 503–504, 513
Pantothenic acid, 509, 536

Papilla (pa-PIL-la): A small, conical projection.
  duodenal, 507
  hair, 115
  renal, 549–550
  on tongue, 283–284
Papillary duct, 550–553
Papillary layer, of dermis, 114
Papillary muscles, 372–374
Pap test, 89
Parafollicular cells, 324–325
Paralysis, 36, 184, 242
Paralytic shellfish poisoning, 270
Paranasal sinuses: Bony chambers lined by respiratory epithelium that open into the nasal cavity; includes the frontal, ethmoidal, sphenoidal, and maxillary sinuses, 9, 144, 460
Paraplegia, 242, 270
Parasitic worm infection, 356
Parasympathetic division: One of the two divisions of the autonomic nervous system; generally responsible for activities that conserve energy and lower the metabolic rate; also called *craniosacral division*, 226–227, 264–267
Parathyroid glands: Four small glands embedded in the posterior surface of the thyroid; responsible for parathyroid hormone secretion, 8, 315, 324–326
Parathyroid hormone (PTH): Hormone secreted by the parathyroid gland when plasma calcium levels fall below the normal range; causes increased osteoclast activity, increased intestinal calcium uptake, and decreased calcium ion loss at the kidneys, 134, 315, 317, 324–326, 333, 513, 556
Paresthesia, 270
Parietal: Referring to the body wall or outer layer.
Parietal bone, 129, 140–142, 145
Parietal cell: Cell of the gastric glands that secretes HCl and intrinsic factor, 497–498, 571
Parietal lobe, 244, 246–247
Parietal pericardium, 19–20, 100, 369, 371
Parietal peritoneum, 100, 490, 496
Parietal pleura, 19–20, 100, 464, 467–468
Parieto-occipital sulcus, 244, 246–247
Parkinson's disease: Progressive motor disorder due to degeneration of the cerebral nuclei, 253, 270
Parotid duct, 492
Parotid glands (pa-ROT-id): Large salivary glands that secrete a saliva containing high concentrations of salivary (alpha) amylase, 89, 492
Partial pressure, 471–472
Parturition (par-tū-RISH-un): Childbirth, delivery, 628–630
Passive immunity, 438
Passive transport, 58, 61
Patella (pa-TEL-ah): The bone of the kneecap, 138–139, 155, 165–166
Patellar ligament, 156, 166, 213
Patellar reflex, 259
Patellar surface, of femur, 155
Paternity test, 354
Pathogenic: Disease-causing, 428
Pathological physiology, 3
Pathologist (pa-THO-lo-jist): A physician specializing in the identification of diseases on the basis of characteristic structural and functional changes in tissues and organs, 104
Pathology, 3
Pathway
  metabolic, 43
  neural, 231
Peak flow meter, 471
Pectineus muscle, 211–213
Pectoral girdle, 139, 150–151
Pectoralis major muscle, 196, 206–207, 597

Pectoralis minor muscle, 205–207, 469
Pedal region, 15–16
Pediatrics, 630, 638
Pedicel (process of podocyte), 553
Pedicles (PE-di-kels): Thick bony struts that connect the vertebral body with the articular and spinous processes, 146–147
Pellagra, 536
Pelvic cavity: Inferior subdivision of the abdominopelvic (peritoneal) cavity; encloses the urinary bladder, the sigmoid colon and rectum, and male or female reproductive organs, 19–20
Pelvic floor, muscles of, 199, 204
Pelvic girdle, 139, 153–155, 165
Pelvic inflammatory disease (PID), 594, 601, 606
Pelvic lymph nodes, 428
Pelvic nerve, 266
Pelvic outlet, 155
Pelvic region, 15–16
Pelvis: A bony complex created by the articulations between the coxae, the sacrum, and the coccyx, 6, 15–16, 154–155
  female vs. male, 154–155
  muscles of, 211–216
Pelvis, renal, 549–550
Penis (PĒ-nis): Component of the male external genitalia; a copulatory organ that surrounds the urethra and serves to introduce semen into the female vagina, 11, 265–266, 582, 588–589, 600
Pepsin, 497, 499, 512–513
Pepsinogen, 497
Peptic ulcer, 497
Peptidase, 511–513
Peptide, 41, 46
Peptide bond, 41
Peptide hormones, 315, 317
Perception, 278
Perforated ulcer, 514
Perforating canal: A passageway in compact bone that runs at right angles to the axes of the osteons, between the periosteum and endosteum, 129–130
Perforin, 440
Performance-enhancing drugs, 332
Perfusion: The blood flow through a tissue.
Pericardial artery, 411–412
Pericardial cavity (per-i-KAR-dē-al): The space between the parietal pericardium and the epicardium (visceral pericardium); covers the outer surface of the heart, 19–20, 368–369, 371, 467
Pericardial effusion, 104
Pericardial fluid, 369
Pericardial sac, 369
Pericarditis, 104, 384
Pericardium (per-i-KAR-dē-um): The fibrous sac that surrounds the heart and whose inner, serous lining is continuous with the epicardium, 19, 100, 368, 370
  parietal, 369, 371
  visceral, 369, 371
Perichondrium (per-i-KON-drē-um): Layer that surrounds a cartilage, consisting of an outer fibrous and an inner cellular region, 96, 131
Perilymph (PER-ē-limf): A fluid similar in composition to cerebrospinal fluid; found in the spaces between the bony labyrinth and the membranous labyrinth of the inner ear, 298, 302–304, 567, A-16
Perimetrium, 594
Perimysium (per-i-MĪS-ē-um): Connective tissue partition that separates adjacent fasciculi in a skeletal muscle, 178–179

Perineum (pe-ri-NĒ-um): The pelvic floor and associated structures, 199, 204, 596

Periodic table, 28, A-11

Periodontal disease, 514

Periodontal ligament, 493–494

Periosteum (pe-rē-OS-tē-um): Layer that surrounds a bone, consisting of an outer fibrous and inner cellular region, 98, 129–130, 158

Peripheral ganglia, 265–266

Peripheral nerve palsy, 258

Peripheral nervous system (PNS): All neural tissue outside the CNS, 7, 226, 254–263

afferent division, 227

anatomical organization of, 231

cranial nerves, 254–256

efferent division, 227

functions of, 226–227

nerve plexuses, 258

reflexes, 258–260

sensory and motor pathways, 260–262

spinal nerves, 163, 231, 240–241, 257

Peripheral neuropathy, 258

Peripheral resistance: The resistance to blood flow primarily caused by friction with the vascular walls, 394–395, 400–402

Peristalsis (per-i-STAL-sis): A wave of smooth muscle contractions that propels materials along the axis of a tube such as the digestive tract, the ureters, or the ductus deferens, 490–491, 495, 501, 510, 514, 586, 593, 601

Peritoneal cavity, 19, 88

Peritoneal fluid, 490

Peritoneum (per-i-tō-NĒ-um): The serous membrane that lines the peritoneal (abdominopelvic) cavity, 19, 100, 562

parietal, 490, 496

visceral, 490

Peritonitis, 104, 514

Peritubular capillaries: A network of capillaries that surrounds the proximal and distal convoluted tubules of the kidneys, 554

Permanent dentition, 493–494

Permeability: Ease with which dissolved materials can cross a membrane; if freely permeable, any molecule can cross the membrane; if impermeable, nothing can cross—most biological membranes are selectively permeable, some materials can cross.

of cell membrane, 58

of epithelium, 84

Permissive effect, of hormones, 333

Pernicious anemia, 447, 536

Peroxisome, 55–56, 65, 69

Perpendicular plate, of ethmoid, 143

Perspiration, 89, 117, 120, 539, 568, A-16

insensible, 538

sensible, 538

in thermoregulation, 12–13, 117

Pes: The foot, 15–16

PET (positron emission tomography) scan, 23

Peyer's patches, 432

pH: The negative exponent of the hydrogen ion concentration, 34–35

of blood, 34–35, 282, 346, 401, 478, 531, 548, 556, 570

buffers and, 34–35, 570–571

of cerebrospinal fluid, 282, 401

of urine, 558

Phagocyte: A cell that performs phagocytosis, 64, 93, 230, 348, 356, 435–436, 444

Phagocytic regulators, 445

Phagocytosis (fa-gō-si-TŌ-sis): The engulfing of extracellular materials or pathogens; movement of extracellular materials into the cytoplasm by enclosure in a membranous vesicle, 63–64, 355

Phalanx/phalanges (fa-LAN-jēz): A bone of the fingers or toes.

of fingers, 138–139, 153

of toes, 138–139, 156–157

Pharmacology: The study of drugs, their physiological effects, and their clinical uses.

Pharyngeal arch, 623

Pharyngeal phase, of swallowing, 494–495

Pharyngeal tonsils, 459–460, 491

Pharyngotympanic tube: A passageway that connects the nasopharynx with the middle ear cavity; also called *Eustachian tube* or *auditory tube. See* **Auditory tube**

Pharynx: The throat; a muscular passageway shared by the digestive and respiratory tracts, 9–10, 458, 460, 489, 494

cancer of, 513

Phenotype (FĒN-ō-tīp): Physical characteristics that are genetically determined, 632, 634

Phenylalanine, 531

Phenylalanine hydroxylase, 531

Phenylketonuria (PKU), 47, 531, 540, 635, 637

Phenylthiocarbamate, ability to taste, 635

Phlebitis, 421

Phlebotomy, 362

Phosphate, absorption of, 133, 325, 331

Phosphate buffer system, 570–571

Phosphate group: PO_4^-, 40, 43–44

Phosphodiesterase, 316

Phospholipid (fos-fō-LIP-id): An important membrane lipid whose structure includes hydrophilic and hydrophobic regions, 38–40, 57, 529

Phospholipid bilayer, 57

Phosphoric acid, 36

Phosphorus, 28, 535

Phosphorylation (fos-for-i-LĀ-shun): The addition of a high-energy phosphate group to a molecule, 316

Photon, 294

Photoreception: Sensitivity to light, 295

Photoreceptors, 289–290, 294–295

Phrenic artery, 411–412

Phrenic nerve, 257–258, 479

Phrenic vein, 415–416

Physical barrier (body defenses), 435–436

Physical conditioning, 193

Physiology (fiz-ē-OL-o-jē): Literally the study of function; considers the ways living organisms perform vital activities, 2–3

Pia mater: The innermost layer of the meninges that surrounds the CNS, 239–240, 246

PID (pelvic inflammatory disease), 594, 601, 606

Pigment: A compound with a characteristic color.

Pigmentation, 112–113

Pigmented part, of retina, 289

Pimple, 116

Pineal gland (PIN-ē-al): Neural tissue in the posterior portion of the roof of the diencephalon, responsible for the secretion of melatonin, 8, 244, 252, 296, 315, 328

Pinkeye, 285

Pinna. *See* **Auricle**

Pinocytosis (pi-nō-sī-TŌ-sis): The introduction of fluids into the cytoplasm by enclosing them in membranous vesicles at the cell surface, 63–64

Pisiform bone, 153

Pitch, of sound, 462

Pituitary gland: Endocrine organ situated in the sella turcica of the sphenoid bone and connected to the hypothalamus by the infundibulum; includes the posterior pituitary and the anterior pituitary glands, 8, 143, 243, 315, 318–323

anterior, 318–323, 589

hormones of, 320–322

posterior, 318–319, 321–323

Pivot joint, 161–163

PKU (phenylketonuria), 47, 531, 540, 635, 637

Placenta (pla-SENT-uh): A complex structure in the uterine wall that permits diffusion between the fetal and maternal circulatory systems; also called *afterbirth*, 418, 620–622

blood supply, 418

hormones and, 620–622

Placental circulation, 620

Placental stage, of labor, 629

Placenta previa, 638

Placentation, 616, 620–622

Planes, sectional, 18–19

Plantar: Referring to the sole of the foot.

Plantar arch artery, 407, 412

Plantar flexion, 161, 195

Plantar reflex, 260

Plantar vein, 413

Plantar venous arch, 413

Plaque, 361, 393, 529

Plasma (PLAZ-mah): The fluid ground substance of whole blood; what remains after the cells have been removed from a sample of whole blood, 94, 96, 344–346, 566–567

Plasma cell: Activated B cells that secrete antibodies, 93, 346, 430, 441, 444, 446

Plasma proteins, 346, 360, 395, 506, 570

Plasmin, 361

Plasminogen, 361

Plateau, of action potential, 376

Platelet factor, 360

Platelet phase, of hemostasis, 359

Platelet plug, 359

Platelets (PLĀT-lets): Small packets of cytoplasm that contain enzymes important in the clotting response; manufactured in the bone marrow by cells called megakaryocytes, 96, 344–346, 357–359

Platysma muscle, 200–201

Pleura (PLOO-ra): The serous membrane lining the pleural cavities, 19–20, 100, 467

parietal, 464, 467–468

visceral, 464, 467–468

Pleural cavities: Subdivisions of the thoracic cavity that contain the lungs, 19–20, 466–468

Pleural effusion, 104

Pleural fluid, 467

Pleuripotent stem cells, 347

Pleurisy, 481

Pleuritis, 104

Plexus (PLEK-sus): A network or braid. *See specific plexuses*

Plica (PLĪ-ka): A permanent transverse fold in the wall of the small intestine, 489, 500

Pneumocystis carinii pneumonia, 466

Pneumonia, 466, 481

Pneumotachometer, 471

Pneumothorax, 467, 481

PNS. *See* **Peripheral nervous system**

Podocyte (PŌ-do-sīt): A cell whose processes surround the glomerular capillaries and assist in filtration, 552–553

Point mutation, 74

Polar body: A nonfunctional packet of cytoplasm containing chromosomes eliminated from an oocyte during meiosis, 591–592, 615

Polar covalent bond: A form of covalent bond in which there is an unequal sharing of electrons, 31

Polio, 217

Pollex: The thumb, 153

Polycythemia, 352, 362, 536
Polydipsia, 322
Polygenic inheritance, 633–635
Polymer: A large molecule consisting of a long chain of subunits.
Polyp, 514
Polypeptide: A chain of amino acids strung together by peptide bonds; those containing over 100 peptides are called proteins, 41
Polysaccharide (pol-ē-SAK-ah-rīd): A complex sugar, such as glycogen or a starch, 37, 46
Polyspermy, 616
Polysynaptic reflex: A reflex with interneurons interposed between the sensory fiber and the motor neuron(s), 259
Polyunsaturated fats: Fatty acids containing carbon atoms linked by double bonds, 38–39
Polyuria, 322, 330, 336
Pons: The portion of the brain anterior to the cerebellum, 243–245, 253
Popliteal (pop-LIT-ē-al): Pertaining to the back of the knee.
Popliteal artery, 398, 407, 412
Popliteal vein, 413, 415
Popliteus muscle, 213–215
Portal system, 319, 416
Position sense, 281
Positive chemotaxis, 355
Positive feedback: Mechanism that increases a deviation from normal limits following an initial stimulus, 12–14
Postcentral gyrus, 244, 247
Posterior: Toward the back; dorsal, 16–17
Posterior cavity, of eye, 286–287, 289–290
Posterior chamber, of eye, 288, 290–291
Posterior column pathway, 260–261
Posterior cruciate ligament, 166
Posterior interventricular branch, of coronary artery, 375
Posterior median sulcus, 240–242
Posterior pituitary, 318–319, 321–323
Posterior section, 18
Postganglionic fiber: The axon of a ganglionic neuron, 264–266
Postnatal development, 614, 630–632
Postovulatory phase: The secretory phase of the menstrual cycle, 595
Postsynaptic neuron, 236
Posture, 178, 253, 258, 299
Potassium, 28, 30, 36, 535, 568–569
  in action potential, 233–234
  in blood, 326, 548, 555, 569
  cardiac output and, 383
  intracellular, 65
  loss of, 326
  in membrane potential, 232–233
  in urine, 555, 569
Potassium channels, 232, 234
Pot belly, 114
Potential difference: The separation of opposite charges; requires a barrier that prevents ion migration, 232
Potential energy, 32
Pott's fracture, 135
Power, muscle, 192–193
Precapillary sphincter, 392–393, 399
Precentral gyrus: The primary motor cortex of a cerebral hemisphere, anterior to the central sulcus, 244, 246
Precipitation, 442–444
Precocious puberty, 335
Prefrontal cortex, 247–248
Prefrontal lobotomy, 248
Preganglionic fiber, 264–266
Preganglionic neuron: Visceral motor neuron inside the CNS whose output controls one or more

ganglionic motor neurons in the PNS, 264–266
Pregnancy. *See also* **Development, prenatal**
  ectopic, 617, 637
  labor and delivery, 628–630
  maternal systems and, 626–627
  mortality rate for, 605
  Rh blood type and, 354
Premature delivery, 630
Premature labor, 630
Premature menopause, 601
Premolars: Bicuspids; teeth with flattened surfaces located anterior to the molar teeth, 493–494
Premotor cortex, 247, 250
Prenatal development. *See* **Development, prenatal**
Preovulatory phase: A portion of the menstrual cycle; period of estrogen-induced repair of the functional zone of the endometrium through the growth and proliferation of epithelial cells in the glands not lost during menses, 595
Prepuce (PRĒ-pūs): Loose fold of skin that surrounds the glans penis in males or the clitoris in females, 582, 588–589, 596
Preputial gland, 589
Presbycusis, 306
Presbyopia, 293, 306
Pressure
  blood. *See* **Blood pressure**
  circulatory, 394
  hydrostatic. *See* **Hydrostatic pressure**
  osmotic. *See* **Osmotic pressure**
  partial, 471–472
Pressure gradient, 468
Pressure point, 398
Pressure receptors, 110
Pressure sensation, 261, 281
Presynaptic membrane: The synaptic surface where neurotransmitter release occurs.
Presynaptic neuron, 236
Prickling pain, 278–279
Primary bronchi, 462–464
Primary center of ossification, 131–132
Primary follicle, 592–593
Primary germ layers, 618–619
Primary immune response, 444
Primary memory, 250
Primary motor cortex, 246–247
Primary sensory cortex, 247, 261, 278–279, 284
Primary spinal curves, 146
Primary taste sensations, 284
Primary tumor, 77
Prime mover: A muscle that performs a specific action, 195
Primitive streak, 618
Primordial follicle, 591–593, 601
Process: A projection or bump of the skeleton, 136–137. *See also specific processes*
Product of reaction, 33, 42–43
Proenzyme, 504
Proerythroblasts, 358
Progesterone (prō-JES-ter-ōn): The most important progestin secreted by the corpus luteum following ovulation, 320, 322, 332, 595, 598–599, 620, 628
Progestins (prō-JES-tinz): Steroid hormones structurally related to cholesterol, 315, 320, 332–333, 590, 598, 601–602, 605
Prognosis: A prediction concerning the possibility or time course of recovery from a specific disease.
Prolactin (prō-LAK-tin): Hormone that stimulates functional development of the mammary gland in females; a secretion of the anterior pituitary gland, 315, 320–323, 333
  placental, 620, 622

Proliferative phase, of uterine cycle, 595
Prolymphocytes, 358
Promonocytes, 358
Promoter, 71–73
Pronation (prō-NĀ-shun): Rotation of the forearm that makes the palm face posteriorly, 152, 160–161, 195
Pronator quadratus muscle, 209–210
Pronator teres muscle, 196, 209–210
Prone, 15
Pronucleus: Enlarged egg or sperm nucleus that forms after fertilization but before amphimixis.
  female, 615–616
  male, 615–616
Prophase (PRŌ-fāz): The initial phase of mitosis, characterized by the appearance of chromosomes, breakdown of the nuclear membrane, and formation of the spindle apparatus.
  meiosis I, 584
  meiosis II, 585
  mitotic, 74, 76
Proprioception (prō-prē-ō-SEP-shun): Awareness of the positions of bones, joints, and muscles, 261, 278
Proprioceptor, 228, 279, 281
Prostaglandin (pros-tah-GLAN-din): A fatty acid secreted by one cell that alters the metabolic activities or sensitivities of adjacent cells; also called *local hormones*, 314–315, 529, 587, 628
Prostatectomy, 606
Prostate gland (PROS-tāt): Accessory gland of the male reproductive tract, contributing roughly one-third of the volume of semen, 11, 562, 582, 587–588, 600
  cancer of, 606, 637
  enlargement of, 573
Prostatitis, 587
Protease, 503–504, 512, 588
Protection
  by connective tissue, 92
  by epithelium, 84
  by integumentary system, 110
  by skeletal system, 128
Protein: A large polypeptide with a complex structure, 36, 40–43, 46
  absorption of, 511–513
  in blood, 555
  in body fluids, A-15, A-16
  caloric content of, 537
  complete, 534
  denaturation of, 42
  digestion of, 499, 511–513
  enzymes. *See* **Enzyme**
  functions of, 40–41, 47
  membrane, 57–58, 68, 316
  metabolism of, 523, 527, 529–532
  modification of, 68, 316
  shape of, 41–42
  structure of, 41–42, 47
  synthesis of, 68, 71–74, 531
Protein buffer system, 570
Protein deficiency disease, 531, 540
Protein hormone, 41
Proteinuria, 573
Prothrombin: Circulating proenzyme of the common pathway of the clotting system; converted to thrombin by the enzyme prothrombinase, 360, 509
Prothrombinase, 360
Proton: A subatomic particle bearing a positive charge, 28
Protraction: To move anteriorly in the horizontal plane, 161, 195
Proximal: Toward the attached base of an organ or structure, 17

Proximal convoluted tubule: The portion of the nephron between Bowman's capsule and the loop of Henle; the major site of active reabsorption from the filtrate, 550–556, 559

Pruritus, 122

Pseudopodia (soo-dō-PŌ-dē-a): Temporary cytoplasmic extensions typical of mobile or phagocytic cells, 64

Pseudostratified epithelium: An epithelium containing several layers of nuclei but whose cells are all in contact with the underlying basement membrane, 89–90

Psoas major muscle, 203, 211–213

Psoriasis, 122, 447

Pterygoid muscle, 199–200

PTH. *See* **Parathyroid hormone**

Puberty: Period of rapid growth, sexual maturation, and the appearance of secondary sexual characteristics; usually occurs at age 10–15, 116, 132, 328, 331–332

   female, 591, 631

   male, 462, 590, 631

   precocious, 335

Pubic angle, 155

Pubic region, 15–16

Pubic symphysis: Fibrocartilaginous amphiarthrosis between the pubic bones of the coxae, 154, 582, 588, 591

Pubis (PŪ-bis): The anterior, inferior component of the coxa, 15–16, 154

Pudendum, 596

Pulled groin, 212

Pulled hamstring, 213

Pulmonary artery, 370, 372, 405–406, 464, 466

Pulmonary circuit: Blood vessels between the pulmonary semilunar valve of the right ventricle and the entrance to the left atrium; the blood circulation through the lungs, 368, 373

Pulmonary circulation, 405–406, 418–419, 466

Pulmonary edema, 397

Pulmonary embolism, 361, 419, 421, 466, 481

Pulmonary function tests, 471

Pulmonary semilunar valve, 372–373

Pulmonary trunk, 370, 372, 375, 390–391, 405–406, 418

Pulmonary vein, 370, 406, 464, 466

Pulmonary ventilation: Movement of air into and out of the lungs, 468–471

Pulp, of spleen, 433

Pulp cavity: Internal chamber in a tooth, containing blood vessels, lymphatics, nerves, and the cells that maintain the dentin, 493

Pulse, 395, 398

Pulse pressure, 395

Punnett square, 634–635

Pupil: The opening in the center of the iris through which light enters the eye, 287–288

   dilation and constriction of, 288

Purkinje cell (pur-KIN-jē): Large, branching neuron of the cerebellar cortex.

Purkinje fibers: Specialized conducting cardiocytes in the ventricles, 377–378

Pus: An accumulation of debris, fluid, dead and dying cells, and necrotic tissue, 437

Putamen, 250

P wave: Deflection of the ECG corresponding to atrial depolarization, 378–379

Pyelitis, 563

Pyelogram, 573

Pyelonephritis, 563, 573

Pyloric sphincter (pī-LOR-ik): Sphincter of smooth muscle that regulates the passage of chyme from the stomach to the duodenum, 497

Pylorus (pī-LOR-us): Gastric region between the body of the stomach and the duodenum; includes the pyloric sphincter, 495–497

Pyramid, renal, 549–550

Pyramidal cells, 262

Pyrexia, 540

Pyridoxine, 529, 536

Pyrogen, 437

Pyruvic acid (pī-RŪ-vik): A three-carbon compound produced by glycolysis, 69, 191, 524–525, 528

Q

QRS complex, 379

Quadrants, abdominopelvic, 16, 23

Quadratus lumborum muscle, 202–203

Quadriceps femoris muscles, 155

Quadriceps muscles, 213, 259

Quadriplegia, 242, 270

Quiet breathing, 470, 476

R

Radial artery, 398, 405, 407, 409

Radial fossa, 152

Radial nerve, 257–258

Radial notch, 152–153

Radial tuberosity, 152

Radial vein, 405, 414–416

Radiation, heat transfer by, 538–539

Radiodensity: Relative resistance to the passage of X-rays, 21

Radioisotope, 29, 47

Radiological procedures, 21–22

Radiologist, 21

Radiology, 23

Radiopharmaceutical, 47

Radius, 138–139, 152–153, 164

Ramus: A branch, 136–137

RAS (reticular activating system), 253, 278

RBC. *See* **Red blood cells**

Reabsorption, by kidney, 555–558

Reactant, 33

Reactions, chemical, 32–35

   decomposition, 33

   endergonic, 34

   energy concepts, 32

   enzymes and, 42–43

   exchange, 33

   exergonic, 34

   reversible, 34

   synthesis, 33

Recall of fluids, 397, 404

Receptor-mediated endocytosis, 63, 443

Receptor protein, 58

Receptor site, 62

Receptors, 5, 12–13, 63, 259, 315–317. *See also specific types of receptors*

   external, 228

   visceral (internal), 228

Recessive gene: An allele that will affect the phenotype only when the individual is homozygous for that trait, 633, 635

Reciprocal inhibition, 260

Recombinant DNA, 77

Recovery period, in muscle, 191

Recruitment, 188

Rectal vein, 417

Rectum (REK-tum): The last 15 cm (6 in.) of the digestive tract, 496, 507–509, 562, 591

   cancer of, 509

Rectus: Straight.

Rectus abdominis muscle, 196, 202, 469

Rectus femoris muscle, 196, 213–214

Rectus muscles, 201–202

Red blood cell count, 347

Red blood cells (RBC), 94, 96, 128, 344–354, 357, 472–475. *See also* **Hemoglobin**

   blood tests, 351–352

   blood type and, 352–354

   crenation, 61

   formation of, 350–351, 358, 402

   hemolysis of, 61, 348, 353

   in isotonic solution, 61

   life span and circulation of, 348–350

   sickling in, 349

   structure of, 347–348

Red bone marrow, 128, 350–351

Red cones, 294

Red muscle, 192

Red pulp, of spleen, 433

Referred pain, 279

Reflex: A rapid, automatic response to a stimulus. *See also specific reflexes*

   aging and, 268

   complex, 259–260

   integration and control of, 260

   monosynaptic, 258

   polysynaptic, 259

   simple, 258–259

   spinal, 258–260

Reflex arc: The receptor, sensory neuron, motor neuron, and effector involved in a particular reflex; may include interneurons, 258–259

Reflex control, of respiration, 477–478

Refraction: The bending of light rays as they pass from one medium to another, 291–292

Refractory period: Period between the initiation of an action potential and the restoration of the normal resting potential; during this period, the membrane will not respond normally to stimulation, 234

Regeneration, 103–104, 437

Regional anatomy, 3

Regulatory hormones, 318

Regulatory T cells, 430

Regurgitation, across heart valves, 373

Relative humidity, 539

Relaxation phase: The period, following a contraction, when the tension in the muscle fiber returns to resting levels, 187

Relaxin: Hormone that loosens the pubic symphysis; a hormone secreted by the placenta, 620, 622, 628

Releasing hormones, 318, 320

Remission, 104

Remodeling, of bone, 133–136

Renal: Pertaining to the kidneys.

Renal artery, 327, 407, 411–412, 549, 553–554, 560

Renal capsule, 549–550, 552

Renal column, 550

Renal compensation, 571–572

Renal corpuscle: The initial portion of the nephron, consisting of an expanded chamber (Bowman's capsule) and the enclosed glomerulus, 550–553, 555

Renal cortex, 549, 554

Renal failure, 560, 573

Renal medulla, 549–550

Renal nerve, 549

Renal papillae, 549–550

Renal pelvis, 549–550

Renal pyramid, 549–550

Renal sinus, 549–551

Renal tubule, 88, 550–551

Renal vein, 327, 413, 415–416, 549, 554

Renin: Enzyme released by the kidney cells when renal blood pressure or O_2 level declines; converts angiotensinogen to angiotensin I, 315, 331, 402–404, 548, 560–561

Renin-angiotensin system, 331, 560–561

Salt: An inorganic compound consisting of a cation other than H$^+$ and an anion other than OH$^-$, 36

Saltatory propagation: Relatively rapid propagation of an action potential between successive nodes of a myelinated axon, 235

Salty taste, 284

SA node (sinoatrial node), 377–378, 401

Saphenous nerve, 257–258

Saphenous vein
great, 413–414
small, 413–414

Sarcolemma: The cell membrane of a muscle cell, 179–180, 182–183, 376

Sarcoma, 217

Sarcomere: The smallest contractile unit of a striated muscle cell, 180–181

Sarcoplasm: The cytoplasm of a muscle cell, 179

Sarcoplasmic reticulum, 179–181, 186

Sartorius muscle, 196–197, 212–214

Saturated fats, 39–40, 530

Saxitoxin, 270

Scab, 118–119

Scala media. *See* **Cochlear duct**

Scala tympani, 299, 301, 303–304

Scala vestibuli, 299, 301, 303–304

Scalp, 200

Scanning electron microscopy (SEM), 54–55, A-14

Scaphoid bone, 153

Scapula, 129, 138–139, 150–151, 164

Scar tissue, 103, 118–119

Schlemm, canal of: Passageway that delivers aqueous humor from the anterior chamber of the eye to the venous circulation, 290

Schwann cells: Glial cells responsible for the neurilemma that surrounds axons in the PNS, 230

Sciatica, 270

Sciatic nerve (sī-A-tik): Nerve innervating the posteromedial portions of the thigh and leg, 257–258

SCID (severe combined immunodeficiency disease), 447

Sclera (SKLER-uh): The fibrous outer layer of the eye, forming the white area of the anterior surface; a portion of the fibrous tunic of the eye, 287–288

Scoliosis, 146, 168

Scopolamine, transdermal, 112

Scrotal cavity, 583

Scrotum (SKRŌ-tum): Loose-fitting, fleshy pouch that encloses the testes of males, 11, 265–266, 582–583, 588

Scurvy, 133, 168, 536

Seasonal affective disorder (SAD), 328

Sebaceous follicle, 116

Sebaceous glands (se-BĀ-shus): Glands that secrete sebum, usually associated with hair follicles, 6, 89, 91, 110, 115–116, 285, 436, 597

Sebum (SĒ-bum): A waxy secretion that coats the surfaces of hairs, 116, 120

Secondary bronchi, 462–464

Secondary center of ossification, 132

Secondary dentition, 493–494

Secondary follicle, 592–593

Secondary immune response, 444

Secondary sex characteristics: Physical characteristics that appear at puberty in response to sex hormones but that are not involved in the production of gametes.
female, 598, 631
male, 590, 631

Secondary spinal curves, 146

Secondary tumor, 77

Second-degree burn, 120

Second heart sound, 381

Second messenger, 315–317

Second trimester, 616, 626–628

Secretin (sē-KRĒ-tin): Duodenal hormone that stimulates pancreatic buffer secretion and inhibits gastric activity, 498, 501–502

Secretion, 86–87, 523. *See also* **Glandular epithelium**
by digestive system, 488
by integumentary system, 110
by kidney, 555–558

Secretory phase, of uterine cycle, 595

Secretory vesicle, 55, 66, 68, 91

Sectional anatomy, 18–21
body cavities, 18–21
clinical technology, 21–22
planes and sections, 18–19

Sectional planes, 18–19

Segmentation, 490–491, 501

Seizure, 251, 270

Selectively permeable membrane, 58, 60

Self antigen, 439, 446

Sella turcica, 142–143, 318–319

SEM (scanning electron microscopy), 54–55, A-14

Semen (SĒ-men): Fluid ejaculate containing spermatozoa and the secretions of accessory glands of the male reproductive tract, 588, A-16

Semicircular canals: Tubular components of the vestibular apparatus responsible for dynamic equilibrium, 297–299

Semicircular ducts, 298–301

Semilunar valve: A three-cusped valve guarding the exit from one of the cardiac ventricles; includes the pulmonary and aortic valves, 380

Semimembranosus muscle, 197, 213–214

Seminal fluid, 588

Seminalplasmin, 587–588

Seminal vesicles (SEM-i-nal): Glands of the male reproductive tract that produce roughly 60 percent of the volume of semen, 11, 582, 587–588, 600

Seminiferous tubules (se-mi-NIF-e-rus): Coiled tubules where sperm production occurs in the testis, 583, 585, 589

Semispinalis capitis muscle, 201–202

Semitendinosus muscle, 197, 213–214

Senescence: Aging, 632. *See also* **Aging**

Senile cataract, 291, 306

Senile dementia, 268

Senility, 268

Sensations, 84, 243, 260, 278

Sensible perspiration, 538

Sensory function, 278–311. *See also specific organs*
adaptation, 278
aging and, 305–306
general senses, 278–282
prenatal and early postnatal development of organs, 624
special senses, 278, 282–305

Sensory neuron, 228, 259–261, 278

Sensory pathways, 231, 260–261

Sensory receptors, 6, 110, 228, 278
somatic, 226, 228
visceral, 226

Sepsis, 122

Septae (SEP-tē): Partitions that subdivide an organ.

Septicemia, 362

Septic shock, 404

SER. *See* **Smooth endoplasmic reticulum**

Serosa
of digestive tract, 489–490
of small intestine, 500
of stomach, 496

Serotonin (ser-ō-TO-nin): A neurotransmitter in the CNS; a compound that enhances inflammation, released by activated mast cells and basophils, 237

Serous cell/secretion: A cell that produces a watery secretion containing high concentrations of enzymes.

Serous fluid, 100, 567

Serous gland, 89, 91–92

Serous membrane: A squamous epithelium and the underlying loose connective tissue; the lining of the pericardial, pleural, and peritoneal cavities, 19, 99–100

Serratus anterior muscle, 196, 203, 205–208, 498

Serratus posterior muscle, 203

Sertoli cells. *See* **Sustentacular cells**

Serum: Blood plasma from which clotting agents have been removed, 346

Severe combined immunodeficiency disease (SCID), 447

Sex chromosome, 632, 635–636

Sex hormones. *See also* **Estrogens; Progesterone; Testosterone**
in bone growth, 132–133

Sex-linked trait, 633, 635

Sexual arousal, 600–601

Sexual function
female, 600–601
male, 600

Sexual intercourse, 588, 596, 600–601

Sexually transmitted disease (STD), 443, 563, 601

Shinbone, 138–139, 156, 166

Shingles, 257, 270

Shivering, 12–13, 192, 539

Shivering thermogenesis, 539

Shock, 270, 421, 449, 556

Short bone, 128–129

Short-term memory, 250

Shoulder blade. *See* **Scapula**

Shoulder girdle, 139, 150–151

Shoulder joint, 150–151, 164
muscles of, 205–211

Sickle-cell anemia, 74, 349, 362, 635, 637

SIDS (sudden infant death syndrome), 478

Sigmoid colon (SIG-moyd): The S-shaped 18-cm portion of the colon between the descending colon and the rectum, 508–509, 591

Simple epithelium: An epithelium containing a single layer of cells above the basement membrane, 87–88
columnar, 87–88
cuboidal, 87–88
squamous, 87–88

Simple inheritance, 634

Simple reflex, 258–259

Simple sugar. *See* **Monosaccharide**

Single covalent bond, 31

Sinoatrial (SA) node, 377–378, 401

Sinus: A chamber or hollow in a tissue; a large, dilated vein, 136–137. *See also specific sinuses*

Sinusoid (SĪ-nū-soyd): An extensive network of vessels found in the liver, adrenal cortex, spleen, and pancreas; similar in histological structure to capillaries, 505

Skeletal muscle: A contractile organ of the muscular system, 7. *See also* **Action potential;** *specific muscles*
atrophy of, 189
contraction of, 182–184
energetics of activity of, 189–192
functions of, 178
gross anatomy of, 178–179
hypertrophy of, 193
innervation of, 227
mechanisms of, 184–189
microanatomy of, 179–182
naming of, 197–198
origins, insertions, and actions of, 195
performance of, 192–193, 216–217
physical conditioning and, 193

in thermoregulation, 12–13, 178, 192
types of muscle fibers, 192–193
in venous return, 397, 403
Skeletal muscle tissue: Contractile tissue dominated by skeletal muscle fibers; characterized as striated, voluntary muscle, 100–102
Skeletal system, 128–175. *See also* **Bone;** *specific bones*
aging of, 136–137
anatomical terms, 136–137
appendicular skeleton, 139, 150–166
axial skeleton, 137, 139–150
components of, 6
functions of, 6
integration with other systems, 166–167
cardiovascular system, 420
digestive system, 515
endocrine system, 337
integumentary system, 121
lymphatic system, 448
muscular system, 218
nervous system, 269
reproductive system, 603
respiratory system, 480
urinary system, 565
prenatal and early postnatal development of, 624
Skill memory, 250
Skin, 100, 110–125. *See also* **Integumentary system**
cancer of, 113–114, 120
color of, 112–113, 633
dermal circulation, 113–114
drug administration through, 112
as physical barrier, 435–436
regeneration after injury, 118
in thermoregulation, 110, 539
thick, 111
thin, 111
Skull, 6, 137–145
of infants and children, 144
Sliding filament theory: The concept that a sarcomere shortens as the thick and thin filaments slide past one another, 181–182, 185–186
Slightly movable joint. *See* **Amphiarthrosis**
Slipped disc, 163, 168
Slow fiber, 192–193
Slow pain, 279
Small intestine: The duodenum, jejunum, and ileum; the digestive tract between the stomach and large intestine, 10, 489, 496, 498–503
absorption in, 499–501
digestion in, 502
hormones of, 501–502
innervation of, ANS, 265–266
intestinal wall, 500–501
movements of, 501
secretions of, 501–502
Smegma, 589
Smell, sense of, 255, 278, 282–283, 458
aging and, 306
Smoking, 67, 104, 393, 466, 472, 477, 479
Smooth endoplasmic reticulum (SER): Membranous organelle where lipid and carbohydrate synthesis and storage occur, 55–56, 67
Smooth muscle, 86, 193–195
blood vessel, 390–392
contraction of, 195
Smooth muscle cells, 54
Smooth muscle tissue: Muscle tissue found in the walls of many visceral organs; characterized as nonstriated, involuntary muscle, 100–102
SNS (somatic nervous system), 226, 261–263
Sodium, 28, 30, 36, 534–535, 568–569
in action potential, 233–234
in blood, 326, 548, 555
in body fluids, 567
intracellular, 65
in membrane potential, 232–233

retention of, 326
in urine, 555, 569
Sodium bicarbonate, 474, 504, 568
Sodium channels, 232, 234
Sodium chloride, 30–31, 36, 61, 568
Sodium hydroxide, 34
Sodium-potassium exchange pump, 62–63, 232–233
Soft palate: Fleshy posterior extension of the hard palate, separating the nasopharynx from the oral cavity, 459, 490–491
Soft spot. *See* **Fontanel**
Sole: The inferior surface of the foot.
Soleus muscle, 187, 196–197, 214–216
Solute: Materials dissolved in a solution, 36, 60
capillary exchange, 396
Solution: A fluid containing dissolved materials, 35–36
Solvent: The fluid component of a solution, 36
Soma (SŌ-ma): Cell body.
Somatic (sō-MAT-ik): Pertaining to the body.
Somatic cells, 74, 614, 632
Somatic motor association area, 247, 250
Somatic motor neuron, 228
Somatic nervous system (SNS), 226, 261–263
Somatic sensory association area, 247
Somatic sensory receptors, 226, 228
Somatomedins: Compounds stimulating tissue growth, released by the liver following GH secretion, 320–322
Somatotropin. *See* **Growth hormone**
Somite, 623
Sound, 303, 462
frequency of, 301, 303
intensity of, 301, 303
pitch of, 303
Sounds of Korotkoff, 398
Sour taste, 284
Spastic colon, 514
Special movements, 160–161
Special physiology, 3
Special senses, 278, 282–305
Specific defenses, 355, 435, 438–447
Specific gravity, of urine, 558
Specificity, immunological, 438
Specific resistance, 435
Speech, 458, 462
Speech center, 248–249
Sperm, 54, 67, 582–585, 588
anatomy of, 586
development of, 583–586
fertilization, 614–616
Spermatic cord: Spermatic vessels, nerves, lymphatics, and the ductus deferens, extending between the testes and the proximal end of the inguinal canal, 586
Spermatids (SPER-ma-tidz): The product of meiosis in the male, cells that differentiate into spermatozoa, 583–585
Spermatocyte (sper-MA-to-sīt): Cells of the seminiferous tubules that are engaged in meiosis, 583–584
primary, 584–585
secondary, 585
Spermatogenesis (sper-ma-to-JEN-e-sis): Sperm production, 584–586, 589–590
Spermatogonia (sper-ma-to-GŌ-nē-uh): Stem cells whose mitotic divisions give rise to other stem cells and spermatocytes, 583–584
Spermatozoa/spermatozoon (sper-ma-to-ZŌ-a): Sperm cells. *See* **Sperm**
Sperm count, 588
Spermiogenesis: The process of spermatid differentiation that leads to the formation of physically mature spermatozoa, 584–585, 589
SPF (sun protection factor), 113
S phase, 74–75

Sphenoidal fontanel, 145
Sphenoidal sinus, 142–144, 458, 460
Sphenoid bone, 140–143, 145, 459
Sphincter (SFINK-ter): Muscular ring that contracts to close the entrance or exit of an internal passageway. *See specific sphincters*
Sphygmomanometer, 398, 421
Spina bifida, 168
Spinal accessory nerve (N XI), 253–256
Spinal cavity, 19–20
Spinal cord, 7, 163, 227, 240–243
gross anatomy of, 240–241
injury to, 242–243
sectional anatomy of, 241–243
Spinal curvature, 146
Spinalis muscle group, 201–202
Spinal meninges, 239
Spinal meningitis, 240, 270
Spinal nerve: One of 31 pairs of peripheral nerves that originate on the spinal cord from anterior and posterior roots, 163, 231, 240–241, 257
Spinal reflex, 240
Spinal shock, 270
Spinal tap, 245
Spindle apparatus: A muscle spindle (intrafusal fibers) and its sensory and motor innervation, 258, 281
Spindle fiber, mitotic, 66, 76
Spine. *See also* **Vertebral column**
muscles of, 199, 201–202
Spine (bone anatomy), 136–137
Spinocerebellar ataxia, 637
Spinocerebellar pathway, 261
Spinothalamic pathway, 261
Spinous process: Prominent posterior projection of a vertebra, formed by the fusion of two laminae, 146–148
Spiral fracture, 135
Spiral ganglion, 302, 304
Spirometer, 471
Splanchnic nerve, 264–265
Spleen: Lymphoid organ important for red blood cell phagocytosis, immune response, and lymphocyte production, 9, 357, 428, 432–434
injury to, 434
innervation of, ANS, 265–266
ruptured, 434
Splenectomy, 434
Splenic artery, 411–412, 433–434, 503
Splenic vein, 417, 433–434, 508
Splenius capitis muscle, 201–202
Splenomegaly, 449
Spongy bone, 129–130, 136, 158
Spontaneous abortion, 630, 632
Sports drinks, 193
Sprain, 168
Squamosal suture, 140, 143, 145
Squamous (SKWĀ-mus): Flattened.
Squamous cell carcinoma, 113, 122
Squamous epithelium: An epithelium whose superficial cells are flattened and platelike, 87–90
Stapedius (stā-PĒ-dē-us): A muscle of the middle ear whose contraction tenses the auditory ossicles and reduces the forces transmitted to the oval window, 298, 305
Stapes (STĀ-pez): The auditory ossicle attached to the tympanum, 297–298, 304
Starch, 37
Static equilibrium, 299
STD (sexually transmitted disease), 443, 563, 601
Stem cells, 91, 93–94
in bone marrow, 430–431
embryonic, 243, 347
in epithelium, 87
lymphoid, 347, 351, 358, 430–431
among muscle fibers, 178

Stem cells *(cont.)*
    myeloid, 347, 351, 356, 358
    neural, 243
    in respiratory tract, 460
    in skin, 118, 120
Stereocilia: Elongate microvilli characteristic of the epithelium of portions of the male reproductive tract and inner ear, 86, 298–299, 304–305
Sterilization, surgical, 604
Sternocleidomastoid muscle, 196–197, 199–201
Sternohyoid muscle, 200–201
Sternothyroid muscle, 200–201
Sternum, 6, 139, 149–150, 200
Steroid: A ring-shaped lipid structurally related to cholesterol, 38–40, 529–530
Steroid hormones, 40, 67, 315–317, 631
Stethoscope, 381
Stimulus: An environmental alteration that produces a change in cellular activities; often used to refer to events that alter the membrane potentials of excitable cells, 258–259
Stomach, 10, 489, 495–499
    cancer of, 498–499, 514
    digestion in, 499
    functions of, 495
    gross anatomy of, 496
    lining of, 496–497
    regulation of activity of, 497–498
    wall of, 496–497
Stool softener, 510
Storage function
    of integumentary system, 110
    of skeletal system, 128, 133
Strabismus, 306
Strain, muscle, 208, 217
Strata, 111
Stratified: Containing several layers.
Stratified epithelium, 87–90
Stratum (STRA-tum)**:** Layer.
Stratum basale, 111–113
Stratum corneum, 111–112
Stratum germinativum, 111–113
Stratum granulosum, 111
Stratum lucidum, 111–112
Stratum spinosum, 111–113
Streptokinase, 361
Stress, 334, 393
    hormones and, 334
    immune response and, 447
Stress incontinence, 564
Stressor, 334
Stretch receptors: Sensory receptors that respond to stretching of the surrounding tissues, 259–260, 281, 477, 479, 563–564
Stretch reflex, 258–259
Striations
    cardiac muscle, 102
    skeletal muscle, 100–102
Stroke, 248, 253, 270, 361, 393, 419, 468, 573
Stroke volume, 381–383, 403
Strong acid, 34
Strong base, 34
Structural protein, 40
Sty, 285
Stylohyoid ligament, 144
Stylohyoid muscle, 200–201
Styloid process, 140–143, 152–153
Subarachnoid space: Meningeal space containing CSF; the area between the arachnoid membrane and the pia mater, 239–240, 245–246
Subclavian (sub-CLA-vē-an)**:** Pertaining to the region under the clavicle.
Subclavian artery, 405–409, 411
Subclavian vein, 405, 413–416, 430
Subclavius muscle, 206–207

Subcutaneous injection, 114
Subcutaneous layer: The layer of loose connective tissue below the dermis; also called *hypodermis,* 6, 110, 114
Subdural hemorrhage, 240
Subdural space, 239–240, 246
Sublingual duct, 492
Sublingual glands (sub-LING-gwal)**:** Mucus-secreting salivary glands situated under the tongue, 89, 492
Submandibular glands: Salivary glands nestled in depressions on the medial surfaces of the mandible; salivary glands that produce a mixture of mucins and enzymes (salivary amylase), 89, 492
Submucosa (sub-mū-KŌ-sa)**:** Region between the muscularis mucosae and the muscularis externa.
    of digestive tract, 488–490
    of small intestine, 500
    of stomach, 496
Submucosal artery, 500
Submucosal glands: Mucous glands in the submucosa of the duodenum, 489
Submucosal plexus, 488–489, 500
Submucosal vein, 500
Subscapularis muscle, 207–208
Substantia nigra, 253
Substrate: A participant (product or reactant) in an enzyme-catalyzed reaction, 42–43
Sucrase, 511–512
Sucrose, 37
Sudden infant death syndrome (SIDS), 478
Sudoriferous gland. *See* **Sweat gland**
Suicide gene, 75
Sulcus (SUL-kus)**:** A groove or furrow, 136–137
Sulfate, 30
Sulfur, 28
Sulfuric acid, 36
Summation, 187
Sunlight exposure, 112–114, 120
Sun protection factor (SPF), 113
Superficial (anatomical direction), 17
Superior: Directional reference meaning *above,* 17
Superior colliculus, 252–253, 296
Superior oblique muscle, 286
Superior orbital fissure, 141
Superior rectus muscle, 286
Superior sagittal sinus, 245–246, 414
Superior section, 18
Superior vena cava: The vein that carries blood from the parts of the body above the heart to the right atrium, 370, 372, 406, 413–416, 430
Supination (su-pi-NA-shun)**:** Rotation of the forearm such that the palm faces anteriorly, 152, 160–161, 195
Supinator muscle, 209–210
Supine (sū-PĪN)**:** Lying face up, with palms facing anteriorly, 15
Support function
    of connective tissue, 92
    of muscular system, 178
    of skeletal system, 128
Supporting connective tissue, 92–93, 96–98
Suppression factor, 441
Suppressor T cells: Lymphocytes that inhibit B cell activation and plasma cell secretion of antibodies, 430, 440–441, 446
Supraorbital foramen, 140
Supraorbital notch, 140
Suprarenal artery, 411–412, 554
Suprarenal gland (soo-pra-RĒ-nal)**:** *See* **Adrenal gland**
Suprarenal vein, 415–416
Supraspinatus muscle, 151, 164, 195, 207–208

Supraspinous fossa, 151
Sural region, 15–16
Surface anatomy, 3, 14–17
Surface tension, 32, 466
Surfactant (sur-FAK-tant)**:** Lipid secretion that coats alveolar surfaces and prevents their collapse, 465–466, 469, 479
Surfactant cells, 465–466
Surgical neck, of humerus, 151–152
Suspensory ligament of the breast, 597
Suspensory ligament of the lens, 287–288, 291–292
Sustentacular cells (sus-ten-TAK-ū-lar)**:** Supporting cells of the seminiferous tubules of the testis, responsible for the differentiation of spermatids and the secretion of inhibin, 331–332, 583–585, 589–590
Suture: Fibrous joint between flat bones of the skull, 157–158
Swallowing, 461, 492, 494–495
Sweat gland, 6, 110, 116–117, 264, 267, 436, 539
Sweat gland duct, 110
Sweat pore, 117
Sweetener, artificial, 37
Sweet taste, 284
Swollen glands, 432–433
Sympathetic chain ganglia, 264–265
Sympathetic division: Division of the autonomic nervous system responsible for "fight or flight" reactions; primarily concerned with the elevation of metabolic rate and increased alertness, 226–227, 264–265, 267, 318
Symphyseal joint, 163
Symphysis: A fibrous amphiarthrosis, such as that between adjacent vertebrae or between the pubic bones of the coxae, 157–158
Symptom: Clinical term for an abnormality of function due to the presence of disease.
Synapse (SIN-aps)**:** Site of communication between a nerve cell and some other cell; if the other cell is not a neuron, the term *neuroeffector junction* is often used.
    structure of, 236
    synaptic function and neurotransmitters, 236–238
Synapsis, 584–585, 592
Synaptic cleft, 182–183, 236
Synaptic delay (sin-AP-tik)**:** The period between the arrival of an impulse at the presynaptic membrane and the initiation of an action potential in the postsynaptic membrane.
Synaptic knob, 183, 227, 236
Synaptic terminal, 102–103, 182–183, 227
Synaptic vesicle, 236
Synarthrosis, 157–158
Synchondrosis, 157–158
Syncytial trophoblast: Multinucleate cytoplasmic layer that covers the blastocyst; the layer responsible for uterine erosion and implantation, 617–619, 621–622
Syncytium (sin-SISH-ē-um)**:** A multinucleate mass of cytoplasm, produced by the fusion of cells or repeated mitoses without cytokinesis.
Syndesmosis, 157–158
Syndrome: A discrete set of symptoms that occur together.
Synergist (SIN-er-jist)**:** A muscle that assists a prime mover in performing its primary action, 195
Synergistic hormones, 333
Synovial cavity (si-NŌ-vē-ul)**:** Fluid-filled chamber in a diarthrosis.
Synovial fluid (si-NŌ-vē-ul)**:** Substance secreted by synovial membranes that lubricates joints, 100, 159, 567, A-16
Synovial joint, 157–159
    movement and structure of, 159–163
    structural classification of, 161–163

Umbilical cord (um-BIL-i-kal): Connecting stalk between the fetus and the placenta; contains the allantois, the umbilical arteries, and the umbilical vein, 418, 620–623

Umbilical ligament, 563

Umbilical stalk, 620–621

Umbilical vein, 418, 620, 622

Umbilicus: The navel, 16

Unicellular gland: Goblet cells, 89, 460

Unipolar neuron: A sensory neuron whose cell body lies in a dorsal root ganglion or a sensory ganglion of a cranial nerve, 228

Universal donor, 354

Universal recipient, 354

Unmyelinated axon: Axon whose neurilemma does not contain myelin and where continuous conduction occurs, 230, 234–235

Unsaturated fats, 38–40

Upper esophageal sphincter, 494

Upper limb

  arteries of, 409

  bones of, 139, 151–153

  joints of, 164

  lymphatics of, 428

  muscles of, 205–211

  veins of, 414

Upper motor neuron, 262

Uracil: One of the nitrogenous compounds in RNA, 44, 73, 531

Urea, 530, 548, 555

Ureteral opening, 562

Ureters (ū-RĒ-terz): Muscular tubes, lined by transitional epithelium, that carry urine from the renal pelvis to the urinary bladder, 10, 548–550, 562–563, 582, 587

Urethra (ū-RĒ-thra): A muscular tube that carries urine from the urinary bladder to the exterior, 10–11, 548–549, 562–563, 582, 587–588, 591, 600

Urethral sphincter

  external, 204–205, 562–564

  internal, 562–564

Urethritis, 563, 573

Uric acid, 531, 548, 555

Urinalysis: Analysis of the physical and chemical characteristics of the urine, 573

Urinary bladder: Muscular, distensible sac that stores urine prior to micturition, 10, 89–90, 265–267, 548–549, 562–563, 582, 587–588, 591

Urinary retention, 573

Urinary system, 548–579. *See also specific organs;* **Urine**

  aging of, 573

  components of, 10

  functions of, 10

  innervation of, ANS, 267

  integration with other systems, 564–565

    cardiovascular system, 420

    digestive system, 515

    endocrine system, 337

    integumentary system, 121

    lymphatic system, 448

    muscular system, 218

    nervous system, 269

    reproductive system, 603

    respiratory system, 480

    skeletal system, 167

  organization of, 548–554

  prenatal and early postnatal development of, 625

  urine production, 555–562

Urinary tract infection, 563, 573

Urination: The voiding of urine; micturition, 281, 563–564

Urine, 550

  composition of, 555, 558, A-15

pH of, 558

production of, 555–562

specific gravity of, 558

transport, storage, and elimination of, 562–564

water loss in, 568

Urokinase, 361

Urology, 573

U.S. system of measurement, A-12

Uterine artery, 594

Uterine cavity, 594

Uterine contractions, 628–629

Uterine cycle, 595, 599–600

Uterine tube, 11, 590, 593–594, 614, 616

Uterine vein, 594

Uterus (Ū-ter-us): Muscular organ of the female reproductive tract where implantation, placenta formation, and fetal development occur, 11, 265–266, 590–591, 594–595, 621–622. *See also* **Pregnancy**

Utricle (Ū-tre-kl): The largest chamber of the vestibular apparatus; contains a macula important for static equilibrium, 298–301

Uvula, 490–491

V

Vaccination. *See* **Immunization**

Vaccine, 438, 449

  mumps, 492

Vagina (va-JĪ-na): A muscular tube extending between the uterus and the vestibule, 11, 590–591, 594–596, 614

Vaginal barriers, 605

Vaginitis, 596, 606

Vagus nerve (N X), 253–256, 266, 282, 284, 382–383, 401, 477, 501

Valine, 531

Valsalva maneuver, 510

Valve

  heart, 372–374

  ileocecal, 499, 507–508

  lymphatic vessels, 429

  venous, 393–394, 419

Valvular heart disease, 373, 384

Variable segment, of antibody, 441–442

Varicose veins, 394, 421

Vasa recta, 554, 559

Vascular: Pertaining to blood vessels.

Vascular phase, of hemostasis, 359

Vascular resistance, 395

Vascular spasm, 359

Vascular tunic, 287–288

Vas deferens. *See* **Ductus deferens**

Vasectomy, 604, 606

Vasoconstriction: A reduction in the diameter of arterioles due to contraction of smooth muscles in the tunica media; elevates peripheral resistance; may occur in response to local factors, through the action of hormones, or to stimulation of the vasomotor center, 391, 399–402, 404

Vasodilation (vaz-ō-dī-LĀ-shun): An increase in the diameter of arterioles due to the relaxation of smooth muscles in the tunica media; reduces peripheral resistance; may occur in response to local factors, through the action of hormones, or following decreased stimulation of the vasomotor center, 103, 391, 399–401, 403

Vasomotion: Alterations in the pattern of blood flow through a capillary bed in response to changes in the local environment, 393

Vasomotor center: Center in the medulla oblongata whose stimulation produces vasoconstriction and an elevation in peripheral resistance, 253, 400, 404

Vastus intermedius muscle, 214

Vastus lateralis muscle, 196, 213–214

Vastus medialis muscle, 196, 213–214

Vastus muscles, 213

Vegetarian diet, 534

Vein: Blood vessel carrying blood from a capillary bed toward the heart, 8, 96, 368, 390, 392. *See also specific veins*

  of abdomen and chest, 415

  anatomy of, 390–391, 393–394

  of head and neck, 414

  large, 393

  medium-sized, 393–394

  structure of vessel walls, 390–391

  systemic, 412–417

  of upper limbs and chest, 414

Venae cavae (VĒ-na KĀ-va): The major veins delivering systemic blood to the right atrium. *See* **Inferior vena cava; Superior vena cava**

Venipuncture, 345, 362

Venous pressure, 394, 397

Venous reserve, 404

Venous return, 382, 397, 403–404

Venous sinusoid, 433

Ventilation: Air movement into and out of the lungs.

Ventral: Pertaining to the anterior surface, 16–17

Ventral body cavity, 19–21

Ventral respiratory group, 476, 479

Ventral root, 240–242

Ventricle (VEN-tri-kl): One of the large, muscular pumping chambers of the heart that discharges blood into the pulmonary or systemic circuit.

  left, 368–370, 372–373, 405, 418

  right, 368–370, 372–373, 405, 418–419

Ventricle of brain, 230, 245

Ventricular diastole, 380–381, 392

Ventricular systole, 380–381, 392

Venule (VEN-ūl): Thin-walled veins that receive blood from capillaries, 390–392

Vermiform appendix, 432, 508

Vertebrae, 6, 129, 138–139, 145–146, 163

  anatomy of, 146

  cervical, 145–147

  lumbar, 145–148

  thoracic, 145–148

Vertebral arch, 146–147

Vertebral artery, 407–411

Vertebral body, 146–147

Vertebral canal: Passageway that encloses the spinal cord, a tunnel bounded by the vertebral arches of adjacent vertebrae, 146

Vertebral column: The cervical, thoracic, and lumbar vertebrae, the sacrum, and the coccyx, 139, 145–146

Vertebral vein, 413–416

Vesicle: A membranous sac in the cytoplasm of a cell.

Vesicular transport, 63

Vestibular branch, of vestibulocochlear nerve, 300–301, 305

Vestibular complex, 298

Vestibular duct, 299, 301, 303–304

Vestibular gland, 596

  greater, 596, 601

Vestibular membrane: The membrane that separates the cochlear duct from the vestibular duct of the inner ear, 302

Vestibular nerve, 256

Vestibular nucleus: Processing center for sensations arriving from the vestibular apparatus; near the border between the pons and the medulla, 301

Vestibule

  of female genitalia, 596

  nasal, 459

  of oral cavity, 490–491

Illustration Credits

FM-01 Frederic H. Martini FM-02 Frederic H. Martini FM-03 Nagamine Photography

Chapter 1 Chapter Opener Mike Timo/Getty Images Inc. 01-10a Science Source/Photo Researchers, Inc. 01-10b Science Source/Photo Researchers, Inc. 01-11b CNRI/Photo Researchers, Inc. 01-11c Photo Researchers, Inc. 01-11d Ben Edwards/Getty Images Inc.

Chapter 2 Chapter Opener Adam Jones/Visuals Unlimited 02-19c Dr. A. Lesk, Laboratory of Molecular Biology/SPL/Photo Researchers, Inc.

Chapter 3 Chapter Opener Prof. P. Motta, Dept. of Anatomy, University La Sapienza, Rome/Science Photo Library/Photo Researchers, Inc. 03-07a David M. Phillips/Visuals Unlimited 03-07c David M. Phillips/Visuals Unlimited 03-14b L. A. Hufnagel, Ultrastructural Aspects of Chemoreception in Ciliated Protists (Ciliophora), Journal of Electron Microscopy Technique, 1991./Photomicrograph by Jurgen Bohmer and Linda Hufnagel, University of Rhode Island. 03-15 CNRI/Science Source/Photo Researchers, Inc. 03-16 Don W. Fawcett, M.D., Harvard Medical School

Chapter 4 Chapter Opener Profs. Motta, Correr, & Nottola, Department of Anatomy, University La Sapienza, Rome/Science Photo Library/Photo Researchers, Inc. 04-03 Custom Medical Stock Photo, Inc. 04-04a Ward's Natural Science Establishment, Inc. 04-04b Pearson Education/PH College 04-04c Frederic H. Martini 04-05a Frederic H. Martini 04-05b left Frederic H. Martini 04-05b right Frederic H. Martini 04-05c Frederic H. Martini 04-09a Science Source/Photo Researchers, Inc. 04-09b Frederic H. Martini 04-09c John D. Cunningham/Visuals Unlimited 04-11a Robert Brons/Biological Photo Service 04-11b Science Source/Photo Researchers, Inc. 04-11c Ed Reschke/Peter Arnold, Inc. 04-12 Frederic H. Martini 04-14a G. W. Willis, M.D./Biological Photo Service 04-14b Phototake NYC 04-14c Pearson Education/PH College 04-15b Pearson Education/PH College

Chapter 5 Chapter Opener Peter Sterling/Getty Images, Inc. 05-02a John D. Cunningham/Visuals Unlimited 05-03 Pearson Education/PH College 05-04a Courtesy of Elizabeth A. Abel, M.D., from the Leonard C. Winograd Memorial Slide Collection, Stanford University School of Medicine. 05-04b Courtesy of Elizabeth A. Abel, M.D., from the Leonard C. Winograd Memorial Slide Collection, Stanford University School of Medicine. 05-05a Manfred Kage/Peter Arnold, Inc. 05-06 Frederic H. Martini ML01-01 John Watney/Science Source/Photo Researchers, Inc. ML01-02 Bettmann/CORBIS

Chapter 6 Chapter Opener Paul Chesley/Getty Images Inc. Paul Chesley/Getty Images Inc. 06-03a © R. G. Kessel and R. H. Kardon, "Tissues and Organs: A Text-Atlas of Scanning Electron Microscopy," W. H. Freeman & Co., 1979. All Rights Reserved. 06-04 Ralph T. Hutchings 06-07a Southern Illinois University/Visuals Unlimited 06-07b Southern Illinois University/Visuals Unlimited 06-07d Southern Illinois University/Peter Arnold, Inc. 06-07e Custom Medical Stock Photo, Inc. 06-07f Scott Camazine/Photo Researchers, Inc. 06-07g Frederic H. Martini 06-07h Southern Illinois University/Visuals Unlimited 06-07i Project Masters, Inc./The Bergman Collection 06-08a Ralph T. Hutchings 06-08b Ralph T. Hutchings 06-17a Ralph T. Hutchings 06-17b Ralph T. Hutchings 06-17c Ralph T. Hutchings 06-19a Ralph T. Hutchings 06-19b Ralph T. Hutchings 06-20a Ralph T. Hutchings 06-21 Ralph T. Hutchings 06-21c Ralph T. Hutchings 06-22a Ralph T. Hutchings 06-22b Ralph T. Hutchings 06-22c Ralph T. Hutchings 06-23a Ralph T. Hutchings 06-23b Ralph T. Hutchings 06-24a Ralph T. Hutchings 06-25 Ralph T. Hutchings 06-28a Ralph T. Hutchings 06-28b Ralph T. Hutchings 06-29 Ralph T. Hutchings 06-30a Ralph T. Hutchings 06-32c left Ralph T. Hutchings 06-32c right Ralph T. Hutchings 06-32d Ralph T. Hutchings 06-34BC Ralph T. Hutchings 06-34BL Ralph T. Hutchings 06-34TC Ralph T. Hutchings 06-34TL Ralph T. Hutchings 06-34TR Ralph T. Hutchings ML02-01 © Salisbury District Hospital/Science Photo Library/Photo Researchers, Inc. ML02-02 Spencer Grant/PhotoEdit

Chapter 7 Chapter Opener Jim Cummins/Getty Images, Inc. 07-04a Don W. Fawcett/Science Source/Photo Researchers, Inc. 07-10a G. W. Willis/Biological Photo Service 07-10b Frederic H. Martini ML03-01 Gary Kufner/CORBIS ML03-02 Hugh S. Rose/Visuals Unlimited

Chapter 8 Chapter Opener Synaptek Scientific Products, Inc./Science Photo Library/Photo Researchers, Inc. 08-02a Ward's Natural Science Establishment, Inc. 08-05a Biophoto Associates/Photo Researchers, Inc. 08-05b Photo Researchers, Inc. 08-10 David Scott/Phototake NYC 08-15b Michael J. Timmons 08-16a Ralph T. Hutchings 08-16b Ralph T. Hutchings 08-16c Ralph T. Hutchings 08-21e Larry Mulvehill/Photo Researchers, Inc. 08-25a Ralph T. Hutchings

Chapter 9 Chapter Opener Prof. P. Motta, Department of Anatomy, University La Sapienza, Rome/Science Photo Library/Photo Researchers, Inc. 09-06b Pearson Education/PH College 09-07a Ralph T. Hutchings 09-11a Ed Reschke/Peter Arnold, Inc. 09-11c Custom Medical Stock Photo, Inc. 09-17 Richmond Products, Inc. 09-25b Ward's Natural Science Establishment, Inc. ML04-01 Suzanne Short/PhotoEdit ML04-02 Getty Images Inc.

Chapter 10 Chapter Opener SW Productions/Getty Images, Inc. 10-05b Manfred Kage/Peter Arnold, Inc. 10-09b Frederic H. Martini 10-11b Frederic H. Martini 10-12c Ward's Natural Science Establishment, Inc. 10-13b Ward's Natural Science Establishment, Inc. ML05-01 AP/Wide World Photos ML05-02 Bob Martin/Getty Images Inc.

Chapter 11 Chapter Opener Lester Lefkowitz Steve Kahn/Getty Images, Inc. 11-01a Martin M. Rotker 11-02a David Scharf/Peter Arnold, Inc. 11-02b Frederic H. Martini 11-03a Stanley Flegler/Visuals Unlimited 11-03b Stanley Flegler/Visuals Unlimited 11-07a Ed Reschke/Peter Arnold, Inc. 11-07b Ed Reschke/Peter Arnold, Inc. 11-07c Ed Reschke/Peter Arnold, Inc. 11-07d Ed Reschke/Peter Arnold, Inc. 11-07e Ed Reschke/Peter Arnold, Inc. 11-09 Custom Medical Stock Photo, Inc.

Chapter 12 Chapter Opener David Arky/CORBIS 12-03 Ralph T. Hutchings 12-05c Peter Arnold, Inc.

Chapter 13 Chapter Opener Biophoto Associates/Photo Researchers, Inc. 13-01 Michael J. Timmons 13-03b Biophoto Associates/Photo Researchers, Inc. 13-07b Jack Star/Getty Images, Inc.

Chapter 14 Chapter Opener Gary Gaugler/Visuals Unlimited 14-02b Frederic H. Martini 14-07c Frederic H. Martini ML06-01 Tony Freeman/PhotoEdit ML06-02 Vittoriano Rastelli/CORBIS

Chapter 15 Chapter Opener Davis Barber/PhotoEdit 15-03b Photo Researchers, Inc. 15-04e Phototake NYC 15-07b Micrograph by P. Gehr, from Bloom & Fawcett, "Textbook of Histology," W. B. Saunders Co. ML07-01 Tony Freeman/PhotoEdit ML07-02 Bill Aron/PhotoEdit

Chapter 16 Chapter Opener Arcimboldo, Giuseppe (1527–93). "La Primavera." Copyright Scala/Art Resource, NY. Pinacoteca Civica, Brescia, Italy. 16-08d Ward's Natural Science Establishment, Inc. 16-13b Frederic H. Martini 16-15c Michael J. Timmons

Chapter 17 Chapter Opener Roger Mear/Getty Images Inc. Sylvain Grandadam/Getty Images Inc. ML08-01 David Young-Wolff/PhotoEdit ML08-02 David Young-Wolff/PhotoEdit

Chapter 18 Chapter Opener Douglas T. Mesney/Corbis/Stock Market 18-03b Ralph T. Hutchings 18-05b David M. Phillips/Visuals Unlimited ML09-01 Michael Newman/PhotoEdit ML09-02 Hot Ideas/Index Stock Imagery, Inc.

Chapter 19 Chapter Opener Mark Richards/PhotoEdit 19-02b Don W. Fawcett, M.D., Harvard Medical School 19-05a Ward's Natural Science Establishment, Inc. 19-05b © R. G. Kessel and R. H. Kardon, "Tissues and Organs: A Text-Atlas of Scanning Electron Microscopy," W. H. Freeman & Co., 1979. All Rights Reserved. 19-10a Frederic H. Martini 19-10b Frederic H. Martini 19-10c Frederic H. Martini 19-10d Frederic H. Martini 19-10e G. W. Willis, M.D./Biological Photo Service ML10-01 David Nunuk/Science Photo Library/Photo Researchers, Inc. ML10-02 Jade Albert Studios/Getty Images, Inc.

Chapter 20 Chapter Opener Petit Format/Nestle/Photo Researchers, Inc. 20-01a Francis Leroy, Biocosmos/Science Photo Library/Custom Medical Stock Photo, Inc. 20-07a Dr. Arnold Tamarin/Arnold Tamarin 20-07b right Photo Lennart Nilsson/Albert Bonniers Forlag 20-07c Photo Lennart Nilsson/Albert Bonniers Forlag 20-07d Photo Lennart Nilsson/Albert Bonniers Forlag 20-08a Photo Lennart Nilsson/Albert Bonniers Forlag 20-08b Photo Researchers, Inc. 20-13 CNRI/Science Photo Library/Photo Researchers, Inc. ML11-01 David Young-Wolff/PhotoEdit ML11-02 Derek Bromhall/Oxford Scientific Films Ltd.

Martini/Bartholomew, Essentials of Anatomy & Physiology Student Tutor CD-ROM, 3rd edition

LICENSE AGREEMENT

YOU SHOULD CAREFULLY READ THE FOLLOWING TERMS AND CONDITIONS BEFORE BREAKING THE SEAL ON THE PACKAGE. AMONG OTHER THINGS, THIS AGREEMENT LICENSES THE ENCLOSED SOFTWARE TO YOU AND CONTAINS WARRANTY AND LIABILITY DISCLAIMERS. BY BREAKING THE SEAL ON THE PACKAGE, YOU ARE ACCEPTING AND AGREEING TO THE TERMS AND CONDITIONS OF THIS AGREEMENT. IF YOU DO NOT AGREE TO THE TERMS OF THIS AGREEMENT, DO NOT BREAK THE SEAL. YOU SHOULD PROMPTLY RETURN THE PACKAGE UNOPENED.

LICENSE

Subject to the provisions contained herein, Prentice-Hall, Inc. ("PH") hereby grants to you a non-exclusive, non-transferable license to use the object code version of the computer software product ("Software") contained in the package on a single computer of the type identified on the package.

SOFTWARE AND DOCUMENTATION

PH shall furnish the Software to you on media in machine-readable object code form and may also provide the standard documentation ("Documentation") containing instructions for operation and use of the Software.

LICENSE TERM AND CHARGES

The term of this license commences upon delivery of the Software to you and is perpetual unless earlier terminated upon default or as otherwise set forth herein.

TITLE

Title, and ownership right, and intellectual property rights in and to the Software and Documentation shall remain in PH and/or in suppliers to PH of programs contained in the Software. The Software is provided for your own internal use under this license. This license does not include the right to sublicense and is personal to you and therefore may not be assigned (by operation of law or otherwise) or transferred without the prior written consent of PH. You acknowledge that the Software in source code form remains a confidential trade secret of PH and/or its suppliers and therefore you agree not to attempt to decipher or decompile, modify, disassemble, reverse engineer or prepare derivative works of the Software or develop source code for the Software or knowingly allow others to do so. Further, you may not copy the Documentation or other written materials accompanying the Software.

UPDATES

This license does not grant you any right, license, or interest in and to any improvements, modifications, enhancements, or updates to the Software and Documentation. Updates, if available, may be obtained by you at PH's then current standard pricing, terms, and conditions.

LIMITED WARRANTY AND DISCLAIMER

PH warrants that the media containing the Software, if provided by PH, is free from defects in material and workmanship under normal use for a period of sixty (60) days from the date you purchased a license to it.

THIS IS A LIMITED WARRANTY AND IT IS THE ONLY WARRANTY MADE BY PH. THE SOFTWARE IS PROVIDED 'AS IS' AND PH SPECIFICALLY DISCLAIMS ALL WARRANTIES OF ANY KIND, EITHER EXPRESS OR IMPLIED, INCLUDING, BUT NOT LIMITED TO, THE IMPLIED WARRANTY OF MERCHANTABILITY AND FITNESS FOR A PARTICULAR PURPOSE. FURTHER, COMPANY DOES NOT WARRANT, GUARANTY OR MAKE ANY REPRESENTATIONS REGARDING THE USE, OR THE RESULTS OF THE USE, OF THE SOFTWARE IN TERMS OF CORRECTNESS, ACCURACY, RELIABILITY, CURRENTNESS, OR OTHERWISE AND DOES NOT WARRANT THAT THE OPERA-

TION OF ANY SOFTWARE WILL BE UNINTERRUPTED OR ERROR FREE. COMPANY EXPRESSLY DISCLAIMS ANY WARRANTIES NOT STATED HEREIN. NO ORAL OR WRITTEN INFORMATION OR ADVICE GIVEN BY PH, OR ANY PH DEALER, AGENT, EMPLOYEE OR OTHERS SHALL CREATE, MODIFY OR EXTEND A WARRANTY OR IN ANY WAY INCREASE THE SCOPE OF THE FOREGOING WARRANTY, AND NEITHER SUBLICENSEE OR PURCHASER MAY RELY ON ANY SUCH INFORMATION OR ADVICE. If the media is subjected to accident, abuse, or improper use; or if you violate the terms of this Agreement, then this warranty shall immediately be terminated. This warranty shall not apply if the Software is used on or in conjunction with hardware or programs other than the unmodified version of hardware and programs with which the Software was designed to be used as described in the Documentation.

LIMITATION OF LIABILITY

Your sole and exclusive remedies for any damage or loss in any way connected with the Software are set forth below. UNDER NO CIRCUMSTANCES AND UNDER NO LEGAL THEORY, TORT, CONTRACT, OR OTHERWISE, SHALL PH BE LIABLE TO YOU OR ANY OTHER PERSON FOR ANY INDIRECT, SPECIAL, INCIDENTAL, OR CONSEQUENTIAL DAMAGES OF ANY CHARACTER INCLUDING, WITHOUT LIMITATION, DAMAGES FOR LOSS OF GOODWILL, LOSS OF PROFIT, WORK STOPPAGE, COMPUTER FAILURE OR MALFUNCTION, OR ANY AND ALL OTHER COMMERCIAL DAMAGES OR LOSSES, OR FOR ANY OTHER DAMAGES EVEN IF PH SHALL HAVE BEEN INFORMED OF THE POSSIBILITY OF SUCH DAMAGES, OR FOR ANY CLAIM BY ANY OTHER PARTY. PH'S THIRD PARTY PROGRAM SUPPLIERS MAKE NO WARRANTY, AND HAVE NO LIABILITY WHATSOEVER, TO YOU. PH's sole and exclusive obligation and liability and your exclusive remedy shall be: upon PH's election, (i) the replacement of your defective media; or (ii) the repair or correction of your defective media if PH is able, so that it will conform to the above warranty; or (iii) if PH is unable to replace or repair, you may terminate this license by returning the Software. Only if you inform PH of your problem during the applicable warranty period will PH be obligated to honor this warranty. You may contact PH to inform PH of the problem as follows:

SOME STATES OR JURISDICTIONS DO NOT ALLOW THE EXCLUSION OF IMPLIED WARRANTIES OR LIMITATION OR EXCLUSION OF CONSEQUENTIAL DAMAGES, SO THE ABOVE LIMITATIONS OR EXCLUSIONS MAY NOT APPLY TO YOU. THIS WARRANTY GIVES YOU SPECIFIC LEGAL RIGHTS AND YOU MAY ALSO HAVE OTHER RIGHTS WHICH VARY BY STATE OR JURISDICTION.

MISCELLANEOUS

If any provision of this Agreement is held to be ineffective, unenforceable, or illegal under certain circumstances for any reason, such decision shall not affect the validity or enforceability (i) of such provision under other circumstances or (ii) of the remaining provisions hereof under all circumstances and such provision shall be reformed to and only to the extent necessary to make it effective, enforceable, and legal under such circumstances. All headings are solely for convenience and shall not be considered in interpreting this Agreement. This Agreement shall be governed by and construed under New York law as such law applies to agreements between New York residents entered into and to be performed entirely within New York, except as required by U.S. Government rules and regulations to be governed by Federal law. YOU ACKNOWLEDGE THAT YOU HAVE READ THIS AGREEMENT, UNDERSTAND IT, AND AGREE TO BE BOUND BY ITS TERMS AND CONDITIONS. YOU FURTHER AGREE THAT IT IS THE COMPLETE AND EXCLUSIVE STATEMENT OF THE AGREEMENT BETWEEN US THAT SUPERSEDES ANY PROPOSAL OR PRIOR AGREEMENT, ORAL OR WRITTEN, AND ANY OTHER COMMUNICATIONS BETWEEN US RELATING TO THE SUBJECT MATTER OF THIS AGREEMENT.

U.S. GOVERNMENT RESTRICTED RIGHTS

Use, duplication or disclosure by the Government is subject to restrictions set forth in subparagraphs (a) through (d) of the Commercial Computer-Restricted Rights clause at FAR 52.227-19 when applicable, or in subparagraph (c) (1) (ii) of the Rights in Technical Data and Computer Software clause at DFARS 252.227-7013, and in similar clauses in the NASA FAR Supplement.

MINIMUM SYSTEM REQUIREMENTS

Windows
- Processor: Intel Pentium processor or comparable class chip-minimum 133 MHz
- RAM: In addition to the RAM required by the operating system, this application requires 32 MB of RAM for Windows 98, ME, and NT.
- Operating System: Windows 98/2000/ME/XP/NT-4.0 or higher
- CD-ROM drive (4× or faster)
- SVGA monitor capable of supporting 800 × 600 resolution at 16-bit
- Sound card
- QuickTime 5 or higher

Macintosh
- Processor: PowerPC processor, minimum 233 MHz
- RAM: In addition to the RAM required by the operating system, this application requires 64 MB of RAM for OS 8.1, 8.5, 9.X.
- Operating System: OS 8.1, 8.5, 9.X
- CD-ROM drive (4× or faster)
- SVGA monitor capable of supporting 800 × 600 resolution at 16-bit
- Sound card
- QuickTime 5 or higher

This CD contains QuickTime movies.

LAUNCHING THE PROGRAM

Windows:
  Insert the CD-ROM.
  Double click on "My Computer."
  Double click on the CD-ROM drive icon.
  Double click Start_EAP3e_CD.exe to launch the program.

NOTE: If you do not currently have QuickTime 5.0.2 or higher installed on your system please run the QuickTime installer found on the CD-ROM.

Macintosh:
  Insert the CD-ROM.
  Double click on the CD icon on your desktop.
  Double click the Start EAP3e CD icon to launch the program.

NOTE: If you do not currently have QuickTime 5.0.2 or higher installed on your system please run the QuickTime installer found on the CD-ROM.

If QuickTime(TM) is not on your system, click the QuickTime(TM) icon in the CD window to install this utility. Simply follow the instructions on screen.

If you want to check which version is currently installed, follow these simple steps:

  Macintosh users: Select the QuickTime(TM) extensions with a single-click. This is found in the Extensions folder in your System Folder. With the extension selected, select "Get Info" from the File menu. This will tell you the version of your QuickTime.

  Windows users: Open the QuickTime(TM) control panel. This is found in your Start Menu, in the Settings: Control Panel folder. With the QuickTime control panel open, the window will be called "QuickTime Settings." Choose "About QuickTime" from the menu at the top of the window. Your version number will be displayed. If you do not see this option, you may have an older version installed. Check the version number displayed in the window.

NOTE: For maximum compatibility, it is common for Windows systems to have multiple versions of QuickTime installed. Carefully search through your Control Panels for other QuickTime control panels.